Handbook of Advanced Ceramics

VOLUME II Processing and their Applications

Elsevier Internet Homepage – http://www.elsevier.com
Consult the Elsevier homepage for full catalogue information on all books, journals and electronic products and services.

Elsevier Titles of Related Interest
ELSSNER *et al.*
Ceramics and Ceramic Composites: Materialographic Preparation
ISBN: 0 444 100 30X

Related Journals:
A sample journal issue is available online by visiting the homepage of the journal (homepage details at the top of this page). Free specimen copy gladly sent on request. Elsevier Ltd, The Boulevard, Langford Lane, Kidlington, Oxford, OX5 1GB, UK

Ceramics International – www.elsevier.com/locate/ceramint
Journal of The European Ceramics Society – www.elsevier.com/locate/jeurceramsoc
Cement and Concrete Research – www.elsevier.com/locate/cemconres
Cement and Concrete Composites – www.elsevier.com/locate/cemconcomp

To contact the Publisher
Elsevier welcomes enquiries concerning publishing proposals: books, journal special issues, conference proceedings, etc. All formats and media can be considered. Should you have a publishing proposal you wish to discuss, please contact, without obligation, the publisher responsible for Elsevier's Composites and Ceramics programme:

Emma Hurst
Publishing Editor
Elsevier Ltd
The Boulevard, Langford Lane
Kidlington, Oxford
OX5 1GB, UK

Phone: +44 1865 843629
Fax: +44 1865 843931
E.mail: e.hurst@elsevier.com

General enquiries, including placing orders, should be directed to Elsevier's Regional Sales Offices – please access the Elsevier homepage for full contact details (homepage details at the top of this page).

books butler

to search for more Elsevier books, visit the Books Butler at http://www.elsevier.com/homepage/booksbutler/

Handbook of Advanced Ceramics

VOLUME II Processing and their Applications

EDITORS

SHIGEYUKI SŌMIYA *(Editor-in-Chief)*
Professor Emeritus, Tokyo Institute of Technology, Tokyo, Japan

FRITZ ALDINGER
Max-Planck-Institut für Metallforschung, Stuttgart, Germany

NILS CLAUSSEN
Technische Universität Hamburg-Harburg, Germany

RICHARD M. SPRIGGS
Alfred University, New York, USA

KENJI UCHINO
The Pennsylvania State University, USA

KUNIHITO KOUMOTO
Nagoya University, Japan

MASAYUKI KANENO
Japan Fine Ceramics Association, Tokyo, Japan

ELSEVIER
ACADEMIC PRESS

Amsterdam • Boston • Heidelberg • London • New York • Oxford
Paris • San Diego • San Francisco • Singapore • Sydney • Tokyo

This book is printed on acid-free paper.

ACADEMIC PRESS
An Imprint of Elsevier
84 Theobald's Road, London WC1X 8RR, UK
http://www.elsevier.com

ACADEMIC PRESS
An Imprint of Elsevier
525 B Street, Suite 1900, San Diego, California 92101-4495, USA
http://www.elsevier.com
ISBN 0–12–654640–1

A catalogue record for this book is available from the Library of Congress

A catalogue record for this book is available from the British Library

Typeset by Newgen Imaging Systems (P) Ltd, Chennai, India
Printed and bound in Great Britain by MPG Books, Bodmin, Cornwall

03 04 05 06 07 MP 9 8 7 6 5 4 3 2 1

To The late Professor Emeritus E. F. Osborn
The Pennsylvania State University
University Park, Pennsylvania, USA

To Professor Emeritus J. A. Pask
University of California, Berkeley, California, USA

To Professor Emeritus Günter Petzow
Max Planck Institut für Metallforschung, Pulvermetallgisches Laboratorium

Shigeyuki Sōmiya

CONTENTS

Part 2: Engineering Ceramics

PREFACE

In 1989 Shigeyuki Sōmiya, the Editor-in-Chief of this book, published *Advanced Technical Ceramics* (Academic Press, Inc.; original publication in Japanese, 1984). Well over a decade has passed without the appearance of an authoritative new title on the ever-changing subject of Advanced Ceramics. The purpose of this book is to provide an up-to-date account of the present status of Advanced Ceramics, from fundamental science and processing to application.

The Handbook of Advanced Ceramics has an internationally renowned group of contributing editors. They are well known throughout the world in their fields of study. These editors discussed the contents and chose the authors of each of the book's chapters very carefully. The chapters consist of review and overview papers written by experts in the field.

Up until about 50 years ago, 'ceramics' were considered to be porcelains, bottle glass, sheet glass, refractory bricks, enamels, cements, lime, gypsum and abrasives. In recent years the field of ceramics has broadened and expanded. Ceramics are now used in new fields of research as well as in the old fields. This handbook describes these developments and the new processes and applications.

The handbook will enable the reader to understand the present status of Ceramics and will also act as an introduction, which may encourage further study, as well as an estimation of the role advanced ceramics may have in the future.

The handbook is a two-volume set. Part I deals with Materials Science and Part II with Processing and Applications.

Part I serves as an introduction to the basic science, raw materials, forming, drying, sintering, innovative processing, single crystal growth, machining, joining, coating, fracture mechanics, testing, evaluation, etc. Part I is intended to provide the reader with a good understanding of the new techniques in advanced ceramics, such as thin films, colloidal processing, active and passive filler, pyrolyses process and precursor derived ceramics, as well as providing a template for the deposition of ceramics from aqueous solutions.

Part II deals with more recent processes and applications and functional and engineering ceramics. The engineering ceramics covered in this book were developed within the last decade. The functional ceramics covered include electro-ceramics, optoelectro-ceramics, superconductive ceramics, etc.

as well as the more recent development of piezoelectric ceramics and dielectric ceramics.

The use of 'Engineering' Ceramics, introduces entirely new fields to be considered. These include mechanical properties, decorative ceramics, environmental uses, energy applications, bioceramics, composites, functionally graded materials, intelligent ceramics and so on.

The term Advanced Ceramics is opposite in meaning to 'Traditional' or 'Classical' Ceramics. In the past, Advanced Ceramics were often confused with New or Newer Ceramics, Modern Ceramics, Special Ceramics and so on. Furthermore, Fine Ceramics, at least in the USA and Europe, is synonymous with Fine Grain Ceramic Products and/or Fine Grain Porcelain; Fine Ceramics in Japan is similar to what we understand as Advanced Ceramics. So for this edition, the term *Advanced Ceramics* was chosen as the most suitable title for a book providing an in-depth survey of the current state of Ceramics Science and its applications.

It is the editors' wish that this book will provide the reader with a detailed understanding of the many applications of Advanced Ceramics in both today's world and in that of the future.

The editors wish to thank all those who participated in the preparation of this book such as authors, publishers and copyright owners in Europe, USA, Asia and the rest of the world.

Fritz Aldinger
Nils Claussen
Masayuki Kaneno
Kunihito Koumoto
Shigeyuki Sōmiya (Editor-in-Chief)
Richard M. Spriggs
Kenji Uchino

ACKNOWLEDGMENTS

First and foremost, I would like to thank all the co-editors, Fritz Aldinger, Nils Claussen, Richard M. Spriggs, Kenji Uchino, Kunihito Koumoto and Masayuki Kaneno for their valuable suggestions with regard to the chapters, authors, and their editorial assistance. Without their help, this book would not have been possible. Especially, Fritz Aldinger, Nils Claussen, and Richard M. Spriggs gave me good advice for this book. Mr Kaneno offered secretarial assistance.

Second, I wish to thank all the publishers who gave permissions to authors to reproduce and use the materials from their original published papers.

Third, many authors wrote their chapters within a short time in spite of their tight schedule. I would like to express my gratitude to all these authors.

I would also like to extend my appreciation to Ms Amanda Weaver and her group at Elsevier, Oxford for their contribution to the publishing works.

Finally, I was able to study abroad in the USA under the Fulbright Exchange Program and in Germany under the scholarship program by Max Planck Institut für Metallforschung, Pulvermetallgisches Laboratorium. Without these experiences, I would not have made it as Editor-in-Chief of this book. I thank all my professors and friends around the world who have made this possible.

Shigeyuki Sōmiya
Editor-in-Chief

CONTRIBUTORS

LIDONG CHEN, Shangai Institute of Ceramics, Academic Sinica, 1295 Ding-Xi Road, Shangai 200050, China

J. FLEIG, Max-Planck-Institut für Festkörperforschung, Heisenbergstrasse 1, 70569 Stuttgart, Germany

MIKIO FUKUHARA, Development of Special Products, Toshiba Tungaloy Ltd., Sugasawa, Tsurumi, Yokohama 230-0027, Japan

TAKASHI GOTO, Institute for Materials Research, Tohoku University, Katahira 2-1-1, Sendai 980-8577, Japan

TOSHIYUKI HAMANAKA, Ceramics Business Group, NGK Insulators Ltd., 2-56 Suda-cho, Mizuho-ku, Nagoya 467-8530, Japan

HAJIME HANEDA, Advanced Materials Laboratory, National Institute of Materials Science, 1-1 Namiki, Tsukuba 305-0044, Japan

ATSUMU ISHIDA, Japan Fine Ceramics Center, 2-4-1 Mutsuno, Atsuta-ku, Nagoya 456-8587, Japan

MASAKAZU KAWASHITA, Graduate School of Engineering, Kyoto University Sakyo-ku, Kyoto 606-8501, Japan

HYUN-MIN KIM, Department of Ceramic Engineering, School of Advanced Materials Engineering, Yonsei University, 134, Shinchon-dong, Seodaemun-gu, Seoul 120-749, Korea

KAZUNORI KOGA, Corporate R&D Group for Components & Devices, Kyocera Corporation, 6 Takeda Tobadono-cho, Fushimi-ku, Kyoto 612-8501, Japan

TADASHI KOKUBO, Graduate School of Engineering, Kyoto University, Sakyo-ku, Kyoto 606-8501, Japan

K. D. KREUER, Max-Plank-Institut für Festkörperforschung, Heisenbergstrasse 1, 70569 Stuttgart, Germany

JOACHIM MAIER, Max-Planck-Institut für Festkörperforschung, Heisenbergstrasse 1, D-70569 Stuttgart, Germany

HIDEAKI MATSUBARA, Japan Fine Ceramics Center, 2-4-1 Mutsuno, Atsuta-ku, Nagoya 456-8587, Japan

KAORU MIYAHARA, Technical Development, Ishikawajima-Harima Heavy Industries Co., Ltd., 1 Shin-nakahara-cho, Isogo-ku, Yokohama 235-8501, Japan

HIROSHI NEMOTO, BIU, NGK Insulators Ltd., 2-56 Suda-cho, Mizuho-ku, Nagoya 467-8530, Japan

HIDEAKI NIIMI, Murata Manufacturing Co. Ltd., Yasu 520-2393, Japan

TAKESHI NOMURA, Materials Research Center, TDK Corporation, Narita 286-8588, Japan

AKIRA OKADA, Japan Fine Ceramics Center, 2-4-1 Mutsuno, Atsuta-ku, Nagoya 456-8587, Japan

YOSHIKI OKUHARA, Japan Fine Ceramics Center, 2-4-1 Mutsuno, Atsuta-ku, Nagoya 456-8587, Japan

YUKIO SAKABE, Murata Manufacturing Co. Ltd., Yasu 520-2393, Japan

TADASHI SASA, Technical Development, Ishikawajima-Harima Heavy Industries Co., Ltd., 1 Shin-nakahara-cho, Isogo-ku, Yokohama 235-8501, Japan

YASUHIRO SHIGEGAKI, Technical Development, Ishikawajima-Harima Heavy Industries Co., Ltd., 1 Shin-nakahara-cho, Isogo-ku, Yokohama 235-8501, Japan

HIROAKI TAKEBAYASHI, EXSEV Engineering Department, Koyo Seiko Co., Ltd., Kokubu Tojyo-machi, Kashihara 582-8588, Japan

MASAYUKI TAKADA, Japan Fine Ceramics Center, 2-4-1 Mutsuno, Atsuta-ku, Nagoya 456-8587, Japan

KAZUMASA TOGANO, Institute for Materials Research, Tohoku University, 2-1-1, Katahira, Aoba-ku, Sendai 980-8577, Japan

TOSHINORI TSURU, Department of Chemical Engineering, Hiroshima University Higashi-Hiroshima 739-8527, Japan

KENJI UCHINO, Materials Research Laboratory, The Pennsylvania State University University Park, PA 16802-4801, USA

TESSHO YAMADA, Sensor Division, Automotive Components Group, Engineering Department, NGK Spark Plug Co. Ltd., Komaki 495-8510, Japan

HIROAKI YANAGIDA, Japan Fine Ceramics Center, 2-4-1 Mutsuno, Atsuta-ku, Nagoya 456-8587, Japan

PART 1

Functional Ceramics

Handbook of Advanced Ceramics
S. Sōmiya *et al.* (Eds.)

CHAPTER 1

1.1 Insulating Ceramics/High Thermal Conductive Ceramics

KAZUNORI KOGA
Corporate R&D Group for Components & Devices, Kyocera Corporation, 6 Takeda Tobadono-cho, Fushimi-ku, Kyoto 612-8501, Japan

1.1.1 GENERAL REMARKS

1.1.1.1 MATERIALS FOR PKG

Because of its high thermal conductivity, high mechanical strength, good insulation characteristics, moderate dielectric properties and high chemical durability, alumina (HTCC: high-temperature co-fired ceramics) is the most popular ceramics material for semiconductor packages.

However, for power devices like power amplifier for base station or for satellites, higher thermal conductivity material is required to dissipate the heat generated in the devices. To meet this requirement, aluminum nitride (AlN), which has high thermal conductivity (TC) and a low thermal expansion coefficient comparable to that of Si, has been adopted for packages requiring high thermal dissipation.

Another market trend, toward higher power, higher working frequencies and lower power consumption, requires reduction of the resistivity of conductors in co-fired packages. To meet this requirement, glass ceramics (LTCC: low-temperature co-fired ceramics) with silver or copper conductors have been developed.

Kyocera has been conducting research and development on AlN and LTCC for more than 10 years, and has produced many kinds of AlN and LTCC products, such as Cer-Quad and multilayer packages, in addition to thin film substrates. Now, we have developed three materials for packages. The first is a novel AlN material (AN75W) that can co-fired at low temperature to reduce cost. The second is a novel LTCC that has a high thermal coefficient of expansion close to that for FR-4. The third is also a novel LTCC that has low permittivity

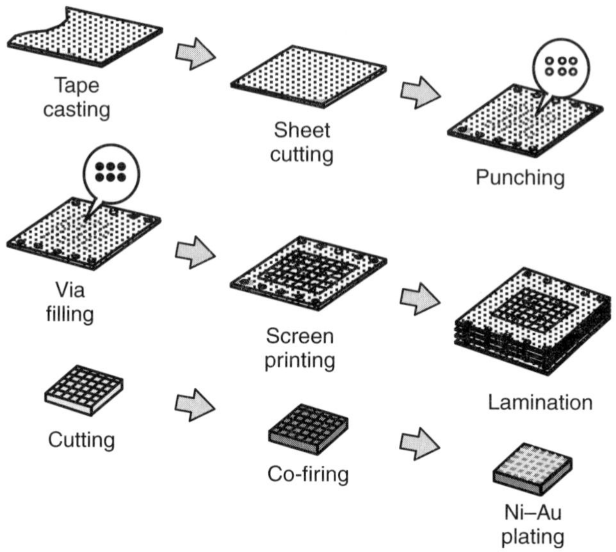

FIGURE 1.1.1 Process flow for multilayer ceramic package fabrication.

and low loss tangent at high frequency. In this section, we describe these new materials.

1.1.1.2 Process Flow

Figure 1.1.1 shows the process flow for a co-fired multilayer ceramic package. There are many steps to produce a multilayer package; however, there are only a few differences among different materials. Of course, material composition, metallize composition and process condition is different for each. Among these steps, metallizing is an especially critical technology for package production.

1.1.2 ALUMINUM NITRIDE

1.1.2.1 Material Properties

Table 1.1.1 shows properties of the new and standard aluminum nitride compositions, compared with multilayer alumina [1]. The main difference between

TABLE 1.1.1 Properties of AlN

	AlN		Al_2O_3
	AN75W	AN242	A-440
Thermal conductivity	75	150	14
Dielectric constant (1 MHz)	8.6	8.7	9.8
$\tan\delta \times 10^{-4}$ (1 MHz)	6	1	24
Volume resistivity (Ω cm at 20°C)	$>10^{14}$	$>10^{14}$	$>10^{14}$
Flexural strength (MPa)	400	400	400
Young's modulus of elasticity (GPa)	310	320	310

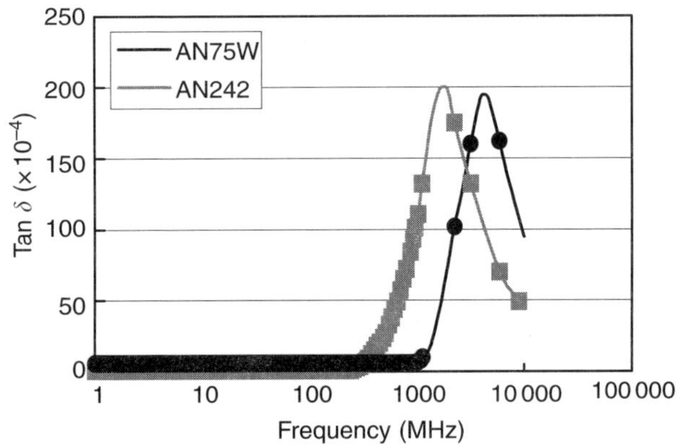

FIGURE 1.1.2 Frequency dependence of dielectric loss of AlN.

these materials is in thermal conductivity, which is 75 W/mK for AN75W and 150 and 14 W/mK for standard AlN and Al_2O_3, respectively. Other fundamental properties are rather similar. AlN, however, has a characteristic dielectric dispersion at high frequencies [2]. Figure 1.1.2 shows the frequency dependence of the dielectric loss ($\tan\delta$) of AN75W and AN242. The dielectric loss of AlN shows a maximum at a few gigahertz. This phenomenon is due to the piezoelectricity of AlN, and the peak frequency inversely depends on crystallite size. As the crystallite size of AN75W is smaller than that of AN242, the peak (dispersion) frequency of AN75W is correspondingly higher.

1.1.2.2 Technology for Multilayer AlN Packages

Concerning the cost, tape casting and co-firing processes of AlN package are particularly expensive compared with those of Al_2O_3 packages, since the cost

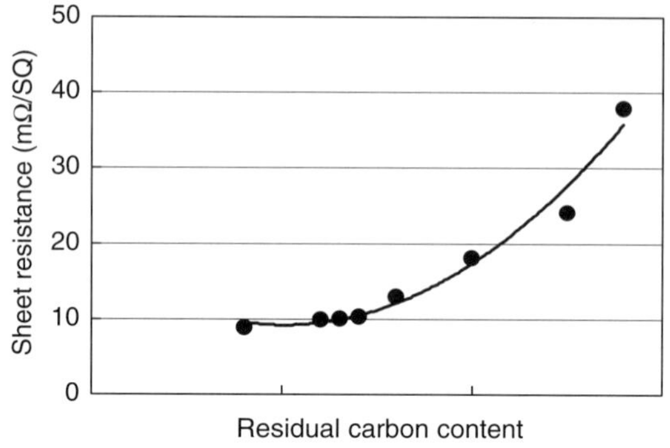

FIGURE 1.1.3 Sheet resistance as a function of residual carbon content in co-fired AlN.

of raw material is high and a significantly higher firing temperature is needed. We have specifically developed AN75W for low-temperature co-firing.

It is not so difficult to sinter AlN below 1700°C by combining rare earth and alkaline-earth compounds as sintering aids. However, as the sintering temperature is reduced, the chemical durability, especially to alkaline solutions, becomes much worse.

Moreover, residual carbon content in the package makes a strong effect on the sheet resistance of the refractory metal conductors (Fig. 1.1.3). In addition, the concentration of residual carbon depends on organic system in tape and sintering condition.

The sintering aids and sintering conditions for AN75W have been developed to optimize the manufacturability, cost, material properties, chemical durability and metallization characteristics. The magnitude of any cost reduction depends on package size and design. In general, however, packages manufactured using AN75W are much cheaper than those produced with standard grades of aluminum nitride.

1.1.2.3 Thermal Resistance

The thermal conductivity of packaging materials will influence its adoption in consumer and other applications. Figure 1.1.4 shows the thermal resistance (θj-a) of a package (OD: 8 mm^2; thickness: 0.762 mm; cavity: 4 mm sq; power consumption: 0.5 W) alone and mounted on a PC board (FR-4: 50 mm^2, 1.5 mm thick) as a function of the thermal conductivity of the ceramic (Al_2O_3, AlN: AN75W and AN242). The thermal resistance of the AN75W package is almost identical to that of the AN242 package especially after mounting onto the PC board.

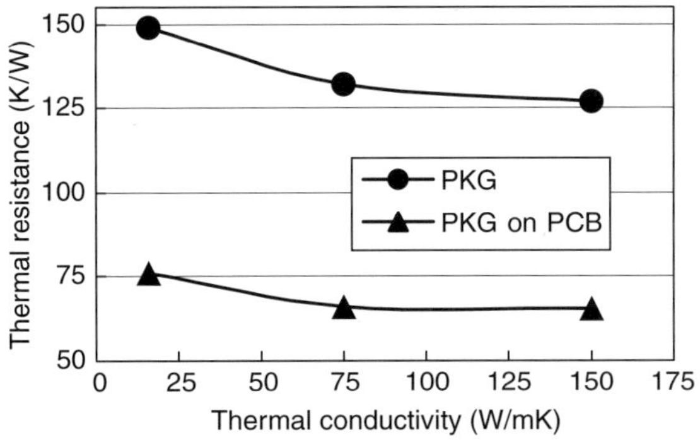

FIGURE 1.1.4 Thermal resistance as a function of ceramic thermal conductivity.

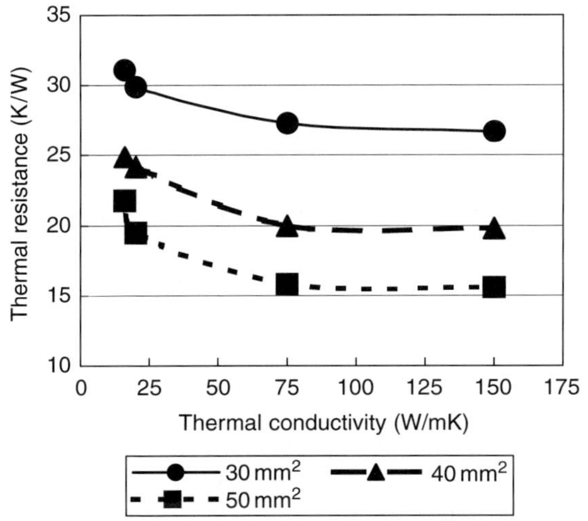

FIGURE 1.1.5 Thermal resistance as a function of thermal conductivity for packages measuring 30, 40 and 50 mm^2.

Figure 1.1.5 shows the thermal resistance (θj-a) of other packages (OD: 30, 40 and 50 mm^2; thickness: 1.0 mm; power consumption: 3 W) as a function of thermal conductivity. In this example, two Al_2O_3 ceramics are included. The thermal resistance of these packages decreases with increasing package size.

This is due to the increased thermal capacity of the package. As in the earlier example, the thermal resistance of the AN75W packages is superior to alumina and comparable to that of the AN242 packages. With increasing package size, however, the effect of the higher thermal conductivity of AN242 is more pronounced than in the smaller outlines. With larger formats, the thermal path

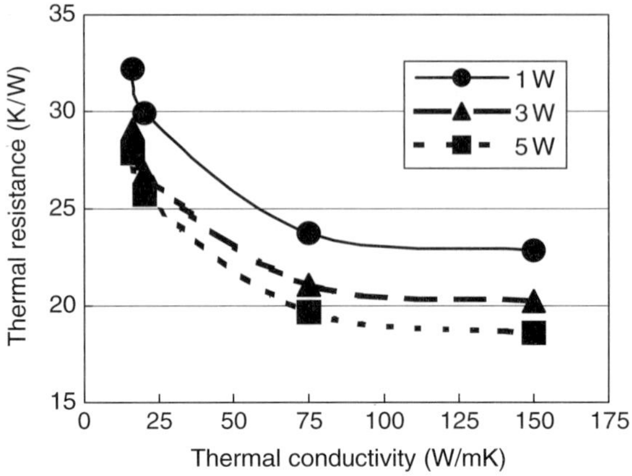

FIGURE 1.1.6 Thermal resistance as a function of ceramic thermal conductivity and input power.

length becomes longer. Thus, thermal conductivity exerts a greater influence on thermal resistance in larger packages.

Figure 1.1.6 shows the effect of package power dissipation, again as a function of ceramic thermal conductivity (OD: 40 mm^2; thickness: 0.762 mm). The thermal resistances of the AN75W and AN242 packages show only a slight difference in the power consumption range from 1 to 5 W.

1.1.2.4 Reliability

We have performed reliability tests on AlN packages produced with both AN75W and AN242. Seal hermeticity, external visual and electric performance, including plating durability, were evaluated after high-temperature storage, low-temperature storage, temperature cycling, and thermal shock. No AlN package has shown any reliability issues. Thus, we confirm that both AN75W and AN242 can be applied for consumer products where high reliability is required.

1.1.2.5 Summary

The characteristics of Kyocera's AlN packages are summarized as follows;

1 AN75W packages have been demonstrated. This material is low temperature co-fireable and can be produced at a cost less than standard AlN.

2 As a result of its fine grain size, AN75W exhibits a low dielectric loss at a few gigahertz. The loss maximum is shifted to higher frequency relative to AN242, which has a coarser crystallite size.
3 The thermal resistance of AN75W packages is comparable to that measured on packages produced from high TC AlN, especially for small outline packages mounted on PWB.
4 Both AN75W and AN242 packages have demonstrated high reliability.

In conclusion, we believe that AN75W is applicable to packages requiring both high thermal dissipation and low cost, for example in consumer products.

1.1.3 LTCC WITH HIGH THERMAL COEFFICIENT OF EXPANSION

1.1.3.1 Introduction

Due to the ever increasing I/O counts for IC devices, packaging trends have been changing to surface mountable area array second-level interconnection, namely BGA and CSP. The driving force for these types of second-level mounting is described as follows [4].

1 Higher wiring density: smaller packages, thinner packages, lighter packages,
2 higher performance: electrical performance, thermal performance, higher I/O counts,
3 lower cost.

A surface mounting technology (SMT) package, such as BGA, has low height interconnection between the substrate and the PWB. When we have a big difference of TCE between the substrate and the PWB, BGA and CSP packages receive more severe shear strain, damaging the reliability of solder joints, compared with PGA. For the second-level mounting of ceramic package on the PWB, this shear strain is a big problem, since alumina ceramics has TCE of 7 ppm/°C while the TCE of a typical PWB, FR-4 board is 12–16 ppm/°C as is illustrated in Figure 1.1.7.

By an FEM analysis for the second-level interconnection, we found that the stress yielded by the TCE mismatch between the substrate and the PWB reaches a minimum value when the TCE is around 11 ppm/°C for a flip chip assembly type packages [5]. We developed a high TCE ceramic material, in which the TCE and the Young's modulus are 11.5 ppm/°C and 114 GPa, respectively.

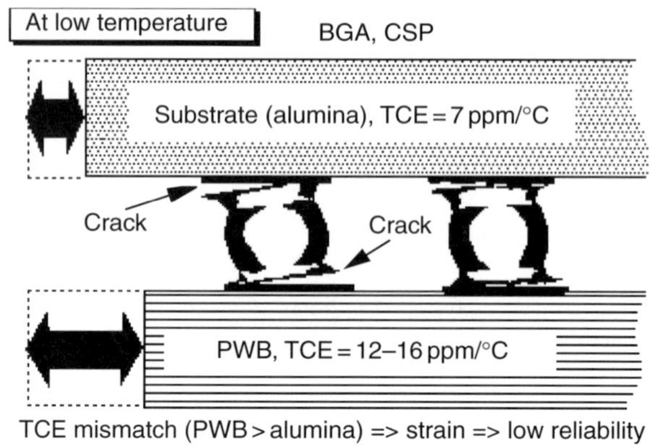

FIGURE 1.1.7 Schematic diagram of TCE mismatch between substrate made of alumina material and PWB.

The FEM analysis has also shown that the stress of the first-level interconnection in the encapsulated flip chip bump is comparably same to that of the alumina package which has high first-level interconnection reliability. We evaluated the solder joint reliability of BGA and CSP by TCT. We confirmed that we could obtain approximately three to ten times longer fatigue life by this high TCE ceramic material package than that of an alumina material package [5].

However, in the case of wire bonded chip assembly type CSP with a potting compound, we found that a sufficient reliability was not achieved compared to life predictions from FEM analyses [6]. Since the shrinkage of the potting compound is larger than that of the ceramic substrate in this assembly, the substrate warped upwards, while the PWB warped downwards. Solder balls at the package corner was most largely displaced due to this warping mechanism. This observation suggests that the reliability largely depends on the TCE mismatch between the substrate and the potting compound. We have developed new high TCE ceramic material according to the FEM results, and evaluated the solder joint reliability between CSP and PWB by using this new material.

1.1.3.2 New High TCE Ceramic Material for Wire Bonded Chip Assembly Type CSP with Potting Compound

According to the target value of TCE, we developed a new material of high TCE ceramics by modifying the 11.5 ppm/°C ceramic material's composition. Table 1.1.2 shows properties of the newly obtained high TCE caramic material.

TABLE 1.1.2 Characteristic Properties of 13 ppm/°C Material, Compared to 11.5 ppm/°C Material and FR-4

Item	Unit	11.5 ppm/°C material	13 ppm/°C material	FR-4
Bulk density	g/cm^3	2.6	2.6	—
Dielectric constant (1 MHz)	—	5.8	5.3	5.5
TCE (40–400°C)	ppm/°C	11.5	13.0	12–16
Flexural strength	MPa	230	200	430
Young's modulus	GPa	114	110	24

The properties of this material are almost same to the 11.5 ppm/°C material with the exception of TCE. The TCE is 13 ppm/°C, which is in the range of PWB's (FR-4: 12–16 ppm/°C) and that of the potting compound's (10–30 ppm/°C). The dielectric constant is 5.3 at 1 MHz, lower than 9.8 of alumina. The Young's modulus is 110 GPa, approximately one-third that of alumina. Also, copper conductor is co-firable.

1.1.3.3 Reliability of the New High TCE Ceramic Material

1.1.3.3.1 Evaluation of Temperature Cycling Test for wire bonded chip assembly type CSP

We evaluated the solder joint reliability by TCT. TCT sample was a wire bonded chip assembly type CSP with potting compound, mounted on a PWB (FR-4). The TCT condition was −40°C (10 min)/125°C (10 min). Size of CSP is 13 mm × 13 mm × 0.4 mm. CSPs were made of the 11.5 ppm/°C material and the 13 ppm/°C material for comparison. The construction of the CSP is 196 pins, 0.8 mm pad. Pitch, 0.35 mm pad diameter and 14 × 14 array.

The construction of the PWB is 1.6 mm thickness, 0.8 mm pad pitch, 0.4 mm pad diameter with two copper conductor layers. Figure 1.1.8 shows a photograph of the TCT sample of the CSP mounted on the PWB. Daisy-chained copper conductors in these package was cofired.

The CSP was mounted on the PWB by using eutectic solder balls (37Pb/63Sn) and paste. The height of solder joints was approximately 0.45 mm.

Figures 1.1.9 and 1.1.10 show cross-sectional SEM photographs of solder joints at a corner of CSP after 1000 cycles, for 13 and 11.5 ppm/°C substrates, respectively. In the case of the 13 ppm/°C material, there is no crack. In the other case of the 11.5 ppm/°C ceramic material, failed solder joints are observed. Figure 1.1.11 shows TCT (−40 to 125°C) results for CSPs made of the 11.5 and the 13 ppm/°C material. The package failure was defined as 50%

FIGURE 1.1.8 Photograph of TCT sample made of the new high TCE material mounted on the PWB.

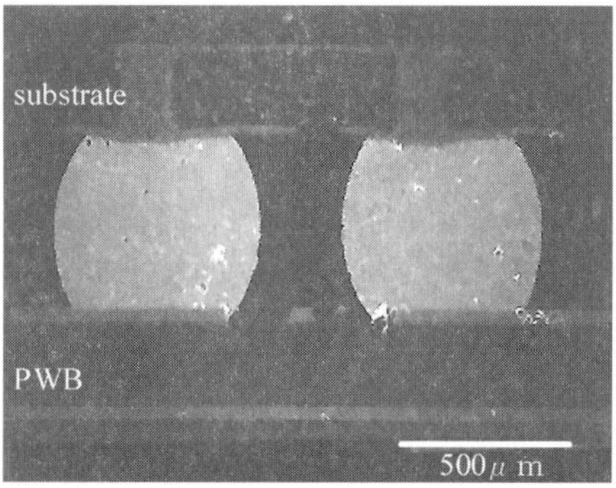

FIGURE 1.1.9 Photograph of cross-sectional SEM of solder joints of TCT (−40 to 125°C) samples of wire bonded chip assembly type CSP made of the 11.5 ppm/°C material after 1000 cycles.

increase of electrical resistance compared to the initial one. The number of samples were 10 for both cases. In the case of 11.5 ppm/°C material CSP, the first failure is observed at 300 cycles. In the case of the 13 ppm/°C material CSP, the first failure occurred at 1750 cycles.

1.1.3.3.2 Evaluation of package reliability for high TCE ceramic CSP

We evaluated a CSP package reliability of the high TCE ceramic material of 11.5 ppm/°C. We evaluated a wiring line resistivity between a chip bonding pad to a land pad, and an insulation resistivity between neighbor wiring lines. We evaluated those changes after package reliability tests of thermal shock, high temperature, high humidity with no bias, high temperature, high humidity with

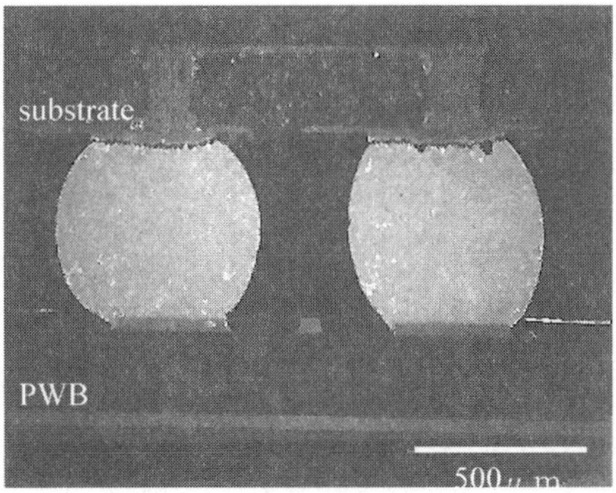

FIGURE 1.1.10 Photograph of cross-sectional SEM of solder joints of TCT (−40 to 125°C) samples of wire bonded chip assembly type CSP made of the 13 ppm/°C material after 1000 cycles.

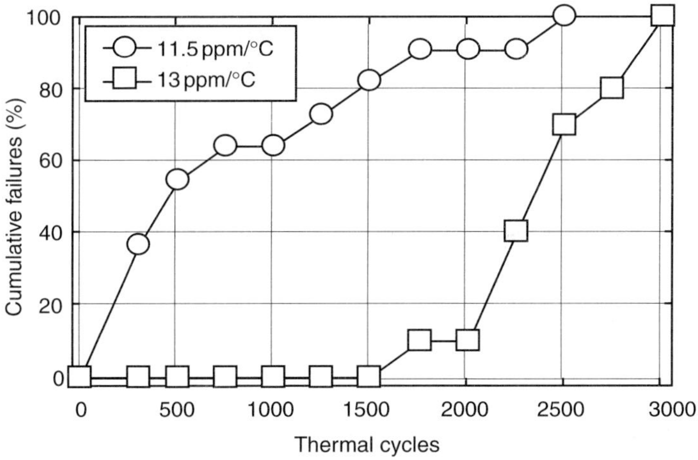

FIGURE 1.1.11 Solder joint reliability by TCT (−40 to 125°C) for wire bonded chip assembly type CSP made of the high TCE ceramic materials.

bias, high temperature and thermal cycling test. Table 1.1.3 shows the condition of package reliability tests. Tables 1.1.4 and 1.1.5 show the results for circuit resistance and insulation resistance, respectively. In all cases, we found very low changes after package reliability tests, and high reliability of CSP comparable to alumina package.

TABLE 1.1.3 Package's Reliability Test Condition

Item	Condition	Number
Thermal shock	−65 to 150°C	20
High temperature, high humidity (1)	85°C, 85% no bias	20
High temperature, high humidity (2)	85°C, 85% 5.5 V	20
High temperature	150°C	20
Thermal cycling test	−65° to 150°C	20

TABLE 1.1.4 Results of Package's Reliability Test for Circuit Resistance

Item	Result
Thermal shock	0 cycle (400 mΩ)
	500 cycles ±1.1%
High temperature high humidity (1)	0 h (400 mΩ)
	1000 h ±3.5%
High temperature high humidity (2)	0 h (400 mΩ)
	1000 h ±4.3%
High temperature	0 h (400 mΩ)
	1000 h ±3.3%
Thermal cycling test	0 cycle (400 mΩ)
	1000 cycles ±4.1%

TABLE 1.1.5 Results of Package's Reliability Test for Insulation Resistance

Item	Result
Thermal shock	0 cycle >1 × 10^{11} Ω
	500 cycles >1 × 10^{11} Ω
High temperature high humidity (1)	0 h >1 × 10^{11} Ω
	1000 h >1 × 10^{10} Ω
High temperature high humidity (2)	0 h >1 × 10^{11} Ω
	1000 h >1 × 10^{10} Ω
High temperature	0 hr >1 × 10^{11} Ω
	1000 h >1 × 10^{11} Ω
Thermal cycling test	0 cycle >1 × 10^{11} Ω
	1000 cycles >1 × 10^{11} Ω

1.1.3.4 Discussion

We have developed 11.5 ppm/°C material [5]. Ikemizu *et al.* [6] reported that by using this material the ceramic fine pitched ball grid array package and

PWB warped into opposite directions due to large shrinkage of the potting compound. TCE of package is smaller than that of the potting compound. Therefore, substrate warps upwards, while the PWB warps downwards with this assembly. The solder ball fracture occurred at 300 cycles. This thermal fatigue life is too short compared to the prediction of an FEM analysis.

We have developed new high TCE ceramic material of 13 ppm/°C. This material and 11.5 ppm/°C material have almost same properties with the exception of TCE. By using the 13 ppm/°C material, the TCE mismatch between the substrate and the potting compound decreases, and the improvement of solder joint reliability is obtained as shown in Figure 1.1.11.

Figure 1.1.12 shows height of solder joints between substrate and PWB along a diagonal direction of CSP. The horizontal axis is the position in the diagonal line of CSP. The vertical axis is the height of solder joints. This distribution of the height represents the warpage of CSP. At the beneath of Si-die, both curves are flat, and at both ends they warp upwards. The 13 ppm/°C material CSP shows a smaller warpage than the 11.5 ppm/°C material CSP. This result verifies the FEM analysis and the evaluation of TCT. The warpages may be caused also by firing distortion of the substrate and/or shrinkages by solidification of the potting compound.

We also evaluated a combination of organic CSP and PWB [6]. The second-level reliability between organic CSP and PWB was sufficient since their warpage occur in the same direction. It shows, however, a larger warpage than the ceramic CSP. A large stress at Si-die may be induced by this warpage. Tsukada has done some simulation and experiments concerning the joint reliability of

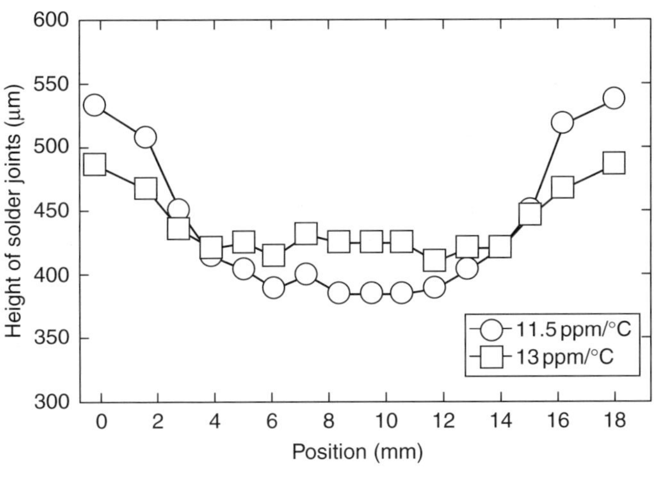

FIGURE 1.1.12 Height of solder joints between substrate and PWB in the diagonal direction of CSP.

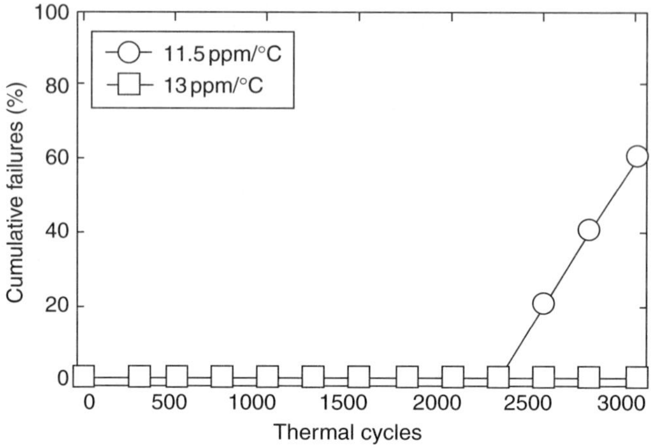

FIGURE 1.1.13 Solder joint reliability by TCT (−40 to 125°C) for flip chip assembly type CSP made of the high TCE ceramic material.

a plastic BGA [7]. A large warpage of the plastic BGA is caused by the TCE mismatch between the Si-die and the substrate, since the organic substrate has high TCE and low Young's modulus. Due to this warpage, thermal fatigue life of a chip assembled BGA is shorter than that of no chip assembled BGA. Young's modulus of the high TCE ceramic material and a BT resin are 110 and 20–25 GPa, respectively. Since the high TCE ceramic material is more rigid than the organic material, this material has an advantage in this subject.

A substrate of flip chip assembly type CSP has to be 11.5 ppm/°C material as was reported [5]. When we evaluated the solder joint reliability of the CSP with no potting compound, the first failure occurs at 2500 cycles, as shown in Figure 1.1.13. Flip chip assembly type CSP has a sufficient reliability with the 11.5 ppm/°C material. Even if the reliability of the second-level interconnection was improved by using 13 ppm/°C material, the TCE mismatch between the Si-die and the substrate shall be increased, and the stress at a flip chip bump becomes large. The first-level interconnection gets weaken by the larger TCE mismatch, and becomes harder to be reinforced by an underfill material. Therefore, the 13 ppm/°C material shall be used for a wire bonded chip assembled type CSP, and the 11.5 ppm/°C material is suitable for a flip chip assembly type CSP.

1.1.3.5 Conclusions

We found that the equivalent plastic strain generated by the TCE mismatch among Si-die, substrate, potting compound and PWB drastically decreases

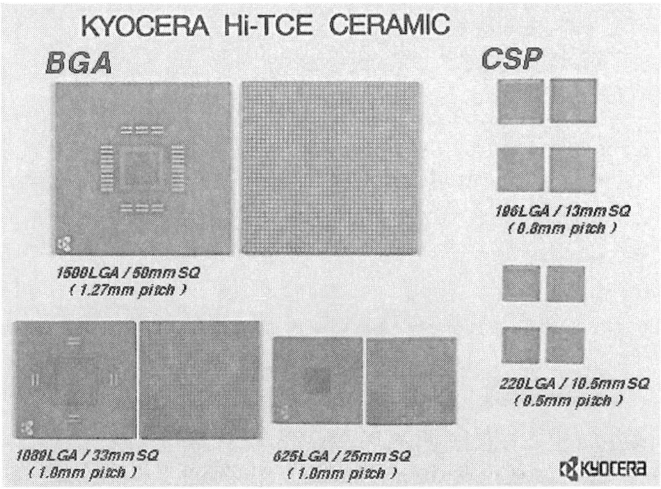

FIGURE 1.1.14 Photograph of BGA and CSP packages co-fired with copper conductors.

as TCE of the substrate increases from 11.5 to 13 ppm/°C. We developed a high TCE material of 13 ppm/°C, and confirmed by TCT that a wire bonded chip assembly type CSP made of this material has sufficient second-level reliability.

1.1.3.6 Future Package Application

Based on the present studies by these two high TCE materials, we believe that we can apply these materials for SMT packages of various shapes. Figure 1.1.14 shows such application samples of BGA, CSP packages cofired with copper conductors at low temperature.

1.1.4 LTCC WITH LOW PERMITIVITY AND LOW LOSS TANGENT AT HIGH FREQUENCY FOR MICROWAVE APPLICATION

1.1.4.1 Introduction

The ceramic package used for microwave applications requires following properties; (a) lower dielectric constant and lower loss tangent in the radio frequency range; (b) lower resistivity conductor; (c) thermal expansion coefficient of the ceramic material close to that of semiconductor chips; and (d) high reliability of hermeticity.

We have developed a new LTCC package that meets these requirements. In this section, we will describe the properties of newly developed LTCC material and copper conductor cofired with LTCC material and the reliability of the package.

1.1.4.2 Characteristics of Material

1.1.4.2.1 LTCC material

A new LTCC material was designed to be able to sinter under 1000°C because of co-firing with copper conductor. The LTCC is composed of lead-free, SiO_2—Al_2O_3—MgO—ZnO—B_2O_3 system glass and ceramic fillers. In order to satisfy electrical and thermal properties, we adjusted the amount of crystalline phases precipitated after sintering [8].

Figure 1.1.15 shows the X-ray diffraction pattern of this LTCC material. The major crystalline phases are $SrAl_2Si_2O_8$, $ZnAl_2O_4$ and SiO_2.

The major characteristics of this material are shown in Table 1.1.6. The coefficient of thermal expansion is 7.5 ppm/°C in the range of 40–300°C. This value is close to that of GaAs chips which are mainly used for microwave applications.

Thermal conductivity and flexural strength and volume resistivity are as good as conventional LTCC material. Because fine leak rate by helium gas through this material is less than 1×10^{-9} Pa m^3/s, it is possible to measure the hermeticity of the package by fine leak method. Dielectric constant and loss tangent were determined by dielectric resonator method using an HP 8757C network

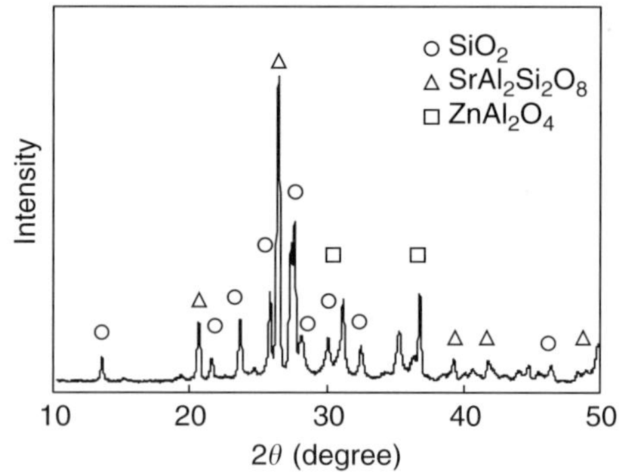

FIGURE 1.1.15 X-ray diffraction pattern of new LTCC.

TABLE 1.1.6 Characteristics of New LTTC

Item	Unit	Developed material
Dielectric constant (30 GHz)	—	6.0
Loss tangent (30 GHz)	—	0.029
Volume resistivity	Ω cm	$\geqq 10^{14}$
Coefficient of thermal expantion (40–300°C)	ppm/°C	7.5
Thermal conductivity	W/mk	1.5
Flexural strength	MPa	200
Fine leak rate	Pa m^3/s	$<10^{-9}$

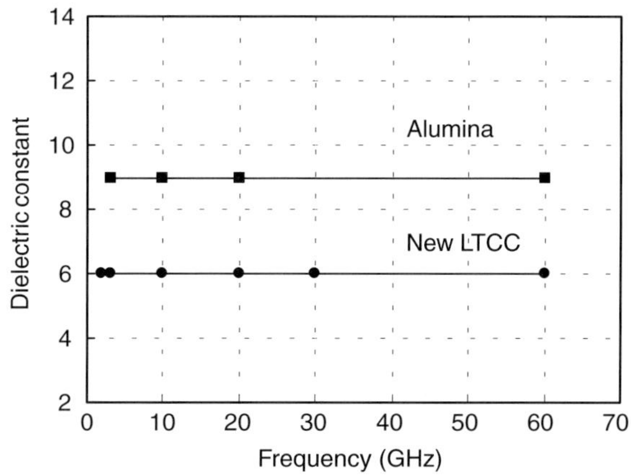

FIGURE 1.1.16 The frequency dependence of dielectric constant.

analyzer at various frequencies [9]. The frequency dependence of dielectric constant of this material is shown in Figure 1.1.16. The dielectric constant is 6.0, which is lower than that of alumina in the range of 2–60 GHz. The frequency dependence of loss tangent is shown in Figure 1.1.17. The loss tangent increases as frequency increases, the value is 0.017 at 2 GHz, 0.0022 at 10 GHz, 0.0029 at 30 GHz, and close to that of alumina, is good for microwave applications.

1.1.4.2.2 Copper metallization

Copper was used for metallization material because of excellent migration resistance. It is important for cofiring process to match the shrinkage behavior of copper metallization to that of LTCC material. Since the shrinkage of copper

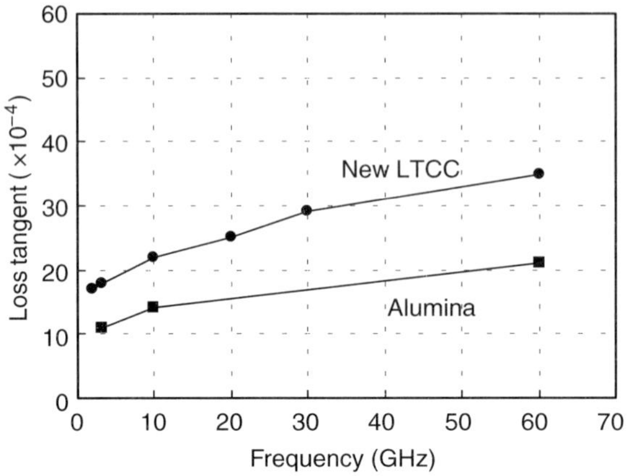

FIGURE 1.1.17 The frequency dependence of loss tangent.

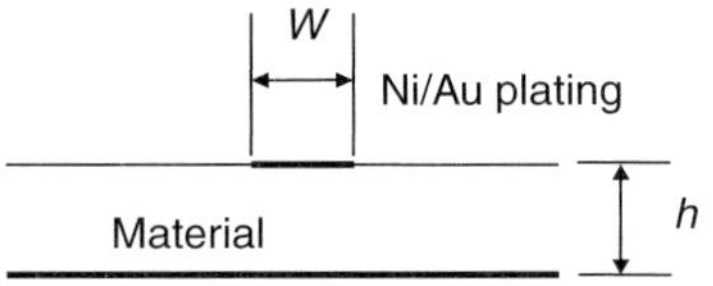

FIGURE 1.1.18 The configuration of a microstrip line. New LTCC ($W = 0.29$ mm, $h = 0.20$ mm); alumina ($W = 0.18$ mm, $h = 0.20$ mm).

starts at lower temperature than that of LTCC material, glass and ceramic fillers are added to copper paste to control the shrinkage behavior of copper metallization [10]. After co-firing, copper metallization is plated LTCC material with nickel–gold or copper–gold.

Adhesion strength of the metallization was measured, and no change was observed after 1000 h aging at 150°C.

Sheet resistance of the metallization was 2.5 mΩ/□ (12 μm thickness), and no increase was observed after 1000 h aging at 150°C. The insulation resistance between lines separated by 100 μm space was more than 10^{12} Ω after 1000 h of HHBT (85°C, 85% RH, 5.5 V).

Transmission characteristics of the LTCC substrate were compared to that of alumina with tungsten metallization. The substrate used in this measurement has a microstrip line configuration (Figure 1.1.18). Figure 1.1.19 shows the measured results of insertion loss (S21). The frequency range was from 0.1 to 50 GHz. The insertion loss of the LTCC was −0.35 dB/cm at 30 GHz, and was

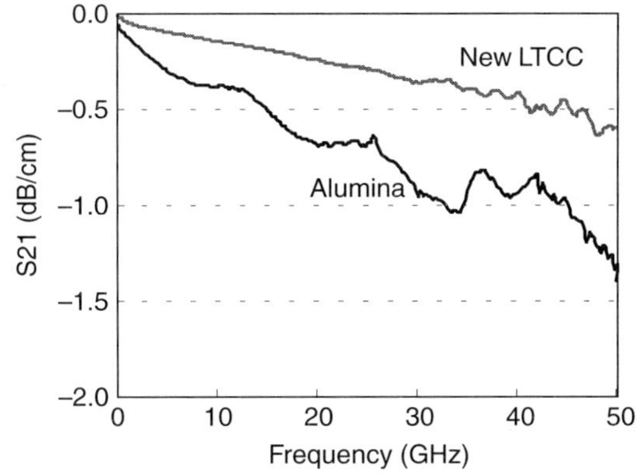

FIGURE 1.1.19 Insertion loss of new LTCC.

less than half of that of alumina. This LTCC material has superior electrical characteristics compared with alumina in microwave frequency range.

1.1.4.3 Reliability of Package

1.1.4.3.1 Hermeticity

Multilayered LTCC package plated with nickel and gold were prepared, and reliabilities of the package were evaluated. The copper sealing pattern was formed by cofiring process. The package was sealed with the Alloy 42 lid (20 mm × 20 mm) by gold–tin eutectic alloy.

The measurement of hermeticity was based on MIL-STD-883E. Fine leak rate by helium gas of the package was less than 5.0×10^{-9} Pa m^3/s and showed no increase after 1000 thermal cycles at −55 to +125°C.

Furthermore, another multilayered package was prepared, and sealed with the iron–nickel–cobalt alloy lid (7 mm × 5 mm) by gold–tin eutectic alloy.

Hermeticity was evaluated after the packages were exposed to several severe conditions. Test conditions and results are listed in Table 1.1.7.

Fine leak rate by helium gas of the packages were maintained less than 5.0×10^{-9} Pa m^3/s after all reliability tests. We confirmed that the LTCC package had excellent hermeticity.

1.1.4.3.2 Resistance change

We measured the sheet resistance of an internal metallization pattern composed of copper lines (0.15 mm width) and copper via (0.15 mm diameter).

TABLE 1.1.7 Reliability of Hermeticity

Contents	Test condition	Result
Mechanical shock	1500 g, 0.5 ms, XYZ-5 times	
Vibration fatigue	20 g, 1.5 ms, XYZ-4 times	$\leq 5 \times 10^{-9}$ Pa m^3/s
Constant acceleration	20 000 g, 1 min	
Salt atmosphere corrosion	0.7%, 35°C, 96 h	

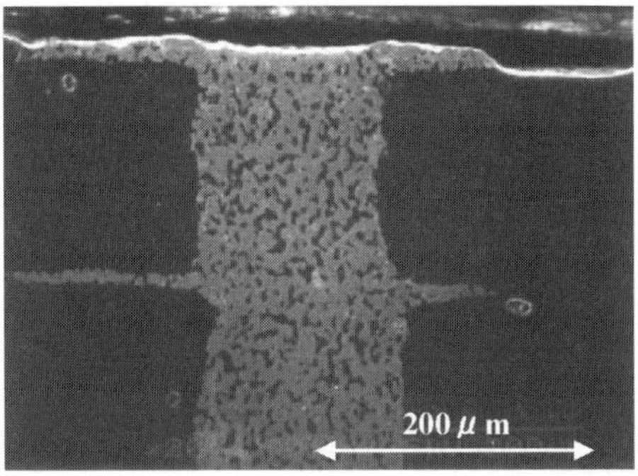

FIGURE 1.1.20 Cross-sectional view of the package.

The resistance variation was less than ±3% after 1000 h aging at 85°C/85% RH, 100 thermal shock cycles at −65 to +150°C, and 1000 h aging at 150°C. Figure 1.1.20 shows the SEM image of cross-sectional view of the package which was used for thermal shock cycle test. No change was observed in microstructure of the inner metallization.

1.1.4.4 Conclusions

We have developed a new LTCC which has excellent electric characteristics in the range of microwave frequency. Sheet resistance of copper metallization cofired with the LTCC was less than 3 m$\Omega/\square$, and the insertion loss of the microstrip line was −0.35 dB/cm at 30 GHz, and was half of that of tungsten metallization.

Multilayered LTCC packages were fabricated using the new material. We have confirmed no change in sheet resistance of metallization pattern of copper trace and via, and in hermeticity of the package after reliability tests.

The new LTCC package has high reliability and is adequate for microwave applications, such as packages for satellite telecommunications and new SAW filter and duplexer packages. Figure 1.1.21 shows a duplexer package using the LTCC material.

1.1.5 FUTURE DEVELOPMENT

Despite many advantages described above, LTCC has some limitations, including low mechanical strength and poor adhesion strength of the metallization. As a result, LTCC packages are sometimes broken during assembly process and the reliability of the metal joints is less than that of HTCC. To address these issues, we have been developing new material for package. This new material has good properties equivalent to those of ordinary HTCC, yet its conductor resistivity is equivalent to that of LTCC (Table 1.1.8).

FIGURE 1.1.21 Duplexer package (5 mm × 5 mm × 1.3 mm).

TABLE 1.1.8 Target Properties of New Material for Package

	New material	A-440	GL-550
Thermal conductivity (W/mk)	15	14	2
Flexural strength (MPa)	400	400	200
Sheet resistance (mΩ/sq.)			
Internal conductors	3	10	<3
Surface conductors	5	10	<3

REFERENCES

1. Kyocera catalog.
2. Hirose, Y., Goto, T., Yamanaka, S., and Tobioka, M. (1997). *Sumitomo Electric Tech. Rev.* **43**: 83–88.
3. Nakayama, A., Nambu, S., Inagaki, M., Miyauchi, M., and Itoh, N. (1996). *J. Am. Ceram. Soc.* **79**: 1453–1456.
4. Marrs, R. (1996). *Electron. Packaging Prod.* **36**: 24–30.
5. Yamaguchi, K., Higashi, M., Shinozaki, M., Hamada, N., Yonekura, H., and Kokubu, M. (1997). "SEMICON Test, Assembly & Packaging 97, Singapore, Proceedings", pp. 225–232.
6. Ikemizu, M., Fukuzawa, Y., Nakano, J., Yokoi, T., Miyajima, K., Funakura, H., and Hosomi, E. (1997). "*IEEE/CPMT International Electronics Manufacturing Technology Symposium*", pp. 447–451.
7. Tsukada, Y. (1997). Japan Institute for Interconnecting and Packaging Electronic Circuits, 2nd symposium on Evaluation Technology, pp. 1–7.
8. Kawai, S. (1999). Development of glass–ceramic material for multilayer substrate in submillmeter-wave application, "Proceedings of 12th Fall Meeting of the Ceramic Society of Japan", p. 397.
9. Terashi, Y. (2000). Development of material for multilayer ceramic substrate with copper wiring in millimeter wave application, "Proceedings of IEMT/IMC Symposium", pp. 382–385.
10. Hamano, S. (1999). Glass ceramic package for mobile communication applications, "Proceedings of IEMT/IMC Symposium", pp. 115–119.

Handbook of Advanced Ceramics
S. Sōmiya *et al.* (Eds.)

2.1 Semiconductive Ceramics

HIDEAKI NIIMI and YUKIO SAKABE
Murata Manufacturing Co., Ltd., Yasu 520-2393, Japan

2.1.1 PTC THERMISTORS

Barium titanate ($BaTiO_3$) is a ferroelectric material with a high dielectric constant and high insulation resistance. Therefore, it has been widely used in the electrical industry for ceramic capacitors since its discovery in 1943. The insulating $BaTiO_3$ ceramic is converted into a semiconductor by adding a small amount of rare earth metal oxide such as Sm_2O_3, CeO_2, Y_2O_3 and La_2O_3. In 1955, unusual temperature dependence of resistance above the Curie temperature of semiconductive $BaTiO_3$ ceramics was discovered.

The resistance of this semiconductor called the positive temperature coefficient (PTC) thermistor drastically increases above the Curie temperature (T_C), up to the temperature (T_n) where the resistance reaches its maximum value. The characterized temperature is divided into three regions (I, II, and III in Figure 2.1.1) according to the resistance behavior.

This drastic increase of resistance above the Curie temperature was a very exciting discovery not only for practical application of thermistors but also for fundamental research of conduction mechanisms.

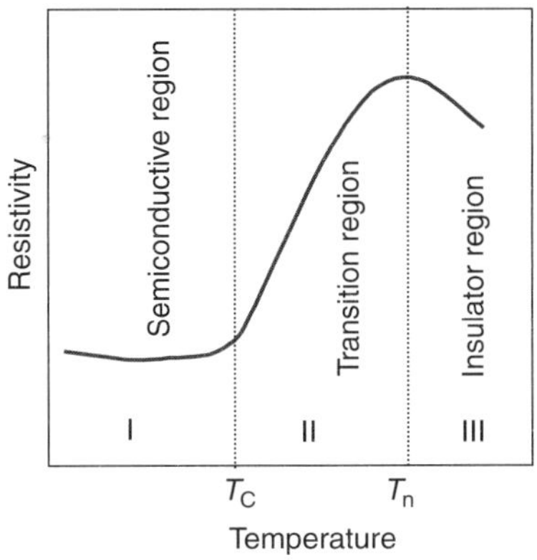

FIGURE 2.1.1 Characterized temperature regions of PTC ceramics.

2.1.1.1 Conduction Mechanisms

The conduction mechanisms in the regions I–III are explained as follows.

In the temperature region I($T < T_C$), the resistivity of PTC thermistor is in the range of 10–10^6 Ω cm. To produce semiconductive $BaTiO_3$, a small amount of rare earth metal ions (e.g. Sm^{3+} or La^{3+}) are substituted at the Ba^{2+} site, or Nb^{5+} and Ta^{5+} ions are substituted at the Ti^{4+} site. These ions provide conductive electrons as shown in the following equations, using the defect notation of Kröger and Vink.

$$\tfrac{1}{2}La_2O_3 \rightarrow La_{Ba} + O_0 + \tfrac{1}{4}O_2 + e \quad (1)$$

$$\tfrac{1}{2}Nb_2O_5 \rightarrow Nb_{Ti} + 2O_0 + \tfrac{1}{4}O_2 + e \quad (2)$$

In the region II above the Curie temperature, resistance across the grain boundary increases exponentially with increasing temperature.

The increase of resistance corresponds to the decrease of spontaneous polarization (P_s) of $BaTiO_3$ due to the phase transition from the ferroelectric tetragonal phase to the paraelectric cubic phase.

Under the influence of deep acceptor states on grain boundary, the double Schottky barrier is generated as shown in Figure 2.1.2. According to Poisson's equation, its barrier height is given by

$$\phi = \frac{e^2 n_D b^2}{2\varepsilon\varepsilon_0} \quad (3)$$

$$b = \frac{n_S}{2n_D} \quad (4)$$

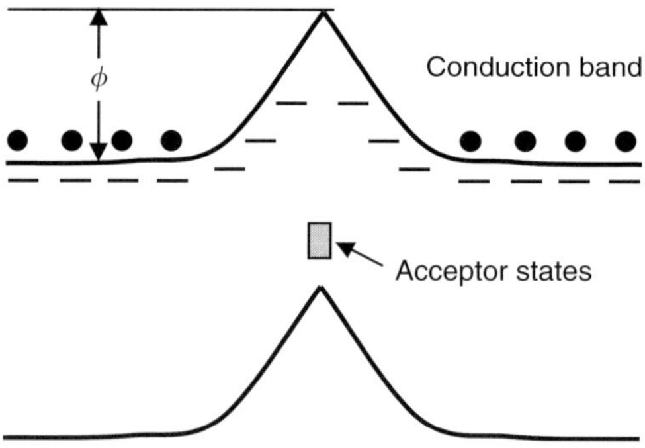

FIGURE 2.1.2 Band model of double Schottky barrier at grain boundary.

where e is the elementary electric charge, n_D the concentration of bulk donor states, b the width of the barrier, n_S the concentration of acceptor states, ε the dielectric constant, and ε_0 the dielectric constant of vacuum.

The acceptor states on the grain boundary are considered to be produced by barium vacancy [1], transition metals [2] or O_2^- [3].

The strong internal field caused by P_s depresses the potential barrier height ϕ in region I. The gradual decrease of P_s and dielectric constant, expressed by the Curie–Weiss law, cause the potential barrier height to recover, which results in an increase of resistance in region II.

According to measurements of resistance–temperature characteristics across single grain boundaries, the PTC characteristic differs from grain boundary to grain boundary, as shown in Figures 2.1.3 and 2.1.4. The drastic resistance

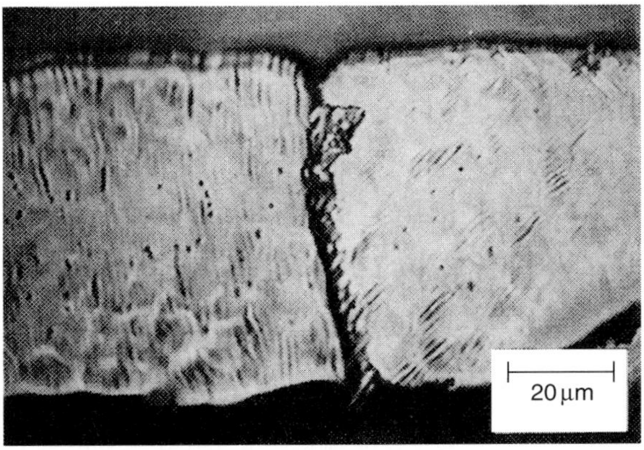

FIGURE 2.1.3 Sample for measurement of PTC characteristics across single grain boundary [4].

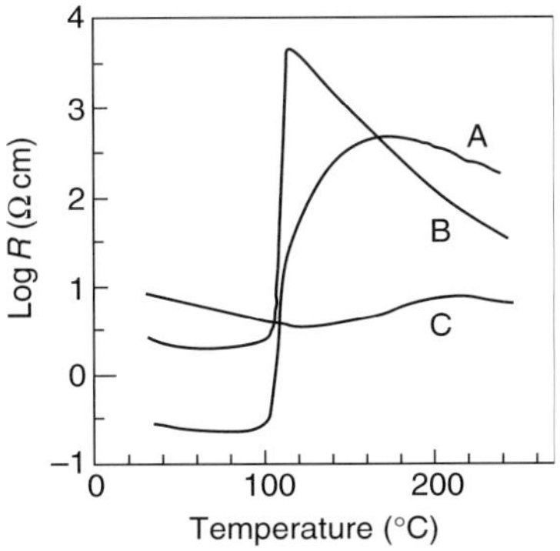

FIGURE 2.1.4 Typical three different PTC characteristics across single grain boundaries [6].

increase above the Curie temperature is dependent on ferroelectric domains and crystal orientations near grain boundary [4–6].

In the temperature region III ($T > T_n$), the electrons that overcome the double Schottky barrier increase with temperature, and resistance decreases from maximum resistance.

2.1.1.2 Manufacturing Process

Additives and their effects on PTC characteristics are listed in Table 2.1.1. The transition temperature T_C can be lowered or elevated from its original value (120°C) by substitution of Sr^{2+} and Pb^{2+} at the Ba^{2+} site. Figure 2.1.5 shows the typical resistivity–temperature curves of PTC thermistors with various T_C.

$BaCO_3$, $SrCO_3$, Pb_3O_4, TiO_2 and donor dopant (e.g. La_2O_3, Sm_2O_3, Y_2O_3 and Nb_2O_5) are used as starting materials. Manufacturing processes of PTC thermistor are almost the same as those in the electronic ceramics industry.

To control the resistance and the temperature coefficient within the exact range, much attention is paid to the firing temperature and ensuing cooling rate. The impurity of rare materials and contamination in manufacturing process must be decreased, because these increase the resistance at room temperature. In particular, Fe and Al strongly affect the resistivity of PTC thermistor.

Non-precious metal electrodes such as Ni, Zn, and Al provide ohmic contact with PTC thermistor, which is n-type semiconductor ceramics. In–Ga alloy also makes ohmic contact, and is used for experimental samples.

TABLE 2.1.1 Additives of PTC Thermistor

Additives	Role
Pb, Sr, Sn	Shifter: Adjusting Curie temperature of $BaTiO_3$
La, Nd, Ce, Sm, Y, Sb, Nb, W	Donor: Semiconduction and control of resistivity
Si, B	Sintering aids: To lowering sintering temperature To suppress abnormal grain growth
Mn, Cr	Acceptor: Enhancement of PTC characteristics

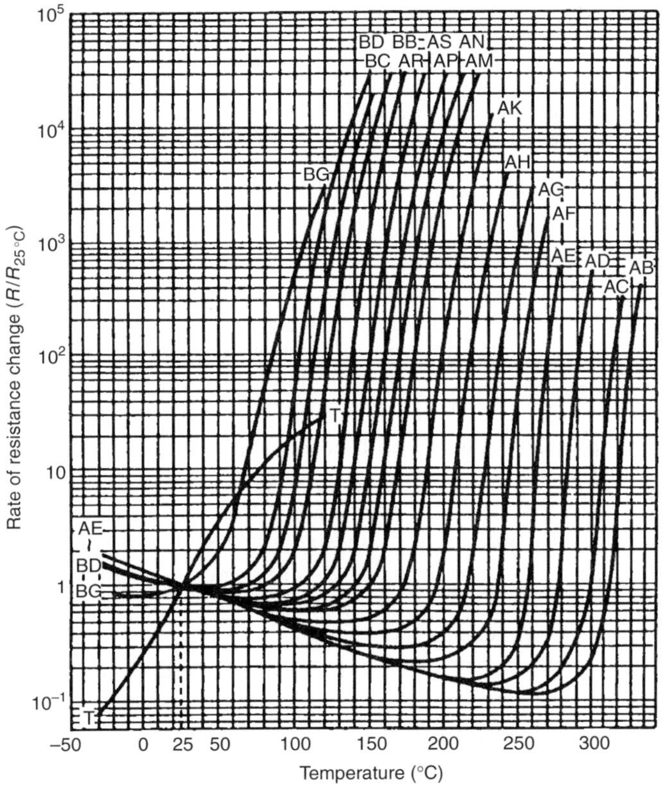

FIGURE 2.1.5 Resistance–temperature characteristics of PTC thermistors with various T_C.

2.1.1.3 Applications

Three basic functions and applications are shown in the Table 2.1.2. PTC thermistors are used in a lot of electric products, such as color televisions, refrigerators, hot-wind heaters, and personal computers.

2.1.1.4 Recent Development and Trend

To meet the requirements of surface mounting technology (SMT) chip-type PTC thermistors have been developed and applied to hybrid circuits. As another

TABLE 2.1.2 Functions and Applications of PTC Ceramics

Functions	Applications
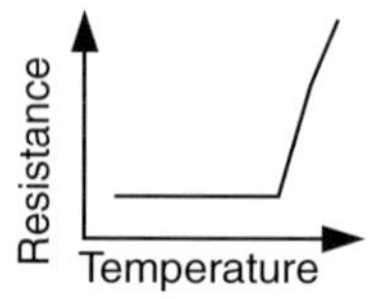 Resistance–temperature	Temperature sensing Temperature indicating Temperature measurement Overheat protection for Power transistor, transformers
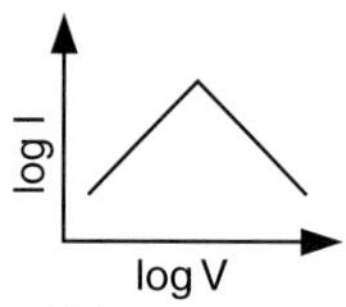 Voltage–current	Protection against overcurrent Constant temperature heating Electrical heater, curing iron
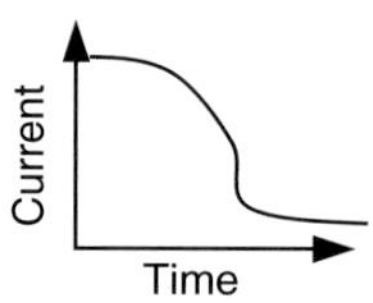 Current–time	Starting motors for compressor Degaussing of color TV Delay time element

trend, the lower resistance at room temperature is required to limit the current in low-voltage circuits. Lowered 0.1 Ω PTC thermistors will be manufactured.

2.1.2 NTC THERMISTORS

The negative temperature coefficient (NTC) thermistors are semiconductive materials whose resistance decreases with increasing temperature as shown in Figure 2.1.6 with other thermistors. Temperature dependence of resistance is given by

$$R = A \exp(B/T) \tag{5}$$

$$B = E/k \tag{6}$$

where A is a constant, B a thermistor constant, E the activation energy, and k Boltzmann's constant.

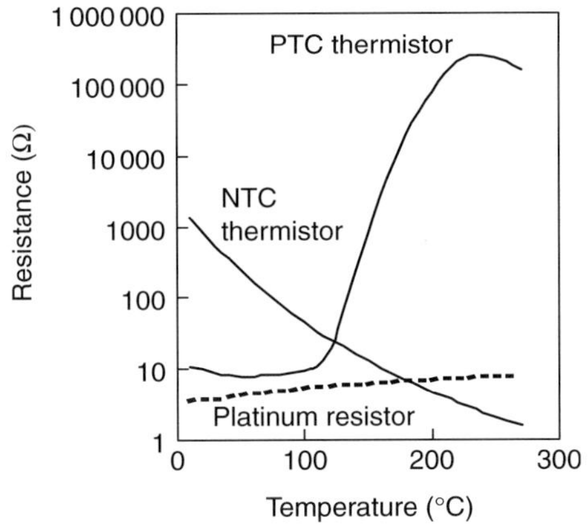

FIGURE 2.1.6 Temperature dependence of different types of thermistors in contrast to a platinum resistor.

2.1.2.1 Conduction Mechanism

NTC thermistor usually consists of transition metals (Cu, Fe, Co, Ni, etc.) spinel manganites. The conductivity is due to the transfer of electrons between Mn^{3+} and Mn^{4+} ions. The resistance and thermistor constant is dependent on the composition, purity, cation distribution, and crystal structures.

2.1.2.2 Applications

NTC thermistors are used as temperature compensation, temperature sensing, and surge current suppression devices. All of these applications are based on the resistance–temperature characteristics of NTC thermistors. Although various thermistor constant B and resistivity are required for many applications, these values are obtained within the limits as illustrated in Figure 2.1.7, because the thermistor constant B is dependent on the resistivity.

The chip-type NTC thermistors have become popular because of their suitability for SMT. In particular, the chip thermistor with inner electrode offers high resistance accuracy and high reliability.

2.1.2.3 Manufacturing Process

Mn_3O_4, NiO, Co_2O_3, and Fe_2O_3 are used as starting materials. NTC thermistors are produced by the general method of the manufacturing of electroceramics.

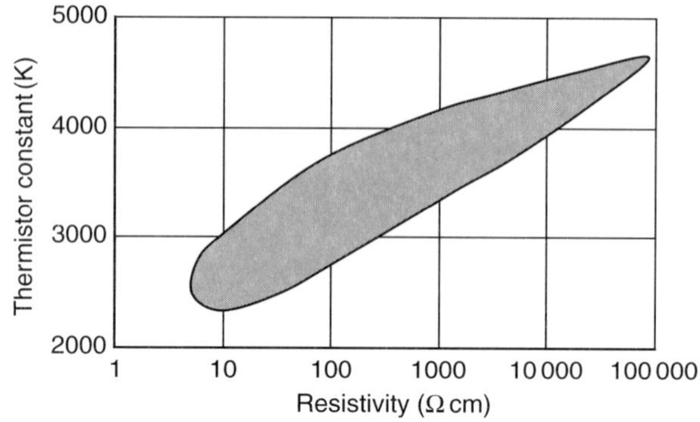

FIGURE 2.1.7 Relationship between thermistor constant and resistivity.

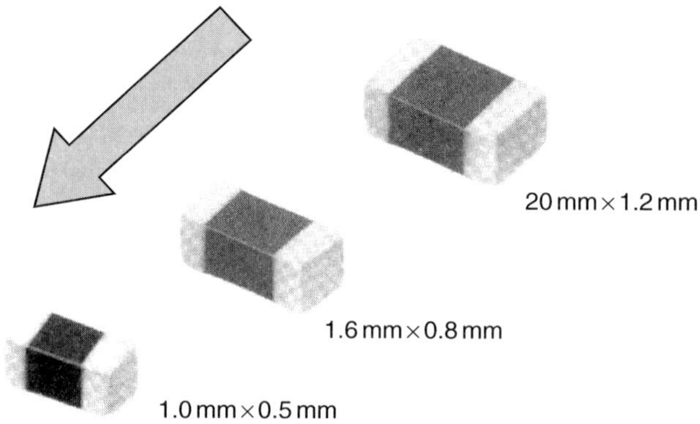

FIGURE 2.1.8 Miniaturization of chip-type NTC thermistor.

Precious metals such as Ag, Pd, and Pt are used for electrodes of NTC thermistor, which is mainly p-type semiconductor ceramics.

2.1.2.4 Recent Development and Trend

The trends in modern NTC ceramics have resulted in thermistor development in three directions. The first trend is the improvement of device stability. The second is the high accuracy of resistance and *B* constant. The third is the miniaturization of the chip-type NTC thermistor. Miniaturized 1.0 mm × 0.5 mm chips have been manufactured, as shown in Figure 2.1.8.

2.1.3 CERAMIC VARISTORS

Metal oxide varistors are ceramic semiconductive devices having highly non-linear current–voltage characteristics, as shown in Figure 2.1.9, expressed as

$$I = (V/C)^{\alpha} \tag{7}$$

$$\alpha = \frac{dI/I}{dV/V} = \frac{\log I_2 - \log I_1}{\log V_2 - \log V_1} \tag{8}$$

where α is the nonlinear exponent, C the constant corresponding to the resistance, and V_1 and V_2 are the voltages at the currents of I_1 and I_2, respectively. C is conveniently given by V_c called varistor voltage, that is, a voltage per unit length (V/mm) when 1 mA/cm^2 of current flows through the body. Thus, the ceramic varistor is characterized by the non-linear exponent α and varistor voltage V_c.

Two types of ceramic varistors are manufactured. Zinc oxide based ceramic varistors were developed in 1970. They exhibit a high non-linearity on voltage–current characteristics. Their α value is in the range of 40–50, and the V_c adjustable to values in the range from 50 to 250 V/mm. Strontium titanate based varistors were developed in 1980. The feature of these varistors is their larger electrostatic capacitance compared with ZnO varistors. The $SrTiO_3$ ceramics are

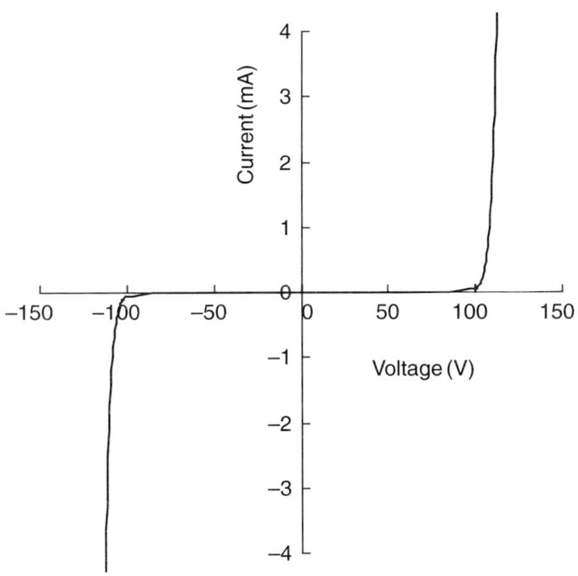

FIGURE 2.1.9 Typical *V–I* characteristic of ceramic.

essentially dielectrics with a die1ectric constant of 320, which is much higher than that of ZnO.

2.1.3.1 High Non-linearity of ZnO Varistor

The ZnO grains are n-type semiconductors. The intergranular traps are formed in this grain boundary due to the presence of Bi_2O_3 or Pr_2O_3 and transition metal oxides, which cause the double Schottky barrier.

Several models have been proposed to explain highly non-linear *V–I* characteristics [7]. One of the most reasonable models is illustrated in Figure 2.1.10 [8].

When a voltage *V* is applied across the double Schottky barrier, the barrier height ϕ decreases with increase of the voltage. A small fraction of electrons can penetrate the depletion region. These electrons, which are accelerated in the depletion region, can excite the valence electrons to conduction band. Therefore, holes recombine with the trapped electrons at the grain boundary, and decrease the barrier height.

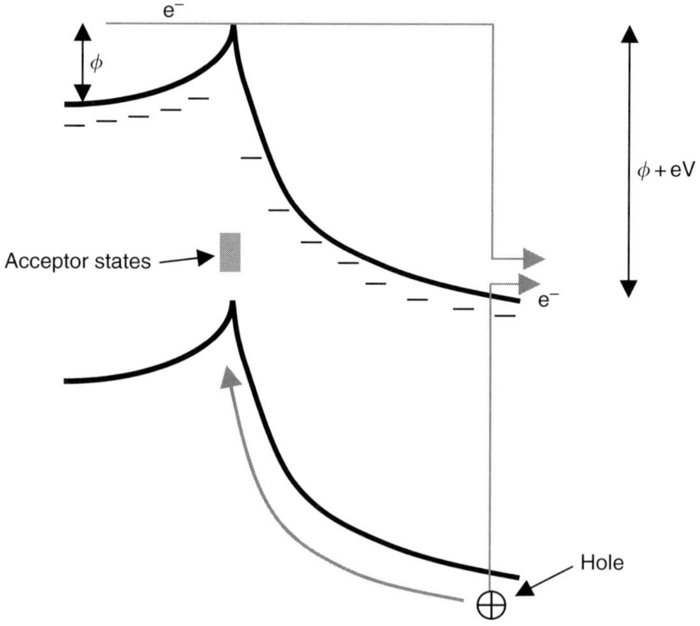

FIGURE 2.1.10 Energy band diagram near a grain boundary under applied voltage *V*.

2.1.3.2 Manufacturing Process

The effect of additives on varistor properties of ZnO varistors are listed in Table 2.1.3. The varistor voltage V_c is dependent on the number of grain boundaries between a couple of electrodes, because the varistor voltage across a single grain boundary is constant value ($\sim$3 V) at each boundary. To obtain the varistors with the various voltages V_c, the grain size are controlled by firing temperature or additives, such as B and Sb.

Strontium titanate based varistors are manufactured by firing in a reducing atmosphere and reoxidized on only grain boundary, like a boundary-layered ceramic capacitor.

2.1.3.3 Applications

Metal oxide varistors are mainly used in circuits for protection against inductive surges, very short spike noise, or power surges. They result to protect circuit simply by inserting between surge entrance line and ground lines shown in Figure 2.1.11. A varistor should be chosen that have a varistor voltage V_c slightly higher than the signal voltage applied to the load to be protected. The varistor is insulator in normal operation where the applied voltage is lower than V_c.

TABLE 2.1.3 Additives of ZnO Varistor

Additives	Role
Bi, Pr, Co, Mn	Basic structure and enhancement of non-linearity
Sb	Suppressing grain growth (= elevating a varistor voltage)
B	Promoting grain growth (= lowering a varistor voltage)
Al, Ga, In	Lowering resistivity of grain

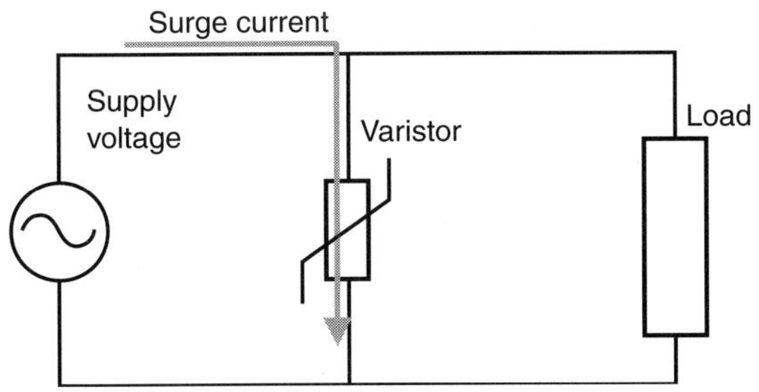

FIGURE 2.1.11 Typical application of ZnO varistor as a transient protective device.

If a transient pulse, whose voltage is higher than V_c, is incident, the current through varistor rapidly increases, resulting in a conducting shunt path for the incident pulse.

ZnO-based varistor have become popular because of the high non-linearity on voltage–current characteristics.

2.1.3.4 Trends

The lower varistor voltage is required in the low-voltage circuits. Varistors with 10 V of varistor voltage have been manufactured. Another trend is a lowering of capacitance for high-frequency circuits.

REFERENCES

1. Daniels, J., and Wernicke, R. (1976). New aspects of an improved PTC model. *Philips Res. Rep.* **31**: 544–559.
2. Hennings, D. (1988). *Surface and Near-Surface Chemistry of Oxide Materials*, pp. 479–505, Nowotny, J., and Dufour, L. C., eds., Amsterdam: Elsevier Science.
3. Takahashi, T., Nakano, Y., and Ichinose, N. (1990). Influence of reoxidation on PTC effects of porous $BaTiO_3$. *J. Jpn. Ceram. Soc.* **98**: 879–884.
4. Matsunaga, T., Shimooka, H., Takahashi, S., and Kuwabara, M. (1995). Organizations and PTCR characteristics of single grain boundaries in barium titanate ceramics. "Proceedings of Fall Meeting of The Ceramic Society of Japan," p. 276.
5. Hayashi, K., Yamamoto, T., and Sakuma, T. (1996). Grain orientation dependnce of the PTCR effect in niobium-doped barium titanate. *J. Am. Ceram. Soc.* **79**: 1669–1672.
6. Matsuda, H., and Kuwabara, M. (1996). PTCR effect across single grain boundary in $BaTiO_3$—in the light of conventional barrier model, "The 16th Electronics Division Meeting Organized by the Ceramic Society of Japan", pp. 99–100.
7. Mahan, G. D. (1979). Theory of conduction in ZnO varistors. *J. Appl. Phys.* **50**: 2799–2813.
8. Pike, G. E., Kurtz, S. R., and Gourley, P. L. (1985). Electroluminescence in ZnO varistors: evidence for hole contributions to the breakdown mechanism. *J. Appl. Phys.* **57**: 5512–5517.

Handbook of Advanced Ceramics
S. Sōmiya *et al.* (Eds.)

3.1 Ionic Conductors/ Oxygen Sensors

TESSHO YAMADA
NGK Spark Plug Co., Ltd., Engineering Department, 2808 Iwasaki, Komaki-shi 495-8510 Aichi, Japan

3.1.1 OXYGEN SENSORS FOR AUTOMOBILES

Oxygen sensors for automobiles are utilized for several kinds of the exhaust gas purification system with three-way catalyst and are very important components to control the combustion for the internal combustion engine.

3.1.1.1 CERAMIC FOR THE OXYGEN SENSORS

As is known, zirconia changes from monoclinic and tetragonal form due to crystal transformation around 1000°C [1]. In order to stabilize the crystal transformation, divalent and trivalent oxides, for example, CaO, MgO, Y_2O_3 are mixed as soluble constituents. Continuous calcination, milling and mixing of compounds, forming and sintering, in sequential order of the production process, are performed to yield stabilized cubic zirconia. This method is used to prevent volume change due to the crystal transformation. But recently zirconia high-strength material are in the process of development. A structure with high mechanical strength was accomplished by partial stabilization composition. Basic performance of sensor element of the oxygen sensor depends on its electrical characteristics, that is, it functions as an oxygen concentration cell. Conditions of thermal resistance requirement on exposure to the high temperature exhaust gas up to 1000°C, gas tightness against the exhaust gas pressure, and intensity warranty to endure mechanical shocks, are some conditions that the sensor must satisfy. According to the circumstance, electrical conduction [2], mechanical toughness, and easily obtainable partial stabilization of zirconia with Y_2O_3, that is the most widely selected sensor.

3.1.1.2 Structure and Operation Principle for the Oxygen Sensors

Zirconia solid electrolyte sensor element is formed in thimble shape. Pt electrode layers as a thin film are adhered inside and outside of the sensor element by plating method. The portion of the sensor element is shown in Figure 3.1.1. The partially stabilized zirconia solid electrolyte sensor element is assumed as one wall. Along the border of the wall, oxygen in ionic form is transferred in a direction that reduces the difference of oxygen partial pressure between chamber A and chamber B. The series construction works as a battery due to the transfer.

At this time, electromotive force is generated. Therefore, it is called an oxygen concentration cell. The electromotive force is given by the Nernst equation:

$$\mathrm{E} = (RT/4F)\ln(P_\mathrm{a}/P_\mathrm{b})$$

where R is the gas constant, T the absolute temperature, F the Faraday constant, P_a the oxygen partial pressure in the high oxygen concentration chamber A (normally, it shows the oxygen partial pressure in the ambient air), and P_b the oxygen partial pressure in the low oxygen concentration chamber B (normally, it shows the oxygen partial pressure in the exhaust gas).

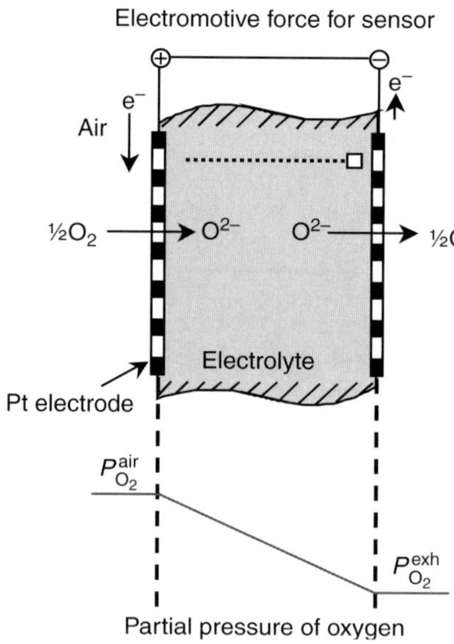

FIGURE 3.1.1 The operation principle for oxygen sensor.

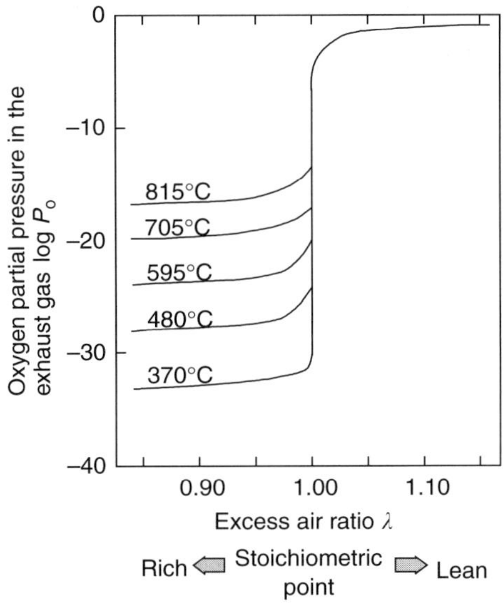

FIGURE 3.1.2 The relationship between oxygen partial pressure and λ.

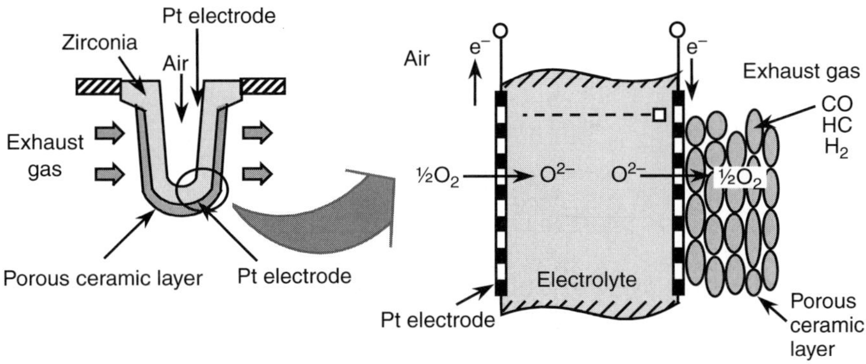

FIGURE 3.1.3 Schematic figure of the Sensor element.

The relationship between the oxygen partial pressure and actual excess air ratio (λ) is shown in Figure 3.1.2.

A schematic figure of the sensor element is shown in Figure 3.1.3. The outer surface is exposed directly to the exhaust gas, whereby plasma jet coating layer (called overcoat layer) is arranged as a ceramic protection layer [3]. The layer has Pt electrode retention for the sensor element surface, Pt sublimation velocity reduction. Further, the layer prevents against the attack of poisoning

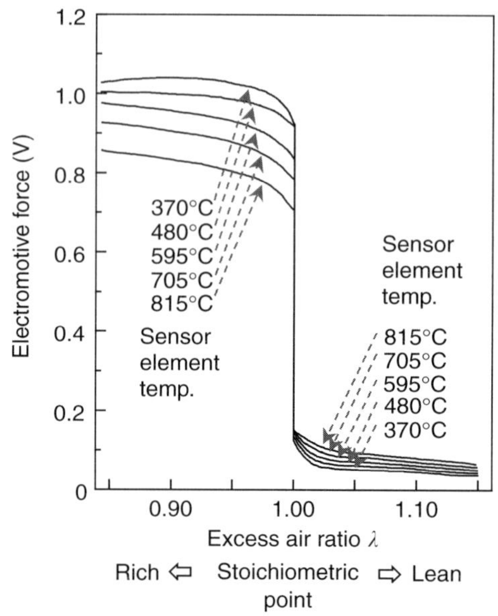

FIGURE 3.1.4 The relationship between EMF and λ.

substances in the exhaust gas, direct shock to the Pt electrode and sensor element. Additionally, it is a diffusion passage for the exhaust gas. It is affected by the response of the important characteristics of the sensor as shown later. Therefore, the layer density, the porosity ratio, and the layer thickness, etc. are controlled.

Exhaust gas is not necessarily burnt completely, even though it is burnt at the stoichiometric point (theoretical air/fuel ratio). The non-equilibrium gases, which are imperfect combustion gases, react and change to the equilibrium status by the catalylic function of the Pt electrode on the sensor element surface. The sensor then detects the remaining oxygen volume. Moreover, it is found that the oxygen partial pressure varies radically, at the border where $\lambda = 1$ as shown in Figure 3.1.2. Due to the change in electromotive force, approximately 1 V output arises and detection control can be $\lambda = 1$ as the oxygen sensor (see Figure 3.1.4.).

3.1.1.3 Oxygen Sensor Structure

The sensor element is formed in a thimble shape and is a dense sintered structure. Pt electrode is adhered inside and outside of the sensor element by plating.

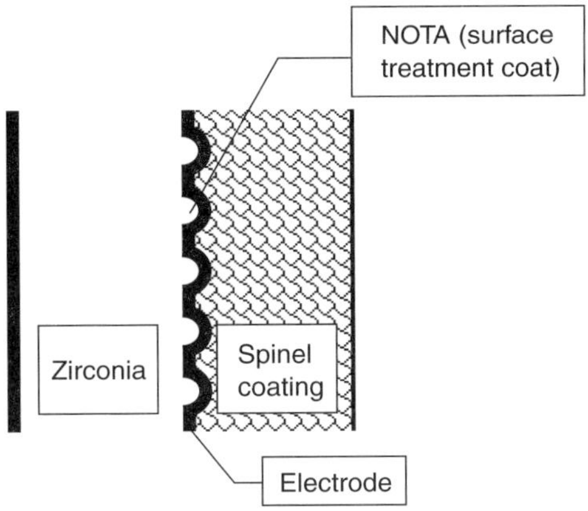

FIGURE 3.1.5 Schematic figure of sensor element cross-section.

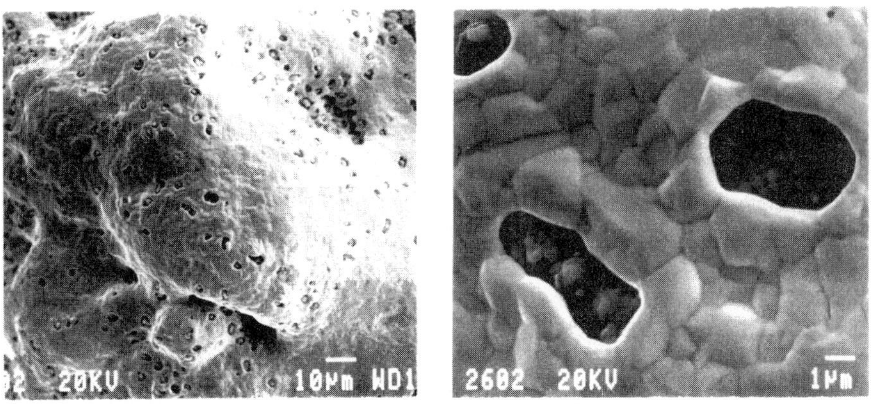

FIGURE 3.1.6 Photographs of element surface without plasma spinel coating (Zirconia ceramic surface is covered by Pt thin film coating layer).

In order to achieve a robust electrode, formed particle is coated with spray dry of approximately 10–100 μm, the same as the sensor element composition, on the outer surface of the sensor element and then sintered (see Figure 3.1.5).

Due to this process, a concavo-convex surface is formed on the sensor element and it is possible that robust adhesion and effective surface area of Pt electrode increases. Pt electrode has several small pores because of heat treatment. A large amount of three-phase boundary, which is called triple point of Pt electrode material, zirconia solid electrolyte, and atmosphere gases, is created, which allows easy electrode reaction (see Figure 3.1.6).

Pt of the electrode material is deposited on the ceramic coating layer so as not to lose the electric characteristics. The coating layer is deposited on the sensor element with plasma jet coating of the spinel powder. It needs to meet the following conditions as an important protection layer: (i) it should not be peeled off easily; and (ii) it should have the porosity such that it is not plugged easily by poisonous substances in the exhaust gas. Therefore, specification are such that it should not be too thick, or too thin, and must be less dense.

The process of the sensor element manufacturing is shown below:

ZrO_2 powder + Y_2O_3 powder
↓
Mixing and milling → Calcination → Milling compounds
→ Mixing compounds → Spray dry → Forming → Grinding
→ Formed particle (NOTA) coating→ Sintering → Plating
→ Heat treatment → Spinel coating

3.1.1.4 Heater Structure and Manufacturing Process for Oxygen Sensors

Sensor element temperature needs to rise to several hundreds degree centigrade in order to operate as oxygen sensor. Zirconia solid electrolyte is an insulator at normal room temperature and it cannot detect the gases. Therefore, gas detection is not possible until the necessary temperature of the sensor element is reached. Because the time of gas detection is short, a heater is embedded in the sensor element to achieve fast light off. The rod type heater is arranged inside of the side of the reference ambient air of the thimble type sensor element, and then the sensor element can be heated. Heating material is printed on the alumina sheet and is laminated. After that, it is wrapped upon the rod type ceramic and then sintered. Extension of the electrode is structured by terminal.

The heater manufacturing process is shown below:

Heating material printing → Laminating → Wrapping → Sintering
→ Terminal brazing

3.1.1.5 Assembling Structure for Oxygen Sensors

In the structure of the zirconia oxygen sensor, the sensor element is stored in the metal shell consisting of a metal with heater resistance. The sensor element is supported by talc in the metal shell by filling press. The gas detection portion which is exposed to the exhaust gas, is covered with a protection tube made up of a metallic material with high temperature durability and is oxidation resistant. Overall, the design must take into consideration high temperature resistance, mechanical shock resistance, thermal shock resistance, water resistance, chemical resistance, and environmental conditions. A passage is built in to lead to the ambient air inside of the thimble type sensor element because the oxygen in ambient air acts as the reference. In order to preserve the ambient air, the filter material, which is water repellent, provides air ventilation (i.e. it allows only gas and not water to pass through), and has good heat resistance, is used on the upper sensor body.

In terms of practical usage, long-term durability is an objective with a warranty for 10 years or more than 100 kmile (160 000 km). Therefore, over time the mechanical strength should not reduce and Pt electrode material should be retained without undergoing sublimation and consumption. These factors help in sustaining the durability of the ceramic coating layer on the zirconia.

The general heat resistance requirements for sensor components are shown in Figure 3.1.7. Sensor layout is transferred further upstream of the catalyst to be a target of fast light off against LEV and ULEV regulations. Moreover,

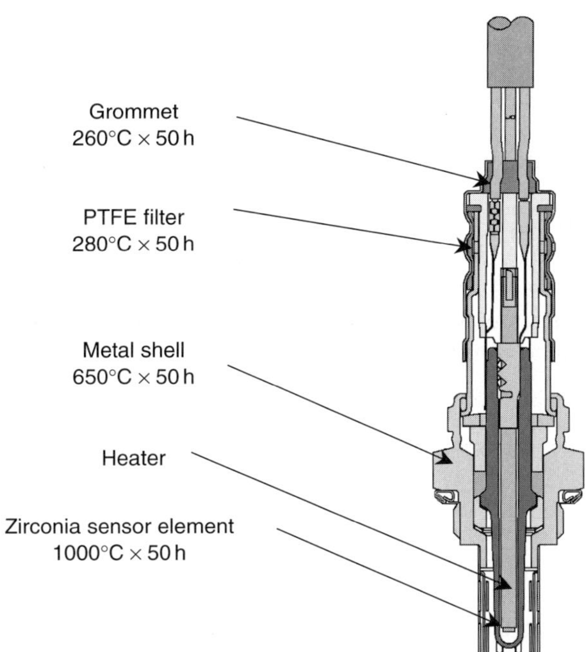

FIGURE 3.1.7 Heat resistance requirements for sensor components.

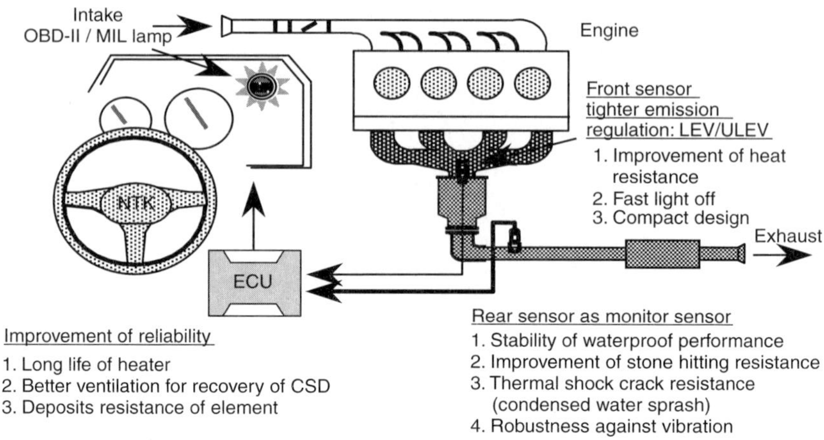

FIGURE 3.1.8 Zirconia sensor design for future requirements.

it must satisfy with requirements regarding installation in the vicinity of the engine exhaust outlet as shown in Figure 3.1.8. Stringent requirements such as water resistance, etc. are applied for installation downstream of the catalyst, which corresponds to the OBD-II regulation.

3.1.1.6 Air/Fuel Ratio Control by Oxygen Sensor

Electromotive force of the oxygen sensor is obtained as the highest output change at the stoichiometric air/fuel ratio. A three-way catalyst (TWC) is used for exhaust gas purification of the automobile internal combustion system. The conversion efficiency for purification is the best at the stoichiometric air/fuel ratio as shown in Figure 3.1.9. Conversion of HC, CO, and NO_x is achieved at the vicinity of the stoichiometric air/fuel ratio, which is called "window", because this region gives the higher output change of the sensor and the fuel injection quantity is controlled.

Zirconia is an insulator at room temperature. The heater is heated up until the sensor element temperature is reached to the region where ion conduction is smooth. Warm up properties of the sensor with 3.3 Ω heater until the output amplitude can be controlled as started is shown in Figure 3.1.10 [4].

Recently on board diagnosis (OBD) system requires the detection of conversion efficiency change of a TWC and an oxygen sensor is installed even in the downstream of the catalyst. Rear oxygen sensor is installed in a cleaner exhaust gas than the front oxygen sensor. Therefore, the sensor itself is hardly deteriorated and it is utilized for lambda point compensation of the front oxygen sensor.

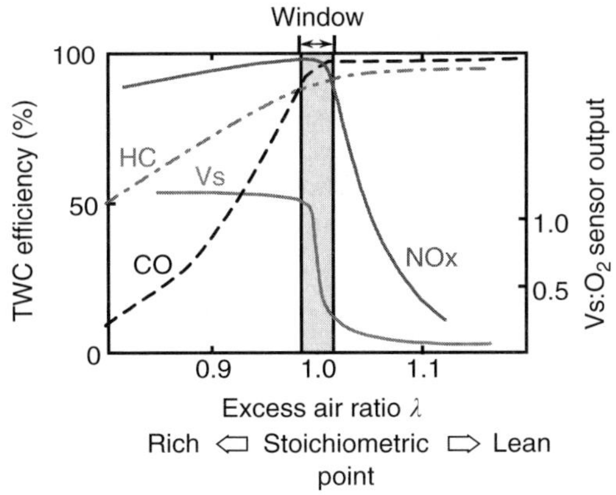

FIGURE 3.1.9 Conversion efficiency for three-way catalyst.

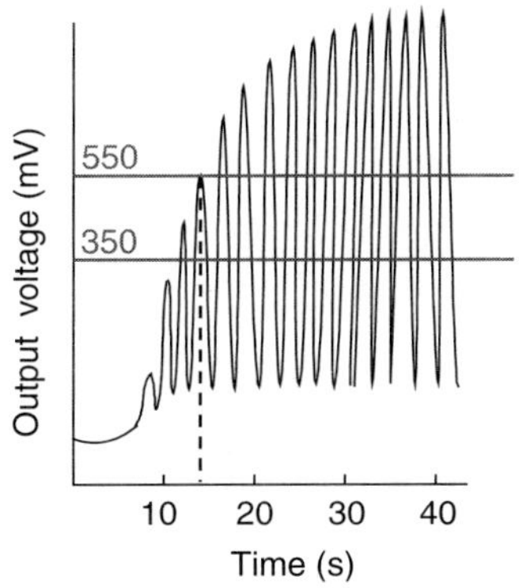

FIGURE 3.1.10 Warm up properties (light off property) with a 3.3 Ω heater.

3.1.2 THICK FILM TYPE OXYGEN SENSOR

Recently, faster activity is a requirement, beside the above mentioned thimble type oxygen sensor. The requirement is resolved due to the more compact sensor element and multilayered co-fired structure of the heater and the sensor element. By way of an example, the element portion of the thick film type oxygen sensor is described [5]. Zirconia powder for the partial stabilization is mixed with an organic binder and a solvent. A sheet is formed and electrode pattern is printed by Pt ink. For the heater, a pattern is printed by Pt ink and the heater is embedded in the alumina or zirconia sheet. Manufacturing process of the thick film type sensor element is shown below:

Zirconia powder + Organic binder mixing → Sheet forming
→ Punching → Zirconia sheet completed → Pt electrode printing
→ Insulation layer printing → Lamination → Dewax → Sintering

It is similar to manufacturing of IC package for electronic components. Typical sensor element structure for the thick film type oxygen sensor is shown in Figure 3.1.11.

The sensor is required to give an output identical to the thimble type oxygen sensor. Therefore, the sensor element as shown in Figure 3.1.11 is created with an oxygen reference in the square shaped chamber. To produce the space, the cavity is filled with carbon material at the sensor element printing process and the chamber is not allowed to collapse at the next lamination process. Finally, the carbon material is burned off on firing and the chamber is retained. However, a problem remains related to the thermal efficiency of the chamber. The ambient air induction passage must be retained on the structure. Recently, a method has been developed in which a small current is flowed to the solid electrolyte and oxygen is obtained from the oxidized gases as water vapor and CO_2, etc. contained in the exhaust gas. The method does not need the ambient air induction passage in particular. This is called as self-generative oxygen reference. Therefore, the sensor element becomes compact and there are no problems caused by ambient air pollution by generation of gases from the sensor component materials. The structure is shown in Figure 3.1.12 with the supplied electric circuit.

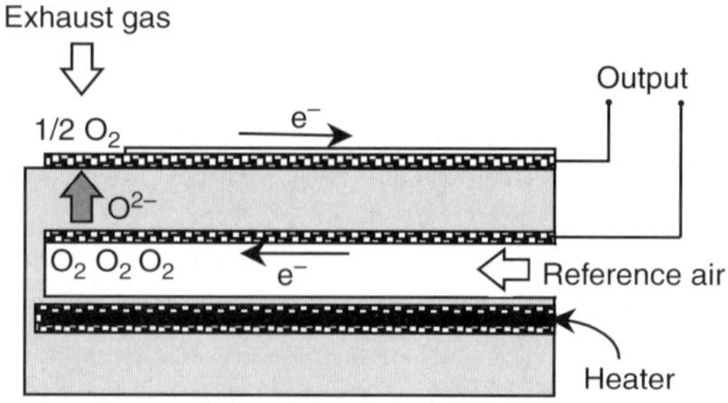

FIGURE 3.1.11 Sensor element structure for thick film type (conventional).

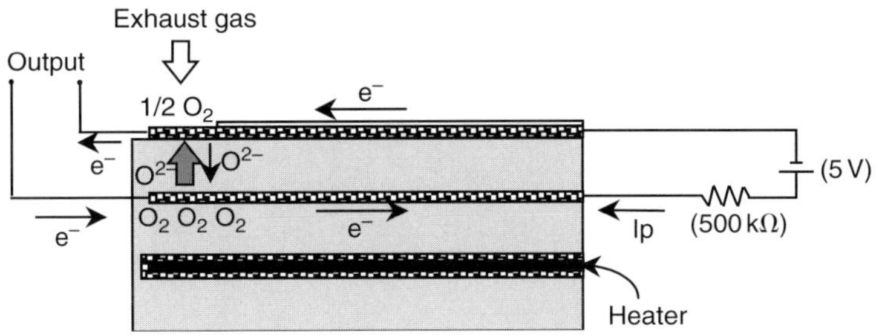

FIGURE 3.1.12 Sensor element structure for thick film type (self-generative oxygen reference).

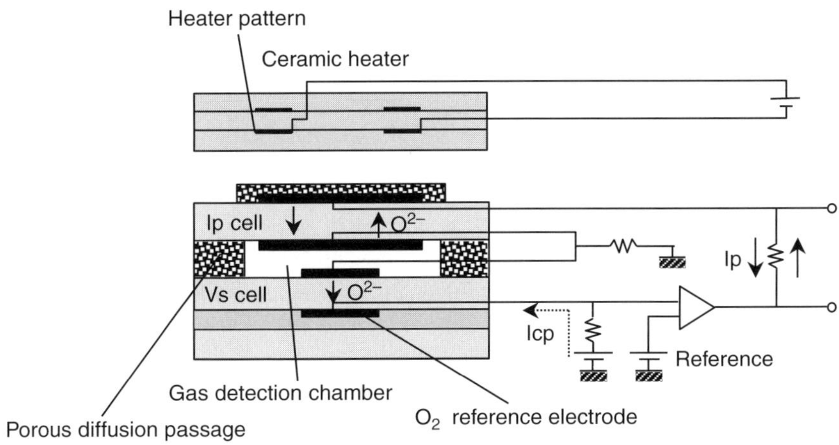

FIGURE 3.1.13 Cross-sectional view of UEGO sensor element.

3.1.3 UNIVERSAL EXHAUST GAS OXYGEN (UEGO) SENSOR

The oxygen sensors, which are central function systems for the automobile exhaust gas purification system, are only used at the stoichiometric point for the three-way catalyst. In order to improve the fuel economy, it needs the combustion control in the lean burn region. Therefore, the sensor is required to have output controlling at the air/fuel ratio in the lean burn region.

3.1.3.1 UEGO Sensor Structure and Principle

A cross-sectional view of the UEGO sensor element is shown in Figure 3.1.13. The element is structured by three substrates of zirconia solid electrolytes. The

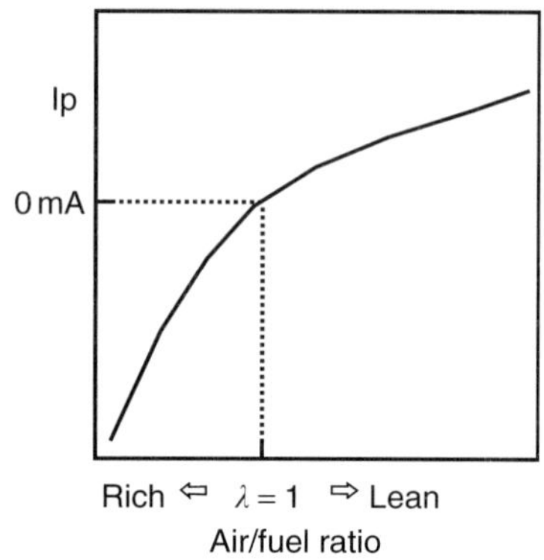

FIGURE 3.1.14 The relationship between Ip signal output and air/fuel ratio.

first substrate is an oxygen pumping cell (called Ip cell). Its top and bottom surfaces are arranged with Pt electrode. The second substrate is an oxygen concentration cell (called Vs cell). Vs cell is also arranged in the same manner as the Ip cell. Vs cell has a self-generative oxygen reference without ambient air induction passage. Connecting cavity (called gas detection chamber) of exhaust gas and porous diffusion layer are structured between the Ip and Vs cells.

Vs cell always monitors the atmosphere in the gas detection chamber. Oxygen pumping current is flowed into the Ip cell due to the constant atmosphere in the detection chamber. Ip signal output can detect the overall air/fuel ratio region, which corresponds to the air/fuel ratio [6]. The relationship between Ip signal output and air/fuel ratio is shown in Figure 3.1.14.

Electric circuit is needed for a constant Vs output signal to control Ip as a series of operations. Actual Vs output is controlled at a constant voltage of 450 mV. As a result, Vs output is equivalent to the electromotive force generated by oxygen concentration cell at the stoichiometric point as is mentioned in the previous section. In other words, gas detection chamber is always maintained at the stoichiometric air/fuel ratio even though exhaust gas is under any atmosphere. Ip current for the retention corresponds to the equation below.

In the lean atmosphere,

$$\mathrm{Ip} = (4FDS/RTL) \times (P_{\mathrm{oe}} - P_{\mathrm{od}})$$

where F is the Faraday constant, D the diffusion coefficient for oxygen molecule, S the area for the diffusion passage, R the gas constant, T the absolute temperature, L the length of gas diffusion passage, P_{oe} the oxygen partial pressure in the exhaust gas, and P_{od} the oxygen partial pressure in the gas detection chamber.

In the rich atmosphere, for example, H_2 and CO are combustible gases

$$\mathrm{Ip} = (2FS/RTL) \times (D_{\mathrm{H_2}} \cdot P_{\mathrm{H_2}} + D_{\mathrm{co}} \cdot P_{\mathrm{co}})$$

where $D_{\mathrm{H_2}}$ is the diffusion coefficient for hydrogen molecule, $P_{\mathrm{H_2}}$ the hydrogen partial pressure in the exhaust gas, D_{co} the diffusion coefficient for the carbon monoxide molecule, and P_{co} the carbon monoxide partial pressure in the exhaust gas.

With regard to hydrocarbons of the combustible gas, if $\mathrm{C}_n\mathrm{H}_m$ exists, then $\mathrm{D}_{\mathrm{C}_n\mathrm{H}_m}$ and $\mathrm{P}_{\mathrm{C}_n\mathrm{H}_m}$ are added as diffusion coefficient and partial pressure respectively, of the hydrocarbon molecule of the above-mentioned equation.

3.1.3.2 UEGO Sensor Manufacturing Process

The sensor element is manufactured with zirconia powder as the partial stabilized composition, which is mixed with organic binder and then is formed into a sheet. The sheet is then punched out and shaped for the printing process. Pt electrode is printed on the top and bottom surface of the Ip sheet. As an electrode protection layer, a ceramic layer is printed in the same manner as the ceramic plasma jet coating layer for thimble-type oxygen sensor. Moreover, Pt electrode is printed on the top and bottom surface of Vs sheet. Between Vs sheet and Ip sheet, insulation alumina layer is printed, which forms the gas detection chamber. Porous diffusion layer is printed at the entrance portion next to the gas detection chamber. The supporting sheet is secured for strength by a material corresponding to the wall for the reference oxygen. Pt wire is inserted and laminated as a lead frame in the spacing between the Ip sheet, Vs sheet, and supporting sheet.

The sensor element manufacturing process is shown as below.

Zirconia powder + Organic binder mixing → Sheet forming → Punching → Zirconia sheet completed → Pt positive and negative electrodes of Ip and Vs sheets printing, protection coat layer printing, insulation layer printing, porous diffusion layer printing → Supporting sheet + Vs sheet + Ip sheet + Pt electrode insertion and lamination → Cutting for separation of each sensor element → Dewax → Sintering

Furthermore, manufacturing process of ceramic heater arranged on the side of the Ip cell is shown as below.

Alumina powder + Organic binder mixing → Sheet forming → Punching → Alumina sheet completed → Heater pattern Pt printing → Pt electrode insertion and lamination → Cutting for separation of each heater element → Dewax → Sintering

3.1.3.3 UEGO Sensor Structure

Sensor element for the thick film type is directly installed in the metal shell and it is different from the thimble-type oxygen sensor in that there is a problem of sealing against the exhaust gas. Sensor element for thick film type is accordingly stored in the thimble-shaped ceramic. The space between sensor element for thick film type and inside of the metal pipe is filled up with talc and glass material. Finally, heat treatment is performed to ensure the sealing. In case of UEGO sensor, the ambient air is not the reference oxygen, but has a self-generative oxygen reference cell. Therefore, the sensor does not have ventilation filter to induce the ambient air as in the thimble-type oxygen sensor.

The thimble-shaped ceramic has a crimped talc ring at the edge of the metal shell, which is same as the thimble-type oxygen sensor element and it ensures the sealing for the outside of the thimble-shaped ceramic.

Typical UEGO sensor structure and features are shown in Figure 3.1.15.

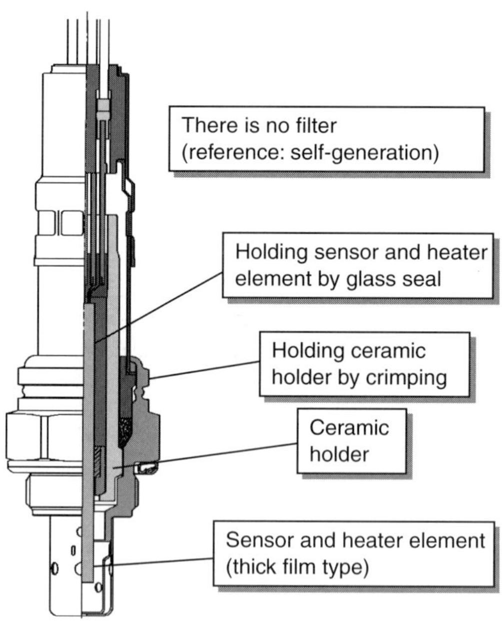

FIGURE 3.1.15 UEGO sensor structure and features.

3.1.3.4 Air/Fuel Ratio Control by UEGO Sensor

Exhaust gas of automobiles will greatly change depending on the driving conditions, air/fuel ratio, temperature, pressure, flow volume, and so on. The sensor is exposed to the exhaust gas and the temperature and pressure effects to achieve the purpose of detecting air/fuel ratio accurately cannot be ignored. As the above mentioned equation, IP $= (4FDS/RTL) \times (P_{oe} - P_{od})$ shows, it is realized that the sensor output will be changed by constituents of porous diffusion passage for the sensor element. Furthermore, diffusion coefficient of gas molecule represented by D depends on temperature and pressure. The diffusion velocity is affected due to the inside diameter of continuous pore for the gas diffusion.

Namely, if the pore diameter is larger than the mean free path of the gas molecule, the gas diffusion is not affected by the pore diameter; this is the so-called molecule diffusion (D_m). Conversely if the pore diameter is smaller than the mean free path, it collides with the pore wall and is diffused; this is the so-called Knudsen diffusion (D_k). The porous diffusion passage follows the mixed diffusion layer [5] that is, both D_m are D_k are followed. Accordingly, the mixed diffusion layer is given by the following equation:

$$\frac{1}{D} = \frac{1}{D_m} + \frac{1}{D_k}$$

The corresponding diffusion coefficient's temperature and pressure dependency are shown in the following:

$$D_m \propto T^{1.75}/P$$

$$D_k \propto T^{0.5}$$

Furthermore, a method has been reported, which selects the pore diameter of the diffusion layer in order to cancel the temperature dependency. Molecule diffusion and Knudsen diffusion are mixed and even though there is a temperature difference of more than 150°C, an example of production applicable to the sensor exports which hardly has any temperature dependency for the output [7].

Consequently, sensor output Ip is substituted by D_m or D_k in the equation IP $= (4FDS/RTL) \times (P_{oe} - P_{od})$ and the effect of the temperature is calculated. Large pore diameter is proportional to 0.75 power of the temperature. Small pore diameter is affected only by temperature and is proportion to -0.5 power of the temperature. Practical porous layer is a mixture of the both diffusion types. Sensor element output change is shown against the pore diameter of mixed diffusion in the practical usage and pressure change of pore diameter for

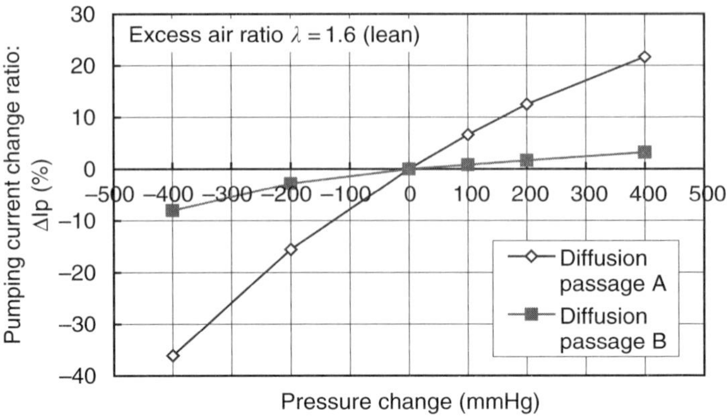

FIGURE 3.1.16 Pressure dependency for UEGO sensor.

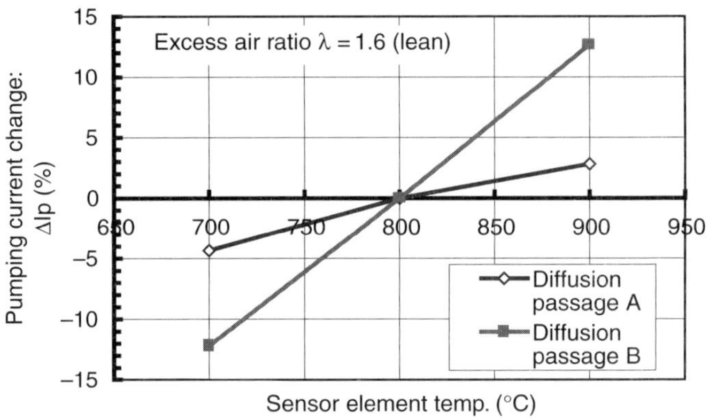

FIGURE 3.1.17 Temperature dependency for UEGO sensor.

molecule diffusion in Figure 3.1.16. Figure 3.1.17 shows the sensor element output change against the temperature change.

Exhaust gas pressure can be compensated by the sensor output change by information of engine revolution, intake pressure, etc. However, the sensor temperature is affected by the temperature change of the environment that is, in cold regions or desert regions and also by the different driving conditions from the starting of the engine to the full-slotted driving. Therefore, the sensor element temperature must be controlled by an accurate detecting method.

A temperature dependency for the electromotive force of the zirconia solid electrolyte sensor element has been reported where Fe_2O_3 as an electric

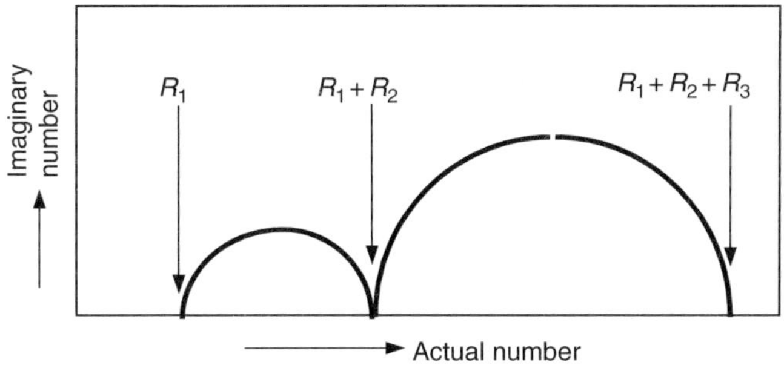

FIGURE 3.1.18 Cole-Cole plot.

conductive material is added to the zirconia to give an oxygen ion conductivity and the solid solution reduces the influence of temperature factor as shown in the Nernst equation [8]. However, this is not a perfect solution and is still not of practical usage. The internal resistance of solid electrolyte is determined by the Cole–Cole plot method as shown in Figure 3.1.18.

Grain internal resistance (R_1) of zirconia particle on the real axis as lead wires and solid electrolyte, zirconia grain boundary matrix resistance (R_2), and interface resistance (R_3) between zirconia Pt electrodes constitute DC component. The sensor element is exposed to the exhaust gas and R_3 resistance suffers effect of the atmosphere. Therefore, if resistance component of the mainstream of grain internal resistance is realized, the sensor element resistance to detect in the high-frequency region, relationship between the sensor element temperature and resistance can be realized without effect of the atmosphere.

Figure 3.1.19 shows the relationship between sensor element temperature and sensor element impedance. Figure 3.1.20 shows the sensor element temperature change to effect the exhaust gas temperature change in each driving region by the sensor element resistance detecting method [9].

The sensor element needs to be maintained at 750°C and is heated up by the heater. However, if heater is operated by a constant voltage, the temperature varies by 242°C in the driving region. Conversely, if the heater is controlled by the sensor element resistance constant, the temperature varies only by 46°C. Furthermore, the temperature drops to 100°C at the constant voltage, when the fuel is cut off but hardly varies as constant control of the sensor element resistance. The purpose of the air/fuel ratio control will be achieved with the control to maintain the sensor output accuracy.

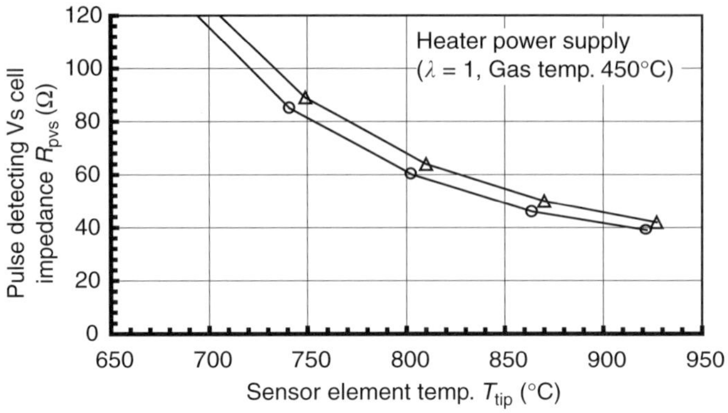

FIGURE 3.1.19 The relationship between sensor element temperature and sensor element impedance.

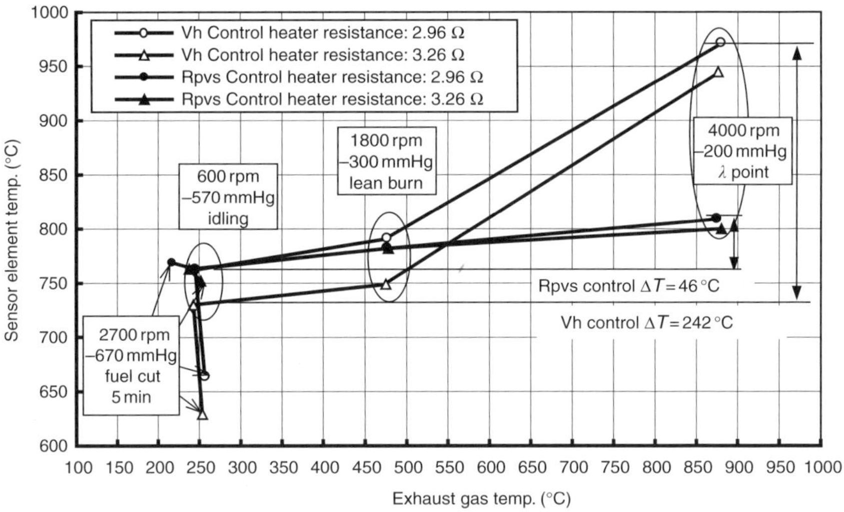

FIGURE 3.1.20 Heater constant voltage control versus Vs cell constant resistance (R_{pvs}) control.

3.1.4 NO_x SENSOR

Recently, developments of direct injection engine and NO_x storage catalyst in the lean region, low emission combustion in the lean burn region in the steady state have been made. Originally, excess oxygen exists in the exhaust gas in the air excess region. Even though NO_x gas is reduced and becomes N_2 gas, it

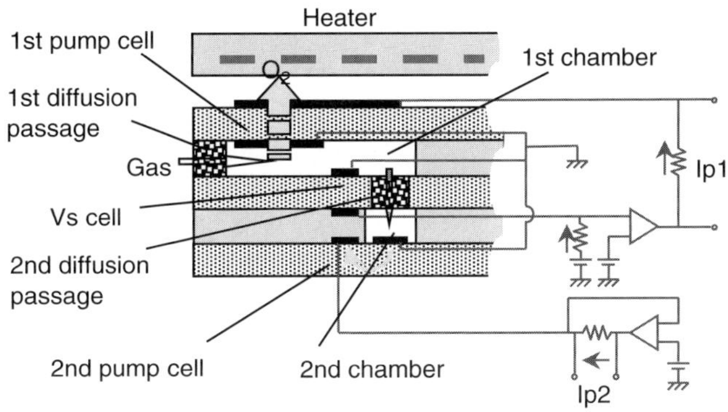

FIGURE 3.1.21 Cross-sectional view of NO_x sensor element.

is oxidized again and reconverted to NO_x. Therefore, NO_x gas is accumulated to the NO_x storage catalyst as nitrate compound. If it reaches more than a prescribed limit, fuel-rich gas is injected for a short time and accumulated NO_x is reduced. On the system, the first purpose is that the timing of the rich-gas injection is decided by NO_x sensor. Secondary deterioration of NO_x storage catalyst should be detected by NO_x sensor when catalyst capability drops because of sulfur attack and general catalyst deterioration. In short, on both of them it is applicable as an OBD deterioration detecting system.

The sensor structure is similar to the one discussed in the previous section (UEGO sensor). Figure 3.1.21 shows the cross-sectional view of the NO_x sensor element. Three zirconia solid electrolyte sheets compose the sensor element. The top and the central sheet are the oxygen pumping cell (Ip1) and oxygen concentration cell (Vs cell), respectively. Additionally, between these cells the gas detection chamber for connecting the exhaust gas and diffusion layer is present. The structure is quite similar to the UEGO sensor.

However, the central portion of the element consists of a second diffusion chamber to induce the gas to the second detection chamber. The bottom sheet composes the second oxygen pumping cell (Ip2 cell). An external voltage is applied to the pumping cell to dissociate NO_x, into N_2 and O_2, and finally transfers the generated O_2 by the pump. At that time, Ip2 pumping current corresponds to the NO_x concentration in the exhaust gas.

Ip1 output is the same as UEGO sensor and pumping current corresponds to the oxygen partial pressure in the exhaust gas. In the first detection chamber, Ip1 cell serves to pump out the excess oxygen in the lean atmosphere. At that time, if oxygen is not pumped out up to the partial NO_x decomposed, a large amount of oxygen affects the NO_x detection. A detection object of NO_x concentration

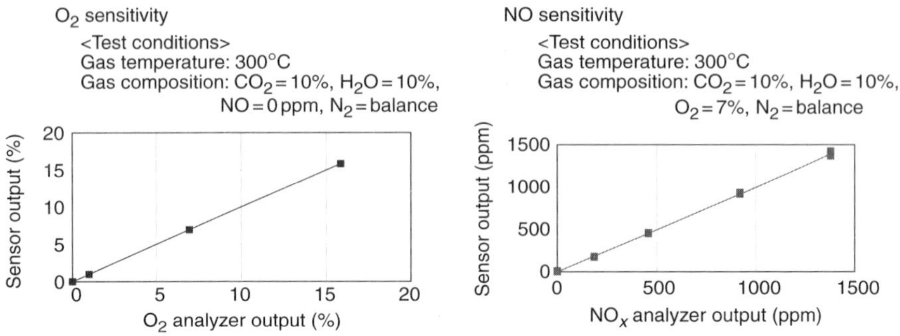

FIGURE 3.1.22 Basic characteristics; O_2 and NO sensibility of NO_x sensor.

in the actual exhaust gas is approximately hundreds of ppm. The oxygen ion current at which NO_x gas is decomposed and flowed, is only of the order of as microamperes. Comparable results of detection output with analyzer output are shown in Figure 3.1.22.

The sensor element is similar in structure to the UEGO sensor element. The process is as follows: printing the Pt electrode on the zirconia sheet, laminating, dewax and finally sintering. As observed from the cross-sectional view, complex printing, two diffusion layers with two oxygen pumping cells and electrode alignment of one oxygen concentration cell are not symmetrical as in the UEGO sensor.

The sensor can simultaneously detect the output corresponding to the oxygen concentration and NO_x concentration in the exhaust gas and therefore can be put to two usages.

3.1.5 OXYGEN SENSORS FOR THE INDUSTRY

In advanced uses of the oxygen sensor in the industry, the function used is the as same as the solid electrolyte sensor element as an oxygen concentration cell. Oxygen partial pressure in the atmosphere is detected from the electromotive force that arises from the oxygen partial pressure ratio based on the Nernst equation:

$$\mathrm{EMF} = RT/4F(ln\, P_a/P_b)$$

where R is the gas constant, T the absolute temperature, F the Faraday constant, P_a the reference oxygen partial pressure (air atmosphere), and P_b the oxygen partial pressure in the measuring atmosphere (e.g. melting metal, flow soldering chamber).

Recently, combustion control by wide range air/fuel ratio sensor with an oxygen pumping function as measurement in the atmosphere of high oxygen volume has been developed, but without much accuracy.

Advanced usage for the industry is shown below:

1 For automatic casting machineries of the copper, etc., oxygen content in the melting metal is measured by the electromotive force of the oxygen concentration cell. Oxygen partial pressure in the chamber is adjusted due to the oxidation protection as the heat treatment. The oxygen sensors are used for measurement of oxygen content during fusing steel in the revolving furnace for steel making in the process of the metal refining, and in steel making as a disposable oxygen detection system. Thimble-type zirconia sensor element is the most widely used. The thimble-type sensor is used for automobiles as oxygen sensors. Moreover, for substrate for the electric circuit, the sensor is also used for the oxygen concentration control in the flow-soldering chamber. Thus, the oxygen sensors are utilized for metal characteristics and performance retention based on the oxidation protection.

2 As an example of the usage in combustion control, there are two combustion control atmospheres, stoichiometric and lean burn for the compressed natural gas combustion control for the dynamo. For these, the normal oxygen sensor and UEGO sensor are applied, respectively. For boiler combustion control, oxygen concentration cell is used for the zirconia oxygen concentration device. Recently, UEGO sensor has been used for incinerator combustion control of flammable refuse during dioxin generation. It measures the oxygen concentration in the combustion gas and air/fuel ratio is controlled to the optimum condition. Consequently, this contributes to the improvement of the combustion efficiency and helps reduce air pollution by purification of the exhaust gas.

3.1.6 SUMMARY

Zirconia solid electrolyte as an oxygen sensor has been used for the combustion control, especially of automobiles, for a quarter of a century. The production of these sensors exceeds 100 million pieces per year worldwide. The zirconia solid electrolyte is also used as a component of the important oxygen concentration control devices for the industry. Improvement and development are under progress with respect to electrical performance and mechanical strength such as improvement of the ceramic solid electrolyte and electrode material. Functional characteristics of the original ceramic material have been accomplished due to establishment of the surrounding technique arrangements in the industry. The product is developed from a tool to sense the oxygen and becomes a central functional component of the exhaust gas purification system for automobiles. In future, it is envisaged that the sensor will be further used for the purpose of contributing towards global environment conservation.

REFERENCES

1. Duwez, P., Odell, F., and Brawn, F. H. Jr. (1952). *J. Am. Ceram. Soc.* **35**: 109.
2. Kvist, A. (1972). *Physics of Electrolyte*, Vol. 1, pp. 319–346. New York: Academic Press.
3. Nakahara, K., and Takami, A. (1991). *Nainenkikan* **30**: 73–82.
4. Noda, Y., Kawahara, K., Yamada, T., Suzuki, S., and Nishio, K. (1992). Annual Report for Overseas Readers, *Fine Ceramics Report*, JFCA, pp. 6–9.
5. Nakahara, K., and Takami, A. (1991). *Nainenkikan* **30**: 73–84.
6. Yamada, T., Hayakawa, N., Kami, Y., and Kawai, T. (1992). SAE Paper No. 920234.
7. Kamo, T., Chujyo, Y., Akatsuka, T., Nakano, J., and Suzuki, M. (1985). Lean mixture sensor, SAE Paper No. 850380.
8. Howarth, D. S., and Wilhelm, R. V. (1978). A zirconia-based lean air fuel ratio sensor, SAE Paper No. 780212.
9. Tsuzuki, M., Kawai, T., Yamada, T., and Nishio, K. (1998). *IEE Jpn.* **118-E**.

Handbook of Advanced Ceramics
S. Sōmiya *et al.* (Eds.)

3.2 Ceramic Fuel Cells

J. FLEIG, K. D. KREUER and J. MAIER
Max-Planck-Institut für Festkörperforschung, Heisenbergstrasse 1, 70569 Stuttgart, Germany

3.2.1 OUTLINE

Fuel cells are composed of galvanic elements in which the reactants and the products are continuously supplied and removed. The basic idea dates back to Schönbein and Grove [1], yet enthusiasm and research activity was never as intense as today.

Even though these electrochemical devices are based on a simple working principle, the requirements are immense as far as materials research is concerned. Fuel cell research and in particular research on ceramic high temperature cells—fields in which milestones were set by Nernst [2], Baur and Preis [3], Kikukkola and Wagner [4]—offers a vivid example of the interdependence of structure, property and performance, and the interplay of optimism and frustration that is characteristic of promising but demanding technologies [5–9].

After a short consideration of why the fuel cell concept is so attractive but difficult to realize, we will present the currently most promising constituents of solid oxide fuel cells, discuss their advantages and shortcomings and then look into the future, that is into perspectives and the potential to develop advanced materials and better devices.

3.2.2 GENERAL ASPECTS

3.2.2.1 ATTRACTIVENESS OF FUEL CELLS

The main attractiveness of fuel cells follows from the definitions given above. It comprises the high theoretical efficiency associated with direct conversion of chemical energy into electrical energy by means of galvanic cells [10]; the selectivity of the electrochemical process; and the advantage of a continuous "metabolism" by using the ambient air to oxidize the steadily supplied fuel.

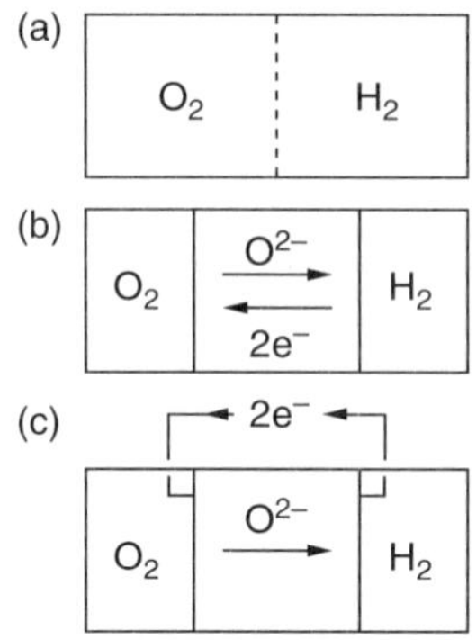

FIGURE 3.2.1 Direct conversion of H_2 and O_2 to water by direct contact (a) or via permeation membrane (b) in contrast to the electrochemical conversion (c).

The increased efficiency of galvanic cells compared to direct combustion is evident from the following consideration. If the fuel, for example, H_2, is brought into direct contact (Fig. 3.2.1a) with air, chemical combustion becomes possible, setting free the associated reaction enthalpy ($\Delta_f H_{H_2O}$), which is -250 kJ/mol at $T \sim 10^3$ K. To convert the heat into electrical energy we might use a heat engine to transform the heat into mechanical and—via a dynamo—into electrical energy.

The efficiency (w) of such a thermal process cannot be higher than Carnot efficiency (w_c)

$$w \equiv \left| \frac{\text{electrical energy}}{\Delta_f H_{H_2O}} \right| \leq w_c = \frac{T_2 - T_1}{T_2}$$

where T_2 and T_1 are the temperatures between which the Carnot process is performed. Typical values for a steam engine are $T_2 = 583$ K (boiling point of water at 100 bar) and $T_1 = 313$ K leading to a maximum efficiency of 46%, real values are distinctly lower. The same consideration is true if the reactants are separated by a mixed conductor and the contact is enabled by a diffusion of ions and electrons through the membrane (Fig. 3.2.1b). If, however, only the ions are allowed to flow internally through a solid electrolyte with negligible electronic conductivity, while the electrons flow through an external circuit, the latter process can be directly used to gain electrical energy (Fig. 3.2.1c). If the process is performed reversibly (current $\longrightarrow 0$) the entire Gibbs energy $\Delta_r G$ ($=\Delta_f G_{H_2O}$ for the H_2/O_2 fuel cell) can be converted into electrical energy:

$$w_g = \left| \frac{\Delta_r G}{\Delta_r H} \right| = \frac{\Delta_r G}{\Delta_r H} = 1 - \frac{T\Delta_r S}{\Delta_r H} = 1 + \frac{T\Delta_r S}{|\Delta_r H|} > w.$$

One recognizes that a positive reaction entropy ($\Delta_r S$) gives rise to a theoretical efficiency greater than 100%, meaning that the process leads to a cooling of the environment (the reaction enthalpy $\Delta_r H$ is negative). If the number of gaseous components is conserved in the reaction, $\Delta_r S$ is small and $w_g \simeq 100\%$. As can be seen in Table 3.2.1 this is the case for the electrochemical combustion of methane to CO_2 and H_2O. The oxidation of hydrogen to water also leads to approximate values of about 80%. Of course, practical efficiencies are distinctly lower owing to the non-zero currents involved (see below). A selection of (overall) anode and cathode reactions of significant fuel cell types is given in Table 3.2.2. The mobile ion in each electrolyte is represented by bold letters. Here, we restrict to high temperature fuel cells composed of ceramic electrolytes and electrodes. Such ceramic fuel cells are, almost without exception, solid oxide fuel cells (SOFCs) based on oxide ion conducting electrolytes (see Table 3.2.2). Besides hydrogen, natural gases and even carbon monoxide can be converted in SOFCs such that H_2O and/or CO_2 are the products in all cases. Besides fuel cell being environmentally benign (concerning pollutant output and noise level), they possess a high flexibility with regard to their size and siting, and particularly the possibility to co-generate electrical power and heat. A very important point in this context is the fact that high efficiency is already obtained at comparatively small power levels.

3.2.2.2 Complications and Demands

The reasons why fuel cells are not yet established and well-developed constituents of our economy lie in the losses caused by current flow, materials problems such as lack of chemical or mechanical strength, and—not independent of these aspects—in the cost of cells, generator and balance-of-plant.

Materials failures as a consequence of mechanical or thermomechanical stresses are inherent problems of the usually brittle high temperature ceramics and have to be kept within tolerable limits by suitable materials selection and engineering. In this section, we will focus on the general electrochemical problems while the reader is referred to later sections for more details on materials.

Voltage losses can be expressed by (generally current dependent) resistivities or overvoltages η and are described by a "voltage efficiency" $w_U = U/E$ (U, real voltage; E, theoretical open circuit voltage). Since the deviation of U from E, the overvoltage, can be usually broken down into contributions from individual processes (see Figure 3.2.2) in series, $\eta = \sum \eta_i$, and since each

TABLE 3.2.1 Thermodynamic Data and Efficiency Factors for Important Fuel Cell Reactions [11]

Cell reaction	Temp (°C)	$\Delta_r H^\circ$ (J/mol)	$\Delta_r S^\circ$ (J/mol/K)	$\Delta_r G^\circ$ (J/mol)	$-U$ (V) (mV^{-1}/K^{-1})	dU/dT	$w_g^\circ = \Delta_r G^\circ/\Delta_r H^\circ$
$H_2 + \frac{1}{2}O_2 \rightarrow H_2O_{(l)}$	25	−285 800	−162.40	−237 400	1.23	+0.840	0.83
	60	−285 050	−159.00	−231 830	1.20	+0.820	0.81
	100	−283 300	−155.00	−220 370	1.17	+0.800	0.78
$H_2 + \frac{1}{2}O_2 \rightarrow H_2O_{(g)}$	25	−241 830	−44.40	−228 580	1.18	+0.230	0.945
	60	−242 180	−45.60	−226 990	1.18	+0.230	0.94
	100	−242 580	−46.60	−225 160	1.17	+0.240	0.93
	500	−246 180	−55.10	−203 530	1.05	+0.280	0.83
$NH_{3(g)} + \frac{3}{4}O_2 \rightarrow \frac{1}{2}N_2 + \frac{2}{3}H_2O_{(l)}$	25	−382 510	−145.50	−339 120	1.17	+0.500	0.89
$NH_{3(g)} + \frac{3}{4}O_2 \rightarrow \frac{1}{2}N_2 + \frac{2}{3}H_2O_{(g)}$	25				1.13		
$N_2H_{4(l)} + O_2 \rightarrow N_2 + 2H_2O_{(l)}$	25	−621 100	+5.10	−622 600	1.61	−0.010	1.00
$2Na + H_2O_{(l)} + \frac{1}{2}O_2 \rightarrow 2NaOH_{(aq)}$	25	−653 210	−174.90	−601 050	3.11	+0.900	0.92

$C_{Gr} + \frac{1}{2}O_2 \rightarrow CO$	25	−110 500	+89.10	−137 080	0.71	−0.460	1.24
	500	110 800	+89.90	−180 300	0.93	−0.460	1.63
$C_{Gr} + O_2 \rightarrow CO_2$	25	−393 500	+2.87	−394 350	1.02	−0.007	1.00
$CO + \frac{1}{2}O_2 \rightarrow CO_2$	25	−283 000	−86.20	−257 300	1.33	+0.440	0.91
$CH_3OH_{(l)} + \frac{2}{3}O_2 \rightarrow CO_2 + 2H_2O_{(l)}$	25	−726 260	−76.50	−703 700	1.21	+0.130	0.97
$CH_4 + 2O_2 \rightarrow CO_2 + 2H_2O_{(g)}$	25	−802 400	−6.00	−800 600	1.04	+0.007	1.00
	60	−802 060	−4.90	−800 420	1.04	+0.006	1.00
	100	−801 700	−3.90	−800 200	1.04	+0.005	1.00
	500	−800 300	−1.70	−798 900	1.03	+0.002	1.00
$CH_4 + 2O_2 \rightarrow CO_2 + 2H_2O_{(l)}$	25	−890 200	−242.60	−817 900	1.06	+0.310	0.92
$C_2H_4 + 3O_2 \rightarrow 2CO_2 + 2H_2O_{(l)}$	25	−1 306 320	−62.10	−1 287 810	1.11	+0.050	0.99
$C_2H_4 + 3O_2 \rightarrow 2CO_2 + 2H_2O_{(g)}$	25				1.09		
$C_3H_8 + 5O_2 \rightarrow 3CO_2 + 4H_2O_{(l)}$	25	−2 218 900	−374.00	−2 107 440	1.09	+0.190	0.95
$C_3H_8 + 5O_2 \rightarrow 3CO_2 + 4H_2O_{(g)}$	25	−2 044 000	+108.00	−2 076 380	1.07	−0.050	1.02
$n\text{-}C_4H_{10} + 6\frac{1}{2}O_2 \rightarrow 4CO_2 + 5H_2O_{(l)}$	25	−2 878 270	−438.00	−2 747 930	1.09	+0.170	0.955

TABLE 3.2.2 Overall Electrode Reactions of Important Fuel Cell Types [9]

Fuel cell types	Anode reaction	Cathode reaction
Phosphoric acid (PAFC) and polymer membrane fuel cell (PEMFC)	$H_2 \longrightarrow 2H^+ + 2e^-$	$\frac{1}{2}O_2 + 2H^+ + 2e^- \longrightarrow H_2O$
Alkali fuel cell (AFC)	$H_2 + 2OH^- \longrightarrow 2H_2O + 2e^-$	$\frac{1}{2}O_2 + H_2O + 2e^- \longrightarrow 2OH^-$
Molten carbonate fuel cell (MCFC)	$H_2 + CO_3^{2-} \longrightarrow H_2O + CO_2 + 2e^-$	$\frac{1}{2}O_2 + CO_2 + 2e^- \longrightarrow CO_3^{2-}$
	$CO + CO_3^{2-} \longrightarrow 2CO_2 + 2e^-$	
Solid oxide fuel cell (SOFC)	$H_2 + O^{2-} \longrightarrow H_2O + 2e^-$	$\frac{1}{2}O_2 + 2e^- \longrightarrow O^{2-}$
	$CO + O^{2-} \longrightarrow CO_2 + 2e^-$ $CH_4 + 4O^{2-} \longrightarrow 2H_2O + CO_2 + 8e^-$	

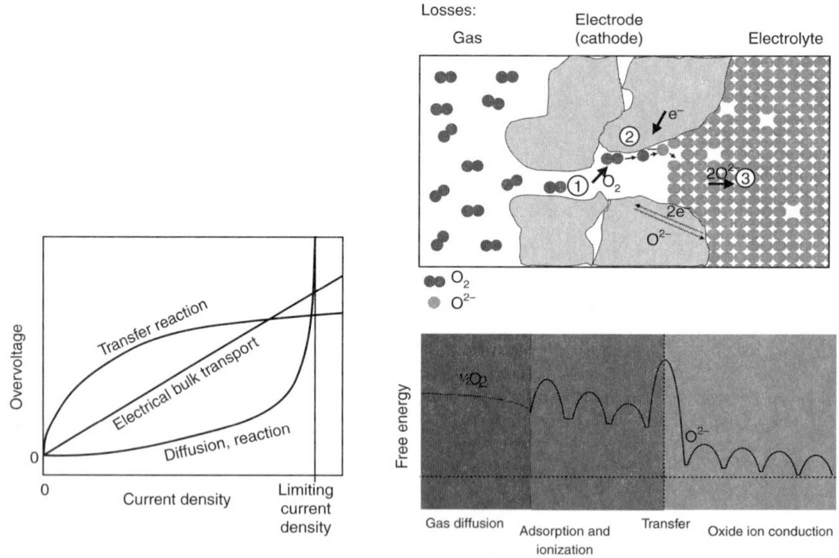

FIGURE 3.2.2 The cathode events for SOFC fuel cells, schematic. The processes actually taking place are usually much more complicated, the same applies to the free enthalpy profile. (Concerning gas diffusion (1) the configuration effect is included in the free enthalpy, while the local standard value (2, 3) is used otherwise; after Ref. [9].) For simplicity and in contrast to reality, it is assumed that the oxygen is completely ionized when it enters the electrolyte.

η_i can be formally written as the product of current and (generally current dependent) resistances R_i, it follows that $w_U = 1 - I\sum_i R_i/E$. Important contributions are voltage losses due to the ohmic resistance of the electrolyte (which is independent of current to a good approximation), the electrode transfer resistance which decreases with increasing current because of the influence of the voltage on the activation thresholds, and the diffusion resistance caused by finite velocity of the gas flow. Proper electrode resistances need not be due to the charge transfer itself but can be also caused by hindered adsorption, dissociation, ionization or space charge transport. Diffusion overvoltage and diffusion resistance increase asymptotically while the current increases towards a limiting value. Such a limiting current results since the transport coefficient hardly depends on potential, and since the driving force for the diffusion cannot be infinitely increased (finite maximum concentration gradient).

Moreover, not all charges transferred lead to the electrochemical reaction under concern (for example incomplete, non-stoichiometric course of reaction). This leads to an additional power loss which is described by a "current efficiency" or more precisely "Faraday efficiency" $w_F = It/nzF$ (n is the number of moles reacting and z the number of electrons transferred in the reaction). If further losses (operational losses, e.g. work $< UIt$ or non-ideal fuel usage, i.e. mole number $n < n_{gas}$) are taken account of by efficiencies w_B and w'_B, the finally usable work is given by $w_B|UIt| = w_B w_F|nzFU| = w_B w_F w_U|nzFE| = w_B w_F w_U w_g n|\Delta_r H_m| = w_B w_F w_U w_g w'_B n_{gas}|\Delta_r H_m|$ and the total efficiency w then by the product of the individual efficiencies (index m refers to 1 mole).

The electrochemical requirements with respect to the cell constituents (cathode, electrolyte, anode) are clear: the electrodes (cathode, anode) should allow for a fast transport to the surface and a fast redox reaction there (consisting of adsorption, dissociation, ionization, transfer into the electrolyte, transport through the space charge layer), the electrolyte should allow for a fast ionic transport (high conductivity and selectivity of the mobile ion; in the usual case of bulk transport this includes high in-grain conductivity, low grain-to-grain resistances including low space charge resistances, negligible electronic contributions) and finally, high resistivities with respect to electronic carriers.

In addition to all this, the structure and composition must be stationary which demands, in addition to the ionic and electronic requirements, sufficient thermal stability (typical performance temperatures of SOFCs are greater than 1000 K), mechanical stabilities (thermal expansion matching, thermoshock resistance, mechanical strength, absence of phase transformations) and chemical stability with respect to reactions of neighboring phases, for example, solid/gas, solid/solid.

Further constraints are imposed by the necessity to connect individual cells together to form high-power stacks. Apart from sealing problems and the specific thermal, chemical, mechanical boundary conditions, leads and interconnects have to be stable and highly electronically conductive.

To illustrate how restrictive such requirements are, let us consider the electrolyte. Not only has it to be stable at the high temperatures with respect to the neighboring solid phases, it must also not corrode under highly oxidizing conditions at the cathode, nor degrade under the highly reducing atmosphere at the anode. Apart from these thermal and chemical conditions and in addition to the mechanical resistance required, the extreme variation of the chemical potential of oxygen from the anode to the cathode is of special significance for the electrical properties. While the ionic conductivity of highly doped or highly disordered materials can be invariant over a wide P_{O_2} range, the electronic conductivities usually pronouncedly change in terms of power laws with the characteristic exponents $-1/4$ for the excess electrons and $+1/4$ for the holes. Negligible electronic contributions then require extended ionic domains which usually implies that the valence states of the cations and anions are extremely stable, as well as reasonably high ionic mobilities and hence sufficient stabilities at high temperatures. Thus, it is not surprising that the oxide electrolyte which is the oldest known [2], stabilized ZrO_2, is still the electrolyte of choice for the SOFCs.

As far as the chemical stability towards the gas phase is concerned, the requirements for the electrode materials are not so harsh. Electrochemical requirements for the anode (in the case of H_2) are less critical than for the cathode. This is because the kinetics of hydrogen oxidation are less demanding than those concerning reduction of oxygen or natural gases. Nonetheless, successful materials engineering managed to reduce cathodic overpotential to approximately the order of magnitude of anodic values. Especially useful are oxides that are appropriately doped mixed conductors, since they offer high catalytic activities, due to presence and mobilities of ionic and electronic defects, and moreover, the possibility of oxygen transport through the solid. In such cases the incorporation is not restricted to a three phase boundary where electrons (from the outer circuit), ionic carriers (from the electrolyte) and the gas phase constituents meet. In this respect, Sr-doped $LaCoO_3$ would be preferable over the cathode of choice consisting of $LaMnO_3$ (Sr-doped), yet it more readily reacts with zirconia. As anode an Ni-YSZ "cermet" (composite of Ni and YSZ) is presently used to guarantee a sufficient density of three phase contacts. While $LaMnO_3$ is almost perfectly matched to the thermal expansion of YSZ, this is not the case for Ni. The YSZ in the cermet is very beneficial in this respect and the Ni content is kept as small as possible (just above the percolation threshold).

3.2.3 STRATEGIES FOR MATERIALS SEARCH AND OPTIMIZATION

As we have discussed, the strategies to realize optimized ceramic fuel cells are essentially materials search strategies.

Since we have to search for candidates which ensure a long-time operation we have to rely on thermodynamically stable structures and compounds, or on metastable situations. Besides the nature of the constituents, the relevant control parameters are temperature, component potential and doping content. Since temperature and partial pressure of oxygen are fixed, the only variable parameters are the major chemical elements as well as the concentrations of dopants. (In the case of multinary oxides usually not all sublattices are in equilibrium and there is additional freedom in fixing the concentration of immobile native constitutents (typically, the A : B ratio in A_xB_yO). This implies identifying the appropriate ground material and finding the optimal nature and content of the dopant. Apart from this homogeneous doping, that is, tuning of the charge carrier concentrations by introducing aliovalent impurities, there is also the tool of heterogeneous doping [12], which makes use of beneficial transport properties brought about by introducing metastable structure elements of higher dimensionality. While using appropriate composites or nanocrystalline solids may be a significant strategy for medium or room temperature applications, this is probably not the case for high-temperature fuel cells: irrespective of morphological stability problems the benefits of interfacially controlled materials are usually lost at sufficiently high temperatures.

To summarize: (i) the most important step is searching for sufficiently stable and mechanically strong structures that exhibit promising potential in terms of electrochemistry, such as fluorites, perovskites, pyrochlores or brownmillerites in the case of electrolyte materials; and (ii) homogeneous doping then is the most important tool to optimize the electrical properties, such as introducing lower valent cations into MO_2 fluorites, which leads to a substantial oxygen conductivity (based on oxygen vacancies). Notwithstanding the specific consequences of adding new components one should recognize the enhanced probability that both thermal and chemical stabilities are expected to suffer if the number of components is increased simply owing to an increased number of thermodynamic and kinetic reaction pathways. Thus, it is not surprising that doped binary or ternary oxides with redox stable cations are the most promising solid electrolytes. Finally, we have to realize the fact that the fuel cell is a heterogeneous materials system implying that the selected components have to be chemically and thermally compatible with each other.

3.2.4 CURRENTLY USED MATERIALS

3.2.4.1 Electrolytes

3.2.4.1.1 Requirements and Commonly Discussed Materials

As already mentioned above, the demands with respect to the electrolyte in a SOFC are enormous. Beside mechanical stability, gas-tightness, compatibility with other cell components—not to mention economic aspects—there are essentially three important requirements to be fulfilled: (i) chemical stability in reducing, oxidizing and CO_2-containing atmosphere; (ii) high ionic conductivity to minimize ohmic losses during operation; and (iii) negligible electronic conductivity to avoid a lowering of the theoretically achievable EMF and fuel losses due to ambipolar permeation of oxygen through the electrolyte.

Doped ZrO_2 [4] and CeO_2 [13] are the only binary oxides which are usually considered as candidates for fuel cell electrolytes; their properties are detailed in the next sections. Bi_2O_3-electrolytes (δ-Bi_2O_3) drew a lot of attention since they exhibit a very high ionic conductivity [14–17]; they become, however, predominantly electronically conducting in reducing atmospheres [17–19] and eventually decompose at oxygen partial pressures which are still higher than the pressures occurring at the anode (e.g. 10^{-13} bar at 600°C [17, 18]). This very limited choice with respect to binary materials led to an intensive search for multinary oxide ion conductors. Many doped $Bi_4V_2O_{11}$ componds (termed BIMEVOX) have considerable ionic conductivities [17, 20–22], are, in contrast to δ-Bi_2O_3, stable at room temperature but suffer from the same redox-problems as Bi_2O_3. Pyrochlores, for example, doped $Gd_2Ti_2O_7$ can exhibit ionic conductivites comparable to that of zirconia [23] and have been mainly discussed for applications in monolithic cells (see section 3.2.6). The most promising non-binary oxide seems to be doped $LaGaO_3$ characterized by rather high ionic conductivities [24, 25]. A discussion of the properties and problems of $LaGaO_3$-based electrolytes is given below. Let us first turn to the binaries ZrO_2 and CeO_2.

3.2.4.1.2 ZrO_2

Pure ZrO_2 exhibits a very low intrinsic point defect concentration and hence negligible ionic conductivity even at 1000°C. Doping with several percent of a lower-valent metal oxide (Y_2O_3, Sc_2O_3, CaO, etc.), however, considerably improves the situation: Since the excess charge introduced by, for example, Y^{3+} onto Zr^{4+}-sites are compensated by oxygen vacancies, the high vacancy concentration leads to an ionic conductivity of ≈0.1 S/cm at 1000°C for a Y_2O_3 content of 8 mol% [26, 27]. Such high doping concentrations also stabilize the

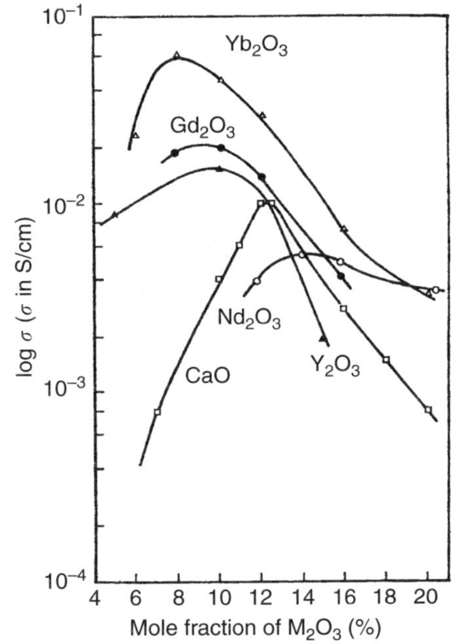

FIGURE 3.2.3 Conductivities of ZrO_2–M_2O_3 systems at 800°C. (From Ref. [34, Fig. 15, p. 1008].)

high temperature cubic phase, and hence the term "yttria stabilized zirconia" (YSZ) is often used for such electrolytes. The requirement of a negligible electronic conductivity is fulfilled in the entire partial pressure range occurring in solid oxide fuel cells [26, 28]. A recent treatment of the defect chemistry has been given by Sasaki and Maier [29]. Lower doping levels (e.g. 2–3 mol% Y_2O_3) lead to the metastable tetragonal zirconia, which exhibits improved fracture toughness and strength, but lower ionic conductivities at temperatures typical for fuel cell applications [30–32].

The conductivity–doping level curves depend on the dopant ion and exhibit maxima at dopant concentrations of typically 8–12 mol% (Fig. 3.2.3) [27, 32–34].

The variations in conductivity with different dopants of the same valence can be understood in terms of defect interactions and lattice relaxation phenomena caused by the dopant: the binding energies of the defect clusters and the effective mobility of the vacancies depend on the dopant size and concentration [35–37]. Doping with Sc^{3+}, the ionic radius of which is almost identical to that of Zr^{4+}, leads to the highest conductivity maximum (ca. 0.3 S/cm at 1000°C) [38, 39]. However, despite the high conductivity of Sc-doped zirconia

8–10 mol% Y_2O_3-doped zirconia is still usually used in SOFCs. One obvious reason to prefer Y over Sc is the price (even though the Sc price has recently fallen). The long-time stability of the ionic conductivity is also an issue [39–41]: serious problems have been reported in this respect but recent studies indicate degradation phenomena to be rather small in the case of ca. 10% Sc doping [40, 41].

There is a trend to reduce the operation temperature of fuel cells to minimize materials problems, particularly in order to gain more flexibility with respect to the choice of the interconnect material (see section titled "Interconnects"). Since the ionic conductivity of doped zirconia is thermally activated (ca. 0.8 eV at 800–1000°C [26, 39]), however, the increase in the resistance accompanying the temperature reduction must be compensated for. The simplest way to do this is to decrease the electrolyte thickness, and there are attempts to employ zirconia films that have thicknesses of a few micrometers or even lower. On the other hand, this approach is limited by the fact that the electrolyte resistance does not only depend on its thickness but also on the contact geometry at the electrode/electrolyte interfaces. At the electrochemically active sites (e.g. three phase boundaries) the current is constricted and the corresponding constriction resistance [42–44] can dominate the overall resistance of the electrolyte and hinder a further resistance reduction with decreasing thickness. This approach together with an optimization of the dopant might lead to cells based on zirconia electrolytes which can be used at temperatures as low as ca. 700–750°C [41] but intermediate temperature SOFCs (500–600°C) certainly require other electrolyte materials. CeO_2 is a promising material in this temperature range as will be discussed in the following section.

3.2.4.1.3 CeO_2

At 1000°C, CeO_2 doped with 10% Gd exhibits a conductivity of ca. 0.25 S/cm which exceeds that of zirconia (of $\sigma \approx 0.1$ S/cm in YSZ); owing to its lower activation energy (ca. 0.65 eV [45, 46]) the difference even increases with decreasing temperature (2.5×10^{-2} S/cm [45] versus ca. 3×10^{-3} S/cm for YSZ [27] at 600°C). Just as in the case of zirconia the influence of defect interactions leads to a maximum in the conductivity–dopant concentration relation [45]. As before, the defect interactions and cluster energies strongly depend on the dopant ion and calculations suggest Gd to be the optimum dopant in order to obtain high bulk conductivity. However, even though high conductivities have indeed been experimentally observed in Gd-doped ceria, the optimum choice of the dopant depends on temperature and is influenced by potential co-dopants [45, 47, 48].

The great disadvantage of ceria is the occurrence of a significant electronic conductivity at low oxygen partial pressures (Fig. 3.2.4), and corresponds to

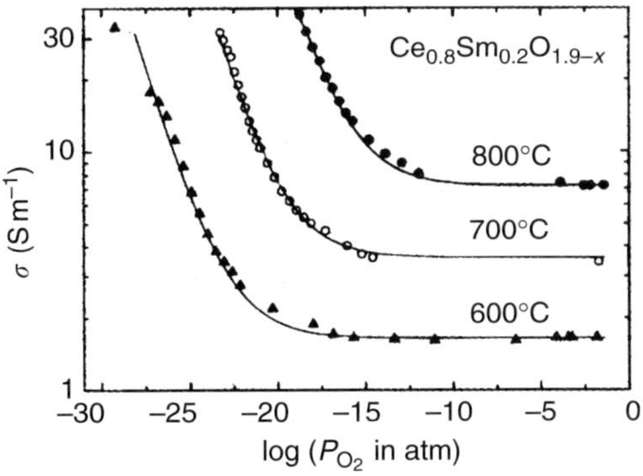

FIGURE 3.2.4 Total electrical conductivity σ_{tot} of doped CeO_2 as a function of oxygen partial pressure. Solid lines show the dependence of $\sigma_{tot} = \sigma_{ion} + kP(O_2)^{-1/4}$. (Reprinted from Gödickemeier, M., and Gauckler, L. J. (1998). *J. Electrochem. Soc.* **145**, 414–421 with permission from The Electrochemical Society, Inc.)

the comparatively easy partial reduction of Ce^{4+} to Ce^{3+}. The partial pressure dependence of this electronic conductivity [13, 45, 49–52] follows from the oxygen exchange reaction

$$\tfrac{1}{2}O_2 + V_O^{\bullet\bullet} + 2e' \rightleftharpoons O_O^{\times} \tag{1}$$

between the gas atmosphere and ceria. Since the oxygen vacancies ($V_O^{\bullet\bullet}$) are majority charge carriers in the entire partial pressure range of interest, their concentration and thus also their activity is determined by the dopant level. From the mass action law for Eq. 1, it follows that the electron concentration is proportional to $p(O_2)^{-1/4}$, and that the electronic conductivity is therefore significantly enhanced in reducing atmospheres. The effect of this electronic conduction on the electrochemical performance of a fuel cell has been simulated [49, 53]; it was found that above ca. 600°C, only a very limited operation window is reasonable from an efficiency and power output point of view. It is also clear from simple defect chemical considerations that under equilibrium conditions the electronic contributions cannot be affected by minor redox-active dopants.

The activation energy of the electronic conductivity is larger than that of the ionic conductivity, and hence the electrolytic domain, that is, the partial pressure regime with predominant ionic conduction, increases with decreasing temperature. The electronic conductivity of ceria therefore plays only a minor role at intermediate temperatures, for example, 500°C. Since the ionic conductivity at these temperatures is still high enough, ceria can be regarded to be an appropriate electrolyte for intermediate temperature fuel cells.

Further information on the properties of ceria and its use in SOFCs was given by [45, 48, 54].

3.2.4.1.4 $LaGaO_3$

The discovery of high ionic conductivity in appropriately doped $LaGaO_3$ [24, 25] has triggered many investigations into its properties. As in the other oxide ion conductors, the ionic conductivity depends not only on the dopant level but also on the kind of dopant ion. In contrast to ceria and zirconia two cations can be substituted leading to a complicated situation that is far from being completely understood. Recent studies have shown that co-doping of Sr on A sites and Mg on B-sites results in a rather high ionic conductivity of ca. 0.12–0.17 S/cm already at 800°C [55–58]. This is similar to doped ceria and considerably exceeds the value of YSZ (ca. 0.03 S/cm at 800°C [26, 27]). The activation energy also varies with composition—it tends, for example, to increase with the Mg content—and can be as low as ca. 0.6 eV [55, 56]. Owing to this relatively low activation energy, the conductivity of $LaGaO_3$ is high enough to be used in fuel cells at about 600–700°C. Below ca. 600°C, however, an activation energy increase has been reported for Sr/Mg-doped $LaGaO_3$ which makes doped ceria the better ionic conductor [56, 57]. The electronic conductivity strongly depends on the dopant ion: Sr/Mg-doped lanthanum gallate exhibits negligible electronic conductivity in the entire partial pressure range occurring in fuel cells. Heavy co-doping with transition metals, on the other hand, can strongly increase the electronic transference number [58, 59]. For certain transition metal concentrations (e.g. $La_{0.8}Sr_{0.2}Ga_{0.8}Mg_{0.115}Co_{0.085}O_3$), however, a further enhancement in ionic conductivity was observed without significant electronic conductivity, and without an increase of the activation energy at lower temperatures [60, 61].

The most severe restriction with respect to the use of $LaGaO_3$ in fuel cells is its limited chemical stability. Volatilization of gallium oxide particularly during sintering has been observed by Yamaji *et al.* [62, 63]; additional phases separate from the parent phase already after heating at 1000°C [64], and during co-firing with typical electrode materials various reaction products have been detected [65]. It is still not clear in how far these problems can be coped with by introducing co-dopant ions, which stabilize the compound, or choosing appropriate fabrication and operation conditions. At any rate, the gallate is not a reasonable candidate for high-temperature fuel cells.

3.2.4.1.5 Influence of Grain Boundaries

As far as the tendency to lower the operation temperatures is concerned, the influence of grain boundaries deserves more detailed considerations. It has

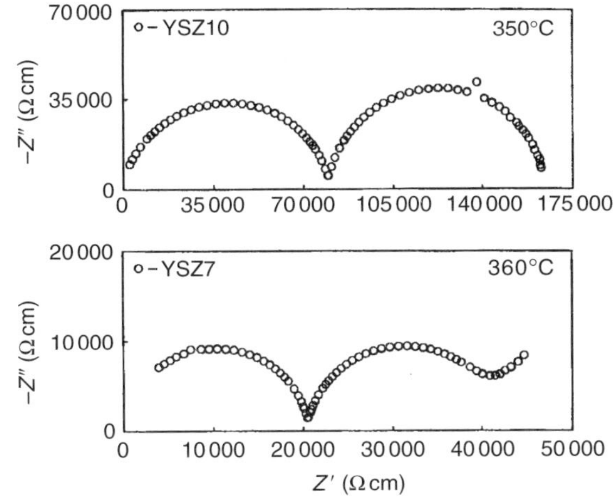

FIGURE 3.2.5 Impedance spectrum measured for 10 and 7 mol% Y_2O_3-doped zirconia. (Reprinted from Badwal, S. P. S. (1995). *Solid State Ionics*, **76**, 67–80 with permission from Elsevier Science.)

long been known [66] that: (i) grain boundaries can increase the resistance of a solid electrolyte; and (ii) impedance spectroscopy can be used to separate grain boundary and bulk resistances in such cases, since highly resistive grain boundaries lead to an additional semicircle in the complex impedance plane (Fig. 3.2.5). Several factors influencing the grain boundary resistance of, for example, zirconia [66–76] and ceria [45, 77–79] have been elucidated and different models are discussed to explain the grain boundary resistances: (i) complete wetting, highly resistive grain boundary phase; (ii) space charge depletion layers; or (iii) current constriction due to only partially contacted grains or because of a partially wetting, ionically blocking grain boundary phase. It has to be emphasized that a current constriction at grain boundaries according to model (iii) occurs in the grain bulk close to the established grain-to-grain contact and hence reflects bulk properties. A quantitative analysis of the corresponding current distributions [80] showed that the size of the resulting "grain boundary" semicircle in the complex impedance plane not only depends on the area of the established grain-to-grain contacts but also on the number of contact spots. The contacted area between the grains therefore cannot be deduced solely from the corresponding resistance [80].

It has been observed that in zirconia and ceria, SiO_2 impurities segregate to the grain boundaries and that, for example, a high silica content increases the grain boundary resistance [45, 70, 71, 73, 74, 76, 77]. It is still under debate, however, whether the corresponding grain boundary resistance is due to the ion transfer across a wetting silica phase or due to the current constriction occurring close to direct grain-to-grain contacts (e.g. pin-holes in the silica). Even in highly pure zirconia a small grain boundary resistance has been measured which might represent a kind of "intrinsic" grain boundary resistance possibly due to charge carrier depleted space charge zones [74, 75]. According to this picture the siliceous phase partially blocks the boundaries while the resistance of the non-blocked passages is determined by Schottky barriers.

In zirconia and ceria, the grain boundary resistance and the bulk resistance frequently exhibit similar activation energies at low temperatures [67–70, 78, 79] (e.g. typically 1.0–1.2 versus 0.9–1.1 eV in stabilized zirconia at temperatures of about 400°C). However, the bulk activation energy lowers towards operation temperature and hence the effect of grain boundaries on the total electrolyte resistance decreases with increasing temperature. In high-temperature SOFCs based on zirconia, electrolyte grain boundaries do not perceptibly influence the ohmic electrolyte resistance. In intermediate temperature fuel cells (based on ceria), however, grain boundaries can significantly affect the effective electrical conductivity [45]. Only recently ceria with very low SiO_2 content has been fabricated which exhibits negligible grain boundary contributions even at about 500°C [45]. In $LaGaO_3$ only few investigations on grain boundary effects exist [81, 82], and a reliable statement on the importance of grain boundaries cannot be given yet. In this context it is mentioned that grain boundaries are not necessarily highly resistive but can also constitute fast current paths [83–85], although for fast oxygen ion conductors such highly conductive grain boundaries have not yet been found to be of relevance.

3.2.5 CATHODES

3.2.5.1 Requirements and General Aspects

Two important requirements to be fulfilled by a cathode material are: (i) a high catalytic activity with respect to the reduction reaction;

$$\tfrac{1}{2}O_2\ (\text{gas}) + 2e^-\ (\text{cathode}) \rightleftharpoons O^{2-}\ (\text{electrolyte}) \tag{2}$$

and (ii) a high electronic conductivity to supply electrons without significant potential drop between the current collector and the electrochemically active site at which the oxygen reduction occurs. Moreover, the cathode material should be chemically compatible with the other cell components and the thermal expansion coefficient of the cathode must be sufficiently close to that of the electrolyte (ca. $10.5 \times 10^{-6}\,K^{-1}$ for YSZ and ca. $12 \times 10^{-6}\,K^{-1}$ for Gd-doped ceria at 600°C [48, 86–88]) in order to avoid cracks and delamination during fabrication or operation.

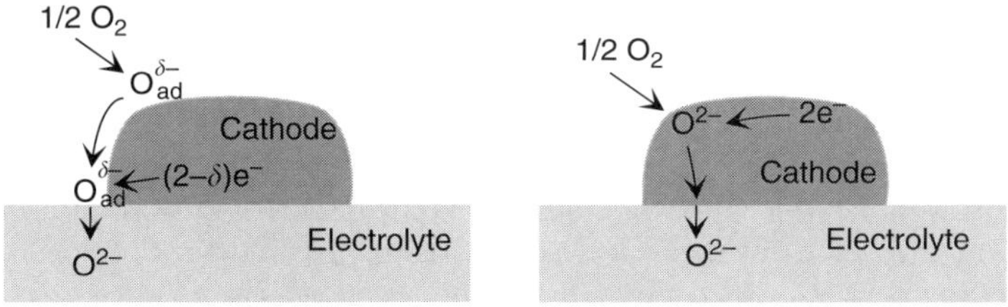

FIGURE 3.2.6 Sketch of possible reduction paths at fuel cell cathodes; (a) surface path; (b) bulk path.

With respect to the mechanism of the oxygen reduction reaction two possible reaction paths are usually discussed (see Figure 3.2.6): (i) The *surface path* includes adsorption of oxygen on the electrode surface, diffusion of a (probably dissociated and partly ionised) oxygen species along the surface towards the three-phase-boundary (TPB) where electrolyte, electrode and gas phase meet, followed by an ionic transfer step into the electrolyte accompanied by complete ionization. The incorporation into the electrolyte does not necessarily occur directly at the TPB; surface or interface diffusion of the ionized species could lead to a certain broadening of the incorporation zone. Modifications of this path include direct adsorption of oxygen at the TPB (without surface diffusion on the cathode surface) or adsorption on the electrolyte plus surface diffusion to the TPB. (ii) The *bulk path* consists of oxygen adsorption and dissociation, ionization and incorporation into the cathode according to Eq. 1, oxide ion transport through the electrode, and the transfer of the ion into the electrolyte. Both paths can contribute to the oxygen reduction, and it depends on the materials properties and the electrode morphology/geometry as to which one dominates the overall reaction rate.

A third path (iii), namely the ionization of the oxygen on the zirconia surface, followed by a direct incorporation into the electrolyte can also not be excluded. In this case, however, the electronic charge carriers, which are required in the oxygen reduction reaction, have to be supplied by the electrolyte. The very low electronic conductivity of zirconia can therefore be expected to restrict the active zone to a region very close to the TPB. Hence, this path is, from a geometrical point of view, similar to the surface path discussed above.

If the surface path dominates, a large TPB length is clearly advantageous and can be achieved if porous cathodes are used. Composite cathodes consisting of electrolyte particles, electrode material and pores even lead to a three-dimensional network of TPBs and hence can further enhance the reaction rate. Simulations on the performance of composite electrodes and suggestions with respect to optimal parameters such as electrode porosity and thickness were given by Sunde [89, 90], Costamagna *et al.* [91] and Tanner *et al.* [92].

If the bulk path dominates the reduction reaction, the effective rate constant for oxygen incorporation into the cathode and the ionic conductivity of the cathode play decisive roles. A slow incorporation reaction, for example, requires a porous cathode with a high surface area. In such a case, the zone involved in oxygen reduction is considerably broadened compared to the narrow TPB region mentioned above. A slow transport within the oxide, on the other hand, demands short transport pathways, and hence a very thin cathode layer would be favorable. Short transport pathway are also accessible at the edges of cathode particles and a large edge length, that is, a porous cathode with a large TPB length, offers another possibility to lower the corresponding reduction resistance. Consequently, the bulk path does not necessarily go hand in hand with a significant broadening of the electrochemically active zone; in the case of a porous cathode and transport in the bulk being the rate determining step, most oxygen is again incorporated close to the TPB [93, 94]. Complex situations arise if both, incorporation and transport, have to be considered as suggested by SIMS experiments on several potential cathode materials [95–97]. Calculations on the optimal geometrical and kinetic properties of such mixed conducting cathodes were presented by Adler *et al.* [98], Deng *et al.* [99], Svensson *et al.* [100] and Liu [101].

With respect to the electrical properties the following conclusions can be drawn: The bulk path requires a certain oxygen ionic conductivity in the cathode; hence mixed ionic and electronic conducting ceramics are particularly promising candidates for ceramic SOFC cathodes. In order to achieve high charge carrier concentrations (e.g. oxygen vacancies and/or holes) again, for a given oxide, aliovalent substitution (e.g. Sr^{2+} on a La^{3+}-site in $LaCoO_3$) is the tool of choice. However, it strongly depends on the oxygen partial pressure, the temperature and in particular on the material under concern whether the charge of, for example, a lower-valent dopant is mainly compensated by oxygen vacancies or holes. In order to achieve or maintain sufficient electronic effects regular constituents with easily changeable valences are necessary (typically transition metal elements). The perovskite structure adopted, for example, by $LaMnO_3$, $LaCoO_3$ or $LaFeO_3$ provides the demanded high stability. In Figure 3.2.7, the typical partial pressure dependences of the defect concentrations are sketched assuming ideal dilution of the defects and partial Schottky disorder (cation and anion vacancies). The indispensable high electronic conductivity and a desired high ionic conductivity (to promote the bulk path) could, for example, be reached in the transition region between regimes III and IV and hence oxides with this transition occurring at ca. 10^{-2}–1 bar oxygen are certainly interesting cathode candidates. However, a simultaneous tailoring of the ionic and electronic conductivity as well as of the thermal and chemical compatibility is rather difficult and has not yet been successful; all known candidates exhibit certain disadvantages, as will be discussed below.

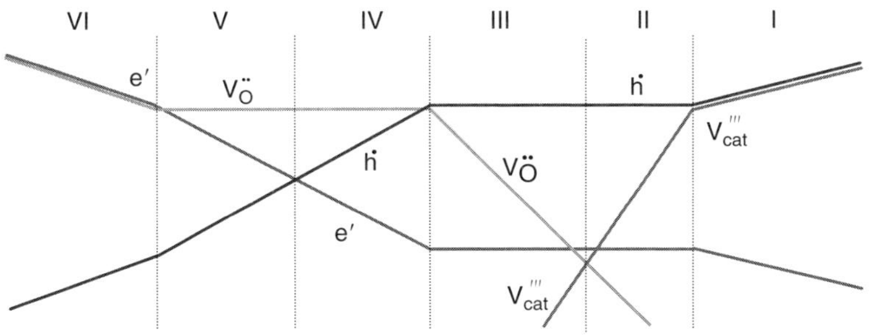

FIGURE 3.2.7 Partial pressure dependence (schematic) of the defect concentrations of an acceptor doped oxide with trivalent cations (cat) exhibiting partial Schottky disorder (cation and anion vacancies).

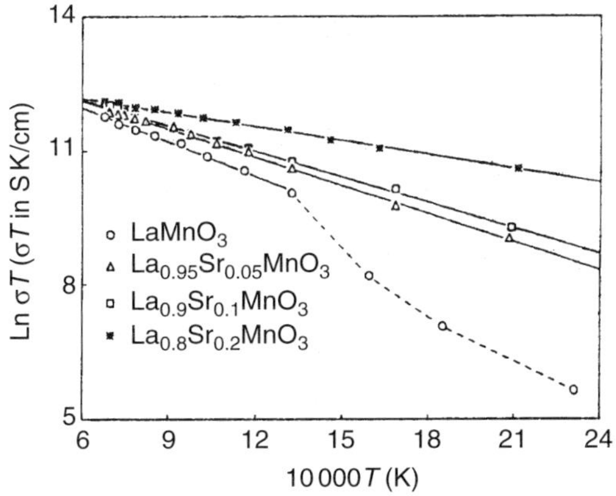

FIGURE 3.2.8 Dopant- and temperature-dependent conductivity of $LaMnO_3$. (From Ref. [104, Fig. 5].)

3.2.5.2 $LaMnO_3$

$LaMnO_3$ is typically used as the cathode material in zirconia-based fuel cells and is also the present material of choice in commercial SOFCs. In order to achieve a high electronic conductivity it is heavily acceptor doped (typically by 10–40% Sr or Ca on the La-site) leading to an enhanced hole concentration corresponding to an enhanced fraction Mn^{4+} (Fig. 3.2.8). The resulting conductivity values of about 100 S/cm at 1000°C [102–104] are not high enough for a complete neglect of the in-plane resistance in the cathode [54, 105], but an adequate geometry and location of the current collector can minimize the corresponding losses. The oxygen content varies with the partial pressure (Fig. 3.2.9), and in air $(La_xSr_{1-x})MnO_{3+\delta}$ (LSM) exhibits positive δ values [106–108]. Owing to

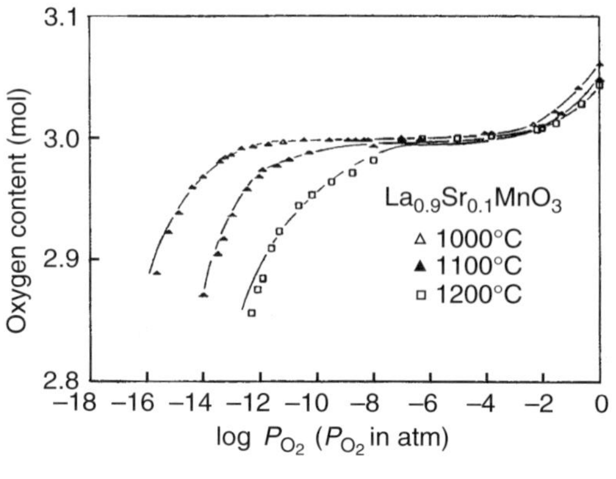

FIGURE 3.2.9 Partial pressure dependence of the oxygen content in $LaMnO_3$. For each isotherm the last point represents the lowest oxygen partial pressure before the oxide decomposed. (From Ref. [106, Fig. 2].)

defect interactions and the possibility of Mn^{3+} to disproportionate, the exact defect model is rather involved [107–110], but essentially confirms the trends of Figure 3.2.7. In other words, $LaMnO_3$ cathodes operate in regime II (Fig. 3.2.7) with high cation vacancy but low oxygen vacancy concentration. Hence, the ionic conductivity is rather low (of the order of 10^{-7}–10^{-8} S/cm at 800°C in air for 10–25% Sr-dopant [97, 111–113]), and it is mostly assumed that the surface path dominates the reaction rate of Eq. 2. However, recent investigations using geometrically well-defined, dense microelectrodes [113, 114] and dense films or large overvoltages [94, 111, 112, 115, 116] showed that it depends on geometry and operating conditions whether the bulk or the surface path is more important: high cathodic overvoltages can increase the oxygen vacancy concentration by shifting the oxygen chemical potential to lower values and thin, broad electrode particles lead to a low TPB-length-to-area ratio. Both factors favor the bulk path while low overvoltages and particles without an extreme aspect ratio promote the surface path. In the case of a dominating bulk path possibly both the oxygen incorporation reaction and the transport within the bulk of the electrode influence the reaction rate [94, 95, 113, 117]. With respect to the rate-determining step of the surface path, many impedance and current–voltage studies have been performed on LSM electrodes, but a generally accepted model has not been achieved yet [94, 111, 112, 114, 115, 118–131].

Despite the rather low ionic conductivity and, hence, the poor utilization of the bulk path, porous LSM cathodes show acceptable polarization resistances in zirconia-based SOFCs at 1000°C. However, the activation energy of the polarization resistance is rather high (ca. 1.8–2.1 eV [94, 95, 119, 120, 131, 132]) and the trend to lower the operation temperature of fuel cells requires an optimization of the LSM cathodes. Several studies [54, 132–135] showed that YSZ/LSM composite cathodes exhibit an improved behavior and might be applicable at temperatures as low as 800°C.

With respect to the chemical and thermal compatibility, LSM cathodes are fairly acceptable although not optimal in YSZ-based fuel cells: the chemical interaction between LSM and YSZ is particularly critical since additional phases such as $La_2Zr_2O_7$ or $SrZrO_3$ can already form at ca. 1000°C leading to enhanced cathodic polarization [120, 136–139]. However, a careful control of the composition and the preparation parameters, as for example, verified for La-deficient $LaMnO_3$, limits these difficulties to an acceptable level [54, 120, 136, 138, 140]. Phase diagram studies [141, 142] indicate a miscibility gap in the $LaMnO_3$–$SrMnO_3$ system below about 1400°C. It is an advantage of $LaMnO_3$ that depending on the dopant its thermal expansion coefficient can be closely matched to the value of zirconia [54, 143–145]. In intermediate temperature SOFCs, on the other hand, the polarization losses of LSM cathodes are certainly too high. Some alternatives to LSM are discussed in the following section.

3.2.5.3 Other Cathode Materials

As already indicated above, from an oxygen permeation point of view, $LaMnO_3$ is unfavorable and several other better ion conducting ceramics have been tested as cathode materials. In particular, doped $LaCoO_3$ exhibits high vacancy concentrations between 10^{-2} and 1 bar oxygen and hence shows high ionic conductivity even under oxidizing conditions [96, 146–151]. The oxygen exchange coefficients are also much higher than in LSM [152, 153] and the electronic conductivity of doped $LaCoO_3$ is excellent (ca. 1000 S/cm at 1000°C in air [154–156]). Nevertheless, due to its high reactivity (formation of $La_2Zr_2O_7$ and $SrZrO_3$ at the interface to YSZ [140, 157]) and its very large thermal expansion coefficient (ca. $20–25 \times 10^{-6}\,K^{-1}$ [154, 156]) YSZ/$LaCoO_3$ interfaces must be avoided and $LaCoO_3$ is not used in YSZ-based fuel cells. Owing to its much lower reactivity with ceria, it is discussed as a promising cathode material for fuel cells based on ceria electrolytes, although the large mismatch in the thermal expansion coefficients is still problematic. Concepts based on bilayer electrolytes (YSZ on the anode side and doped ceria on the cathode side) are being investigated [158–165], but solid state reactions and interdiffusion between these two materials appear to be critical in such systems.

Alternatives to $LaCoO_3$ with high ionic conductivity are doped $LaFeO_3$ and compounds based on their solid solutions (La–Fe–Ca–Co–O system or La–Fe–Sr–Co–O system). Compositions of the latter type are often favored as electrode materials in ceria-based intermediate temperature fuel cells since combining excellent ionic conductivity (in some cases even higher than in YSZ) and favorable thermal expansion coefficients (at least for certain compositions) [45, 98, 147, 154, 156, 166–168]. In addition, $PrMnO_3$, $YMnO_3$, $CaMnO_3$ or $SmCoO_3$ have been tested and the search for optimized electrode materials is still going on. Investigations on cathodes for $LaGaO_3$-based fuel cells (e.g. Sr-doped $SmCoO_3$ [169], doped $LaCoO_3$ [170, 171] or $La_{0.9}Sr_{0.1}Ga_{0.5}Ni_{0.5}O_3$ [172]) have been reported only recently, and there are still many open questions particularly in view of chemical compatibility and long-term stability.

3.2.5.4 Anodes

Owing to the reducing atmosphere on the fuel side, it is possible to use non-noble metals in fuel cell anodes. The anodic reaction requires electrons from the metal electrode, fuel and oxygen ions from the electrolyte. It can therefore be assumed that the electrochemical reaction occurs close to the anodic TPB [173] and porous electrodes are again required. Ni exhibits excellent catalytic properties with respect to the hydrogen oxidation, but Ni particles tend to sinter and coarsen at high temperatures leading to a reduction of the porosity and TPB length. By adding YSZ particles the aggregation of the Ni particles can be avoided. Moreover, such composite anodes (cermets) yield an extended three-dimensional network of TPBs which can substantially lower the polarization resistance if the Ni as well as the YSZ particles form percolating paths. Usually, one starts with NiO/YSZ cermets and the NiO is then reduced under fuel conditions. Calculations on the optimal morphology, porosity and thickness of such cermet anodes were presented, for example, by Sunde [89, 90], Costamagna *et al.* [91], Tanner *et al.* [92], Abel *et al.* [174] and Divisek *et al.* [175]. The thermal expansion coefficient of Ni/YSZ cermets can be tailored such that it fits to the other cell components. A cross-section of a fuel cell set-up with a typical Ni anode is shown in Figure 3.2.10.

A number of oxides such as $LaCrO_3$ or TiO_2 have been examined with respect to their performance as anodes, but a material with sufficient electronic conductivity, high catalytic activity and chemical stability under fuel conditions has not yet been found [95, 176, 177]. There are also attempts to use two-phase systems, the oxide constituents of which shows high catalytic activity with respect to the oxidation of, for example, CH_4 while the second phase functions as a current collector [95].

For intermediate temperature fuel cells based on ceria the adequate anode would be a Ni/ceria cermet. However, at temperatures as low as 500–600°C and high fuel conversion levels, the relatively high oxygen partial pressure on

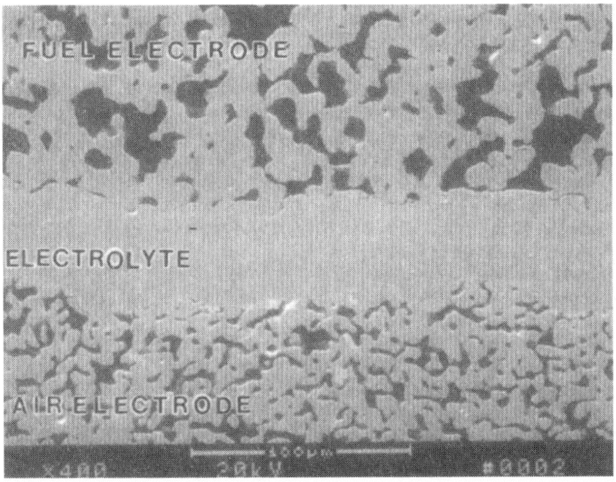

FIGURE 3.2.10 Microstructure of a cross-section of a Siemens Westinghouse solid oxide fuel cell. (Reprinted from Singhal, S. C. (2000). *Solid State Ionics*, **135**, 305–313 with permission from Elsevier Science.)

the anode side can lead to the formation of NiO and hence an alternative to Ni might be required [167]. In the case of $LaGaO_3$ electrolytes different anode compositions (e.g. Ni doped ceria) are under investigation. There the situation is additionally aggravated by the fact that NiO and $LaGaO_3$ easily react [178]. Hence, at least in specific cases, the choice of a Ni-cermet as the anode material is under debate.

3.2.5.5 Interconnects

The theoretical electromotive force of a single electrochemical cell is too low for most applications, and hence several cells are usually connected in series to form stacks. In such stacks interconnects are required to: (i) connect the anode and the cathode of adjoining cells, and if necessary; (ii) separate the fuel at the anode of one cell from the oxygen at the cathode of the next cell. Therefore, in addition to the need of matching thermal expansion coefficients, providing gas-tightness (no connected porosity) and chemical compatibility with the anode and cathode materials, three additional key requirements have to be fulfilled by the interconnect: it has to exhibit; (i) high electronic conductivity to avoid ohmic losses; (ii) negligible ionic conductivity to avoid permeation of oxygen and hydrogen via ambipolar transport of ions and electrons; and (iii) chemical stability in oxygen and fuel atmospheres. Doped $LaCrO_3$ is used as the interconnect in high temperature fuel cell stacks based on YSZ electrolytes

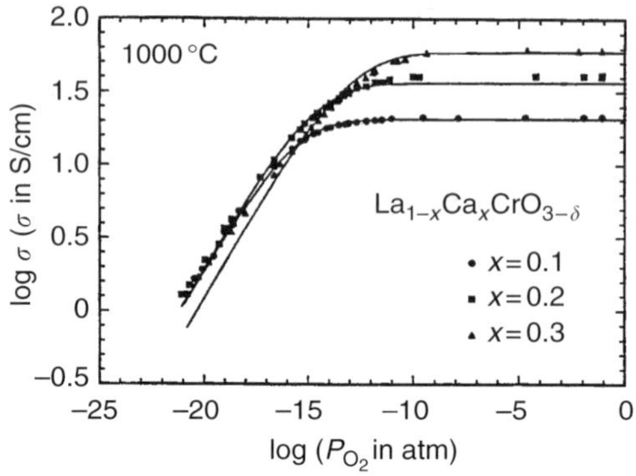

FIGURE 3.2.11 Conductivity of Ca-doped $LaCrO_3$ versus oxygen partial pressure at 1000°C. (Reprinted from Yasuda, I., and Hikita, T. (1993). *J. Electrochem. Soc.* **140**, 1699–1704 by permission of The Electrochemical Society, Inc.)

and in the following it is discussed in how far this material fulfills the above-mentioned requirements.

In terms of maximized electronic and minimized ionic conductivity a material is desirable which is—in the entire partial pressure range occurring in fuel cells—characterized by regime I or II (Fig. 3.2.7). $LaCrO_3$ doped with several percent of for example, Sr, Mg or Ca, however, lies in the transition region between regimes III and IV [179–181] and hence exhibits an oxygen partial pressure dependent electronic conductivity (Fig. 3.2.11).

While on the cathode side, conductivity values of typically 20–50 S/cm are achieved at 1000°C, this value can be more than one order of magnitude lower on the fuel side [182–184] (Fig. 3.2.11). The ionic conductivity, on the other hand, is enhanced at the anode side [181, 185, 186]. Owing to this non-negligible ionic conductivity oxygen permeation can occur through the interconnect [186, 187], and an appropriate interconnect geometry and doping has to be chosen to suppress significant leakage. $LaCrO_3$ is chemically stable in fuel as well as in oxygen atmosphere although a reaction of the dopant with CO_2 has to be kept in mind. Phase diagram studies were performed by Peck *et al.* [188, 189]. Under operation conditions $LaCrO_3$ is chemically compatible with $LaMnO_3$ and Ni/YSZ, yet co-firing can lead to an interaction, for example, due to migration of liquid phases if sintering aids are admixed to improve the very poor sinterability of $LaCrO_3$ [140, 190]. The thermal expansion coefficient of $LaCrO_3$ is not very far from the values of the other components and can be adjusted by co-doping on both A and B sites [145, 184, 190].

Thus, doped $LaCrO_3$ fulfills the electrical, thermal and chemical requirements only partially. Therefore, and particularly due its high costs and its shortcomings with respect to the fabrication process, there are many efforts to replace $LaCrO_3$ interconnects by metallic alloys based on low-corrosive constituents.

Alloys, however, require operation temperatures significantly below 1000°C, and hence numerous activities are going on to adjust the electrolyte and the electrodes to temperatures lower than 800°C as already discussed in the previous sections.

Closely connected with the choice of the interconnect material and its geometry is the design of fuel cell stacks consisting of several cells. The largest SOFC power plants are built by Siemens Westinghouse: a 100-kW SOFC is running since 1997 and a 250-kW hybrid SOFC gas turbine system is presently (2001) being fabricated. This company uses a stack design which is based on cathode supported tubular single cells (Fig. 3.2.12a). In this design rather small interconnects (9-mm wide strips) are used and no further sealing is needed [86, 145, 191, 192]. The long electronic current path in tubes as long as 150 cm as well as the low power density are disadvantages of this design. Mainly two further stack designs are discussed and tested: The planar approach as sketched in Figure 3.2.12b [193] and the monolithic design (see Figure 3.2.12c) using a corrugated structure [145, 191, 192]. Both concepts yield higher power densities than the tubular cells but require high-temperature sealing materials. The success of these designs depends on how challenges such as the need for seals or the fabrication of the corrugated structure in the case of monolithic cells can be satisfactorily responded to.

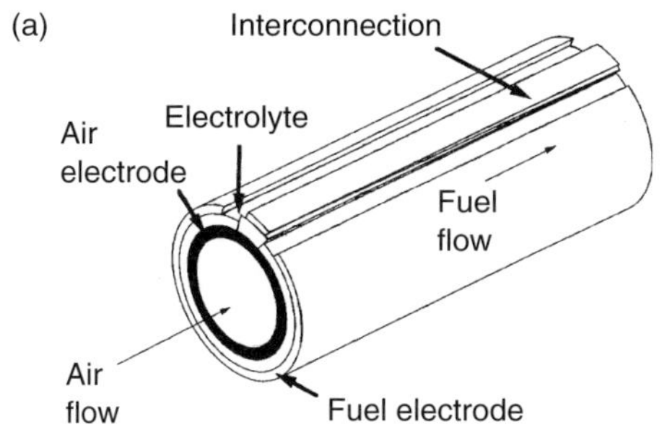

FIGURE 3.2.12 (a) Tubular cell used by Siemens-Westinghouse. (Reprinted from Singhal, S. C. (2000). *Solid State Ionics*, **135**, 305–313 with permission from Elsevier Science.) (b) Sketch of a simple planar design. (From Ref. [193, Fig. 12.27].) (c) A cross-flow monolithic stack design. (*J. Am. Ceram. Soc.* by N. Q. Minh. Copyright 1993 by Am. Ceram. Soc. Inc. Reproduced with permission of Am. Ceram. Soc. Inc. in the format Other Book via Copyright Clearance Center.)

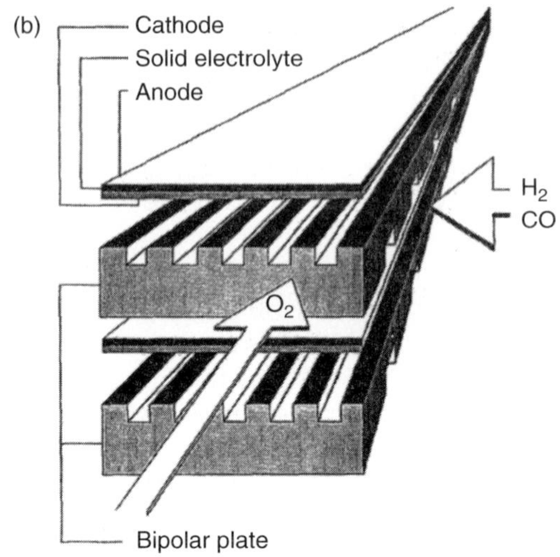

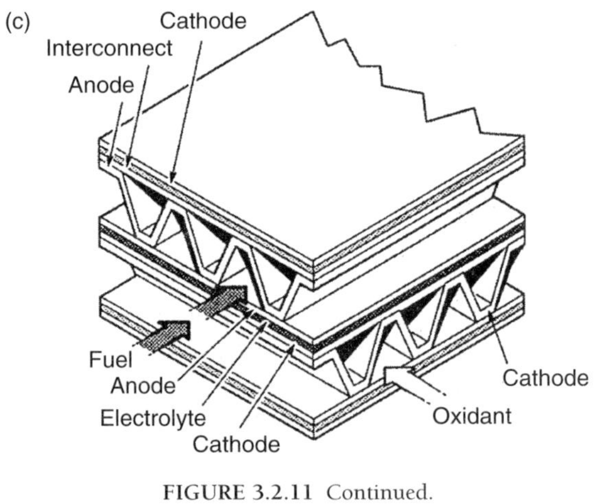

FIGURE 3.2.11 Continued.

3.2.6 ALTERNATIVE APPROACHES

The well-established ceramic fuel cell concepts discussed above comprise oxide ion conducting oxides as solid electrolyte separator material, distinct electrocatalytically active electrodes made from metals or mixed conducting oxides and well separated gas chambers. Alternative approaches are based on electrolytes conducting other species (essentially protons), the combination of the functions of electrolyte and electrodes in one and the same phase (monolithic fuel cell) and the abundance of well-separated gas compartments. Although some of these approaches seem to be exotic at first sight, they may have the potential to solve some of the problems inherent to conventional SOFC concepts.

3.2.6.1 Proton and Mixed Proton/Oxide Ion Conducting Separator Materials

While ceramics with high oxide ion conductivity (especially zirconia based) have been known for almost a century (see above), protons were first shown to exist as minority charge carriers in oxides by Stotz and Wagner [194] in the 1960s. The systematic search of Takahashi and Iwahara [195] later demonstrated that acceptor doped perovskite-type oxides (e.g. doped $LaAlO_3, LaYO_3, SrZrO_3$), which were already known for their moderate oxide ion conductivity [196] could become proton conductors in water containing atmospheres. Unfortunately, the observed conductivities were still far too low to compete with the high oxide ion conductivity of yttria stabilized zirconia, the standard electrolyte material for SOFCs at that time. But soon, related compounds (in particular acceptor doped $SrCeO_3$ [197] and $BaCeO_3$ [198]) with higher proton conductivities were discovered and tested in different kind of electrochemical cells. For laboratory fuel cells, power densities up to 0.2 W/cm^2 have been achieved for a temperature of 800°C [199–204]. Although these cells were not optimized, for example, with respect to the thickness of the electrolyte (typically cut disks of 0.5-mm thickness were used) and the choice of the electrode materials (generally porous platinum or nickel for the anode; platinum, silver, or a cobaltate for the cathode) these early results already demonstrated that the use of proton-conducting oxides may provide the possibility to reduce the operation temperature of SOFCs. While this may help to overcome some compatibility problems related to electrode and interconnect materials (see above), the use of proton-conducting oxides also raises new problems.

In a conventional SOFC, carbon- and hydrogen-containing species (e.g. H_2, CO, CH_4) may be oxidized at high temperatures, while the use of proton-conducting separators only allows the de-hydrogenation of the fuel and hence limits the choice of fuels to hydrogen-rich gases. On the other hand a separation of exhaust (essentially water produced at the air side) from fuel (fuel processing) may be avoided. The very low stability of the cerates turns out to be a more severe problem. Several independent stability tests clearly demonstrated that both $SrCeO_3$ and $BaCeO_3$: (i) are only marginally thermodynamically stable with respect to the decomposition into the binary oxides [205]; (ii) react with low levels of CO_2 to form carbonates [206]; and (iii) form hydroxides at high water activities [207].

Recently, there have been several attempts to increase the stability of $BaCeO_3$-based proton conductors by forming solid solutions with $BaZrO_3$, which is thermodynamically much more stable [208]. For Gd- and Nd-doped materials, a decrease of the proton conductivity with increasing Zr content is generally observed. This is, however, only true for the bulk conductivity of Ce-rich compositions [209, 210]. For Ce-poor compositions, however, the observed decrease of the total conductivity [211] most likely reflects the behavior of grain boundaries rather than the bulk of the solid solutions, as discussed by Kreuer [212] for $BaZrO_3$-based oxides.

Apart from such empirical approaches, there has been a tremendous effort to understand the fundamentals of proton conduction in oxides in more detail rather than developing an SOFC technology based on available proton-conducting oxides. This has led to the discovery of high proton conductivity in other perovskite-based oxides (e.g. Ca-doped $Ba_3Nb_2CaO_{9-\delta}$ [213] or $Ba_2YSnO_{5.5}$ [214]) and the moderate conductivity in diverse, very stable rare earth oxides [215]. The initial results seemed to confirm that high proton conductivity and stability are mutually exclusive properties. But improved insight into the relation between the formation and mobility of protonic defects and the thermodynamic stability of oxides on the one hand, and structural and chemical parameters on the other hand side, eventually led to the development of stable oxides with high proton conductivity [212]. A few aspects of the fundamentals underlying this development will therefore be summarized in the following subsections.

3.2.6.1.1 Formation of Protonic Defects

For large band gap oxides, the most important reaction leading to the formation of protonic defects at moderate temperatures is the dissociative absorption of water, which requires the presence of oxide ion vacancies. The latter may be intrinsically present in the structure (e.g. in $Ba_2YSnO_{5.5}$) or they may be formed extrinsically by acceptor doping. In the case of perovskite-type oxides, one generally employs a substitution of up to 25% of the B-site cation by a lower-valent cation (e.g. in $Ba(Ce_{1-x}Y_x)O_{3-\delta}$). In order to form protonic defects, water from the gas phase dissociates into a hydroxide ion and a proton, the hydroxide ion filling an oxide ion vacancy and the proton forming a covalent bond with a lattice oxide ion. In Kröger–Vink notation this reaction is written as:

$$H_2O + V_O^{\bullet\bullet} + O_O^{\times} \rightleftharpoons 2OH_O^{\bullet} \tag{3}$$

by which two hydroxide ions substituting for oxide ions, that is, positively charged protonic defects ($OH_O^{\bullet}$), are formed. Of course such charged defects can diffuse into the bulk of the oxide only when accompanied by the counter diffusion of oxide ion vacancies ($V_O^{\bullet\bullet}$) [216]. This implies that such oxides must show some oxide ion conductivity in the dry state, as reported by Iwahara *et al.* [217]. Since a water molecule is eventually split into a hydroxide ion and a proton, this reaction may be considered an amphoteric reaction, where the oxide acts as an acid (absorption of hydroxide ions by oxide ion vacancies) as well as a base (protonation of lattice oxide ions). There has been some controversy about the extent to which structural and chemical parameters determine the enthalpy of that reaction [218, 219], but the growing data base now seems to confirm that the enthalpy of the hydration reaction tends to become more exothermic with decreasing electronegativity of the cations interacting with the lattice oxygen, that is, with decreasing Brønsted basicity of the oxide. This is illustrated for some acceptor-doped perovskite-type oxides (ABO_3) in Figure 3.2.12. The equilibrium constant of the hydration reaction decreases in the order cerate, zirconate, stannate, niobate, titanate, that is, with increasing electronegativity of the B-site cation. Similar trends are also observed for the choice of the acceptor dopant on the B-site [220] and the A-site cation, that is, protonic defects are stabilized in the order A = Ca, Sr, Ba. It should be mentioned that the solubility limit indeed approaches the dopant concentration for cubic perovskites [212], that is, structures with only one crystallographic site for oxygen. Any reduction in symmetry leads not only to a decrease of the solubility limit but also to a drop in the mobility of protonic defects (see below).

3.2.6.1.2 Stability versus Concentration of Protonic Defects

The observation that preferentially basic oxides stabilize protonic defects seems to suggest that the formation of stable protonic defects is not compatible with high stability of the oxide with respect to the reaction with acidic gases such as CO_2, SO_2 or SO_3. Indeed, the stability of the perovskite-type oxides shown in Figure 3.2.12 roughly decreases with increasing stability of protonic defects [208, 212]. Besides the oxide basicity various other parameters are of influence, which leaves significant freedom for material optimization. For example, a perfect chemical matching between the acceptor dopant and its environment appears to stabilize protonic defects entropically to higher temperatures [212, 220]. Just as observed for any reduction of the crystallographic symmetry [221], any local symmetry reduction (e.g. by doping) tends to destabilize protonic defects at high temperatures [222]. With respect to the stability of the oxide, not only geometrical parameters, such as the perovskite tolerance factor, but also the nature of the chemical interactions must be considered.

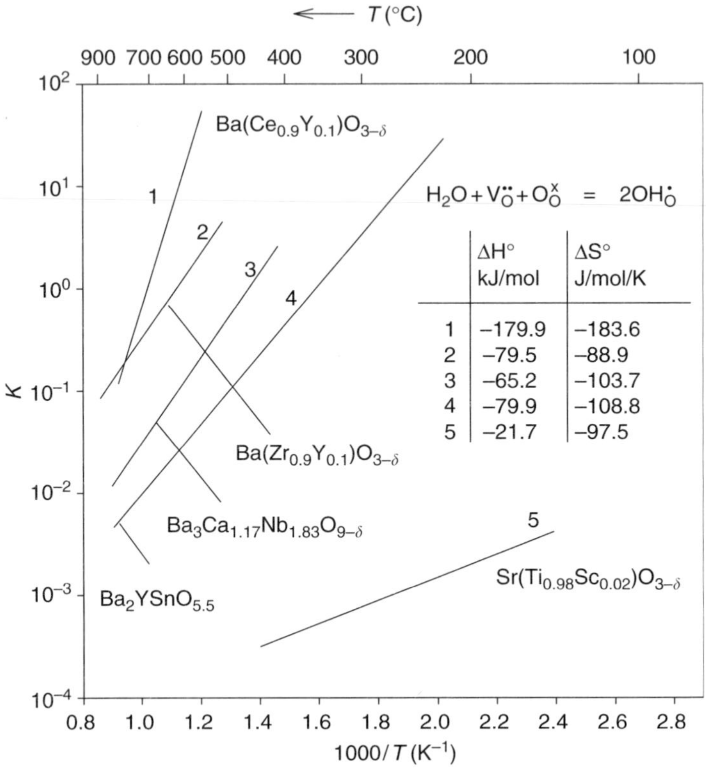

FIGURE 3.2.12 Equilibrium constant for the formation reaction of protonic defects (dissociative absorption of water) in different perovskite-type oxides (calculated from literature data [212]). The standard reaction enthalpies and entropies are given in the insert.

3.2.6.1.3 Mobility of Protonic Defects

The elementary reactions underlying the mobility of protonic defects have been investigated in great detail, experimentally as well as by computer simulations, and their contributions to the total activation enthalpy have been estimated.

The mobility of protonic defects essentially comprise the reorientation and transfer of a proton to a neighboring oxygen, either on the same or a neighboring octahedron. An important feature of this mechanism is the formation of short high-energy hydrogen bonds between the protonic defect and its 8–12 oxygen neighbors in the perovskite structure [223, 224]. In oxides with high proton mobility the free energy the system gains by hydrogen bonding, and the free energy it has to spend to aquire an adequate lattice deformation, are balanced over a wide range in configuration space, in particular over a wide range of OH/O separations (local lattice softening). This leads not only to strong bond length fluctuations, but also to rapid bond forming and breaking processes, which are the elementary reactions leading to defect reorientation.

During the lifetime of the hydrogen bonds, very short OH/O bond lengths may also occur with high probability, providing potential paths for proton transfer. As a result of the repulsive interaction between the B-site cation and the proton, however, the hydrogen bonds are strongly bent most of the time, which prevents the proton from being transfered. In the transition state configurations of proton transfer, one generally observes strongly elongated B–O bonds, which reduces the bending of the hydrogen bonds. Nevertheless, the remaining hydrogen bond bending leads to some remaining proton transfer barrier in the transition state configuration as indicated by pronounced H/D-isotope effects [225, 226].

Except for perovskite-type oxides with small lattice constants, that is, short oxygen separations, reorientation of protonic defects (rotational diffusion) is generally fast, and proton transfer is the rate limiting process. According to the above described interactions, the activation enthalpy of the latter exhibits contributions from the compression of the OH/O separation and elongation of the B/O and O/H bond. Not only symmetry reduction of the average structure, but also local symmetry perturbations, for example, by acceptor dopants, may significantly reduce the mobility of protonic defects [212].

3.2.6.1.4 The Case of Y-doped $BaZrO_3$

At the moment, highly Y-doped $BaZrO_3$ seems to be the only proton-conducting oxide, which meets the essential requirements for SOFC electrolyte materials. Surprisingly, the development of this material required very little compromising, which can be understood from the above considerations.

Its high symmetry is essential for the high solubility limit of protonic defects and for the high isotropic proton mobility. The large lattice constant reduces the Zr/H-repulsive interaction and, therefore, the activation enthalpy of the mobility of protonic defects. Only $Ba_3CaNb_2O_9$ has a similarly favorable lattice constant, but Ca/Nb ordering on the B-site leads to a symmetry reduction [227]. Less densely packed $BaCeO_3$ already shows a significant orthorhombic lattice distortion.

Another key feature is the availability of a nearly perfect acceptor dopant (i.e. a dopant which does not change the electronic structure of the oxygen). While in all other reported cases, the increase of the acceptor dopant concentration leads to a reduction of the proton mobility and an entropic destabilization of protonic defects [222], both the proton mobility and the thermodynamics of hydration are practically unchanged for dopant levels up to 20% Y in $BaZrO_3$ (Fig. 3.2.13). High proton mobility and entropically stabilized protonic defects even at high dopant concentrations and the high-solubility limit lead to the enormous proton

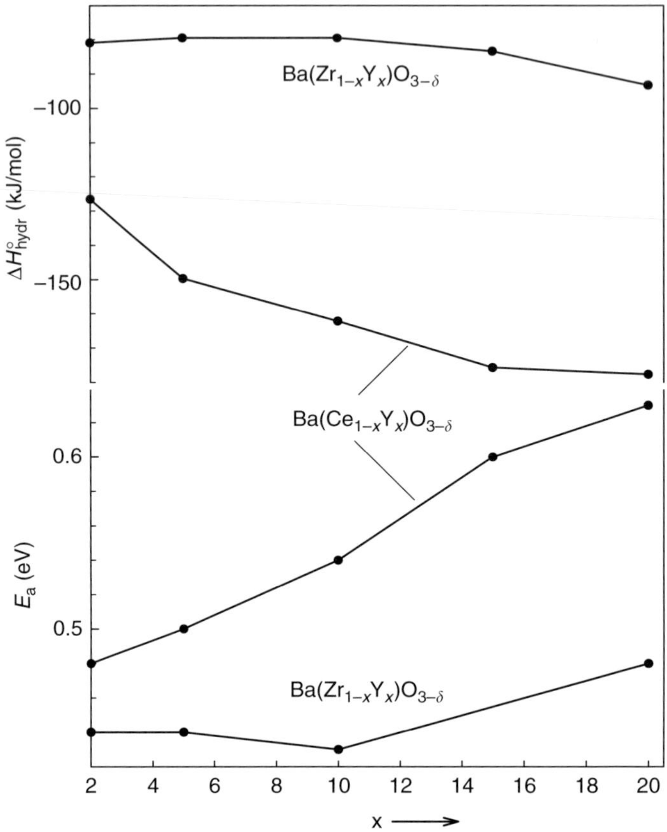

FIGURE 3.2.13 Standard hydration enthalpy ΔH°_{hydr} and activation enthalpy E_a of proton mobility for Y-doped $BaZrO_3$ and $BaCeO_3$ as a function of the dopant concentration [220, 222].

conductivity of this material. For temperatures below about 700°C and a water partial pressure of 23 hPa, this exceeds the oxide ion conductivity of the best oxide ion conductors (see section titled "Electrolytes" and Figure 3.2.14). Like $BaCeO_3$-based electrolytes, $BaZrO_3$-based oxides show some additional oxide ion conductivity above about 500°C [212, 220] which even renders possible the oxidation of traces of CO when being used as electrolyte in an SOFC. Although the conductivity of Y-doped $BaZrO_3$ is even slightly higher than the proton conductivity of $BaCeO_3$-based oxides (Fig. 3.2.14), the chemical stability is far more advantageous, as expected from the higher electronegativity of Zr compared to Ce. For the CO_2 partial pressure of air (380 ppm), pure $BaZrO_3$ is stable above 300°C, which is only slightly higher than for $BaTiO_3$ and $SrTiO_3$, which are known for their superior stabilities [208].

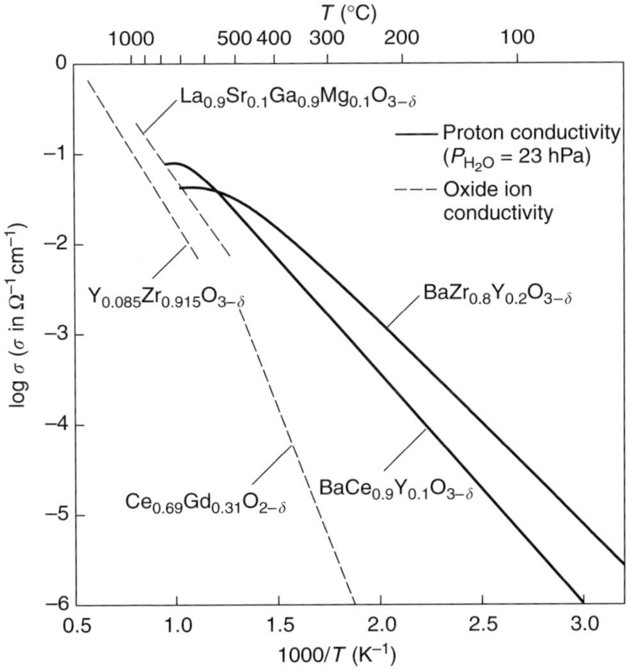

FIGURE 3.2.14 Proton conductivity of Y-doped $BaZrO_3$ [220] and $BaCeO_3$ [222] compared to the oxide ion conductivity of the best oxide ion conductors (see section titled "Currently used materials").

High proton conductivity, high stability and a wide ionic domain [228] make Y-doped $BaZrO_3$ an interesting alternative as SOFC electrolyte material. Even the formation of thin films has been proven to be possible, for example by sol–gel techniques [229]. Not much effort has been spent yet on the development of compatible electrode materials, but the promising properties of Y-doped $BaZrO_3$ and the encouraging performance of $BaCeO_3$-based electrolytes in fuel cells operating with pure hydrogen as a fuel (see above) provide a profound reason for such developments in the near future. For Y-doped $BaZrO_3$, the remaining grain boundary resistivities, the low thermal expansion coefficient ($\alpha = 6 \times 10^{-6}\,K^{-1}$), brittleness and the slight tetragonal distortion for highly doped materials require and leave some space for further optimization.

3.2.6.2 MONOLITHIC FUEL CELLS

Apart from the primary requirements for SOFC materials (see sections titled "Complications and demands" and "Requirements and commonly discussed materials"), the lack of their mutual compatibility frequently prevents them from being used in fuel cells. A well known example is $(La_{1-x}Sr_x)CoO_{3-\delta}$, which shows superior cathode properties in freshly prepared SOFCs, but its progressive reaction with YSZ leads to the formation of a growing, blocking layer of La_2ZrO_7. Also changes in sample dimensions due to temperature and P_{O_2} or even P_{H_2O} variations must be minimized, which puts additional constraints on the materials selection. A conceptually interesting approach to reduce some of these internal compatibility problems is based on using a single, crystalline structure/phase, whose composition is then modulated to achieve the desired functionality for the different parts of the cell. The center section of the material thus serves as a solid electrolyte, while the ends exposed to the fuel and the air may serve as anode and cathode, respectively.

One major problem to realize such a monolithic fuel cell is the choice of an adequate base material. Apart from very high ionic conductivity, additionally a very high electronic conductivity must be introduced for highly reducing and oxidizing environments by comparatively small compositional changes. So far, this has only been systematically investigated for oxides with the pyrochlore structure (general formula $A_2B_2O_7$) [230]. Especially for $Gd_2Ti_2O_7$, Ca doping on the A site leads to reasonably high oxygen ion conductivity ($\sim 10^{-2}$ S/cm at 1000°C) while Mo and Ru doping on the B site gives rise to electronic conductivity and catalytic activity under reducing [231] and oxidizing conditions [232], respectively. Unfortunately the choice of the dopant level for the part operating as an anode requires some compromise between electronic conductivity and stability at low external oxygen partial pressures [233, 234], which has prevented monolithic fuel cell concept based on $Gd_2Ti_2O_7$ from being verified. Nevertheless, it is definitely worth considering this concept not only for other oxide ion conductors but also for proton conducting oxides. In particular perovskite-type oxides should offer a high flexibility in this respect.

3.2.6.3 Single Chamber Fuel Cells

In conventional fuel cell designs, the electrolyte not only mediates the electrochemical reactions taking place at the anode and cathode, but also separates the fuel from the oxidant to prevent direct combustion. It was already recognized by Gool [235] that: (i) many fuels do not directly react with oxygen at typical fuel cell temperatures; and (ii) the catalytic activity of anode and cathode materials are quite selective to certain type of reactions. This renders possible single chamber fuel cells, that is, fuel cells without separation of anode and cathode compartment (sometimes called mixed gas fuel cells). Anode and cathode may then be placed either on the two sides of the electrolyte like in conventional designs or even on the same side generating electrical power in a flow of mixtures of fuel and oxidant. A quantitative relation between the open circuit voltage and the cell parameters, such as the selectivities of the electrodes, has been derived by Riess *et al.* [236].

Apart from the advantage of avoiding gas separation problems, surprisingly high power densities have been achieved experimentally with single chamber fuel cells. Hibino *et al.* [237] in particular have pushed forward the idea of single chamber fuel cells during the last few years. Initial designs were based on Gd-doped $BaCeO_3$ electrolytes with gold and platinum electrodes for selective reduction and oxidation of oxygen and methane, respectively. Later, they demonstrated, that YSZ-based fuel cells with conventional electrodes (Ni and LSM) may also operated as single chamber fuel cells even at high temperature ($T = 950°C$) [238]. Recently, a peak power density of $0.4\,W/cm^2$ has been demonstrated for a ceria-based single chamber fuel cell with Ni and $Sm_{0.5}Sr_{0.5}CoO_3$ electrodes operating in mixtures of ethane or propane and air at 500°C [239]. Since this power density clearly exceeds those being achieved with conventional designs, it may be suspected however that some direct combustion may locally increase the temperature and therefore the reaction rates for this type of fuel cell.

REFERENCES

1. Pehnt, M. (2002). Energierevolution Brennstoffzelle? pp. 31–34, Weinheim, Germany: Wiley-VCH.
2. Nernst, W. (1899). Über die elektrolytische Leitung fester Körper bei sehr hohen Temperaturen (On electrolytic conduction in solids at very high temperatures). *Z. Elektrochem.* **6**: 41–43.
3. Baur, E., and Preis, H. (1937). Über Brennstoffketten mit Festkörpern (On fuel chains with solids). *Z. Elektrochemie* 43: 727–732.
4. Kiukkola, K., and Wagner, C. (1957). Measurements on galvanic cells involving solid electrolytes. *J. Electrochem. Soc.* **104**: 379–387.
5. Kordesch, K., and Simader, G. (1996). Fuel Cells and their Applications, Weinheim, Germany: VCH.
6. Steele, B. C. H. (2001). Materials science and engineering: the enabling technology for the commercialisation of fuel cell systems. *J. Mater. Sci.* **36**: 1053–1068.
7. Singhal, S. C. (1996). Status of solid oxide fuel cell technology. High Temperature Electrochemistry: Ceramics and Metals, pp. 123–138, Poulsen, F. W., Bonanos, N., Linderoth, S., Mogensen, M., and Zachau-Christiansen, B., eds., Roskilde, Denmark: Risø National Laboratory.
8. Möbius, H.-H. (1997). On the history of solid electrolyte fuel cells. *J. Solid State Electrochem.* **1**: 2–16.
9. Kreuer, K. D., and Maier, J. (1995). Physikalisch-chemische Aspekte von Festelektrolyt-Brennstoffzellen (Physicochemical aspects of solid electrolyte fuel cells). *Spektrum der Wissenschaft* 7: 92–97.

10. Ostwald, W. (1894). Die wissenschaftliche Elektrochemie der Gegenwart und die technische der Zukunft (The contemporary scientific and the future technical electrochemistry). *Z. Elektrochem.* **1**: 122–125.
11. von Sturm, F. (1969). Elektrochemische Stromerzeugung (Electrochemical current generation), 1st edn, Weinheim, Germany: VCH.
12. Maier, J. (1995). Ionic conduction in space charge regions. *Prog. Solid State Chem.* **23**: 171–263.
13. Tuller, H. L., and Nowick, A. S. (1979). Defect structure and electrical properties of nonstoichiometric CeO_2 single crystals. *J. Electrochem. Soc.* **126**: 209–217.
14. Takahashi, T., Iwahara, H., and Nagai, Y. (1972). High oxide ion conduction in sintered Bi_2O_3 containing SrO, CaO or La_2O_3. *J. Appl. Electrochem.* **2**: 97–104.
15. Cahen, H. T., van den Belt, T. G. M., de Wit, J. H. W., and Broers, G. H. J. (1980). The electrical conductivity of δ-Bi_2O_3 stabilized by isovalent rare-earth oxides R_2O_3. *Solid State Ionics* **1**: 411–423.
16. Verkerk, M. J., and Burggraaf, A. J. (1981). High oxygen ion conduction in sintered oxides of the Bi_2O_3–DY_2O_3 system. *J. Electrochem. Soc.* **128**: 75–82.
17. Shuk, P., Wiemhöfer, H.-D., Guth, U., Göpel, W., and Greenblatt, M. (1996). Oxide ion conducting solid electrolytes based on Bi_2O_3. *Solid State Ionics* **89**: 179–196.
18. Takahashi, T., Esaka, T., and Iwahara, H. (1977). Conduction in Bi_2O_3-based oxide ion conductor under low oxygen pressure. 2. Determination of partial electronic conductivity. *J. Appl. Electrochem* **7**: 303–308.
19. Bouwmeester, H. J. M., Kruidhof, H., Burggraaf, A. J., and Gellings, P. J. (1992). Oxygen semipermeability of erbia-stabilized bismuth oxide. *Solid State Ionics* **53–56**: 460–468.
20. Abraham, F., Boivin, J. C., Mairesse, G., and Nowogrocki, G. (1990). The BIMEVOX series—a new family of high performances oxide ion conductors. *Solid State Ionics* **40–41**: 934–937.
21. Kendall, K. R., Navas, C., Thomas, J. K., and zur Loye, H.-C. (1996). Recent developments in oxide ion conductors: Aurivillius phases. *Chem. Mater.* **8**: 642–649.
22. Yan, J., and Greenblatt, M. (1995). Ionic conductivities of $Bi_4V_{2-x}M_xO_{11-x/2}$ (M = Ti, Zr, Sn, Pb) solid-solutions. *Solid State Ionics* **81**: 225–233.
23. Kramer, S., Spears, M., and Tuller, H. L. (1994). Conduction in titanate pyrochlores—role of dopants. *Solid State Ionics* **72**: 59–66.
24. Ishihara, T., Matsuda, H., and Takita, Y. (1994). Doped $LaGaO_3$ perovskite-type oxide as a new oxide ionic conductor. *J. Am. Ceram. Soc.* **116**: 3801–3803.
25. Feng, M., and Goodenough, J. B. (1994). A superior oxide-ion electrolyte. *Eur. J. Solid State Inorg. Chem.* **31**: 663–672.
26. Park, J. H., and Blumenthal, R. N. (1989). Electronic transport in 8 mole percent Y_2O_3–ZrO_2. *J. Electrochem. Soc.* **136**: 2867–2876.
27. Casselton, R. E. W. (1970). Low-field dc-conduction in yttria stabilized zirconia. *Phys. Stat. Sol. (a)* **2**: 571–585.
28. Weppner, W. (1977). Electronic transport properties and electrically induced p–n-junction in ZrO_2+10m/o Y_2O_3. *J. Solid State Chem.* **20**: 305–314.
29. Sasaki, K., and Maier, J. (2000). Re-analysis of defect equilibria and transport parameters in Y_2O_3-stabilized ZrO_2 using EPR and optical relaxation. *Solid State Ionics* **134**: 303–321.
30. Birkby, I., and Stevens, R. (1996). Applications of zirconia ceramics. *Key. Eng. Mater.* **122–124**: 527–552.
31. Weppner, W. (1992). Tetragonal zirconia polycrystals—a high performance solid oxygen conductor. *Solid State Ionics* **52**: 15–21.
32. Badwal, S. P. S. (1992). Zirconia-based solid electrolytes: microstructure, stability and ionic conductivity. *Solid State Ionics* **52**: 23–32.

33. Baumard, J. F., and Abelard, P. (1984). Defect structure and transport properties of ZrO_2-based solid electrolytes. Advances in Ceramics, Vol. 12, pp. 555–571, Claussen, N., Rühle, M., and Heuer, A. H., eds., Columbus, OH: The American Ceramic Society.
34. Takahashi, T. (1972). Solid electrolyte fuel cells (theoretics and experiments), *Physics of Electrolytes*, Vol. 2, pp. 980–1052, Hladic, J., ed., London: Academic Press.
35. Zacate, M. O., Minervini, L., Bradfield, D. J., Grimes, R. W., and Sickafus, K. E. (2000). Defect cluster formation in M_2O_3-doped cubic ZrO_2. *Solid State Ionics* **128**: 243–254.
36. Yamamuru, Y., Kawasaki, S., and Sakai, H. (1999). Molecular dynamics analysis of ionic conduction mechanism in yttria-stabilized zirconia. *Solid State Ionics* **126**: 181–189.
37. Meyer, M., Nicoloso, N., and Jaenisch, V. (1997). Percolation model for the anomalous conductivity of fluorite-related oxides. *Phys. Rev. B* **56**: 5961–5966.
38. Spiridonov, F. M., Popova, L. N., and Ya, R. (1970). On the phase relation and the electrical conductivity in the system ZrO_2–Sc_2O_3. *J. Solid State Chem.* **2**: 430–438.
39. Badwal, S. B. S., Ciacchi, F. T., Rajendran, S., and Drennan, J. (1998). An investigation of conductivity, microstructure and stability of electrolyte compositions in the system 9 mol% (Sc_2O_3–Y_2O_3)–ZrO_2(Al_2O_3). *Solid State Ionics* **109**: 167–186.
40. Nomura, K., Mizutani, Y., Kawai, M., Nakamura, Y., and Yamamoto, O. (2000). Aging and Raman scattering study of scandia and yttria doped zirconia. *Solid State Ionics* **132**: 235–239.
41. Badwal, S. B. S., Ciacchi, F. T., and Milosevic, D. (2000). Scandia–zirconia electrolytes for intermediate temperature solid oxide fuel cells. *Solid State Ionics* **136–137**: 91–99.
42. Fleig, J., and Maier, J. (1997). The influence of inhomogeneous potential distributions on the electrolyte resistance of SOFC. Vol. 97–40, pp. 1374–1383, "Proc. of the 5th Int. Symp. on Solid Oxide Fuel Cells" Stimming, U., Singhal, S. C., Tagawa, H., and Lehnert, W., eds., Pennington, NJ: The Electrochemical Society.
43. Fleig, J., and Maier, J. (1997). The influence of laterally inhomogeneous contacts on the impedance of solid materials: a three-dimensional finite-element study. *J. Electroceram.* **1**: 73–89.
44. van Heuveln, F. H., Bouwmeester, H. J. M., and van Berkel, F. P. F. (1997). Electrode properties of Sr-doped $LaMnO_3$ on yttria-stabilized zirconia. 1. Three-phase boundary area. *J. Electrochem. Soc.* **144**: 126–133.
45. Steele, B. C. H. (2000). Appraisal of $Ce_{1-y}Gd_yO_{2-y/2}$ electrolytes for IT-SOFC operation at 500 degrees C. *Solid State Ionics* **129**: 95–110.
46. Huang, K., Feng, M., and Goodenough, J. B. (1998). Synthesis and electrical properties of dense $Ce_{0.9}Cd_{0.1}O_{1.95}$ ceramics. *J. Am. Ceram. Soc.* **81**: 357–362.
47. Minervini, L., Zacate, M. O., and Grimes, R. W. (1999). Defect cluster formation in M_2O_3-doped CeO_2. *Solid State Ionics* **116**: 339–349.
48. Mogensen, M., Sammes, N. M., and Tompsett, G. A. (2000). Physical, chemical and electrochemical properties of pure and doped ceria. *Solid State Ionics* **129**: 63–94.
49. Gödickemeier, M., and Gauckler, L. J. (1998). Engineering of solid oxide fuel cells with ceria-based electrolytes. *J. Electrochem. Soc.* **145**: 414–421.
50. Lübke, S., and Wiemhöfer, H.-D. (1999). Electronic conductivity of Gd-doped ceria with additional Pr-doping. *Solid State Ionics* **117**: 229–243.
51. Mogensen, M., Lindegaard, T., and Hansen, U. R. (1994). Physical properties of mixed conductor solid oxide fuel-cell anodes of doped CeO_2. *J. Electrochem. Soc.* **141**: 2122–2128.
52. Yahiro, H., Eguchi, K., and Arai, H. (1986). Ionic conduction and microstructure of the ceria-strontia system. *Solid State Ionics* **21**: 37–47.

53. Riess, I., Gödickemeier, M., and Gauckler, L. J. (1996). Characterization of solid oxides fuel cells based on solid electrolytes or mixed ionic electronic conductors. *Solid State Ionics* **90**: 91–104.
54. Mogensen, M., Primdahl, S., Jorgensen, M. J., and Bagger, C. (2000). Composite electrodes in solid oxide fuel cells and similar solid state devices. *J. Electroceram.* **5**: 141–152.
55. Stevenson, J. W., Armstrong, T. R., McCready, D. E., Pederson, L. R., and Weber, W. J. (1997). Processing and electrical properties of alkaline earth-doped lanthanum gallate. *J. Electrochem. Soc.* **144**: 3613–3620.
56. Huang, K., Tichy, R. S., and Goodenough, J. B. (1998). Superior perovskite oxide-ion conductor; strontium- and magnesium-doped $LaGaO_3$: I, phase relationships and electrical properties. *J. Am. Ceram. Soc.* **81**: 2565–2575.
57. Huang, P., and Petric, A. (1996). Superior oxygen ion conductivity of lanthanum gallate doped with strontium and magnesium. *J. Electrochem. Soc.* **143**: 1644–1648.
58. Trofimenko, N., and Ullmann, H. (1999). Transition metal doped lanthanum gallates. *Solid State Ionics* **118**: 215–227.
59. Kharton, V. V., Viskup, A. P., Yaremchenko, A. A., Baker, R. T., Gharbage, B., Mather, G. C., Figueiredo, F. M., Naumovich, E. N., and Marques, F. M. B. (2000). Ionic conductivity of $La(Sr)Ga(Mg, M)O_{3-\delta}$ (M = Ti, Cr, Fe, Co, Ni): effects of transition metal dopants. *Solid State Ionics* **132**: 119–130.
60. Ishihara, T., Akbay, T., Furutani, H., and Takita, Y. (1998). Improved oxide ion conductivity of Co doped $La_{0.8}Sr_{0.2}Ga_{0.8}Mg_{0.2}O_3$ perovskite type oxide. *Solid State Ionics* **113–115**: 585–591.
61. Trofimenko, N., and Ullmann, H. (1999). Co-doped LSGM: composition–structure–conductivity relations. *Solid State Ionics* **124**: 263–270.
62. Yamaji, K., Horita, T., Ishikawa, M., Sakai, N., and Yokokawa, H. (1999). Chemical stability of the $La_{0.9}Sr_{0.1}Ga_{0.8}Mg_{0.2}O_{2.85}$ electrolyte in a reducing atmosphere. *Solid State Ionics* **121**: 217–224.
63. Yamaji, K., Negishi, H., Horita, T., Sakai, N., and Yokokawa, H. (2000). Vaporization process of Ga from doped $LaGaO_3$ electrolytes in reducing atmospheres. *Solid State Ionics* **135**: 389–396.
64. Tao, S. W., Poulsen, F. W., Meng, G. Y., and Sorensen, O. T. (2000). High-temperature stability study of the oxygen-ion conductor $La_{0.9}Sr_{0.1}Ga_{0.8}Mg_{0.2}O_{3-x}$. *J. Mater. Chem.* **10**: 1829–1833.
65. Nguyen, T. L., and Dokiya, M. (2000). Electrical conductivity, thermal expansion and reaction of $(La, Sr)(Ca, Mg)O_3$ and $(La, Sr)AlO_3$ system. *Solid State Ionics* **132**: 217–226.
66. Bauerle, J. E. (1969). Study of solid electrolyte polarization by a complex admittance method. *J. Phys. Chem. Solids* **30**: 2657–2670.
67. Badwal, S. P. S. (1995). Grain boundary resistivity in zirconia-based materials: effect of sintering temperatures and impurities. *Solid State Ionics* **76**: 67–80.
68. van Dijk, T., and Burggraaf, A. J. (1981). Grain boundary effects on ionic conductivity in ceramics $Gd_xZr_{1-x}O_{2-(x/2)}$ solid solutions. *Phys. Stat. Sol. (a)* **63**: 229–240.
69. Verkerk, M. J., Middelhuis, B. J., and Burggraaf, A. J. (1982). Effect of grain boundaries on the conductivity of high-purity ZrO_2–Y_2O_3 ceramics. *Solid State Ionics* **6**: 159–170.
70. Badwal, S. P. S., and Drennan, J. (1987). Yttria–zirconia: effect of microstructure on conductivity. *J. Mater. Sci.* **22**: 3231–3239.
71. Kleitz, M., Bernard, H., Fernandez, E., and Schouler, E. (1981). Impedance spectroscopy and electrical resistance measurements on stabilized zirconia. Advances in Ceramics. Science and Technology of Zirconia, Vol. 3, 310–336, Heuer, A. H., and Hobbs, L. W., eds., Washington, DC: The American Ceramic Society.

72. Steil, M. C., Thevenot, F., and Kleitz, M. (1997). Densification of yttria-stabilized zirconia. *J. Electrochem. Soc.* **144**: 390–398.
73. Badwal, S. P. S., and Rajendran, S. (1994). Effect of micro- and nano-structures on the properties of ionic conductors. *Solid State Ionics* **70–71**: 83–95.
74. Aoki, M., Chiang, Y.-M., Kosacki, I., Jong-Ren Lee, L., and Tuller, H. Liu, Y. (1996). Solute segregation and grain boundary impedance in high-purity stabilized zirconia. *J. Am. Ceram. Soc.* **79**: 1169–1180.
75. Guo, X., and Maier, J. (2001). Grain boundary blocking effect in zirconia: a Schottky barrier analysis. *J. Electrochem. Soc.* **148**: E121–E126.
76. Gödickemeier, M., Michel, B., Orliukas, A., Bohac, P., Sasaki, K., Gauckler, L., Heinrich, H., Schwander, P., Kostorz, K., Hofmann, H., and Frei, O. (1994). Effect of intergranular glass films on the electrical conductivity of 3Y-TZP. *J. Mater. Res.* **9**: 1228–1240.
77. Gerhardt, R., and Nowick, A. S. (1986). Grain-boundary effect in ceria doped with trivalent cations: I. Electrical measurements. *J. Am. Ceram. Soc.* **69**: 641–646.
78. Christie, G. M., and van Berkel, F. P. F. (1996). Microstructure–ionic conductivity relationship in ceria–gadolinia electrolytes. *Solid State Ionics* **83**: 17–27.
79. Adham, K. E., and Hammou, H. (1983). Grain boundary effect on ceria based solid solutions. *Solid State Ionics* **9–10**: 905–912.
80. Fleig, J., and Maier, J. (1999). Finite-element calculations on the impedance of electroceramics with highly resistive grain boundaries: I. Laterally inhomogeneous grain boundaries. *J. Am. Ceram. Soc.* **82**: 3485–3493.
81. Huang, K., Tichy, R. S., and Goodenough, J. B. (1998). Superior perovskite oxide-ion conductor; strontium- and magnesium-doped $LaGaO_3$: II, ac impedance spectroscopy. *J. Am. Ceram. Soc.* **81**: 2576–2580.
82. Lybye, D., Poulsen, F. W., and Mogensen, M. (2000). Conductivity of A- and B-site doped $LaAlO_3$, $LaGaO_3$, $LaScO_3$ and $LaInO_3$ perovskites. *Solid State Ionics* **128**: 91–103.
83. Maier, J. (1996). On the conductivity of polycrystalline materials. *Ber. Bunsenges. Phys. Chem.* **90**: 26–33.
84. Fleig, J., Rodewald, S., and Maier, J. (2000). Spatially resolved measurements of single grain boundaries using microcontact impedance spectroscopy. *Solid State Ionics* **135**: 905–912.
85. Moya, E. G. (1994). Some apects of grain boundary diffusion in oxides. Science of Ceramics Interfaces II, 277–309, Novotny, J., ed., Amsterdam: Elsevier Science.
86. Singhal, S. C. (2000). Advances in solid oxide fuel cell technology. *Solid State Ionics* **135**: 305–313.
87. Hayashi, H., Kanoh, M., Quan, C. J., Inaba, H., Wang, S., Dokiya, M., and Tagawa, H. (2000). Thermal expansion of Gd-doped ceria and reduced ceria. *Solid State Ionics* **132**: 227–232.
88. Mori, M., Hiei, Y., Yamamoto, T., and Itoh, H. (1999). Lanthanum alkaline-earth manganites as a cathode material in high-temperature solid oxide fuel cells. *J. Electrochem. Soc.* **146**: 4041–4047.
89. Sunde, S. (1996). Monte Carlo simulations of polarization resistance of composite electrodes for solid oxide fuel cells. *J. Electrochem. Soc.* **143**: 1930–1939.
90. Sunde, S. (2000). Simulations of composite electrodes in fuel cells. *J. Electroceram.* **5**: 153–182.
91. Costamagna, P., Costa, P., and Antonucci, V. (1998). Micro-modelling of solid oxide fuel cell electrodes. *Electrochim. Acta* **43**: 375–394.

92. Tanner, C. W., Fung, K.-Z., and Virkar, A. V. (1997). The effect of porous composite electrode structure on solid oxide fuel cell performance. 1. Theoretical analysis. *J. Electrochem. Soc.* **144**: 21–30.
93. Fleig, J. (2003). On the width of the electrochemically active region in mixed conducting solid oxide fuel cell cathodes. *J. Power Sources.*
94. Siebert, E., Hammouche, A., and Kleitz, M. (1995). Impedance spectroscopy analysis of $La_{1-x}Sr_xMnO_3$ yttria-stabilized zirconia electrode kinetics. *Electrochim. Acta* **40**: 1741–1753.
95. Steele, B. C. H. (1996). Survey of materials selection for ceramic fuel cells. 2. Cathodes and anodes. *Solid State Ionics* **86–88**: 1223–1234.
96. Steele, B. C. H. (1997). Behaviour of porous cathodes in high temperature fuel cells. *Solid State Ionics* **94**: 239–248.
97. De Souza, R. A., and Kilner, J. A. (1998). Oxygen transport in $La_{1-x}Sr_xMn_{1-y}Co_yO_{3+\delta}$ perovskites—Part I. Oxygen tracer diffusion. *Solid State Ionics* **106**: 175–187.
98. Adler, S. B., Lane, J. A., and Steele, B. C. H. (1996). Electrode kinetics of porous mixed-conducting oxygen electrodes. *J. Electrochem. Soc.* **143**: 3554–3564.
99. Deng, H., Zhou, M., and Abeles, B. (1994). Diffusion reaction in mixed ionic–electronic solid oxide membranes with porous electrodes. *Solid State Ionics* **74**: 75–84.
100. Svensson, A. M., Sunde, S., and Nisancioglu, K. (1998). Mathematical modeling of oxygen exchange and transport in air–perovskite–yttria-stabilized zirconia interface regions—II. Direct exchange of oxygen vacancies. *J. Electrochem. Soc.* **145**: 1390–1400.
101. Liu, M. (1998). Equivalent circuit approximation to porous mixed-conducting oxygen electrodes in solid state cells. *J. Electrochem. Soc.* **145**: 142–154.
102. Kertesz, M., Riess, I., Tannhauser, D. S., Langpape, R., and Rohr, F. J. (1982). Structure and electrical conductivity of $La_{0.84}Sr_{0.16}MnO_3$. *J. Solid State Chem.* **42**: 125–129.
103. Ahlgren, E. O., and Poulsen, F. W. (1996). Thermoelectric power and electrical conductivity of strontium-doped lanthanum manganite. *Solid State Ionics* **86–88**: 1173–1178.
104. Kuo, J. H., Anderson, H. U., and Sparlin, D. M. (1990). Oxidation reduction behavior of undoped and Sr-doped $LaMnO_3$ defect structure, electrical conductivity, and thermoelectric power. *J. Solid St. Chem.* **87**: 55–63.
105. Sasaki, K., Wurth, J.-P., Gschwend, R., Gödickemeier, M., and Gauckler, L. J. (1996). Microstructure–property relations of solid oxide fuel cell cathodes and current collectors—cathodic polarization and ohmic resistance. *J. Electrochem. Soc.* **143**: 530–543.
106. Kuo, J. H., Anderson, H. U., and Sparlin, D. M. (1989). Oxidation reduction behavior of undoped and Sr-doped $LaMnO_3$ nonstoichiometry and defect structure. *J. Solid State Chem.* **83**: 52–60.
107. Mizusaki, J., Tagawa, H., Naraya, K., and Sasamoto, T. (1991). Nonstoichiometry and thermochemical stability of the perovskite type $La_{1-x}Sr_xMnO_{3-\sigma}$. *Solid State Ionics* **49**: 111–118.
108. Mizusaki, J., Mori, N., Takai, H., Yonemura, Y., Minamiue, H., Tagawa, H., Dokiya, M., Inaba, H., Naraya, K., Sasamoto, T., and Hashimoto, T. (2000). Oxygen nonstoichiometry and defect equilibrium in the perovskite-type oxides $La_{1-x}Sr_xMnO_{3+\delta}$. *Solid State Ionics* **129**: 163–177.
109. van Rosmalden, J. A. M., and Cordfunke, E. H. P. (1994). The defect chemistry of $LaMnO_{3+\delta}$. 4. Defect model for $LaMnO_{3+\delta}$. *J. Solid State. Chem.* **110**: 109–112.
110. Poulsen, F. W. (2000). Defect chemistry modelling of oxygen stoichiometry, vacancy concentrations, and conductivity of $(La_{1-x}Sr_x)_yMnO_{3+\delta}$. *Solid State Ionics* **129**: 145–162.
111. Mizusaki, J., Saito, T., and Tagawa, H. (1996). A chemical diffusion-controlled electrode reaction at the compact $La_{1-x}Sr_xMnO_3$-stabilized zirconia interface in oxygen atmospheres. *J. Electrochem. Soc.* **143**: 3065–3073.

112. Endo, A., Ihara, M., Komiyama, H., and Yamada, K. (1996). Cathodic reaction mechanism for dense Sr-doped lanthanum manganite electrodes. *Solid State Ionics* **86–88**: 1191–1195.
113. Brichzin, V., Fleig, J., Habermeier, H.-U., Cristiani, G., and Maier, J. (2003). The geometry dependence of the polarization resistance of Sr-doped $LaMnO_3$ microelectrodes on yttria stabilized zirconia. *Solid State Ionics* **152–153**: 499–507.
114. Brichzin, V., Fleig, J., Habermeier, H.-U., and Maier, J. (2000). Geometry dependence of cathode polarization in SOFCs investigated by defined Sr-doped $LaMnO_3$-microelectrodes. *Electrochem. Solid State Lett.* **3**: 403–406.
115. Gharbage, B., Pagnier, T., and Hammou, A. (1994). Oxygen reduction at $La_{0.5}Sr_{0.5}MnO_3$ thin film yttria stabilized zirconia interface studied by impedance spectroscopy. *J. Electrochem. Soc.* **138**: 2118–2121.
116. Ioroi, T., Hara, T., Uchimoto, Y., Ogumi, Z., and Takehara, Z. (1997). Preparation of perovskite-type $La_{1-x}Sr_xMnO_3$ films by vapor-phase processes and their electrochemical properties. *J. Electrochem. Soc.* **144**: 1362–1370.
117. Bouwmeester, H. J. M., Kruidhof, H., and Burggraaf, A. J. (1994). Importance of the surface exchange kinetics as rate-limiting step in oxygen permeation through mixed-conducting oxides. *Solid State Ionics* **72**: 185–194.
118. Mizusaki, J., Tagawa, H., Tsuneyoshi, K., and Sawata, A. (1991). Reaction-kinetics and microstructure of the solid oxide fuel-cells air electrode $La_{0.6}Ca_{0.4}MnO_3$/YSZ. *J. Electrochem. Soc.* **138**: 1867–1873.
119. van Herle, J., McEvoy, A. J., and Thampi, K. R. (1996). A study on the $La_{1-x}Sr_xMnO_3$ oxygen cathode. *Electrochim. Acta* **41**: 1447–1454.
120. Mitterdorfer, A., and Gauckler, L. J. (1998). $La_2Zr_2O_7$ formation and oxygen reduction kinetics of the $La_{0.85}Sr_{0.15}Mn_yO_3$, O_2(g) vertical bar YSZ system. *Solid State Ionics* **111**: 185–218.
121. Østergård, M. J. L., and Mogensen, M. (1993). AC-impedance study of the oxygen reduction-mechanism on $La_{1-x}Sr_xMnO_3$ in solid oxide fuel-cells. *Electrochim. Acta* **38**: 2015–2020.
122. Lee, H. Y., Cho, W. S., Oh, S. M., Wiemhöfer, H.-D., and Göpel, W. (1995). Active reaction sites for oxygen reduction in $La_{0.9}Sr_{0.1}MnO_3$/YSZ electrodes. *J. Electrochem. Soc.* **142**: 2659–2664.
123. Fukunaga, H., Ihara, M., Sakaki, K., and Yamada, K. (1996). The relationship between overpotential and the three phase boundary length. *Solid State Ionics* **86–88**: 1179–1185.
124. Kamata, H., Hosaka, A., Mizusaki, J., and Tagawa, H. (1998). High temperature electrocatalytic properties of the SOFC air electrode $La_{0.8}Sr_{0.2}MnO_3$/YSZ. *Solid State Ionics* **106**: 237–245.
125. Hammouche, A., Siebert, E., Hammou, A., Kleitz, M., and Caneiro, A. (1991). Electrocatalytic properties and nonstoichiometry of the high-temperature air electrode $La_{1-x}Sr_xMnO_3$. *J. Electrochem. Soc.* **138**: 1212–1216.
126. Odgaard, M., and Skou, E. (1996). SOFC cathode kinetics investigated by the use of cone shaped electrodes: the effect of polarization and mechanical load. *Solid State Ionics* **86–88**: 1217–1222.
127. van Berkel, F. P. F., van Heuveln, F. H., and Huijsmans, J. P. P. (1994). Characterization of solid oxide fuel-cell electrodes by impedance spectroscopy and I–V characteristics. *Solid State Ionics* **72**: 240–247.
128. Wang, S. Z., Jiang, Y., Zhang, Y. H., Yan, J. W., and Li, W. Z. (1998). The role of 8 mol% yttria stabilized zirconia in the improvement of electrochemical performance of lanthanum manganite composite electrodes. *J. Electrochem. Soc.* **146**: 1932–1939.

129. van Heuveln, F. H., and Bouwmeester, H. J. M. (1997). Electrode properties of Sr-doped $LaMnO_3$ on yttria-stabilized zirconia. 2. Electrode kinetics. *J. Electrochem. Soc.* **144**: 134–140.
130. Horita, T., Yamaji, K., Ishikawa, M., Sakai, M., Yokokawa, H., Kawada, T., and Kato, T. (1998). Active sites imaging for oxygen reduction at the $La_{0.9}Sr_{0.1}MnO_{3-x}$/yttria-stabilized zirconia interface by secondary ion mass spectrometry. *J. Electrochem. Soc.* **145**: 3196–3202.
131. Matsuzaki, Y., and Yasuda, I. (1999). Relationship between the steady-state polarization of the SOFC air electrode, $La_{0.6}Sr_{0.4}MnO_{3+\delta}$/YSZ, and its complex impedance measured at the equilibrium potential. *Solid State Ionics* **126**: 307–313.
132. Juhl, M., Primdahl, S., Manon, C., and Mogensen, M. (1996). Performance/structure correlation for composite SOFC cathodes. *J. Power Sources* **61**: 173–181.
133. Kenjo, T., and Nishiya, M. (1992). $LaMnO_3$ air cathodes containing ZrO_2 electrolyte for high-temperature solid oxide fuel-cells. *Solid State Ionics* **57**: 295–302.
134. Østergård, M. J., Clausen, C., Bagger, C., and Mogensen, M. (1995). Manganite–zirconia composite cathodes for SOFC—influence of structure and composition. *Electrochim. Acta* **40**: 1971–1981.
135. Murray, E. P., Tsai, T., and Barnett, S. A. (1998). Oxygen transfer processes in (La, Sr)MnO_3/Y_2O_3-stabilized ZrO_2 cathodes: an impedance spectroscopy study. *Solid State Ionics* **110**: 235–243.
136. van Roosmalen, J. A. M., and Cordfunke, E. H. P. (1992). Chemical reactivity and interdiffusion of (La, Sr)MnO_3 and (Zr, Y)O_2, solid oxide fuel-cell cathode and electrolyte materials. *Solid State Ionics* **52**: 303–312.
137. Yokokawa, H., Sakai, T., Kawada, T., and Dokiya, M. (1991). Thermodynamic analysis of reaction profiles between $LaMO_3$ (M = Ni, Co, Mn) and ZrO_2. *J. Electrochem. Soc.* **138**: 2719–272.
138. Wiik, K., Schmidt, C. R., Faaland, S., Shamsili, S., Einarsrud, M.-A., and Grande, T. (1999). Reactions between strontium-substituted lanthanum manganite and yttria-stabilized zirconia: I. Powder samples. *J. Am. Ceram. Soc.* **82**: 721–728.
139. Lee, H. Y., and Oh, S. M. (1996). Origin of cathodic degradation and new phase formation at the $La_{0.9}Sr_{0.1}MnO_3$/YSZ interface. *Solid State Ionics* **90**: 133–140.
140. Kawada, T., and Yokokawa, H. (1997). Materials and characterization of solid oxide fuel cells. *Key Eng. Mater.* **125–126**: 187–248.
141. Majewski, P., Epple, L., and Aldinger, F. (2000). Phase-diagram studies in the La_2O_3–SrO–CaO–Mn_3O_4 system at 1200 degrees C in air. *J. Am. Ceram. Soc.* **83**: 1513–1517.
142. Majewski, P., Epple, L., Rozumek, M., Schluckwerder, H., and Aldinger, F. (2000). Phase diagram studies in the quasi binary systems $LaMnO_3$–$SrMnO_3$ and $LaMnO_3$–$CaMnO_3$. *J. Mater. Res.* **15**: 1161–1166.
143. Minh, N. Q., and Takahashi, T. (1995). Cathode—thermal expansion. Science and Technology of Ceramic Fuel Cells, pp. 135–137, Minh, N. Q., and Takahashi, T., eds., Amsterdam: Elsevier Science.
144. Hammouche, A., Siebert, E., and Hammou, A. (1989). Crystallographic, thermal and electrochemical properties of the system $La_{1-x}Sr_xMnO_3$ for high-temperature solid electrolyte fuel-cells. *Mater. Res. Bull.* **2**: 367–380.
145. Minh, N. Q. (1993). Ceramic fuel cells. *J. Am. Ceram. Soc.* **76**: 563–588.
146. van Hassel, B. A., Kawada, T., Sakai, N., Yokokawa, H., Dokiya, M., and Bouwmeester, H. J. M. (1993). Oxygen permeation modeling of perovskites. *Solid State Ionics* **66**: 295–305.
147. Teraoka, Y., Nobunaga, T., Okamoto, K., Miura, N., and Yamazoe, N. (1991). Influence of constituent metal-cations in substituted $LaCoO_3$ on mixed conductivity and oxygen permeability. *Solid State Ionics* **48**: 207–121.

148. Chen, C. H., Bouwmeester, H. J. M., van Doorn, R. H. E., Kruidhof, H., and Burggraaf, A. J. (1997). Oxygen permeation of $La_{0.3}Sr_{0.7}CoO_{3-\delta}$. *Solid State Ionics* **98**: 7–13.
149. Figueiredo, F. M., Marques, F. M. B., and Frade, J. R. (1998). Electrical properties and thermal expansion of $LaCoO_3/La_2(Zr, Y)_2O_7$ composites. *Solid State Ionics* **111**: 173–182.
150. Zipprich, W., Waschilewski, S., Rocholl, F., and Wiemhöfer, H.-D. (1997). Improved preparation of $La_{1-x}Me_xCoO_{3-\delta}$ (Me = Sr, Ca) and analysis of oxide ion conductivity with ion conducting microcontacts. *Solid State Ionics* **101–103**: 1015–1023.
151. Zipprich, W., and Wiemhöfer, H. D. (2000). Measurement of ionic conductivity in mixed conducting compounds using solid electrolyte microcontacts. *Solid State Ionics* **135**: 699–707.
152. De Souza, R. A., and Kilner, J. A. (1999). Oxygen transport in $La_{1-x}Sr_xMn_{1-y}Co_yO_{3+\delta}$ perovskites. Part II. Oxygen surface exchange. *Solid State Ionics* **126**: 153–161.
153. Carter, S., Selcuk, A., Chater, R. J., Kajda, J., Kilner, J. A., and Steele, B. C. S. (1992). Oxygen transport in selected nonstoichiometric perovskite-structure oxides. *Solid State Ionics* **53–56**: 597–605.
154. Tai, L.-W., Nasrallah, M. M., Anderson, H. U., Sparlin, D. M., and Sehlin, S. R. (1995). Structure and electrical properties of $La_{1-x}Sr_xCo_{1-y}Fe_yO_3$. 1. The system $La_{0.8}Sr_{0.2}Co_{1-y}Fe_yO_3$. *Solid State Ionics* **76**: 259–271.
155. Mizusaki, J., Tabuchi, J., Matsuura, T., Yamauchi, S., and Fueki, K. (1989). Electrical conductivity and Seebeck coefficient of nonstoichiometric $La_{1-x}Sr_xCoO_{3-\delta}$. *J. Electrochem. Soc.* **136**: 2082–2088.
156. Petric, A., Huang, P., and Tietz, F. (2000). Evaluation of La–Sr–Co–Fe–O perovskites for solid oxide fuel cells and gas separation membranes. *Solid State Ionics* **13**: 719–725.
157. Minh, N. Q., and Takahashi, T. (1995). Cathode—other materials, Science and Technology of Ceramic Fuel Cells, p. 139, Minh, N. Q., and Takahashi, T., eds., Amsterdam: Elsevier Science.
158. Inoue, T., Setoguchi, T., Eguchi, K., and Arai, H. (1989). Study of a solid oxide fuel cell with a ceria-based solid electrolyte. *Solid State Ionics* **35**: 285–291.
159. Eguchi, K., Setoguchi, T., Inoue, T., and Arai, H. (1992). Electrical properties of ceria-based oxides and their application to solid oxide fuel cells. *Solid State Ionics* **52**: 165–172.
160. Virkar, A. V. (1991). Theoretical analysis of solid oxide fuel cells with 2-layer, composite electrolytes—Electrolyte stability. *J. Electrochem. Soc.* **138**: 1481–1487.
161. Tsoga, A., Naoumidis, A., and Stöver, D. (2000). Total electrical conductivity and defect structure of ZrO_2–CeO_2–Y_2O_3–Gd_2O_3 solid solutions. *Solid State Ionics* **135**: 403–409.
162. McEvoy, A. J. (2000). Thin SOFC electrolytes and their interfaces—a near-term research strategy. *Solid State Ionics* **132**: 159–165.
163. Tsai, T., Perry, E., and Barnett, S. (1997). Low-temperature solid-oxide fuel cells utilizing thin bilayer electrolytes. *J. Electrochem. Soc.* **144**: L130–L132.
164. Marques, F. M. B., and Navarro, L. M. (1997). Performance of double layer electrolyte cells. 2. GCO/YSZ, a case study. *Solid State Ionics* **100**: 29–38.
165. Uchida, H., Arisaka, S., and Watanabe, M. (1999). High performance electrode for medium-temperature solid oxide fuel cells—La(Sr)CoO_3 cathode with ceria interlayer on zirconia electrolyte. *Electrochem. Solid State Lett.* **2**: 428–430.
166. Chen, C. C., Nasrallah, M. M., and Anderson, H. U. (1995). Immittance response of $La_{0.6}Sr_{0.4}Co_{0.2}Fe_{0.8}O_3$ based electrochemical cells. *J. Electrochem. Soc.* **142**: 491–496.
167. Steele, B. C. S. (2000). Materials for IT-SOFC stacks. 35 years R&D: the inevitability of gradualness? *Solid State Ionics* **134**: 3–20.
168. Worell, W. L. (1992). Electrical properties of mixed-conducting oxides having high oxygen-ion conductivity. *Solid State Ionics* **52**: 147–151.

169. Ishihara, T., Honda, M., Shibayama, T., Minami, H., Nishiguchi, H., and Takita, Y. (1998). Intermediate temperature solid oxide fuel cells using a new $LaGaO_3$ based oxide ion conductor—I. Doped $SmCoO_3$ as a new cathode material. *J. Electrochem. Soc.* **145**: 3177–3183.
170. Feng, M., Goodenough, J. B., Huang, K. Q., and Milliken, C. (2000). Fuel cells with doped lanthanum gallate electrolyte. *J. Power Sources* **63**: 47–63.
171. Horita, T., Yamaji, K., Sakai, N., Yokokawa, H., Weber, A., and Ivers-Tiffée, E. (2000). Stability at $La_{0.6}Sr_{0.4}CoO_{3-\delta}$ cathode/$La_{0.8}Sr_{0.2}Ga_{0.8}Mg_{0.2}O_{2.8}$ electrolyte interface under current flow for solid oxide fuel cells. *Solid State Ionics* **133**: 143–152.
172. Lecarpentier, F., Tuller, H. L., and Long, N. (2000). Performance of $La_{0.9}Sr_{0.1}Ga_{0.5}Ni_{0.5}O_3$ as a cathode for a lanthanum gallate fuel cell. *J. Electroceramics* **5**: 225–229.
173. Mizusaki, J., Tagawa, H., Saito, T., Yamamura, T., Kamitani, K., Hirano, K., Ehara, S., Takagi, T., Hikita, T., Ippommatsu, M., Nakagawa, S., and Hashimoto, K. (1994). Kinetic studies of the reaction at the nickel pattern electrode on YSZ in H_2–H_2O atmospheres. *Solid State Ionics* **70–71**: 52–58.
174. Abel, J., Kornyshev, A. A., and Lehnert, W. (1996). Correlated resistor network study of porous solid oxide fuel cell anodes. *J. Electrochem. Soc.* **144**: 4253–4259.
175. Divisek, J., Wilkenhöner, R., and Volfkovich, Y. (1999). Structure investigations of SOFC anode cermets—Part I: Porosity investigations. *J. Appl. Electrochem.* **29**: 153–163.
176. Vernoux, P., Guindet, J., and Kleitz, M. (1998). Gradual internal methane reforming in intermediate-temperature solid oxide fuel cells. *J. Electrochem. Soc.* **145**: 3487–3492.
177. Pudmich, G., Boukamp, B. A., Gonzales-Cuenca, M., Jungen, W., Zipprich, W., and Tietz, F. (2000). Chromite/titanate based perovskites for application as anodes in solid oxide fuel cells. *Solid State Ionics* **135**: 433–438.
178. Huang, P., Horky, A., and Petric, A. (1999). Interfacial reaction between nickel oxide and lanthanum gallate during sintering and its effect on conductivity. *J. Am. Ceram. Soc.* **82**: 2402–2406.
179. Mizusaki, J., Yamauchi, S., Fueki, K., and Ishakawa, A. (1984). Nonstoichiometry of the perovskite-type oxide $La_{1-x}Sr_xCrO_{3-\delta}$. *Solid State Ionics* **12**: 119–124.
180. Mizusaki, J. (1992). Nonstoichiometry, diffusion, and electrical properties of perovskite-type oxide electrode materials. *Solid State Ionics* **52**: 79–91.
181. Yasuda, I., and Hikita, T. (1993). Electrical conductivity and defect structure of calcium-doped lanthanum chromites. *J. Electrochem. Soc.* **140**: 1699–1704.
182. Yasuda, I., and Hishinuma, M. (1995). Electrical conductivity and chemical diffusion coefficient of Sr-doped lanthanum chromites. *Solid State Ionics* **80**: 141–150.
183. Weber, W. J., Griffin, C. W., and Bates, J. L. (1987). Effects of cation substitution on electrical and thermal transport properties of $YCrO_3$ and $LaCrO_3$. *J. Am. Ceram. Soc.* **70**: 265–270.
184. Anderson, H. U. (1992). Review of p-type doped perovskite materials for SOFC and other applications. *Solid State Ionics* **52**: 33–41.
185. Suzuki, M., Sasaki, H., and Kajimura, A. (1997). Oxide ionic conductivity of doped lanthanum chromite thin film interconnectors. *Solid State Ionics* **96**: 83–88.
186. Yasuda, I., and Hishinuma, M. (1996). Electrochemical properties of doped lanthanum chromites as interconnectors for solid oxide fuel cells. *J. Electrochem. Soc.* **143**: 1583–1590.
187. Sakai, N., Horita, T., Yokokawa, H., Dokiya, M., and Kawada, T. (1996). Oxygen permeation measurement of $La_{1-x}Ca_xCrO_{3-\delta}$ by using an electrochemical method. *Solid State Ionics* **86–88**: 1273–1278.
188. Peck, D. H., Miller, M., and Hilpert, K. (1999). *Solid State Ionics* **123**: 59–65.
189. Peck, D. H., Miller, M., and Hilpert, K. (1999). Phase diagram study in the CaO–Cr_2O_3–La_2O_3 system in air and under low oxygen pressure. *Solid State Ionics* **123**: 47–57.

190. Minh, N. Q., and Takahashi, T. (1995). Interconnect—thermal expansion. Science and Technology of Ceramic Fuel Cells, pp. 181–183, Minh, N. Q., and Takahashi, T., eds., Amsterdam: Elsevier Science.

191. Minh, N. Q., and Takahashi, T. (1995). Stack design and fabrication. Science and Technology of Ceramic Fuel Cells, pp. 233–306, Minh, N. Q., and Takahashi, T., eds., Amsterdam: Elsevier Science.

192. Badwal, S. P. S., and Foger, K. (1996). Solid oxide electrolyte fuel cell review. *Ceram. Int.* **2**: 257–265.

193. Hammou, A., and Guindet, J. (1997). Solid oxide fuel cells, The CRC Handbook of Solid State Electrochemistry, pp. 432–443. Gellings, P. J., and Bouwmeester, H. J. M., eds., New York: CRC Press.

194. Stotz, S., and Wagner, C. (1966). Die Löslichkeit von Wasserdampf und Wasserstoff in festen Oxiden. *Ber. Bunsenges. Phys. Chem.* **70**: 781–788.

195. Takahashi, T., and Iwahara, H. (1980). Solid-state ionics: protonic conduction in perovskite type oxide solid solutions. *Revue de Chimie minérale* **17**: 243–253.

196. Browall, K. W., Muller, O., and Doremus, R. H. (1976). Oxygen ion conductivity in oxygen-deficient perovskite-related oxides. *Mater. Res. Bull.* **11**: 1475–1482.

197. Iwahara, H., Uchida, H., and Tanaka, S. (1983). High-temperature type proton conductor based on $SrCeO_3$ and its application to solid electrolyte fuel cells. *Solid State Ionics* **9–10**: 1021–1025.

198. Iwahara, H., Uchida, H., Ono, K., and Ogaki, K. (1988). Proton conduction in sintered oxides based on $BaCeO_3$. *J. Electrochem. Soc.* **135**: 529–533.

199. Iwahara, H. (1988). High temperature proton conducting oxides and their applications to solid electrolyte fuel cells and steam electrolyzer for hydrogen production. *Solid State Ionics* **28–30**: 573–578.

200. Iwahara, H., Uchida, H., and Morimoto, K. (1990). High temperature solid electrolyte fuel cells using perovskite type based on $BaCeO_3$. *J. Electrochem. Soc.* **137**: 462–465.

201. Bonanos, N., Ellis, B., and Mahmood, M. N. (1991). Conduction and operation of fuel cells based on the solid electrolyte $BaCeO_3$:Gd. *Solid State Ionics* **44**: 305–311.

202. Taniguchi, N., Hatoh, K., Niikura, J., and Gamo, T. (1992). Proton conductive properties of Gd-doped barium cerates at high temperatures. *Solid State Ionics* **53–56**: 998–1003.

203. Iwahara, H., Yajima, T., Hibino, T., and Ushida, H. (1993). Performance of solid oxide fuel cell using proton and oxide ion mixed conductors based on $BaCe_{1-x}Sm_xO_{3-\alpha}$. *J. Electrochem. Soc.* **140**: 1687–1691.

204. Bonanos, N., Knight, K. S., and Ellis, B. (1995). Perovskite solid electrolytes: structure, transport properties and fuel cell applications. *Solid State Ionics* **79**: 161–170.

205. Scholten, M. J., Schoonman, J., van Miltenburg, J. C., and Cordfunke, E. H. P. (1995). The thermodynamic properties of $BaCeO_3$ at temperatures from 5 to 940 K. *Thermochim. Acta* **268**: 161–168.

206. Scholten, M. J., Schoonman, J., van Miltenburg, J. C., and Oonk, H. A. J. (1993). Synthesis of $SrCeO_3$, $BaCeO_3$, $SrZrO_3$ and $BaZrO_3$ and their reaction with CO_2. Lithium Batteries, Vol. PV 93-4, pp. 146–155, Doddapaneni, V. N., and Landgrebe, A. R., eds., Pennington, NJ: The Electrochemical Society.

207. Tanner, C. W., and Virkar, A. V. (1996). Instability of $BaCeO_3$ in H_2O-containing atmospheres. *J. Electrochem. Soc.* **143**: 1386–1389.

208. Kreuer, K. D. (1997). On the development of proton conducting materials for technological applications. *Solid State Ionics* **97**: 1–15.

209. Wienströer, S., and Wiemhöfer, H. D. (1997). Investigation of the influence of zirconium substitution on the properties of neodymium-doped barium cerates. *Solid State Ionics* **101–103**: 1113–1117.

210. Kwang, H. R., and Haile, S. M. (1999). Chemical stability and proton conductivity of doped $BaCeO_3$–$BaZrO_3$ solid solutions. *Solid State Ionics* **125**: 355–367.
211. Katahira, K., Kohchi, Y., Shimura, T., and Iwahara, H. (2000). Protonic conduction in Zr-substituted $BaCeO_3$. *Solid State Ionics* **138**: 91–98.
212. Kreuer, K. D. (1999). Aspects of the formation and mobility of protonic charge carriers and the stability of perovskite-type oxides. *Solid State Ionics* **125**: 285–302.
213. Liang, K. C., and Nowick, A. S. (1993). High-temperature protonic conduction in mixed perovskite ceramics. *Solid State Ionics* **61**: 77–81.
214. Murugaraj, P., Kreuer, K. D., He, T., Schober, T., and Maier, J. (1997). High proton conductivity in barium yttrium stannate $Ba_2YSn_{5.5}$. *Solid State Ionics* **98**: 1–6.
215. Norby, T., Dyrlie, O., and Kofstad, P. (1992). Protonic conduction in acceptor-doped cubic rare-earth sesquioxides. *J. Am. Ceram. Soc.* **75**: 1176–1181.
216. Kreuer, K. D., Schönherr, E., and Maier, J. (1994). Proton and oxygen diffusion in $BaCeO_3$ based compounds: a combined thermal gravimetric analysis and conductivity study. *Solid State Ionics* **70–71**: 278–284.
217. Iwahara, H., Yajima, T., and Ushida, H. (1994). Effect of ionic radii of dopants on mixed ionic conduction ($H^+ + O^{2-}$) in $BaCeO_3$-based electrolytes. *Solid State Ionics* **70–71**: 267–271.
218. Larring, Y., and Norby, T. (1995). Protons in rare-earth oxides. *Solid State Ionics* **77**: 147–151.
219. Kreuer, K. D., Dippel, T., Baikov, Y. M., and Maier, J. (1996). Re-evaluation of water solubility, proton and oxygen diffusion in acceptor doped $BaCeO_3$ single crystals. *Solid State Ionics* **86–88**: 613–620.
220. Kreuer, K. D., Adams, S., Fuchs, A., Klock, U., Münch, W., and Maier, J. (2003). Proton conducting alkaline earth zirconates and titanates for high-drain electrochemical applications. *Solid State Ionics*.
221. Münch, W., Kreuer, K. D., Adams, S., Seifert, G., and Maier, J. (1999). The relation between crystal structure and the formation and mobility of protonic charge carriers in perovskite-type oxides: a case study of Y-doped $BaCeO_3$ and $SrCeO_3$. *Phase Transitions* **68**: 567–586.
222. Kreuer, K. D., Münch, W., Ise, M., Fuchs, A., Traub, U., and Maier, J. (1997). Defect interactions in proton conducting perovskite-type oxides. *Ber. Bunsenges. Phys. Chem.* **101**: 1344–1350.
223. Kreuer, K. D. (2000). On the complexity of proton transport phenomena. *Solid State Ionics* **136**: 149–160.
224. Münch, W., Kreuer, K. D., Seifert, G., and Maier, J. (2000). Proton diffusion in perovskites: comparison of cerates, zirconates and titanates using quantum molecular dynamics. *Solid State Ionics* **136**: 183–189.
225. Kreuer, K. D., Fuchs, A., and Maier, J. (1995). H/D-isotope effect and proton conduction mechanism in oxides. *Solid State Ionics* **77**: 157–162.
226. Scherban, T., Nowick, A. S., Boatner, L. A., and Abraham, M. M. (1992). Protons and other defects in Fe-doped $KTaO_3$. *Appl. Phys. A (Mater. Sci. and Process.)* **55**: 324–331.
227. Du, Y., and Nowick, A. S. (1995). Structural transitions and proton conduction in nonstoichiometric perovskite-type oxides. *J. Am. Ceram. Soc.* **78**: 3033–3039.
228. He, T., Kreuer, K. D., Murugaraj, P., Baikov, Y. M., and Maier, J. (1997). Defect chemistry and mass transport in proton conducting perovskite oxides. "Proc. EUROSOLID 4", pp. 60–67, Negro, A., and Montanaro, L., eds., Turin, Italy: Politecnico di Torino, Italy.
229. Veith, M., Mathur, S., Lecerf, N., Huch, V., and Decker, T. (2000). Sol-gel synthesis of nano-scaled $BaTiO_3$, $BaZrO_3$ and $BaTi_{0.5}Zr_{0.5}O_3$ oxides via single-source alkoxide precursors and semi-alkoxide routes. *J. Sol-Gel Sci. Technol.* **15**: 145–158.

230. Tuller, H. L. (1996). Ionic and mixed conductors: materials design and optimization, High Temperature Electrochemistry: Ceramics and Metals, pp. 139–153, Poulsen, F., Bonanos, N. S., Linderoth, S., Mogensen, M., and Zachau-Christiansen, B., eds., Roskilde, Denmark: Risø National Laboratory.
231. Tuller, H. L., Kramer, S., and Spears, M. A. (1993). Flexible control of ionic and electronic conduction in pyrochlore oxides. *Proc. 14th Risø Int. Symposium Mater. Sci.*, pp. 151–173, Poulsen, F. W., Bentzen, J. J., Jacobson, T., Skou, E., and Østergård, M. J. L., eds., Roskilde, Denmark: Risø National Laboratory.
232. Spears, M. A., and Tuller, H. L. (1994). Electrical conductivity and phase stability of ruthenium substituted gadolinium titanate, Ionic and Mixed Conducting Ceramics III, Vol. PV 97-24, p. 94, Ramanarayanan, T. A., Worrell, W. L., Tuller, H. L., Mogensen, M., and Khandkar, A. C., eds., Pennington, NJ: The Electrochemical Society.
233. Porat, O., Heremans, C., and Tuller, H. L. (1997). Phase stability and electrical conductivity in $Gd_2Ti_2O_7$–$Gd_2Mo_2O_7$ solid solutions. *J. Am. Ceram. Soc.* **80**: 2278–2284.
234. Porat, O., Heremans, C., and Tuller, H. L. (1997). Stability and mixed ionic electronic conduction in $Gd_2(Ti_{1-x}Mo_x)_2O_7$ under anodic conditions. *Solid State Ionics* **94**: 75–83.
235. van Gool, W. (1965). The possible use of surface migration in fuel cells and heterogeneous catalysis. *Philips Res. Rep.* **20**: 81–93.
236. Riess, I., Vanderput, P. J., and Schoonman, J. (1995). Solid oxide fuel cells operating on uniform mixtures of fuel and air. *Solid State Ionics* **82**: 1–4.
237. Hibino, T., Ushiki, K., and Kuwahara, Y. (1996). New concept of simplifying sofc system. *Solid State Ionics* **91**: 69–74.
238. Hibino, T., Kuwahara, Y., and Wang, S. (1999). Effect of electrode and electrolyte modification on the performance of one-chamber solid oxide fuel cells. *J. Electrochem. Soc.* **146**: 2821–2826.
239. Hibino, T., Hashimoto, A., Inoue, T., Tokuno, J., Yoshida, S., and Sano, M. (2000). A low-operating temperature solid oxide fuel cell in hydrocarbon–air mixtures. *Science* **288**: 2031–2033.

Handbook of Advanced Ceramics
S. Sōmiya *et al.* (Eds.)

4.1 Piezoelectric Ceramics

KENJI UCHINO
International Center for Actuators and Transducers,
134 Materials Research Laboratory, Pennsylvania State University,
University Park, PA 16802-4801, USA

Certain materials produce electric charges on their surfaces as a consequence of applying mechanical stress. The induced charges are proportional to the mechanical stress. This is called the direct piezoelectric effect and was discovered in quartz by Piere and Jacques Curie in 1880. Materials showing this phenomenon also conversely have a geometric strain proportional to an applied electric field. This is the converse piezoelectric effect. The root of the word "piezo" means "pressure"; hence the original meaning of the word piezoelectricity implied "pressure electricity".

Piezoelectricity is extensively utilized in the fabrication of various devices such as transducers, actuators, surface acoustic wave devices, frequency control and so on. In this chapter we describe the piezoelectric materials that are used, and various potential applications of piezoelectric materials [1–4].

4.1.1 PIEZOELECTRIC MATERIALS AND PROPERTIES

4.1.1.1 Piezoelectric Figures of Merit

There are five important figures of merit in piezoelectrics: the piezoelectric strain constant d, the piezoelectric voltage constant g, the electromechanical coupling factor k, the mechanical quality factor Q_M, and the acoustic impedance Z. These figures of merit are considered in this section.

4.1.1.1.1 Piezoelectric Strain Constant *d*

The magnitude of the induced strain x by an external electric field E is represented by this figure of merit (an important figure of merit for actuator applications):

$$x = dE. \tag{1}$$

4.1.1.1.2 Piezoelectric Voltage Constant g

The induced electric field E is related to an external stress X through the piezoelectric voltage constant g (an important figure of merit for sensor applications):

$$E = gX. \tag{2}$$

Taking into account the relation, $P = dX$, we obtain an important relation between g and d:

$$g = d/\varepsilon_0\varepsilon \quad (\varepsilon\text{: permittivity}) \tag{3}$$

4.1.1.1.3 Electromechanical Coupling Factor *k*

The terms, electromechanical coupling factor, energy transmission coefficient, and efficiency are sometimes confused [5]. All are related to the conversion rate between electrical energy and mechanical energy, but their definitions are different [6].

(a) *The electromechanical coupling factor k*

$$k^2 = (\text{stored mechanical energy/input electrical energy}) \tag{4}$$

or

$$k^2 = (\text{stored electrical energy/input mechanical energy}) \tag{5}$$

Let us calculate Eq. 4, when an electric field E is applied to a piezoelectric material. Since the input electrical energy is $(1/2)\varepsilon_0\varepsilon E^2$ per unit volume and the stored mechanical energy per unit volume under zero external stress is given by $(1/2)x^2/s = (1/2)(dE)^2/s$, k^2 can be calculated as

$$\begin{aligned} k^2 &= \left[(1/2)(dE)^2/s\right]/\left[(1/2)\varepsilon_0\varepsilon E^2\right] \\ &= d^2/\varepsilon_0\varepsilon \cdot s. \end{aligned} \tag{6}$$

(b) *The energy transmission coefficient* λ_{max}

Not all the stored energy can be actually used, and the actual work done depends on the mechanical load. With zero mechanical load or a complete clamp (no strain) zero output work is done.

$$\lambda_{max} = (\text{output mechanical energy/input electrical energy})_{max} \tag{7}$$

or

$$\lambda_{max} = (\text{output electrical energy/input mechanical energy})_{max} \tag{8}$$

Let us consider the case where an electric field E is applied to a piezoelectric under constant external stress X (<0, because a compressive stress is necessary to work to the outside). As shown in Figure 4.1.1, the output work can be calculated as

$$\int (-X)\,\mathrm{d}x = -(dE + sX)X, \tag{9}$$

while the input electrical energy is given by

$$\int E\,\mathrm{d}P = (\varepsilon_0\varepsilon E + dX)E. \tag{10}$$

We need to choose a proper load to maximize the energy transmission coefficient. From the maximum condition of

$$\lambda = -(dE + sX)X/(\varepsilon_0\varepsilon E + dX)E, \tag{11}$$

we can obtain

$$\begin{aligned}\lambda_{\max} &= [(1/k) - \sqrt{(1/k^2) - 1}]^2 \\ &= [(1/k) + \sqrt{(1/k^2) - 1}]^{-2}\end{aligned} \tag{12}$$

Notice that

$$k^2/4 < \lambda_{\max} < k^2/2 \tag{13}$$

depending on the k value. For a small k, $\lambda_{\max} = k^2/4$, and for a large k, $\lambda_{\max} = k^2/2$.

It is also worth noting that the maximum condition stated above does not agree with the condition which provides the maximum output mechanical energy. The maximum output energy can be obtained when the load is half of the maximum generative stress: $-(dE - s(dE/2s))(-dE/2s) = (dE)^2/4s$. In this case, since the input electrical energy is given by $(\varepsilon_0\varepsilon E + d(-dE/2s))E$,

$$\lambda = 1/2[(2/k^2) - 1], \tag{14}$$

which is close to the value $\lambda_{\max}$, but has a different value that is predicted theoretically.

(c) *The efficiency* η

$$\eta = \text{(output mechanical energy)/(consumed electrical energy)} \tag{15}$$

or

$$\eta = \text{(output electrical energy)/(consumed mechanical energy)}. \tag{16}$$

In a work cycle (e.g. an electric field cycle), the input electrical energy is transformed partially into mechanical energy and the remaining is stored

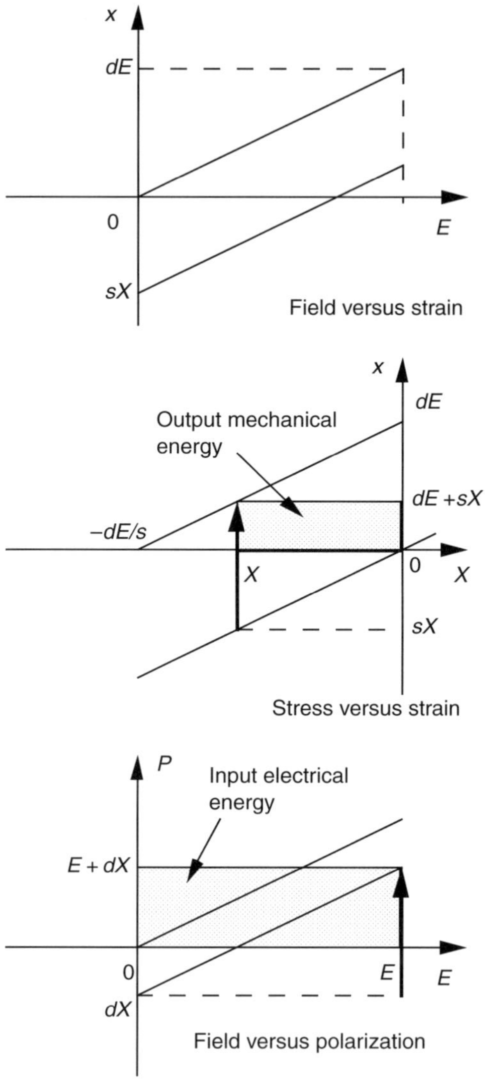

FIGURE 4.1.1 Calculation of the input electrical and output mechanical energy.

as electrical energy (electrostatic energy like a capacitor) in an actuator. In this way, the ineffective energy can be returned to the power source, leading to near 100% efficiency, if the loss is small. Typical values of dielectric loss in PZT are about 1–3%.

4.1.1.1.4 Mechanical Quality Factor Q_M

The mechanical quality factor, Q_M, is a parameter that characterizes the sharpness of the electromechanical resonance spectrum. When the motional admittance Y_m is plotted around the resonance frequency ω_0, the mechanical quality factor Q_M is defined with respect to the full width at $Y_m/\sqrt{2}[2\Delta\omega]$ as:

$$Q_M = \omega_0/2\Delta\omega. \tag{17}$$

Also note that Q_M^{-1} is equal to the mechanical loss ($\tan\delta_m$). The Q_M value is very important in evaluating the magnitude of the resonant strain. The vibration amplitude at an off-resonance frequency (dEL, L: length of the sample) is amplified by a factor proportional to Q_M at the resonance frequency. For a longitudinally vibration rectangular plate through d_{31}, the maximum displacement is given by $(8/\pi^2)Q_M d_{31}EL$.

4.1.1.1.5 Acoustic Impedance Z

The acoustic impedance Z is a parameter used for evaluating the acoustic energy transfer between two materials. It is defined, in general, by

$$Z^2 = \text{(pressure/volume velocity)}. \tag{18}$$

In a solid material,

$$Z = \sqrt{\rho c}, \tag{19}$$

where ρ is the density and c the elastic stiffness of the material.

In more advanced discussions, there are three kinds of impedances; specific acoustic impedance (pressure/particle speed), acoustic impedance (pressure/volume speed) and radiation impedance (force/speed). See Ref. [6] for the details.

4.1.1.2 Piezoelectric Materials [7]

This section summarizes the current status of piezoelectric materials: single-crystal materials, piezoceramics, piezopolymers, piezocomposites and piezofilms. Table 4.1.1 shows the material parameters of some of the piezoelectric materials [8].

4.1.1.2.1 Single Crystals

Although piezoelectric ceramics are widely used for a large number of applications, single-crystal materials retain their utility, being essential for applications

TABLE 4.1.1 Piezoelectric Properties of Representative Piezoelectric Materials [7, 8]

Parameter	Quartz	$BaTiO_3$	PZT 4	PST5H	(Pb, Sm)TiO_3	PVDF-TrFE
d_{33} (pC/N)	2.3	190	289	593	65	33
g_{33} (10^{-3} V m/N)	57.8	12.6	26.1	19.7	42	380
k_t	0.09	0.38	0.51	0.50	0.50	0.30
k_p	—	0.33	0.58	0.65	0.03	—
$\varepsilon_3^T/\varepsilon_0$	5	1700	1300	3400	175	6
Q_M	$>10^5$	—	500	65	900	3–10
T_C (°C)	—	120	328	193	355	—

such as frequency stabilized oscillators and surface acoustic devices. The most popular single-crystal piezoelectric materials are quartz, lithium niobate ($LiNbO_3$), and lithium tantalate ($LiTaO_3$). The single crystals are anisotropic, exhibiting different material properties depending on the cut of the materials and the direction of bulk or surface wave propagation.

Quartz is a well-known piezoelectric material. α-Quartz belongs to the triclinic crystal system with point group 32 and has a phase transition at 537°C to its β-form, which is not piezoelectric. Quartz has a cut with a zero temperature coefficient. For instance, quartz oscillators, operated in the thickness shear mode of the AT-cut, are used extensively for clock sources in computers, frequency stabilized ones in TVs and VCRs. On the other hand, an ST-cut quartz substrate with X-propagation has a zero temperature coefficient for surface acoustic wave and so is used for SAW devices with high-stabilized frequencies. The another distinguished characteristic of quartz is an extremely high mechanical quality factor $Q_M > 10^5$.

$LiNbO_3$ and $LiTaO_3$ belong to an isomorphous crystal system and are composed of oxygen octahedron. The Curie temperatures of $LiNbO_3$ and $LiTaO_3$ are 1210 and 660°C, respectively. The crystal symmetry of the ferroelectric phase of these single crystals is $3m$ and the polarization direction is along c-axis. These materials have high electromechanical coupling coefficients for surface acoustic wave. In addition, large single crystals can easily be obtained from their melt using the conventional Czochralski technique. Thus both materials occupy very important positions in the surface acoustic wave (SAW) device application field.

4.1.1.2.2 Polycrystalline Materials

Barium titanate ($BaTiO_3$) is one of the most thoroughly studied and most widely used piezoelectric materials. Just below the Curie temperature (120°C), the vector of the spontaneous polarization points in the [001] direction (tetragonal phase), below 5°C it reorients in the [011] (orthrhombic phase) and below −90°C in the [111] direction (rhombohedral phase). The dielectric and piezoelectric properties of ferroelectric ceramic $BaTiO_3$ can be affected by its own stoichiometry, microstructure, and by dopants entering onto the A or B site in solid solution. Modified ceramic $BaTiO_3$ with dopants such as Pb or Ca ions have been developed to stabilize the tetragonal phase over a wider temperature range and are used as commercial piezoelectric materials. The initial application was for Langevin-type piezoelectric vibrators.

Piezoelectric $Pb(Ti,Zr)O_3$ solid solutions (PZT) ceramics have been widely used because of their superior piezoelectric properties. The phase diagram for the PZT system ($PbZr_xTi_{1-x}O_3$) is shown in Figure 4.1.2. The crystalline symmetry of this solid-solution system is determined by the Zr content. Lead titanate also has a tetragonal ferroelectric phase of perovskite structure. With increasing Zr content, x, the tetragonal distortion decreases and at $x > 0.52$ the structure changes from the tetragonal $4mm$ phase to another ferroelectric phase of rhombohedral $3m$ symmetry. The line dividing these two phases is called the *morphotropic phase boundary* (MPB). The boundary composition is considered to have both tetragonal and rhombohedral phases coexisting together. Figure 4.1.3 shows the dependence of several piezoelectric d constants on composition near the MPB. The d constants have their highest values near

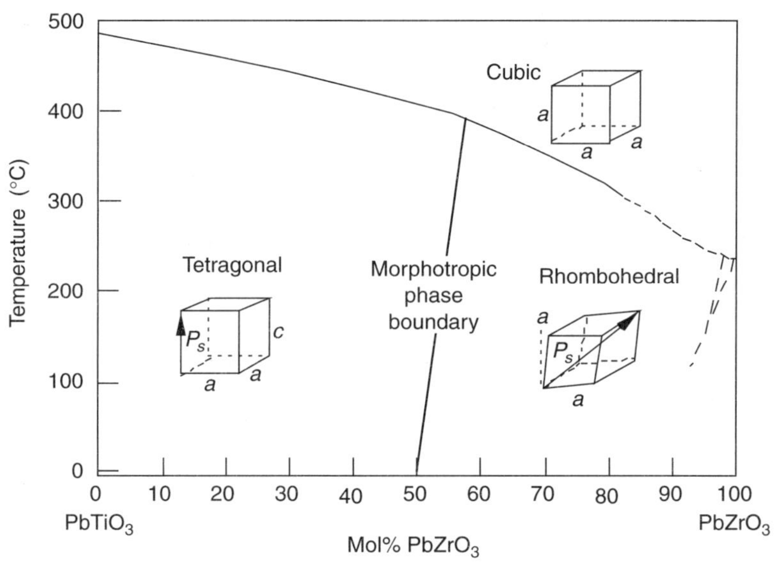

FIGURE 4.1.2 Phase diagram of lead zirconate titanate (PZT).

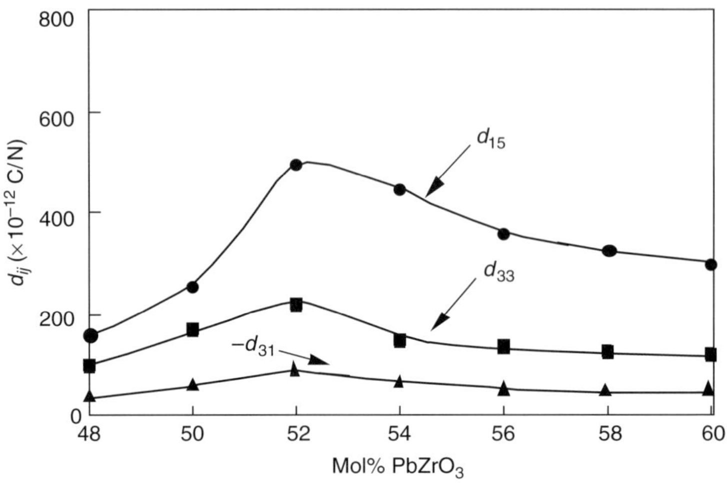

FIGURE 4.1.3 Dependence of several *d* constants on composition near the morphotropic phase boundary in the PZT system.

the MPB. This enhancement in piezoelectric effect is attributed to the increased ease of reorientation of the polarization under an applied electric field.

Doping the PZT material with donor or acceptor ions changes its properties dramatically. Donor doping with ions such as Nb^{5+} or Ta^{5+} provides soft PZTs, like PZT-5, because of the facility of domain motion due to the resulting Pb-vacancies. On the other hand, acceptor doping with Fe^{3+} or Sc^{3+} leads to hard PZTs, such as PZT-8, because the oxygen vacancies will pin domain wall motion.

Subsequently, PZT in ternary solid solution with another perovskite phase has been investigated intensively. Examples of these ternary compositions are: PZTs in solid solution with $Pb(Mg_{1/3}Nb_{2/3})O_3$, $Pb(Mn_{1/3}Sb_{2/3})O_3$, $Pb(Co_{1/3}Nb_{2/3})O_3$, $Pb(Mn_{1/3}Nb_{2/3})O_3$, $Pb(Ni_{1/3}Nb_{2/3})O_3$, $Pb(Sb_{1/2}Sn_{1/2})O_3$, $Pb(Co_{1/2}W_{1/2})O_3$, $Pb(Mg_{1/2}W_{1/2})O_3$, all of which are patented by different companies.

The end member of PZT, lead titanate has a large crystal distortion. $PbTiO_3$ has a tetragonal structure at room temperature with its tetragonality $c/a = 1.063$. The Curie temperature is 490°C. Densely sintered $PbTiO_3$ ceramics cannot be obtained easily, because they break up into a powder when cooled through the Curie temperature due to the large spontaneous strain. Lead titanate ceramics modified by adding a small amount of additives exhibit a high piezoelectric anisotropy. Either $(Pb, Sm)TiO_3$ [9] or $(Pb, Ca)TiO_3$ [10] exhibits an extremely low planar coupling, that is, a large k_t/k_p ratio. Here,

k_t and k_p are thickness-extensional and planar electromechanical coupling factors, respectively. Since these transducers can generate purely longitudinal waves through k_t associated with no transverse waves through k_{31}, clear ultrasonic imaging is expected without "ghost" caused by the transverse wave. (Pb,Nd)(Ti,Mn,In)O_3 ceramics with a zero temperature coefficient of SAW delay have been developed as superior substrate materials for SAW device applications [11].

4.1.1.2.3 Relaxor Ferroelectrics

Relaxor ferroelectrics can be prepared either in polycrystalline form or as single crystals. They differ from the previously mentioned normal ferroelectrics in that they exhibit a broad phase transition from the paraelectric to ferroelectric state, a strong frequency dependence of the dielectric constant (i.e. dielectric relaxation) and a weak remanent polarization. Lead-based relaxor materials have complex disordered perovskite structures.

Relaxor-type electrostrictive materials, such as those from the lead magnesium niobate–lead titanate, $Pb(Mg_{1/3}Nb_{2/3})O_3$–$PbTiO_3$ (or PMN–PT), solid solution are highly suitable for actuator applications. This relaxor ferroelectric also exhibits an induced piezoelectric effect. That is, the electromechanical coupling factor k_t varies with the applied DC bias field. As the DC bias field increases, the coupling increases and saturates. Since this behavior is reproducible, these materials can be applied as ultrasonic transducers which are tunable by the bias field [12].

Recently, single-crystal relaxor ferroelectrics with the MPB composition have been developed which show tremendous promise as ultrasonic transducers and electromechanical actuators. Single crystals of $Pb(Mg_{1/3}Nb_{2/3})O_3$ (PMN), $Pb(Zn_{1/3}Nb_{2/3})O_3$ (PZN) and binary systems of these materials combined with $PbTiO_3$ (PMN–PT and PZN–PT) exhibit extremely large electromechanical coupling factors [13, 14]. Large coupling coefficients and large piezoelectric constants have been found for crystals from the morphotropic phase boundaries of these solid solutions. PZN–8% PT single crystals were found to possess a high k_{33} value of 0.94 for the (001) crystal cuts; this is very high compared to the k_{33} of conventional PZT ceramics of around 0.70–0.80.

4.1.1.2.4 Polymers

Polyvinylidene difluoride, PVDF or PVF2, is piezoelectric when stretched during fabrication. Thin sheets of the cast polymer are then drawn and stretched in the plane of the sheet, in at least one direction, and frequently also in the perpendicular direction, to transform the material to its microscopically polar phase. Crystallization from the melt forms the non-polar α-phase, which can be

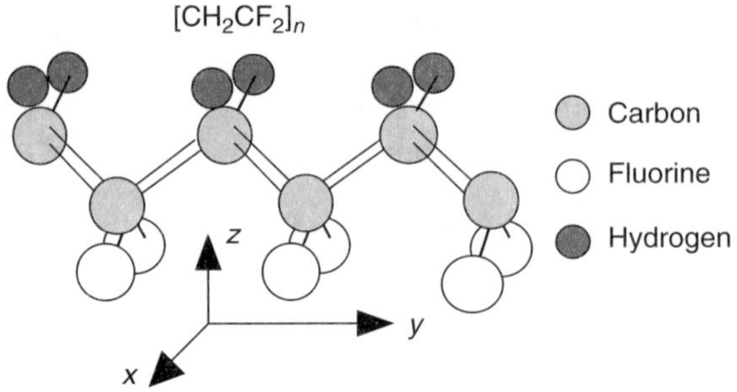

FIGURE 4.1.4 Structure of polyvinylidene diflouride (PVDF).

converted into the polar β-phase by a uniaxial or biaxial drawing operation; the resulting dipoles are then reoriented through electric poling (see Figure 4.1.4).

Large sheets can be manufactured and thermally formed into complex shapes. The copolymerization of vinilydene difluoride with trifluoroethylene (TrFE) results in a random copolymer (PVDF–TrFE) with a stable, polar β-phase. This polymer need not be stretched; it can be poled directly as formed. A thickness-mode coupling coefficient of 0.30 has been reported. Piezoelectric polymers have the following characteristics: (a) small piezoelectric *d* constants (for actuators) and large *g* constants (for sensors); (b) light weight and soft elasticity, leading to good acoustic impedance matching with water or the human body; and (c) a low mechanical quality factor Q_M, allowing for a broad resonance band width. Such piezoelectric polymers are used for directional microphones and ultrasonic hydrophones.

4.1.1.2.5 Composites

Piezocomposites comprised of a piezoelectric ceramic and a polymer phase are promising materials because of their excellent and readily tailored properties. The geometry for two-phase composites can be classified according to the dimensional connectivity of each phase into 10 structures; 0–0, 0–1, 0–2, 0–3, 1–1, 1–2, 1–3, 2–2, 2–3 and 3–3 [15]. A 1–3 piezocomposite, such as the PZT-rod/polymer composite is a most promising candidate. The advantages of this composite are high coupling factors, low acoustic impedance, good matching to water or human tissue, mechanical flexibility, broad bandwidth in combination with a low mechanical quality factor and the possibility of making undiced arrays by structuring the electrodes. The thickness-mode electromechanical coupling of the composite can exceed the k_t (0.40–0.50) of the constituent ceramic, approaching almost the value of the rod-mode electromechanical coupling, k_{33} (0.70–0.80) of that ceramic [16]. Acoustic impedance is the square root of the product of its density and elastic stiffness. The acoustic match to tissue or water (1.5 Mrayls) of the typical piezoceramics (20–30 Mrayls) is significantly improved by forming a composite structure, that is, by replacing some of the heavy, stiff ceramic with a light, soft polymer. Piezoelectric composite materials are especially useful for underwater sonar and medical diagnostic ultrasonic transducer applications.

4.1.1.2.6 Thin Films

Both zinc oxide (ZnO) and aluminum nitride (AlN) are simple binary compounds with a Wurtzite-type structure, which can be sputter-deposited as a *c*-axis oriented thin film on a variety of substrates. ZnO has large piezoelectric coupling and thin films of this material are widely used in bulk acoustic and surface acoustic wave devices. The fabrication of highly oriented (along the *c*-axis) ZnO films have been studied and developed extensively. The performance of ZnO devices is limited, however, due to their low piezoelectric coupling (20–30%). PZT thin films are expected to exhibit higher piezoelectric properties. At present the growth of PZT thin films is being carried out for use in microtransducers and microactuators.

4.1.2 PRESSURE SENSORS/ACCELEROMETERS/ GYROSCOPES

One of the very basic applications of piezoelectric ceramics is a gas igniter. The very high voltage generated in a piezoelectric ceramic under applied mechanical stress can cause sparking and ignite the gas (Fig. 4.1.5). There are two means to apply the mechanical force, either by a rapid, pulsed application or by a more gradual, continuous increase.

Piezoelectric ceramics can be employed as stress sensors and acceleration sensors, because of the *direct piezoelectric effect*. Figure 4.1.6 shows a three-dimensional (3D) stress sensor designed by Kistler. By combining an appropriate number of quartz crystal plates (extensional and shear types), the multilayer device can detect 3D stresses [17].

Figure 4.1.7 shows a cylindrical gyroscope commercialized by Tokin (Japan) [18]. The cylinder has six divided electrodes, one pair of which are used to excite the fundamental bending vibration mode, while the other two pairs are used to detect the acceleration. When the rotational acceleration is applied about the axis of this gyro, the voltage generated on the electrodes is modulated by the *Coriolis force*. By subtracting the signals between the two sensor electrode pairs, a voltage directly proportional to the acceleration can be obtained.

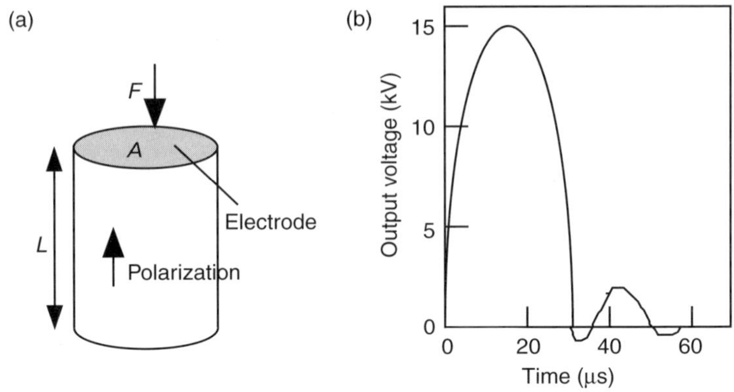

FIGURE 4.1.5 (a) Gas igniter and (b) output voltage.

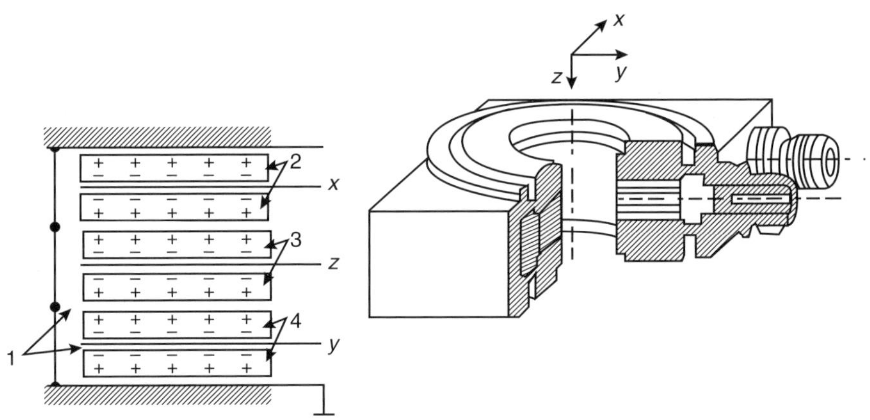

FIGURE 4.1.6 3D stress sensor (by Kistler).

The *converse electrostrictive effect*—the stress dependence of the permittivity—is also used in stress sensors [19]. A bimorph structure provides superior stress sensitivity and temperature stability. A measuring system with a bimorph structure, which subtracts the static capacitances of two dielectric ceramic plates, has been proposed [19]. The capacitance changes of the top and bottom plates have opposite signs for uniaxial stress and the same sign for temperature deviation. The response speed is limited by the capacitance measuring frequency to about 1 kHz. Unlike piezoelectric sensors, electrostrictive sensors are effective in the low-frequency range, especially DC.

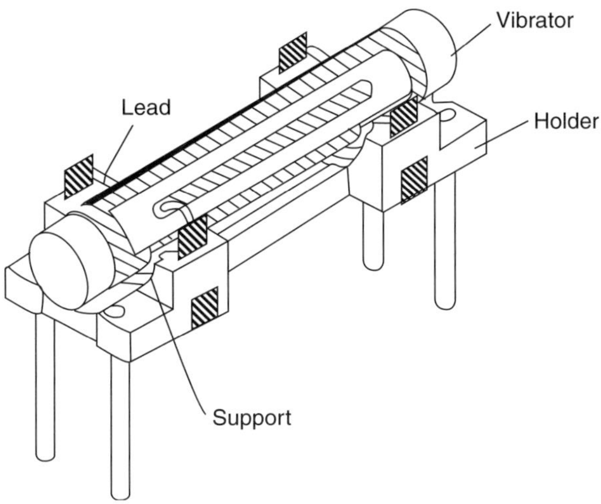

FIGURE 4.1.7 Cylindrical gyroscope (by Tokin).

4.1.3 PIEZOELECTRIC VIBRATORS/ULTRASONIC TRANSDUCERS

4.1.3.1 Piezoelectric Resonance

4.1.3.1.1 The Piezoelectric Equations

When an electric field is applied to a piezoelectric material, deformation (ΔL) or strain ($\Delta L/L$) arises. When the field is alternating, mechanical vibration is caused, and if the drive frequency is adjusted to a mechanical resonance frequency of the device, large resonating strain is generated. This phenomenon can be understood as a strain magnification due to accumulating input energy, and is called *piezoelectric resonance*. Piezoelectric resonance is very useful for realizing energy trap devices, actuators, etc. The theoretical treatment is as follows.

If the applied electric field and the generated stress are not large, the stress X and the dielectric displacement D can be represented by the following equations:

$$x_i = s_{ij}^E X_j + d_{mi} E_m, \tag{20}$$

$$D_m = d_{mi} X_i + \varepsilon_{mk}^X E_k \tag{21}$$

where $i, j = 1, 2, \ldots, 6;\ m, k = 1, 2, 3.$

These are called the *piezoelectric equations*. The number of independent parameters for the lowest symmetry trigonal crystal are 21 for s_{ij}^E, 18 for d_{mi}

and 6 for ε^X_{mk}. The number of independent parameters decreases with increasing crystallographic symmetry. Concerning the polycrystalline ceramics, the poled axis is usually denoted as the z-axis and the ceramic is isotropic with respect to this z-axis (Curie group $C_{\infty v}(\infty m)$). The number of non-zero matrix elements in this case is 10 ($s^E_{11}, s^E_{12}, s^E_{13}, s^E_{33}, s^E_{44}, d_{31}, d_{33}, d_{15}, \varepsilon^X_{11}$, and ε^X_{33}).

4.1.3.1.2 Electromechanical Coupling Factor

Next let us introduce the *electromechanical coupling factor* k, which corresponds to the rate of electromechanical transduction. The internal energy U of a piezoelectric vibrator is given by summation of the mechanical energy U_M ($= \int x\,dX$) and the electrical energy U_E ($= \int D\,dE$). U is calculated as follows, when linear relations Eqs 20 and 21 are applicable:

$$U = U_M + U_E = \left[\frac{1}{2}\sum_{i,j} s^E_{ij} X_j X_i + \frac{1}{2}\sum_{m,i} d_{mi} E_m X_i\right] + \left[\frac{1}{2}\sum_{i,j} d_{mi} X_i E_m + \frac{1}{2}\sum_{k,m} \varepsilon^X_{mk} E_k E_m\right]. \tag{22}$$

The s and E terms represent purely mechanical and electrical energies (U_{MM} and U_{EE}), respectively, and the d term denotes the energy transduced from electrical to mechanical energy or vice versa through the piezoelectric effect. The coupling factor k is defined by:

$$k = U_{ME}/\sqrt{U_{MM} \cdot U_{EE}}. \tag{23}$$

The k value varies with the vibrational mode (even in the same ceramic sample), and can have a positive or negative value. Note that this definition is equivalent to the definition provided in section on "Piesoelectric figures of merit":

$$k^2 = (\text{stored mechanical energy/input electrical energy})$$

or

$$k^2 = (\text{stored electrical energy/input mechanical energy}).$$

4.1.3.1.3 Longitudinal Vibration Mode

Let us consider the longitudinal mechanical vibration of a piezoceramic plate through the transverse piezoelectric effect (d_{31}) as shown in Figure 4.1.8. If the polarization is in the z-direction and x–y planes are the planes of the electrodes,

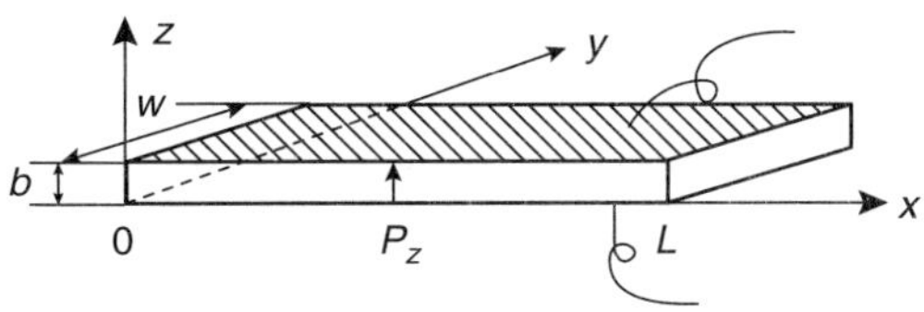

FIGURE 4.1.8 Longitudinal vibration through the transverse piezoelectric effect (d_{31}) in a rectangular plate.

the extentional vibration in the x-direction is represented by the following dynamic equation:

$$\rho(\partial^2 u/\partial t^2) = F = (\partial X_{11}/\partial x) + (\partial X_{12}/\partial y) + (\partial X_{13}/\partial z), \qquad (24)$$

where u is the displacement of the small volume element in the ceramic plate in the x-direction. The relations between stress, electric field (only E_z exists) and the induced strain are given by:

$$\begin{aligned}
x_1 &= s_{11}^E X_1 + s_{12}^E X_2 + s_{13}^E X_3 + d_{31} E_z,\\
x_2 &= s_{12}^E X_1 + s_{11}^E X_2 + s_{13}^E X_3 + d_{31} E_z,\\
x_3 &= s_{13}^E X_1 + s_{13}^E X_2 + s_{33}^E X_3 + d_{33} E_z,\\
x_4 &= s_{44}^E X_4,\\
x_5 &= s_{44}^E X_5,\\
x_6 &= 2(s_{11}^E - s_{12}^E) X_6.
\end{aligned} \qquad (25)$$

When the plate is very long and thin, X_2 and X_3 may be set equal to zero through the plate. Since shear stress will not be generated by the electric field E_z, Eq. 25 is reduced to:

$$X_1 = x_1/s_{11}^E - (d_{31}/s_{11}^E)E_z. \qquad (26)$$

Introducing Eq. 26 into Eq. 24, and allowing for $x_1 = \partial u/\partial x$ and $\partial E_z/\partial x = 0$ (due to the equal potential on each electrode), leads to a harmonic vibration equation:

$$-\omega^2 \rho s_{11}^E u = \partial^2 u/\partial x^2. \qquad (27)$$

Here, ω is the angular frequency of the drive field, and ρ is the density. Substituting a general solution $u = u_1(x)e^{j\omega t} + u_2(x)e^{-j\omega t}$ into Eq. 26, and with the boundary condition $X_1 = 0$ at $x = 0$ and L (sample length), the following solution can be obtained:

$$\partial u/\partial x = x_1 = d_{31}E_z[\sin\omega(L-x)/v + \sin(\omega x/v)/\sin(\omega L/v). \qquad (28)$$

Here, v is the *sound velocity* in the piezoceramic which is given by

$$v = 1/\sqrt{\rho s_{11}^{E}}. \tag{29}$$

When the specimen is utilized as an electrical component such as a filter or a vibrator, the electrical impedance [(applied voltage/induced current) ratio] plays an important role. The current flow into the specimen is described by the surface charge increment, $\partial D_3/\partial t$, and the total current is given by:

$$i = j\omega w \int_0^L D_3 \, dx = j\omega w \int_0^L [(\varepsilon_{33}^{X} - d_{31}^2/s_{11}^{E})E_z + (d_{31}/s_{11}^{E})x_1] \, dx. \tag{30}$$

Using Eq. 28, the admittance for the mechanically free sample is calculated to be:

$$\begin{aligned}(1/Z) &= (i/V) = (i/E_z t) \\ &= (j\omega wL/t)\varepsilon_{33}^{LC}[1 + (d_{31}^2/\varepsilon_{33}^{LC}s_{11}^{E})(\tan(\omega L/2v)/(\omega L/2v)],\end{aligned} \tag{31}$$

where w is the width, L the length, t the thickness of the sample, and V the applied voltage. ε_{33}^{LC} is the permittivity in a longitudinally clamped sample, which is given by

$$\varepsilon_{33}^{LC} = \varepsilon_{33}^{X} - (d_{31}^2/s_{11}^{E}). \tag{32}$$

The piezoelectric resonance is achieved where the admittance becomes infinite or the impedance is zero. The resonance frequency f_R is calculated from Eq. 31, and the fundamental frequency is given by

$$f_R = v/2L = 1/(2L\sqrt{\rho s_{11}^{E}}). \tag{33}$$

On the other hand, the antiresonance state is generated for zero admittance or infinite impedance:

$$(\omega_A L/2v)\cot(\omega_A L/2v) = -d_{31}^2/\varepsilon_{33}^{LC}s_{11}^{E} = -k_{31}^2/(1 - k_{31}^2). \tag{34}$$

The final transformation is provided by the definition,

$$k_{31} = d_{31}/\sqrt{s_{11}^{E}\varepsilon_{33}^{X}}. \tag{35}$$

The resonance and antiresonance states are described by the following intuitive model. In a high electromechanical coupling material with k almost equal to 1, the resonance or antiresonance states appear for $\tan(\omega L/2v) = \infty$ or 0 [i.e. $\omega L/2v = (m - 1/2)\pi$ or $m\pi$ (m: integer)], respectively. The strain amplitude x_1 distribution for each state (calculated using Eq. 28) is illustrated in

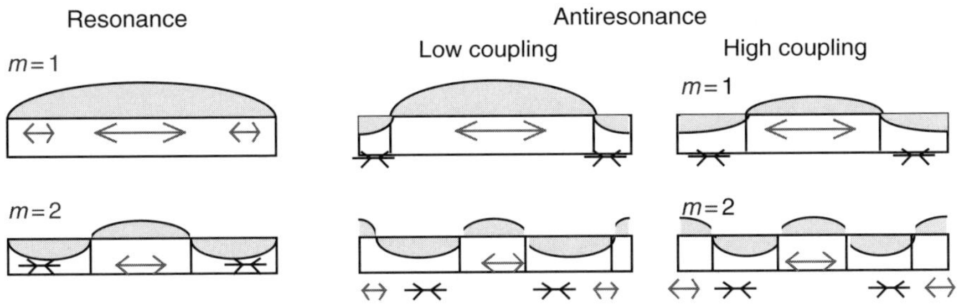

FIGURE 4.1.9 Strain generation in the resonant or antiresonant state.

Figure 4.1.9. In the resonance state, large strain amplitudes and large capacitance changes (called *motional capacitance*) are induced, and the current can easily flow into the device. On the other hand, at antiresonance, the strain induced in the device compensates completely, resulting in no capacitance change, and the current cannot flow easily into the sample. Thus, for a high k material the first antiresonance frequency f_A should be twice as large as the first resonance frequency f_R.

In a typical case, where $k_{31} = 0.3$, the antiresonance state varies from the previously mentioned mode and becomes closer to the resonance mode. The low-coupling material exhibits an antiresonance mode where capacitance change due to the size change is compensated completely by the current required to charge up the static capacitance (called *damped capacitance*). Thus, the antiresonance frequency f_A will approach the resonance frequency f_R.

The general processes for calculating the electromechanical parameters (k_{31}, d_{31}, s_{11}^E, and ε_{33}^X) are described below:

1. The sound velocity v in the specimen is obtained from the resonance frequency f_R (refer to Figure 4.1.10), using Eq. 33.
2. Knowing the density ρ, the elastic compliance s_{11}^E can be calculated.
3. The electromechanical coupling factor k_{31} is calculated from the v value and the antiresonance frequency f_A through Eq. 34. Especially in low-coupling piezoelectric materials, the following approximate equation is available:

$$k_{31}^2/(1 - k_{31}^2) = (\pi^2/4)(\Delta f/f_R) \quad (\Delta f = f_A - f_R) \qquad (36)$$

4. Knowing the permittivity ε_{33}^X, the d_{31} is calculated through Eq. 35.

Figure 4.1.10 shows observed impedance curves for a typical k material (PZT 5H, $k_{33} = 0.70$) and a high-k material (PZN–PT single crystal, $k_{33} = 0.90$). Note a large separation between the resonance and antiresonance peaks in the high-k material, leading to the condition $f_A = 2f_R$.

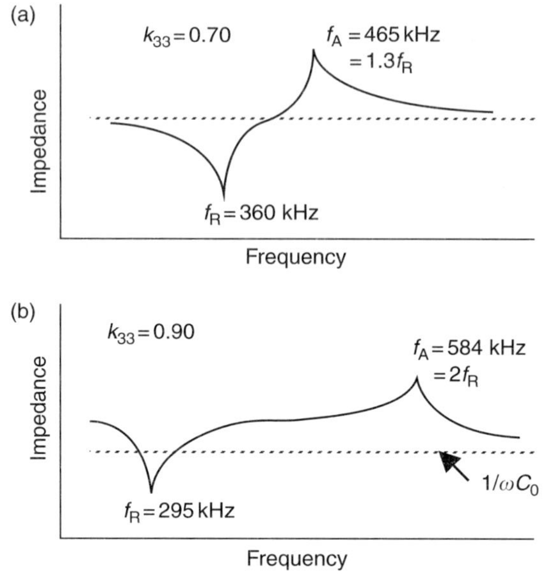

FIGURE 4.1.10 (a) Impedance curves for a reasonable k material (PZT 5H, $k_{33} = 0.70$); and (b) a high-k material (PZN–PT single crystal, $k_{33} = 0.90$).

4.1.3.2 Equivalent Circuits of Piezoelectric Vibrators

The *equivalent circuit* for the piezoelectric acuator is represented by a combination of L, C and R. Figure 4.1.11a shows an equivalent circuit for the resonance state, which has a very low impedance. C_d corresponds to the electrostatic capacitance, and the components L_A and C_A in a series resonance circuit are related to the piezoelectric motion. For example, in the case of the longitudinal vibration of the above rectangular plate through d_{31}, these components are represented by

$$L_A = (\rho/8)(Lb/w)(s_{11}^{E2}/d_{31}^2), \tag{37}$$

$$C_A = (8/\pi^2)(Lw/b)(d_{31}^2/s_{11}^E). \tag{38}$$

The component R_A corresponds to the mechanical loss.

In contrast, the equivalent circuit for the antiresonance state of the same actuator is shown in Figure 4.1.11b, which has high impedance.

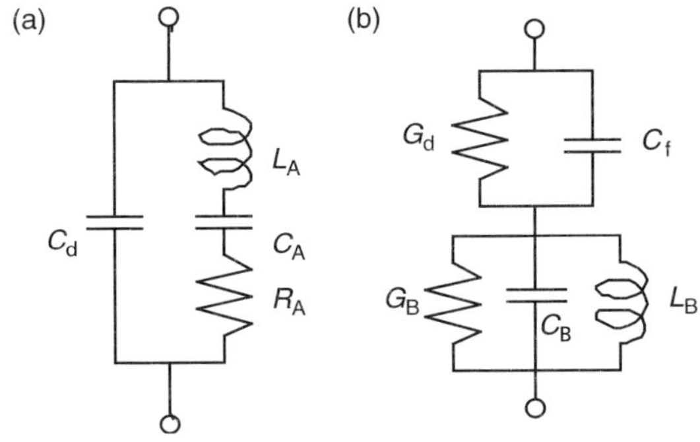

FIGURE 4.1.11 Equivalent circuit of a piezoelectric device for: (a) the resonance: and (b) the antiresonance states.

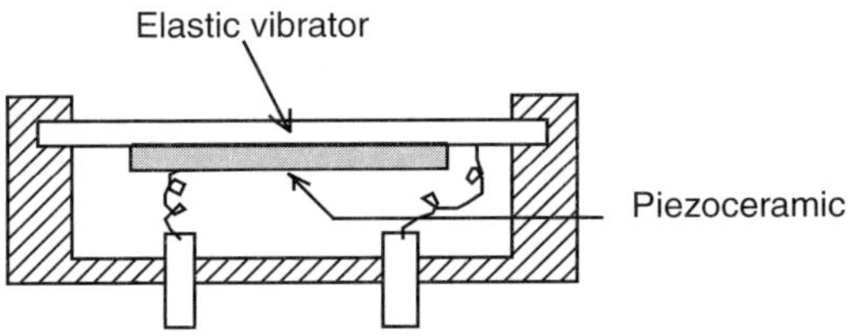

FIGURE 4.1.12 Piezoelectric buzzer.

4.1.3.3 Piezoelectric Vibrators

In the use of mechanical vibration devices such as filters or oscillators, the size and shape of a device are very important, and both the vibrational mode and the ceramic material must be considered. The resonance frequency of the bending mode in a centimeter-size sample ranges from 100 to 1000 Hz, which is much lower than that of the thickness mode (100 kHz). For these vibrator applications the piezoceramic should have a high mechanical quality factor (Q_M) rather than a large piezoelectric coefficient d; that is, hard piezoelectric ceramics are preferable.

For speakers or buzzers, audible by humans, devices with a rather low resonance frequency are used (kilohertz range). Examples are a bimorph consisting of two piezoceramic plates bonded together, and a piezoelectric fork consisting of a piezodevice and a metal fork. A piezoelectric buzzer is shown in Figure 4.1.12, which has merits such as high electric power efficiency, compact size and long life.

4.1.3.4 Ultrasonic Transducers

Ultrasonic waves are now used in various fields. The sound source is made from piezoelectric ceramics as well as magnetostrictive materials. Piezoceramics are generally superior in efficiency and in size to magnetostrictive materials. In particular, hard piezoelectric materials with a high Q_M are preferable. A liquid medium is usually used for sound energy transfer. Ultrasonic washers, ultrasonic microphones for short-distance remote control and underwater detection, such as sonar and fish finding, and non-destructive testing are typical applications. Ultrasonic scanning detectors are useful in medical electronics for clinical applications ranging from diagnosis to therapy and surgery.

One of the most important applications is based on ultrasonic echo field [20, 21]. Ultrasonic transducers convert electrical energy into mechanical form when generating an acoustic pulse and convert mechanical energy into an electrical signal when detecting its echo. The transmitted waves propagate into a body and echoes are generated which travel back to be received by the same transducer. These echoes vary in intensity according to the type of tissue or body structure, thereby creating images. An ultrasonic image represents the mechanical properties of the tissue, such as density and elasticity. We can recognize anatomical structures in an ultrasonic image since the organ boundaries and fluid-to-tissue interfaces are easily discerned. The ultrasonic imaging process can also be done in real time. This means we can follow rapidly moving structures such as the heart without motion distortion. In addition, ultrasound is one of the safest diagnostic imaging techniques. It does not use ionizing radiation like X-rays and thus is routinely used for fetal and obstetrical imaging. Useful areas for ultrasonic imaging include cardiac structures, the vascular systems, the fetus and abdominal organs such as liver and kidney. In brief, it is possible to see inside the human body without breaking the skin by using a beam of ultrasound.

Figure 4.1.13 shows the basic ultrasonic transducer geometry. The transducer is mainly composed of matching, piezoelectric material and backing layers [22]. One or more matching layers are used to increase sound transmissions into tissues. The backing is added to the rear of the transducer in order to damp the acoustic backwave and to reduce the pulse duration. Piezoelectric materials are used to generate and detect ultrasound. In general, broadband transducers should be used for medical ultrasonic imaging. The broad bandwidth response corresponds to a short pulse length, resulting in better axial resolution. Three factors are important in designing broad bandwidth transducers; acoustic impedance matching, a high electromechanical coupling coefficient of the transducer, and electrical impedance matching. These pulse echo transducers operate based on thickness mode resonance of the piezoelectric thin plate. Further, a low planar mode coupling coefficient, k_p, is beneficial for limiting

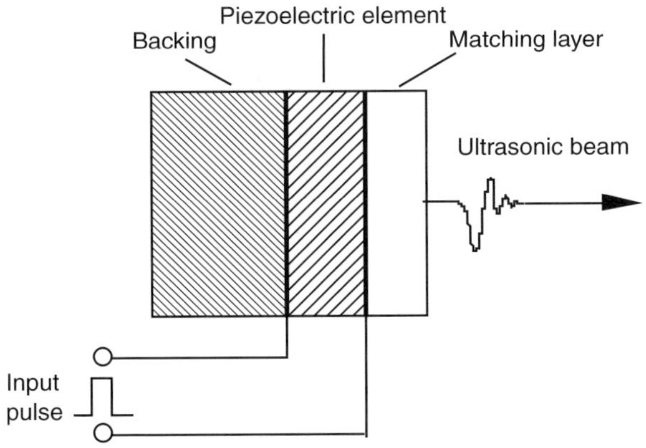

FIGURE 4.1.13 Basic transducer geometry for acoustic imaging applications.

energies being expended in non-productive lateral mode. A large dielectric constant is necessary to enable a good electrical impedance match to the system, especially with tiny piezoelectric sizes.

There are various types of transducers used in ultrasonic imaging. Mechanical sector transducers consist of single, relatively large resonators and can provide images by mechanical scanning such as wobbling. Multiple element array transducers permit discrete elements to be individually accessed by the imaging system and enable electronic focusing in the scanning plane to various adjustable penetration depths through the use of phase delays. Two basic types of array transducers are linear and phased (or sector). A linear array is a collection of elements arranged in one direction, producing a rectangular display (see Figure 4.1.14). A curved linear (or convex) array is a modified linear array whose elements are arranged along an arc to permit an enlarged trapezoidal field of view. The elements of these linear type array transducers are excited sequentially group by group with the sweep of the beam in one direction. These linear array transducers are used for radiological and obstetrical examinations. On the other hand, in a phased array transducer, the acoustic beam is steered by signals that are applied to the elements with delays, creating a sector display. This transducer is useful for cardiology applications where positioning between the ribs is necessary.

4.1.3.5 Resonators/Filters

When a piezoelectric body vibrates at its resonant frequency, it absorbs considerably more energy than at other frequencies resulting in a dramatic decrease

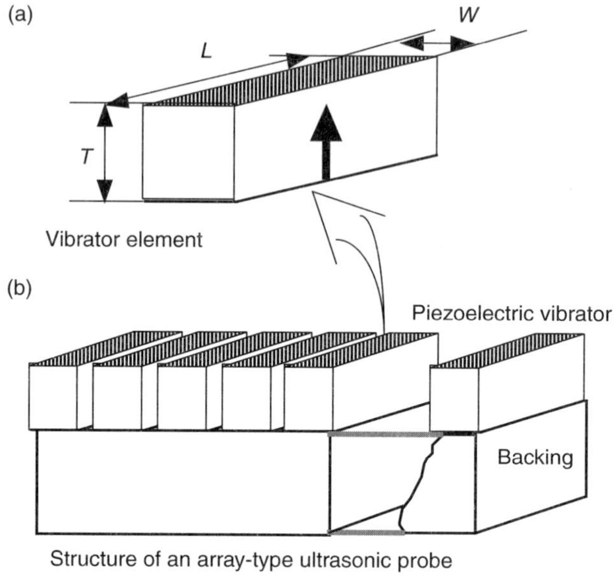

FIGURE 4.1.14 Linear array type ultrasonic probe.

in the impedance. This phenomenon enables piezoelectric materials to be used as a wave filter. A filter is required to pass a certain selected frequency band or to block a given band. The band width of a filter fabricated from a piezoelectric material is determined by the square of the coupling coefficient k, that is, it is nearly proportional to k^2. Quartz crystals with a very low k value of about 0.1 can pass very narrow frequency bands of approximately 1% of the center resonance frequency. On the other hand, PZT ceramics with a planar coupling coefficient of about 0.5 can easily pass a band of 10% of the center resonance frequency. The sharpness of the passband is dependent on the mechanical quality factor Q_M of the materials. Quartz also has a very high Q_M of about 10^6, which results in a sharp cut-off to the passband and a well-defined oscillation frequency.

A simple resonator is a thin disk type, electroded on its plane faces and vibrating radially, for filter applications with a center frequency ranging from 200 kHz to 1 MHz and with a bandwidth of several percent of the center frequency. For a frequency of 455 kHz, the disk diameter needs to be about 5.6 mm. However, if the required frequency is higher than 10 MHz, other modes of vibration such as the thickness extensional mode are exploited, because of its smaller size. The trapped-energy type filters made from PZT ceramics have been widely used in the intermediate frequency range for applications such as the 10.7 MHz FM radio receiver and transmitter. When the trapped-energy phenomena are

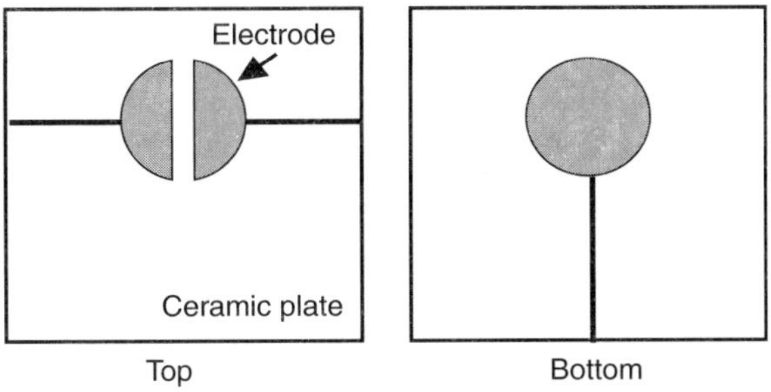

FIGURE 4.1.15 Schematic drawing of a trapped-energy filter.

utilized, the overtone frequencies are suppressed. The plate is partly covered with electrodes of a specific area and thickness. The fundamental frequency of the thickness mode of the ceramic beneath the electrode is less than that of the unelectroded portion, because of the extra inertia of the electrode mass. The lower-frequency wave of the electroded region cannot propagate into the unelectroded region. The higher-frequency overtones, however, can propagate away into the unelectroded region. This is called the *trapped-energy principle*. Figure 4.1.15 shows a schematic drawing of a *trapped-energy filter*. In this structure the top electrode is split so that coupling between the two parts will only be efficient at resonance. More stable filters suitable for telecommunication systems have been made from single crystals such as quartz or $LiTaO_3$.

4.1.4 SURFACE ACOUSTIC WAVE DEVICES

A *surface acoustic wave* (SAW), also called a *Rayleigh wave*, is essentially a coupling between longitudinal and shear waves. The energy carried by the SAW is confined near the surface. An associated electrostatic wave exists for a SAW on a piezoelectric substrate, which allows electroacoustic coupling via a transducer. The advantages of SAW technology are [23, 24]:

1. The wave can be electroacoustically accessed and tapped at the substrate surface and its velocity is approximately 10^4 times slower than an electromagnetic wave.
2. The SAW wavelength is on the same order of magnitude as line dimensions produced by photolithography and the lengths for both short and long delays are achievable on reasonably sized substrates.

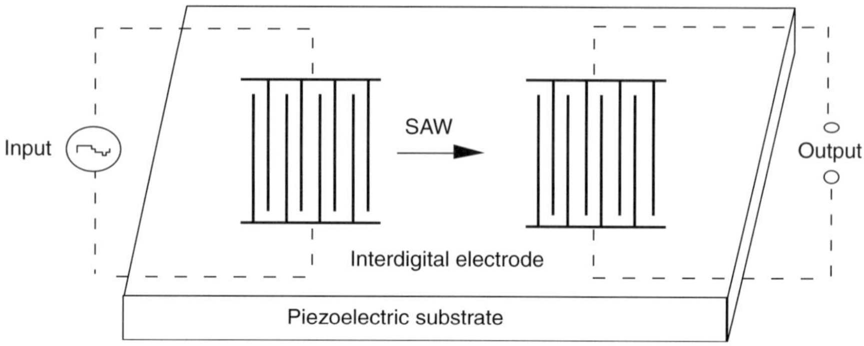

FIGURE 4.1.16 Fundamental structure of a SAW device.

There is a very broad range of commercial system applications which include front-end and intermediate frequency (IF) filters, community antenna television (CATV) and video cassette recorder (VCR) components, synthesizers, analyzers and navigators. In SAW transducers, finger (*interdigital*) electrodes provide the ability to sample or tap the wave and the electrode gap gives the relative delay. A SAW filter is composed of a minimum of two transducers. A schematic of a simple SAW bidirectional filter is shown in Figure 4.1.16. A bidirectional transducer radiates energy equally from each side of the transducer. Energy which is not associated with the received signal is absorbed to eliminate spurious reflection.

Various materials are currently being used for SAW devices. The most popular single-crystal SAW materials are $LiNbO_3$ and $LiTaO_3$. The materials have different properties depending on the cut of the material and the direction of propagation. The fundamental parameters considered when choosing a material for a given device applications are SAW velocity, temperature coefficients of delay (TCD), electromechanical coupling factor and propagation loss. Surface acoustic waves can be generated and detected by spatially periodic, interdigital electrodes on the plane surface of a piezoelectric plate. A periodic electric field is produced when an RF source is connected to the electrode, thus permitting piezoelectric coupling to a traveling surface wave. If an RF source with a frequency, f, is applied to the electrode having periodicity, d, energy conversion from an electrical to mechanical form will be maximum when

$$f = f_0 = v_s/d, \tag{39}$$

where v_s is the SAW velocity and f_0 is the center frequency of the device. The SAW velocity is an important parameter determining the center frequency. Another important parameter for many applications is temperature sensitivity. For example, the temperature stability of the center frequency of SAW bandpass filters is a direct function of the temperature coefficient for the velocity and the delay for the material used. The first-order temperature coefficient of delay is given by:

$$(1/\tau) \cdot (d\tau/dT) = (1/L) \cdot (dL/dT)(1/v_s) \cdot (dv_s/dT), \tag{40}$$

where $\tau = L/v_s$ is the delay time and L is the SAW propagation length. The surface wave coupling factor, k_s^2, is defined in terms of the change in SAW velocity which occurs when the wave passes across a surface coated with a thin massless conductor, so that the piezoelectric field associated with the wave is effectively short-circuited. The coupling factor, k_s^2, is expressed by:

$$k_s^2 = 2(v_f v_m)/v_f, \tag{41}$$

where v_f is the free surface wave velocity and v_m the velocity on the metallized surface. In actual SAW applications, the value of k_s^2 relates to the maximum bandwidth obtainable and the amount of signal loss between input and output, which determines the fractional bandwidth as a function of minimum insertion loss for a given material and filter. Propagation loss is one of the major factors that determines the insertion loss of a device and is caused by wave scattering at crystalline defects and surface irregularities. Materials which show high electromechanical coupling factors combined with small temperature coefficients of delay are generally preferred. The free surface velocity, v_0, of the material is a function of cut angle and propagation direction. The TCD is an indication of the frequency shift expected for a transducer due to a temperature change and is also a function of cut angle and propagation direction. The substrate is chosen based on the device design specifications which include operating temperature, fractional bandwidth, and insertion loss.

Piezoelectric single crystals such as 128°*Y*–*X* (128°-rotated-*Y*-cut and *X*-propagation)—$LiNbO_3$ and *X*–112°*Y* (*X*-cut and 112°-rotated-*Y*-propagation)—$LiTaO_3$ have been extensively employed as SAW substrates for applications in VIF filters. A *c*-axis oriented ZnO thin film deposited on a fused quartz, glass or sapphire substrate has also been commercialized for SAW devices. Table 4.1.2 summarizes some important material parameters for these SAW materials.

A delay line can be formed from a slice of glass such as PbO or K_2O doped SiO_2 glass in which the velocity of sound is nearly independent of temperature. PZT ceramic transducers are soldered on two metallized edges of the slice of glass. The input transducer converts the electrical signal to a shear acoustic wave which travels through the slice. At the output transducer, the wave is reconverted into an electrical signal delayed by the length of time taken to travel around the slice. Such delay lines are used in color TV sets to introduce a delay of approximately 64 μs and are also employed in videotape recorders.

TABLE 4.1.2 SAW Material Properties

Material	Cut–propagation direction	k^2(%)	TCD (ppm/C)	v_0 (m/s)	ε_r
Single crystal					
Quartz	ST–X	0.16	0	3158	4.5
$LiNbO_3$	128°Y–X	5.5	−74	3960	35
$LiTaO_3$	X112°–Y	0.75	−18	3290	42
$Li_2B_4O_7$	(110)–⟨001⟩	0.8	0	3467	9.5
Ceramic					
$PZT–In(Li_{3/5}W_{2/5})O_3$		1.0	10	2270	690
$(Pb, Nd)(Ti, Mn, In)O_3$		2.6	<1	2554	225
Thin film					
ZnO/glass		0.64	−15	3150	8.5
ZnO/sapphire		1.0	−30	5000	8.5

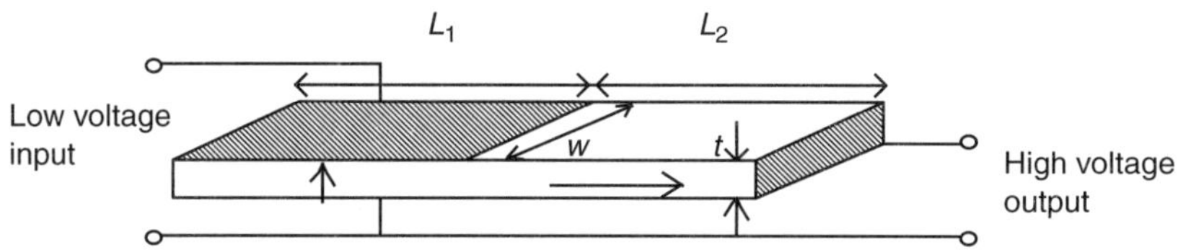

FIGURE 4.1.17 Piezoelectric transformer proposed by Rosen [25].

4.1.5 PIEZOELECTRIC TRANSFORMERS

When input and output terminals are fabricated on a piezo device and input/output voltage is changed through the vibration energy transfer, the device is called a *piezoelectric transformer*. Piezoelectric transformers were used in color TVs because of their compact size in comparison with the conventional electromagnetic coil-type transformers. Since serious problems were found initially in the mechanical strength (collapse occurred at the nodal point!) and in heat generation, the development approach was the same as that used for fabricating ceramic actuators. Recent lap-top computers with a liquid crystal display require a very thin, no electromagnetic-noise transformer to start the glow of a fluorescent back-lamp. This application has recently accelerated the development of the piezotransformer.

Since the original piezotransformer was proposed by Rosen [25], there have been a variety of such transformers investigated. Figure 4.1.17 shows a fundamental structure where two differently poled parts coexist in one piezoelectric plate. A standing wave with a wavelength equal to the sample length is excited,

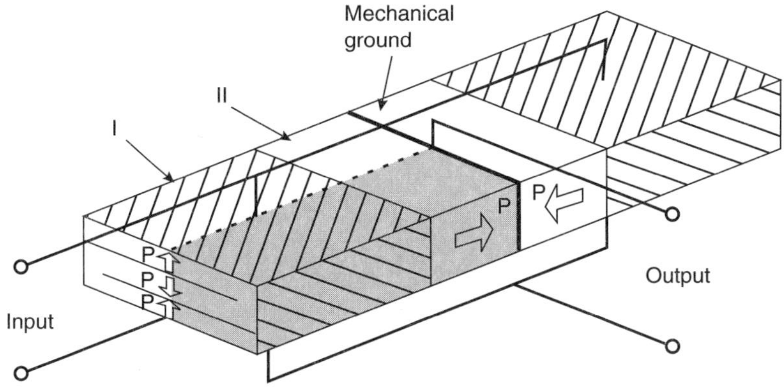

FIGURE 4.1.18 Multilayer type transformer by NEC.

a half wavelength existing on both the input (L_1) and output (L_2) parts. The voltage rise ratio r (*step-up ratio*) is given for the unloaded condition by:

$$r = \left(4/\pi^2\right) k_{31}k_{33}Q_M(L_2/t)\left[2\sqrt{s_{33}^E/s_{11}^E}\Big/\left(1+\sqrt{s_{33}^D/s_{11}^E}\right)\right]. \qquad (42)$$

The r ratio is increased with an increase of (L_2/t), where t is the thickness.

NEC proposed a multilayer type transformer (Fig. 4.1.18) in order to increase the voltage rise ratio [26]. Usage of the third-order longitudinal mode is another idea to distribute the stress concentration.

4.1.6 PIEZOELECTRIC ACTUATORS

Piezoelectric and electrostrictive devices have become key components in smart actuator systems such as precision positioners, miniature ultrasonic motors and adaptive mechanical dampers. This section reviews the developments of piezoelectric and related ceramic actuators with particular focus on the improvement of actuator materials, device designs and applications of the actuators.

Piezoelectric actuators are forming a new field between electronic and structural ceramics [27–30]. Application fields are classified into three categories: positioners, motors and vibration suppressors. The manufacturing precision of optical instruments such as lasers and cameras, and the positioning accuracy for fabricating semiconductor chips, which must be adjusted using solid-state actuators, are generally on the order of 0.1 μm. Regarding conventional electromagnetic motors, tiny motors smaller than 1 cm^3 are often required in office or factory automation equipment and are rather difficult to produce with sufficient energy efficiency. Ultrasonic motors whose efficiency is insensitive to size are considered superior in the mini-motor area. Vibration suppression in space structures and military vehicles using piezoelectric actuators is another promising field of application.

New solid-state displacement transducers controlled by temperature (shape memory alloy) or magnetic field (magnetostrictive alloy) have been proposed, but are generally inferior to the piezoelectric/electrostrictive ceramic actuators because of current technological trends aimed at reduced driving power and miniaturization [30]. The shape memory actuator is too slow in response with a very low energy efficiency, while the magnetostrictor requires a driving coil which is very bulky and generates magnetic noise.

4.1.6.1 Ceramic Actuator Materials

Actuator materials are classified into three categories; piezoelectric, electrostrictive and phase-change materials. Modified lead zirconate titanate [PZT, $Pb(Zr,Ti)O_3$] ceramics are currently the leading materials for piezoelectric applications. The PLZT [$(Pb,La)(Zr,Ti)O_3$] 7/62/38 compound is one such composition [31]. The strain curve is shown in Figure 4.1.19a (left). When the applied field is small, the induced strain x is nearly proportional to the field E ($x = dE$, where d is called the piezoelectric constant). As the field becomes larger (i.e. greater than about 1 kV/cm), however, the strain curve deviates from this linear trend and significant hysteresis is exhibited due to polarization reorientation. This sometimes limits the use of such materials for actuator applications that require non-hysteretic response.

An interesting new family of actuators has been fabricated in Germany from the barium stannate titanate system [$Ba(Sn,Ti)O_3$] [32]. The useful property of $Ba(Sn_{0.15}Ti_{0.85})O_3$ is its unusual strain curve, in which the domain reorientation occurs only at low fields, and there is then a long linear range at higher fields (Figure 4.1.19a, right); that is, the coercive field is unusually small. Moreover, this system is particularly intriguing since it contains no Pb ions, an essential feature as ecological concerns grow in the future.

The second category of actuators is based on electrostriction as exhibited by PMN [$Pb(Mg_{1/3}Nb_{2/3})O_3$] based ceramics. Although it is a second-order phenomenon of electromechanical coupling ($x = ME^2$, where M is called the electrostrictive coefficient), the induced strain can be extraordinarily large (more than 0.1%) [33]. An attractive feature of these materials is the near absence of hysteresis (Fig. 4.1.19b). The superiority of PMN to PZT was demonstrated in a scanning tunneling microscope (STM) [34]. The STM probe was

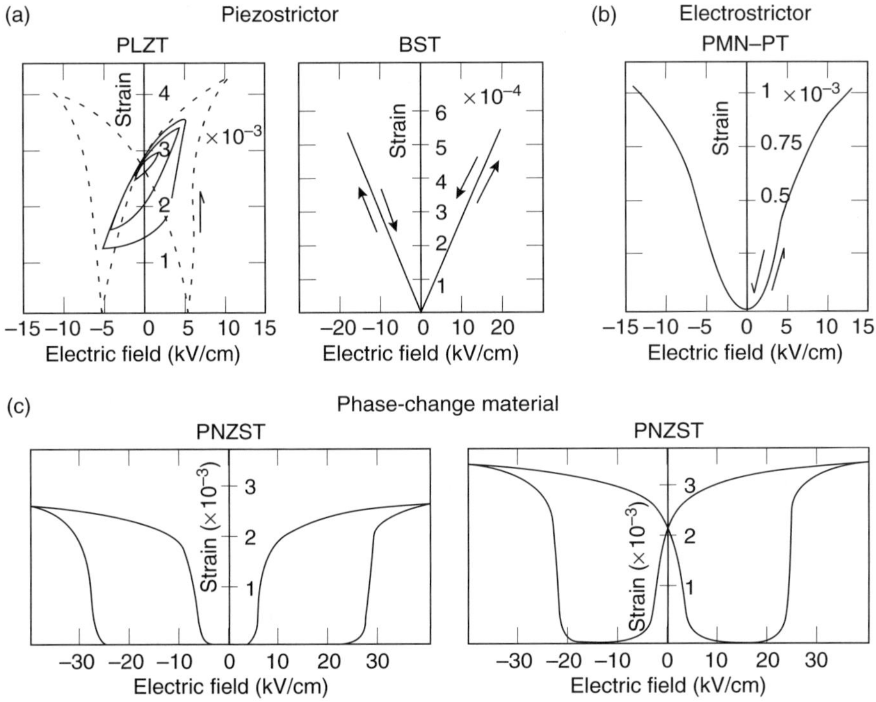

FIGURE 4.1.19 Electric field-induced strains in ceramics: (a) piezoelectric $(Pb,La)(Zr,Ti)O_3$ and $Ba(Sn,Ti)O_3$; (b) electrostrictive $Pb(Mg_{1/3}Nb_{2/3},Ti)O_3$; (c) phase-change material $Pb(Zr,Sn,Ti)O_3$.

scanned mechanically by a PMN actuator, which produces only extremely small distortion of the image, even when the probe was scanned in the opposite direction due to a negligibly small hysteresis.

The third category is based on phase-change-related strains, that is, polarization induced by switching from an antiferroelectric to a ferroelectric state [35]. Figure 4.1.19c shows the field-induced strain curves taken for the lead zirconate stannate based system $[Pb_{0.99}Nb_{0.02}((Zr_xSn_{1-x})_{1-y}Ti_y)_{0.98}O_3]$. The longitudinally induced strain reaches more than 0.3%, which is much larger than that exhibited by normal piezostrictors or electrostrictors. A rectangular-shaped hysteresis in Figure 4.1.19c (left) characterizes the response of these devices which are referred to as "digital displacement transducers" because of the two on/off strain states. Moreover, this field-induced transition is accompanied by a shape memory effect for appropriate compositions (Figure 4.1.19c, right). Once the ferroelectric phase has been induced, the material "memorizes" its ferroelectric state even under zero-field conditions, although it can be erased with the application of a small reverse bias field [36]. This shape memory ceramic is used in energy saving actuators. A latching relay was composed of a shape memory ceramic unimorph and a mechanical snap action switch, which was driven by a pulse voltage of 4 ms duration. Compared with the conventional electromagnetic relays, the new relay was much simpler and more compact in structure with almost the same response time.

4.1.6.2 Actuator Designs

Two of the most popular actuator designs are the multilayers [37] and bimorphs (see Figure 4.1.20). The multilayer, in which roughly 100 thin piezoelectric/electrostrictive ceramic sheets are stacked together, has the advantages of low driving voltage (100 V), quick response (10 μs), high generative force (1000 N), and high electromechanical coupling. But the displacement, on the order of 10 μm, is not sufficient for some applications. This contrasts with the characteristics of the bimorph which consists of multiple piezoelectric and elastic plates bonded together to generate a large bending displacement of several hundred μm, but has relatively low response time (1 ms) and generative force (1 N).

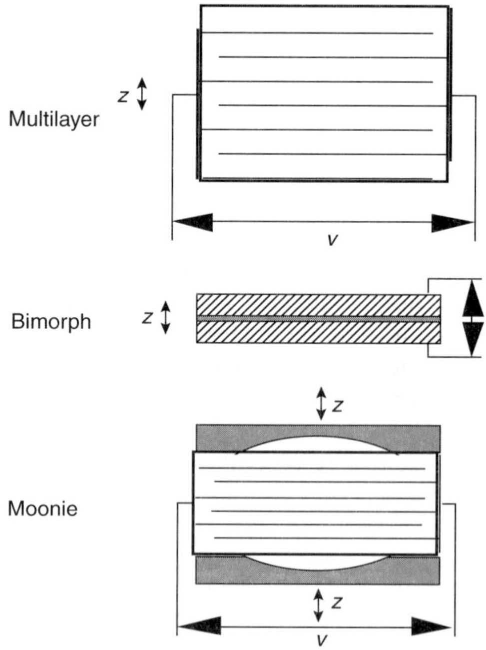

FIGURE 4.1.20 Typical designs for ceramic actuators: multilayer, bimorph and moonie.

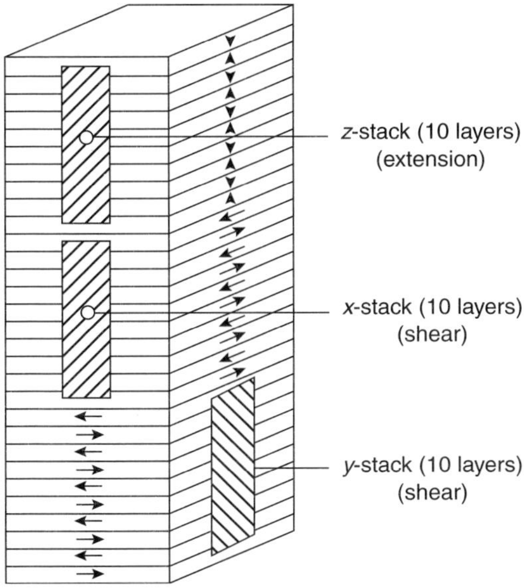

FIGURE 4.1.21 3D positioning multilayer actuator. Notice that the *x*- and *y*-stacks are using shear mode with the spontaneous polarization perpendicular to the applied electric field direction.

A 3D positioning actuator with a stacked structure as pictured in Figure 4.1.21 was proposed by a German company, where shear strain was utilized to generate the *x* and *y* displacements [38]. Polymer-packed PZT bimorphs have been commercialized by ACX for vibration reduction/control applications in smart structures [39].

A composite actuator structure called the "moonie" (or "cymbal") has been developed to provide characteristics intermediate between the multilayer and bimorph actuators; this transducer exhibits an order of magnitude larger displacement than the multilayer, and much larger generative force with quicker response than the bimorph [40]. The device consists of a thin multilayer piezoelectric element and two metal plates with narrow moon-shaped cavities bonded together as shown in Figure 4.1.20. The moonie with a size of $5 \times 5 \times 2.5\,\text{mm}^3$ can generate a $20\,\mu\text{m}$ displacement under 60 V, eight times as large as the generative displacement produced by a multilayer of the same size [41]. This new compact actuator has been utilized in a miniaturized laser beam scanner.

4.1.6.3 Drive/Control Techniques

Piezoelectric/electrostrictive actuators may be classified into two categories, based on the type of driving voltage applied to the device and the nature

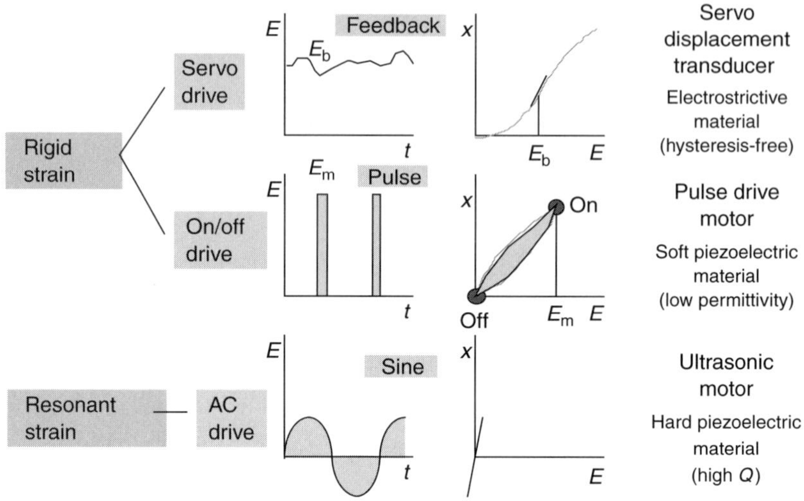

FIGURE 4.1.22 Classification of piezoelectric/electrostrictive actuators.

of the strain induced by the voltage (Fig. 4.1.22): (a) rigid displacement devices for which the strain is induced unidirectionally along the direction of the applied DC field; and (b) resonating displacement devices for which the alternating strain is excited by an AC field at the mechanical resonance frequency (ultrasonic motors). The first can be further divided into two types: servo displacement transducers (positioners) controlled by a feedback system through a position-detection signal, and pulse drive motors operated in a simple on/off switching mode, exemplified by dot-matrix printers.

The material requirements for these classes of devices are somewhat different, and certain compounds will be better suited to particular applications. The ultrasonic motor, for instance, requires a very hard piezoelectric with a high mechanical quality factor Q_M, to suppress heat generation. Driving the motor at the antiresonance frequency, rather than at resonance, is also an intriguing technique to reduce the load on the piezoceramic and the power supply [42]. The servo displacement transducer suffers most from strain hysteresis and, therefore, a PMN electrostrictor is used for this purpose. The pulse drive motor requires a low permittivity material aimed at quick response with a certain power supply rather than a small hysteresis, so soft PZT piezoelectrics are preferred rather than the high-permittivity PMN for this application.

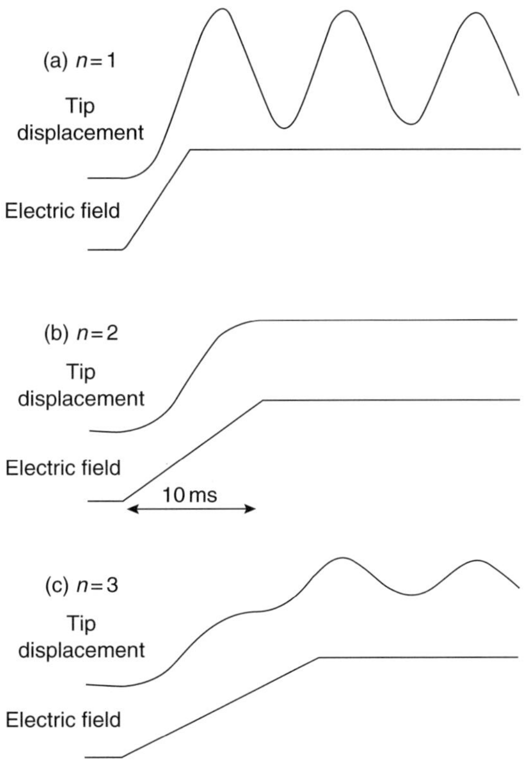

FIGURE 4.1.23 Transient vibration of a bimorph excited after a pseudo-step voltage applied. n is a time scale with a unit of half of the resonance period, that is, $2n$ = resonance period.

Pulse drive techniques for ceramic actuators are very important for improving the response of the device [43, 44]. Figure 4.1.23 shows transient vibrations of a bimorph excited after a pseudo-step voltage is applied. The rise time is varied around the resonance period (n is the time scale with a unit of $T_0/2$, where T_0 stands for the resonance period). It is concluded that the overshoot and ringing of the tip displacement is completely suppressed when the rise time is precisely adjusted to the resonance period of the piezodevice (i.e. for $n = 2$) [43]. A flight actuator was developed using a pulse drive piezoelectric element and a steel ball. A 5 μm rapid displacement induced in a multilayer actuator can hit a 2 mm steel ball up to 20 mm in height. A dot-matrix printer head has been developed using a flight actuator as shown in Figure 4.1.24 [45]. By changing the drive voltage pulse width, the movement of the armature was easily controlled to realize no vibrational ringing or double hitting.

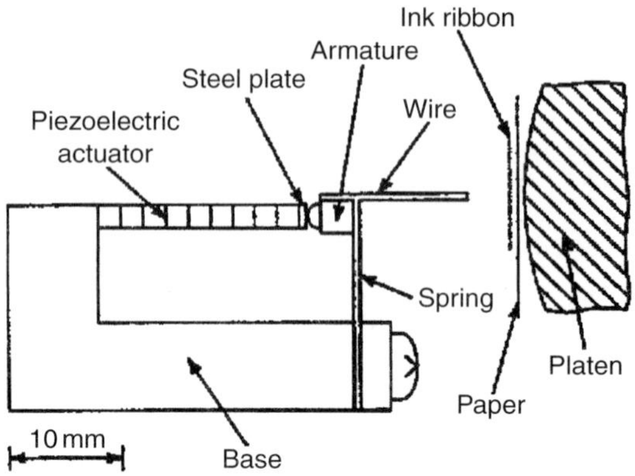

FIGURE 4.1.24 Dot-matrix printer head using a flight actuator mechanism.

4.1.6.4 Device Applications

4.1.6.4.1 Servo Displacement Transducers

A typical example is found in a space truss structure proposed by the Jet Propulsion Laboratory [46]. A stacked PMN actuator was installed at each truss nodal point and operated so that unnecessary mechanical vibration was suppressed immediately. A “Hubble” telescope has also been proposed using multilayer PMN electrostrictive actuators to control the phase of the incident light wave in the field of optical information processing (Fig. 4.1.25) [47]. The PMN electrostrictor provided superior adjustment of the telescope image because of negligible strain hysteresis.

The US Army is interested in developing a rotor control system in helicopters. Figure 4.1.26 shows a bearingless rotor flexbeam with attached piezoelectric strips [48]. Various types of PZT-sandwiched beam structures have been investigated for such a flexbeam application and for active vibration control [49].

Concerning home appliance applications, there is already a large market in VCR systems. The requirement for high-quality images has become very stringent for VCRs especially when played in still, slow or quick mode. As illustrated in Figure 4.1.27, when the tape is running at a speed different from the normal speed, the head trace deviates from the recording track depending on the velocity difference. Thus, the head traces on the guard band, generating *guard band noise* [50]. The auto tracking scan system by Ampex operates with a piezoelectric actuator so that the head follows the recording track. The piezoelectric device generates no magnetic noise.

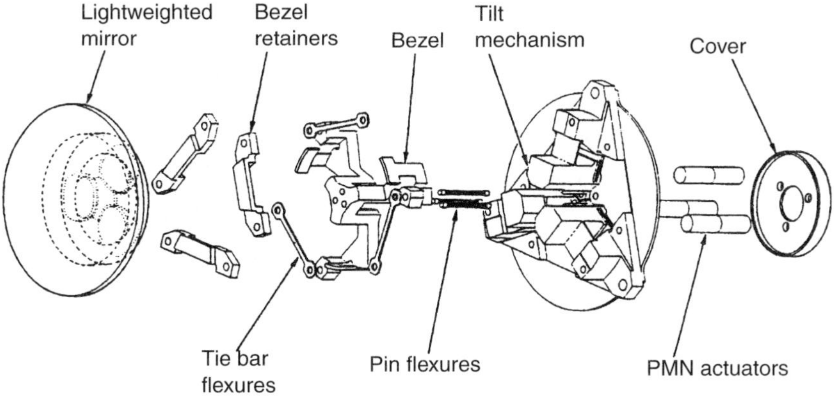

FIGURE 4.1.25 "Hubble" telescope using three PMN electrostrictive actuators for optical image correction.

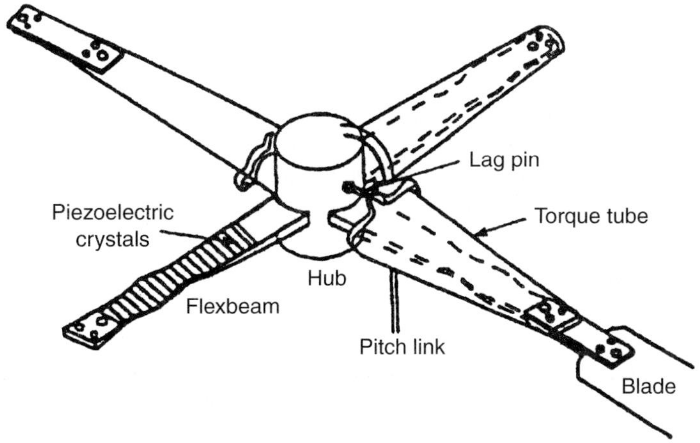

FIGURE 4.1.26 Bearingless rotor flexbeam with attached piezoelectric strips. A slight change in the blade angle provides for enhanced controllability.

Bimorph structures are commonly used for this tracking actuator because of their large displacement. However, special care has been taken not to produce a *spacing angle* between the head and the tape, because a single bimorph exhibits deflection with slight rotation. Various designs have been proposed to produce a completely parallel motion.

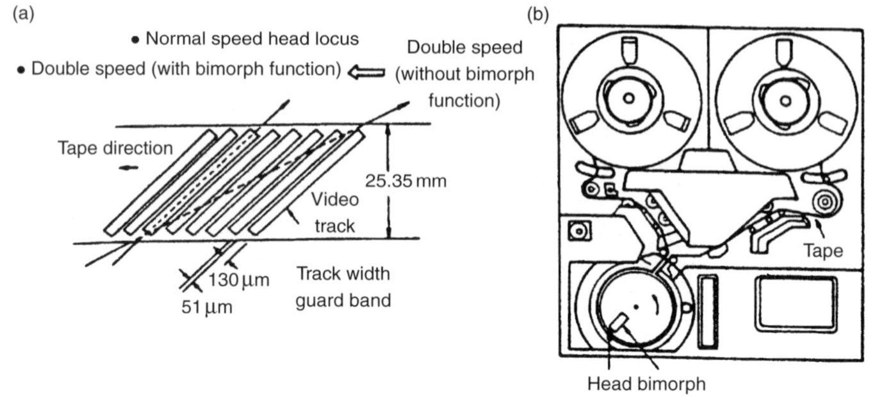

FIGURE 4.1.27 Locus of the video head and the function of the piezoactuator.

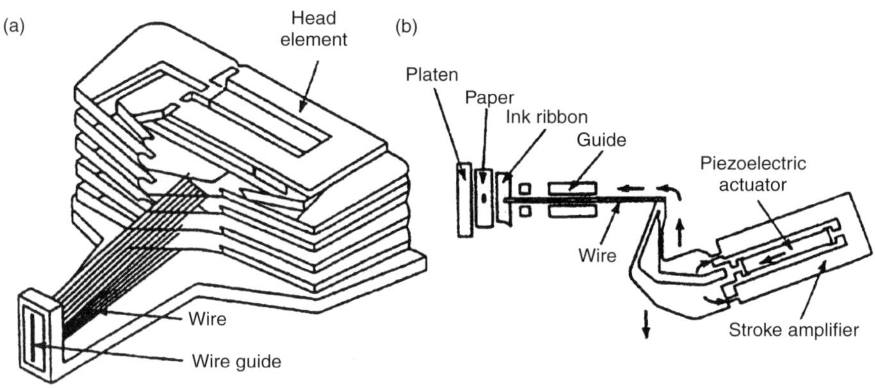

FIGURE 4.1.28 Structure of a printer head (a), and a differential-type piezoelectric printer-head element (b). A sophisticated monolithic hinge lever mechanism amplifies the actuator displacement by 30 times.

4.1.6.4.2 Pulse Drive Motors

A dot-matrix printer is the first widely commercialized product using ceramic actuators. Each character formed by such a printer is composed of a 24 × 24 dot matrix. A printing ribbon is subsequently impacted by a multiwire array. A sketch of the printer head appears in Figure 4.1.28a [51]. The printing element is composed of a multilayer piezoelectric device, in which 100 thin ceramic sheets 100 μm in thickness are stacked, together with a sophisticated magnification mechanism (Fig. 4.1.28b). The magnification unit is based on a monolithic

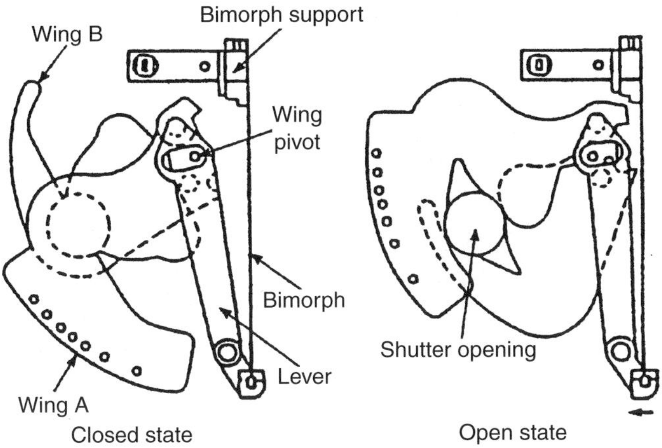

FIGURE 4.1.29 Camera shutter mechanism using a piezoelectric bimorph actuator.

hinge lever with a magnification of 30, resulting in an amplified displacement of 0.5 mm and an energy transfer efficiency greater than 50%.

A piezoelectric camera shutter is currently the largest production item (Fig. 4.1.29). A piece of piezoelectric bimorph can open and close the shutter in a milli-second through a mechanical wing mechanism [52].

Toyota developed a Piezo TEMS (Toyota Electronic Modulated Suspension), which is responsive to each protrusion on the road in adjusting the damping condition, and installed it on a "Celcio" (equivalent to Lexus, internationally) in 1989 [53]. In general, as the damping force of a shock absorber in an automobile is increased (i.e. "hard" damper), the controllability and stability of a vehicle are improved. However, comfort is sacrificed because the road roughness is easily transferred to the passengers. The purpose of the electronically controlled shock absorber is to obtain both controllability and comfort simultaneously. Usually the system is set to provide a low damping force ("soft") so as to improve comfort, and the damping force is changed to a high position according to the road condition and the car speed to improve the controllability. In order to respond to a road protrusion, a very high response of the sensor and actuator combination is required.

Figure 4.1.30 shows the structure of the electronically controlled shock absorber. The sensor is composed of five layers of 0.5 mm thick PZT disks. The detecting speed of the road roughness is about 2 ms and the resolution of the up–down deviation is 2 mm. The actuator is made of 88 layers of 0.5 mm thick disks. Applying 500 V generates a displacement of about 50 μm, which is magnified by 40 times through a piston and plunger pin combination. This

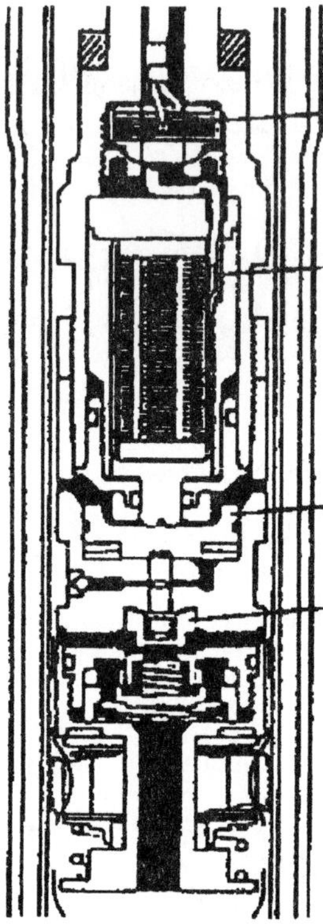

FIGURE 4.1.30 Electronic modulated suspension by Toyota.

stroke pushes the change valve of the damping force down, then opens the bypass oil route, leading to the decrease of the flow resistance (i.e. "soft").

Figure 4.1.31 illustrates the operation of the suspension system. The up–down acceleration and pitching rate were monitored when the vehicle was driven on a rough road. When the TEMS system was used (top figure), the up–down acceleration was suppressed to as small as the condition fixed at "soft", providing comfort. At the same time, the pitching rate was also suppressed to as small as the condition fixed at "hard", leading to better controllability.

Figure 4.1.32 shows a walking piezomotor with four multilayer actuators [54]. The two shorter actuators function as clamps and the longer two provide the movement by an inchworm mechanism.

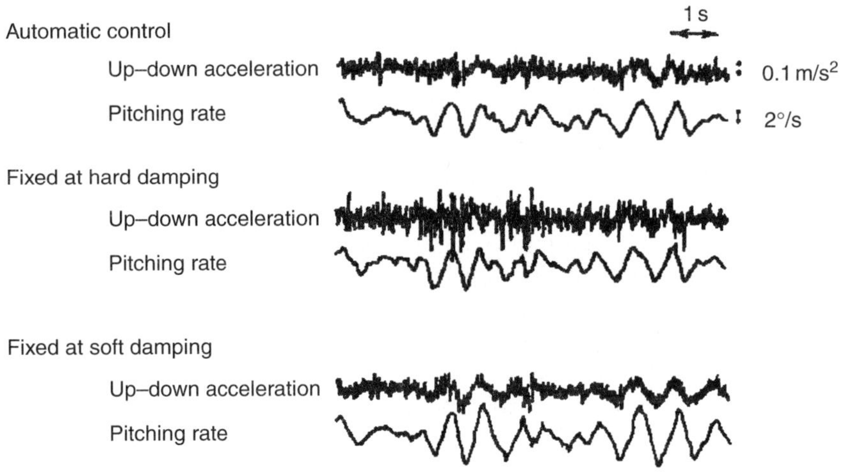

FIGURE 4.1.31 Function of the adaptive suspension system.

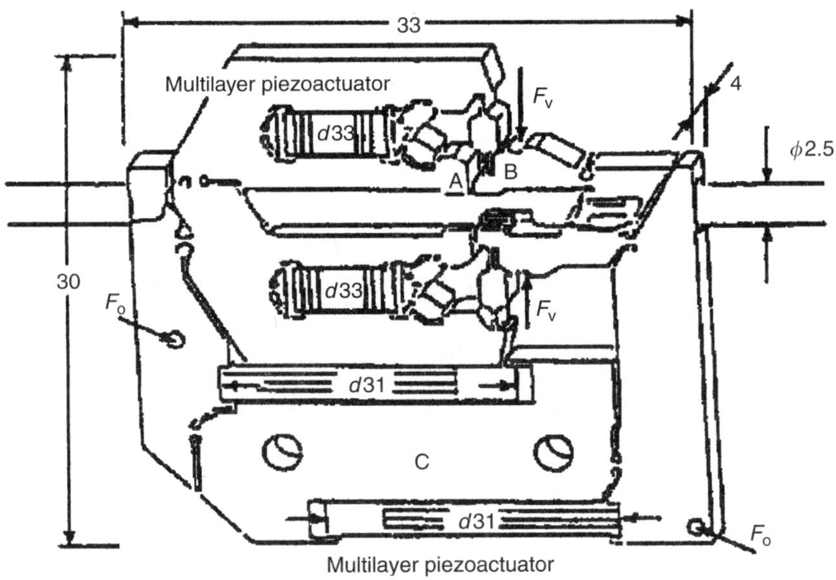

FIGURE 4.1.32 Walking piezomotor using an inchworm mechanism with four multilayer piezoelectric actuators by Philips.

4.1.6.4.3 Ultrasonic Motors

4.1.6.4.3.1 *Background*

Electromagnetic motors were invented more than a 100 years ago. While these motors still dominate the industry, a drastic improvement cannot be expected except through new discoveries in magnetic or superconducting materials. Regarding conventional electromagnetic motors, tiny motors smaller than 1 cm^3 are rather difficult to produce with sufficient energy efficiency. Therefore, a new class of motors using high power ultrasonic energy, the ultrasonic motor, is gaining widespread attention. Ultrasonic motors made with piezoceramics whose efficiency is insensitive to size are superior in the mini-motor area. Figure 4.1.33 shows the basic construction of most ultrasonic motors, which consist of a high-frequency power supply, a vibrator and a slider. The vibrator is composed of a piezoelectric driving component and an elastic vibratory part, and the slider is composed of an elastic moving part and a friction coat.

Although there had been some earlier attempts, the first practical ultrasonic motor was proposed by H. V. Barth of IBM in 1973 [55]. The rotor was pressed against two horns placed at different locations. By exciting one of the horns, the rotor was driven in one direction, and by exciting the other horn, the rotation direction was reversed. Various mechanisms based on virtually the same principle were proposed by Lavrinenko [56] and Vasiliev [57] in the former USSR. Because of difficulty in maintaining a constant vibration amplitude with temperature rise, wear and tear, the motors were not of much practical use at that time.

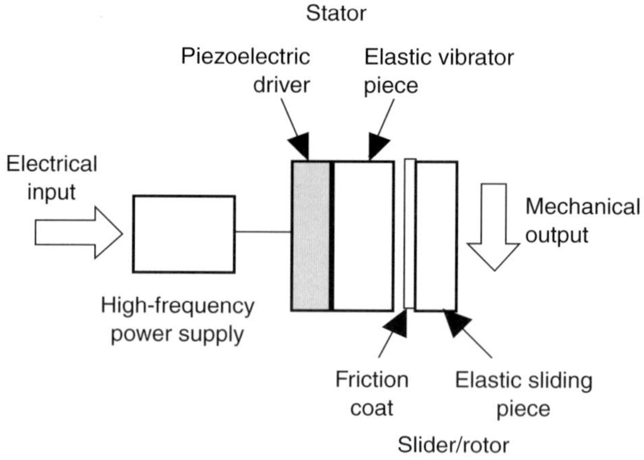

FIGURE 4.1.33 Fundamental construction of an ultrasonic motor.

In the 1980s, with increasing chip pattern density, the semiconductor industry began to demand much more precise and sophisticated positioners which would not generate magnetic field noise. This urgent need accelerated the development of ultrasonic motors. Another advantage of ultrasonic motors over conventional electromagnetic motors with expensive copper coils is the improved availability of piezoelectric ceramics at reasonable cost. Japanese manufacturers are currently producing piezoelectric buzzers at about 30–40 cents per unit.

Let us summarize the merits and demerits of the ultrasonic motor:

Merits

1. Low speed and high torque
   direct drive
2. Quick response, wide velocity range, hard brake and no backlash
   excellent controllability
   fine position resolution
3. High power/weight ratio and high efficiency
4. Quiet drive
5. Compact size and light weight
6. Simple structure and easy production process
7. Negligible effect from external magnetic or radioactive fields, and also no generation of these fields.

Demerits

8. Necessity for a high-frequency power supply
9. Less durability due to frictional drive
10. Drooping torque versus speed characteristics

4.1.6.4.3.2 *Classification and principles of ultrasonic motors*

From a customer's point of view, there are rotary and linear type motors. If we categorize them according to the vibrator shape, there are rod type, π-shaped, ring (square) and cylinder types. Two categories are being investigated for ultrasonic motors from a vibration characteristic viewpoint: a standing-wave type and a propagating-wave type. Refresh your memory on the wave formulas. The standing wave is expressed by

$$u_s(x,t) = A\cos kx \cdot \cos \omega t, \tag{43}$$

while the propagating wave is expressed as

$$u_p(x,t) = A\cos(kx - \omega t). \tag{44}$$

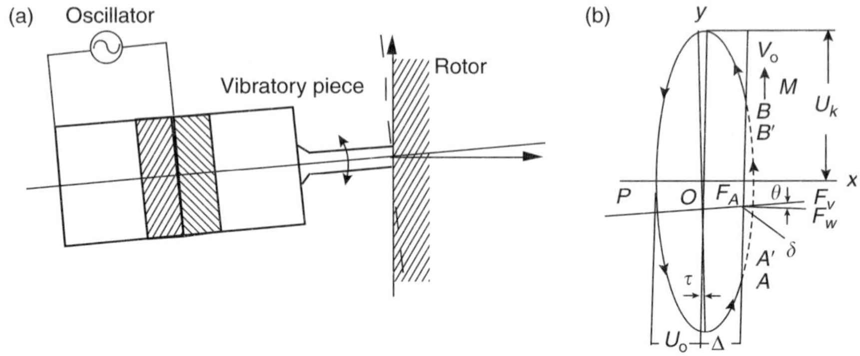

FIGURE 4.1.34 Vibratory coupler type motor (a) and its tip locus (b).

Using a trigonometric relation, Eq. 44 can be transformed as

$$u_p(x,t) = A\cos kx \cdot \cos \omega t + A\cos(kx - \pi/2) \cdot \cos(\omega t - \pi/2). \qquad (45)$$

This leads to an important result, a propagating wave can be generated by superimposing two standing waves whose phases differ by 90° both in time and in space. This principle is necessary to generate a propagating wave on a limited volume/size substance, because only standing waves can be excited stably in a solid medium of finite size.

The standing-wave type is sometimes referred to as a vibratory-coupler type or a "woodpecker" type, where a vibratory piece is connected to a piezoelectric driver and the tip portion generates a flat-elliptical movement. Figure 4.1.34 shows a simple model proposed by Sashida [58]. A vibratory piece is attached to a rotor or a slider with a slight cant angle. When a vibration is excited at the piezoelectric vibrator, the vibratory piece generates bending because of restriction by the rotor, so that the tip moves along the rotor face between A → B, and freely between B → A. If the vibratory piece and the piezovibrator are tuned properly, they form a resonating structure, which is an elliptical locus. Therefore, only the duration A → B provides a unidirectional force to the rotor through friction, and, therefore, an intermittent rotational torque or thrust. However, because of the inertia of the rotor, the rotation speed ripple is not observed to be large. The standing-wave type, in general, is low in cost (one vibration source) and has high efficiency (up to 98% theoretically), but lacks control in both the clockwise and counterclockwise directions.

By comparison, the propagating-wave type (a surface-wave or "surfing") combines two standing waves with a 90° phase difference both in time and in space. The principle is shown in Figure 4.1.35. A surface particle of the elastic body draws an elliptical locus due to the coupling of longitudinal and transverse

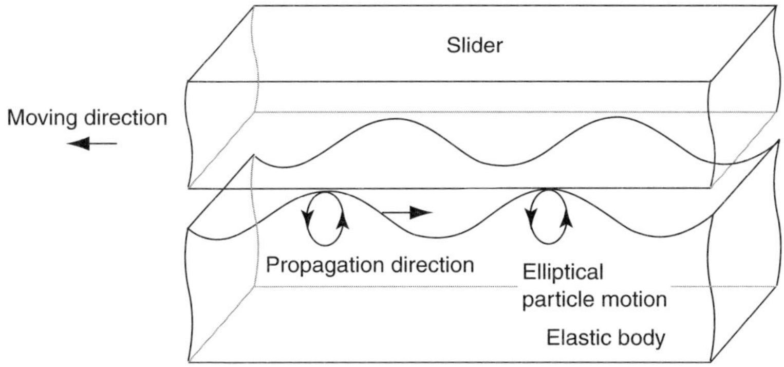

FIGURE 4.1.35 Principle of the propagating-wave type motor.

waves. This type requires, in general, two vibration sources to generate one propagating wave, leading to low efficiency (not more than 50%), but it is controllable in both rotational directions.

4.1.6.4.3.3 *Various ultrasonic motors*

Sashida developed a rotary-type motor similar to the fundamental structure [58]. Four vibratory pieces were installed on the edge face of a cylindrical vibrator, and pressed onto the rotor. This is one of the prototypes which triggered the present active development of ultrasonic motors. A rotation speed of 1500 rpm, torque of 0.08 N m and an output of 12 W (efficiency 40%) are obtained under an input of 30 W at 35 kHz. This type of ultrasonic motor can provide a speed much higher than the inchworm types, because of its high operating frequency and amplified vibration displacement at the resonance frequency.

Hitachi Maxel significantly improved the torque and efficiency by using a torsional coupler replacing Sashida's vibratory pieces (Fig. 4.1.36), and by increasing the pressing force with a bolt [59]. The torsional coupler looks like an old-fashioned TV channel knob, consisting of two legs which transform longitudinal vibration generated by the Langevin vibrator to a bending mode of the knob disk, and a vibratory extruder. Notice that this extruder is aligned with a certain cant angle to the legs, which transforms the bending to a torsional vibration. This transverse moment coupled with the bending up–down motion leads to an elliptical rotation on the tip portion, as illustrated in Figure 4.1.36b. The optimum pressing force to get the maximum thrust is obtained, when the ellipse locus is deformed roughly by half. A motor 30 mm × 60 mm in size and with a 20–30° in cant angle between leg and vibratory piece can generate torques as high as 1.3 N m with an efficiency of 80%. However, this type provides only unidirectional rotation. Notice that even though the drive of the

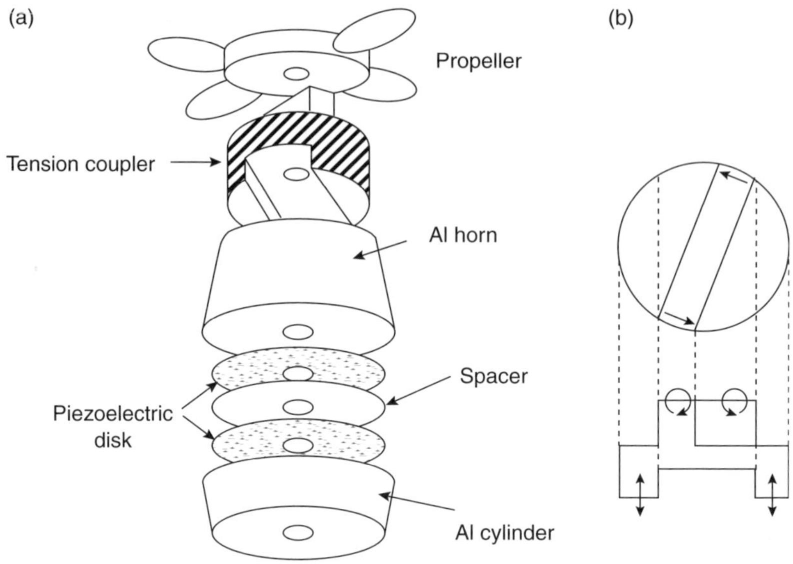

FIGURE 4.1.36 Torsional coupler ultrasonic motor (a) and the motion of the torsional coupler (b).

motor is intermittent, the output rotation becomes very smooth because of the inertia of the rotor.

A compact ultrasonic rotory motor, as tiny as 3 mm in diameter, has been developed at the Pennsylvania State University. As shown in Figure 4.1.37, the stator consists of a piezoelectric ring and two concave/convex metal endcaps with "windmill" shaped slots bonded together, so as to generate a coupling of the up–down and torsional vibrations [60]. Since the number of components is reduced and the fabrication process is much simplified, the fabrication price is decreased remarkably, and a disposable design becomes feasible. When driven at 160 kHz, a maximum revolution of 600 rpm and a maximum torque of 1 mN m were obtained for a 11 mm diameter motor.

Tokin developed a piezoelectric ceramic cylinder for a torsional vibrator [61]. Using an interdigital type electrode pattern printed with a 45° cant angle on the cylinder surface, torsion vibration was generated, which is applicable for a simple ultrasonic motor.

Ueha proposed a two-vibration-mode coupled type (Fig. 4.1.38), that is, a torsional Langevin vibrator was combined with three multilayer actuators to generate larger longitudinal and transverse surface displacements of the stator, as well as to control their phase difference [62]. The phase change can change the rotation direction.

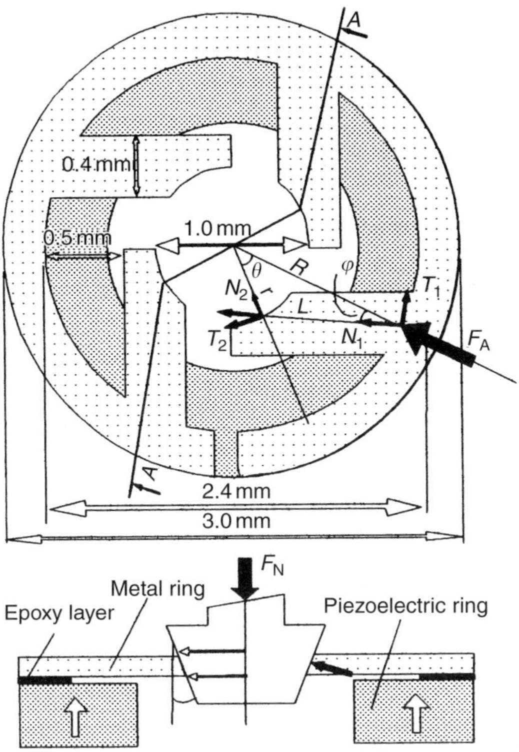

FIGURE 4.1.37 "Windmill" motor with a disk-shaped torsional coupler.

Uchino invented a π-shaped linear motor [63]. This linear motor is equipped with a multilayer piezoelectric actuator and fork-shaped metallic legs as shown in Figure 4.1.39. Since there is a slight difference in the mechanical resonance frequency between the two legs, the phase difference between the bending vibrations of both legs can be controlled by changing the drive frequency. The walking slider moves in a way similar to a horse using its fore and hind legs when trotting. A test motor, $20 \times 20 \times 5\ \text{mm}^3$ in dimension, exhibits a maximum speed of 20 cm/s and a maximum thrust of 0.2 kgf with a maximum efficiency of 20%, when driven at 98 kHz at 6 V (actual power = 0.7 W). This motor has been employed in a precision X–Y stage.

Tomikawa's rectangular plate motor is also intriguing [64]. As shown in Figure 4.1.40, a combination of the two modes of vibration forms an elliptical displacement. The two modes chosen were the first longitudinal mode (L_1 mode) and the eighth bending mode (B_8), whose resonance frequencies were almost the same. By applying voltages with a phase difference of 90° to

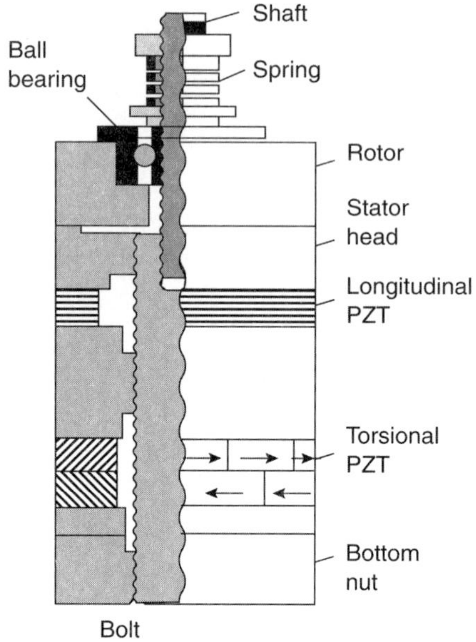

FIGURE 4.1.38 Two-vibration-mode coupled type motor.

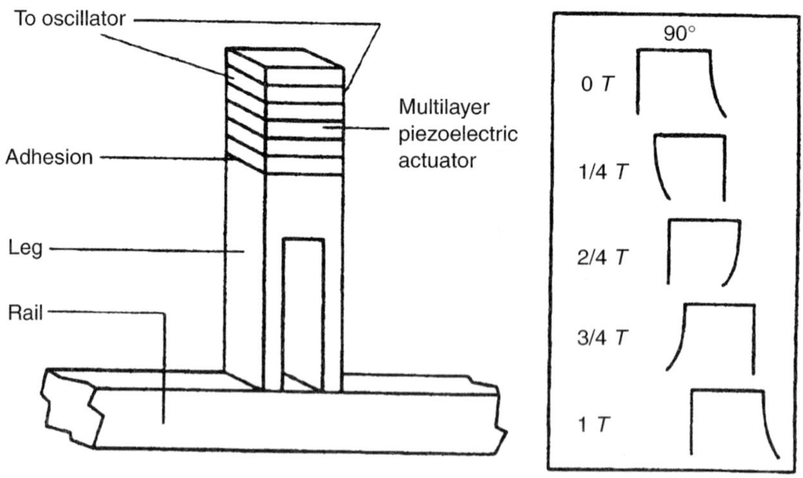

FIGURE 4.1.39 π-shaped linear ultrasonic motor: (a) construction and (b) walking principle. Note the 90° phase difference of two legs similar to that associated with human walking.

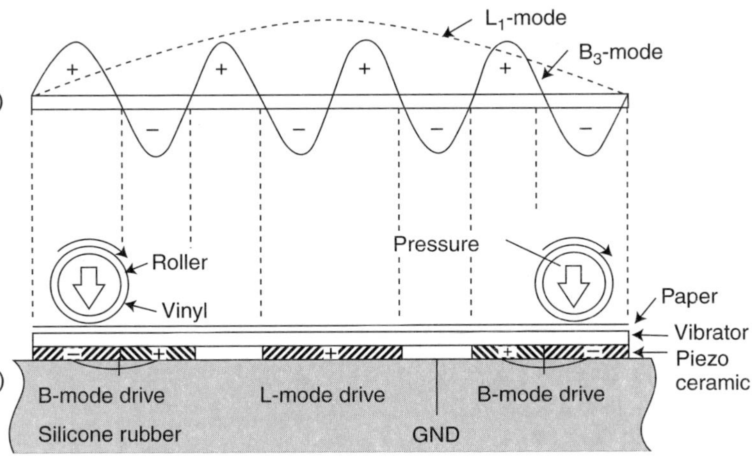

FIGURE 4.1.40 L_1 and B_8 double-mode vibrator motor.

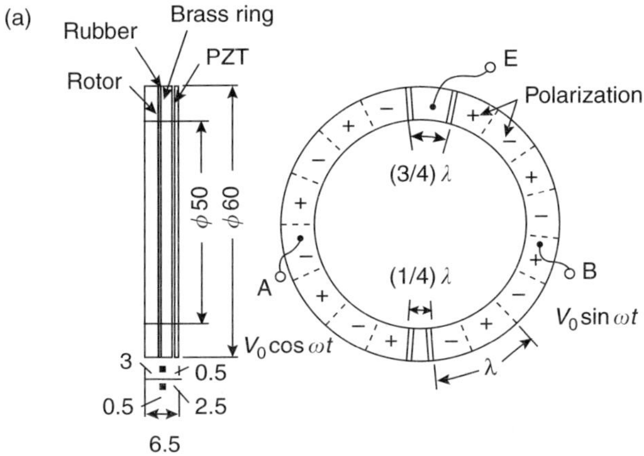

FIGURE 4.1.41 Stator structure of Sashida's motor.

the L- and B-mode drive electrodes, elliptical motion in the same direction can be obtained at both ends of this plate, leading to rotation of the rollers in contact with these points. Anticipated applications are paper or card senders.

Figure 4.1.41 shows the famous Sashida motor [65]. By means of the traveling elastic wave induced by a thin piezoelectric ring, a ring-type slider in contact with the "rippled" surface of the elastic body bonded onto the piezoelectric is driven in both directions by exchanging the sine and cosine voltage inputs. Another advantage is its thin design, which makes it suitable for installation in cameras as an automatic focusing device. Eighty percent of the exchange lenses in Canon's "EOS" camera series have already been replaced by the ultrasonic motor mechanism.

The PZT piezoelectric ring is divided into 16 positively and negatively poled regions and two asymmetric electrode gap regions so as to generate a ninth mode propagating wave at 44 kHz. A prototype was composed of a brass ring of 60 mm in outer diameter, 45 mm in inner diameter and 2.5 mm in thickness, bonded onto a PZT ceramic ring of 0.5 mm in thickness with divided electrodes on the back-side. The rotor was made of polymer coated with hard rubber or polyurethane. Figure 4.1.42 shows Sashida's motor characteristics.

Canon utilized the "surfing" motor for a camera automatic focusing mechanism, installing the ring motor compactly in the lens frame. It is noteworthy that the stator elastic ring has many teeth, which can magnify the transverse elliptical displacement and improve the speed. The lens position can be shifted back and forth with a screw mechanism. The advantages of this motor over the conventional electromagnetic motor are:

1. Silent drive due to the ultrasonic frequency drive and no gear mechanism (i.e. more suitable for video cameras with microphones).
2. Thin motor design and no speed reduction mechanism such as gears, leading to space saving.
3. Energy saving.

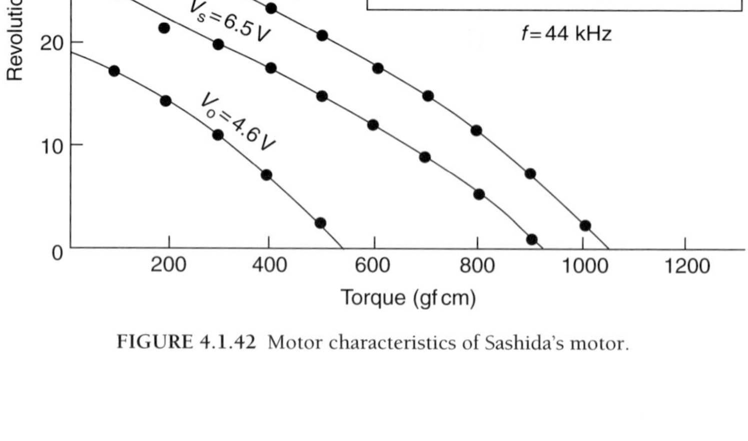

FIGURE 4.1.42 Motor characteristics of Sashida's motor.

A general problem encountered for these traveling-wave type motors is the support of the stator. In the case of a standing-wave motor, the nodal points or lines are generally supported; this causes minimum effects on the resonance vibration. A traveling wave, however, does not have such steady nodal points or lines. Thus, special considerations are necessary. In Figure 4.1.41, the stator is basically supported very gently along the axial direction on felt so as not to suppress the bending vibration. It is important to note that the stop pins which latch onto the stator teeth only provide high rigidity against the rotation.

Matsushita Electric proposed a nodal line support method using a higher-order vibration mode [66]. A stator wide ring is supported at the nodal circular line and "teeth" are arranged on the maximum amplitude circle to get larger revolution.

Seiko Instruments miniaturized the ultrasonic motor to dimensions as tiny as 10 mm in diameter using basically the same principle [67]. Figure 4.1.43 shows the construction of one of these small motors with a 10 mm diameter and a 4.5 mm thickness. A driving voltage of 3 V and a current of 60 mA produces 6000 rpm (no-load) with a torque of 0.1 mN m. Allied Signal developed ultrasonic motors similar to Shinsei's, which are utilized as mechanical switches for launching missiles [68].

It is important to note that the unimorph (a piezoceramic plate and a metal plate bonded together) bending actuation cannot provide high efficiency theoretically, because the electromechanical coupling factor k is usually less than 10%. Therefore, instead of the unimorph structure, a simple disk was

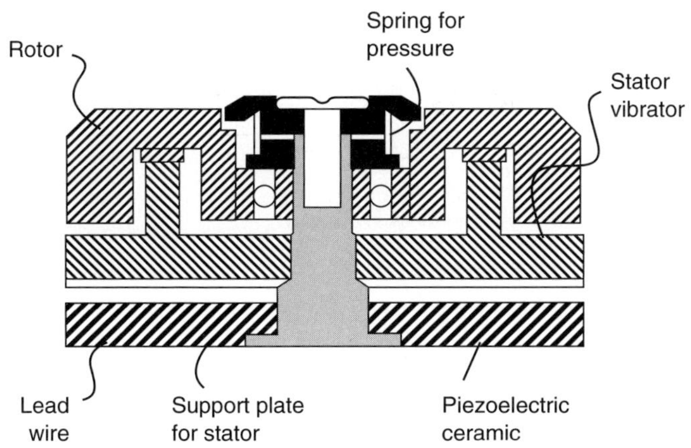

FIGURE 4.1.43 Construction of Seiko's motor.

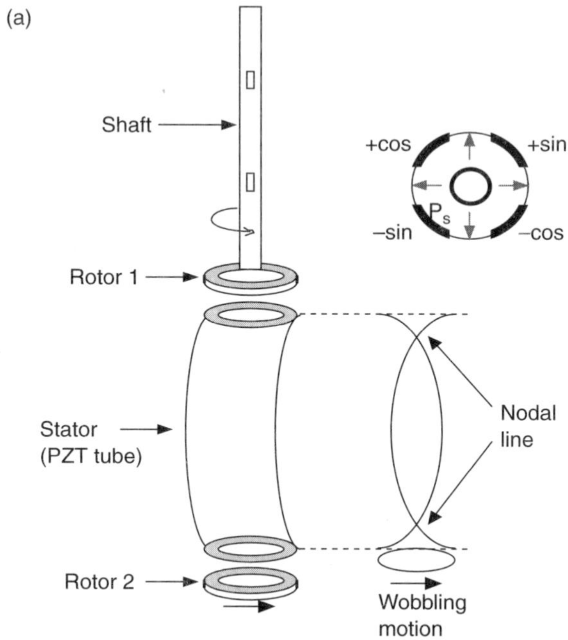

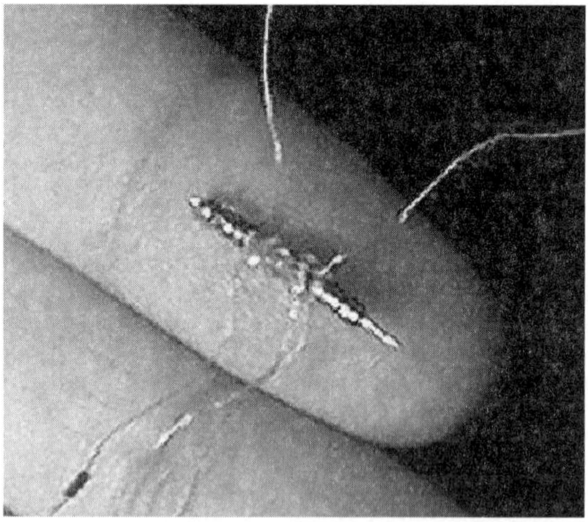

FIGURE 4.1.44 "Plate-spinning" type motor by Penn State and IMRE (1.5 mm in diameter).

directly used to make motors [69, 70]. The (1,1), (2,1) and (3,1) modes of a simple disk, which are axial-asymmetric modes, are proposed to use. Both the inner and outer circumferences can provide a rotation like a "hula hoop."

Another intriguing design is a "plate-spinning" type proposed by Tokin [71]. Penn State and IMRE miniaturized the size down to 1.5 mm in diameter [72]. Figure 4.1.44 shows its principle of operation and a photograph. A rotary bending vibration is excited on a PZT rod by a combination of sine and cosine voltages, then to rotors are brought into contact with the "spinning" rod at the end faces to produce the rotation.

In summary, the standing-wave type, in general, is low in cost (one vibration source) and has high efficiency (up to 98% theoretically), but lack of control in both the clockwise and counterclockwise directions is a problem. In comparison, the propagating-wave type combines two standing waves with a 90° phase difference both in time and space. This type requires, in general, two vibration sources to generate one propagating wave, leading to low efficiency (not more than 50%), but it is controllable in both rotational directions.

REFERENCES

1. Jaffe, B., Cook, W., and Jaffe, H. (1971). *Piezoelectric Ceramics*, London: Academic Press.
2. Cady, W. G. (1964). *Piezoelectricity*, New York: McGraw-Hill, Revised Edition by Dover Publications.
3. Lines, M. E., and Glass, A. M. (1977). *Principles and Applications of Ferroelectric Materials*, Oxford: Clarendon Press.
4. Uchino, K. (1996). *Piezoelectric Actuators and Ultrasonic Motors*, Boston, MA: Kluwer Academic Publishers.
5. Uchino, K. (1986). *Piezoelectric/Electrostrictive Actuators*, Tokyo: Morikita Publishing.
6. Ikeda, T. (1984). *Fundamentals of Piezoelectric Materials Science*, Tokyo: Ohm Publishing Co.
7. Ito, Y., and Uchino, K. (1999). Piezoelectricity, *Wiley Encyclopedia of Electrical and Electronics Engineering*, Vol. 16, p. 479, New York: John Wiley & Sons.
8. Smith, W. A. (1992). *Proc. SPIE—The International Society for Optical Engineering*, 1733.
9. Takeuchi, H., Jyomura, S., Yamamoto, E., and Ito, Y. (1982). *J. Acoust. Soc. Am.* **74**: 1114.
10. Yamashita, Y., Yokoyama, K., Honda, H., and Takahashi, T. (1981). *Jpn. J. Appl. Phys.* **20**: 183.
11. Ito, Y., Takeuchi, H., Jyomura, S., Nagatsuma, K., and Ashida, S. (1979). *Appl. Phys. Lett.* **35**: 595.
12. Takeuchi, H., Masuzawa, H., Nakaya, C., and Ito, Y. (1990). "Proc. IEEE 1990 Ultrasonics Symposium", p. 697.
13. Kuwata, J., Uchino, K., and Nomura, S. (1982). *Jpn. J. Appl. Phys.* **21**: 1298.
14. Shrout, T. R., Chang, Z. P., Kim, N., and Markgraf, S. (1990). *Ferroelectric Lett.* **12**: 63.
15. Newnham, R. E., Skinner, D. P., and Cross, L. E. (1978). *Mater. Res. Bull.* **13**: 525.
16. Smith, W. A. (1989). "Proc. 1989 IEEE Ultrasonic Symposium", p. 755.
17. Kistler, Stress Sensor, Production Catalog, Switzerland.
18. Tokin, Gyroscope, Production Catalog, Japan.
19. Uchino, K., Nomura, S., Cross, L. E., Jang, S. J., and Newham, R. E. (1981). *Jpn. J. Appl. Phys.* **20**: L367. Uchino, K. Proc. Study Committee on Barium Titanate, XXXI-171-1067 (1983).
20. Auld, B. A. (1990). *Acoustic Fields and Waves in Solids*, 2nd edn, Melbourne: Robert E. Krieger.
21. Kino, G. S. (1987). *Acoustic Waves: Device Imaging and Analog Signal Processing*, Englewood Cliffs, NJ: Prentice-Hall.

22. Desilets, C. S., Fraser, J. D., and Kino, G. S. (1978). *IEEE Trans. Sonics Ultrason.* **SU-25**: 115.
23. Campbell, C. (1989). *Surface Acoustic Wave Devices and Their Signal Processing Applications*, San Diego, CA: Academic Press.
24. Matthews, H. (1977). *Surface Wave Filters*, New York: Wiley Interscience.
25. Rosen, C. A. (1957). "Proc. Electronic Component Symp.", p. 205.
26. Kawashima, S., Ohnishi, O., Hakamata, H., Tagami, S., Fukuoka, A., Inoue, T., and Hirose, S. (1994). "Proc. IEEE Int'l Ustrasonic Symp. '94", France.
27. Uchino, K. (1986). *Bull. Am. Ceram. Soc.* **65**: 647.
28. Uchino, K. (1993). *MRS Bull.* **18**: 42.
29. Uchino, K., Editor in Chief (1994). *Handbook on New Actuators for Precision Position Control*, Tokyo: Fuji Technosystem.
30. Uchino, K. (1996). Recent developments in ceramic actuators, "Proc. Workshop on Microsystem Technologies in the USA and Canada, Germany, mst news, special issue, VDI/VDE", pp. 28–36.
31. Furuta, K., and Uchino, K. (1986). *Adv. Ceram. Mater.* **1**: 61.
32. von Cieminski, J., and Beige, H. (1991). *J. Phys. D* **24**: 1182.
33. Cross, L. E., Ang, S. J. J., Newnham, R. E., Nomura, S., and Uchino, K. (1980). *Ferroelectrics* **23**: 187.
34. Uchino, K. (1988). Ceramic actuators, *Ceramic Data Book '88*, Tokyo: Inst. Industrial Manufacturing Tech., Tokyo.
35. Uchino, K., and Nomura, S. (1983). *Ferroelectrics* **50**: 191.
36. Furuta, A., Oh, K. Y., and Uchino, K. (1992). *Sensors Mater.* **3**: 205.
37. Takahashi, S., Ochi, A., Yonezawa, M., Yano, T., Hamatsuki, T., and Fujui, I. (1993). *Ferroelectrics* **50**: 181.
38. Bauer, A., and Moller, F. (1994). "Proc. 4th Int'l Conf. New Actuators", p. 128.
39. Active Control Experts, Inc. (1996). Catalogue "PZT Quick Pack".
40. Sugawara, Y., Onitsuka, K., Yoshikawa, S., Xu, Q. C., R. Newnham, E., and Uchino, K. (1992). *J. Am. Ceram. Soc.* **75**: 996.
41. Goto, H., Imanaka, K., and Uchino, K. (1992). *Ultrasonic Technol.* **5**: 48.
42. Kanbe, N., Aoyagi, M., Hirose, S., and Tomikawa, Y. (1993). *J. Acoust. Soc. Jpn. (E)* **14**: 235.
43. Sugiyama, S., and Uchino, K. (1986). "Proc. Int'l. Symp. Appl. Ferroelectrics '86", p. 637, IEEE.
44. Kusakabe, C., Tomikawa, Y., and Takano, T. (1990). *IEEE Trans. UFFC* **37**: 551.
45. Ota, T., Uchikawa, T., and Mizutani, T. (1985). *Jpn. J. Appl. Phys.* **24** (Suppl. 24-3): 193.
46. Dorsey, J. T., Sutter, T. R., and Wu, K. C. (1992). "Proc. 3rd Int'l Conf. Adaptive Structures", p. 352.
47. Wada, B. (1993). JPL Document D-10659, p. 23.
48. Straub, F. K. (1996). *Smart Mater. Struct.* **5**: 1.
49. Chen, P. C., and Chopra, I. (1996). *Smart Mater. Struct.* **5**: 35.
50. Ohgoshi, A., and Nishigaki, S. (1981). *Ceramic Data Book '81*, p. 35, Tokyo: Inst. Industrial Manufacturing Technology.
51. Yano, T., Sato, E., Fukui, I., and Hori, S. (1989). "Proc. Int'l Symp. Soc. Information Display", p. 180.
52. Tanaka, Y. (1994). *Handbook on New Actuators for Precision Control*, p. 764. Fuji Technosystem.
53. Yokoya, Y. (1991). *Electron. Ceram.* **22**: 55.
54. Koster, M. P. (1994). "Proc. 4th Int'l Conf. New Actuators", Germany, p. 144.
55. Barth, H. V. (1973). *IBM Tech. Disclosure Bull.* **16**: 2263.
56. Lavrinenko, V. V., Vishnevski, S. S., and Kartashev, I. K. (1976). *Izvestiya Vysshikh Uchebnykh Zavedenii, Radioelektronica* **13**: 57.
57. Vasiliev, P. E. *et al.* (1979). UK Patent Application GB 2020857 A.

58. Sashida, T. (1982). *Oyo Butsuri* **51**: 713.
59. Kumada, A. (1985). *Jpn. J. Appl. Phys.* **24** (Suppl. 24–2): 739.
60. Koc, B., Bouchilloux, P., and Uchino, K. (2000). *IEEE Trans. UFFC* **47**: 836.
61. Fuda, Y., and Yoshida, T. (1994). *Ferroelectrics* **160**: 323.
62. Nakamura, K., Kurosawa, M., and Ueha, S. (1993). *Proc. Jpn. Acoustic Soc.* No.1-1-18, 917.
63. Uchino, K., Kato, K., and Tohda, M. (1988). *Ferroelectrics* **87**: 331.
64. Tomikawa, Y., Nishituka, T., Ogasawara, T., and Takano, T. (1989). *Sensors Mater.* **1**: 359.
65. Sashida, T. (1983). *Mech. Automation Jpn.* **15**: 31.
66. Ise, K. (1987). *J. Acoust. Soc. Jpn.* **43**: 184.
67. Kasuga, M., Satoh, T., Tsukada, N., Yamazaki, T., Ogawa, F., Suzuki, M., Horikoshi, I., and Itoh, T. (1991). *J. Soc. Precision Eng.* **57**: 63.
68. Cummings, J., and Stutts, D. (1994). Design for manufacturability of ceramic components, *Am. Ceram. Soc. Trans.*, p. 147.
69. Kumada, A. (1989). *Ultrasonic Technol.* **1**: 51.
70. Tomikawa, Y., and Takano, T. (1990). *Nikkei Mechanical* (Suppl.) 194.
71. Yoshida, T. (1989). "Proc. 2nd Memorial Symp. Solid Actuators of Japan: Ultra-precise Positioning Techniques and Solid Actuators for Them", p. 1.
72. Dong, S., Uchino, K., and Lim, L. C. (2003). *IEEE Trans. UFFC* **50**(4): 361.

Handbook of Advanced Ceramics
S. Sōmiya *et al.* (Eds.)

5.1 Dielectric Ceramics

YUKIO SAKABE
Murata Manufacturing Co., Ltd., Yasu 520-2393, Japan

A material with high electric resistivity is categorized as an insulator material. When we pay attention to their dielectric polarization and apply the materials to the electronics circuits, we usually call them "dielectrics". Dielectric ceramics are essential electrical materials for today's advanced electronics devices. Production quantity of the dielectric ceramic is the largest among the other electronics ceramics such as magnetic, semiconductors, insulators, resistors and piezoelectric, and electro-optic materials. Main applications are for ceramic capacitors and microwave resonators. The dielectric ceramics is classified into two groups based on their dielectric properties.

5.1.1 CLASSIFICATION OF THE DIELECTRIC CERAMICS

5.1.1.1 HIGH-*Q* MATERIALS

The dielectrics of this group are called "temperature compensating dielectrics", because they can compensate the temperature dependence of other components. Dielectric constant changes linearly with temperature. Ceramic capacitors with these materials stabilized the resonant circuits in which a high quality factor (*Q* value) and a resonant frequency are extremely important. In some cases, the ceramics also called "linear dielectrics", because the polarization changes linearly with an applied electric field. Dielectric constant of this group ranges from about 4 to 400. The temperature coefficient is in the range from +120 to −4700 ppm/°C. *Q* value (defined as a reciprocal number of a dissipation factor tan δ) is in the range from 1000 to 100 000. These characterized values are intrinsically given by the compositions, and are modified with kinds and contents of the composed elements. Typical dielectrics and their properties are shown in Table 5.1.1. Titanate-based materials are dominant compositions, which sinter at normally higher than 1100°C. Some of the high *Q* materials for microwave application need very high soaking temperature (>1400°C).

TABLE 5.1.1 Dielectric Properties of the Typical High-Q Materials

Dielectric	ε_r (1 MHz)	TCC (10^{-6}/°C)	Q (1 MHz)
TiO_2	90	N750	>5000
$MgTiO_3$	17	P100	>5000
$CaTiO_3$	150	N1500	>5000
$SrTiO_3$	240	N3300	>1500
La_2O_3–$2TiO_2$	35–38	P60	>5000
ZnO–TiO_2	35–38	N60	>1500
Bi_2O_3–$2TiO_2$	104–110	N150	>2000
$MgTiO_3$–$CaTiO_3$	17–45	P100–N150	>5000
BaO-$4TiO_2$–TiO_2	35–65	N15–N500	>3000
2MgO-SiO_2–SrO–BaO–TiO_2	6–13	P100–N1000	>5000
$BaTiO_3$–Nd2O_3–TiO_2	35–87	P100–N330	>2500
$CaTiO_3$–La_2O_3–TiO_2	100–150	N470–N1000	>3000
$SrTiO_3$–$CaTiO_3$–Bi_2O_3–TiO_2	240–300	N1000–N2000	>1500
$CaTiO_3$–La_2O_3–Bi_2O_3–TiO_2	145–210	N750–N1500	>2500
$BaTiO_3$–$SrTiO_3$–La_2O_3–TiO_2	360–650	N3300–N4700	>1500

ε_r, dielectric constant; TCC, temperature coefficient of *K*; *Q*, quality factor.

Today, on the other hand, glass ceramics are widely used for ceramic multilayer substrate with Ag and Cu as an inner conductor. About 40–50% of grass elements such as Al_2O_3, SiO_2, MgO and alkali-earth elements compose the dielectrics, which can sinter at relatively lower temperature (<900°C).

5.1.1.2 High-ε_r Materials

Barium titanate ($BaTiO_3$) is the main dielectric in this category, which provides the dielectric constant ε_r higher than 1000. Many $BaTiO_3$-based dielectrics are developed to have the composition with the other titanate such as $SrTiO_3$, $CaTiO_3$, $BaTiO_3$ and zirconate such as $BaZrO_3$, $CaZrO_3$. Wide variety of the dielectric properties has been developed to perform the high volumetric efficiency. With increasing the dielectric constant at room temperature, the capacitance change increases at rated temperature range. Stabilizing the high dielectric constant at wider temperature range has been the principal work for the capacitor engineers. Lead-based relaxor dielectrics have been also developed, which have much higher ε_r of 30 000 at room temperature, and provide better temperature and bias voltage performances than $BaTiO_3$-based dielectrics. The representative composition is Pb ($Mg_{1/3}Nb_{2/3}$) O_3–$PbTiO_3$, which is used for ceramic capacitors providing high volumetric efficiency.

5.1.2 CERAMIC CAPACITORS

With the advent of advanced electronic devices such as handy phone and personal computer, the demand for surface mountable chip components continues to increase. Under this circumstance, the principal developments in ceramic capacitor industry are miniaturization, improvement of volumetric efficiency, cost reduction, improvement in reliability and the design of new products with high performance.

Chip-type capacitors are the main products due to their superior frequency performance and volumetric efficiency, which meet the requirements from the advanced electronics devices. A schematic diagram of the multilayer ceramic capacitor (MLC) is shown in Figure 5.1.1. The capacitor is composed of many thin plate capacitors in parallel connection. Thinning the dielectric layer, and stacking them maximum number into the limited thickness achieve maximum capacitance. The capacitance of the MLC of electrode area S, dielectric thickness t and number of dielectric layers n is given by

$$C = \varepsilon_\mathrm{r} \times S \times n/(k \times t)$$

where ε_r is the specific dielectric constant of the dielectrics, and k is a constant.

The volume efficiency C/V is given by

$$C/V = \varepsilon_\mathrm{r}/(k \times t^2).$$

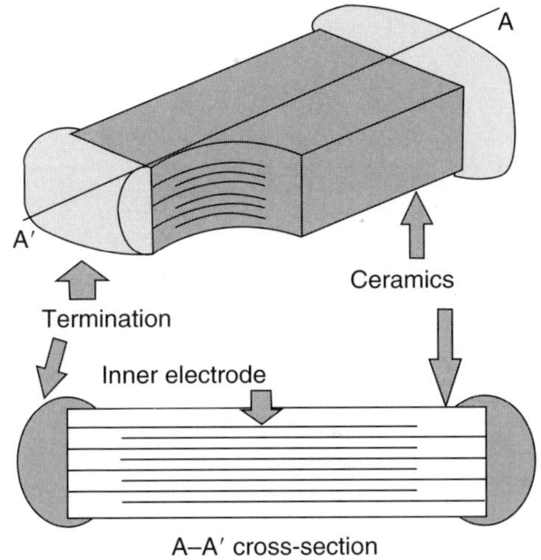

FIGURE 5.1.1 Schematic diagram of a multilayer ceramic capacitor construction.

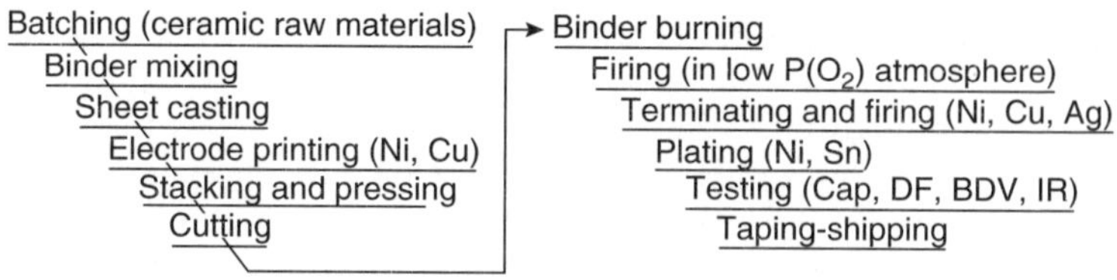

FIGURE 5.1.2 Manufacturing process flowchart of MLCs.

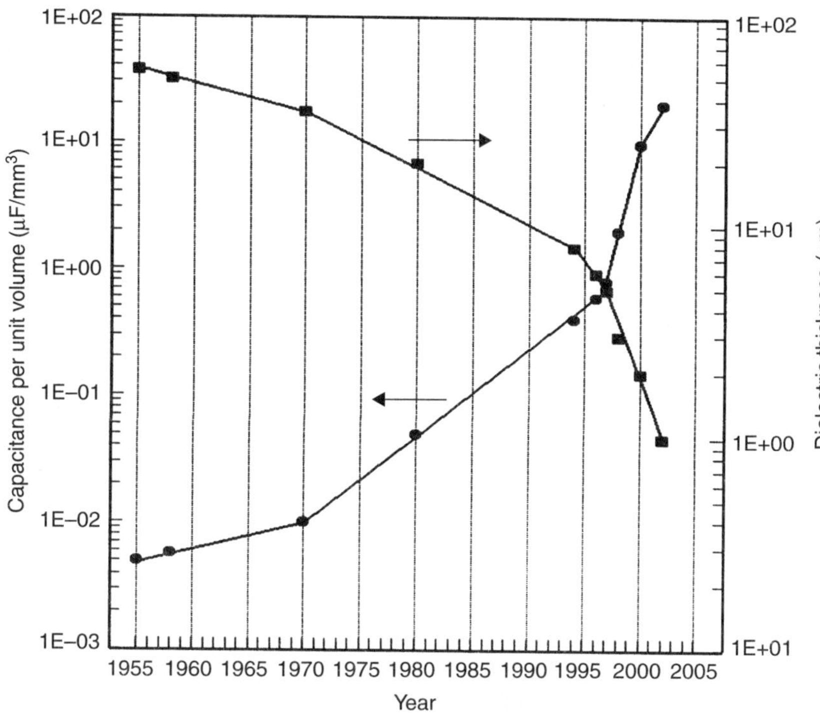

FIGURE 5.1.3 Historical trends of volumetric efficiency and dielectric thickness of MLCs (X7R).

The manufacturing process flowchart of the MLCs is shown in Figure 5.1.2. Green sheet process is suitable to prepare the defect-free thin ceramic film. Thickness reduction of the dielectric layer is a most effective method to design the higher capacitance capacitor with given dielectrics and chip dimension.

Figure 5.1.3 shows the historical trend of volumetric efficiency and dielectric thickness of X7R (EIA-code) MLCs. Capacitance per unit volume (C/V) has increased more rapidly during last 5 years. The advanced technologies of fine powder synthesis, thin film preparation and stacking the high layer count more than 600 have realized the high volumetric efficiency comparable to tantrum electrolytic capacitors.

In the year 2000, because of the rapid popularization of the telecommunication equipment, the production of MLC showed production units growth more than 40% world wide. The total production quantity in the world reached to five billion units. Today MLCs have large capacitances up to 400 μF which is comparable in performance to the tantalum electrolytic capacitors. Replacement of the Ta capacitors with the compatible MLCs are in progress for de-coupling capacitors at high power and high-frequency circuits. This trend will continue from now on, because the MLC has higher volumetric efficiency, lower ESR and lower ESL at high frequency than the electrolytic capacitors. These advantages meet the all needs from today's telecommunication equipment's and personal computer.

On the other hand, the price of the standard MLC products is steadily decreasing at a rate of more than 10% a year. The average price of MLC is equal to or less than 1 cent and is the cheapest among capacitors.

5.1.2.1 Nickel Electrode MLCs

Since dramatic increase of palladium price during 1995 to 1997, the MLC manufactures had extreme pressure to reduce the electrode cost, and so accelerated the expansion of MLCs production with base metal electrode such as nickel and copper to substitute the palladium. Nickel metal is easily oxidized in air at elevated temperature. Therefore, the capacitor with nickel and copper inner electrode must be fired in a low oxygen pressure atmosphere. Conventional ceramics, however, have poor insulation resistance after heat treatment in such reducing atmosphere.

To prevent the $BaTiO_3$-based dielectric ceramics from reduction, essential points of the composition designing are [1–3]:

1 Control of the molar ratio $(Ba+Ca)O/(Ti+Zr)O_2$ larger than unity.
2 Substitution of Ca ions to Ba lattice site of $BaTiO_3$.
3 Doping the acceptor as Mn ion.

The resistivity changes of the Ca-doped and undoped $BaTiO_3$ as a function of mole ratio *m* are shown in Figure 5.1.4. All the samples which were sintered in air exhibited high insulation resistance and were independent of Ca content and molar ratio *m*. However, the samples, which had a molar ratio less than 1.00, were considerably reduced, and became semiconducting when fired under a low $P(O_2)$ atmosphere.

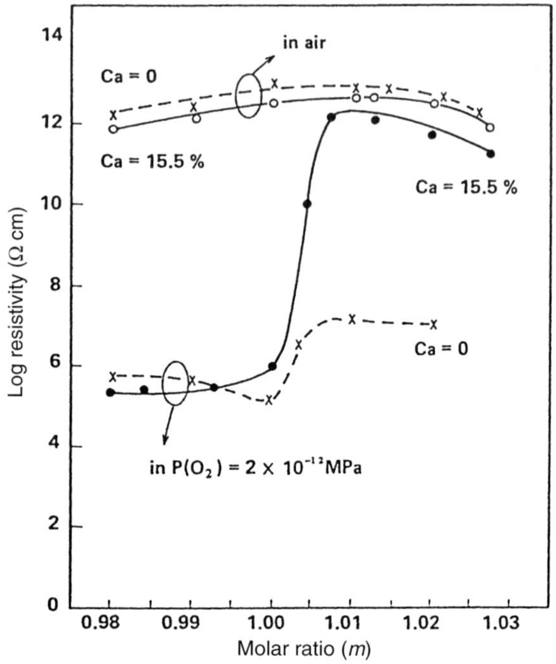

FIGURE 5.1.4 Resistivity of $[(Ba_{1-x}Ca_x)O]_m$ TiO_2 ceramics sintered in air and in an atmosphere of 2×10^{-12} MPa (O_2) at 1400°C for 2 h as a function of molar ratio *m*.

Mn ion and some of the Ca ions incorporate into Ti lattice site and act as acceptor ions compensating the electrons resulting from dissociation of oxygen [4, 5]. The composition has excellent initial characteristics, but stability of the capacitance and reliability at highly accelerated life conditions are not as high as that of conventional Pd electrodes. The effects of rare earth oxide on dielectric properties and reliability have been discussed in recent papers which concluded that rare earth oxides play an important role on the properties of X7R dielectrics and are considerably effective in prolonging the lifetime. Dy and Y ions provide stable temperature dependence of capacitance. Voltage dependence of insulation resistance is the smallest for Dy-doped dielectric and largest for La-doped one. Dy doped dielectric is much more reliable than the others. Microstructures of the dielectrics are characterized with these rare earth oxides. Dy exists uniformly in the dielectric layer, while La and Y are segregated. Dy and Y ions can corporate into the grain to form a core–shell structure, while La-doped dielectric does not have core–shell structured grain [6].

The dielectric doped with rare earth oxide (Dy, Ho and Er) which has smaller ionic radius showed a lower aging rate of capacitance and a longer lifetime [7].

Doping of donors such as Y_2O_3 or V_2O_5, which compensate the acceptor levels, is also effective to improve the capacitance stability and reliability at accelerated life testing [8]. Base metal electrode MLCs have expanded the product line-up utilizing the nickel compatible X7R dielectrics, and fine Ni and Cu powder. Fabrication of nickel paste reached 80% in weight of inner electrode paste consumption. Over the next few years, the usage in electronics is likely to fall as nickel electrode MLC technology has improved so much so as to be adopted for higher specific products of large capacitance with X7R materials.

Till 5 years ago, high-Q materials have been used only for the MLCs of relatively small capacitance capacitors, because the electrode cost of high-Q capacitors with Pd were too high to compete with other capacitors. Highly reliable nickel compatible high-Q dielectrics were also developed to replace the film capacitors. Copper is an ideal electrode material of MLC because of its high conductivity and relatively low price. Low fire dielectric with high reduction resistance was developed to utilize the copper as an inner electrode. The base metal electrode MLCs yield high-Q and lower equivalent series resistance (ESR) than Pd electrode MLCs at high frequency.

5.1.2.2 Fine Grained $BaTiO_3$ Ceramics

Further miniaturization of chip capacitors with thinning the dielectric layer and electrode laydown is also effective for the production cost. By incorporating thinner dielectric layers with nickel compatible dielectrics and nickel electrode, large capacitance MLCs comparable to tantalum electrolytic capacitors have been developed. To reduce the dielectric thickness less than few micrometers, further development of ultrafine grain $BaTiO_3$ is indispensable.

Extensive work on the particle size effects of $BaTiO_3$ on dielectric properties has been carried out, and several models for critical size of ferroelectricity have been proposed [9–21]. Arlt *et al.* reported that there was an optimal value of grain size at about 0.8 μm for the maximum dielectric constant [11]. Results from reliability study claimed that dielectric layer should be pore-free ceramics with grain of smaller in sized, and at least five grains are required for the ceramic layer. To design a capacitor of large capacitance per unit volume, the dielectric is expected to have large dielectric constant with smaller grain size. To realize the ideal $BaTiO_3$ powder of fine and good crystallinity, the $BaTiO_3$ powder of 70 nm in size was synthesized by the hydrolysis method [22, 23]. Dense dielectric ceramics of grain size ranging from 150 to 900 nm were prepared with calcined $BaTiO_3$ powder doping the grain growth inhibitors and glass powder as sintering agent. MLC test samples were prepared with 1.6 μm thick green sheets and Ni paste for inner electrode. The chip capacitors of size 2.0 mm × 1.25 mm

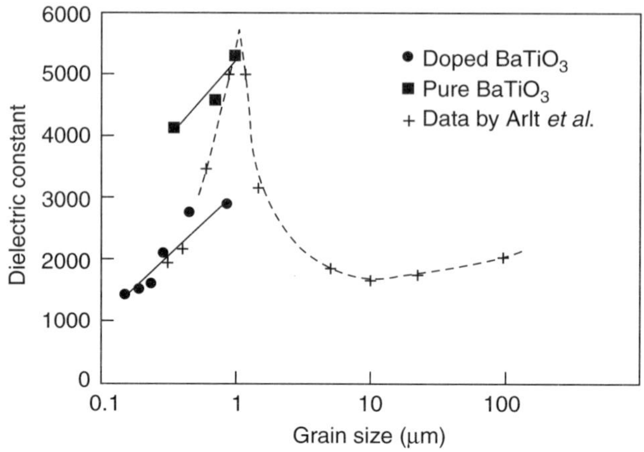

FIGURE 5.1.5 Grain size dependence of dielectric constant.

were fired in a reducing atmosphere at various temperature ranging from 1000 to 1250°C for 2 h. Ferroelectric domain structure was still observed in these 0.15 μm sized grains. Grain boundary layer of 1 nm thickness was recognized for the doped $BaTiO_3$ ceramics. It was confirmed by TEM analysis that Si, Mg, Ti and Ba ions located in the boundary layer and triple point of the grains.

Grain size dependence of dielectric constant is shown in Figure 5.1.5. Dielectric constant at room temperature decreased with grain size at a region smaller than 0.8 μm. The doped $BaTiO_3$ has smaller dielectric constant than the pure $BaTiO_3$, because of the formation of the heterogeneous grain boundary layer. The doped $BaTiO_3$, however, provided relatively high dielectric constant of 1000 even with small grain size of 100 nm.

Temperature dependence of dielectric constant is shown in Figure 5.1.6. Dielectric constant of the doped specimens decreased with grain size, resulting in stable temperature dependence. Curie temperature was linearly related to the reciprocal of the grain size, accordingly, Curie temperature decreased with grain size. D–E hysteresis curve of the pure and doped $BaTiO_3$ ceramics was measured. Non-linear D–E curve was still observed in doped $BaTiO_3$ ceramics with a grain size of 150 nm.

Electrical properties of the MLC with dielectric layer of 1 μm thickness are shown in Table 5.1.2. High-ε ceramics with stable characteristics under high AC and DC voltage stress are indispensable for designing the MLC with thinner layers. The ultra fine $BaTiO_3$ was rather stable under voltage stress than conventional dielectrics with coarse grain. This is a big advantage for high volumetric capacitors designing with thinner layer. It was supposed that the high surface tension in the ultra fine grain provided the small voltage dependence.

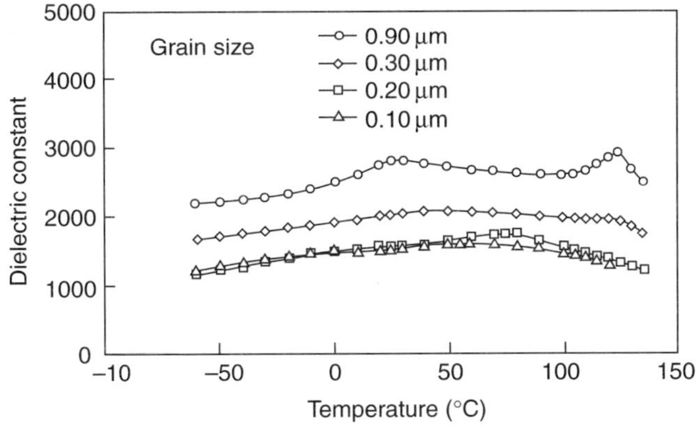

FIGURE 5.1.6 Temperature dependence of dielectric constant of the ultrafine grained $BaTiO_3$.

TABLE 5.1.2 Electrical Properties of MLC with Ultrafine Grain $BaTiO_3$ Ceramics (Average Grain Size of 150 nm)

Chip size (mm)	2.0 × 1.25 × 0.4
Dielectric thickness (μm)	1.0
Active layer's number	5
Inner electrode	Ni
Firing temperature (°C)	1150
Capacitance (nF) at 1 kHz/1 V_{rms}	89
Dielectric constant	1560
Dissipation factor (%)	2.1
Resistivity (Ω cm) log R at 10 V	12.5
Breakdown voltage (kV/mm)	115

One of the most critical processing parameters is the degree of homogeneous mixing of additives and binder in the slurry. High density and defect-free layer is realized by preparing the homogeneous slurry for sheet casting. The surface roughness becomes a serious problem with thinning of the dielectric sheet. The roughness of 3 μm thick sheet must be controlled to less than 0.3 μm to insure a smooth surface contact with the inner nickel electrode. This is also very important factor to avoid the concentration of electric field at asperities, where the charge emission from the inner electrode is accelerated, resulting in short failure.

For the same reason, the nickel metal powder for the electrode paste must also be very fine, typically less than 0.4 μm, and well dispersed in the paste. Very fine Ni powder (<0.4 μm) prepared by CVD and/or chemical synthesis

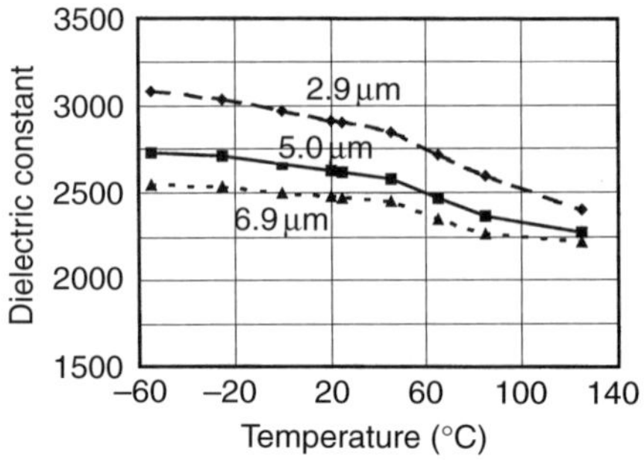

FIGURE 5.1.7 Thickness dependence of dielectric constant at 0.5 V_{rms} 1 kHz.

has been supplied. The powder provides MLCs with higher capacitance because of the higher coverage of the electrode at the ceramic interface.

5.1.2.3 Performance Changes with Thinning Layer

Voltage stress increases with thinning the dielectric layer, resulting in the serious changes in capacitance and dissipation factor of $BaTiO_3$ ceramics. This tendency is remarkable for the dielectrics having higher permittivity. Figure 5.1.7 shows the temperature dependence of dielectric constant of X7R material of various thicknesses [24]. Applied AC voltage was 0.5 V_{rms} at 1 kHz. Dielectric constant of ferroelectric phase increases with decreasing thickness of the layer, resulting in the clockwise rotation in the *T–C* curve. New dielectric materials, which provide high dielectric constant and stable characteristics under high voltage stress, are required to realize the large capacitance MLCs with thinner layer and higher layer count.

The thickness of the inner electrode also has effects on the temperature dependence of dielectric constant as shown in Figure 5.1.8. Dielectric constant increases due to an increase of horizontal compression from the thick metal electrode layer.

5.1.2.4 Large Capacitance MLCs (X5R 100 μF)

A ceramic chip capacitor of 100 μF was developed with nickel electrode system to meet the condition of decoupling capacitors for VLSI and switching mode power supply. The fundamental characteristics of the newly developed X5R 100 μF MLC are shown in Table 5.1.3. A large volumetric efficiency

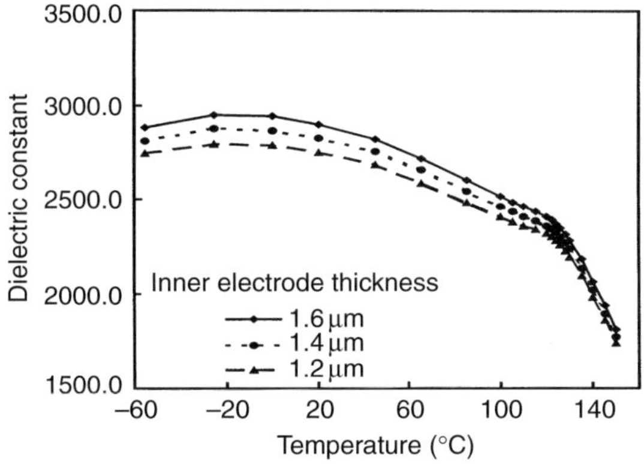

FIGURE 5.1.8 Dielectric constant changes with thickness of the electrode. Dielectric thickness was 2.9 mm. Applied voltage was 0.5 V_{rms}, 1 kHz.

TABLE 5.1.3 Characteristics of MLC of X5R 100 μF 6.3 V

Chip size (mm)	5.7 × 5.0 × 3.0
Dielectric thickness (μm)	3.3
Inner electrode	Ni
Active layer number	525
Capacitance (1 kHz, 0.5 V_{rms}) (μF)	108
Dissipation factor (1 kHz, 0.5 V_{rms})	3.3%
TCC	X5R
Insulation resistance (Ω)	1.0×10^7
Breakdown voltage (V_{DC})	197

of 1.17 μF/mm^3 was achieved by chip size of 5.7 × 5.0 × 3.0 mm^3 and 3.3 μm thick dielectrics of 525 layers. Used X5R dielectrics had dielectric constant of 3900. The frequency response of the MLC was excellent for the applications due to the low ESR and low ESL characteristics as shown in Figure 5.1.9.

The large capacitance MLCs has been realized by reliable Ni compatible dielectrics and established technologies of manufacturing process, handling the thinner sheets and firing in a low $P(O_2)$ atmosphere.

The capacitors have large potential to replace Ta and Al electrolytic capacitors in the range from 10 to 100 μF. The major technical advantages over Ta and Al capacitors are higher breakdown voltage, higher reliability and lower ESR at high frequency.

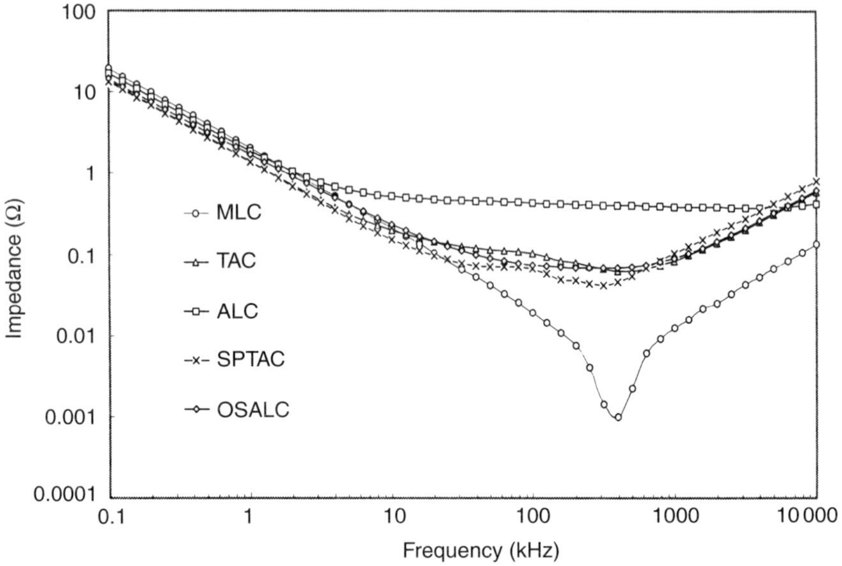

FIGURE 5.1.9 Frequency dependence of impedance of the several 100 μF capacitors. MLC, multilayer ceramic capacitor; TAC, Ta electrolytic capacitors; ALC, Al electrolytic capacitors; SPTAC, Ta electrolytic capacitors with polymer electrode; OSALC, Al electrolytic capacitors with polymer electrode.

5.1.2.5 MLCs WITH MOCVD

Thinner dielectric layer of MLC less than few micrometers is formed by sheet casting machines such as reverse roll coater, lip coater and die coater with homogeneous mixture of fine ceramic powder, additives and binder. It is believe that these conventional sheet methods have technical limits of the dielectric thickness of around 1 μm. Therefore, new techniques are required to prepare the dielectric layer of sub-micron thickness for higher volumetric efficiency. (BaSr)TiO_3 (BST) thin film was prepared by metal-organic chemical vapor deposition (MOCVD), and investigated the microstructure and dielectric properties. The MLCs was designed to demonstrate the new technologies for next generation of capacitor manufacturing [25, 26].

A schematic diagram of the MOCVD system used to deposit the BST layers is shown in Figure 5.1.10. ($Ba(C_{11}H_{19}O_2)_2(C_8H_{23}N_5)_2$), ($Sr(C_{11}H_{19}O_2)_2(C_8H_{23}N_5)_2$) or ($Sr(C_{11}H_{19}O_2)_2(C_6H_{18}N_4)_2$), and ($Ti(i\text{-}OC_3H_7)_4$) were used as Ba, Sr, and Ti sources respectively. Oxygen was used as an oxidant, and argon as a carrier gas. The deposition conditions for the BST layers are summarized in Table 5.1.4. The structure of the MLC with BST

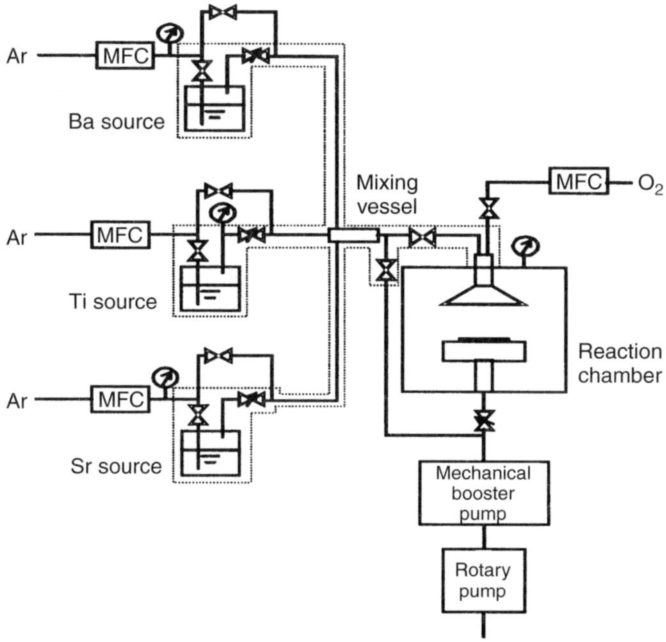

FIGURE 5.1.10 Schematic diagram of the MOCVD apparatus for BST thin films.

TABLE 5.1.4 Deposition Conditions of BST Layers

Source temperature (°C)	
Ba source	130–158
Sr source	107–128
Ti source	40
Substrate temperature (°C)	650
Chamber pressure (kPa)	13.3
Deposition time (min)	65–120

thin layers is schematically shown in Figure 5.1.11. MgO single crystal was used as a substrates and Pt as the electrode of the MLC with BST thin layers. RF magnetron sputtering formed the Pt electrode at temperature lower than 100°C.

SEM image of the cross-sectional view of the MLC with $(Ba_{0.7},Sr_{0.3})TiO_3$ thin layers is shown in Figure 5.1.12. The capacitor was constructed with 15 BST dielectric layers and 16 Pt electrode layers. The average thickness of dielectric and electrode were 0.22 and 0.23 μm, respectively, giving a total thickness of 7.0 μm. The MLC of 0.4 × 0.4 mm^2 in size yielded capacitance of 34 nF

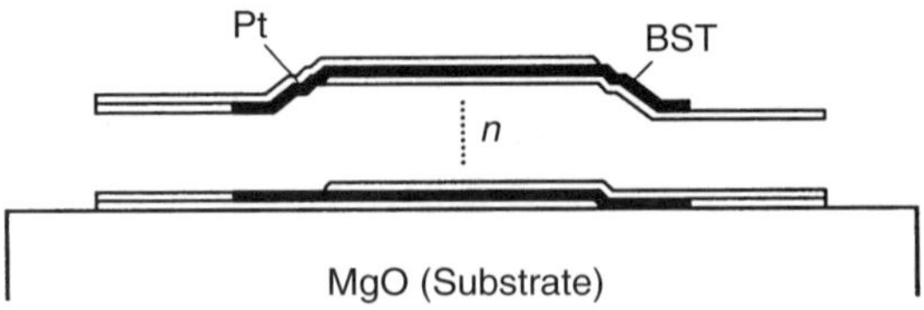

FIGURE 5.1.11 Structure of the MLC on MgO substrate.

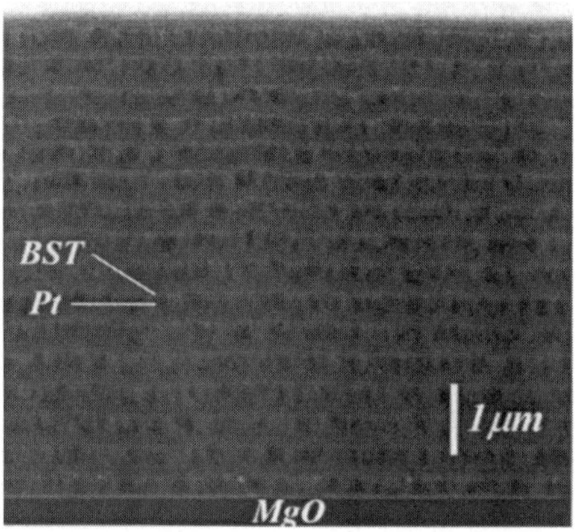

FIGURE 5.1.12 Cross-sectional SEM image of multilayer capacitor with 15 $(Ba_{0.7},Sr_{0.3})TiO_3$ dielectric layers.

and dissipation factor (tan δ) of 2.6% at 1 kHz and 100 mV. Capacitance per unit volume was 30 μF/mm^3 which was 10 times larger than MLCs of today. The dielectric constant was evaluated about 350, which is comparable to that of the other reports [27]. Figure 5.1.13 shows the temperature dependence of the single- and multilayer capacitor with $(Ba_{0.7},Sr_{0.3})TiO_3$ dielectrics. The maximum capacitance value was obtained at −10 and −30°C, respectively.

The capacitance of this capacitor changes considerably with DC biasing field. It decreased 27 and 34% under DC biasing voltage of 1.5 and 2.0 V, respectively. This is one of the important subjects in future works. The leakage current at 1 V was about 0.2 nA, and the breakdown voltage was about 6 V. The acceptable *CR* product of 121 MΩ μF was obtained. But short failure rate was still high, around 50%, which may be due to the inhomogeneity of the deposited ceramic layer. Surface of the single BST film was homogenous and smooth. With increasing the layer count, however, the surface mohology had change to the rather poor structure. Further development on multistacking technologies of

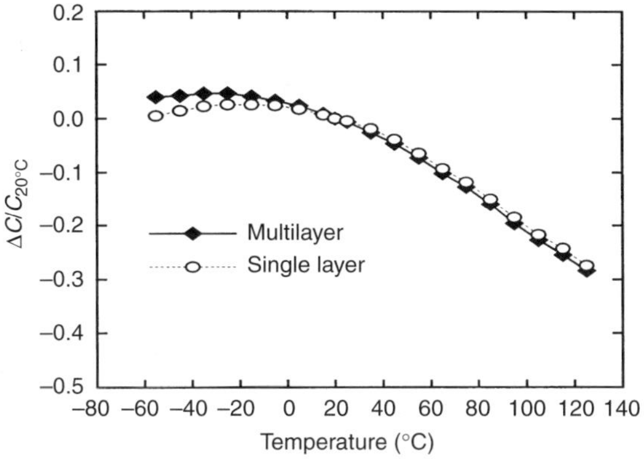

FIGURE 5.1.13 Capacitance change of MLC with $(Ba_{0.7},Sr_{0.3})TiO_3$ dielectric layer by MOCVD.

the homogenous dielectrics of nanometer thickness with base metal electrode is expected for advanced circuits.

5.1.3 DIELECTRIC RESONATOR

Dielectric resonators have gained a position as key elements in microwave components for size reduction in microwave filters and as frequency stabilizing elements in oscillator circuits. Dielectric resonators reduce the physical size of resonant systems because the electromagnetic wavelength is shortened in dielectrics to $1/\sqrt{\varepsilon_r}$ of its value in free space, where ε_r is the dielectric constant of the resonator.

The required properties for dielectric resonator materials are: (a) high dielectric constant; (b) low dielectric loss tangent $\tan\delta$; and (c) low temperature coefficient of resonant frequency τ_f. Many kinds of dielectric resonator materials have been developed since the 1970s [28–31]. Table 5.1.5 shows the dielectric properties of some materials that are commercially available now. In the table, the quality factor Q is reciprocal of dielectric loss tangent; $Q = l/\tan\delta$. As $\tan\delta$ is proportional to frequency for ionic paraelectric materials, the product of Q and frequency is the value inherent to each material. Some materials have high Q value equal to copper cavity and some have the temperature coefficient as stable as Inver cavity. The material with lower ε_r generally has higher Q value.

It is known that the complex dielectric constant of the resonator materials follows the dielectric dispersion equation, which is expressed as the superposition of electronic and ionic polarization:

$$\dot{\varepsilon}(\omega) - \varepsilon(\infty) = \frac{\omega_T^2(\varepsilon(0) - \varepsilon(\infty))}{\omega_T^2 - \omega^2 - j\gamma\omega} \tag{1}$$

where ω_T and γ are the resonance frequency and damping constant of the infrared active lattice vibration modes, and $\varepsilon(\infty)$ is the permittivity due to the electronic polarization.

As the condition $\omega^2 \ll \omega_T^2$ is reasonably adopted at microwave and millimeter wave frequencies, following equations are derived from Eq. 1:

$$\varepsilon'(\omega) - \omega(\infty) = \varepsilon'(0) - \varepsilon(\infty) \tag{2}$$

$$\tan\delta = \frac{\varepsilon''(\omega)}{\varepsilon'(\omega)} = \frac{\gamma}{\omega_T^2}\omega \tag{3}$$

TABLE 5.1.5 Microwave Properties of Dielectric Resonator Materials

Material	ε_r	$Q \times f$ (GHz)	τ_f (ppm/°C)
$MgTiO_3$–$CaTiO_3$ system	21	60 000	0–6
$Ba(Mg,Ta)O_3$ system	24	200 000–350 000	0–6
$Ba(Zn,Ta)O_3$ system	30	100 000–180 000	0–6
$Ba_2Ti_9O_{20}$	38	50 000	4
$(Zr,Sn)TiO_4$	38	60 000	0–6
$CaTiO_3$–$NdAlO_3$	43	45 000	0–6
$Ba(Sm,Nd)_2Ti_4O_{12}$	80	100 000	0–6
$(Ba,Pb)Nd_2Ti_4O_{12}$	92	5 000	0–6
$(Ba,Pb)(Nd,Bi)_2Ti_4O_{12}$	110	2 500	0–6

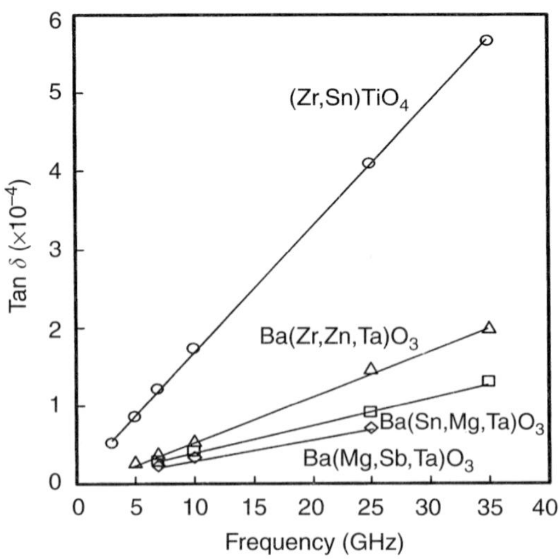

FIGURE 5.1.14 Frequency dependence of tan δ from 1 to 35 GHz.

These equations show that the dielectric constant is independent of frequency and the dielectric loss tangent increases proportionately to frequency at microwave frequency. Frequency dependence of tan δ from 1 to 35 GHz for the microwave dielectric materials is shown in Figure 5.1.14. As predicted from

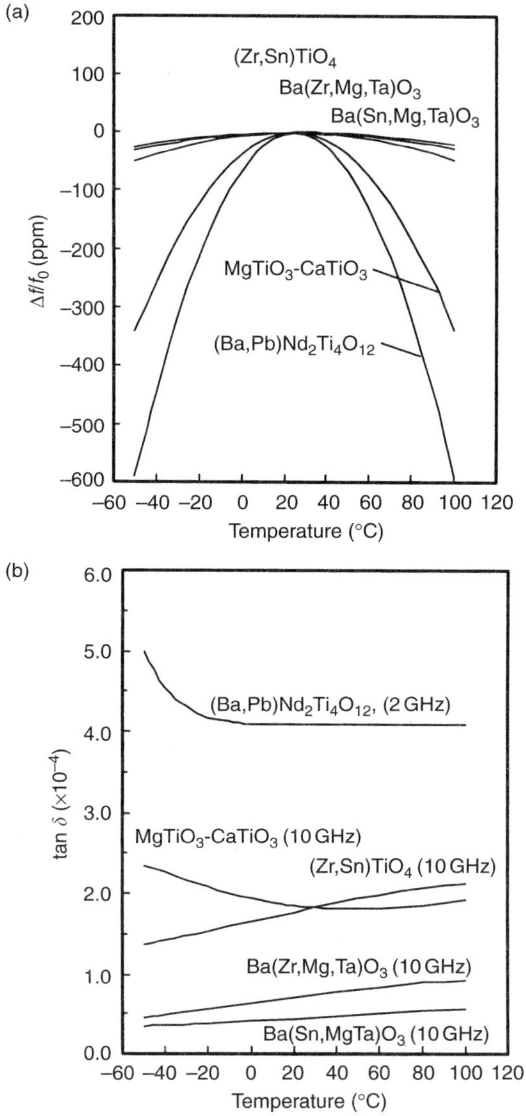

FIGURE 5.1.15 Temperature dependence of: (a) resonant frequency; and (b) tan δ.

Eq. 3, $\tan\delta$ increases proportionately to frequency. Temperature dependence of the resonance frequency and $\tan\delta$ is shown in Figure 5.1.15.

Superior temperature stability of resonant frequency is obtained by selecting the composition of each material. The temperature coefficient of resonant frequency τ_f is defined by the following equation.

$$\tau_f = -\tfrac{1}{2}\tau_\varepsilon - \alpha \tag{4}$$

where τ_ε is the temperature coefficient of dielectric constant and α is the linear thermal expansion coefficient of the dielectric specimen.

The intrinsic origin of damping constant γ in Eq. 3 is the inharmonic terms in the crystal's potential energy by which the phonons of infrared active modes decay into the thermal phonons. As the phonon energy follows the Planck distribution, γ, and $\tan\delta$, increase proportionately to temperature when the abnormality such as phase transition does not occur in the observed temperature range. Figure 5.1.16 shows the temperature dependence of $\tan\delta$ from 30 to 300 K.

The main applications for dielectric resonator materials are the antenna duplexers of cellular mobile phones, filters of cellular base stations, oscillators of DBS-TV converter, and filters at millimeter-wave frequencies.

The materials $(Pb,Ba)(Bi,Nd)_2Ti_4O_{12}$ system with high ε_r are popularly used for antenna duplexers of cellular mobile phones. The materials $(Zr,Sn)TiO_4$ and $CaTiO_3$–$NdAlO_3$ with high Q and high ε_r are used for filters of cellular

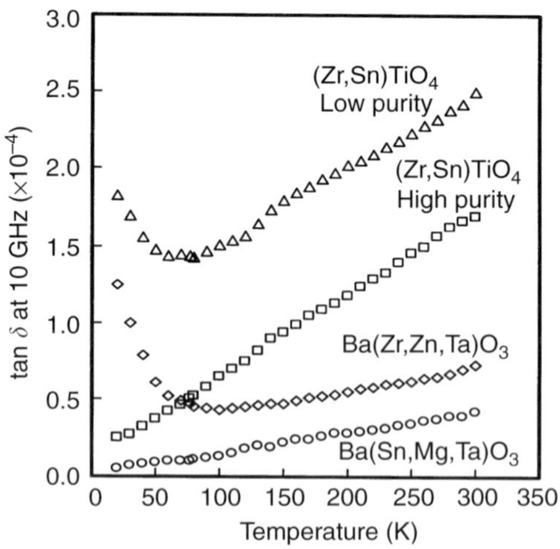

FIGURE 5.1.16 Temperature dependence of $\tan\delta$ from 30 to 300 K.

base stations. The complex perovskite materials $Ba(Zn,Ta)O_3$ and $Ba(Mg,Ta)O_3$ systems with very high Q value are used for applications higher than 10 GHz.

REFERENCES

1. Herbert, J. M. (1963). *Trans. Br. Ceram. Soc.* **62**: 645–653.
2. Burn, I., and Maher, G. H. (1975). *J. Mater. Sci.* **10**: 633–638.
3. Sakabe, Y., Minai, K., and Wakino, K. (1981). *Jpn. J. Appl. Phys.* **20**: 147–152.
4. Sakabe, Y. (1987). *Ceram. Bull.* **66**: 1338–1341.
5. Hennings, D. F. K., and Schreinemacher, H. (1995). *J. Eur. Ceram. Sci.* **15**: 795–800.
6. Hamaji, Y., Sano, H., Wada, H., and Tomono, K. (1995). "Proceedings of 7th *US–Japan Seminar on Dielectric and Piezoelectric Ceramics.*" *November 14–17* Tukuba, Japan **IV-5**, pp. 273–276.
7. Okino, Y., Shizuno, H., Kusumi, S., and Kisi, H. (1994). *Jpn. J. Appl. Phys.* **33**: 5393–5396.
8. Nomura, T., Kawano, N., Yamamatu, J., and Arashi, T. (1995). *Jpn. J. Appl. Phys.* **34**: 5398.
9. Bussen, W. R., Cross, L. E., and Goswami, A. (1966). *J. Am. Ceram. Soc.* **49**: 33.
10. Yamaji, A., Enomoto, E., Kinosita, K., and Murakami, T. (1977). *J. Am. Ceram. Soc.* **60**: 97–101.
11. Arlt, G., Hennings, D., and de With, G. (1985). *J. Appl. Phys.* **58**: 1619.
12. Uchino, K., Sadanaga, E., and Hirose, T. (1989). *J. Am. Ceram. Soc.* **72**: 1555.
13. Uchino, K., Sadanaga, E., Oohashi, K., Morohasi, T., and Yamamura, H. (1989). *Ceram. Trans.* **8**: 107.
14. Frey, M. H., and Payne, D. A. (1993). *Appl. Phys. Lett.* **63**: 2753.
15. Wada, S., Suzuki, T., and Noma, T. (1996). *J. Ceram. Soc.* **104**: 383.
16. Hsiang, H., and Yen, F. (1996). *J. Am. Ceram. Soc.* **79**: 1053.
17. Frey, M. H., and Payne, D. A. (1996). *Phys. Rev. B*, **54**: 3125.
18. Li, X., and Shih, W. (1997). *J. Am. Ceram. Soc.* **80**: 2844.
19. Viswanath, R. N., and Ramasamy, S. (1997). *Nano-Struct. Mater.* **8**: 155.
20. Zhong, W. L., Wang, Y. G., Zhang, P. L., and Qu, B. D. (1994). *Phys. Rev. B*, **50**: 698.
21. Frey, M. H., and Payne, D. A. (1996). *Phys. Rev. B*, **54**: 3153.
22. Sakabe, Y., Wada, N., and Hamaji, Y. (1998). *J. Korean. Phys. Soc.* **32**: 260.
23. Sakabe, Y., Wada, N., and Hamaji, Y. (1998). "Proceedings, 11th. IEEE, ISAF," p. 565.
24. Yoneda, Y., Hosokawa, T., Ohmori, N., and Takeuchi, S. (1996). *CARTS-EUROPE'* **96**: p. 11.
25. Takesima, Y., Shiratuyu, K., Takagi, H., and Sakabe, Y. (1997). *J. Jpn. Appl. Phys.* **36**: 5870.
26. Sakabe, Y., Takesima, Y., and Tanaka, K. (1999). *J. Electroceram.* **3**: 115.
27. Watt, M. M., Woo, P., Rywak, T., McNeil, L., Kassam, A., Joshi, V., Cuchiaro, J. D., and Melnick, B. M. (1998). "Proceedings, 11th. IEEE, ISAF," p. 11.
28. O'Bryan, H. M. Jr., Thomson, J. Jr., and Plourde, J. K. (1974). *J. Am. Ceram. Soc.* **57**: 450–453.
29. Kawashima, S., Nishida, M., Ueda, I., and Ouchi, H. (1983). *J. Am. Ceram. Soc.* **66**: 421–423.
30. Tamura, H., Konoike, T., and Wakino, K. (1984). *J. Am. Ceram. Soc.* **67**: C-59–C-61.
31. Wakino, K., Minai, K., and Tamura, H. (1984). *J. Am. Ceram. Soc.* **67**: 278–281.

Handbook of Advanced Ceramics
S. Sōmiya *et al.* (Eds.)

6.1 Magnetic Ceramics

TAKESHI NOMURA
TDK Corporation, 570-2 Matsugashita, Minamihatori, Narita-shi, Chiba-Prefecture 286-8588, Japan

6.1.1 MAGNETIC FUNCTION AND MATERIALS

In addition to the electronics and information communication fields, magnetic materials play an active part in the personal sphere of modern life. Nowadays, magnetic materials are used in various fields. Although magnetic functions are required for each purpose, it can be roughly classified into soft and hard from the viewpoint of magnetic property. If classifications of hard and soft are described simply, soft can be attracted to a permanent magnet; on the contrary, hard can become a permanent magnet. Although both are aggregates of micromagnet particles, as for 'soft', large magnetic fields cannot be generated to the outside, on the other hand, in the case of 'hard'; it is possible that it generates magnetic fields.

The complex oxide, which contains trivalent iron ion as the main ingredient, is generally called ferrite. It is the substance group, which generally exhibits ferrimagnetism, and is widely used in industry. Magnetic materials are roughly divided into metal magnetic materials and oxide magnetic materials (ceramics). Although initial permeability and magnetic flux density of metal and alloy magnetic materials are high, loss by eddy current is large at high frequencies because of their lower electrical resistivity. For this reason, it is generally used in the form of a multilayer core of rolled thin plates. When permeability can be sacrificed in order to make the electrical resistivity higher and to reduce loss, it may be used in the form of dust core. Recently, high-frequency characteristics, which exceed ferrites by multiplying the thin films, have also been acquired. On the other hand, there is no worry about oxidation for oxide magnetic materials (ferrites). The soft ferrites, whose characteristics at high frequency is excellent because of their higher electrical resistivity, are abundantly used for inductors or core materials of transformer. The hard ferrite is also used abundantly as permanent magnets for speakers and motors.

From the viewpoint of an applied field, soft ferrite is used into an alternating magnetic field. Magnetic property is excellent in high frequency as compared with metal magnetic materials since ferrite shows higher electrical resistively and smaller eddy current loss. Therefore, in the high-frequency band, ferrites are widely used. Since a magnetic field is alternating, it can be expressed by $H = H_0 \exp(j\omega t)$. On the other hand, since magnetic flux density, B, generally cannot follow in H but has phase difference δ, it is expressed as $B = B_0 \exp j(\omega t - \delta)$. Permeability is $\mu = B/H = (B_0/H_0)\exp(-j\delta) = (B_0/H_0)\cos\delta - j(B_0/H_0)\sin\delta = \mu' - j\mu''$ and it is divided into complex permeability μ' and μ''. In addition, μ'' is a loss ingredient and is presented as $\mu''/\mu' = \tan\delta(= 1/Q)$.

The following are mainly mentioned as magnetic characteristics required for high-frequency materials:

1. high permeability,
2. low loss (high effective permeability),
3. high permeability at high frequency,
4. stable permeability against temperature,
5. stable permeability against time,
6. higher magnetic flux density.

It is important to achieve stable permeability. Permeability is strongly affected by not only the composition but also the microstructure because it is a structure sensitive characteristic. Nomura and Ohta [1] describe the factors affecting permeability as follows:

1. Chemical composition and a small amount of ingredient: high permeability can be obtained by the composition of which magnetocrystalline anisotropy and magnetostriction constant are set to zero, respectively (Fig. 6.1.1).
2. Sintered density (pore): pore becomes a pinning site of domain wall that prevents domain wall motion (Fig. 6.1.2).
3. Grain size: the number of domain walls increase when grain size is larger, and permeability becomes higher due to a superior magnetizing process by the movement of domain walls (Fig. 6.1.3).
4. Inclusions: a precipitate becomes a pinning site of domain wall, and prevents its movement (Fig. 6.1.2).
5. Completeness of a crystal: the permeability is reduced by the imperfection of the crystal, that is, a defect, dislocation, etc. because it can lower the mobility of domain wall or increase the induced magnetic anisotropy.

The causes of magnetic loss have been studied for a long time, and are divided into: (a) hysteresis loss; (b) eddy current loss; and (c) residual loss. The contribution of each loss toward high-frequency loss of metal magnetic materials and ferrite is typically shown in Figure 6.1.4. Hysteresis loss is equivalent to

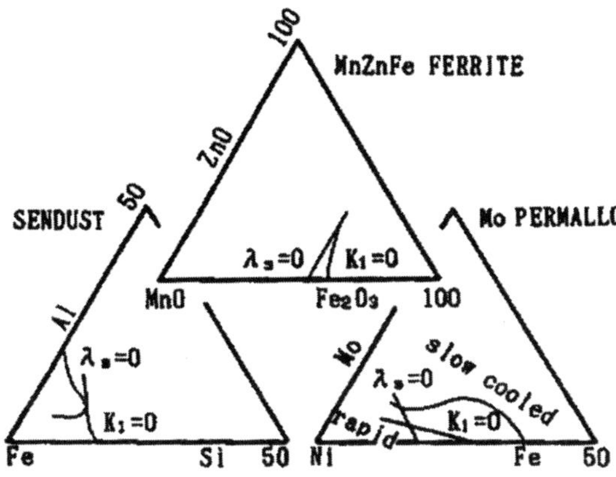

FIGURE 6.1.1 Magnetocrystalline anisotropy and magnetostriction of magnetic materials represented in miscibility diagram.

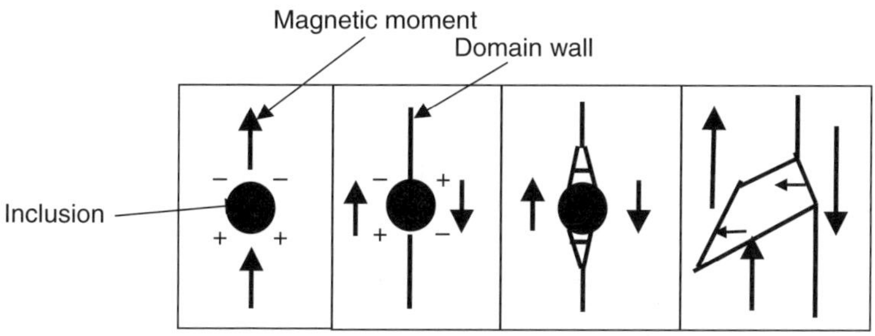

FIGURE 6.1.2 Domain structure around the inclusion.

the area of *B–H* curve, and increases in proportion to driving frequency. Eddy current loss is generated to reduce the change of magnetic field, and increases in proportion to the second power of driving frequency. In the case of metal magnetic materials, since the electrical resistivity is very low, at higher frequencies, contribution of eddy current loss becomes very large, and consequently are not available for practical use. On the other hand, ferrites have been considered ideal materials for high frequency since electrical resistivity is high, that is, eddy current loss is very low. Permeability is, however, not extended surprisingly to high frequency because residual loss suddenly expands from around a few megahertz.

Therefore, it is effective to enlarge anisotropic magnetic field, to extend permeability to higher frequencies. On the other hand, in order to obtain high permeability, it is also important to choose composition of small magnetocrystalline anisotropy. That is, permeability becomes higher as the anisotropy

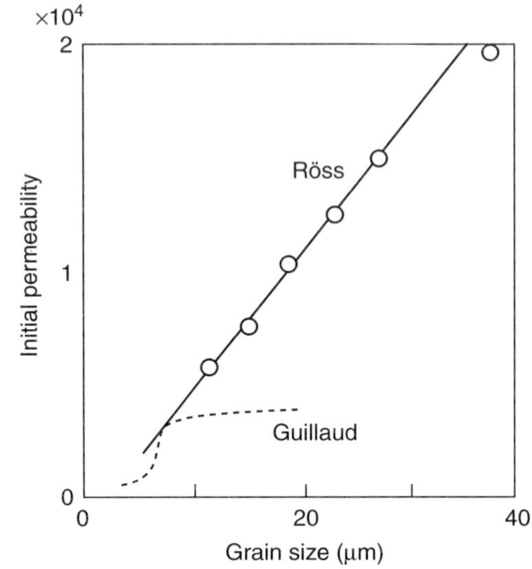

FIGURE 6.1.3 Effect of grain size on the initial permeability.

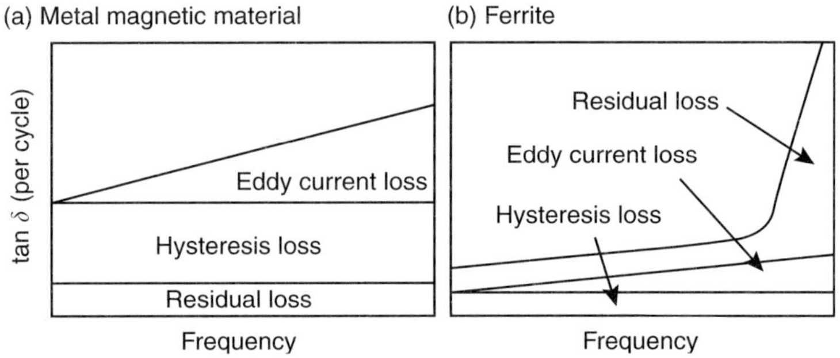

FIGURE 6.1.4 Dependence on loss factor at high frequency.

decreases. Conversely, available frequency decreases with increase of permeability. This relationship was pointed out by Snoek [2]. The relationship between permeability of cubic crystal and anisotropic magnetic field, is given by $\mu' - 1 = (2I_s/3\mu_0)(1/H_A)$.

If it combines with $f_r = (\gamma/2\pi)H_A$, it becomes $f_r(\mu' - 1) = (\gamma/3\pi\mu_0)I_s = 2 \times 10^{10} I_s$, and μf is a constant when I_s is considered to be constant. The frequency characteristic of initial permeability as for NiZn ferrites is shown in Figure 6.1.5. As for permeability, it becomes smaller for ferrites with higher resonance frequency. As a result, the straight line which shows the limit of μ' is obtained and this is called a limit of Snoek. However, there is a hexagonal crystal ferrite, which exceeds the relation of an upper formula and can be used

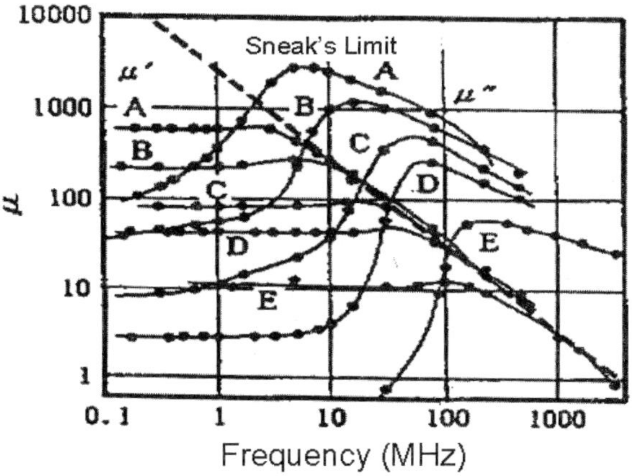

FIGURE 6.1.5 Frequency dependence of permeability of NiZn ferrite.

to increase frequency. This anisotropy fields of the vertical direction and the horizontal direction, to C-plane, differ drastically from these materials. Consequently, isotropy or uniaxial magnetic anisotropy ferrite that has the same μ_i can have high resonance frequency.

When magnetic field changes, magnetization change is delayed. The phenomenon is called magnetic aftereffect, which is shown in the following equation.

$$I_t = I_\infty[1 - \exp(-t/\tau)]$$
$$\tau = \tau_0 \exp(Q/kT)$$

where I_t is the magnetization at time t, I_∞, the magnetization after infinite time, T, the time, and τ, the relaxation time.

When τ is close to the periodic time of measurement frequency, relaxation type loss is observed because magnetization change cannot follow that of an alternating magnetic field, and τ cannot be observed whether it is very long or short. There are two kinds of magnetic aftereffect, which are the diffusion aftereffect and the heat fluctuation aftereffect. Diffusion aftereffect is also called Richter type and is a relaxation phenomenon by the diffusion or migration of the atom or vacancy. This is observed in soft magnetic materials. On the other hand, heat fluctuation aftereffect is also called Jordan type and it is a relaxation phenomenon by spontaneous magnetization changing its direction at the probability by heat vibration, and is observed in hard magnetic materials. As a concrete example of diffusion aftereffect, carbon in iron is well known. Among the magnetic aftereffects, the permeability change with passage of time, especially the change after demagnetization is called disaccommodation (DA). Example of MgZn ferrite, which was demonstrated by Ohta [3] is shown in Figure 6.1.6. Such demagnetizing curve of μ is called Richter-type relaxation curve, and is obtained when τ continuously distributes over between τ_1 and τ_2, not a constant τ. DA at near room temperature of MnZn ferrite is dissolved by the Po_2 (oxygen partial pressure) controlled firing. In order to control the formation of cation vacancy, small DA ferrite is obtained by controlling the oxygen partial pressure during firing.

The operating frequency range of typical soft ferrite is shown in Figure 6.1.7. Moreover, the example of use is shown in Table 6.1.1. In the frequency

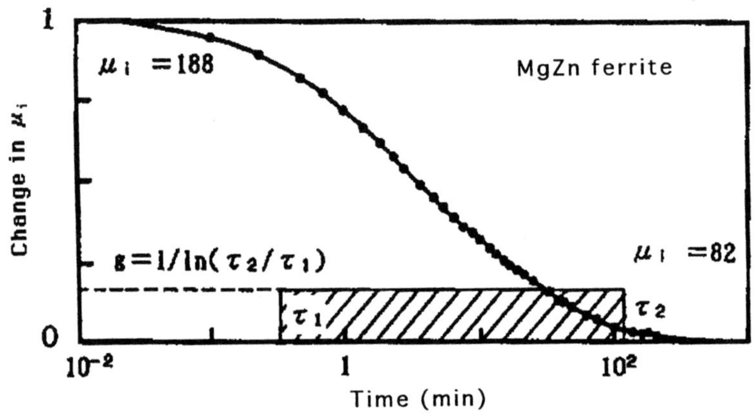

FIGURE 6.1.6 Disaccommodation of MnZn ferrite.

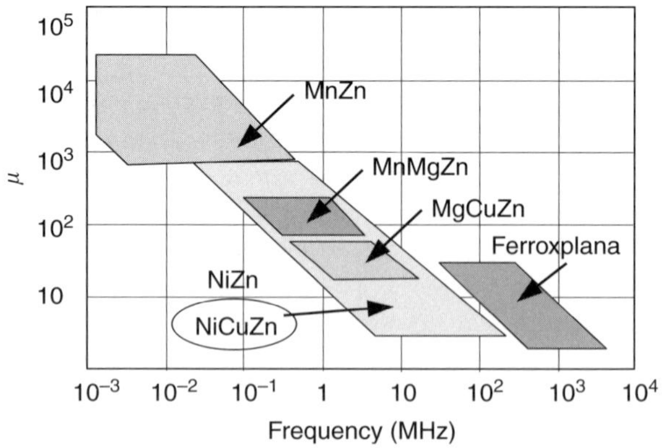

FIGURE 6.1.7 Relationship between driving frequency and initial permeability of ferrite.

TABLE 6.1.1 Main uses of soft ferrites

Use	Frequency	Ferrites	Property
Coil (communication)	1 kHZ ~ 1 MHz	MnZn	Low loss Low temperature coefficient
	0.5 ~ 80MHz	NiZn	Inductance adjustment
Pulse transformer		MnZn	High permeability Low loss
		NiZn	Low temperature coefficient
Transformers	~300kHz	MnZn	High permeability High magnetic flux density Low loss
Flyback transformer	15.75 kHz	MnZn	High permeability High magnetic flux density Low loss
Deflection yoke	15.75kHz	MnZn MnMgZn NiZn	Precise shape High permeability High resistivity
Antenna	0.4 ~ 50 MHz	NiZn	Large μQ Temperature characteristics
RF transformer	0.3 ~ 200MHz	NiZn	Large μQ Temperature characteristics Inductance adjustment
Magnetic head	1kHz ~ 10MHz	MnZn	High magnetic flux density High permeability High anti-wearability
Isolator Circulator Latching	30MHz ~ 30GHz	MnMgAl YIG YIG	Tensor permeability Magnetic flux density Resonance half-value width
Temperature Switch		MnCuZn	Curie Temperature

range of several megahertz or less, MnZn system ferrite is mainly used. As shown in Figure 6.1.1, there are some compositions where magnetocrystalline anisotropy and magnetostriction become zero or nearly zero, which show high permeability. Moreover, magnetic flux density is also high and the flexibility of composition selection is large to meet with the purpose of use. In a spinel-type ferrite, although MnZn ferrite has the highest permeability and magnetic flux density, the electrical resistivity is very low compared with other spinel ferrites and its operating frequency is limited to several megahertz. The high permeability materials used for noise filters, wide band transformers, pulse transformers, a transformers for communication, etc, are prepared with this material. In addition, the high magnetic flux density materials for transformers of switched mode power supply or choke coil is also manufactured using MnZn ferrite. With such use, when driving frequency is comparatively low, the greatest demand is that magnetic flux density is large, and many metal magnetic materials, such as silicon iron plate, molybdenum permalloy, and iron powder are used. However, core loss increases at high frequency area so that MnZn ferrite is used. High-frequency driving is effective for miniaturizing a power supply, and it is thought that the substitution to a ferrite material will be accelerated in the near future.

Although the higher-frequency band against soft ferrite is between several megahertz and gigahertz, the main use in this range is intermediate frequency transformer, which is used in TV, radio and VTR, inductor, balun transformer and a bar antenna and so on. The typical requisite characteristics of ferrite are low loss and stability for temperature. The losses, which easily become a problem in the high-frequency range, are eddy current loss and residual loss. In order to decrease eddy current loss, the ferrite with the high electrical resistivity is advantageous.

In the case of ferrite with a microwave band, it is used where the condition that internal spin magnetic moment of magnetic materials is completed toward the same direction, by applying an external magnetic field. When a high-frequency magnetic field is applied in the vertical direction to the magnetic field, the spin magnetic moment slightly leans and rotates. This phenomenon is useful for microwave applications. Such a phenomenon is called the gyromagnetic effect. In this movement, applied magnetostatic field and materials have the settled characteristic cycle, and a resonance phenomenon happens when a high-frequency magnetic field corresponds with this cycle. This resonance phenomenon is called as ferromagnetic resonance. Since the ferrite used with a microwave range differs from usual high-permeability materials, the required magnetic characteristic are also unique. Usually, saturation magnetization, magnetic resonance half-value width, spin resonance half-value width, specific dielectric constant, dissipation factor and Curie temperature, are of prime importance. Some spinel-type ferrites such as Mg, Ni, and Li ferrites can be used for microwave application. The garnet type ferrite is a microwave magnetic material used most widely at present. Typical yttrium iron garnet (YIG) is the material with the smallest loss of magnetic materials for microwave known now, and magnetic resonance half-value width is below 40 A/m (0.5 e). In order to obtain a suitable value of saturation magnetization, partial substitution of iron with aluminum is performed, and this is used with VHF and UHF bands. Moreover, there are also materials with partial substitution of yttrium with gadolinium, which improved the flatness of temperature dependence of saturation magnetization. Ferrites for microwave band are also used for isolator, circulator, and latching circuit devices.

Recently, the noise problem attracted much attention because of digitalization of electrical devices, high-frequency driving of CPU, new apparatus of a high-speed information transmission interface (IEEE1394, USB), progress of the mobile communications which is further seen as a rapid expansion of cellular phone use, etc. Ferrite materials are effectively used as noise suppressors regarding these electronic devices. As for line noise, MnZn or NiZn ferrites of high permeability are used. Nickel ferrite that can correspond to each frequency with composition against a radiation noise of the broad band, are abundantly used. Moreover, ferroxplana-type ferrite materials are beginning to be used for the noise reduction in the gigahertz band.

6.1.2 SOFT FERRITE

In soft ferrite, there are two chemical formulas, spinel type ($MeFe_2O_4$) and the garnet type ($Me_3Fe_5O_{12}$), are shown in Table 6.1.2. Here, the spinel-type ferrite, which is the most useful material, is described. Soft ferrite has small coercive force of a hysteresis loop, and generally permeability is of prime importance. Magnetic flux density is large and magnetocrystalline anisotropy and magnetostriction are small so that permeability increases. With spinel type structure, as magnetocrystalline anisotropy is comparatively smaller, the permeability is large. The requisite characteristics of ferrite are high Curie temperature, high permeability, and high stability, but it cannot satisfy all these requirements so that various spinel types of ferrites are being used depending upon the purpose. Practical materials are a solid solution of a ferrimagnetic single ferrite

TABLE 6.1.2 Magnetic properties of single ferrites

Crystal structure	Chemical formula	Magnetic flux density at room temperature I_s [G]	Curie temperature T_c [K]	specific gravity
Spinel	$MnFe_2O_4$	400	570	5.00
Spinel	$FeFe_2O_4$	480	860	5.24
Spinel	$CoFe_2O_4$	425	790	5.29
Spinel	$NiFe_2O_4$	270	860	5.38
Spinel	$CuFe_2O_4^{*1}$	170	455	5.35
Spinel	$MgFe_2O_4^{*1}$	230	700	4.52
Spinel	$Li_{0.5}Fe_{2.5}O_4$	310	940	4.75
Garnet	$Y_3Fe_5O_{12}$	135	550	5.17
Ferrox planar	$Ba_2Co_2Fe_{12}O_{22}$	185	613	5.40
Ferrox planar	$Ba_3Co_2Fe_{24}O_{41}$	270	683	5.33

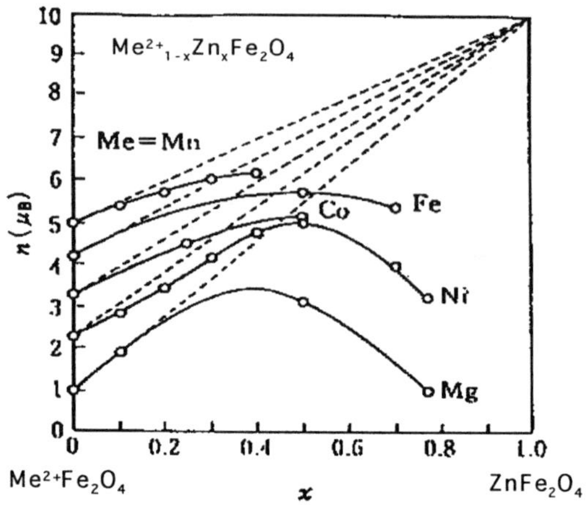

FIGURE 6.1.8 Effect of Zn ferrite on the number of Bohr magneton of single ferrite (Gorter).

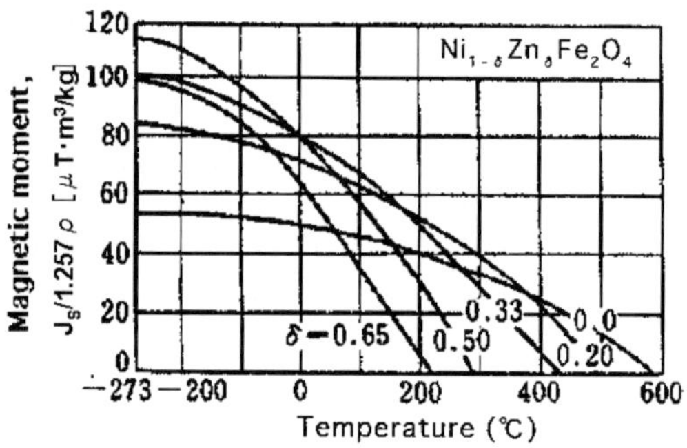

FIGURE 6.1.9 Temperature dependence of saturation magnetic moment of NiZn ferrites.

and an antiferromagnetic Zn ferrite. In this solid solution, Zn ions preferentially occupy the A-site, and make magnetic moment of A-site small, and it makes the difference of magnetic moment larger between A- and B-sites. This behavior that was demonstrated by Gorter [4] is shown in Figure 6.1.8. In the region where the content of $ZnFe_2O_4$ is in minority, magnetic moment gradually increases with the increase in $ZnFe_2O_4$. As a result, magnetic flux density increases. However, A–O–B super exchange interaction becomes weak because the magnetic moment in A-site becomes smaller with increase of $ZnFe_2O_4$. As a result, it becomes weak against heat turbulence, and Curie temperature, which is the transition temperature from ferrimagnetism to parmagnetism, decreases as is shown in Figure 6.1.9 according to the findings of Smit and Wijn [5].

MnZn ferrite, which has lower electrical resistivity but has larger saturation magnetic flux density, initial permeability, and lower core loss compared with NiZn and MgZn ferrites, is widely used as core material for transformers. Moreover, the Fe_2O_3-excess compositions, rather than stoichiometric composition, is widely employed because of their high magnetic performances. The excessive Fe_2O_3 forms solid solution with $MnFe_2O_4$ and $ZnFeO_4$ as γ-Fe_2O_3 or Fe_3O_4. In order to control the valence of excess iron oxide, especially strict control of oxygen partial pressure of ambient atmosphere during firing is required. According to the data of Ceramic Data Book [6], a production process of MnZn ferrite is shown in Figure 6.1.10. Metal oxides or carbonates are generally employed as raw materials. They are well mixed and then calcined followed by comminution. In order to acquire the desired sintering property, powder properties are controlled mainly by the calcination and comminution conditions. Calcination is usually performed at 800–1000°C in the air. Therefore, calcined powder of MnZn ferrites is not a single phase of spinel. It is generally a mixture of α-Fe_2O_3(ss), α-MnO_3(ss), $ZnFe_2O_4$, $MnFe_2O_4$, and $ZnMn_2O_4$. A discharge of oxygen is spread out by the reactions of $Mn_2O_3 \rightarrow 2MnO + (1/2)O_2 \uparrow$ and $Fe_2O_3 \rightarrow (2/3)Fe_3O_4 + (1/6)O_2 \uparrow$. Control of oxygen partial pressure during firing is of prime importance not only for the control of microstructure but also for ferritization reactions.

According to Morineau and Paulus [7], representative phase diagram of MnZn ferrite is shown in Figure 6.1.11. Spinel one phase is stable between the upper and lower boundaries. Spinel and wustite phases are stable below the lower boundary, on the contrary, spinel and α-Fe_2O_3 are stable above the upper line. Excess iron oxide can be changed between Fe_3O_4 and γ-Fe_2O_3, so that

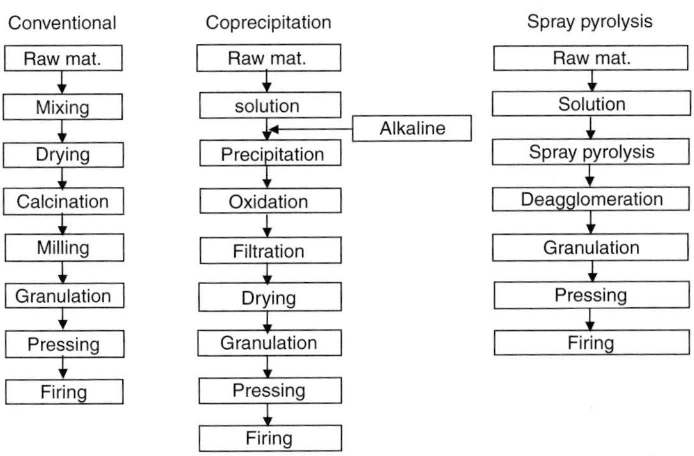

FIGURE 6.1.10 Producing process of MnZn ferrite.

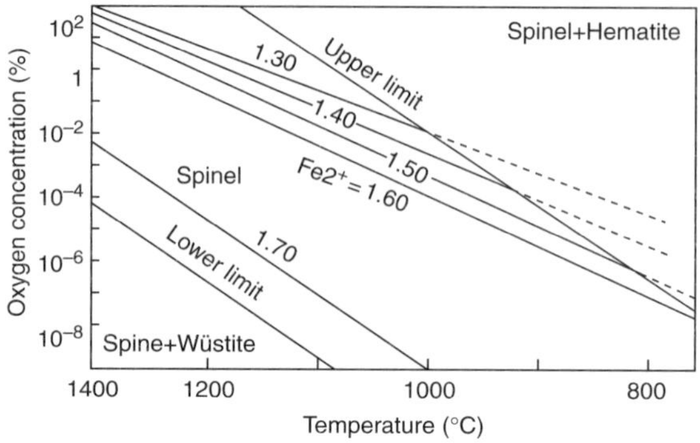

FIGURE 6.1.11 Diagram for MnZn ferrite (Fe_2O_3: 54.6, MnO: 29.3, ZnO: 16.1 mol%).

the spinel one phase region spreads out. In order to maintain constant oxygen content, it is necessary to choose an appropriate partial pressure of oxygen corresponding to temperature. For example, in order to keep Fe^{2+} content to 1.7 wt%, partial pressure of oxygen is set to the 1.7 wt%-line, and should be continuously changed with temperature during the cooling step. The relationship between oxygen content and magnetic property of ferrite sintered body is shown in Figure 6.1.12 [8]. In spinel one phase region, initial permeability increases with oxygen content. Moreover, the quality factor (Q) at high frequency and electrical resistivity has a peak at γ-Fe_2O_3 side of Fe_3O_4–γ-Fe_2O_3 pseudo-binary system. Additionally, the change of initial permeability with time become larger by formation of cation vacancy in side of γ-Fe_2O_3 composition. As a result, the balance of various magnetic properties is considered and composition is chosen. It is important to maintain the oxygen content with cooling. If the hematite phase precipitates during cooling, it causes an air gap in the magnetic circuit.

MnZn ferrite can be used for amplifying fine signals as initial permeability is high, and it is an excellent material. According to the findings of Ross *et al.* [9], the larger grain size is preferable for higher initial permeability and effect of grain size on the initial permeability is shown in Figure 6.1.3. Fe^{2+} is required to control to obtain higher initial permeability because the excess Fe_2O_3 has to become the spinel phase. The initial permeability when MnZn ferrite was developed was about 2000, but it materials whose initial permeability is over 20 000 were successfully obtained according to the findings of Yasuhara *et al.* [10]. That is, Bi_2O_3 was used for the additive, and firing conditions were precisely controlled, ensuring homogeneous distribution of oxygen in the ferrite core. The effect of Bi_2O_3 addition on the distribution of oxygen content in ferrite is shown in Figure 6.1.13.

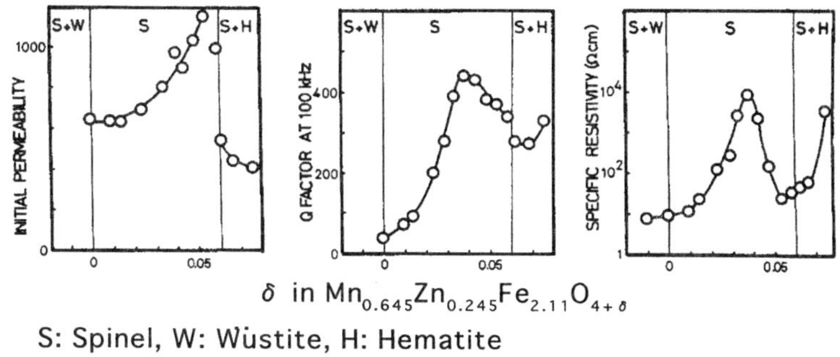

FIGURE 6.1.12 Effect of oxygen content on the magnetic properties of the MnZn ferrite.

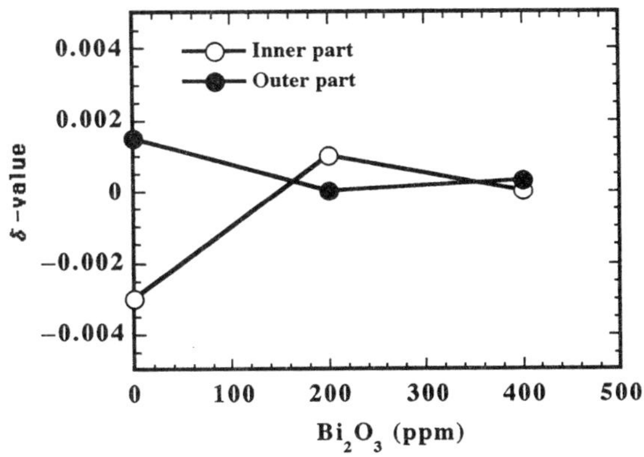

FIGURE 6.1.13 Effect of Bi_2O_3 doping on the nonstoichiometric oxygen content of MnZn ferrite.

Low loss materials are being used in transformers for power supply, which handles large electric power. Therefore, the most important requirement for these materials is a low loss with high saturation magnetic flux density. This is because magnetic losses increase remarkably by applying larger magnetic field compared with high-permeability materials. The driving frequency moves from around 10 to 100–500 kHz, and ferrite is required to have low loss at higher frequencies. A low loss material was developed by the optimization of small amount of additives, the control of impurities in raw materials, and the precise control of firing conditions [11, 12]. The transition of loss is shown in Figure 6.1.14. Super low loss, which was 199 kW/m^3 (100 kHz, 200 mT) demonstrated by Ceramic Data Book [6], is realized at present by SnO_2 substitution. Such materials will aid the development of an electric car which will address environmental concerns. Some requirements toward ferrite materials

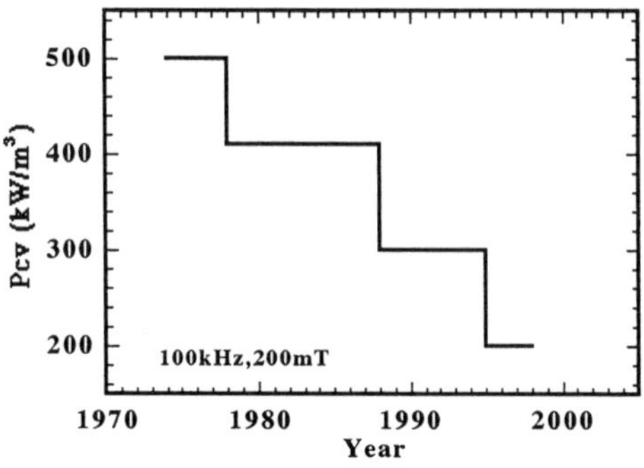

FIGURE 6.1.14 A change in the loss of MnZn ferrite.

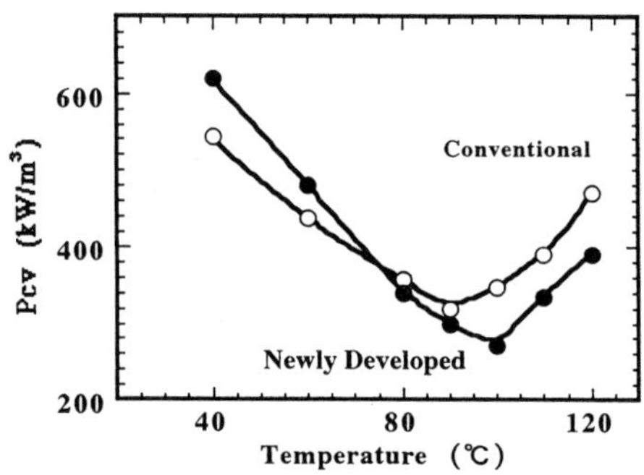

FIGURE 6.1.15 Temperature dependence of core loss of MnZn ferrite for the electric car.

is a low loss and to have high magnetic flux density under high temperature. Temperature dependence of loss and saturation magnetic flux density are shown in Figures 6.1.15 and 6.1.16, respectively. High performance MnZn ferrite for electric car usage has been developed by Murase *et al.* [13].

Mg ferrite was put to practical use for the first time as electronic materials in microwave applications. Ferrites for microwave are classified into spinel and garnet types. These ferrites were discovered, and were put to practical use for the first time in MnMg ferrite. The requisite characteristics of ferrite for microwave use are small magnetocrystalline anisotropy and uniform microstructure. Many studies have been done for Mg Ferrite from the viewpoint of composition, additives, and production process. According to Paladino [14], it is known that it is hard to obtain a single phase ferrite for $FeO–MgO–Fe_2O_3$. Fe^{2+} is formed when equilibrium partial pressure of oxygen is higher than that of Fe_2O_3, and especially MgO easily segregates when it exceeds limits of solid solution [15]. Mg ferrite has the cation distribution of mixed type of normal spinel and inverse spinel. Moreover, Bertaut [16] has reported that cation distribution of spinel

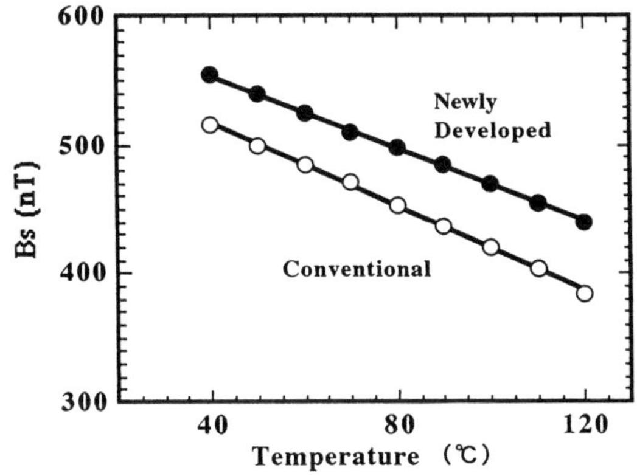

FIGURE 6.1.16 The temperature dependence of saturation flux density of MnZn ferrite for electric car.

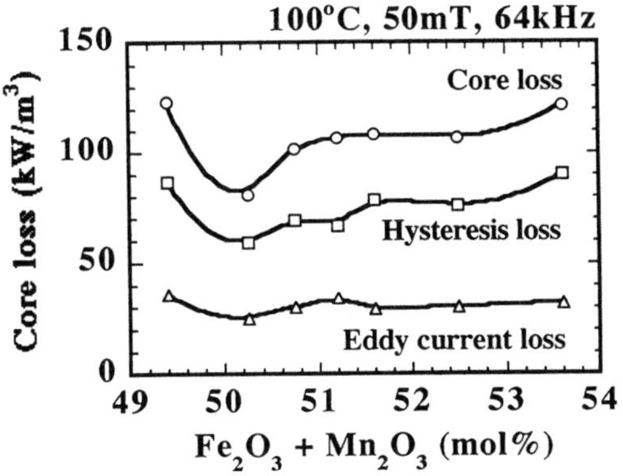

FIGURE 6.1.17 Effect of chemical composition on the core loss of MgZn ferrite.

is changed by cooling step after sintering at high temperature. The result of an enormous amount of research on Mg ferrite for microwave use is applied to present ferrite materials for deflection yoke core. Recently, the deflection yoke was put to use at higher frequencies and higher electric current, because wide, high vision, and high definition display televisions are in demand. For MnMgZn ferrites also high saturation magnetic flux density and low magnetic loss are required. Effect of the composition on magnetic loss has been reported from the viewpoint of stoichiometry by Murase *et al.* [17]. Magnetic loss of MgZn ferrite fired at 1300°C in the air is shown in Figure 6.1.17. At the composition where the sum of Fe_2O_3 and Mn_2O_3 is ca. 50 mol%, hysteresis loss becomes the smallest so that core loss also becomes the smallest. As for composition less than 50 mol% of Fe_2O_3 and Mn_2O_3, the segregation of some (MgZnFe)O is observed and this causes larger core loss by non-magnetic phase. As for composition over 50 mol% of Fe_2O_3 and Mn_2O_3, cation vacancies are formed, which prevent the movement of domain walls and cause higher hysteresis loss. Many studies have been performed by Sawai *et al.* [18] with special reference to composition, additives, and process to develop a high-performance MnMgZn ferrite. As a result, substitution of Mg for Cu is an effective way to achieve high performance.

As for the trend of electronic instruments in recent years, it proceeds with high density mounting of electronic components on printed circuit boards. As for ferrite components, it was thought to be hard to manufacture a chip that compared to multilayer chip capacitor or chip resistor, but in 1980 a multilayer chip inductor, which is multilayer ferrite chip component (MLFC) and reported by Takaya [19] had been developed by technologies of low temperature sintering and thick film printing. Silver internal conductor is used in order to lower the dc resistance. In order to fire inner Ag and ferrite materials at same time, low temperature sinterable NiCuZn ferrites have been developed by selected composition and control of powder properties. It was reported by Nakano *et al.* [20] that previously this was sintered at ca. 1200°C, but is now sintered at below 900°C. The printing technology of MLFC is shown in Figure 6.1.18. A three-dimensional coil is formed in the direction of the multilayer and it layers the pattern for a half turn coil conductor and magnetic body alternately according to Nomura and Takaya [21]. The size has been reduced from 3216 (3.2 mm × 1.6 mm) to 1005 (1.0 mm × 0.5 mm). Moreover, an excellent point of MLFC is its ability to obtain high inductance with small size and low-level electromagnetic interference to electronic devices because of the closed magnetic circuit, and additionally, MLFC is suitable for high-density

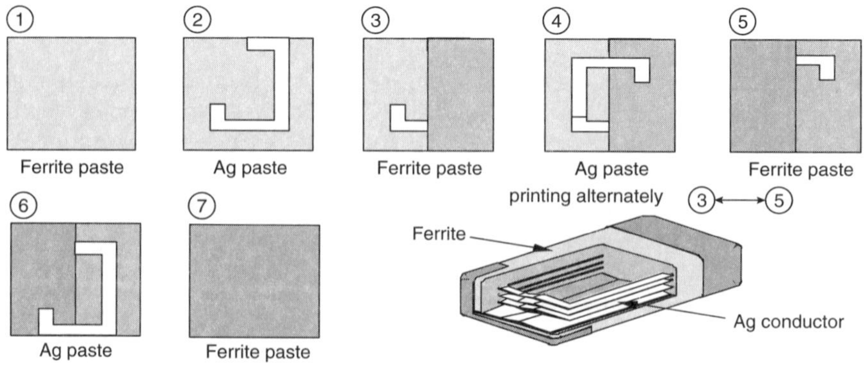

FIGURE 6.1.18 Printing method for multilayer ferrite chip inductor.

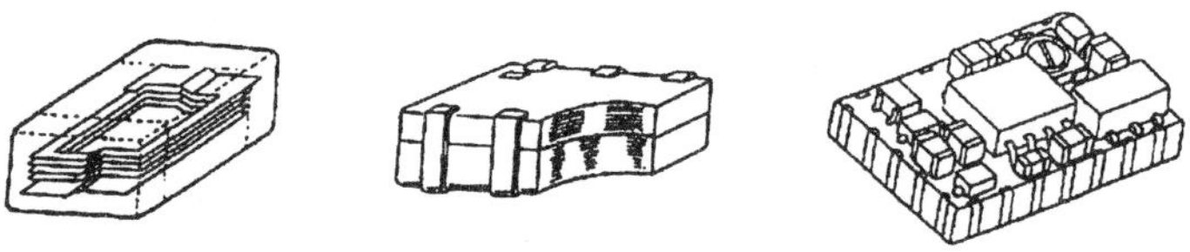

FIGURE 6.1.19 Schematic models of representative multilayer ferrite chip components (Multilayer chip inductor, Multilayer LC filter, and Multilayer hybrid component).

mounting. Technical development of MLFC is shown in Figure 6.1.19. A multilayered chip bead, which uses magnetic loss of ferrite, absorbs the in-line noise, and the production has shown abrupt increase in recent years. A multilayer LC filter is a composite component made to combine magnetic and dielectric materials, and is substantial miniaturization is being achieved by building up several inductors and capacitors in one chip. Furthermore, a multilayer hybrid circuit device (MHD) is a functional IC block module composed of IC on a one-chip transformer. These MLFCs are widely used for electronic equipment and their demand has risen because of their excellent properties.

REFERENCES

1. Nomura, T., and Ohta, K. (1991). *Seminar Text of Magnetic Materials*, Jpn. Inst. Metal, p. 13.
2. Snoek, J. L. (1948). *Physica* **14**: 207.
3. Ohta, K. (1954). *J. Phys. Soc. Jpn.* **16**: 295.
4. Gorter, E. W. (1954). *Philips Res. Rep.* **9**: 295.
5. Smit, J., and Wijn, H. P. J. (1959). Ferrite. *Philips Tech. Lib.*
6. Nomura, T. (1986). *Ceramic Data Book*, Kogyo Seihin Gijutu Kyoukai, p. 376.
7. Morineau, R., and Paulus, M. (1973). *Phys. Stat. Solidi (a)* **20**: 373.
8. Morita, A., and Okamoto, A. (1980). "Proc. Int. Conf. on Ferrites," p. 313.
9. Ross, E., Hanke, I., and Moser, E. (1961). *Z. Angew. Phys.* **13**: 247.
10. Yasuhara, K., Horino, K., Nomura, T., and Sano, T. (1995). *J. Magn. Soc. Jpn.* **19**: 417.
11. Iimura, T. (1976). *J. Jpn. Soc. Powder Powder Metall.* **23**: 253.
12. Nomura, T. (1989). "Proc. of MRS International Meeting on Advanced Materials," vol. 11, p. 233.
13. Murase, T., Sato, N., Nakano, A., and Nomura, T. (1999). *KINZOKU*. **69**: 707.
14. Paladino, A. E. (1960). *J. Am. Ceram. Soc.* **43**: 4.
15. Richard, R. G., and White, J. (1954). *Trans. Br. Ceram. Soc.* **53**: 422.
16. Bertaut, E. (1951). *J. Phys. et Radium* **12**: 252.
17. Murase, T., Igarashi, K., Yamazaki, T., Nomura, T., and Ochiai, T. (1995). *J. Magn. Soc. Jpn.* **19**: 413.
18. Sawai, J., Murase, T., and Nomura, T. (1997). *J. Magn. Soc. Jpn.* **21**: 915.
19. Takaya, M. (1983). *Denshi-Gijutu* **25**: 42.
20. Nakano, A., Suzuki, T., Kanagawa, Y., Watanabe, H., and Nomura, T. (1991). "Proc. of 10th Takei Seminar," p. 1.
21. Nomura, T., and Takaya, M. (1987). *HYBRIDS* **3**: 15.

Handbook of Advanced Ceramics
S. Sōmiya *et al.* (Eds.)

CHAPTER 7

7.1 Optoelectroceramics

HAJIME HANEDA
Advanced Materials Laboratory, National Institute of Materials Science,
1-1 Namiki, Tsukuba, Tsukuba 305-0044, Japan

7.1.1 INTRODUCTION

In the modern information age, electronic technology based on semiconductor materials has become indispensable in all fields from advanced science to daily life. But as society continues to move forward, the need has arisen for new technologies that can perform even better than electronics. Steps have already been taken in this direction by combining optics and electronics to produce the new field of optoelectronics. In this field, work is under way to find functions that can be performed by exploiting phenomena to which electrons and photons both contribute, and to develop new systems based on these functions.

The use of advanced ceramics in optoelectronic systems (optoelectroceramics) first attracted serious interest in 1970, when G.H. Heartling at the Sandia Research Laboratory in the United States achieved the first successful synthesis of the polycrystalline transparent perovskite PLZT [1, 2]. Then, in 1973, Greskovich *et al.* [3] at General Electric succeeded in developing a transparent sintered solid-state laser (NDY) by sintering a system of Y_2O_3 including 10% ThO_2 and 1% Nd_2O_3 at 2170°C in hydrogen atmosphere. Since then, developments in the field of optoelectroceramics have taken place at an ever-increasing rate. Recently, $Y_3Al_5O_{12}$ (YAG) transparent sintered materials are being produced with characteristics that surpass those of single-crystal ceramics in some respects, and have been successfully used in laser devices [4].

In this section, the basic properties of optoelectroceramics and there evaluation methods are described, and the production methods and applications of such devices are summarized.

The term "ceramics" refers to a broad range of materials including not only polycrystalline materials, but also powdered materials, thin films and single crystals, and glassy inorganic materials. In this section, we will use a slightly narrower definition of ceramics as polycrystalline sintered materials.

7.1.2 THE BASIC PROPERTIES OF OPTOELECTROCERAMICS

7.1.2.1 Reflection, Refraction, Birefringence and Polarization

Light propagates in a straight line through a vacuum or a homogeneous medium, and is reflected and refracted at the boundaries between different media. In the plane including the normal vector of an interfacial surface and the incident light rays, the light is reflected by an angle equal to the angle of incidence θ as shown in Figure 7.1.1.

In Figure 7.1.1, at the interfacial plane between medium 1 and medium 2, some of the incident light with the angle of incidence θ is reflected back into medium 1, and the rest is refracted into medium 2. If medium 1 is a vacuum, then the following relationship holds between the angle of incidence θ and the angle of refraction φ:

$$\frac{\sin\theta}{\sin\varphi} = n \tag{1}$$

where n is the refractive index. The refractive index and birefringence values of some typical solids are listed in Table 7.1.1.

According to Fresnel's law, the energy reflection factor R and the proportion of refracted light T for a perpendicular incident light ray are expressed as follows:

$$R = \frac{(n_1 - n_2)^2}{(n_1 + n_2)^2} \quad T = \frac{4n_1n_2}{(n_1 + n_2)^2} \quad (= 1 - R) \tag{2}$$

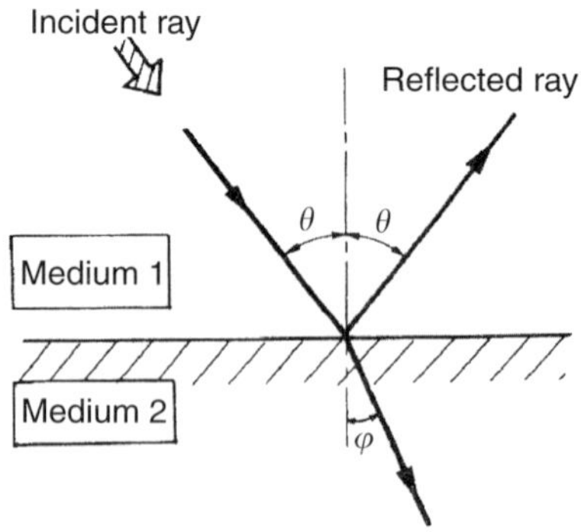

FIGURE 7.1.1 Reflection and refraction of light.

TABLE 7.1.1 Median Refractive Index and Birefringence of Some Solid Materials

Materials	Median refractive index	Birefringence
Oxide crystals		
SiO_2	1.55	0.009
BeO	1.73	0.014
MgO	1.74	—
Al_2O_3	1.76	0.008
Gd_2O_3	1.96	—
Y_2O_3	1.92	—
ZrO_2	2.20	0.070
PbO	2.61	0.130
TiO_2	2.71	0.287
Double oxide crystals and solid solution		
$3Al_2O_3 \cdot 2SiO_2$	1.64	0.010
$Al_2O_32SiO_2$	1.65	0.021
$MgAl_2O_4$	1.72	—
$ZrSiO_4$	1.95	0.055
(Pb, La)(Zr, Ti)O_3	2.50	0.055
$LiNbO_3$	2.31	−0.086
$BaTiO_3$	2.40	−0.070
$Gd_3Ga_5O_{12}$	2.03	—
Glasses		
Silicate glass	1.46	—
Vycor glass	1.46	—
Pyrex glass	1.47	—
Soda lime glass	1.51	—
Flint glass	1.70	—
Others		
SiC	2.38	—
ZnSe	2.68	0.043
CdS	2.62	−0.017
Si	3.49	—
Ge	1.43	—
CaF_2	4.00	0.172

where n_1 and n_2 are the refractive indices of medium 1 and medium 2, respectively.

Birefringence (double refraction) is an optical phenomenon—exhibited by certain transparent materials—whereby double images are seen when nearby objects are viewed through them. As Figure 7.1.2 shows, when light is incident on a material that exhibits birefringence, the light is split into two waves with orthogonal electric field vectors. Of these two waves, the one whose electric field vector oscillates at right angles to the optical axis of the material is referred to as the "ordinary ray", and the other is referred to as the "extraordinary ray".

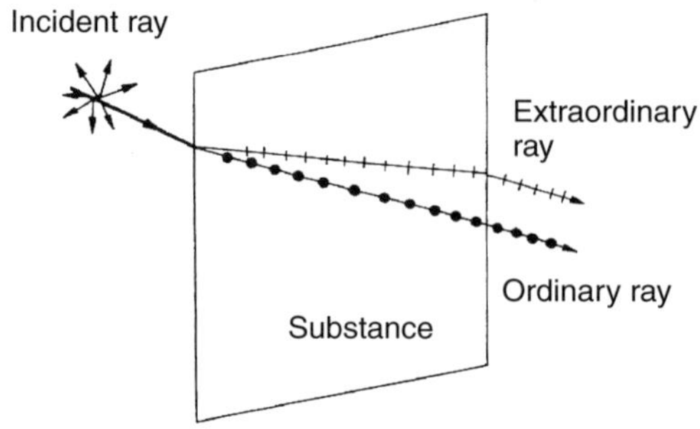

FIGURE 7.1.2 Birefringence.

If n_0 and n_e are the refractive indices for the ordinary and extraordinary rays, then the birefringence Δn is given by the difference between the two:

$$\Delta n = n_e - n_0 \tag{3}$$

A crystal is said to exhibit positive birefringence when Δn is positive, and negative birefringence when Δn is negative.

When linearly polarized light is incident on an anisotropic material that exhibits birefringence, the light propagates as two plane waves with orthogonal electric field vectors as shown in Figure 7.1.2. The electric field components E_x and E_y of these plane waves can be expressed as shown in Eq. 4:

$$\begin{aligned} E_x &= A \exp\left\{ i\left(\omega t - \frac{2\pi}{\lambda} n_x z\right)\right\} \\ E_y &= A \exp\left\{ i\left(\omega t - \frac{2\pi}{\lambda} n_y z\right)\right\} \end{aligned} \tag{4}$$

Here, for the sake of simplicity, it is assumed that $\delta_x = \delta_y = 0$. Also, n_x and n_y are the refractive indices of light having electric field components oscillating in the x and y directions, respectively.

The ωt terms can be eliminated from Eq. 4 to yield Eq. 5:

$$\left(\frac{E_x}{A_x}\right)^2 + \left(\frac{E_y}{A_y}\right)^2 - 2\left(\frac{E_x}{A_x}\right)\left(\frac{E_y}{A_y}\right)\cos(\Delta\beta z) = \sin^2(\Delta\beta z) \tag{5}$$

where $\Delta\beta (= \beta_y - \beta_x)$ is the difference between the propagation constants $\Delta\beta_y (= 2\pi n_y/\lambda)$ and $\Delta\beta_x (= 2\pi n_x/\lambda)$ of the two electric field components.

Therefore, depending on the magnitude of $\Delta\beta z$, an incident ray of linearly polarized light may become elliptically polarized or linearly polarized. In this case, the polarization state differs depending on the thickness z of the medium and the refractive index difference $\Delta n (= n_y - n_x)$.

7.1.2.2 The Optical Properties of Ceramics

The optical properties of ceramics are naturally affected either directly or indirectly by factors such as the constituent elements constituting the ceramic, the crystalline symmetry, the microstructure of the ceramic, and changes in the external field. Table 7.1.2 lists some of the typical factors that affect the optical properties of ceramics. Even in single crystals—which are the simplest form of ceramics under the broad definition—these factors rarely act individually on the optical properties. And in a polycrystalline material, the situation is obviously much more complicated.

The change in optical properties Δp resulting from a change Δs in the external field can be expressed in general terms by the following formula:

$$\Delta p = a_1 \Delta s + a_2(\Delta s)^2 + a_3(\Delta s)^3 + \cdots \quad (6)$$

where the set of values a_n are physical constants that represent the change in a physical quantity for a change in the external field. These constants are expressed in the form of tensor quantities. Since only small changes in the external field are normally dealt with, it is usual to consider only the first one or two terms of Eq. 6.

Taking optical transmittance as an example of the optical and physical properties of a transparent ceramic, this depends on the surface roughness due to

TABLE 7.1.2 Some Factors Reacting on the Optical Properties of Ceramics

Factors	On single grain	On aggregation of grains
Structure		
Crystal structure (symmetry)	×	×
Electronic structure	×	
Lattice defect structure	×	
Microstructure		×
(pore, grain, surface, grain boundary, interface, etc.)		
Composition		
Constituting elements	×	
Compositional inhomogeneity	×	×
Impurity	×	×
Stoichiometry	×	
External field		
Electric field	×	×
Magnetic field	×	×
Stress	×	×
Temperature, light, etc	×	×
Flint glass	×	×

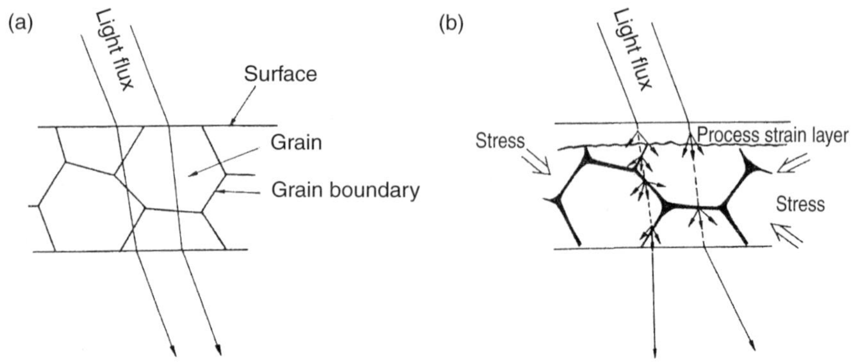

FIGURE 7.1.3 Interaction between the light and the isotropic polycrystal ceramics. (a) Ideal homogenous polycrystal with theoretical density. Light is not scattered about grain boundaries. (b) Polycrystal with internal strain and external stress. Light is scattered at inhomogenous region.

the degree of polishing of the ceramic surface, and the optical transmittance changes when the scattering of light is different. Furthermore, local distortion and strained layers resulting from surface processing give rise to localized changes of refractive index, causing light to be scattered and affecting the optical transmittance. In particular, distortion often builds up close to the surface of a ceramic and at grain boundaries, causing changes in refractive index as shown in Figure 7.1.3. (schematic diagram). If Eq. 6 is applied to the change in refractive index due to this distortion, then the change $\Delta p(= \Delta(1/n_{ij}^2)$, where n_{ij} are the refractive index tensor components) resulting from a change of Δs in the distortion occurring in the ceramic, is related by a quaternary tensor quantity called the *distorted optical constant* (a_1). This is known as the photoelastic effect, and changes in the refractive index tensor are closely related to the crystalline structure (crystalline symmetry) of the ceramic. So even when we consider the single property of optical transmittance, we can see that its effects on the optical and physical properties are mutually interconnected with the complex structure in the ceramics.

Below, taking an electric field as a typical example of an external field, we will discuss the electro-optic effects that result from the effect of this field on the optical properties. Electro-optic effects are effects such as changes to the deformation and rotation of the index ellipsoid caused by the action of an electric field (E). When the field is applied, the change in the refractive index tensor $\Delta(1/n_{ij}^2)$, which is an optical parameter, can be expressed as follows by expanding the electric field terms:

$$\Delta(1/n_{ij}^2) = \sum_k r_{ijk} E_k + 1/2 \sum_k \sum_l R_{ijkl} E_k E_l \tag{7}$$

Here, the first term is referred to as the first-order electro-optic effect (*Pochels effect*), and the second term is referred to as the second-order electro-optic effect (*Kerr effect*). The coefficients r_{ijk} and R_{ijkl} are ternary and quaternary tensor quantities known as the Pochels constant (first-order electro-optic constant) and Kerr constant (second-order electro-optic constant), respectively. As Table 7.1.3 shows, a second-order electro-optic effect is present in materials, including isotropic materials such as glass, whereas first-order electro-optic effects are only observed in piezoelectric crystals. In Table 7.1.3, electro-optic effects are present in crystals belonging to point groups.

Electro-optic effects in ferroelectric materials can also be dealt with by similar arguments to those used so far. For example, above the Curie temperature (about 120°C), $BaTiO_3$ belongs to a cubic system ($m3m$), and since it has a center of symmetry does not exhibit piezoelectric or first-order electro-optic effects. Accordingly, the electro-optic effect in this paraelectric phase is the Kerr effect. Using the polarization optical constant R^* in Eq. 7 instead of an electric field, it can also be expressed in terms of polarization as follows:

$$\delta(\Delta n) = -\tfrac{1}{2}(R_{11}^* - R_{12}^*)n_0^3\, P_3 \tag{8}$$

In the ferroelectric phase below the Curie point, it becomes a tetragonal system and spontaneous polarization P_s occurs. The polarization P_3 when an electric field is applied in the direction of spontaneous polarization can be expressed using the electric susceptibility χ as follows:

$$P_3 = P_s + \chi E_3 \tag{9}$$

By substituting Eq. 9 into Eq. 8, we find that the spontaneous birefringence in the absence of an electric field is proportional to the square of P_s, so it results only in a spontaneous Kerr effect (see Table 7.1.4).

Next, we will discuss electro-optic effects in polycrystalline materials. In the case of a PLZT ceramic ($(Pb,La)(Zr,Ti)O_3$), the phase diagram in Figure 7.1.4 shows that it exhibits a first-order electro-optic effect in the constituent domains of a tetragonal crystal with a large resisting field, and exhibits a second-order electro-optic effect at boundary phase regions between the ferroelectric and paraelectric phases (the region where the *P–E* curve describes a slim loop: ⑤) [2]. However, an electro-optic ceramic in which the crystallites are sintered with random orientations does not exhibit first-order electro-optic effects, just as piezoelectric ceramics do not exhibit piezoelectric effects. When polarization processing is performed, it belongs to the $C_{\infty v}$ point group—which is isotropic around the polarization axis (like a piezoelectric ceramic)—and here it exhibits a first-order electro-optic effect for the first time. In this case, the matrix elements of the electro-optic constants r_{ijk} and R_{ijkl} have the same form as those of C_{6v} ($6mm$) and C_{4v} ($4mm$).

TABLE 7.1.3 Crystal Symmetry and Related Optical Properties

Crystal system	Point group		Without external field	With external field		
	International index	Schoenfles symbol	Optical anisotropy (birefringence)	Electro-optic effect		Photoelastic effect
				First order	Second order	
Triclinic	1	c_1	Existent (two axis)	Existent	Existent	Existent
	1	$C_i(S_2)$	Existent (two axis)	Non		
Monoclinic	2	C_2	Existent (two axis)	Existent		
	m	$C_s(C_{ih})$	Existent (two axis)	Existent	Existent	Existent
	$2/m$	C_{2h}	Existent (two axis)	Non		
Orthorhombic	222	$D_2(V)$	Existent (two axis)	Existent		
	$mm2$	C_{2v}	Existent (two axis)	Existent	Existent	Existent
	mmm	$D_{2h}(V_h)$	Existent (two axis)	Non		
Tetragonal	4	C_4	Existent (one axis)	Existent		
	4	S_4	Existent (one axis)	Existent		
	$4/m$	C_{4h}	Existent (one axis)	Non		
	422	D_4	Existent (one axis)	Existent	Existent	Existent
	$4mm$	C_{4v}	Existent (one axis)	Existent		
	$42m$	$D_{2d}(V_d)$	Existent (one axis)	Existent		
	$4/mmm$	D_{4h}	Existent (one axis)	Non		

Trigonal	3	C_3	Existent (one axis)	Existent		
	3	$C_{3i}(S_6)$	Existent (one axis)	Non		
	32	D_3	Existent (one axis)	Existent	Existent	Existent
	3*m*	C_{3v}	Existent (one axis)	Existent		
	3*m*	D_{3d}	Existent (one axis)	Non		
Hexagonal	6	$C6$	Existent (one axis)	Existent		
	6	C_{3h}	Existent (one axis)	Existent		
	6/*m*	C_{6h}	Existent (one axis)	Non		
	622	D_6	Existent (one axis)	Existent	Existent	Existent
	6*mm*	C_{6v}	Existent (one axis)	Existent		
	6*m*2	D_{3h}	Existent (one axis)	Existent		
	6/*mmm*	D_{6h}	Existent (one axis)	Non		
Cubic	23	T	Non	Existent		
	3m	T_h	Non	Non		
	432	O	Non	Non	Existent	Existent
	43m	T_d	Non	Existent		
	333	O_h	Non	Non		
Isotropic			Non	Non	Existent	Existent

TABLE 7.1.4 Electro-optic Coefficients

Materials	Symmetry	Linear		Quadratic	
		ij	$10^{12} r_{ij}$ (m/V)	ij	R^*_{ij} (m^4/C^2)
Bi_2GeO_{20}	23	41	9.7		
		11	0.23		
		11	-0.47^a		
SiO_2 (quartz)	32	41	0.1		
		41	0.2^a		
$LiNbO_3$	$3m$	13	8.6	11	0.16
		33	30.8	12	0.04
		51	28	14	0.11
		22	3.4		
		c^{*2}	21		
		c	19^a		
$BaTiO_3$	$4mm$	51	820		
		c	23		
		c	19^a		
$BaTiO_3$	$m3m$			c^b	0.13
$Sr_{0.75}Ba_{0.25}Nb_2O_6$	$4mm$	33-13	1400		
ZnO	$6mm$	33	2.6		
		13	1.4		
$SrTiO_3$	$m3m$			c	0.14
$KTaO_3$	$m3m$			c	0.16
				44	0.12
$KTa_{0.65}Nb_{0.35}O_3$	$m3m$			c	0.17
$Ba_2NaNb_5O_{15}$	$mm2$			c	0.12

[a]Values under constant stress.
[b] $R^*_c = R^*_{11} - R^*_{12}$.

Accordingly, in the polarization processed state, if we choose the polarization direction and applied electric field direction to be along the z axis as in Figure 7.1.5, then the effective birefringence Δn_{eff} (i.e. the average birefringence of the bulk polycrystalline material, as opposed to the value for each individual crystal grain) brought about by the electric field is given by the following formula:

$$\begin{aligned} \Delta n_{\text{eff}} &= -r_c n_0^3 E_3 \qquad \text{(Pochels effect)} \\ &\left(r_c = r_{33} - \frac{n_0^3}{n_e} r_{13} \right) \\ \Delta n_{\text{eff}} &= -\tfrac{1}{2} R_c n_0^3 E_3^2 \qquad \text{(Kerr effect)} \\ &(R_c = R_{11} - R_{12}) \end{aligned} \tag{10}$$

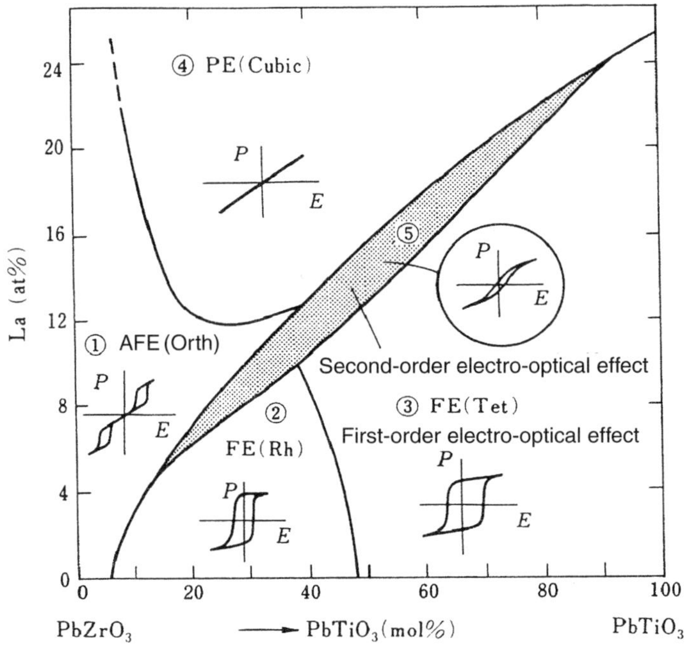

FIGURE 7.1.4 Phase diagram and relation between electric field (E) and polarization.

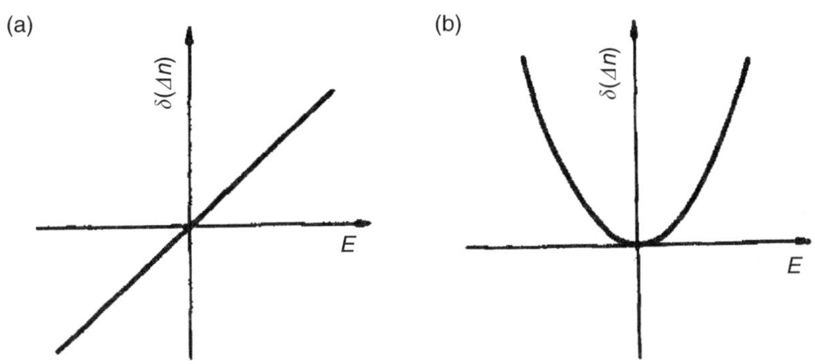

FIGURE 7.1.5 Birefringence induced by the electro-optic effect and the electric field. (a) Linear electro-optic effect. Linear electro-optic coefficient: gradient. (b) Quadratic electro-optic effect. Coefficient of E^2 is proportional to $(R_{11} - R_{12})$ value.

where r_c and R_c are the first- and second-order transverse electro-optic constants, respectively. Several PLZT ceramics (compositions of the form $(Pb_{1-x}, La_x)(Zr_{1-y}, Ti_y)O_3$; abbreviated to PLZT $- x/(1-y)/y$) have attracted interest due to their having larger lateral electro-optic effects than single crystals.

7.1.3 METHODS FOR MEASURING OPTICAL PROPERTIES

7.1.3.1 Optical Transmittance, Absorption Coefficient, and Reflectance

When light travels through a ceramic containing a large number of light-scattering bodies, the light rays spread out due to optical scattering. Accordingly, the transmittance of a transparent ceramic may be measured either in terms of the overall amount of transmitted light including scattered light components, or in terms of just the light that passes straight through it, with scattered components excluded (the central ray in Figure 7.1.6). In some cases it is appropriate to measure the overall amount of transmitted light including scattered light components, for example, when evaluating transparent ceramics used in fluorescent light tubes. To separate the absorption coefficient and scattering coefficient expressing the amount of attenuation of transmitted light, measurements must be made by both methods.

As an example of how these optical characteristics are measured, Figure 7.1.7 shows a system that can be used to measure the optical transmittance. In the optical system, the light emitted from a light source (S) consisting of a xenon lamp or a laser is passed through a system of lenses (L_1, L_2) and a pinhole aperture (A_1) to obtain a well-collimated beam which is passed through a second pinhole (A_2) measuring about 1 mm across to produce a narrow beam that is directed toward the measurement sample. Light that passes through the sample

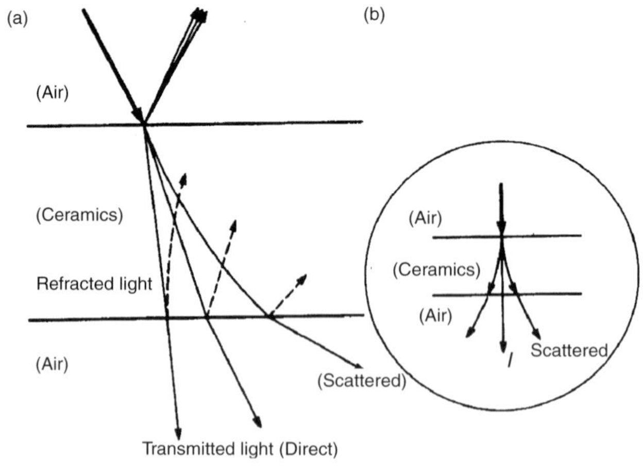

FIGURE 7.1.6 Schematic drawing of light transmission in ceramics: (a) general case; (b) normal incident case.

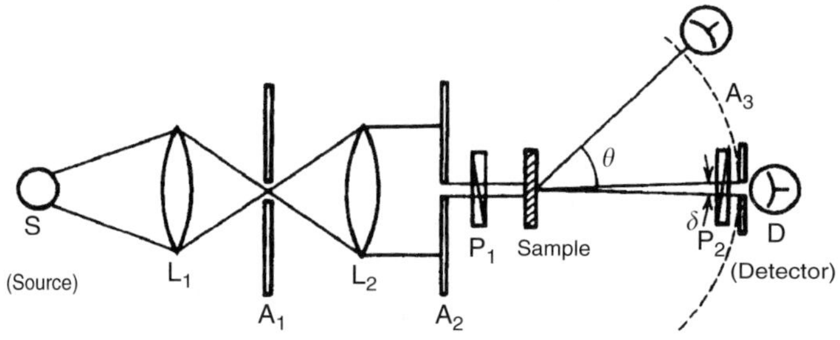

FIGURE 7.1.7 Optical system for transmittance measurement.

is detected by passing it through another pinhole (A_3) situated in front of a photodetector (D). If the diameter of (A_3) is made sufficiently small—that is, if the detector aperture angle δ is sufficiently small—then the proportion of scattered light components in the transmitted light can be made negligibly small and the true transmittance is obtained as a result.

In actual measurements, the measured quantities are the amount of incident light I_0 and the amount of light I transmitted when the sample is inserted into the optical system. These two quantities are related as follows:

$$\frac{I}{I_0} = (1 - R^2)e^{-\beta t} \tag{11}$$

The transmittance is either given by I/I_0 or (when corrected for reflectance) by $I/I_0(1-R)^2$. The light attenuation coefficient β can be determined by measuring the transmittance of a number of samples of different thickness and determining the gradient of a graph of $\ln(I)$ against thickness t. A simple method for making this measurement is to calculate it from the following formula based on the transmittance values of two samples with different thicknesses.

$$\begin{aligned}\beta &= \frac{\ln(I_2/I_1)}{t_1 - t_2} \\ &= \alpha_0 + S_{im} + S_{op}\end{aligned} \tag{12}$$

where α_0 is the linear absorption coefficient, and S_{im} and S_{op} are coefficients due to the crystal imperfection and the optical anisotropy among constituent grains, respectively.

Next, we will briefly discuss methods for separating the linear absorption coefficient α_0 and the scattering coefficient $S_0(= S_{im} + S_{op})$ from the overall attenuation coefficient β (Eq. 11). In principle, this is done by measuring the total amount of transmitted light including scattered components (the overall transmittance) and expressing it as a fraction of the amount of incident light produce an expression in which the attenuation coefficient β is expressed only in terms of α_0. The total amount of scattered light can be measured by methods involving the use of an integrating sphere, for example. When the ceramic includes a light-absorbing material that absorbs light at particular wavelengths, the dependence of β on the incident light wavelength can be measured to allow α_0 and S_0 to be separated.

The reflectance R can also be measured by methods similar to those used to measure the transmittance. Specifically, from Eq. 11, the amount of light transmitted at the limit where the sample thickness t approaches zero becomes $(1 - R)^2 I_0$, and the reflectance can be calculated by extrapolating a plot of $\ln(I)$ against t to the point where $t \rightarrow 0$.

7.1.3.2 Refractive Index and Birefringence

Although measurements of refractive index can be inferred by applying the Fresnel relationship to the measured reflectance values, this is an imprecise method. Instead, methods such as immersion in liquids and measurement of the Brewster angles are generally used.

The liquid immersion method allows measurements to be made with comparative ease even with small samples such as coarse powder grains. The sample is placed in a liquid with a fixed refractive index, and is observed under a microscope while the refractive index of the liquid is gradually changed until it matches that of the sample. When the refractive indices of the liquid and sample are different, bright lines called Becke lines can be seen around the outside of the sample. But when the refractive indices are the same, the Becke lines disappear and the interface between the liquid and the sample becomes invisible. This is because no refraction or reflection occurs at the interface between the materials when their refractive indices are identical, and light behaves as if it is propagating through the same medium. This method is usually capable of making measurements with an accuracy of three places of decimals. Refractive indices that can be measured in this way lie in the range of approximately 1.4–1.8.

The measurement of refractive indices based on Brewster angle measurements is described with reference to Figure 7.1.8a. This method is based on the principle that the reflected ray produced by light incident on a transparent material of refractive index n with an incidence angle θ derived from the formula $n = \tan\theta$ will be completely linearly polarized at right angles to the plane of incidence. This angle is known as the Brewster angle, and obeys the relationship whereby the sum of the angles of incidence and refraction is equal to $\pi/2$. Such measurements can be made as shown in Figure 7.1.8b, where an optical system is mounted on a goniometer. Light from a laser or xenon tube that

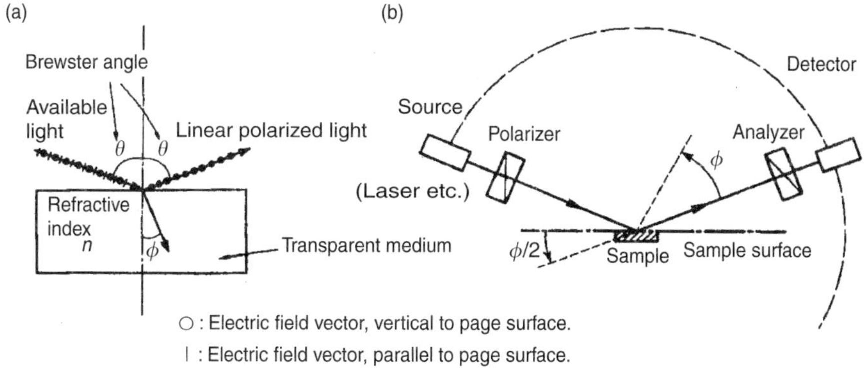

FIGURE 7.1.8 Brewster angle description (a) and its measurement method (b).

has been fully collimated is passed through a polarizer and is incident on the sample. The reflected light is passed through an analyzer and measured. When the polarization planes of the polarizer and analyzer both match the plane of incidence, the amount of light reaches a minimum at an angle corresponding to the Brewster angle. The precision of measurements made in this way is of a similar order to that of the liquid immersion method.

7.1.3.3 Electro-Optic and Magneto-Optic Effects

The electro-optic constants (Pochels constant r_{ijk} and Kerr constant R_{ijkl}) can be measured using more or less the same optical system as that of Figure 7.1.7, except that a polarizer, compensator and analyzer are used in the sample part as shown in Figure 7.1.9. Linearly polarized light is directed toward the sample, and the phase change that occurs as it passes through the sample is measured. Figure 7.1.9 shows where the electrodes are placed when measuring: (a) the lateral effect, and (b) the longitudinal effect.

Measurements are made by adjusting the compensator so as to minimize the amount of light transmitted through the polarizer and analyzer (which is at right angles to the polarizer), thereby measuring the change in the phase of the light after it has passed through the sample. Here, to improve the extinction ratio of light passing through the system, the polarization direction of the polarizer is arranged by rotating it through $\pi/4$ radians relative to the optical axis of the crystal as shown in Figure 7.1.9a. In this case, the positional difference Γ (rad) is related to the sample thickness t and birefringence Δn as follows:

$$\Gamma = \frac{2\pi}{\lambda}\Delta nt \tag{13}$$

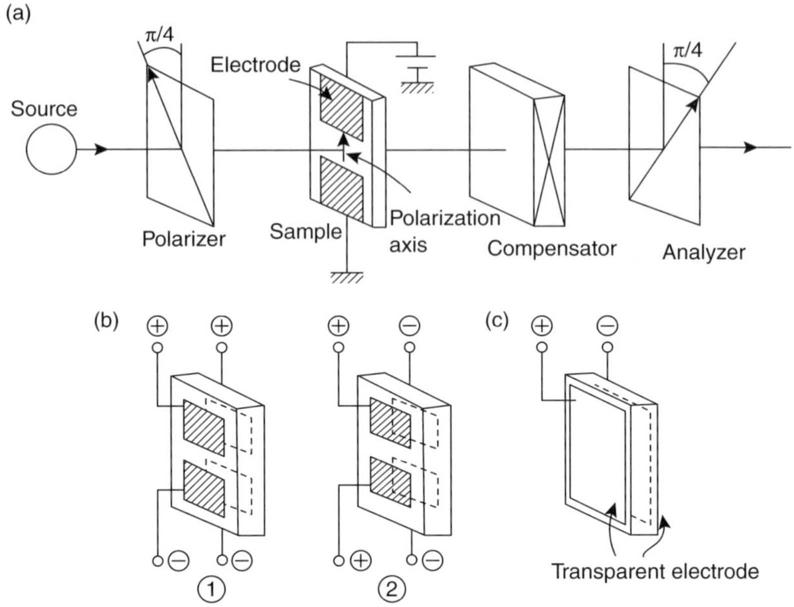

FIGURE 7.1.9 Measuring method for electro-optical effect. (a) System of measurement for the lateral effect. (b) Electrode for the longitudinal effect measurement. ①, polarization; ②, electric filed impressing during measurement. (c) Sample configuration.

Measuring Γ therefore allows Δn to be determined, and the electro-optic constants r_c and R_c can then be found from Eq. 10.

7.1.4 PROCESSES FOR PREPARATION OF OPTOELECTROCERAMICS

7.1.4.1 Introduction

Mechanisms for making ceramics transparent have been actively studied since the appearance of transparent alumina polycrystalline materials—which dispelled the myth that all ceramics are opaque. As a result of this research, many ceramics have been reclassified as transparent materials, and the mechanisms that make them transparent have also gradually become clearer. Here, we discuss the conditions necessary for making ceramic polycrystalline materials transparent, and we describe a few examples of processes for used to achieve this.

First, the formula expressing the attenuation to which light is subjected when it propagates through a ceramic polycrystalline material is qualitatively expressed by the Lambert–Beer formula as shown in Eq. 11 above. β can be expressed in terms of the linear absorption coefficient α_0, the optical scattering coefficient S_{im} resulting from structural imperfections, and the optical scattering coefficient S_{op} resulting from the optical anisotropy of each individual crystal grain constituting the polycrystalline ceramic. If the value of β can be reduced, then it should be possible to obtain a transparent ceramic. Table 7.1.5 lists the correspondence between these attenuation factors and the basic parameters of the transparent ceramic production process.

TABLE 7.1.5 Relation between Factors for Transparent Ceramics and the Basic Items of Processes

Factor for transparent ceramics	Process	Basic items
Reducing of absorption coefficient α_0	Purify	Purification of precursor materials Usage of ultra high purity materials Inhibition of contamination during synthesis (mixing, pulverization, calcination, sintering)
	Composition control	Identification of precursor material composition (ignition loss, crystallization water) Inhibition of compositional fluctuation during process (solubility to solvent, evaporation during calcination and sintering) Atmosphere (oxidation, reduction)
Reducing of scattering coefficient S_{im}	Reduction of pores	Usage of fine and homogeneous powders Inhibition of abnormal grain growth during sintering Reduction of large void (bridge) in green compact Usage of gas with high diffusivity (elimination of trap in pores)
	Elimination of precipitates	Optimization of blending composition Inhibition of compositional fluctuation during process Homogenizing of microstructure
Reducing of scattering coefficient S_{op}	Reduction of optical anisotropy	Choice of materials with low anisotropy
	Composition control	Control of anisotropy with additives

A fundamental requirement is that the ceramic materials should not include—either as constituent parts or as impurities—any elements whose characteristic absorption lies in the required range of optical wavelengths. In particular, allowing impurities to become mixed in at any stage in the entire process from adjustment of the raw material powder to final sintering will result in increased α_0. Since changes in composition also cause the lattice defect structure of the material to change, this can also lead to increased α_0. Since sintering involves high temperatures, compositions that include materials with a high vapor pressure should be checked for changes in composition after firing. Since α_0 is also increased by the reduction of specific components by the sintering atmosphere, it is also important to control the oxygen potential in the atmosphere. For example, TiO_2 readily becomes black in a reducing atmosphere.

Any inclusions in the parent material that have a different refractive index—such as air holes and precipitate particles—are accounted for by the S_{im} term. In processes for making ceramic polycrystalline materials transparent, reducing this term is of particular importance. Since transparent ceramic polycrystalline materials are synthesized at high temperature in the solid state, diffusion phenomena in the material constitute the basic process for eliminating air holes from the polycrystalline material. The sintering parameters include the molding conditions, sintering temperature, time, atmosphere, pressure and additives, and must be precisely controlled.

S_{op} is a coefficient that represents the scattering and attenuation of light due to refraction and reflection at interfacial surfaces where there are discontinuities in refractive index due to the optical anisotropy of each crystal constituting the ceramic polycrystalline material. To reduce S_{op}, it is thus necessary to change the composition of the ceramic to reduce the optical anisotropy. Since light scattering effects occur due to birefringence in the crystal grains of a ceramic that exhibits optical anisotropy, another way of reducing S_{op} is to reduce the number of crystal grains. Specifically, this involves increasing the size of the crystal grains as much as possible while trying to suppress anisotropic grain growth and avoid leaving air holes behind during sintering.

In ceramic polycrystalline materials such as Al_2O_3 and BeO, the optical transmittance can be raised by making the values of α_0 and S_{im} in Eq. 12 very small, but even so these polycrystalline materials do not become completely transparent and only end up being translucent. This is because of the large values of S_{op} resulting from their large birefringence values—0.008 for Al_2O_3, and 0.014 for BeO. On the other hand, compounds such as MgO, Y_2O_3, $Y_3Al_5O_{12}$ (YAG) and $MgAl_2O_4$ are isotropic crystals and have no optical anisotropy. As a result, they form transparent materials because the grain boundaries do not have discontinuous refractive indices and scattering does not occur.

Table 7.1.6 lists the principal ceramics that have so far been made transparent by paying attention to these points. This table also shows their optical transmittance values and summarizes the production methods of these ceramics. Although the translucent ceramics have different production methods, their basic concepts are shown in Table 7.1.5.

In the following, we will present a detailed discussion of PLZT ceramics, which are a typical example of optoelectroceramics, and we will also briefly discuss yttrium aluminum garnet (YAG) ceramics, which have recently attracted interest.

7.1.4.2 PLZT Ceramics

Solid solutions of lead zirconate ($PbZrO_3$) and lead titanate ($PbTiO_3$) are denoted by the chemical formula $Pb(Zr,Ti)O_3$, and are unique materials that exhibit both piezoelectricity and ferroelectricity. They are often referred to by the abbreviation PZT.

PZT has a perovskite crystalline structure of the form ABO_3, and by varying the types and quantities of the elements in solid solution at the A sites and B sites, the PZT can be given a variety of characteristics.

To start with, by substituting some of the Pb sites in PZT with Bi, a slightly translucent ceramic polycrystalline material can be obtained. Next, by using La instead of Bi, a material with exceptionally high optical transmittance is produced. In addition to clarifying the factors affecting transparency, numerous ceramic polycrystalline materials have been developed by substituting the A and B sites in PZT with various elements.

Transparent ceramic polycrystalline materials of the form $(Pb,La)(Zr,Ti)O_3$ [1, 2, 5–7] were developed by Heartling and are referred to by the abbreviation PLZT. The chemical formula of PLZT is as follows:

$$(Pb_{1-x}, La_x)(Zr_y, Ti_{1-y})_{1-x/4}O_3 \tag{14}$$

The A sites in an ABO_3 type perovskite structure are all filled with Pb and La, but the B sites are not necessarily filled with Zr and Ti, and the PLZT material as a whole is electrically compensated by the presence of vacancies. Accordingly, for a composition with 4 at% La, a single vacancy exists at the B sites.

Oxide materials can be used to produce the solid solution raw material powders, and are sometimes made by a wet precipitation method. Here, we will discuss methods for adjusting the raw materials by an ethanol/oxalic acid method and a multistage wet method, and various properties of polycrystalline ceramics produced using these as raw materials.

TABLE 7.1.6 Optical Transmittance and Sintering Conditions of Ceramic Materials

Materials	Additives	Optical transmittance (%)	Wavelength (μm)	Sample thickness (mm)	Crystal system	Sintering condition temperature (K) × duration (h), pressure	Sintering atmosphere	Reference
Al_2O_3	MgO (0.25 wt%)	40–60	0.3–2	1	Hexagonal	2123–2173 × 16	H_2	8
	Y_2O_3 (0.1 wt%), MgO (0.05 wt%)	70	3–11	0.5	Hexagonal	1973 × 5	H_2	9
	MgO (0.05 wt%)	85–90	Visible	0.75	Hexagonal	2000–2073 × 17–30	H_2	10
ALON	AlN (30 mol%)	Translucent	Visible		Cubic	1473 × 24 → 250–2300 × 1	Air → N_2	11
MgO	LiF, NaF (1 wt%)	80–85	1–7	5	Cubic	1273 × 0.25, 1500 psi	Vacuum	12–14
	NaF (0.25 wt%)	Transparent	Visible		Cubic	1873 × 111	O_2	15, 16
$MgAl_2O_4$		Transparent	Visible		Cubic	1673 × 1–2, 800 kg/cm^2 → 1473 × 1	Vacuum → air	17–19
		Transparent	Visible		Cubic	1623–1673 × 16 → 1953 × 6	H_2	20
	CaO (0.25 wt%)	20–80	0.3–65		Cubic	1773 × 1 → 2123 × 8	Vacuum → air	17
Y_2O_3		60–80	2–7	2.5	Cubic	1573–1773 × 1–2, 5000–7000 psi	Vacuum	21
		80	0.55	0.76	Cubic	2573	H_2	22
	LiF (3–5 wt%)	80–85	1–7	1	Cubic	1223 × 48, 10 000–12 000 psi	Vacuum	23
	ThO_2(10 wt%), Nd_2O_3(1 wt%)	75	Visible		Cubic	2443 × 58–125	H_2	3, 24
	2.6% HfO_2, 0.5%Nd_2O_3	80	2.3–6.3	1	Cubic	1923, 1 h and 1973, 3 h 196 MPa	Vacuum → Hiped	25
	Non-doped	20–50	0.4–1.2	1	Cubic	1973, 2 h	Vacuum	26
	0.5% CaO	45–70	0.4–1.2	1	Cubic	1973, 2 h	Vacuum	27
ZrO_2	CaO, Y_2O_3 (15 mol%)	10–15	Visible	2.5	Cubic	1973 × 1, 10 kbar		28
	Y_2O_3 (6 mol%)	10	Visible	1	Cubic	1723 × 16	Air	29
SIALON	Li, Na, K	40	4.5		Cubic	1873–2073 × 2, 100–500 kg/cm^2	N_2	30
PLZT	La: 9/65/35	70	0.6	1	Pseudocubic	1473 × 60 min	PbO, O_2	6, 7
	La: 8/65/35	60–70	0.5–6	1	Pseudocubic	1100 × 16, 2000 psi	O_2	2, 5
	La 9/65/35	50–70	0.5–0.8	1	Pseudocubic	1100 × 20 min 2000 psi	O_2	31, 32
	La 9/55/45	75	0.5–0.8	1	Pseudocubic	1473 × 60 min	PbO, O_2	35
K(Ta,Nb)O_3		60–70	0.4–1	0.14		1473 × 1, 3700 psi	O_2	36
$Y_3Al_5O_{12}$		70	0.3–1	1	Cubic	2073 × 24	H_2	37, 38
	SiO_2 (0.02–0.05 wt%)	70	0.3–1	10	Cubic	1973 × 24	Vacuum	4, 39, 40
		80	0.3–1	10	Cubic	2023 × 20, 1673 × 20	Vacuum, O_2	41–43

By concentrating on the fact that oxalic acid is soluble in ethanol whereas the oxalates of Ti and alkali earth metals such as Sr, Ca and Pb are not, Yamamura *et al.* [31–34]. have developed a new method involving the use of oxalic acid and ethanol. In this method, the starting raw materials are nitrates of the constituent metals. A quantity of oxalic acid 10% greater than that necessary to precipitate out all the metal ions as oxalate salts is measured out and dissolved in ethanol to produce oxalic acid ethanol. A pre-prepared aqueous solution containing the constituent metal ions is then added dropwise to this mixture, and aqueous ammonia is also added dropwise until it reaches pH 9.0. The resulting precipitate is washed several times in ethanol, dried, and heat-treated in air at 8000°C for 2 h, resulting in a single-phase perovskite. A flow diagram of this procedure of PLZT is shown in Figure 7.1.10.

Also, Shirasaki *et al.* [35] have proposed a "multistage wet method" as a simple and easy method for the synthesis of PLZT which uses ordinary inexpensive titanium tetrachloride as its starting material, and have produced a flow diagram for this process. Initially, as the first-stage precipitate formation process, a mixed nitric acid solution of three components (excluding the titanium tetrachloride) is adjusted and added dropwise to 6 N aqueous ammonia while stirring to bring about simultaneous precipitation of the hydroxides. Then, as

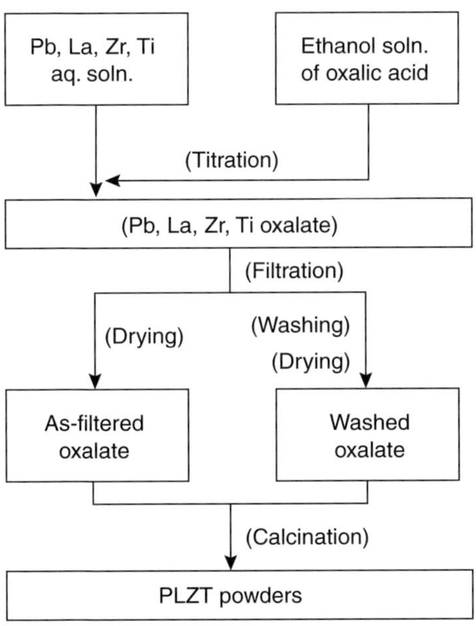

FIGURE 7.1.10 Flow diagram for PLZT powder preparation with ethanol–oxalate method.

the second-stage precipitate formation process, an acidic solution of titanium tetrachloride in hydrochloric acid is added dropwise to this slurry to obtain a mixed hydroxide of PLZT. By separating the titanium hydroxide formation process and the lead hydroxide formation process into two stages, it is possible to form a precipitate of hydroxides without producing any lead chloride. The product is washed several times, and after the chlorine components have been eliminated it is dried at 1100°C to obtain a dry hydroxide. This is fired at 700–1000°C to obtain powdered PLZT.

The resulting powder can be used to obtain a transparent ceramic by hot press (HP) sintering or by atmospheric sintering. Sufficient transparency is obtained by hot press sintering at 1100°C for 20 min. Figure 7.1.11 shows the wavelength dependence of transmittance in PLZT obtained under these sintering conditions. At 600 nm, the transmittance values resulting from the HP and atmospheric sintering processes are 67 and 66%, respectively. These results compare well with the theoretical value of 68%.

One possible use of PLZT is in ceramic actuators, where the most important factors are the amount of deformation and hysteresis. Figure 7.1.12 shows the relationship between the electric field and deformation for PLZT materials having various different grain sizes, and Figure 7.1.13 summarizes the grain sizes and the amounts of deformation. In general, the amount of deformation is found to increase as the grain size increases, but the hysteresis also tends to become larger. Figure 7.1.14 shows the dependence of hysteresis degree on the grain size.

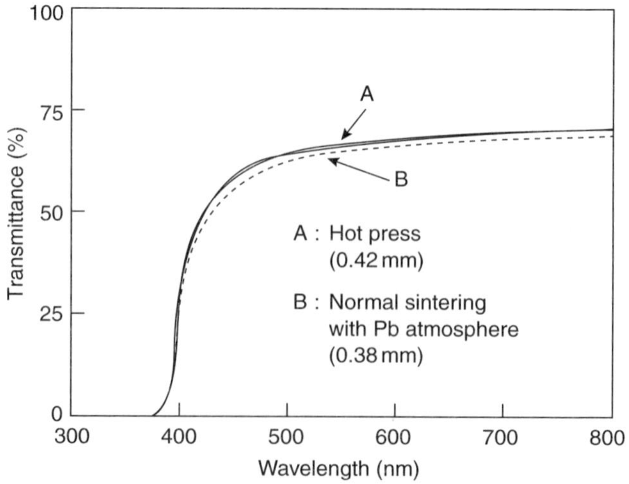

FIGURE 7.1.11 Transmittance of PLZT ceramics.

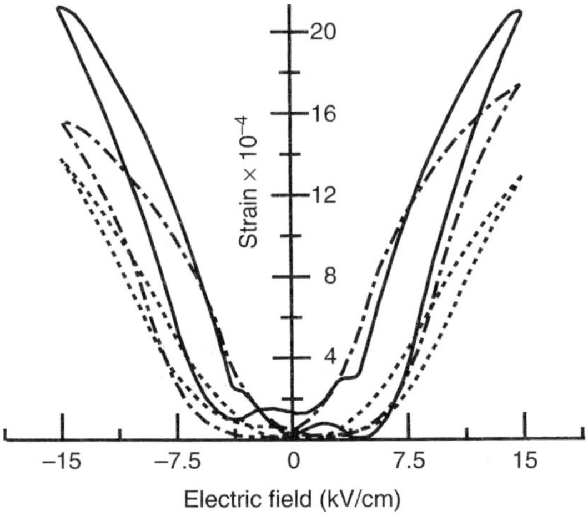

FIGURE 7.1.12 Hysteresis strain curves versus electric field for PLZT ceramics with various grain sizes. Solid line, with 4.5 μm diameter. Dot-dashed line, with 2.4 μm. Broken line, with 1.1 μm.

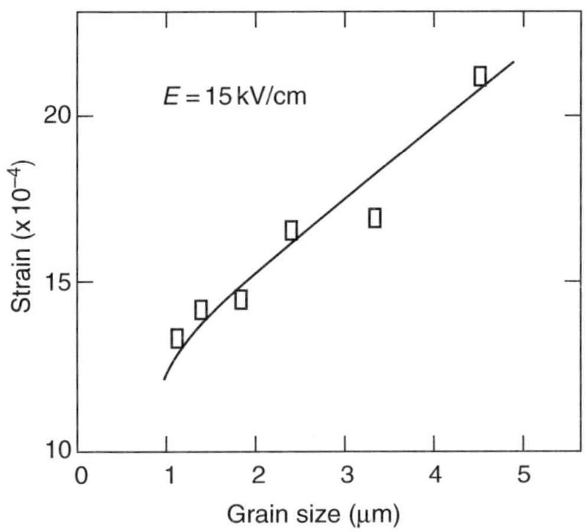

FIGURE 7.1.13 Grain size dependence of strain amount.

Furthermore, Figure 7.1.15 shows how the quadratic electro-optic coefficient *R*—which is the most important physical property for optoelectroceramics—varies with grain size. As this figure demonstrates, the *R* coefficient also depends strongly on grain size, and decreases sharply below a grain size of 1.7 μm.

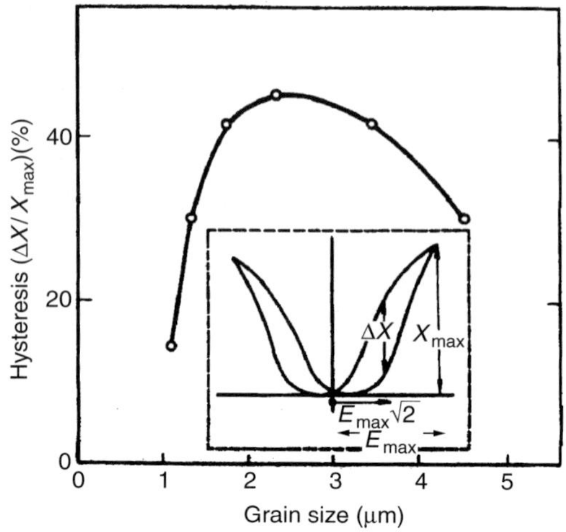

FIGURE 7.1.14 Grain size dependence of strain hysteresis.

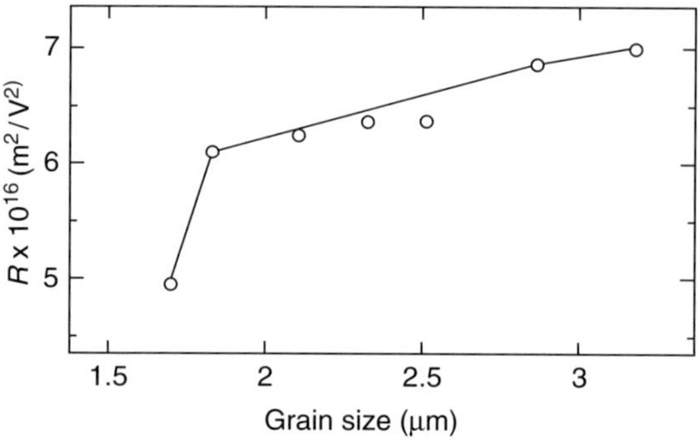

FIGURE 7.1.15 Grain size dependence of the quadratic electro-optic coefficient.

7.1.4.3 YAG Ceramics

Of the numerous oxide ceramics that have successfully been made transparent, attention has recently been focused on transparent ceramics made from yttrium aluminum garnet ($Y_3Al_5O_{12}$; commonly referred to as YAG). Since it was successfully made transparent by hot press production methods [44], there have also been reports of it being made transparent by ordinary sintering. Furthermore, DeWith's group [37, 38] and Haneda *et al.* [4, 39, 40] have made progress in the development of transparent YAG ceramics based on their discovery that silicon ions are effective for making it transparent.

Meanwhile, it has come to light that ytterbium iron garnet $Yb_3Fe_5O_{12}$ (YbIG)—which has the same garnet structure—can be sintered almost as far as the theoretical density by using raw materials derived from the urea method, which is a type of homogenous precipitation method [45]. The advantages of this method are that the pH of the solution can be uniformly controlled, and that it is easy to obtain a powder that is readily sintered. Recently, by applying this method to the production of YAG sintered materials, it has become possible to obtain transparent YAG ceramics having optical characteristics equivalent to those of single crystals [4]. This material is expected to find applications as a substitute for transparent alumina ceramics or as a window material with superior specular transmittance. It is also a promising material for use in high-temperature windows since it has superior strength [46] and creep resistance [47] at high temperature. The applications with the closest bearing on optoelectroceramics seem to be those that involve using it as a fluorescent material that is optically activated by a rare earth element. In fact, yttrium-based transparent ceramics were developed for use in scintillators before the arrival of YAG, and are now being put to practical use [3, 24]. Of course, these materials are expected to be particularly useful in polycrystalline laser applications.

Figure 7.1.16 shows a flow diagram of the process for synthesizing YAG sintered material using urea method powder as a raw material. This powder has a size of about 20 nm when fired. It is thought that secondary particles are formed

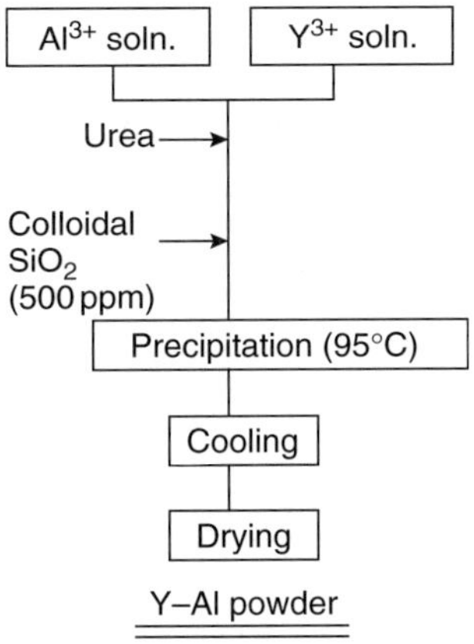

FIGURE 7.1.16 Flow diagram for YAG powder preparation with the urea homogeneous precipitation method.

FIGURE 7.1.17 Dy-doped YAG ceramics sintered in vacuum at 1973 K, 1 cm diameter and 1.5 cm height.

when $YAlO_3$ perovskite, alumina and YAG are uniformly mixed together, and the homogeneity of the composition is greatly sacrificed at this stage. But it is easy to end up with a homogeneous composition after sintering in a vacuum furnace or in a hydrogen atmosphere, so this is not a problem. The sintering can be assisted by the addition of silica components, but only small quantities should be added due to their effects on the optical characteristics. It is currently possible to get by with as little as 100 ppm. Optically active ceramics can be obtained by using a solution in which some of the yttrium ions are substituted with other rare earth elements. Compared with single crystals, one of the advantages of ceramics is that it is possible to avoid problems associated with the distribution coefficient of rare-earth elements between the liquid and solid phases during the production of a single crystal, and as a result the concentration of rare earth elements can be controlled with a certain degree of freedom. Figure 7.1.17 shows a YAG ceramic produced in this way. It appears to be in no way inferior to a single crystal.

It is important to evaluate the optical characteristics of these materials in terms of their application to optical materials. In YAG ceramics used as fluorescent materials, it is also important to measure the rare-earth light emission spectrum. Examples of optical characteristics that should be measured include the optical absorption spectrum, the light emission spectrum, and the light emission lifetime. Table 7.1.7 shows the results of measuring the absorption coefficient and α in samples with no added rare earth ions. Figure 7.1.18 shows the optical spectrum of a sample to which 1 at% Nd ions was added, along with a breakdown of the results for each spectral component. The spectral width is equivalent to that of a single crystal, showing that the ceramic exhibits good crystalline properties. These results show that the optical properties of Nd ions

TABLE 7.1.7 Spectral parameters for transitions between the $^4F_{3/2}$ and $^4I_{9/2,11/2}$ manifolds, from the decomposition of the absorption and emission spectra. N_0 denotes the N_d concentration, α the absorption coefficients, L the sample thickness, I the peak intensities, ν the central energy of the Lorentzian shape emission, and $\Delta\nu$ the full-width at half-maximum of the peak intensity

	Ceramic	Single crystal[a]	Unit
N_0	1.4×10^{20}	1.4×10^{20}	cm^{-3}
α_{R_1}	0.952	0.74	cm^{-1}
α_{R_2}	3.76	2.98	cm^{-1}
L	0.1556	0.3114	cm
I_{R_1}	5.76	4.29	
ν_{R_1}	11432.24	11 423	cm^{-1}
$\Delta\nu_{R_1}$	12.96	14.00	cm^{-1}
I_{R_2}	9.56	6.71	
ν_{R_2}	11518.36	11 508	cm^{-1}
$\Delta\nu_{R_2}$	17.57	18.90	cm^{-1}
I_{I_1}	0.886	0.583	
ν_{I_1}	9400.1	9394	cm^{-1}
$\Delta\nu_{I_1}$	5.27	4.53	cm^{-1}
I_{I_2}	2.05	1.61	
ν_{I_2}	9403.6	9397	cm^{-1}
$\Delta\nu_{I_2}$	5.02	4.95	cm^{-1}

[a]Reference [39].

in ceramics are more or less the same as those of single crystals, and it can thus be considered that the Nd ions are placed in a similar environment to that of a single crystal.

With regard to laser activity, it is meaningful to evaluate the stimulated emission cross-section σ_1^{in} of Nd ions in this ceramic. As a result, it is found that the value of σ_1^{in} in a YAG ceramic is 4.9×10^{-19} cm^2, while the values reported for single crystals lie in the range from 2.7×10^{-19} to 8.8×10^{-19} cm^2 [39]. The above results were obtained in the initial stages of the development of YAG ceramics. Recently, further progress has been made in powder synthesis methods, and as Figure 7.1.19 shows, it is now becoming possible to achieve laser activity with high output power [40]. The absorption and emission spectra of samples doped with other rare-earth ions (Pr, Eu and Er) are also being evaluated [41].

When such materials are considered as a replacement for the translucent alumina ceramics used in fluorescent tubes, the reactivity and corrosiveness in a metal sodium environment are also important characteristics. We

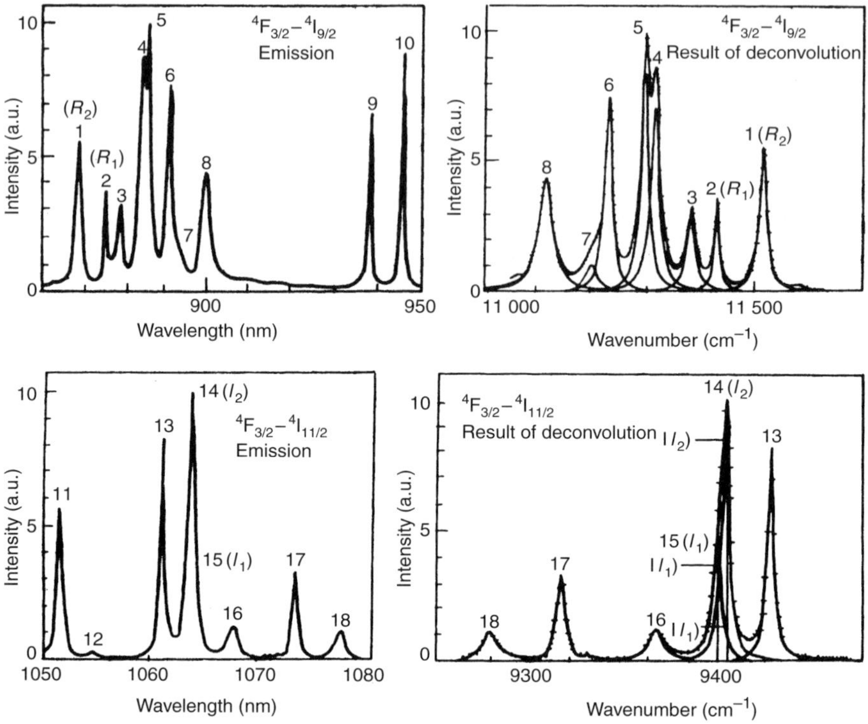

FIGURE 7.1.18 Emission spectra (left) of Nd 1%-doped YAG ceramics and their deconvolutions (right). (a) Transition between $^4F_{3/2}$ and $^4I_{9/2}$ (860–950 nm). (b) Transition between $^4F_{3/2}$ and $^4I_{11/2}$ (1050–1080 nm).

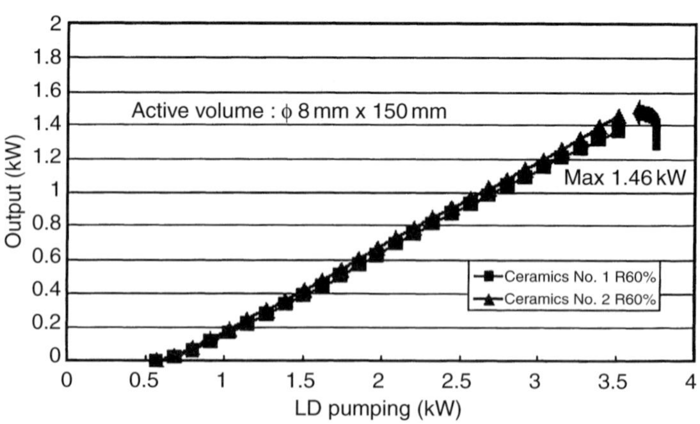

FIGURE 7.1.19 High power laser action of Nd-doped YAG ceramics.

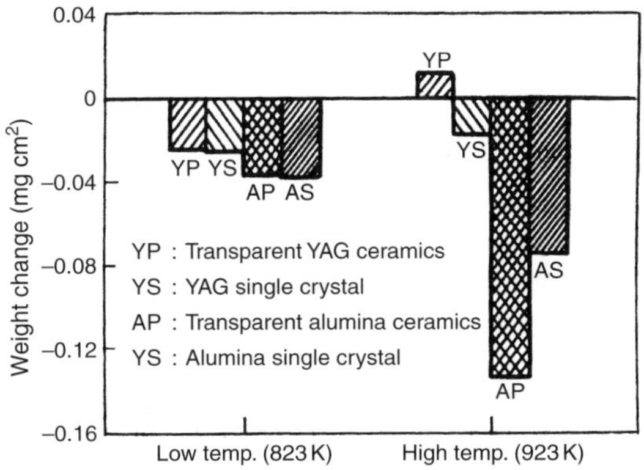

FIGURE 7.1.20 Corrosion of alumina and YAG in liquid sodium metal.

will therefore briefly discuss the results of corrosion tests in liquid metal sodium [48]. Figure 7.1.20 illustrates the mass reduction caused by corrosion after immersion in sodium for 1000 h. The data for a transparent alumina sintered material is also shown for comparison. The transparent alumina sintered material is a ceramic material with excellent corrosion resistance to liquid sodium, but the transparent YAG ceramic has even better corrosion resistance.

To apply the HID lamp and ceramic laser, the creep processing has to be controlled (see next section). High-temperature creep, as well as other kinetic processes (sintering, sinter-forging, solid-state reaction, and grain growth) are controlled by the transport of matter, which requires the motion of the various species (cations and anions) in the crystal. Thus, the diffusivities of the constituent ions in YAG are essential properties for its industrial applications. The oxygen self-diffusion coefficient has already been determined in YAG, both in volume (lattice) and along the grain boundaries [50, 51]. For cation, diffusion data had been taken by Haneda *et al.* with a secondary mass spectrometry [52]. These results were compared with the apparent diffusivities [53] deduced independently from the analysis of high-temperature deformation data reported in the literature [54–58] (see Figure 7.1.21). According to these results, the cation activation energies (~550 kJ/mol) are much larger than those for oxygen (300–350 kJ/mol). The effective dlffusion coefficient deduced from high-temperature deformation data reported in the literature for YAG polycrystals, assuming grain-boundary sliding accommodated by volume diffusion, is in excellent agreement, both in magnitude and activation energy, with the cation diffusion data.

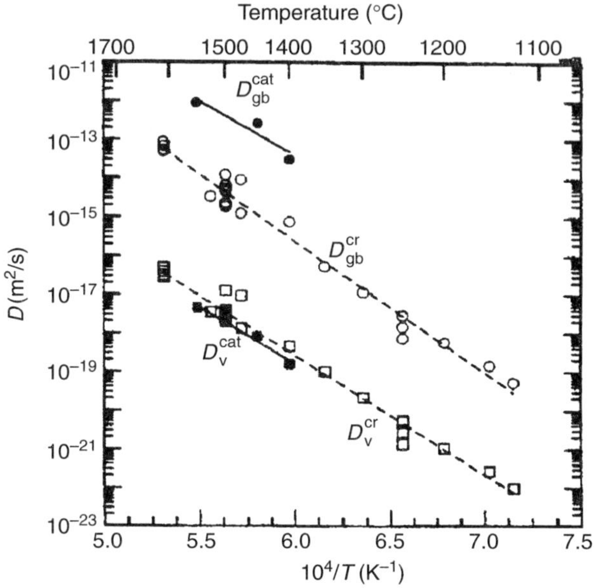

FIGURE 7.1.21 Comparison of diffusion coefficients derived from the creep data and cation diffusion experiments.

7.1.5 THE APPLICATIONS OF OPTOELECTROCERAMICS

7.1.5.1 INTRODUCTION

In this section, we will summarize the applications of optoelectroceramics in various broad categories and discuss how the various optical phenomena exhibited by these ceramics can be put to effective use. Table 7.1.8 classifies the applications of optoceramics in each category.

The optical behavior of ceramics often results from a combination of several different phenomena. For example, most applications of optoelectroceramics are impossible to achieve without controlling their composition and structure. Specifically, the light-emission efficiency of fluorescent materials and the electro-optic effects in optical ceramics are naturally governed very strongly by the composition and crystal grain size of the ceramic, and by the structure of the crystal grains.

It should be possible to use polycrystalline ceramic materials with good transparency in the same applications as glass and quartz. Specific examples include various lighting windows, lenses and prisms. Since the thermal resistance

TABLE 7.1.8 Application of Opto-electro Ceramics Using Some Optical Properties

Classification of optical phenomena	Typical optical property	Application
Optical with control of composition and structure	High transparency	Window for infrared light using high temperature
	Light guide property	Reflection reducing film
		Light source
		Optical wave guide (optical fiber)
		Optical circuit device (passive device)
		Optical fiber sensor
Optical phenomena based on electron transition	Optical adsorption (electrochromism, photochromism)	EC display
		Chromatic glasses
		Optical switch
	Luminescence (fluorescence, electroluminescence)	Phosphors for CRT, lump, etc.
		Solid state laser
		EL display
Optical phenomena depending on crystal structure	Electro-optic effect	Optical switch
	Magneto-optic effect	Display
	Photoelastic effect	Modulator, optical deflector
		Optical isolator
		Optical circuit device (active device)
		Optical fiber sensor

and mechanical properties of ceramic polycrystalline materials (Table 7.1.9) compare very favorably with those of alumina (melting point 2050°C) and yttria (2400°C), they are ideal for use in high-temperature viewports where thermal resistance and a high degree of transparency are both required. Applications of this sort include scientific apparatus and equipment for use in various industrial processes such as high temperature chemistry, iron manufacture and ceramics manufacture. Thesematerials are particularly useful in association with automatic control systems.

7.1.5.2 High Intensity Discharged (HID) Lamp Tubes

Although tungsten lamps, fluorescent lights, mercury tubes and the like are widely used as lighting sources, demand for more efficient lighting means has grown in recent years in order to save energy. High-pressure sodium discharge

TABLE 7.1.9 Heat-resisting Transparent Ceramics

Materials	Composition	Melting point (K)	Hardness, *Hv* (kg/mm^2)	Bending strength (kg/mm^2)
Dysprosium oxide	Dy_2O_3–10%HfO_2	≈ 2573	≈ 600	≈ 20
Erbium oxide	Er_2O_3–10%ThO_2	≈ 2673	≈ 600	≈ 20
Hafnia	HfO_2–7%Y_2O_3	≈ 2973	≈ 765	≈ 20
Lithium aluminate	$LiAl_5O_8$	—	—	—
Magnesia	MgO–1%LiF	3073	650	24
Spinel	$MgAl_2O_4$	2408	1650	17.5
Scandium oxide	Sc_2O_3	2678	900	≈ 20
Thoria	ThO_2–5%Y_2O_3	≈ 3273	970	11
Yttria	Y_2O_3	2673	700	≈ 20
	Y_2O_3	≈ 2400	≈ 700	≈ 20

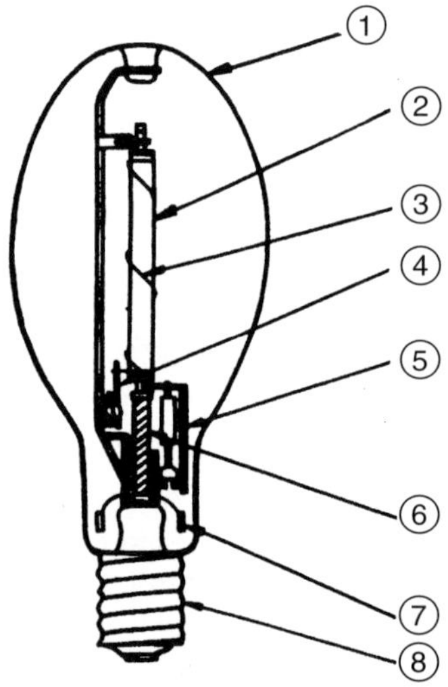

FIGURE 7.1.22 HID lamp structure. (1): outer tube (hard glass), (2): arc tube (transparent ceramics), (3): conductor for starting (heat-resistant metal), (4): bimetal switch protecting arc tube, (5): starter (heater and bimetal switch), (6): resistor generating pulse voltage, (7): getter keeping up the vacuum level, (8): cap.

lamps are highly efficient lamps that have been developed for this purpose, and useful lamps of this sort have been developed with light-emitting tubes made of transparent alumina. Other materials that can be used for light-emitting tubes besides alumina include Y_2O_3, MgO, $MgAl_2O_4$ and YAG. Figure 7.1.22 shows the structure of a discharge lamp with a light-emitting tube made of a transparent ceramic [59].

When the lamp is lit, the temperature at the outer wall of the central part of the light-emitting tube can reach as high as 1200°C, while a sodium vapor pressure of 100–300 Torr is present inside the light-emitting tube. Consequently, the alumina tube should be resistant to corrosion at high temperatures [60–62]. In general, Al_2O_3 either reacts with Na to produce β-alumina ($Na_2O \cdot 11Al_2O_3$), or is reduced by Na to produce Al and a lower oxide of aluminum. However, these reactions are very minor under the normal operating conditions of the lamp—that is, when the wall of the light-emitting tube is subjected to a load of approximately 20 W/cm^2 and the tube wall temperature is approximately 1150–1200°C.

7.1.5.3 Laser Host Materials

When a substance is irradiated with light energy, specific ions in this substance are promoted to a higher energy level. When these ions subsequently fall back to a lower energy level, laser light is emitted.

In ruby lasers, which are well known, a single crystal of alumina is doped with a small quantity of Cr^{3+} ions and laser light with a wavelength of 0.694 μm is emitted. Yttrium aluminum garnet (YAG) doped with Nd^{3+} ions emits laser light at a wavelength of 1.06 μm. Since YAG ceramic lasers were discussed in the section on synthesis, here we will describe consisting of yttria ceramics.

A laser rod that uses an yttria ceramic may have a composition such as Y_2O_3–10 mol% ThO_2–1Nd_2O_3 (NDY). This is obtained by doping a highly translucent Y_2O_3–10ThO_2 ceramic with a small quantity of Nd_2O_3, and is a solid-state laser having characteristics close to those of an Nd glass laser [3]. The optical transmittance of NDY is about 80% for a thickness of 6.3 mm at a wavelength of about 0.65 μm, which is approximately 10% lower than that of an Nd glass laser, but as a ceramic polycrystalline material it is exceptionally transparent. The wavelength of light from an NDY laser is 1.074 μm, the spectral line width is 3.3 nm, and the light emission efficiency is about 0.1%, which is lower than that of an Nd glass laser (about 0.44%). However, the thermal conductivity of NDY is about six times higher than glass, and the impurity content is also very low. It is therefore expected to be a useful material for producing long-life solid state lasers.

The transparent YAG ceramic had been developed for the high-power laser application [40], which is described in details in the previous section.

7.1.5.4 Applications with Electro-Optical Effects

As mentioned above, electro-optical ceramics such as PLZT have composition regions that exhibit: (a) memory effects; (b) primary electro-optical effects; and

(c) secondary electro-optical effects. It is therefore possible to change the characteristics by varying the composition, allowing these materials to be applied to a variety of devices. Moreover, by changing structural parameters such as the crystal grain size of the ceramic polycrystalline material, the optical scattering effects can also be made more pronounced in addition to the above effects, allowing devices to be implemented with a wide variety of configurations. These devices include image memories, display devices, optical modulators, spectral filters, optical shutters, optical gates and variable density filters. The principal applications of ceramic polycrystalline materials are discussed below.

Optical devices that use the birefringence and optical scattering effects of electro-optical ceramics can be operated in either of two basic modes. One is the transverse mode, and the other is the longitudinal mode. As Figure 7.1.23a shows, the transverse mode corresponds to the case where the electric field is applied to the electro-optical ceramic perpendicular to the direction in which light travels through it. This mode normally allows effective use to be made of changes in birefringence. Positive and negative electrodes are formed separated by suitable gaps from the surfaces of a polished ceramic plate, which is set between an analyzer and a polarizer. When light shines on the ceramic, it travels through the gap between the electrodes, so the amount of light passing

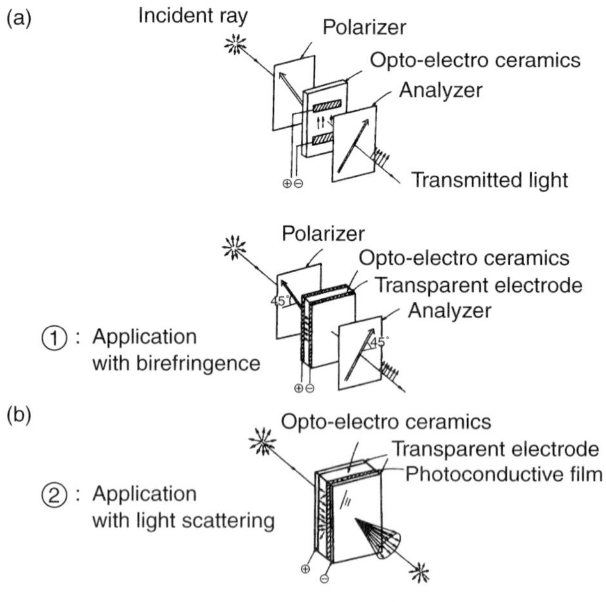

FIGURE 7.1.23 Basic operation mode of opto-electro ceramics: (a) horizontal mode; (b) vertical mode.

through it is related to the size of the gap. This mode is used in materials that exhibit primary and secondary electro-optical effects.

As Figure 7.1.23b shows, the longitudinal mode corresponds to the case where the electric field is applied in the same direction as the direction in which the light travels. Transparent electrodes such as indium oxide and tin oxide (In_2O_3–SnO_2) are formed on both sides of a polished electro-optical ceramic, and when a voltage is applied between them, polarization occurs perpendicular to the electro-optic ceramic plate. When polarizing plates are placed on both sides, the intensity of light that passes through the electro-optical ceramic plate is varied according to the electric field. Unlike the transverse mode, the intensity of the transmitted light in this mode is unaffected by the surface area of the electrodes. The mode shown in Figure 7.1.23b② is also longitudinal, and is suitable for optical memories and display elements that use electro-optical ceramics having a composition and structure that produce a large optical scattering effect.

In general, natural light such as white light sources drops in intensity by about 50% as it passes through a polarizer, so optical devices that use two polarizers tend to be rather dim. However, devices that operate based on the wide angle scattering of light do not require the use of a polarizer, making it possible to produce brighter devices.

7.1.5.4.1 Optical shutters

An optical shutter is a device that exploits the ability to change the scattering characteristics and birefringence of PZLT by applying an electric field. Figure 7.1.24 shows the configuration of an optical shutter [63]. As the figure shows, arch-shaped (interdigital) electrodes are formed on a PLZT plate, which is sandwiched between a pair of crossed polarizers. The polarization state of light that enters the PLZT by passing through the polarizer is changed according to a pulsed voltage applied to the electrodes on the PLZT plate, causing the

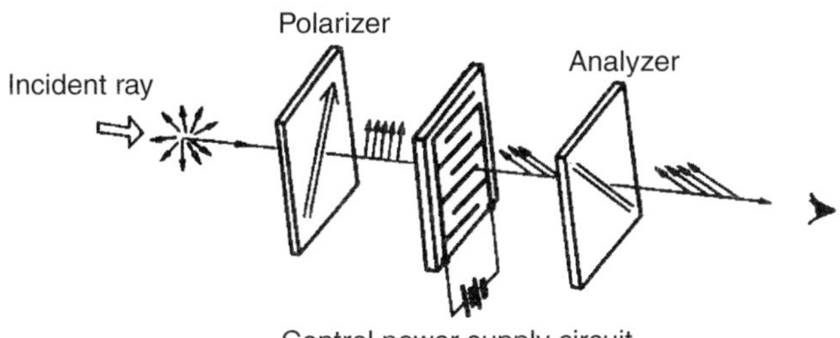

FIGURE 7.1.24 Configuration of optical shutter (on-state).

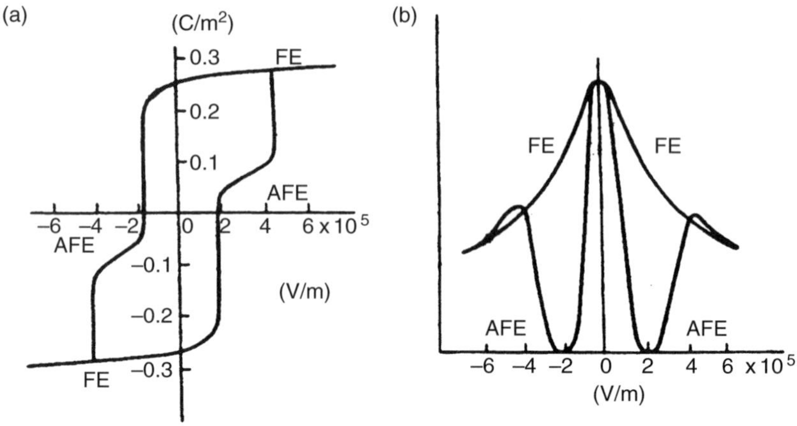

FIGURE 7.1.25 The scattering property changing due to the phase transition from AFE to FE PLZT with 8.4/65/35 composition. (a) Double hysteresis curve. (b) Electric field dependence of light scattering.

light that passes through the analyzer to be switched on and off, thereby acting as an optical shutter.

When a voltage is applied to the PLZT plate, a ferroelectric phase is induced as shown in Figure 7.1.25a, resulting in a crystal structure with a large degree of optical anisotropy which exhibits optical scattering. If the microstructure of the PLZT is observed at this time, it is found to undergo a forced phase transition from an anti-ferroelectric phase (AFE) to a ferroelectric phase (FE), causing large numbers of domain boundaries to develop inside the PLZT crystal grains.

The changes caused by this forced phase transition are reversible, and correspond to the double hysteresis curve shown in Figure 7.1.25a, with the relationship between the electric field and dispersed light intensity varying as shown in (b). If PLZT having characteristics of this sort is sandwiched between two crossed polarizers, then when it is in the AFE phase the PLZT does not scatter light and the light passing through the element is blocked by the two polarizers, corresponding to the "off" state. But when a voltage is applied and the PLZT is induced into the FE phase, it exhibits scattering and the device changes to the "on" state. The optical scattering characteristics of the AFE and FE phases depend on the size of the crystal grains in the PLZT ceramic. As Figure 7.1.26 shows, the scattering in the FE phase increases as the grains get larger, so when a PLZT ceramic with this composition is used in an optical shutter, it is advantageous to use a PLZT ceramic with as large a grain size as possible.

One example of an application for this optical shutter is in three-dimensional goggles [64, 65]. When humans look at an object, the left and right eyes see the

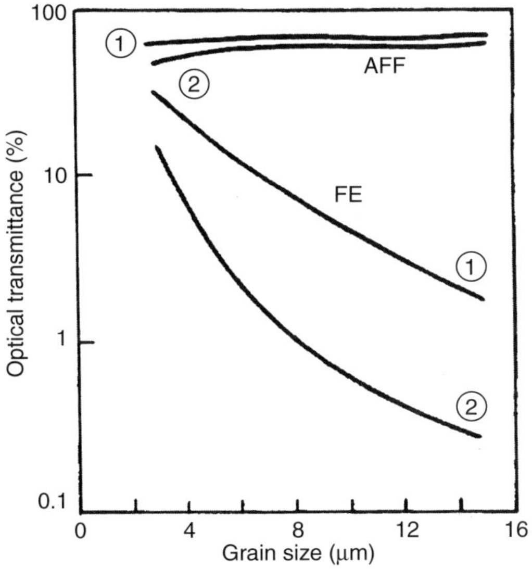

FIGURE 7.1.26 Relation between the optical transmittance and grain size in AFE and FE PLZT with 7.6/70/30 composition.

object from different angles, allowing the object to be viewed in three dimensions. This principle is applied to 3D goggles, where PLZT optical shutters are attached to the left and right eyepieces and are opened and closed alternately to allow viewing in 3D. There are a great many examples of applications for 3D goggles configured in this way.

Variable density filters have also been proposed that use a similar action to allow the intensity of transmitted light to be arbitrarily changed. For example, a PLZT welding mask has been proposed [65]. This mask uses the transverse mode shown in Figure 7.1.23a, and involves operating a shutter according to the intensity of the workplace so that when the welding starts, the optical shutter is closed according to the intensity of the sparks as soon as these sparks enter the optical shutter. The input to the PLZT is controlled according to the output from a photodiode situated close to the PLZT element. A similar principle could be employed in goggles for aircraft pilots, which would prevent them from being dazzled by intense sunlight.

7.1.5.4.2 Memory devices

PLZT has also been considered for use in image memory elements (see Figure 7.1.27) [66]. These elements use the longitudinal mode. A PLZT plate ①

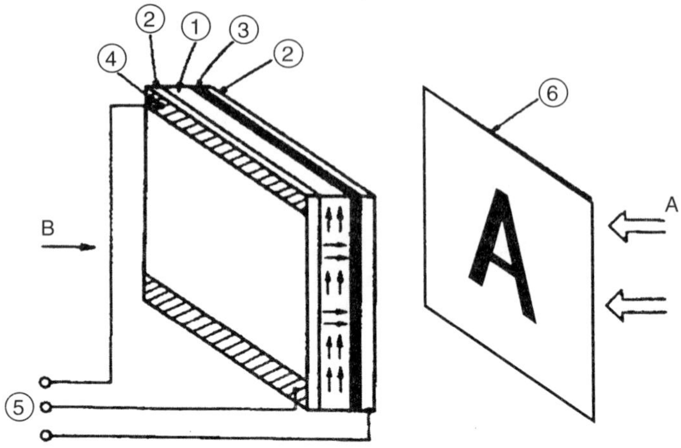

FIGURE 7.1.27 Configuration of image memory and display system with PLZT ceramics. ①, PLZT ceramics; ②, transparent electrode; ③, photoconductive film; ④, metal electrode; ⑤, lead terminal; ⑥, image mask.

with an 8/65/35 composition is optically polished on both sides to a thickness of 200 μm. A transparent electrode ② made of ($In_2O_3-SnO_2$) with a sheet resistance of about 10Ω/□ is then formed on one side, and a spinner is used to form a 1.2 μm film of polyvinyl carbazole (PVCz) on the other side. Another transparent electrode is formed on top of this film, lead terminals ② are attached to both electrodes, and the whole assembly is fixed to a secure insulating frame.

There are several ways in which this element can be used to store images. One involves shining light from direction A in the figure onto an image mask ± placed to the right of the element, and forming a silhouette of the image mask on the element. When a voltage of about 200 V is applied to the transparent electrodes, a photocurrent flows in the parts of the photoconductive film that are lit by the masked image, and at the same time a voltage is applied to the PLZT plate from both sides. The PLZT plate is subjected beforehand to polarization processing, and the direction of spontaneous polarization is aligned parallel to the PLZT plate. The parts of the PLZT plate that are struck by the light and to which a voltage is applied undergo polarization inversion corresponding to the image of the mask as shown in the figure, and the image is thereby stored for about 0.1 s. The write speed depends on the characteristics of the photoconductive film, and can be increased by using materials such as cadmium sulfide (CdS).

A second approach involves using PLZT with the same composition as that of the optical shutter discussed above, and using an electric field to induce FE–AFE phase transitions. This method uses the same element structure as in Figure 7.1.27, but instead of using light exposure to write a pattern on the ceramic plate, a ferroelectric phase pattern corresponding to the image is produced on the ceramic plate by exposing it to an electric field.

The methods used to read out and display the image stored in the memory are as follows. In the first method, the PLZT plate is inserted between a polarizer and an analyzer, and the light passing through due to birefringence (which is varied by the electric field) is projected onto a screen. The second method involves exploiting the difference between the optical scattering characteristics of the FE and AFE phases (see Figure 7.1.25b). If a parallel beam of light is directed at the element and the light that passes through the element is projected onto a screen via an aperture or the like, then an image of arbitrary size can be reproduced. Figure 7.1.28 shows an optical system for displaying this reproduced image. The purpose of the aperture is to block out the scattered light that has passed through the PLZT element, whereby a clear image can be effectively produced on the screen. This sort of image memory/display element has a maximum resolution of several tens of lines per millimeter, and is capable of producing a contrast ratio (i.e. the difference in brightness between the light and dark parts of the displayed image) of roughly 20–100.

To clear the stored image from the PLZT element, a voltage is applied to the metal electrode ④ in Figure 7.1.27, causing a current to flow in the plane of the transparent electrode and resulting in Joule heating of the PLZT element while an erasure voltage is applied. In Figure 7.1.28, a reverse phase pulse current with a period of 10 ms is made to flow in alternating directions about 10 times between gate A and gate B of the metal electrodes, and a reverse bias voltage is applied at the same time. In this way, the image can be completely erased in about 0.1 s.

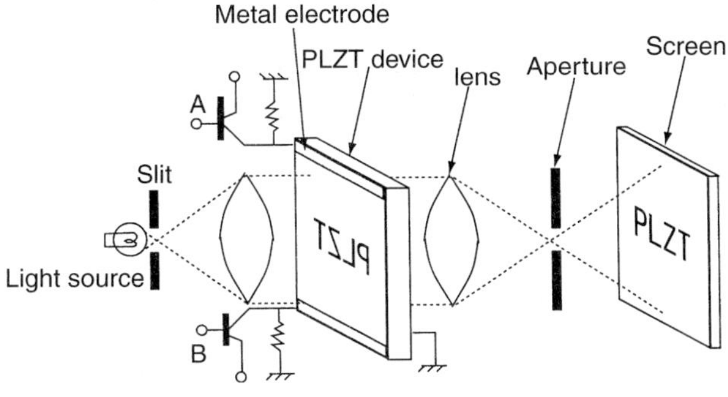

FIGURE 7.1.28 Image display and clearing system with the difference of optical scattering between the AFE and FE PLZT ceramics.

Another memory-related product that has been considered is an analog space modulator. Laser holographic memory (LHM) is a fast, high-density and low-cost optical data storage means that is expected to play a major role in the next generation of file storage devices. This sort of memory would be very useful if could record and play back image data as seen in ordinary printed material, like a sort of photocopier. However, the light reflected from printed material has no coherency, and therefore cannot be used directly in LHM systems, which require coherent signals. It is therefore necessary to somehow convert incoherent data into coherent data. A device that performs this sort of conversion function is called an analog space modulator.

Figure 7.1.29 shows the basic configuration of the modulator. Printed documents containing material such as photographs, pictures and text are lit with a source of white light, and the light reflected from these documents, which contains the information held on them, is guided to the analog space modulator that stores the image in specific parts of a PLZT element. Light from a laser is then directed at the memory element, and the transmitted light is projected onto a single point in the hologram film together with a reference light sent directly from the laser light source. This forms a hologram of the information corresponding to a single page of the printed material. Successive pages of the printed material can be stored by forming further holograms. Here, the PLZT element acts as a buffer memory that provides a temporary storage means for the formation of holograms. The PLZT element used in an analog space modulator has the same structure as shown in Figure 7.1.23b②.

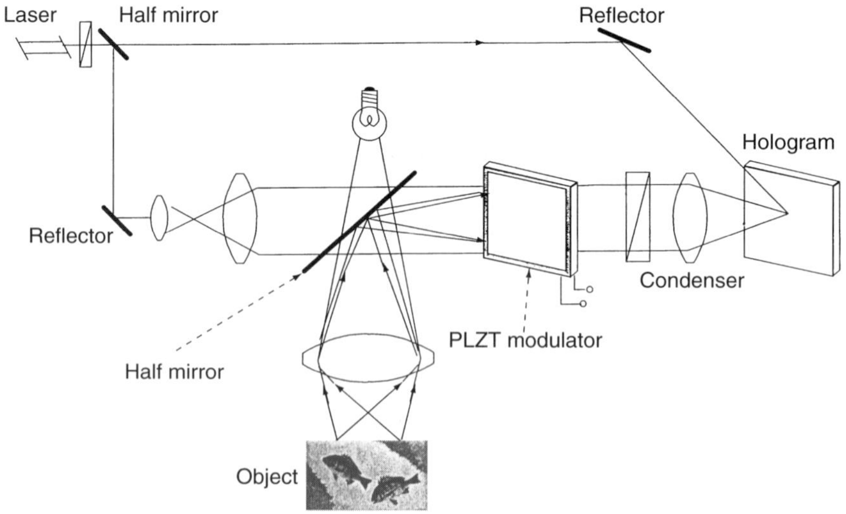

FIGURE 7.1.29 Schematic drawing for crating the hologram with PLZT modulator.

REFERENCES

1. Heartling, G. H. (1970). *Am. Ceram. Bull.* **49**: 564.
2. Heartling, G. H., and Land, C. E. (1971). *J. Am. Ceram. Soc.* **54**: 1.
3. Greskovich, C., and Chernoch, J. P. (1973). *J. Appl. Phys.* **44**: 4599.
4. Haneda, H., Yanagitani, T., Sekita, S., Okamura, F., and Shirasaki, S. (1991). *Mater. Sci. Monogr.* **66D**: 2401–2409.
5. Heartling, G. H. (1971). *J. Am. Ceram. Soc.* **54**: 303.
6. Snow, G. S. (1973). *J. Am. Ceram. Soc.* **56**: 91.
7. Snow, G. S. (1973). *J. Am. Ceram. Soc.* **56**: 479.
8. Coble, R. L. (1962). US Patent 3026210.
9. Toda, G., Noro, T., and Muta, A. (1974). *J. Jpn. Soc. Powder Powder Metall.* **21**: 76.
10. Bates, C. H., Drew, C. J., and Kell, R. C. (1971). *Trans. Br. Ceram. Soc.* **70**: 128.
11. MaCauley, J. M., and Corbin, N. D. (1979). *J. Am. Ceram. Soc.* **62**: 476.
12. Miles, G. D., Sambell, R. A. J., Rutherford, J., and Stephenson, G. W. (1961). *Trans. Br. Ceram. Soc.* **60**: 319.
13. Rice, R. W. (1971). *J. Am. Ceram. Soc.* **54**: 205.
14. Ramakrishnan, P. (1968). *Trans. Br. Ceram. Soc.* **68**: 135.
15. Smethurst, E., and Budworth, D. W. (1972). *Trans. Br. Ceram. Soc.* **71**: 45.
16. Smethurst, E., and Budworth, D. W. (1972). *Trans. Br. Ceram. Soc.* **71**: 51.
17. Bratton, R. J. (1971). *J. Am. Ceram. Soc.* **57**: 283.
18. Hamano, K., and Kanzaki, S. (1977). *J. Ceram. Soc. Jpn.* **85**: 225.
19. Kanzaki, S., Saito, K., Nakagawa, Z., and Hamano, K. (1978). *J. Ceram. Soc. Jpn.* **86**: 485.
20. Hing, P. (1976). *J. Mater. Sci.* **11**: 1919.
21. Dutta, S. K., and Gazza, G. E. (1969). *Mater. Res. Bull.* **4**: 791.
22. Tsukuda, Y., and Muta, A. (1976). *J. Ceram. Soc. Jpn.* **84**: 585.
23. Lefever, R. A., and Matsko, J. (1967). *Mater. Res. Bull.* **2**: 865.
24. Greskovich, C., and Wood, K. N. (1972). GE Rept. No. 72 CRD243.
25. Ikesue, A., Kamata, K., and Yoshida, K. (1996). *J. Am. Ceram. Soc.* **79**: 359.
26. Saito, N., Matsuda, S., and Ikegami, T. (1998). *J. Am. Ceram. Soc.* **81**: 2023.
27. Saito, N., Haneda, H., Sakaguchi, I., and Ikegami, T. (2001). *J. Mater. Res.* **16**: 2362.
28. Mazdiyasni, K. S., Lynch, C. T., and Smith II, J. S. (1967). *J. Am. Ceram. Soc.* **50**: 532.
29. Vahldiek, F. W. (1967). *J. Less-Common Met.* **13**: 530.
30. Mitomo, M. (1982). *J. Mater. Sci. Let.* 25.
31. Yamamura, H., Shirasaki, S., and Tanada, M. (1984). Japanese Patent S59-35711.
32. Yamamura, H., Tanada, M., Song, B. M., Kim, D. Y., and Shirasaki, S. (1985). *Jpn. J. Appl. Phys.* Supplement 24: 439.
33. Yamamura, H., Kuramoto, S., Haneda, H., Watanabe, A., and Shirasaki, S. (1986). *J. Ceram. Soc. Jpn.* **94**: 18.
34. Yamamura, H., Kuramoto, S., Haneda, H., Watanabe, A., and Shirasaki, S. (1986). *J. Ceram. Soc. Jpn.* **94**: 545.
35. Shirasaki, S., Nakanishi, H., and Kakegawa, K. (1984). Japanese Patent S-59-228760.
36. Debely, P. E., Gunter, P., and Arend, H. (1979). *Am. Ceram. Bull.* **58**: 606.
37. deWith, G., and van Dijk, H. J. A. (1984). *Mater. Res. Bull.* **19**: 1669.
38. Mulder, C. M., and deWith, G. (1985). *Solid State Ionics*, **16**: 81.
39. Sekita, M., Haneda, H., Yanagitani, T., and Shirasaki, S. (1990). *J. Appl. Phys.* **67**: 453.
40. Yanagitani, T., and Yagi, H. (2001). *Optronics* **20**: 168–173 (in Japanese).
41. Sekita, M., Haneda, H., Shirasaki, S., and, Yanagitani, T. (1991). *J. Appl. Phys.* **69**: 3709.
42. Ikesue, A., Yoshida, K., and Kamata, K. (1995). *J. Am. Ceram. Soc.* **78**: 2545.

43. Ikesue, A., Yoshida, K., and Kamata, K. (1996). *J. Am. Ceram. Soc.* **79**: 507.
44. Ikesue, A., Yoshida, K., Yamamoto, T., and Yamaga, I. (1997). *J. Am. Ceram. Soc.* **80**: 1517.
45. Gazza, G. E., and Dutta, S. K. (1973). US Patent 3767745.
46. Haneda, H., Yanagitani, T., Watanabe, A., and Shirasaki, S. (1990). *J. Ceram. Soc. Jpn.* **98**: 285. (in Japanese).
47. Keller, K., Mah, T., and Parthasarthy, T. A. (1990). *Ceram. Eng. Sci. Proc.* **11**: 1122.
48. Parthasarthy, T. A., Mah, T., and Keller, K. (1992). *J. Am. Ceram. Soc.* **75**: 1756.
49. Mitsuhashi, T., Haneda, H., Otani, S., Kano, S., and Yoshida, E. (1992). "Proceedings of the International Symposium on Material Chemistry in Nuclear Environment", pp. 109–116.
50. Haneda, H., Miyazawa, Y., and Shirasaki, S. (1984). Oxygen diffusion in single crystal yttrium alumiman garnet. *J. Cryst. Growth* **63**: 581–588.
51. Sakaguchi, I., Haneda, H., Tanaka, I., and Yanagitani, T. (1996). Effect of composition on the oxygen tracer diffusion in transparent yttrium aluminum garnet (YAG) ceramics. *J. Am. Ceram. Soc.* **79**: 1627–1632.
52. Haneda, H., Miyazawa, Y., Sakaguchi, I., Nozawa, H., Yanaginatni, T., and Jimenz-Melendo, M. (2001). Sc, Ga, Yb diffusion in single and polycrystalline YAG. *J. Ceram. Soc. Jpn.* **109**: 115–122.
53. Corman, G. S. (1993). Creep of yttrium aluminum garnet single crystals. *J. Mater. Sci. Lett.* **12**: 379–382.
54. Jimenez-Melendo, M., Haneda, H., and Nozawa, H. (2001). *J. Am. Ceram. Soc.* **84**: 2356–2360.
55. Blumenthal, W. R., and Paillips, D. S. (1996). High-temperature deformation of single-crystal yttrium aluminum garnet (YAG). *J. Am. Ceram. Soc.* **79**: 1047–1052.
56. Parthasarathy, T. A., Mali, T.-I., and Keller, K. (1992). Creep mechanism of polycrystalline yttrium aluminum garnet. *J. Am. Ceram. Soc.* **75**: 1756–1759.
57. Karato, S., Wang, Z., and Fujino, K. (1994). High-temperature creep of yttrium-aluminum garnet single crystals. *J. Mater. Sci.* **29**: 6458–6462.
58. French, J. D., Thao, J., Harmer, M. P., Chan, H. M., and Miller, G. A. (1994). Creep of duplex microstructures. *J. Am. Ceram. Soc.* **77**: 2857–2865.
59. Louden, W. C., and Schmitt, K. (1965). *Illum. Eng.* **60**: 696.
60. Hing, P. (1981). *J. Illum. Eng.* **10**: 194.
61. Hing, P. (1976). *J. Mater. Sci.* **11**: 1919.
62. Campbell, W. R. (1972). *J. Illum. Eng.* **1**: 281.
63. Toda, G. (1978). *Kagaku Kogyou* **29**: 160.
64. Hertling, G. H. (1981). *Am. Sci. Symp. Ser.* No. 164, 265.
65. Kato, K., Fukuhara, S., and Komada, T. (1977). *Scanning Electron Microsco.* **1**: 41.
66. Kumada, A., Suzuki, K., and Kitta, K. (1975). *Jpn. J. Appl. Phys.* **44**: 87.

Handbook of Advanced Ceramics
S. Sōmiya *et al.* (Eds.)

8.1 Superconductive Ceramics

KAZUMASA TOGANO
Institute for Materials Research, Tohoku University, Sendai 980-8577, Japan

8.1.1 INTRODUCTION

Since the discovery of superconductivity in mercury in 1911 by Kamerlingh Onnes, superconductivity was found in many other metals and in an extremely large number of metallic alloys and compounds. The highest temperature of the transition to the superconducting state (critical temperature T_c) increased in steps with year as shown in Figure 8.1.1 and had reached 23 K for Nb_3Ge by 1973. Stable cooling to achieve superconductivity for these so-called conventional metallic superconductors is possible only by using liquid helium (4.2 K), which makes the technology complex and costly. As a result, superconductivity application was limited to special ones such as the generation of strong magnetic fields, which could not be attained by using conventional copper wire.

In early 1986, Bednorz and Mueller [1] made the amazing and unexpected discovery of high-temperature superconductivity at around 35 K in La–Ba–Ca–Cu–O system. Immediately after this discovery, a group at Tokyo University succeeded to identify the superconducting phase responsible for superconductivity in Bednorz and Mueller's mixed phase samples as $(La, Ba)_2CaCu_4O_{4-x}$ (214 phase) [2]. This was an entirely new class of layered-perovskite copper oxide and the discovery initiated a rapid development in the researches in ceramic oxide superconductors. The subsequent discovery in early 1987 by Wu and Chu of the $YBa_2Cu_3O_y$ (123 phase), which was superconducting at 93 K, caused further great excitement in the scientific and commercial community [3]. In addition to this 123 material family, compounds based on bismuth ($Bi_2Sr_2Ca_2Cu_3O_x$) and thallium ($Tl_2Ba_2Ca_2Cu_3O_x$), with transition temperatures near 110 and 125 K, respectively, were discovered in rapid succession in January and February of 1988 [4, 5]. The maximum critical temperature T_c has now reached to 135 K for mercury system ($HgBa_2Ca_2Cu_3O_x$) [6].

The most important feature of these oxide ceramic superconductors is that superconductivity can be achieved in liquid nitrogen (77 K). This is very

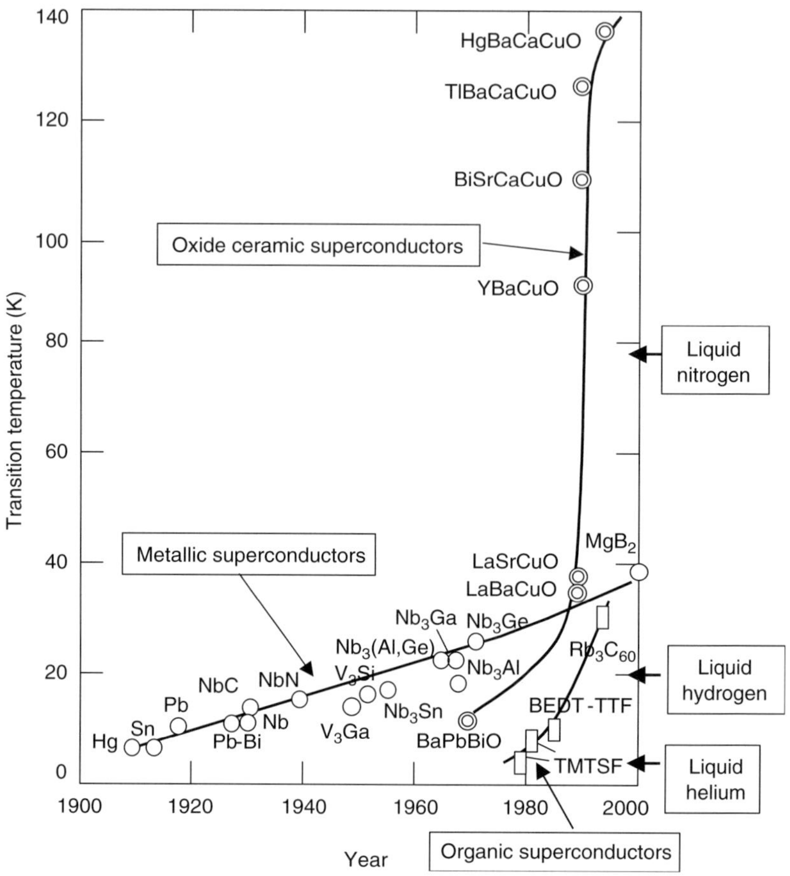

FIGURE 8.1.1 Change of superconducting transition temperature (T_c) with year.

advantageous from the viewpoint of cooling cost. Nitrogen is relatively much more abundant in the atmosphere than helium, while the major source of the world's supply of helium is limited to natural gas wells in the United States. In addition, the availability and reliability of liquid-nitrogen-cooled equipment are considerably higher than those for liquid helium. Another advantage of the ceramic high temperature superconductors is that the superconductivity can be maintained at much higher magnetic fields than the conventional metallic superconductors when they are used at the temperatures below 30 K. Therefore, the use of the ceramic superconductors offers the prospects of operating at higher magnetic fields than those used for metallic superconductors. Until now, some ceramic high temperature superconducting materials can be made into useful conductor shapes of long wires and tapes, permitting engineering scale prototype electrical machines to be demonstrated.

8.1.2 STRUCTURES

8.1.2.1 Cubic Perovskite and its Derivatives

Before the discovery of high-temperature superconductors in 1986, there already existed oxide ceramic superconductors such as $SrTiO_3$ (T_c = 0.4 K) [7], $BaPbO_3$ (T_c = 0.4 K) and $Ba(Pb, Bi)O_3$ (T_c = 12 K). They have a cubic perovskite structure. In those, the $Ba(Pb, Bi)O_3$ attracted much interest from scientists due to its high T_c despites of relatively small carrier density. In 1998, Cava *et al.* discovered that $(Ba, K)BiO_3$ with same cubic perovskite structure has an extremely higher T_c of 40 K [8] in cubic perovskite superconductors. Figure 8.1.2 shows the crystal structure of cubic perovskite. The unit cell is

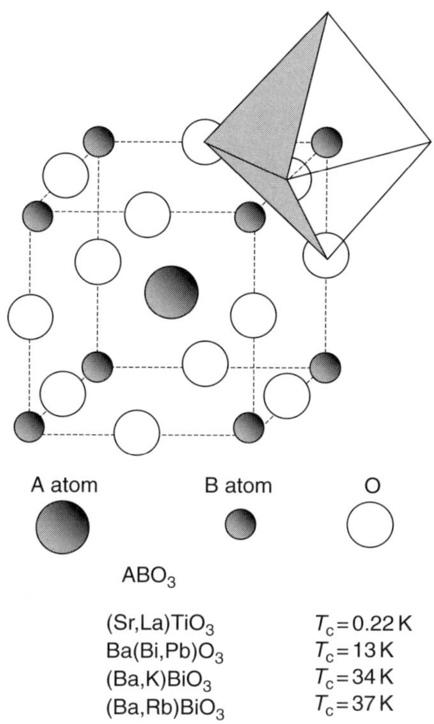

FIGURE 8.1.2 Structure of cubic perovskite.

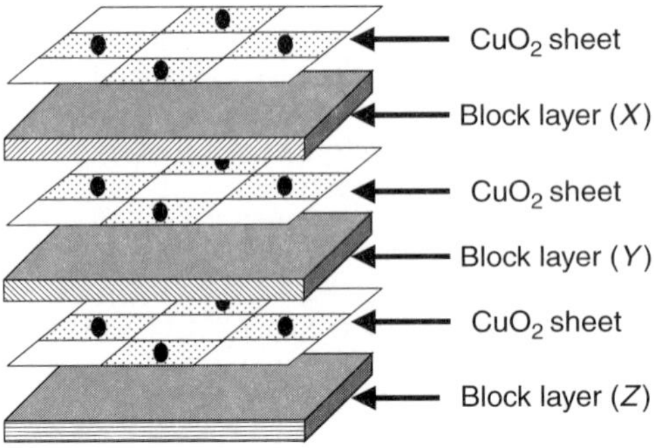

FIGURE 8.1.3 Basic structure of high-temperature oxide ceramics.

composed of eight A atoms at the corner, one B atom at the body center and six oxygen atoms at the face center. Therefore, the chemical formula is ABO_3. The A atom is also located at the center of octahedron, which is composed of six oxygen atoms at the corner.

Since the discovery of the La–Ba–Cu–O superconductor by Bednorz and Mueller, more than a dozen of copper oxide ceramics with high T_c have been synthesized. Structurally, all those copper oxide ceramics are layered perovskite structure. Basic structures composing the layered structure are cubic perovskite or its derivatives as shown in Figure 8.1.3. Important building blocks are CuO_6 octahedral, which may degenerate to CuO_5 pyramids, CuO_4 squares or Cu–O chain as a result of oxygen deficiency. All of the octahedral, pyramid and square have two-dimensional CuO_2 sheets. The crystal structure of the copper oxide high-T_c superconductors can be described as alternating stack of these CuO_2 sheets and blocks along the *c*-axis. For example, stacking of $(La, Ba)_2O_2$ block and CuO_2 sheets results in the first high-T_c compound $(La, Ba)_2CuO_4$. There are also many other structures composed of two or more kinds of blocks: the first above-90-K superconductor $YBa_2Cu_3O_{7-y}$ is composed of two blocks of the oxygen-vacant Y layer and the BaO–CuO_{1-y}–BaO block layer and a one-dimensional CuO chain. Those CuO_2 planes are crucial determinants of the physical properties of the materials, and hence, of their superconductivity.

8.1.2.2 $YBa_2Cu_3O_7$

$YBa_2Cu_3O_7$ has an oxygen-deficient complex perovskite structure with an orthorhombic crystal structure, which is shown in Figure 8.1.4. The lattice

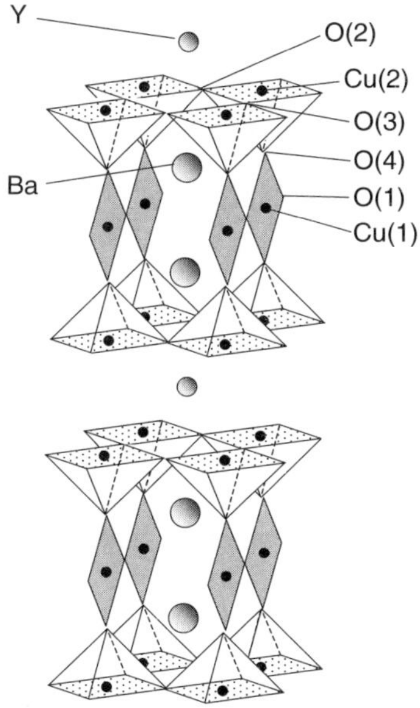

FIGURE 8.1.4 Crystal structure of $YBa_2Cu_3O_7$ superconductor.

parameters are: $a = 3.821$ Å, $b = 3.885$ Å, $c = 11.676$ Å, space group = *Pmmm*. The structure is a variant of the perovskite unit cell, in which with Y and Ba cations are ordered along the *c*-axis to give the tripled unit cell. The unit cell has two different sites of copper atom, one Cu(1) and two Cu(2) and four different kinds of oxygen atoms from O(1) to O(4) in the structure. Due to the oxygen deficiency, Cu(1) and O(1) atoms are linked in linear chains along the *b*-axis, arising orthorhombic symmetry. The Cu(2) atoms are five coordinate and are located at the apex of a pyramid, with four neighboring O-atoms (O(2) and O(3)) at a distance of ~1.9 Å forming a square CuO_2 plane along the *a–b* plane. Due to the oxygen deficiency in Y-123, there exists mixed valency in the Cu and the nominal chemical formula can be written as $YBa_2Cu_2^{2+}Cu_1^{3+}O_7$.

Although metal atom ratio in the $YBa_2Cu_3O_{7-\delta}$ structure is well defined as 1 : 2 : 3, the oxygen stoichiometry varies largely depending on the synthesizing condition. δ value in $YBa_2Cu_3O_{7-\delta}$ can vary from $\delta \sim 0.0$ to as high as 1.0 and in the stoichiometric Y-123, δ is always positive. δ values range from 0.05 to 0.35 in the materials synthesized by conventional methods such as solid-state reaction. In order to have optimum superconducting properties, δ value must be maintained at lower value. To do so, an oxygen treatment in a flowing oxygen gas at temperatures ranging from 500 to 650°C is required. The structure is orthorhombic for the δ values of $0.0 < \delta < 0.6$, while it is tetragonal for the δ

values of $0.6 < \delta < 1.0$, due to the destruction of the Cu–O chains and/or the disorder in the occupancy of the O(1) and O(5) positions.

8.1.2.3 Bismuth Cuprates

Superconductivity in a bismuth-containing cuprate was first discovered in the Bi–Sr–Cu–O system by Michel *et al.* [9], whose T_c was as low as around 10 K. However, Maeda *et al.* [4] discovered extremely high T_c value above 100 K in the Bi–Sr–Ca–O system, attracting wide attention as a new high-temperature superconducting cuprate. Subsequent work has revealed the existence of a homologous series of superconducting oxides of the ideal formula $Bi_2Sr_2Ca_{n-1}Cu_nO_{2n+4}$ with $n = 1, 2, 3, \ldots$. Figure 8.1.5 shows the structures of the homologous series. The structure is orthorhombic with $a \sim b \sim 3.8$ Å and $c \sim 18.6 + n(6.1)$ Å. The structure consists of rocksalt-like BiO-bilayers that alternate with perovskite-like $Sr_2Ca_{n-1}Cu_nO_{2n+2}$ slabs. The orthorhombic structure arises due to the unusual feature of this material of an incommensurate modulated superstructure along one of the basal plane axes.

A more common structural feature is the presence of CuO_2 sheets oriented in the *a–b* plane, which is common to all high-T_c superconducting cuprates discovered so far. The number of CuO_2 sheets in unit cell corresponds to the *n* value in the formula for the homologous series, $Bi_2Sr_2Ca_{n-1}Cu_nO_{2n+4}$. For the $n = 1$ member (Bi-2201), there is one CuO_2 sheet where each Cu coordinates to

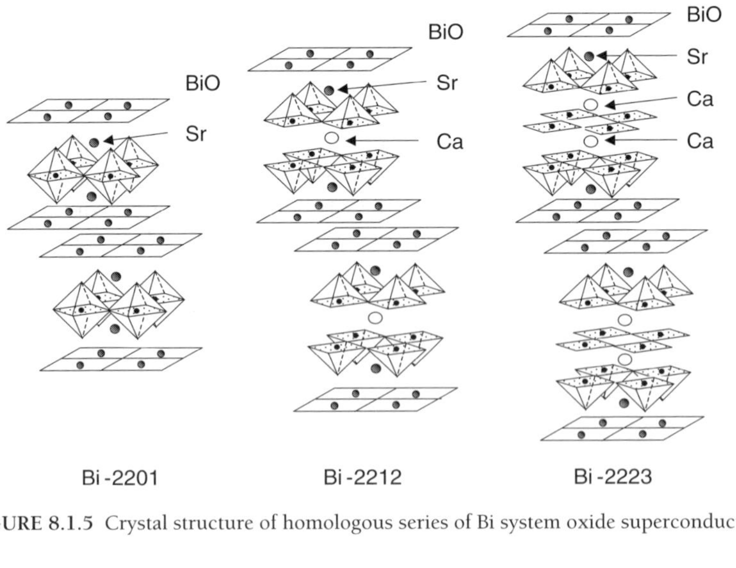

FIGURE 8.1.5 Crystal structure of homologous series of Bi system oxide superconductor.

two additional oxygen atoms above and below the sheet to form an axially elongated octahedron. The $n = 2$ member (Bi-2212) possesses double CuO_2 sheets separated by calcium atoms, the separation between CuO_2 sheets being ~3.2 Å. The $n = 3$ member (Bi-2223) has the highest T_c of 110 K and three consecutively stacked CuO_2 sheets separated by Ca. The middle layer contains planar CuO_4 units, while the two outer ones contain pyramidally coordinated Cu.

8.1.2.4 Thallium Cuprates

High-temperature superconductivity in thallium-based cuprates was discovered by Sheng and Hermann [5]. Subsequent works revealed the presence of two kinds of homologous series of $m = 1$ and 2 for $Tl_mBa_2Ca_{n-1}Cu_nO_{2n+m+2}$. Up to $n = 3$, single pahse can be easily prepared by conventional solid-state reaction. The m value corresponds to the number of rocksalt-like TlO layer and the n value corresponds to the number of CuO_2 sheet. T_c increases with the increase of n number up to $n = 3$ for both $m = 1$ and 2. The highest T_c of the Tl-based superconductors is 125 K, which was recorded for the $m = 1$ and $n = 3$ structure, $Tl_2Ba_2Ca_2Cu_3O_{10}$.

8.1.3 SUPERCONDUCTING PROPERTIES

8.1.3.1 Critical Temperature and Critical Field

The essential property of superconductors is the complete absence of electrical resistance below the transition temperature T_c. Table 8.1.1 shows the T_c values for typical high-temperature oxide superconductors. The $HgBa_2Ca_2Cu_3O_x$ shows the highest T_c of 135 K [6]. It was also reported that the T_c of $HgBa_2Ca_2Cu_3O_x$ was increased to 160 K by applying high pressure. In addition to T_c, there are two more important factors that are critical for superconductivity to occur: the critical magnetic field (H_c) and the critical current density (J_c). Of these, T_c and H_c are intrinsic properties of the material while J_c can be altered by processing procedures. As shown in Figure 8.1.6, these parameters represent a three-dimensional surface of maximum values above which superconductivity does not exist. This clearly shows that the critical value of each parameter (T, H, and J) for superconductivity varies with the other two parameters. For example, the critical current density (J_c) declines with increasing H and T.

Depending on the response to applied magnetic field, the superconductors are divided to two categories: type I and type II as shown in Figure 8.1.7. The type II superconductors show more complicated magnetic behavior, however,

TABLE 8.1.1 T_c Values for Typical High-Temperature Oxide Superconductors

Copper oxide	T_c (K)
$(La, Ba)_2Cu_4$	35
$(La, Sr)_2Cu_4$	40
$YBa_2Cu_3O_7$	92
$(RE)Ba_2Cu_3O_7$	62–94
(RE = La, Nd, Sm, Eu, Gd, Dy, Ho, Er, Tm, Yb, Lu)	
$Bi_2Sr_2CaCu_2O_x$	80
$(Bi, Pb)_2Sr_2Ca_2Cu_3O_x$	110
$TlBa_2CuO_x$	95
$TlBa_2CaCu_2O_x$	82
$TlBa_2Ca_2Cu_3O_x$	118
$Tl_2Ba_2CuO_x$	87
$Tl_2Ba_2CaCu_2O_x$	114
$Tl_2Ba_2Ca_2Cu_3O_x$	125
$HgBa_2CuO_x$	95
$HgBa_2CaCu_2O_x$	128
$HgBa_2Ca_2Cu_3O_x$	135

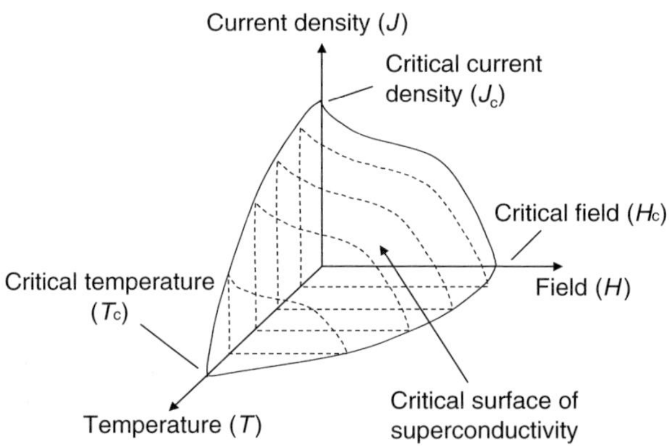

FIGURE 8.1.6 Three-dimensional surface of superconductivity.

and are more practically important. When a type II superconductor is cooled below its T_c and subjected to an increasing magnetic field, the field is initially expelled from the superconductor by the Meissner effect up to lower critical field H_{c1}. When the applied field exceeds H_{c1}, the flux begins to penetrate the superconductor as quantized fluxoid or vortex, which is essentially a microscopic thread-like region of normal state through which the magnetic field is channeled. This state of the coexistence of the normal regions of flux

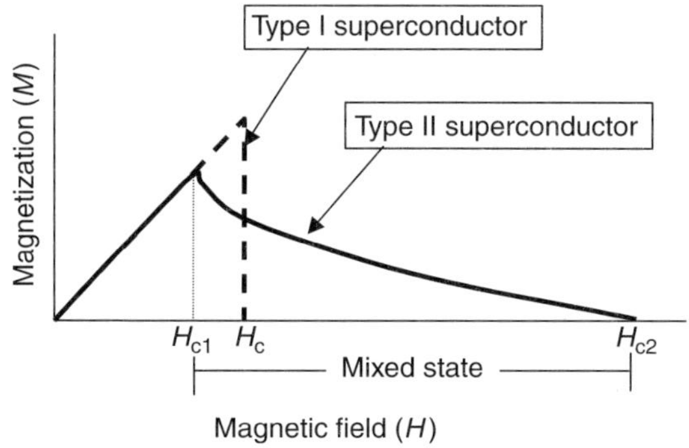

FIGURE 8.1.7 Type I and type II superconductivities.

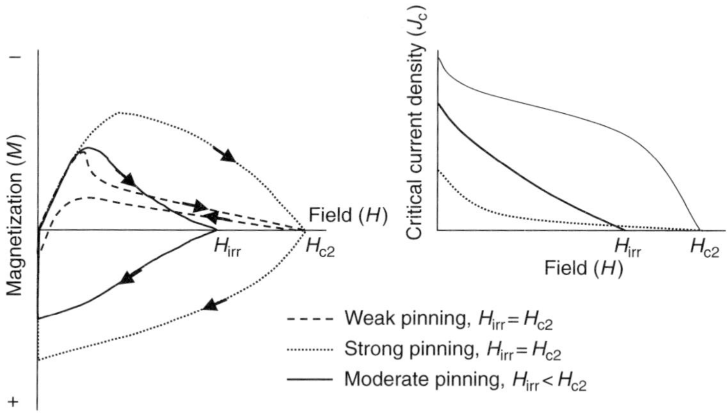

FIGURE 8.1.8 Schematic diagrams showing the relation of magnetization hysteresis ΔM and critical current density J_c.

lines and a superconducting matrix region is called as mixed state and continues up to upper critical field H_{c2}, where the whole region becomes normal state.

The technologically important feature of type II superconductors is that upper critical field H_{c2} is much larger than the thermodynamic critical field H_c. When current is passed along the type II superconductors in the mixed state, the flux lines are forced to move by Lorentz force. The movement of flux causes the appearance of voltage and hence energy dissipation. However, the zero resistance state can be achieved when the motion of flux lines is completely prevented by pinning centers against the Lorentz force. Therefore, larger pinning force yields larger critical current density. The strength of the pinning force can be also evaluated by the hysteresis curve in the magnetization measurement in increasing and decreasing external fields as shown in Figure 8.1.8.

If the flux can move freely without any pinning force, the magnetization curve is completely reversible. On the other hand, if the material has any pinning force, the curve becomes irreversible showing the hysteresis, which is caused by the difference in flux distribution inside the specimen between in the increasing and decreasing applied fields. The hysteresis becomes larger for larger pinning force. The pinning sites in superconductors are the various types of defects such as grain boundary, inclusions, lattice defects and so on. Therefore, J_c can be varied by the microstructure of the material and hence is not a intrinsic property of the material.

8.1.3.2 Irreversibility Field

In case of conventional metallic superconductors, the materials usually have any pinning force effective up to the field close to H_{c2} and hence the magnetization curve is irreversible almost up to H_{c2}. However, in case of oxide high-temperature superconductors, the hysteresis becomes almost zero far below H_{c2}. The magnetic field that separates reversible and irreversible region

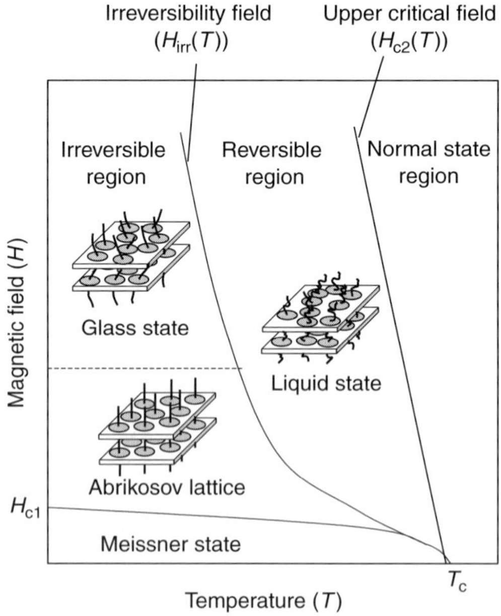

FIGURE 8.1.9 Schematic H–T phase diagram of Cu oxide high-temperature superconductor.

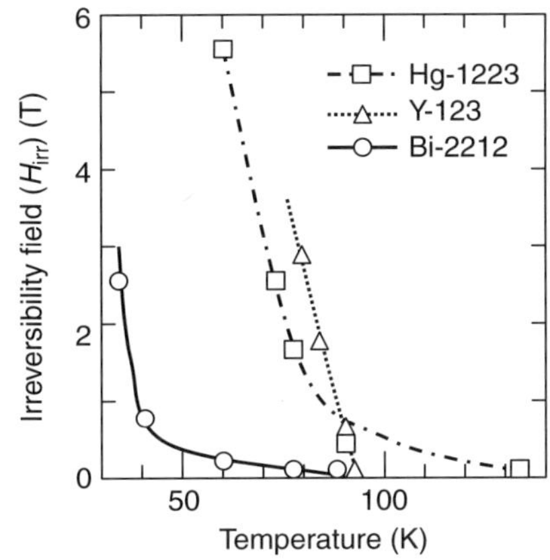

FIGURE 8.1.10 H_{irr}–T curves for three high-temperature Cu oxide superconductors.

is called as irreversibility field H_{irr}. In the H–T phase diagram, H_{irr} is the line as shown in Figure 8.1.9. Since type II superconductors cannot be used beyond this line, the H_{irr} is much more practically important than H_{c2}, particularly in the case of high-temperature oxide superconductors.

Figure 8.1.10 shows the H_{irr}–T curves for various types of high-temperature superconductors. The H_{irr} of Bi-based superconductors is much lower than that of Y-123 superconductors. This is due to the larger anisotropy in Bi-based superconductors. The Cu–O planes in the Bi-based superconductors are more weakly coupled, making the flux line movement much easier. Therefore, the Bi-based superconductors are not applicable for magnetic field generation in liquid nitrogen. The improvements of H_{irr} for Bi-base superconductors are now being attempted by various methods such as irradiation, oxygen control, Pb-doping and so on.

8.1.4 FABRICATION METHODOLOGY

8.1.4.1 Wires and Tapes

8.1.4.1.1 Bi-2223 and Bi-2212

Wire production process has been well developed for Bi-based high-temperature superconductors, Bi-2223 and Bi-2212. There are two major techniques for the production of Bi-2223 and Bi-2212 long conductors; powder-in-tube (PIT) method and ceramic coating methods.

The PIT process was one of the first to be developed for making high-temperature superconductors [10]. The PIT process is now in use in many laboratories and companies around the world. It is sometimes used for Bi-2212 and Tl-1223 and is almost always used for processing Bi-2223 into a conductor. Silver or silver alloys are used for the tube of PIT process. This is because silver is permeable to oxygen and nonreactive with the superconducting oxide in the core. Silver also lowers the melting point of Bi-based materials during thermal processing and forms a template upon which the crystal of Bi-based superconducting phase can grow. Figure 8.1.11 shows the standard PIT process. The tube is filled with oxide powder, then extruded or drawn to wire about 1–2 mm diameter. For the fabrication of multifilamentary conductor, the wire is drawn in a hexagonal shape, cut into shorter lengths, and formed into a stack. The stack is then inserted in another tube, and the composite is extruded or drawn to wire. In case of Bi-2223, the wire must be flattened into a tape shape by rolling process with several times intermediate annealings at around 845°C. During the rolling process, the Bi-2223 grains with flat shape are forced to be oriented parallel to the tape surface; that is, parallel to the current flow. The final aspect ratio is typically 10 : 1. Tl-1223 is processed similarly to Bi-2223.

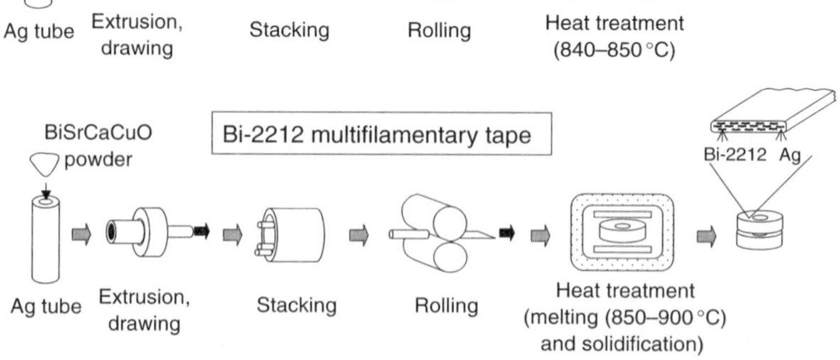

FIGURE 8.1.11 Standard PIT processes for Bi-2223 and Bi-2212 multifilamentary superconducting tapes.

Bi-2212 is subjected to a partial melting process at 850–900°C at the final stage of the process [11]. During the partial melting, Bi-2212 grains with flat shape grow from the liquid phase in the direction parallel to the interface with the metal sheath. This results in a nice texturing of *a–b* plane parallel. Both Bi-2223 and Bi-2212 are highly anisotropic materials, and the J_c within the *a–b* plane may be several orders of magnitude larger than J_c along the *c*-axis. Therefore, obtaining good *c*-axis orientation of the grains in these materials is necessary to achieve high J_c. Using the PIT process, kilometer-length wires and tapes are now fabricated by many companies and being used for various prototype application systems. Figure 8.1.12 shows the cross-sections of the Bi-2223 and Bi-2212 tapes and wires.

Another process used for manufacturing superconductors employs typical ceramic techniques such as dip coating or doctor-blade coating of the Bi-2212 material mixed into a slurry with an organic binder and a solvent onto

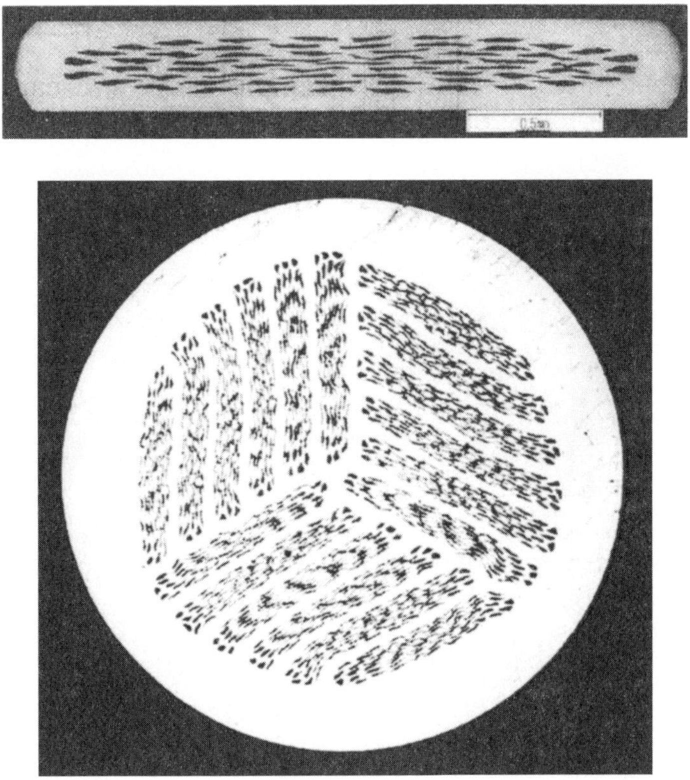

FIGURE 8.1.12 Cross-sections of PIT processed Bi-2223 (upper: Sumitomo Electric) tape and (lower: Hitachi Cable) Bi-2212 round wire in which 18 multifilamentary tapes are assembled.

a substrate, typically silver or silver alloy [12]. After the coating, the dip-coated tape is subjected to heat treatment to remove the organic materials and subsequently the heat treatment of partial melting and slow cooling to form large grain and *c*-axis oriented Bi-2212 on the substrate tape.

The highest J_cs at 77 K and 0 T obtained so far are 7×10^4 and 4×10^4 A/cm^2 for Bi-2223 and Bi-2212 tapes, respectively. However, the J_cs rapidly decrease in applied magnetic field, making the application of high-field generation in liquid nitrogen very difficult. However, the Bi-2212 and Bi-2223 conductors show very nice J_c characteristics in applied magnetic field at the temperatures below 30 K as shown in Figure 8.1.13. The J_c value of Bi-2212 tapes is well over 10^5 A/cm^2 at 4.2 K in strong magnetic fields higher than 20 T. This makes the Bi-2212 wires attractive for the applications of superconducting magnet generating above 23 T, which cannot be attained by conventional metallic superconductors.

8.1.4.1.2 Y-123 Coated Conductor

In the fabrication of Y-123 conductor, both *c*-axis and in-plane alignments are required to achieve high transport J_c of the wire, making the fabrication

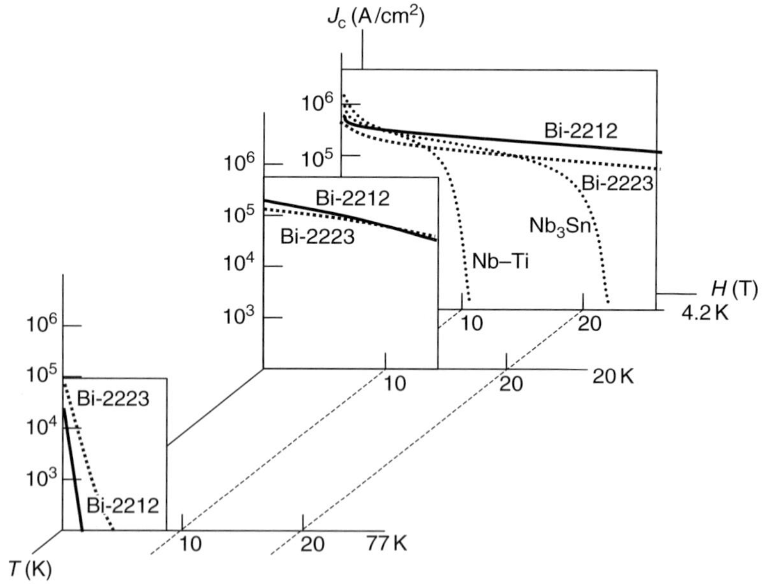

FIGURE 8.1.13 J_c–H curves of PIT processed Bi-2223 and Bi-2212 superconducting tapes at temperatures of 4.2, 20, and 77 K. J_c–H curves of conventional Nb–Ti and Nb_3Sn wires are also included for comparison.

process of long Y-123 wire more complicated. This is largely different from Bi-based superconductors for which only *c*-axis alignment is required. The classic experiment by Dimos *et al.* [13] showed that the J_c drops by an order of magnitude for current transfer between grains with parallel *c*-axes, but with the *a*-axis misoriented by only 14°.

Although the best J_c values for Y-123 are obtained for thin films grown epitaxially on single crystal substrates, such as $SrTiO_3$ or $LaAlO_3$, they are impractical for long conductors fabrication due to high cost, small size, and poor mechanical properties. There are two major techniques to produce long tapes with biaxial grain alignment as shown in Figure 8.1.14, which are intensively being developed in the world. The first process successfully examined is using a buffer layer of yttrium-stabilized zirconia (YSZ) deposited on a commercially available stainless steel tape [14]. IBAD technique was successfully used to achieve a biaxial texturing for YSZ buffer layer on which a biaxially textured Y-123 was grown epitaxially by conventional deposition techniques such as pulse laser deposition (PLD), sputtering and chemical vapor deposition (CVD). A J_c of 10^5 A/cm^2 has been achieved for 1 m long Y-123 tape processed by IBAD technique. The other process is the epitaxial deposition on a textured metallic substrate. The metallic substrate tape such as Ni and Ag can be biaxially textured by thermomechanical processing, typically a very large rolling reduction followed by heating to obtain preferred grain orientation. Y-123 film is then deposited directly on the

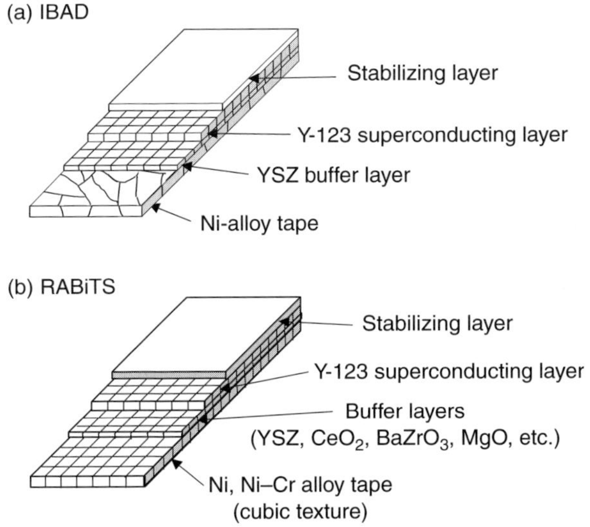

FIGURE 8.1.14 Conductor structures of IBAD processed and RABiT processed YBCO-coated conductors.

metallic substrate or on any intermediate buffer layers formed on the metallic substrate. This is the so-called RABiT process [15]. A protective coating may be made over the high-temperature superconductor film to protect it from the environment and to facilitate attachment of electrical contacts.

8.1.4.2 Bulk Materials

The idea of using bulk superconductors for practical applications, such as flywheels, bearings, current leads and so on, is unique for high-temperature superconductors. Bulk materials of Bi-2212 and Bi-2223 superconductors are already used for current leads for several application systems. The materials are prepared either by sintering a powder compact of rod or cylinder shape or by subjecting a partial melting process to the sintered rod or cylinder. The product is not nearly as highly textured as that from the PIT or dip coating method, but since it has a much larger cross-sectional area, it is able to carry a substantial current.

For Y-123 bulk material, it is important to grow a large grain (>1 cm) in order to achieve larger shielding current pass and hence larger remnant magnetic field [16, 17]. The technique is basically same with those of semiconductor industry; the crystal growth from the liquid in thermal gradient. Seeding with single crystals of Sm-123 has also been effective in nucleating the growth of large Y-123 grains. Various additions, such as Pt and Y_2BaCuO_5 (Y-211), result in enhanced pinning force and hence stronger levitation force for Y-123. Bulk YBCO material approximately 10 cm in diameter and 1–2 cm thick can grown by this technique.

In addition to Y-123, rare earth-123 (RE-123) is also a promising material for bulk applications. However, the material must be grown under controlled low oxygen pressure to prevent site exchange of Nd and Ba that would results in low T_c. Processing to control site exchange and to produce precipitates of Nd-422 phase yields higher J_c values at 1–3 T and 77 K than can be achieved with Y-123.

8.1.4.3 Thin Films

Since oxide high-temperature superconductors have a strong crystalline and electronic anisotropy, the materials must be synthesized in the form of oriented, virtually single crystal films. Furthermore, due to the extremely short coherence length of those materials, the quality of the films must be well controlled at the surface or interface as well as inside the film for device applications. The critical current density is one of the indications of the quality and the highest J_c obtained so far is $>5 \times 10^6$ A/cm^2 at 77 K in 0 T.

Most works on the thin films of high-temperature superconductors have been carried out for Y-123 and their families of RE-123, because of relative easiness of synthesizing high-quality films. Various deposition techniques have been examined on various substrate such as MgO, ZrO_2, $SrTiO_3$, $LaAlO_3$, and $NdGaO_3$. Substrate wafers 4 in. in diameter can now be obtained. Heteroepitaxial multiplayer structures of Y-123 and semiconducting PrBCO or insulating perovskite ($SrTiO_3$, $LaAlO_3$) have also been made. At present time, Y-123 seems to be most promising for microelectronic device applications. Very recently, Y-123 film with extremely high quality for wide area was successfully synthesized by triphase epitaxy method.

Advances have been also made in the fabrication of BiSrCaCuO films. Bi-2212 is better than Bi-2223 for the fabrication of high-quality thin films due to its thermodynamic stability. The most important issue in the fabrication of high-quality Bi-2212 films is to avoid the intergrowth of homologous structures such as Bi-2201 and Bi-2223.

8.1.5 APPLICATIONS

The most important reason for evaluating the new high-T_c ceramic superconductors for practical application in electric power systems and electronic devices is that the critical temperature of these materials is above the boiling point of liquid nitrogen (77 K). At temperatures of 77 K or higher, considerable simplification and cost savings for the refrigeration system using liquid nitrogen are obtained compared with refrigeration using liquid helium. Another feature of the high-T_c ceramic superconductor is the higher upper critical field, H_{c2}, over those of conventional practical superconductors of Nb–Ti and Nb_3Sn. This makes the high-T_c superconductors attractive for the generation of magnetic field even when they are used in lower temperatures of liquid helium or refrigerator operation.

8.1.5.1 Power Application

Huge amounts of energy are lost in various electric power applications due to the presence of electrical resistance of conventional wires such as copper wire. Electric power systems made from the coils wound by high temperature superconducting wires are expected to have reduced volume and mass, enhanced performance and improved operating efficiency as a result of their larger currents, generated higher fields and smaller resistance losses. Therefore, various kinds of electric power systems are being developed by using high temperature superconductors as shown in Figure 8.1.15.

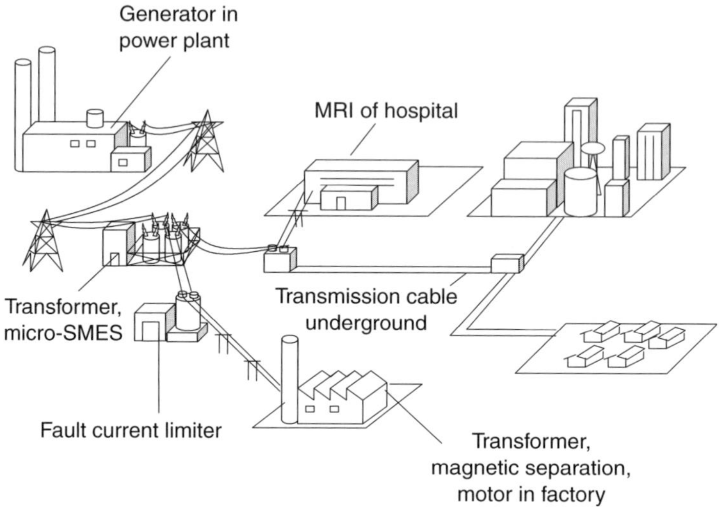

FIGURE 8.1.15 Various kinds of power applications being developed by using high-temperature oxide superconductors.

Major advantages of motors and generators made from the high-temperature ceramic superconducting wires are reduced weight and size as well as higher efficiency. However, alternating current loss in the superconductor may limit the use of high-temperature superconducting wire in certain motor types. Transformers using high-temperature superconducting wires are a more realistic application. In addition to higher efficiency, less flammability and less environmental impact are also great advantages of the high-temperature transformers. Figure 8.1.16 shows the 500 kVA transformer developed by the group of Kyushu University using Bi-2223 multifilamentary tape [19].

Transmission and distribution cables made from the superconductors are expected to have increased current capacities. The application of high-temperature transmission cables is particularly attractive in dense urban area where the demand for electric power is rapidly increasing while the space is extremely limited and most underground ducts are filled to capacity. The replacement of conventional cable with high-temperature superconducting cable is planned in Japan and the United States. Figure 8.1.17 shows the 7-m prototype high-temperature transmission cable developed by Sumitomo Electric Co. and TEPCO using Bi-2223 wire [20].

High-temperature ceramic superconductors are also promising for the application of fault current limiter, which protects the power transmission system from the damages caused by unanticipated power disturbances. The high-temperature superconducting current limiter has tremendous benefits of

FIGURE 8.1.16 500 kVA transformer made of Bi-2223 coils [19].

enhanced operating capacity, increasing safety, reliability and power quality, because it can limit the fault current without adding impedance to the circuit during normal operation. Figure 8.1.18 shows the schematic drawing of inductive type fault current limiter, in which the high-temperature superconducting wire is used for the winding of a coil. The use of high-temperature superconducting films such as YBaCuO film is also being developed for the application of fault current limiter.

Bulk superconductors of $YBa_2Cu_3O_7$ are being used for flywheel energy storage. Conventional flywheel has a problem of the energy loss in the bearing. One solution of this problem is to use the repulsive force between the high-temperature superconductor and permanent magnet. Figure 8.1.19 shows a conception of the flywheel energy storage using high-temperature bulk superconductors. Superconducting magnet bearings have demonstrated losses of 10^{-2}–10^{-3} W/kg for a 2000 rpm rotor.

8.1.5.2 High Magnetic Field Generation

High-temperature superconductors have an important role to play in the development of superconducting magnet. Since Bi-2212 and Bi-2223 tapes show

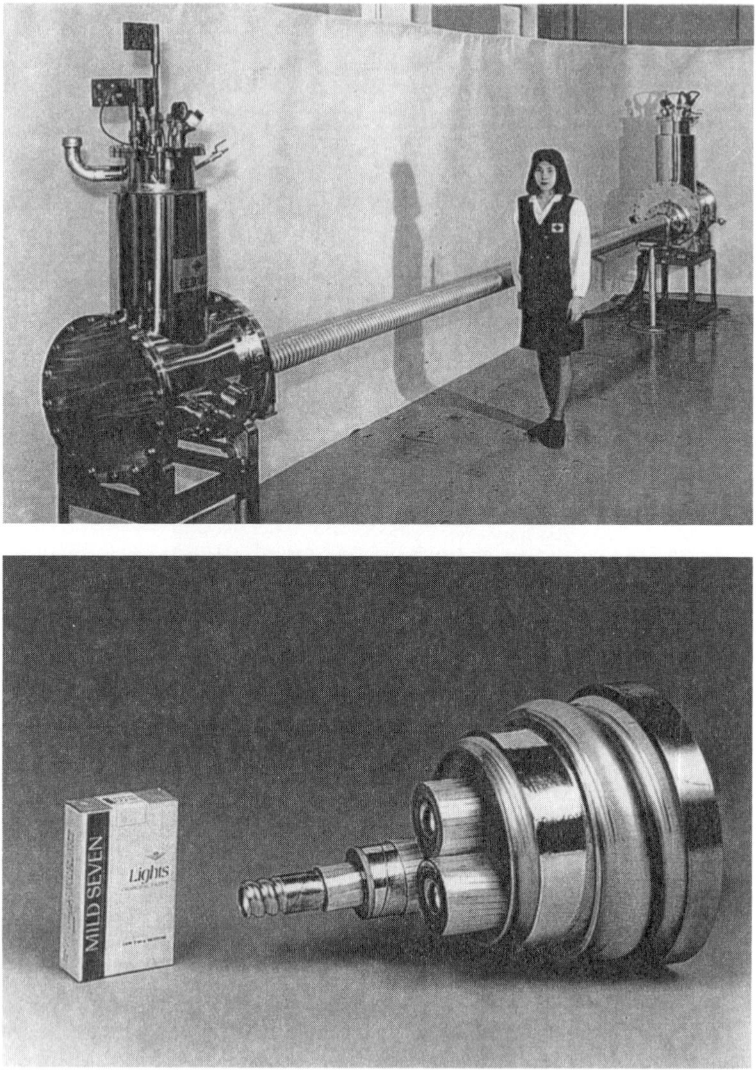

FIGURE 8.1.17 A 7-m prototype high-temperature transmission cable system and cut model of the cable made by using Bi-2223 tapes [20].

excellent high-field performance at the temperatures below 20 K, they are being used for the construction of "cryogen-free" conduction-cooled superconducting magnet operated by refrigerator. Figure 8.1.20 shows a 10 T superconducting magnet made by using Bi-2212 multifilamentary wire for windings. The current

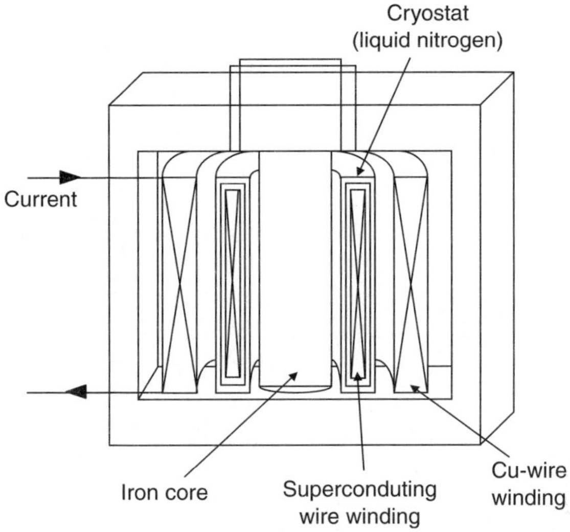

FIGURE 8.1.18 Schematic drawing of inductive type high-temperature superconducting fault current limiter.

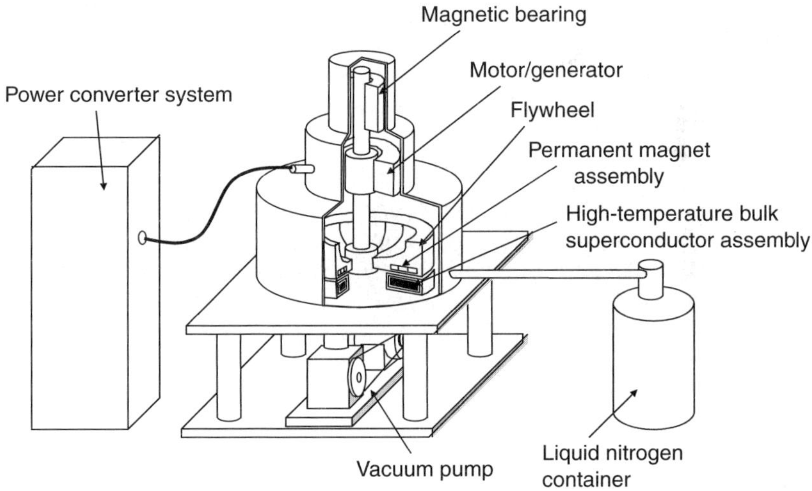

FIGURE 8.1.19 Conception of flywheel energy storage using high-temperature bulk superconductors.

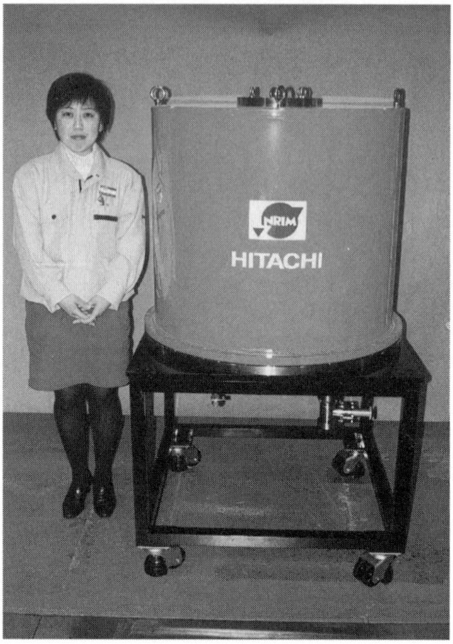

FIGURE 8.1.20 Cryocooled 10 T superconducting magnet made by using Bi-2223 tapes (by courtesy of Hitachi and Hitachi Cable).

leads of high-temperature superconductors also play an important role in the development of helium-free conduction cooled superconducting magnets.

There are several attempts to generate extremely high magnetic field in liquid helium using BiSrCaCuO superconducting wires. Their idea is to use conventional metallic superconducting coils of Nb–Ti and Nb_3Sn as the outer magnets and the BiSrCaCuO coil as the insert magnet as shown in Figure 8.1.21. In this case, the insert magnet plays the role of booster to generate the magnetic field well over 23 T, which can never be attained by conventional metallic superconductors. Projects using 23.5 T superconducting magnets for 1 GHz nuclear magnetic resonance analyzing system (1 GHz NMR) are under progress at the Tsukuba Magnet Laboratory of NIMS [21], Japan and HFML of Florida State University, USA.

8.1.5.3 Electronic Devices

The application of high-temperature oxide superconductors for electronic devices is being successfully performed for superconducting quantum

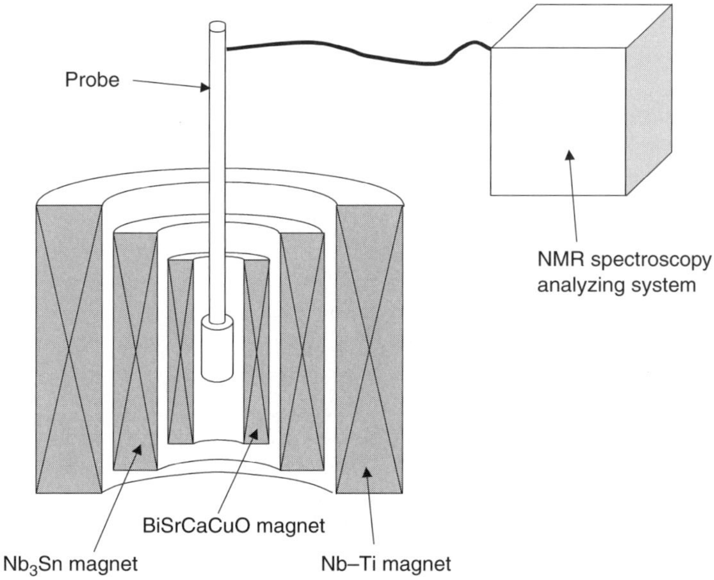

FIGURE 8.1.21 Configuration of high field superconducting magnet using Bi-2212 insert magnet. The magnet is being used for 1 GHz NMR system.

FIGURE 8.1.22 Chip of high-T_c SQUID [22].

interference device (SQUID) magnetometers and filters for wireless telecommunications. These systems are being developed by using $YBa_2Cu_3O_7$ thin films. Figure 8.1.22 shows the chip of high-T_c SQUID [22]. The high-T_c SQUID has a great potential for applications in medical diagnostics, nondestructive tests and geological survey.

The so-called "intrinsic Josephson junction" in BiSrCaCuO superconductors has also potential to be used for new type of superconducting device [23]. The crystal structure of BiSrCaCuO superconductors is the alternative stacking of superconducting oxide layer containing CuO sheet and nonsuperconducting (or weakly superconducting) oxide layer. This means the structure naturally contains the stacked series arrays of SIS junction. This intrinsic Josephson junction would be useful for high-frequency applications such as Josephson voltage standards, mixers, and oscillators. For such applications, large and high-quality single crystal is required.

REFERENCES

1. Bednorz, J. G., and Mueller, K. A. (1986). *Z. Phys. B* **64**: 189.
2. Uchida, S., Takagi, H., Kitazawa, K., and Tanaka, S. (1987). *Jpn. J. Appl. Phys.* **26**: L1.
3. Wu, M. K., Ashburn, J. R., Torng, C. J., Hor, P. H., Meng, P. L., Gao, L., Haang, Z. J., Wang, Y. Q., and Chu, C. W. (1987). *Phys. Rev. Lett.* **58**: 908.
4. Maeda, H., Tanaka, Y., Fukutomi, M., and Asano, T. (1988). *Jpn. J. Appl. Phys.* **27**: L209.
5. Shen, Z. Z., and Hermann, A. M. (1988). *Nature* **332**: 138.
6. Shilling, A., Cantoni, M., Gao, D., and Ott, H. R. (1993). *Nature* **365**: 56.
7. H. Suzuki *et al.* (1996). *J. Phys. Soc. Jpn*, **65**: 1529.
8. Cava, R. J., Batlogg, B., Krajewski, J. J., Farrow, R., Rupp, L. W. Jr, White, A. E., Peck, W. F. Jr, and Kometani (1988). *Nature* **332**: 814.
9. Michel, C., Hervieu, M., Borel, M. M., Grandin, A., Deslands, F., Provost, J., and Raveau, B. (1987). *Z. Phys. B* **68**: 421.
10. Yamada, Y., Fukushima, N., Nakayama, S., Yoshino, H., and Murase, S. (1987). *Jpn. J. Appl. Phys.* **26**: L865.
11. Kase, J., Irisawa, N., Morimoto, T., Togano, K., Kumakura, H., Dietderich, D. R., and Maeda, H. (1990). *Appl. Phys. Lett.* **56**: 970.
12. Kase, J., Togano, K., Kumakura, H., Dietderich, D. R., Irisawa, N., Morimoto, T., and Maeda, H. (1990). *Jpn. J. Appl. Phys.* **29**: L1096.
13. Dimos, D., Chaudhari, P., Manhart, J., and Legoues, F. K. (1988). *Phys. Rev. Lett.* **61**: 219.
14. Iijima, Y., Tanabe, N., Kohno, O., and Ikeno, Y. (1992). *Appl. Phys. Lett.* **60**: 769.
15. Goyal, A., Norton, D. P., Budai, J. D., Paranthanman, M., Specht, E. D., Kroeger, D. M., Christen, D. K., He, Q., Saffian, B., List, F. A., Lee, D. F., Martin, P. M., Klabunde, C. E., Hartfield, E., and Sikka, K. (1996). *Appl. Phys. Lett.* **69**: 1795.
16. S. Jin *et al.* (1989). *Phys. Rev. B* **37**: 7859.
17. M. Murakami *et al.* (1989). *Jpn. J. Appl. Phys.* **28**: 1189.
18. Y. Shiohara *et al.* (1997). *Mater. Sci. Eng.* **R19**: 1.
19. Funaki, K., Iwakuma, M., Takeo, M., Yamafuji, K., Suehiro, J., Hara, M., Konno, M., Kasagawa, Y., Itoh, I., Nose, S., Ueyama, M., Hayashi, K., and Sato, K. (1997). *IEEE Trans. Appl. Supercond.* 7: 824.
20. Fujikami, J., Saga, N., Ohmatsu, K., Shibata, T., Isojima, S., Sato, K., Ishii, H., and Hara, T. (1996). *Advances in Superconductivity VIII*, p. 862, Tokyo: Springer-Verlag.
21. K. Inoue *et al.* (1996). Proceedings of the 16th CEC/ICMC, Kitakyushu, p. 1103.
22. Sumitomo Electric Hightechs, High T_c SQUID, Catalog http://www.shs.co.jp.
23. Kleiner, R., Steinmeyer, F., Kunkel, G., and Mueller, P. (1992). *Phys. Rev. Lett.* **68**: 2394.

PART 2

Engineering Ceramics

Handbook of Advanced Ceramics
S. Sōmiya *et al.* (Eds.)

9.1 High-Temperature High-Strength Ceramics

KAORU MIYAHARA, YASUHIRO SHIGEGAKI
and TADASHI SASA
Ishikawajima-Harima Heavy Industries Co., Ltd., Shin-nakahara-cho, Isogo-ku, Yokohama 235-8501, Japan

9.1.1 INTRODUCTION

The energy crisis in the 1970s made the industrialized countries aware of the importance of the effective utilization and conversion of energy for the first time. Since the late 1980s, the efficient utilization of resources and the protection of environments have been widely recognized as the most crucial issues in the global scale. As one of the most promising solutions to the problems, attention has been given to ceramics with high temperature durability due to their highly probable contribution to the heat-engine efficiency improvements. From this point of view, tremendous efforts have been carried out in the area of science and engineering of structural ceramics for high-temperature applications through the late 1970s to 1990s.

In the present chapter, the developments in science and technology of high-temperature high-strength ceramics for heat-engine applications, especially those related with silicon nitride and silicon carbide materials, are summarized with special emphasis on certain typical examples.

9.1.2 SILICON-BASED CERAMICS AS HIGH-TEMPERATURE HIGH-STRENGTH MATERIALS

The improvements in the efficiency of heat-engines generally require the structural materials to be utilized under severer operation conditions. Typically in the case of gas turbine engines, the turbine inlet temperature (TIT) has continued to increase for the improved efficiency, which has concurrently required structural materials with higher temperature durability. The TIT of the most advanced large-scale gas turbines has already far exceeded the melting point of the super-alloys, which are currently utilized for these high-temperature components with tremendous air cooling.

TABLE 9.1.1 Silicon Nitride and Carbide as High-Temperature High-Strength Ceramics

Characteristics as substance	Characteristics as material	Characteristics for heat engine components
Compound between Si and N/C	Abundance as element materials	Abundance as resources
		Potential low cost
	Oxidation to form SiO_2 layer with low O permeability	High oxidation resistance
	Low specific gravity	Light weight
Covalent bonding	High-temperature deformation resistance	High fracture strength
		High creep resistance
		High fatigue resistance
	High Young's modulus	High rigidity
	Low thermal expansion	Low thermal deformation
		Thermal shock resistance
	High hardness	Wear resistance

The general requirements to the high-temperature structural materials for heat-engines such as gas turbines and diesel engines are as follows:

- high fracture strength from ambient to high temperatures, especially high strength per density;
- high fatigue strength from ambient to high temperatures;
- high thermal shock and thermal fatigue resistance;
- high creep resistance to high temperatures;
- high oxidation and corrosion resistance;
- high wear resistance;
- high impact resistance.

Although many kinds of materials are classified as ceramics, only limited materials are capable of simultaneously satisfying the conditions listed above. Silicon nitride and silicon carbide are the most promising ceramic materials from this point of view. Table 9.1.1 summarizes the characteristics of silicon nitride and silicon carbide as high-temperature high-strength ceramics.

9.1.3 FABRICATION AND MICROSTRUCTURE CONTROL OF SILICON-BASED MONOLITHIC CERAMICS

9.1.3.1 Silicon Nitride

Silicon nitride is a highly covalent material, which means bulk diffusion rate is too low to give densification. Therefore, sintering additives are used to obtain the fully desified material. The added metal oxides (MgO, Al_2O_3, Y_2O_3, rare earth oxides, etc.) and metal nitride form a liquid phase by the reaction with Si_3N_4 and surface SiO_2 on Si_3N_4 powder at high temperatures. Liquid phase sintering is considered to be composed of the three stages, namely, rearrangement, solution–reprecipitation and grain growth [1]. Silicon nitride has two crystal structures, α and β. α-Silicon nitride, which is a stable phase at lower temperatures is generally used as a raw powder, where the amount of initial β-Si_3N_4 nuclei and the particle size largely influence the sintering behavior and the final microstructure. At the first stage, the raw powder rearranges to give more close packing by a capillary force of liquid phase. Shrinkage rate depends on the characteristics of the raw powder and the liquid phase. Second, small Si_3N_4 particles dissolve into the liquid phase and then solutes diffuse and reprecipitate on the larger particles. α-Si_3N_4 transforms to β-Si_3N_4 during this stage. The rate of shrinkage is controlled by either interface reaction or diffusion through the liquid phase. The final stage is a grain growth, where residual pores are eliminated and the further grain growth proceeds. The grain boundary energy drives the grain growth, which is called Ostwald ripening.

The application of mechanical and gas pressure during sintering is also effective to increase density and control the microstructure [2]. Hot-pressing (HP), where an uniaxial pressure is applied, is one of the most effective sintering method of simple shape parts. Complex shape parts can be densified by hot-isostatic-pressing (HIP), which utilizes isostatic pressure of gas. The other important point is the temperature limitation due to the decomposition of Si_3N_4, which depends on the temperature. Serious decomposition occurs at a temperature of 1800°C under ambient nitrogen atmosphere. Gas pressure sintering (GPS) is often applied to suppress the decomposition at higher temperatures.

Silicon nitrides have composite microstructure, consisting of rod-like large grains and equiaxial small grains. Final grain size and morphology also are affected by the sintering additives. Most additives remain at the grain boundary after sintering usually as glassy phase, which strongly affect the thermal and mechanical properties of the sintered body. In general, glassy phase is located at triple points, which are called glass pockets, and along a grain boundary between two grains. The glass pockets can be changed to refractory crystallized phases by post-sintering heat treatment.

Figure 9.1.1 shows SEM micrograph of the typical microstructure of Si_3N_4 obtained by using the sintering additives: (a) Y_2O_3–Al_2O_3, (b) Yb_2O_3–SiO_2. The sintered body containing Yb_2O_3 has a relatively coarser microstructure compared to that containing Y_2O_3 owing to the higher melting point of eutectic systems (Y(Yb)–Al–Si–O–N).

Linear fracture mechanics predicts that strength and fracture toughness have a proportional relationship in the case of the same size of the fracture origin. However, high strength and high fracture toughness cannot be attained

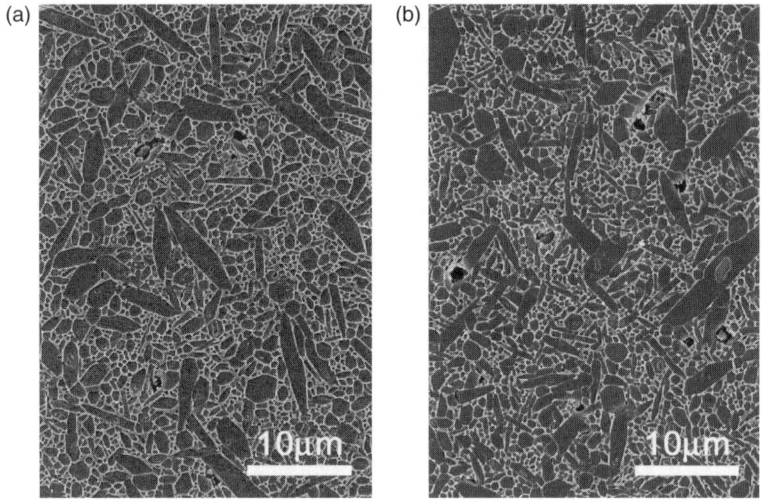

FIGURE 9.1.1 SEM micrographs of sintered silicon nitrides: (a) Y_2O_3–Al_2O_3; (b) Yb_2O_3–SiO_2.

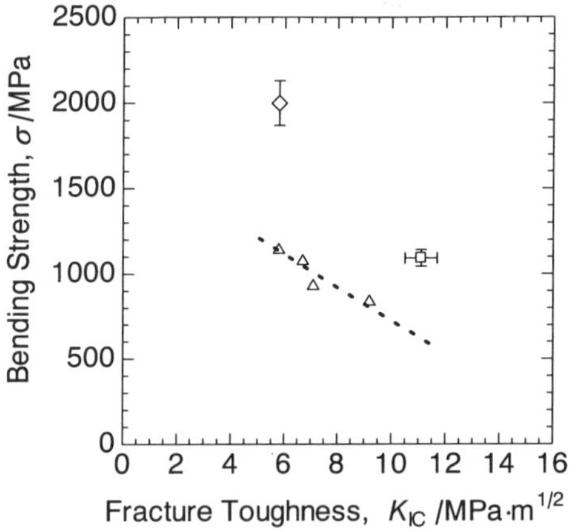

FIGURE 9.1.2 Relationship between strength and fracture toughness of typical silicon nitride.

concurrently in a single material made by a conventional processing technique. The dotted line in Figure 9.1.2 shows this limited relationship between strength and toughness [3, 4]. Recently, the new strategy of microstructural design has been proposed to attain the high strength and the high fracture toughness. Hirao *et al.* [5, 6] reported that the well-controlled microstructure was obtained by adding β-seeds as nuclei, and tape casting or extrusion technique introduced

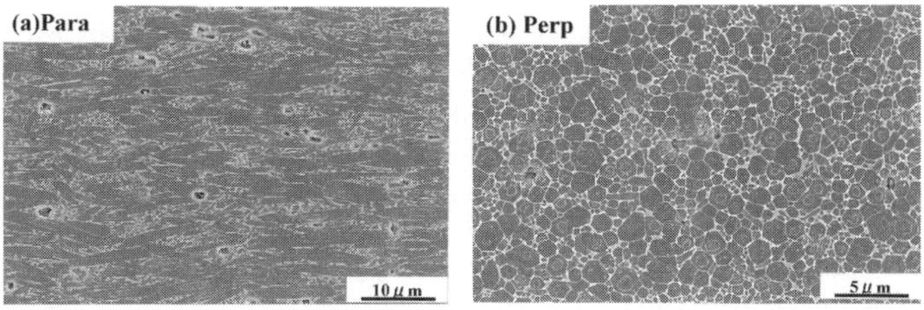

FIGURE 9.1.3 SEM micrographs of the microstructure of the highly anisotropic silicon nitride: (a) parallel; (b) perpendicular.

high orientation into the sintered body. Figure 9.1.3 shows the anisotropic microstructure, which clearly indicates the aligned rod-like grains. These materials enabled high strength and high fracture toughness concurrently.

9.1.3.2 Silicon Carbide

Although silicon carbide has a covalent bond, SiC can be densified by solid state sintering. A combination of boron and carbon is used as typical sintering additives [7]. The added carbon is considered to react with SiO_2 on the surface of silicon carbide particles, and boron is thought to increase grain boundary diffusion rate. The liquid phase sintering is also applied by using the similar additives as Si_3N_4 [8]. Silicon carbide has two crystal structures, α and β type, where β-SiC is the low-temperature phase. The transformation during sintering and the particle size of the powder largely affect the microstructure and the sintering behavior.

SiC produced by solid state sintering has no heterogeneous phase at the grain boundary, which leads to better oxidation resistance and high temperature strength than SiC produced by liquid phase sintering. However, SiC by the solid state sintering has lower toughness and lower thermal shock resistance, compared to SiC or Si_3N_4 having the heterogeneous phase at the grain boundary.

9.1.4 MECHANICAL PROPERTIES OF Si-BASED MONOLITHIC CERAMICS

9.1.4.1 Spontaneous Fracture

Monolithic ceramic materials, as typical brittle materials, possess very limited plastic deformation capability. The lack of the capability of release of local stress concentration results in extremely high sensitivity of the materials to the microstructure defects. Besides the typical defects, that is, voids, inclusions or surface flaws, the heterogeneities in the grain microstructure such as elongated grains also have to be carefully taken into consideration.

According to linear fracture mechanics, the fracture stress of the material of containing a defect with size a is described by the following equation:

$$\sigma_f = K_{IC}/Y\sqrt{a} \tag{1}$$

where K_{IC} is the critical stress intensity factor or fracture toughness and Y the geometrical factor.

The dependence of the fracture stress on the size and the geometry of the defect gives statistical nature to the fracture behavior. The statistical distribution of the spontaneous fracture strength of many ceramic materials is reasonably well described by the one-parameter Weibull distribution described below.

$$P_f = 1 - \exp\left[-(\sigma_f/\sigma_0)^m\right] \tag{2}$$

where P_f is the accumulative failure probability, σ_f the fracture strength, σ_0 the characteristic stress parameter, and m the statistical parameter (Weibull parameter).

9.1.4.2 Time-Dependent Deformation and Fracture

Many ceramic materials also show time-dependent mechanical behavior. Glasses are known to show sub-critical crack growth (SCG) in the presence of humidity. Silicon nitrides with metal oxide additives also show SCG at elevated temperatures where the crack growth occurs along the grain boundaries. SCG under static stress is often called as static fatigue. While sintered silicon nitrides show noticeable static fatigue (Fig. 9.1.4), silicon carbides with B and C show little SCG even at high temperatures (Fig. 9.1.5) [9].

SCG rate is dependent on the stress intensity factor K_I. The typical behavior is shown in Figure 9.1.6. SCG is found to start above a certain threshold value of K_I called 'fatigue limit', and reaches a certain level of crack growth rate which increases as a function of K_I described by the following equation:

$$da/dt = AK_I^n \tag{3}$$

where A and n are constants.

The combinations of Eqs. 1–3 give the following equation as the life of a ceramic component under steady state static fatigue as a function of applied stress.

$$t_f = \frac{2}{\left[(n-2)AY^n\sigma^n a_i^{(n-2)/2}\right]} \tag{4}$$

where a_i is the size of the initial crack.

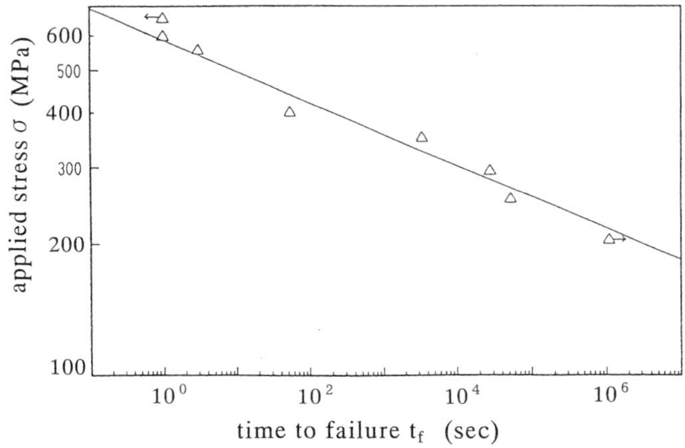

FIGURE 9.1.4 Static fatigue strength of sintered silicon nitride (four-point bending, 1250°C). Reprinted with kind permission of Fine Ceramics Technical Research Association.

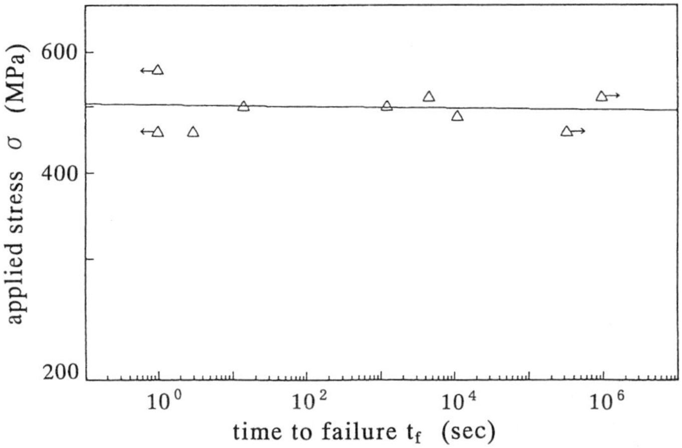

FIGURE 9.1.5 Static fatigue strength of sintered silicon carbide (four-point bending, 1400°C). Reprinted with kind permission of Fine Ceramics Technical Research Association.

SCG rate steeply increases as K_I approaches the critical value K_{IC}, leading to spontaneous fracture. In the case of sintered silicon nitrides at high temperatures, the fatigue limit is considered as the threshold between the crack healing and the crack growth.

Many ceramic materials also show creep deformation at elevated temperature. Among the several creep mechanisms studied or proposed for ceramic materials, sintered silicon nitride with metal oxide additives is considered to show creep with grain boundary sliding mechanism. At the final stage of the

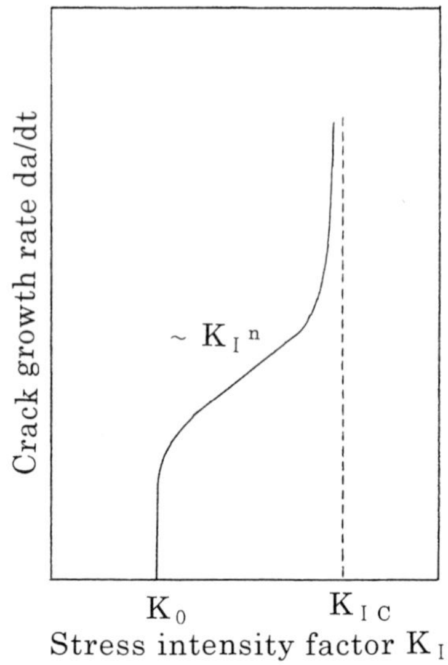

FIGURE 9.1.6 Sub-critical crack growth in silicon nitrides.

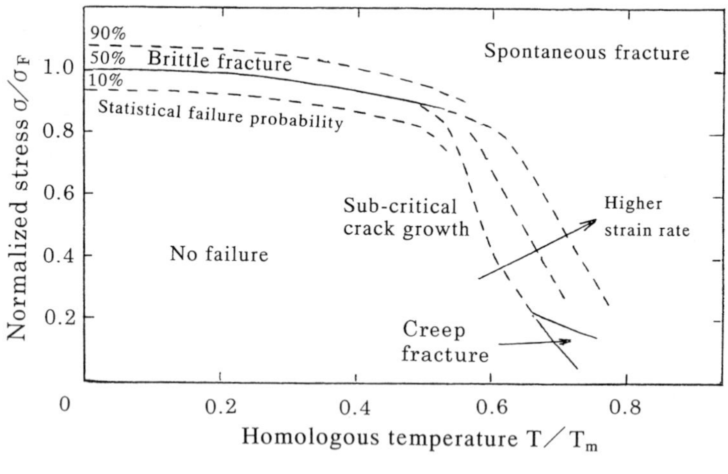

FIGURE 9.1.7 Fracture map of sintered silicon nitrides.

creep, the formations of cavities at grain triple points and their coalescence lead to the accelerated deformation rate typical to the final stage. The time-dependent deformation or fracture behavior of the ceramics described above are further described by the fracture map. A typical example of the fracture map of sintered silicon nitride is shown in Figure 9.1.7.

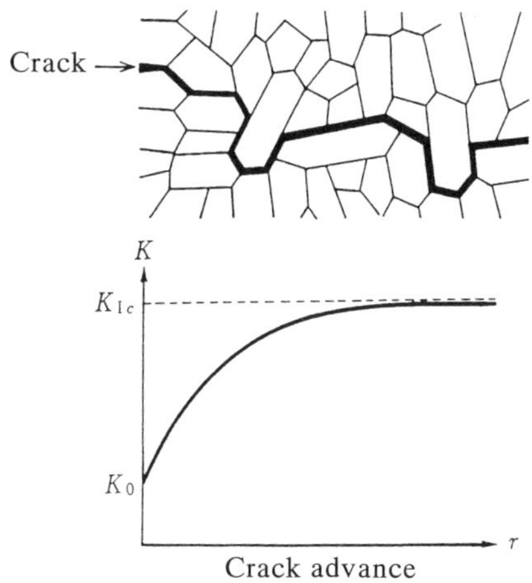

FIGURE 9.1.8 *R*-curve behavior of sintered silicon nitride with enhanced fracture toughness.

9.1.4.3 Anelasticity

In the case of fracture of polycrystalline materials, grain boundaries generally play a role of crack branching or bowing, which results in higher fracture toughness of polycrystalline materials than that of single crystals. Especially in the case of polycrystalline silicon nitride with grains of high aspect ratios, process-zones with a mechanism of grain pull-out and bridging play an important role to give *R*-curve behavior and enhanced fracture toughness as shown in Figure 9.1.8. The material developments with higher toughness will be discussed in the following sections.

9.1.4.4 Oxidation

Silicon nitride and silicon carbide are thermodynamically unstable in oxygen-rich atmosphere, although they are exceedingly oxidation resistant compared with super-alloys. Their oxidation resistance basically relies on the oxidation product layer on the surface. The oxidation of silicon-based ceramics under relatively high oxygen partial pressure is called passive oxidation, where silica or silica-based oxide layer formed on the surface plays the role of oxidation protection and fairly high durability of the materials are expected. The compositions of the grain boundary phase have to be carefully considered since they strongly influence on the oxidation resistance of the silicon nitride or carbide materials. On the other hand, the exposure of silicon nitride and carbide to low oxygen partial pressure atmosphere has to be carefully avoided, since volatile SiO is formed and poor oxidation resistance of the materials called active oxidation takes place.

The oxidation is more enhanced in high-velocity streams as are encountered in gas turbines, where high-velocity gas retards the formation of surface protection films. The effect of aggressive atmosphere has to be also taken into considerations in the evaluation of the time-dependent fracture behavior such as fatigue or creep.

9.1.5 TOUGHENING OF Si-BASED CERAMICS BY FIBER REINFORCEMENT

Si-based ceramics have a remarkable potential for various structural applications at high temperatures as already described. However, their catastrophic fracture behavior sometimes limit further use of Si-based ceramic materials. Therefore, considerable efforts have been made in order to improve the toughness of Si-based ceramics especially by whisker and fiber reinforcement.

Silicon nitride reinforced with silicon carbide whisker is one of the material systems that has been most extensively investigated, because superior mechanical properties and heat resistance can be expected. An example of the fracture toughness of hot-pressed silicon nitride with SiC whisker reinforcement is shown in Table 9.1.2. These materials, which have a different orientation of whiskers achieved by cold-pressing or extrusion in powder forming, show a improved fracture toughness by whisker reinforcement [11]. Silicon nitride can be toughened by the enhanced acicular grain growth caused by heat treatment at relatively high temperature. These heat treatments often cause strength degradation, because elongated grains tend to act as fracture origins.

As whisker reinforcement usually inhibits the growth of the matrix grains, this approach has a potential to simultaneously achieve higher strength and toughness.

TABLE 9.1.2 Fracture Toughness of Whisker-reinforced Si_3N_4 and Unreinforced Si_3N_4 (after Ref. [11])

Whisker orientation	Forming process	Hot-pressing temperature (°C)	Fracture toughness (MPa m$^{1/2}$)
Two-dimensional	Cold-pressing	1700	8.2 ± 0.4
Unidirectional	Extrusion	1700	10.0 ± 0.2
Two-dimensional	Cold-pressing	1750	8.0 ± 0.3
Unidirectional	Extrusion	1750	10.2 ± 0.3
Unreinforced	Cold-pressing	1750	7.5 ± 0.6

Since toughening can be achieved by bridging of whiskers and crack deflection at whiskers, various microstructural aspects, namely morphology, distribution and orientation of whiskers and so on, are considered to affect the toughening performance. Among these microstructural aspects, a property of the interface between the whiskers and matrix is primary factor to determine the toughness. Effect of interface is obviously shown in the fracture surface of silicon carbide reinforced with silicon carbide whiskers (Fig. 9.1.9). When carbon coating was introduced on the whiskers, carbon interface was formed between the whiskers and the matrix. Consequently, a remarkable interaction between crack and whiskers is observed on the fracture surface, while the materials without carbon-coating show smooth typically brittle fracture surface.

Figure 9.1.10 shows the fracture resistance also obtained in silicon carbide reinforced with the carbon-coated SiC whiskers. These materials have

(a)
(b)
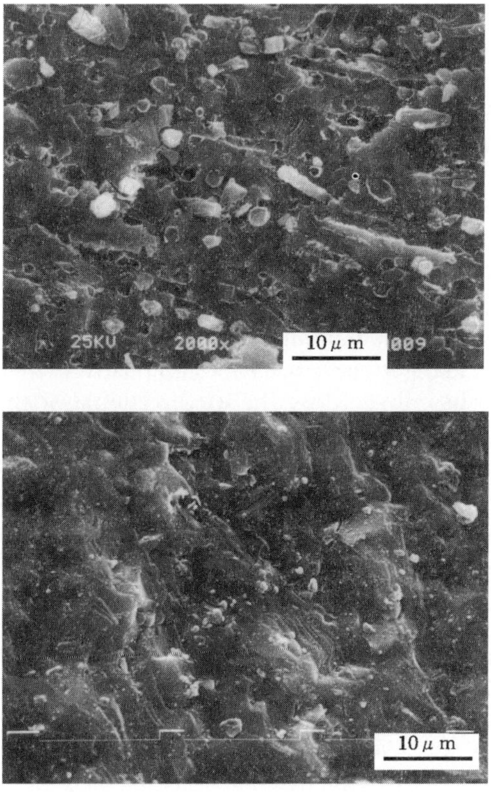

FIGURE 9.1.9 Fracture surface of SiC whisker-reinforced SiC ceramics: (a) with carbon coating; (b) without carbon coating.

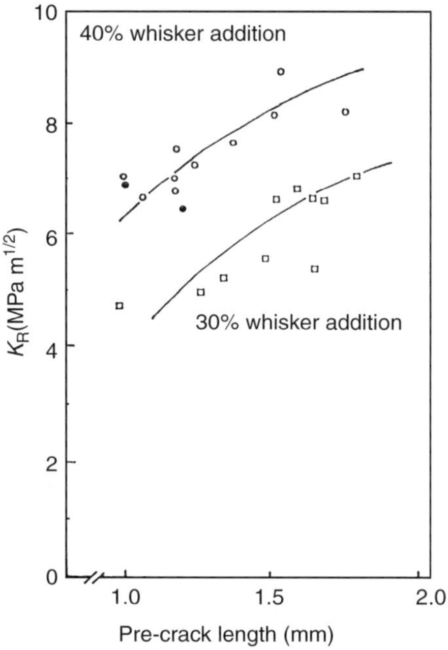

FIGURE 9.1.10 Rising *R*-curve behavior of whisker-reinforced SiC ceramics fabricated by extrusion process.

an unidirectional orientation of whiskers, which was achieved by an extrusion forming of compounds containing SiC powder and whiskers. These materials show an obvious rising *R*-curve behavior and relatively high toughness, which is considered to be enhanced by the unidirectional orientation of whiskers. Figure 9.1.11 shows strength degradation of a monolithic and a whisker-reinforced silicon carbide after a pin-on-disk test, in which a test contact load was applied with a ceramic pin, and then the specimen was slid at a constant speed. Reinforced material shows little strength degradation, while monolithic material degrades steeply at relatively low or load. As contact stress or foreign object damage is most probable in the actual applications, whisker reinforcement is expected to improve the performance.

For the further increase of toughness, continuous fiber reinforcement has been employed. In this category, silicon carbide reinforced with silicon carbide fibers, which is usually fabricated by the densification process using chemical vapor infiltration (CVI) or impregnation and pyrolysis of organosilicon polymers, is one of the most attractive materials. In these materials, the interface is also essentially important. As shown in Figure 9.1.12, appropriate thickness of carbon interface introduced by CVI process leads to a nonlinear fracture.

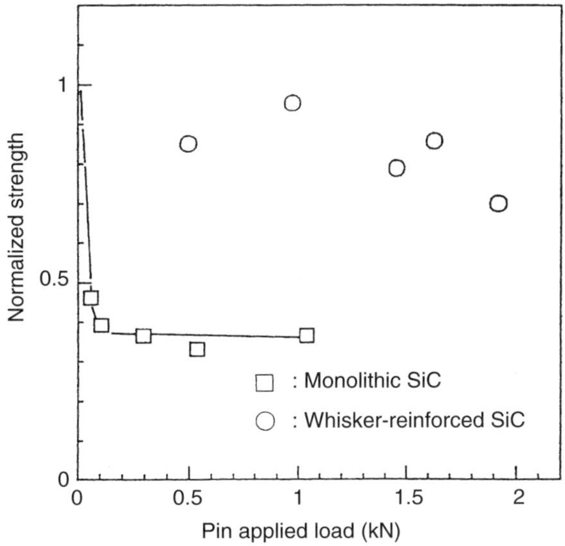

FIGURE 9.1.11 Normalized strength of a monolithic SiC and a whisker-reinforced SiC after pin-on-disk test (four-point bending strength after pin-on-disk test was normalized by the initial four-point bending strength).

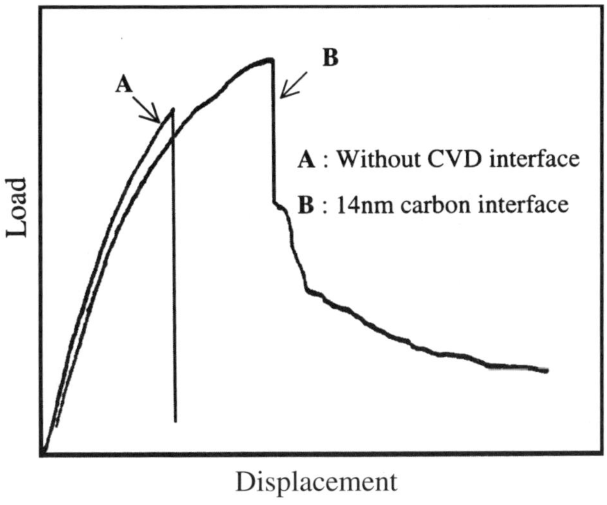

FIGURE 9.1.12 Flexural stress–strain behavior of continuous SiC fiber-reinforced SiC.

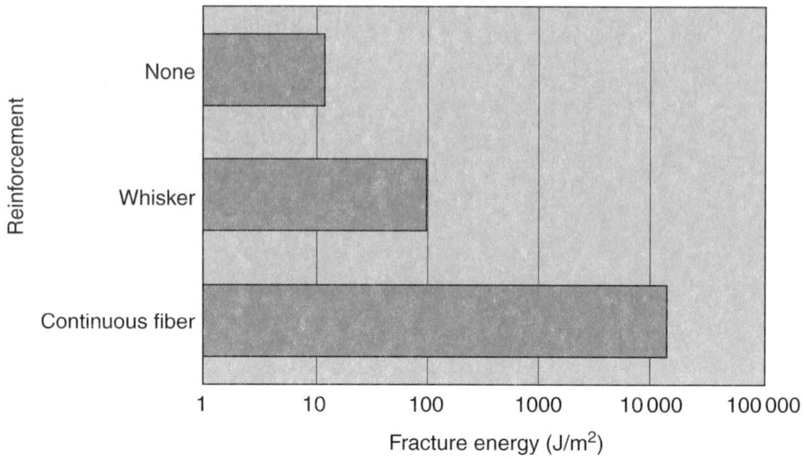

FIGURE 9.1.13 Fracture energy of monolithic, whisker- and fiber-reinforced SiC ceramics.

Fracture energy of the aforementioned silicon carbide materials are compared in Figure 9.1.13. Significant toughening is thus possible in fiber-reinforced materials.

9.1.6 LAMINATED COMPOSITE STRUCTURE WITH ENHANCED FRACTURE RESISTANCE

Multilayered ceramic composites have attracted attention in recent years to overcome the brittleness of ceramic materials. An alternate layered composite composed of different materials has enhanced fracture resistance and/or damage tolerance, as reported in ZrO_2/Al_2O_3, SiC/C systems and so forth. The large difference in the layer properties leads to improved mechanical properties. On the other hand, the difference of shrinkage and thermal expansion coefficient causes the subsequent delamination or cracking through the sintering stage. If the aimed properties are distinguished between each layer, strengthening and toughening mechanism should arise even in the same material system. From this point of view, the multilayered silicon nitride composed of dense and porous layers has been developed. Porous silicon nitride has lower elastic moduli, compared to dense silicon nitride, where a distinction of elastic property between the layers is realized.

The whisker was used, both for anisotropic grain growth and for retarding the densification of porous layer by percolation. Figure 9.1.14 shows the SEM micrographs for the typical microstructure incorporating 70 vol% whisker for

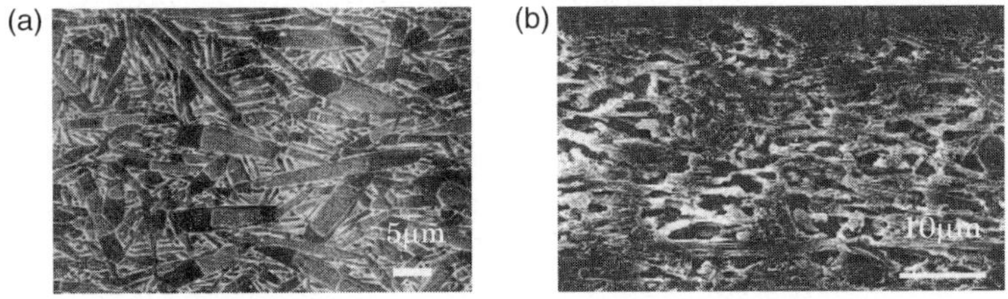

FIGURE 9.1.14 SEM micrographs of multilayered silicon nitrides: (a) dense layer; (b) porous layer.

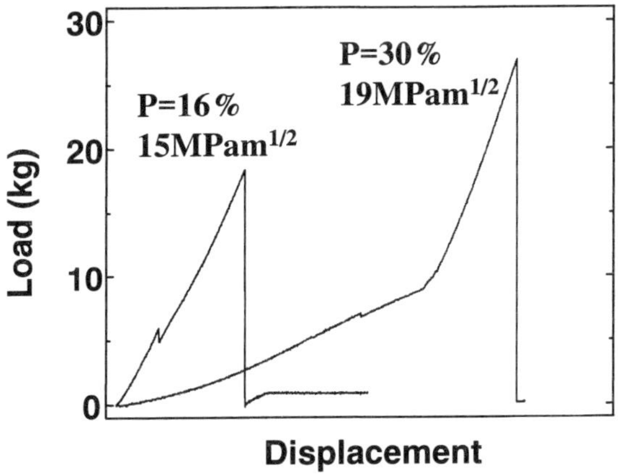

FIGURE 9.1.15 Load–displacement curves in SENB test.

the porous layer and 5 vol% for the dense layer. The duplex layered structure was successfully fabricated and the anisotropic microstructure was clear in the both layers. The layers with sintering additives were sintered to nearly full density and no cracks or delaminations was observed, neither within the layers nor at the interfaces. The porous layer showed unique microstructure, where rod-like grains were aligned in the casting direction and anisotropic pores were generated by tightly tangled elongated grains. The microstructure and layered structure can be controlled by modifying the distribution of the sintering additives and the content of whisker.

The fracture of the layered composite originated in the dense layer on the tensile surface owing to lower strain to failure, compared to the porous layer. Actually, a fracture origin was always confirmed in the dense layer. The strength behavior strongly depended on the thickness of the dense layer. The material with the thin dense layer presented high strength, over 1 GPa.

FIGURE 9.1.16 SEM micrograph of the crack path (side view).

The difference in the elastic properties between the layers caused the zigzag *R*-curve behavior. The porosity in the porous layer directly affected the fracture behavior. Figure 9.1.15 shows the load–displacement curves in the SENB test for the materials with porosity in the porous layer of 30 and 16%. The high porosity clearly contributed to the higher fracture resistance due to the enhanced delamination in the porous layer (Fig. 9.1.16).

9.1.7 FABRICATION TECHNOLOGY OF MONOLITHIC CERAMIC COMPONENTS FOR HEAT ENGINES

9.1.7.1 Fabrication Technology of Ceramic Components

The applications of structural materials to various utilizations such as heat engines require the forming technology developments of the materials to desired size and geometry as the components. A forming process is generally composed of three elements: a certain raw material with shaping capability, a shaping process, and a consolidation process of the raw material. In the case of the metallic components, the most popular forming process is casting, where the raw material is liquid metal, the shaping process is pouring into molds, and the consolidation process is solidification of the molten metal. The casting process can be also applied to ceramic materials except that such materials as silicon nitride or silicon carbide sublimate at high temperatures instead of melting. The application of the casting process is, however, quite limited in ceramic materials due to the fact that the melting of ceramic materials usually requires extremely high temperatures and also that the microstructures of the materials obtained by this process generally show considerably coarse grain structures, which lead to poor mechanical properties. The most popular forming process for ceramics is powder forming followed by sintering, which enables the processing of ceramics at reasonable temperatures and also the formation of reasonably fine grain structures.

Figure 9.1.17 shows the general flow sheet of the fabrication process of ceramic materials for machinery components. The shaping process of the raw material powder is especially crucial to obtain high-quality and low-cost components. Different component geometries require different forming technologies. Major forming technologies for typical component geometries are summarized in Table 9.1.3.

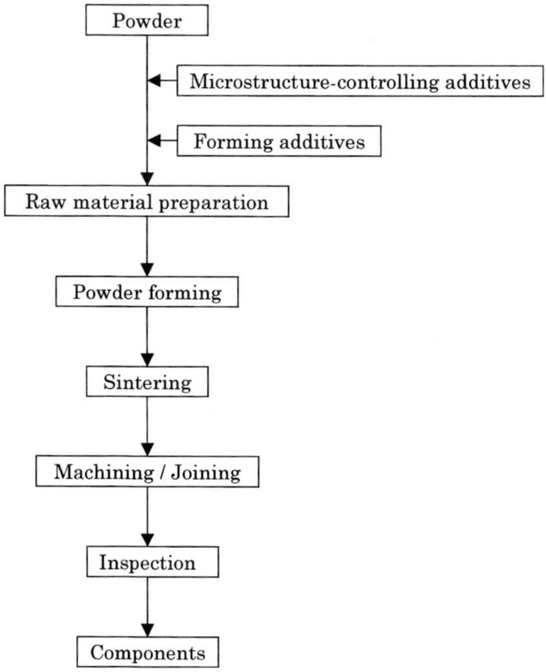

FIGURE 9.1.17 Fabrication process of ceramic components.

TABLE 9.1.3 Major forming technologies of ceramic components

Forming technology	Component geometry
Isostatic pressing + (machining)	Two-dimensional shape or simple semi-two-dimensional shape
	Thick or large shape
Slip casting (drain casting)	Three-dimensional hollow shape
Slip casting (solid casting)	Three-dimensional solid shape
Injection molding	Three-dimensional precise shape
Die pressing	Plate-like shape
	Small simple shape
Extrusion	Tubular or planar shape
Tape casting	Thin plate shape

9.1.7.2 Applications of Si-based Ceramics to Heat Engines

The initial target of the application of silicon-based ceramics to heat engines was automotive engine turbochargers, which was started in the 1970s, and silicon nitride turbocharger rotors for diesel engines and gasoline engines were commercialized in the 1980s (Fig. 9.1.18). The injection molding method has been developed and applied to the fabrication of the ceramic turbocharger rotors. Further, several components of automotive diesel engines have been commercialized successively, including igniter plugs, swirl chambers, cams, and valves. The cold isostatic pressing method followed by machining has been applied to these components. The major hot-section components in diesel engines such as piston-head, piston-cylinder are still in the process of development.

There have been a number of programs in the United States, the European Union, and Japan for the development of gas turbine engines with ceramic components since the 1970s. Many of the programs have been for automotive engine applications. From late 1980s to 1990s, the development of larger-size gas turbine engines for co-generation applications have been started in the United States and in Japan.

Figure 9.1.19 shows the ceramic components utilized in a gas turbine engine of 300 kW-class developed for co-generation applications, which is a single-shaft engine with a two-stage axial-flow type turbine and a single-can type combustion chamber installed with a shell-tube type heat exchanger. Especially, the turbine rotor is composed of ceramic blades inserted to a metallic disk (Fig. 9.1.20), and the turbine nozzle assembly is composed of individual vanes supported by a retainer ring. The ceramic turbine blades and the turbine nozzle vanes have been fabricated by the injection molding method. The combustion

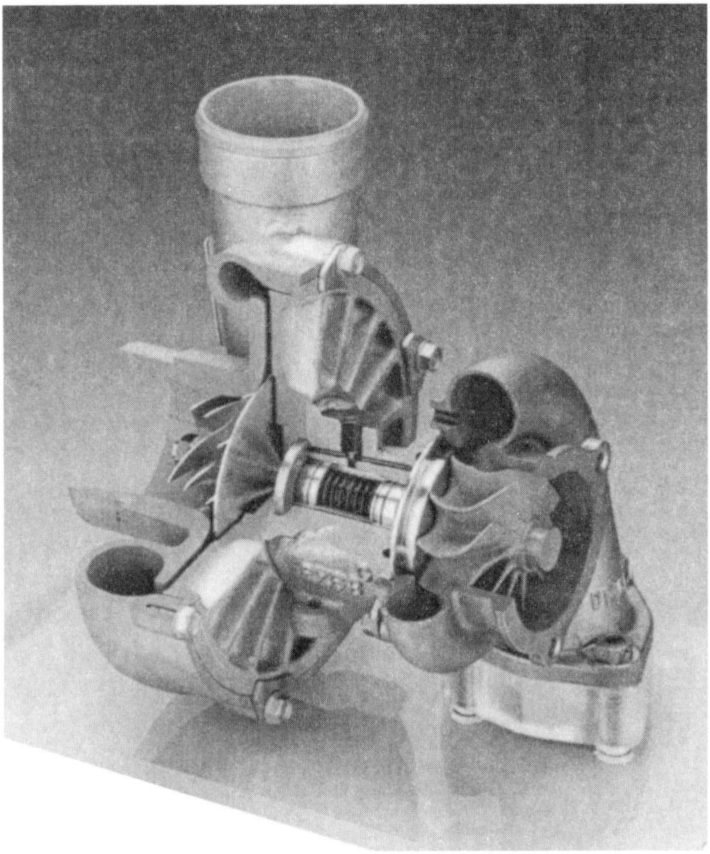

FIGURE 9.1.18 Automotive turbocharger with a silicon nitride rotor.

chambers have been fabricated by the cold isostatic pressing method followed by machining, and especially in case of the turbine nose cones (Fig. 9.1.21), the joining technology by cold isostatic pressing combined with machining has been developed and applied. The heat-exchanger tube assemblies shown in Figure 9.1.22 have been fabricated by the extrusion method combined with joining technology.

Improvements in the properties and reliability of silicon-based ceramic components as well as those in the technologies of designing and assembling of them have enabled the extremely high engine efficiency of 35% or more with the 300 kW-class gas turbine systems installed with ceramic components, which usually show efficiencies of around 20% with metallic components.

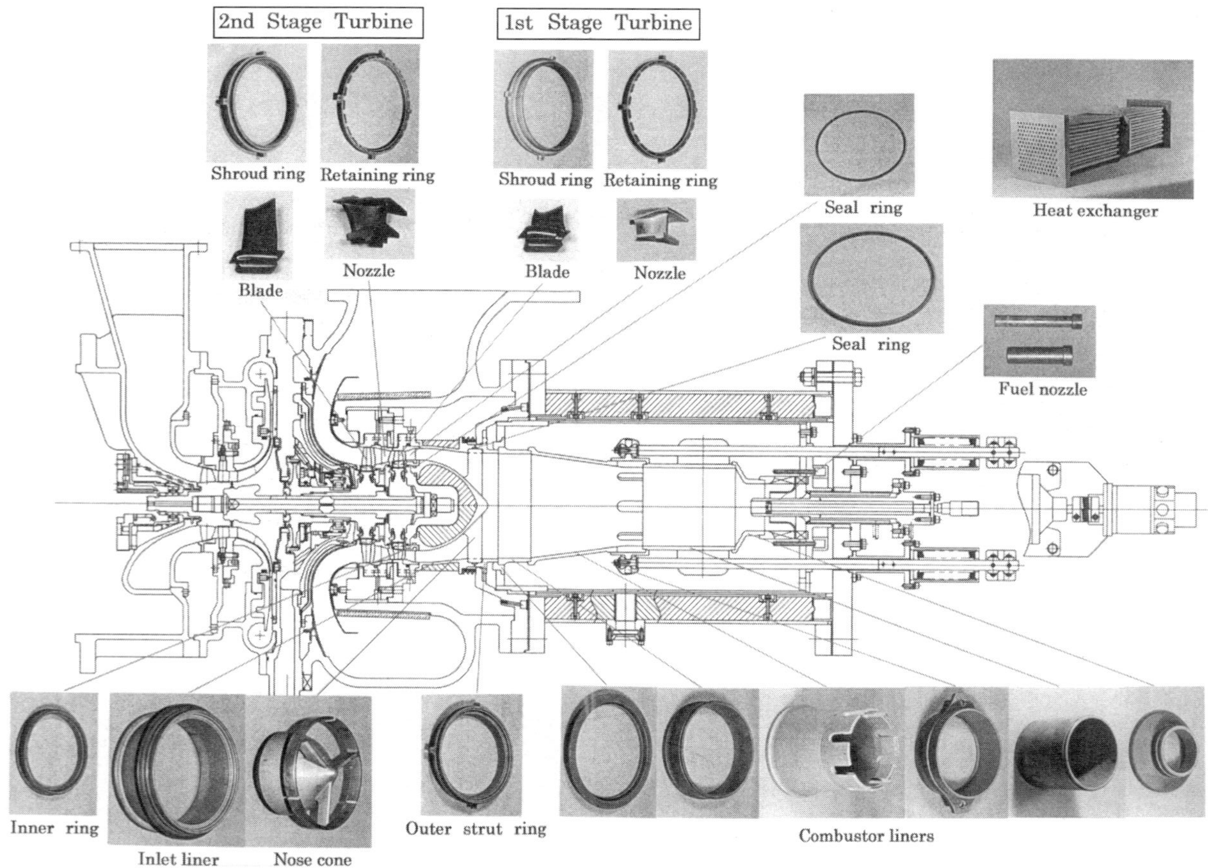

FIGURE 9.1.19 300 kW-class gas turbine CGT301 for co-generation applications and ceramic components utilized in the engine system.

FIGURE 9.1.20 Ceramic turbine blades inserted to a metallic disk.

FIGURE 9.1.21 Ceramic turbine nose-cone.

FIGURE 9.1.22 Ceramic heat-exchanger tube assembly.

9.1.8 CONCLUSIONS

Silicon-based ceramics, especially sintered silicon nitride materials have shown extraordinary advancements in the past decades. Not only the improvements in the high-temperature mechanical properties of the materials but also the technologies for the applications to machinery components have shown considerable advancements, that is, the forming technologies of precise complex shapes, the evaluation and the quality control technologies of defect-sensitive ceramic components, the design technologies suited to brittle ceramic components. Further improvements in the technologies and developments of newer applications are expected to proceed steadily hereafter in the world of engineering ceramics, although the extensive commercialization of structural ceramics for heat engines, which many ceramists had once dreamed of, has not yet come to fruition.

REFERENCES

1. Jack, K. H. (1976). Review: Sialons and related nitrogen ceramics. *J. Mater. Sci.* **11**: 1135–1158.
2. Mitomo, M., and Tajima, Y. (1991). Sintering, properties and application of silicon nitride and Sialon ceramics. *J. Ceram. Soc. Jpn.* **99**: 1014–1025.
3. Kawashima, H., Okamoto, H., Ymamoto, H., and Kitamura, A. (1991). Grain size dependence of the fracture toughness of silicon nitride. *J. Ceram. Soc. Jpn.* **99**: 320–323.
4. Yoshimura, M., Nishioka, T., Yamakawa, A., and Miyake, M. (1995). Grain size cotrolled high-strength silicon nitride ceramics. *J. Ceram. Soc. Jpn.* **103**: 407–408.
5. Hirao, K., Ohashi, M., Brito, M. E., and Kanzaki, S. (1995). Processing strategy for producing highly anisotropic silicon nitride. *J. Am. Ceram. Soc.* **78**: 1687–1690.
6. Kondo, N., Suzuki, Y., and Ohji, T. (1999). Superplastic sinter-forging of silicon nitride with anisotropic microstructure formation. *J. Am. Ceram. Soc.* **82**: 1067–1069.
7. Prochazka, S., and Charles, R. J. (1973). *Am. Ceram. Soc. Bull.* **52**: 885–888.
8. Suzuki, K., and Sasaki, M. (1986). *Fundamental Structure Ceramics.* Trra Sci. Pub. Co.
9. Inamura, T., Suzuki, A., Hayashi, S., and Shigegaki, Y. (1992). Static and cyclic fatigue. Final Report of Fine Ceramics for Future Industries Research Program, Fine Ceramics Research Association, pp. 753–775.
10. Sakida, T., Tanaka, S., Mikami, T., Tatsuzawa, M., and Taoka, T. (1999). Development of the Ceramic Gas Turbine Engine System (CGT301). ASME 99-GT-104, pp. 1–7.
11. Goto, Y., and Tsuge, A. (1993). Mechanical properties of unidirectionally oriented SiC-whisker-reinforced Si_3N_4 fabricated by extrusion and hot-pressing. *J. Am. Ceram. Soc.* **76**: 1420–1421.

Handbook of Advanced Ceramics
S. Sōmiya *et al.* (Eds.)

CHAPTER 10

10.1 Porous Ceramics for Filtration

TOSHINORI TSURU
Department of Chemical Engineering, Hiroshima University, Higashi-Hiroshima 739-8527, Japan

10.1.1 INTRODUCTION

A general definition of a membrane is that it is "a selective barrier between two phases" [1]. Therefore, using membranes, the feed is separated into two streams, that is, the retentate and permeate streams, as shown in Figure 10.1.1. Either the retentate or the permeate could be product stream, depending upon types of membranes used and the feed stream. The permeate stream is the product stream, if the solvent is purified by removing solutes using a membrane which allows the permeation of solvent and retains the permeation of solutes, such as in the desalination of seawater. If the purpose of the separation process is the concentration of solutes, then the retentate becomes the product.

Membrane separation, which is a relatively new separation process, has been commercialized in the last two decades. The majority of membrane materials which have been commercialized thus far are polymeric. Porous ceramic membranes have great potential for opening up new types of applications to which polymeric membranes cannot be applied. This review will summarize the present status and a potential application of porous ceramics as materials for membrane separation.

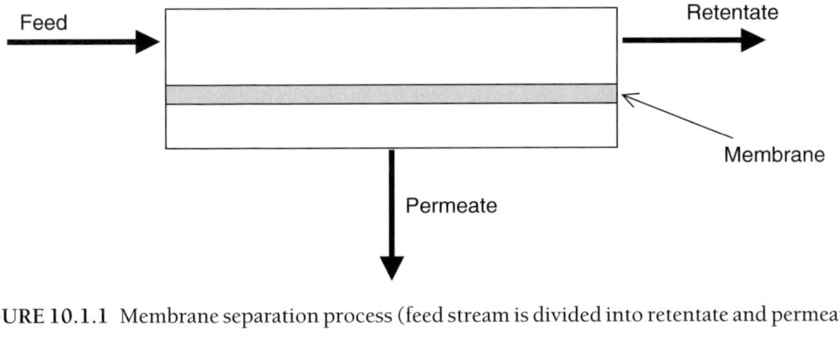

FIGURE 10.1.1 Membrane separation process (feed stream is divided into retentate and permeate).

10.1.2 MEMBRANE AND MEMBRANE SEPARATION PROCESS

Membrane separation processes can be categorized based on the phases of the feed and permeate, as well as the types of driving force for the separation, as shown in Table 10.1.1 [1]. For the case where both the feed and permeate streams are a liquid phase and the driving force is a pressure difference between the two phases, the separation process is referred to a filtration process such as microfiltration (MF), ultrafiltration (UF), nanofiltration (NF) and reverse osmosis (RO), depending upon the pore sizes. It should be noted that membranes, the pore sizes of which are larger than 10 μm, are usually categorized as filters. Gas separation is conducted in the gas phase for both feed and permeate stream. For the case of pervaporation, the feed stream is a liquid phase, while the permeate stream is a gas phase by evacuation. Other separation processes (electrodialysis, dialysis, membrane distillation) are also conducted in the liquid phase, but ceramic porous membranes have not yet been applied to these types of separation processes. The separation mechanism is mainly controlled by the sieving effects, where solutes which are smaller than the pore sizes of membranes permeate through the porous membranes. Another mechanism is the affinity of solutes for membrane materials, but this mechanism plays an important role in cases where the pore size is relatively small.

Most applications have been carried out using polymeric membranes, while inorganic membranes, including ceramic membranes, have been utilized to

TABLE 10.1.1 Membrane Processes Based on Phase and Driving Force

Membrane process	Phase		Driving force	Pore size	Note
	Feed	Permeate			
Microfiltration	L	L	ΔP	0.1–10 μm	Separation of particles
Ultrafiltration	L	L	ΔP	2–100 nm	Separation of macromolecules
Nanofiltration	L	L	ΔP	1–2 nm	Separation of low MW solutes (MWCO 200–1000)
Reverse osmosis	L	L	ΔP	<1 nm	Reject electrolytes
Gas separation	G	G	ΔP	<0.5–1 nm	
Vapor permeation	G	G	ΔP	<0.5–1 nm	
Pervaporation	L	G	ΔP	0.5–2 nm	
Electrodialysis	L	L	$\Delta \Phi$		
Dialysis	L	L	ΔC		
Membrane distillation	L	L	ΔT		

L, liquid phase; G, gas phase; ΔP, (partial) pressure difference; $\Delta \Phi$, electrical potential difference; ΔC, concentration difference; ΔT, temperature difference.

TABLE 10.1.2 Advantages and Disadvantages of Inorganic Membranes

	Applications and comments
Advantages	
Thermal stability	Separation at high temperature, steam sterilization
Resistance to organic solvents	Separation of non-aqueous systems, separation of oil
Resistance to chemicals, acidic and basic pH	Chemical cleaning, recovery of acid/basic
Resistance to peroxide	Chemical cleaning, application of textile processing
Mechanical strength	Back-washing
Long life time	
Uniform pore size	Dependent upon preparation methods
Disadvantages	
Difficult sealing and module construction	
Cost	Expensive sauce chemicals, complex processing

some extent in practical applications. For example, inorganic micro- and ultrafiltration membranes, which are currently commercially available, have been applied in areas of food, beverage, and biotechnology, and comprise approximately 12% of the membranes used in these applications [2]. As shown in Table 10.1.2, inorganic materials have the advantages of thermal stability, resistance to solvents and chemicals. It should be noted that the advantages are dependent on the types of materials used. Mechanical strength and long-life are also one of the advantages of ceramic over polymeric materials [2, 3]. Inorganic membranes have the disadvantage of difficulty in sealing between the inorganic membranes and metal housings, and constructing modules, and also are considered to have a relatively high cost because of the expensive source materials, the complex processing of membrane fabrication, and low membrane surface area per volume of a membrane module (low packing density compared with polymeric membranes). The selectivity of inorganic membranes is considered to be rather low compared with polymeric membranes, but recently it has been made clear several inorganic membranes show superior separation characteristics, compared to conventional polymeric membranes, which will be described in a later section.

10.1.3 PREPARATION OF POROUS CERAMIC MEMBRANES

Porous membranes must have pores which are connected continuously from the feed stream to permeate stream, otherwise no permeation through the

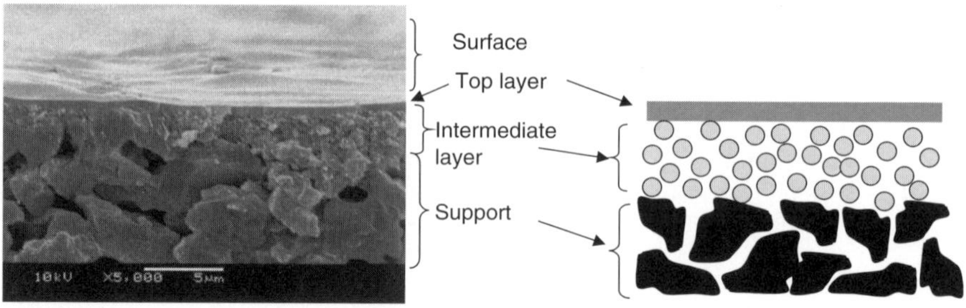

FIGURE 10.1.2 SEM photo (a) and a schematic representation (b) of an asymmetric, composite membrane.

membranes is possible. Figure 10.1.2 shows a SEM photo and a schematic representation of a typical structure of porous membranes: asymmetric-composite membranes. Porous supports, which are usually manufactured using powders by extrusion or slip-casting, have large pore sizes in excess of 1 μm and a thickness of the order of millimeters, and are designated for purposes of mechanical strength. The intermediate layer is coated on the support layer to reduce pore sizes for the further coating of the top layer. The separation top layer, which has separation ability and needs to have controlled pore sizes suitable for the specific separation, is formed as thin as possible on the intermediate layer. In this way, the pore size of asymmetric composite membranes shows a gradient structure from porous supports to the separation layer, so as to minimize the resistance to permeation across a membrane.

Table 10.1.3 summarizes the preparation processes of inorganic membranes and the pore size limits [2, 3]. Membranes can be classified into two groups based on differences in their morphology: porous or nonporous membranes. In powder sintering methods, ceramic powders, which have been pre-ground to several micrometers in diameter, can be formed in various types of membrane shapes such as tubes, plates and monoliths, by extrusion, tape-casting, and slip-casting. Since there is a low limit of particle diameters at approximately several hundred nm, the pore sizes of the membranes are approximately 0.1 μm at a minimum, which is in the microfiltration range, and are not small enough for molecular separation. Therefore, membranes prepared by powder sintering have been used as supports for preparing membranes of small pore sizes by further coating.

Porous membranes have been prepared by the sol–gel process from a variety of metal oxides and composite oxides. The sol–gel process is divided into two main routes: the polymeric sol–gel route and the colloidal sol–gel route [3]. A metal alkoxide or inorganic salt is hydrolyzed and a simultaneous condensation reaction occurs to form polymeric or colloidal sols. In the colloidal sol route,

TABLE 10.1.3 Preparation Methods of Inorganic Membranes

Type	Process	Materials	Pore diameter
Porous membranes	Powder sintering (extrusion or tape-casting)	α-Al_2O_3, ZrO_2, TiO_2	100 nm–
	Sol–gel	SiO_2, γ-Al_2O_3, ZrO_2, TiO_2, Fe_2O_3	1–50 nm
	CVD	SiO_2	<1 nm
	Pyrolysis	C, SiC, Si_3N_4	<1 nm
	Hydrothermal treatment	Silicalite, NaA, NaY	<1 nm
	Anodic oxidation	Al_2O_3 (amorphous)	10 nm–
	Phase separation/leaching	SiO_2	4 nm–
	Dynamic membranes	Amorphous ZrO_2	
	Sintered metal	Stainless steel	0.1 μm–
Non-porous membranes	Solid electrolyte	YSZ, perovskite	
	Metal	Pd, Pd/Ag	
Composite membranes	Modification with organic component	Silane coupling agent	
	Modification with inorganic component	MgO, Ag	

the hydrolysis and condensation reaction is fast, in comparison with the case of the polymeric sol route, which is achieved by adjusting the reaction conditions (type of alkoxide, solvent, catalyst, composition of reactants, molar ratio of water, temperature, etc.), resulting in a fully hydrolyzed alkoxide (highly branched polymer). The rapid condensation reaction causes particulate growth and/or the formation of precipitates. Particulate sols can be obtained by the precise control of reaction conditions and/or the peptization of the precipitate by adding acid. In the polymeric sol route, the hydrolysis reaction is kept slower and typically achieved by adding a small amount of water, resulting in a partially hydrolyzed alkoxide and the formation of a linear inorganic polymer. Through the subsequent gellation process, polymeric sols form a gel network. In both routes, the pore size of membranes can be controlled by the size of the sols; pore sizes can be controlled by the void spaces among the packed colloidal particles (i.e. interparticle pore) for the case of the colloidal sol route and the size of the gel network for the case of the polymeric gel route, respectively. Since the size of the gel network is smaller than the size of colloidal sols, the polymeric sol route is thought to be appropriate for the preparation of microporous materials. Pore sizes by the sol–gel process can be controlled from less than 1 nm to approximately 50 nm. Materials prepared have been silica, γ-alumina, titania, zirconia, and composite oxides such as silica-zirconia, and will be discussed extensively in a later section.

Chemical vapor deposition (CVD) of metal alkoxides such as tetraethoxysilane (TEOS) has been carried out to prepare microporous membranes, the pore sizes of which are controlled less than 0.5 nm, which is suitable especially for hydrogen separation [4–6]. A pyrolysis method is also employed for the preparation of microporous membranes [2, 3, 7]. Polyimide, polyvinylidene chloride, cellulose, polyacrylonitrile, and phenolic resin, which are thermosetting polymers or have high melting points, have been pyrolyzed in the absence of oxygen at high temperature in a range from 600 to 800°C, in a form of supported membranes or self-standing fibers such as hollow fibers. The pore size is less than 1 nm, and shows gas separation ability such as oxygen/nitrogen. Zeolite or zeolitic membranes have been prepared by hydrothermal treatment. To date, MFI (silicalite and ZSM-5), A, and Y-type zeolite membranes have been reported to be successfully manufactured [8–11], and some are currently commercially available. The pore sizes of zeolite membranes are basically determined by the crystalline intrapores (zeolitic pores), and therefore usually less than 1 nm. It should be noted that the separation mechanism of zeolite membranes remains controversial, as to whether the permeation is controlled by zeolitic intrapores or intercrystalline pores. Anodic oxidation of aluminum in acid solutions gives uniform cylindrical pores as shown in Figure 10.1.3 [12], which is different in structure from membranes prepared by other methods, but the pore size is normally larger than 10 nm.

In phase separation and leaching method [13–15], a homogeneous glass such as ternary mixtures of SiO_2–B_2O_3–Na_2O is melt at high temperature (1000–1500°C), separates into two phases: a silica-rich and -poor phase, after leaving at a specified temperature, based on spinodal demixing. After leaching

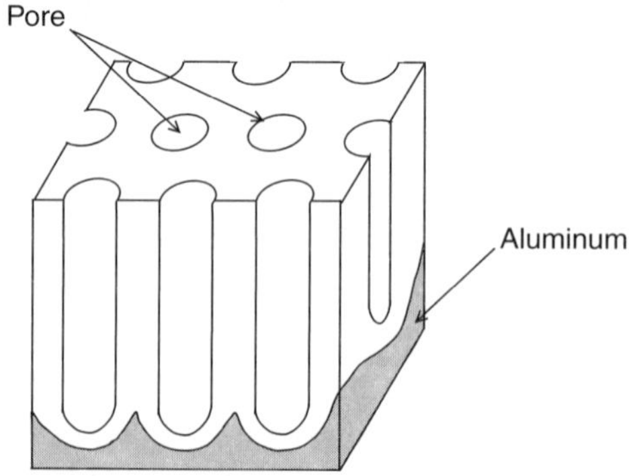

FIGURE 10.1.3 Pore structure of an anodic oxide aluminum membrane.

the silica-poor phase by acid, a porous silica membrane can be obtained. The typical membrane is a Vycor glass porous membrane, which has pore size of 4 nm. A pore size by this method can be reduced to approximately 0.5 nm by the precise control of the conditions used in the phase separation process.

Concerning non-porous membranes, these are categorized as dense ceramic electrolytes such as yttria-stabilized zirconia (YSZ) and perovskite membranes [16], which allow only the permeation of ionic oxygen. Permeation through metal membranes such as palladium and a palladium alloy is based on the selective dissolution of hydrogen and diffusion through the metal membrane.

10.1.4 LIQUID PHASE SEPARATION

Membrane performance for liquid phase separation can be evaluated by two parameters: permeability (L_p) and rejection (R). The permeability dimension is $m^3\,m^{-2}\,s^{-1}\,Pa^{-1}$ in SI units, and indicates how much solvent, such as water, permeates through the unit membrane area per second under a pressure difference of 1 Pa. Rejection, R, is the ratio of the concentration retained in feed stream over the feed concentration as defined in Eq. 1 with the concentration of permeate, C_p, and feed, C_f.

$$R = 1 - (C_p/C_f) \tag{1}$$

$R = 0$ indicates that no separation between the feed and permeate, while only solvent permeates with no permeation of solutes for the case of $R = 1$. Rejection is largely determined by the pore size and the pore size distribution. The pore size distribution of porous membranes must be sharp for selective separation. Therefore, it is important to understand the current status between pore size and inorganic materials for filtration materials. Figure 10.1.4 shows a schematic diagram of pore sizes obtained by several materials which have been used for liquid phase separation. α-Alumina, which is a common material used for the preparation of a porous ceramic membrane, has excellent stability in acidic as well as basic pH. However, there is a limitation in pore size, which is in a range of microfiltration (larger than 100 nm). On the other hand, γ-alumina, which is prepared by coating boehmite sols and subsequent heat-treatment, is reported to have a pore size in a range of ultrafiltration membranes (the pore size can be controlled down to approximately 4 nm), but is not as stable as in the case of α-alumina. Titania and zirconia have an excellent stability in aqueous solutions, and the pore size in the NF range is under development. Amorphous silica, which is mainly prepared via the hydrolysis and condensation of TEOS, has a great advantage in controllability of the pore size in a wide range from RO/NF to UF, but is not stable in aqueous solutions. Therefore,

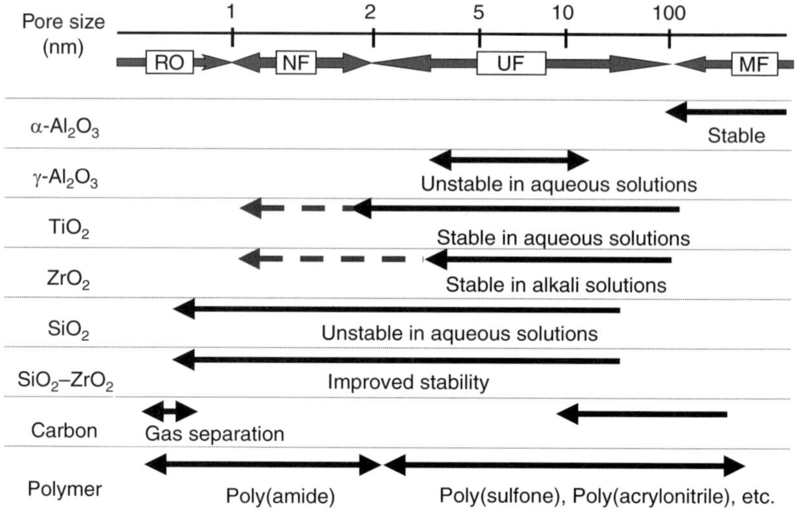

FIGURE 10.1.4 Materials and pore sizes for use in liquid phase separation.

the incorporation of other metal oxides such as zirconia into silica has been investigated. Carbon has excellent stability in aqueous solution in acidic and alkali pH, and therefore, has potential for use as materials for filtration. Carbon membranes which have been commercialized are in a range of mainly MF and partly UF. Concerning polymeric membranes, materials for MF and UF are polyacrylonitrile, polysulfone, cellulose, and cellulose triacetate, which can be used at relatively high temperatures (≈80°C). For the case of polymeric RO and NF membranes, which have pore sizes of less than 2 nm, have been manufactured mostly from poly(amide). Polyamide can be controlled to give pore size less than 2 nm, but the utilization is limited normally in the pH range from 3 to 11 and at temperatures from 20 to 40°C [17].

In summary, inorganic MF and UF membranes have now been commercialized from α-alumina, titania, zirconia and carbon, and applied to a variety of applications. Extensive investigation has been devoted to the development of porous membranes having pore sizes in the NF range. New trends in membrane development are summarized below, followed by reviewing practical applications.

10.1.4.1 Sol–Gel Membranes

In the sol–gel process, porous ceramic membranes are manufactured by sol-coating on porous substrates, and drying for gelling, followed by firing process.

The pore sizes of porous membranes by the sol–gel process can be controlled by colloidal diameters and firing temperature, since the pores are considered to be formed as spaces among packed colloidal particles, that is, interparticle pores.

10.1.4.1.1 Alumina Membranes

Alumina porous membranes have been prepared by the Yoldas method. That is, boehmite sol is prepared by the hydrolysis of aluminum butoxide or aluminum propoxide in hot water (higher than 80°C), followed by peptization by acids such as nitric acid [18, 19]. γ-Alumina membranes can be obtained after firing boehmite membranes at 400–600°C. Figure 10.1.5 shows a schematic representation of γ-alumina membranes [19], indicating that plate-shaped particles (boehmite and γ-alumina after firing) are packed to form slit-shape pores, in which distance between the slits is approximately several nanometers. Figure 10.1.6 shows molecular weight cut-off (MWCO) curves, that is, the rejection of various molecules plotted as a function of molecular weight [18, 20]. The MWCOs, which are defined as a molecular weight showing an approximate rejection of 90%, are in the approximate range of 2000–10 000, depending on the firing temperature. MWCO fired at higher temperature increases because of the growth in particle size. A further increase in firing temperature above 1000°C causes a phase transition of γ-alumina to α-alumina, resulting in the conversion of the UF membranes to MF membranes. On the other hand, Larbot *et al.* [21] reported the successful preparation of colloidal sol solutions which gave a pore size of 0.6 nm after firing at 450°C. γ-Alumina membranes having MWCO of 350 and 450 were prepared by firing at 400 and 650°C, respectively. As mention above, γ-alumina is not stable, especially in aqueous solution. At present, no γ-alumina membranes have been commercialized. Another approach to utilize γ-alumina is to apply it to non-aqueous solutions. Persin *et al.* [22] applied γ-alumina membranes fired at 450°C (MWCO $\approx$ 400) to toluene/cyclohexane separation in pervaporation and reported a separation factor of 3 for toluene over cyclohexane. No separation occurred through γ-alumina membranes fired at 650°C.

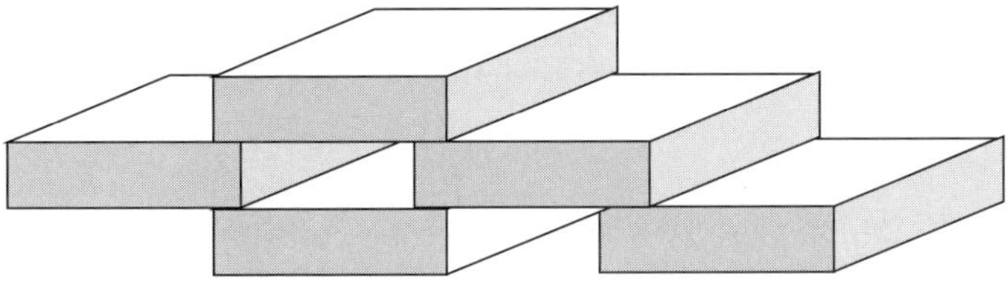

FIGURE 10.1.5 Pore structure of γ-alumina (boehmite) membrane proposed by Leenaars and Burggraaf [19].

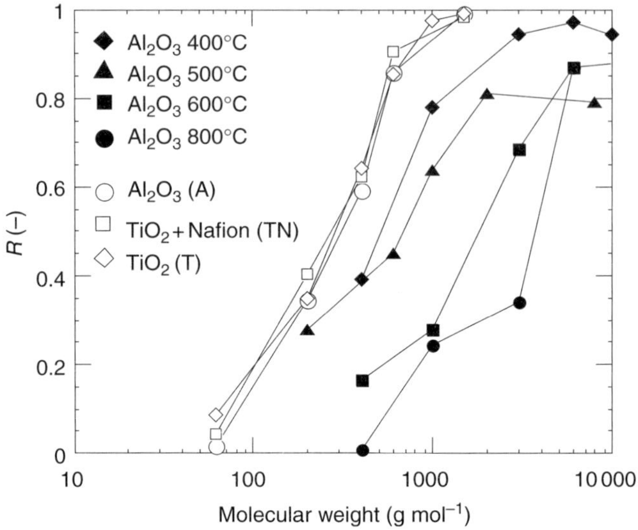

FIGURE 10.1.6 Molecular weight cut-off curves for alumina and titania porous membranes: filled symbols [18]; open symbols [27].

10.1.4.1.2 Titania Membranes

Titania membranes show excellent chemical resistance, and can be used in both acidic and basic pH; and moreover, they show interesting photocatalytic activity. Titania UF membranes have been commercialized by several companies. At present, extensive efforts have focused on the preparation of porous membranes having small pore sizes in the NF range (1–2 nm). Since pore sizes are believed to be controlled by the sizes of the packed particles, controlling the sol size is a crucial process for membrane processing. Anderson *et al.* [23] prepared nanosized TiO_2 and ZrO_2 particles (3–5 nm) by carrying out hydrolysis and condensation reactions of metal *tert*-amyloxide with a small amount of water (molar ratio of $H_2O/Ti = 3–5$) in *tert*-amyl alcohol solutions. Hydrolysis and condensation reactions of metal alkoxides are written as follows:

$$Ti(OR)_4 + H_2O \rightarrow Ti(OR)_3(OH) + ROH \qquad \text{(hydrolysis)}$$

$$Ti(OR)_4 + Ti(OR)_3(OH) \rightarrow (RO)_3TiOTi(OR)_3 + ROH \qquad \text{(condensation)}$$

Since both reactions take place by nucleophilic substitution, the electronegativity and the size of alkoxy groups is important. By using a large branched alkoxy group (*tert*-amyl alkoxide), the reaction rate can be reduced, resulting in small-sized sols. An amorphous phase predominated in the xerogel

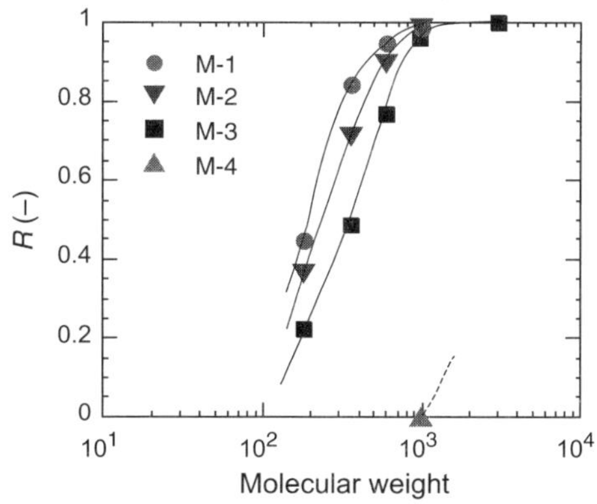

FIGURE 10.1.7 Molecular weight cut-off curves for titania porous membranes (solutes: sugars, 500 ppm) [25].

dried at room temperature and fired at 300°C, and phase transition from amorphous to anatase and from anatase to rutile occurred in the range 300–400 and 400–600°C, respectively. Therefore, it is important to control firing temperature in order to obtain controlled pore sizes of membranes. Titania membranes which were coated on porous alumina membranes and fired at 200°C, showed approximately a 100% rejection of polyethylene glycol 200 (PEG200, MW 200), depending to operating conditions [24]. Another strategy for controlling MWCOs is the appropriate choice of sol diameters, which is used at the final coating for the separation top layer. Tsuru *et al.* [25] successfully prepared a variety of MWCOs for TiO_2 membranes at the same firing temperature of 450°C using sol solutions having different sizes of sol diameters, as shown in Figure 10.1.7. M1, M2, and M3 show MWCOs approximately 500, 600, and 800, while M4 shows nearly no rejection for α-dextrin (MW 972).

On the other hand, Soria and Cominottim claimed that they had commercialized TiO_2 NF membranes having a MWCO of 1000 [26]. Colloidal sols with a diameter of 15 nm, which showed an anatase phase and which were stable upto 600°C, were coated on porous alumina supports and fired at 500°C. The MWCO curves of titania (T), alumina (A), and titania–nafion composite (TN) membranes prepared by this research group are also shown in Figure 10.1.6 [27]. Table 10.1.4 shows MWCOs and water permeability along with recent reports [28, 29]. Table 10.1.4 indicates that titania membranes show a relatively large water permeability, in comparison with those of polymeric NF membranes [17].

TABLE 10.1.4 Nanofiltration Performances of Ceramic Inorganic Membranes

Membrane	MWCO ($R_{obs} = 0.9$)	$L_p \times 10^{11}$ ($m^3\ m^{-2}\ s^{-1}\ Pa^{-1}$)	Pore radius (nm)	Reference
γ-Al_2O_3 (A)	≈600	0.68	0.6	27
TiO_2 + Nafion (TN)	≈600	0.53	0.6	27
TiO_2 (T)	800–1000	1.7	0.9	27
TiO_2	≈1000	33	1–2	29
TiO_2	500	5.5	0.9	28

10.1.4.1.3 Zirconia Membranes

Zirconia, which is stable as well as titania, especially, in alkali solution, is also one of the promising materials for separation membranes. Several manufacturers have commercialized zirconia UF membranes, but not NF membranes. Zirconium propoxide [30] or butoxide [31], which are mainly used as precursors of zirconia sols, are highly sensitive to water to form suspensions (the reaction rate with water is much faster than titanium alkoxide), the preparation of nanosized sols is difficult. Therefore, very few reports have appeared on the successful preparation of zirconia nanofiltration membranes. Vacassy *et al.* [30] added acetylacetone to zirconia propoxide to prevent hydrolysis, and reported a successful preparation of porous zirconia membranes showing a water permeability of $3.4 \times 10^{-11}\ m\ s^{-1}\ Pa^{-1}$ and a rejection of 55% towards saccharose (MW = 342).

10.1.4.1.4 Composite Oxide Membranes

Amorphous silica, which is an acidic metal oxide, is not stable in aqueous solutions, especially in neutral and alkaline pHs, but has a great advantage in terms of pore-size controllability. Therefore, the incorporation of zirconia into silica has been investigated. TEOS and zirconium propoxide or zirconium butoxide were applied to the composite precursor for reverse osmosis membranes [32] and nanofiltration membranes [33]. As shown in Figure 10.1.8, the MWCOs of silica-zirconia membranes (molar ratio: 9 Si–1 Zr), were controlled in a range from 200 to 1000, showing a smaller MWCO compared with those of γ-alumina and titania. Tsuru *et al.* applied silica–zirconia NF membranes to separation in aqueous solutions (neutral solutes and electrolytes) [33, 34], and non-aqueous solutions [35], and found them to be stable in aqueous solutions after long-term permeation experiments.

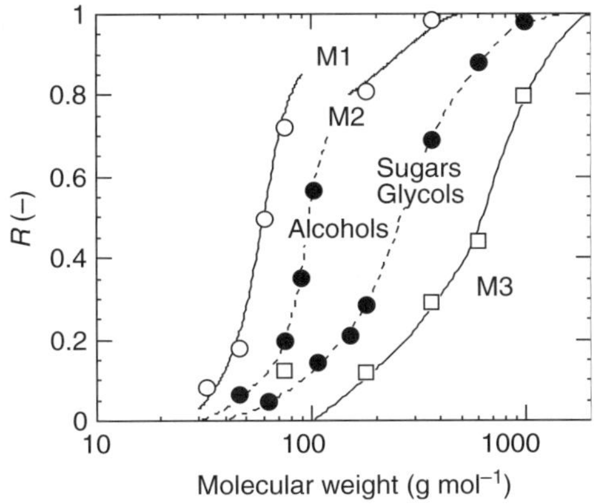

FIGURE 10.1.8 Molecular weight cut-off curves for silica–zirconia membranes [33].

10.1.4.1.5 Other Materials

Hafnia, which can be sintered up to 1850°C without transition of the monoclinic form, was used for preparing NF membranes, showing a MWCO of 420 and a pore size close to 1 nm by sintering at 450°C [36]. Various types of metal oxides such as SnO_2 [37], $ZnAl_2O_4$ [38], which were prepared by the sol–gel process, have been proposed for the fabrication of porous membranes, although the pore size is in the UF range.

10.1.4.2 Other Preparation Processes

Nakashima *et al.* [39] developed tubular glass membranes of different compositions from a so-called Vycor composition, by a phase separation/leaching method of $CaO–Al_2O_3–B_2O_3–SiO_2$ glass. The pore sizes, which were controlled in a range larger than 50 nm, showed a sharp distribution. The glass membranes were used for filtration in aqueous solutions. Another interesting application is membrane emulsification by taking advantage of the sharp pore size distribution.

Carbon, which has stability in aqueous as well as non-aqueous solutions, is also a candidate for porous inorganic membranes. At present, a few manufacturers have commercialized carbon UF membranes, the pore sizes of which are larger than 10 nm. No carbon nanofiltration membranes have been reported, while carbon gas separation membranes, which have pore sizes less than 1 nm, have been reported by several groups [7]. Nomura *et al.* [40] coated poly (vinylidene chloride) on α-alumina supports and pyrolyzed them at 825°C in a N_2 atmosphere. The porous carbon membranes consisted of two types of pores: mesopores of 10 nm (interparticle) and micropores of 0.7 nm (intraparticle), showing a MWCO of 10 000 and a adsorption ability based on the micropores.

Organic–inorganic composite membranes have been investigated for their excellent stability by inorganic materials and high selectivity of organic membranes. Nafion, which is a negatively charged polyelectrolyte membrane, was used for a composite membrane with a mesoporous titania membrane [26]. γ-Alumina membranes were modified with silane coupling agents to reduce the pore size and enhance the selectivity of toluene over a toluene/lubricant mixture [41]. Ohya *et al.* [42] modified the pores of α-alumina membranes by the molecular-wise growth in liquid and gas-phase copolymerization of polyimide to control pores sizes for UF and gas separation membranes.

Several types of zeolite membranes such as A-type, Y-type, silicalite, ZSM-5, etc. have been developed, and have been applied mainly to gas and pervaporation separations. Kumakiri *et al.* [43] prepared A-type zeolite membranes by hydrothermal synthesis with seed growth, and applied these to the reverse osmosis separation of water/ethanol mixtures. The zeolite A membrane showed a rejection of 40% and a permeate flux of 0.06 kg m^{-2} h^{-1} for 10 wt% ethanol at a pressure difference of 1.5 MPa, while a permeate flux of 0.8 kg m^{-2} h^{-1} and a separation factor of 80 were obtained in PV.

One of the features of inorganic membranes is their controlled pore structure. Anodic aluminum oxide membranes have uniform cylindrical pores, and were applied to an investigation of the analysis of transport mechanism [12]. Another route involves the application of a micelle template to membrane preparation [44]. Cubic mesoporous silica (MCM48) membranes were prepared on a stainless steel supports [45] to possible applications for filtration membranes and membrane reactions.

Porous membranes were prepared by using clays such as montmorillonite and sepiolite, and the dynamic filtration of solutions having fine particles of hydrous zirconium oxide, although their pore sizes are in a range of ultrafiltration membranes.

10.1.4.3 Applications

Inorganic microfiltration and ultrafiltration membranes have been used in a wide variety of processings. Typical applications were reported by Hsieh [2] and Bhave [3], and are summarized in Table 10.1.5. MF, which has pore sizes larger than 100 nm, can be applied to remove or concentrate particles or microorganisms, while UF membranes have been used for the separation of components, the sizes of which are from 2 to 100 nm, such as proteins and colloidal solutes. In this section, potential applications will be reviewed.

TABLE 10.1.5 Applications of Liquid Phase Separations

Application area	Application examples
Water treatment	Drinking water, wastewater treatment
Waste oil treatment	Oil–water separation, purification of used oils
Dairy industry	Bacteria removal of milk, concentration of milk, concentration of protein
Beverage	Clarification of fruit juice, clarification of wine, bacteria removal, microorganism separation from fermented
Biotechnology	Microorganism separation and cell debris filtration, plasma separation
Petrochemical processing	Removal of precipitated heavy metals and solids

10.1.4.3.1 Water Treatment

One of the largest applications is water treatment, including drinking water as well as wastewater. For the case of water treatment using porous membranes, the surface charge can play a very important role in determining ion separation and flux based on electrokinetic effects. The surface charge influences several aspects of membrane separation. Ion separation is possible based on electrostatic interaction between ions and surface charge; coions, which have the same charge as the membrane surface, receive repulsive force from the membrane surface, which is referred to Donnan exclusion. The repulsive force for divalent coions is greater than for monovalent coions, and, therefore, ion separation can be successfully carried out using nanofiltration membranes, the pore size of which is even much greater than ion sizes [46, 47].

In terms of practical applications, fouling, which occurs via formation of a cake/gel layer on the membrane surface and pore blocking by colloidal particulates and adsorption of dissolved solutes, reduces permeate flux during filtration, and is an important problem to be solved. Modification of membrane surface with a helix geometry configuration, which promoted turbulence, was found to be effective in recovering permeate flux by a factor of 6, compared with non-modified membranes [48].

10.1.4.3.2 Application for Non-aqueous Systems

Membranes have been developed for applications in aqueous systems such as desalination and drinking water treatment. Recently, it has been expected that membrane separation will be expanded to various applications such as separation of non-aqueous solutions and filtration in non-aqueous solutions. Petrochemical processing is one of the applications to which organic membranes cannot be applied. Ceramic membranes: ZrO_2/C (pore size 140–3.7 nm), silica

(10 nm), and γ-alumina (5 nm) membranes were applied for an oil deasphalting process at temperatures in the range of 115–180°C [49]. ZrO_2/carbon membrane showed the best results, while no permeation occurred through silica and γ-alumina membranes which had a strong affinity for the feed solution. ZrO_2, silica and γ-alumina membranes were applied for cleaning up used oils at temperatures from 150 to 260°C. A γ-alumina membrane showed a gradual decrease in flux as a result of fouling, which was caused by affinity between components in the feed and the membrane surface, while the other two membranes showed a steady flux. The treated oils, which were separated from heavy metals, ash, and suspended solids, are reported to be considered as clean oils [50].

Affinity between feed solutions with the membrane surface plays an important role in determining permeation properties through porous membranes. The effects of pore sizes on permeation performances through porous membranes were investigated based on the dependency of the types of solvents and temperature [51]. It has been made clear that the transport mechanism through inorganic porous membranes in the NF range is an activated process and does not obey the viscous flow mechanism which is an ordinal flow mechanism through UF and MF membranes having relatively large pore sizes. Another interesting application is the use of inorganic membranes for separation and concentration in non-aqueous solutions such as ethanol and hexane [52, 53] and in supercritical fluids [54].

10.1.5 GAS PHASE SEPARATION/PERVAPORATION

In comparison with liquid phase separation, gas phase separation has not been so widely applied from the viewpoints of market size as well as application varieties, but has great potential for future development. This section will briefly describe gas phase separation and pervaporation using ceramic membranes.

10.1.5.1 Particulate Filtration in Gas Phase

Particulate filtration from a gas stream is also possible using porous ceramic membranes whose pore size is in the MF range. Ceramic porous membranes can be used for clean room applications based on chemical stability, while polymeric fabric filters could be degraded by reaction with reactive gases [2]. Particle filtration from hot exhaust gases is another promising application because of the excellent thermal stability of these types of membranes. The removal of particulates from flue gases of diesel engine automobiles is also a promising application. Larbot *et al.* [55] reported more than 99.99% of polydispersed NaCl particles (10–70 nm) could be removed using porous membranes prepared by an extrusion method (pore size ≈ 10 μm). Several materials such as cordielite, α-alumina, and mullite, have been commercialized/under development for particulate filtration at high temperatures.

10.1.5.2 Separation of Gaseous Mixtures

The separation of gaseous mixtures by ceramic membranes is summarized in Table 10.1.6. Various types of ceramic membranes such as SiO_2, carbon, and metals such as palladium, have been prepared for the separation of a wide variety of gaseous mixtures.

To date, hydrogen semipermeable membranes have been prepared by the sol–gel processes, the phase separation/leaching method, the chemical vapor deposition (CVD) technique and metal such as palladium. Generally, the permeabilities of gases prepared by the sol–gel process are relatively high due to the thin top layer compared to membranes prepared by other methods. Silica represents a typical material, and the composite metal oxide such as silica–zirconia has been recently investigated [56]. As shown in Figure 10.1.9, SiO_2–ZrO_2 membranes have permselectivity for hydrogen with separation factors of approximately 500 over nitrogen and a relatively high permeance of $2 \times 10^{-5}\,m^3(STP)m^{-2}\,s^{-1}\,kPa^{-1}$. Silica glasses prepared by a phase separation/leaching method have been applied to microporous hollow fiber

TABLE 10.1.6 Typical Applications of Gas Separation Using Inorganic Membranes

Application area	Separation mixtures	Membranes	Applications
Hydrogen	$H_2/N_2, H_2/CH_4$, $H_2/CO, H_2/CO_2$, H_2/organic gases	Amorphous SiO_2 (sol–gel, CVD, phase separation/leaching)	Recovery of hydrogen, refinery hydrogen recovery
Inorganic gaseous mixtures	$CO_2/N_2, O_2/N_2$	SiO_2, NaY, carbon	Removal of CO_2 from flue gas, air separation
Inorganic/organic gaseous mixtures	CO_2/CH_4, hydrocarbon/air, He/hydrocarbon	Silica, carbon	Removal of CO_2 from natural gases, hydrocarbon recovery, pollution control, recovery of helium
Organic/organic gaseous mixtures	C_3H_6/C_3H_8, C_2H_4/C_2H_6 n-C_4H_{10}/i-C_4H_{10}	SiO_2, MFI zeolite (ZSM, silicalite)	Olefin/paraffin separation, isomer separation

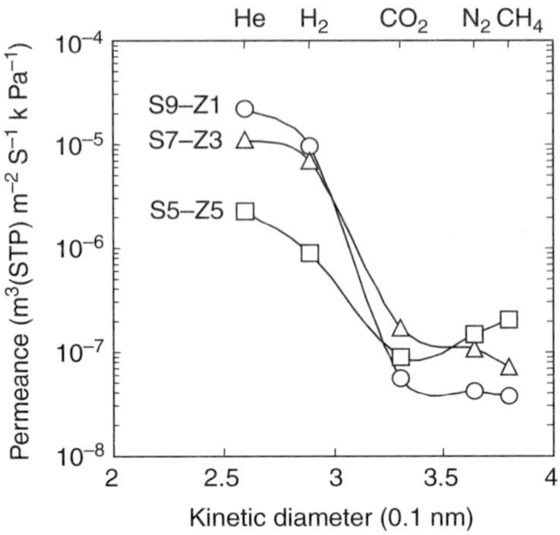

FIGURE 10.1.9 Permeance of silica–zirconia composite membranes as a function of kinetic diameter at 500°C (S9–Z1, S7–Z3 and S5–Z5 have molar compositions of SiO_2–ZrO_2 of 9–1, 7–3, and 5–5, respectively) [56].

membranes [14, 15]. Silica layers can be prepared by CVD on porous substrates [4–6]. Morooka *et al.* [57] reported the forced cross-flow CVD to fabricate hydrogen-selective silica layer inside pores of the substrates; in this method the silica layer is thought to be maintained in a mechanically stable state. However, the permeability is generally small.

Silica membranes prepared by the sol–gel process, the pore size of which can be controlled by the colloidal size, have been applied for CO_2/N_2 separations [58] and olefin/paraffin separations [59]. For example, silica microporous membranes showed a separation factor of propylene over propane as high as 75 with an approximate permeance of 0.2×10^{-5} m^3 m^{-2} s^{-1} kPa^{-1} at 35°C. However, it was pointed out that silica membranes needs to be improved with respect to their stability in humid air, because microporous silica becomes densified, resulting in a decrease in permeance [56].

Zeolite membranes have relatively large pore sizes for the separation of inorganic gaseous mixtures, and, therefore, have been applied for the separation of H_2/organic gases and organic/organic gaseous mixtures such as mixtures of *n*-/*i*-butane [10, 60].

10.1.5.3 Pervaporation

In pervaporation, the feed stream is kept in the liquid phase, which means that the separation mixtures are relatively large molecules compared with gaseous molecules, while the permeate stream is kept in gas phase by evacuation. Therefore, pore sizes for PV separation membranes need to be larger than those for gas separation, that is, approximately in a range from 0.5 to 2 nm. Silica and the composite oxide prepared by the sol–gel process have been employed for this purpose. Since the water molecule is the smallest and has hydrophilic properties, silica and silica composite membranes were found to be effective for dehydration such as organic acidic aqueous solutions and alcohol aqueous solutions. Organic/organic mixtures such as methanol/benzene and ethanol/cyclohexane can be separated using silica-zirconia membranes prepared by the sol–gel process [61].

Another type of inorganic membranes used to the PV separation is a zeolite membrane. Na-type zeolite membranes have been applied for dehydration of aqueous alcohol. Kita *et al.* [9] reported that a permeation flux of $3\,kg\,m^{-2}\,h^{-1}$ and separation factor (α) over 10 000 isopropyl alcohol aqueous solution (90 wt% isopropyl alcohol), which corresponds to much larger flux and selectivity compared with polymeric membranes (normally $\alpha \approx 1000$; flux $<0.1\,kg\,m^{-2}\,h^{-1}$.) On the other hand, a silicalite membrane, which is hydrophobic, preferentially permeates alcohol over water, showing a selectivity of 60 and flux of $0.8\,kg\,m^{-2}\,h^{-1}$ at 5 wt% of ethanol at 60°C [8].

10.1.6 CONCLUDING REMARKS

Porous ceramic membranes have been reviewed from the viewpoint of membrane preparation methods and applications for separation. These new classes of porous ceramic membranes hold considerable promise in applications such as separation at high temperatures. A membrane reaction where separation and reaction is combined in one system, will be realized using porous ceramic membranes, since most chemical reactions occur at high temperature where polymeric membranes cannot be applied. The preparation of porous ceramic membranes, which need to have uniform pore sizes, to be as thin as possible without defects, seems to represent a different strategy from conventional preparation of ceramic bulk bodies. This new research field of ceramic processing will contribute much to the development of membrane science and technology.

REFERENCES

1. Mulder, M. (1996). *Basic Principles of Membrane Technology*, Dordrecht: Kluwer Academic Publishers.
2. Hsieh, H. P. (1996). *Inorganic Membranes for Separation and Reaction*, Amsterdam: Elsevier.

3. Bhave, R. R. (1991). *Inorganic Membranes: Synthesis, Characterization, and Applications*, New York: Van Nostrand Reinhold.
4. Gavalas, G. R., Megiris, C. E., and Nam, S. W. (1989). Deposition of H_2-permselective SiO_2 films. *Chem. Eng. Sci.* **44**: 1829–1835.
5. Tsapatsis, M., Gavalas, G. R., and Xomeritakis, G. (2000). Chemical vapor deposition, *Recent Advances in Gas Separation by Microporous Ceramic Membranes*, pp. 397–416, Kanellopoulos, N. K., ed., Amsterdam: Elsevier.
6. Nakao, S., Suzuki, T., Sugawara, T., Tsuru, T., and Kimura, S. (2000). Preparation of microporous membranes by TEOS/O_3 CVD in the opposing reactants geometry. *Microporous Mesoporous Mater.* **37**: 145–152.
7. Morooka, S., Kusakabe, K., Kusuki, Y., and Tanihara, N. (2000). Microporous carbon membranes, *Recent Advances in Gas Separation by Microporous Ceramic Membranes*, pp. 323–334, Kanellopoulos, N. K., ed., Amsterdam: Elsevier.
8. Sano, T., Yanagishita, H., Kiyozumi, Y., Mizukami, F., and Haraya, K. (1994). Separation of ethanol/water mixture by silicalite membrane on pervaporation. **95**: 221–228.
9. Kita, H., Hori, K., Ohtoshi, Y., Tanaka, K., and Okamoto, K. (1995). Synthesis of a zeolite NaA membrane for pervaporation of water/organic liquid mixtures. **14**: 206–208.
10. Coronas, J., and Santamaria, J. (1999). Separation using zeolite membranes. *Sep. Purif. Methods* **28**: 127–177.
11. Kusakabe K., Kuraoka T., Uchino, K., Hasegawa, Y., and Morooka, S. (1999). Gas permeation properties of ion-exchanged faujasite zeolite membranes. *AIChE J.* **45**: 1220–1226.
12. Ichimura, S., Tsuru, T., Nakao, S., and Kimura, S. (2000). Analysis of linear macromolecule transport through aluminum anodic oxide membranes by pore model. *J. Chem. Eng. Jpn.* **33**: 141–151.
13. Elmer, T. H. (1992). Porous and reconstructed glass, *Engineered Materials Handbooks*, Vol. 4: *Ceramics and Glasses*, pp. 427–432, ASM International.
14. Shelekhin, A. B., Dixon, A. G., and Ma, Y. H. (1995). Theory of gas diffusion and permeation in inorganic molecular-sieve membranes. *AIChE J.* **41**: 58–67.
15. Kuraoka, K., Amakawa, R., Matsumoto, K., and Yazawa, T. (2000). Preparation of molecular-sieving glass hollow fiber membranes based on phase separation. *J. Membr. Sci.* **175**: 215–223.
16. Bouwmeester, H. J. M., and Burggraaf, A. J. (1996). Dense ceramic membranes for oxygen separation, *Fundamentals of Inorganic Membrane Science and Technology*, pp. 435–528, Burggraaf A. J., and Cot, L., eds., Amsterdam: Elsevier.
17. Tsuru, T., Shutoh, T., Nakao, S., and Kimura, S. (1994). Peptide and amino acid separation by nanofiltration membranes. *Sep. Sci. Technol.* **29**: 971.
18. Uhlhorn, R. J. R., Huisin'tveld, M. H., Keizer, K., and Burggraaf, A. J. (1992). Synthesis of ceramic membranes. *J. Mater. Sci.* **27**: 527.
19. Leenaars, A. F. M., Keizer, K., and Burggraaf, A. J. (1984). The preparation and characterization of alumina membranes with ultra-fine pores. *J. Mater. Sci.* **19**: 1077.
20. Peterson, R. A., Anderson, M. A., and Hill, C. G. (1990). Permselectivity characteristics of supported ceramic alumina membranes. *Sep. Sci. Technol.* **25**: 1281.
21. Larbot, A., Alami-Younssi, S., Persin, M., Sarrazin, J., and Cot, L. (1994). Preparation of a γ-Alumina nanofiltration membrane. *J. Membr. Sci.* **97**: 167.
22. Persin, M., Larbot, A., Alami-Younssi, S., Elbaz-Poulichet, F., Sarrazin, J., and Cot, L. (1993). The behavior in nanofiltration and pervaporation experiments of gamma alumina membranes elaborated by sol gel process, "Proceedings of Third International Conference on Inorganic Membranes," pp. 525–528.
23. Xu, Q., and Anderson, M. (1994). Sol-gel route to synthesis of microporous ceramic membranes: preparation and characterization of microporous TiO_2 and ZrO_2 xerogels. *J. Am. Ceram. Soc.* 77: 1939–1945.

24. Wildman, D. L., Peterson, R. A., Anderson, M. A., and Hill. C. G. (1993). Investigation of titania membranes for nanofiltration, "Proceedings of Third International Conference on Inorganic Membranes," pp. 111–117.
25. Tsuru, T., Hironaka, D., Yoshioka, T., and Asaeda, M. (2003). Titania membranes for liquid phase separation: effect of surface charge on flux. *Sep. Purif. Technol.* 25(2001): 307–314.
26. Soria R., and Cominottim, S. (1996). Nanofiltration ceramic membranes, "Proceedings of Fourth International Conference on Inorganic Membranes," pp. 194–197.
27. Sarrade, S. J., Rios, G. M., and Carles, M. (1994). Dynamic characterization and transport comparision of three nanofiltration membranes, "Proceedings of Fourth International Conference on Inorganic Membranes," pp. 157–165.
28. Puhlfürß, P., Voigt, A., Weber, R., and Morbé, M. (2000). Microporous TiO_2 membranes with a cut off <500 Da. *J. Membr. Sci.* **174**: 123–133.
29. Benfer, S., Popp, U., Richter, H., Siewert, C., and Tomandl, G. (2001). Development and characterization of ceramic nanofiltration membranes. *Sep. Purif. Technol.* **22–23**: 231–237.
30. Vacassy, V., Guizard, C., Thoraval, V., and Cot, L. (1997). Synthesis and characterization of microporous zirconia powders: application in nanofilters and nanofiltration characteristics. *J. Membr. Sci.* **132**: 109–118.
31. Wu, J. C., and Cheng, L. (2000). An improved synthesis of ultrafiltration zirconia membranes via the sol-gel route using alkoxide precursor. *J. Membr. Sci.* **167**: 253–261.
32. Yazawa, T., Tanaka, H., Nakamichi, H., and Yokoyama, T. (1991). Preparation of water and alkali durable porous glass membrane coated on porous alumina tubing by sol-gel method. *J. Membr. Sci.* **60**: 307–317.
33. Tsuru, T., Wada, S., Izumi, S., and Asaeda, M. (1998). Preparation of microporous silica–zirconia membranes for nanofiltration. *J. Membr. Sci.* **149**: 127–135.
34. Tsuru, T., Takezoe, H., and Asaeda, M. (1998). Ion separation by porous silica–zirconia nanofiltration membranes. *AIChE J.* **44**: 765–768.
35. Tsuru, T., Sudoh, T., Yoshioka, T., and Asaeda, M. (2001). Nanofiltration in non-aqueous solutions by inorganic porous membranes. *J. Membr. Sci.* **185**: 253–261.
36. Blanc, P., Larbot, A., Palmeri, J., Lopez, M., and Cot, L. (1998). Hafnia ceramic nanofiltration membranes. Part 1: Preparation and characterization. *J. Membr. Sci.* **149**: 151–161.
37. Santos, L. R. B., Santilli, C. V., Larbot, A., Persin, M., and Pulcinelli, S. H. (2001). Influence of membrane–solution interface on the selectivity of SnO_2 ultrafiltration membranes. *Sep. Purif. Technol.* **22–23**: 17–22.
38. Elmarraki, Y., Cretin, M., Persin, M., Sarrazin, J., and Larbot, A. (2001). Elaboration and properties of TiO_2–$ZnAl_2O_4$ ultrafiltration membranes. *Mater. Res. Bull.* **36**: 227–237.
39. Nakashima, T., Shimizu, M., and Kawano, M. (1987). Articles of porous glass and process for preparing the same. US Patent 4657875.
40. Sakoda, A., Nomura, T., and Suzuki, M. (1996). Activated carbon membrane for water treatments: application to decolorization of coke furnace wastewater. *Adsorption* **3**: 93–98.
41. Miller, J. R., and Koros, W. J. (1990). The formation of chemically modified γ-Alumina microporous membranes. *Sep. Sci. & Technol.* **25**: 1257–1280.
42. Sawamoto, S., Ohya, H., Yanase K., Semenova, S., Aihara, M., Takeuchi, T., and Negishi, Y. (2000). Nanotechnological method to control pore diameter of organic-inorganic composite membrane. Part 2. Molecular-wise vapor polymerization. *J. Membr. Sci.* **174**: 151–159.
43. Kumakiri, I., Yamaguchi, T., and Nakao, S. (2000). Application of a zeolite A membrane to reverse osmosis process. *J. Chem. Eng. Jpn.* **33**: 333–336.
44. Guizard, C. (1996). Sol-gel chemistry and its application to porous membrane processing, *Fundamentals of Inorganic Membrane Science and Technology*, pp. 227–258, Burggraaf A. J., and Cot, L., eds., Amsterdam: Elsevier.
45. Nishiyama, N., Koide, A., Egashira, Y., and Ueyama, K. (1998). Mesoporous MCM-48 membrane synthesized on a porous stainless steel support, *Chem. Commun.* 2147–2148.

46. Tsuru, T., Takezoe, H., and Asaeda, M. (1998). Ion separation by porous silica–zirconia nanofiltration membranes. *AIChE J.* **44**: 765–768.
47. Rios, G. M., Joulie, R., Sarrade, S. J., and Carlès, M. (1996). Investigation of ion sepration by microporous nanofiltration membranes. *AIChE J.* **42**: 2521–2528.
48. Broussous, L., Ruiz, J. C., Larbot, A., and Cot, L. (1998). Stamped ceramic porous tubes for tangential filtration. *Sep. Purif. Technol.* **14**: 53–57.
49. Guizard, C., Rambault D., Urhing, D., Dufour, J., and Cot, L. (1993). Deasphalting of long residue using ultrafiltration inorganic membranes, "Proceedings of Third International Conference on Inorganic Membranes," pp. 345–354.
50. Higgins, R., Bishop, B., and Goldsmith, R. (1993). Reclamation of waste lubricating oil using ceramic membranes, "Proceedings of Third International Conference on Inorganic Membranes," pp. 447–456.
51. Tsuru, T., Sudoh, T., Kawahara, S., Yoshioka, T., and Asaeda, M. (2000). Permeation of liquids through inorganic nanofiltration membranes, *J. Colloid Interf. Sci.* **228**: 292–296.
52. Tsuru, T., Sudoh, T., Yoshioka, T., and Asaeda, M. (2001). Nanofiltration in non-aqueous solutions by inorganic porous membranes. *J. Membr. Sci.* **185**: 253–261.
53. Wu, J., and Lee, E. (1999). Ultrafiltration of soybean/hexane extract by porous ceramic membranes. *J. Membr. Sci.* **154**: 251–259.
54. Sarrade, S., Rios, G. M., and Carlès, M. (1996). Nanofiltration membrane behavior in a supercritical medium. *J. Membr. Sci.* **114**: 81–91.
55. Ruiz, J. C., Larbot A., Prouzet, E., Blanc. P., Laffont P., and Coryn, P. (1998). New porous ceramic for air filtration, "Proceedings of the Fifth International Conference on Inorganic Membranes," pp. 232–235.
56. Yoshida, K., Hirano, Y., Fujii, H., Tsuru, T., and Asaeda, M. (2001). Hydrothermal stability and performance of silica-zirconia membranes for hydrogen separation in hydrothermal condition, *J. Chem. Eng. Jpn.* **34**: 523–530.
57. Morooka, S., Yan, S., Kusakabe, K., and Akiyama, Y. (1995). Formation of hydrogen-permselective SiO_2 membrane in macropore of α-Alumina support tube by thermal deposition of TEOS. *J. Membr. Sci.* **101**: 89–98.
58. Yoshioka, T., Nakanishi, E., Tsuru, T., and Asaeda, M. (2001). Experimental and analytical studies of gas permeation through microporous silica membranes. *AIChE J.* **47**: 2052–2063.
59. Asaeda, M., Yamamichi, A., Satoh, M., and Kamakura, M. (1993). Preparation of porous silica membranes for separation of propylene/propane gaseous mixtures, "Proceedings of Third International Conference on Inorganic Membranes," pp. 315–323.
60. Coronas, J., Noble, R., and Falconer, J. (1998). Separation of C_4 and C_6 isomers in ZSM tubular membranes. *Ind. Eng. Chem. Res.* **37**: 166–176.
61. Asaeda, M., and Tsuru, T. (1996). Porous silica–zirconia membranes for separation of organic molecular mixtures by pervaporation, "Proceedings of Fouth International Conference on Inorganic Membranes," pp. 68–78.

Handbook of Advanced Ceramics
S. Sōmiya *et al.* (Eds.)

11.1 Ceramic Bearing

HIROAKI TAKEBAYASHI
EXSEV Engineering Department, Koyo Seiko Co., Ltd., Kokobu Tojyo-machi, Kashihara 582-8588, Japan

11.1.1 INTRODUCTION

As a result of recent technological progress, the environments and condition under which rolling bearings are used are becoming severe as well as diverse. Therefore, bearings made of bearing steel are not operating in severe conditions.

Engineering ceramics such as silicon nitride (Si_3N_4), zirconia (ZrO_2), silicon carbide (SiC) and alumina (Al_2O_3) are excellent at heat resistance, corrosion resistance and wear resistance. Accordingly, engineering ceramics are being researched and developed for applications of bearings [1–4].

This chapter will present ceramic bearings. First, the applicability of all types of ceramic materials for ceramic bearings will be shown, and next, silicon nitride, which shows characteristic of being the most superior bearing material among various ceramic materials, will be focused on, and its performance for application to bearings will be presented. Specifically, such subjects as static load carring capacity, rolling contact fatigue life, fitting and application examples will be discussed.

11.1.2 APPLICABILITY OF CERAMICS TO BEARINGS

Table 11.1.1 shows the features of various ceramic materials [5]. Ceramic materials include silicon nitride (Si_3N_4), zirconia (ZrO_2), silicon carbide (SiC) and alumina (Al_2O_3). Mechanical strength is in the order alumina $<$ silicon carbide $<$ silicon nitride $<$ zirconia, and corrosion resistance is in the order silicon nitride $<$ zirconia $<$ silicon carbide $=$ alumina. Silicon nitride and zirconia are sintered by hot isostatic pressing (HIP).

Figure 11.1.1 shows the results of rolling life testing, in which a thrust type bearing tester was used to evaluate the applicability of various types of ceramics materials for bearings [5]. Life is indicated in the figure as that point in time when damage occurred on the ceramic flat plate as test load was increased each

TABLE 11.1.1 Characteristics of Fine Ceramics

Item	Ceramic material			
	Silicon nitride Si_3N_4	Zirconia ZrO_2	Silicon carbide SiC	Alumina Al_2O_3
Density (g/cm^3)	3.2	6.0	3.1	3.8
Linear expansion coefficient, (1/°C)	3.2×10^{-6}	10.5×10^{-6}	3.9×10^{-6}	7.1×10^{-6}
Vickers hardness, HV	1 500	1 200	2 200	1 600
Modulus of longitudinal elasticity (GPa)	320	220	380	350
Poisson's ratio	0.29	0.31	0.16	0.25
Three-point bending strength (MPa)	1 100	1 400	500	300
Fracture toughness ($MPa\,m^{1/2}$)	6	5	4	3.5

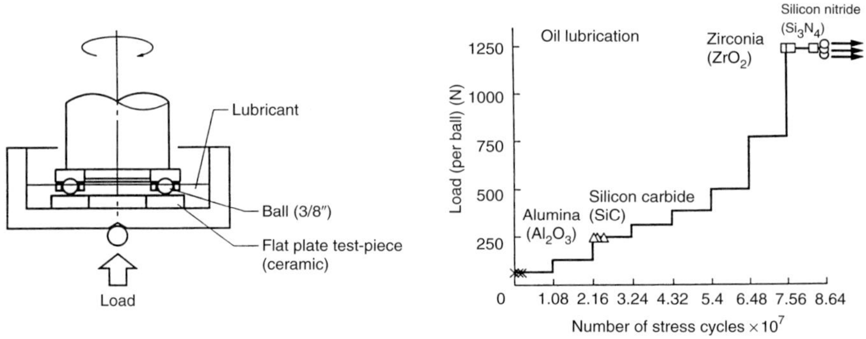

FIGURE 11.1.1 Life test results of ceramics.

1.08×10^7 stress cycles. Test results showed load resistance and rolling life to be in the following order (best to worst): silicon nitride, zirconia, silicon carbide, and then alumina.

Table 11.1.2 shows the suitability of each type of ceramic for use in bearings based on the characteristics of each type of ceramic materials and the results of the rolling life test shown in Figure 11.1.1 [6]. This table is the standard when selecting ceramic materials for bearings. Because the load capacity and rolling life of silicon nitride are equal to or better than those of high-carbon chrome bearing steel, silicon nitride is commonly used as a bearing material.

Silicon nitride, however, has a problem with corrosion resistance in high-temperature corrosive environments such as acid or alkali, and therefore zirconia or silicon carbide is the best material for such applications. However,

TABLE 11.1.2 Applicability of Ceramics to Bearings

	Application to rolling bearings		
	Judgment	Performance/application	Features
Silicon nitride Si_3N_4	◎	Load capacity and life equal to or better than those of bearing steels, suitable for applications requiring high performance	High-speed rotation, suitable for high vacuum, corrosion resistance, heat resistance, non-magnetic, high rigidity
Zirconia ZrO_2	○	Load must be limited, usable in strong corrosive chemicals	High corrosion resistance
Silicon carbide SiC	○	Load must be limited, usable in strong corrosive chemicals	High corrosion resistance, usable in ultrahigh temperatures
Alumina Al_2O_3	×	Not suitable for rolling bearings	—

◎, Suitable; ○, suitable for some applications; ×, not suitable.

it is necessary in such cases to limit bearing load because the load capacity and rolling life of zirconia and silicon carbide are inferior to those of silicon nitride. Because the load capacity (strength) of alumina is low, it cannot be used basically as a bearing material.

11.1.3 STATIC LOAD RATING OF CERAMIC BEARINGS

The allowable load that can be statically loaded on rolling bearings is specified as a basic static load rating. JIS B1519 (1989), specification has defined it as "The static radial load in the center of the contact area between the rolling element and the raceway receiving the maximum load, corresponding to the calculated contact stresses shown as follows:

Self-aligning ball bearings	4600 MPa
Other radial ball bearings	4200 MPa
Radial roller bearings	4000 Mpa"

This is because, with the amount of permanent deformation of the rolling element and the raceway, under their contact stress, being approximately 0.0001 times the diameter of the rolling element, the normal revolution of the bearings becomes impossible. In other words, in steel bearings, the allowable load can be determined from the amount of permanent deformation of the bearings.

In the case of bearings manufactured with brittle materials like silicon nitride, where plastic deformation cannot be expected, the JIS B1519 static load rating specification cannot be applied. It is well known that if a ball is pushed against the flat plate of a brittle material such as silicon nitride, ring cracks called as Hertz cracks will occur around the boundary of the contact area. This is caused by tensile stresses in the radial direction of the contact circle on the surfaces of the contact point. In this section, the static load rating of all ceramic bearings from the load at which these cracks appear and the static load rating of hybrid ceramic bearing based on JIS B1519 will be discussed.

11.1.3.1 Static Load Rating of All-Ceramic Bearings

In this section, noticing the occurrence of cracks in silicon nitride, the static load rating was studied for an all ceramic bearing, in which silicon nitride is used for the inner and outer rings and balls. (This is because if silicon nitride is loaded, cracks appear usually without the occurrence of permanent deformation.) [7, 8].

Figure 11.1.2 shows the measuring method of the cracking load [7]. A silicon nitride flat plate and a silicon nitride ball were used as test materials. The cracking load was decided by detecting acoustic emissions while gradually adding a load to the test pieces. Three kinds of balls, 5/32, 5/16 and 3/8 in., were used in this test.

Figure 11.1.3 is a comparison of the measuring results of the cracking load of silicon nitride balls, and the static load rating for conventional steel bearing calculated at the maximum contact stress of 4600 MPa [9]. From these results, it can be seen that the allowable loads, using cracking loads silicon nitride rolling bearings, is considerably larger than static load ratings of steel rolling bearings.

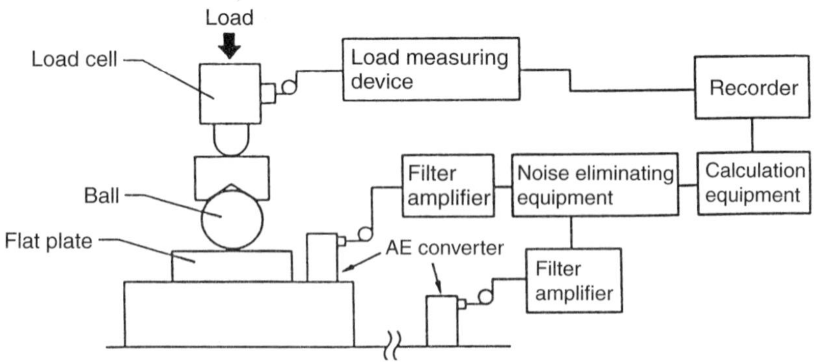

FIGURE 11.1.2 Measuring method of the cracking loads.

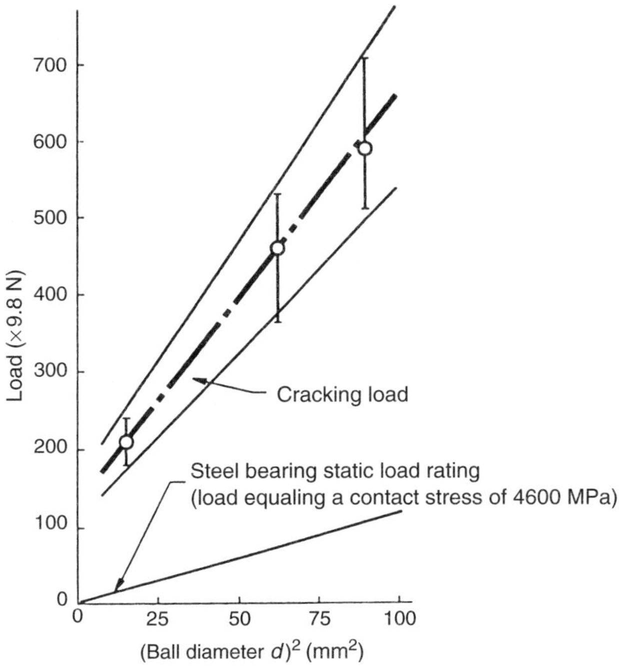

FIGURE 11.1.3 Comparison of cracking load and static load rating.

In the case of silicon nitride ball being pushed on a silicon nitride flat plate, all cracks occurred in the silicon nitride flat plate.

Consequently, it can be seen that an all-ceramic bearing made of silicon nitride can well withstand the same load as the static load rating of a conventional steel bearing, and that there are no problems at all in their practical use in terms of their static load rating.

11.1.3.2 Static Load Rating of Hybrid Ceramic Bearing

In the case of the static load rating of hybrid ceramic bearings (inner and outer rings made of high-carbon chromium bearing steel and rolling elements of ceramics), because the inner and outer rings of steel bearings are permanently deformed, the philosophy of static load rating for steel bearings can be adopted.

The results of measuring permanent deformation (depth indentation) when high-carbon chromium bearing steel balls and ceramic balls, respectively, are pressed against a plate of high-carbon chromium bearing steel by the method

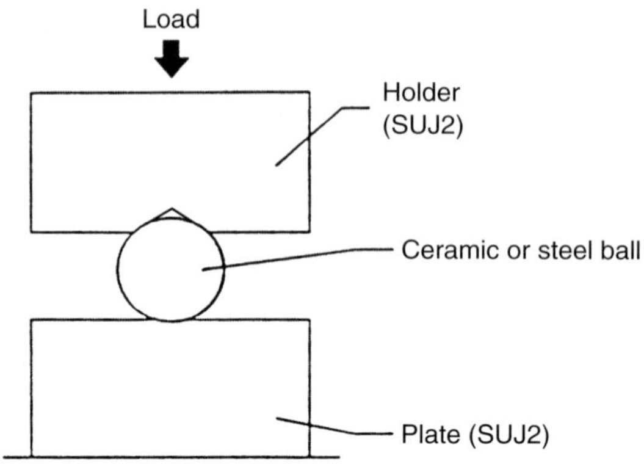

FIGURE 11.1.4 Method of making indentation.

TABLE 11.1.3 Results of Deformation Test

Load (kN)	Permanent deformation (average) (mm)		Permanent deformation (summation) (mm)
	Plate (bearing steel)	Ball	
Ceramic ball			
0.65	0.5	—	0.5
1.3	1.9	—	1.9
2.6	5.2	—	5.2
3.9	9.3	—	9.3
Steel ball			
0.65	0.4	—	0.4
1.3	1.3	0.11	1.41
2.6	4.0	0.41	4.41
3.9	6.8	1.18	7.98

shown in Figure 11.1.4 are given in Table 11.1.3 [10]. The results showed that permanent deformation was not observed in the ceramic balls, and that the permanent deformation produced on the steel plate when ceramic balls were used was approximately 1.2 times of the combined permanent deformation produced on the ball and the plate when steel balls were used.

Thus, the static load rating of hybrid ceramic bearings is limited by the deformation of steel races. Therefore, the static load rating of hybrid ceramic bearings is 0.85 times the static load rating of steel bearings.

TABLE 11.1.4 Static Load Rating of Ceramic Rolling Bearings

Bearing	View on static load rating	Static load rating
All ceramic bearing	Cracking load	Same as that of steel bearings (SUJ2)
Hybrid ceramic bearing balls: silicon nitride inner/outer rings: SUJ2	Permanent deformation	0.85 times that of steel bearings (SUJ2)

11.1.3.3 Views on the Static Load Rating of Ceramic Bearings

Silicon nitride bearings include all-ceramic bearings, which use silicon nitride for inner and outer rings and balls, and hybrid ceramic bearings, in which inner and outer rings are made of bearing steel and balls are made of silicon nitride.

Table 11.1.4 shows the views on the static load rating of ceramic bearings [9]. At the moment, there are no standards established by ISO or JIS, regarding the static load rating of all-ceramic bearings. As mentioned in the previous section, the static load rating of all-ceramic bearings has the same value as that of conventional steel bearing (SUJ2) based on the results of the cracking load of silicon nitride balls. In addition, for a hybrid ceramic bearing, the views on the static load rating of a conventional steel bearing (JIS B1519) can be used, because the inner and outer rings made of conventional bearing steel undergo plastic deformation. As a result, the static load rating of a hybrid ceramic bearing is 0.85 times that of a conventional steel bearing.

11.1.4 ROLLING FATIGUE LIFE OF CERAMIC BEARINGS

In this section, the life test results for ceramic bearings will be presented. Specifically, life test results for three types of bearings will be discussed; all-ceramic bearings, hybrid ceramic bearings, and steel bearings.

11.1.4.1 Test Bearings and Test Methods

Figure 11.1.5 shows the dimensions of the test bearing and Table 11.1.5 shows the configurations of the test bearings [11]. The test bearings are equivalent

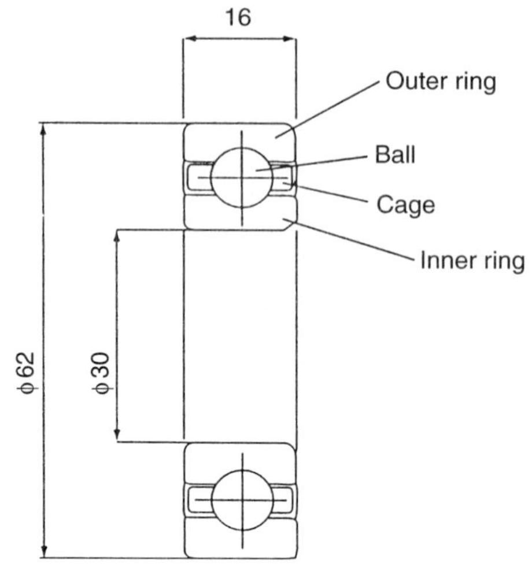

FIGURE 11.1.5 Dimensions of test bearings.

TABLE 11.1.5 Configurations of Test Bearings

Item	Type		
	All ceramic bearing NC6206	Hybrid ceramic bearing 3NC6206	Steel bearing M50 6206
Material			
Inner and outer ring	Silicon nitride Si_3N_4	AISI M50	AISI M50
Ball	Silicon nitride Si_3N_4	Silicon nitride Si_3N_4	AISI M50
Cage	AMS6414 silver plating	AMS6414 silver plating	AMS6414 silver plating

to 6206 deep-groove ball bearings, and the three types of bearings shown in Table 11.1.5 were used. The three types of bearings are all-ceramic bearings NC 6206, hybrid ceramic bearings 3NC 6206, and steel bearings 6206. The cage used for the three types of bearings is silver-plated AMS 6414. Nine balls with a diameter of 9.525 mm are used in each bearing. The three types of bearings were manufactured so that the inner- and outer-ring raceway curvature, the roughness of the inner- and outer-ring raceways and of the balls' surface, the internal clearance of the bearings, and other features were essentially the same.

Figure 11.1.6 shows the test equipment and Table 11.1.6 shows the test conditions [11]. For the test, a radial life test apparatus is used. A load is added

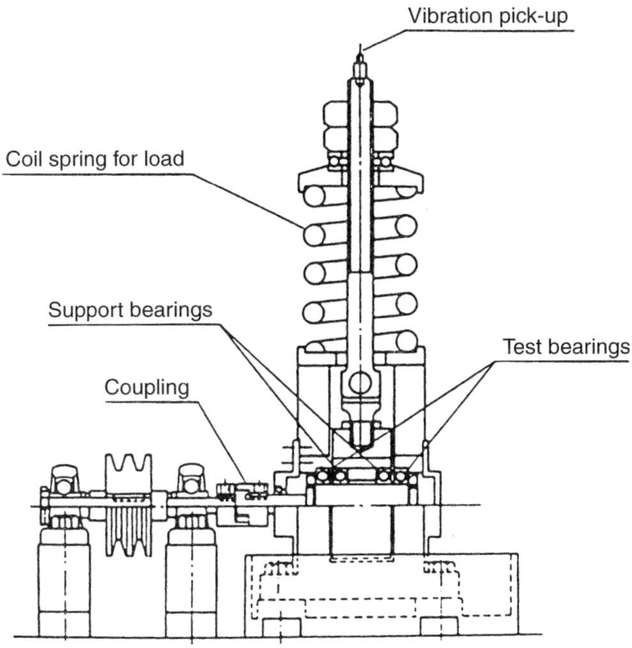

FIGURE 11.1.6 Test method.

TABLE 11.1.6 Test Conditions

Item	Condition
Load (N)	5800
Number of revolutions (r/m)	8000
Oil	Aero-Shell turbine oil #500
Temperature (°C)	70 ± 2

by using a coil spring, and the lubricating oil temperature is 70°C. Four bearings are used at one time, and as shown in Figure 11.1.6, the test bearings are two of these bearings at the outside edges. A failure of test bearings are detected using vibration pick-up, and the equipment is designed so that if the vibration value reaches twice that of its starting value, the test equipment will shut down. In addition, the interference between the inner ring of the all-ceramic bearing and the steel shaft is about 14 μm and the circumferential stress of the bore surface of ceramic inner ring becomes about 110 MPa.

11.1.4.2 Test Results

Figure 11.1.7 shows the life test results of three types bearings, all-ceramic bearings NC 6206, hybrid ceramic bearing 3NC 6206, and steel bearing 6206 [11]. Two test bearings are used in one test, and the results are plotted on Weibull probability papers using the sudden death method. With a test load of 5880 N, the maximum contact stress P_{max} occurring in the three types of bearings is as follows: $P_{max} = 4.3$ GPa for all-ceramic bearings, $P_{max} = 3.8$ Gpa for hybrid ceramic bearing, and $P_{max} = 3.3$ GPa for steel bearings. From the results in Figure 11.1.7, it can be seen that, although all-ceramic bearings NC 6206 and hybrid ceramic 3NC 6206 have a comparatively higher maximum contact stress than steel ball bearing 6206 under the same load, they have a rolling contact fatigue life equal to or longer than that of steel bearings. Moreover, the damage

Test bearing	Load (N)	Number of revolutions (r/m)	L_{10} life (h)	L_{50} life (h)	Weibull slope
○ NC6206	5880	8000	82.6	589.2	0.95
× 6206	5880	8000	46.2	269.5	1.06
✱ 3NC6206	5880	8000	49.4	294.6	1.05

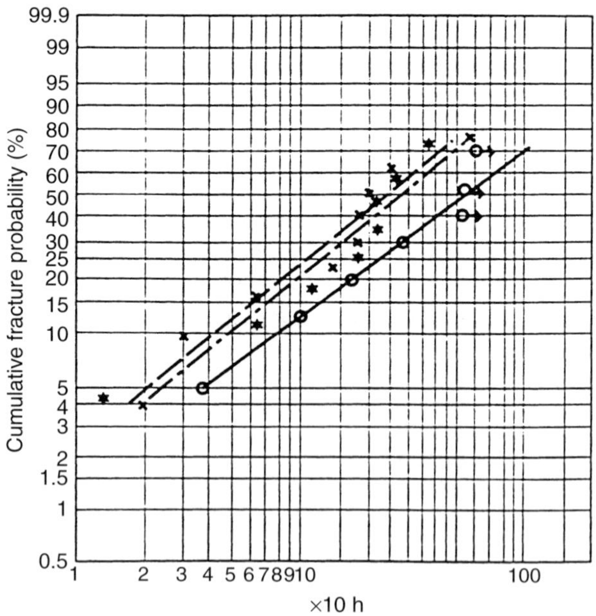

FIGURE 11.1.7 Life test results.

form of all ceramic bearing and hybrid ceramic bearing due to rolling contact fatigue is identical in form to that of the rolling contact fatigue flaking observed with bearing steel [7, 12]. Consequently, the life of all-ceramic bearings and hybrid ceramic bearings can be detected by the vibration pick up in the same manner as for steel bearings.

Using the above results, in terms of the life of all-ceramic bearings and hybrid ceramic bearings, the dynamic load rating is the same value as that of steel bearings, and therefore life prediction can be conducted using the life calculation formula for steel bearings [7].

11.1.5 FITTING OF CERAMIC BEARINGS

It is necessary to think about two kinds of fitting: fitting between the inner ring and the shaft; and fitting between the outer ring and the housing. For all-ceramic bearing, in the case of fitting between the silicon nitride inner ring and the steel shaft, the linear expansion coefficient of steel is four to five times greater than that of silicon nitride. Consequently, if temperature rises, the shaft expands, and there is a possibility of the silicon nitride inner ring being failed [13]. Meanwhile, in the case of fitting between the silicon nitride outer ring and the steel housing, with an expansion of the housing an excessive clearance occurs between the outer ring and the housing, and it is possible that this will adversely affect the rotation performance of the bearing.

This section will introduce the results of examining the fitting between the silicon nitride inner ring and the steel shaft, which is possibly linked to fatal damage of the bearing. First, using a silicon nitride ring and a steel shaft, the static limit of interference fit for silicon nitride ring damage will be shown, and next, using all-ceramic bearing, the dynamic limit of interference fit for the silicon nitride ring and the steel shaft during operation will be shown.

11.1.5.1 Static Limit of Interference Fit [14]

A silicon nitride ring and a SUS303 steel shaft were used for fitting tests. Table 11.1.7 shows the physical properties of the silicon nitride and of the SUS303 used for the shaft. The linear expansion coefficient of the SUS303 is approximately five times that of silicon nitride. Figure 11.1.8 shows the dimensions of the silicon nitride ring, and Figure 11.1.9 shows those of the steel

TABLE 11.1.7 Physical Properties of Ring and Shaft

Item	Ring	Shaft
Material	HIP-Si_3N_4	SUS303
Young's modulus (MPa)	31×10^4	21×10^4
Poisson's ratio	0.29	0.3
Linear expansion coefficient (1/°C)	3.2×10^6	17.2×10^6

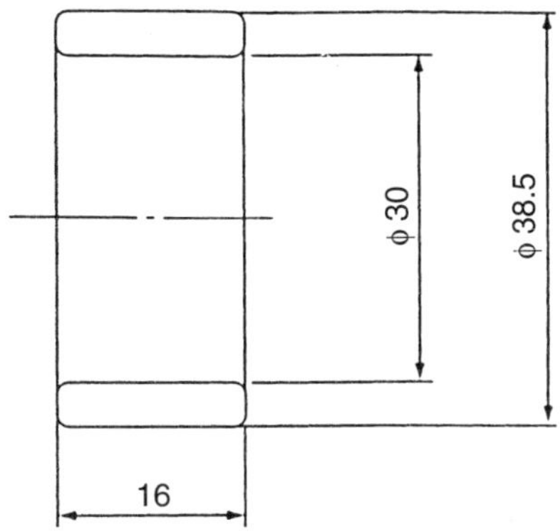

FIGURE 11.1.8 Dimensions of Si_3N_4 ring.

shaft. Three types of shaft are used: a solid shaft, a hollow shaft, and a spline shaft. For the spline shaft, the test is designed so that the contact area between this shaft and the bore surface of the silicon nitride ring will be 30% of the surface area.

Figure 11.1.10 shows the test method. In the test, the silicon nitride ring is fitted to the steel shaft, is heated from both ends by heaters. Due to the difference in the linear expansion coefficients, the interference between the two components is produced, and with the rise in temperature the interference increases, finally resulting in the fracture of the silicon nitride ring.

Table 11.1.8 shows the influence of the shaft types on silicon nitride ring fracture. The tests were conducted 15 times for each type of shafts. If fracture interference values at a cumulative fracture probability of 50% are compared for the three types of shafts, it can be seen that the interference values of the hollow and spline shafts are approximately 1.3 times greater than that of the solid shaft. Consequently, it can be said that compared with a solid shaft, a hollow or a spline shaft contributes to the relief of the circumferential stress generating in the bore surface of the silicon nitride ring, and makes the fracture

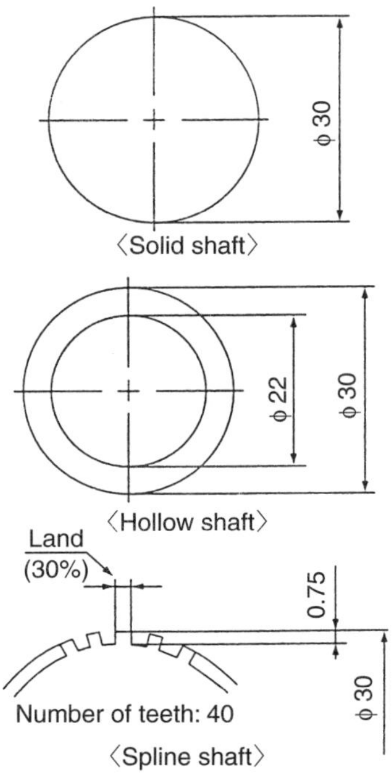

FIGURE 11.1.9 Dimensions of steel shaft.

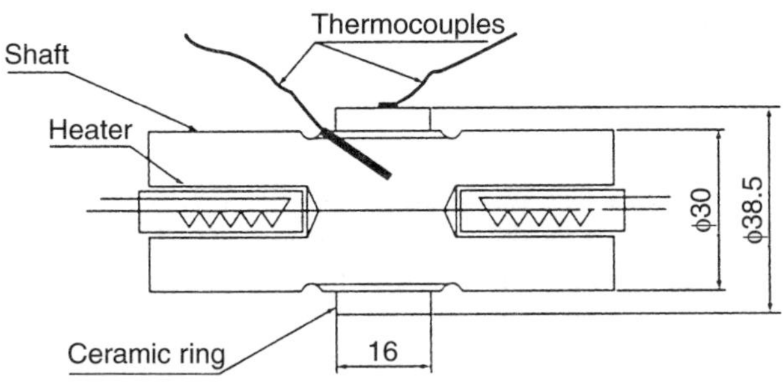

FIGURE 11.1.10 Test method for fitting.

interference increase. For the solid shaft, the fracture interference at a cumulative fracture probability of 10% is approximately 50 μm, and the circumferential stress generating in the bore surface of the silicon nitride ring at that time is approximately 400 MPa.

TABLE 11.1.8 Test Results of Ring Fracture

	B10 (μm)	B50 (μm)	Minimum value of fracture stress (MPa)	Weibull coefficient	Interference ratio
Solid shaft	49.8	58.4	399	11.8	1
Hollow shaft	68.1	77.3	332	14.9	1.3
Spline shaft	66.6	74.6	—	16.5	1.3

11.1.5.2 Dynamic Limit of Interference Fit [14]

The test bearings used are all-ceramic bearings NC6206 equivalent to 6206 (see Figure 11.1.5 and Table 11.1.5). Figure 11.1.9 (see previous section) shows the details of the fitting areas of the steel shafts. The shaft material is SUJ2 and three types of shaft are used: a solid shaft, a hollow shaft and a spline shaft.

Figure 11.1.6 (see section titled "Test bearings and test methods") shows the test method and Table 11.1.6 shows the test conditions. Four bearings are used for the test, one of which is the test bearing. From the difference between the linear expansion coefficient of silicon nitride and that of bearing steel (SUJ2 : $12.5 \times 10^{-5}/°C$), when an all-ceramic bearing is mounted, 10°C temperature rise causes the interference to increase by 2.8 μm. If a test is carried out under the conditions shown in Table 11.1.6, the shaft temperature becomes 80°C and the interference increase is 16.8 μm during the test.

Table 11.1.9 shows the test results for dynamic fitting. Interference tests are conducted using three types of shafts, in other words, interference tests were conducted with three kinds of interference. Life tests are conducted under each type of interference condition in the table, and if there are no failures, the tests are suspended after 100 h. Bearing damage occurs in the either form of inner ring fractures or of inner and outer ring fractures. In addition, most bearings failures occur within just one hour after the test starts. It can be considered that a fracture of the inner ring occurs first and then that of the outer ring occurs. Figure 11.1.11 shows the typical appearance of fracture of the inner ring. The fracture appears in one area. From the test results in Table 11.1.9, it can be seen that fracture of the inner ring does not occur within an interference of 31 μm in the solid shaft, of 43 μm in the hollow shaft, and of 39 μm in the spline shaft. It is thus clear that in dynamic fitting tests too, compared to the solid shaft, the hollow and spline shafts can make the interference increase until fractures occur.

Table 11.1.10 is a summary of the interference for the solid and hollow shafts, which do not cause fracture of the inner ring, and the circumferential stress occurring in the bore surface of the silicon nitride inner ring at that time. The spline shaft is omitted because circumferential stress cannot be calculated

TABLE 11.1.9 Test results for Dynamic Fitting

n	Interference (μm)		
	Solid shaft	Hollow shaft	Spline shaft
1	33 (inner ring fracture)	51 (inner and outer ring fracture)	53 (inner and outer ring fracture)
2	33 (inner ring fracture)	51 (inner ring fracture)	53 (inner and outer ring fracture)
3	33 (inner ring fracture)	51 (inner and outer ring fracture)	47 (inner ring fracture)
4	39 (inner ring fracture)	51 (inner and outer ring fracture)	48 (inner and outer ring fracture)
5	38 (inner ring fracture)	49 (inner ring fracture)	47 (inner ring fracture)
6	35 (inner ring fracture)	43 (no abnormalities)	39 (no abnormalities)
7	34 (inner ring fracture)	41 (no abnormalities)	39 (no abnormalities)
8	15 (no abnormalities)	41 (no abnormalities)	38 (no abnormalities)
9	31 (no abnormalities)	34 (no abnormalities)	38 (no abnormalities)
10	29 (no abnormalities)	35 (no abnormalities)	38 (no abnormalities)
11	29 (no abnormalities)	29 (no abnormalities)	30 (no abnormalities)
12	31 (no abnormalities)	34 (no abnormalities)	29 (no abnormalities)

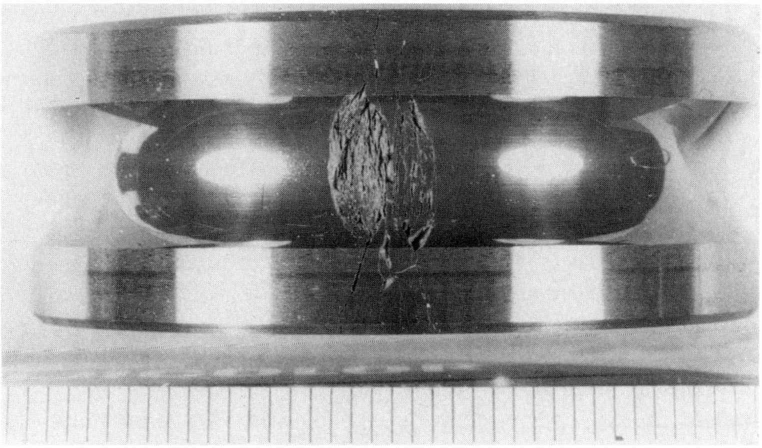

FIGURE 11.1.11 Appearance of silicon nitride inner ring.

using the formula in the strength of materials. Looking at Table 11.1.10, it can be seen that when an all-ceramic bearing is mounted in a steel shaft and then used, the circumferential stress occurring in the silicon nitride inner ring must be designed to be less than 200 MPa.

TABLE 11.1.10 Limit of Fitting

	Interference (μm)	Circumferential stress (MPa)
Solid shaft	31	243
Hollow shaft	43	204
Spline shaft	39	—

11.1.6 APPLICATION EXAMPLE OF CERAMIC BEARINGS

11.1.6.1 Characteristics of Ceramic Bearings

The characteristics of the ceramics (silicon nitride, Si_3N_4) used in ceramic bearings are compared to those of high carbon chrome bearing steel (SUJ2) used for conventional types of bearings. The result is given in Table 11.1.11, together with the advantages of ceramic bearings [15]. Because of the preferable characteristics of ceramic material (silicon nitride), ceramic bearings are considered useful in a variety of applications.

For example, the higher heat resistance of ceramic bearings allows them to be used in high-temperature environments [16, 17]. Further, their lower density greatly reduces the mass of the bearings and contributes to the reduction in centrifugal force generated by the rolling elements (balls or rollers) when the bearings run at high speeds [18–20]. In addition, covalent bonding of ceramic material gives higher resistance to seizure caused by discontinued oil film during high-speed rotation.

11.1.6.2 Composition of Ceramic Bearings

The composition of ceramic bearings, as well as some examples of their applications, are presented in Table 11.1.12 [6].

Ceramic bearings can be divided into all ceramic type and hybrid ceramic type. For the all ceramic type bearings, all the outer rings, inner rings, and rolling elements are made of ceramics. For the hybrid type bearings, only the rolling elements are made of ceramics, and the outer and inner rings are made of high-carbon chrome bearing steel or other special steel.

Some applications require that the hybrid type bearings run at high speeds. In such applications, problems may occur because of inner ring expansion due to the centrifugal force produced by high-speed running and hence the interference between the inner ring and shaft will loosen. In such cases, rolling elements and inner rings made of ceramics are used.

The retainers are made of metallic material or plastic material, depending upon the intended use of the bearings.

TABLE 11.1.11 Comparison of Characteristics between Ceramic Material (Silicon Nitride) and High-Carbon Chrome Bearing Steel (SUJ2) and Advantages of Ceramic Bearings

Item (unit)	Ceramic Material (Si_3N_4)	Bearing Steel (SUJ2)	Advantage of Ceramic Bearings
Heat resistance (°C)	800	180	Higher load durability maintained in high-temperature ranges
Density (g/cm^3)	3.2	7.8	Reduction of centrifugal force induced by rolling elements (balls or rollers) →Increased service life and restricted increase in temperature
Linear expansion coefficient (1/°C)	3.2×10^{-6}	12.5×10^{-6}	Smaller change of internal clearance caused by temperature rise → Reduced vibration, small change of preload
Vickers hardness (HV)	1 500	750	
Module of longitudinal elasticity (GPa)	320	208	Smaller change of deformation at rolling contact point →High rigidity
Poisson's ratio	0.29	0.3	
Corrosion resistance	Good	Not good	Can be used in acid solutions, alkali solutions, and other special environments
Magnetism	Non-magnetic material	Ferromagnetic material	Smaller speed fluctuation caused by magnetism in intense magnetic field
Conductivity	Insulating material	Conductive material	Eliminates electric pitting (applicable to electric motors, etc.)
Bonding of raw material	Covalent bonding	Metallic bonding	Minimized seizure (or cohesion) at contact points, usually resulting from discontinued oil film

11.1.7 CONCLUSIONS

The physical as well as mechanical characteristics of ceramics are very different from those of steels. Consequently, to expand the applications of ceramic

TABLE 11.1.12 Composition of Ceramic Bearings and Some Examples of Their Use: Ceramics

(1) High-speed Rotation	(2) For Use in a Vacuum Environment	(3) For Corrosion Resistance
Specific gravity 40% of bearing steel Suitable for high-speed rotation because lower centrifugal force is produced by the rolling elements.	Can be used in a vacuum of 1–10^{-10} Pa Lubricating method should be selected according to use.	Can be used in acid, alkali, salt water, and molten metal.
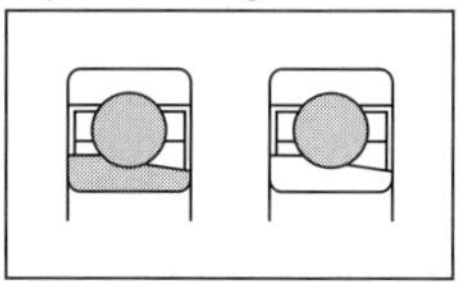	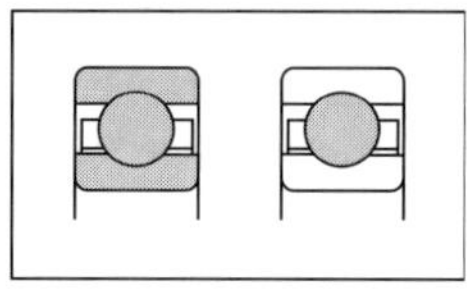	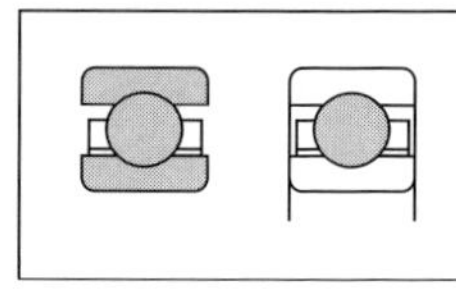
■**Example of use** Main spindle of machine tools, turbo chargers for automobiles, and industrial equipment (spin testers, etc.)	■**Example of use** Semi-conductor production facilities and vacuum equipment (turbo molecular pump, etc.)	■**Example of use** Chemical equipment, steel production facility, and textile machinery
(4) For High-temperatures	(5) Non-magnetism	(6) Insulation
Ceramic is heat resistant up to 800°C. Lubricating method should be selected according to the temperature.	Can be used in magnetic fields.	Ceramics are insulating materials, and can be used in applications where electric leakage may occur.
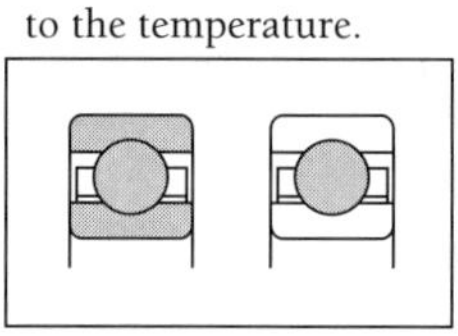	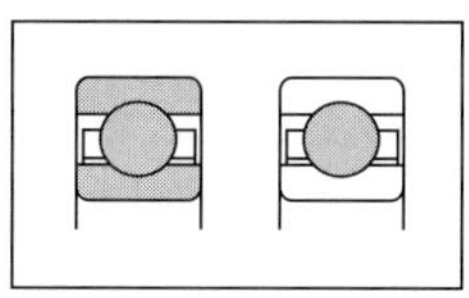	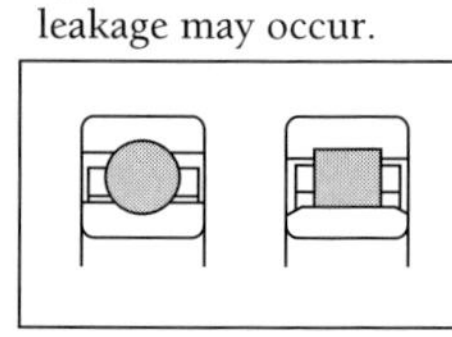
■**Example of use** Steel production facilities, industrial equipment, and automotive diesel engines	■**Example of use** Semi-conductor production facilities, superconductivity-related equipment, and nuclear power generators	■**Example of use** Railway rolling stock and electric motors

Remarks: There are two types of ceramic bearings, all ceramic type and hybrid ceramic type. All ceramic type: all outer rings, inner rings, and rolling elements are made of ceramics. Hybrid ceramic type: Only the rolling elements, or both rolling elements and inner rings are made of ceramics.

bearings, it is very important to grasp their characteristics thoroughly and to understand their merits and demerits as a bearing material. To further understand the basic properties of ceramic bearings, we will continue to vigorously promote further research and development in this field.

REFERENCES

1. Baumgartner, H. R. (1978). Ceramic bearings for turbine applications, "5th Army Mater. Technol. Conf." p. 423.
2. Miner, J. R., Grace, W. A., and Valori, R. (1980). Demonstration of High-speed gas Turbine bearings using silicon nitride rolling elements, ASLE Preprint, 80-AM-3C-3.
3. Kitamura, K., Takebayashi, H., and Ikeda, M. (1997). Development of Ceramic cam follower for engine application, SAE Technical Paper 972774.
4. Shiratori, M., Kano, K., Takebayashi, H., and Okuda, K. (1993). Characteristics of ceramic spherical sliding bearing for severe application, "Advanced Materials '93", Preprint, p. 417.
5. Hattori, T., Kitamura, K., and Takebayashi, H. (1998). "Tribology Conference (Japanese Society of Tribologists)" p. 124.
6. EXSEV Bearings Series: Ceramics Bearings and EXSEV bearings, CAT. No. 208E, Koyo Seiko Co., Ltd.
7. Fujiwara, T., Yoshioka, T., Kitahara, T., Koizumi, S., Takebayashi, H., and Tada, T. (1988). Study on load rating property of silicon nitride of rolling bearing materials, *J. JSLE* **33**: 301.
8. Yoshioka, T., Kitahara, T., Takebayashi, H., and Yuine, T. (1989). A new method for static load rating of ceramic rolling bearing, *Wear* **133**: 373.
9. (1994). Performance and application of ceramic bearing. *Koyo Eng. J.* **145**: 24.
10. Ichikawa, Y., and Tabata, S. (1989). *Koyo Eng. J.* **135**: 62.
11. Takebayashi, H. (1998). Study of ceramic bearing for the basic performance and applications, Doctoral Dissertation, p. 31.
12. Takebayashi, H., Johms, T. M., Rokkaku, K., and Tanimoto, K. (1990). Performance of ceramic bearing in high speed turbine application, SAE Technical Paper 901629.
13. Hamburg, G., Cowley, P., and Valori, R. (1980). Operation of an all-ceramic mainshaft roller bearing in a J-402 gas-turbine Engine, ASLE Preprint, 80-AM-3C-1.
14. Takebayashi, H., Kitamura, K., and Hattori, T. (1999). Study on fitting of silicon nitride bearings. *J. Jpn. Soc. Tribologists* **44**: 61.
15. Rokkaku, K., and Nishida, K. (1988). Ceramic roller bearings. *J. JSPE* **54**: 1240.
16. Takebayashi, H., Yuine, T., and Yoshioka, T. (1993). Performance of ceramic ball bearings at high temperature (Part 1). *J. Jpn. Soc. Tribologists* **38**: 935.
17. Yoshioka, T., Fujita, K., Takebayashi, H., and Yuine, T. (1993). Performance of ceramic ball bearings at high temperature (Part 2). *J. Jpn. Soc. Tribologists* **38**: 1077.
18. Takebayashi, H., Tanimoto, K., and Hattori, T. (1998). Performance of hybrid ceramic bearing at high speed condition (Part 1). *J. Gas Turbine Soc. Jpn.* **26**: 55.
19. Takebayashi, H., Tanimoto, K., and Hattori, T. (1998). Performance of hybrid ceramic bearing at high speed condition (Part 2). *J. Gas Turbine Soc. Jpn.* **26**: 61.
20. Tanimoto, K., Kajihara, K., and Yanai, K. (2000). Hybrid ceramic ball bearings for turbochargers, SAE Technical Paper 2000-01-1339.

Handbook of Advanced Ceramics
S. Sōmiya *et al.* (Eds.)

11.2 Cutting Tools

MIKIO FUKUHARA
Toshiba Tungaloy Ltd., Sugasawa, Tsurumi, Yokohama 230-0027, Japan

11.2.1 INTRODUCTION

Cutting method has been widely used as precession machining operation. Significant increases in manufacturing productivity can be achieved by increasing metal-removal rate. However, when cutting hard or abrasive materials at high speed, productivity is severely limited by cutting tool edge wear. The rate determining factor in the chip making process has been the cutting tool material itself. Machine tools and procedures have always been designed around the maximum capabilities of new tool materials as they were developed. It may readily be seen from Figure 11.2.1 that since about 1900 we have experienced an exponential increase in productivity capability as measured by the cutting speeds available [1]. There is no reason to believe that this tend will not continue in the future. Zsolnay [2] has discussed the use of ceramic tools in the machining of steel and cast iron at speeds up to 3000 m/min (50 m/s). Not only were no problem encountered at the high temperature generated under these

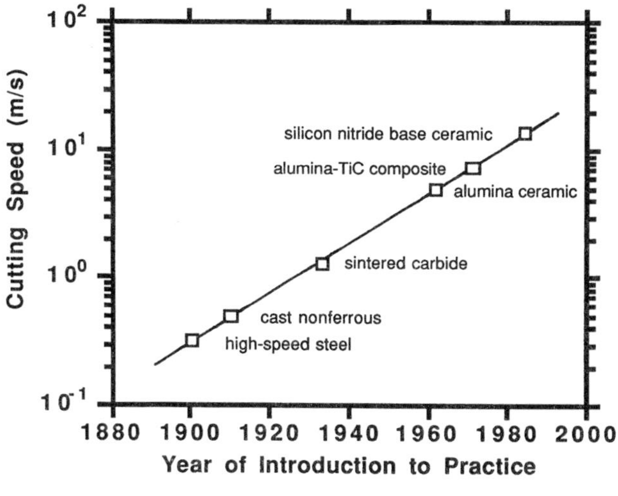

FIGURE 11.2.1 Improvement of tool materials.

conditions, but there was even less wear on the oxide cutting tool per unit of metal cut.

Favorable properties that promote increased metal removal rates include good hot hardness, low coefficient of friction, high wear resistance, chemical inertness, and low coefficient of thermal conductivity. Thus materials for cutting tools are classified into three categories by virtue of their cutting speed; usage of high-speed steels is comfortable for speed below 100 m/min (1.67 m/s), cemented carbides such as WC–(Ti,Ta)C–Co and Ti(C,N)–Mo_2C–Ni are mainly used in cutting speed between 100 and 300 m/min (i.e. between 1.67 and 5 m/s) and intrinsic ceramic tools are suitable for higher speed than 300 m/min (5 m/s). For high-speed cutting, inorganic materials play the leading part.

Since inorganic compounds of III, IVa, Va and VIa group elements in Periodic Table have high hardness, low thermal expansion and high thermal conductivity and excellent oxidation and corrosion resistance, they are candidate for high speed cutting tool materials. Among these compounds, alumina Al_2O_3, silicon nitride Si_3N_4, titanium carbide TiC or titanium carbonitride Ti(C,N), diamond and cubic boron nitride cBN are actually used as the ceramic tool. Silicon carbide SiC, which have high hardness in high temperature, is not suitable material for cutting of metals, especially steels or ferrous alloys, because of thermal reaction with the work metals during cutting process. However, silicon carbide whiskers are used as additive of alumina composite tool for cutting of nickel based superalloys such as Inconel and Waspaloy. Low thermal conductivity of zirconia ZrO_2 controls usage as main cutting tool material.

Fundamental to tool development is the ability to control microstructure in tool materials so as to maximize those physical and mechanical properties responsible for optimum machining performance.

11.2.2 PHYSICAL PROPERTIES

Some physical properties of $Al_2O_3 + TiC$, $Al_2O_3 + Ti(C,N)$, $Al_2O_3 + SiC_W$ (SiC whisker), Si_3N_4, sialon $(Si,Al)_3(N,O)_4$, diamond and cBN based tool materials are listed in Table 11.2.1 [3, 4]. The material conditions required for cutting tool materials are deformation, thermal crack and oxidation resistance at the high-temperature region. The thermal shock parameter [5]

$$R = \sigma\kappa(1 - \nu)/\alpha E, \tag{1}$$

is an important parameter for thermal crack control, where σ is the transverse rupture strength, κ the thermal conductivity, ν the Poisson's ratio, α the thermal expansion coefficient and E the Young's modulus. The thermal crack is come from alternating thermal stress which occurs with interrupted cutting

TABLE 11.2.1 Some Physical Properties of $\mathbf{Al_2O_3 + TiC, Al_2O_3 + Ti(C,N), Al_2O_3–SiC_w, Si_3N_4}$, Sialon, cBN and diamond based tool materials

Tool material	Density (mg/m)	Hardness			Transverse rupture strength (MPa)	Fracture toughness ($MN\,m^{-3/2}$)	Young's modulus (GPa)	Poisson's ration	Thermal expansion (10^{-6}/K)	Thermal conductivity (W/mK)	Thermal shock parameter
		HRA (298 K)	Hv 298 K	Hv 1273 K							
$Al_2O_3 + TiC$	4.24	93.8	2130	770	760	4.3	373	0.219	7.6	22.1	11.5
$Al_2O_3 + Ti(C,N)$	4.30	94.0	2050	670	890	5.7	405	0.219	7.4	21.9	12.5
$Al_2O_3 + SiC_W$	3.74	94.4	2100	—	700	6.0	392	—	—	—	100
Si_3N_4	3.27	92.8	1820	1170	941	9.4	287	0.233	3.6	54.9	91.5
Sialon	3.33	93.5	1650	1300	1300	6.8	290	0.273	3.6	45.0	80.1
cBN	3.66	—	3200	2670	1071	6.2	—	—	—	65.7	—
Diamond	3.52	—	9000	—	—	—	—	—	—	104.3	—

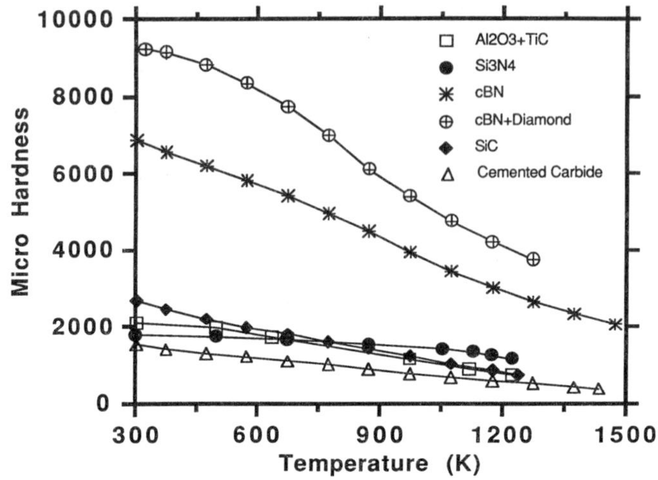

FIGURE 11.2.2 Temperature dependence of micro hardness for Al_2O_3 + TiC, Si_3N_4, SiC, cBN, cBN–25 mass% diamond and cemented carbide.

operations (e.g. milling) and when machining with coolant. The *R* parameter above 10 is desirable for the tool material in high-speed cutting. For cutting tools, fracture toughness is also another important material parameter, because it presents resistance against propagation of cracks occurred at the tool edge during the cutting process.

Hardness is a measure evaluating deformation. Temperature dependence of hardness for these materials is shown in Figure 11.2.2. Although diamond and cBN possess high hardness at room temperature, their hardness drops steeply at elevated temperature above 1373 K due to diamond/graphite and cBN/hBN (hexagonal boron nitride) transitions, respectively. Thus diamond tools are not suitable for the machining of ferrous metals, titanium alloys or the high nickel based alloys in that diamond disintegrates oxidation and graphitization at the high temperatures generated. In other words, the diamond tools are most suited for the machining of copper, aluminum and other non-ferrous abrasive materials such as phenolic plastics, fused silica, graphite, cemented carbon, presintered tungsten carbide, ceramics and abrasive rubber base materials [1]. In this regard the cBN is superior to diamond for oxidation resistance and transition temperature. On the other hand, hardness of other ceramics decreases slowly with increasing temperature, but they have not phase transition accompanied by vanishing of hardness. Especially, silicon nitride keeps high hardness above 1000 at 1273 K.

The deformation behavior at elevated temperature is assumed by using the deformation mechanism map [6, 7] proposed by Ashby. In order to estimate the deformation pattern of representative Al_2O_3, Al_2O_3 + TiC and Si_3N_4 ceramic tools during cutting practice, deformation mechanism maps for each component (α-Al_2O_3 [3], TiC [8] and β-Si_3N_4 [3]) can be constructed, provided that the strain rate is $10^{-2}\,s^{-1}$ and the edge stress is mainly compressive. These maps are shown in Figure 11.2.3a–c, respectively, where the region of the broken line indicates the temperature and the pressure of a tool edge in cutting practice.

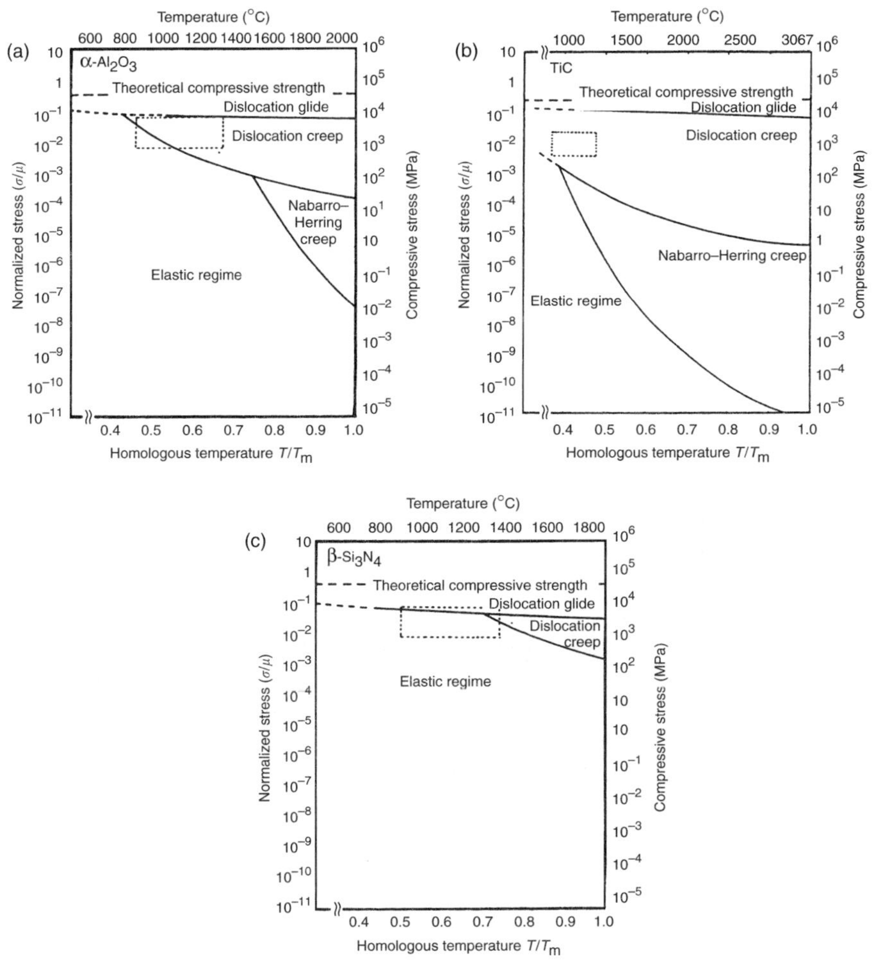

FIGURE 11.2.3 Deformation mechanism map for: (a) α-Al_2O_3 (grain size, 1 μm) [3]; (b) TiC (grain size, 1.4 μm) [8]; and (c) β-Si_3N_4 (grain size, 0.4 μm) [3] for a critical strain rate of $10^{-2}\,s^{-1}$.

The tool edge during cutting is under severe conditions such as temperature region from 1073 to 1573 K and compressive strength region from 0.1 to 1 GPa.

For α-Al_2O_3 and TiC (Fig. 11.2.3a and b), deformation of the tool edge in cutting depends on dislocation creep, while dislocation creep dominates for β-Si_3N_4 (Fig. 11.2.3c) having crystalline boundary phases. Thus, it can be expected that the chipping resistance of the silicon nitride ceramic tool is superior to those of Al_2O_3 and TiC in cutting. In addition, gain boundary sliding is generally accepted as a creep mechanism in silicon nitride having the glassy boundary phases [9, 10]. The exploitation of heat generated in cutting process contributes to elongation of tool life.

As can be seen from Table 11.2.1, the Al_2O_3 + TiC and Al_2O_3 + Ti(C,N) ceramics have moderate thermal conductivity compared with that of Al_2O_3 one ($\approx 10^{10}\ \Omega$ m). This is due to formation of continuous skeleton of TiC or Ti(C,N) in Al_2O_3 matrix [11]. It is very important for cutting tool materials to posses excellent thermal conductivity or diffusivity [12].

The microstructures of the representative tool materials are presented in Figure 11.2.4. The microstructure of Al_2O_3 + Ti(C,N), cBN and diamond tool materials are characterized by homogeneous distribution of fine grains, while silicon nitride, Al_2O_3+SiC_W and Al_2O_3+TiC materials have elongated and large grains in their matrices, respectively. Grain shapes of alumina and silicon nitride matrices are granular and angular, respectively. For materials with elongated grains such as the Al_2O_3 + SiC_W and silicon nitride materials, it is proposed that propagation of cracks dominates the pulling out of elongated grains from their matrices, and consequently the thoughness of the material increases [13]. Contrast to Al_2O_3 + TiC composite, usage of Ti(C,N) in Al_2O_3 + Ti(C,N) material controls to grain growth of Ti(C,N).

11.2.3 CUTTING PERFORMANCE

In dry turning of the carbon steel S48C (HB 214), the Al_2O_3+TiC was compared with a silicon nitride based tool and a alumina coated silicon nitride one. The result after 600 s cutting is illustrated in Figure 11.2.5. The flank wear width of the Al_2O_3 + TiC tool was one-tenth that of the silicon nitride one. This means that bonding strength between silicon nitride grains and intergranular phases is not yet high compared with that between Al_2O_3 and TiC grains, because flank wear is the result of abrasion and grooving process [14]. This flank wear problem of silicon nitride can be gotten around by coating of alumina, titanium carbide or titanium nitride layers on the silicon nitride tool. However, this treatment makes weaken the interrupted cutting performance and increase the production cost.

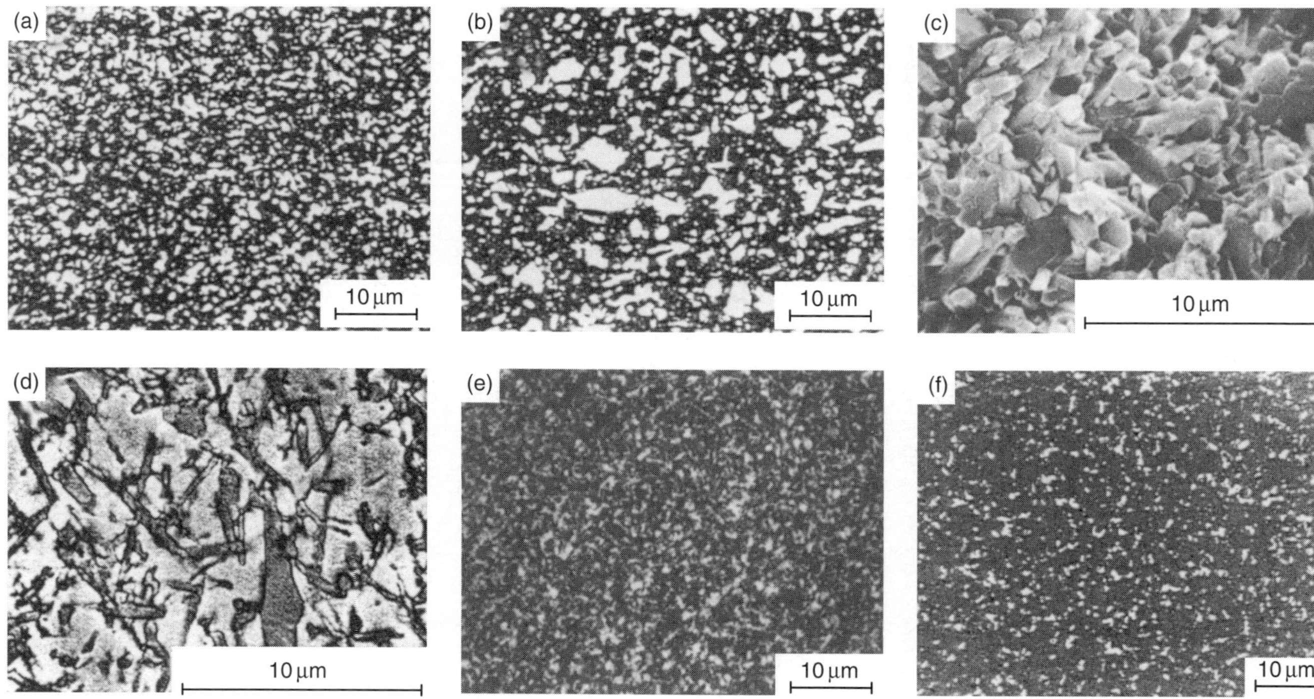

FIGURE 11.2.4 Microstructures of representative ceramic tools: (a) $Al_2O_3 + Ti(C,N)$; (b) $Al_2O_3 + TiC$; (c) Si_3N_4; (d) $Al_2O_3 + SiC_w$; (e) cBN; and (f) diamond.

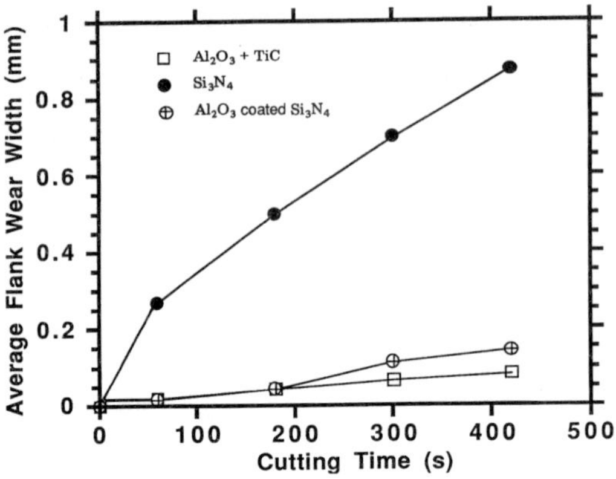

FIGURE 11.2.5 Relation between average flank wear width and cutting time in dry turning of S48C (cutting speed $V = 1.67\,m/s$, feed rate $f = 0.1\,mm/rev$; depth of cut $d = 1.5\,mm$; SNCN 432; machine tool, NC lathe (55 kW)).

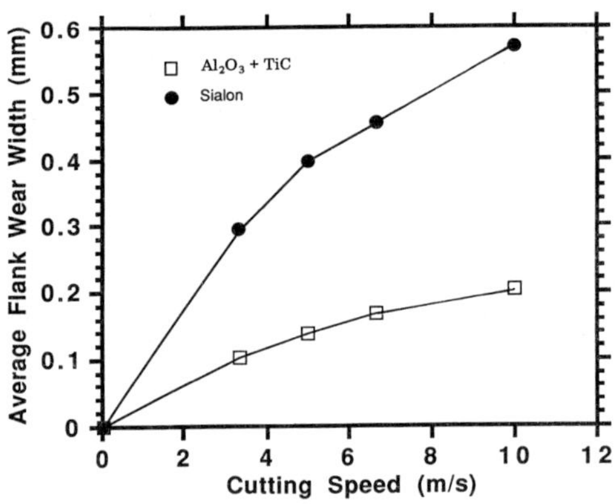

FIGURE 11.2.6 Relation between average flank wear width and cutting time in wet turning of S55C (cutting speed $V = 3.3, 5, 6.7\,m/s$, feed rate $f = 0.2\,mm/rev$; depth of cut $d = 1.5\,mm$; SNCN 432; machine tool, NC lathe (55 kW)).

In above cutting tests on mild carbon steel, no coolant was used. Additional turning tests on the carbon steel S55C (HB 254) were evaluated with coolant. The cutting speed was varied from 3.3 to 10 m/s (200–600 m/min) at a fixed feed rate of 0.2 mm/rev. The results for the Al_2O_3 + TiC tool together with the results for Sialon tools were plotted as the average flank wear width after 1200 s of the cutting versus cutting speed and are shown in Figure 11.2.6.

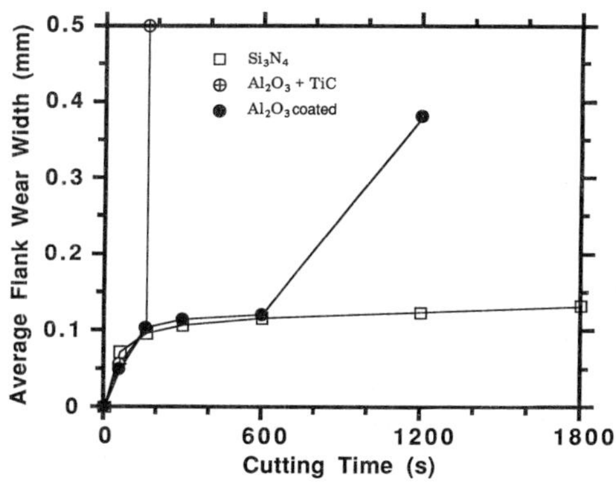

FIGURE 11.2.7 Relation between average flank wear width and cutting time in wet turning of FC35 (cutting speed $V = 10\,m/s$, feed rate $f = 0.7\,mm/rev$; depth of cut $d = 1.5\,mm$; SNGN 432; machine tool, NC lathe (55 kW)).

The flank wear width of the Al_2O_3 + TiC tool was one-half that of the sialon one all over the cutting speed. Thus, in wet turning of the mild steel with high hardness, the Al_2O_3 + TiC tool showed excellent performance in high speed and heavy cutting. This indicates that there is less chemical reaction between Al_2O_3 + TiC material and medium carbon steel in cutting, because solution of the tool material in the material being cut is primarily responsible for creating wear [15]. Especially, alumina is the best material for inertness with steel and ferrous alloys.

Wet turning of cast iron FC35 (HB 179) under condition of high speed (10 m/s) and feed rate (0.7 mm/rev) was performed by silicon nitride tools [3]. The relation between average flank wear width and cutting time of the silicon nitride tool is shown in Figure 11.2.7, together with data for inserts of the Al_2O_3 + TiC and Al_2O_3 coated cemented carbide tool. The Al_2O_3 + TiC tool failed catastrophically in the first 180 s of cutting, while the value of V_B of the Al_2O_3 coated tool rapidly increased over 600 s and chipping occurred after 1200 s. In contrast, the silicon nitride tool demonstrated better wear resistance without failure up to 1800 s. Consequently the silicon nitride is regarded as a good tool material with long life in cast iron cutting.

An engine block of FC25 (HB 217) cast iron was milled with a single sialon insert in the conventional negative-type face mill. The relation between cutting time and cutting length is shown in Figure 11.2.8. The flank wear width of the Al_2O_3 + TiC tool was twice that of the sialon. Thus it is found that the silicon nitride base ceramic is the most effective material for high speed milling of cast iron.

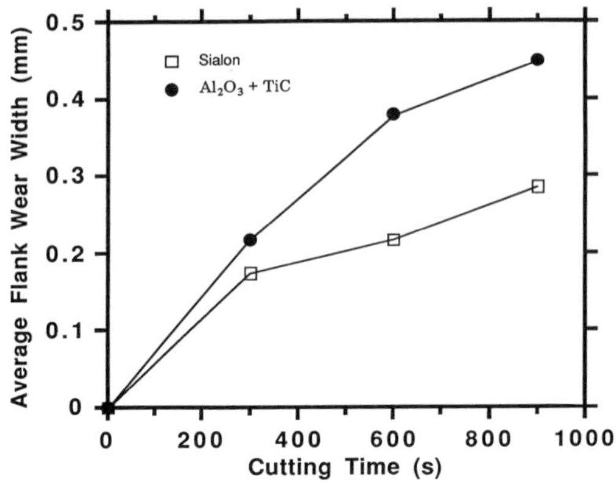

FIGURE 11.2.8 Comparison of flank wear curves of dry milling for FC 25 using sialon and $Al_2O_3 + TiC$ ceramic tools (cutting speed $V = 16.75\,m/s$; feed rate $f = 0.1\,mm/tooth$; depth of cut $d = 1.0\,mm$; machine tool, vertical milling machine (7.5 kW)).

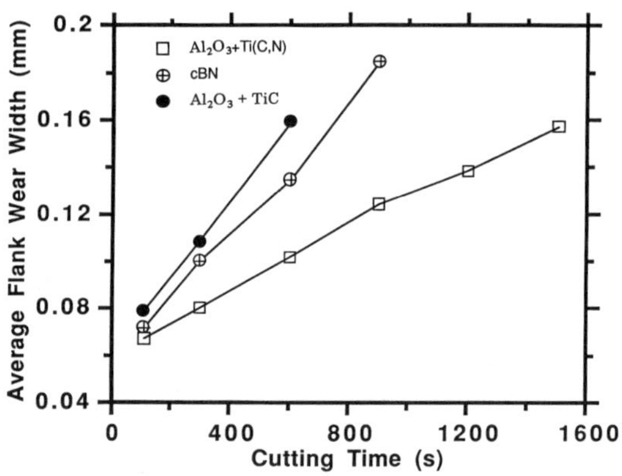

FIGURE 11.2.9 Relation between average flank wear width and cutting time in dry turning of SKD11 (cutting speed $V = 2.5\,m/s$, feed rate $f = 0.1\,mm/rev$; depth of cut $d = 0.5\,mm$; SNGN 120408; machine tool, NC lathe (55 kW)).

Dry turning performance of tool steel SKD11 with high hardness (HRC60) by $Al_2O_3 + Ti(C,N)$ tool is presented in Figure 11.2.9, along with those of cBN and $Al_2O_3 + TiC$ tools. The $Al_2O_3 + Ti(C,N)$ tool could perform cutting life up to 1500 s, but the cBN and $Al_2O_3 + TiC$ ones failed catastrophically at 923 and 649 s, respectively. The long life of the $Al_2O_3 + Ti(C,N)$ tool comes from fine microstructure, can be seen from Figure 11.2.4.

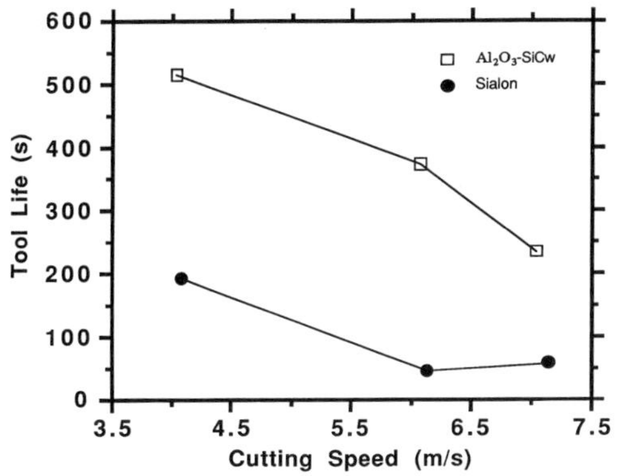

FIGURE 11.2.10 Tool life of Al_2O_3–SiC_W and sialon cutting tool materials in wet machining Ni-base superalloy Inconel 718 at 4.0–7.1 m/s with SNCN 644 insert (cutting feed rate $f = 0.2$ mm/rev and $d = 2.54$ mm).

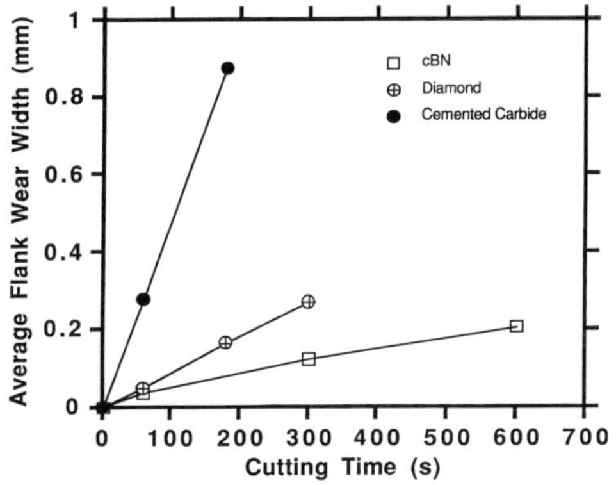

FIGURE 11.2.11 Relation between average flank wear width and cutting time in wet turning of Waspaloy (cutting speed $V = 3$ m/s, feed rate $f = 0.2$ mm/rev; depth of cut $d = 0.3$ mm; TNGN 160412; machine tool, NC lathe (55 kW)).

Cutting life test for Ni-based superalloy Inconel 718 (HRC 42.5) was performed with $Al_2O_3 + SiC_W$ and sialon tool materials to analyze their performances at different cutting speeds. This result is shown in Figure 11.2.10 [16]. At high speed of 4.2–7.0 m/s (200–450 m/min), the $Al_2O_3 + SiC_W$ tool has more than double the tool life of the sialon tool material. The behavior shows that SiC-whisker-reinforcement of Al_2O_3 exhibits a superior combination of wear resistance and fracture resistance [17].

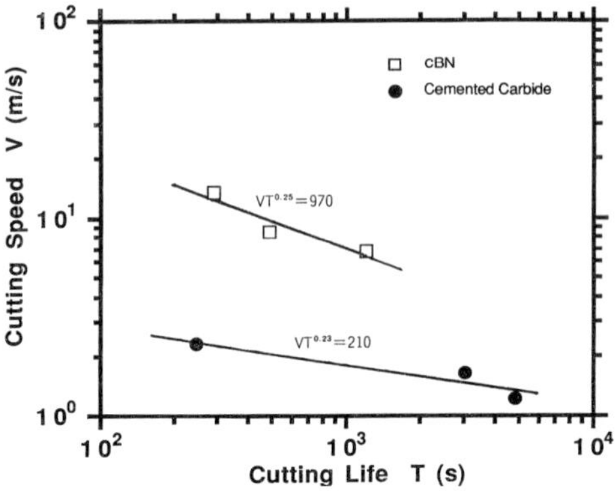

FIGURE 11.2.12 Flank wear tool-life curve based on $V_B = 0.2$ mm in dry turning of diamond and cemented carbide tool materials for Al–20 mass% Al (cutting feed $f = 0.1$ mm/rev; cut of depth $d = 0.5$ mm; SPGN120308; machine NC lathe (55 kW)).

Figure 11.2.11 shows the average flank wear versus cutting time of the cBN tool for wet turning of Ni-based superalloy Waspaloy (HB 388), containing tough γ' metallic compound Ni_3Al phases, in comparison with cemented carbide and diamond ones [18]. The cBN tool shows excellent cutting result compared with other two tool materials. This result is attributed to good hot hardness of cBN, as can be seen from Figure 11.2.2.

The flank wear's tool-life curve of in dry turning of diamond tool for Al–20 mass% Si alloy is presented in Figure 11.2.12 [19]. In comparison with the cemented carbide, the diamond tool shows superior cutting performance. This is due to extremely high hardness and superior thermal conductivity of diamond. The diamond is suitable material for machining such non-ferrous alloys.

11.2.4 RESULTS

From the above-mentioned cutting performances and discussions, the recommended cutting condition of all kinds of ceramic tool is graphically illustrated in Figure 11.2.13 [20]. This figure clearly reflects mechanical and thermal properties of tool materials. Generally speaking, alumina is suitable for steel cutting, while silicon nitride shows superior performance for cutting of cast irons. The applications of the cBN and diamond tools are limited to hard steels and non-ferrous alloys, and non-ferrous mild metals and alloys, respectively.

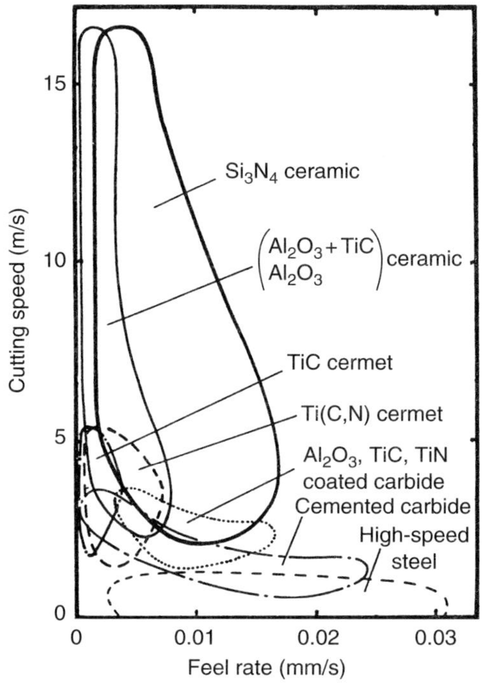

FIGURE 11.2.13 The recommended cutting condition of all kinds of ceramic tool.

REFERENCES

1. Whitney, E. D. (1978). *Powder Metall. Int.* **10**: 16–21.
2. Zsolnay, L. M. (1976). *Ceram. Ind. Mag.* **August**: 26–27.
3. Fukuhara, M., Fukazawa, K., and Fukawa, A. (1985). *Wear* **102**: 195–210.
4. Wakatsuki, M. (1974). *Kagaku* **44**: 481–486.
5. Davidge, R. W. (1979). *Mechanical Behaviour of Ceramics*, pp. 122–124, London: Cambridge University Press.
6. Ashby, M. F. (1972). *Acta Metall.* **20**: 887–897.
7. Fukuhara, M. (1984). *Tungaloy* **31**: 41–47.
8. Katsumura, Y., and Fukuhara, M. (1986). High tech ceramics, "Proc. World Congress on High Tech Ceramics", 6th CIMTEC, Milan, Italy, 24–28 June, Pt. C, pp. 2735–2745, Vincenzini, P. ed., Amsterdam: Elsevier Science.
9. Weston, J. E., and Pratt, P. L. (1978). *J. Mater. Sci.* **13**: 2147–2156.
10. Katz, K. N., and Gazza, G. E. (1978). Processing of crystalline ceramics. *Mater. Sci. Res.* **11**: 547–560.
11. Fukuhara, M. (1989). *J. Am. Ceram. Soc.* **72**: 236–242.
12. Matta, J. E., Roper, W. L., Hasselman, D. P. H., and Kane, G. E. (1976). *Wear* **37**: 323–331.
13. Evans, A. G., Heuer, A. H., and Porter, D. L. (1977). *Fracture* **1**: 527.
14. Suh, N. P., and Fillion, P. D. (1980). *Wear* **62**: 123.

15. Kramer, B. M., and Suh, N. P. (1980). *J. Eng. Ind.* **102**: 303.
16. Billman, E. R., Mehrotra, P. K., Shuster, A. F., and Beeghly, C. W. (1988). *Ceram. Bull.* **67**: 1016–1019.
17. Vignean, J., Bordel, P., and Leonard, A. (1987). *CIRP Ann.* **36**: 13–16.
18. Uehara, K., Onizuka, M., and Suzuki, M. (1994). *Tungaloy* **34**: 51–58.
19. Kiso, H. (1994). *Tungaloy* **34**: 30–37.
20. Drozda, T. J. (1985). *Manufac. Eng.* 34–40.

Handbook of Advanced Ceramics
S. Sōmiya *et al.* (Eds.)

11.3 Decorative Ceramics

MIKIO FUKUHARA
Toshiba Tungaloy Ltd., Sugasawa, Tsurumi, Yokohama 230-0027, Japan

11.3.1 INTRODUCTION

Refractory hard carbides, nitrides and oxides of the transition elements in Group IV–VI of the Periodic Table are characterized by having a beautiful golden color, high corrosion resistance and hardness. Among these compounds, titanium, zirconium and hafnium nitrides (TiN, ZrN and HfN, respectively) and titanium monooxide TiO have typical golden color. Since one of these nitrides, TiN has a low density and moderately cheapness in addition to its gold tone, it has been used as decorative scratch-proof sintered bodies and decorative coated ceramics. For other golden compounds, TaC-, TaN-, HfC- and HfN-based sintered alloys are unsuitable for portable decoration because of their high cost, high density and week golden tone. On the other hand, TiO has an advantage for decorative ceramics as well as TiN, but it is not generally available, because of metastable phase and brittleness. Here, we note two compounds, TiN and TiO as decorative ceramics.

11.3.2 PHYSICAL PROPERTIES OF TiN AND TiO

11.3.2.1 TiN

Titanium mononitride δ-TiN_x ($0.42 < x < 1.0$) has wide solubility from 10.93 to 22.63% of interstitial nitrogen [1]. The TiN is stable golden compound of face-centered cubic (B1) structure. Strictly speaking, stoichiometric composition $TiN_{1.0}$ cannot exist under atmospheric pressure, but substoichiometric composition $TiN_{0.97}$ is stable. For this reason, the atomic vacancies [2] in sublattices of Ti and N reduce height of Fermi level based on valence electron concentration (VEC) of transition metal compounds [3], resulting in stabilization of band structure. It was experimentally reported that the VEC value of the transition metal compounds is stable at 8.8 under atmospheric pressure [4]. In case of titanium nitride, $TiN_{0.96}$ is the most stable composition.

TABLE 11.3.1 Physical Properties of TiN, TiO and Au at room temperature

Material	TiN	TiO	Au
Structure	B1	B1	B1
Latice constant (nm)	0.4240	0.4182	0.4079
Density (Mg/mm^3)	5.39	4.98	19.3
Melting point (K)	3222	2023	1336
Hardness	2000	1140	—
Tone (chroma)			
x	0.413	0.373	0.399
y	0.409	0.377	0.387
Electric conductivity ($\mu\Omega$ cm)	53.9	285	2.2
Thermal conductivity (W/mK)	19.3	—	296.2
Thermal coefficient $\times 10^{-6}$(1/K)	9.35	—	14.43

There is an interesting phenomenon about a relation between nitrogen solubility and hardness, concerning of atomic vacancy; hardness of TiN_x shows mountain-like curve with maximum value at $x = 0.78$ [5]. This means that the stoichiometric composition does not show maximum hardness. It can be explained that nitrogen atoms on the sublattices select regular replacement, as well as carbon atom in vanadium carbide VC_{1-x} [6]. Thus, we use TiN hereafter as titanium nitride in place of $TiN_{0.97}$.

Physical properties for TiN are presented in Table 11.3.1 [7]. Assuming from difference of electronegativity of Ti and N, TiN is the compound of 58% covalency and 42% ionicity. However, since it has electric conductivity as same order as Si [8], it has also nature of metallic bonding in some measure. The thermal characteristics of TiN are fairly different from ordinal typical metallic elements. The thermal conductivity strongly depends on N/Ti molar ratio and temperature; it increases with increasing the ratio and temperature. The room temperature conductivity lies in between metallic and covalent bonds [9]. The thermal conducting mechanism mainly arises from lattice vibration, that is, phonon conduction. The conductivity decreases by contribution of electron scattering being proportion to T^2 [10]. Superior thermal conductivity of TiN in high temperature is an important parameter for heat-resistant application such as cutting tools.

The corrosion resistance of TiN can be presented by anodic polarization curves. The polarization result of $TiN_{0.97}$ sintered body determined at room temperature is shown in Figure 11.3.1 [11]. Since its electrode potential directly after immersion in dilute sulfuric acid is positive (+0.016 V), it is not soluble for the dilute sulfuric acid. The current density increases rapidly with increasing the voltage, and then decreases reversibly from 0.1 to 0.5 V, showing passivation of TiN. This behavior resembles to anodic polarization curve of titanium. This would be due to substoichiometric composition TiN_x with excess amount of titanium. Dissolution of TiN is possible in hot fluoric nitric acid solution alone.

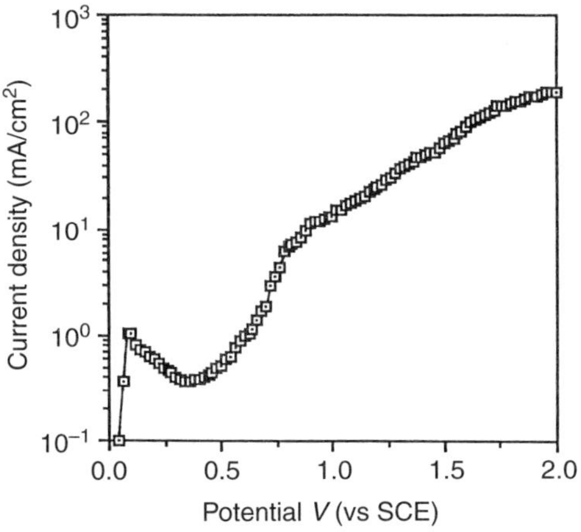

FIGURE 11.3.1 Anodic polarization curve of TiN in dilute sulfuric acid.

11.3.2.2 TiO

The physical properties for TiO are presented in Table 11.3.1. The titanium monoxide has wide range of composition range from $TiO_{0.9}$ to $TiO_{1.25}$ at temperatures below about 1263 K [12]. The structure is monoclinic, but is similar to that of NaCl (B1). The structure has an ordered array of vacant lattice sites [13]. An important feature of this phase is that its crystal structure has varying properties of titanium and oxygen vacancies. The color tone of TiO is also golden, but it is not as well known as that of TiN. However, TiO shows also poor sinterability, because of deoxidation and the resulting formation of other compounds during sintering.

11.3.3 APPLICATIONS

We classify golden decoration ceramics into two categories, sintered bodies and coated ceramics.

11.3.3.1 Sintered Bodies

Sinterability of TiN and TiO in liquid sintering under vacuum suffers from denitrification [14] and deoxidation [5], respectively. To improve poor sinterability of both TiN and TiO compounds, the TiO and double carbide of

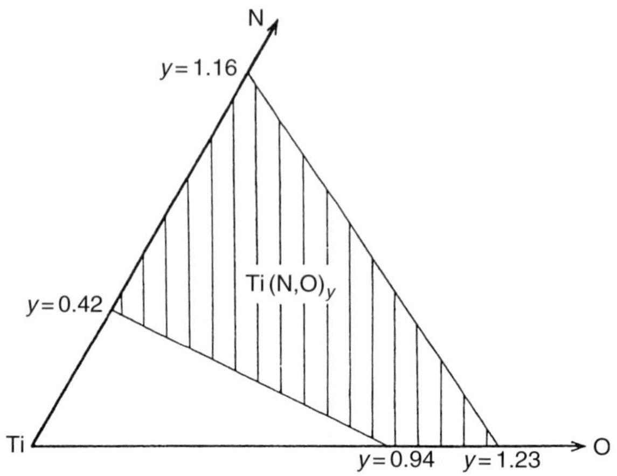

FIGURE 11.3.2 Isothermal diagram of the Ti–N–O system.

the group Va element (V,Ta)C were added to the TiN. The solid solution Ti(N,O) of TiN and TiO stabilizes the TiO phase, as can be seen from isothermal diagram (Fig. 11.3.2) of the Ti–N–O system [15]. Hardness of sintered compounds $Ti(N_{1-x}, O_x)_y$–z wt% (V,Ta)C can be described as a half ellipsoid (Fig. 11.3.3) in which the highest hardness (91.5 in HRA) can be found at or near the point when $x = 0.4$, $y = 0.78$ and $z = 18$ [5]. Some properties of the sintered material are shown in Table 11.3.2. Although the transverse rupture strength and the fracture toughness of the compound is unsatisfactory, in comparison with zirconia and silicon nitride based ceramics, it has the characteristics of low density, high hardness, good corrosion resistance and a golden color. To obtain full density, isostatically hot-pressed method is used after pressure-less sintering. The hot-pressed method is impractical, because degradation of golden color due to formation of Ti(C,N) and Ti(O,C) phase coming from diffusion of carbon from carbon mold in the hot press furnace. Furthermore, it is not available for production of complicated parts.

We show golden ceramic reels (line rollers) for fishing, as the representative decoration parts. The reels have been used in one million pieces during the past 10 years all over the world. The sliding resistance of fishing strings abrasing the reel groove immersed in artificial sweat is shown in Figure 11.3.4 as a function of durability number. The Ti(N,O) ceramic reel has superior sliding resistance comparable with those of other materials such as Si_3N_4, SiC and Al_2O_3. This would be due to excellent solid lubricant between TiN material and the string during sliding.

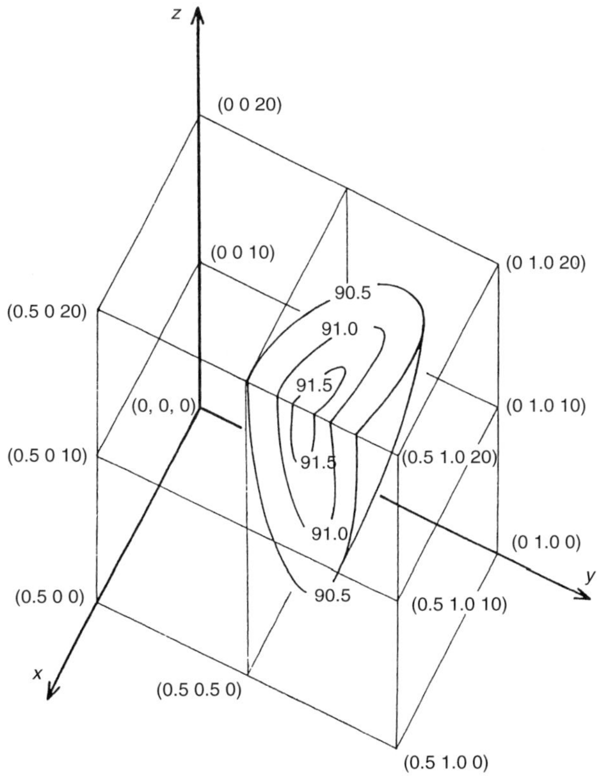

FIGURE 11.3.3 Hardness of sintered compacts of $Ti(N_{1-x}, N_x)_y$–z wt% (V,Ta)C.

The Ti(N,O) has also used as golden spikes for golf shoes. The wear resistance and toughness are important material points for the spikes. In this case, toughness is reinforced with small amount of nickel binder.

11.3.3.2 Coated Ceramics

TiN-coated decorative ceramics are used on wrist watches, because they have lightness, scratch proof, corrosion resistance and golden color. They are held in esteem all over the world as high-class goods. Although the luxury goods depend on the times, one million pieces a year were produced at the height of its popularity. The cemented carbide and Ni–Al–Cr alloy were selected as their substrates in view of their scratch-proof properties. Chemical vapor

TABLE 11.3.2 Some Properties of a Ti(N,O)–(V,Ta)C Compound

Density(Mg/m^3)	5.38
Hardness	
HRA	91.5
Hv (500 g)	1382
Transverse rupture strength(MPa)	540.0
Fracture toughness $K_{Ic}(MN/m^{3/2})$	3.0
Corrosion resistance artificial (pH 2.5–5.0)	No color changed and no corrosion
Tone (chroma)	Golden color ($x = 0.363$, $y = 0.367$)
Young's modulus (GPa)	311.0
Thermal expansion ($10^{-6}K^{-1}$)	9.2
Thermal conductivity (W/mK)	1.16

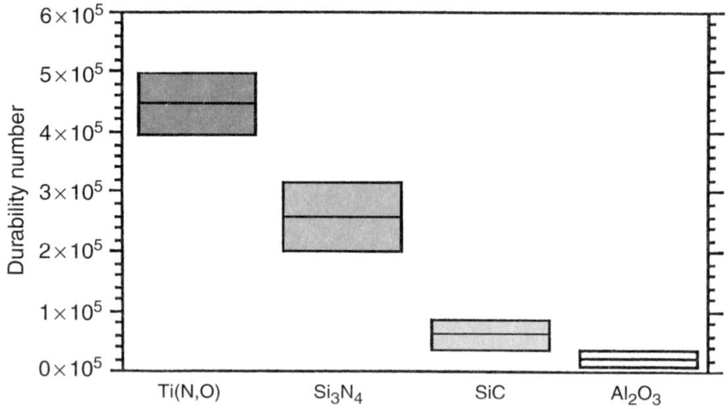

FIGURE 11.3.4 Sliding resistance durability number of fishing strings for four kinds of ceramics.

deposition (CVD) and physical vapor deposition (PVD) methods are used. The former provides high adhesion strength and mass production for complicated goods, whereas the latter is suitable for middle-scale production of simple form goods with adhesion strength less than the former. The products by the former method are somewhat reddish golden in tone, while the latter's ones are light golden color due to shortage of nitrogen ($TiN_x(x < 1)$).

REFERENCES

1. Palty, A. E., Margolin, H., and Nielsen, J. P. (1954). *Trans. Am. Soc., Metals* **46**: 312.
2. Brauer, G., and Kirner, H. (1964). *Z. Anorg. Chem.* **328**: 34.
3. Bilz, H. (1958). *Z. Phys.* **153**: 338–358.
4. Fukuhara, M., Mitsuda, T., Katsumura, Y., and Fukawa, A. (1985). *J. Mater. Sci.* **20**: 710–717.
5. Fukuhara, M., and Mitani, H. (1978). *Trans. Jpn. Inst. Metals* **21**: 211–218.
6. Venables, J. D., Kahn, D., and Lye, R. G. (1968). *Phil. Mag.* **18**: 177.
7. Metals and Ceramics Information Center Report (1976). Engineering Property Data on Selected Ceramics, Vol. 1, Nitrides, Battelle, Columbus, OH, 5.3.4-104.9.
8. Passall, N., Hulm, J. K., and Walker, M. S. (1967). Westinghaus, Research Laboratories, Final Rep. AF33 (615)-272.
9. Samsonov, G. V., and Verkhoglyadova, T. S. (1962). Physical properties of the nitrides of the transition metals. *Dolkalady Akad. Nauk USSR* **142**: 608–611.
10. Radosevich, L. G., and Williams, W. S. (1969). *Phys. Rev.* **181**: 110.
11. Fukuhara, M. (1999). *Nitride Ceramics*, pp. 119–126, Tokyo: Nikkan Kogyo.
12. Hansen, M. (1958). *Constitution of Binary Alloys*, p. 990, New York: McGraw-Hill.
13. Watanabe, D., Catles, J. R., Jostsons, A., and Malin, A. S. (1967). *Acta Crystallogr.* **23**: 307–313.
14. Fukuhara, M. (1979). Study on sintering of TiN–Ni system, PhD thesis, Osaka University.
15. Granier, B., Chatillon, C., and Allibert, M. (1982). *J. Am. Ceram. Soc.* **65**: 465–469.

Handbook of Advanced Ceramics
S. Sōmiya *et al.* (Eds.)

12.1 Ceramic Materials for Energy Systems

HIROSHI NEMOTO
NGK Insulators Ltd., 2-56 Suda-cho, Mizuho, Nagoya 467-8530, Japan

12.1.1 Li-ION BATTERY

Ceramic materials also play an important role in the field of battery technology. The Li-ion battery is a typical case in which ceramic materials are applied. In Li-ion batteries, lithium oxides are used for a positive active material, and carbons for a negative active material. Both of the active materials are considered to be ceramics prepared by normal ceramic production processes. They are used in powder form in Li-ion batteries.

One of the important parameters in evaluating a battery is its capacity to store electric energy. Coulomb capacity is normally expressed as a unit of ampere hour. One ampere hour capacity delivers 1 A current for 1 h (3600 s); that is, it has energy of 3600 C. Therefore, 1 molar active materials can give 26.8 A h capacity (= 96 500/3600), if they have monovalence. When the mean voltage of a battery is in volts V energy capacity is expressed to be V × A h, which is W h. Therefore, a higher voltage gives higher energy, and also higher power because power is proportional to V squared.

Another important parameter for battery is *C*. *C*-rate can normalize current density for a battery. One *C*-rate current for a battery with 1 A h capacity is 1 A. Battery engineers often compare performance of batteries with *C*-rate when the batteries have the same chemistry of electrode active materials.

There are many books on Li-ion battery technologies that can be referred to for further information [1, 2].

12.1.1.1 LITHIUM AND LITHIUM-ION

An important aspect of Li-ion battery is that lithium is the lightest metal element. In practical use, a lightweight battery is preferable to a heavier one if the stored energy of the battery is the same. Furthermore, when active material in a battery is metal, it is beneficial for electrode materials because metal provides electron conductivity. So far, lithium has been regarded as the best active material for a battery.

Lithium has high ionization potential, which means that lithium is also the best material for a battery from the viewpoint of energy density. This is why so many scientists and engineers have been trying to utilize lithium as the active material for many years. However, lithium metal is very reactive and unstable to handle.

Metallic lithium has first been applied to primary batteries (non-rechargeable batteries), but it has not been applicable to secondary batteries (rechargeable batteries) up to now. During recharging of the lithium secondary battery, lithium metal tends to deposit as dendrites on the lithium surface. This lithium metal dendrites build up until they eventually cause an internal short circuit in the battery. Consequently, this often leads to a fire accident, as the battery continues to heat up.

To avoid this accident, for Li-ion batteries, carbon is utilized as the negative active material instead of lithium metal, and Li-containing transition metal oxides are used for the positive active material. The combination of these positive and negative materials gives more than 3 V potential. An organic liquid or polymer electrolyte is used for Li-ion batteries. Water-based electrolytes decompose to oxygen and hydrogen at more than 1.23 V in such batteries.

12.1.1.2 Structure of Li-ion Battery

Typical cylindrical cell structure and electrode configuration of Lithium-ion battery are illustrated in Figure 12.1.1.

12.1.1.2.1 Positive Electrode (Cathode)

The most popular positive electrode material for Li-ion batteries is $LiCoO_2$. This material is a kind of ceramic (discovered by Mizushima *et al.* in 1988) that delivers 4 V potential. During charging, Li-ion/electron pairs are extracted electrochemically from $LiCoO_2$ ceramic particles. Electrons come to the negative electrode through an external circuit, while Li-ions also come to the negative electrode through the electrolyte inside the battery. The valence of trivalent Co ions in the particles increases tetra-valence to maintain electrostatic neutrality after the extraction of Li-ions and electrons. The equation of this reaction can be expressed as follows:

$$\mathrm{Li}^{+}\mathrm{Co}^{+3}\mathrm{O}_2 \rightarrow n\mathrm{Li}^{+} + ne + \mathrm{Li}^{+}_{(1-n)}\mathrm{Co}^{+3}_{(1-n)}\mathrm{Co}^{+4}_{n}\mathrm{O}_2$$

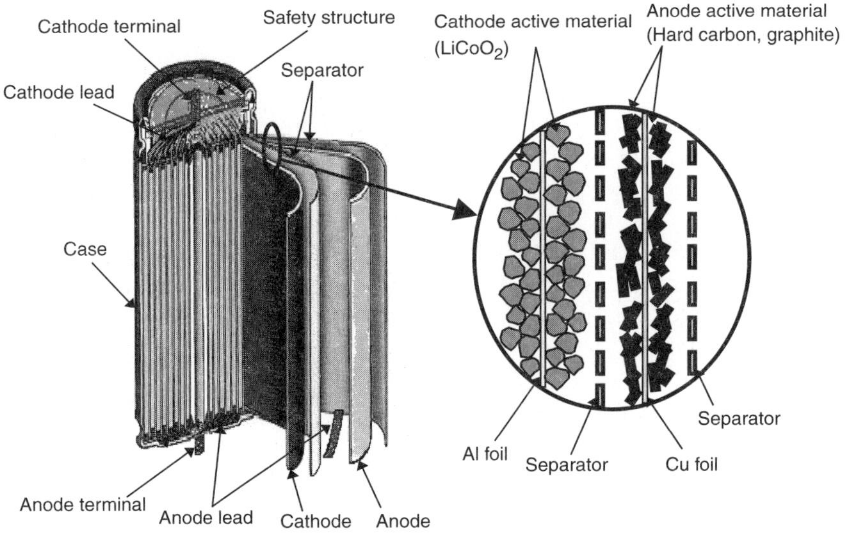

FIGURE 12.1.1 Cell structure of lithium-ion battery.

where, *n* is the number of ions and electrons. This valence change in the positive electrode is so-called solid state reduction–oxidation (redox) reaction. For this reaction, positive electrode active material must contain a transition element. This reaction is characteristic for Li-ion batteries. Oxygen deficiency in the $LiCoO_2$ affects the Co valence and keeps electrical neutrality. Therefore, control of oxygen partial pressure is very important for producing $LiCoO_2$ ceramic material.

As electrons are extracted from $LiCoO_2$ ceramic particles via external circuit during charging, the particles have to make electrical contact with the metal current collector. At the same time, as Li-ions are extracted from $LiCoO_2$ ceramic particles through the electrolyte during charging, the particles have to contact the liquid electrolyte as well. Therefore, $LiCoO_2$ ceramic particles are normally mixed with Acetylene Black as an electron conducting agent and polyvinyl di-fluoride (PVDF) as a binder, and then, are coated onto the current collector of aluminum thin foil, as shown in Figure 12.1.1.

$LiCoO_2$ ceramic powder is synthesized by a solid state reaction using Li_2CO_3 and Co_2O_3 raw materials, and firing at around 900°C. Controlling the oxygen atmosphere during firing is important because it affects stoichiometry of $LiCoO_2$. $LiCoO_2$ ceramic is normally used in Li-ion batteries not as sintered body, but as powder with a particle size range of 5–20 μm. Scanning electron micrograph (SEM) of typical $LiCoO_2$ ceramic powder is shown Figure 12.1.2.

FIGURE 12.1.2 SEM of $LiCoO_2$ ceramic powder for positive electrode active material.

When all of the Li is extracted from $LiCoO_2$, the final residual composition would be CoO_2. This CoO_2 is chemically unstable, and tends to dissociate to oxygen. This oxygen oxidizes the organic electrolyte at elevated temperatures in the Li-ion battery; this chemical reaction is unsafe for the Li-ion battery. To avoid this oxygen dissociation, extraction of Li is normally stopped at the point of $Li_{0.5}CoO_2$ composition during charging. This composition is normally controlled by the charging voltage of the battery. In case of $LiCoO_2$ and graphite, the limit of this charging voltage is 4.1 V. For other positive electrode materials, $LiNiO_2$ and $LiMn_2O_4$ are utilized. They also give 4 V.

The theoretical capacity of 1 molar metal lithium is very large, delivering 3.86 A h/g. The theoretical capacity of $LiCoO_2$ and $LiNiO_2$ is relatively small, delivering 0.274 A h/g because of oxides. As half of the Li in these oxides is utilized, as discussed above, real capacity is actually 0.173 A h/g in a practical Li-ion battery. In the case of $LiMn_2O_4$, all of the Li can be extracted because the final residual composition of MnO_2 is chemically stable. Therefore, $LiMn_2O_4$ is appreciated as a stable positive material during over-charging in Li-ion batteries. $LiMnO_2$ also acts as positive active material, but the potential between $Li_{0.5}MnO_2$ ($=LiMn_2O_4$) and $LiMnO_2$ gives 3 V, and this potential degrades during charge–discharge cycling. Therefore, the reaction between $LiMn_2O_4$ and Mn_2O_4 ($=MnO_2$) is normally utilized for 4 V battery.

$LiFeO_2$ and $LiFePO_4$, both iron compounds, have potential as a positive active material. These materials are currently being studied because Fe is an abundant natural resource.

In order to improve the charge–discharge cycle capability, above transition metals (Co, Ni and Mn) of the positive electrode materials are often substituted with foreign elements such as Ti, Cr, Al and so on.

12.1.1.2.2 Negative Electrode (Anode)

Carbon is normally used as the negative electrode active material of Li-ion batteries. It is widely known that the use of carbon for negative electrode generated great advances in Li-based battery technologies. As discussed before, lithium metal is too reactive for practical use.

During charging, electrons extracted from positive electrode transport to the negative carbon electrode through an external circuit. At the same time, Li-ions also extracted from positive electrode transport to the negative through liquid electrolyte. These two particles form a pair and exist in the carbon material in charged state. The equation of this reaction can be expressed as follows:

$$\mathrm{C} + n\mathrm{Li}^{+} + ne \rightarrow \mathrm{CLi}_{n}$$

Negative electrode carbon is normally powder with particle size range of 5–20 μm. As carbon particles must receive electrons from external circuit and Li-ions from liquid electrolyte, the carbon particles have to contact both the metal current collector and the electrolyte liquid. Carbon powder is mixed with PVDF binder and then coated by a machine onto a thin metal copper foil, as shown in Figure 12.1.1.

Carbon materials are classified into two kinds: graphitic carbon, which has a normally layered, ordered crystal structure; and hard carbon, which has an amorphous, disordered crystal structure.

Two graphitic carbons, natural graphite and synthetic graphite, are used for Li-ion batteries. Graphitic carbon can be intercalated with Li up to LiC_6 composition. The theoretical capacity of LiC_6 is 0.372 A h/g. Excess Li must be deposited as metal lithium on the graphite surface when Li ions are intercalated beyond this theoretical capacity. This lithium metal may cause an internal short circuit in the battery, so in practical use intercalation must be stopped at the point of around 0.3 A h/g. Synthetic graphite is produced by heating up carbon raw materials to about 2000°C in an electric furnace. Oil pitch or coal pitch is a good source of raw materials.

When Li ions are intercalated to graphite, the Li ions penetrate not into basal plane but into the edge plane of the graphite crystal structure. The Li ions do not cross over the interlayer of the graphite. Graphite normally has a plate like crystal structure, and consequently the edge plane is small and narrow compared with the basal plane.

FIGURE 12.1.3 SEM of hard carbon ceramic powder for negative electrode active material.

Hard carbons are synthesized by heating polymer raw materials, epoxy resin, for example, at a temperature range of 900–1000°C in an inert gas atmosphere. SEM of typical hard carbon is shown in Figure 12.1.3. As they have an amorphous crystal structure, the structure is considered to have a random arrangement of molecular size graphite crystals. Intercalated Li ions can occupy the space among the molecular size graphites, which makes the Li ion capacity of hard carbons larger than that of graphitic carbons. A capacity over 0.6 A h/g can often be obtained, but it depends on the raw materials and other synthetic conditions. When Li ions are de-intercalated from hard carbons during discharging, some ions are trapped in the carbon microstructure disordered, and therefore, actual discharge capacity normally decreases to around 0.3 A h/g. These trapped ions cause low efficiency of movable Li-ions contributing to discharging battery.

Compared with that of graphitic carbons, the charge–discharge curve of hard carbons is not steep. Therefore, the average voltage of battery reaction for hard carbons is lower than that for graphitic carbons. As the temperature to synthesize hard carbons is much lower than that needed to synthesize graphitic carbons, action cost of hard carbons is generally lower than that of synthetic graphitic carbons.

Because hard carbons have an amorphous structure rather than a layered structure, Li ions can intercalate or de-intercalate throughout the entire surface of carbon particles.

12.1.1.2.3 Organic Electrolyte and Separator

A Li-ion battery generally has a voltage range of 3.5–4 V. Aqueous electrolytes cannot be applied to electrolyte solvents, because water decomposes at a voltage range above 1.23 V. Therefore, alkyl carbonates such as propylene carbonate (PC) and ethylene carbonate (EC) having high dielectric constants are used for a basic electrolyte. To optimize the viscosity of the electrolyte solution, chain carbonates such as diethyl carbonate (DEC) and dimethoxyethane (DME) are mixed with PC and EC. These organic compounds are stable in normal battery reaction. But some components in the electrolyte mixture begin to decompose in the battery at around 4.5 V, and evolve into a gas of free radicals at 5 V. When over-charging by applying more than 5 V, this gas raises the internal pressure inside the battery case, and then, gives a tendency to explode it.

The so-called Lewis acids such as $LiPF_6$ and $LiBF_4$ are normally used for salts in Li-ion batteries. These salts tend to decompose under water, and to form hydrofluoric acid (HF). The HF is very corrosive to the metal cell-case and to the positive active materials. For this reason, the level of water contamination in electrolyte solution as delivered from electrolyte suppliers is guaranteed to be less than 20 ppm.

Polypropylene (PP) and polyethylene (PE) microporous separators (e.g. with 20 μm thickness and 50% porosity) are used for electrically separating the positive electrode and negative electrode. SEM of microporous separator is shown in Figure 12.1.4. As organic solvents are wettable to PP and PE, the solvents can penetrate into such micropores. The pore size of the separator is normally less than 0.5 μm, in order to ensure that fine active ceramic particles of electrodes do not pass through the separator. A PP/PE/PP layered separator is often used for practical Li-ion batteries because of a "shut down" effect. When battery temperature approaches the melting point of PE (130°C), micropores of only PE are suddenly closed, and the battery reaction coming from Li-ion transportation is stopped by the separator.

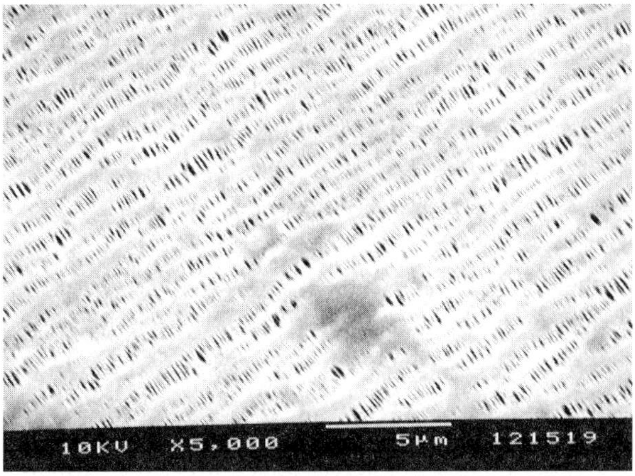

FIGURE 12.1.4 SEM of PP/PE/PP microporous separator.

12.1.1.2.4 Battery Reaction

Battery reaction of Li-ion battery is illustrated in Figure 12.1.5. Electrochemical reactions in the positive and negative electrodes during charging are expressed by the following equations:

$$\text{Positive electrode: } LiCoO_2 \rightarrow nLi + ne + Li_{(1-n)}CoO_2$$

$$\text{Negative electrode: } C + nLi + ne \rightarrow CLi_n$$

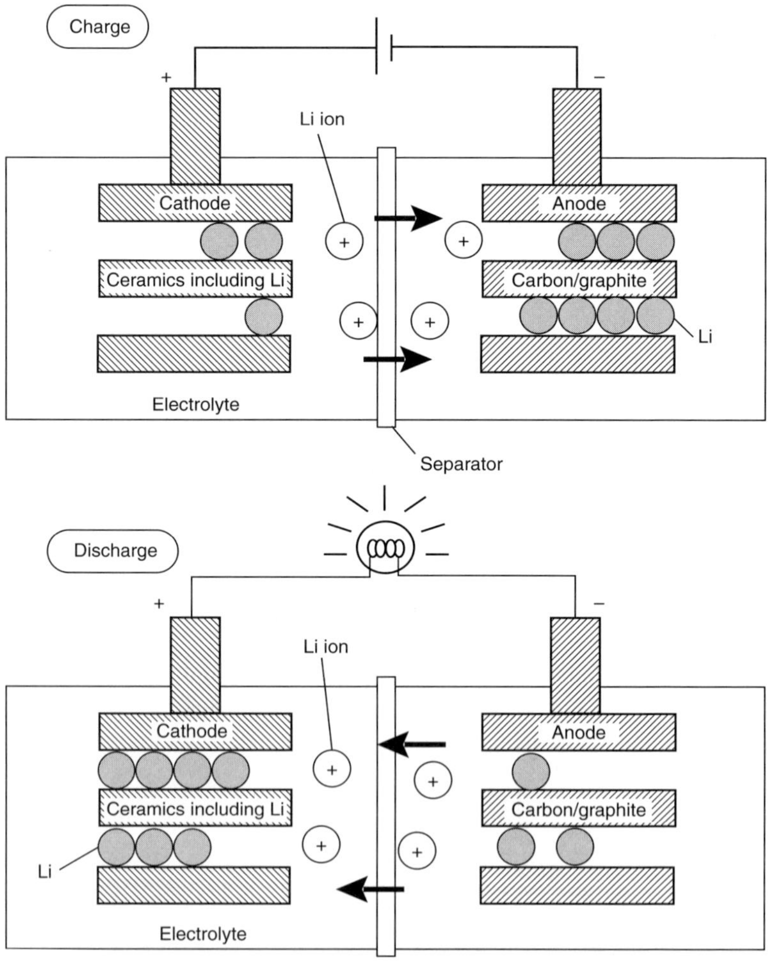

FIGURE 12.1.5 Charge–discharge reaction of Li-ion battery.

Overall reaction is as follows.

$$LiCoO_2 + C \rightarrow Li_{(1-n)}CoO_2 + CLi_n \quad \text{(charging reaction)}$$

One of advantages in this battery system is no side reaction during charging and discharging. As can be seen in these reactions, a Li-ion battery is manufactured under a state of discharge. About 10% of capacity is lost in the first charging, because 10% of the Li ions are residual in negative carbon. As the Li-ion battery has low internal resistance, and has no side reaction, the efficiency of discharging energy versus charging energy is very high; normally able to deliver more than 98%. During charging and discharging, Li ions go and return between the positive and negative electrodes. Because of this, the Li-ion battery is nicknamed the "rocking-chair" battery. Charge–discharge cycle of Li-ion batteries reaches more than 1000 cycles at 0.2 C current.

12.1.1.3 Production Process of Li-ion Battery

The production process of Li-ion battery is illustrated in Figure 12.1.6.

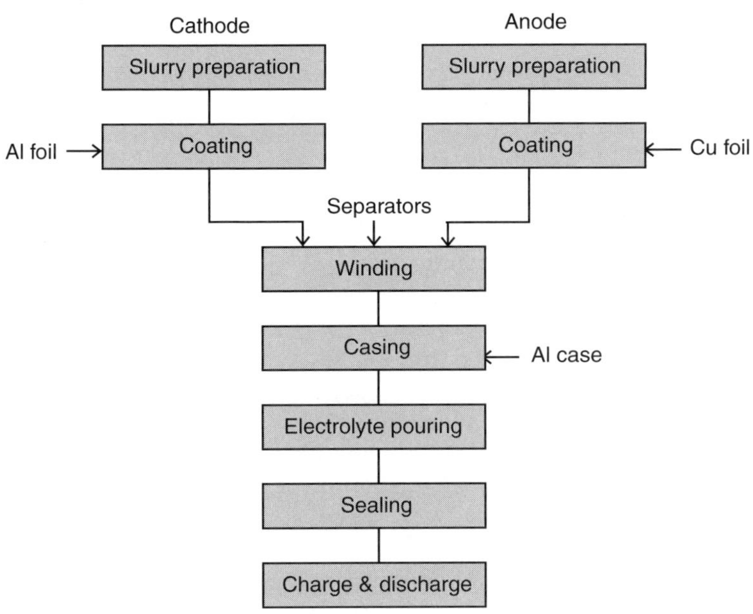

FIGURE 12.1.6 Production process of Li-ion battery.

The positive electrode is produced as follows: the $LiCoO_2$ powder of the positive active material and carbon black conducting agent are mixed with an *N*-methyl pyrrolidone (NMP) solution of PVDF binder to make a paste of appropriate viscosity. A mixture of water and carboxyl methylcellulose (CMC), with styrene butadiene rubber (SBR, used as a binding agent), is also used for making such paste.

The paste is coated on both sides of the aluminum metal thin foil by a coating machine. Normally, the thickness of the coated layer has a range of 50–80 μm. The coated foil is dried in a drier at about 150°C to vaporize the NMP solvent.

Similarly, the carbon powder of the negative active material is mixed with an NMP solution of PVDF binder to create a paste of appropriate viscosity. This paste is coated on both sides of the copper metal thin foil by a coating machine. The thickness of the coated layer has a range of 50–80 μm. The coated foil is similarly dried in a drier at 150°C to vaporize the NMP solvent.

After drying, the coated foils of the electrodes have low density to achieve loose contact with the powder particles. Of course, particles electrically apart from the metal foil can not act as active material. In order to make the particles contact tightly, the foils have to be roll-pressed to increase apparent density.

In making a cylindrical cell, the positive and negative foils are wound with two separators by a winding machine to make cell body. In making a prismatic cell, the films are just stacked with separators. The cell body is inserted into a cell case after the body is connected to electrode terminals with lead tabs, and then the cell body is filled with the liquid electrolyte under an inert gas atmosphere. Because Li-ion cells are made under discharged conditions, safety is well maintained. After being filled with electrolyte, the cells are immediately charged with 0.2 C current, for example. If they are not charged immediately, the cells come to a state of over-discharge.

12.1.1.4 Applications of Li-ion Batteries

Compared with conventional batteries such as lead acid, Ni–Cd, and Ni–MH, Li-ion batteries provide higher energy density and higher power density. Since the Li-ion battery was introduced in 1992, a huge worldwide market has developed for portable cell phones and laptop computers. In 2000, Li-ion batteries represented half of the worldwide rechargeable battery market with a value of $3 billion, and still growing.

12.1.1.4.1 Small Size Li-ion Battery

Small size Li-ion batteries have a typical capacity of 2 A h. Cylindrical and prismatic battery shapes are valued for applications in cell phones and personal computers. The average cell voltage of Li-ion batteries is normally about 3.6 V, so the batteries can drive an integrated circuit on only one cell. For personal computers, four Li-ion cells are connected in series to drive the display. To avoid serious accidents caused by over-charge, an electronic circuit and a fuse are normally attached to the four cells.

12.1.1.4.2 Large Size Li-ion Battery

Large size cells with a capacity of about 100 A h, for example, are now being developed for the electric energy storage and electric vehicle (EV) fields. For energy storage applications, the energy density of the battery is important to reduce space needed for installation. A Japanese national project called New Energy Development Organization (NEDO) is leading the development of this application. The target cell energy density is 150 W h/kg. This energy density is the highest practical value available among practical rechargeable batteries.

In EV applications, power density is the factor needed to attain good acceleration in the vehicles. NEDO is also targeting EV applications. In the United States, the Partnership of New Generation Vehicle national project is also targeting EV applications. So far power density is reaching 2000 W/kg, making this the highest value among rechargeable batteries.

REFERENCES

1. Wakihara, M. *et al.* (1998). *Lithium Ion Batteries*, New York: Wiley-VCH.
2. Pistoia, G. *et al.* (1994). *Lithium Batteries: New Materials, Developments and Perspectives*, Amsterdam: Elsevier.

Handbook of Advanced Ceramics
S. Sōmiya *et al.* (Eds.)

CHAPTER 13

13.1 Extruded Cordierite Honeycomb Ceramics for Environmental Applications

TOSHIYUKI HAMANAKA
NGK Insulators Ltd., Automotive and Industrial Ceramics Division, 2-56 Suda-cho, Mizuho, Nagoya 467-8530, Japan

13.1.1 INTRODUCTION

Extruded cordierite ceramic honeycombs were developed as substrates for automotive catalytic converters to solve the environmental pollution problems caused by exhaust gas from gasoline engines [1, 2]. Since the 1970s, US and Japanese auto-manufacturers have used ceramic honeycomb catalytic converters to eliminate toxic substances such as HC, CO, and NO_x in accordance with emission regulations. In the 1980s, Australian, Korean, and many European auto-manufacturers also began using ceramic honeycomb catalytic converters. During the 1990s, many newly industrialized countries such as Taiwan, Mexico, and Brazil adopted automotive emission regulations. Ceramic honeycomb converters have been used in highly motorized regions all over the world as shown in Table 13.1.1.

Extruded cordierite honeycombs also have applications in other fields because of their unique material and structural properties such as high porosity, low thermal expansion, high geometric surface area, and low gas flow restriction [2]. Utilizing their porous ceramic wall as filters, extruded honeycombs can be used as trap oxidizers to eliminate toxic particulate matter from diesel engine exhaust.

This paper describes the properties and applications of extruded cordierite honeycombs. Future tasks to upgrade performance of extruded cordierite honeycombs are also discussed.

TABLE 13.1.1 Worldwide Autocatalyst Use

Countries	Adoption year	Countries	Adoption year
US	1973	Mexico	1992
Japan	1978	Hong Kong	1992
Australia	1986	Thailand	1993
Korea	1988	Argentina	1994
Europe[a]	1988	Malaysia	1995
Brazil	1991	Chile	1995
Taiwan	1992	India	1996

[a]Region.

13.1.2 EXTRUDED CORDIERITE HONEYCOMBS

13.1.2.1 Manufacturing Process

Extrusion is the most widely used method of forming cordierite honeycombs. Other methods, such as corrugated paper dipped in ceramic slurry, comb-shaped panel lamination, and press forming have been investigated. However, from the view point of mass production, extrusion is the best method for forming honeycombs.

Figure 13.1.1 shows the manufacturing process flow chart for cordierite honeycomb. The raw materials for cordierite are kaolin, alumina, and talc, which are pulverized to suitable sizes for extrusion. These raw material powders are mixed with organic binders and water, then kneaded. The kneaded clay mixture is extruded through a die, dried and cut to specified length. These dried bodies are then fired at around 1400°C.

Figure 13.1.2 shows the structure of the extrusion die. The die slit width and cell pitch are determined by the desired wall thickness and cell pitch of the cut honeycomb product. During extrusion, inorganic raw material powders become favorably positioned after passing through a narrow slit, as shown in the next section.

13.1.2.2 Material Properties

Table 13.1.2 shows the material properties of standard extruded cordierite honeycombs. Extruded cordierite honeycombs have an extremely low coefficient of thermal expansion from 40 to 800°C, hence, high thermal shock resistance is expected. The softening temperature is above 1410°C. Porosity is 35%, and mean pore diameter is 5 μm (typical value).

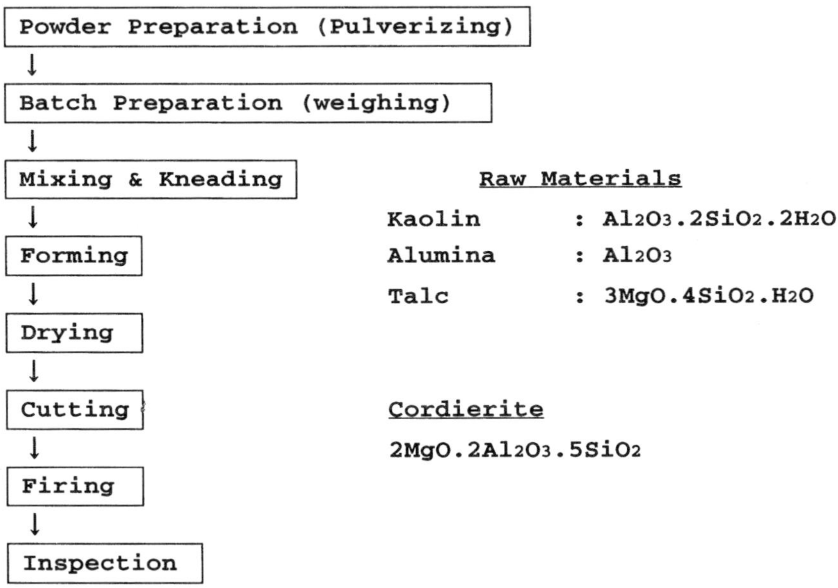

FIGURE 13.1.1 Manufacturing process Flow chart of cordierite honeycombs.

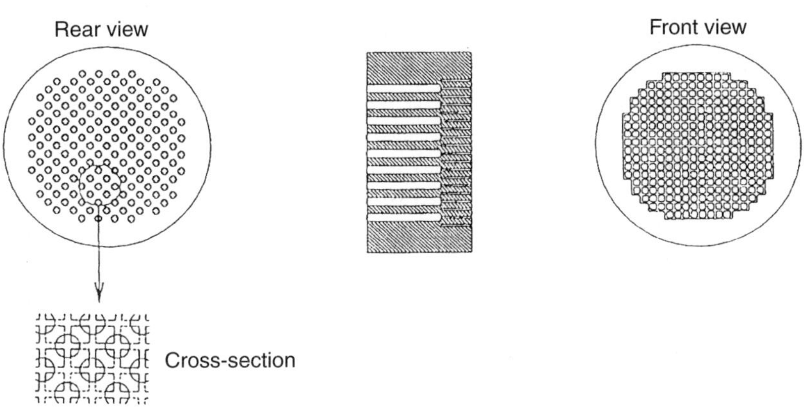

FIGURE 13.1.2 Structure of extruding die.

Figure 13.1.3a shows the low thermal expansion scheme of extruded cordierite due to the orientation of the crystals. The *c*-axes of flat-shaped kaolinite crystals are oriented vertical to the honeycomb wall by the shearing force generated during extrusion through the narrow slit of the die. The *c*-axis of

TABLE 13.1.2 Material Properties of Extruded Cordierite Honeycombs

Items	Properties
Crystalline phase	Cordierite
Coefficient of thermal expansion (40–800°C)	0.6×10^{-6}/°C
Softening temperature	1410°C
Porosity	35%
Mean pore diameter	5 μm
Thermal shock resistance	>700°C
Compressive strength (extruding direction)	>100 kg/cm^2

cordierite crystals, which have negative thermal expansion due to anisotropy, as shown in Figure 13.1.3b, grow vertically to the *c*-axis of kaolinite. As a result, thermal expansion of extruded cordierite body except in the wall thickness direction is reduced, and high thermal shock resistance is achieved [3]. Figure 13.1.3c is a scanning microscope photograph of cordierite crystals. Along the cell wall, hexagonal crystal growth is observed.

Figure 13.1.4 shows the thermal expansion of extruded cordierite from 40 to 800°C, in comparison with other ceramics. No material is equal to cordierite in thermal expansion properties. Hence, extruded cordierite honeycombs having extremely low thermal expansion, are widely used as automotive catalyst substrates, diesel particulate filters, and heat exchangers, where high thermal shock resistance is required.

13.1.2.3 Structural Properties

Table 13.1.3 shows the structural properties of extruded cordierite honeycombs with typical cell structures. Cell structures of 17 mil/100 cpi^2 and 12/200 cpi^2 are combined with high porosity material for use in diesel particulate filters, where low flow restriction is required for the cell walls. Cell structure of 6 mil/400 cpi^2 combined with normal porosity material is used for automotive catalyst substrates as the standard. Cell structures of 4 mil/600 cpi^2, 4 mil/400 cpi^2, 3 mil/600 cpi^2, 3 mil/400 cpi^2, and 2 mil/900 cpi^2 are used for automotive catalyst substrates, especially for the tightened emission regulations of United States, Europe, and Japan after 1995.

Extruded honeycomb structures have large geometric surface areas (GSA) per unit volume, as shown in Figure 13.1.5. In diesel particulate filters, large

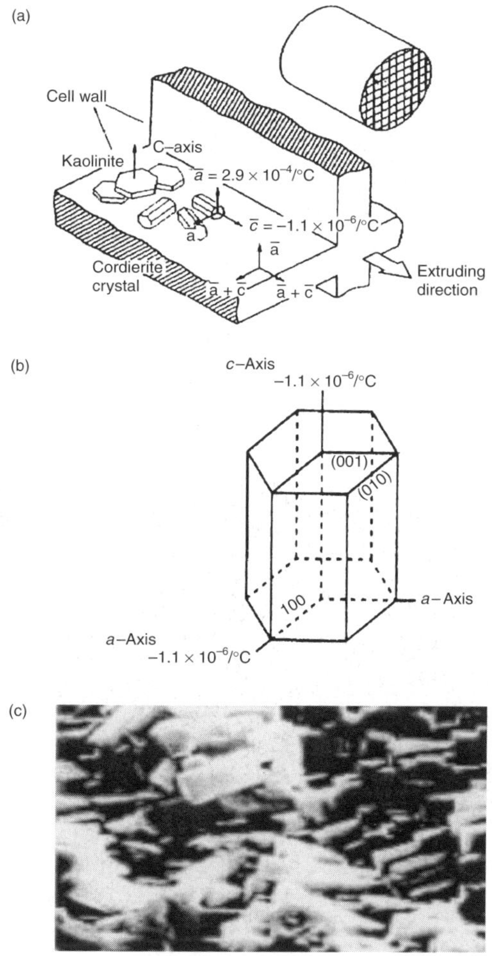

FIGURE 13.1.3 (a) Orientation of cordierite crystals. (b) Anisotropy of cordierite crystals. (c) SEM photograph of cordierite crystals.

GSA provides high accumulation capacity of particulate during engine operation. In substrates for automotive catalysts, large GSA gives a high catalytic reaction area between the catalyst and the exhaust gas.

Extruded honeycomb structures have low pressure drop compared to gas flow, as shown in Figure 13.1.6. As substrates for automotive catalytic converters, honeycomb with low back pressure drop has less effect on engine output and fuel penalty.

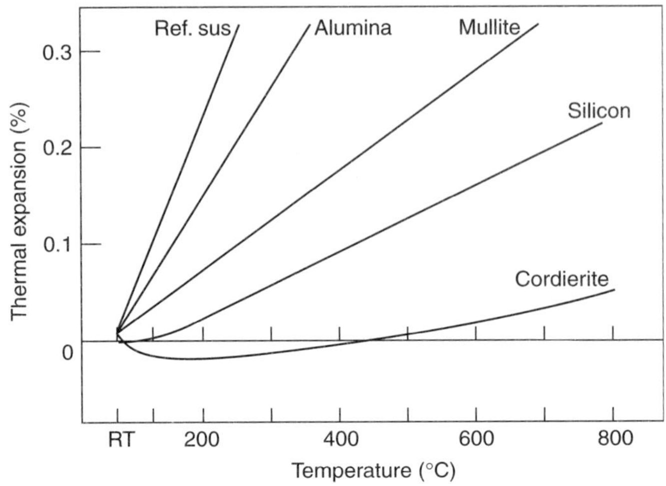

FIGURE 13.1.4 Thermal expansion of extruded cordierite.

TABLE 13.1.3 Structural Properties of Extruded Cordierite Honeycombs

Cell structure (mil/cpi^2)	17/100	12/200	6/400
Cell shape	Square	Square	Square
Wall thickness (mm)	0.43	0.31	0.17
Geometrical surface area (m^2/l)	1.3	1.8	2.7
Bulk density (g/cm^3)	0.36	0.36	0.55
Porosity (%)	54	54	35
Cell Structure (mil/cpi^2)	4/400	3/600	2/900
Cell shape	Square	Square	Square
Wall thickness (mm)	0.15	0.75	0.50
Geometrical surface area (m^2/l)	3.0	3.5	4.5
Bulk density (g/cm^3)	0.25	0.25	0.20
Porosity (%)	35	35	35

13.1.3 APPLICATION OF EXTRUDED CORDIERITE HONEYCOMBS

13.1.3.1 SUBSTRATES FOR AUTOMOTIVE CATALYTIC CONVERTERS

Due to durability requirements of catalysts for long-term vehicle use such as 120 000 miles or so, extruded cordierite honeycombs with low thermal

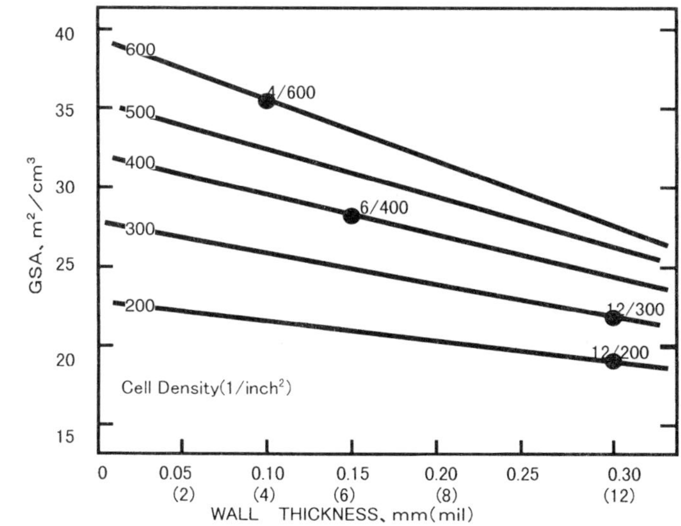

FIGURE 13.1.5 Geometric surface area of honeycombs.

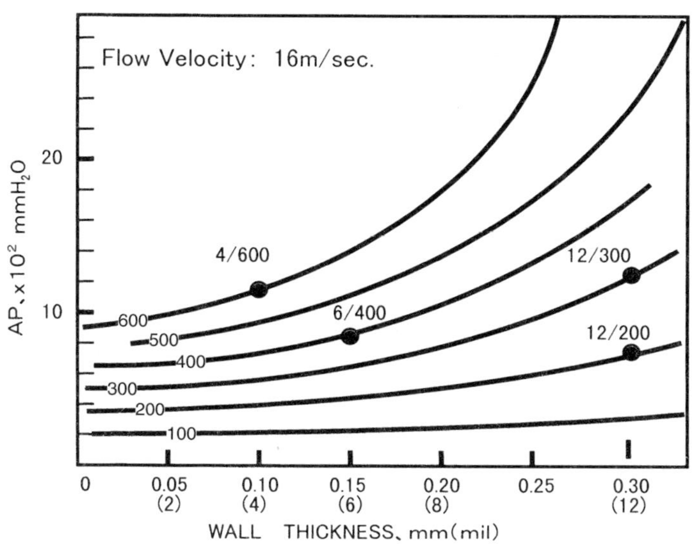

FIGURE 13.1.6 Pressure drop of honeycombs.

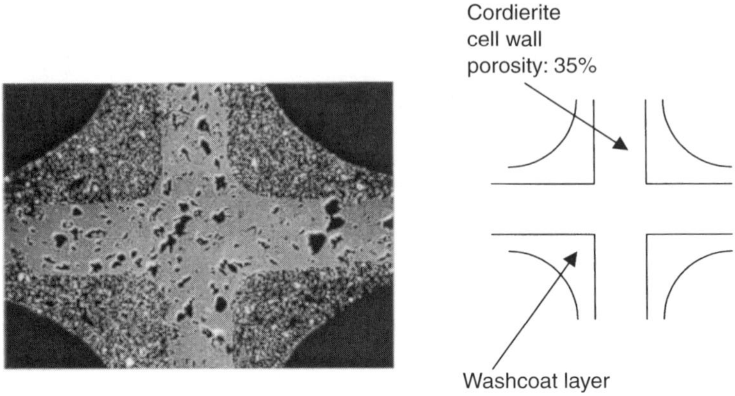

FIGURE 13.1.7 SEM photograph of washcoat layer of catalyst.

expansion have been used as substrates for automotive catalytic converters. In this section, coatability, catalytic efficiency at cold start of engines, and thermal shock resistance of cordierite honeycomb substrates are discussed.

13.1.3.1.1 Coatability

To maintain catalytic performance during long-term vehicle operation, porosity of the cordierite material is important to fix the catalyst layer on the cell wall of the honeycomb substrates. Figure 13.1.7 shows the washcoat layer, which contains γ-alumina composite carrier and precious metal catalyst, on the porous cell wall surface. This photograph shows a 6 mil/400 cpi^2 standard cell structure catalyst made of 35% porosity cordierite.

13.1.3.1.2 Conversion Efficiency of the Catalysts [4–7]

13.1.3.1.2.1 *Experiments*

Two tests to determine the conversion efficiency of the catalyst are conducted. One is the light-off test to examine the conversion efficiency of the catalyst itself by slowly raising exhaust gas temperature at the rate of 10°C/ min, as shown in Figure 13.1.8. The other test is the warm-up test to examine the conversion efficiency of the catalyst including its heat mass factors in order to simulate the cold-start of engines. Hence, the exhaust gas temperature is raised quickly at the rate of 160°C/ min, also shown in Figure 13.1.8.

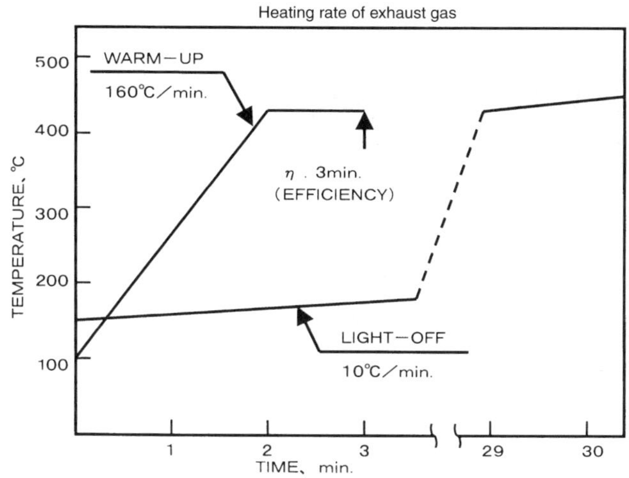

FIGURE 13.1.8 Conversion efficincy tests of catalyst.

TABLE 13.1.4 Test Substrates for Conversion Efficiency Test

	Cell structure (mil/cpi^2)		
	4/400	6/400	8/400
Wall thickness (mm)	0.10	0.17	0.20
Cell pitch (mm)	1.27	1.27	1.27
Porosity (%)	24	35	35
Bulk density (g/cm^3)	0.35	0.44	0.50
(with catalyst[a])	(0.60)	(0.68)	(0.74)

[a]Three-way catalyst: Pt/Rh = 10/1, 35 g/ft^3.

13.1.3.1.2.2 *Test substrates for light-off and warm-up tests*

Table 13.1.4 shows cell structures of the three test substrates. Wall thickness is 4 mil (0.10 mm), 6 mil (0.17 mm), and 8 mil (0.20 mm); cell density is 400 cells per square inch (1.27 mm pitch).

High-density cordierite material with 24% porosity is used for 4 mil substrates to improve mechanical strength. Normal cordierite with 35% porosity is used for 6 and 8 mil substrates. Bulk density of substrates varies by wall thickness and porosity; however, the amount of washcoat per unit volume of the catalyst is kept equal in this test to compare the effect of heat mass of substrates

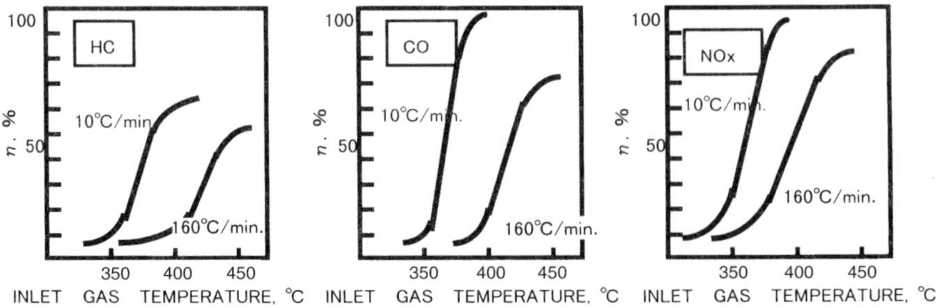

FIGURE 13.1.9 Effect of heating rate of exhaust gas on conversion efficiency.

on warm-up characteristics. All substrates are coated with the same amount of Pt/Rh catalyst, using a ratio of 10/1 and concentration of 35 g/ft^3.

13.1.3.1.2.3 *Effect of heating rate of exhaust gas on conversion efficiency*

Figure 13.1.9 compares the conversion efficiencies of HC, CO, and NO_x of 6 mil/400 cpi^2 catalysts in the light-off and warm-up tests, after engine aging at 750°C for 50 h. In the warm-up tests where the heating rate of exhaust gas is 160°C/ min, HC, CO, and NO_x conversion efficiencies are lower than in the light-off tests, when the heating rate of exhaust gas is 10°C/ min. To improve conversion efficiency of the catalysts, especially at the cold start of engines, warm-up characteristics should be improved.

13.1.3.1.2.4 *Effect of bulk density of the catalysts on warm-up characteristics*

Figure 13.1.10 shows the effects of bulk density of 400 cpi^2 catalysts on warm-up characteristics after engine aging. To examine the effects of heat mass of catalysts, 4, 6, and 8 mil catalysts are used. As catalyst bulk density increase, inlet gas temperature at 50% conversion efficiency ($T_{50\%}$) increases, and conversion efficency during the initial 3 min ($\eta_{3\,\mathrm{min}}$) decreases in the HC, CO, and NO_x warm-up tests. Consequently, reducing the heat mass of catalysts presumably improves warm-up characteristics.

13.1.3.1.2.5 *Test substrates for HC emission in the FTP cycle*

Table 13.1.5 shows the cell structures of the six test substrates. Wall thickness is 2 mil (0.06 mm), 3 mil (0.09 mm), 4 mil (0.11 mm), and 6 mil (0.16 mm); cell density is 400 cells per square inch (1.27 mm pitch), 600 cells per square inch (1.04 mm pitch), and 900 cells per square inch (0.85 mm pitch).

High-density cordierite material with 28% porosity is used for 2, 3, and 4 mil substrates to improve mechanical strength. Normal cordierite with 35%

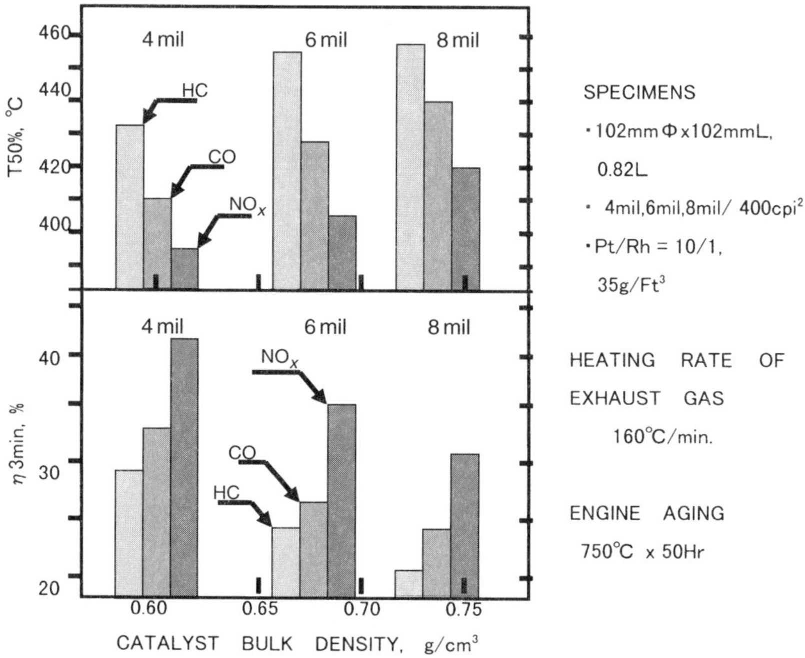

FIGURE 13.1.10 Effect of bulk density on warm-up characteristics.

TABLE 13.1.5 Test Substrates for HC Emission in the FTP Cycle

	Cell Structure (mil/cpi^2)					
	6/400	4/400	4/600	3/600	3/900	2/900
Wall thickness (mm)	0.16	0.11	0.11	0.09	0.09	0.06
Cell pitch (mm)	1.27	1.27	1.04	1.04	0.85	0.85
Porosity (%)	35	28	28	28	28	28
Bulk density (g/cm^3)	0.42	0.32	0.38	0.30	0.37	0.28
GSA[a] (cm^2/cm^3)	27.3	28.8	34.3	35.3	42.3	43.7

[a]Geometric surface area.

porosity is used for 6 mil substrates. The amount of washcoat per unit volume of the catalyst is kept equal as 90 kg/m^3 in this test. All substrates are coated with the same amount of Pd/Pt/Rh catalyst, using a ratio of 15/1/1 and concentration of 120 g/ft^3.

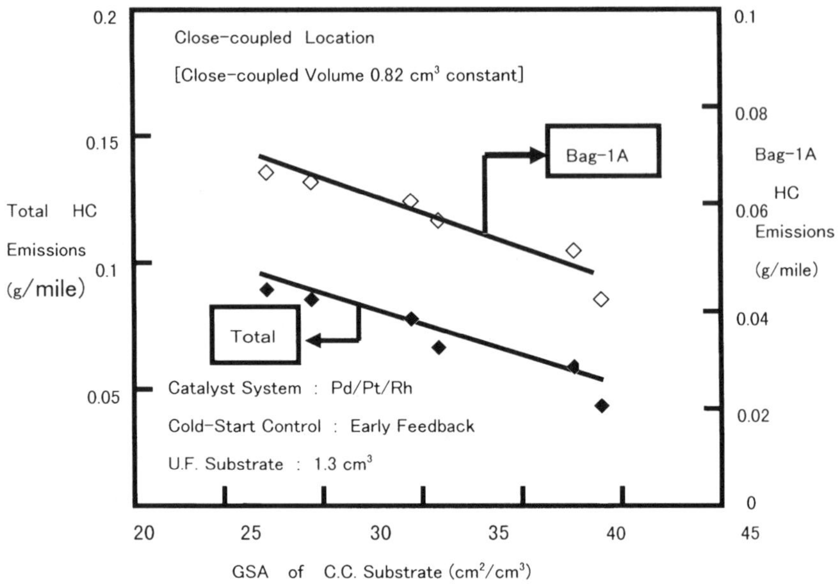

FIGURE 13.1.11 Effect of GSA on HC emissions.

13.1.3.1.2.6 *Effect of GSA on HC emissions*

Figure 13.1.11 compares the total HC emissions of various catalysts in the FTP-75 engine test cycle, after engine aging at 750°C for 100 h. The catalyst location is in so called close-coupled converter installed 400 mm downstream from the engine.

These results indicate the GSA and bulk density of honeycombs have an influence on the HC emissions. High GSA and low bulk density of 2 mil/900 cpi^2 catalysts shows very low HC emissions, and has the potential to reduce the total HC emissions as low as 40% compared with 6 mil/400 cpi^2 catalysts.

13.1.3.1.3 Thermal Shock Resistance

Since exhaust gas temperature changes rapidly during engine operation, ceramic honeycomb substrates must have thermal shock resistance. Thermal shock resistance of ceramic honeycombs is determined by electric furnace or gas burner testing. Thermal shock resistance of ceramic material is generally represented by the following equation [8]. As the coefficient of thermal expansion of extruded cordierite is extremely low, high thermal shock resistance is expected.

$$\Delta T = K \times (1/\alpha) \times (1 - \mu) \times (\sigma/E)$$

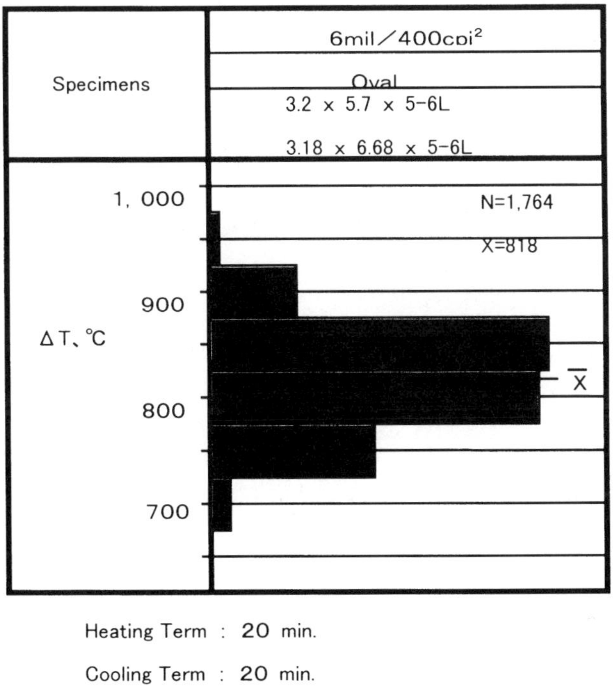

FIGURE 13.1.12 Thermal shock resistance of cordierite honeycombs.

where ΔT is the Thermal shock resistance, K a constant, α the coefficient of thermal expansion, μ the Poisson's ratio, σ the tensile strength, and e the Young's modulus.

Figure 13.1.12 shows the electric furnace test results for thermal shock resistance of extruded cordierite honeycombs for automotive catalyst substrates. The withstanding temperature of the substrates is more than 700°C.

13.1.3.2 Diesel Particulate Filters (DPF)

Due to their structural and material properties, wall flow type cordierite honeycomb filters DPF have good features such as high trapping efficiency, high soot accumulation capability relative to small volume, and high thermal shock resistance [2]. Hence, cordierite honeycomb DPFs have been tested in city buses and utility vehicles in Europe and the United States to reduce toxic particulate matter from diesel engine emissions [9–10]. In this section, structure, material properties in relation to soot accumulation, and some regeneration test results of DPF are described.

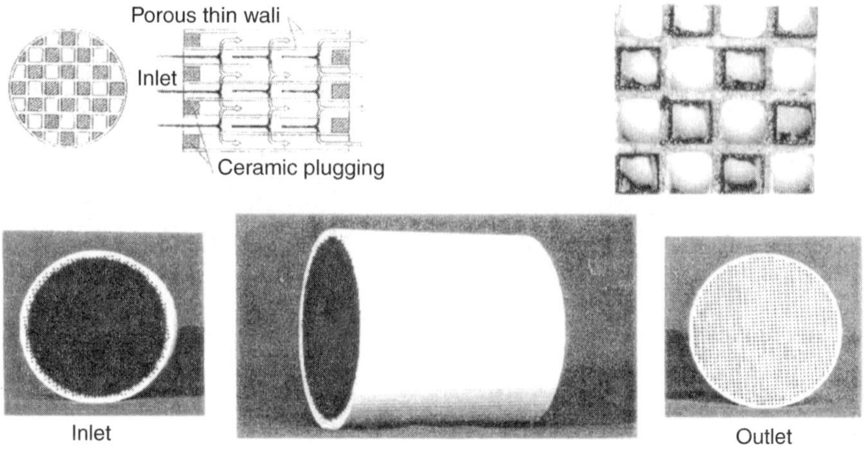

FIGURE 13.1.13 DPF schematic.

13.1.3.2.1 Structure of DPF

Figure 13.1.13 shows a DPF schematic. Cell channel openings are plugged in a checker flag pattern at one end and alternatively at the other. Thus, the exhaust gas is forced through the porous thin cell walls, which serve as filters.

13.1.3.2.2 Material Properties

Figure 13.1.14 shows the pore size distribution of DPF materials with 54 and 60% porosity. High trapping efficiency DHC-558 material has a 15 μm mean pore diameter, and most of pores are distributed from 5 to 50 μm. Low pressure drop DHC-611 material has a 25 μm mean pore diameter and higher porosity of 60%.

Figure 13.1.15 shows the pressure drop of the two DPF materials during soot accumulation, using 17 mil/100 cpi^2 and 12 mil/300 cpi^2 cell structures. DHC-558 exhibits a rapid increase in pressure drop, while DHC-611 has a low pressure drop especially in the cell structure of 12 mil/300 cpi^2 due to the higher GSA for soot accumulation.

Figure 13.1.16 shows the trapping efficiency of the two DPF materials during soot accumulation, using 17 mil/100 cpi^2 cell structure. Both DHC-558 and DHC-611 shows more than 90% trapping efficiency.

13.1.3.2.3 Regeneration of DPF

Figure 13.1.17 shows the soot accumulation and regeneration process of DPF with 17 mil/100 cpi^2 cell structure and DHC-558 material using a diesel burner,

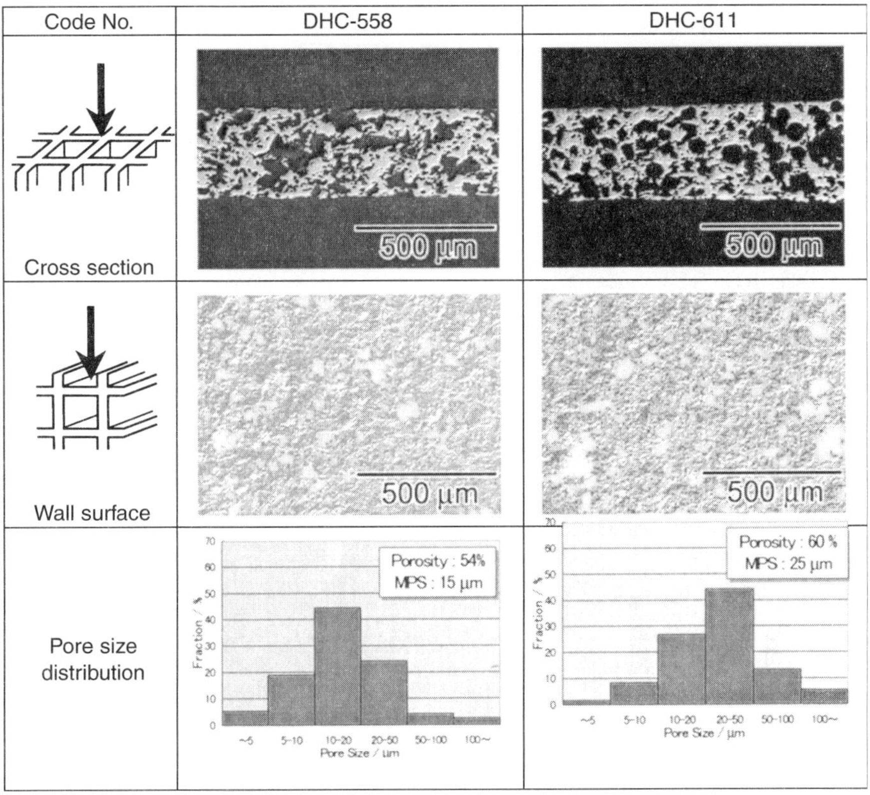

FIGURE 13.1.14 Pore size distribution of DPF materials.

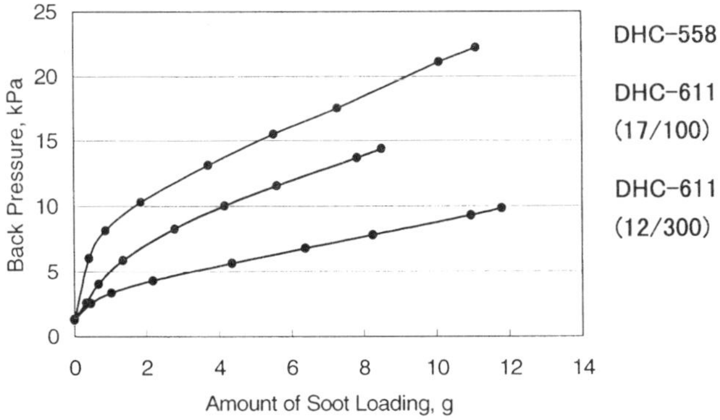

FIGURE 13.1.15 Pressure drop of DPF materials.

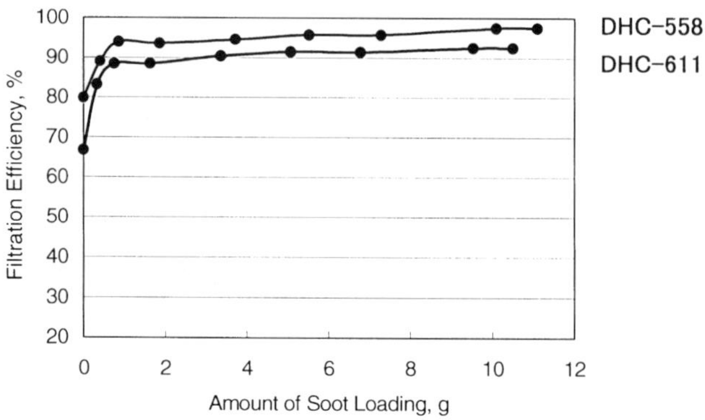

FIGURE 13.1.16 Trapping efficiency of DPF materials.

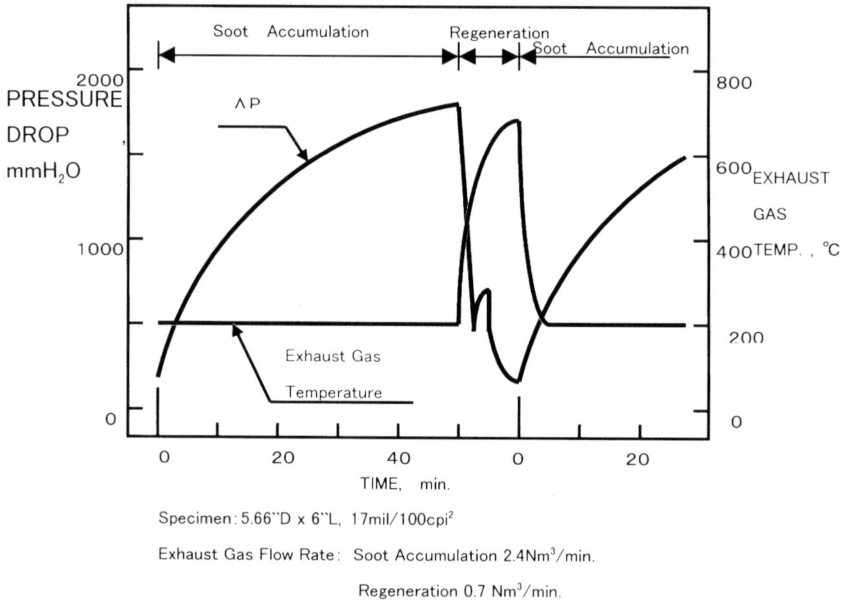

FIGURE 13.1.17 Soot accumulation and regeneration process of DPF.

which generates exhaust gas of 200°C and soot concentration of 20 g/h [11]. During 50 minutes of operation, the pressure drop of DPF increases from 100 to 2000 mmH_2O due to soot accumulation on the cell wall. At this point, exhaust gas temperature is raised from 200 to 700°C in 10 min by reducing secondary air, and the pressure drop of the DPF returns to the original level by burning the accumulated soot.

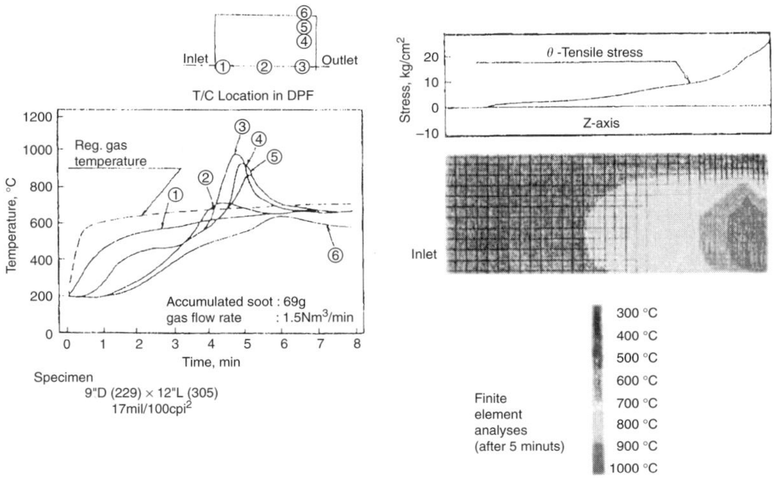

FIGURE 13.1.18 Temperature profile of regeneration and finite element analysis.

Figure 13.1.18 shows the temperature profile of regeneration. A maximum temperature of 950°C is observed at the central portion of the outlet end. The DPF withstands regeneration of accumulated soot of 69 g per 12.5 l volume at a gas flow rate of 1.5 Nm^3/ min [12]. Figure 13.1.18 also shows the temperature profile and finite element analysis at 5 min after regeneration from the point where maximum temperature was observed [13, 14]. Due to the high temperature of the outlet portion, maximum tensile stress of 20 kg/cm^2 is calculated by the finite element analysis. DPF with extremely low thermal expansion cordierite withstands high temperature regeneration.

13.1.4 FUTURE TASKS

To develop the application of extruded cordierite honeycombs continually, upgrading the structural and material properties is essential. As substrates for automotive catalytic converters, cell walls thinner than the standard 6 mil improve conversion efficiency at the cold start of engines which will be important to comply with the further tightening of emission regulations, and will also reduce component weight and pressure drop. For DPF, improvement of pore size distribution will reduce pressure drop during soot accumulation while maintaining high trapping efficiency. To meet the demand to protect the environment from motor vehicle emission pollutants, continuous improvement of extruded cordierite honeycombs is increasingly becoming important.

REFERENCES

1. Howitt, J. (1980). Thin wall ceramic as monolithic catalysts supports, SAE Paper No. 800082.
2. Kitagawa, J. (1990). Ceramic honeycombs for automotive catalysts. *Kogyo-Zairyo (Jap)* **38**: p. 49.
3. Lachman, I. M. *et al.* (1975). US Patent 3,885,977.
4. Yamamoto, H. *et al.* (1990). Reduction of wall thickness of ceramic substrates for automotive catalysts, SAE Paper No. 900613.
5. Yamamoto, H. *et al.* (1991). Warm-up characteristics of thin wall honeycomb catalysts, SAE Paper No. 910611.
6. Kitagawa, J. *et al.* (1991). Characteristics of thin wall honeycomb substrates for automotive catalysts, "24th International Symposium on Automotive Technology and Automation," Florence, Italy, May 20–24.
7. Umehara, K. *et al.* (1997). Advanced ceramic substrate: catalytic performance by high GSA and low heat capacity, SAE Paper No. 971029.
8. Hesselmann, A. (1969). United theory of thermal shock fracture initiation and crack propagation in brittle ceramics. *J. Am. Ceram. Soc.* **52**: 900.
9. Balzotti, A. *et al.* (1990). Italian city buses with particulate traps, SAE Paper No. 900114.
10. Barris, M. (1990). Durability studies of trap oxdizer systems, SAE Paper No. 900108.
11. Higuchi, N. *et al.* (1983). Optimized regeneration conditions of ceramic honeycomb diesel burner systems, SAE Trans. 830078.
12. Kitagawa, J. *et al.* (1989). Effect of DPF volume on thermal shock failures during regeneration, SAE Paper No. 890173.
13. Kitagawa, J. *et al.* (1990). Analyses of thermal shock failure on large volume DPF, SAE Trans. 900113.
14. Hijikata, T. *et al.* (1991). Regeneration condition of ceramic honeycomb diesel particulate filters, SAE Paper No. 911018.

Handbook of Advanced Ceramics
S. Sōmiya *et al.* (Eds.)

14.1 Ceramics for Biomedical Applications

TADASHI KOKUBO[1], HYUN-MIN KIM[2] and
MASAKAZU KAWASHITA[1]
[1]*Department of Material Chemistry, Graduate School of Engineering, Kyoto University, Yoshida, Sakyo-ku, Kyoto 606-8501, Japan*
[2]*Department of Ceramic Engineering, School of Advanced Materials Engineering, Yonsei University, 134, Shinchon-Dong, Seodaemun-gu, Seol 120-749, Korea*

14.1.1 INTRODUCTION

Ceramics are generally defined as being materials based on inorganic substances, and inorganic substances are substances that are not related to living organisms. Nobody, therefore, believed that ceramics could play an important role in repairing living tissues and organs. However, it has been shown over the last three decades, that some ceramics can promote the regeneration of neighbouring tissue, can spontaneously bond to living tissues, and that some ceramics can locally destroy cancer cells so that normal tissue regeneration can occur after treatment. In this chapter, some ceramics which can play an important role in repairing living tissue and organs will be described.

14.1.2 CERAMICS FOR ARTIFICIAL JOINTS

Our body is supported by bones from the top of our head to the bottom of our feet. Important organs such as the brain, heart, and lungs are protected from external forces by these bones. We can walk, bend, and grasp objects because our skeletons are comprised of 206 bones, connected through joints.

To achieve these functions, our bones are composed of 99 vol% of an extracellular matrix, and by only 1 vol% of living cells. The extracellular matrix is composed of 50 vol% (or 70 wt%) of inorganic apatite in the form of small, nanometer-sized crystallites, and 50 vol% (or 30 wt%) of organic collagen fibres that are fabricated into a three-dimensional structure, as shown in Figure 14.1.1 [1]. At the joints, bones are covered with a soft cartilage so that they can move smoothly.

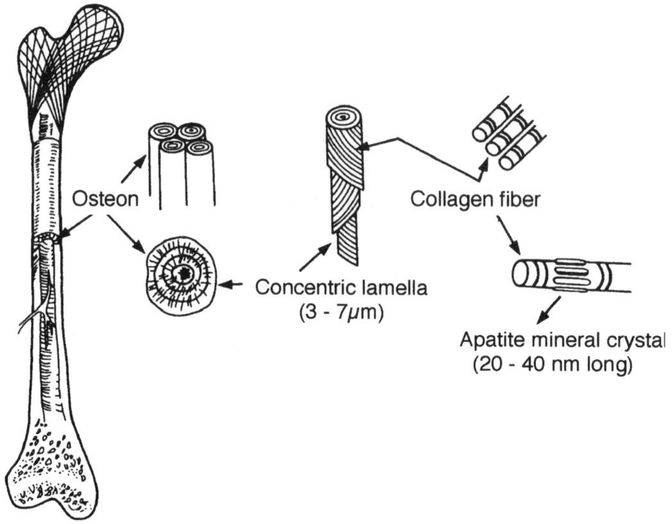

FIGURE 14.1.1 Structure of typical human bone [1].

Bones, and their joints, are sometimes damaged by accidents, disease, aging, etc. When hip and knee joints are damaged, patients sometimes feel severe pain, and have much difficulty in walking, since these joints must slide smoothly at least a few thousand times every day under a load over several times body weight. In such cases, the joints must be replaced with artificial ones. The present number of hip and knee joints replaced with artificial ones total about 100 000 in Japan, and 600 000 in the United States.

The first successful hip joint was developed by Charnley, a British orthopedic surgeon, in 1960. It consisted of a head and stem made of 316 stainless steel, and a cup socket made of ultrahigh molecular weight polyethylene. These were fixed to the surrounding bones by filling the gap between each component and the surrounding bone with a bone cement consisting of polymethylmethacrylate powder and its monomer liquid, and solidifying it *in situ*, as shown in Figure 14.1.2. This was expected to work well for over 20 years. In practice, a loosening of the joint fixation to the surrounding bone was observed within 5–10 years in some cases, and the components had to be retrieved. The reason for this loosening was attributed to wear of the cup and head, and an unstable fixation of the joint components with the bone cement.

Our body fluid contains the same types of ions as those found in seawater [2], although their concentrations are almost a quarter of those of seawater, and our body temperature is higher than that of seawater. Stainless

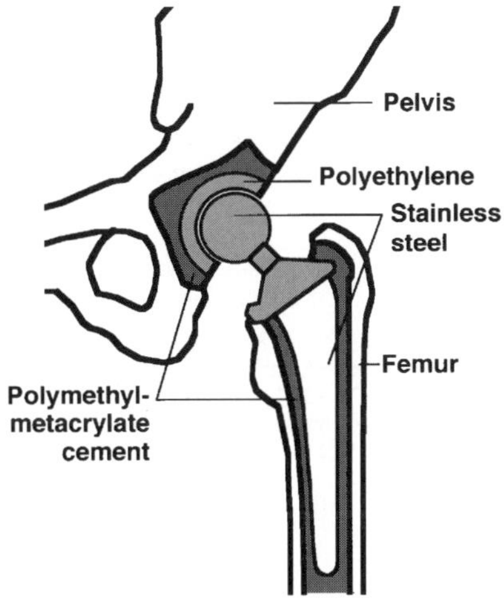

FIGURE 14.1.2 Artificial hip joint system in early 1960s.

steel is liable to corrosion in such circumstances, and so can impart a roughness on the articulating surface of the implant head. As a result, wear debris from the stainless steel head and polyethylene cup were produced, and the coefficient of friction of the articulation increased. The wear debris was phagocytized by macrophages which produced cytokines to induce the resorption of the surrounding bone. In addition, fixation of the joint components to the surrounding bone by the bone cement was not intrinsically stable. The bone cement was encapsulated by a thick membrane of collagen fibres, and had to be isolated from the surrounding bone, because it generated a large amount of heat on solidification from the polymerization of its monomers. The unreacted monomers that were released had an adverse effect on the cardiovascular system. The increased friction gave rise to the extraordinarily large stress at the interface between the components and the surrounding bone, and induced the loosening of the fixed joint.

In 1970, Bautin, a French orthopedic surgeon, replaced both the cup and head of the hip joint with high-density, high-purity sintered alumina [3]. As a result, production of the polyethylene and stainless steel debris was eliminated. The sintered alumina was superior in terms of its mechanical strength, hardness, chemical durability, and hydrophilicity. As a result, the smooth surfaces of the head and cup were maintained for a longer period, and hence, the coefficient of the friction of the articulation was also maintained at a low level over a long period, as shown in Figure 14.1.3 [4]. Later, it was found that even when only the head was made of sintered alumina and the cup was made of polyethylene, as shown in Figure 14.1.4, the production of polyethylene debris and the increase

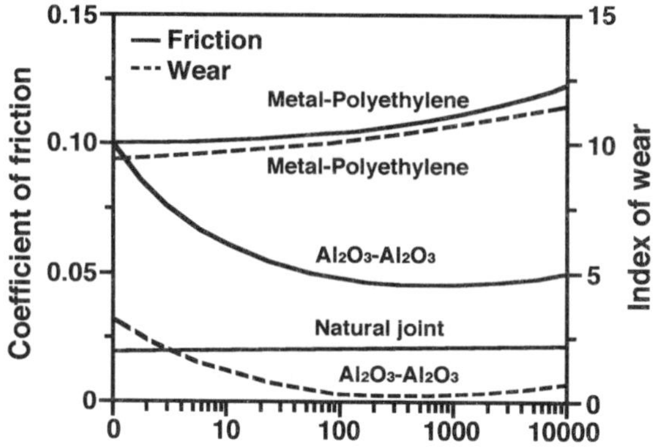

FIGURE 14.1.3 Time dependence of coefficient of friction and wear of alumina-alumina versus metal-polyethylene hip joint in vitro [4].

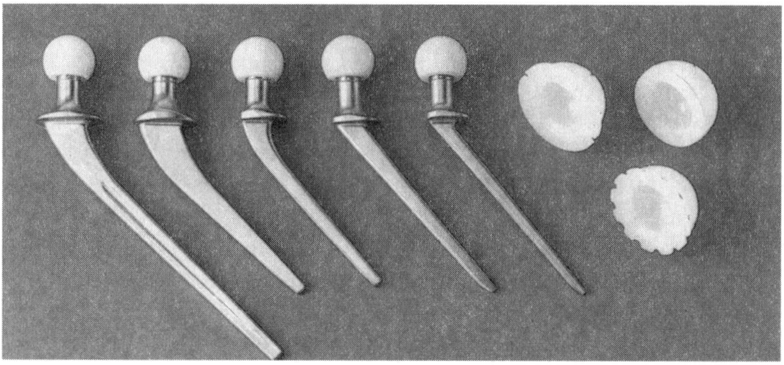

FIGURE 14.1.4 Alumina head (left) and polyethylene cup (right) in hip joint system (photograph courtesy of Kyocera Co., Kyoto, Japan).

in the coefficient of friction of the articulation was remarkably suppressed. Fitting of the head to the cup, both in size and shape, was much easier for this combination than for the ceramic-ceramic combination. Therefore, the alumina–polyethylene combination has been widely used. This combination has also been applied to knee joints.

Recently, however, it has been shown that polyethylene debris produced by this combination can also induce bone resorption over a long period, and these are gradually being replaced with the ceramic–ceramic combination. The mechanical strength of sintered alumina for this use has been improved by suppressing grain growth to achieve a fully dense sintered body, as shown in Table 14.1.1. For this purpose, a small amountof MgO (<0.5%) is added, and

TABLE 14.1.1 Properties of alumina and zirconia ceramics used in surgical implants

Properties	Al_2O_3[1]	Y-ZrO_2[2]	Ce-ZrO_2/Al_2O_3[3]
Density (g/cm^3)	3.98	6.05	5.57
Average Grain Size (μm)	3.6	0.2–0.4	—
Bending Strength (MPa)	595	1000	1000
Compressive Strength (MPa)	4250	2000	—
Young's Modulus (GPa)	400	150	250
Hardness (HV)	2400	1200	1200
Fracture Toughness ($MPa{\cdot}m^{1/2}$)	5	7	18

[1] Purity >99.7 wt% [3].
[2] 3Y-TZP ($3Y_2O_3{\cdot}97ZrO_2$ in mol%) [3].
[3] 70 vol% of 10Ce-TZP ($10CeO_2{\cdot}90ZrO_2$ in mol%) and 30 vol% of Al_2O_3 [5].

the total amount of SiO_2 and Na_2O is suppressed to below 0.1%. The amount of CaO is suppressed to below 0.1% in order to obtain a high resistance to static fatigue.

Other types of ceramics with higher mechanical strength and fracture toughness have also been tested for use for this purpose. Sintered partially stabilized zirconia is one of these materials. It is, however, liable to exhibit a decrease in its mechanical strength due to a transformation of the metastable tetragonal phase to the stable monoclinic phase in an aqueous environment, such as is found in the living body. Sintered zirconia generally shows a lower hardness value than sintered alumina. In addition, radioactive elements such as thorium and uranium emitting α- and γ-rays with a very long half-life, are liable to be contaminants in the zirconium oxide. The phase-transformation problem in zirconia was easily solved by suppressing the grain growth, and the radioactivity of commercially available zirconia ceramics was confirmed to be negligible. Sintered yttrium or magnesium partially stabilized zirconia is already clinically used as the head of hip joints in combination with a polyethylene cup. The use of these zirconia heads in combination with cups made from the same zirconia is controversial with respect to the wear rate of the articulation. A ceria-stabilized zirconia–alumina nanocomposite [5] shows a higher mechanical strength, fracture toughness, hardness, and stability against phase transformation. This ceramic is now being tried for use both as the joint cup and the head. The properties of these zirconia ceramics are listed in Table 14.1.1.

It is believed that the problems of debris and friction in artificial joints could be solved in practical terms in the near future. The remaining problem to be solved is the stable fixation of the components of the artificial joints to the surrounding bone.

14.1.3 CERAMICS FOR ARTIFICIAL BONE

Artificial materials implanted into bone defects are generally encapsulated by a tissue of collagen fibres and are isolated from the surrounding bone, as shown in Figure 14.1.5. This is a normal reaction against a foreign presence by the protection mechanism of our bodies. Owing to this encapsulation, artificial material cannot be fixed rigidly to the surrounding bone.

In the early 1970s, Hench showed that some glasses in the Na_2O–CaO–SiO_2–P_2O_5 system spontaneously bonded to living bone without forming any fibrous tissue around them, as shown in Figure 14.1.6 [6, 7]. These glasses were the first man-made materials that had been found to bond to living tissue. They were named Bioglass®. Their typical composition (45S5) is $SiO_2 = 45, Na_2O = 24.5, CaO = 24.5$, and $P_2O_5 = 6$ wt%. Their tensile strength is 42 MPa, whereas that of the human cortical bone is a maximum of 150 MPa, as shown in Table 14.1.2 [8]. Consequently, they are used clinically only in areas under reduced loads, such as periodontal fillers, as shown in Figure 14.1.7 [9].

In 1976, Jarcho *et al.* [10] showed that sintered hydroxyapatite ($Ca_{10}(PO_4)_6(OH)_2$) also bonds to living bone. Although it forms a highly dense body that shows a maximum bending strength around 200 MPa, bodies in practical use generally show a bending strength at around 115 MPa, which is lower than that of human cortical bones, as shown in Table 14.1.2 [11]. This ceramic is widely used clinically in the fields of orthopaedics and dentistry, as a bone

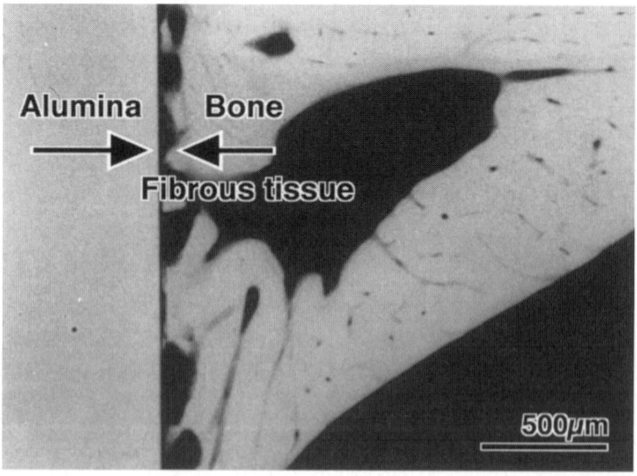

FIGURE 14.1.5 Contact microradiograph of the interface between alumina and rabbit tibial bone (12 weeks after implantation).

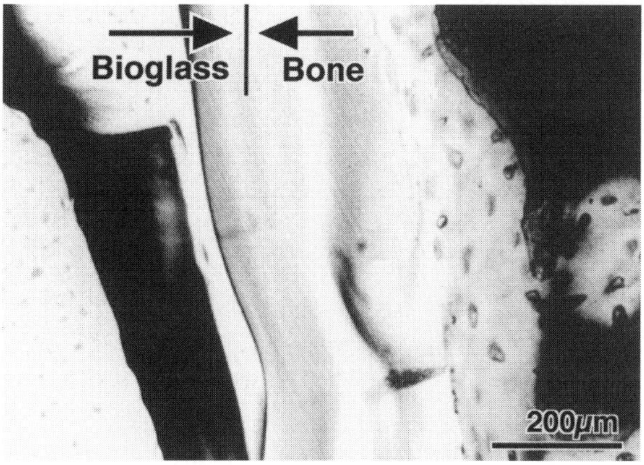

FIGURE 14.1.6 Optical micrograph of the interface between Bioglass 45S5 and rat bone (1 year after implantation) [7].

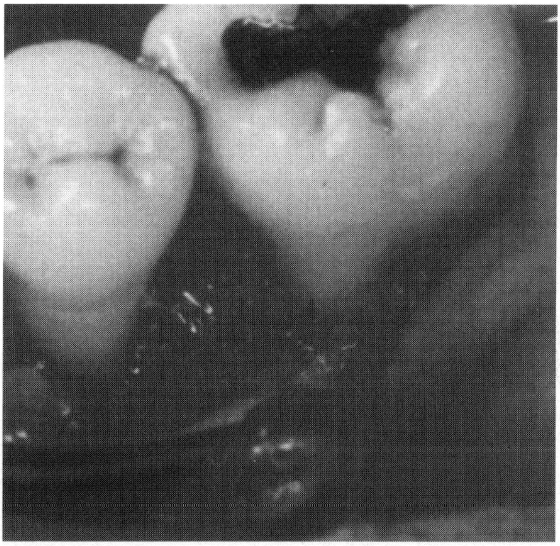

FIGURE 14.1.7 Clinical use of Bioglass as periodontal filler (photograph courtesy of US Biomaterials, Baltimore, MD, USA).

filler in a bulk, or granular form that has a dense, or porous, structure, as shown in Figure 14.1.8 [12].

In 1981, Kokubo *et al.* [13] developed a glass-ceramic containing 38 wt% of crystalline oxyfluoroapatite ($Ca_{10}(PO_4)_6(O, F_2)$) and 34 wt% of β-wollastonite ($CaSiO_3$) 50–100 nm in size, in a MgO—CaO—SiO_2 glassy matrix by the sintering and crystallization of a glass powder compact with composition

TABLE 14.1.2 Properties of bioactive ceramics and human bone

Properties	Bioglass 45S5	Sintered hydroxyapatite	Glass-ceramic A-W	Human cortical bone
Composition (wt%)				
Na_2O	24.5	$Ca_{10}(PO_4)_6(OH)_2$	0	
MgO	0		4.6	
CaO	24.5		44.7	
SiO_2	45.0		34.0	
P_2O_5	6.0		16.2	
CaF_2	0		0.5	
Phase	Glass	Apatite	Apatite ß-Wollastonite Glass	
Density (g/cm^3)	2.66	3.16	3.07	1.6–2.1
Vicker's hardness (HV)	458	600	680	
Compresive Strength (MPa)		500–1000	1060	100–230
Bending strength (MPa)	42 (Tensile)	115–200	215	50–150
Young's Modulus (GPa)	35	80–110	118	7–30
Fracture Toughness K_{IC} ($MPa \cdot m^{1/2}$)		1.0	2.0	2–6
Slow crack growth, *n*		12–27	33	

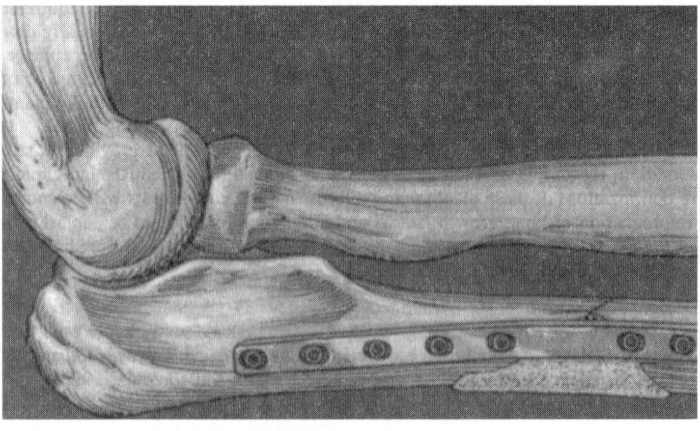

FIGURE 14.1.8 Clinical use of porous sintered hydroxyapatite in forearm fracture treatment [12].

MgO = 4.6, CaO = 44.7, SiO_2 = 34.0, P_2O_5 = 16.2, and CaF_2 = 0.5 wt%. This glass-ceramic was named A-W, after the names of crystalline phases. It also directly bonds to living bone. The period of time required for the glass-ceramics A-W to bond to living bone was longer than that for Bioglass 45S5, but shorter than that for sintered hydroxyapatite. Its bending strength, 215 MPa, was higher than that of human cortical bones, as shown in Table 14.1.2. It was estimated from the dependence of the bending strength of the glass-ceramic upon stressing rate in a simulated body fluid, that it could withstand usage over 10 years under continuous loading of bending stresses of 65 MPa, where the sintered hydroxyapatite may be broken within 1 min [14]. Glass-ceramic A-W has been used clinically in Japan as artificial vertebrae, intervertebral disks, and iliac crests, etc. since 1991. More than 60 000 patients have received it as a bone substitute. Figure 14.1.9 shows an X-ray photograph of an artificial vertebra which was substituted for the lumbar vertebra of a sheep, and bonded to the surrounding

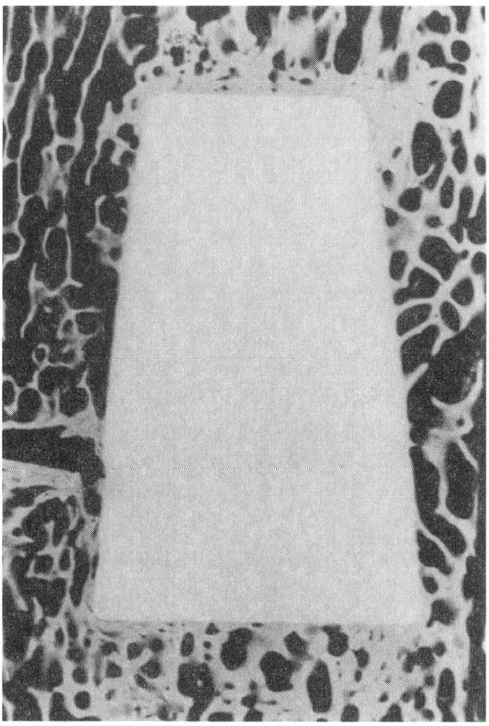

FIGURE 14.1.9 X-ray photograph of artificial vertebra of glass-ceramic A-W which substituted for a lumbar vertebra of a sheep and bonded to the surrounding cancellous bone (2 years after implantation: photograph courtesy of Professor T. Yamamuro, Kyoto University, Japan).

cancellous bone [15]. These bone-bonding ceramics such as Bioglass®, sintered hydroxyapatite, and glass-ceramic A-W, are now called bioactive ceramics.

However, even glass-ceramic A-W cannot replace highly loaded bones such as the femoral and tibial bones, since its fracture toughness of 2 MPa $m^{1/2}$ is not so high as that of the human cortical bones at 6 MPa $m^{1/2}$. For these applications, metallic materials such as Ti–6Al–4V alloy coated with hydroxyapatite using a plasma spray method are mainly utilized. In the above technique, however, hydroxyapatite powders are momentarily heated in a flame above 10 000°C to be partially molten and decomposed. As a result, a porous layer of calcium phosphate, different from the hydroxyapatite, is weakly bonded to the metallic substrate. Such a layer is not stable in a living body over a long period [16]. Metallic materials with a high fracture toughness are desired to exhibit bioactivity by themselves, without any coating of a foreign material. How is this possible?

14.1.4 REQUIREMENT FOR ARTIFICIAL MATERIAL TO BOND TO LIVING BONE

Using a transmission electron microscope, it can be observed that all the bioactive ceramics described above spontaneously form an apatite layer on their surfaces in the living body and bond to living bone through this apatite layer, as shown in Figure 14.1.10 [17]. This apatite layer can be reproduced on their surfaces even in an acellular simulated body fluid (SBF) with ion concentrations nearly equal to those of the human blood plasma [2], given in Table 14.1.3 [18] and as shown in Figure 14.1.11 [19]. According to detailed analyses using thin film X-ray diffraction and Fourier transform infrared reflection spectra, this surface layer consists of apatite composed of small crystallites with a composition such that the Ca/P atomic ratio (1.65) is smaller than that of stoichiometric apatite (1.67); small concentrations of Na^+, Mg^{2+}, Cl^-, and CO_3^{2-} ions are also included. These structural and compositional characteristics are very similar to those of bone mineral. Therefore, bone-producing cell, called an osteoblast, can preferentially proliferate and differentiate on its surface to produce apatite and collagen, rather than the fibrous tissue-producing cell called a fibroblast, similar to the surface of fractured bone [20, 21]. Consequently, the surrounding bone comes into direct contact with the surface apatite as shown in Figure 14.1.10. When this occurs, a tight chemical bond is formed between the bone mineral and the surface apatite in order to reduce the interfacial energy between them. It can be concluded from these findings that the essential requirement for an artificial material to bond to bone is the formation of a bone-like apatite layer on its surface in the living body [22, 23]. What kind of materials forms the bonelike apatite layer on their surfaces in the living body?

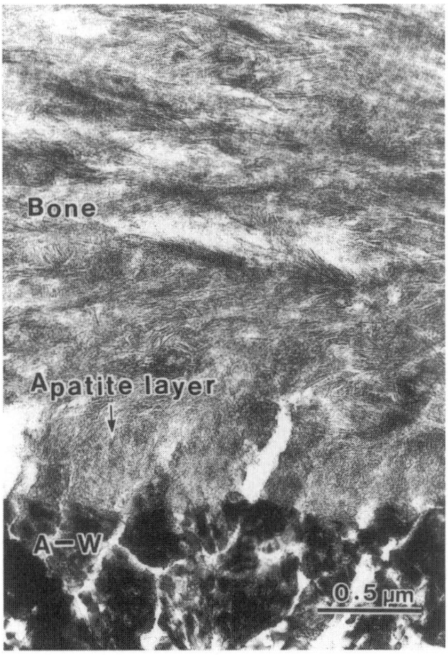

FIGURE 14.1.10 Transmission electron micrograph of the interface between glass-ceramic A-W and rat tibial bone (8 weeks after implantation).

TABLE 14.1.3 Ion concentrations of human blood plasma and simulated body fluid (SBF)

	Concentration/mM							
	Na^+	K^+	Mg^{2+}	Ca^{2+}	Cl^-	HCO_3^-	HPO_4^{2-}	SO_4^{2-}
Blood plasma	142.0	5.0	1.5	2.5	103.0	27.0	1.0	0.5
SBF*	142.0	5.0	1.5	2.5	148.8	4.2	1.0	0.5

*Buffered at 7.4 with tris-hydroxymethyl-aminomethane ($(CH_2OH)_3CNH_3$) and 1M-HCl.

14.1.5 REQUIREMENT FOR ARTIFICIAL MATERIAL TO FORM APATITE

Our body fluid is already supersaturated with respect to apatite, even under normal conditions. It can therefore form apatite anywhere in the body. Nevertheless, it is not normally formed except in bone tissue, since the energy barrier for apatite nucleation in the body fluid is generally very high, but is reduced in bone tissue by some cell reactions. Even artificial material, therefore, can form

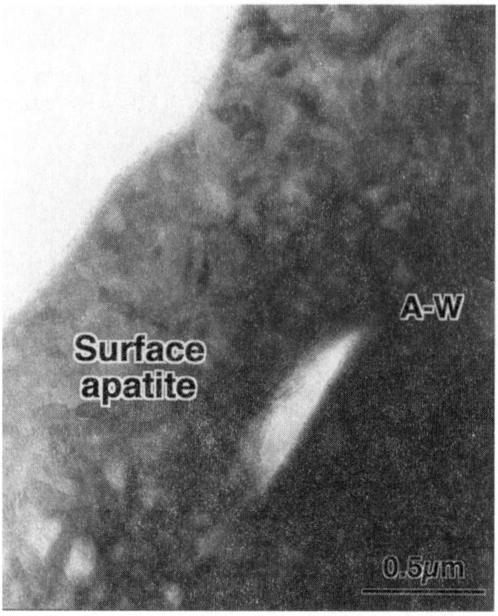

FIGURE 14.1.11 Transmission electron micrograph of the apatite formed on glass-ceramic A-W after soaking in SBF for 3 days.

apatite on its surface, if it has a functional group that can reduce the energy barrier for apatite nucleation, that is, it can induce the heterogeneous nucleation of apatite on its surface.

In the simple ternary system $CaO-SiO_2-P_2O_5$, apatite is formed on the surface of CaO, SiO_2-based glasses in an SBF, whereas CaO, P_2O_5-based glasses do not form an apatite layer, as shown in Figure 14.1.12 [24]. Apatite formation on the former surface can be interpreted as follows. The CaO, SiO_2-based glasses release Ca^{2+} ions into the SBF via ion exchange with the H_3O^+ ion in the SBF to form Si–OH groups, as shown in Figure 14.1.13. The water molecule in the SBF reacts with the Si–O–Si bonds in the glass, and breaks the bonds to form Si–OH groups. The thus-formed Si–OH groups on the surface of the glass then induce apatite nucleation. The released Ca^{2+} ions accelerate apatite nucleation by increasing the ionic activity product of the apatite in the SBF, since the Ca^{2+} ion is a component of apatite [24]. Once the apatite nuclei are formed, they spontaneously grow by consuming the calcium and phosphate ions from SBF, since the SBF is already supersaturated with respect to apatite. It can be seen from these findings, that an essential requirement for artificial material to form apatite on its surface in the body environment is the presence of functional groups that are effective for apatite nucleation on its surface. Now our problem is, "What kind of functional group is effective for apatite nucleation?"

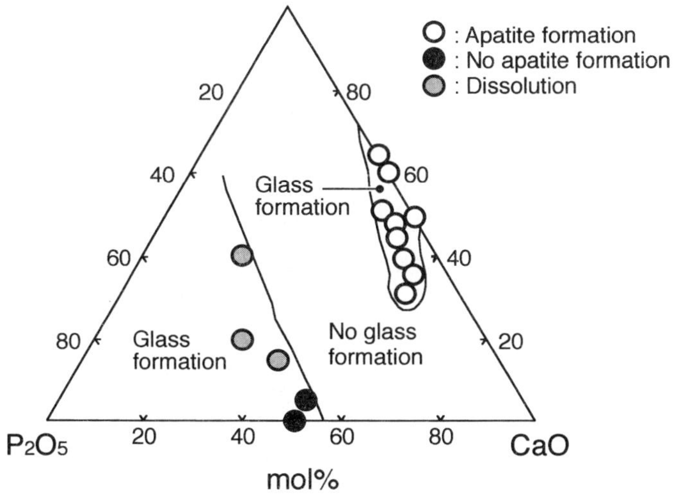

FIGURE 14.1.12 Compositional dependence of apatite formation on $CaO-P_2O_5-SiO_2$ glasses after soaking in SBF for 28 days.

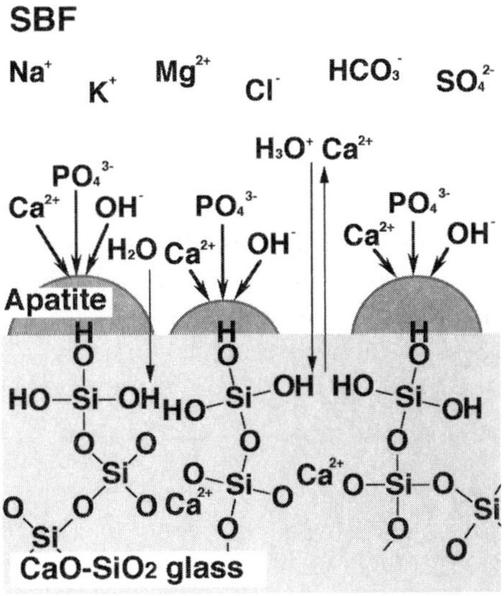

FIGURE 14.1.13 Schematic representation of the mechanism of apatite formation on a $CaO-SiO_2$ glass in SBF.

14.1.6 FUNCTIONAL GROUPS EFFECTIVE FOR APATITE NUCLEATION

Pure metal oxide gels of SiO_2 [25], TiO_2 [26], ZrO_2 [27], Nb_2O_5 [28], and Ta_2O_5 [29], were all prepared by the sol–gel method, and were found to form a bonelike apatite layer on their surfaces in an SBF, as shown in Figure 14.1.14. However, Al_2O_3 gels [26, 30] did not form an apatite layer. This means that functional groups such as Si—OH, Ti—OH, Zr—OH, Nb—OH, and Ta—OH that are abundant on the surface, are effective for apatite nucleation, but the Al—OH group is not. All the former functional groups are negatively charged at values around pH = 7, whereas the latter is positively charged.

Not all the gels with the compositions described above, are equally effective for apatite nucleation. For example, an SiO_2 gel prepared by the hydrolysis and polycondensation of tetraethoxysilane in an aqueous solution containing polyethylene glycol, is more effective than that prepared from pure water, although the structural differences between them cannot be detected using X-ray diffraction and FT-IR [31]. A TiO_2 gel with the anatase structure, prepared using the sol–gel process and subsequent heat treatment, is much more effective than that with an amorphous or a rutile structure [32]. Tetragonal, or monoclinic, ZrO_2 gels which were prepared from the sol-gel process and subsequent heat treatments, are much more effective than those with an amorphous or cubic structure [33].

Self-assembled monolayers terminated with COOH and H_2PO_4 groups, form bone-like apatite on their surfaces in an SBF [34]. This means that COOH and H_2PO_4 groups are also effective for apatite nucleation. These functional

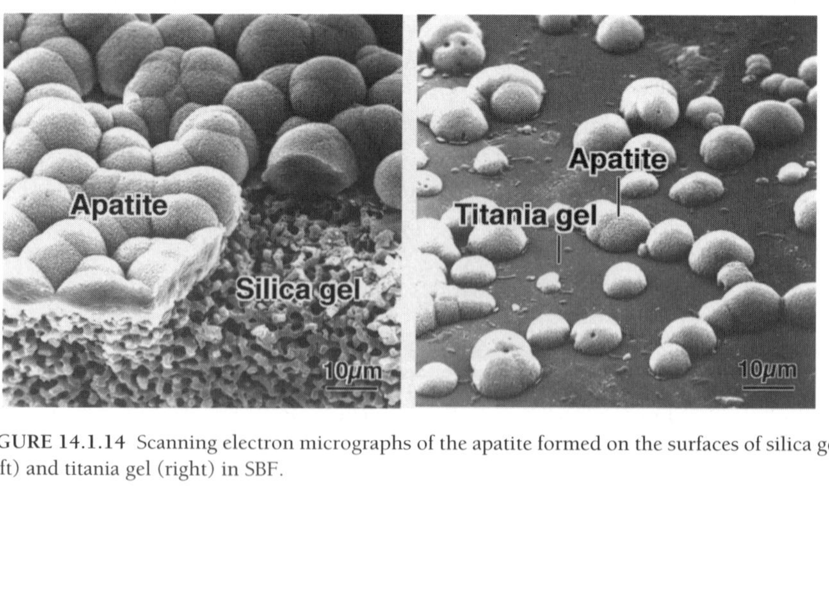

FIGURE 14.1.14 Scanning electron micrographs of the apatite formed on the surfaces of silica gel (left) and titania gel (right) in SBF.

groups also form negatively charged groups in an aqueous solution around pH = 7.

On the basis of these findings, we can provide a guide for the apatite-forming ability of various materials, that is, not only ceramics, but also metals and organic polymers that posses the functional groups effective for apatite nucleation on their surfaces, and hence we can obtain bioactive materials with different mechanical properties. In the following sections, some examples of these are described.

14.1.7 APATITE-FORMING METALS

Pure titanium metal is generally chemically durable, since it is covered with a passive thin TiO_2 layer, as shown in Figure 14.1.15a. Even this TiO_2 layer can react with an NaOH solution to form a sodium titanate hydrogel layer, as shown in Figure 14.1.15b [35, 36]. Although this gel layer is very soft, it can be stabilized as an amorphous sodium titanate layer, as shown in Figure 14.1.15c [35, 36]. It should be noted here, that a graded structure, where the Na_2O content gradually decreases towards the interior, while the Ti content gradually increases towards the interior, is formed within a 1 μm thick layer at the surface. It is confirmed that the mechanical properties of the titanium metal, including its dynamic fatigue in saline solution, is not adversely affected by the NaOH and heat treatment [37].

It is observed from X-ray photoelectron spectra [38], and from transmission electron microscopy [39], that when the NaOH- and heat-treated titanium metal is soaked in an SBF, it releases Na^+ ions via ion exchange with the H_3O^+ ions in the SBF to form abundant Ti–OH groups on its surface within a short period, as shown in Figure 14.1.16. The thus-formed Ti–OH groups combine immediately with the Ca^{2+} ions in the SBF to form an amorphous calcium titanate. Later, these combine with the phosphate ion in the SBF to form an amorphous calcium phosphate with a low Ca/P ratio. The Ca/P ratio increases, to eventually reach a value of 1.65, and transforms into crystalline apatite, as shown in Figure 14.1.16. This apatite formation process is interpreted in terms of the electrostatic reaction of the negatively charged Ti–OH groups on the surface, and the positively charged Ca^{2+} ions and the negatively charged phosphate ions in the SBF [40].

The thus-formed apatite nuclei spontaneously grow by consuming the calcium and phosphate ions in the SBF to form a dense, uniform apatite layer, as shown in Figure 14.1.17a. At the interface between the apatite and the titanium metal, a graded structure is formed, where the apatite content gradually decreases, while the Ti content increases toward the interior, as shown

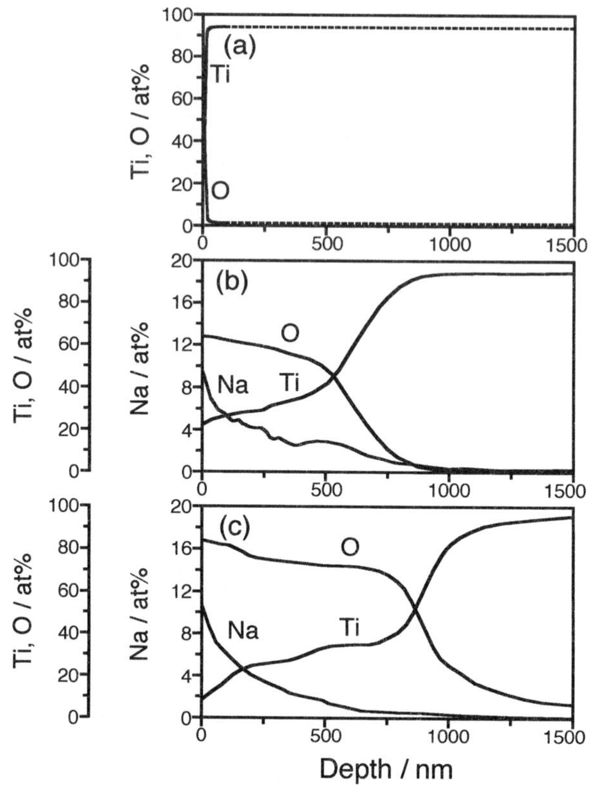

FIGURE 14.1.15 Depth profiles of Auger electron spectroscopy of the surface of titanium metal before (a) and after soaking in 5.0M—NaOH solution for 24 h (b) and subsequent heat treatment at 600°C for 1 h (c).

in Figure 14.1.17b [36]. Consequently, the thus-formed apatite is tightly integrated with the titanium substrate.

The NaOH-and heat-treated titanium metal forms the same apatite layer on its surface even *in vivo*, and tightly bonds to living bone through the apatite layer [41, 42]. For example, when an NaOH-and heat-treated titanium metal rod, 5 mm in diameter and 25 mm in length, was implanted into the intramedullar canal of a rabbit femur, it formed an apatite layer on its surface within 3 weeks, as shown in Figure 14.1.18 [43], and was completely surrounded by living bone within 12 weeks as shown in Figure 14.1.18b [43], At a period of 12 weeks after implantation, the rod was pulled out from the intramedullar canal, being accompanied by a fragment of the surrounding bone.

It has already been confirmed that similarly treated titanium-based alloys such as, Ti—6Al—4V, Ti—6Al—2Nb—Ta, and Ti—15Mo—5Zr—3Al also form

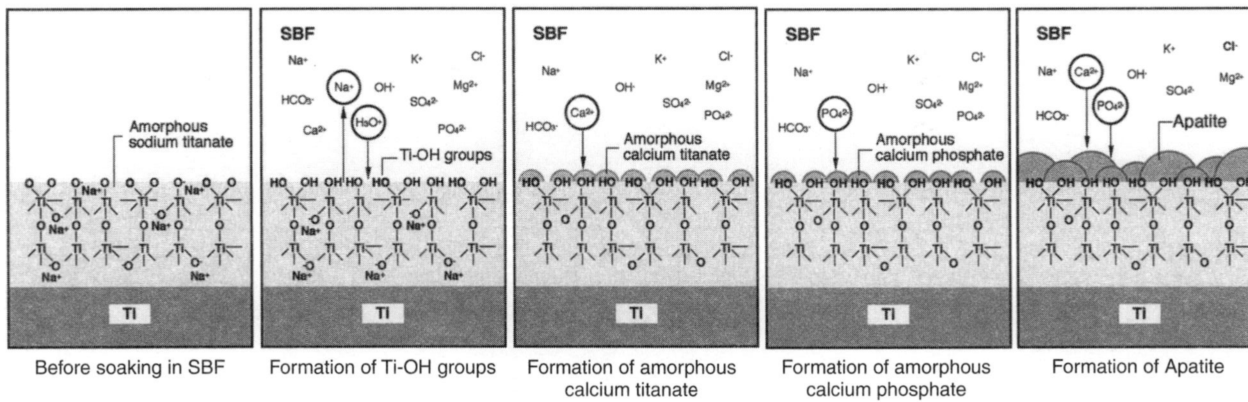

FIGURE 14.1.16 Schematic representation of the mechanism of apatite formation on the NaOH- and heat-treated titanium metal in SBF.

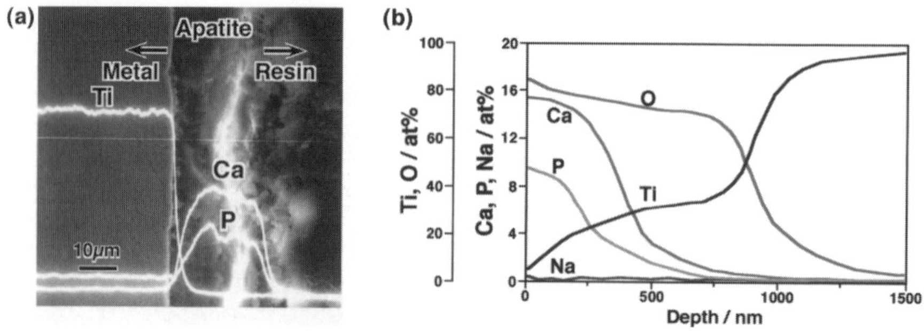

FIGURE 14.1.17 Scanning electron micrograph and energy-dispersive X-ray microanalysis (a) and profile of Auger electron spectroscopy (b) of the cross-section of apatite formed on the NaOH- and heat-treated titanium metal in SBF.

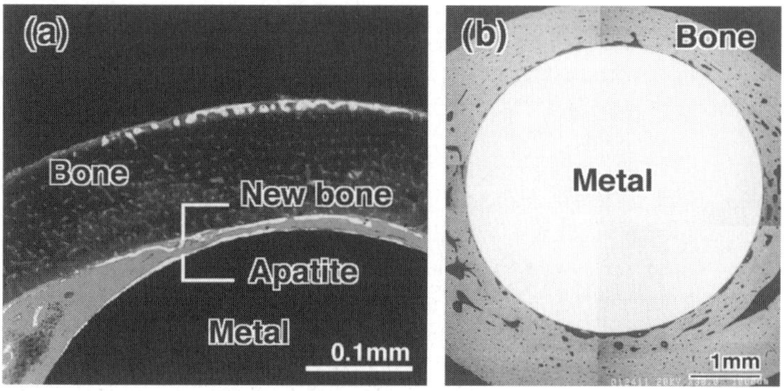

FIGURE 14.1.18 Confocal laser scanning micrograph (a) and scanning electron micrograph (b) of the cross-section of the NaOH- and heat-treated titanium metal rod implanted in intramedullar canal of rabbit femur (3 weeks (a) and 12 weeks (b) after implantation).

bone-like apatite on their surfaces in an SBF, as well as in the living body, and bond to living bone [35, 44].

Tantalum metal also forms an amorphous sodium tantalate layer with a graded structure [45] on treatment with a 0.5 M NaOH solution at 60°C for 24 h [46] and heat treatment at 300°C for 1 h [47]. The thus-treated tantalum metal also forms a bone-like apatite layer on its surface in an SBF as well as in the living body, and bonds to living bone [48].

These bone-bonding metals can be called bioactive metals, since they also form bone-like apatite on their surfaces in the living body, and bond to the living bone, similarly to the bioactive ceramics. These bioactive metals are useful as bone substitutes even under load-bearing conditions, such as hip and knee joints, vertebrae, and dental implants, since they exhibit bioactivity as well as high fracture toughness. Clinical trials assessing their suitability for application to hip joints are now being conducted. They are expected to be rigidly fixed to the surrounding bone over a long period.

Not only these metals, but also nanocomposites of ceria-stabilized zirconia with alumina show a high mechanical strength as well as high fracture toughness, as shown in Table 14.1.1. The Zr–OH groups here are effective for apatite nucleation, as described in Section 14.1.6. This indicates that ZrO_2-based composites also can show bioactivity, if a large number of Zr–OH groups are formed on its surface by some suitable chemical treatment. When the zirconia–alumina nanocomposite was treated with a 5 M H_3PO_4 solution at 95°C for 4 days, many Zr–OH groups were formed on its surface. The thus-treated composite formed a bone-like apatite layer on its surface in an SBF within 7 days [49].

Young's moduli of the metals and zirconia ceramics described above are 100–200 GPa. They are higher than that of human cortical bone. In such cases, the surrounding bone is liable to be resorbed, since the stress is mainly borne by these materials, but not by the surrounding bones. The contemporary desire is to develop bioactive materials with low elastic moduli and ductility.

14.1.8 APATITE–POLYMER COMPOSITE

In order to obtain a bioactive material with a lower elastic modulus and ductility, Bonfield prepared a hydroxyapatite–polyethylene composite [50]. Hydroxyapatite powders can be dispersed in a polyethylene matrix up to 45 vol% without losing any ductility of the polymer. The resultant composite shows a Young's modulus value of about 3 GPa, an ultimate tensile strength of 22–26 MPa, and a fracture toughness, K_{IC}, of 2.9 MPa m$^{-1/2}$ [51]. This composite is already used clinically as an artificial middle ear bone, etc.

14.1.9 APATITE-FORMING INORGANIC–ORGANIC HYBRIDS

$CaO–SiO_2$ glasses form a bone-like apatite layer on their surfaces in the body, and bond to living bone [52]. It is expected that a ductile bioactive material with a low elastic modulus can be obtained, if some organic molecules can be incorporated into the structure of these glasses. In practice, when an organic molecule such as polydimethylsiloxane (PDMS; $HO{-}[Si(CH_3)_2{-}O]_n{-}H$) or 3-isocyanatopropyltriethoxysilyl-terminated polytetramethylene oxide (Si-PTMO; $(C_2H_5O)_3Si(CH_2)_3NHCOO{-}((CH_2)_4O)_n{-}CONH(CH_2)_3Si(OC_2H_5)_3$), is incorporated into the structure of a $CaO{-}SiO_2{-}TiO_2$ glass by the sol–gel

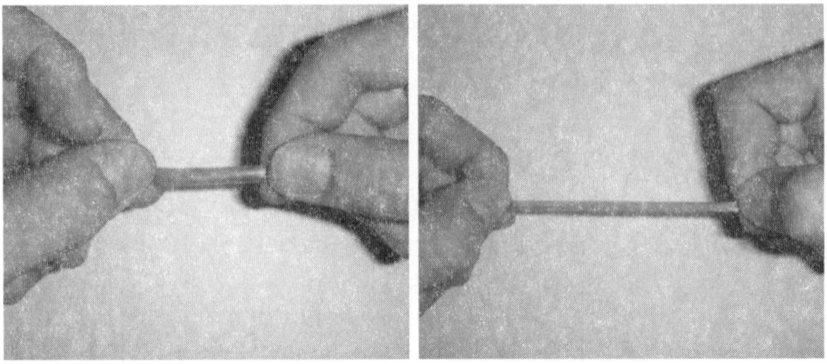

FIGURE 14.1.19 Photographs of PDMS—TiO_2 hybrid with a high deformability (as-prepared (left) and elongated (right)).

process, the resultant hybrids show an apatite-forming ability in an SBF as well as a high ductility [53, 54]. Their mechanical strength is, however, liable to decrease in the body environment.

When a Ca-free PDMS–TiO_2 hybrid was prepared using the sol–gel process, and treated in pure water around 80°C, a hybrid containing anatase-type TiO_2 nanoparticles was obtained. The resultant hybrid showed an apatite-forming ability in an SBF, as well as high deformability, as shown in Figure 14.1.19 [55, 56]. Its mechanical strength was, however, lower than that of human cortical bone.

14.1.10 APATITE–POLYMER FIBER COMPOSITES

In order to obtain a bioactive material with a high mechanical strength, high fracture toughness, low elastic modulus and ductility similar to natural bone, an apatite–polymer fiber composite with a structure analogous to that of the natural bone given in Figure 14.1.1 needs to be fabricated. Such a composite could be obtained, if a synthetic organic polymer fiber was fabricated into a three-dimensional structure analogous to that of the fiber collagen (found in natural bone), and modified with a functional group on its surface known to be effective for apatite nucleation, and then soaked in an SBF, as shown in Figure 14.1.20.

It is already known that a fabric of synthetic organic polymer fiber in a three-dimensional structure can show various mechanical properties by itself [57]. Several attempts have been made to form effective functional groups for apatite nucleation on organic polymers [58, 59]. Most of the resultant polymers, however, form apatite only in solutions that are more highly supersaturated with respect to apatite than SBF, and do not form apatite in

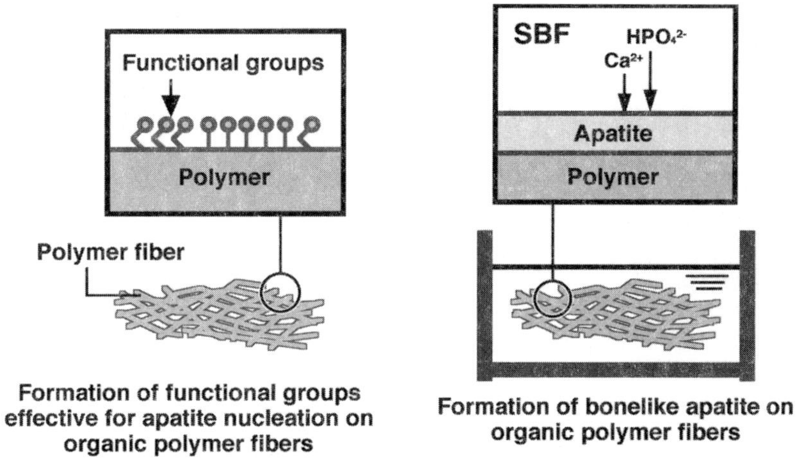

FIGURE 14.1.20 Schematic representation of fabrication process of apatite-polymer composite with analogous three dimensional structure to that of natural bone.

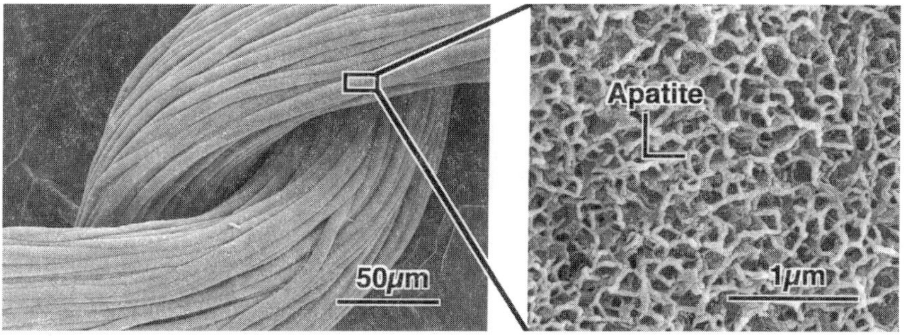

FIGURE 14.1.21 Scanning electron micrograph of EVOH fibers constituting a fabric, which were treated with silane coupling agent and calcium silicate solution, and then soaked in SBF for 2 days.

SBF. Apatite formed in solutions different from an SBF in ion concentrations, is different from bone mineral in its composition and structure [60].

Recently, it was shown that ethylene–vinyl alcohol copolymer (EVOH) fibers constituting a fabric, formed small bone-like apatite crystallites on their surfaces in an SBF, when they were modified with a silane coupling agent and a $CaO-SiO_2$ gel on their surfaces, as shown in Figure 14.1.21 [61]. The same polymer also formed bonelike apatite on its surface in an SBF, when it was modified by a titania gel on its surface, and treated in hot water around 80°C to precipitate anatase [62].

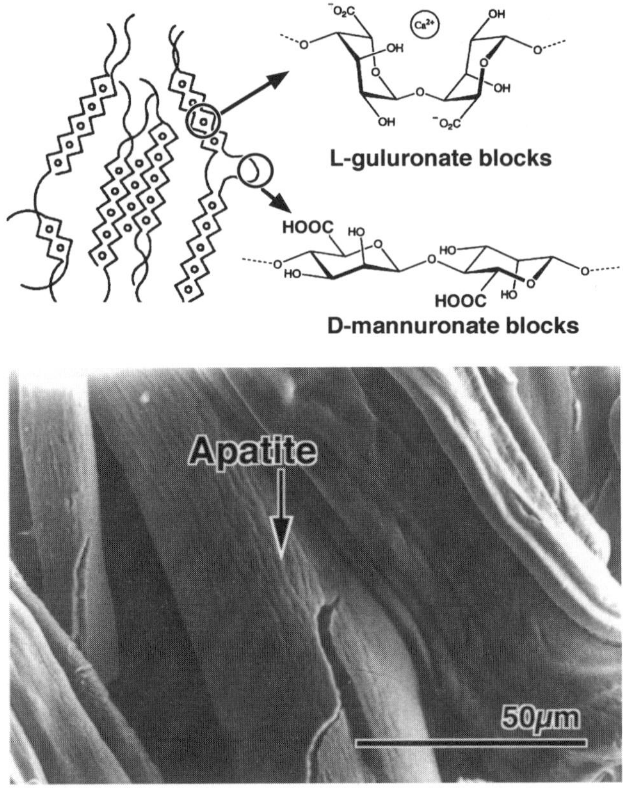

FIGURE 14.1.22 Structure and calcium alginate fiber (top) and scanning electron micrograph of apatite formed on the fiber (left).

The COOH group is also effective for apatite nucleation, as described in Section 14.1.6. Alginic acid possesses the COOH group in its structure, as shown in Figure 14.1.22. Therefore, its fibres form bone-like apatite on its surface in an SBF within 7 days, when they are previously treated in an aqueous saturated $Ca(OH)_2$ solution, as shown in Figure 14.1.22 (Kokubo, unpublished). Chitin fibres can be modified to form COOH groups on their surfaces by carboxymethylation. They also form apatite on their surfaces in an SBF within 7 days, when they are previously treated in an aqueous saturated $Ca(OH)_2$ solution [63].

Homogeneous deposition of the apatite on the individual fibers constituting a fabric in a three-dimensional structure is being attempted. The resultant composite is expected to exhibit analogous mechanical and biological properties to those of the natural bone.

TABLE 14.1.4 Some bioactive cements

Powder	Liquid	Reference
Tetracalcium phosphate ($Ca_4(PO_4)_2O$) Dicalcium phosphate anhydrate ($CaHPO_4$)	Sodium phosphate aqueous solution	Chow, 1998
α-Tricalcium phosphate (α-$Ca_3(PO_4)_2$) Calcium carbonate ($CaCO_3$) Monocalcium phosphate monohydrate ($Ca(H_2PO_4)_2 \cdot H_2O$)	Sodium phosphate aqueous solution	Chow, 1998
α-Tricalcium phosphate (α-$Ca_3(PO_4)_2$) Dicalcium phosphate dihydrate ($CaHPO_4 \cdot 2H_2O$) Tetracalcium phosphate ($Ca_4(PO_4)_2O$)	Sodium succimate aqueous solution Sodium chondroitin sulfate aqueous solution	Asaoka, 1999

14.1.11 BIOACTIVE CEMENTS

Bone defects sometimes exhibit complex shapes. Bioactive cements are useful for repairing such defects, and are usually composed of powder and liquid. When the powder and liquid are mixed in an appropriate ratio, they show fluidity, and in a few minutes, they solidify, forming bone-like apatite, and later bond to the surrounding living bone. They can be injected into the bone defects as a viscous liquid, or filled into the bone defects as a paste. Typical examples of these are given in Table 14.1.4 [64, 65]. All these cements set within 5–10 min after being mixed, forming bone-like apatite, and later bond to the living bone. Their compressive strengths after setting are in the range 60–90 MPa. They are already widely used in clinical applications. Various attempts to improve their mechanical strengths are still being carried out.

14.1.12 CERAMICS FOR *IN SITU* RADIOTHERAPY OF CANCERS

The most popular cancer treatment is the excision of the diseased part by surgery. Once an organ is excised, however, its full function is hardly recovered. The development of a cancer treatment, in which only the cancer cells are destroyed locally, so that the normal tissue can regenerate after treatment, is desired.

Radiotherapy is one treatment with such potential. Conventionally, however, irradiation has been performed externally. As a result, an inadequate radiation dosage could be given to deep-seated cancers, and the normal tissues near the surface of the body could be damaged.

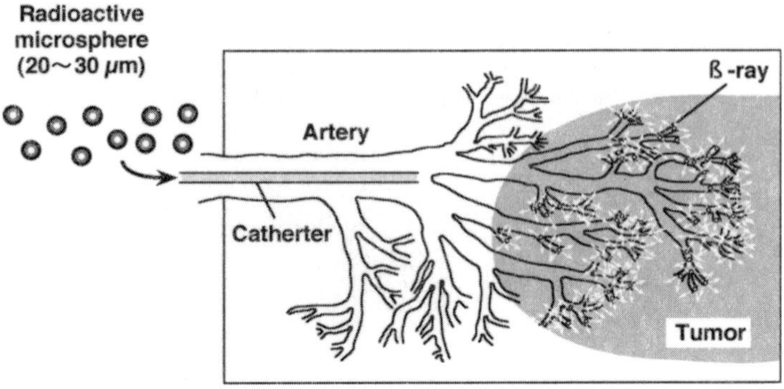

FIGURE 14.1.23 *In situ* cancer treatment using a radioactive microsphere.

In 1987, Erhardt *et al.* [66] showed that $17Y_2O_3{-}19Al_2O_3{-}64SiO_2$ (mol%) glass microspheres were useful for *in situ* radiotherapy. These glass microspheres can be prepared using the conventional melting technique, and are not radioactive as prepared. Yttrium-89 (^{89}Y) present in this glass, can be activated by neutron bombardment to form ^{90}Y, which is a β-emitter with a half-life of 64.1 h. When these glass microspheres, 20–30 μm in size, are dispersed into a saline, and injected into a liver tumour through the hepatic artery by a catheter, as shown in Figure 14.1.23, they are entrapped in a capillary bed in the tumor. As a result, they shut off the blood supply to the tumour and directly, and locally, irradiate the surrounding tumour with the β-rays to destroy the cancer cells. Since β-rays transmit only 5–10 mm in living tissue, it does not have any adverse effect on the normal tissue. This glass has a high chemical durability, and hence releases hardly any of the radioactive ^{90}Y into the living tissue, and does not damage the normal tissue. The radioactivity of this glass decays to a normal level within 21 days after the neutron bombardment. These microspheres are already used clinically for the treatment of liver cancer in the United States, Canada, and China [67]. The Y_2O_3 content in the microsphere is limited to only 17 mol% when the microsphere is prepared by the conventional melting technique for glasses. In addition, the half-life of ^{90}Y is too short. Consequently, the radioactivity of the microspheres may substantially decay even before the cancer treatment.

Recently, Kawashita *et al.* [68] successfully prepared pure Y_2O_3 crystalline microspheres 20–30 μm in size by using a high-frequency induced thermal plasma melting technique, as shown in Figure 14.1.24. Phosphorus-31 (^{31}P) is also not usually radioactive, but can be activated by neutron bombardment to form ^{32}P, which is a β-emitter with a half-life of a little over 14.3 days. It is expected that microspheres containing P_2O_5 will be more effective for *in situ* radiotherapy. The high-frequency induced thermal plasma melting technique has also been used to prepare YPO_4 crystalline microspheres, 20–30 μm in

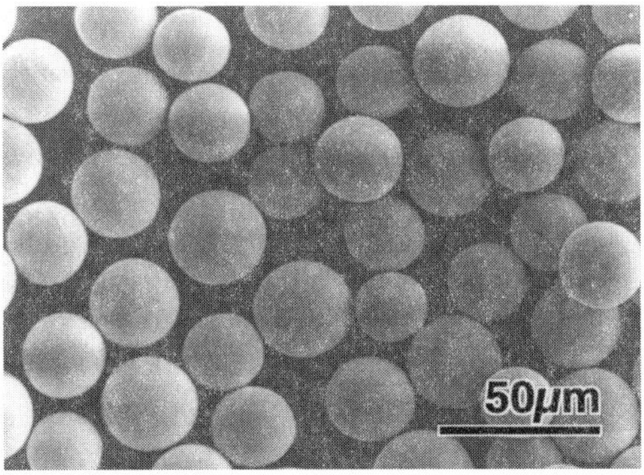

FIGURE 14.1.24 Scanning electron micrograph of Y_2O_3 microspheres.

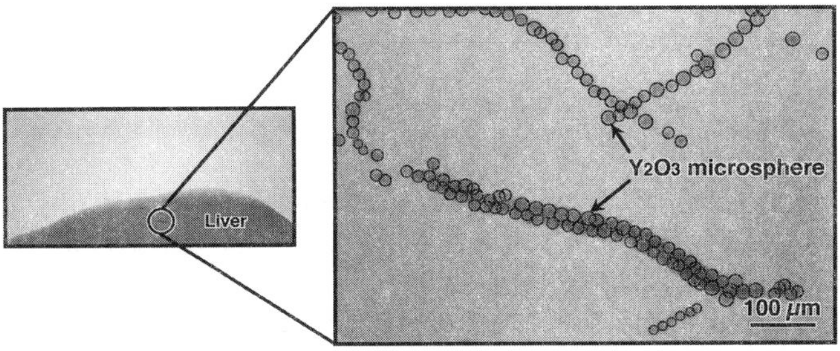

FIGURE 14.1.25 X-ray photograph of Y_2O_3 microspheres which were entrapped in the capillary bed of a rabbit liver.

size [68]. Both of these microspheres have already been confirmed to be highly chemically durable. They are now the subjects of animal experiments for *in situ* radiotherapy of liver cancer. Figure 14.1.25 shows an X-ray photograph of Y_2O_3 microspheres that were entrapped in the capillary bed of the liver of a rabbit.

Liver cancer is the third main cause of death by cancer in Japan. More than 30 000 patients die from liver cancer each year. The development of effective treatments for liver cancer without any side effects is most desirable.

14.1.13 CERAMICS FOR *IN SITU* HYPERTHERMIA THERAPY OF CANCER

Cancer cells are destroyed when they are heated up to about 43°C, and that is only 6°C above the normal body temperature, whereas normal cells are not damaged up to 48°C. In addition, a tumour is preferentially heated, as nerve and blood systems are not fully developed in a tumour. Therefore, hyperthermia therapy can also be an effective non-invasive treatment for cancer. Conventionally, however, heat treatment has been performed externally by using hot water, infrared rays, ultrasonic waves, and electromagnetic microwave radiation, etc. It is difficult to heat deep-seated cancers both effectively and locally. The normal tissue near to the surface of the body can be damaged.

Magnetic fields can penetrate into living tissue without being absorbed by it. Ferro-or ferrimagnetic materials generate heat under an alternating magnetic field in an amount proportional to the area of magnetic hysteresis loop and frequency of the magnetic field. Therefore, when ferro- or ferrimagnetic materials with high chemical durabilities are implanted around tumors and placed under an alternating magnetic field, the tumors can be locally heated up to 43°C and be destroyed by the magnetic hysteresis loss.

Ebisawa *et al.* [69] developed a glass-ceramic containing 36 wt% of magnetite (Fe_3O_4) crystal particles 200 nm in size, in a $CaO–SiO_2$-based matrix by crystallization of a $19.50Fe_2O_3–40.25CaO–40.25SiO_2–3.35B_2O_3–1.65P_2O_5$ (molar ratio) glass. This showed ferrimagnetism with saturation magnetization of 32 emu/g, and a coercive force of 120 Oe. Heat generation of this glass-ceramic under an alternating magnetic field of 300 Oe at a frequency of 100 kHz, was estimated to be 10 W/g [70]. A pin, 5 cm in length and 3 mm in diameter, was inserted into the medullary canal of a rabbit tibia transplanted with bone tumour, and subjected to an alternating magnetic field of 300 Oe at 100 kHz for 50 min. It was confirmed 3 weeks later that the cancer cells in the medullary canal were completely killed, and that the shape and function of the bone were recovered, as shown in Figure 14.1.26 [71].

Such treatment, however, cannot be applied to humans, since cancers metastasize by this kind of treatment. For humans, ferro-or ferrimagnetic materials must be implanted around the tumors through the vascular system in a type of microsphere, similar to that seen for the radioactive materials. In these cases, the efficiency of the heat generation of the magnetic material must be further increased. Pure magnetite microspheres, 20–30 μm in size, can also be prepared by the high-frequency induced thermal plasma melting technique described above. These show a saturation magnetization as high as 92 emu/g, but show a small coercive force of only 50 Oe, since they consist of magnetite particles as large as 1 μm. Consequently, their heat generation capacity under the same alternating magnetic field as described above, is 10 W/g. When goethite (FeOOH) was deposited on silica microspheres 12 μm in size, in an aqueous solution, and then heat-treated at 600°C in an atmosphere of $70CO_2 \cdot 30H_2$, microspheres 20–30 μm in size deposited with magnetite were obtained [72]. They show a little smaller saturation magnetization of 53 emu/g, but a large coercive force of 156 Oe, since they consist of magnetite particles as small as 50 nm. Consequently, these show heat generation as large as 41 W/g. Their hysteresis curves under a magnetic field of 300 Oe are shown in Figure 14.1.27

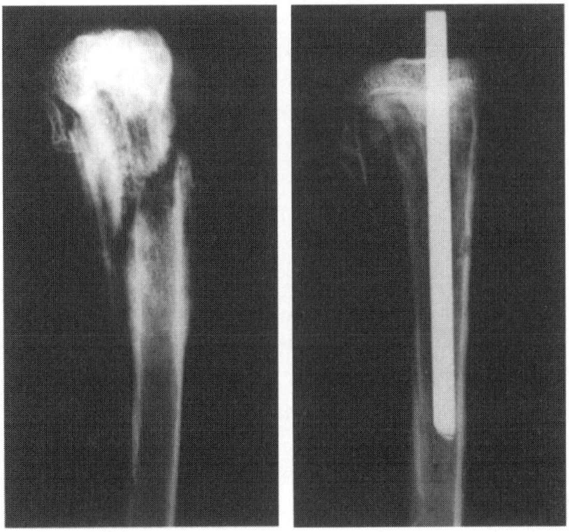

FIGURE 14.1.26 X-ray photograph of rabbit tibial bone transplanted with bone tumor, 5 weeks after the transplantation, (a) No treatment, (b) A ferrimagnetic glass-ceramic pin was inserted into the medullary canal and placed under an alternating magnetic field for 50 min, 2 weeks after the transplantation.

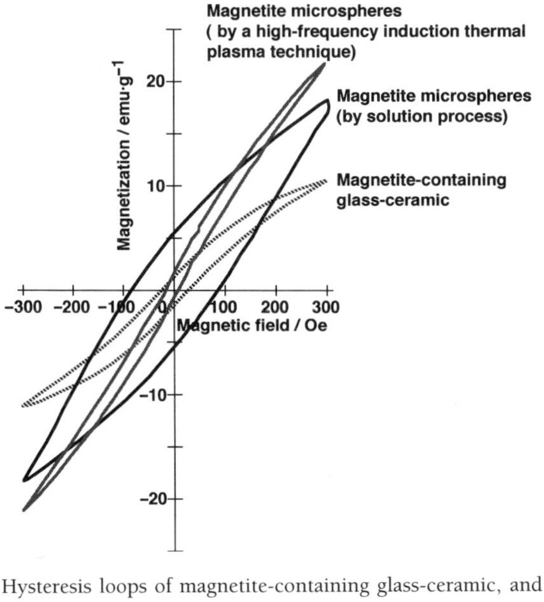

FIGURE 14.1.27 Hysteresis loops of magnetite-containing glass-ceramic, and magnetite microspheres prepared using a high-frequency induction thermal plasma technique and a solution process.

for comparison. The preparation of microspheres consisting of pure magnetite around 50 nm in size is being attempted, to obtain thermoseeds effective for *in situ* hyperthermia therapy.

14.1.14 SUMMARY

Various types of novel materials based on inorganic substances have been developed for biomedical applications over the last three decades. Some of them already play an important, and indispensable role in repairing bone defects, and in cancer treatments. New, advanced ceramic-based materials are expected to be developed for minimally invasive medical treatments in the future.

REFERENCES

1. Park, J. B., and Lakes, R. S. (1992). *Biomaterials: An Introduction*, 2nd edn, New York: Plenum.
2. Gamble, J. (1967). *Chemical Anatomy, Physiology and Pathology of Extracellular Fluid*, 6th edn, Harvard University Press.
3. Hulbert, S. F. (1993). The use of alumina and zirconia in surgical implants, *An Introduction to Bioceramics*, pp. 25–40, Hench, L. L., and Wilson, J., eds., Singapore: World Scientific.
4. Hench, L. L. (1991). Bioceramics from concept to clinic. *J. Am. Ceram. Soc.* **74**: 1487–1510.
5. Nawa, M., Nakamoto, S., Sekino, T., and Niihara, K. (1998). Tough and strong Ce-TZP/alumina nanocomposites doped with titania. *Ceram. Int.* **24**: 497–506.
6. Hench, L. L., Splinter, R. J., Allen, W. C., and Greenlee, T. K. (1972). Bonding mechanisms at the interface of ceramic prosthetic materials. *J. Biomed. Mater. Res. Symp.* **2**: 117–141.
7. Hench, L. L., and Clark, A. E. (1982). Adhision to bone, *Biocompatibility of Orthopaedic Implants*, Vol. II, pp. 129–170, Boca Raton, FL: CRC Press.
8. Hench, L. L., and Andersson, Ö. (1993). Bioactive glass, *An Introduction to Bioceramics*, pp. 41–62, Hench, L. L., and Wilson, J., eds., Singapore: World Scientific.
9. Wilson, J., Yli-Urpo, A., and Happonen, R.-P. (1993). Bioactive glasses: clinical applications, *An Introduction to Bioceramics*, pp. 63–73, Hench, L. L., and Wilson, J., eds., Singapore: World Scientific.
10. Jarcho, M., Kay, J. F., Drobeck, H. P., and Doremus, R. H. (1976). Tissue, cellular and subcellular events at bone-ceramic hydroxyapatite interface. *J. Bioeng.* **1**: 79–92.
11. LeGeros, R. Z., and LeGeros, J. P. (1993). Dense hydroxyapatite, *An Introduction to Bioceramics*, pp. 139–180. Hench, L. L., and Wilson, J., eds., Singapore: World Scientific.
12. Shors, E. C., and Holmes, R. E. (1993). Porous hydroxyapatite, *An Introduction to Bioceramics*, pp. 181–198, Hench, L. L., and Wilson, J., eds., Singapore: World Scientific.
13. Kokubo, T., Shigematsu, M., Nagashima, Y., Tashiro, M., and Higashi, S. (1982). Apatite-and wollastonite-containing glass-ceramic for prosthetic application. *Bull. Inst. Chem. Res. Kyoto Univ.* **60**: 260–268.
14. Kokubo, T. (1993). A/W glass-ceramic: processing and properties, *An Introduction to Bioceramics*, pp. 75–88, Hench, L. L., and Wilson, J., eds., Singapore: World Scientific.
15. Yamamuro, T. (1993). A/W glass-ceramic: clinical applications, *An Introduction to Bioceramics*, pp. 89–104, Hench, L. L., and Wilson, J., eds., Singapore: World Scientific.

16. Klein, C. P. A., Wolke, J. G. C., and de Groot, K. (1993). Stability of calcium phosphate ceramics and plasma sprayed coating, *An Introduction to Bioceramics*, pp. 199–222, Hench, L. L., and Wilson, J., eds., Singapore: World Scientific.
17. Neo, M., Kotani, S., Nakamura, T., Yamamuro, T., Ohtsuki, C., Kokubo, T., and Bando, Y. (1992). A comparative study of ultrastructures of the interface between four kinds of surface-active ceramic and bone. *J. Biomed. Mater. Res.* **26**: 1419–1432.
18. Kokubo T., Kushitani, H., Sakka, S., Kitsugi, T., and Yamamuro, T. (1990). Solutions able to reproduce in vivo surface-structure changes in bioactive glass-ceramic. *J. Biomed. Mater. Res.* **24**: 721–734.
19. Ohtsuki, C., Aoki, Y., Kokubo, T., Bando, Y., Neo, M., and Nakamura, T. (1995). Transmission electron microscopic observation of glass-ceramic A-W and apatite layer formed on its surface in a simulated body fluid. *J. Ceram. Soc. Jpn.* **103**: 449–454.
20. Loty, C., Sautier, J. M., Boulekbache, H., Kokubo, T., Kim, H.-M., and Forest, N. (2000). In vitro bone formation on a bonelike apatite layer prepared by a biomimetic process on a bioactive glass-ceramic. *J. Biomed. Mater. Res.* **49**: 423–434.
21. Neo, M., Nakamura, T., Ohtsuki, C., Kokubo, T., and Yamamuro, T. (1993). Apatite formation on three kinds of bioactive materials at an early stage in vivo: a comparative study by transmission electron microscopy. *J. Biomed. Mater. Res.* **24**: 331–343.
22. Kokubo, T. (1991). Bioactive glass ceramics: properties and applications. *Biomaterials* **12**: 155–163.
23. Kokubo, T. (1992). Bioactivity of glasses and glass-ceramics. *Bone-bonding Biomaterials*, pp. 31–46, Ducheyne, P., Kokubo, T., and Blitterswijk, C. A., eds., Netherlands: Reed Healthcare Communications.
24. Ohtsuki, C., Kokubo, T., and Yamamuro, T. (1992). Mechanism of apatite formation on $CaO–SiO_2–P_2O_5$ glasses in a simulated body fluid. *J. Non-Cryst. Solids* **143**: 84–92.
25. Li, P., Ohtsuki, C., Kokubo, T., Nakanishi, K., Soga, N., Nakamura, T., and Yamamuro, T. (1992). Apatite formation induced by silica gel in a simulated body fluid. *J. Am. Ceram. Soc.* **75**: 2094–2097.
26. Li, P., Ohtsuki, C., Kokubo, T., Nakanishi, K., Soga, N., Nakamura, T., and Yamamuro, T. (1992). A role of hydrated silica, titania and alumina in forming biologically active apatite on implant, "Trans. Fourth World Biomater. Cong." p. 4.
27. Uchida, M., Kim, H.-M., Kokubo, T., Miyaji, F., and Nakamura, T. (2001). Bonelike apatite formation induced on zirconia gel in a simulated body fluid and its modified solutions. *J. Am. Ceram. Soc.* **84**: 2041–2044.
28. Miyazaki,T., Kim, H.-M., Kokubo, T., Ohtsuki, C., and Nakamura, T. (2001). Apatite-forming ability of niobium oxide gels in a simulated body fluid. *J. Ceram. Soc. Japan.* **109**: 929–933.
29. Miyazaki, T., Kim, H.-M., Kokubo, T., Kato, H., and Nakamura, T. (2001). Induction and acceleration of bonelike apatite formation on tantalum oxide gel in simulated body fluid. *J. Sol-Gel Sci. Tech.* **21**: 83–88.
30. Li, P., Ohtsuki, C., Kokubo, T., Nakanishi, K., Soga, N., and Nakamura, T. (1994). A role of hydrated silica, titania and alumina in forming biologically active bone-like apatite on implant. *J. Biomed. Mater. Res.* **28**: 7–15.
31. Cho, S. B., Nakanishi, K., Kokubo, T., Soga, N., Ohtsuki, C., and Nakamura, T. (1996). Apatite formation on silica gel in simulated body fluid: its dependence on structures of silica gels prepared in different media. *J. Biomed. Mater. Res.: Appl. Biomter.* **33**: 145–151.
32. Uchida, M., Kim, H.-M., Kokubo, T., Fujibayashi, S., and Nakamura, T. (2003). Structural dependence of apatite formation on titania gels in a simulated body fluid. *J. Biomed. Mater. Res.* **64A**: 164–170.
33. Uchida, M., Kim, H.-M., Kokubo, T., Tanaka, K., and Nakamura, T. (2002). Structural dependence of apatite formation on zirconia gels in a simulated body fluid. *J. Ceram. Soc. Japan.* **110**: 710–715.

34. Tanahashi, M., and Matsuda, T. (1997). Surface functional group dependence on apatite formation on self-assembled monolayers in a simulated body fluid. *J. Biomed. Mater. Res.* **34**: 305–315.
35. Kim, H.-M., Miyaji, F., Kokubo, T., and Nakamura, T. (1996). Preparation of bioactive Ti and its alloys via simple chemical treatment. *J. Biomed. Mater. Res.* **32**: 409–417.
36. Kim, H.-M., Miyaji, F., Kokubo, T., Nishiguchi, S., and Nakamura, T. (1999). Graded surface structure of bioactive titanium metal prepared by chemical treatment. *J. Biomed. Mater. Res.* **45**: 100–107.
37. Kim, H.-M., Sasaki, Y., Suzuki, J., Fujibayashi, S., Kokubo, T., Matsushita, T., and Nakamura, T. (2000). Mechanical properties of bioactive titanium metal prepared by chemical treatment, *Bioceramics*, Vol. 13, pp. 227–230, Giannini, S., and Moroni, A., eds., Switzerland: Trans Tech Publications Ltd.
38. Takadama, H., Kim, H.-M., Kokubo, T., and Nakamura, T. (2001). An X-ray photoelectron spectroscopy study of process of apatite formation on bioactive titanium metal. *J. Biomed. Mater. Res.* **55**: 185–193.
39. Takadama, H., Kim, H.-M., Kokubo T., and Nakamura, T. (2001). TEM-EDX study of mechanism of bonelike apatite formation on bioactive titanium metal in simulated body fluid. *J. Biomed. Mater. Res.* **57**: 441–448.
40. Himeno, T., Kawashita, M., Kim, H.-M., Kokubo, T., and Nakamura, T. (2001). Zeta-potential variation of bioactive titanium metal during apatite formation on its surface in simulated body fluid, Bioceramics, Vol. 14, pp. 641–644, Brown, S., Clarke, I., and Williams, P., eds., Switzerland: Trans Tech Publications Ltd.
41. Yan, W.-Q., Nakamura, T., Kobayashi, M., Kim, H.-M., Miyaji, F., and Kokubo, T. (1997). Bonding of chemically treated titanium implant to bone. *J. Biomed. Mater. Res.* **37**: 265–275.
42. Nishiguchi, S., Nakamura, T., Kobayashi, M., Kim, H.-M., Miyaji, F., and Kokubo, T. (1999). The effect of heat treatment on bone-bonding ability of alkali-treated titanium. *Biomaterials* **20**: 491–500.
43. Kokubo, T., Kim, H.-M., Nishiguchi, S., and Nakamura, T. (2000). In vivo apatite formation induced on titanium metal and its alloys by chemical treatment, *Bioceramics*, Vol. 13, pp. 3–6, Giannini, S., and Moroni, A., eds., Switzerland: Trans Tech Publications Ltd.
44. Nishiguchi, S., Kato, H., Fujita, H., Kim, H.-M., Miyaji, F., Kokubo, T., and Nakamura, T. (1999). Enhancement of bone-bonding strengths of titanium alloy implants by alkali and heat treatments. *J. Biomed. Mater. Res.: Appl. Biomater.* **48**: 689–696.
45. Miyazaki, T., Kim, H.-M., Kokubo, T., Ohtsuki, C., Kato, H., and Nakamura, T. (2002). Enhancement of bonding strength by graded structure at interface between apatite layer and bioactive tantalum metal. *J. Mater. Sci.: Mater. Med.* **11**: 651–655.
46. Miyazaki, T., Kim, H.-M., Miyaji, F., Kokubo, T., and Nakamura, T. (2000). Bioactive tantalum metal prepared by NaOH treatment. *J. Biomed. Mater. Res.* **50**: 35–42.
47. Miyazaki, T., Kim, H.-M., Kokubo, T., Miyaji, F., Kato H., and Nakamura, T. (2001). Effect of thermal treatment on apatite-forming ability of NaOH-treated tantalum metal. *J. Mater. Sci: Mater. Med.* **12**: 683–687.
48. Kato, H., Nakamura, T., Nishiguchi, S., Matsusue, Y., Kobayashi, M., Miyazaki, T., Kim, H.-M., and Kokubo, T. (2000). Bonding of alkali-and heat-treated tantalum implant to bone. *J. Biomed. Mater. Res.: Appl. Biomater.* **53**: 28–35.
49. Uchida, M., Kim, H.-M., Kokubo, T., Nawa, M., Asano, T., Tanaka, K., and Nakamura, T. (2002). Apatite-forming ability of a zirconia/alumina nano-composite induced by chemical treatment. *J. Biomed. Mater. Res.* **60**: 277–282.
50. Bonfield, W. (1993). Design of bioactive ceramic-polymer composites. *An Introduction to Bioceramics*, pp. 299–305, Hench, L. L., and Wilson, J., eds., Singapore: World Scientific.
51. Hench, L. L. (1998). Bioceramics. *J. Am. Ceram. Soc.* **81**: 1705–1728.

52. Ohura, K., Yamamuro, T., Nakamura, T., Kokubo, T., Ebisawa, Y., Kotoura, Y., and Oka, M. (1991). Bone-bonding ability of P_2O_5-free $CaO–SiO_2$ glasses. *J. Biomed. Mater. Res.* **25**: 357–365.
53. Chen, Q., Miyata, N., Kokubo, T., and Nakamura, T. (2000). Bioactivity and mechanical properties of PDMS-modified $CaO–SiO_2–TiO_2$ hybrids prepared by sol-gel process. *J. Biomed. Mater. Res.* **51**: 605–611.
54. Miyata, N., Fuke, K., Chen, Q., Kawashita, M., Kokubo, T., and Nakamura, T. (2000). Bioactivity and mechanical behavior of PTMO-modified $CaO–SiO_2$ hybrids prepared by sol-gel process, *Bioceramics*, Vol. 13, pp. 681–684, Giannini, S., and Moroni, A., eds., Switzerland: Trans Tech Publications Ltd.
55. Kamitakahara, M., Kawashita, M., Miyata, N., Kokubo, T., and Nakamura, T. (2001). Apatite-forming ability and mechanical properties of polydimethylsiloxane (PDMS)-TiO_2 hybrid treated with hot water, Bioceramics, Vol. 14, pp. 633–636, Brown, S., Clarke, I., and Williams, P., eds., Switzerland: Trans Tech Publications Ltd.
56. Kamitakahara, M., Kawashita, M., Miyata, N., Kokubo, T., and Nakamura, T. (2003). Apatite-forming ability and mechanical properties of CaO-free poly(tetramethylene oxide) (PTMO)-TiO_2 hybrids treated with hot water. *Biomaterials.* **24**: 1357–1363.
57. Shikinami, Y., and Kawarada, H. (1998). Potential application of a triaxial three-dimensional fabric (3-DF) as an implant. *Biomaterials* **19**: 617–635.
58. Tretinnikov, O. N., Kato, K., and Ikada, Y. (1994). In vitro hydroxyapatite deposition onto a film surface-grafted with organo phosphate polymer. *J. Biomed. Mater. Res.* **28**: 1365–1373.
59. Mucalo, M. R., Yokogawa, Y., Toriyama, M., Suzuki, T., Kawamoto, Y., Nagata, F., and Nishizawa, K. (1995). Growth of calcium phosphate on surface-modified cotton. *J. Mater. Sci.: Mater. Med.* **6**: 597–605.
60. Kim, H.-M., Kishimoto, K., Miyaji, F., Kokubo, T., Yao, T., Suetsugu, Y., Tanaka, J., and Nakamura, T. (2000). Composition and structure of apatite formed on organic polymer in SBF with a high content of carbonate ion. *J. Mater. Sci.: Mater. Med.* **11**: 421–426.
61. Oyane, A., Kawashita, M., Nakanishi, K., Kokubo, T., Minoda, M., Miyamoto, T., and Nakamura, T. (2003). Bonelike apatite formation on ethylene-vinyl alcohol copolymer modified with silane coupling agent and calcium silicate solutions. *Biomaterials.* 24: 1729–1735.
62. Oyane, A., Kawashita, M., Kokubo, T., Minoda, M., Miyamoto, T., and Nakamura, T. (2002). Bonelike apatite formation on ethylene-vinyl alcohol copolymer modified with a silane coupling agent and titania solution. *J. Ceram. Soc. Japan.* **110**: 248–254.
63. Kawashita, M., Nakao, M., Minoda, M., Kim, H.-M., Beppu, T., Miyamoto, T., Kokubo, T., and Nakamura, T. (2003). Apatite-forming ability of carboxyl group-containing polymer gels in a simulated body fluid. *Biomaterials.* **24**: 2477–2484.
64. Chow, L. C. (1998). Calcium phosphate cements: chemistry and applications. *Bioceramics*, Vol. 11, pp. 45–49, LeGeros, R. Z., and LeGeros, J. P., eds., Singapore: World Scientific.
65. Asaoka, N., Misago, M., Hirano, M., and Takeuchi, H. (1999). Mechanical and chemical properties of the injectable calcium phosphate cement. *Bioceramics*, vol. 12, pp. 525–528, Ohgushi, H., Hastings, G. W., and Yoshikawa, T., eds., Singapore: World Scientific.
66. Ehrhardt, G. J., and Day, D. E. (1987). Therapeutic use of ^{90}Y microspheres. *Nucl. Med. Biol.* **14**: 233–242.
67. Day, D. E., and Day, T. E. (1993). Radiotherapy glass. *An Introduction to Bioceramics*, pp. 305–317, Hench, L. L., and Wilson, J., eds., Singapore: World Scientific.
68. Kawashita, M., Shineha, R., Kim, H.-M., Kokubo, T., Inoue, Y., Araki, N., Nagata, Y., Hiraoka, M., and Sawada, Y. (2003). "Preparation of ceramic microspheres for in situ radiotherapy of deep-seated cancer. *Biomaterials.* 24: 2955–2963.
69. Ebisawa, Y., Sugimoto, Y., Hayashi, T., Kokubo, T., Ohura, K., and Yamamuro, T. (1991). Crystallization of (FeO, Fe_2O_3)–$CaO–SiO_2$ glasses and magnetic properties of their crystallized products. *J. Ceram. Soc. Jpn* **99**: 7–13.

70. Kokubo, T., Ebisawa, Y., Sugimoto, Y., Kiyama, M., Ohura, K., Yamamuro, T., Hiraoka, M., and Abe, M. (1992). Preparation of bioactive and ferromagnetic glass-ceramic for hyperthermia, *Bioceramics*, Vol. 3, pp. 213–224, Hulbert, J. E., and Hulbert, S. F., eds., Terra Haute: Rose-Hulman Institute of Technology.
71. Ikenaga, M., Ohura, K., Yamamuro, T., Kotoura, Y., Oka, M., and Kokubo, T. (1993). Localized hyperthermic treatment of experimental bone tumors with ferromagnetic ceramics. *J. Orthop. Res.* **11**: 849–855.
72. Kawashita, M., Tanaka, M., Kokubo, T., Yao, T., Hamada, S., and Shinjo, T. (2001). Preparation of magnetite microspheres for hyperthermia of cancer, Bioceramics, Vol. 14, pp. 645–648, Brown, S., Clarke, I., and Williams, P., eds., Switzerland: Trans Tech Publications Ltd.

Handbook of Advanced Ceramics
S. Sōmiya *et al.* (Eds.)

15.1 Ceramic-Matrix Composites

AKIRA OKADA
Japan Fine Ceramics Center, 2-4-1 Mutsuno, Atsuta-ku, Nagoya 456-8587, Japan

15.1.1 INTRODUCTION

It is widely known that controlling the microstructure of materials can modify their physical properties. Porous materials having cellular or fibrous structures, for example, are typically lightweight, thermally insulating, and easy to deform elastically, compared with dense materials. The strength of steel is obviously due to the dispersed hard particles which obstruct the movement of dislocations, and a suitable microstructure can be achieved by the heat treatment, such as tempering and annealing. Glass ceramics are produced by annealing the metastable glassy phase to yield numerous precipitates, and the suitably controlled microstructure results in high strength.

Fiber-reinforced plastics are the most successful composite materials. In spite of the poor load-bearing ability of polymeric materials, excellent mechanical properties are achieved by using fiber architectures of glass and carbon fibers in a manner similar to the reinforcement of concrete with steel rods and frames. For example, toughened polymeric materials with dispersed rubber particles in a polymer matrix exhibit high fracture energy. In composite materials, introduction of secondary materials into the matrix can improve the mechanical properties considerably.

In the case of polymeric materials, the major advantage of using composite systems is the improvement in strength. For ceramics, however, the most serious problem hindering use in structural applications is their brittle failure. Accordingly, numerous attempts have been conducted to improve the brittle nature of ceramics, and a variety of toughened ceramics have been made, including *in situ* reinforced silicon nitrides and composite ceramics reinforced with particulates, whiskers and continuous fibers. These composite ceramics are thus attracting considerable attention due to their high fracture toughness while at the same time maintaining excellent mechanical properties at high temperature.

15.1.2 TOUGHENING MECHANISMS IN CERAMICS

Ceramic materials are in general brittle, and, according to the fracture mechanics, the strength is governed by the flaw size and the fracture toughness. General approaches for producing strong ceramics are to reduce the maximum size of processing flaws or to enhance the fracture toughness. The former approach, however, is limited by the nature of the microstructure of ceramics, since a grain boundary itself can be a flaw responsible for brittle fracture, and surface flaws generated in use may also reduce the strength of ceramics. Considerable effort has been made to develop toughened ceramics with high fracture toughness.

The fracture toughness of ceramics is improved by introduction of secondary phases into matrix materials when the secondary phases are chosen to act as barriers to crack propagation. Whiskers introduced into a ceramic matrix, for example, can retard the crack propagation because the stresses in a whisker spanning the crack plane will tend to pull the crack shut. This phenomenon, known as "crack bridging", leads to higher fracture toughness due to the additional stress required for further propagation of the crack. Moreover, continuous fiber composites exhibit quasi-ductile fracture behavior resulting from extensive fiber bridging.

Ceramic matrix composites may be classified into two categories. One is a group of toughened ceramics reinforced with particulates and whiskers, and these materials exhibit brittle behavior in spite of considerable improvements in fracture toughness and strength. The maximum in fracture toughness is around 10 MPa $m^{1/2}$ or more. The second consists of continuous-fiber composites exhibiting quasi-ductile fracture behavior accompanied by extensive fiber pull out. The fracture toughness of this class of materials can be higher than 20 MPa $m^{1/2}$ when produced with weak interfaces between the fibers and matrix. Toughening mechanisms in ceramics explored in recent years may be summarized as follows.

15.1.2.1 Crack Bowing

Brittle materials containing secondary phase dispersions can possess higher strength than those of homogeneous materials. The strength increases with an increase in the volume fraction of dispersed particles and decreases with dispersed particle size [1]. The strength therefore depends on the particle spacing. This effect is not only found in systems of hard particle dispersions but also in void dispersed systems due to the localized blunting of the crack tip at the voids [2, 3]. In the case of single crystals such as MgO, the position of the crack front

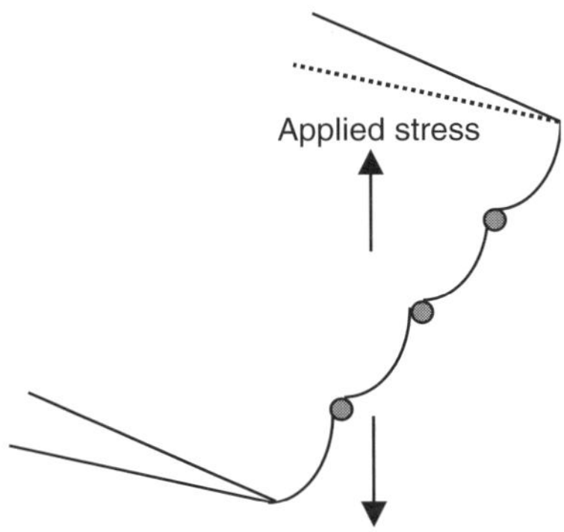

FIGURE 15.1.1 A tensile crack bowing between dispersed particles (after Ref. [2]).

is estimated from cleavage step patterns formed on the fracture surface during cleavage fracture. This reveals that the crack front is pinned by the dispersed particles and voids, and that cleavage crack bowing occurs around a pair of dispersoids (Fig. 15.1.1). As a result, the particle spacing determines the fractional increase in crack front length per unit extension. Assuming that the fracture energy depends not only on the new surface area but also on the newly formed length of the crack front, the strengthening is achieved in a manner analogous to the bowing of a dislocation between pinning points [3, 4].

15.1.2.2 Crack Deflection

According to calculations of the strain energy release rate for a deflected crack, the crack deflection can enhance the fracture toughness because the crack propagation mode is a mixture of $K_{\rm I}$, $K_{\rm II}$ and $K_{\rm III}$. In this case, a greater tensile stress is required than in mode I crack propagation [5]. The most effective morphology for deflecting a propagating crack has been found to be rod-shaped grains having a high aspect ratio, and this can account for fourfold increases in fracture toughness. High fracture toughness values in ceramics consisting of rod-shaped grains have been demonstrated in silicon nitride and a lithium-alumino-silicate glass ceramic [6].

Another crack deflection mechanism for toughening ceramics is as a result of the existence of the local residual stress in the vicinity of dispersed secondary particles. Composites of 25 vol% TiC particles in a matrix of SiC have 60% higher fracture toughness and 40% higher strength than the matrix material alone. The improved flexural strength and fracture toughness in this system is thought to result from crack deflection due to residual stress [7]. The mismatch

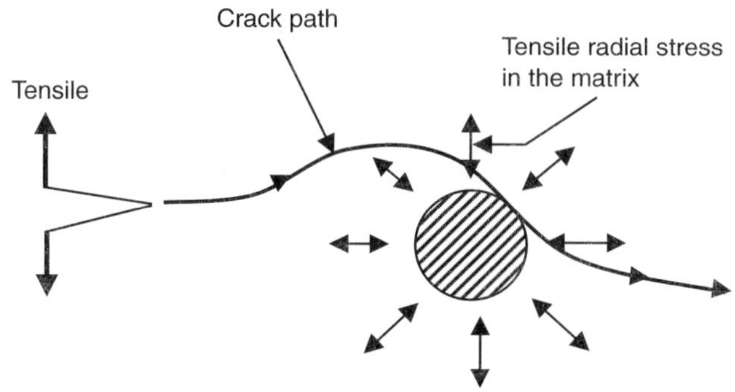

FIGURE 15.1.2 Crack deflection by a dispersed particle and associated matrix stress due to thermal expansion mismatch. Note that a tensile radial stress appears in the matrix when the thermal expansion coefficient of the particle is greater than that of the matrix. As the crack moves around the particle it can be attracted back to the particle interface (after Ref. [7]).

between the linear thermal expansion coefficients of the SiC matrix and TiC particles results in the generation of residual stresses in the particles and surrounding matrix during cooling after fabrication. The resultant tensile stress generated in the matrix causes crack deflection around the TiC particles, leading to high fracture toughness and strength (Fig. 15.1.2).

15.1.2.3 Microcracking

In composites containing secondary particles, the local residual stress due to mismatch of thermal expansion coefficients is generated during cooling from the processing temperature. The residual stresses cause spontaneous microcracking if they exceed a certain size. However, when the residual stress is less than the local strength of the material, the internal stress remains in the material. In this case, applying a stress causes microcracks ahead of the crack tip, where huge local stresses are generated, because the applied stress reduces the critical size for microcracking below that for spontaneous microcracking [8]. Toughening due to microcracking is expected to occur according to the following mechanism. First, a microcracking zone is generated ahead of the crack tip. The incorporation of the microcracking zone into the crack as it extends produces microcracked layers surrounding the crack. The volume expansion of the cracked layers leads to the generation of compressive stresses to the crack interface, and this results in an increase in the fracture toughness [9].

The fracture toughness of Al_2O_3 is considerably enhanced by incorporation of fine monoclinic ZrO_2 particles. For example, a hot-pressed composite containing 15 vol% ZrO_2 has a K_{Ic} value of approximately 10 MPa $m^{1/2}$, twice of that of the Al_2O_3 matrix alone [10, 11]. This high value results from the high density of matrix microcracks, formed by the expansion of ZrO_2 during the tetragonal to monoclinic transformation.

Ceramics with non-cubic crystal structures have anisotropic elastic and thermal expansion coefficients, and exhibit a large fracture energy within a particular grain size range. The maximums in the fracture energy of Al_2O_3 and $MgTi_2O_5$ were reported to occur for a grain size of approximately 100 and 20 μm, respectively [12]. This is explained by the counter-trend between increasing numbers of microcracks and decreasing energy absorption per microcrack as grain size increases [13].

15.1.2.4 Transformation Toughening

The microstructure of partially stabilized zirconia (PSZ) in a system such as $CaO–ZrO_2$ consists of pure monoclinic ZrO_2 grains distributed within a matrix of fully stabilized material [14]. However, when the diameter of the second phase is maintained at less than 100 nm, the second phase becomes metastable tetragonal zirconia and the flexural strength of this tetragonal phase containing PSZ is 650 MPa, much greater than that of PSZ containing a purely monoclinic phase (250 MPa) [15]. Furthermore, a high fracture toughness is achieved by controlling the size of the precipitates [16].

The stress-induced transformation of tetragonal zirconia to monoclinic symmetry contributes to the toughening because all particles within several microns of the cracks become monoclinic while all other particles remain tetragonal [16]. The volume and morphology of the particles ahead of the crack tip change due to the phase transformation, and this modifies the stress distribution. As a result, a compressive stress is generated at the crack interface due to the volume expansion accompanying phase transformation, in a similar way to microcracking. Accordingly, a high fracture toughness results from generation of residual compressive stresses at the crack tip [16].

ZrO_2 ceramics doped with around 3 mol% Y_2O_3 exhibit high strength when fabricated from an ultrafine powder to produce fine-grained materials. Since Rieth *et al.* reported a flexural strength of 890 MPa in 3.9 mol% Y_2O_3-doped ZrO_2 [17], materials with strengths exceeding 1 GPa have been produced using pressure sintering methods, such as hot-pressing and hot-isostatic pressing [18, 19]. The crystalline phases found in the sintered products are tetragonal in 2 mol% Y_2O_3-doped ZrO_2 and cubic in 6 mol% Y_2O_3-doped ZrO_2, and intermediate compositions have a mixture of the two crystalline phases [20].

15.1.2.5 Bridging

By aligning strong fibers along the tensile stress, high fracture strength and fracture energy are obtained in continuous fiber/glass composites [21–23]. High fracture energy is in general obtained when long fibers protrude from the fracture surfaces, because the frictional stress between the fiber and the matrix resists crack propagation in the pullout process. A weak interface leading to debonding of fibers from the matrix is essential for producing toughened composites [23].

A similar stress transfer between crack interfaces also occurs in polycrystalline ceramics. For example, when a crack passes through a coarse-grained alumina, the crack occasionally propagates discontinuously and grain-localized interfacial bridging is formed behind the advancing crack tip [24]. This leads to an increase in fracture toughness because of the bridging grains transferring a stress that resists crack advancement [25]. The bridging stresses are generated by a variety of micro-processes, such as frictional interlocking, fiber bridging and sliding pullout.

Toughening with silicon carbide whiskers has been realized in ceramics, such as alumina and silicon nitride, and it has been shown that a larger whisker diameter leads to greater toughness [26]. High fracture toughness in whisker composites is achieved by introducing a weak interface between whisker and matrix, so that the debonded whiskers can transfer a large amount of bridging stress [27].

15.1.3 FIBERS AND WHISKERS FOR REINFORCEMENT

Continuous fibers and whiskers are widely used for reinforcement of ceramic matrices. In the case of continuous fibers, fine diameters are preferred as this gives flexibility for fabricating complex fiber architectures. Methods of producing continuous fibers may be classified into two types. One is produced from liquid, similar to the organic fiber process: either directly produced from a viscous melt, such as glass fibers, or by an indirect process through precursor fibers. Carbon and SiC fibers are usually produced in a single step as organic fibers, such as polyacrylonitrile and polycarbosilane, and converted to their final states by breaking of the organic bonds. The precursor method can produce fibers of small diameter, typically around 10 μm. The second type is chemical vapor deposition on tungsten filaments or carbon fibers. In this case, fiber diameters are usually large and more than 100 μm.

SiC whiskers and continuous fibers that have small diameters and are durable at high temperatures are often selected for fabrication of ceramic matrix composites. Tables 15.1.1 and 15.1.2 list the properties of continuous fibers and whiskers, respectively, used for ceramic matrix composites.

15.1.3.1 Carbon Fibers

Carbon fibers are able to withstand very high temperatures without strength loss up to and above 3000°C only in the absence of oxygen [28, 29]. The most common carbon fibers are produced from polyacrylonitrile (PAN). PAN fibers are partially oxidized at a temperature of 200–300°C in air to be cross-linked, and the fibers are then heat-treated to convert them to carbon in a non-oxidizing atmosphere around 800–1500°C. Their mechanical properties depend on the final heat treatment temperature because the crystallization of carbon occurs intensively at elevated temperatures. Since the crystallization of graphite occurs during heat-treatment at a temperature of 2000–3000°C, high-modulus graphite fibers are produced. Carbon fibers with less crystallization generally have a large elongation to failure and high tensile strength. Carbon fibers produced from mesophase pitch can possess an extremely high Young's modulus with values up to 900 GPa.

15.1.3.2 SiC Fibers

Yajima *et al.* have reported the synthesis of continuous silicon carbide fibers from an organic polymer [30]. SiC fibers have been commercially produced by Nippon Carbon Co. under the trade name "Nicalon", and similar SiC fibers containing Ti are produced by UBE Industries under the trade name "Tyranno fiber" [31].

SiC fibers are produced from polycarbosilane, $(\text{–Si–C–})_n$, synthesized from dimethyldichlorosilane, $(CH_3)_2SiCl_2$. Viscous polycarbosilane is melt-spun to produce preceramic fibers, and the fibers are cured in air to be cross-linked with oxygen. The polycarbosilane fibers are then heat-treated in vacuum in order to break the organic bonds, such as Si–CH_3 and C–H, and are thereby converted into continuous SiC fibers containing oxygen. The oxygen contained in the fibers is introduced during the curing process. Nicalon fiber, composed of ultrafine β-SiC crystallites and an amorphous mixture of silicon, carbon and oxygen, contains 11.7% oxygen.

An electron beam curing process can reduce the oxygen content by an oxygen free cross-linking process [32]. The fiber, composed of fine β-SiC crystallites

TABLE 15.1.1 Continuous Fibers Used for Fabricating Ceramic Matrix Composites

Material		Trade name	Tensile strength (GPA)	Tensile modulus (GPA)	Elongation to break (%)	Diameter (μm)	Density $(g/cm)^3$	Manufacturer
Carbon	PAN-based	—	3–7	200–700	0.5–2	4–7	1.8	—
	Pitch-based	—	3–4	600–900	0.50	7–10	2.2	—
SiC	56.6Si–31.7C–11.7O	Nicalon	3.0	220	1.40	14	2.55	Nippon Carbon
	62.4Si–37.1C–0.5O	Hi-Nicaron	2.8	270	1.00	14	2.74	Nippon Carbon
	68.9Si–30.9C–0.2O	Hi-Nicalon S	2.6	420	0.60	12	3.1	Nippon Carbon
	51.0Si–27.9C–17.7O–3.1Ti	Tyranno	3.3	195	—	8.5–15	—	Ube Industry
	66.6Si–28.5C–0.8O–2.1Ti–2.3B–0.4N	Sylramic	3.4	386	—	10	3.1	Dow Corning
Alumina	99.5% α–Al_2O_3	Almax	1.8	330	—	10	3.6	Mitsui Mining Materials
	>99% Al_2O_3	Nextel 610	2.93	373	—	10–12	3.96	3M
Oxides	85% Al_2O_3–15% SiO_2	Nextel 720	2.1	260	—	10–12	3.4	3M

TABLE 15.1.2 Whisker Materials Used for Fabricating Whisker Composites

Material	Trade name	Length (μm)	Diameter (μm)	Density (g/cm^3)	Manufacturer
β-SiC	TWS-100	5–15	0.3–0.6	3.20	Tokai Carbon
β-SiC	SCW	5–200	0.05–1.5	3.18	Tateho Chemical
α-Si_3N_4	SNW	5–200	0.1–1.6	3.18	Tateho Chemical

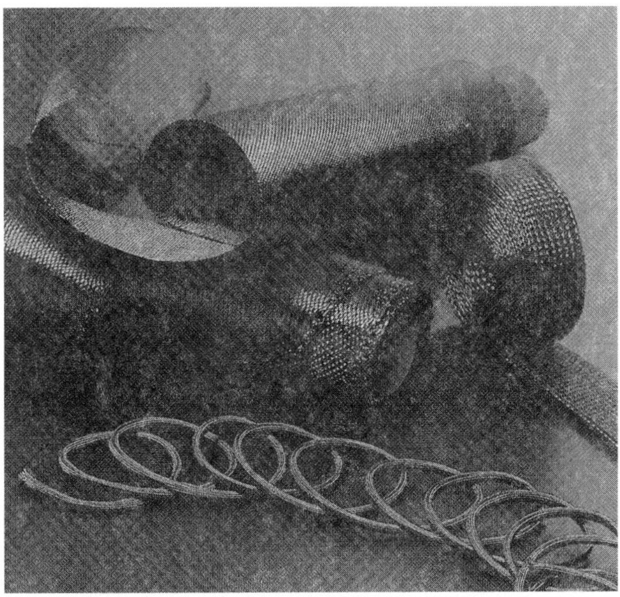

FIGURE 15.1.3 A photograph of Tyranno fiber fabrics (courtesy of Ube Industry).

and carbon, has higher thermal stability than Nicalon fiber, and is commercially produced by Nippon Carbon Company under the trade name "Hi-Nicalon" with a typical oxygen content of 0.5%, and typically the fiber has a diameter of 14 μm, a tensile strength of 2.8 GPa and a tensile modulus of 270 GPa. Type S, created using decarbonation pyrolysis in a hydrogen atmosphere, is a stoichiometoric β-SiC fiber having high modulus, high creep resistance, and high resistance against oxidation at elevated temperature.

Tyranno fiber, shown in Figure 15.1.3, is produced from pyrolysis of polytitanocarbosilanes containing 1.5–4.0 mass% titanium, and has a diameter of 8.5–15 μm, a tensile strength of 3.3 GPa and an elastic modulus of 195 GPa. The fiber is durable up to a temperature of 1500°C.

Sylramic fiber (by Dow Corning) is an SiC fiber containing a small amount of TiB_2 and B_4C, and produced using polytitanocarbosilane precursors containing a small amount of titanium [33].

15.1.3.3 Oxide Fibers

Oxide fibers include glass fibers, mullite fibers, zirconia fibers and alumina fibers. Of these, α-alumina-based fibers have been used intensively for ceramic matrix composites. Fiber FP, manufactured by Du Pont in 1979, was the first wholly α-alumina fiber produced [34]. At present, Almax (Mitsui Mining Material Co. Ltd., Japan) and Nextel 610 (3M Co., USA) are commercially available α-alumina fibers. Almax contains 99.5% alumina and has an elastic modulus of 330 GPa, and Nextel 610 has a tensile strength of 2.4 GPa and an elastic modulus of 380 GPa [35].

The addition of a secondary phase to α-alumina can improve the mechanical properties at high temperatures. Nextel 720 consists of α-Al_2O_3 and mullite, and has a chemical composition of 85% Al_2O_3 and 15% SiO_2. The fiber has a tensile strength of 2.1 GPa and an elastic modulus of 260 GPa. The high-temperature properties such as tensile strength and creep resistance are improved in comparison with the α-alumina fiber of Nextel 610 [36]. PRD-166 is a continuous multifilament yarn with a polycrystalline alumina-zirconia composition, developed by Du Pont but commercially unavailable. The fiber has a tensile strength of 2.1 GPa and a modulus of 380 GPa [37].

15.1.3.4 Whiskers

Silicon carbide whiskers, usually produced from silica and carbon, are widely used for ceramic matrix composites [38]. The vapor–liquid–solid (VLS) process is accepted as the mechanism for SiC whisker production (Fig. 15.1.4). This process requires formation of a liquid phase of a transition metal at the reaction sites. Elemental sources of Si and C are usually supplied in vapor phase form as SiO and CH_4, respectively [39]. The vapors are deposited on the surface of the liquid metal and dissolved into the liquid metal droplets. Since the whiskers are precipitated from the dissolved components, metal droplets are often observed at the tip of the whiskers [39].

SiC whiskers having a diameter of 4–6 μm and a length of 5 mm have been reported to possess an averaged tensile strength of 8.40 GPa and an average elastic modulus of 581 GPa [40]. Commercially available SiC whiskers are usually smaller, and typically have a diameter of approximately 1 μm and a length of several tens of microns or more (see Figure 15.1.5).

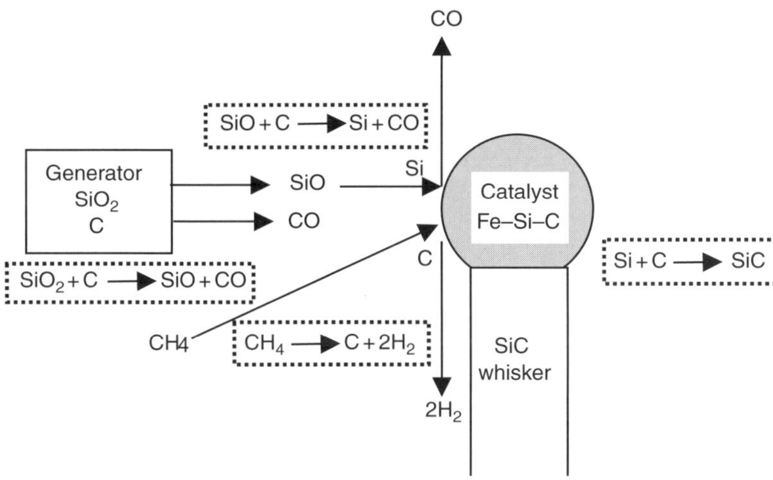

FIGURE 15.1.4 VLS process in SiC whisker production (after Ref. [39]).

FIGURE 15.1.5 A scanning electron micrograph of SiC whiskers (courtesy of Tokai Carbon).

15.1.4 PROCESSING OF CERAMIC COMPOSITES

When fabricating ceramic composites, the reaction between the reinforcement and matrix must be minimized. Consequently, a suitable low processing temperature should be selected to minimize degradation of the reinforcing phases. Use of continuous fibers can modify the mechanical properties if suitable fiber orientations are chosen. A well-designed fiber architecture is first prepared, and the matrix material introduced into the voids of the structure. Infiltration into the fiber architecture is often performed using chemical vapor infiltration or pre-ceramic polymer infiltration-pyrolysis techniques. These techniques have the advantage of requiring low temperatures in comparison with sintering processes. Textile production techniques, such as weaving, stitching, knitting and braiding are used for producing three-dimensional composite structures [41], and the stacked sheets of fabrics are used for manufacturing panel structures. Hot pressing is used for fabricating glass matrix composites due to the relatively low processing temperatures, and this is also used for the fabrication of whisker composites.

15.1.4.1 Hot Pressing

Uniaxial and two-dimensional fiber composites with glass matrices can be produced by hot pressing [42]. The fibers are immersed in a slurry of matrix particles and then dried. The fiber/matrix powder preform is then cut into suitable dimensions, and stacked for hot pressing in order to produce dense composites. The fabricated composites are usually in the shape of flat disks or rectangular plates.

Densification of whisker composites is usually performed by hot pressing. Careful control is required both to avoid damage to the whiskers and to achieve a homogeneous dispersion of whiskers.

15.1.4.2 Chemical Vapor Impregnation

Chemical vapor impregnation (CVI) is a method of infiltrating fiber architectures with matrix particles via the vapor phase. Although this is a similar process to chemical vapor deposition (CVD) in terms of gas reaction, decomposition conditions in CVI are chosen for in-depth decomposition rather than coating the surface of the substrate.

Several techniques have been developed to introduce the reaction gases into the fiber architecture, such as temperature gradient, pressure gradient and pulse CVI [43]. Figure 15.1.6 shows the equipment used for the temperature gradient technique [44]. CVI has been used for carbon and silicon carbide matrix formation. Hydrocarbon gases such as CH_4 are used for manufacturing carbon matrix composites, and the decomposition of methyltrichlorosilane is used in the production of silicon carbide matrix composites [43], according to the reaction given in Eq. 1.

$$CH_3SiCl_3 + H_2 \rightarrow SiC + 3HCl + H_2 \quad (1)$$

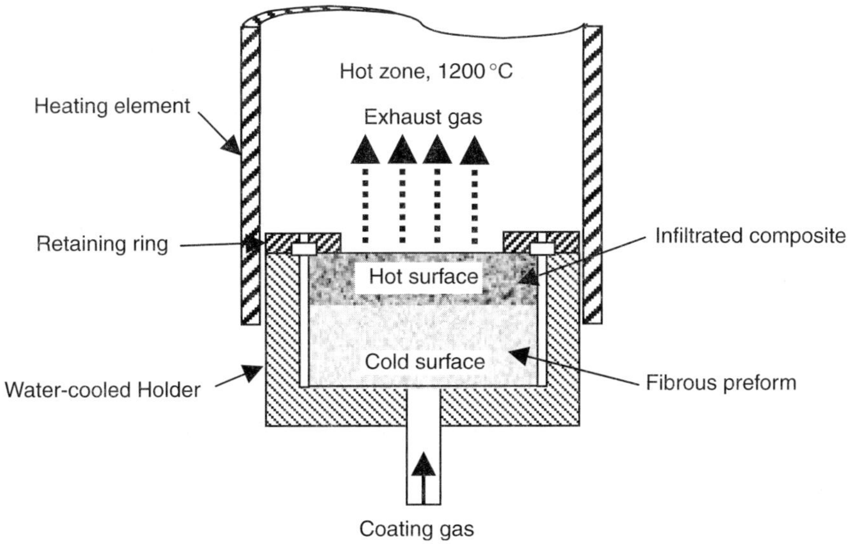

FIGURE 15.1.6 CVI equipment for the temperature gradient technique (after Ref. [44]).

15.1.4.3 Liquid Impregnation and Pyrolysis

Preceramic polymer can fill a preform with liquid polymers, either molten or in solution, which are then pyrolized to make a ceramic matrix. Polymer impregnation and thermal decomposition are repeated several times. Resins, such as phenol and pitch, are used for producing carbon matrix composites, and organosilicon compounds, such as polycarbosilane and polyvinylsilane, are used for the impregnation of silicon carbide [45–48]. Sols have also been used for infiltrating preforms to produce oxide matrices, such as alumina, silica, zirconia and mullite. Following infiltration, they are gelled by drying or by adjusting the temperature, and the matrix is formed after heat treatment.

15.1.4.4 Novel Techniques

Melt infiltration into fibrous preforms combined with oxidation of the metal matrix can produce ceramic matrix composites. Since this type of process was developed by Lanxide Corporation, it is called the Lanxide process [49]. For example, a Nicalon SiC fiber and alumina matrix composite was produced as follows [50]: stacked fibrics of Nicalon fiber were coated by CVD. The major purposes of the coating were to protect the fiber from aluminum alloys during the matrix growth process and to provide a weak fiber–matrix interface. The fiber architectures were then placed on molten aluminum to allow the matrix to grow in the fiber preform by direct oxidation of the aluminum alloy.

Eutectic consolidation using a single crystal production technique can also be used in the production of two-phase ceramic composites. Aligned composite structures can be produced by unidirectional solidification of binary eutectics. The $BaFe_{12}O_{19}$–$BaFe_2O_4$ eutectic has been, for example, produced from powder mixtures of Fe_2O_3 and $BaCO_3$. The melt was consolidated by moving the specimen gradually in the furnace, in a similar way to the Bridgman technique [51]. Using a similar technique, a variety of eutectic systems were investigated [52–56]. The microstructure revealed that fibrous structures are developed along the longitudinal direction, and the flexural strength is greater when the crack propagates across the fibrous structure.

15.1.5 PARTICULATE COMPOSITES

Ceramic composites containing particulates are similar to multiphase ceramics, and these materials can be fabricated through traditional ceramic processes.

15.1.5.1 Al_2O_3–TZP Composites

Fine-grained ZrO_2 typically doped with 3 mol% Y_2O_3 is known to exhibit flexural strengths exceeding 1 GPa, and the strength can be enhanced further by hot pressing or hot isostatic pressing. Additional strength increases have been observed in composites of Y_2O_3-doped ZrO_2 (tetragonal zirconia polycrystals, TZP) with Al_2O_3. The addition of alumina to TZP greatly improves the strength. For example, Tsukuma *et al.* reported a flexural strength of 2.4 GPa for a composite containing 20 mass% Al_2O_3 [57], although the strength decreased with temperature. Two mole percent Y_2O_3-doped ZrO_2/Al_2O_3 composites have been reported to maintain a strength of 1.0 GPa up to 1000°C [58]. In addition, Shikata *et al.* reported a flexural strength of 3.0 GPa for Y-TZP/Al_2O_3 composites (Fig. 15.1.7) [59]. A flexural strength of 3 GPa was achieved for alumina contents of 20–40 mass% in 2 mol% Y_2O_3-doped TZP matrix composites, and also for a 40 mass% alumina content in 3 mol% Y_2O_3-doped TZP. Both TZP powders of 2 mol% Y_2O_3 and 3 mol% Y_2O_3 were prepared by a hydro-decomposition method, and the alumina powder was 99.9% pure with a mean particle size of 0.54 μm. The powder mixtures were ball-milled in ethanol for 120 h, dried and die-pressed at 39 MPa, and subsequently cold isostatically pressed at 98 MPa.

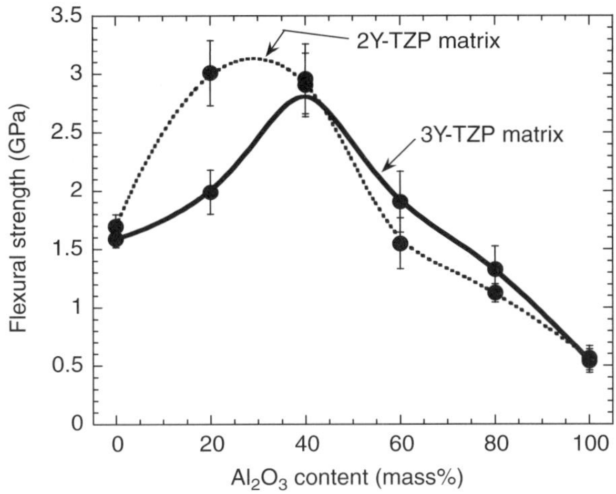

FIGURE 15.1.7 Flexural strength of HIPed composites as a function of Al_2O_3 content. A flexural strength of 3 GPa was achieved for an alumina content of 20–40 mass% in a 2 mol% Y–TZP matrix and for a 40 mass% alumina content in a 3 mol% Y–TZP matrix (after Ref. [59]).

The powder compacts were sintered at a temperature between 1400 and 1600°C for 3 h, followed by HIPing at 1450°C in argon gas under a pressure of 98 MPa for 1 h.

15.1.5.2 SiC–Si_3N_4 Composites

Addition of SiC particulates to a silicon nitride matrix is expected to improve the mechanical properties of silicon nitride. The flexural strength at 1400°C is considerably improved with the addition of 5 μm SiC particles to a silicon nitride matrix using MgO as a sintering aid [60].

Niihara *et al.* fabricated a Si_3N_4–32% SiC nanocomposite from an amorphous Si–C–N powder produced from a vapor phase reaction, and 8 mass% Y_2O_3 powder was added for densification [61]. In this material, nanometer-sized SiC particulates are mainly dispersed in the silicon nitride grains, and no reactive layer or amorphous layer was found between Si_3N_4 and SiC. The flexural strength of the composite was 970 MPa at 1400°C, and this is considerably greater than the 500 MPa of monolithic silicon nitride containing the same additive. This nanocomposite also exhibits excellent tensile creep resistance [62].

15.1.6 WHISKER COMPOSITES

The green density of the mixture of whiskers and matrix powders is generally low due to a high aspect ratio of whiskers. Pressure sintering, such as hot pressing, is therefore used for densification. Another problem is due to the agglomeration of whiskers, and careful mixing of whiskers with matrix forming powders is required to avoid inducing serious damage to the whiskers. A typical procedure for this is to disperse the whiskers in liquid followed with successive filtration to remove agglomerated whiskers. The whiskers are then prepared for composite fabrication by blending with powders of the matrix material [63–68].

15.1.6.1 SiC Whisker–Al_2O_3 Matrix Composites

Fully dense, fine-grained (less than 4 μm) alumina matrix composites have been obtained by hot-pressing a mixture of alumina powder containing 20 vol% SiC whiskers at 1850°C [63]. Hot-pressing causes the whiskers to become oriented, with the length of the whisker essentially randomly oriented in a plane perpendicular to the hot-pressing axis because of the high aspect ratio of the whiskers. The materials exhibit a fracture toughness of 8.7 MPa $m^{1/2}$ and flexural strength of 800 MPa. The dispersion of SiC whiskers is improved by ultrasonic dispersion techniques.

Al_2O_3–SiC whisker composites are commercially used for cutting-tool materials, and the materials contain 30–40 vol% SiC whiskers [69]. The advantage of using Al_2O_3–SiC whisker composites for cutting tools is the substantially increased metal-removal rates in machining Ni-based superalloys.

15.1.6.2 SiC Whisker–Si_3N_4 Matrix Composites

Ueno *et al.* have reported fabricating silicon carbide whisker–silicon nitride matrix composites [64]. SiC whiskers were dispersed in water and filtered to remove the agglomerates. Silicon nitride powder was ball-milled with sintering aids of Y_2O_3 and La_2O_3 in ethanol. The SiC whiskers and silicon nitride powder were then mixed and dispersed in water. Thin sheets were produced by filtration of the mixed slurry. These sheets were stacked in a carbon mold for hot pressing, carried out at 1800°C. The composites maintained a flexural strength of 590–680 MPa up to 1300°C with a small reduction in strength at high temperature. A whisker content greater than 40% leads to difficulty in full densification and a porosity of around 10%. Statistical analysis of the flexural strengths revealed a high Weibull modulus of 25, much greater than for usual silicon nitride.

FIGURE 15.1.8 A burner diffuser made of an SiC whisker reinforced silicon nitride composite (courtesy of JMC New Materials [71]).

SiC whisker reinforced silicon nitride has excellent thermal shock resistance, and has been successfully used in burner diffusers for boilers in thermal power plants, as illustrated in Figure 15.1.8 [70, 71].

15.1.7 CONTINUOUS FIBER COMPOSITES

The mechanical behavior of continuous fiber composites is very different to that of other brittle ceramics. In tensile loading, a change in the linear stress–strain relation occurs after matrix cracking, and sliding pullout contributes to the load bearing ability afterward [72, 73]. Moreover, shear failure and compression failure are often observed in flexural tests, resulting from delamination due to shear stresses and fiber buckling due to compression [74]. Such a failure mode is obviously derived from the weak interface between the fiber and matrix. Therefore, the presence of lubricant carbon and boron nitride at the interfaces is preferred. Although excess carbon on the surface of SiC fibers acts as a lubricant, the carbon layer may oxidize in air. Boron nitride coating on fibers has been carried out to maintain weak interfaces at high temperatures [75].

15.1.7.1 Glass Matrix Composites

Continuous fibers such as carbon fibers and SiC fibers can be used to reinforce a glass matrix. The strengthening mechanism is similar to that in resin matrix composites, and the fibers carry most of the load due to their much higher Young's modulus compared to the matrix. Of all the continuous fiber composites, glass matrix composites are particularly dense as they are produced by impregnation of a glass melt. The mechanical properties are characterized by high strength and large fracture energy. The large fracture energy is explained in terms of intensive pullout of fibers from the matrix glass [76]. This indicates that the fiber to matrix bond is poor due to the presence of lubricant carbon layers on the surfaces of both graphite and silicon carbide fibers. In contrast,

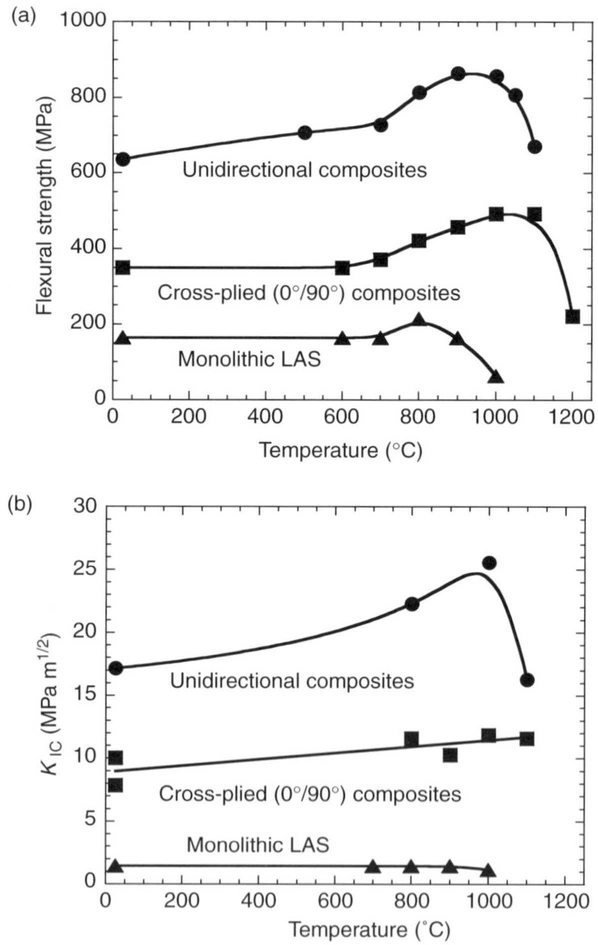

FIGURE 15.1.9 Mechanical properties of Nicalon/LAS glass composites: (a) Flexural strength and (b) fracture toughness (after Ref. [76]).

glass matrix composites using oxide fibers exhibit low flexural strength due to the high bonding strength between fiber and glass matrix [42].

Carbon fiber glass matrix composites were intensively studied in the 1970s [77, 78] and SiC fiber composites in the 1980s [79, 80]. In both cases the fabrication method is essentially the same. Fiber tows or fabrics are first immersed in a glass powder suspension, and the powder-containing fiber sheets are stacked ready to be hot-pressed into laminates.

Figure 15.1.9 shows the flexural strength and fracture toughness of Nicalon/LAS glass as a function of temperature [76]. The unidirectionally reinforced composites exhibit a flexural strength of over 620 MPa at room temperature, which increases to over 830 MPa between 900 and 1000°C. Cross-piled samples (0°/90°) exhibit strengths of over 350 MPa at room temperature and 480 MPa at 1000°C.

15.1.7.2 Carbon/Carbon Composites

The development of carbon/carbon composites began in 1958, and they have been applied to the hot parts of missiles and the Space Shuttle, such as nose caps and leading edges [81]. Carbon/carbon composites can withstand temperatures higher than 3000°C in a vacuum and in an inert atmosphere, without losing strength as the operating temperature is increased. However, they oxidize and sublime when in an oxygen atmosphere at 600°C. Silicon carbide coatings are therefore coated with a layer of glass to protect them in high temperature applications. The protection mechanism is as follows. When the part is cooled down from the coating temperature, microcracks develop in the silicon carbide layer, resulting from thermal expansion mismatch between carbon and silicon carbide. These cracks might cause oxidation of the substance if exposed to the air, but are immediately impregnated with the overcoated glass.

Carbon/carbon composites have successfully replaced metallic brake discs in racing cars and aircraft because of their lightweight. Civilian aircraft, such as the Concorde supersonic jet and Boeing 767 use carbon/carbon composite brakes. In comparison to steel brakes, a 40% weight saving is achieved using carbon/carbon composites due to their large heat capacity (2.5 times that of steel) and high strength (twice that of steel) at elevated temperatures [82].

Carbon/carbon composites are produced by pyrolyzing an organic matrix or by CVI. CVI of carbon from hydrocarbon gas is normally accomplished at 1100°C, and pyrolytic carbon is obtained. Carbonized organic composites are typically produced from graphite fabrics pre-impregnated with phenolic resin. The fabrics are laid up as a laminate and cured. They are then pyrolyzed to form a matrix of glassy carbon around the graphite fibers. As a result of repeated impregnations of resins and pyrolyzations, densification is achieved.

15.1.7.3 SiC/SiC Composites

Since the oxidation resistance of SiC is much better than that of carbon, SiC/SiC composites have been developed for aerospace application such as propulsion and high velocity systems. Similar to carbon/carbon composites, the SiC/SiC continuous fiber composites consist of a fiber architecture made of silicon carbide fibers in a matrix of silicon carbide. The matrix is usually produced by CVI or preceramic polymer impregnation and pyrolysis.

The Société Européenne de Propulsion (SEP) has commercially produced SiC/SiC composites through a CVI process [83, 84]. Typical SiC/SiC composites by SEP consist of two-dimensional flat laminates with Nicalon fibers, and have a density of about 2.4 Mg/m^3 with a remaining open porosity near 10%. Their four-point flexural strength at room temperature is about 300 MPa, slightly increases up to 1300°C, and exhibits a progressive decrease with some strength remaining at a temperature of 2000–2200°C. The fracture toughness is as high as 25 $MPa\,m^{1/2}$ and remains constant up to 1400°C. In oxidation resistance tests conducted at 1100°C in air, no degradation in flexural strength was observed after 500 h. The SiC/SiC composites can be used up to 1200°C as long as Nicalon fibers retain enough strength. In oxidizing atmospheres, SiC/SiC are self-healed at high temperatures [84]. Thermostructural parts have been developed by SEP for engines and rockets, as well as turbojets and for space-plane thermal protection systems. For example, an air–kerosene ramjet chamber with an SiC/SiC wall has been tested for 2000 s at a temperature higher than 2000 K and the resultant weight loss of the total chamber was only 0.2% with no local erosion [84].

The fabrication of a one-piece-SiC composite heat exchanger using a CVI technique has been reported [85]. A 500 kW CVD reactor was used and this reactor was also used for producing a corrugated Nicalon panel of 0.8 m × 0.8 m × 1 cm with a maximum cross-sectional thickness of 5 mm.

Nicaloceram products are produced from Nicalon fabrics impregnated with polycarbosilane solution and SiC power slurry (Fig. 15.1.10). The impregnated fabrics are piled to make a laminate. After curing, the laminate is pyrolyzed to convert it to an SiC matrix. Impregnation of polycarbosilane solution to the pyrolized material is repeated several times and finally fired at a temperature higher than 1200°C. Table 15.1.3 shows typical properties of Nicaloceram.

Figure 15.1.11 shows an SiC/SiC exhaust nozzle flap produced using combined precursor impregnation and pyrolysis (PIP) and CVI processes, and this

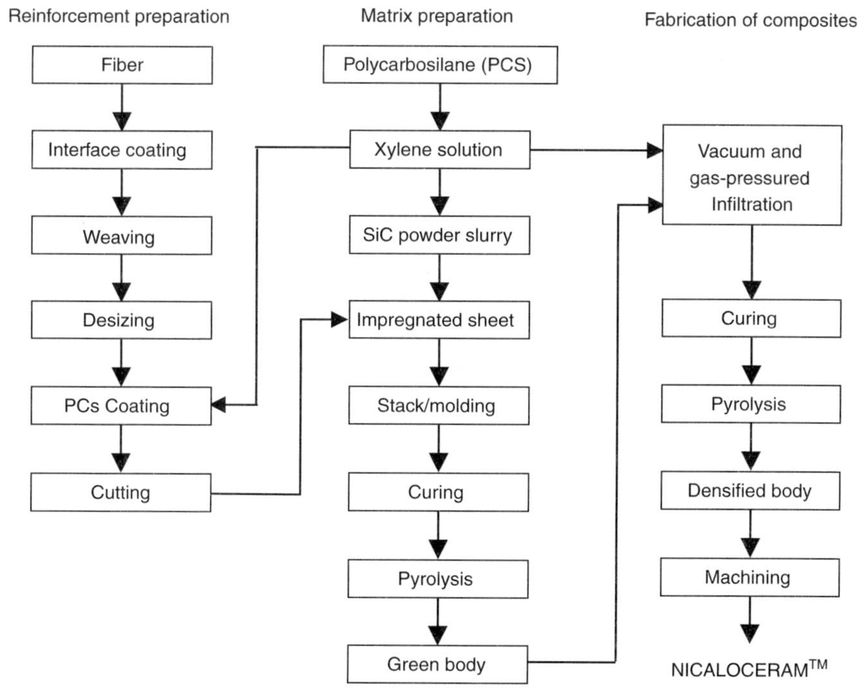

FIGURE 15.1.10 Typical manufacturing process flow chart for Nicaloceram.

TABLE 15.1.3 Typical properties of Nicaloceram

Grade	Interface material	Reinforcement (type of Fiber)	Volume fraction (vol%)	Tensile strength (MPa)	Tensile modulus (GPa)	Flexural strength (MPa)	Density (Mg/m^3)
Nicaloceram	Carbon	Nicalon	30	110	60	110	2.0
Hi-Nicaloceram	Boron Nitride	Hi-Nicalon	30–40	330	110	550	2.2–2.3

part is used for an aero-engine [86]. The advantages of using SiC/SiC composites in advanced aero-engines are their light weight and durability at elevated temperatures. The engine test was conducted successfully for a total of 15 h and no damage was found.

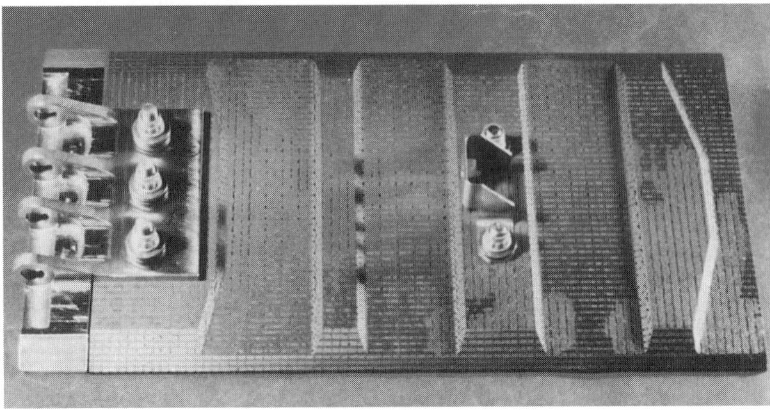

FIGURE 15.1.11 An SiC/SiC exhaust nozzle flap for an aero-engine (courtesy of Ishikawajima-Harima Heavy Industries [86]).

15.1.7.4 Oxide/Oxide Composites

In continuous fiber composites, a weak interface between fiber and matrix is preferred because a large fracture energy is generated as a result of frictional slip at the fiber/matrix interface after matrix cracking. In the case of SiC continuous fibers, the presence of excess carbon on the surface of the SiC fibers contributes to their sliding ability, although the excess carbon may disappear during long-term use due to oxidation and the oxide reaction product formed at the cracked matrix may bond with the interface. As a result, embrittlement occurs. It is therefore important to maintain weak interfaces at high temperatures in oxidative atmospheres, and successful results have been obtained for SiC continuous fiber/Al_2O_3 composites using SiC fibers coated with BN layers [75].

Since oxide materials have no oxidation problems, development of all-oxide composites has been a major goal of recent research. Such composites have an interface configuration which allows a crack to propagate along the interface after matrix cracking. There are several microstructural design strategies. The first is to use fugitive layers, the second is to use stable oxide interfaces with suitably low fracture toughness, and the third is to use a porous matrix because the porous interlayers act as crack deflection paths [87, 88].

Monazite ($LaPO_4$) provides a weakly bonded interface between alumina fibers and the alumina matrix because monazite is thermochemically stable with alumina and mullite, and, moreover, has a low interface fracture toughness. Al_2O_3 fiber/Al_2O_3 composites having a monazite interphase have been fabricated by manually dip coating on Al_2O_3 fibers in a monazite slurry, embedding the coated fibers in alumina powder, and hot-pressing at 1400°C [89].

Due to the weakly bonded monazite interface, the non-linear deformation is enhanced and the composites are less sensitive to the presence of sharp notches [90].

15.1.8 EUTECTIC COMPOSITES

Recently, Al_2O_3–$Y_3Al_5O_{12}$ eutectic composites have been attracting considerable attention. Firstly, the creep rates of Al_2O_3–$Y_3Al_5O_{12}$ eutectic composites are considerably lower due to the excellent creep resistance of $Y_3Al_5O_{12}$ single crystals, and meet the design guidelines for use in gas turbines [91]. Furthermore, these composites have greater fracture toughnesses than single crystals [92], and maintain their high flexural strengths up to 1700°C [93, 94].

REFERENCES

1. Hasselman, D. P. H., and Fulrath, R. M. (1966). Proposed fracture theory of a dispersion-strengthened glass matrix. *J. Am. Ceram. Soc.* **49**: 68–72.
2. Ahlqusit, C. N. (1975). On the interaction of cleavage cracks with second phase particles. *Acta Metall.* **23**: 239–243.
3. Lange, F. F. (1970). The interaction of a crack front with a second-phase dispersion. *Phil. Mag.* **22**: 983–992.
4. Evans, A. G. (1972). The strength of brittle materials containing secondary phase dispersions. *Phil. Mag.* **26**: 1327–1344.
5. Faber, K. T., and Evans, A. G. (1983). Crack deflection process—I. Theory. *Acta Metall.* **31**: 565–576.
6. Faber, K. T., and Evans, A. G. (1983). Crack deflection process—II. Experiment. *Acta Metall.* **31**: 577–584.
7. Wei, C., and Becher, P. F. (1984). Improvements in mechanical properties in SiC by the addition of TiC particles. *J. Am. Ceram. Soc.* **67**: 571–574.
8. Green, D. J. (1981). Stress-induced microcracking at second-phase inclusions. *J. Am. Ceram. Soc.* **64**: 138–141.
9. Evans, A. G., and Faber, K. T. (1981). Toughening of ceramics by circumferential microcracking. *J. Am. Ceram. Soc.* **64**: 394–398.
10. Claussen, N. (1976). Fracture toughness of Al_2O_3 with an unstabilized ZrO_2 dispersed phase. *J. Am. Ceram. Soc.* **59**: 49–51.
11. Claussen, N., Steeb, J., and Pabst, P. F. (1977). Effect of induced microcracking on the fracture toughness of ceramics. *Am. Ceram. Soc. Bull.* **56**: 559–562.
12. Rice, R. W., Freiman, S. W., and Becher, P. F. (1981). Grain-size dependence of fracture energy in ceramics: I. Experiment. *J. Am. Ceram. Soc.* **64**: 345–350.
13. Rice, R. W., and Freiman, S. W. (1981). Grain-size dependence of fracture energy in ceramics: II. Model for noncubic materials. *J. Am. Ceram. Soc.* **64**: 350–354.
14. Garvie, R. C., and Nicholson, P. S. (1972). Structure and thermomechanical properties of partially stabilized zirconia in the CaO–ZrO_2 system. *J. Am. Ceram. Soc.* **53**: 152–157.

15. Garvie, R. C., Hannink, R. H., and Pascoe, R. T. (1975). Ceramic steel? *Nature* **258**: 703–704.
16. Porter, D. L., and Heuer, A. H. (1977). Mechanisms of toughening partially stabilized zirconia (PSZ). *J. Am. Ceram. Soc.* **60**: 183–184.
17. Rieth, P. H., Reed, J. S., and Naumann, A. W. (1976). Fabrication and flexural strength of ultrafine-grained yttria-stabilized zirconia. *Am. Ceram. Soc. Bull.* **55**: 717–721, 727.
18. Masaki, T., and Kobayashi, K. (1981). Mechanical properties of hot-pressed ZrO_2–Y_2O_3 ceramics, "Proceeding of the Japanese Ceramic Society Meeting," pp. 2–3 (in Japanese).
19. Tsukuma, K., and Shimada, M. (1985). Hot-isostatic pressing of Y_2O_3-partially stabilized zirconia. *Am. Ceram. Soc. Bull.* **64**: 310–313.
20. Tsukuma, K., Kubota, Y., and Nobugai, K. (1984). Thermal and mechanical properties of Y_2O_3-partially stabilized zirconia. *Yogyo-Kyokai-shi* **92**: 233–241 (in Japanese).
21. Sambell, R. A. J., Briggs, A., Phillips, D. C., and Bowen, D. H. (1972). Carbon fiber composites with ceramic and glass matrix. Part 2: continuous fiber. *J. Mater. Sci.* **7**: 676–681.
22. Philips, D. C. (1972). The fracture energy of carbon fiber reinforced glass and glass-ceramics. *J. Mater. Sci.* **7**: 1175–1191.
23. Philips, D. C. (1974). Interfacial bonding and the toughness of carbon fiber reinforced glass and glass-ceramics. *J. Mater. Sci.* **9**: 1847–1854.
24. Swanson, P. L., Fairbanks, C. J., Lawn, B. R., Mai Y. W., and Hockey, B. J. (1987). Crack–interface grain bridging as a fracture resistance mechanism in ceramics: I. Experimental study on alumina. *J. Am. Ceram. Soc.* **70**: 279–289.
25. Mai, Y. W., and Lawn, B. R. (1987). Crack–interface grain bridging as a fracture resistance mechanism in ceramics: II. Theoretical fracture mechanics model. *J. Am. Ceram. Soc.* **70**: 289–294.
26. Becher, P. F., Heueh, C., Angellita, P., and Tiegs, T. N. (1988). Toughening behavior in whisker-reinforced ceramic matrix composites. *J. Am. Ceram. Soc.* **71**: 1051–1061.
27. Homeny, J., Vaughn, W. L., and Ferber, M. K. (1990). Silicon carbide whisker/alumina matrix composites: effect of whisker surface treatment on fracture toughness. *J. Am. Ceram. Soc.* **73**: 394–402.
28. Warren, R. (1992). *Ceramic-Matrix Composite*, London: Blackie and Son.
29. Chawla, K. K. (1993). *Ceramic Matrix Composites*, London: Chapman & Hall.
30. Yajima, S., Okamura, K., Hayashi, J., and Omori, M. (1976). Synthesis of continuous SiC fibers with high tensile strength. *J. Am. Ceram. Soc.* **59**: 324–327.
31. Yamamura, T., Ishikawa, T., Shibuya, M., and Okamura, K. (1988). Development of a new continuous Si-Ti-C-O fibre using an organometallic polymer precursor. *J. Mater. Sci.* **23**: 2589–2594.
32. Takeda, M., Imai, Y., Ichikawa, H., and Ishikawa, T. (1991). Properties of the low oxygen content SiC fiber on high temperature heat treatment. *Ceram. Eng. Sci. Proc.* **12**: 1007–1018.
33. Lipowitz, J., and Rabe, J. A. (1997). Structure and properties of Sylramic silicon carbide fiber—a polycrystalline, stoichiometric β-SiC composition. *Ceram. Eng. Sci. Proc.* **18**: 145–157.
34. Dhingra, A. K. (1980). Alumina fiber FP. *Phil. Trans. R. Soc.* **A294**: 411–417.
35. Wilson, D. M., Lueneburg, D. C., and Leider, S. L. (1993). High temperature properties of Nextel 610 and alumina-based nanocomposite fibers. *Ceram. Eng. Sci. Proc.* **14**: 609–621.
36. Wilson, D. M., Lueneburg, D. C., and Leider, S. L. (1995). Microstructure and high temperature properties of Nextel 720 fibers. *Ceram. Eng. Sci. Proc.* **16**: 1005–1014.
37. Romine, J. C. (1987). New high-temperature ceramic fiber. *Ceram. Eng. Sci. Proc.* **8**: 755–765.
38. Lee, J.-G., and Culter, I. B. (1975). Formation of silicon carbide from rice hulls. *Am. Ceram. Soc. Bull.* **54**: 195–198.
39. Hollar, Jr. W. E., and. Kim, J. J. (1991). Review of SiC whisker growth technology. *Ceram. Eng. Sci. Proc.* **12**: 979–991.

40. Perrovic, J. J., Milewski, J. V., Rohr, D. L., and Gac, F. D. (1985). Tensile mechanical properties of SiC whiskers. *J. Mater. Sci.* **20**: 1167–1177.
41. Ko, F. K. (1989). Preform fiber architecture for ceramic-matrix composites. *Am. Ceram. Soc. Bull.* **68**: 401–414.
42. Prewo, K. M., Brennan, J. J., and Layden, G. K. (1986). Fiber reinforced glass and glass-ceramics for high performance applications. *Am. Ceram. Soc. Bull.* **65**: 305–313, 322.
43. Lackey, W. J. (1989). Review, status, and future of the chemical vapor infiltration process for fabrication of fiber-reinforced ceramic composites. *Ceram. Eng. Sci. Proc.* **10**: 577–584.
44. Stinton, D. P., Caputo, A. J., and Lowden, R. A. (1986). Synthesis of fiber-reinforced SiC composites by chemical vapor infiltration. *Am. Ceram. Soc. Bull.* **65**: 347–350.
45. Boisvert, R. P., and Diefendorf, R. J. (1988). Polymeric precursor SiC matrix composites. *Ceram. Eng. Sci. Proc.* **9**: 873–880.
46. Schilling, Jr. C. L., Wesson, J. P., and Williams, T. C. (1983). Polycarbosilane precursors for silicon carbide. *Am. Ceram. Soc. Bull.* **62**: 912–915.
47. Walker, Jr. B. E., Rice, R. W., Becher, P. F., Bender, B. A., and Coblenz, W. S. (1983). Preparation and properties of monolithic and composite ceramics produced by polymer pyrolysis. *Am. Ceram. Soc. Bull.* **62**: 916–923.
48. Sato, K., Morizumi, H., Tezuka, A., Furuyama, O., and Isoda, T. (1995). Interface and mechanical properties of ceramic fiber reinforced silicon nitride composites prepared by a preceramic polymer impregnation method, *High-Temperature Ceramic-Matrix Composite II: Manufacturing and Materials Development*, pp. 199–203, Evans, A. G. and Naslain, R. eds., Westerville, OH: The American Ceramic Society.
49. Chiang, Y. C., Haggerty, J. S., Messner, R. P., and Demetry, C. (1989). Reaction-based processing methods for ceramic-matrix composites. *Am. Ceram. Soc. Bull.* **68**: 420–428.
50. Fareed, A. S., Sonuparlak, B., Lee, C. T., Fortini, A. J., and Schiroky, G. H. (1990). Mechanical properties of 2-D Nicalon^{TM} fiber-reinforced LAXIDE^{TM} aluminum oxide and aluminum nitride matrix composites. *Ceram. Eng. Sci. Proc.* **11**: 782–294.
51. Galasso, F. S., Darby, W. L., Douglas, F. C., and Batt, J. A. (1967). Unidirectional solidification of the $BaFe_{12}O_{10}$–$BaFe_2O_4$ eutectic. *J. Am. Ceram. Soc.* **50**: 333–334.
52. Kennard, F. L., Bradt, R. C., and Stubican, V. S. (1973). Eutectic solidification of MgO–$MgAl_2O_4$. *J. Am. Ceram. Soc.* **56**: 566–569.
53. Kennard, F. L., Bradt, R. C., and Stubican, V. S. (1974). Eutectic solidification of the ZrO_2–MgO eutectic. *J. Am. Ceram. Soc.* **57**: 428–431.
54. Kennard, F. L., Bradt, R. C., and Stubican, V. S. (1976). Mechanical properties of the directionally solidified MgO–$MgAl_2O_4$ eutectic. *J. Am. Ceram. Soc.* **59**: 160–163.
55. Ashbrook, R. L. (1977). Directionally solidified ceramic eutectics. *J. Am. Ceram. Soc.* **60**: 428–435.
56. Sorrell, C. C., Stubican, V. S., and Bradt, R. C. (1986). Mechanical properties of ZrC–ZrB_2 and ZrC–TiB_2 directionally solidified eutectics. *J. Am. Ceram. Soc.* **69**: 317–321.
57. Tsukuma, K., Ueda, K., and Shimada, M. (1985). Strength and fracture toughness of isostatically hot-pressed composites of Al_2O_3 and Y_2O_3-partially-stabilized ZrO_2. *J. Am. Ceram. Soc.* **68**: C4–C5.
58. Tsukuma, K., Ueda, K., and Shimada, M. (1985). High temperature strength and fracture toughness of Y_2O_3-partially-stabilized ZrO_2/Al_2O_3 composites. *J. Am. Ceram. Soc.* **68**: C56–C58.
59. Shikata, R., Urata, Y., Shiono, T., and Nishikawa, T. (1990). Strengthening of Y-PSZ–Al_2O_3 composite cereamics, *Funtai-Oyobi-Funmatsu-Yakin* **37**: 357–361 (in Japanese).
60. Lange, F. F. (1973). Effect of microstructure on strength of Si_3N_4–SiC composite system. *J. Am. Ceram. Soc.* **56**: 445–450.

61. Niihara, K., Izak, K., and Kawakami, T. (1990). Hot-pressed Si_3N_4–32% SiC nanocomposite from amorphous Si–C–N powder with improved strength above 1200°C. *J. Mater. Sci. Lett.* **10**: 112–114.
62. Hirano, T., Niihara, K., Ohji, T., and Wakai, F. (1996). Improved creep resistance of Si_3N_4/SiC nanocomposites fabricated from amorphous Si–C–N precursor powder. *J. Mater. Sci. Lett.* **15**: 505–507.
63. Becher, P. F., and Wei, G. C. (1984). Toughening behavior in SiC-whisker-reinforced alumina. *J. Am. Ceram. Soc.* **67**: C267–C269.
64. Ueno, K., and Toibana, Y. (1983). Mechanical properties of silicon nitride ceramic composite reinforced with silicon carbide whisker. *Yogyo-Kyokai-Shi* **91**: 491–497 (in Japanese).
65. Singh, J. P., Goretta, K. C., Kupperman D. S., and Roubort, J. L. (1988). Fracture toughness and strength of SiC-whisker-reinforced Si_3N_4 composites. *Adv. Ceram. Mater.* **3**: 357–360.
66. Lundberg, R., Kahlman, L., Pompe, R., and Carlson, R. (1987). SiC-whisker-reinforced Si_3N_4 composites. *Am. Ceram. Soc. Bull.* **66**: 330–333.
67. Buljan, S. T., Baldoni, J. G., and Huckbee, M. L. (1987). Si_3N_4–SiC composites. *Am. Ceram. Soc. Bull.* **66**: 347–352.
68. Shalek, P. D., Petrovic, J. J., Hurley, G. F., and Gac, F. D. (1986). Hot pressed SiC whisker/Si_3N_4 matrix composites. *Am. Ceram. Soc. Bull.* **65**: 351–356.
69. Billman, E. R., Mehrotra, P. K., Shuster, A. F., and Beeghly, C. W. (1988). Machining with Al_2O_3–SiC-whisker cutting tools. *Am. Ceram. Soc. Bull.* **67**: 1016–1019.
70. Yonezawa, T. (1994). Pressureless sintering of silicon-nitride components. *Compos. Sci. Technol.* **51**: 265–269.
71. Yonezawa, T., Saito, S., Matsuda, T., Sugita, Y., Itoh, H., and Kazama, K. (1997). 6th International Symposium on Ceramic Materials and Components for Engines, pp. 189–192, Niihara, K., Kanzaki, S., Komeya, K., Hirao, S. and Morinaga, K. eds., Tokyo, Japan: Technoplaza Co., Ltd.
72. Sbaizero, O., and Evans, A. G. (1986). Tensile and shear properties of laminated ceramic matrix composites. *J. Am. Ceram. Soc.* **69**: 481–486.
73. Mah, T., Mediratta, M. G., Katz, A. P., Ruh, R., and Mazdiyasni, K. S. (1985). Room-temperature mechanical behavior of fiber-matrix composites. *J. Am. Ceram. Soc.* **68**: C27–C30.
74. Marshall, D. B., and Evans, A. G. (1985). Failure mechanisms in ceramic-fiber/ceramic-matrix composites. *J. Am. Ceram. Soc.* **68**: 225–231.
75. Fareed, A. S., Schiroky, G. H., and Kennedy, C. R. (1993). Development of BN/SiC duplex fiber coatings for fiber-reinforced aluminia matrix composites fabricated by directed metal oxidation. *Ceram. Eng. Sci. Proc.* **18**: 794–801.
76. Brennan, J. J., and Prewo, K. M. (1982). Silicon carbide fibre reinforced glass–ceramic matrix composites exhibiting high strength and toughness. *J. Mater. Sci.* **17**: 2371–2383.
77. Sambell, R. A. J., Bowen, D. H., and Phillips, D. H. (1972). Carbon fiber composites with ceramic and glass matrix part 1 discontinuous fiber. *J. Mater. Sci.* **7**: 663–675.
78. Sambell, R. A. J., Briggs, A., Phillips, D. C., and Bowen, D. H. (1972). Carbon fiber composites with ceramic and glass matrix part 2 continuous fiber. *J. Mater. Sci.* **7**: 676–681.
79. Prewo, K. M., and Brennan, J. J. (1980). High-strength silicon carbide fiber-reinforced glass-matrix composites. *J. Mater. Sci.* **15**: 463–468.
80. Prewo, K. M., and Brennan, J. J. (1982). Silicon carbide yarn reinforced glass matrix composites. *J. Mater. Sci.* **17**: 1201–1206.
81. Buckley, J. D. (1988). Carbon–carbon, an overview. *Am. Ceram. Soc. Bull.* **67**: 364–368.
82. Awasthi, S., and Wood, J. L. (1988). C/C composite materials for aircraft brakes. *Adv. Ceram. Mater.* **3**: 449–451.

83. Lamicq, P. J., Bernhart, G. A., Dauchier, M. M., and Mace, J. G. (1986). SiC/SiC composite ceramics. *Am. Ceram. Soc. Bull.* **65**: 336–338.
84. Janet, J. F., and Lamicq, P. (1993). Composite thermostructures: an overview of the French experience, *High Temperature Ceramic Matrix Composites*, pp. 735–742, Naslain, R., Lamon, J., and Documeingts, D. eds., Cambridge: Woodhead Pub. Ltd.
85. Reagan, P., Ross, M. F., and Huffman, F. N. (1988). CVI/CVD for SiC composites. *Adv. Ceram. Mat.* **3**: 198–201.
86. Ishizaki, M., Shioda, M., Miyahara, K., Sasa, T., Araki, T., Masaki, S., Imamura, R., and Ohnabe, H. (1996). Study on fabrication process of fiber-reinforced ceramic composites and application to engines. *Ishikawajima-Harima Eng. Review* **34**: 418–422 (in Japanese).
87. Levi, C. G., Yang, J. Y., Dalgleish, B. J., Zok, F. Z., and Evans, A. G. (1998). Processing and performance of an all-oxide ceramic composite. *J. Am. Ceram. Soc.* **81**: 2077–2086.
88. Evans, A. G., Marshall, D. B., Zok, F., and Levi, C. (1999). Recent advance in oxide–oxide composite technology. *Adv. Composite Mater.* **8**: 17–23.
89. Morgan, P. E. D., and Marshall, D. B. (1995). Ceramic composites of monazite and alumina. *J. Am. Ceram. Soc.* **78**: 1555–1563.
90. Davis, J. B., Marshall, D. B., and Morgan, P. E. D. (2000). Monazite-containing oxide/oxide composites. *J. Eur. Ceram. Soc.* **20**: 583–587.
91. Parthasarathy, T. A., Mah, T., and Matson, L. E. (1990). Creep behavior of an Al_2O_3–$Y_3Al_5O_{12}$ eutectic composite. *Ceram. Eng. Sci. Proc.* **11**: 1628–1637.
92. Mah, T., Parthasarathy, A., and Matson, L. E. (1990). Processing and mechanical properties of an Al_2O_3–$Y_3Al_5O_{12}$ (YAG) eutectic composite. *Ceram. Sci. Eng. Proc.* **11**: 1628–1637.
93. Waku, W., Ohtsubo, H., Nakagawa, N., and Kohtoku, Y. (1996). Sapphire matrix composites reinforced with single crystal YAG phases. *J. Mater. Sci.* **31**: 4663–4670.
94. Waku, W., Nakagawa, N., Wakamoto, T., Ohtsubo, H., Shimizu, K., and Kohtoku, Y. (1998). High-temerature strength and thermal stability of a unidirectionally solidified Al_2O_3/YAG eutectic composite. *J. Mater. Sci.* **33**: 1217–1225.

Handbook of Advanced Ceramics
S. Sōmiya *et al.* (Eds.)

CHAPTER 16

16.1 Functionally Graded Materials

LIDONG CHEN[1] and TAKASHI GOTO[2]
[1]*Shangai Institute of Ceramics, Academia Sinica, 1295 Ding-Xi Road, Shangai 200050, China*
[2]*Institute for Materials Research, Tohoku University, Katahira 2-1-1, Sendai 980-8577, Japan*

16.1.1 INTRODUCTION

In the development of new materials, attainment of homogeneity of the material characteristics is often attempted. The efforts in pursuit of uniformity in chemical composition, structure and texture have resulted in the continuing progress of material science and technology. In recent years, however, as the environments in which materials are used become more demanding, there are frequently cases in which the conventional homogeneous materials cannot withstand severe environments. For example, in aerospace applications such as turbines and combustion chambers, there is no approved homogeneous material that can withstand the prevailing temperatures (up to 1800°C in combustion chambers) at a usable stress level. To meet the increasing requirement for industrial materials, studies have also been conducted to design inhomogeneous composites, such as coated and joined materials. These inhomogeneous composites are characterized by having different characteristics on separate surfaces or parts, and therefore have two or more different functions within the given material. Unfortunately, however, these materials possess sharp boundaries, the existence of which often results in various undesirable behavior caused by the discontinuities in the physical and chemical characteristics at the boundary. The boundary separation in a coated material due to thermal stress is one typical example of such disadvantages.

In the 1980s, in an effort to develop thermal stress-resistant materials for aerospace applications, a new concept of materials was proposed to deal with the boundary problem [1, 2]. That is, a ceramic coating on metal or a ceramic/metal joined material with continuous texture was developed in order to increase the adhesion strength and minimize the thermal stress near the ceramic/metal boundary. This new type of material is termed as "functionally graded materials" (FGMs). Although the conceptual idea of gradient structure was initially advanced for composites and polymeric materials in the early 1970s [3, 4], there was no systematic experimental investigation of the design, fabrication and evaluation of the graded structure until the 1980s. In a general definition, an FGM is a composite consisting of two or more phases, structures or textures, which vary gradually in a certain direction [5]. As the result of variation in composition and/or structure, the functions also vary along with the graded direction within the material. For example, a ceramic/metal FGM consists of pure ceramic at the hotter end and pure metal at the cooler end, so that the different parts within a material are endowed with better resistance of ceramics to high temperatures and the better mechanical and heat-transfer properties of metal according to the different environments. A schematic of a ceramic/metal FGM for the relaxation of thermal stress is shown in Figure 16.1.1.

In this chapter, we briefly review the modeling and design, processing and fabrication, and applications of functionally graded materials. In the next section, after reviewing the general modeling and design procedure of FGMs,

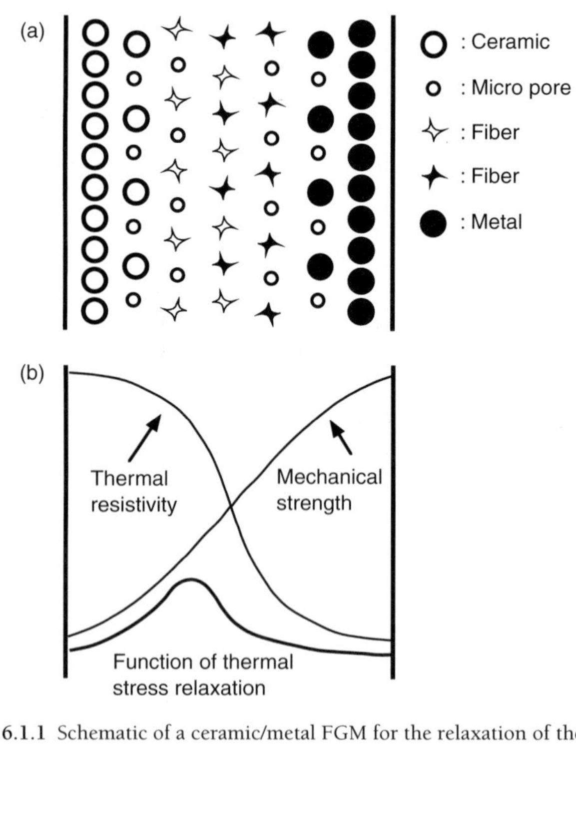

FIGURE 16.1.1 Schematic of a ceramic/metal FGM for the relaxation of thermal stress.

we discuss the microstructural characteristics of composites and the functions of composition distribution in an FGM. Next, the processes for fabricating FGMs are reviewed with emphasis on the chemical vapor deposition and powder processes. New results for powder configuration are also described. In the penultimate section, recent achievements in developing functionally graded materials including structural and functional materials, are reviewed. The recent expansion of FGM application to electrical, magnetic, optical and other fields is especially significant. Finally, in the last section some comments on the present problem and future developments in FGM research are given.

16.1.2 MODELING AND DESIGN

16.1.2.1 General Approach and Design Procedure

In the design of an FGM, two important problems have to be solved aside from the selection of the material system. One is how to predict the characteristics of an FGM for a given composition profile. The fact that the microstructure is strongly dependent on the composition and therefore also varies with position within an FGM means that the description of characteristics is very complex. Although many models to calculate microstructure-dependent thermophysical properties have been developed, there is no one all-around model for different materials and properties. Another problem is how to determine the optimum spatial dependence for a certain composition based on the description of characteristics. This can be regarded as the composition profile which best accomplishes the intended purpose of the material. Although the design of FGMs should deal with a wide range of issues such as thermal, electrical, and chemical properties, the contents of this section are limited to thermophysical properties because research in this field has been conducted extensively over the years.

Limiting the design of the material to the relaxation of thermal stress, the final purpose of the design is to determine the optimum composition distribution within a given material for a given thermal and mechanical environment. A useful approach to obtain such an optimum distribution function is based on a systems-analysis strategy [6–9], which has been developed by many researchers. Figure 16.1.2 illustrates an "inverse design procedure" developed on the basis of such a systems-analysis strategy [8, 9]. In this approach, the geometric structure and boundary condition are initially specified. Then, the material ingredients (constituents) are selected and the initial dependent mixture ratio is assumed. The temperature and thermal stress distributions are then calculated by employing the known thermophysical properties of materials. The

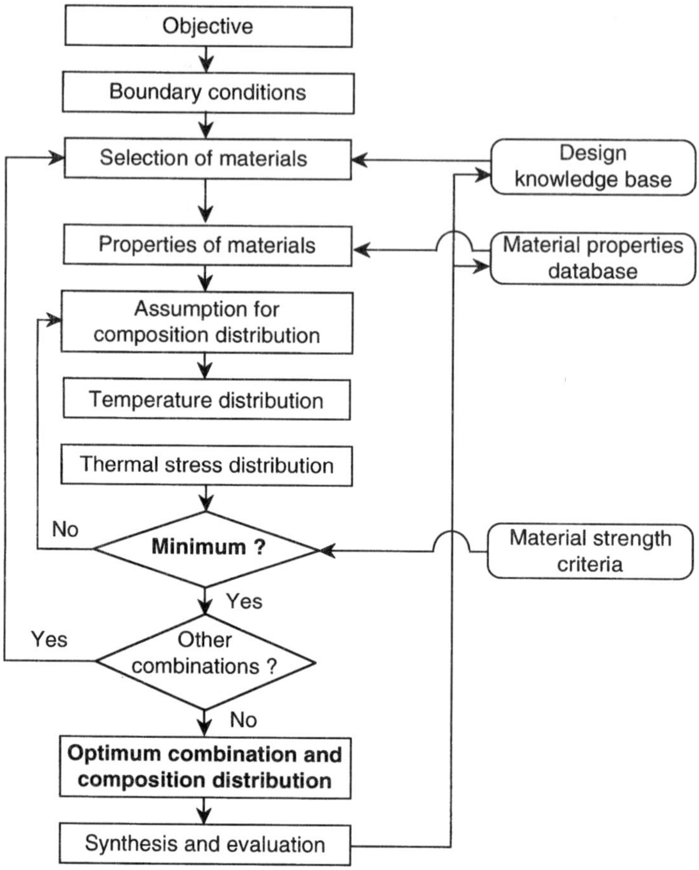

FIGURE 16.1.2 Flow chart for the "inverse design procedure" used for designing thermal stress relaxation FGM.

results are then compared with the material strength criteria to judge whether the optimum conditions have been obtained. If they have not, a modified calculation is re-performed by changing the mixture ratio profile. Such calculation routing is repeated until the optimum condition has been reached. In each routing, the composition distribution is modified based on the result of the former step of calculation. In this design procedure, the use of appropriate models for the microstructure-dependent material properties and the use of flexible compositional distribution functions are the key factors essential to obtain reliable results. These two subjects are discussed in the next subsections.

16.1.2.2 Distribution Functions of Composition

An FGM is characterized by having a graded composition and/or microstructure. Considering a general case of composition gradation, an assumption of the spatial distribution of their constituent phases is required. For an FGM having two constituents, which are denoted as A and B here, let us assume that the geometry is one-dimensional with the x-direction being the direction of the compositional gradient, and that the inner and outer surfaces are pure phase A and pure phase B with border regions of x_A and x_B, respectively. As one typical example, the local volume fraction of phase A, $f_A(x)$, can be presented as a continuous function as follows [6]:

$$f_A(x) = \left(\frac{x_B - x}{x_B - x_A}\right)^n \qquad (1)$$

where n is a variable parameter. Figure 16.1.3 illustrates the $f_A(x)$ function for several selected values of n with $x_A = 0$, and $x_B = 1$. It can be seen from Figure 16.1.3 that the curvature of the compositional distribution can be changed from concave upward to concave downward in a wide range by proper choice of n. In the "inverse design procedure" described above, "n" is the main parameter to be changed to search for the optimum conditions. Besides Eq. 1, several other basic functions have also been applied to describe the compositional function in an FGM. For example, a quadratic function was used by Markworth and Saunders [10]:

$$f_1(x) = a_0 + a_1x + a_2x^2 \qquad (2)$$

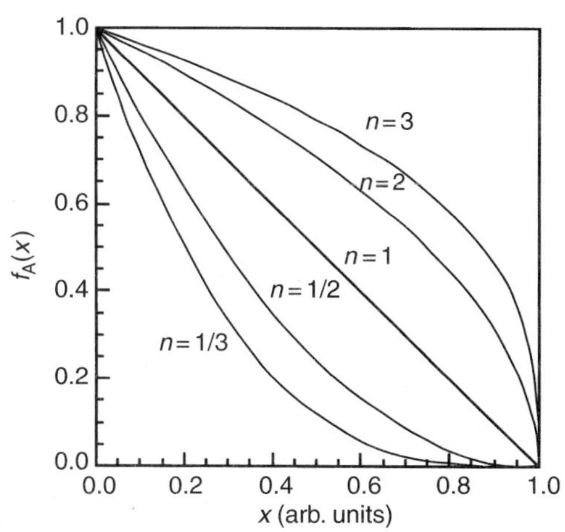

FIGURE 16.1.3 Plot of $f_A(x)$ versus x (Eq. 1) for several selected n values.

From an analytical point of view, Eq. 2 is much simpler to deal with than Eq. 1. However, Eq. 1 offers a greater range of composition profiles than Eq. 2.

Another frequently used approach to modeling the spatial variation of composition is to select $f_A(x)$ such that it changes discontinuously, that is, in a finite number of steps across the gradient direction. This would be appropriate for describing an FGM which is fabricated by bonding several layers of material that differ in composition from one to the next. This is a practical approach for fabrication experiments because step-wise gradation is more easily to be realized than the continuous texture approach from the viewpoint of process technology.

16.1.2.3 Models for Thermophysical Properties

After choosing the distribution function of composition, it becomes essential to predict the characteristics of an FGM with a given composition profile. Firstly, it is necessary to know the material characteristics for each composition (non-FGM composites). Many rules of mixtures have been developed to calculate the effective thermophysical properties of heterogeneous materials [11–16]. These rules are based on the structure and dispersion state.

However, these mixture rules are valid for only a limited number of typical and special cases of microstructures. The real structures are usually more complex. Figure 16.1.4 shows the typical manner of microstructure change in

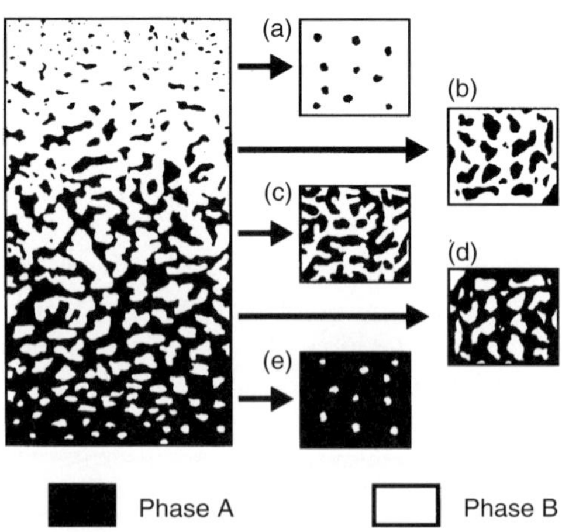

FIGURE 16.1.4 Typical manner of the microstructural change in an FGM.

an FGM. This particular FGM is shown to consist of two pure components (A and B), which are at opposite ends. Between these ends, there are five typical structures: (a) phase A randomly and discretely disperses in phase B; (b) A disperses in B with an aggregated structure; (c) A and B disperse in each other in intertwined networks, that is, in a percolation-like structure; (d) B conversely disperses in A with an aggregated structure; and (e) B randomly disperses in A.

To apply the mixture rules to the design of FGMs, much effort has been made to modify the mixture rules. Here, an effective medium theory [6, 17, 18] is described because of its moderately successful application to the modeling of FGMs. Consider a material having two components, denoted as A and B. Let P_A and P_B be the values of some particular property for pure A and pure B, respectively, and let their respective volume fractions be f_A and f_B, where $f_B = 1 - f_A$ if the material is fully dense. For an FGM, these volume fractions are dependent upon their position along the gradient direction. The well-known Voigt-type estimate for the effective value (P) can be represented by Eq. 3 for a typical case of longitudinal fiber dispersion:

$$P = f_A P_A + f_B P_B \tag{3}$$

which is an arithmetic mean. On the other hand, for the dispersion of cross fibers (another typical case), the Reuss-type estimate is given by

$$\frac{1}{P} = \frac{f_A}{P_A} + \frac{f_B}{P_B} \tag{4}$$

Unfortunately, these two expressions have only limited validity for the two dispersion states described above. In an effective medium theory, a more general expression for the randomly dispersed structure has been proposed:

$$P = f_A P_A + f_B P_B + f_A f_B Q_{AB} \tag{5}$$

where Q_{AB} is a function that depends on P_A, P_B, f_A, f_B and the microstructure. For the case in which the microstructure consists of spherical particles of constituent B embedded in a matrix phase A, several properties (e.g. thermal conductivity, thermal expansion coefficient, bulk modulus, and shear modulus) have been examined and the functional form of Q_{AB} has been given. For example, thermal conductivity, λ, can be represented as

$$\lambda = f_A \lambda_A + f_B \lambda_B + f_A f_B \frac{\lambda_A - \lambda_B}{\{3/[(\lambda_B/\lambda_A) - 1]\} + f_A} \tag{6}$$

The effective-medium approach is valid only for the random-dispersion structure including the cases in which phase B disperses in matrix phase A and phase A conversely disperses in matrix phase B. However, for the percolation-like structure, in which the identification of dispersion phase and matrix phase is difficult to determine, the effective-medium theory cannot be used directly. To deal with such a transition area, a newly developed type of fuzzy logic [19, 20] may be useful for describing the complex microstructure and thermophysical properties.

16.1.3 FABRICATION PROCESS

The fabrication process is one of the key items in FGM research. One of the important objectives in fabricating an FGM is to realize a well-controlled distribution of composition, texture, structure, and other elements according to the design. In general, such a gradient distribution can be realized by the use of appropriate preparation conditions, such as nominal composition and other processing parameters. The existing method of preparation for conventional homogeneous materials can be applied to the fabrication of FGMs but only after introducing necessary modifications. Here, some typical fabrication methods for FGMs are described.

16.1.3.1 Vapor Deposition Method

The vapor deposition (VD) method includes mainly chemical vapor deposition (CVD) and physical vapor deposition (PVD). In this method, a film or a plate-like material can be fabricated layer by layer through the process of deposition. The composition gradient along with the growth direction is comparatively easily obtained by directly changing the source composition during the deposition. For controlling the gradient microstructure, it may be helpful to change the deposition conditions such as gas pressure and/or substrate temperature because the microstructure of vapor-deposited materials is usually sensitive to such deposition conditions.

16.1.3.1.1 Chemical Vapor Deposition

The CVD method is a well-studied process, in which materials are yielded by the decomposition or chemical reactions of source gases at high temperatures. Hydride, bromide, chloride, and organometallic compounds are generally used for the source materials. When the source materials are in a liquid or solid state at room temperature, they are usually vaporized by heating and transported into a reactor (reaction chamber) by carrier gas. By continuously changing the mixture ratio of the source gases or by controlling the CVD conditions, such as deposition temperatures and gas pressures, an FGM with graded composition and/or graded microstructure can be obtained. In order to realize such continuous control, it is necessary to know the relationship between the deposition conditions and the composition or microstructure of the deposited materials. The microstructures are also dependent on the deposited composition, which makes the control of the gradient in thermophysical properties more difficult. So far, many systems of FGMs, such as SiC/C [21], TiC/C [22], BN/Si_3N_4 [23], and SiC/TiC [24], have been developed by CVD. Figure 16.1.5 shows a typical microstructure of SiC/C FGM prepared by CVD at 1500°C and 1.3 kPa on a graphite plate [21]. In this synthesis process, $SiCl_4$ and CH_4 were used as the source gases. During the process, the $SiCl_4/CH_4$ ratio was changed during the deposition to obtain a deposit with a graded composition.

Besides the CVD method, a similar method, so-called chemical vapor infiltration (CVI), is also used to fabricate FGMs. CVI is derived from CVD for use in surface coating. This process generally uses porous materials such as ceramic bodies, ceramic cloths having many residual pores as the matrix. Decomposition and/or chemical reaction of the source gases deposit a coating on the surface of the open pores or inside the porous substrate. A typical example is a two-dimensional (2D) fabric carbon fiber composite coated with SiC/C FGM [25]. The gas-tightness and oxidation-resistance of C/C composite are improved by the SiC/C FGM coating [26].

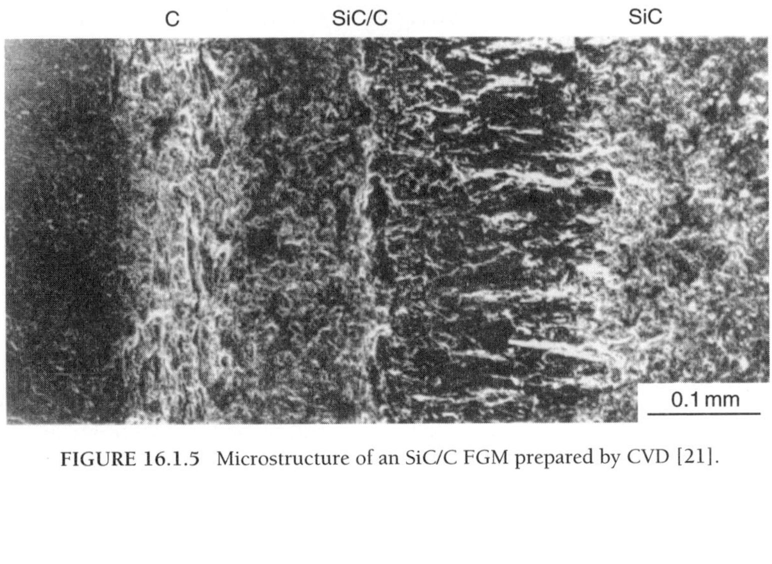

FIGURE 16.1.5 Microstructure of an SiC/C FGM prepared by CVD [21].

16.1.3.1.2 Physical vapor deposition

The PVD method is also a well-known technology for preparing thin films. Many PVD techniques, such as electron-beam PVD, ion plating, activated reactive evaporation and sputtering, have been applied to fabricate FGMs. Similar to CVD, a gradient composition and/or gradient structure can be realized by directly changing the vapor composition and/or other deposition conditions. Up to now, Ti/TiN, Ti/TiC FGMs have been prepared by PVD [27, 28]. Recently, a new sputtering technique called helicon plasma deposition has been used to fabricate graded TiO_2/SiO_2 multilayer films for an optical filter application [29, 30]. Figure 16.1.6 shows a schematic of helicon plasma sputtering equipment. This sputtering system utilizes a pair of helicon cathodes in which fused silica and sintered TiO_2 targets are set. In this film, the refractive index is graded within the thickness direction, and the graded refractive index is realized by changing the TiO_2/SiO_2 ratio. The TiO_2/SiO_2 ratio in each layer is obtained by changing the RF power applied to the TiO_2 and SiO_2 targets during deposition. In the PVD technique, it is possible to deposit not only thermodynamically stable phases but also metastable phases depending on the deposition conditions. Therefore, a wider range of composite states can be obtained by PVD than by CVD.

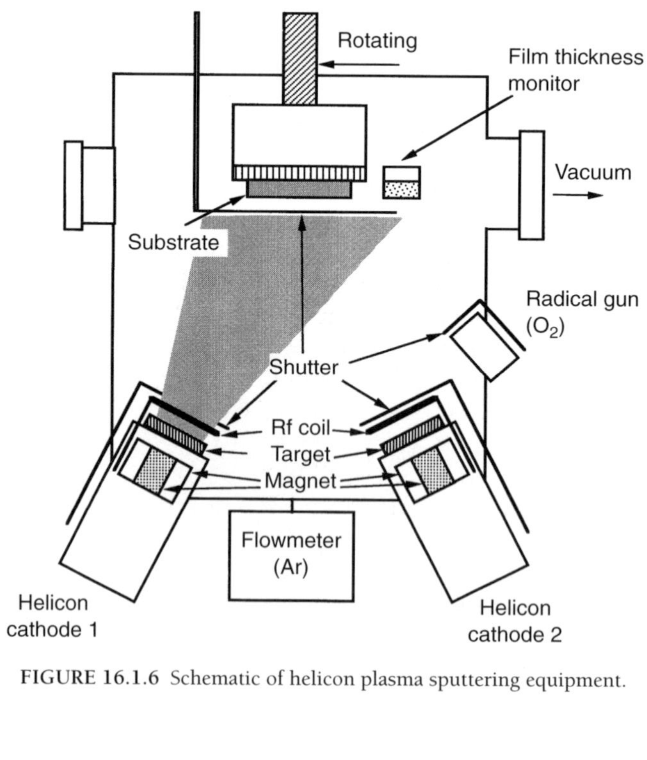

FIGURE 16.1.6 Schematic of helicon plasma sputtering equipment.

16.1.3.2 Solid Phase Method

16.1.3.2.1 Fabrication Methods for Green Bodies with Compositional Gradient

A solid phase technique for fabricating ceramic- or metal-based FGMs consists of two important processes: fabricating a green body with graded composition and then sintering the green body to a dense bulk. Because the compositional profile for most cases are determined by the first process, study of the fabrication method of the green body is a key issue in FGM research.

A powder stacking technique is a typical one used for preparing a graded green body. In this technique, a graded powder compact is produced by sequentially layering the powder mixture with various compositions, which are then pressed. Most of the powder stacking reported to date has been carried out by hand, by which only a laminated type of FGM can be fabricated and purely artificial control of the material's functions is difficult. To overcome this difficulty, many new techniques such as powder spray-stacking technology, centrifugal technology, and computer-assisted technology have been developed. In the powder spray-stacking technique [31], a powder suspension having a varied mixture ratio is sprayed using a roller pump or a compressed air nozzle onto the preheated substrate. The resulting deposits are then dried to obtain a green body with a graded composition. The compositional distribution can be controlled by changing the suspension composition using a computer-aided controller system. In the centrifugal technique [32], a mixture of source powder is supplied to the center of a rapidly rotating centrifuge. The mixture ratio of the powder is computer regulated. The mixed powder is deposited on the inner wall by the centrifugal force and the deposited layer-like compact serves as a green compact for sintering. Recently, a particle configuration process [33], in which individual single particles can be positioned at the allotted sites according to the theoretical design, has also been developed. After this technique is fully developed, accurate control of the composition will become easy to realize.

A slurry technique is also a popular method used to fabricate FGM green bodies. One typical, well-studied technique using slurry is the so-called film layering method. In this process, slurries with different powder compositions are converted to thin films by slip casting. An FGM green body is obtained by layering these films. The FGMs of ZrO_2/Ni and other ceramic/metal systems have been prepared by the film layering and pressureless sintering [34]. Besides the thin film layering method, a process using slurry to form thick green bodies based on filtration and sediment operation has also been developed [35]. In this process, composition-graded cake layers can be obtained by continuously changing the mixture ratio of two kinds of slurries that are fed into the tank and filtrated.

16.1.3.2.2 Sintering Method

Although all of the conventional sintering methods such as pressureless sintering, hot press and hot isotactic pressing (HIP) can be used for the fabrication of FGMs, the FGM sintering process is more difficult than the sintering of homogeneous materials. Typical common problems are the appearance of cracks, the deformation of the sintered body, and the difficult densification of all parts. These difficulties are associated with the differences in the sintering characteristics and thermophysical properties of the different parts, such as sintering temperature, shrinkage, density and thermal expansion coefficients. Many efforts have been made to solve these process problems. For example, the shrinkage rate of the powder compact has been controlled by blending fine and coarse particles. The heating rate is also a sensitive parameter for use in controlling the deformation and/or cracking.

In order to densify an FGM having different parts with different sintering temperatures, a special temperature gradient sintering technique has been developed. That is, the portion containing more of the higher sintering temperature component is sintered at a higher temperature and the portion containing more of the lower sintering temperature component is sintered at a lower temperature. Sintering is carried out under a prescribed temperature gradient. One way to achieve this temperature gradient is to apply additional heat energy by directing a laser beam or infrared beam onto one side of the powder compact [36]. Another newly developed temperature gradient sintering technique is spark plasma sintering (SPS) [37]. In this method, a pulse electric current is applied to the subject powder compact and a specially shaped graphite die is used. Figure 16.1.7 shows a schematic of the SPS equipment. The diameter of the graphite die is varied from the top to the bottom, and a gradient temperature profile within the powder compact can be obtained. The gradient temperature enables preparation of a dense FGM body, in which the component materials have different sintering temperatures. Many FGMs such as polyimide/Cu and PSZ/Ni have been successfully prepared by this technique [5, 37].

16.1.3.3 Liquid Phase Method and Others

Since the early stage of FGM research, many studies on the liquid phase method, such as sol–gel, electrodeposition, and plasma spray for the preparation of FGMs have been done because the liquid phase method is convenient and facilitates control of the graded composition. Cu/CuZn and Cu/CuNi FGM films with a thickness of 50–200 μm have been prepared on a Cu substrate by the electrodeposition technique [38]. A glass rod having a radially graded refractive index was

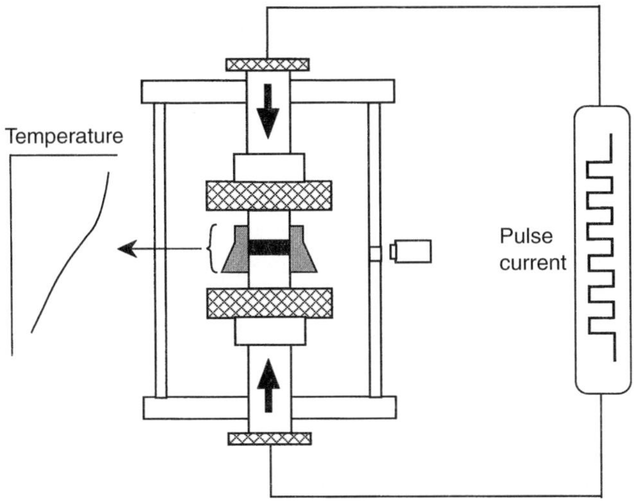

FIGURE 16.1.7 Schematic of spark plasma sintering equipment.

prepared using a sol–gel process [39]. This process utilizes metal alkoxides of the $Si(OCH_3)_4-Ge(OC_2H_5)_4$ system and the $Si(OCH_3)_4-Ti(O\text{-}n\text{-}C_4H_9)_4$ system as the source materials of SiO_2-GeO_2 and SiO_2-TiO_2 glasses, respectively. Immersion of the rod-shaped wet gels in an acidic solution results in leaching out of the dopant (Ge and Ti). The dopants concentration gradient in the leached gel contributes to the formation of the concentration gradient in the densified glasses (SiO_2-GeO_2 and SiO_2-TiO_2).

A plasma spray method is also a well-studied method for FGM fabrication. In this method, the source powder (metal and/or ceramic) is transported to the plasma jet by a torch nozzle. Then the source materials in a molten state by plasma are sprayed onto a substrate to form a coating. The composition gradient within the sprayed coating can be realized by changing the composition of the supply source. To date, various kinds of ceramic/metal, ceramic/ceramic and metal/metal FGMs (such as ZrO_2/NiCrAl, NiCr/Al and NiCrAlY/YSZ) have been fabricated by the plasma spray technique [40, 41].

Some other new processes for fabricating bulk, coating, thin film FGMs have also been reported. For example, eutectic bonding has been developed for making metal–intermetallic compound FGMs [42]. A dissolution–diffusion method has been employed to prepare polymer FGM [43]. A copolymerization method has been developed for preparing polymer optical fiber with a graded refractive index [44]. The electrophoretic deposition method has been used for forming graded ceramic green bodies by controlling the pH values of the suspension [45].

Although many studies have been reported on the FGM fabrication process, at present it is still difficult to prepare an ideal FGM with a continuous distribution of composition and/or structure in a designed profile. It is expected that new developments in the FGM process will greatly stimulate the investigation and application of FGMs.

16.1.4 APPLICATIONS AND FUTURE DEVELOPMENT

16.1.4.1 Structural Materials

Since the concept of FGMs was first proposed, much research has been done, especially focusing on the effective relaxation of thermal stress [46–49]. Nowadays, FGM research is being carried out in a wide variety of materials such as structural materials, electronic materials, optical materials and biomaterials [49–52].

Studies on heat-shielding materials, including thermal barrier coatings, have been the most prevalent in the past 10 years or more. One remarkable achievement is the development of CVD-SiC/C FGM (Fig. 16.1.5) [21, 53]. CVD-SiC/C FGM was first reported by Sasaki and Hirai in 1990. The SiC/C FGM (0.2–0.8 mm in thickness) was deposited on a graphite substrate by using an $SiCl_4-CH_4-H_2$ system. The gradient composition was easily achieved by controlling the $SiCl_4$ and CH_4 gas flow rates. This CVD SiC/C FGM showed superior thermal barrier characteristics and thermal shock resistance compared with homogeneous SiC. SiC/C FGM coatings were also applied to the three-dimensional C/C composites. The SiC/C FGM coating on a hemispherical C/C composite was evaluated as to exposure to a supersonic gas flow at a temperature of approximately 1627°C, a simulated re-entry environment of the nose-cone part of the spaceplane. In this evaluation experiment, an ordinary SiC-coated sample broke on only one exposure to this environment. However, the SiC/C FGM-coated sample showed no discernible change in its structure even after 10 exposure cycles [26]. SiC/C FGMs were also coated on a plate-like C/C composite by Nippon Oil Company. These samples were evaluated together with several non-FGM SiC/C and SiC/C–C composites by the high performance material experiment (HIPMEX) under an actual re-entry environment of spacecraft. In the HIPMEX experiment, the SiC/C and SiC/C–C specimens were mounted on the surface of the re-entry capsule, which was launched out of the atmosphere, then re-entered and finally landed. It was found that the SiC/C FGMs had the best performance in heat resistance, mechanical shock resistance and mechanical strength among all the specimens [54].

16.1.4.2 Functional Materials

Recently, basic research has been conducted for creating new functions through FGM technology. Some research has emphasized the clarification of the mechanism of new functions made possible by FGMs and the discovery of new properties by the formation of graded compositions and/or graded structures in metallic, ceramic, polymeric, plant and biological materials.

Dielectric materials have a wide range of industrial applications as the electrical resistant substrate for the transportation of high frequency waves. Nishida [55] designed and fabricated a new composite dielectric substrate, in which the external part was designed with a low conductivity while the central part was conceived with a high-conductivity area. Such graded substrates allow minimization of the loss of high-frequency signals and at the same time ensure the availability of a large substrate width for circuit print. Yamada *et al.* [56] designed and prepared a transducer with a piezoelectrically graded structure, and found that the graded-type transducer performed better than the conventional one.

A new concept to control electronic transport behavior by FGMs has been reported for perovskite-type $(La_{1-x}Ca_x)MnO_3$. For the electronic device application of $(La_{1-x}Ca_x)MnO_3$, it is necessary to independently control the metal–insulator transition temperature (T_t) and $d\rho/dT$. However, this is difficult for the homogeneous oxide. Taguchi [57] investigated the electrical properties of $(La_{1-x}Ca_x)MnO_3$ with graded composition and found that the relationship of T_t–x and $(d\rho/dT)$–x can be independently controlled by the composition gradation. Figure 16.1.8 shows the relationship between $d\rho/dT$ and composition (x) of graded and homogeneous $(La_{1-x}Ca_x)MnO_3$. In Figure 16.1.8 the x value of graded $(La_{1-x}Ca_x)MnO_3$ refers to the average composition of a bulk sample. The $d\rho/dT$ of a step-wise gradient $LaMnO_3$-$CaMnO_3$ sintered body can be controlled by changing the gradient of composition profile.

In the field of optical materials, it was well known that an optical fiber with a continuously changing refractive index in the radius direction, in which the refractive index at the core is higher than that near the sheath, has a higher transmission capacity than a homogeneous fiber [58]. Organic and inorganic optical fibers with graded refractive indices have been developed and applied industrially [44, 59]. Another novel effectiveness of refractive index gradation was found in optical filter design [29, 30]. Figure 16.1.9 compares the designs of a conventional TiO_2/SiO_2 multilayer film (a) and a novel graded multilayer film (b) for an optical filter, and Figure 16.1.10 compares their transmittance spectra. They have the same reflection band with the center at about 730 nm. However, the transmittance spectrum of the usual alternative multilayer film has many sidelobes outside the reflection band (Fig. 16.1.10a), which causes the transport efficiency to greatly decrease. The calculation results indicated that

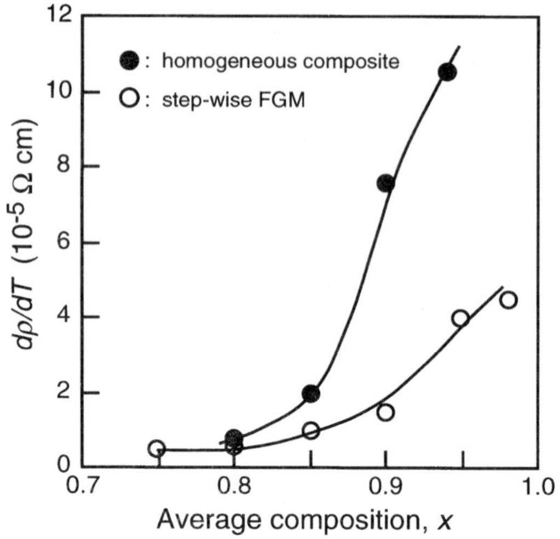

FIGURE 16.1.8 Relationship between $d\rho/dT$ and Ca content (x) for $(La_{1-x}Ca_x)MnO_3$.

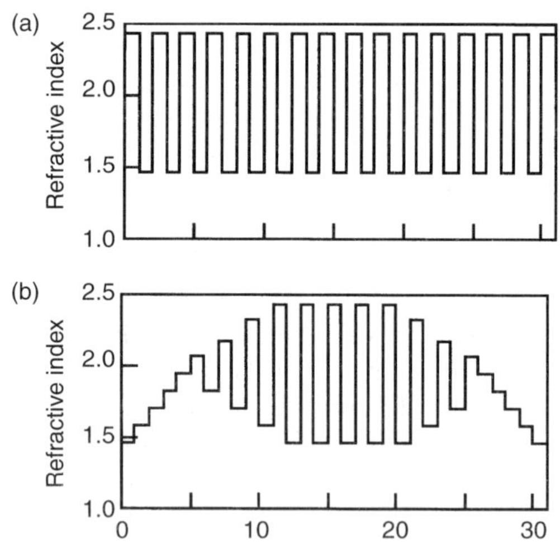

FIGURE 16.1.9 Refractive index profiles for TiO_2/SiO_2 multilayer filter (a) and TiO_2/SiO_2 graded multilayer filter (b).

the introduction of the layers with graded refractive index into the multilayer film can significantly eliminate the sidelobes (Fig. 16.1.10b). A TiO_2/SiO_2 FGM film has been prepared by helicon plasma deposition and its transmittance spectrum was found to agree well with the calculated results.

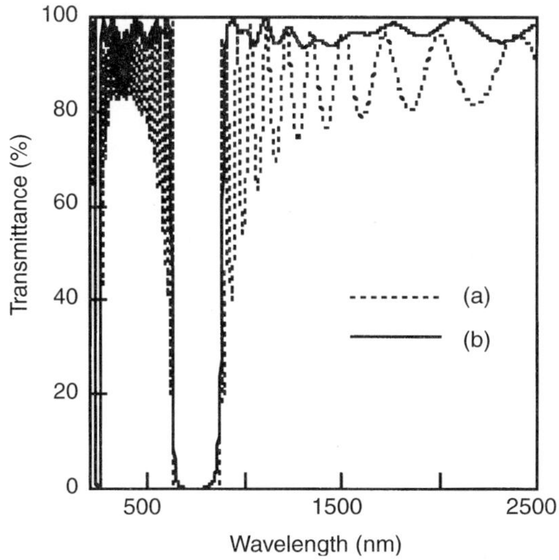

FIGURE 16.1.10 Comparison of the transmittance spectra of TiO_2/SiO_2 multilayer filter (a) and TiO_2/SiO_2 graded multilayer filter (b).

In the field of biomaterials, an FGM technique is being applied to combine the mechanical properties and biocompatibility within a single material. A bioactive graded Ti–6Al–4V alloy was developed through NaOH treatment and subsequent heat treatment. A bonelike apatite with graded structure can be formed on the alloy substrate in the environment of the human body. This graded structure provides a strong interfacial bonding strength between the apatite layer and the alloy substrate, thereby making the alloy tightly bonded to and integrated with living bone through the apatite layer [60].

Another Ti-based graded implant, Ti–Ti/20HAP (20% hydroxyapatite), has also been developed by the SPS sintering technique [61]. In this implant, one end is pure Ti and the other consists of 20% hydroxyapatite; between them, the hydroxyapatite content changes gradually. This implant has a higher mechanical strength at the pure Ti end and a higher biocompatibility at the Ti/20HAP end. This type of implant is expected to apply to the jawbone and teeth.

16.1.5 FINAL COMMENTS

The FGM concept was originated in an effort to reduce thermal stress for thermal barrier materials. Subsequently, much research has been done focusing on the thermal and mechanical aspects of FGMs. Recent research has revealed that the concept of FGM can also be advantageously extended to electronic, optical, biomedical and other fields. Many interesting and unique concepts and experimental results have been forthcoming to appear not only in structural materials but also in other various fields.

Generally, the most popular applications of the FGM concept are based on combining two or more incompatible functions into a given material (such as high thermal resistivity and excellent mechanical properties) in order to suit a particular circumstance. However, many phenomena and experimental results, such as the control of electrical properties of perovskite oxide and the optical filter characteristics of graded TiO_2/SiO_2 film mentioned above, cannot be explained by such simple combination of two functions. In addition, recent research has revealed that natural plants and living bodies have many intelligent properties due to their graded structures. Although the details of the mechanism involved remain to be clarified, the results to date encourage us to create new functions through gradation techniques. Especially, the establishment of FGM design theory for various functions (including the effects of graded structure on the properties of materials), and the development of new FGM processes are most important for the development of new FGMs.

REFERENCES

1. Niino, M., Kumakawa, A., Hirano, T., Sumiyoshi, K., and Watanabe, R. (1985). "Proc. 36th Congress of the International Astronautical Federation", Stockholm, p. 1.
2. Niino, M., Hirai, T., and Watanabe, R. (1987). *J. Jpn. Soc. Compos. Mater.* **13**: 257.
3. Bever, M. B., and Duwez, D. E. (1972). *Mater. Sci. Eng.* **10**: 1.
4. Shen, M., and Bever, M. B. (1972). *J. Mater. Sci.* **7**: 741.
5. Hirai, T. (1996). *Materials Science and Technology, Processing of Ceramics, Part 2*, p. 293, Brook, R. J. ed., Weinheim: VCH Verlagsgesellschaft mbH.
6. Kawashima, K., Hirano, T., and Niino, M. (1990). *Space Applications of Advanced Structural Materials*, European Space Agency.
7. Watanabe, R., and Kawasaki, A. (1990). "Proc. 1st International Symposium on FGM", p. 107, FGM Forum, Japan.
8. Hirano, T., Yamada, T., Teraki, J., Niino, M., and Kumakawa, A. (1988). "Proc. 16th International Symposium on Space Tech. and Sci.", p. 375, AGNE, Tokyo.
9. Niino, M., and Maeda, S. (1990). *ISIJ Int.* **30**: 699.
10. Markworth, A. J., and Saunders, J. H. (1995). *Mater. Lett.* 22: 103.
11. Torquato, S. (1991). *Appl. Mech. Rev.* **44**: 37.
12. Becker, R., and Richmond, O. (1994). *Model. Simul. Mater. Sci. Eng.* **2**: 439.
13. Bergman, D. J., and Stroud, D. (1992). *Solid State Phys.* **46**: 147.
14. Nan, C. W. (1993). *Prog. Mater. Sci.* **37**: 1.
15. Chernikov, A. A., and Rogalsky, A. V. (1994). *Chaos* **4**: 35.
16. Wimmer, J. M., Graham, H. C., and Tallan, N. M. (1974). *Electrical Conductivity in Ceramics and Glass*, p. 169, Tallan, N. M. ed., New York: Marcel Dekker.

17. Fan, Z., Tsakiropoulos, P., and Miodownik, A. P. (1994). *J. Mater. Sci.* **29**: 141.
18. Christensen, R. M. (1979). *Mechanics of Composite Materials*, p. 316, New York: Wiley.
19. Tanaka, K., Tanaka, Y., Watanabe, H., Poterasu, V. F., and Sugano, Y. (1993). *Computer Mech. Appl. Mech. Eng.* **109**: 377.
20. Hirano, T., Teraki, J., and Yamada, T. (1990). Proc. 1st Inter. Symp. FGM", p. 5, FGM Forum, Japan.
21. Hirai, T., and Sasaki, M. (1991). *Ceram. Int.* **17**: 275.
22. Sasaki, M., Hiratani, T., and Hirai, T. (1993). *Ceramic Transactions, Vol. 34—Functionally Gradient Materials*, p. 369. Holt, J. B., Koizumi, M., Hirai, T., and Munir, Z. A. eds., American Ceramic Society.
23. Kowbel, W. (1993). *Ceramic Transactions, Vol. 34—Functionally Gradient Materials*, p. 237, Holt, J. B., Koizumi, M., Hirai, T., and Munir, Z. A. eds., American Ceramic Society.
24. Kawasaki, A., and Watanabe, R. (1990). *J. Jpn. Soc. Powder Powder Metall.* **37**: 287.
25. Suemitsu, T., Matsuzaki, Y., Fujioka, J., Uchida, M., Sohda, Y., Kude, Y., Uemura, S., Kuroda, Y., Ueda, S., and Moro, A. (1993). *Ceramic Transactions, Vol. 34—Functionally Gradient Materials*, p. 315, Holt, J. B., Koizumi, M., Hirai, T., and Munir, Z. A. eds., American Ceramic Society.
26. Sohda, Y., Kude, Y., Uemura, S., Saitoh, T., Wakamatsu, Y., and Niino, M. (1993). *Ceramic Transactions, Vol. 34—Functionally Gradient Materials*, p. 125, Holt, J. B., Koizumi, M., Hirai, T., Munir, Z. A. eds., American Ceramic Society.
27. Niino, M., and Chen, L. (1993). *Hyomen* (in Japanese) **31**: 519.
28. Shinohara, Y., Imai, Y., Ikeno, S., Shiota, I., and Fukushima, T. (1993). *Ceramic Transactions, Vol. 34—Functionally Gradient Materials*, p. 255, Holt, J. B., Koizumi, M., Hirai, T., Munir, Z. A. eds., American Ceramic Society.
29. Wang, X. R., Masumoto, H., Someno, Y., and Hirai, T. (1998). *Appl. Phys. Lett.* **72**: 3264.
30. Wang, X. R., Masumoto, H., Someno, Y., Chen, L. D., and Hirai, T. (2000). *J. Mater. Res.* **15**: 274.
31. Watanabe, R., Kawasaki, A., and Takahashi, H. (1991). *Mechanics and Mechanisms of Damage in Composites and Multi-Materials*, p. 285, Baptiste, D. ed., Mater. Eng. Publisher.
32. Ilschner, B. (1990). "Proc. 1st International Symp. FGM", p. 101, FGM Forum, Japan.
33. Watanabe, R., Ichiki, K., Hayama, T., and Kawasaki, A. (1999). *Mater. Sci. Forum* **308–311**: 19.
34. Takabe, H., Teshima, T., Nakashima, M., and Morinaga, K. (1992). *J. Ceram. Soc. Jpn.* **100**: 387.
35. Iwata, M., Yi, W. D., Nakamura, M., and Toyama, S. (1992). *J. Soc. Powder Technol. Jpn.* **39**: 762.
36. Kawasaki, A., and Watanabe, R. (1990). *J. Soc. Powder Technol. Jpn.* **37**: 287.
37. Omori, M., Kawahara, M., Sakai, H., Okubo, A., and Hirai, T. (1994). *J. Jpn Soc. Powder Powder Metall.* **41**: 649.
38. Merk, N., Ding, X., Guo, X., and Ilschner, B. (1993). *Ceramic Transactions, Vol. 34—Functionally Gradient Materials*, p. 279, Holt, J. B., Koizumi, M., Hirai, T., Munir, Z. A. eds., American Ceramic Society.
39. Konishi, S., Shingyouchi, K., and Makishima, A. (1988). *J. Non-Cryst. Solids.* **100**: 511.
40. Eroglu, S., Birla, N. C., Demirci, M., and Baykara, T. (1993). *J. Mater. Sci.* **12**: 1099.
41. Fukushima, T., Kurioda, S., and Kitahara, S. (1993). *Proc. 1st International Symp. FGM*, p. 145, FGM Forum, Japan.
42. Kirihara, S., Tomota, Y., and Tsujimoto, T. (1997). *Mater. Trans. JIM* **38**: 650.
43. Agari, Y. (1998). *Kagaku To Kogyo* (in Japanese) **72**: 205.
44. Koike, Y., Tanio, N., Nihei, E., and Ohtsuka, Y. (1989). *Polym. Eng. Sci.* **29**: 1200.
45. Zhao, C., Vandeperre, L., Vleugels, J., and Van Der Biest, O. (2000). *Brit. Ceram. Trans.* **99**: 284.

46. Hirai, T. (1993). *Ceramic Transactions, Vol. 34—Functionally Gradient Materials*, Holt, J. B., Koizumi, M., Hirai, T., and Munir, Z. A. eds., Westerville, OH: American Ceramic Society.
47. Suresh, S. (1994). Proc. 3rd International Symp. Struct. Funct. Gradient Mater., p. 711, Ilschner, B., and Cherradi, N. eds., Presses Polytechniques et Universitaires Romandes.
48. Rodel, J. (1994). Proc. 3rd International Symp. Struct. Funct. Gradient Mater., p. 712, Ilschner, B., and Cherradi, N. eds., Presses Polytechniques et Universitaires Romandes.
49. Miyamoto, Y., Niino, M., and Koizumi, M. (1997). *Functionally Graded Materials 1996*, p. 1, Shiota, I., and Miyamoto, Y. eds., Amsterdam: Elsevier.
50. Rodel, J., and Neubrand, A. (1997). *Functionally Graded Materials 1996*, p. 9, Shiota, I., and Miyamoto, Y. eds., Amsterdam: Elsevier.
51. Niino, M., and Chen, L. (1994). "Proc. 12th International Conf. Thermoelectrics", p. 527, Matsuura, K. ed., IEE of Japan.
52. Niino, M., and Koizumi, M. (1994). "Proc. 3rd International Symp. Struct. Funct. Gradient Mater.", p. 601, Ilschner, B. and Cherradi, N. eds., Presses Polytechniques et Universitaires Romandes.
53. Sasaki, M., and Hirai, T. (1991). *J. Physique IV, Collogue C2* **1**: 649.
54. Kiuchi, N., Kude, Y., Ido, Y., and Sohda, Y. (1997). "Proc. FGM'97 Jpn", p. 95 FGM Forum.
55. Nishida, T. (1999). *Mater. Integra.* **12**: 21.
56. Yamada, K., Sakamura, J., Nakamura, K., and Hongo, H. (1997). Technical Report of Inst. Electron Commun. Eng. Jpn, US97–49, p. 31.
57. Taguchi, H. (1997). Research Report 1997: Chemistry and Physics of Functionally Graded Materials, Grant-in-Aid for Scientific Research on Priority Area, No. 274, MESSC, p. 157.
58. Kawakami, S., and Nishizawa, J. (1965). *Proc. IEEE* **53**: 2148.
59. Koike, Y., Ishigure, T., and Nihei, E. (1995). *J. Lightwave Technol.* **13**: 1475.
60. Yan, W. Q., Nakamura, T., Kobayashi, M., Kim, H. M., Miyaji, F., and Kokubo, T. (1997). *J. Biomed. Mater. Res.* **37**: 265.
61. Watari, F., Yokoyama, A., Saso, F., Uo, M., Matsuno, H., and Kawasaki, T. (1997). Research Report 1997: Chemistry and Physics of Functionally Graded Materials, Grant-in-Aid for Scientific Research on Priority Area, No. 274, MESSC, p. 225.

Handbook of Advanced Ceramics
S. Sōmiya *et al.* (Eds.)

17.1 Intelligent Ceramics—Design and Development of Self-Diagnosis Composites Containing Electrically Conductive Phase

HIDEAKI MATSUBARA, YOSHIKI OKUHARA, ATSUMU ISHIDA, MASAYUKI TAKADA and HIROAKI YANAGIDA
Japan Fine Ceramics Center, 2-4-1 Mutsuno, Atsuta-ku, Nagoya 456-8587, Japan

17.1.1 INTRODUCTION

Recently, the study of "intelligent materials" has grown enormously all around the world. Means of introducing self-diagnosis or self-monitoring functions into modern materials are wide-ranging and various. For example, piezoelectric ceramics, shape memory alloys and optical fibers can all be used to give materials a certain degree of "intelligence". In heavy construction, however, large volumes of material are required, so it is essential to avoid high complexity and high cost. The use of electrical conductivity or resistance is perhaps the most straightforward way of obtaining a relatively simple material system as compared with other high-tech materials.

Yanagida *et al.* have reported fracture detection using fiber-reinforced plastics (FRP) embedded in structural materials as a reinforcement [1]. The electrical resistance change of carbon-fiber glass-fiber reinforced plastics (CFGFRP) indicates the local or partial damage in a composite prior to its fatal fracture. Although the CFGFRP showed appreciable resistance change in the strain range above 0.7–1.5% (due to fracture of the carbon fiber) the detectable strain level is too large for diagnosing local damage in structural materials such as concrete (where typical strains are about 0.05%). It is therefore necessary to develop a self-diagnosis material with higher sensitivity in the low strain region if we are to achieve simple health monitoring system.

In our recent work [2–7], composites containing a continuous structure of electrically conductive particles, or a so-called "percolation structure", have been designed and fabricated. We have successfully the FRP composites and the ceramics matrix composites (CMCs) with high performances in the low-strain region due to "percolation structure" of conductive particles in a matrix phase. These self-diagnosis functions were evaluated from the changes in resistance with applied strain.

In the present paper, the electrically conductive composites having continuous structure of conductive particles with the continuous or network structure, were designed and fabricated in the FRP and the CMC. These self-diagnosis functions were evaluated from the measurement of resistance changes with applied strain in normal or cyclic loading tests. The practicability of the function was examined in bending tests for mortar specimen embedding FRP. Percolation phenomena with the second phase of various aspect ratios were also studied by two-dimensional computer simulation.

17.1.2 SELF-DIAGNOSIS FUNCTION OF FRP

Figure 17.1.1 is the schematic drawing of the structural design for conductive FRP which basically consists of vinyl ester resin (Showa High Polymer Co., Ltd. RIPOXY R-804) and glass fiber (Asahi Glass Fiber Co, Ltd. ER2220). The carbon fiber (pitch-based CF, Toho Rayon Co., Ltd. BESFIGHT UM63) introduced in replacement of a part of glass fiber as shown in Figure 17.1.1a forms conductive path and enhances strength of the composite in the longitudinal direction. The carbon particles (graphite, SEC Co., Ltd. SPG5) dispersed in a part of matrix

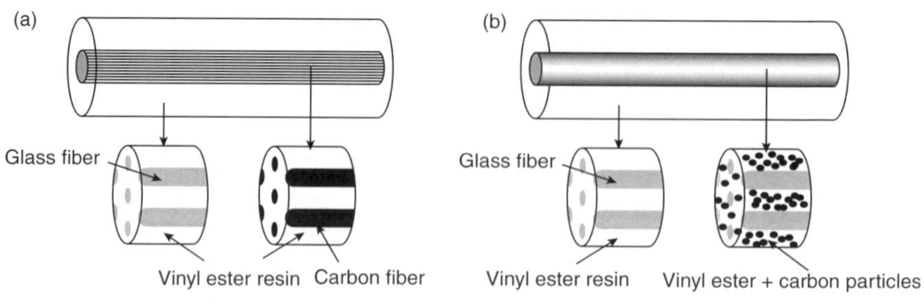

FIGURE 17.1.1 Schematic drawings of the structural design for CFGFRP (a) and CPGFRP (b).

create conductivity due to the formation of percolation structure as shown in Figure 17.1.1b. The composite containing carbon particles was indicated by carbon-particles glass-fiber reinforced plastics (CPGFRP).

The self-diagnosis functions of these materials were evaluated by simultaneous measurements of stress and electrical resistance change as a function of applied strain in tensile loading tests. The resistance change was defined as relative change in resistance $(R - R_0)/R_0$, indicated by $\Delta R/R_0$ in which R_0 denotes initial resistance before loading. The types of loading were in two ways: (a) a normal tensile test until specimen fracture; and (b) cyclic loading–unloading test below the maximum stress level.

Figures 17.1.2 shows the changes in electrical resistance and applied stress for CFGFRP and CPGFRP as a function of applied strain in tensile test. The stresses in both specimens increased linearly in proportion to the strains until fracture of carbon fiber or glass fiber. The CFGFRP indicated slight change in resistance below 0.5% strain and tremendous change around 0.7% strain; namely, the resistance of CFGFRP exhibited non-linear response to applied strain as shown in Figure 17.1.2a. The initial resistance R_0 for CPGFRP was higher than that for CFGFRP because of slight electrical contact between carbon particles in percolation structure. As can be seen from Figure 17.1.2b, the CPGFRP indicated linear increase in resistance with increasing tensile strain. The response of the resistance to applied strain appeared around the strain of 0.01% (100 μ strain) or less. Comparing Figure 17.1.2a and b shows that the CPGFRP possesses higher sensitivity at the small strain level and wider detectable strain range than CFGFRP. These results mean that the percolation structure formed with carbon particles enables more sensitive and adaptable diagnosis of damage than the structure consisting of carbon fiber. The fine response of resistance for CPGFRP were attributed to local break in

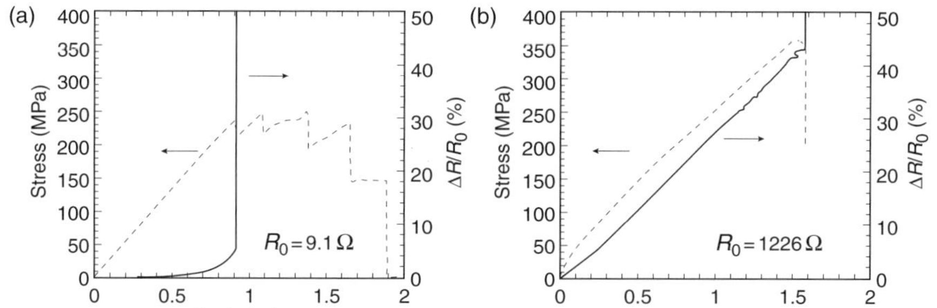

FIGURE 17.1.2 Changes in electrical resistance (solid line) and applied stress (dashed line) as a function of applied strain in tensile tests for CFGFRP (a) and CPGFRP (b).

electrical contact between carbon particles because of microcrack formation in the matrix or of rearrangement of the percolation structure under tensile stress.

Figure 17.1.3 shows resistance change and applied strain as a function of time in cyclic loading test for CPGFRP. Specimens were loaded and unloaded cyclically under a progressive increase in stress. It can be seen that the resistance change corresponded well with variation in strain. It is worthy of notice that a part of resistance change in the elongated composites remained after unloading despite elastic deformation. The residual resistance appeared after the application of 0.2% strain, and then increased with increase in maximum applied strain. The maximum resistance change during loading indicated by ΔR_{max} and the residual resistance change after unloading denoted by ΔR_{res} were arranged with respect to maximum strain applied in the past as shown in Figure 17.1.4. The change in residual resistance correlated closely with previous maximum strain, suggesting that the CPGFRP has the ability to memorize maximum applied through the measurement of residual resistance. It seems likely that this phenomenon is due in part to the irreversible change in percolation structure. Although the elongation of CPGFRP removed elastically after unloading, the percolation structure must have not returned reversibly to initial state because of the crack formation in the matrix or the rearrangement of percolation structure.

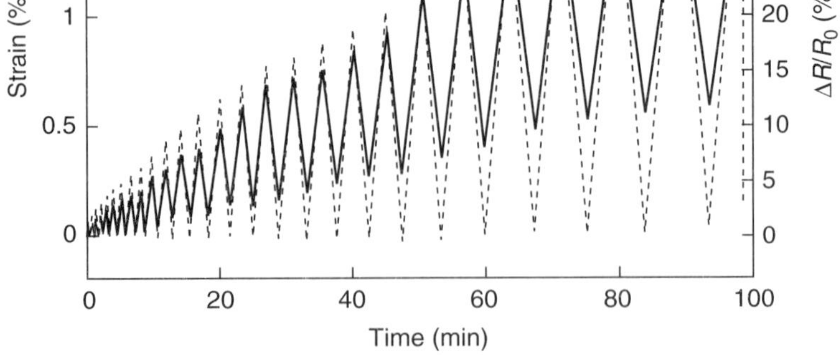

FIGURE 17.1.3 Change in resistance (solid line) and applied strain (dashed line) as a function of time in cyclic loading tests for the CFGFRP.

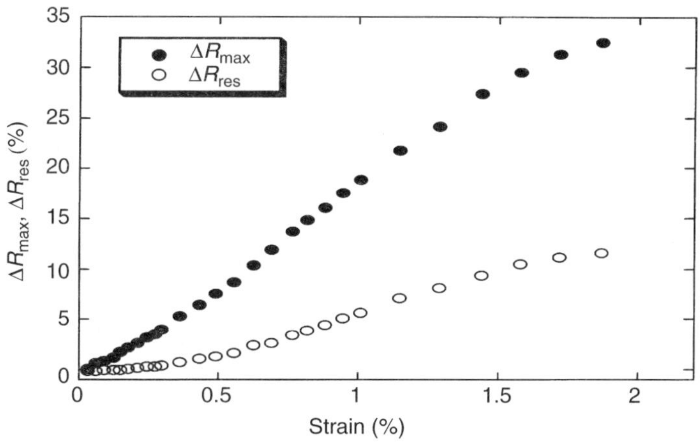

FIGURE 17.1.4 Maximum resistance change at loading state and residual resistance change at unloading state as a function of applied strain in cyclic loading tests for the CFGFRP.

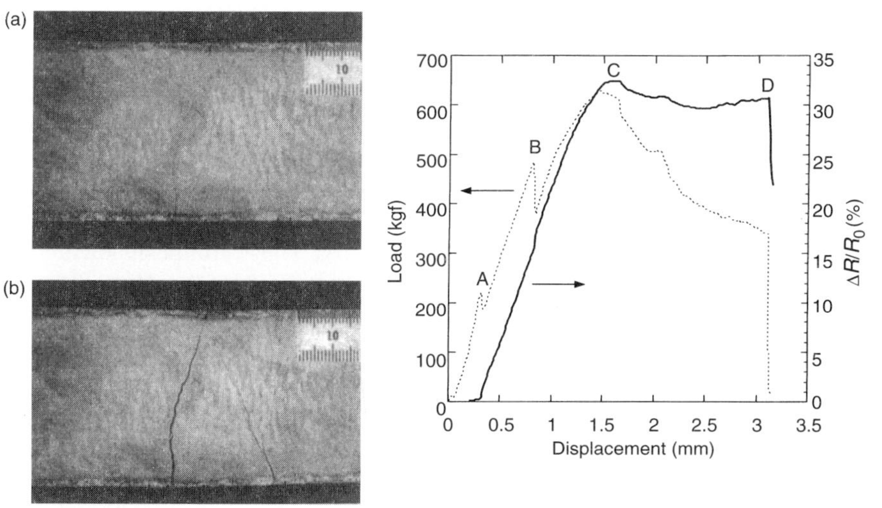

FIGURE 17.1.5 Changes in resistance (solid line) and applied load (dashed line) in a bending test for CPGFRP embedded in mortar block. These points A and B on the graph correspond with the left marked photographs of mortar block.

17.1.3 APPLICATION OF SELF-DIAGNOSIS FRP

The FRP containing carbon particles was embedded in tensile side of mortar blocks in order to demonstrate the self-diagnosis function. Figure 17.1.5 shows applied load and resistance change of CPGFRP as a function of displacement in bending test. The embedded CPGFRP was located 8 mm apart from tensile surface of mortar block of $40 \times 40 \times 160\,mm^3$. The load–displacement curve indicated discontinuous change at the points of A and B, which corresponded to crack formation and propagation in the mortar block, respectively. The crack formation and propagation were shown in photographs of mortar block. The resistance of CPGFRP began to increase slightly before crack formation. Noteworthy is that the obvious increase in resistance appeared simultaneously with microcrack formation and discontinuous resistance change generated in response to crack propagation. The residual resistance was observed in the FRP material after unloading at point D. The resistance change of embedded CPGFRP corresponded well with propagation of damage inflicted on mortar block. These results demonstrated that the embedded CPGFRP had the ability to diagnose microcrack formation/propagation and loading history in cement-based structural materials.

17.1.4 SELF-DIAGNOSIS FUNCTION OF CMC

Figure 17.1.6 is the schematic drawings of structural design for CMC materials. The composites were fabricated by the filament winding method using Si_3N_4 particles (Ube Industries Co., Ltd. SN-COA) as the matrix and SiC fiber (Nippon Carbon Co., Ltd. NL-401) as the reinforcement for strengthening or toughening. A portion of the fibers was replaced with W wire (Nippon Tungsten Co., Ltd. 30 μm). The conductive particles of Si_3N_4–40%TiN (Japan New Metals Co., Ltd.) were dispersed in a part of the matrix. These composites were hot pressed under 40 MPa at 1773 K in N_2 atmosphere for 1 h. The sintered specimens were cut into bars of $3 \times 4 \times 45\,mm^3$ for bending test pieces, with either conductive wire or particles located 0.5 mm from the tensile side.

The self-diagnosis functions of the CMC were evaluated by simultaneous measurements of stress and electrical resistance change ΔR as a function of

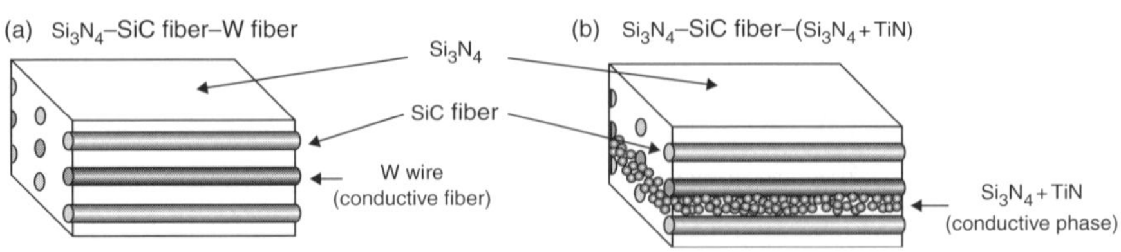

FIGURE 17.1.6 Schematic drawings of the structural design for CMC containing W wire (a) and TiN particles (b).

applied strain in four-point bending tests. The types of loading were in two ways: (a) a normal bending test until specimen fracture; (b) cyclic loading–unloading test below the maximum stress level.

The dependence of applied stress and change in resistance on strain for CMC are shown in Figure 17.1.7. Both composites indicated non-linear response

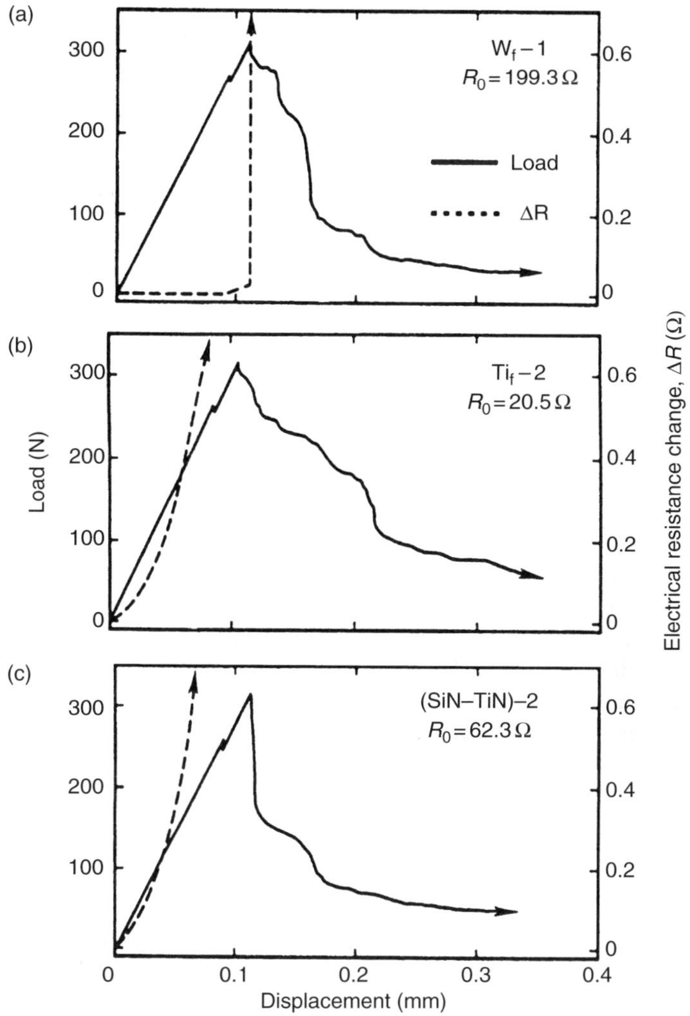

FIGURE 17.1.7 Load versus displacement curves and electrical resistance change versus displacement curves of the specimens. R_0 is the initial value of electrical resistance.

of resistance change to applied strain and fractured at about 0.2% strain. The CMC with W wire showed slight change in resistance in a small strain level, and then drastic change were accompanied by own fracture as shown in Figure 17.1.7a. The CMC containing TiN particles exhibited distinct change in resistance from a small strain level to fracture in the composite as shown in Figure 17.1.7b. These results suggest that the monitoring of the resistance for CMC with percolation structure is also advantageous to diagnose damages to the composites.

Figure 17.1.8 presents the variation of resistance for the CMC with TiN particles in cyclic bending test as a function of number of repetitions. The stress applied at loading state was kept constant at 50 and 70% of maximum stress (250 MPa) for the CMC. The residual resistance after unloading rapidly increased up to 10 cycles. It should be noted that the residual resistance proportionally increased with increasing number of repetitions after 20 cycles, which was distinct for cyclic loading under 70% of maximum stress. This result suggests that the CMC containing TiN particles have the ability to diagnose a cumulative fatigue for the composite by estimation of the residual resistance.

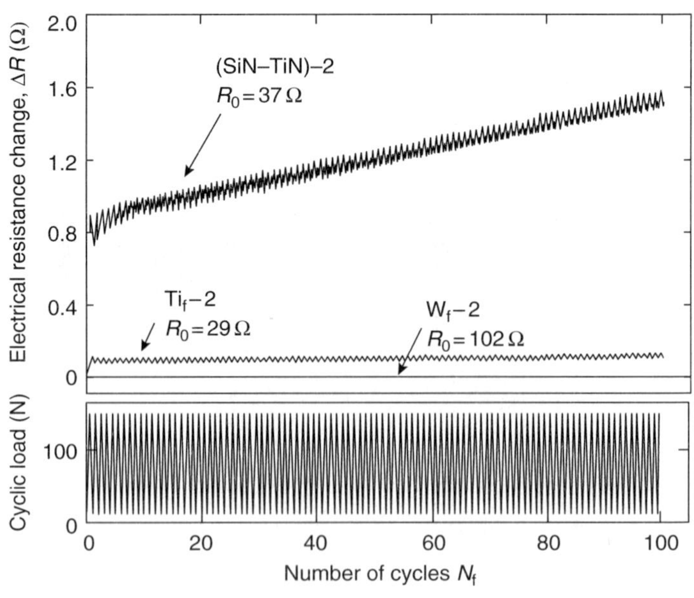

FIGURE 17.1.8 Change of electrical resistance under cyclic load. Applied load is 150 N.

17.1.5 COMPUTER SIMULATION OF PERCOLATION STRUCTURE IN COMPOSITES

For the optimization of microstructure and a material system in formation of the percolation structure, many experimental works on observation of microstructure and evaluation of properties in composites are necessary. In this sense, computer simulation is considered to be a promising method for the estimation of properties and functions from design of microstructure in composites. The purpose of this study is the investigation of two-dimensional computer simulation to design percolation structure of the second phases with different aspect ratios. That is to clarify the relation between formation of percolation structure, the morphology and the configuration of the second phase.

Figure 17.1.9 shows a schematic of an array for the computer simulation which demonstrates structures of the dual phases of a matrix and dispersoids (second phase). This type of the array is called as a triangular (or hexagonal) lattice in two dimension. The lattice size is 1000 × 1000 and the dark and white phases in the figure indicate the second and matrix phases, respectively. The construction of the simulation program has been started from the process how are the second phase arranged in the matrix phase. The arrangement of the second phase was given by the following two different computational programs. In the first one, when second phases are dispersed randomly in the matrix, an overlap among second phases is prohibited; that is, when a certain trial results

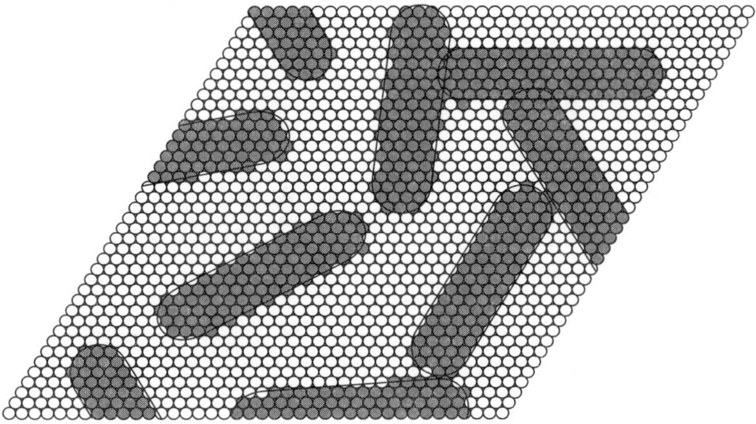

FIGURE 17.1.9 Schematic drawing of the array of the two dimensional triangular (or hexagonal) lattice for the computer simulation. Dark and white lattices indicate second and matrix phases, respectively.

in overlapping, the trial is canceled and the next trial is carried on. Trials are continued until the second phase cannot be arranged anymore. In the second program, an overlap among second phases is allowed so that the percolation structure forms beyond a certain amount of the second phase. These programs are written in FORTRAN, one of the popular programming languages in the field of science and engineering.

In the simulated microstructures arranged by the above two ways, the formation of the percolation structure was evaluated in the following points. In the first program, the threshold fraction of percolation was predicted by the saturation of the second phases. In the second program, a continuous path of the second phases in the array was evaluated by measuring the length of X-direction and the number of second phases forming the continuous path. Many parameters in the microstructure imaginable for the percolation structure. This study has treated the aspect ratio of the second phase as the parameter of great importance for the design of the percolation structure.

Figure 17.1.10 shows an area fraction of the second phase arranged in the array (f) as a function of trial number of arranging the second phase with various aspect ratios. In this simulation, f increase rapidly with trial number in the early stage, but it saturates to a certain value, which is lower in case of higher aspect ratio. Graphic expressions of the computer simulation with the second phase of different aspect ratios in a nearly saturated stage are shown in Figure 17.1.11. It is clearly seen that the second phases with higher aspect ratio

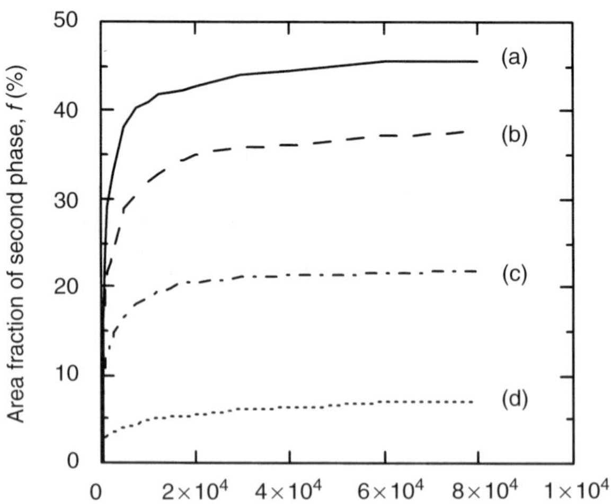

FIGURE 17.1.10 Area fraction of the second phase (f) as a function of trial number of arranging the second phase with various aspect ratios (a.r.) in the computer simulation prohibiting second phase overlap. a.r. = (a) 1, (b) 5, (c) 20, and (d) 100.

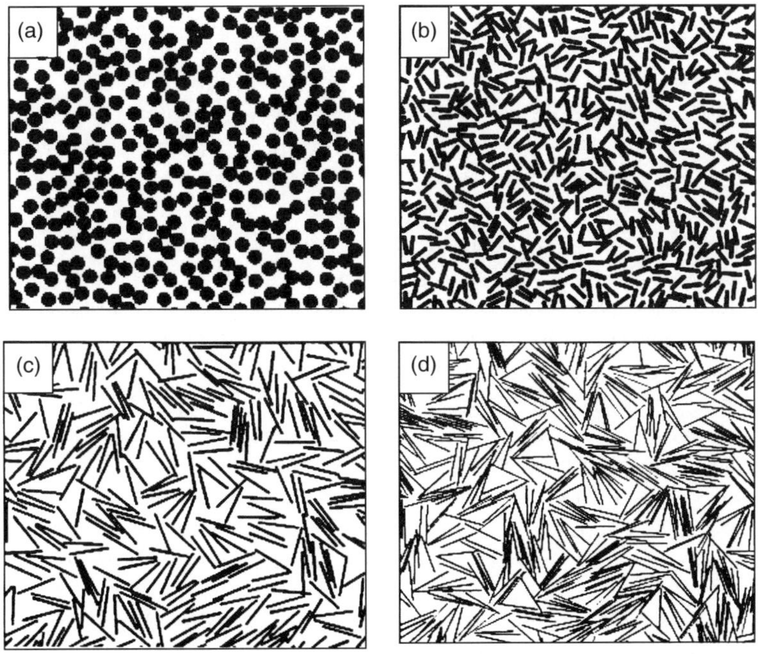

FIGURE 17.1.11 Computer graphics of the simulation prohibiting the overlap of second phases with various aspect ratios (a.r.). Area fraction of the second phase (f) is nearly at saturated stage in Fig. 1. (a) a.r. = 1, f = 46%; (b) 5, 39%; (c) 20, 21%; and (d) 100, 7.8%.

occupy the space in lower content of second phases. However, the feature of the simulated structure is coercive and unnatural especially in case of high aspect ratios. From these results of prohibiting an overlap among second phases, it can be predicted that an increase in aspect ratios of the second phase is advantageous for the formation of the continuous path with a small amount of the second phase. For instance, the second phase with the aspect ratio of 100 saturated in the array by f of around 7%.

Figure 17.1.12 shows the graphic expressions in the simulation allowing an overlap among second phases with four grades of aspect ratios. The overlapped area in the array is not double counted in measuring the area fraction. In this simulation, the second phases are freely dispersed in the array (or matrix), so that the simulated structures seem to be relatively smooth and natural even in case of higher aspect ratio. As shown in Figure 17.1.12a and b, a continuous path which spans from one side of the array to the other (this is called as a complete path, hereafter) was not formed when the second phase with aspect ratio of 1 and 5 were arranged by f of 44 and 36%, respectively. On the other

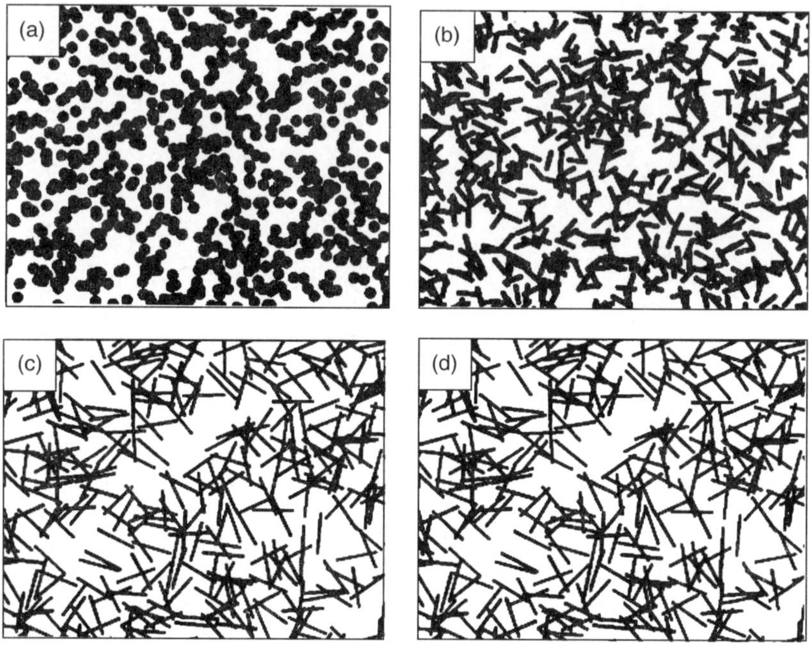

FIGURE 17.1.12 Computer graphics of the simulation allowing the second phases overlap with various aspect ratios (a.r.). (a) a.r. = 1, f = 44%; (b) 5, 36%; (c) 20, 20%; and (d) 100, 5.6%.

hand, as shown in Figure 17.1.12c and d, a complete path was formed when the second phase with aspect ratio of 20 and 100 were arranged by f of 20 and 5.6%, respectively. It is clear that the second phase with higher aspect ratio can form a complete path with a small amount of the second phase. A complete path of percolation (continuous distance = 1000) is observed by f of over 4%. Graphic expressions of the selected second phase which forms a complete path at f of 4 and 5% are shown in Figure 17.1.13.

17.1.6 CONCLUSIONS

The self-diagnosis functions for several types of conductive composites have been investigated. Compared with the composites including conductive fiber or wire, the composites containing percolation structure were found to possess the advantages to diagnose deformation or damage in the composites by themselves. The FRP containing carbon particles in particular diagnosed microcrack formation and propagation in cement based materials, showing that the self-diagnosis function of the FRP has the ability to monitor the health condition of concrete structures. Moreover, the FRP composites indicated the ability to

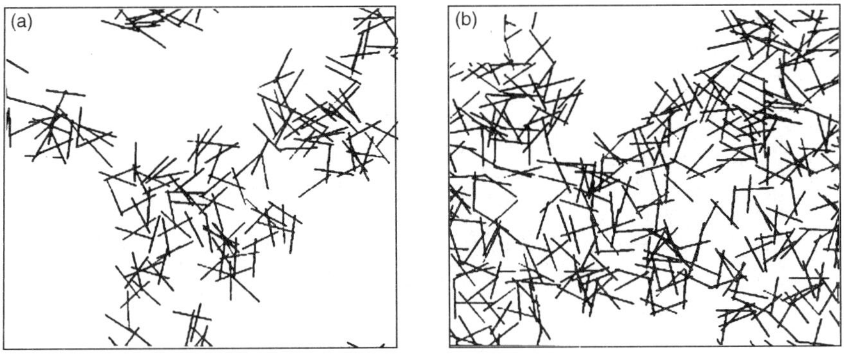

FIGURE 17.1.13 Computer graphics of the selected second phase forming a complete path in the simulation allowing an overlap among second phases (aspect ratio of 100). (a) $f = 4.0\%$, (b) 5.0%.

memorize maximum applied through the measurement of residual resistance. The CMC materials also showed the ability to diagnose a cumulative fatigue for the composite by estimation of the residual resistance. It should be noted that these self-diagnosis functions are able to be easily obtained by simple measurement of electrical resistance. It is expected that the percolation structure has the ability to be applied in other composite systems and the range of possible applications for self-diagnosis can be extended.

Percolation phenomena with the second phase of various aspect ratios were studied by two-dimensional computer simulation. The computer simulation prohibiting second phase overlap gives a certain saturated value in the amount of the second phase. It was predicted that an increase of aspect ratios of the second phase is advantageous for a formation of a continuous path with a small amount of the second phase.

REFERENCES

1. Muto, N., Yanagida, H., Nakatsuji, T., Sugita, M., and Ohtsuka, Y. (1993). *J. Am. Ceram. Soc.* **76**: 875–879.
2. Takada, M., Shin, S.-G., Matsubara, H., and Yanagida, H. (1999). *J. Jpn. Soc. Compos. Mater.* **25**: 225–230.
3. Okuhara, Y., Shin, S.-G., Matsubara, H., and Yanagida, H. (2000). *Trans. MRS-J* **25**: 581–584.
4. Takada, M., Matsubara, H., Shin, S.-G., Mitsuoka, T., and Yanagida, H. (2000). *J. Ceram. Soc. Jpn* **108**: 397–401.
5. Okuhara, Y., Shin, S.-G., Matsubara, H., Yanagida, H., and Takeda, N. (2000). *Proc. SPIE* **4238**: 314–322.
6. Nishimura, H., Sugiyama, T., Okuhara, Y., Shin, S.-G., Matsubara, H., and Yanagida, H. (2000). *Proc. SPIE* **3985**: 335.
7. Ishida, A., Matsubara, H., Furukawa, K., Miyayama, M., and Yanagida, H. (1995). *J. Ceram. Soc. Jpn* **103**: 996–999.

INDEX

Elsevier Internet Homepage – http://www.elsevier.com
Consult the Elsevier homepage for full catalogue information on all books, journals and electronic products and services.

Elsevier Titles of Related Interest
ELSSNER *et al.*
Ceramics and Ceramic Composites: Materialographic Preparation
ISBN: 0 444 100 30X

Related Journals:
A sample journal issue is available online by visiting the homepage of the journal (homepage details at the top of this page). Free specimen copy gladly sent on request. Elsevier Ltd, The Boulevard, Langford Lane, Kidlington, Oxford, OX5 1GB, UK

Ceramics International – www.elsevier.com/locate/ceramint
Journal of The European Ceramics Society – www.elsevier.com/locate/jeurceramsoc
Cement and Concrete Research – www.elsevier.com/locate/cemconres
Cement and Concrete Composites – www.elsevier.com/locate/cemconcomp

To contact the Publisher
Elsevier welcomes enquiries concerning publishing proposals: books, journal special issues, conference proceedings, etc. All formats and media can be considered. Should you have a publishing proposal you wish to discuss, please contact, without obligation, the publisher responsible for Elsevier's Composites and Ceramics programme:

Emma Hurst
Publishing Editor
Elsevier Ltd
The Boulevard, Langford Lane
Kidlington, Oxford
OX5 1GB, UK

Phone: +44 1865 843629
Fax: +44 1865 843931
E.mail: e.hurst@elsevier.com

General enquiries, including placing orders, should be directed to Elsevier's Regional Sales Offices – please access the Elsevier homepage for full contact details (homepage details at the top of this page).

Handbook of Advanced Ceramics

VOLUME I Materials Science

EDITORS

SHIGEYUKI SŌMIYA (*Editor-in-Chief*)
Professor Emeritus, Tokyo Institute of Technology, Tokyo, Japan

FRITZ ALDINGER
Max-Planck-Institut für Metallforschung, Stuttgart, Germany

NILS CLAUSSEN
Technische Universität Hamburg-Harburg, Germany

RICHARD M. SPRIGGS
Alfred University, New York, USA

KENJI UCHINO
The Pennsylvania State University, USA

KUNIHITO KOUMOTO
Nagoya University, Japan

MASAYUKI KANENO
Japan Fine Ceramics Association, Tokyo, Japan

ELSEVIER
ACADEMIC
PRESS

Amsterdam • Boston • Heidelberg • London • New York • Oxford
Paris • San Diego • San Francisco • Singapore • Sydney • Tokyo

This book is printed on acid-free paper.

ACADEMIC PRESS
An Imprint of Elsevier
84 Theobald's Road, London WC1X 8RR, UK
http://www.elsevier.com

ACADEMIC PRESS
An Imprint of Elsevier
525 B Street, Suite 1900, San Diego, California 92101-4495, USA
http://www.elsevier.com
ISBN 0–12–654640–1

A catalogue record for this book is available from the Library of Congress

A catalogue record for this book is available from the British Library

Typeset by Newgen Imaging Systems (P) Ltd, Chennai, India
Printed and bound in Great Britain by MPG Books, Bodmin, Cornwall

03 04 05 06 07 MP 9 8 7 6 5 4 3 2 1

To The late Professor Emeritus E. F. Osborn
The Pennsylvania State University
University Park, Pennsylvania, USA

To Professor Emeritus J. A. Pask
University of California, Berkeley, California, USA

To Professor Emeritus Günter Petzow
Max Planck Institut für Metallforschung, Pulvermetallgisches Laboratorium

Shigeyuki Sōmiya

CONTENTS

PREFACE

In 1989 Shigeyuki Sōmiya, the Editor-in-Chief of this book, published *Advanced Technical Ceramics* (Academic Press, Inc.; original publication in Japanese, 1984). Well over a decade has passed without the appearance of an authoritative new title on the ever-changing subject of Advanced Ceramics. The purpose of this book is to provide an up-to-date account of the present status of Advanced Ceramics, from fundamental science and processing to application.

The Handbook of Advanced Ceramics has an internationally renowned group of contributing editors. They are well known throughout the world in their fields of study. These editors discussed the contents and chose the authors of each of the book's chapters very carefully. The chapters consist of review and overview papers written by experts in the field.

Up until about 50 years ago, 'ceramics' were considered to be porcelains, bottle glass, sheet glass, refractory bricks, enamels, cements, lime, gypsum and abrasives. In recent years the field of ceramics has broadened and expanded. Ceramics are now used in new fields of research as well as in the old fields. This handbook describes these developments and the new processes and applications.

The handbook will enable the reader to understand the present status of Ceramics and will also act as an introduction, which may encourage further study, as well as an estimation of the role advanced ceramics may have in the future.

The handbook is a two-volume set. Part I deals with Materials Science and Part II with Processing and Applications.

Part I serves as an introduction to the basic science, raw materials, forming, drying, sintering, innovative processing, single crystal growth, machining, joining, coating, fracture mechanics, testing, evaluation, etc. Part I is intended to provide the reader with a good understanding of the new techniques in advanced ceramics, such as thin films, colloidal processing, active and passive filler, pyrolyses process and precursor derived ceramics, as well as providing a template for the deposition of ceramics from aqueous solutions.

Part II deals with more recent processes and applications and functional and engineering ceramics. The engineering ceramics covered in this book were developed within the last decade. The functional ceramics covered include electro-ceramics, optoelectro-ceramics, superconductive ceramics, etc.

as well as the more recent development of piezoelectric ceramics and dielectric ceramics.

The use of 'Engineering' Ceramics, introduces entirely new fields to be considered. These include mechanical properties, decorative ceramics, environmental uses, energy applications, bioceramics, composites, functionally graded materials, intelligent ceramics and so on.

The term Advanced Ceramics is opposite in meaning to 'Traditional' or 'Classical' Ceramics. In the past, Advanced Ceramics were often confused with New or Newer Ceramics, Modern Ceramics, Special Ceramics and so on. Furthermore, Fine Ceramics, at least in the USA and Europe, is synonymous with Fine Grain Ceramic Products and/or Fine Grain Porcelain; Fine Ceramics in Japan is similar to what we understand as Advanced Ceramics. So for this edition, the term *Advanced Ceramics* was chosen as the most suitable title for a book providing an in-depth survey of the current state of Ceramics Science and its applications.

It is the editors' wish that this book will provide the reader with a detailed understanding of the many applications of Advanced Ceramics in both today's world and in that of the future.

The editors wish to thank all those who participated in the preparation of this book such as authors, publishers and copyright owners in Europe, USA, Asia and the rest of the world.

Fritz Aldinger
Nils Claussen
Masayuki Kaneno
Kunihito Koumoto
Shigeyuki Sōmiya (Editor-in-Chief)
Richard M. Spriggs
Kenji Uchino

ACKNOWLEDGMENTS

First and foremost, I would like to thank all the co-editors, Fritz Aldinger, Nils Claussen, Richard M. Spriggs, Kenji Uchino, Kunihito Koumoto and Masayuki Kaneno for their valuable suggestions with regard to the chapters, authors, and their editorial assistance. Without their help, this book would not have been possible. Especially, Fritz Aldinger, Nils Claussen, and Richard M. Spriggs gave me good advice for this book. Mr Kaneno offered secretarial assistance.

Second, I wish to thank all the publishers who gave permissions to authors to reproduce and use the materials from their original published papers.

Third, many authors wrote their chapters within a short time in spite of their tight schedule. I would like to express my gratitude to all these authors.

I would also like to extend my appreciation to Ms Amanda Weaver and her group at Elsevier, Oxford for their contribution to the publishing works.

Finally, I was able to study abroad in the USA under the Fulbright Exchange Program and in Germany under the scholarship program by Max Planck Institut für Metallforschung, Pulvermetallgisches Laboratorium. Without these experiences, I would not have made it as Editor-in-Chief of this book. I thank all my professors and friends around the world who have made this possible.

Shigeyuki Sōmiya
Editor-in-Chief

CONTRIBUTORS

FRITZ ALDINGER, Max-Planck-Institut für Metallforschung, Heisenbergstrasse 3, D-70569 Stuttgart, Germany

B. BALZER, ETH Zurich, Institute for Nonmetallic Materials, CH-8092 Zurich, Switzerland

J. BILL, Max-Planck-Institut für Metallforschung, Heisenbergstrasse 3, D-70569 Stuttgart, Germany

NILS CLAUSSEN, Advanced Ceramics Group, Technische Universitat Hamburg-Harburg, D-21073 Hamburg, Germany

LUTGARD C. DE JONGHE, Department of Materials Science and Engineering, University of California at Berkeley, and the Lawrence National Laboratory, Materials Science Division, Berkeley, CA 94720-1760, USA

P. EHRHART, IFF/EKM, FZ-Jülich, D-52425 Jülich, Germany

J. FETT, Forschungszentrum Karlsruhe and Universität Karlsruhe, D-76021 Karlsruhe, Germany

T. FUCHS, Max-Planck-Institut für Metallforschung, Heisenbergstrasse 3, D-70569 Stuttgart, Germany

TSUGUO FUKUDA, Institute of Multidisciplinary Research for Advanced Materials, Tohuku University, Sendai 980–8577, Japan

L. J. GAUCKLER, ETH Zurich, Institute for Nonmetallic Materials, CH-8092 Zurich, Switzerland

PETER GREIL, Department of Materials Science, University of Erlangen-Nuernberg, Martensstrasse 5, D-91058 Erlangen, Germany

R. HOFFMAN, Max-Planck-Institut für Metallforschung, Heisenbergstrasse 3, D-70569 Stuttgart, Germany

NOBUYA IWAMOTO, Ion Engineering Laboratory, Hirakata 573-0128, Japan

TOSHIO KASAI, School of Engineering, Tokyo Denki University, Tokyo 101-8457, Japan

SUNITA KAUSHIK, Department of Materials Science and Engineering, University of Florida, Gainesville, FL 32611-6400, USA

SEONGTAE KWON, Materials Research Institute, The Pennsylvania State University, University Park, PA 16802, USA

ERIC LAARZ, Department of Materials Science and Engineering, University of Florida, Gainesville, FL 32611-6400, USA

GARY L. MESSING, Department of Materials Science and Engineering, College of Earth and Mineral Sciences, The Pennsylvania State University, University Park, PA 16802-5005, USA

ERIK MUELLER, Department of Materials Science and Engineering, University of Florida, Gainesville, FL 32611-6400, USA

D. MUNZ, Institut fur Keramik im Maschienenbau und Forschungszentrum Karlsruhe, Institut fur Materialforschung II, Universitat Karlsruhe, D-76021 Karlsruhe, Germany

MAKIO NAITO, JWRI, Osaka University, 11-1, Mihogaoka, Ibaraki, Osaka 567-0047, Japan

TATSUKI OHJI, AIST, Synergy Materials Research Center, National Institute of Advanced Industrial Science and Technology, Moriyama-ku, Nagoya 463-8687, Japan

HITOSHI OHMORI, Institute of Physical and Chemical Research, Wako 351-0198, Japan

MOHAMED N. RAHAMAN, Department of Ceramic Engineering, University of Missouri-Rolla, Rolla, MO 65409-0330, USA

EDWARD M. SABOLSKY, Materials Research Institute, The Pennsylvania State University, University Park, PA 16802, USA

KEN H. SANDHAGE, Department of Materials Science and Engineering, The Ohio State University, Columbus, OH, USA

KIYOSHI SHIMAMURA, Kagami Memorial Laboratory for Materials Science and Technology, Waseda University, Tokyo 169-0051, Japan

NOBUHIRO SHINOHARA, Asahi Glass Co. Ltd., 1150 Hazawa-cho, Kanagawa-ku, Yokohama 221-8755, Japan

WOLFANG SIGMUND, Department of Materials Science and Engineering, University of Florida, PO Box 11640, Gainesville, FL 32611-6400, USA

SHIGEYUKI SŌMIYA, 3-7-19 Seijo, Setagaya-ku, Tokyo 157-0066, Japan

KEIZO UEMATSU, Department of Chemistry, Nagaoka University of Technology, Kamitomioka-cho, Nagaoka 940-2188, Japan

R. WASER, IFF/EKM, FIZ-Jülich, D-52425 Jülich, Germany

MARKUS WEINMANN, Max-Planck-Institut für Metallforschung, Heisenbergstrasse 3, D-70569 Stuttgart, Germany

WILLIAM B. WHITE, Materials Research Institute, The Pennsylvania State University, University Park, PA 16802, USA

YUNPENG YANG, Department of Materials Science and Engineering, University of Florida, Gainesville, FL 32611-6400, USA

Basic Science of Ceramics

Handbook of Advanced Ceramics
S. Sōmiya *et al.* (Eds.)

CHAPTER 1

1.1 Basic Science of Advanced Ceramics

WILLIAM B. WHITE
Materials Research Institute, The Pennsylvania State University, University Park, PA 16802, USA

1.1.1 INTRODUCTION

The definition of what constitutes a ceramic has broadened in recent decades. Instead of a handful of oxides and silicates making up the subject matter of ceramics there are now sulfides, nitrides, and many other classes of compounds. In addition to the traditional structural ceramics, there are ceramics used as dielectric, magnetic, and optical materials. The subject of this chapter might better be defined by what it is not. Excluded are polymers, metals (although some ceramics have metallic conductivity, even superconductivity), semiconductors (although the boundary between II–VI and III–V compounds as ceramics and as semiconductors is exceedingly fuzzy), and glasses. Glasses, of course, are traditionally a subset of ceramics and are excluded here only because of a deliberate choice to limit the discussion to crystalline materials. Ceramics may, however, take many forms including single crystals, sintered bodies, films, fibers, coatings, and unconsolidated powders in a great range of particle sizes and particle morphologies.

The basic science of advanced ceramics is approached in three steps. First, the major classes of ceramic materials are introduced and their crystal structures and crystal chemistry described. Then, the various models for chemical bonding in ceramic compounds are briefly outlined. The third step is to relate physical properties to crystal chemistry and crystal structure. Structure–property relations are, of necessity, rather abbreviated without much discussion of the subtleties. Tabular data for specific properties are illustrative rather than comprehensive. In each of the areas, enough references are provided to lead the reader into more detail on the subject. Where possible, reference is made to specialized books and reviews.

1.1.2 CRYSTAL CHEMISTRY

A key concept in crystal chemistry is that of the *structure type*. One example of a given structure is taken as a prototype for the structure type. If the prototype compound occurs in nature, the mineral name of the naturally occurring compound is used as a label for all compounds belonging to the structure type. Thus, CaF_2 is taken as a prototype for all compounds with the same atomic arrangement. We speak of the fluorite structure for compounds such as $PbF_2, UO_2, GeMg_2$, or even Na_2S (the latter has the antifluorite structure), which have the same atomic arrangement although entirely different chemical compositions.

One could, of course, simply prepare a catalog of all known structures as in Wyckoff's compilations [1]. However, an objective of crystal chemistry is to organize a very large number of inorganic compounds into a much smaller number of structure types. In the strict sense, compounds belonging to the same structure type will have the same atomic arrangement of ions within the unit cell, the same cation and anion coordination numbers, the same linkages between the cation–anion polyhedra that make up the structure, and belong to the same space group. In a less strict sense, one can allow for distorted structures, ordered structures, and superstructures by relaxing the requirement of identical symmetry and require only that the compounds have a similar atomic arrangement.

1.1.2.1 Oxide Ceramics

Oxides are the dominant ceramic compounds. In part, this is because of their generally refractory character and, in part, because they are chemically stable at high temperatures in the presence of an oxygen-containing atmosphere.

The oxide ion, O^{2-}, has the electronic configuration $1s^22s^22p^6$, giving it a filled shell electronic configuration. Electron transfer in most oxides of ceramic interest is essentially complete, making the bonding in oxides strongly ionic. The radius of the oxide ion for O^{2-} in six-fold coordination is 126 pm on the Shannon–Prewitt crystal radii scale or 140 pm on the Shannon–Prewitt ionic radii scale [2, 3]. The oxide ion radius is relatively insensitive to changes in coordination, ranging from 121 pm for O^{2-} in two-fold coordination to 128 pm for O^{2-} in eight-fold coordination. Crystal radii have a better theoretical foundation but ionic radii fit better with the Goldschmidt or Pauling radii widely used in the older literature. Crystal radii will be used exclusively in this review.

The fluoride ion, F^-, has the same electronic structure as the oxide ion and is also a tightly bound, closed shell configuration. For this reason, fluoride compounds for the most part take the same structures as the corresponding oxides.

Fluorides generally have lower melting points and are more chemically reactive than the corresponding oxides. The sulfide, S^{2-}, ion is also isoelectronic with the oxide ion but the bonding p-orbitals have a greater spatial extension, thus greater overlap with cation orbitals often leading to a covalent or metallic bonding character. For this reason, the sulfides (and also selenides and tellurides) are discussed separately although many oxides and sulfides belong to the same structure types.

Many oxide compounds can be approximated as a closest packing of oxide ions with the cations occupying tetrahedral or octahedral interstices. Two alternatives are possible: the cubic closest packing and the hexagonal closest packing (Fig. 1.1.1). Depending on cation sizes and cation bonding requirements, many structures occur that are distortions of the basic packing arrangements. Compounds with strongly directed bonding (i.e. more covalent than ionic) often have anion arrangements that are not closest packed. Many silicates deviate strongly from closest packed structures.

1.1.2.1.1 Binary Oxides

Useful parameters for describing the crystal chemistry of the binary oxides are the bulk composition and the ionic radius of the cation (Fig. 1.1.2). The cation coordination increases with increasing cation radius in a systematic way.

ZnO and BeO crystallize in the wurtzite structure, space group $P6_3mc$, with the cation in four-fold coordination [4]. This structure is better represented among AX chalcogenides and pnictides and will be discussed later. Considerations of size alone would place ZnO in the NaCl structure but the tetrahedral bonding scheme is stabilized by overlap of the filled d-shell of Zn^{2+} with the p-orbitals of the oxide anions. Pressures in excess of 10 GPa are required to invert ZnO into the NaCl structure [5].

MgO, with a melting point of 2832°C, is widely used as a highly refractory ceramic. MgO, along with most other oxides with the MO stoichiometry, is isostructural with NaCl. The NaCl structure is cubic, space group $Fm3m$, with all cations in six-fold coordination. The structure consists of a cubic closest packing of anions with cations in the octahedral interstices leading to an array of edge-sharing octahedra (Fig. 1.1.3). The transition metal oxides TiO, VO, CrO, MnO, FeO, CoO, and NiO all take the NaCl structure. Only CuO has a uniquely different structure. However, TiO, VO, and FeO are defect structures with large concentrations of cation and anion vacancies depending on temperature and the oxygen fugacity of the coexisting gas phase. TiO and VO are metallic and FeO and CoO are semiconductors. In the corresponding halide structures there exists the CsCl structure taken by compounds with the largest cations. The cation in the CsCl structure is in eight-fold coordination. Oxides do not take the CsCl structure.

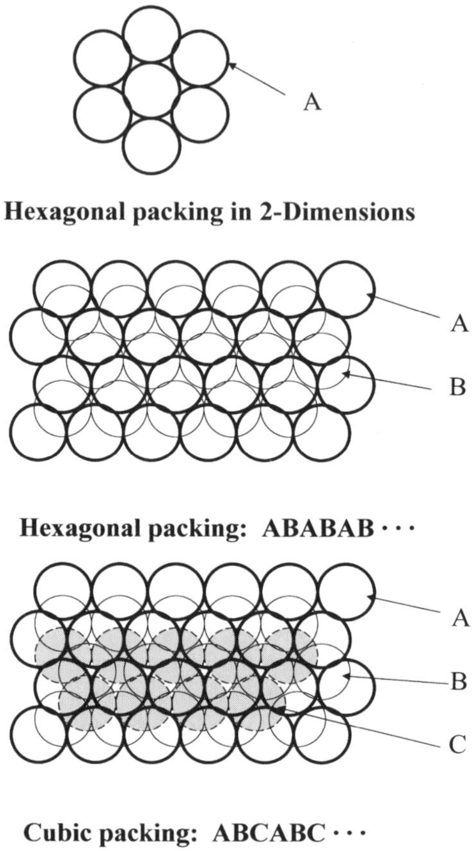

FIGURE 1.1.1 Sketch showing closest packing of equal-diameter spheres. In two dimensions only a hexagonal arrangement is possible. In three dimensions, both a hexagonal (AB—[0001] stacking direction) and a cubic (ABC—[111] stacking direction) packing are possible.

Compounds with the M_2O_3 stoichiometry occur in the corundum structure and the rare earth oxide structures. B_2O_3 is in a class by itself with boron in three-fold coordination. Although B_2O_3 is a glass-former and an essential component of borate and borosilicate glasses, the end-member oxide has little application as a ceramic.

Sesquioxides of intermediate size cations take the corundum structure. Corundum, Al_2O_3, is trigonal, space group $R\bar{3}c$, with the cation in six-fold coordination. The structure is based on a hexagonal closest packing of anions with systematic filled and vacant octahedral interstices (Fig. 1.1.4a). The AlO_6 octahedra are face-sharing, giving corundum structure compounds an

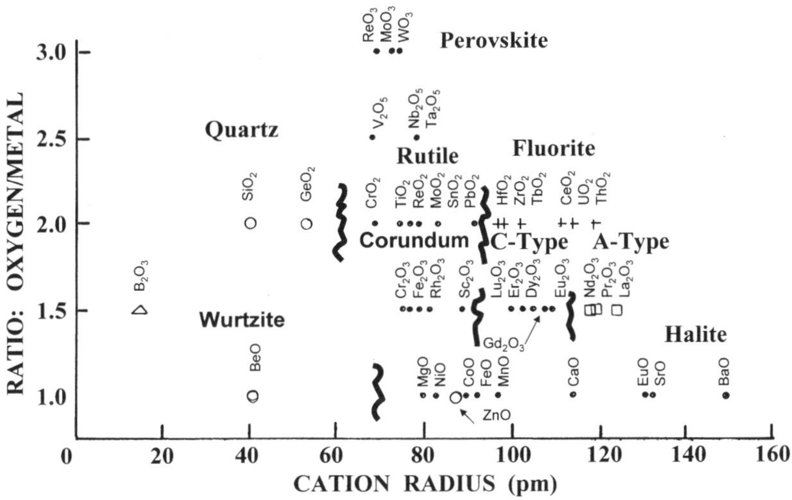

FIGURE 1.1.2 Distribution of binary oxide structures as a function of composition and ionic radius. Structures according to Wyckoff [1]. Radii are Shannon–Prewitt [2] crystal radii taken for the appropriate coordination number.

exceptionally high density. Compounds with the corundum structure include α-Ga_2O_3, Ti_2O_3, Cr_2O_3, and Fe_2O_3. The trivalent cations can be replaced by other ions if charge balance is maintained. A II–IV valence pair, for example, $FeTiO_3$ or $MgTiO_3$, goes into the corundum structure in an ordered way to produce the ilmenite structure with derivative space group $R\bar{3}$. A I–V valence pair will produce a different ordering scheme, the $LiNbO_3$ structure, with space group $R3c$.

Sesquioxides of larger cations, particularly the trivalent rare earths, take on one of three rare earth oxide structures known, rather unimaginatively, as the A, B, and C types [6] (Fig. 1.1.5). The A-type structure is trigonal, space group $C3m$. The A-type structure is taken by the oxides of the larger rare earths, La_2O_3 through Nd_2O_3, and the actinides such as Ac_2O_3 and Pu_2O_3. There is a single cation site with four short cation–oxygen distances and three somewhat longer cation–oxygen distances giving overall distorted seven-fold coordination polyhedra. There are two anion sites so that the formula could be written $La_2O(I)O(II)_2$. Sulfur can substitute for oxygen on O(I) but not on O(II), which produces a series of oxysulfides, La_2O_2S through Lu_2O_2S, all with the A-type structure. The intermediate size rare earths oxides, Sm_2O_3 and Gd_2O_3, take the highly distorted monoclinic B-type structure. The oxides of the smaller rare earths take the C-type structure (also known as the bixbyite structure after Mn_2O_3). The structure is cubic, space group $Ia3$, with a large

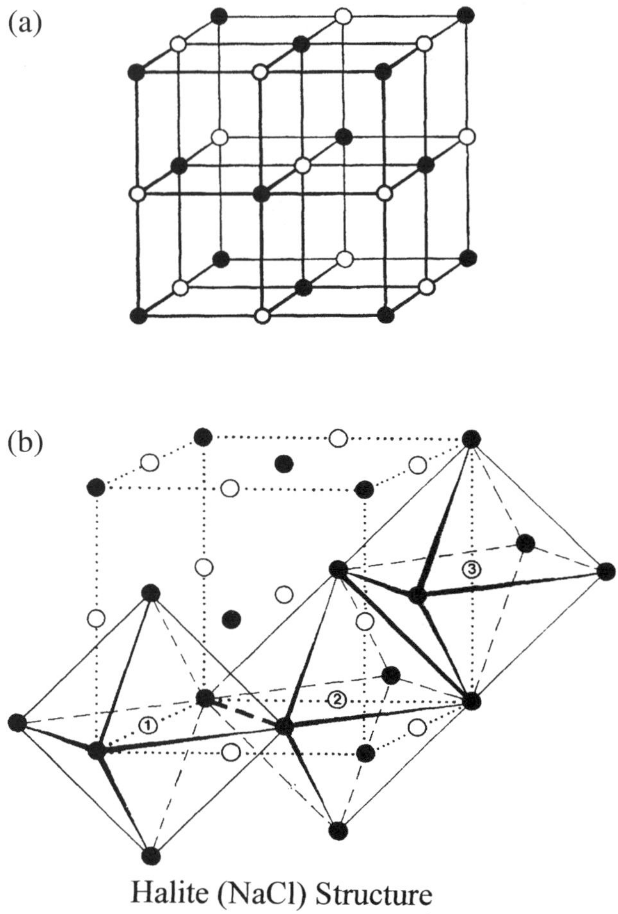

FIGURE 1.1.3 The halite or NaCl structure: (a) the usual face-centered cell; (b) an expanded cell showing the arrangement of edge-sharing octahedra.

cell, about 1000 pm on edge. The C-type structure is related to the fluorite structure. The cations occupy similar positions to the cations in fluorite but with only three-fourths of the anion positions occupied, thus giving six-fold rather then eight-fold coordination.

Oxides with the AO_2 stoichiometry take a series of structures of which the quartz structure, the rutile structure, and the fluorite structures represent four-, six-, and eight-fold coordination. Quartz, SiO_2, is a three-dimensional framework of corner-sharing silica tetrahedra, and so is not strictly a packing structure. Further discussion is deferred to the section on silicates.

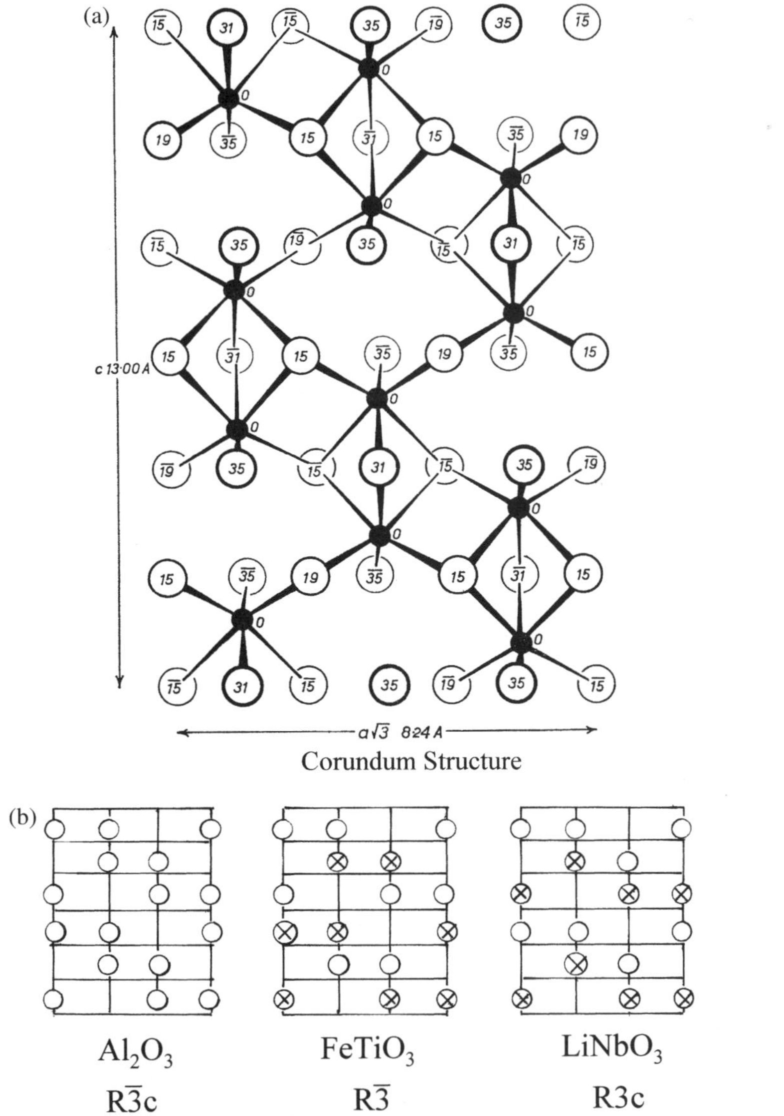

FIGURE 1.1.4 (a) The corundum structure shown as a projection on (2110). Solid circles are Al-ions which all lie in the projection plane. Numbers in open circles give heights of oxygens above or below the plane of the aluminum. (b) Comparison of corundum, ilmenite, and $LiNbO_3$ structures.

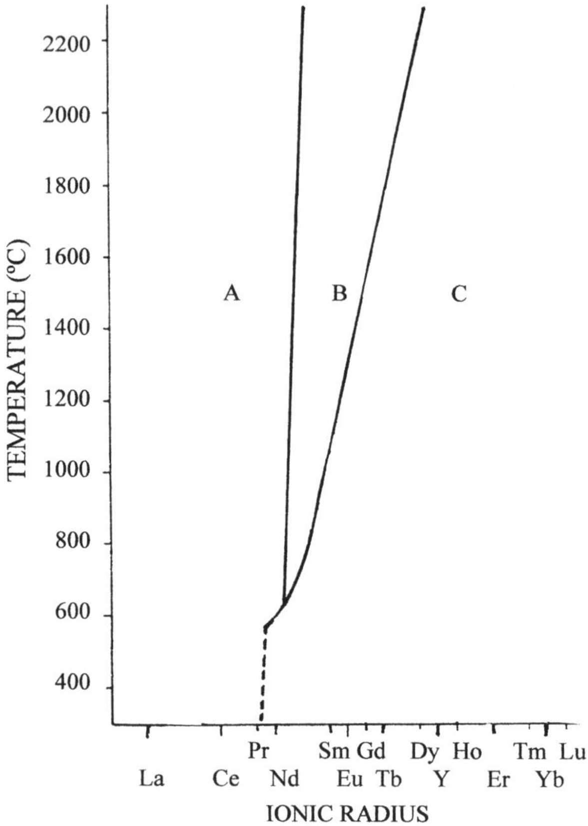

FIGURE 1.1.5 Stability ranges of the rare earth oxide structures. Adapted from Warshaw and Roy [6].

Oxides of intermediate size tetravalent cations take the rutile structure. The structure is tetragonal, space group $P4_2/mnm$. The single cation site is six-coordinated. The octahedra share edges to form chains along the *c*-axis so that the structure is not closest packed (Fig. 1.1.6). TiO_2 is the prototype. Other rutile structure compounds include CrO_2, SnO_2, and PbO_2. The oxides of tetravalent transition metal ions such as VO_2, MoO_2, and WO_2 are distorted variants of the rutile structure. The rutile structure lends itself to several distorted and ordered variants including the $AlNbO_4$ structure—distorted rutile, the wolframite ($FeWO_4$) structure—ordered and distorted rutile, the columbite ($FeNb_2O_6$) structure 1:2 ordered and distorted rutile, and the trirutile ($ZnSb_2O_6$) structure—another 1:2 ordered but undistorted rutile. The orthorhombic columbite structure also

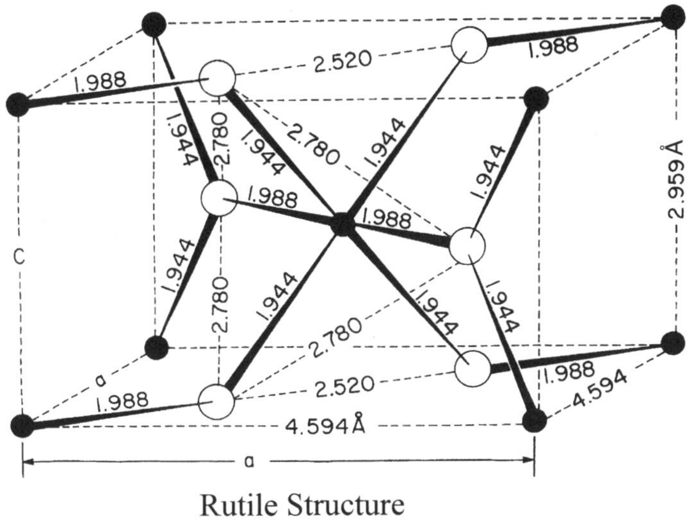

FIGURE 1.1.6 The rutile structure. Bond lengths are shown for the specific example of TiO_2.

appears as a high-pressure polymorph of binary rutile oxides. All of these structural variants are reflections of the different ways of linking octahedra [7].

Oxides of large tetravalent metal ions take the fluorite structure. The prototype is CaF_2, a cubic structure, space group *Fm*3*m*, with a unit cell of about 500 pm. Examples are ZrO_2, CeO_2, ThO_2, UO_2, and PuO_2, a good collection of important ceramic materials. The structure can be considered to be composed of octants, one half of which are occupied by cations in eight-fold coordination with the other half unoccupied (Fig. 1.1.7). Because of the open cation sites, the fluorite structure is easily subject to interstitial substitutions, an aspect that becomes important in the zirconia ceramics. The C-type rare earth oxide structure can be considered to be constructed from eight fluorite cells with one-fourth of the anions missing. The empty anion sites form an ordered array, thus preserving the cubic cell, but lowering the space group symmetry from *Fm*3*m* to *Ia*3.

Other stoichiometries for binary compounds exist, for example, A_2O_5, AO_3, and AO_4. The oxides V_2O_5, Nb_2O_5, and Ta_2O_5 are of interest to ceramic science but only a few other ceramic oxides fit into these stoichiometries. As the cation valence state increases, the bonding becomes progressively more covalent and simple packing models become less and less applicable. Likewise, as the bonding becomes more covalent, the compounds become less refractory.

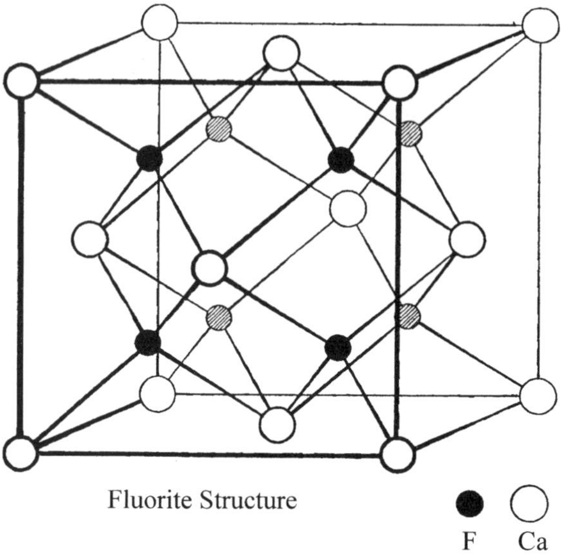

FIGURE 1.1.7 The standard unit cell of the fluorite structure. The four-coordination of the anions is clear but the eight-coordination of the cations is not well displayed in a single unit cell.

1.1.2.1.2 Alumina Ceramics

A great variety of ceramic bodies are prepared from aluminum oxide with properties depending strongly on particle size and processing conditions. Although the stable oxide is α-Al_2O_3 with the corundum structure, other modifications of Al_2O_3 exist and may appear depending on processing conditions [8, 9].

Al_2O_3 precursors are hydroxides or oxyhydroxides. The hydroxides are chemically $Al(OH)_3$ but occur in three crystallographically distinct forms known by the mineral names of bayerite, gibbsite, and nordstrandite. All are closely related hexagonal layer structures made up of $Al(OH)_6$ octahedra sharing edges, thus producing layers of OH^- ions alternating with layers of Al^{3+} ions perpendicular to the hexagonal *c*-axis [10]. Of these, bayerite has the simplest structure, with one cation layer ABAB repeat distance. Gibbsite has two cation layers with a more complex ABBAAB stacking sequence. Nordstrandite is a still larger polytypic stacking with a four-cation repeat. There are two oxyhydroxides, boehmite and diaspore, both orthorhombic but with different space groups and a different arrangement of oxygen and hydroxyl polyhedra.

Depending on dehydration conditions a series of different Al_2O_3 structures can be formed, most related to the spinel structure with Al^{3+} ions in both four- and six-fold coordination. All of the intermediate aluminas appear to be metastable phases. Some are defect structures and some may contain essential

OH^- groups. Some of the structural data are summarized in Table 1.1.1. The reaction schemes from various precursors to the end product corundum are shown in Figure 1.1.8. Because of the structural similarity of many of the transitional aluminas, there remains some uncertainty in the number of phases and in their structural relationships. One structure, κ-Al_2O_3, has been deduced by band theory calculations [11]. The transitional aluminas are important because

TABLE 1.1.1 The Transitional Oxides of Aluminum

Compound	Symmetry	Space group	Comments
γ-Al_2O_3	Tetragonal		Defect spinel
δ-Al_2O_3	Tetragonal		Triple spinel block
η-Al_2O_3	Cubic	*Fd3m*	Spinel structure
θ-Al_2O_3	Monoclinic	*C2/m*	Isostructural with β-Ga_2O_3
χ-Al_2O_3	Hexagonal		six-coordinated layer structure
κ-Al_2O_3	Hexagonal		Large cell
ι-Al_2O_3	Orthorhombic	*Pbam*	Mullite-like

Data summarized from Wefers and Misra [8].

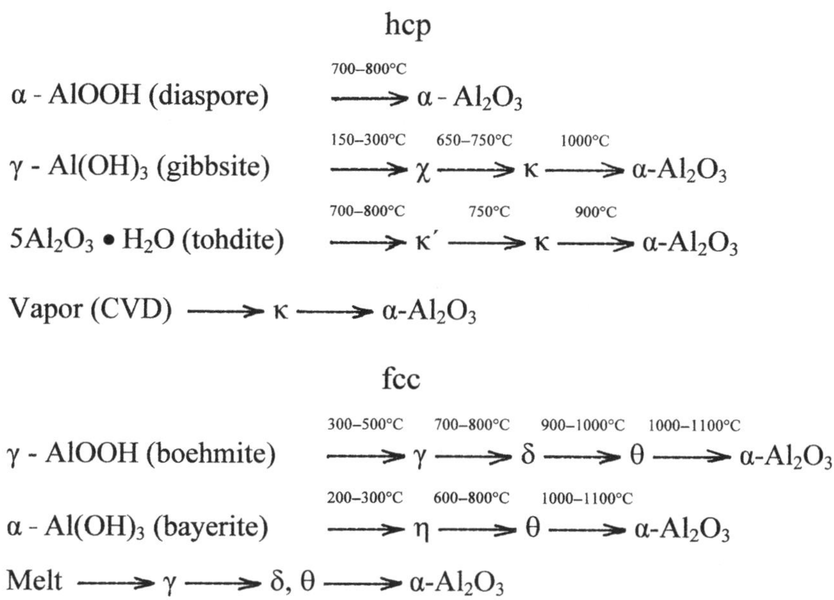

FIGURE 1.1.8 Temperature relationships of the transitional aluminas. Data drawn largely from Ref. [8].

the final microstructures of alumina ceramics depends on details of precursors and firing schedules.

1.1.2.1.3 Zirconia Ceramics

Zirconium appears only as the Zr^{4+} ion so there is a single oxide, ZrO_2. At atmospheric pressure, ZrO_2 can exist in three polymorphic forms. The low-temperature phase is monoclinic (baddeleyite), space group P_1/c [12]. Between 1170 and 2370°C, the stable phase is tetragonal, space group $P4_2/nmc$ [13]. The tetragonal phase is a distorted fluorite structure with eight-coordinated Zr^{4+} ions but with the oxygens in a distorted tetrahedral environment. Above 2370°C, ZrO_2 has the cubic fluorite structure that persists to the melting point at 2680°C. At pressures up to 50 GPa, monoclinic ZrO_2 transforms into a series of three orthorhombic phases [14].

The transition from the high-temperature cubic structure to tetragonal or monoclinic forms takes place rapidly on cooling. Large volume changes result in fracturing or shattering ceramic bodies so that pure ZrO_2 is of little value as a ceramic material. However, solid solutions of ZrO_2 with such oxides as CaO or Y_2O_3 stabilize the high-temperature form and lead to the important class of zirconia ceramics. The charge imbalance caused by replacing Zr^{4+} by Ca^{2+} is compensated by an equal number of oxygen vacancies giving a defect solid solution of the form $Zr_{1-x}Ca_xO_{2-x}\Box_x$. Yttria-stabilized zirconia produces similar solid solutions but with the possibility of structural ordering of Y^{3+} and vacancies. Phase equilibria studies of the system CaO–ZrO_2 [15] show that the solid solution extends to $x = 0.2$ where an ordered compound forms. The lowest temperature at which the cubic phase is stable is 850°C at $x = 0.15$. Yttria-stabilized zirconia has a wider composition range for the cubic phase (Fig. 1.1.9) as well as a lower equilibrium temperature [16]. The Y^{3+} ion is more effective than the Ca^{2+} ion in destabilizing the tetragonal phase.

Zirconia is one of the most important refractory ceramics. There has been a series of five international conferences on its processing and properties [17–21].

1.1.2.1.4 Spinels and Related Structures

The prototype for the spinel structure is Al_2MgO_4, the mineral spinel. The structure is based on a cubic closest packing of anions with space group $Fd3m$. The large face-centered cell contains eight A_2BX_4 formula units. The cell can be divided into octants with alternating tetrahedral and octahedral groupings

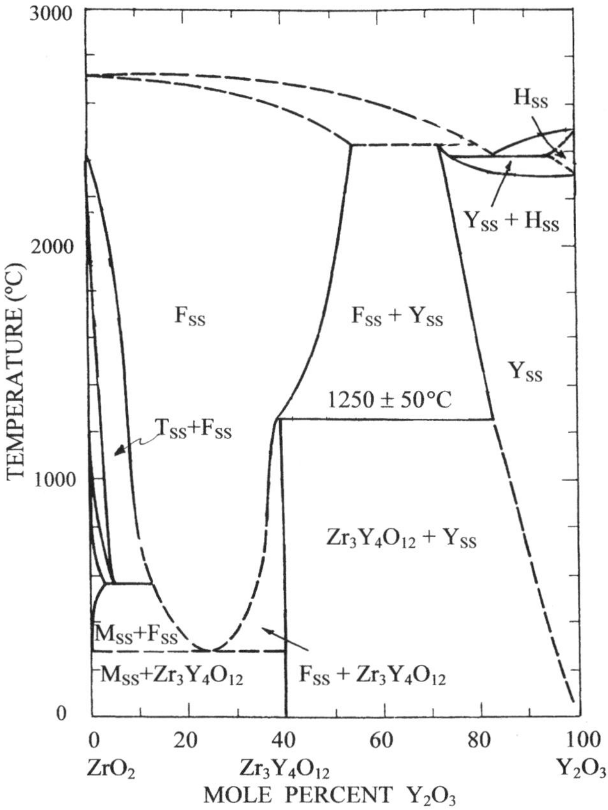

FIGURE 1.1.9 Phase diagram for the system ZrO_2–Y_2O_3. Adapted from Ref. [16]. F_{ss}, T_{ss}, M_{ss}, and Y_{ss} are fluorite zirconia solid solution, tetragonal zirconia solid solution, monoclinic zirconia solid solution, and yttrium oxide solid solution, respectively. Reprinted with permission of the American Ceramic Society.

so that two octants form a primative unit cell containing two formula units (Fig. 1.1.10) [22]. The four-coordinated ion has full tetrahedral symmetry while the local symmetry of the octahedral ion is trigonal, producing a slightly distorted octahedron. The exact positioning is described by a single anion parameter, *u*. In the ideal structure, known as normal spinel, the coordination arrangement is $Al_2^{VI}Mg^{IV}O_4$. Inverse spinels are those with an arrangement such as $[GaMg]^{VI}Ga^{IV}O_4$. Many spinels have mixed cation distributions between the normal and inverse arrangements. The exact distribution is a function of temperature. The spinel structure is very stable with many possible cation and also anion substitutions possible. There are several hundred known compounds with the spinel structure including, in addition to oxides, chalcogenide, nitride, and even cyanide spinels [23–25].

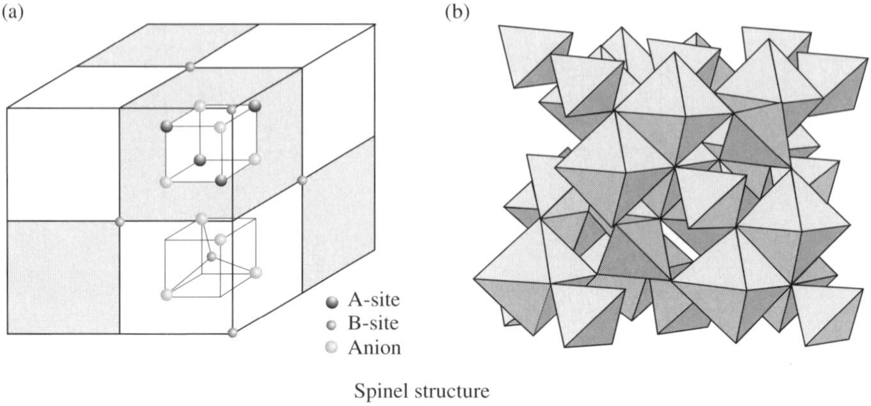

FIGURE 1.1.10 The unit cell of spinel: (a) octahedral and tetrahedral sub-cells drawn from Ref. [22]; (b) the polyhedral packing.

Spinels can be broadly categorized as III–II spinels, II–IV spinels, and I–VI spinels:

III–II spinels	II–IV spinels	I–VI spinels
Al_2MgO_4	Mg_2TiO_4	Na_2WO_4
Cr_2FeO_4	Mg_2GeO_4	Li_2SO_4
Fe_2CoO_4		

Spinel structure compounds have a variety of ceramic applications. Hot-pressed Al_2MgO_4 is an optical window material, the ferrite spinels are an important family of magnetic materials, and the chromite spinels are important refractories and also chromium ores.

The spinel structure will tolerate a substantial concentration of defects. Two examples are γ-Al_2O_3 and γ-Fe_2O_3 , both of which are spinel structures with a vacant tetrahedral cation site and a charge-compensating vacant oxygen site.

The spinel structure lends itself to a variety of ordering schemes that generally result in derivative space groups and may result in superstructure cells. Examples are $Cr_4[GaLi]O_8$, 1:1 order on the tetrahedral sites; $[LiNb]^{VI}Zn^{IV}O_4$, 1:1 order on the octahedral sites; $[LiAl_3]^{VI}[Al_2]^{IV}O_8$, 1:3 order on the octahedral sites.

Two important superstructures are composed of spinel building blocks. The magnetoplumbite structure (prototype $PbFe_{12}O_{19}$) consists of two spinel blocks separated by an anion layer with one site containing Pb^{2+} (Fig. 1.1.11b). The magnetoplumbite structure can be built with large divalent ions such as Ca^{2+}, Sr^{2+}, and Ba^{2+} with Al_2O_3, Fe_2O_3, or Ga_2O_3. The iron compounds form an

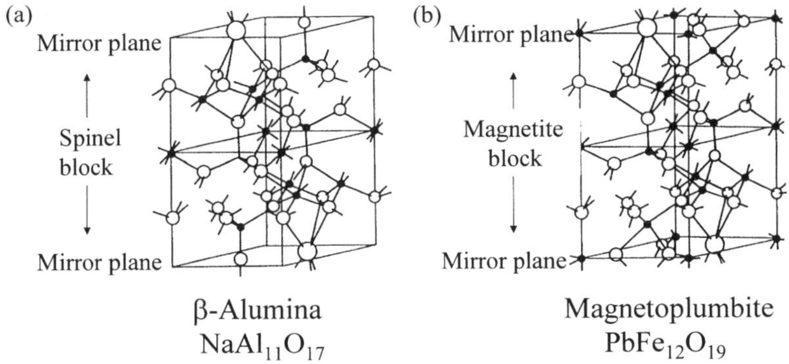

FIGURE 1.1.11 Superstructures based on spinel: (a) the β-alumina structure; (b) the magnetoplumbite structure.

important family of ferrimagnets. The aluminum compounds are used as phosphor hosts particularly with rare earth substitutions of the form $LaMgAl_{11}O_{19}$ in which charge compensation for the trivalent rare earth is provided by replacing one of the Al^{3+} ions in the spinel block with Mg^{2+}.

A second superstructure constructed of spinel blocks is the β-Al_2O_3 structure, so named because it was originally thought to be a polymorph of corundum. The prototype is $NaAl_{11}O_{17}$. The structure consists of two spinel blocks separated, like magnetoplumbite, by an anion layer with the Na^+ ion in one of the anion positions (Fig. 1.1.11a). One of the other anion positions in the central mirror plane is vacant allowing an easy diffusion path for the Na^+ ion in the plane. As a result, β-Al_2O_3 is the prototype for a family of fast ion conductors. The Na^+ ion can be replaced by other alkali ions. There are also a large number of substitutions possible by replacing Al^{3+} in the spinel blocks with Mg^{2+} and other ions. These compounds are important as electrolytes in high-energy-density batteries such as the sodium–sulfur battery.

1.1.2.1.5 Perovskites and Related Structures

The generic formula for compounds with the perovskite structure can be written $A^{XII}B^{VI}O_3$, with a large cation in the 12-fold site and an intermediate size cation in the six-fold site. The space group is cubic, *Pm3m*. The structure consists of a three-dimensional network of corner-sharing BO_6 octahedra (Fig. 1.1.12). The mineral perovskite itself, $CaTiO_3$, is an orthorhombic distortion of the basic structure and there exists the situation where the mineral perovskite does not have exactly the perovskite structure. $SrTiO_3$ is often used as the prototype.

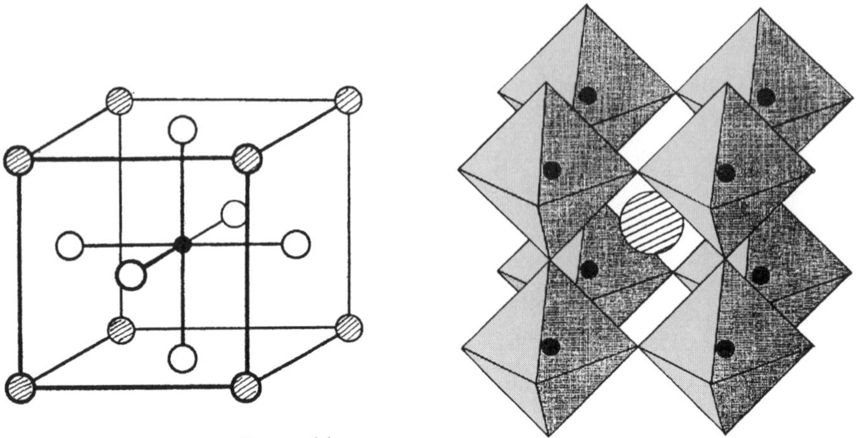

FIGURE 1.1.12 The structure of perovskite showing the primitive cell and the corner-sharing octahedral linkages.

The stability of the perovskite structure is described by a tolerance factor defined as

$$t = \frac{r_A + r_O}{\sqrt{2}(r_B + r_O)}$$

where r is the ionic radii of the A, B, and O ions, respectively. The tolerance factor should be between 0.75 and 1.0 for a stable perovskite structure. Few perovskite compounds actually form the ideal cubic structure. Rotation of the octahedra or ordering of substituted ions reduce the symmetry to tetrahedral and trigonal forms (Fig. 1.1.13).

In addition to the II–IV valence combination, there are III–III perovskites such as $LaAlO_3$ and $GdFeO_3$, which have, respectively, trigonal and orthorhombic distortions of the perovskite structure, I–V perovskites such as $NaNbO_3$, and perovskites that might be called 0–VI in which the 12-fold site is vacant. ReO_3 is a cubic, undistorted 0–VI perovskite while MoO_3 and WO_3 are slightly distorted variants. There are also a range of oxygen-deficient perovskites, some of which have ordered arrangements of vacancies [26]. There is great flexibility in custom designing perovskites with optimum properties because of the great variety of possible substitutions. Some substitutions lead to ordered derivative structures [27]. Some examples of what are known as "half-breed" derivatives are $[KNd]^{XII}[Ti_2]^{VI}O_6$, $[Pb_3]^{XII}[MgNb_2]^{VI}O_9$ (known as PMN—lead magnesium niobate), and $[BaNa]^{XII}[NbTi]^{VI}O_6$. One example of 1:1 ordering on the B-sites is $BaSc_{0.5}Ta_{0.5}O_3$, in which the structure becomes face-centered cubic, space group *Fm3m*, with a perovskite sub-cell.

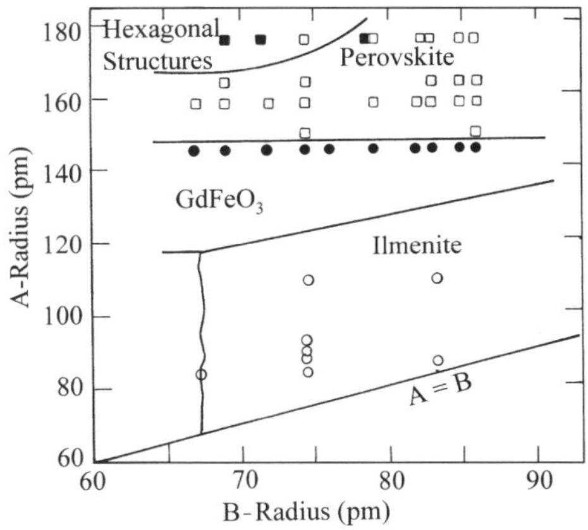

FIGURE 1.1.13 Structure field map for II–IV perovskites. Data taken from Ref. [24]. Such plots do not display the full complexity of these compounds because many of them take on more than one crystal structure.

Perovskites of importance as dielectric materials include such isovalent substitutions as $BaTiO_3$, $PbTiO_3$, and $PbZrO_3$. Solid solutions permit the tuning of the dielectric properties such as $PbZr_{1-x}Ti_xO_3$ (PZT). Transition metal perovskites are important catalyst materials [28]. Because the perovskite structure is stabilized by high pressures, silicate perovskites are thought to be stable phases deep in the Earth's mantle [29].

The perovskite structure also lends itself to the building of superstructures. The Ruddlesden–Popper phases [30] (Fig. 1.1.14a) are a series of structures usually represented by Sr_2TiO_4, $Sr_3Ti_2O_7$, and $Sr_4Ti_3O_{10}$. These can be described as $SrO.nSrTiO_3$ and consist of sequences of n perovskite blocks separated by an NaCl-structure block. The Ruddlesden–Popper phases cannot be extended much beyond $n = 3$; ordered domains appear in an otherwise disordered matrix [31, 32].

The first member of the Ruddlesden–Popper series is an example of the breakdown of the scheme of classifying compounds entirely by stoichiometry. Sr_2TiO_4 is an A_2BX_4 compound and is isostructural with K_2NiF_4, the prototype for the K_2NiF_4 structure. There can be produced half-breed derivatives such as $BaLaAlO_4$ with both Ba^{2+} and La^{3+} on the nine-coordinated A site. Carrying this one step farther produces La_2CuO_4. Now go back and replace some of the La^{3+} by Ba^{2+}. The mismatched charges can be compensated by increasing the valence state of copper giving the composition $La_{2-x}Ba_xCu^{2+}_{1-x}Cu^{3+}_xO_4$, the first of the copper-based high-temperature superconductors [33]. For the structures of many other oxide superconductors see review by Cava [34].

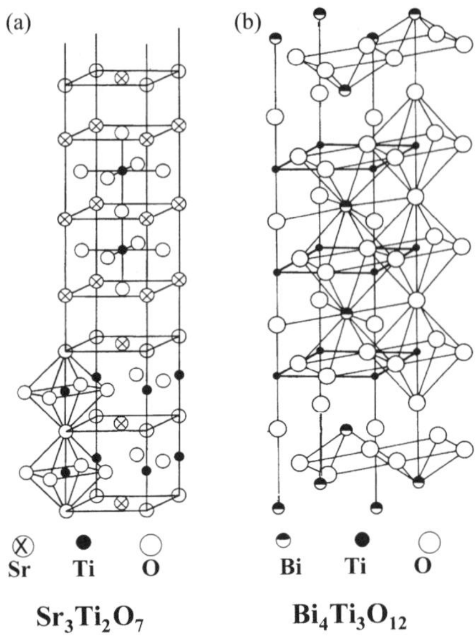

FIGURE 1.1.14 Superstructures based on perovskite: (a) a Ruddlesden Popper phase, $Sr_3Ti_2O_{12}$. (b) an Aurivillius phase, $Bi_4Ti_3O_{12}$.

A second family of superstructures are the Aurivillius phases of which $Bi_4Ti_3O_{12}$ is the prototype. Again, the superstructure is composed of perovskite blocks but separated by Bi_2O_2 layers. These layers consist of sheets of BiO_4 edge-sharing square pyramids (Fig. 1.1.14b).

1.1.2.1.6 Garnets and Related Structures

The generic formula for compounds with the garnet structure is $A_3^{VIII}B_2^{VI}C_3^{IV}O_{12}$. The structure is cubic, space group *Ia*3, with a large unit cell containing 16 formula units (Fig. 1.1.15). Because of the availability of three coordination sites, the garnet structure accepts a great variety of ionic substitutions. The common silicate garnets fall naturally into two families:

$Ca_3Al_2Si_3O_{12}$	grossular	$Mg_3Al_2Si_3O_{12}$	pyrope
$Ca_3Fe_2Si_3O_{12}$	andradite	$Fe_3Al_2Si_3O_{12}$	almandine
$Ca_3Cr_2Si_3O_{12}$	uvarovite	$Mn_3Al_2Si_3O_{12}$	spessartine

Germanate analogs of the silicate garnets can be prepared as can garnets with other substitutions on the eight- and six-fold sites [35]. The coordination polyhedra in the garnet structure are all somewhat distorted. The tetrahedral site has S_4 symmetry and the octahedral site S_6 symmetry although the distortions are

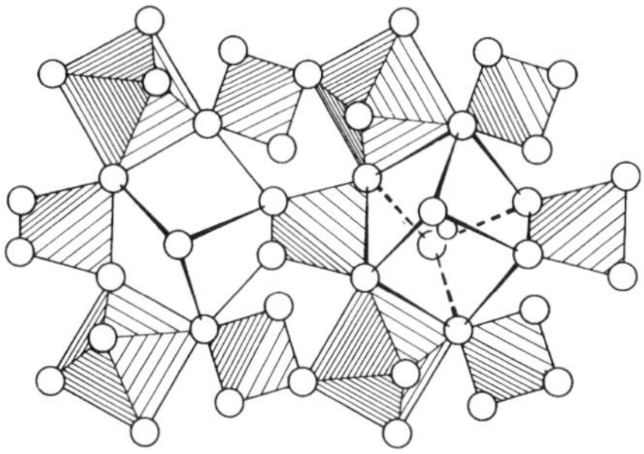

FIGURE 1.1.15 The structure of garnet showing tetrahedral, octahedral, and cubal sites.

not large. The cube site, although eight-coordinated, is a highly distorted cube, D_2 symmetry, which can better be described as a triangular dodecahedron [36].

The tetrahedral site in the garnet structure will accept a variety of trivalent ions such as Al^{3+}, Ga^{3+}, and Fe^{3+}. The loss of one positive charge on the tetrahedral site is compensated by substituting a large trivalent ion on the eight-fold site. The result is the important family of rare earth garnets. Examples are:

$Y_3^{3+}Al_2^{3+}Al_3^{3+}O_{12}$, yttrium aluminum garnet (YAG);
$Y_{3-x}Nd_xAl_5O_{12}$, Nd-YAG an important laser material;
$Y_3Fe_2^{3+}Fe_3^{3+}O_{12}$, yttrium iron garnet (YIG), a ferrimagnet;
$Gd_3Ga_5O_{12}$, gadolinium gallium garnet (GGG), a ferrimagnet.

The first two examples are important optical materials. The latter two are examples of an important class of microwave ferrites. These compounds are good insulators but are magnetically ordered and thus can be used in devices that operate at very high frequencies. Garnet structures can be prepared from all of the rare earths and many combinations of intermediate size trivalent ions on the six- and four-fold sites.

1.1.2.1.7 Silica and Silicate Ceramics

Historically, silicates are the oldest ceramics. Pottery, composed of a mix of silicate compounds, dates back thousands of years. The basic building block of all silicate structures is the SiO_4 tetrahedron. The bonds of the silica tetrahedron have a strong covalent character. As a result, the tetrahedra occur in polymerized form, sharing various numbers of corners, but very rarely edges and never faces. The degree of polymerization serves as a basis for the classification of silicate compounds [37, 38]. One can count the number of

TABLE 1.1.2 Classification of Silicate Structures

Structure	Si/O ratio	NBO/T	*Q*	Example
Nesosilicates (orthosilicates)	1:4	4	0	Mg_2SiO_4
Sorosilicates (pyrosilicates)	2:7	3	1	$Ca_2MgSi_2O_7$
Cyclosilicates	1:3	2	2	$Be_3Al_2Si_6O_{18}$
Inosilicates (metasilicates)	1:3	2	2	$CaMgSi_2O_6$
Phyllosilicates (sheets)	2:5	1	3	$Mg_3Si_4O_{10}(OH)_2$
Tectosilicates (frameworks)	1:2	0	4	SiO_2

oxygens with a formal negative charge (non-bridging oxygens) per tetrahedron (NBO/T) or one can count the number of oxygen that are shared between adjacent tetrahedra (bridging oxygens) and express this number on the *Q*-scale (Table 1.1.2).

Structures containing isolated SiO_4 tetrahedra include olivine, $(Mg_{1-x}Fe_x)_2SiO_4$, zircon, $ZrSiO_4$, mullite, $Al_6Si_2O_{13}$, the aluminum orthosilicates kyanite, andalusite, and sillimanite, Al_2SiO_5, and the silicate garnets. Tetrahedra sharing two corners form chain structures producing the pyroxene and pyroxenoid families of minerals of relatively less importance to ceramics. Cross-linking the chains produces the double chain amphibole family of silicates including the asbestiform minerals. Tetrahedra sharing three corners produces the two-dimensional sheets of the clay and mica minerals. Silicate tetrahedra sharing all corners form the many polymorphs of silica as well as aluminosilicates such as the feldspars and zeolites.

The olivine structure is orthorhombic, space group *Pbnm*. The end member composition $Mg(I)Mg(II)SiO_4$, forsterite, is used in steatite ceramics. The structure is a distorted hexagonal packing with two distinct six-coordinated cation sites [39]. At pressures in the range of 10 GPa, the olivine structure converts to the spinel structure by a rearrangement to cubic closest packing. Analogous germanates can be prepared. These have much lower olivine-to-spinel transition pressures. Many substitutions are possible of intermediate-sized divalent cations on the Mg-sites. Aluminate and borate olivines are also known, for example, Al_2BeO_4, chrysoberyl, and $MgAlBO_4$, sinhalite.

Zircon, $Zr^{VIII}SiO_4$, is tetragonal, space group $I4_1/amd$ (Fig. 1.1.16). Other compounds with large tetravalent cations such as $USiO_4$ and $ThSiO_4$ also have the zircon structure. The structure is extremely stable. Natural zircons survive episodes of mountain building, erosion, sedimentation, and new mountain building. Radiometric dating of zircons has given some of the best estimates for the age of the Earth. Because of its stability, $ZrSiO_4$ is used in glazes. Some Zr^{4+} replaced by V^{4+} produce a highly stable bright blue,

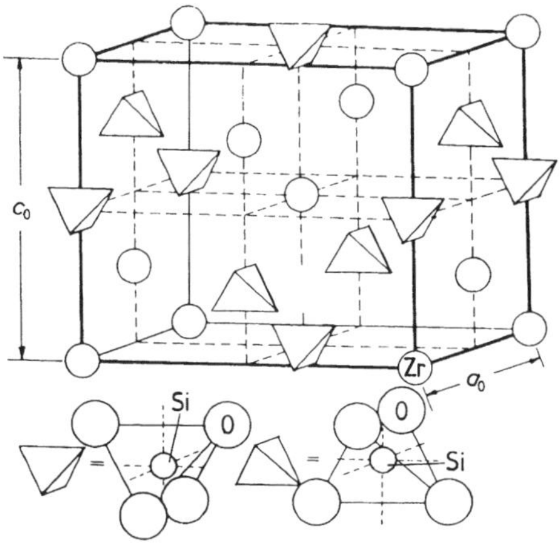

FIGURE 1.1.16 The structure of zircon, $ZrSiO_4$.

Pr^{3+} substitution gives a yellow, and Fe^{3+} a pink. There are a group of closely related structures with an isolated tetrahedral unit and an eight-coordinated large cation:

Zircon	$ZrSiO_4$	$I4_1/amd$
Scheelite	$CaWO_4$	$I4_1/a$
Monazite	$CePO_4$	$P2_1/n$
Fergusonite	$YTaO_4$	$I2$

The monazite structure is chemically resistant and takes many rare earth and actinide elements into solid solution. It has been considered as a phase for nuclear waste immobilization ceramics.

The aluminosilicates present something of a puzzle. To the ceramist, there is one dominant compound, mullite. Mullite is an incommensurate structure with a range of compositions that can be described as $Al_2(Al_{2+2x}Si_{2-2x})O_{10-x}$ where a typical value is $x = 0.4$. An often quoted composition is $3Al_2O_3 \cdot 2SiO_2$ or $Al_6Si_2O_{13}$ but the composition varies. The structure is orthorhombic, space group *Pbam*. It consists of columns of edge-sharing AlO_6 octahedra that form chains parallel to the *c*-axis. These columns are cross-linked by SiO_4 and AlO_4 tetrahedra [40]. Mullite is one of the oldest and most widely used refractory materials. It forms readily by firing in the range of 1000–1400°C at atmospheric pressure and is stable in the ambient environment to a congruent melting point at 1850°C. To the mineralogist there are three aluminosilicate polymorphs, all with composition $Al_2SiO_5$and none of which appear on the Al_2O_3–SiO_2 phase diagram at atmospheric pressure. The structures consist of AlO_6 octahedra combined with silica tetrahedra and a second aluminum of varying coordination number [39, 41]:

Kyanite	$Al^{VI}Al^{VI}Si^{IV}O_5$	$P\bar{1}$
Andalusite	$Al^{VI}Al^{V}Si^{IV}O_5$	*Pnnm*
Sillimanite	$Al^{VI}Al^{IV}Si^{IV}O_5$	*Pbnm*

The exact *P–T* equilibria between the three phases and the invariant point where the *P–T* boundaries intersect have been subject of a great many experimental investigations [42]. The phase diagram (Fig. 1.1.17) shows the difficulty in reconciling the atmospheric pressure "ceramic" part of the diagram with the high-pressure "mineralogy" part of the diagram.

The amphiboles are an extremely complex family of silicate structures. The basic building block is a double chain or ribbon of silica tetrahedra with

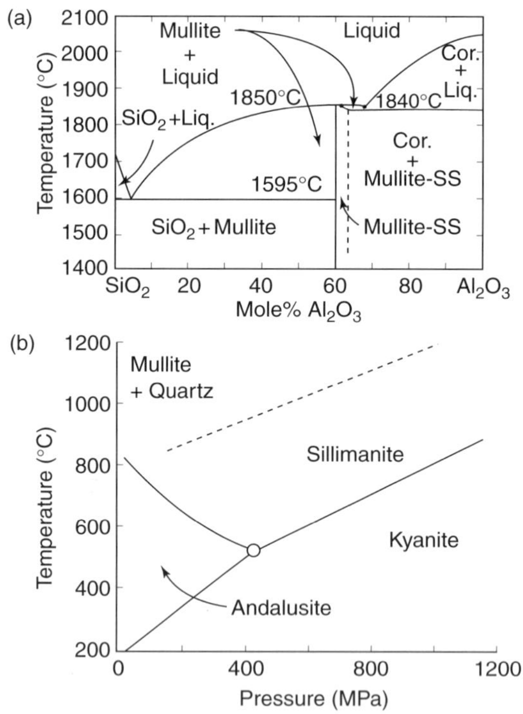

FIGURE 1.1.17 Phase diagram for the system Al_2O_3–SiO_2. (a) high temperature, atmospheric pressure "ceramic" phase diagram; (b) *P–T* diagram at Al_2SiO_5 composition showing high-pressure Al_2O_3:SiO_2 phases and their decomposition into mullite and silica at temperatures below that of the "ceramic" diagram. Diagram constructed from data in Kerrick [42].

composition Si_4O_{11}. The prototype is often taken as the mineral tremolite, $Ca_2Mg_5Si_8O_{22}(OH)_2$. There are four cations sites that allow many substitutions of intermediate-size ions. Al substitution into the silicate ribbon is also possible leading to still more structural complexity. Of most interest to ceramists are a subgroup of these fibrous silicates known collectively as asbestos. These were widely used as refractory insulation until the health hazards associated with asbestos fibers became recognized in the 1970s.

Linking three corners of the SiO_4 tetrahedra produces a two-dimensional sheet. The three Si–O–Si linked bonds form a hexagonal array with the three bridging oxygens forming the base and the bond with the single non-bridging oxygen with its formal negative charge perpendicular to the plane of the sheet. The fundamental silicate grouping can be written $Si_2O_5^{2-}$. Alkali salts, for example, $Na_2Si_2O_5$, have a pure silicate layer structure although the hexagonal rings are organized differently in the various polymorphs [37, 43]. The clay minerals contain these silica sheets—referred to as the tetrahedral layer—with the free charges, one per tetrahedron, matched by superposition of an octahedral layer of $Al(OH)_6$ (gibbsite layer) or $Mg(OH)_6$ (brucite layer). The gibbsite (or brucite) layer is accommodated by replacing a portion of the hydroxy groups by the non-bridging oxygen, thus making the combined 700-pm layer electrically neutral (Fig. 1.1.18). The layers are bound only by weak van der Waals forces. The aluminum prototype is kaolinite, $Al_2Si_2O_5(OH)_4$. The corresponding magnesium compound is serpentine, $Mg_3Si_2O_5(OH)_4$. It is also possible to insert a second tetrahedral layer "upside down" on the octahedral layer producing a 1000-pm sheet with composition $Al_2Si_4O_{10}(OH)_2$ (pyrophyllite) and $Mg_3Si_4O_{10}(OH)_2$ (talc). Many variants in stacking of these basic structures are possible.

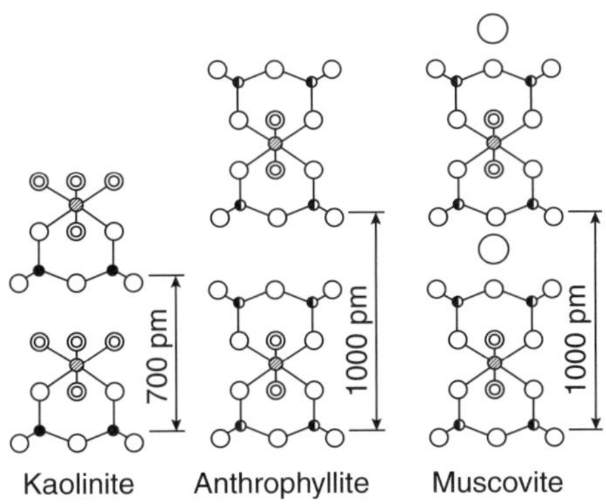

FIGURE 1.1.18 The 700-pm kaolinite structure, the 1000-pm two-layer (anthrophyllite) structure, and the structure of muscovite mica showing arrangement of tetrahedral and octahedral layers.

In the clays and related minerals, all Al^{3+} are in six-fold coordination. There is a small mismatch between the octahedral and tetrahedral layers, which results in increasing strain as the layers become larger. Eventually, the internal strain destabilizes the layer, resulting in the very small crystal sizes for the clay minerals. Al^{3+} can be substituted for Si in the tetrahedral layer. In this case, replacing Si by Al results in a deficiency of one positive charge for each tetrahedral aluminum. Charge balance is maintained by placing large cations in the interlayer position (Fig. 1.1.18). The resulting composition is $KAl_2(Si_3Al)O_{10}(OH)_2$, muscovite. This and other possible substitutions form the mica family of structures. The larger Al^{3+} ion on the Si-site relieves the strain so that mica crystals are among the largest found in nature. Synthetic layer silicates can be prepared by replacing silicon by germanium or Al^{3+} by Ga^{3+} or Fe^{3+}.

Silicon dioxide, silica, SiO_2, is one of the most important of ceramic materials and also one of the most structurally complicated. Silica is used as refractory brick, as feed stock for preparing other ceramic materials, as glass, as high-quality single crystals for the electronics industry, and as insulator films in the manufacture of semiconductor chips to name a few. Some of the silica polymorphs are listed in Table 1.1.3. Most of what follows is drawn mainly from Heaney *et al.* [44]. The stable low-temperature form, α-quartz is trigonal, space group $P3_121$, and its enantiomorphic partner, $P3_221$. Crystals of quartz occur in both left- and right-handed varieties. The structure consists of

TABLE 1.1.3 The Polymorphs of SiO_2

Structure	Space group	Comments
α-Quartz	$P3_121$	Left handed
	$P3_221$	Right handed
β-Quartz	$P6_422$	Left handed
	$P6_222$	Right handed
HP tridymite	$P6_3/mmc$	
Low tridymite		At least five variants
α-Cristobalite	$P4_12_12$	
β-Cristobalite	$Fd3m$	
Coesite	$C2/c$	High-pressure phase
Stishovite	$P4_2/mnm$	Very high pressure phase
Moganite	$I2/a$	
Melanophlogite	$P4_232$	Clathrate structure
Silica K	$P4_32_12$	
Silica W		Edge-sharing tetrahedra
Silica glass	—	
Densified glass	—	
Silica aerogel	—	

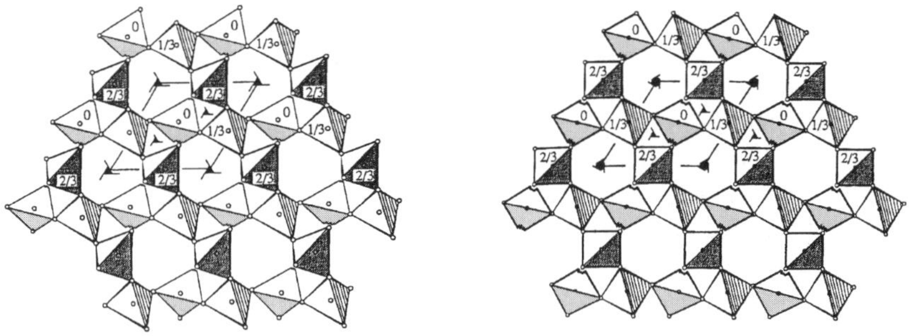

α-Quartz β-Quartz

FIGURE 1.1.19 The trigonal low- (or α-) quartz structure compared with the hexagonal high- (or β-) quartz structure. Adapted from Heaney *et al.* [44].

a three-dimensional framework of corner-sharing silica tetrahedra. The tetrahedra form spirals along the *c*-direction. The possibility of left- and right-handed spirals gives the quartz structure its handedness. At 573°C, quartz undergoes a fast, displacive phase transition. The bond angles relax and the structure becomes the hexagonal, space group $P6_422$, β-quartz structure (Fig. 1.1.19). At 867°C, the β-quartz form converts to tridymite and at 1470°C tridymite converts to cristobalite. The conversion to tridymite and cristobalite are reconstructive and kinetically very sluggish. At a pressure of 3.35 GPa and at a temperature of 500–800°C, silica converts into the dense but four-coordinated high-pressure phase, coesite. At pressures in the range of 12 GPa is a phase change to a dense six-coordinated rutile structure known as stishovite. At still higher pressures, 35–40 GPa, a more dense form of silica with the $CaCl_2$ structure has been reported.

In many silicate structures, AlO_4 tetrahedra can be substituted for SiO_4 tetrahedra providing that charge balance can be established. Consider these replacements in the three-dimensional silica framework:

$$Na + Al \rightarrow 2SiO_2 \rightarrow NaAlSiO_4$$, nepheline, prototype feldspathoid

$$Na + Al \rightarrow 3SiO_2 \rightarrow NaAlSi_2O_6 \cdot H_2O$$, analcime, protype zeolite

$$Na + Al \rightarrow 4SiO_2 \rightarrow NaAlSi_3O_8$$, albite, prototype feldspar

The feldspars occur in nature as solid solutions of $NaAlSi_3O_8$ (albite) and $CaAl_2Si_2O_8$ (anorthite) and as $KAlSi_3O_8$ (orthoclase). These have been used for centuries as feedstocks for the manufacture of porcelain ceramics. There is a huge literature on the detailed structure and chemistry of the feldspar family [45].

$NaAlSi_2O_6 \cdot H_2O$ is a prototype of a large family of compounds, both natural and synthetic, known as the zeolites. These are frameworks of AlO_4 and SiO_4 tetrahedra that contain structural tunnels and cages of various sizes. Synthetic zeolites can also be prepared with PO_4^{3-} tetrahedra substituting for SiO_4 as in the $AlPO_4$ (ALPO) family of structures [46]. Because of the ability to control tunnel and cage sizes, synthetic zeolites are used as filters, adsorbants, and as catalyst substrates.

The feldspathoids, like the feldspars, are examples of stuffed structures. The large alkali or alkaline earth ions required for charge balance occupy a cavity within the Si–Al–O framework. The feldspathoids have less application other than pollucite, $CsAlSi_2O_6$, which has been considered as a host phase for immobilization of ^{137}Cs in nuclear waste.

1.1.2.2 Chalcogenide Ceramics

1.1.2.2.1 Binary Sulfides, Selenides, and Tellurides

From simple packing principles, one might expect binary chalcogenides to form the same structures as the oxides only with larger cations to maintain radius ratios with the larger sulfide, selenide, and telluride ions. This only partially true. Some oxide structures, particularly the NaCl structure, appear among the chalcogenides. However, rutile and fluorite structures rarely appear and new structures take their place.

Binary chalcogenides of the alkaline earth metals such as MgS and CaS take the NaCl structure. However, these compounds react readily with water vapor and are unstable in ambient environments. Sulfides of the divalent rare earths and also compounds such as LaS also have the NaCl structure; the former are semiconductors while the latter are metals.

Binary chalcogenides of many of the transition metals take the niccolite (or NiAs) structure, which contains face-sharing octahedra. The face-sharing structure is stabilized by direct metal–metal orbital overlap and thus metallic bonding (Fig. 1.1.20). The NiAs structure appears in transition metal compounds such as CoSe, FeS, IrSb, NiAs, NiTe, and PdSn to name only a few. The d-orbital electrons are necessary to stabilize the NiAs structure; it does not appear in the compounds of filled shell ions. The structure is hexagonal, space group $P6_3/mmc$, with the cations in six-fold coordination.

Most important are the binary chalcogenides of Zn^{2+} and Cd^{2+}. Hybridization of the filled d-orbitals stabilizes four-coordinated structures. ZnS, the prototype, appears as the cubic sphalerite (also called zincblende) structure and the hexagonal wurtzite structure (Fig. 1.1.21). The sphalerite structure consists of a three-dimensional framework of corner-sharing ZnS_4 (or SZn_4) tetrahedra.

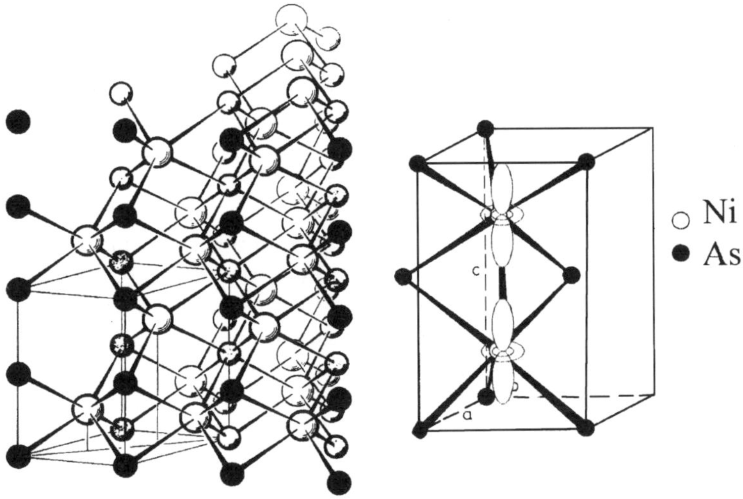

FIGURE 1.1.20 The NiAs (niccolite) structure with sketch showing the direct orbital overlap across the shared faces.

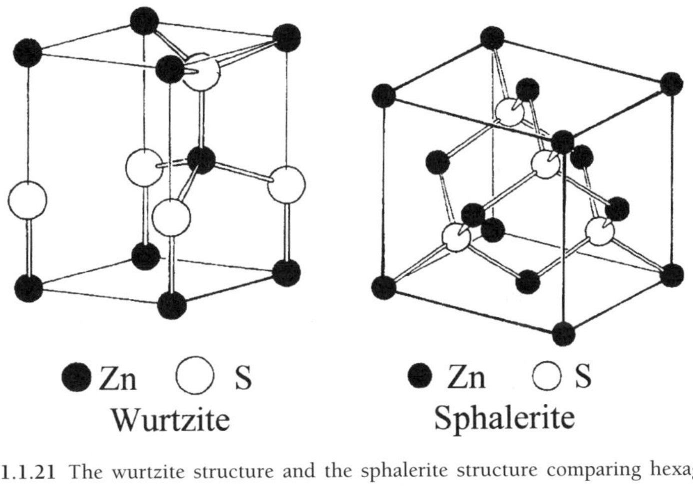

FIGURE 1.1.21 The wurtzite structure and the sphalerite structure comparing hexagonal and cubic packing.

The space group is $F\bar{4}3m$. The wurtzite structure is also a framework of corner-sharing tetrahedra but based on a hexagonal rather than cubic packing. The space group is $P6_3mc$. In addition to the Zn^{2+} and Cd^{2+} chalcogenides, the sphalerite and wurtzite structures are taken by the III–V semiconductors such as GaP, GaAs, and InSb.

ZnS, usually in the sphalerite form and activated with copper and/or aluminum, is one of the oldest phosphors for cathode ray tube screens. Sphalerite structure III–V compounds are used as semiconductors, as light emitting diodes, and as infrared detectors. Hot-pressed ZnS and ZnSe are used as infrared window materials.

AX_2 chalcogenide compositions do not follow the oxide pattern. The rutile and fluorite structures do not appear. Many transition metal sulfides, selenides, and tellurides along with some arsenides and antimonides take the pyrite structure. Pyrite itself, FeS_2, is cubic, space group $Pa3$, with Fe^{2+} in six-fold coordination. The chalcogens have a tendency to form dimers, S_2^{2-}, so that pyrite consists of a packing of Fe^{2+} ions and S_2^{2-} molecules (Fig. 1.1.22). Most of the pyrite and related structures, such as marcasite and arsenopyrite, are semiconductors or metals with limited technological application.

Some AX_2 chalcogenides such as ZrS_2 and MoS_2 form layer structures. MoS_2, the molybdenite structure, is hexagonal, space group $P6_3/mmc$, with strong bonding within the layers and only weak van der Waals bonding between the layers. It finds application as a solid-state lubricant.

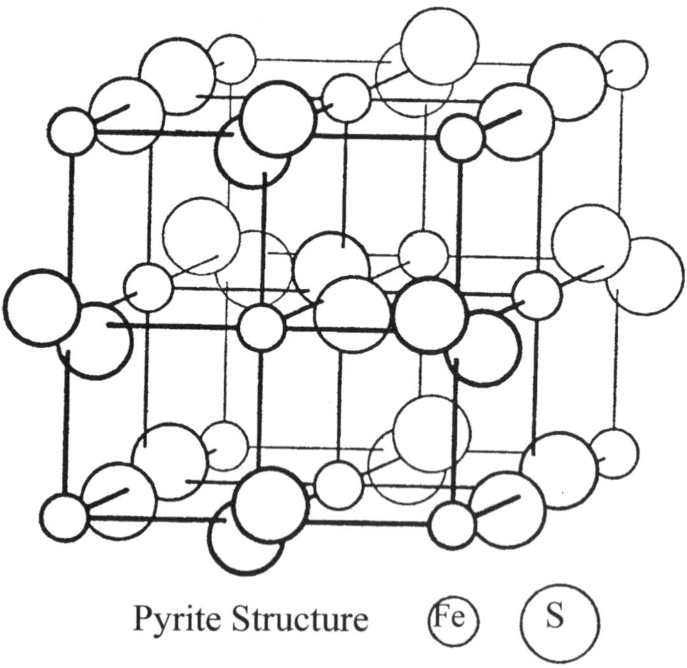

FIGURE 1.1.22 The pyrite structure. Note dumbbell-shaped pairs of S_2 molecules.

TABLE 1.1.4 The Polymorphic Form of the Rare Earth Sesquichalocogenides

Phase	Structure type	Space group	Z	c.n. (cation)	c.n. (anion)
α	Gd_2S_3	*Pnma*	4	7–8	5
β	$Pr_{10}S_{14}O$	$I4_1/acd$	8	7–8	4
γ	Defect Th_3P_4	$I\bar{4}3d$	6	8	5.33
δ	Ho_2S_3	$P2_1/m$	6	7–6	4.33
ε	Corundum	$R\bar{3}c$	6	6	4

The rare earth sesquichalcogenides form a series of structures, generally different from the corresponding oxides. Following Flahout [47], these phases are given a Greek letter notation (Table 1.1.4). The rare earth sesquichalcogenides are mainly high-gap semiconductors and most are highly refractory compounds with melting points from 1600 to 2000°C.

1.1.2.2.2 Ternary Sulfides, Selenides, and Tellurides

The chalcopyrite structure is a superstructure consisting of two stacked sphalerite-type cells. The structure is tetragonal, space group $I\bar{4}2d$. Like the parent sphalerite structure, chalcopyrite is polar and as such compounds with the chalcopyrite structure have been of some interest as electrooptic materials.

There are chalcogenide analogs of the olivine and spinel structures. There also exists the Cr_3S_4 structure taken by many transition metal sulfides with the A_2BX_4 composition. With the larger rare earths there are formed new families of structures that have no oxide analogs (Fig. 1.1.23). The Th_3P_4 structure is cubic, space group $I\bar{4}3d$. Both cations are in eight-fold coordination. Compounds with this structure such as $CaLa_2S_4$ are highly refractory with melting points near 2000°C. There has been interest in hot-pressed or hot-isostatic-pressed ceramics of this composition as infrared optical materials that would be transparent in the 8–14-μm atmospheric window [48].

1.1.2.3 Nitride and Oxynitride Ceramics

1.1.2.3.1 Binary Nitrides

The ambient pressure form of BN, sometimes known as "white graphite", is hexagonal, space group, $C6/mmc$. It is a layer structure containing edge-sharing hexagonal rings of alternating boron and nitrogen atoms (Fig. 1.1.24). There is strong covalent bonding within the sheets but only weak van der Waals bonding

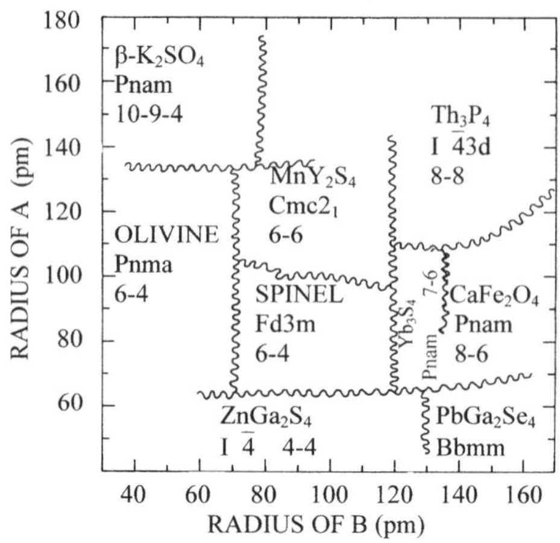

FIGURE 1.1.23 Structure field map for A_2BX_4 chalcogenides. Compounds listed are the prototypes for each structure type.

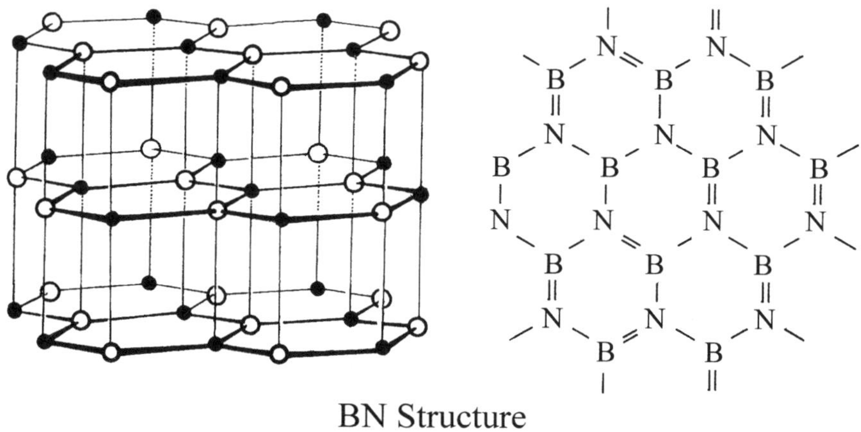

FIGURE 1.1.24 Structure of ambient pressure BN showing the layer structure and the double bond character.

between them. Like graphite, low-pressure BN is a soft material and can be used as a lubricant. At high temperatures and at pressures in excess of 8.5 GPa, BN converts to a sphalerite structure, extremely hard material often abbreviated as c-BN [49].

Other four-coordinated nitrides, with the sphalerite or wurtzite structures, form from trivalent ions: AlN, GaN, ScN, InN, and others. These along with the corresponding phosphides, arsenides, and antimonides form the important family of III–V semiconductors. Development of a synthesis method for ScN, in particular, made possible the construction of an intense white light emitting diode.

1.1.2.3.2 Silicon and Aluminum Nitrides

The silicon and aluminum ternary nitrides were extensively investigated as possible light, high-temperature, tough ceramics for use in turbines—the concept of a ceramic engine [50]. Processing and fabrication difficulties have largely frustrated these applications.

Consider as a starting point Si_3N_4. Si_3N_4 is known to exist as two polymorphs, α and β, both with all ions in four-fold coordination. β-Si_3N_4 is isostructural with phenakite, Be_2SiO_4. A cubic high-pressure form of Si_3N_4 with the spinel structure has been identified in which silicon is in six-fold coordination [51]. Band structure calculations predict an entire family of cubic spinel-structure nitrides [52].

Si_3N_4 consists of a three-dimensional framework of SiN_4 tetrahedra. Oxygen will substitute for nitrogen giving silicon oxynitride, Si_2N_2O. This composition can be thought of as a 1:1 compound between Si_3N_4 and SiO_2. Another starting point is AlN. Reactions between AlN and Al_2O_3 leads to the complex sequence of aluminum oxynitrides known as the ALONs, structural ceramics of light weight and high hardness and toughness. Putting it all together as a phase diagram (Fig. 1.1.25) [53] shows some of the ternary solid solutions and ordered compound known as the sialons.

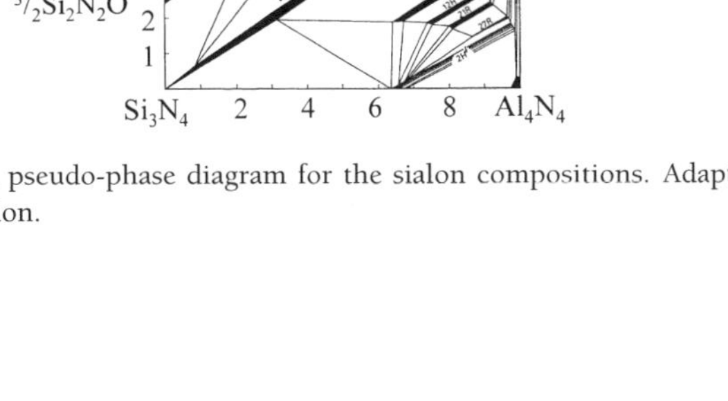

FIGURE 1.1.25 A pseudo-phase diagram for the sialon compositions. Adapted from Ref. [53]. Used with permission.

1.1.2.4 Carbides and Borides

1.1.2.4.1 Diamond and Silicon Carbide

Diamond is cubic, space group *Fd3m*, with eight carbon atoms in the face-centered cell. The structure consists of a three-dimensional framework of corner sharing CC_4 tetrahedra. Because of the strong covalent bonding in three dimensions, diamond is the hardest substance known. Because of the saturated carbon–carbon bonds, diamond is an insulator, band gap 5.5 eV. Diamond has a single Raman-active zone-center phonon at $1332\,cm^{-1}$, which gives diamond an exceptionally high thermal conductivity. In addition to the cubic packing of carbon atoms in diamond, a hexagonal packing is also known (Fig. 1.1.26). Hexagonal diamond has been synthesized [54] and occurs in nature as the rare mineral lonsdaleite.

The stable form of carbon at atmospheric pressure is graphite, a layer structure with π-bonds within the layers and weak van der Waals bonding between the layers. Diamond is stable only at pressures in excess of 4–10 GPa at temperatures in the range of 2000–3000°C [55]. The phase transition from graphite to diamond requires complete reconstruction of the bonds so that the kinetics are extremely sluggish. As a result the direct synthesis of diamond requires high pressures, high temperatures, and usually the presence of a metal catalyst [56]. The discovery, in the late 1980s, that diamond films could be metastably synthesized by chemical vapor deposition reactions was a great breakthrough [57–59].

Silicon carbide is related to diamond in the sense that one of the two carbons in diamond is replaced by silicon while retaining the three-dimensional framework of corner-sharing tetrahedra. However, the arrangement of alternating CSi_4 and SiC_4 tetrahedra removes elements of symmetry so that SiC has the structure of sphalerite rather than diamond. A hexagonal form of SiC isostructural with wurtzite is also known. SiC displays an extremely complex sequence of polytypes ranging from 2H, the basic hexagonal stacking, to at least 393 R,

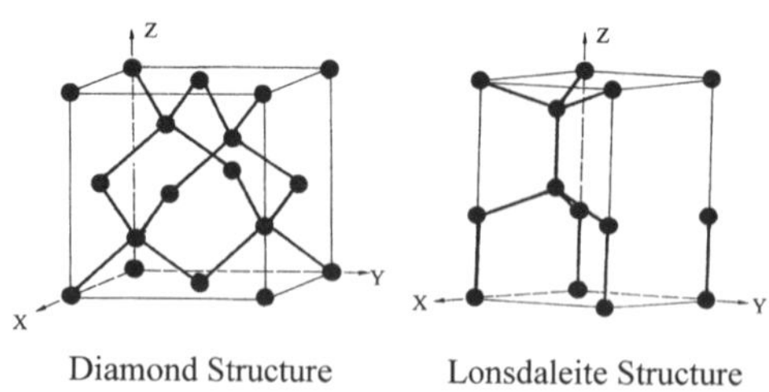

FIGURE 1.1.26 Structure of cubic and hexagonal diamond. SiC has the same arrangement with Si and C atoms alternating.

with a hexagonal c-axis of 9896 pm. SiC is a hard material, useful as an abrasive, and also as a high-temperature semiconductor.

1.1.2.4.2 Heavy Metal Carbides

Heavy metal carbides for the most part are considered to be intermetallic compounds and so outside the scope of this review. The structures consist of a packing of metal atoms with the small carbon atoms occupying an octahedral interstice within the metal structure [60]. Of importance is tungsten carbide, WC, a cutting tool material, an especially simple structure with a hexagonal cell with $a = 290$ pm and $c = 284$ pm. Other carbides such as the Ti_3SiC_2family of machinable carbides [61] occupy the borderland between ceramics and alloys.

1.1.2.4.3 Borides

Borides are in large part unique structures because of the tendency of boron to bond with itself, often forming octahedra and icosahedra consisting entirely of boron. Because of the strong B–B bonding, many of these compounds are metallic and are very hard materials.

B_4C is the third hardest substance known, the first and second being diamond and cubic boron nitride, respectively. B_4C has a rhombohedral cell, space group $R\bar{3}m$. It can be visualized as a three-dimensional network of B_{12} icasohedra with C_3 groups filling the largest holes.

Among the borides of interest to ceramists are the diborides with the AlB_2 structure. The unit cell is hexagonal, space group $C\bar{3}m$. It is a layer structure with alternating aluminum layers and boron layers perpendicular to the c-axis. Two compounds are of interest. TiB_2, extremely hard (18–27 GPa) and with a high melting point (2970°C), has application as armor, nozzles, and wear parts. Another compound with the AlB_2 structure, MgB_2, has been found to be a superconductor with a transition temperature of 39 K [62].

The diversity of boride structures can be illustrated with the rare earth borides [63]. There are LnB_2 compounds with the AlB_2 structure. The LnB_4 compound is tetragonal, space group $P4/mbm$, with two kinds of boron, boron with sp^2 bonding and B_6 octahedra. LnB_6 is cubic, space group $Pm3m$. The lanthanide ion occupies the center of the cell. Each corner is occupied by cross-bonded B_6 octahedra (Fig. 1.1.27). Compounds such as LaB_6 are metallic and have a high thermionic emission rate so that the compounds find some application as filaments. The concept of three-dimensional structures composed entirely of boron is extended in the cubic ($Fm3m$) LnB_{12} structure. The boron framework is composed of cross-bonded B_{12} dodecahedra. In LnB_{66}, the framework consists of “super icosahedra” with composition $B_{12}(B_{12})_{12}$.

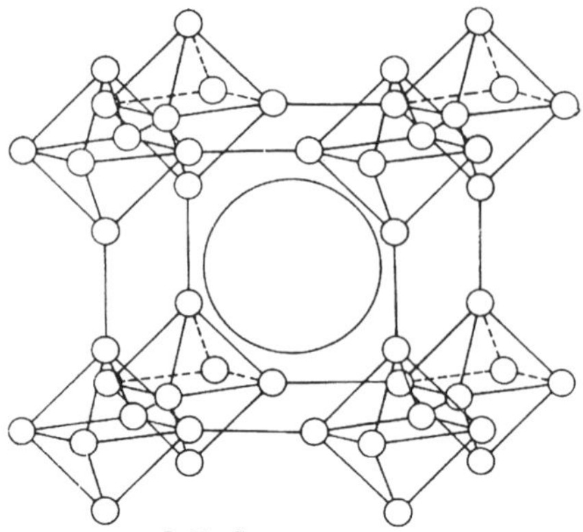

FIGURE 1.1.27 The LaB_6 structure. Note arrangement of cross-bonded (not corner-sharing) boron octahedra.

1.1.3 CHEMICAL BONDING IN CERAMICS

There are three main models that can be applied to chemical bonds in ceramics and each has its own theoretical structure. These are the classical electrostatic model, augmented by crystal field theory, for ionic compounds, the molecular orbital model for covalent compounds, and the band model for semiconductors and metals. All have their advantages and drawbacks.

1.1.3.1 The Electrostatic (Ionic) Model

If the internal ionization is complete, that is, the electron transfer from cation to anion leaves no residual electron density between the ions, both cation and anion can be treated as charged spheres. The bonding between them then results from their mutual electrostatic attraction. The electrostatic attraction is counter-balanced by the repulsive forces that arise when the outer electron shells of the ions are pulled together. The attractive forces are Coulomb forces that fall off with the reciprocal of the distances between the ions. The repulsive forces rise as reciprocal of the eighth or ninth power of the distance as the ions are brought closer together (or exponentially with a Born–Mayer potential). The balance between these oppositely acting forces defines the equilibrium cation–anion distance. The potential function for a cation–anion pair takes the form shown in Figure 1.1.28.

The potentials for pairs of ions can be summed over the entire crystal to obtain an expression for the lattice energy. Adding two additional

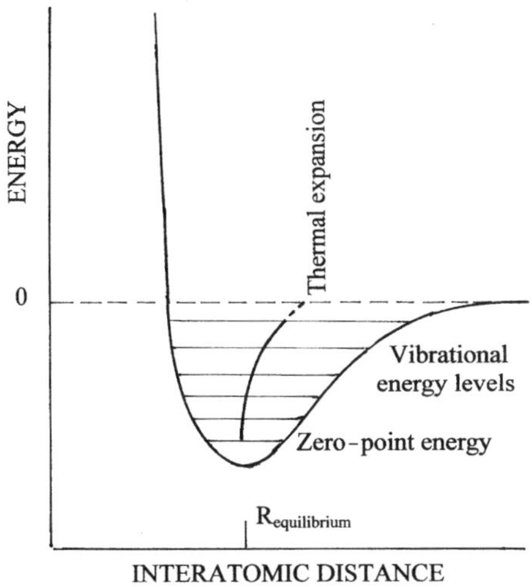

FIGURE 1.1.28 Potential function for the classical ionic model. Horizontal lines represent vibrational states. The lowest of these is the zero-point energy.

embellishments, the van der Waals interaction between the ions and the zero point energy (which have opposite signs and partially cancel), the lattice energy is given by

$$U = -\frac{e^2 z_+ z_- N_0 A}{r} + BN_0\, e^{-r/\rho} - \frac{CN}{r^6} + 2.25 N_0 h\nu$$

The four terms are the Coulomb attraction, the Born–Mayer repulsion, the van der Waals attraction, and the zero point energy. In this equation, U is the lattice energy in kJ/mol, e is the electron charge, z_+ and z_- are the formal changes on the ions, N_0 is Avagadro's number, B and ρ are repulsion coefficients, CN the coordination number, and ν the vibrational frequency of the crystal. Because the Coulomb attraction falls off very slowly with distances, the bonding between any ionic pair also contains contributions from next nearest neighbors (cation–cation—repulsive), next next nearest neighbors (cation–second shell anions—attractive), and on outward from the target bond pair. This effect gives a slowly converging series that sums to the Madelung constant, the parameter A in the lattice energy equation. The Madelung constant depends on the crystal structure type but not the specific compound.

Considering the rough approximations, the classical electrostatic model gives lattice energies for binary oxide compounds that often lie within 5% of measured values, a quite remarkable result. The pure ionic model becomes less and less accurate as the covalency of the bonds increases.

1.1.3.2 Crystal Field Theory

Although the classical electrostatic model produces a reasonably good approximation to the lattice energy of insulator crystals, the potential function describes only the ground state. There is no provision for the excited states responsible for color, luminescence, other optical properties, or magnetic properties or electrical conductivity. Crystal field theory takes into account the bound states arising from partially filled d- and f-orbitals in insulating crystals. Electrons in the d-orbitals make a small contribution to bond energies but a large contribution to optical and magnetic properties. Electrons in the f-orbitals contribute little to bonding but also determine optical and magnetic properties.

1.1.3.2.1 The Crystal Field Model

Crystal field theory is a quantum mechanical approach that is quite precise within the constraints of its assumptions. Details may be found in a number of textbooks such as Ballhausen [64], Sugano *et al.* [65], and Henderson and Imbusch [66], for transition metals, and Wybourne [67] for the rare earth ions. Also of value is the mineralogical text of Burns [68], which contains many spectra and calculations of crystal field stabilization energies for compounds of interest to ceramists.

The fundamental assumption is that the anions surrounding the site of a transition or rare earth ion in a crystal act strictly as point electrostatic charges. Their interaction with the d- or f-orbital electrons is through the Coulomb field with no provision for overlap or exchange. Considering only the d-orbitals, the orbital lobes are oriented within a six-fold coordination site as shown in Figure 1.1.29. The orbitals d_{xy}, d_{xz}, and d_{yz} are oriented between the anions. The other two, $d_{x^2-y^2}$, and d_{z^2}, point directly toward the anions. The resulting energy level diagram is sketched in Figure 1.1.30. The core states represent the minimum in the potential function for the electrostatic model and represent the balance between Coulomb attraction and Born repulsion without the effect of the d-electrons. Introducing electrons into the d-orbitals actually increases the average repulsion. However, the differing orientations of the d-orbitals splits the orbital states into two sets with the three orbitals oriented between the anions becoming stabilized with respect to the mean value, and the other two becoming proportionally destabilized. The electrostatic interaction between the d-electrons and the negative charges on the anions can be treated by perturbation theory using hydrogen-like orbitals for the d-orbital states. In a crystal field with perfect octahedral symmetry, the angular parts of the wavefunctions integrate exactly to small integers. The integrals of the radial wavefunctions cannot be calculated analytically, but can be expressed in terms of a single

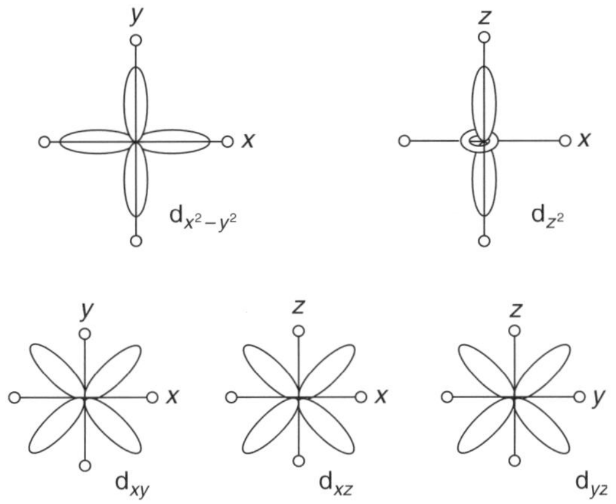

FIGURE 1.1.29 Arrangement of d-orbitals in an octahedral site.

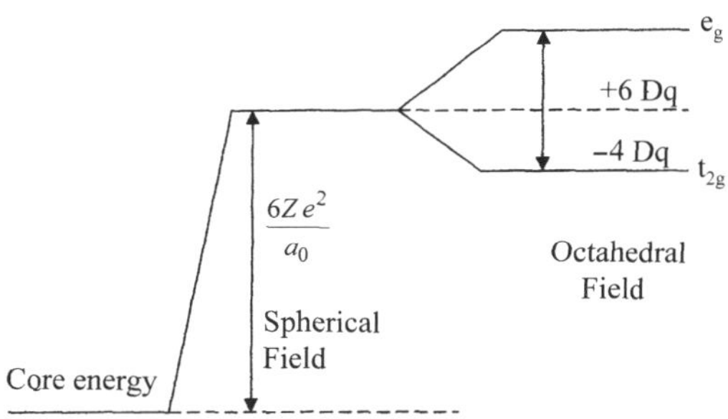

FIGURE 1.1.30 The energy level diagram for the d-orbitals scaled in terms of the crystal field strength parameter *Dq*.

parameter, *q*. The overall crystal field interaction for a single electron in an octahedral crystal field can be expressed by a single parameter, *Dq*, as shown in Figure 1.1.30. *D* is the crystal field potential. *Dq* is treated as an empirical parameter to be determined experimentally from optical absorption spectra.

When more than one electron occupies the d orbitals there arises another interaction due to the mutual repulsion of the d-electrons. Because the d-electron interactions with the crystal field and with each other are of comparable magnitude, they must be calculated together. In addition to the crystal field and the interelectron repulsion, there is an interaction between the electron spin and the electron orbital momentum. The full Hamiltonian for the interactions of both d- and f-electrons is

$$H = \sum_i \frac{h}{\pi m}\nabla_i^2 + \sum_i \frac{Ze^2}{r_i} + \sum_i \frac{e^2}{r_{ij}} + V_{cf} + \sum_i \xi_i \vec{s}_i \cdot \vec{\ell}_i$$

For the free ion, $V_{cf} = 0$. If the spin–orbit interaction is taken to be small compared with the interelectron repulsion, the free ion energy states can also be calculated with the integrals of the radial wave functions expressed in terms of the Racah parameters *A*, *B*, and *C*. Like *Dq*, the Racah parameters are treated as empirical parameters to be determined from spectroscopic data.

The relative magnitudes of the three perturbation terms in the Hamiltonian lead to three different interactions of the d- and f-electrons. For such compounds as oxides and halides of the transition metals, the interelectron repulsion dominates the crystal field:

$$\frac{e^2}{r_{ij}} \geq V_{cf} \geq \xi_i \vec{s}_i \cdot \vec{\ell}_i$$

In these compounds, the unpaired spins will be a maximum. This is referred to as the "high spin" or "weak field" case. Certain ligands that occur in transition metal complexes such as NH_3 and CN^- produce sufficiently high crystal fields to quench the spins and produce the "low spin" or "strong field" case:

$$V_{cf} \geq \frac{e^2}{r_{ij}} \geq \xi_i \vec{s}_i \cdot \vec{\ell}_i$$

Some sulfides are also low-spin compounds. The distinction between high- and low-spin arrangements is illustrated in Figure 1.1.31. The spin arrangement strongly affects the optical and magnetic properties and also the thermodynamics and crystal structures of the compounds. Ionic radii for high-spin ions are greater than ionic radii for low-spin ions.

The strength of the crystal field, described by the parameter, *Dq*, varies continuously. There is an abrupt transition where certain energy levels cross and the behavior shifts from weak field to strong field. The complete range

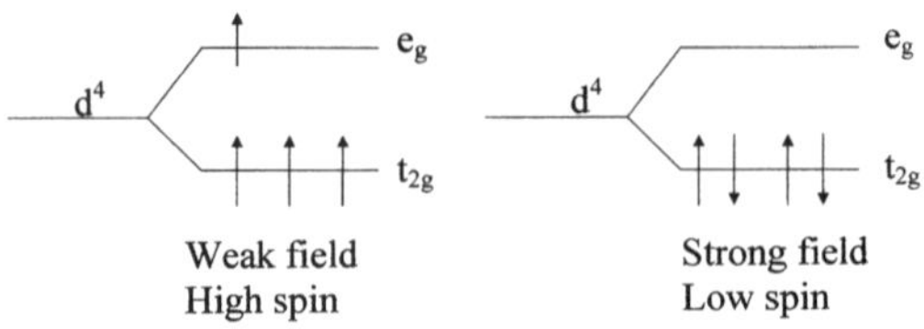

FIGURE 1.1.31 High spin and low spin electron configurations for a d^4 ion in an octahedral field.

of behavior is described by the Tanabe–Sugano matrices that contain both *Dq* and the Racah parameters *B* and *C*. Diagonalizing these matrices produces complete energy level diagrams for the transition metal ions (ignoring spin–orbit coupling). These, in turn, allow the assignments and interpretation of optical absorption spectra, luminescence spectra, and magnetic behavior of the transition metal ions.

In the rare earth ions, the f-electrons are well shielded and the interaction with the crystal field is very small. Likewise, the strength of the spin–orbit coupling is higher in the rare earths than in the transition metals. The result is what may be called the "feeble field":

$$\frac{e^2}{r_{ij}} \geq \xi_i \vec{s}_i \cdot \vec{\ell}_i \geq V_{cf}$$

Because of the small influence of the crystal field, optical spectra and magnetic properties of rare earth compounds are to a first approximation independent of the host structure. A single-energy-level diagram, known as the Dieke–Crosswhite diagram, describes the main features of rare earth ion spectra (see Blasse and Grabmaier [69] for the diagram and its application to luminescent materials).

The energies associated with the crystal field states are in the range of 0.5–3 eV so that the transitions between these states fall into the visible range of the electromagnetic spectrum. Crystal field transitions are between states of the same parity and thus are forbidden in the first order. Crystal field absorptivities are in the range of 1–1000 cm^{-1}, compared to 10^4–10^6 cm^{-1} for band gap transitions. The familiar colors of transition metal compounds are for the most part due to these transitions. Cr_2O_3 is a deep green color as are glasses colored by Cr^{3+}. Ruby and spinel are pink with the colors also arising from Cr^{3+}. Ni^{2+} produces more of a yellow-green. Co^{2+} in four-ccordination gives an intense deep blue whereas Co^{2+} in six-coordination gives a pale pink. Both coordination number and interatomic distance play a role. Co^{2+} in solid solution in ZnO gives a deep green. In contrast to the deep colors of the transition metal ions, the rare earth ions produce weak pastel colors: pale purple for Nd^{3+}, apple green for Pr^{3+}, and pink for Er^{3+}. The crystal field model accounts rather well for the positions and intensities of the absorption bands and their relation to host structure. Likewise, with the Tanabe–Sugano matrices, the optical absorption spectra allow the calculation of *Dq*, *B* and *C* for transition ions in specific host structures.

Crystal field parameters are to a great extent portable. The spectra of the hexaquo complexes of the transition metal ions are often used as standards from which are calculated reference values for *Dq* [70] (Table 1.1.5). These can then be scaled for other ligands by *Dq* (ligand) = *fDq* (ref). The scaling

TABLE 1.1.5 References Values for *Dq* in Hexaquo Complexes

M^{2+}	*Dq* (cm^{-1})	M^{3+}	*Dq* (cm^{-1})
V^{2+}	1180	Ti^{3+}	2030
Cr^{2+}	1400	V^{3+}	1800
Mn^{2+}	750	Cr^{3+}	1760
Fe^{2+}	1000	Mn^{3+}	2100
Co^{2+}	1000	Fe^{3+}	1400
Ni^{2+}	860	Co^{3+}	1910
Cu^{2+}	1260		

Data compiled from Holmes and McClure [70].

parameter, $f = 1.0$ for the oxide or hydroxide ion, 0.9 for fluoride, 0.8 for chloride, and 0.7 for sulfide. *Dq* is scaled for coordination number by noting that *Dq* (tetrahedral) $= -4/9Dq$ (octahedral) and *Dq* (cube) $= -8/9Dq$ (octahedral) (for equal interatomic distances). *Dq* scales with the inverse fifth power of the interatomic distance.

If the crystal field assumption of a purely electrostatic interaction between cation and anion held exactly, the Racah parameters for ions in crystals would be identical to the values calculated for the free ions from atomic emission spectra. In reality, the crystal values of Racah *B* and *C* are smaller by 10–20% compared to the free ions, an effect due to expansion of the d-orbitals caused by mixing the d-electron states of the cation with the ligand p-electron states. The quantity $(1 - B_{\text{cryst.}}/B_{\text{free ion}})$ has been proposed as a measure of covalency in transition metal ion bonds [71].

1.1.3.2.2 Crystal Field Stabilization Energies

Crystal field effects play a role in the relative stability and site distribution of transition metal ions in crystals [72]. Referring back to Figure 1.1.30, there is an additional stabilization energy for electrons added to the t_{2g} orbitals and a destabilization energy for electrons occupying the e_g orbitals. The crystal field stabilization energy is given by

$$E_{\text{cfs}}\ (\text{oct}) = 11.962[4n_t - 6n_e]Dq\ (\text{oct})$$

E_{cfs} is in J/mol and *Dq* is in wavenumbers. The crystal field stabilization energy is a maximum when there are three or eight electrons in the d-orbitals. The common ions with these configurations are Cr^{3+} and Ni^{2+}. As a trivalent ion, *Dq* for Cr^{3+} is larger (1760 cm^{-1}) than *Dq* for the divalent Ni^{2+} (860 cm^{-1}). Cr^{3+} and Ni^{2+} show the largest crystal chemical anomalies and the largest deviations from the expected linear dependence of thermodynamic properties on atomic number across the transition metal series. Crystal field stabilization energies range from zero for ions with half-filled shells, to several hundred kilojoules/mole, large enough to make a significant contribution to bulk thermodynamic properties.

The crystal field stabilization energy scales with Dq. Dq, in turn, scales with coordination number and with interatomic distance so that ions may show a preference for one site over another in a crystal. This was first proposed for the spinel structure as the octahedral site preference energy, $E_{oct} = E_{cfs}$ (oct) $- E_{cfs}$ (tetr), which provided a partial explanation for why some spinels, such as the chromite spinels, were always normal, while other spinels were inverse [73]. This concept has been extended to other competitions between sites in complex crystals [74].

1.1.3.3 Molecular Orbital Approaches to Chemical Bonding

The molecular orbital model for chemical bonding has its origins in the explanation of molecular structures, particularly strongly covalent molecules such as most organic compounds. The application to ceramic materials has been slow in developing. Although the basic concepts were set down in 1932, it was not until the widespread availability of large mainframe computers in the 1960s and later supercomputers, that high accuracy calculations could be made.

The basic concept is that the electrons that make up chemical bonds move in orbitals composed of linear combinations of the atomic orbitals of the atoms forming the bonds. When the atomic orbitals are mixed they split into bonding orbitals with energies much less that that of the atoms individually and antibonding orbitals that are destabilized with respect to the individual atoms (Fig. 1.1.32). Electrons are paired into these orbitals thus forming electron pair bonds. Non-degenerate orbitals can hold two electrons and are known

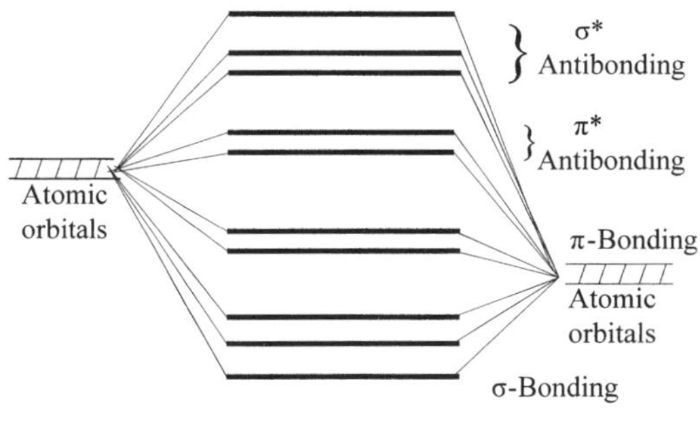

FIGURE 1.1.32 Schematic molecular orbital diagram.

as σ-orbitals. Doubly degenerate orbitals, π-orbitals, can hold four electrons. In general, σ-orbitals provide the greatest stabilization energy and thus the strongest bonds. The π-orbitals provide less stabilization energy but tend to control properties.

The core calculations are, in principle, simple. For the hydrogen molecule, for example, each hydrogen atom with the $1s^1$ configuration provides one orbital and one electron. The bonding and antibonding orbitals are linear combinations of the atomic orbitals

$$\Psi_B = \frac{1}{\sqrt{2}}(\phi_1 + \phi_2) \qquad \Psi_{AB} = \frac{1}{\sqrt{2}}(\phi_1 - \phi_2)$$

However, hidden in the simplicity of the hydrogen molecule is the source of a huge problem in computation when the same concepts are applied to more complex structures. To get a complete and accurate description of the chemical bonds in a complex molecule, one must, at least, in principle, include all possible orbitals (the basis) and all electrons that might participate in bonding. These all interact with each other. Each of the hydrogen-like atomic orbitals contains angular and radial functions and the calculation of orbital and thus bond energies requires integrals of these functions. Angular functions can often be integrated analytically while the radial functions are approximated with functions of gaussian form. Early molecular orbital calculations required simplifications and assumptions that greatly limited both accuracy and predictive power of the results. There are a great variety of these calculations in the literature, usually labeled with an acronym indicating the assumptions and the degree of simplification. Current generation calculations are *ab initio* because they follow from first principles—meaning that wavefunctions, mostly gaussian, are calculated, not replaced by empirical parameters. These have been reduced to computer programs such as Gaussian 98 that are widely available and used as practical tools for modeling chemical bonds [75].

Anions are electron-rich. These electrons can do some combination of the following:

1. Pair with cation electrons to form σ-bonds. This is usually the most energetically favorable process and forms the primary bonds of the molecular unit. There is then the question of what happens to electrons in excess of those needed to form the primary bonds.
2. The remaining electrons can occupy π-orbitals and so lead to double bond formation.
3. The excess electrons can remain localized on the anions to produce a negative charge, thus producing a molecular anion rather than a neutral molecule.

4. The excess electrons can participate in bonding with adjacent molecules to form polymeric structures such as the silicates.

These choices are not mutually exclusive. The energy differences involved are small and easily divert the bonding scheme into different structures. Consider carbon and silicon, with the same electronic structure. With hydrogen, when only one pairing electron per ligand is needed, both CH_4 and SiH_4 are molecular gases. With oxygen and carbon, the extra electrons form double bonds so that CO_2, O=C=O, is a molecular gas. With oxygen and silicon, O=Si=O does not exist. Instead, the extra electrons become formal negative charges on the SiO_4^{4-} tetrahedral unit or bond with additional tetrahedra to form the three-dimensional framework of quartz. In contrast, a CO_4^{4-} tetrahedral unit does not exist. The small size of carbon compared with silicon requires a lower coordination number to produce the carbonate, CO_3^{2-}, ion which, however, retains substantial double bond character and does not form polymerized structures.

Molecular orbital models and associated energy level diagrams are useful for the interpretation of some optical and other properties. The early empirical calculations were useful for this purpose because they provided an energy level structure that could be empirically fitted to spectroscopic data. The next step in complexity was to rationalize bond lengths and bond angles for known structures and likewise to account for observed thermodynamic data. A milestone was the estimation of the Si—O bond length by computing the energy of H_4SiO_4 as a neutral surrogate for SiO_4^{4-} as a function of bond length, and then selecting the bond length that minimized the energy (Fig. 1.1.33) [76]. Advances in computation and improved molecular orbital computer codes permitted the computations of increasingly complex structures [77, 78]. The final step in molecular orbital approaches has been to rationalize the entire crystal structure. The spinel structure is a useful test case (Fig. 1.1.34) where the positions of all atoms are specified by the lattice parameter and by a single oxygen position parameter, u. The calculation gives a rather complex energy surface with a distinct minimum that defines both the lattice parameter and the oxygen parameter [79].

The various *ab initio* families of molecular orbital models can provide much useful information [80] although these models specifically assume bound states. They can describe the bonding in molecular structures quite well but offer some difficulty when applied to crystals. The difficulty in applying these calculations to ceramic materials is the difficulty in defining a "molecule". Extracting a cluster of atoms that carries the essentials of the crystal structure has been used with success with the SCF – $X\alpha$ models [81]. These types of calculations have been especially applicable to sulfide and other semiconductor materials and

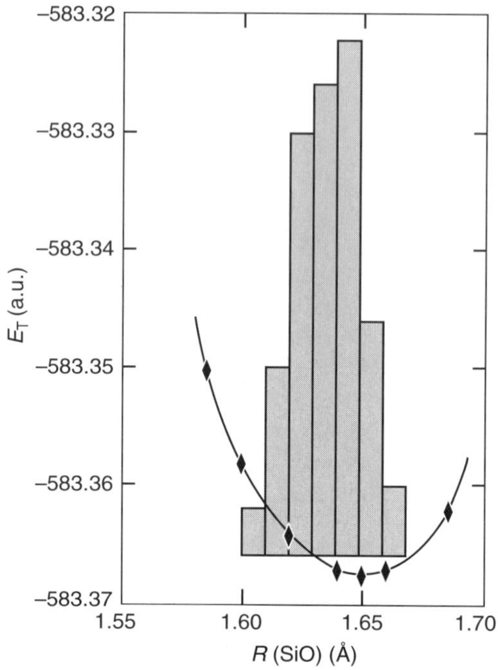

FIGURE 1.1.33 Energy of the H_4SiO_4 molecule as a surrogate for SiO_2. Energy is plotted as a function of Si–O bond length with the equilibrium distance defined by the minimum in the potential curve. Adapted from Gibbs [76]. Used with permission.

provide a better representation of the excited states needed to interpret spectra and transport properties.

1.1.3.4 Band Structure of Ceramics

The third of the theoretical models for the chemical bond is the most comprehensive and most difficult from both conceptual and computational points of view. Crystal field theory and molecular orbital theory contain only bound states. Electrons are fixed either on their parent atoms in crystal field theory or localized within molecular orbitals in molecular orbital theory. Each state has a unique energy and the concept of electron momentum does not apply Band theory allows for delocalized states and thus can account for transport properties such as electrical conductivity but at the expense of requiring calculation of both energies and momenta for all states.

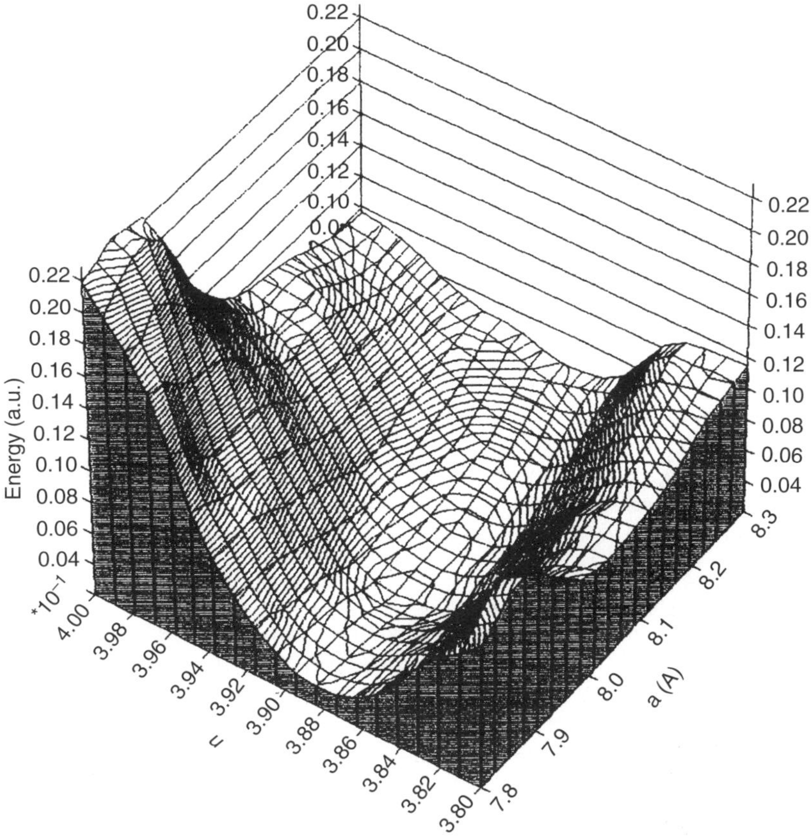

FIGURE 1.1.34 Energy surface for Al_2MgO_4 as a function of cell edge and oxygen parameter. Adaped from D'Arco *et al.* [79]. Reproduced by permission of American Geophysical Union.

1.1.3.4.1 The Band Model

The underlying framework of band structure calculations is an array of atomic nuclei (actually nuclei plus core electrons) that define a periodic potential (Fig. 1.1.35). The potential conforms to the space group symmetry of the crystal. The wave functions for the outer electrons overlap to form a continuum of states throughout the crystal. Because of the Pauli exclusion principle, these states must not have exactly the same energy so the assemblage of states broadens and is referred to as a "band". Typical widths of energy bands range from fractions of an eV to 5 eV. Because band states are not bound, electrons have, in addition to an energy, also a momentum, expressed as a wave vector, **k**. Energy band diagrams are plotted as energy versus wave vector. The wave vector has

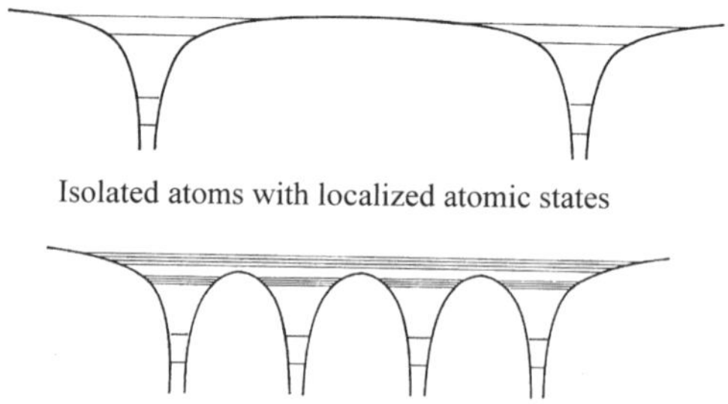

FIGURE 1.1.35 Schematic periodic potential with core and band states.

units of reciprocal length, typically cm^{-1}, so that the appropriate plotting space is not the unit cell but the reciprocal call, which, with a scaling factor, is known as the Brillouin zone. Because of the translational symmetry of the crystal lattice, the Brillouin zone contains all information about the band structure of the entire crystal [82].

In principle, constructing band diagrams consists of solving the Schrodinger equation for electrons moving in the periodic potential defined by the symmetry and interatomic distances of the crystal structure to arrive at wave functions of the form

$$\Psi = Ae^{-ik \cdot r}u(r)$$

In practice, all atomic states must be taken into account and the calculation of energy band diagrams is an intensive effort in computer programming. As a result, both form and interpretation of energy band diagrams lack the intuitive insight of crystal field and molecular orbital calculations.

As with molecular orbital calculations, as computational capability has increased, band structure calculations have evolved more into a working tool for rationalization of crystal properties. Calculations and property interpretations have been published for such ceramic materials as MgO [83], Al_2O_3 [84, 85], SiO_2 [86, 87], and TiO_2 [88–91]. Figures 1.1.36 and 1.1.37 show the energy band diagrams for TiO_2, a moderately ionic compound, and quartz, a partially covalent compound. Band diagrams are drawn as a series of sections along specified directions through the Brillouin zone with Γ indicating the zone center. Maxima and minima in the bands indicate high densities of states. The relatively flat bands in quartz indicate delocalization in **k**-space, thus localized electrons in real space, characteristic of covalent bonds. The more advanced band structure models allow calculation of electron density maps and thus insight into the bond character of such oxides as MgO [92].

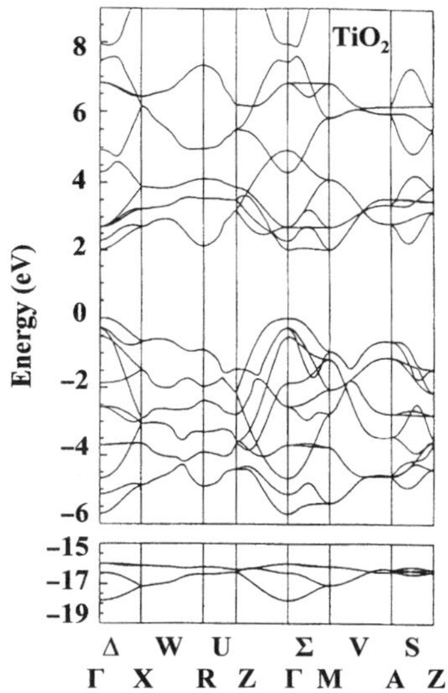

FIGURE 1.1.36 Energy bands for selected directions in the Brillouin zone of TiO_2 from Ref. [89]. Capital letters mark points on the surface of the Brillouin zone. Greek letters label directions within the zone. See Koster [93] for description of the notation. Used with permission.

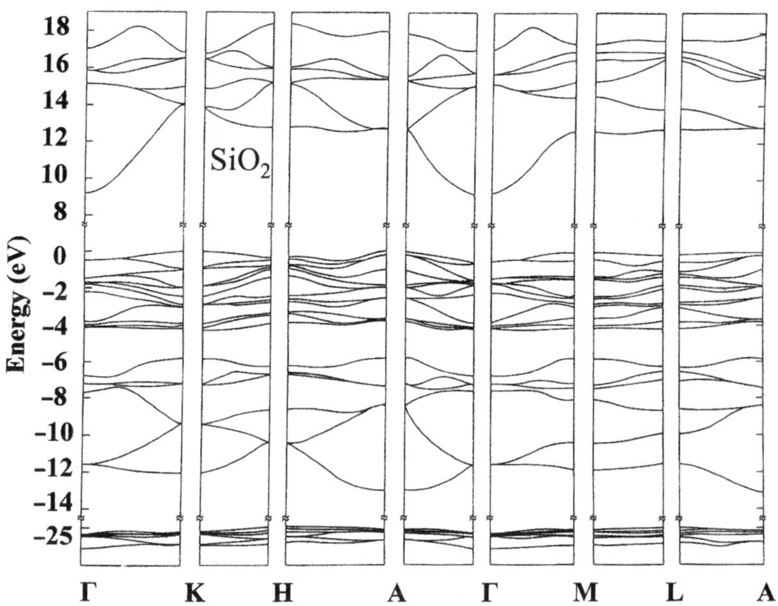

FIGURE 1.1.37 Energy band diagram for quartz from Ref. [86]. Similar notation to Figure 1.1.36. Used with permission.

The most distinguishing feature of the energy band diagram is the band gap, a region of energy in which there are no band states. In the band structure of metals, bands would everywhere overlap and there would be no band gap. In semiconductors and insulators, there is a finite gap. The valence band, below the gap, is composed of filled states; the conduction band above the gap consists of empty states. For small gaps, a few multiples of kT, electrons can be thermally excited across the gap producing intrinsic semiconducting behavior. For gaps of all widths, electrons can be optically excited across the gap to produce the optical absorption edge. Compounds with gaps less than the energy of visible light are strongly absorbing at visible wavelengths and have a metallic luster. Compounds with gaps in excess of 3.1 eV are transparent at visible wavelengths. Compounds with gaps corresponding to visible wavelengths are usually brightly colored.

Some information on the structure of the conduction band in ceramic insulators can be extracted from vacuum UV spectra (Fig. 1.1.38) [94]. The sharp absorption edge appears at 7.5, 7.8, and 9.1 eV for MgO, $MgAl_2O_4$, and Al_2O_3, respectively. The spectra are plotted as the imaginary part of the dielectric permittivity. The peak values correspond to absorptivities in the range of 10^5–10^6 cm^{-1}. The structure in the spectra relates to transitions to specific

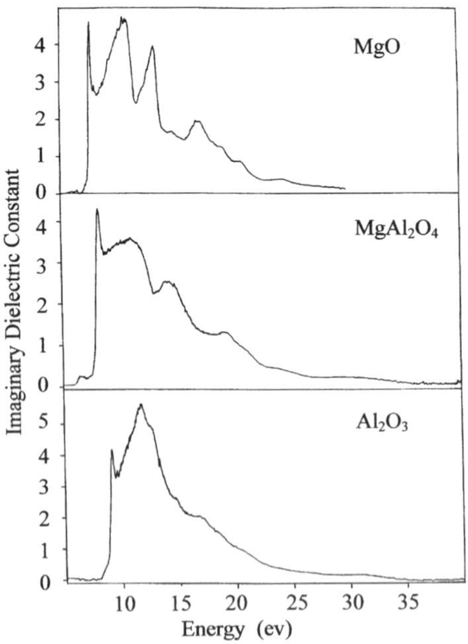

FIGURE 1.1.38 Ultraviolet spectra of MgO, $MgAl_2O_4$, and Al_2O_3. From Bortz *et al.* [94]. Used with permission.

bands within the conduction band but specific assignments are usually difficult. The overall width of the complete set of conduction bands is seen to be about 25 eV.

Pressure in the GPa range forces atoms closer together, decreases interatomic distances, and also usually decreases the band gap. If the gap is squeezed completely closed, the material undergoes a semiconductor–metal transition. Such behavior has been observed in wide gap semiconductors such as silicon, germanium, and the zinc chalcogenides and also in such molecular materials as iodine [95].

The band gap is clearly related to chemical bonding. One measure of bond strength is the heat of sublimation, which, within any family of compounds, is linearly related to the band gap (Fig. 1.1.39) [96]. The various linear plots can be expressed as

$$E_g = A(\Delta G_s - B)$$

The slope and intercept parameters vary with the specific family of structures.

1.1.3.4.2 Narrow d-Band Models

The band model works best with homodesmic crystals in which the valence and conduction bands are made up mainly of s- and p-orbitals. Bands arising

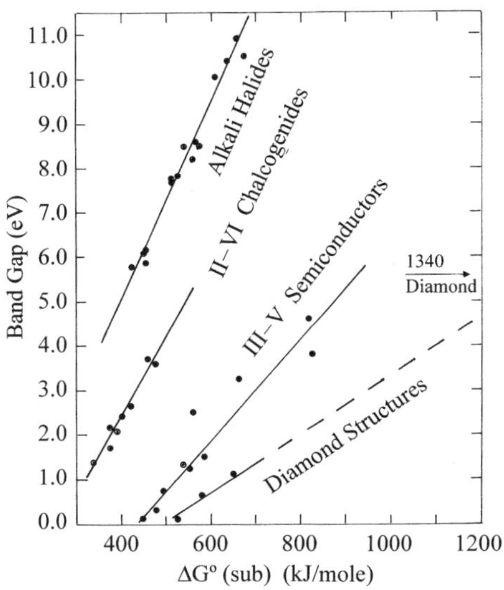

FIGURE 1.1.39 Plot of band gap as a function of heat of sublimation from various homodesmic binary compounds. Numerical data taken from Bailly and Manca [96].

from d-orbital states are more complicated. Consider these oxides. MnO, NaCl structure, is an insulator with a wide band gap. It has the pale pink color expected from tightly bound Mn^{2+} ions in an octahedral field. VO, NaCl structure, is a metal. TiO_2, rutile structure, is an insulator with a band gap of about 3 eV. VO_2, rutile structure, is a semiconductor that undergoes a semiconductor–metal phase transition. ReO_3 has a defect perovskite structure but is a metal with both a coppery metallic luster and the electrical conductivity of copper.

A useful conceptual model, proposed long ago by Goodenough [97, 98], combines features of band theory with crystal field theory and molecular orbital theory. There are three possible interactions for the d-electrons: (a) they can remain completely localized, in which case crystal field theory is the appropriate model; (b) the d-orbitals can directly overlap between next-nearest neighbor cations so that the d-orbitals form narrow bands that are largely decoupled from the broad bands formed from cation and anion s- and p-orbitals; (c) the d-orbitals can mix with anion s- and p-orbitals to form what would be the equivalent of π-molecular orbitals. The test to determine which mechanism applies is the number of orbitals required for σ bonding

$$n\sigma = [(4p^3 + 4s + 3e_g^2)(\text{number of cations})]/(\text{number of anions})$$

If $n\sigma > 4$, the compound will exhibit Type I behavior in which the d-orbitals will either be localized or directly overlap to form cation d-bands. If $n\sigma < 3$, there will be orbitals available to form π cation–anion–cation bands.

Type I narrow d-bands depend on a critical cation–cation separation, R_c. If the cation–cation distance is less than R_c, the d-orbitals will overlap and narrow d-bands will be formed. If the cation–cation distance is greater than R_c, the d-orbitals will be localized and described by crystal field theory. The critical distance has been fitted by an empirical equation

$$R_c = 3.20 - 0.05m - 0.03(Z - Z_{Ti}) - 0.04S_i(Si_i + 1)$$

R_c is given in angstroms, m is the formal charge on the cation, Z is atomic number, and S_i is a spin correction term.

For the MO compounds, $n\sigma = 6$ so they fall into Type I. VO has a cation–cation distance of 289 pm, smaller than the calculated $R_c = 292$ pm, so metallic narrow d-band behavior is predicted. In contrast, the cation–cation distance in MnO is 314 pm, greater than $R_c = 266$ pm, so that localized Mn^{2+} crystal field states are predicted (Fig. 1.1.40).

ReO_3 is an example of Type II behavior with $n\sigma = 2$ [99]. In this case, the oxygen s- and p-orbitals are not completely used up in the construction of σ-orbitals. The unused oxygen p-orbitals mix with the cation t_{2g} orbitals to form a pair of π-bonding and π* antibonding orbitals. The π-bonding

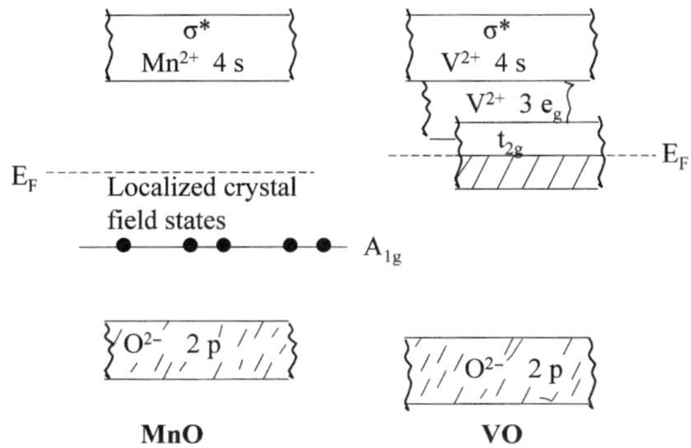

FIGURE 1.1.40 Schematic energy diagram for MnO, an insulator, and VO, a narrow d-band metal, according to the Goodenough model.

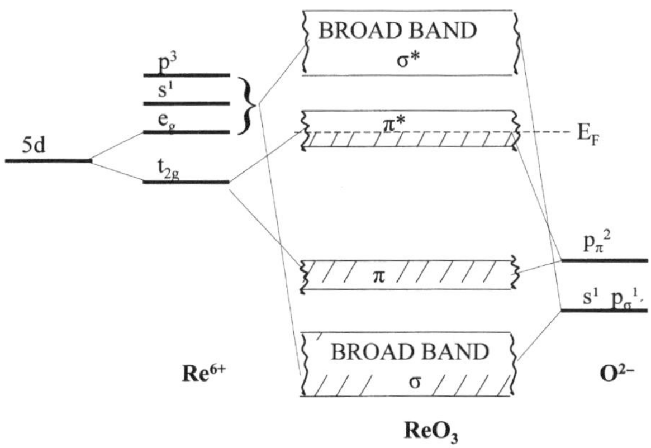

FIGURE 1.1.41 Schematic energy band diagram for ReO_3, a cation–anion–cation overlap system according to the Goodenough model.

bands are filled and indeed Re_2O_7 is a covalently bonded molecular solid. The extra d-electron in Re^{6+}, however, goes into the π^* band making it $\frac{1}{6}$ filled. ReO_3 then becomes metallic with the Fermi surface located in the π^* band (Fig. 1.1.41).

The narrow d-band models are useful for conceptualizing structure and properties of many transition metal compounds including oxides, chalcogenides, and others [100].

1.1.4 STRUCTURE–PROPERTY RELATIONS

Properties of ceramic materials are a composite of properties intrinsic to the crystalline compounds and properties that derive from the polycrystalline ceramic. Single crystal properties are tensors of various ranks (taking scalars as tensors of rank zero). In general, these properties vary with crystal direction and must be specified in terms of components, the number of which is determined by crystal symmetry. When crystals are combined as the grains of bulk ceramics, single crystal properties are modified by grain interfaces and porosity within the ceramic and the orientational dependence is usually lost.

Ceramic properties divide into two broad categories. Thermomechanical properties depend mainly on interatomic distances and ionic packing. Various optical, magnetic, and electronic properties depend both on crystal structure and on the electronic structure of specific ions that make up the compounds. The sections that follow summarize some important ceramic properties and their relation to crystal structure and chemical bonding. It should be remembered that these are only some of the first order properties. In addition, there are properties that depend on higher-order terms in the various equations. There are mixed properties (e.g. magnetoelasticity). There is the dependence of the various properties on temperature and pressure.

1.1.4.1 Thermomechanical Properties

1.1.4.1.1 Thermal Expansion

The crystal potential function as described by the classical ionic model (Fig. 1.1.28) is asymmetric. The vibrational modes of the crystal are discrete energy states stacked within the potential function. When the temperature is increased, more and more of these vibrational levels become populated. As a result, the equilibrium interatomic distance shifts to larger values with increasing temperature, thus providing the classical explanation for the thermal expansion observed in most materials. Thermal expansion is important both in terms of dimensional changes on heating and cooling of ceramic materials and as an important component of thermal shock resistance.

Either linear thermal expansion, for example, of a ceramic rod, or volume thermal expansion can be measured directly and is usually expressed as fractional change per degree Celsius:

$$\alpha_\ell = \frac{1}{d}\left(\frac{\partial d}{\partial T}\right)_P \qquad \alpha_V = \frac{1}{V}\left(\frac{\partial V}{\partial T}\right)_P$$

On an atomic scale in heterodesmic crystals, the thermal expansion is distributed among the polyhedra that make up the structure. By determining the crystal structure at a sequence of temperatures it has been found that the thermal expansion of individual polyhedra is largely determined by chemical bonding and by coordination number, and is essentially independent of interatomic distance or the specific chemical composition of the material [101]. Most of the thermal expansion is taken up by high coordination polyhedra with tetrahedral units expanding relatively little. The mean linear expansion of a polyhedron is given by the relation

$$\alpha_{1000} = 4.0 \times 10^{-6} \frac{\text{CN}}{S^2 z_c z_a}$$

where CN is the coordination number, z_c, z_a are formal charge on cation and anion, respectively, and S is an ionicity factor with numerical values of 0.50 for oxides and silicates, 0.75 for halides, 0.40 for chalcogenides, 0.25 for phosphides and arsenides, and 0.20 for nitrides and carbides.

Some representative data for isodesmic compounds are given in Table 1.1.6. Silica occupies a unique position with a near-zero thermal expansion, one of the properties that makes silica glass so important. Diamond has a low coefficient but most other materials have moderate to high thermal expansion coefficients. Low thermal expansion in heterodesmic compounds is achieved by compensating polyhedral expansions within the structure. The thermal expansion of non-cubic single crystals may be highly anisotropic and indeed negative thermal expansions may be found in certain crystallographic directions.

TABLE 1.1.6 Linear Coefficient of Thermal Expansion (in Units of 10^{-6}/K)

Compound	Structure	c.n.	α_{1000}
MgO	NaCl	6	13.0
ZnO	Wurtzite	4	7.3
Al_2O_3	Corundum	6	8.6
ThO_2	Fluorite	8	9.0
ZrO_2	Baddeleyite	7	8.2
β-SiO_2	High Quartz	4	0.0
ZnS	Sphalerite	4	9
GaAs	Sphalerite	4	6.7
BN	Sphalerite	4	13.0
TiC	NaCl	6	8
C	Diamond	4	3.5

Date selected from much larger compilation by Hazen and Finger [101].

1.1.4.1.2 Thermal Conductivity

In an isotropic solid, the flow of heat is driven by the temperature gradient and is described by

$$\mathbf{h} = -k \text{ grad } T$$

where the proportionality constant, k, is the thermal conductivity. Heat flow is a vector and the temperature gradient is a vector so in a single crystal the thermal conductivity becomes a symmetric tensor of rank 2:

$$h_i = -k_{ij} \frac{\partial T}{\partial x_j}$$

In polycrystalline materials, both metals and ceramics, the thermal conductivity is taken as a single numerical constant.

In metals, heat is carried mainly by free electrons. In insulators heat is carried mainly by phonons, that is, by the low wavenumber lattice vibrations. Acoustic phonons are the main contributors in low coordination structures but low wavenumber optic modes may also contribute particularly in high coordination structures. In principle, it should be possible to calculate thermal conductivity from a knowledge of the lattice vibrations. The calculation turns out to be difficult because phonons from all points in the Brillouin zone contribute to the thermal conductivity [102]. Infrared and Raman spectra give only zone center phonons. Further, the phonon mean free path is a contributing parameter and is difficult to extract from infrared and Raman spectra. Hofmeister [103] has devised an approximation for thermal conductivity at ambient pressure and temperature calculated from spectroscopic data:

$$k = \frac{1}{3} \frac{\rho}{ZM} \frac{\langle u \rangle^2}{\langle \Gamma \rangle} C_v$$

where ρ is the density, Z the number of formula units per primitive unit cell, M the molecular weight, $\langle u \rangle$ the mean sound velocity $= (u_s + u_p)/2$, Γ the average full-width at half-maximum for infrared modes, and C_v = heat capacity at constant volume.

Most ceramics have low thermal conductivities compared with metals, in part, because of phonon scattering at defects, dislocations, and grain boundaries. High thermal conductivity is important in semiconductor substrates and other device applications because of the need to dissipate waste heat effectively. Comparison of data for a variety of crystal structures [104] led to the conclusion that compounds with high thermal conductivity have low atomic masses, strong bonding, small number of atoms in the unit cell, and low anharmonicity. Some illustrative data are given in Table 1.1.7. These illustrate the effect of more complex structures, longer bond lengths, and higher coordination numbers on lowering the thermal conductivity.

TABLE 1.1.7 Selected Values for Thermal Conductivity

Compound	Structure	$k(\mathrm{w\,m^{-1}\,K^{-1}})$	Reference
Ag	A1 (fcc)	430	Slack [104]
Cu	A1 (fcc)	400	Slack [104]
C	Diamond	2000	Srivastava [102]
c-BN	Sphalerite	1300	Srivastava [102]
SiC	Sphalerite	490	Srivastava [102]
AlN	Sphalerite	320	Srivastava [102]
Si	Diamond	160	Srivastava [102]
Al_2O_3	Corundum	24	Savage [48]
Al_2MgO_4	Spinel	10.2	Hofmeister [103]
Cr_2MgO_4	Spinel	7.5	Hofmeister [103]
Mg_2SiO_4	Olivine	5.2	Hofmeister [103]
Fe_3O_4	Spinel	5.1	Hofmeister [103]
ZnS	Sphalerite	17.2	Savage [48]
$CaLa_2S_4$	Th_3P_4	1.7	Savage [48]

1.1.4.1.3 Strength and Elastic Properties

The volume of a bulk solid, single crystal or ceramic, in general, decreases when subject to pressure. The compressibility is defined as

$$\beta = \frac{1}{V}\left(\frac{\partial V}{\partial P}\right)_T$$

where V can be molar, specific, or bulk volume because the units of volume cancel. The inverse of the compressibility is known as the bulk modulus.

If, instead of a uniform pressure, the solid is subject to a directional stress, σ, there will be a deformation, ε, called a strain. In one dimension, and for deformations that do not exceed the elastic limit,

$$\varepsilon = s\sigma \qquad \sigma = c\varepsilon$$

where s and c are, respectively, the elastic compliance and the elastic stiffness. These equations describe the elastic deformation of wires or rods that can be treated as one-dimensional objects.

For isotropic materials—glasses or polycrystalline ceramics—the elastic properties are described by Young's modulus, E, and the modulus of rigidity, G [105]:

$$\sigma = E\varepsilon$$

$$\tau = G\gamma$$

TABLE 1.1.8 Representative Elastic Parameters of Some Ceramic Materials

Compound	Structure	E (GPa)	G (GPa)	ν
Copper	A1, fcc	129.8	48.3	0.342
Iron	A2, bcc	211.8	81.6	0.293
Al_2O_3	Corundum	414	—	—
C	Diamond	965	—	—
SiC	Sphalerite	470	—	—
WC	Hex. WC Str.	77.5	31.8	0.22

Data selected from Hertzberg [106].

where τ is the shear stress and γ the shear strain. When a solid is extended under tension, the thickness must decrease. The ratio of thickness decrease to length increase is Poisson's ratio

$$\nu = \frac{\Delta d/d}{\Delta l/l}$$

These bulk parameters can be related to the elastic moduli (see, e.g. Hertzberg [106]). Some representative values are given in Table 1.1.8.

In three dimensions, both stress and strain have a dimensional dependence that makes them second-rank tensors. The compliance and stiffness constants then become fourth-rank tensors [107] and the equations become

$$\varepsilon_{ij} = s_{ijkl}\,\sigma_{kl}$$

$$\sigma_{ij} = c_{ijkl}\,\varepsilon_{kl}$$

If written out in full, each of the tensor equations stands for nine algebraic equations, each of which has nine terms on the right-hand side. In the general case, the stiffness tensor and the compliance tensor each has 81 components. In the linear case, the stiffness and compliance constants are simply the inverses of each other, $c = s^{-1}$. The same relation holds in the full tensor representation but the inverses are mathematically much more complicated.

Taking into account the internal symmetry of the tensors, it is possible to reduce these expressions to a matrix (not tensor) form

$$\varepsilon_i = s_{ij}\,\sigma_j$$

$$\sigma_i = c_{ij}\,\varepsilon_j$$

where s_{ij} and c_{ij} are 6×6 matrices. Application of the crystal symmetry further reduces the number of independent components with some of them becoming

TABLE 1.1.9 Representative Elastic Moduli for Some Ceramic Materials

Compound	Structure	c_{11}	c_{12}	c_{44}	s_{11}	s_{12}	s_{44}
Copper	A1, fcc	16.84	12.14	7.54	1.50	−0.63	1.33
Iron	A2, bcc	23.70	14.10	11.60	0.80	−0.28	0.86
MgO	Halite	29.3	9.2	15.5	0.401	−0.096	0.648
NaCl	Halite	4.87	1.26	1.27	2.29	−0.47	7.85
Al_2MgO_4	Spinel	27.9	15.3	15.3	0.585	−0.208	0.654
TiC	Halite	51.3	10.6	17.8	0.21	−0.036	0.561
ZnS	Sphalerite	10.29	2.22	4.12	2.0	−0.802	2.43

Data selected from Hertzberg [106].

equal and many becoming zero for high symmetry crystals. For a cubic crystal, the compliance matrix becomes

$$\begin{pmatrix} s_{11} & s_{12} & s_{12} & 0 & 0 & 0 \\ s_{12} & s_{11} & s_{12} & 0 & 0 & 0 \\ s_{12} & s_{12} & s_{11} & 0 & 0 & 0 \\ 0 & 0 & 0 & s_{44} & 0 & 0 \\ 0 & 0 & 0 & 0 & s_{44} & 0 \\ 0 & 0 & 0 & 0 & 0 & s_{44} \end{pmatrix}$$

The stiffness matrix has the same form. Only three independent parameters need be determined. More parameters are needed for lower symmetries: hexagonal—5, trigonal and tetragonal—6 or 7, orthorhombic—9, monoclinic—13, and triclinic—21. Some representative values for cubic materials are given in Table 1.1.9.

The elastic moduli are related in complicated ways to both structural arrangement and strength of bonding in crystals. The longitudinal elastic stiffness coefficient, c_{11}, scales with interatomic distance for structures with the same structure and the same formal ionic charges [108]:

$$c_{11} = AR_{c\text{–}a}^{-4}$$

where $R_{c\text{–}a}$ is the cation–anion distance and A is related to the formal charges on the ions.

1.1.4.1.4 Hardness

Hard materials have applications as cutting tools, abrasives, and armor. Exceptionally hard materials must have strong chemical bonds with a three-dimensional structure. The prototype is diamond with a three-dimensional framework of corner-sharing carbon tetrahedra. One family of hard materials is diamond-related and includes such materials as cubic BN and Si_3N_4. Others are based on borides with strong boron–boron bonding such as B_4C. Hardness scales with bulk modulus [109], predicting a third family of hard materials derived from high-pressure polymorphs such as stishovite, the rutile structure polymorph of SiO_2 and RuO_2 with the cubic cotunnite structure.

Hardness is usually measured by an indenter technique giving either a Knoop or Vickers hardness. Indenters are made from diamond, the hardest substance presently known with a Knoop hardness of 90 GPa. Use of spherical indenters provides a useful tool for the study of fracture characteristics of ceramics [110].

According to a theoretical prediction [111], C_3N_4, structurally equivalent to Si_3N_4, if it could be synthesized, should be harder than diamond. To date, attempts to synthesize this hypothetical superhard material have not met with much success [112, 113].

1.1.4.1.5 Fracture Toughness

Hardness is resistance to scratching or indenting. Toughness is resistance to fracture. Tough materials are not necessarily hard materials. Fracture toughness is described by K_{Ic}, the resistance of the material to unstable crack propagation. Because fracture toughness is related to crack propagation, it is more determined by microstructure than by structure at the atomic scale. Materials with fibrous microstructures are generally tougher than materials with granular microstructures. Polycrystalline materials are usually tougher than single crystals of the same materials. Composites are often tougher than single phase materials.

1.1.4.2 Optical Properties

1.1.4.2.1 Refractive Index, Dispersion, Scattering, and Absorption

The optical properties of an insulating solid are described by a complex refractive index

$$N = n - ik$$

where n is the real refractive index and k the extinction coefficient. The refractive index is a characterizing parameter for the propagation of electromagnetic waves through the solid and k is a damping term that describes the energy loss. These parameters are related to the dielectric permittivity and the dielectric loss in the material:

$$\varepsilon' = n^2 - k^2$$

$$\varepsilon'' = 2nk$$

In the general case, all of the above quantities are symmetric tensors of rank 2. The number of independent components is determined by the crystal symmetry.

The transmission of electromagnetic radiation through a solid is determined by measuring the intensity of light emerging from the solid in comparison to the intensity of light entering the solid. The intensity of light in an absorbing solid falls off exponentially along the transmission path according to the Lambert–Beer law:

$$I = I_0 \mathrm{e}^{-\alpha t}$$

The absorption coefficient, α, which is a function of wavenumber (or wavelength), is an intrinsic property of the material. However, the usual measurement does not distinguish between loss of light intensity due to the intrinsic absorption by the material and intensity loss due to scattering. Intrinsic absorption is due to electronic transitions on an atomic scale. Scattering is due to refractive index mismatches usually on the scale of the wavelength of the electromagnetic radiation. The experimental absorptivity can be considered to be the sum of the intrinsic absorptivity and the scattering coefficient.

The intrinsic absorptivity is directly related to imaginary part of the refractive index

$$\alpha = 4\pi k \nu$$

The intensity of electromagnetic radiation reflected at normal incidence from the surface of a material, $R = I(\text{reflected})/I_0$, is also determined by the optical parameters

$$R = \frac{(n-1)^2 + k^2}{(n+1)^2 + k^2}$$

The absorption across the electromagnetic spectrum is divided into three regions in insulating solids (Fig. 1.1.42). At wavelengths shorter than the optical band edge, the material is strongly absorbing due to dipole-allowed interband transitions. At wavelengths longer than those corresponding to the highest energy lattice vibrations, the material is strongly absorbing due to infrared-allowed crystal vibrations. Between the strongly absorbing regions is a window region where absorption is determined by electronic transitions in ions with unfilled d- and f-orbitals, by trapped charges (color centers), and by miscellaneous impurities. The window region is widest in high gap insulators, narrows in semiconductors, and disappears in metals.

Optical components can be prepared from glass, dense ceramics, or single crystals. Components cut from single crystals can make use of the directional properties of the refractive index and also the higher-order terms. In polycrystalline ceramics, the optical characteristics are averaged. Theoretically dense hot-pressed or hot-isostatic-pressed cubic crystals such as spinel, ZnS, ZnSe, the alkali halides, and others are used as window materials in the ultraviolet,

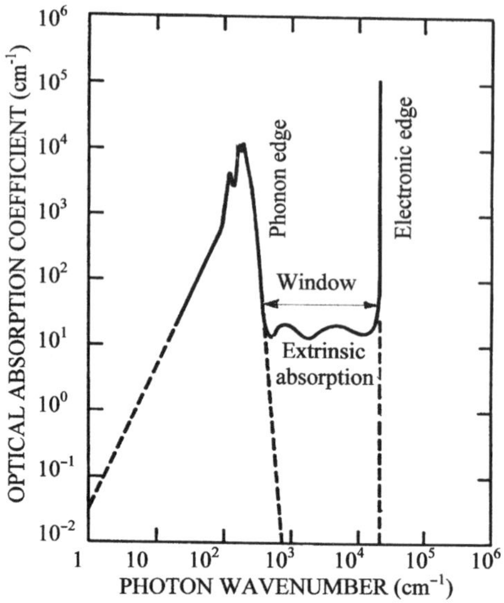

FIGURE 1.1.42 Schematic absorption spectrum showing lattice vibration region, window region, and interband absorption region.

visible, and infrared regions of the spectrum depending on the window region of the material [48].

1.1.4.2.2 Color

The phenomenon called color results from the interaction of three distinct systems: a source, an object, and the human visual system. The latter is important. In a very real sense, color is something that has its origins in the human brain; it is not an intrinsic property of a material. To observe color, there must be light in the wavelength range from roughly 400 to 700 nm, the range to which the human visual system responds. An ideal white light would have all wavelengths represented at equal intensity. Sunlight is usually taken as the standard although its spectral energy curve is not flat. Artificial sources have been prepared with the complete range of spectral energy distributions from near-white to monochromatic. The wavelength response of the human eye is also wavelength dependent with maximum visual acuity in the green region of the spectrum. Visual sensitivity falls off, first gradually and then rapidly as the wavelength approaches either the violet or the deep red. The aspect of color, which is an intrinsic property, depends on the wavelength-dependent absorption (or reflection) of light. The intrinsic absorption spectrum (or reflection spectrum) is then combined with source and visual system to produce an actual color.

This review is concerned only with colors produced by the absorption characteristics of ceramic materials. A complete tabulation gives 15 possible sources of color according to Nassau [114], but these include incandescence, dispersion, interference, and other optical effects. Also excluded from this review are metals and organic compounds.

High absorption coefficients produce deep colors; small absorption coefficients produce pastel colors. The purity of the color depends on the shape of the absorption curve. Sharp edges or sharp bands produce the purest colors. The types of electronic transitions giving rise to color include:

1. Band gap transitions with the optical absorption edge in the visible region of the spectrum have very high absorption coefficients. Examples are such compounds as HgS (red-orange), CdS–CdSe (deep yellow), and ternary sulfides of various compositions. Most compounds with colors produced by band gap transitions are chalcogenides. Because of limited stability at high temperatures and concern over toxicity, these compounds have limited application as ceramic pigments.
2. Transitions between molecular orbital states occur in chromates. Absorption coefficients are high, resulting in intense absorption bands. Yellow, orange, and red pigments have been based on chromate compositions but now have limited use because of toxicity considerations.
3. Intervalence electron transfer transitions. Electron exchange in compounds containing mixed valence ions such as Fe^{2+}/Fe^{3+}, Ti^{3+}/Ti^{4+}, Fe^{2+}/Ti^{4+}, and many others produces intense absorption bands usually in the visible region of the spectrum. Deep colors such as the blue of sapphire ($Al_{2-x}Ti_xO_3$) and the black of magnetite (Fe_3O_4) result from this process.
4. Transitions between unfilled d-orbitals in transition metal ions are relatively weak but are responsible for the color of many ceramic compounds. These transitions are described by crystal field theory, which can account for the Cr greens, Co blues, Cu greens, and other colors.
5. Transitions between unfilled f-orbitals in rare earth ions are very weak and produce only pastel colors. The absorption spectra consist of very complex patterns of sharp lines. These spectra can also be interpreted by crystal field theory.

The main technological application of strongly colored ceramic compounds is as pigments and colored structural ceramics. Optically, a pigment must be capable of imparting color of the desired hue, chroma, and value. In addition, pigments intended for use in glazes and other ceramic applications must be

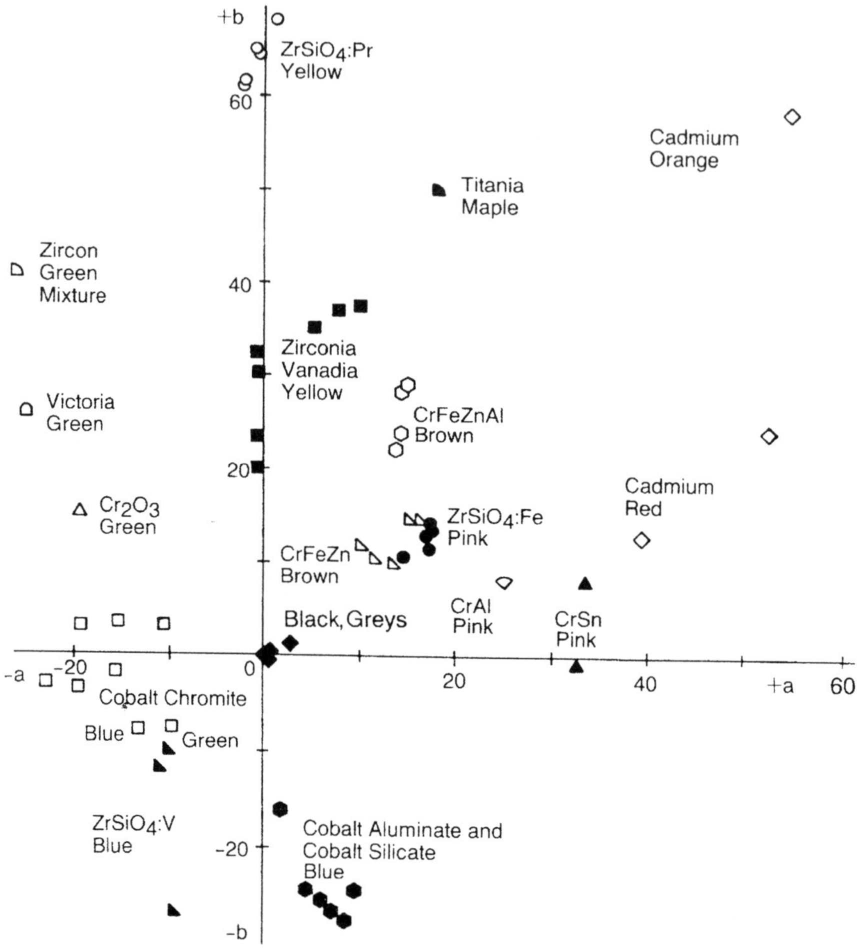

FIGURE 1.1.43 Chroma diagram showing many of the common ceramic pigments. From Ref. [115]. Used with permission.

stable at the firing temperatures. The number of widely useful pigments is limited and most derive their color from either crystal field or intervalence electron transfer transitions (Fig. 1.1.43) [115].

1.1.4.2.3 Luminescence

Certain materials will emit light when excited by some higher-energy source such as ultraviolet radiation or electron beams. Light emission in excess of black body radiation (incandescence) is known as luminescence. Materials that emit light efficiently are known as phosphors. Phosphors, as ceramic materials, are used mainly as fine-grained powders, occasionally as thin films. Preparation requires considerations of particle size, crystal perfection, defects, and surface states as well as carefully controlled concentrations of activators and co-activators.

Phosphors can be broadly divided into three classes: organic phosphors, semiconductor phosphors, and insulator phosphors [116]. Organic phosphors are based on molecular orbital transitions between bound states in molecular structures and have applications such as fluorescent paints, dyes, and, more recently, organic light-emitting diodes (OLEDs). There are some inorganic systems that exhibit molecular orbital luminescence, for example, the uranyl ion (UO_2^{2+}), and the tungstate and molybdate ions. Luminescence from semiconductors arises from the recombination of electron–hole pairs formed near the bottom of the conduction band. The emission wavelength and luminescent efficiency is determined by the band structure of the material as modified by defects and dopants. Important examples are ZnS:Ag (blue) and ZnS:Cu,Cl (green), which are used in cathode ray tubes. Other semiconductor phosphors include GaAs–GaP solid solutions and other III–V compounds that are used as light emitting diodes (LEDs). Insulator phosphors are high band gap crystals in which the luminescence arises from substituted ions known as activators [69]. The emission wavelength and luminescent efficiency is determined by the electronic structure of the activators and by their nearest neighbor environment in the crystal structure.

The energy flow in an insulator phosphor is shown in Figure 1.1.44. The simplest possible system requires an electronic ground state and an electronic excited state, both represented by parabolas. Within the parabolas are sketched horizontal lines representing vibrational energy levels. The x-axis is a configurational coordinate. If the system were a diatomic molecule, the configurational coordinate would be the actual bond length. In a complex system, such as a crystal, the configurational coordinate is more schematic although it is related to the normal coordinates that describe crystal vibrations. Electronic transitions are fast compared to ionic relaxation so that these transitions are represented by vertical arrows on the diagram. Absorption of ultraviolet radiation places the system in both an excited electronic and an excited vibrational state. The vibrations relax on a time scale of nanoseconds and the nuclei move to increase the configurational coordinate. Emission from the electronic excited state occurs on time scales from nanoseconds to seconds (or sometimes longer). Emission of radiation terminates in a vibrationally excited level of the electronic ground state from which the system must relax again to the original ground state. Because of the energy loss to lattice vibrations, the emission wavelength is longer than the excitation wavelength, an energy loss known as the Stokes shift.

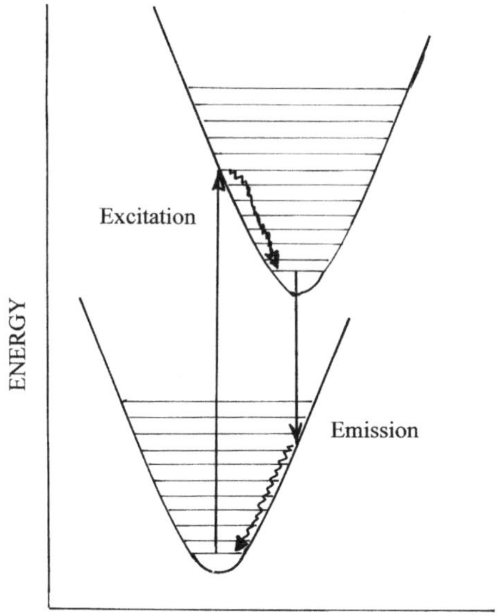

FIGURE 1.1.44 Schematic energy level diagram showing the energy flow in a two-level phosphor.

Transition metal ions that serve as activators include Cr^{3+}, Mn^{4+}, Mn^{2+}, and Fe^{3+}. The energy level structure is acceptably described by crystal field theory. Of these, Mn^{2+} is most important. The emission is very sensitive to the crystal structure. The emission is green if Mn^{2+} is in four-fold coordination, for example, $Zn_2SiO_4:Mn^{2+}$, and orange red if Mn^{2+} is in six-fold coordination. The white phosphor used in fluorescent lamps is an apatite structure halophosphate, $Ca_5(PO_4)_3(F,Cl):Sb^{3+}, Mn^{2+}$. The white color is achieved by combining a blue emission from Sb^{3+} with an orange from Mn^{2+}.

Rare earth ions that serve as activators include Ce^{3+}, Nd^{3+}, Tb^{3+}, Eu^{2+}, Eu^{3+}, and Er^{3+} among others. Transitions between excited d-orbital states and f-orbital states in Ce^{3+} and Eu^{2+} result in a broad band emission that can be tuned over a wide wavelength range by changing coordination number, interatomic distances, and site distortion in the host crystals. The luminescence of Nd^{3+}, Tb^{3+}, Eu^{3+}, and Er^{3+} is due to f–f transitions that produce the same emission wavelengths largely independent of the crystalline host. Tb^{3+} activated phosphors emit in the green while Eu^{3+} activated phosphors such as $Y_2O_3:Eu^{3+}$ are red emitters. Nd^{3+} and Er^{3+} emit in the infrared and have application as lasers, for example, Nd-YAG, and in fiber optic systems, respectively.

1.1.4.3 Dielectric Properties

1.1.4.3.1 Principles

Dielectric properties are those that result from the interaction of an electromagnetic field with an insulating material and thus are particularly significant for ceramic materials. When an electric field is applied to an insulating solid, the material becomes polarized with the polarization proportional to the field strength:

$$\vec{P} = \alpha\vec{E}$$

Both polarization, $\vec{P}$, and electric field, $\vec{E}$, are vectors. In the general case, the polarizability, α, is a symmetric tensor of rank 2. There are four distinct contributions to the polarizability

$$\alpha = \alpha_{\mathrm{e}} + \alpha_{\mathrm{i}} + \alpha_{\mathrm{d}} + \alpha_{\mathrm{s}}$$

where the terms are, respectively, the electronic polarizability, the ionic polarizability, the dipolar polarizability, and the space charge polarizability.

The electronic polarizability arises from the distortion of the electron clouds of the atoms under the influence of the electric field. Electronic polarizability increases with increasing atomic number so that ions such as Cs^{+} or Ba^{2+} have a large polarizability while ions such as Al^{3+} have a small polarizability. Ions with non-bonding electrons such as Pb^{2+} and Bi^{3+} have exceptionally large polarizabilities. Ionic polarizability arises from the field-induced displacement of ions in ionic crystals. Movable dipoles and drifting charge carriers produce the dipolar and space charge polarizabilities.

Each component of the polarizability has a different frequency response. Dipolar and space charge polarizabilities are most important at low frequencies at which the orientation of dipoles and the drift of charges can respond to the alternating electromagnetic field. Ionic polarizability extends to infrared frequencies where resonances produce infrared absorption spectra. Only the electronic polarizability extends to visible and ultraviolet regions.

1.1.4.3.2 Dielectrics: Bulk and Particulate

Like the refractive index, the permittivity of a ceramic material is a complex quantity:

$$\varepsilon = \varepsilon' - \mathrm{i}\varepsilon''$$

Materials to be used as dielectrics must be good insulators. The imaginary part of the permittivity is a measure of the dielectric loss and is related to the conductivity. In the absence of long-range effects, materials are paraelectric and the permittivity is determined in large part by the polarizability of individual ions with the crystal structure playing a secondary role.

1.1.4.3.3 Pyroelectric Materials

The pyroelectric effect is the development of a macroscopic electric polarization due to a change in temperature or the modification of an existing polarization due to change in temperature:

$$\Delta P_i = p_i \Delta T$$

where p_i are the three pyroelectric coefficients that describe the magnitude of the polarization. Because the change in polarization is a vector and the temperature change is a scalar, the pyroelectric coefficient must also be a vector.

By what is called Neumann's principle, any real physical property of a crystal must conform to the point group symmetry of the crystal. Since the pyroelectric effect is a vector, it must be zero for centrosymmetric crystals because the operation of the center of symmetry converts $p_i \rightarrow -p_i$ and this can only be true if p_i vanishes. Thus, only crystals belonging to 21 non-centrosymmetric crystal classes can exhibit a pyroelectric effect.

1.1.4.3.4 Piezoelectric Materials

The piezoelectric effect is the spontaneous polarization that develops in certain crystals in response to applied stress. It is defined by the tensor relationship

$$P_i = d_{ijk}\sigma_{jk}$$

The polarization is a vector and the stress is a symmetric second-rank tensor. The piezoelectric modulii, d_{ijk}, are represented by a third-rank tensor. Because d_{ijk} is a tensor of odd rank, the components must change sign under the operation of a center of symmetry. Thus, the piezoelectric effect does not appear in crystals belonging to centrosymmetric point groups. For the non-centrosymmetric crystals, the number of non-zero components of the piezoelectric modulus tensor depends on details of the crystal symmetry. Nye [107] gives a table showing the relationships of the piezoelectric modulii for the 21 non-centrosymmetric crystal classes.

The closely related converse piezoelectric effect is the appearance of a mechanical strain from the application of an electric field. The same modulii apply:

$$\varepsilon_{jk} = d_{ijk}E_i$$

The piezoelectric effect and its converse provide the coupling between mechanical deformations and electrical signals and thus can be applied to a great variety of transducer and actuator devices. These range from needle cartridges for old-time phonographs to microphones, hydrophones, and most recently to a class of applications known as "smart materials". A smart material may use the piezoelectric effect to generate a signal and then use the converse piezoelectric effect to produce a response. Others use higher-order effects such as electrostriction or magnetostriction in their operation. Most smart materials are composites of various architectures. Newnham [117] summarizes the variety of smart materials that have been developed.

1.1.4.3.5 Ferroelectric Ceramics

Ferroelectric materials are characterized by the presence of spontaneous polarizations that are coupled to form domains. Ferroelectric domains can be re-positioned by the application of external electric fields. This leads to hysteresis between the applied field and the resulting electric polarization. Ferroelectric behavior is usually temperature sensitive. As temperature is increased, there appears the Curie temperature at which the material undergoes a displacive phase transformation and the spontaneous polarization and domain structure disappear. Some materials undergo a series of phase transformations from one ferroelectrically ordered structure to another before reaching the Curie temperature and the final transformation into a paraelectric state. There is a second category of materials known as relaxor ferroelectrics, which display broad, diffuse phase transitions.

Ferroelectricity is most often associated with the perovskite structure. The best-known prototype is $BaTiO_3$. At temperatures above 120°C, $BaTiO_3$ has the ideal cubic perovskite structure. At lower temperatures, the Ti^{4+} ion moves slightly off center in its coordination polyhedron, generating a strong electric dipole. These dipoles couple to form ferroelectric domains and the bulk structure distorts into a tetragonal form. Two other phase transformations into an orthorhombic and a trigonal structure occur near 0 and −100°C, respectively.

The large internal polarization of ferroelectric materials gives them a high dielectric permittivity, which is a sensitive function of temperature. Mostly, the permittivity peaks near the phase transition temperature but relaxor ferroelectrics have a high permittivity over a long temperature range. The great variety of possible substitutions into the perovskite structure make possible the fine tuning of the dielectric properties for particular applications. Many of the applications are in the construction of capacitors, where high permittivity makes possible high capacitances in small volumes. Because many ferroelectrics are also piezoelectric there are further applications to transducer and actuator devices. Some of the most useful materials are solid solutions such as $PbZr_{1-x}Ti_xO_3$ (PZT) and $Pb_3MgNb_2O_9$ (PMN).

1.1.4.4 Magnetic Properties

1.1.4.4.1 Principles

Magnetic interactions in materials require the presence of magnetic dipoles that can interact with external magnetic fields. At the atomic scale, magnetic dipoles arise from electrons moving in orbitals or from electron spin. Of these the most important is electron spin. If electrons in orbital or band states are paired, the spins cancel and there is no net magnetic moment. Materials with unpaired spins are generally those with partially filled d-orbitals or partially filled f-orbitals.

The magnetic moment of a single electron is known as the Bohr magneton and has the theoretical value

$$\mu_\beta = \frac{eh}{4\pi m_e c} = 9.274 \times 10^{-24}\,\mathrm{J/T}$$

The induced magnetization is proportional to the magnetic field

$$I = kH$$

where the proportionality constant is the magnetic susceptibility and is a property of the material. Usually, susceptibility is expressed as specific susceptibility, $\chi = k/\rho$, which may be multiplied by atomic weight to give atomic susceptibility or by molecular weight to give molecular susceptibility. The magnetic permeability, u, of electromagnetic theory is simply related to susceptibility by

$$u = 4\pi k + 1$$

Diamagnetic materials are those in which the magnetic field lines are repelled from the material. Susceptibilities are in the range of -1 to $-10 \times 10^{-6}\,\mathrm{cm^3/mol}$. Paramagnetic materials are those in which the magnetic field is attracted to the material. Typical susceptibilities are in the range of $100 \times 10^{-6}\,\mathrm{cm^3/mol}$.

At low fields and moderate temperatures, to a first approximation the susceptibility is proportional to the reciprocal of the temperature, a relationship known as the Curie law, which relates the susceptibility to the magnetic moment. For a system with only one spin

$$\chi = \frac{I}{H} = \frac{Ng^2\mu_\beta^2}{2kT}$$

where g is the Landé factor with numerical value of approximately 2 for one electron. Each unpaired spin and each quantum of angular momentum contributes to the magnetic moment. However, angular momentum is quenched in compounds of transition metal ions so that the magnetic moment is to be expressed in terms of the number of Bohr magnetons, p, which, in turn, gives the number of unpaired spins:

$$p = g[S(S+1)]^{1/2}$$

$$\chi = \frac{Np^2\mu_\beta^2}{3kT}$$

The measured susceptibility can be used to calculate the number of unpaired spins and thus determine whether a particular compound is in a high- or low-spin state. Likewise, the measurement will reveal the valence state of transition metal ions.

There is a long-range field, the Weiss field, that couples the spins and forces them into alignment to form magnetic domains. If there is a single crystallographically equivalent set of ordered spins, the material is *ferromagnetic*. If there is more than one set of spins, these may point in different directions resulting in a partial cancellation of the magnetic moment giving a *ferrimagnetic* material. If the spins cancel exactly, but remain ordered in domain states, the material is said to be *antiferromagnetic*. Small particles, on the order of 15–60 nm may consist of single domains producing a phenomenon known as *superparamagnetism*. In unmagnetized material, the domains are randomly aligned and there is no net moment in the bulk material. When a ferro- or ferrimagnetic material is placed in a magnetic field, the domain walls move, the domains align with the field, and there is a net magnetization. If the field is cycled, there is a characteristic hysteresis loop (Fig. 1.1.45). The hysteresis loop defines most of the important properties of magnetic materials. For soft magnetic applications such as transformer cores, H_c and B_R should be small. For permanent magnets, both should be large. Most hard ferromagnets are alloys of transition metals such as iron and nickel, or of rare earth metals. Ionic, that is, ceramic, compounds tend to be soft magnets.

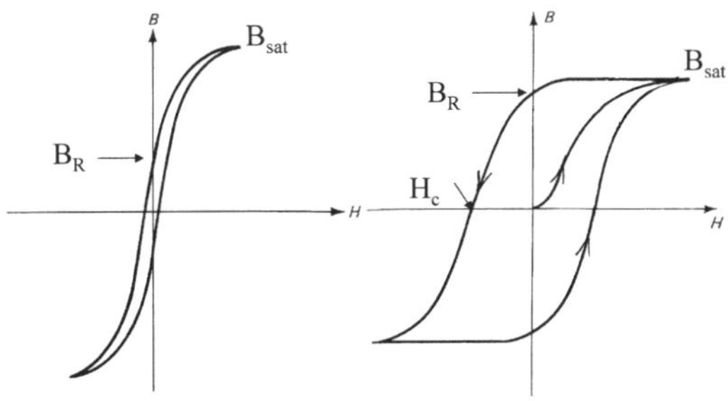

FIGURE 1.1.45 Schematic hysteresis loops for soft and hard magnetic materials showing characterizing parameters.

As the temperature rises, the domains become disoriented. Eventually, thermal motion will overcome the Weiss field, the individual spins will randomize, the domain structure is lost, and the ferromagnetic or ferrimagnetic material will become paramagnetic. The temperature at which ferro- or ferrimagnetic ordering is lost is the Curie temperature.

1.1.4.4.2 Spinel-based Ferrimagnets

Only a few binary compounds, those that contain only one cation sublattice, exhibit ferromagnetic ordering. The best-known example is CrO_2 with the rutile structure, which is widely used in magnetic recording tape. Much more complex magnetic behavior is possible when multiple cation sites are available. One widely used family are compounds with the spinel structure. Fe_3O_4, magnetite, has the spinel structure with Fe^{3+} on the tetrahedral site and Fe^{3+} and Fe^{2+} on the octahedral sites. Likewise, γ-Fe_2O_3, a strong ferrimagnet, is a defect spinel structure in spite of its nominal binary composition.

Spinels containing Fe^{3+} with some divalent cation such as Fe_2MgO_4, Fe_2MnO_4, or $(NiFe)FeO_4$ have Curie temperatures in the range of 300–580°C while spinels containing Cr^{3+} with some divalent cation are magnetically ordered only in the range of liquid nitrogen temperature. Chromite spinels are all normal. Ferrite spinels generally have a mixed cation distribution, the details of which depend on synthesis temperatures and other preparation conditions [118].

A second closely related family of magnetic structures are the magnetoplumbite ferrites. $PbFe_{12}O_{19}$ is the prototype. The magnetic interactions occur in the spinel blocks. Substitution of other large divalent ions is possible as is the substitution of medium-size trivalent ions. Magnetoplumbite structures are sometimes called the hexagonal ferrites in contrast to the cubic ferrites, the spinels. Although most of the spinel and magnetoplumbite ferrites are hopping semiconductors, the conductivity is sufficiently low to allow application as inductor cores at radio frequencies.

1.1.4.4.3 Garnet-based Ferrimagnets

The garnet structure provides a useful host for a distinct family of ferrimagnets. The large eight-coordinated site can be occupied by rare earth ions which have from 0 to 7 unpaired spins. The tetrahedral site can be occupied by Fe^{3+} or by other moderately small trivalent ions. $Y_3Fe_5O_{12}$, yttrium iron garnet (YIG), is a pale green compound and a very good insulator. Because of the low electrical conductivity, the garnet ferrimagnets can be used in microwave applications where hysteresis losses are very important. The magnetic properties can be fine-tuned by varying the substitutions on the three cation sites. $Gd_3Fe_5O_{12}$,

(GGG), takes advantage of seven unpaired spins on Gd^{3+} and five unpaired spins on Fe^{3+}. Many combinations are possible.

1.1.4.5 Electrical Properties

1.1.4.5.1 Models for Electrical Conductivity in Ceramics

Ceramics are generally regarded as insulators and certainly most traditional oxide and silicate ceramics have low electrical conductivities. For those compounds, MgO, alumina, and silicates, with no electronic states to contribute to the conductivity, charge movement is by means of vacancies and interstitials. Charged ions under the influence of an electric field diffuse into nearby vacancies or take on interstitial positions. The net result is a flow of ions one way and a flow of vacancies the other way. Defect charge transport is highly temperature dependent with the conductivity generally following an Arrhenius relation such as

$$\sigma = \frac{\sigma_0}{T} \exp - \frac{E_a}{kT}$$

E_a is the activation energy for ion/vacancy diffusion, typically in the range of 0.5–2.5 eV. Conductivities are usually very low and become appreciable only at temperatures close to the melting point.

There are some combinations of ions and structures that permit much higher ionic conductivities, and there are ions with electronic structures that permit electron conductivities either through loosely bound states or band states.

1.1.4.5.2 Fast Ion Conductors

Ionic conductivities at appropriate temperatures can reach values 1–100 S/cm, values comparable to that of some metals (for reference, the conductivity of copper at 25°C is 6×10^5 S/cm). These fall into three broad categories: anion conductors, silver halides, and alkali ion conductors.

Anion conductors are materials such as $Zr_{1-x}Ca_xO_{1-x}\square_x$. Oxygen is the diffusing ion. Lime-doped zirconia at high temperatures acts as an oxygen permeable membrane. With rare earth substitution, conductivities become sufficient for these materials to act as ceramic heating elements.

The Ag^+ ion is exceptionally mobile in many ionic crystals. The silver halides, particularly AgI, have high conductivities. The conductivity of $RbAg_4I_5$ reaches 0.2 S/cm. Limitations on the stability of silver salts constrain the application of these materials as ceramic conductors.

Some of the highest alkali ion conductivities are found in compounds with the beta alumina structure. The prototype is $NaAl_{11}O_{17}$. The structure is composed of spinel blocks but with an anion vacancy in the mirror plane holding the alkali ion. Because of the vacancy structure, the activation energy for migration of the Na^+ ion is very low. There are two families of these structures: β-alumina and a Mg^{2+} substituted structure known as β″ alumina. Many complex substitutions are possible leading to different electrical properties [119]. There was great interest in these structures because the β-aluminas are chemically stable and refractory and have application as electrolytes in high-energy-density sodium–sulfur batteries.

1.1.4.5.3 Conductivity in Mixed Valence Ceramics

If a compound contains mixed valence elements there can be an exchange of electrons with the concurrent reversal of the valence state of the two ions. If there are two cation sites occupied by iron, for example, the exchange is

$$Fe^{2+}(I) + Fe^{3+}(II) \rightarrow Fe^{3+}(I) + Fe^{2+}(II)$$

If the cation–cation distance is such that the coupling takes place only in an optically excited state, the process produces intervalence electron transfer absorption and the development of strong colors but little electrical conductivity. If the cation–cation distance is smaller, the exchanged electron can migrate through the structure under an electric field to produce what is sometimes called "hopping" semiconduction. The conductivity of many iron, titanium, vanadium, manganese, and nickel compounds arises from this mechanism.

1.1.4.5.4 Metallic Conductors

Some ceramic materials are true metallic conductors in the sense that electrons move in continuous bands under the influence of electric fields. There tend to be few oxides with metallic conductivity, mostly those with conductivity arising from narrow d-bands. Examples are ReO_3 and compounds such as TiO and VO_2 that undergo semiconductor/metal phase transitions. Chalcogenides, arsenides, and related anions have extended p-orbitals so that orbital overlap to form a conduction band is more common.

Metallic conductivity can also arise from delocalized electrons in compounds where charge balance is not satisfied by conventional valence states. Examples are the rare earth chalcogenides with the NaCl structure such as LaS. Lanthanum occurs only as La^{3+} and sulfur as S^{2-} but regardless, there are stable LnS compounds (and also LnSe and LnTe) for all of the rare earths. Local charge balance is satisfied by an electron loosely bound to the rare earth ion with a nominal radius of 1–5 nm. Such electrons overlap to form a broad conduction band and complete metallic character.

1.1.4.5.5 Ceramic Superconductors

Superconductivity with a transition temperature higher than occurs in intermetallic compounds (e.g. Nb_3Sn) was first found in the K_2NiF_4 structure compound $La_{2-x}Ba_xCuO_4$ and its Sr^{2+} analog. Other copper-containing compounds related to perovskite have been found to be superconducting with superconducting transition temperatures above that of liquid nitrogen. The prototype compound first to be discovered was $YBa_2Cu_3O_7$. Numerous other compounds have been discovered with more complex structures as reviewed by Cava [34]. Because these compounds are ceramics and very brittle, as well as having limited thermodynamic stability, development of useful applications has not proceeded as well as originally hoped.

1.1.5 CONCLUSIONS

Modern ceramics are a very diverse group of materials with widely varying physical properties that can be adapted to many technological needs. The properties depend on crystal structure, on the electronic structure of the individual ions that make up the structure, and by the chemical bonding within the structure. Structures are determined in large part by ratios of the atomic constituents and by ionic radii. Bonding is a function of the collective interaction (or lack of interaction) of the outer shell electrons with each other.

ACKNOWLEDGEMENTS

I am indebted to many researchers in many parts of the world for reprints of their crystal chemical research. I am especially indebted to my mentor and colleague, Professor Rustum Roy, from whom I first learned crystal chemistry many decades ago. Special thanks are due to Dr Elizabeth L. White for her assistance in preparing the illustrations.

REFERENCES

1. Wyckoff, R. W. G. (1982). *Crystal Structures*, three volumes, Malabar, FL: R.E. Krieger Publishing Co.
2. Shannon, R. D., and Prewitt, C. T. (1969). Effective ionic radii in oxides and fluorides. *Acta Crystallogr. B* **25**: 925–946.
3. Shannon, R. D. (1976). Revised effective ionic radii and systematic studies of interatomic distances in halides and chalcogenides. *Acta Crystallogr. A* **32**: 751–767.
4. Abrahams, S. C., and Bernstein, J. L. (1969). Remeasurement of the structure of hexagonal ZnO. *Acta Crystallogr. A* **25**: 1233–1236.
5. Bates, C. H., White, W. B., and Roy, R. (1962). New high-pressure polymorph of zinc oxide. *Science* **137**: 993.
6. Warshaw, I., and Roy, R. (1961). Polymorphism of the rare earth sesquioxides. *J. Phys. Chem.* **65**: 2048.
7. Hyde, B. G., and Andersson, S. (1989). *Inorganic Crystal Structures*, New York: John Wiley.
8. Wefers, K., and Misra, C. (1987). *Oxides and Hydroxides of Aluminum*, Alcoa Tech. Paper 19, Pittsburgh.
9. Levin, I., and Brandon, D. (1998). Metastable alumina polymorphs: crystal structures and transition sequences. *J. Am. Ceram. Soc.* **81**: 1995–2012.
10. Ramos-Gallardo, A., and Vegas, A. (1996). The cation array in aluminum oxides, hydroxides and oxihydroxides. *Zeits. Krist.* **211**: 299–303.
11. Yourdshahyan, Y., Ruberto, C., Halvarsson, M., Bengtsson, L., Langer, V., Lundqvist, B. I., Ruppi, S., and Rolander, U. (1999). Theoretical structure determination of a complex material: κ-Al_2O_3. *J. Am. Ceram. Soc.* **82**: 1365–1380.
12. Smith, D. K., and Newkirk, H. W. (1965). The crystal structure of baddeleyite (monoclinic ZrO_2) and its relation to the polymorphism of ZrO_2. *Acta Crystallogr.* **18**: 983–991.
13. Teufer, G. (1962). The crystal structure of tetragonal ZrO_2. *Acta Crystallogr.* **15**: 1187.
14. Leger, J. M., Tomaszewski, P. E., Atouf, A., and Pereira, A. S. (1993). Pressure-induced structural phase transitions in zirconia under high pressure. *Phys. Rev. B* **47**: 14075–14083.
15. Stubican, V. S., and Ray, S. P. (1977). Phase equilibria and ordering the system ZrO_2–CaO. *J. Am. Ceram. Soc.* **60**: 534–537.
16. Stubican, V. S., Hink, R. C., and Ray, S. P. (1978). Phase equilibria and ordering in the system ZrO_2–Y_2O_3. *J. Am. Ceram. Soc.* **61**: 17–21.
17. Heuer, A. H., and Hobbs, L. W. (eds.) (1981). *Science and Technology of Zirconia*, Advances Ceram. 3, Columbus, OH: Amer. Ceram. Soc.
18. Claussen, N., Rühle, M., and Heuer, A. H. (eds.) (1984). *Science and Technology of Zirconia II*, Advances Ceram. 12, Columbus, OH: Amer. Ceram. Soc.
19. Somiya, S., Yamamoto, N., and Yanagida, H. (eds.) (1988). *Science and Technology of Zirconia III*, Advances Ceram. 24A,B, Columbus, OH: Amer. Ceram. Soc.
20. Meriani, S., and Palmonari, C. (eds.) (1989). *Zirconia '88—Advances in Zirconia Science and Technology*, New York: Elsevier.
21. Badwal, S. P. S., Banniser, M. J., and Hannink, R. H. J. (eds.) (1993). *Science and Technology of Zirconia V*, Lancaster, PA: Technomic.
22. Sickafus, K. E., Wills, J. M., and Grimes, N. W. (1999). Structure of spinel. *J. Am. Ceram. Soc.* **82**: 3279–3292.
23. Blasse, G. (1964). Crystal chemistry and some magnetic properties of mixed metal oxides with the spinel structure. *Philips Res. Rep. Suppl.* **3**: 1–139.
24. Muller, O., and Roy, R. (1974). *Crystal Chemistry of Non-Metallic Materials: The Major Ternary Structural Families*, Berlin: Springer-Verlag.

25. Hill, R. J., Craig, J. R., and Gibbs, J. V. (1979). Systematics of the spinel structure type. *Phys. Chem. Mineral.* **4**: 317–339.
26. Anderson, M. T., Vaughey, J. T., and Poeppelmeier, K. R. (1993). Structural similarities among oxygen-deficient perovskites. *Chem. Mater.* **1993**: 151–165.
27. Galasso, F. S. (1969). *Structure, Properties and Preparation of Perovskite-Type Compounds*, Oxford: Pergamon Press.
28. Peña, M. A., and Fierro, J. L. G. (2001). Chemical structures and performance of perovskite oxides. *Chem. Rev.* **101**: 1981–2017.
29. Navrotsky, A., and Weidner, D. J. (1989). *Perovskite: A Structure of Great Interest to Geophysics and Materials Science*, Washington, DC: American Geophysical Union.
30. Sharma, I. B., and Singh, D. (1998). Solid state chemistry of Ruddlesden–Popper type complex oxides. *Bull. Mater. Sci.* **21**: 363–374.
31. Tian, W., Pan, X. Q., Haeni, J. H., and Schlom, D. G. (2001). Transmission electron microscopy study of n = 1-5$Sr_{n-1}Ti_nO_{3n+1}$ epitaxial thin films. *J. Mater. Res.* **16**: 2013–2026.
32. Šturm, S., Rečnik, S. C., and Čeh, M. (2000). Formation of Ruddlesden–Popper faults and polytype phases in SrO-doped $SrTiO_3$. *J. Mater. Res.* **15**: 2131–2139.
33. Bednorz, J. G., and Müller, K. A. (1986). Possible high-T_c superconductivity in the La–Ba–Cu–O system. *Zeits. Phys. B: Condens. Matter* **64**: 188.
34. Cava, R. J. (2000). Oxide superconductors. *J. Am. Ceram. Soc.* **83**: 5–28.
35. Geller, S. (1967). Crystal chemistry of the garnets. *Zeits. Krist.* **125**: 1–47.
36. Novak, G. A., and Gibbs, J. V. (1971). The crystal chemistry of silicate garnets. *Am. Miner.* **56**: 791–825.
37. Liebau, F. (1985). *Structural Chemistry of Silicates*, Berlin: Springer-Verlag.
38. Smyth, J. R., and Bish, D. L. (1989). *Crystal Structures and Cation Sites of the Rock-Forming Minerals*, London: Allen and Unwin.
39. Papike, J. J. (1987). Chemistry of the rock-forming silicates: ortho, ring, and single chain structures. *Rev. Geophys.* **25**: 1483–1526.
40. Angel, R. J., and Prewitt, C. T. (1986). Crystal structure of mullite: a re-examination of the average structure. *Am. Mineral.* **71**: 1476–1482.
41. Papike, J. J., and Cameron, M. (1976). Crystal chemistry of silicate minerals of geophysical interest. *Rev. Geophys. Space Phys.* **14**: 37–80.
42. Kerrick, D. M. (1990). The Al_2SiO_5 polymorphs. *Rev. Mineral.* **22**: 1–406.
43. Fleet, M. E., and Henderson, G. S. (1995). Epsilon sodium disilicate: a high-pressure layer structure [$Na_2Si_2O_5$]. *J. Solid State Chem.* **119**: 400–404.
44. Heaney, P. J., Prewitt, C. T., and Gibbs, G. V. (eds.). (1994). *Silica: Physical Behavior, Geochemistry and Materials Applications, Rev. Mineral, Vol. 29*, Washington, DC: Mineral. Soc. Amer.
45. Smith, J. V. (1974). *Feldspar Minerals*, vols 1 and 2, New York: Springer-Verlag.
46. Bennett, J. M., Dytrych, W. J., Pluth, J. J., Richardson, Jr., J. W., and Smith, J. V. (1986). Structural features of aluminophosphate materials with Al/P = 1. *Zeolites* **6**: 349–360.
47. Flahaut, J. (1979). Sulfides, selenides and tellurides, *Handbook of the Physics and Chemistry of Rare Earths*, pp. 1–88 (Gschneidner, Jr., K. A., and Eying, L., eds.), vol. IV, Chapter 31, Amsterdam: North Holland.
48. Savage, J. A. (1990). Crystalline optical materials for ultraviolet, visible and infrared applications, *Optical Materials*, pp. 323–419 (Musikant, S., ed.), Chapter 3, New York: Marcel Dekker.
49. Wentorf, Jr., R. H. (1961). Synthesis of the cubic form of boron nitride. *J. Chem. Phys.* **34**: 809–812.

50. Riley, F. L. (1977). *Nitrogen Ceramics*, NATO Advanced Study Institute Series, Leyden: Noordhoff.
51. Zerr, A., Miege, G., Serghious, G., Schwarz, M., Kroke, E., Riedel, R., Fuess, H., Kroll, P., and Boehler, R. (1999). Synthesis of cubic silicon nitride. *Nature* **400**: 340–342.
52. Ching, W.-Y., Mo, S.-D., Ouyang, L., Rulis, P., Tanaka, I., and Yoshiya, M. (2002). Theoretical prediction of the structure and properties of cubic spinel nitrides. *J. Am. Ceram. Soc.* **85**: 75–80.
53. Jack, K. H. (1976). Sialons and related nitrogen ceramics. *J. Mater. Sci.* **11**: 1135–1158.
54. Bundy, F. P., and Kasper, J. S. (1967). Hexagonal diamond—a new form of carbon. *J. Chem. Phys.* **46**: 3437–3446.
55. Bundy, F. P. (1980). The P, T phase and reaction diagram for elemental carbon, 1979. *J. Geophys. Res.* **85**: 6930–6936.
56. Bundy, F. P. (1963). Direct conversion of graphite to diamond in a static pressure apparatus. *J. Chem. Phys.* **38**: 631–643.
57. Angus, J. C., and Hayman, C. C. (1988). Low-pressure, metastable growth of diamond and "diamondlike" phases. *Science* **241**: 913–921.
58. Ashfold, M. N. R., May, P. W., Rego, C. A., and Everitt, N. M. (1994). Thin film diamond by chemical vapour deposition methods. *Chem. Soc. Rev.* 21–30.
59. Spear, K. E. (1989). Diamond—ceramic coating of the future. *J. Am. Ceram. Soc.* **72**: 171–191.
60. Parthe, E., and Yvon, K. (1970). On the crystal chemistry of the close packed transition metal carbides. II. A proposal for the notation of the different structures. *Acta Crystallogr. B* **26**: 153–163.
61. Barsoum, M. W. (2000). The $M_{N+1}AX_N$ phases. A new class of solids: thermodynamically stable nanolaminates. *Progr. Solid State Chem.* **28**: 201–281.
62. Nagamatsu, J., Nakagawa, N., Muranaka, T., Zenitani, Y., and Akimitsu, J. (2001). Superconductivity at 39 K in magnesium diboride. *Nature* **410**: 63–64.
63. Spear, K. E. (1976). Phase behavior and related properties of rare earth borides, *Phase Diagrams*, pp. 91–159 (Alper, A. M., ed.), New York: Academic Press.
64. Ballhausen, C. J. (1962). *Introduction to Ligand Field Theory*, New York: McGraw Hill.
65. Sugano, S., Tanabe, Y., and Kamimura, H. (1970). *Multiplets of Transition-Metal Ions in Crystals*, New York: Academic Press.
66. Henderson, B., and Imbusch, G. F. (1989). *Optical Spectroscopy of Inorganic Solids*, Oxford: Oxford University Press.
67. Wybourne, B. G. (1965). *Spectroscopic Properties of Rare Earths*, New York: John Wiley.
68. Burns, R. G. (1993). *Mineralogical Applications of Crystal Field Theory*, Cambridge: Cambridge University Press.
69. Blasse, G., and Grabmaier, B. C. (1994). *Luminescent Materials*, Berlin: Springer Verlag.
70. Holmes, O. G., and McClure, D. S. (1957). Optical spectra of hydrated ions of the transition metals. *J. Chem. Phys.* **26**: 1686–1694.
71. Jorgensen, C. K. (1962). The nephelauxetic series. *Prog. Inorg. Chem.* **4**: 73–124.
72. George, P., and McClure, D. S. (1959). The effect of inner orbital splitting on the thermodynamic properties of transition metal compounds and coordination complexes. *Progr. Inorg. Chem.* **1**: 381–463.
73. McClure, D. S. (1965). The effects of inner-orbitals on thermodynamic properties, *Some Aspects of Crystal Field Theory*, pp. 77–95 (Dunn, T. M., McClure, D. S., and Pearson, R.G., eds.), Chapter 4, New York: Harper and Row.
74. Burns, R. G. (1985). Thermodynamic data from crystal field spectra. *Rev. Mineral.* **14**: 277–316.
75. Hinchliffe, A. (2000). *Modelling Molecular Structures*, 2nd edn, Chicester: Wiley.

76. Gibbs, G. V. (1982). Molecules as models for bonding in silicates. *Am. Mineral.* **67**: 421–450.
77. Tossell, J. A., and Gibbs, G. V. (1977). Molecular orbital studies of geometries and spectra of minerals and inorganic compounds. *Phys. Chem. Mineral.* **2**: 21–57.
78. Geisinger, K. L., Gibbs, G. V., and Navrotsky, A. (1985). A molecular orbital study of bond length and angle variations in framework structures. *Phys. Chem. Mineral.* **11**: 266–283.
79. D'Arco, P., Silvi, B., Roetti, C., and Orlando, R. (1991). Comparative study of spinel compounds: a pseudopotential periodic Hartree–Fock calculation of Mg_2SiO_4, Mg_2GeO_4, Al_2MgO_4, and Ga_2MgO_4. *J. Geophys. Res.* **96**: 6107–6112.
80. Simons, J. (1991). An experimental chemist's guide to *ab initio* quantum chemistry. *J. Phys. Chem.* **95**: 1017–1029.
81. Vaughan, D. J., and Tossell, J. A. (1983). Electronic structures of sulfide minerals—theory and experiment. *Phys. Chem. Minerals* **9**: 253–262.
82. Jones, H. (1962). *The Theory of Brillouin Zones and Electronic States in Crystals*, Amsterdam: North-Holland.
83. McCarthy, M. I., and Harrison, N. M. (1994). *Ab initio* determination of the bulk properties of MgO. *Phys. Rev. B* **49**: 8574–8582.
84. French, R. H. (1990). Electronic band structure of Al_2O_3, with comparison to AlON and AlN. *J. Am. Ceram. Soc.* **73**: 477–489.
85. French, R. H., Jones, D. J., and Loughin, S. (1994). Interband electronic structure of α-alumina up to 2167 K. *J. Am. Ceram. Soc.* **77**: 412–422.
86. Chelikowsky, J. R., and Schlüter, M. (1977). Electron states in α-quartz: a self-consistent pseudopotential calculation. *Phys. Rev. B* **15**: 4020–4029.
87. Pantelides, S. T., and Harrison, W. A. (1976). Electronic structure, spectra, and properties of 4:2-coordinated materials. I. Crystalline and amorphous SiO_2 and GeO_2. *Phys. Rev. B* **13**: 2667–2691.
88. Daude, N., Gout, C., and Jouanin, C. (1977). Electronic band structure of titanium dioxide. *Phys. Rev. B* **15**: 3229–3235.
89. Glassford, K. M., and Chelikowsky, J. R. (1992). Optical properties of titanium dioxide in the rutile structure. *Phys. Rev. B* **45**: 3874–3877.
90. Glassford, K. M., and Chelikowsky, J. R. (1992). Structural and electronic properties of titanium dioxide. *Phys. Rev. B* **46**: 1284–1298.
91. Mo, S.-D., and Ching, W. Y. (1995). Electronic and optical properties of three phases of titanium dioxide: Rutile, anatase, and brookite. *Phys. Rev. B* **51**: 13023–13032.
92. Schwarz, K. (1987). Band theoretical studies of the electronic structure of oxides. *Phys. Chem. Minerals* **14**: 315–319.
93. Koster, G. F. (1957). Space groups and their representations, *Solid State Physics*, pp. 1–84 (Seitz, F., and Turnbull, D., eds.), New York: Academic Press.
94. Bortz, M. L., French, R. H., Jones, D. J., Kasowski, R. V., and Ohuchi, F. S. (1990). Temperature dependence of the electronic structure of Al_2O_3, $MgAl_2O_4$, and MgO. *Phys. Scripta* **41**: 537–541.
95. Drickamer, H. G., and Frank, C. W. (1973). *Electronic Transitions and High Pressure Chemistry and Physics of Solids*, London: Chapman and Hall.
96. Bailly, F., and Manca, P. (1966). Correlation entre largeur de bande interdite et l'energie libre d'atomisation des materiaux semi-conducteurs. *J. Phys. Chem. Solids* **27**: 783–793.
97. Goodenough, J. B. (1965). Transition-metal oxides with metallic conductivity. *Bull. Soc. Chim. France 1965*, 1200–1207.
98. Goodenough, J. B. (1967). Narrow-band electrons in transition-metal oxides. *Czech. J. Phys. B* **17**: 304–336.

99. Ferretti, A., Rogers, D. B., and Goodenough, J. B. (1965). The relation of the electrical conductivity in single crystals of rhenium trioxide to the conductivities of Sr_2MgReO_6 and Na_xWO_3. *J. Phys. Chem. Solids* **26**: 2007–2011.
100. Vaughan, D. J. (1985). Spectroscopy and chemical bonding in the opaque minerals, *Chemical Bonding and Spectroscopy in Mineral Chemistry*, pp. 251–292 (Berry, F. J., and Vaughan, D. J., eds.), London: Chapman and Hall.
101. Hazen, R. M., and Finger, L. W. (1982). *Comparative Crystal Chemistry*, New York: John Wiley.
102. Srivastava, G. P. (2001). Theory of thermal conduction in nonmetals. *MRS Bull.* **26**: 445–450.
103. Hofmeister, A. M. (2001). Thermal conductivity of spinels and olivines from vibrational spectroscopy: ambient conditions. *Am. Mineral.* **86**: 1188–1208.
104. Slack, G. A. (1973). Nonmetallic crystals with high thermal conductivity. *J. Phys. Chem. Solids* **34**: 321–335.
105. Kingery, W. D., Bowen, H. K., and Uhlmann, D. R. (1976). *Introduction to Ceramics*, New York: John Wiley.
106. Hertzberg, R. W. (1983). *Deformation and Fracture Mechanics of Engineering Materials*, 2nd edn, New York: John Wiley.
107. Nye, J. F. (1957). *Physical Properties of Crystals*, Oxford: Oxford University Press.
108. Newnham, R. E. (1974). Elastic properties of oxides. *Ceram. Bull.* **53**: 821–829.
109. Haines, J., and Leger, J. M. (1998). The search for superhard materials: a new approach. *J. Superhard Mater.* **20**: 1–7.
110. Lawn, B. R. (1998). Indentation of ceramics with spheres: a century after Hertz. *J. Am. Ceram. Soc.* **81**: 1977–1994.
111. Liu, A. Y., and Cohen, M. L. (1989). Prediction of new low compressibility solids. *Science* **245**: 841–842.
112. Sung, C.-M., and Sung, M. (1996). Carbon nitride and other speculative superhard materials. *Mater. Chem. Phys.* **43**: 1–18.
113. Matsumoto, S., Xie, E.-Q., and Izumi, F. (1999). On the validity of the formation of crystalline carbon nitrides, C_3N_4. *Diamond Relat. Mater.* **8**: 1175–1182.
114. Nassau, K. (1983). *The Physics and Chemistry of Color*, New York: John Wiley.
115. Eppler, R. A. (1987). Selecting ceramic pigments. *Ceram. Bull.* **66**: 1600–1604.
116. White, W. B. (1975). Luminescent materials. *Trans. Am. Crystallogr. Assoc.* **18**: 31–49.
117. Newnham, R. E. (1997). Molecular mechanisms in smart materials. *Mater. Res. Soc. Bull.* **22**: 20–33.
118. Broese van Groenou, A., Bongers, P. F., and Stuyts, A. L. (1968). Magnetism, microstructure and crystal chemistry of spinel ferrites. *Mater. Sci. Eng.* **3**: 317–392.
119. Farrington, G. C., and Briant, J. L. (1979). Fast ionic transport in solids. *Science* **204**: 1371–1379.

Handbook of Advanced Ceramics
S. Sōmiya *et al.* (Eds.)

CHAPTER 2

2.1 Raw Materials

MAKIO NAITO[1], NOBUHIRO SHINOHARA[2] and KEIZO UEMATSU[3]
[1]*JWRI, Osaka University, Ibaraki, Japan*
[2]*Asahi Glass Co. Ltd., Hazawa, Kanagawa-ku, Yokohama, Japan*
[3]*Department of Chemistry, Nagaoka University of Technology, Nagaoka, Japan*

2.1.1 INTRODUCTION

Raw material is very important in ceramics in various aspects. It is one of the most decisive factors for the quality of a product [1]. Roughly speaking, half of the quality of a product is determined by the characteristics of raw materials in the current production technology. Use of proper raw material is critical for producing ceramics of high quality at a minimal cost. Technologically, the raw powder sets a fair starting point for industries. Raw materials of high quality are in the market, and are equally available for all users for their disposal. Proper selection of raw powders and their post-treatment is critical for successful competition in the market.

A variety of factors affects the characteristics of raw powders. Information on the product specification as well as their production method is very important in the proper selection of raw materials. Post-treatments such as grinding, mixing and classification also have critical effects on the characteristics. Their effects must be considered carefully to control and also to understand the characteristics of raw powders.

General criteria for good raw powders are shown in Table 2.1.1. They may vary, however, with application and the technical level of processing. For example, fine powders are desirable for many applications, at least, in principle, for their low sintering temperature, the resultant fine grains after sintering and high quality of products. In practice, however, their processing is often very difficult without high skill, and the quality of ceramics made from them may be lower than that made from coarse powders, if they are inadequately processed. Our skill often has not reached the high level needed for handling extremely fine powders.

In this chapter, we tentatively define good raw materials as follows: it can be formed into a chemically and physically uniform green compact of both macro- and microscopic levels with proper treatment within the current technical level. The skill of processing can affect this definition itself. The effect of

TABLE 2.1.1 Criteria for Good Raw Powder

Chemistry	
Composition	Homogeneous
Impurity	None
Stoichiometry	Stoichiometric
Phase	Stable
Particle shape	Uniform and spherical for most applications
Particle size	Small
Distribution	Adequate

processing can even be strong enough to reverse the conclusion. For example, fine powders can be very good raw materials, provided that there are means to disperse them. In many practical cases, fine powders are very difficult to handle properly, and their use may seriously reduce the uniformity of green compacts.

The quality of current raw materials needs to be critically evaluated. In an optimistic view, our advanced powders have reached the perfect level except for the price. With common analytical tools, we cannot find any technical problem in these advanced raw powders. This may not be true, however.

The apparent absence of a detrimental structure in green compacts is clearly due to our poor ability of characterization, especially with modern tools. A direct evidence of fair quality of green compact is noted in Figure 2.1.1. It shows an example of a green compact examined [2] by the optical method, which has very high sensitivity for detecting minor defects in green compacts [3]. The green compact contains a variety of structures and is far from uniform, although it should be one of the most homogeneous green body structures.

In this critical view, certain problems are noted in any raw powder currently available even in advanced powder systems. There is a clear evidence that the above criteria for good raw powder have not been met in virtually any system in reality. The quality of current raw materials is far from perfect. They contain a variety of imperfections such as aggregates and coarse particles. The unpredictable variation also occurs from batch to batch. Clearly, the non-uniformity in green compact develops defects during sintering and causes reduction and variation of mechanical strength. The imperfection of raw powder is one of the major causes that makes the quality of ceramic products, especially for structural application, far below the theoretical prediction.

The following sections describe the characteristics of raw powder to be considered, the detailed description on important powders and post-treatment such as grinding, mixing and classification.

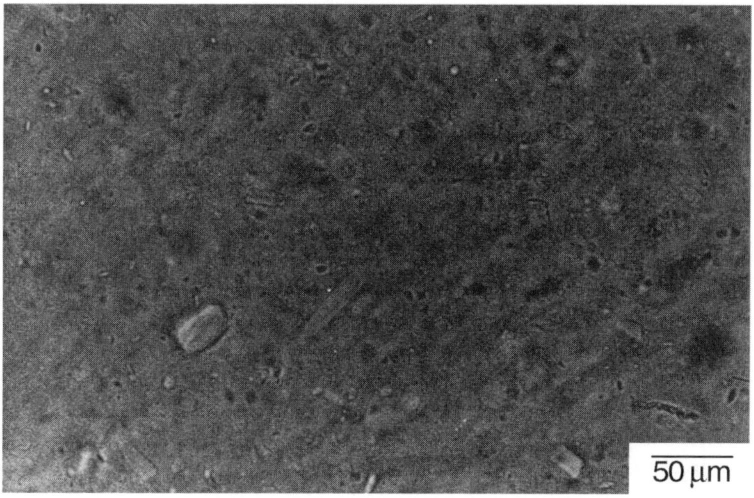

FIGURE 2.1.1 Structure of green compact examined by the liquid immersion method.

2.1.2 CHARACTERISTICS

Chemical and physical properties affect the characteristics of raw powders as explained in detail below.

2.1.2.1 Chemical Composition

The chemical composition in a powder should be fixed and constant. In simple oxides made of typical elements like alumina, the quantities of constituent elements are expressed by a simple ratio, for example, 2:3 for aluminum to oxygen in alumina. This stoichiometry is seldom met in practical powders that contain transition metal elements of variable valency. The non-stoichiometry can happen in the ratios of both cation/anion and cation/cation, affecting the property significantly. In iron oxide, the iron to oxygen ratio can vary very widely. In barium titanate, the ratio of titanium to barium and the ratio for cation to anion can vary. The stoichiometry affects the sintering of barium titanate for capacitors and uranium oxide for nuclear fuel.

2.1.2.2 Impurity

Impurity is a minor element that is not intentionally added to the powder. Certain elements govern the production of ceramics and/or limit the application.

A typical example is sodium in alumina powder. Sodium affects the dispersion of the powder in slurry and the sintering behavior, exerting critical influence on the quality of products. It also limits the application of the products for ceramic package for semiconductors, since this is one of the most detrimental elements for silicon semiconductors. A minor element can be also very important in ceramics for electronics. In alumina again, a trace of radioactive thorium is crucial. Thorium is present in bauxite, and a small fraction is preserved in the final powder. Gamma rays from thorium causes malfunction in CPUs and semiconductor memories. Calcia and silica in alumina tend to be segregated at the grain boundaries and affects the chemical and physical properties of the grain boundary significantly [4]. Their minor quantity causes anisotropic grain growth of alumina in sintering. High-purity alumina (>99.99%) shows much less tendency for abnormal grain growth, if proper caution is taken to avoid impurity pick-up during heating [5].

Oxygen in non-oxide powder is important. It affects the sintering behavior as well as the quality of final products. Oxygen can be present as oxide, oxynitride, hydroxide, etc. in the nitride powders. It contributes to the formation of the liquid phase and thus to sintering at low temperatures.

Certain elements do not have markedly detrimental effect on processing and properties, like oxygen and carbon in silicon semiconductors. Elements of the same column in the table of elements often have minor effects.

2.1.2.3 PHASE

Phase may significantly affect the characteristic of raw materials with nominally the same chemical composition. Important examples are noted in rutile and anatase in titania, alpha- and gamma-alumina, and monoclinic, tetragonal and cubic zirconia. In each material, the former corresponds to the low-temperature phase and the latter to the high-temperature phase. In general, the activity of powder is higher in the low-temperature phase than in the high-temperature phase. The sintering starts at lower temperature in the low-temperature phase than in the high-temperature phase. With a proper heating schedule, titania can be sintered to high density [6]. However, the phase transformation associated with heating may cause damage in sintered materials. The specific volume is larger in a low-temperature phase than in a high-temperature phase. An exception is zirconia, in which cracks are developed due to a large volume change associated from high-temperature cubic phase to low-temperature monoclinic phase transformation during cooling from sintering temperature.

2.1.2.4 Particle Size

Particle size directly affects rate and temperature of sintering significantly. Figure 2.1.2 shows the density reached for various alumina powders after sintering for fixed time at various temperatures. Simple sintering theory can predict the densification temperatures needed for these powders. In Figure 2.1.2, high density was reached after sintering at 1850 and 1550 K for alumina powders with particle sizes 0.5 and 0.1 μm, respectively. According to the simple theory of sintering, the sintering rate varies roughly minus 3–4th powder of particle size for the mechanism involving grain-boundary and bulk diffusion [7]. In the above example, the sintering rate should be approximately 125–625 times faster for the finer powder at a given condition. Meanwhile, the mass transport rate decreases rapidly with decreasing temperature due to the high activation energy for solid-state diffusion; approximately 1/4–1/6 with each decrease of temperature by 100 K or 1/64–1/216 by 300 K. This slow mass transport is well compensated by the fast sintering rate in the fine powder.

The handling of small particles is difficult. The dispersion in slurry is especially difficult. The dispersant needed is proportional to the surface area, since a given fraction of total surface must be covered by the molecules of dispersant to generate repelling interaction among particles. Unrealistically large quantities of dispersant may be required to disperse very fine powders. Dispersion through pH adjustment and/or dispersant with low molecular weight can be effective in dispersing fine powders.

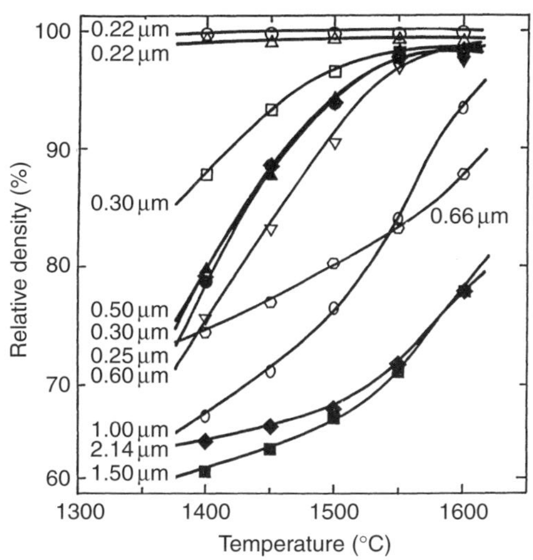

FIGURE 2.1.2 Effect of particle size on the density after sintering at various temperatures.

Improperly handled, small particles tend to form strong aggregates. Aggregates tend to form fracture origins and are very detrimental in the production of ceramics.

2.1.2.5 Particle Size Distribution

Proper size distribution is important for high packing density. The highest density achievable is 74% for mono-dispersed particles in theory. It can be much higher for a well-designed mixture of mono-dispersed powders [8]. In the latter system, the coarsest particles are packed closely and the space between them is filled with the second coarsest particles. The void space left is filled with the third coarsest particles, and so on.

In reality, it is very difficult to attain the high density predicted. The typical density achievable is around 80 and 60% for multi- and mono-dispersed particles, respectively.

A system with broad particle size distribution is prone to exaggerated grain growth in sintering [9]. Few large particles behave as nuclei for grain growth. The resultant abnormally large particles have a very detrimental effect on various properties of ceramics.

Coarse particles are present in almost all powders, even though they are apparently absent in the ordinal examination with a particle size analyzer. A special tool, such as the liquid immersion method [2, 3], is needed to show their presence. Figure 2.1.3 shows an example of coarse particles noted in alumina by this tool. Their fraction is much below 1%, which is under the detection limit of most types of analyzers. Yet, they have a very important effect in microstructure development of ceramics. Coarse particles can be either a single crystal or aggregates consisting of small particles.

Accurate measurement of particle size distribution is difficult. Figure 2.1.4 shows the particle size distribution determined with commercial equipment of the laser diffraction type [10]. The results vary significantly with the equipment; the mean particle size differs more than 50% and it is very difficult to obtain the true value. The distribution varies more significantly when other types of equipment are included. The results also depend on operator significantly.

2.1.2.6 Particle Shape

Particle shape exerts an important effect on the packing structure in a compact. High packing density is readily attained with particles of near spherical shape.

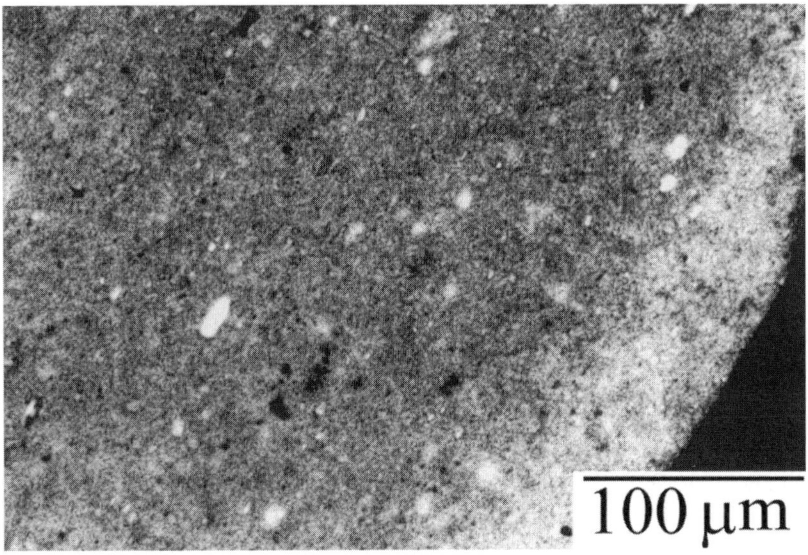

FIGURE 2.1.3 Large particles in alumina green compact examined by the liquid immersion method with the crossed polarized light microscope. White features are alumina particles.

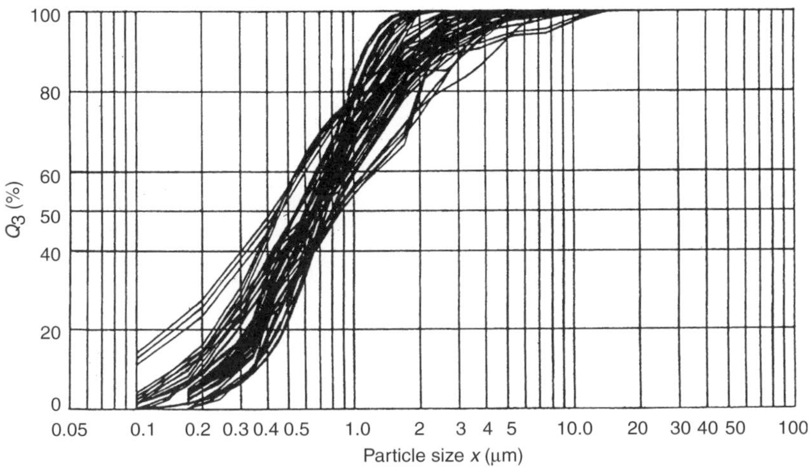

FIGURE 2.1.4 Particle size distribution of a silicon carbide powder examined by various particle size analyzers of laser diffraction type.

It is very difficult to get high packing density with particles of platelet or needle shape if they are randomly oriented. They tend to leave large voids between them. Regularly packed, their packing density can be fairly high.

Shear stress in forming tend to orient particles of non-spherical shapes. Platelets and/or needles readily form oriented structures [11]. They form a unique microstructure and thus have anisotropic properties after sintering. A ceramic of interesting characteristics can be made through doctor-blade processing, which allows uniform shear stress field in the processing. It should be very difficult to achieve uniform structure with other forming methods such as extrusion and injection molding where the shear stress varies with the distance from the die wall.

Particle shape also exerts considerable effect on sintering. Two types of compacts were formed with particles that were of spherical (aspect ratio: 1.1) and near-spherical shapes (aspect ratio: 1.5) by uniaxial pressing followed by cold isostatic pressing. They have very uniform density, yet a significant anisotropy was noted in sintering the compact made with the particle of near-spherical shape. Isotropic shrinkage was found in the compact made with the spherical particles [12].

2.1.2.7 Agglomerate

Agglomerates are formed by weak attractive interactions between fine primary particles. The weak interactions include Van der Waals force, capillary force due to condensed liquid and electrostatic force between oppositely charged particles. Agglomerates can be made into primary particles with gentle mechanical force once the attractive interaction is compensated adequately, for example, by the addition of a dispersant in a liquid media. An agglomerate is formed again easily, however, if adequate care is not taken.

Agglomerates can have a detrimental effect on the structure of green compact and thus on the microstructure of ceramics after sintering. They form a local region with density different from that of the surrounding matrix [13, 14].

2.1.2.8 Aggregates

In aggregates, a group of particles are bonded tightly by strong solid bonding formed by sintering, etc. Gentle stirring cannot break these particles into primary particles. Rigorous grinding, such as ball milling, is needed to break them. The increased sinterability by grinding was once explained by the increased

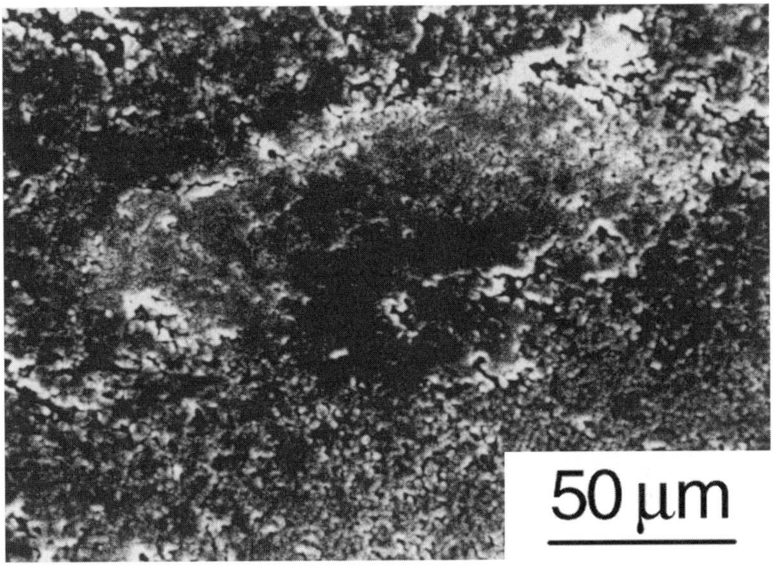

FIGURE 2.1.5 SEM micrograph of aggregates in alumina powder.

activity of the powder due to stored mechanical energy. This explanation, however, is not accepted today. The increased activity is ascribed to the elimination of aggregates by grinding [15]. Aggregates are very detrimental for sintering.

Aggregates are formed by a variety of ways, such as partial sintering during a thermal process in powder synthesis and/or precipitation of salts at the neck of particles in drying of slurry. They are also formed by the topotactic reaction in the thermal decomposition of a mother salt. The crystallites formed in this reaction are oriented, since their crystal axes have a fixed relationship to that of the mother crystal. Figure 2.1.5 shows the SEM micrograph noted in very fine alumina made through the thermal decomposition route [16]. The aggregates consisting of alumina particles of 0.1 μm have a size of about 30 μm. The orientation is noted for the crystal axis of these fine alumina particles.

The aggregate may be detected in the particle size measurement. However, they are not found in the surface area measurement. Its surface area is approximately equal to the sum of the surface area of each primary particle. The neck between the primary particles has only a small effect on the specific surface area.

The aggregate behaves like a single particle with the size of outer dimension in many operations. The aggregates have a very detrimental effect on the processing of ceramics. It forms defects in powder compacts, and thus in sintered ceramics. If its content is high in green compact, even the sintering behavior is affected. They retard the densification in sintering. Their elimination may improve the sintering characteristics as well as the quality of product significantly [17, 18].

2.1.3 IMPORTANT POWDERS

2.1.3.1 Alumina (Al_2O_3)

Alumina [19–25] has excellent physical and chemical properties, and is used in a variety of industrial fields. Various types of high-purity grades of powders such as hydrated, activated, calcined, low-soda, reactive, tabular and fused aluminas are available in the market.

2.1.3.1.1 Production of Alumina

Alumina occurs abundantly in nature, most often as impure hydroxides that are the essential constituents of bauxites. Bauxite is converted to alumina through the Bayer process. In the Bayer process, the Al_2O_3 component in bauxite dissolves in NaOH solution to form $NaAlO_2$, subsequently followed by the precipitation of $Al(OH)_3$ by the hydrolysis of $NaAlO_2$. Al_2O_3 can be produced by calcining $Al(OH)_3$ at $>1000°C$. Through the process, impurities (e.g. SiO_2, Fe_2O_3 and TiO_2) are removed and a nominal 99.5% Al_2O_3 can be produced with Na_2O ($\sim$0.5 to $<$0.05%) as the dominant impurity. Low-soda calcined alumina contains Na_2O of $<$0.1% and is applied to spark plugs, substrates, electronic components, and various engineering ceramics parts. Submicron-sized alumina powders with purity 99.8–99.9% are available.

High purity (99.99% or more) and ultra-fine powders are also synthesized through advanced processes: hydrolysis and heat treatment of aluminum alkoxide, and pyrolysis of ammonium alum, $(NH_4)_2SO_4 \cdot Al_2(SO_4)_3 \cdot 12H_2O$, or ammonium dawsonite, $NH_4AlCO_3(OH)_2$, through the following reactions:

- Hydrolysis and heat treatment of aluminum alkoxide

$$2Al + 6ROH \rightarrow 2Al(OR)_3 + 3H_2$$

$$2Al(OR)_3 + 4H_2O \rightarrow Al_2O_3 \cdot H_2O + 6ROH$$

$$Al_2O_3 \cdot H_2O \rightarrow \text{Calcination} \rightarrow Al_2O_3 + H_2O$$

- Pyrolysis of ammonium alum

$$Al_2(SO_4)_3 \cdot 18H_2O + (NH_4)_2SO_4 \rightarrow (NH_4)_2SO_4 \cdot Al_2(SO_4)_3 \cdot 12H_2O + 6H_2O$$

$$(NH_4)_2SO_4 \cdot Al_2(SO_4)_3 \cdot 12H_2O \rightarrow \text{Pyrolysis} \rightarrow Al_2O_3 + 2NH_3 + 4SO_3 + 13H_2O$$

- Pyrolysis of ammonium dawsonite

$$(\text{Aluminum salt}) + NH_4HCO_3 \rightarrow NH_4AlCO_3(OH)_2$$

$$2NH_4AlCO_3(OH)_2 \rightarrow Al_2O_3 + 2NH_3 + 2CO_2 + 3H_2O$$

In these production methods, aluminum intermediate resources made from the Bayer process are used as a starting material. Virtually all commercial alumina products are made through the Bayer process.

2.1.3.1.2 Powder Characteristics of Alumina

2.1.3.1.2.1 *Activated alumina*

Activated alumina has a very high surface area and is used as a variety of adsorbents and catalysts. It may include the transition alumina such as ρ-, η-, δ-phases other than boehmite (AlOOH) and γ-Al_2O_3. Activated alumina normally has a large pore volume and a high surface area of 100–400 m^2/g. The products are produced through a special flash calcination of aluminum trihydrate, or by pyrolysis of hydrolyzed aluminum salts or aluminum alkoxides. After activation, the products typically are formed into granules or pellets and sold as desiccants, adsorbents, bed supports, raw materials for catalyst, etc. Products of various grades are commercially available. The powder characteristics are adjusted according to the applications.

2.1.3.1.2.2 *Calcined alumina*

Calcined aluminas have a highly stable crystalline form, α-Al_2O_3. They are manufactured by calcination in a rotary kiln of aluminum trihydroxide (gibsite, $Al(OH)_3$), which is made through the Bayer process. A variety of calcined aluminas are produced and classified by soda content, particle size and degree of calicination to meet the requirements for the refractories, ceramics and glass industries. Calcined alumina, which has been extensively ground by ball-milling to produce a relatively fine product, is called reactive alumina. Milling conditions are precisely controlled to yield the products with desired powder characteristics.

2.1.3.1.2.3 *Low-soda alumina*

Low-soda aluminas have soda levels of less than 0.1%. These aluminas are calcined alumina, from which the residual soda had been removed to make it better suited to ceramic processing. Reduction of the soda content is achieved by adding boric acid or silica to aluminum trihydroxide during the calcination process. Low-soda alumina is widely used to help in sintering of electronic components and various structural ceramics. Low-soda alumina is also used for abrasives and polishing compounds. A variety of grades are commercially available. Typical examples of the powders are presented in Table 2.1.2.

For the production of ceramics, the characteristics of the powder must be well controlled to achieve stable properties in both green and sintered bodies. In the market, there are many types of powders available. Their characteristics, such as purity, specific surface area, particle size and size distribution, particle morphology, agglomerate or aggregate structures, are appropriately controlled to meet the demands for various applications. Average particle size of commercially available low-soda alumina ranges from 0.4 to several tens of micrometers. Although the purity of the products are usually 99.7–99.9%, high-purity powders with a purity of 99.95% or more are also produced.

2.1.3.1.2.4 *High-purity alumina*

High-purity aluminas, whose purity is 99.99% or more, are synthesized through the processes of the hydrolysis and calcination of aluminum alkoxide, the pyrolysis of ammonium alum, etc. The powders manufactured with these methods have general characteristics such as high purity, small particle size of submicron scale and uniform size distribution. Their typical application includes high-strength ceramics (cutting tools, bearings, etc.), optical materials (translucent parts, YAG, etc.) and electronic materials (IC substrates, ceramic parts for semiconductor manufacturing, etc.). Typical characteristics of the commercial high-purity powders are shown in Table 2.1.3.

2.1.3.2 Zirconia (ZrO_2)

Pure zirconia [22, 26–31] has a high melting temperature (2700°C) and a low thermal conductivity. Its application includes raw materials for lead–zirconia–titanate electronic ceramics, solid electrolyte, structural components and refractories. The crystal structure of pure zirconia is monoclinic at room temperature, transforming to tetragonal at 1170°C and cubic at ~2370°C. The transformation of the monoclinic structure is accompanied by a large volume change and causes many cracks within the sintered structures.

Zirconia can be stabilized by the addition of some oxides called stabilizers, such as CaO, MgO and Y_2O_3. They form a solid solution with zirconia and make the cubic and/or tetragonal structure stable for all temperatures, eliminating the detrimental phase transformation during heating and cooling. Addition of more than 16 mol% of CaO (7.9 wt%), 16 mol% MgO (5.86 wt%) or 8 mol% of Y_2O_3 (13.75 wt%) into the zirconia structure is needed to form a fully stabilized zirconia. This solid solution is termed as stabilized zirconia. Stabilized zirconia,

TABLE 2.1.2 Typical Characteristics of Commercial Activated, Calcined, Low-Soda and Reactive Alumina Powders

	Classification	Activated	Activated	Calcined	Calcined	Low-soda	Low-soda	Reactive	Reactive	Reactive	Reactive
	Supplier[a]	Alcan[b]	Sumitomo[c]	Sumitomo[c]	Alcoa[b]	Sumitomo[c]	Sumitomo[c]	Showa[d]	Alcoa[b]	Sumitomo[c]	Alcan[b]
	Grade	AA-100	A-11	A-21	A-13	AL-41	AL-M41	AL-160SG	A-16SG	AES-11	RAC45B
	Crystalline phase	χ	γ	α	α	α	α	α	α	α	α
Chemical properties (typical)	Al_2O_3, %	—	99.6	99.6	99.6	99.9	99.9	99.9	99.8	99.8	—
	Na_2O, %	0.4	0.33	0.25	0.15–0.22	0.04	0.03	0.05	0.06	0.04	0.04
	SiO_2, %	0.02	0.02	0.01	<0.025	0.05	0.06	0.03	0.03	0.06	0.01
	CaO, %	—	—	—	<0.04	—	—	—	0.02	—	0.01
	Fe_2O_3, %	0.015	0.03	0.01	<0.025	0.01	0.01	0.02	0.02	0.01	0.01
	MgO, %	—	—	—	<0.002	—	—	—	0.03	0.1	0.05
Physical properties (typical)	d_{50}; μm	70–90	40–50	50	−325 mesh	40–60	1.6	0.5	0.4	0.5	0.45
	Crystallite size, μm	—	—	2–4	0.4	1–2	1–2	0.4	—	0.3	—
	SSA, m^2/g	260–280	150	—	11–14	—	—	8	9.5	—	8

[a]Alcan: Alcan Inc. Sumitomo: Sumitomo Chemical Co., Ltd. Alcoa: Alcoa Inc. Showa: Showa Denko K. K.
[b]Data are quoted from the product data presented in the homepage of the company.
[c]Kazuhiko Nakano (2001). Alumina powders and their prices, *Bull. Ceram. Soc. Jpn* **36**: 248–253.
[d]Data are quoted from *Fine Ceramkic Catalog* published by Japan Fine Ceramics Association (1987).

TABLE 2.1.3 Typical Characteristics of Commercial High Purity Alumina Powders[a]

Raw material for synthesis of alumina		Alkoxide			Ammonium alum		Dawsonite	
	Supplier[b]	Sumitomo	Sumitomo	Sumitomo	Showa	Showa	Taimei	Taimei
	Grade	AKP-30	AKP-3000	AKP-G015	UA5055	UA5805	TM-DA	TM-100J
	Crystalline phase	α	α	γ	α	γ	α	θ
Chemical properties (typical)	Al_2O_3, %	>99.99	>99.995	>99.995	>99.99	>99.99	>99.99	>99.99
	Si, ppm	<40	<20	<8	10	10	10	10
	Na, ppm	<10	<10	<3	15	15	8	8
	Mg, ppm	<10	<10	<3	1	1	2	2
	Cu, ppm	<10	<10	<3	—	—	1	1
	Fe, ppm	<20	<10	<8	4	4	8	8
Physical properties (typical)	d_{50}, μm	~0.3	0.4–0.7	0.1[c]	0.4	0.03[c]	0.1[c]	—
	SSA, m^2/g	5–10	4–8	150	5	85	13.5	120

[a]Kazuhiko Nakano (2001). Alumina powders and their prices, *Bull. Ceram. Soc. Jpn* **36**: 248–253.
[b]Sumitomo: Sumitomo Chemical Co., Ltd. Showa: Showa Denko K. K. Taimei: Taimei Chemicals Co., Ltd.
[c]The size of primary particles.

which is a solid electrolyte, has high oxide ion conductivity and is used in applications such as electrodes, oxygen sensors and solid oxide fuel cells.

Partially stabilized zirconia is obtained with a smaller addition of stabilizer. In this material, very fine precipitates of metastable tetragonal and/or monoclinic phases are dispersed in a cubic zirconia matrix. Typical amounts of additives to form partially stabilized zirconia are more than 8 mol% (2.77 wt%) for MgO, 8 mol% (3.81 wt%) for CaO or 3–4 mol% (5.4–7.1 wt%) for Y_2O_3.

Partially stabilized zirconia has excellent fracture strength and high toughness at room temperature. It can be used for various structural applications such as grinding media, extrusion dies, ceramic liners, optical fiber connectors and medical implants. The toughening mechanisms in partially stabilized zirconia are microcracking and induced stress.

2.1.3.2.1 Preparation of Zirconia Powders

Zirconia powders are manufactured by electric fusion processes (90–99% purity), chemical precipitation processes (99–99.9% purity), such as co-precipitation, hydrolysis of zirconium salts or alkoxides, and hydrothermal processes.

In fusion processes, zircon and carbon are heated in an electric arc furnace to >2000°C. Silica (SiO_2) is reduced to SiO and evaporates leaving behind zirconia. The compounds containing stabilizer, such as $CaCO_3$, are added in the melt when stabilized zirconia are produced.

Chemical precipitation processes are commonly used in the production of sinterable, fine and high purity zirconia powders. A typical process for the production of zirconia powder through the precipitation process is as follows:

$$\begin{array}{c}\text{Zircon } (ZrSiO_4) + NaOH\\ \downarrow \text{Solution}\\ Na_2ZrO_3\\ \downarrow HCl\\ ZrOCl_2 \cdot 8H_2O\\ \downarrow NH_4OH\\ Zr(OH)_4\\ \downarrow \text{Calcination}\\ ZrO_2\end{array}$$

Zircon is first converted to zirconium oxychloride ($ZrOCl_2 \cdot 8H_2O$) through the solution-precipitation process. Zirconium hydroxide, $Zr(OH)_4$, is precipitated by the addition of NH_4OH solution into zirconium oxychloride solution. After filteration and washing, the precipitated $Zr(OH)_4$ is calcined to form zirconia. Powder characteristics including grain size, particle shape, agglomerate size, and specific surface area can be adjusted within a certain degree by controlling the conditions of precipitation and calcination.

Stabilized or partially stabilized zirconia can be obtained by the co-precipitation process. During the chemical processing, a specific amount of the salt for stabilizers such as YCl_3 can be added into the zirconium oxychloride solution and mixed. Addition of NH_4OH solution into the mixed solution results in co-precipitation of $Zr(OH)_4$ and the hydroxides of stabilizer, such as $Y(OH)_3$, as mixtures. A cubic or tetragonal phase zirconia can be formed during calcination of the co-precipitated mixture. The powders obtained have chemically high uniformity and can be used in applications such as refractories, engineering ceramics and thermal barrier coatings.

The hydrolysis process is used to produce ZrO_2 or (ZrO_2+ oxides of stabilizer) powders. The precipitated sol is prepared through the hydrolysis of the solution of zirconium oxychloride and/or its mixture with the salt of stabilizers, such as YCl_3, with heating, and then calcined.

2.1.3.2.2 Characteristics of Commercial Zirconia, Stabilized Zirconia and PSZ Powders

Table 2.1.4 lists the typical characteristics of commercial zirconia powders made through various production processes. Well-controlled precipitated zirconia and PSZ powders have fine and fairly uniform particle size. Average particle size of the commercial powders ranges from 0.2 to several micrometers, and each particle is generally constituted from fine crystallites with the size of 20–150 nm. Small particle size, large specific surface area and well-balanced particle size distribution enhance the sintering kinetics, contributing to reduced time and temperature for sintering. In the selection of powders for specific application, the physical properties that need to be considered include composition, impurity, particle size and its distribution and specific surface area.

2.1.3.3 Silicon Nitride (Si_3N_4)

Silicon nitride [18, 32–36] is a man-made inorganic compound. It does not exist naturally and is one of the typical artificially synthesized substances. Pure silicon nitride stays solid at temperatures up to 1800–1900°C and sublimates without forming a liquid phase at higher temperatures under an atmospheric pressure due to its high covalent bonding nature. This nature also interdicts densification of powder compacts in solid-state sintering. Addition of appropriate

TABLE 2.1.4 Typical Characteristics of Zirconia Powders

Production process of zirconia		Electric fusion		Co-precipitation	Hydrolysis		
	Supplier[a]	Saint-Gobain[b]		Daiichi Kigenso[c]	TOSOH[b]		Osaka[b]
	Grade	NDZ	NSZ	HSY-3	TZ-0	TZ-3Y	OZC-3YFA
Chemical properties (typical)	ZrO_2, wt%	99.17	95.64	94	99.9	94.7	<94
	Y_2O_3, wtl%	—	—	<5.3	0	5.2	5.1–5.5
	CaO, wt%	0.23	3.82	—	—	—	—
	MgO, wt%	0.15	0.15	—	—	—	—
	SiO_2, wt%	0.14	0.1	0.1	—	—	<0.01
	Fe_2O_3, wt%	0.04	0.07	0.05	—	—	<0.01
	TiO_2, wt%	0.15	0.15	0.15	—	—	<0.01
	Al_2O_3, wt%	0.12	0.07	—	<0.1	<0.1	0.25
Physical properties (typical)	d_{50}, μm	~350 mesh	~350 mesh	0.4–0.7	0.4	0.3	0.6
	Crystallite size, nm	—	—	—	27	24	—
	SSA, m^2/g	—	—	—	14	16	15.5

[a]Saint-Gobain: Saint-Gobain Ceramic Materials K.K. Daiichi Kigenso: Daiichi Kigenso Kagakukogyo Co., Ltd. TOSOH: TOSOH Corporation. Osaka: Sumitomo Osaka Cement Co., Ltd.

[b]Data are quoted from the product data presented in the homepage of the company.

[c]Data are quoted from Fine Ceramic Catalog published by Japan Fine Ceramics Association (1987).

oxides is required for the densification of silicon nitride ceramics by liquid-phase sintering. The microstructure of sintered silicon nitride ceramics is often composed of elongated grains with high aspect ratio. The high covalent nature and the unique microstructure of sintered silicon nitride provide a variety of attractive properties for its applications, that is, high hardness, low thermal expansion coefficient, high strength over a wide temperature range, excellent thermal shock resistance, high fracture toughness, high chemical resistance against acids and molten metals, etc.

These attractive features of silicon nitride make the sintered silicon nitride ceramics applicable for various high-temperature and high-stress applications. Typical applications of silicon nitride ceramics are cutting tools, bearing balls and rollers, refractory parts such as the containers for sintering, crucibles and various parts for molten metals, and metal tube forming rolls and dies. Engine wear and moving-parts such as cam followers and valves are also promising applications of silicon nitride ceramics, although the use has been limited mainly due to the high costs. Other uses include high-temperature parts of heat engine such as blades and vanes of gas turbine and turbocharger rotors.

Silicon nitride has demonstrated its excellent properties required for many high-temperature and high-stress applications. However, high costs have continued to restrict its growth. Recent development is concentrated for improved and cheaper processing to provide low cost, reliable parts.

2.1.3.3.1 Characteristics of Silicon Nitride Powders

The required characteristics of the silicon nitride powders are generally as follows, although they depend on the forming and sintering processes used for the production of the parts for the individual application:

1. submicron-sized powders with narrow particle size distribution;
2. small agglomerate size;
3. powder particles with equiaxed or spherical shapes;
4. less impurities;
5. high ratio of α-phase;
6. improved powder compaction behavior.

These are important factors that can affect performance in the final products. For example, particle size and size distribution, and also agglomerate size, have a great impact on the final properties, because they affect the sintering kinetics and final grain size. Fine and uniform particle size sometimes causes negative effect on the compaction of powder particles in forming. The desired particle size and size distribution of raw powders should be determined by considering the forming processes and sintering conditions for the specific applications.

Coarse particles or agglomerates are very detrimental for the final properties. Even a few coarse particles, or agglomerates, can cause a significant reduction in the strength of sintered bodies. The powders often have to be milled before further processing in order to remove them.

A small amount of metallic impurities, including Fe, Ca, Al, and Mg, are desirable in order to improve the high-temperature properties of sintered bodies. Oxygen content in the bulk and on the surface of the powder particles can affect the densification kinetics and final properties. It is important to use the silicon nitride powders of suitable oxygen content for the required properties of sintered bodies.

The phase composition, α/β ratio, is also important in the selection of the desired powder. A high ratio of α-phases in powder particles generally results in better final properties, because it enhances the formation of an interlocked microstructure of elongated grains with high aspect ratio during liquid-phase sintering.

2.1.3.3.2 Preparation of Silicon Nitride Powders

Direct nitridation of silicon and silicon imide reductions are the most common methods for producing silicon nitride powders. Direct nitridation of Si metal powders involves the heating process at 1200–1500°C in nitrogen atmosphere. The nitridation of Si metal is accompanied by a large amount of exothermic heat. The purity and the particle size of Si metal powders must be controlled appropriately before the nitridation process. The reaction must be well controlled, especially when a high content of α-phase is needed. Many types of powders are produced and are commercially available to meet the required characteristics such as the crystalline phase, purity, specific surface area, particle size and size distribution. Typical characteristics of the powders are presented in Table 2.1.5.

Production of silicon nitride powder by silicon imide reductions includes three processes: (a) synthesis of silicon diimide from silicon tetrachloride and ammonia: (b) thermal decomposition of silicon diimide; and (c) crystallization of silicon nitride. The reactions are as follows:

$$SiCl_4 + 6NH_3 \rightarrow Si(NH)_2 + 4NH_4Cl \tag{1}$$

$$3Si(NH)_2 \rightarrow Si_3N_4 + 2NH_3 \tag{2}$$

In this process, silicon diimide, $Si(NH)_2$, is first synthesized with the gas-phase reaction between silicontetrachloride and ammonia gases. Silicon diimide obtained is heat-treated at ~1000°C in nitrogen or ammonia atmosphere to form amorphous silicon nitride through the thermal decomposition process. Amorphous silicon nitride powders are then crystallized by heating at 1200–1500°C in a nitrogen atmosphere. A very careful control is crucial in the crystallization

TABLE 2.1.5 Typical Characteristics of Commercial Silicon Nitride Powders[a]

Production process of silicon nitride		Direct nitridation			Silicon imide reduction		
	Supplier[b]	Denka			Ube		
	Grade	SN-9FWS	SN-9	SN-E	SN-E10	SN-E05	SN-ESP
Physical	α-ratio	92	92	<20	97	100	98
properties	d_{50}, μm	0.7	5	5	0.55	0.7	0.7
(typical)	SSA, m^2/g	11	5	3	11	5	7
Chemical	C, %	0.15	0.2	0.3	0.1	0.1	0.12
properties	O, %	0.7	1.2	0.2	1.3	1.2	1.2
(typical)	Fe, %	100	2000	7000	<50	<50	<50
	Al, %	500	2000	3000	<20	<20	<20
	Ca, %	1000	2000	2000	<10	<10	<10

[a]Takashi Kawasaki (2001). Silicon nitride powder, *Bull. Ceram. Soc. Jpn* **36**: 260–261.
[b]Denka: Denka Co., Ltd. Ube: Ube Industries, Ltd.

process of silicon nitride, because it governs the characteristics of the powders such as particle shape, particle size, α-ratio and remaining chlorine amount. Silicon nitride powders produced with this process are fine and contain small amounts of metallic impurities. The powders are generally more expensive in comparison with the powders made by the direct nitridation of Si metal powders.

Costs vary considerably, from as low as $30 per kg to as high as $150 per kg, depending on processing and powder characteristics. There are several powder manufacturers, primarily in Europe and Japan. Total estimated usage of silicon nitride powders is about 400 tons, with Japan being the primary market.

2.1.3.4 Aluminum Nitride (AlN)

Aluminum nitride [18, 36, 37–42] is a non-oxide ceramic material having wurtzite structure (hexagonal). Aluminum nitride is the only stable compound in the binary system Al–N and does not occur in nature. Pure aluminum nitride has a density of 3.26 g/cm^3 and dissociates under an atmospheric pressure above 2500°C. Dense aluminum nitride ceramics exhibits a variety of unique properties:

- high thermal conductivity equal to metal Al;
- high electrical resistivity;
- moderate low dielectric constant and loss;
- thermal expansion coefficient close to Si semiconductor.

These properties make the material very useful for electronic applications. The material is especially valuable when careful thermal management is needed, such as the case when large amounts of waste heat must be quickly removed. Ceramic substrates, packages, heat-sinks or coolers are typical examples of the applications of aluminum nitride ceramics. Aluminum nitride is recently being applied to the components for semiconductor production equipment, such as plasma etcher, because the material has excellent erosion resistance to the gas plasma of various halogen compounds.

Aluminum nitride is a material having both covalent and ionic bond natures and can be sintered more easily in comparison with silicon nitride or silicon carbide ceramics. Although the density of >90% can be achieved without any sintering aids in aluminum nitride ceramics, the addition of certain oxides is generally required to promote the densification in industrial production purposes. Typical additives for sintering of aluminum nitride include rare-earth and alkaline-earth oxides. Yttrium compounds are primarily used as additives to achieve high thermal conductivity. The formation and favorable distribution of yttrium aluminum garnet in the microstructure promote both densification and thermal properties. The sintering temperatures depend much on the powder characteristics and additives, ranging from 1600 to 1900°C.

2.1.3.4.1 Production Processes of Aluminum Nitride Powders

Three primary routes are used in the production of aluminum nitride powders: carbothermal reduction, direct nitridation and CVD methods. Carbothermal reduction is a primary process to produce aluminum nitride powders and about 70% of the total amount of commercial aluminum nitride powders is produced in this method.

In direct nitridation process, aluminum powder is heat-treated at 1200–1500°C in nitrogen atmosphere after the purity and particle size are appropriately controlled. Control of the reaction temperature is very important because the reaction is accompanied by the large amount of exothermic heat.

Aluminum nitride powders that are made from triethylaluminum, $Al(C_2H_5)_3$, by the CVD process are also commercialized. The specific characteristics of the powders produced by this method are the narrow particle size distribution and high purity.

In the carbothermal reduction process, alumina and carbon react in nitrogen atmosphere to form aluminum nitride:

$$Al_2O_3 + 3C + N_2 \rightarrow 2AlN + 3CO$$

This process is endothermic and the temperature can be controlled more precisely in comparison with the direct nitridation method. The powders obtained are generally composed of fine particles with narrow size distribution. Excess carbon is oxidized and removed by heat-treating the powders in air. Because aluminum nitride is oxidized concurrently during the process, a thin oxide layer is formed on the surface of the powder particles. The oxide layer contributes to minimize the susceptibility to hydrolysis of aluminum nitride powders by water and humidity. The thickness of the oxide layer can be controlled by adjusting the heat-treatment conditions.

2.1.3.4.2 Characteristics of Aluminum Nitride Powders

Aluminum nitride is generally used as dense, sintered materials in order to utilize the characteristic high thermal conductivity and high electrical resistivity. The characteristics of the powder, such as the size, size distribution, morphology and the agglomerate structure, need to be appropriately controlled to meet the demands for efficient powder packing as well as for improved sinterability. Small size, narrow size distribution, spherical shape and the small agglomerate size may be the preferable characteristics of aluminum nitride powders. Typical characteristics of commercial aluminum nitride powders are shown in Table 2.1.6.

Minimum amounts of impurities, especially for Fe, O, and Si, are also crucial requirements in order to maximize the unique properties of aluminum nitride, because they have detrimental effects on thermal conductivity. The oxide layer with an appropriate thickness may be required on the surface of aluminum

TABLE 2.1.6 Typical Characteristics of Commercial AlN Powders

	Grade	Supplier[a]			
		Tokuyama[b]		H. C. Starck[c]	
		F	H	A	C
Chemical properties (typical)	Al, wt%	65.4	65.4	—	—
	N, wtl%	33.6	33.6	>32	>30
	O, wt%	0.9	0.9	<1.5	<2.5
	C, wt%	0.06	0.06	<0.1	<0.1
	Ca, ppm	70	300	—	—
	Si, ppm	<30	<100	—	—
	Fe, ppm	<10	<20	<50	<50
	Ni, ppm	<10	<20	—	—
Physical properties (typical)	d_{50}, μm	1.4	1.8	6–10	1.7–2.6
	SSA, m^2/g	3.4	2.7	<2	4–8

[a]Tokuyama: Tokuyama Corporation. H. C. Starck: H. C. Starck GmbH & Co. KG.
[b]Data are quoted from the product data presented on homepage.
[c]Data are quoted from the product data presented in *Am. Ceram. Soc. Bull.* **80**: 120 (2001).

nitride powder particles in order to prevent the reaction of aluminum nitride with water and/or moisture. Aluminum nitride rapidly forms hydroxide with the moisture in the atmosphere by hydrolysis, and develops the characteristic ammonia smell ($AlN + 3H_2O \rightarrow Al(OH)_3 + NH_3$). Aluminum nitride powders should be carefully preserved to avoid hydrolysis.

2.1.3.5 Silicon Carbide (SiC)

Silicon carbide [22, 29, 30, 36, 41, 43–47] was discovered in the 1890s by Acheson who was studying the synthesis of diamond. The material occurs in nature but only a very small amount. Virtually all of the commercially used silicon carbide is man-made and is manufactured mainly via the reaction between silica and carbon by the Acheson process.

Silicon carbide decomposes without melting at >2500°C under an atmospheric pressure. Industrial application of silicon carbide began in abrasives, and expands into refractories and metallurgical fields. Electric heating elements have also long been familiar as an application of silicon carbide ceramics. Today, silicon carbide is used in a range of high-performance applications including structural, semiconductor, optics, wear and electronic fields. The properties that have led to its widespread use are the high hardness, high electrical conductivity, high oxidation resistance and excellent thermal shock resistance. Another factor that makes silicon carbide especially attractive for various applications is the unique combination of low values in density and thermal expansion coefficient, and high values of thermal conductivity, elastic modules and flexural strength.

Silicon carbide is applied in the sintered and powder forms. Sintering processes such as hot-pressing, presureless sintering, siliconizing or reaction-bonding are used for the production of silicon carbide ceramics. The products such as abrasives are applied in a powder form. Chemical vapor deposition (CVD) has in recent years been attracting great interest as a process for producing silicon carbide ceramics of very high purity.

2.1.3.5.1 Characteristics of Silicon Carbide Powders

Silicon carbide exists in many polytypic forms, all based on various stacking sequences of the planar units of Si and C in tetrahedral coordination. Several nomenclatures have been proposed to describe these polytypes. Among them, the Ramsdell notation is widely used. This is a number/letter notation, in which the number and letter refer to the number of close-packed Si–C double layers in the unit cell and the crystal symmetry, respectively. The only cubic polytype with the zinc blend structure is 3C, which has three repeated layers in the unit cell. This structure is also known as β-SiC and appears preferentially when silicon carbide is made through a reaction of silicon and carbon sources. The other polytypes have either hexagonal or rhombohedral symmetries and are collectively known as α-SiC. It is known that β-SiC transforms to a mixture of several polytypes such as 6H and 15R in α-SiC when heated at high temperatures.

Both α- and β-SiC powders can be used for sintering purpose. The particle size is reduced to submicron range to improve the sinterability of the material. In the production of reaction-bonding silicon carbide, various grades of silicon carbide powders are mixed to promote efficient powder packing and to achieve high green density.

α-SiC powders are mass-produced by the Acheson process. They are ground and refined extensively to produce raw powders for ceramics. Various grades of α-SiC powders are commercially available. In the synthesis of β-SiC powders, several methods have been developed and applied. However, there are small number of grades available for β-SiC powders in the market. Table 2.1.7 shows typical characteristics of silicon carbide powders now in the market. As noted in Table 2.1.7, the purity of the commercial silicon carbide powders is typically 97–98% with oxygen as a dominant impurity. Most oxygen is present on the surface of the powders as silica, which has been formed by the oxidation during grinding operation to form fine powders. Typical metallic impurities in the powder include Si, Al, and Fe. The kinds, distribution and form of impurities exert a crucial influence on the densification and grain growth behaviors in sintering of silicon carbide, and thus on the properties of final products. Excessively fine powders with extremely large specific surface area is not favorable in many applications. They often cause difficulty of achieving high green density, producing an excessive shrinkage of the sintered bodies. Specific surface area ranges from 10 to 20 m^2/g for silicon carbide powders generally available for the sintering purpose.

In the selection of proper powder, their characteristics in purity, particle size and particle size distribution, etc., as well as the processes used for the manufacturing of the products should be considered to fulfill the required properties for the specific applications.

2.1.3.5.2 Production Processes of Silicon Carbide Powders

The Acheson process is the primary process for the mass production of silicon carbide powders. In the process, silica and carbon are reacted to produce silicon carbide at temperatures in the range of 1700–2500°C:

$$SiO_2 + 3C = SiC + 2CO \quad - 618.5\,kJ$$

TABLE 2.1.7 Typical Characteristics of Commercial Silicon Carbide Powders

Grade		Supplier[a]							
		Showa[b]		Yakushima[b]	H. C. Starck[c]		Ibiden[b]	Pacific[c]	
		A-1	A-2	OY-15	UF-15	BF-12	UF	H	S
Production process of silicon carbide									
Chemical properties (typical)	Crystalline phase	α	α	α	α	β	β	α	α
	SiC, %	97	98	97	—	—	0.12	98.5	96
	Free C, %	0.8	0.5	0.9	—	—	1.2	0.33	0.34
	SiO_2, %	0.7	0.3	—	—	—	<50	—	—
	Al, %	0.01	0.01	—	—	—	—	13 ppm	0.02
	Fe, %	0.04	0.04	0.08	—	—	<20	50 ppm	0.25
	Total, %	0.11	0.08	—	—	—	<10	—	—
Physical properties (typical)	d_{50}, μm	0.4	0.7	0.4	0.6	0.75	0.7	—	—
	SSA, m^2/g	15	9	15	14–16	11–13	7	6, 8, 12, 15	6, 8, 12, 15

[a]Showa: Showa Denko K. K. Yakushima: Yakushima denko Co., Ltd. H. C. Starck: H. C. Starck GmbH & Co. KG. Ibiden: Ibiden Co., Ltd. Pacific: Pacific Rundum Co., Ltd.

[b]Data are quoted from Fine Ceramic Catalog published by the Japan Fine Ceramics Association (1987).

[c]Data are quoted from the product data presented in the homepage of the company.

The raw materials are silica sand and carbon, and their mixture is loaded around a central graphite core in the furnace. The graphite core is heated by supplying electric energy. Silicon carbide is formed as an ingot of solid cylindrical shape around the core. Coarse crystals of α-SiC are formed in the inner region of the product ingot, where the temperature is the highest. Fine crystals containing α-SiC are formed in the outer layer of the ingot. In the colder part of the outermost layer, unreacted silica and carbon are included. The produced α-SiC is sorted accurately from the ingot and then ground and refined to produce powders for ceramics.

The quality of silicon carbide is conveniently divided into two, that is, black and green silicon carbides. The green silicon carbide has a slightly higher purity and is preferred for many applications in abrasives and ceramics.

Two main methods for the production of β-SiC powders are carbothermal reduction of silica and plasma CVD processes. In the carbothermal reduction process, the mixture of silica and carbon powders are heated below 2000°C in an inert atmosphere to form β-SiC. Unreacted silica and carbon in the produced powders are removed by the subsequent refining processes. Although the β-SiC powders manufactured with this process are composed of fine particles, agglomerates or small amount of coarse particles are generally included in as-produced powders. Those agglomerates and coarse particles should be removed by milling or other appropriate methods prior to the subsequent forming and sintering processes. They exert a crucial influence on the densification and the properties of the final products.

In the plasma CVD process, β-SiC powder is produced by the reaction of silane and hydrocarbon gases, such as SiH_4 and C_2H_4, in the high-frequency plasma of argon. The powders produced by this process have very small particle size, that is, often $<0.1\ \mu m$, and contain very few amount of metallic impurities.

2.1.4 GRINDING OF POWDERS

2.1.4.1 Introduction

Grinding is an unit operation to break solid particles mechanically into fine particles. It is one of the most fundamental and the oldest unit operations in our history. A wide variety of raw powder materials with varying sizes is produced with this method. Many kinds of grinding machines have been developed to fulfill a variety of needs. Table 2.1.8 summarizes the classification and features of existing grinding machines. Grinding machines are classified into primary crusher, secondary crusher and fine grinding machines. Rapid advance of technology is still going on, and the current research is focused on the development of fine grinding machines.

In this section, the recent progress of grinding technology is reviewed from three viewpoints. The first subject covers the progress in submicron grinding technique for producing submicron powders in dry as well as in wet phases. The second topic is the quality control of the ground powders. It covers the control of coarse particles contained in a very small amount in raw powder materials, contamination control in the product and the discussion relevant to mechano-chemistry. The third topic covers the issues on grinding aids to improve the grinding operation. The application of grinding aids for the grinding operation will be explained.

2.1.4.2 Grinding Techniques for Submicron Powders

In general, there are two approaches to produce fine powders. The first is called the breakdown process; the grinding technology belongs to this category. The second is called the build-up process, in which the particles are synthesized from atoms and molecules in a medium. There has been a limitation of minimum particle size in the breakdown process; it was capable of producing powders of sizes down to 1 μm. In contrast, there is a limit of maximum size in the latter process at around 0.1 μm. With very intensive efforts, several grinding machines have been recently developed to produce submicron powders as shown in Table 2.1.8. Now the problem that was once called a "valley" in the size range of submicron powders no longer exists.

The grinding mechanisms to obtain fine particles are classified into the following two categories [48]: the first is called volume grinding, in which solid particles are broken into small pieces of fine particles by impaction. A typical particle size distribution of the ground powder is schematically shown in Figure 2.1.6a. The second one corresponds to the surface grinding. In this case, very fine particles are produced from the surface of solid particles by the friction forces between the surfaces of solid particles. This latter mechanism is more effective for the production of submicron size powders, and is dominantly applied in a variety of milling machines, including roller milling, frictional grinding milling, ball milling and agitation ball milling. They are also promising methods to produce submicron powders in high efficiency. A very interesting point to be emphasized is that it is now possible to produce submicron powders continuously by the dry-phase method.

Figure 2.1.7 compares the grinding capacity in dry phase for various mills in a continuous grinding mode with talc as a model raw material [49]. The vertical axis indicates the grinding capacity measured in the amount

TABLE 2.1.8 Classification and Features of Existing Grinding Machines

Classification	Types of machines	Examples	Applications
Primary crusher	Compressive crusher	Jaw crusher	Ore, etc.
		Cone crusher	
	Shearing crusher	Cutter mill	Soft material, such as synthetic resin, fibrous material and wastes
	Impact crusher	Hammer crusher	Primary crushing of soft materials, such as limestone and agglomerates
Secondary crusher	Roller mill	Roll crusher	Ore, agricultural products
		Edge runner mill	Raw materials of pottery, hard materials
		Cage mill	Grinding and disintegration for very brittle material and fibrous material
	Hammering type	Stamp mill	Ore, foods, etc. (small capacity)
	High-speed impact mill	Hammer mill	Soft material, fibrous material, etc. Can be used as a secondary crusher
		Pin mill	Minerals, foods, etc.
		Disc mill	
Fine grinding mill		Centrifugal classifying type mill	Soft inorganic materials, organic materials
	Roller mill	Roller mill	Grinding for materials, such as limestone, cement and ceramics in dry phase
		High-pressure roller mill	
	Frictional grinding mill	Angmill	Grinding for minerals in dry phase
	Autogeneous mill	Aerofall mill	Mainly used for ore
	Ball mill	Tumbling ball mill	The most popular type. Widely used for both dry and wet phase grinding
		Vibration ball mill	Easy to control the temperature and ambient gas conditions
		Planetary ball mill	Mainly used for small scale production
		CF mill	Dry phase grinding
		UF mill	Can be used for both dry and wet phase grinding
	Agitating ball mill	Tower mill	Ore, etc.
		Mixing vessel type	Originally used for dispersion of the pigment. Wide variety of applications: mechanical alloying, cell disintegration, etc.; must pay attention for temperature control
		Tube type	
		Annular type	
		Annular agitating type	
	Jet mill	Jet mill	Toner, etc. Used for dispersing fine powder
		Liquid jet mill	Emulsification, dispersion and grinding can be operated simultaneously
	Wet grinding mill	Colloid mill	Fine grinding and dispersion of the powder in liquid
	Others	Mortar	
		Stone mill	Cereals, foods, etc.

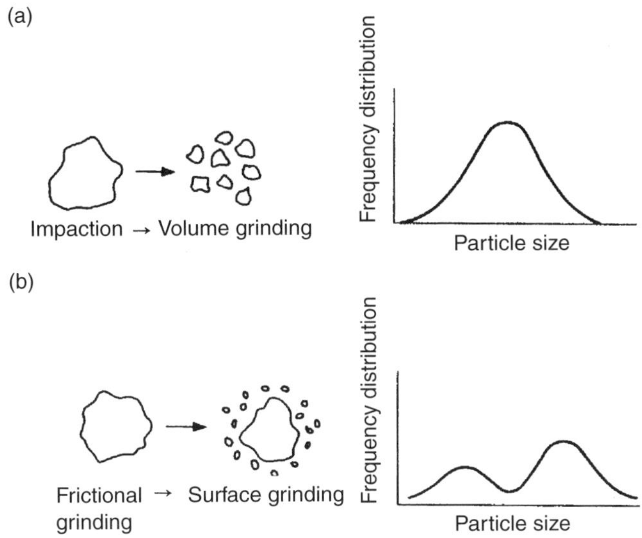

FIGURE 2.1.6 Grinding mechanism of particles: (a) volume grinding; (b) surface grinding [48].

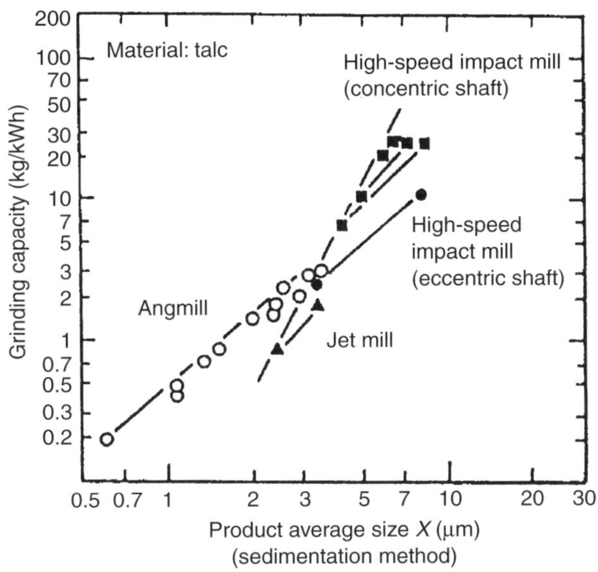

FIGURE 2.1.7 Comparison of grinding performance between four kinds of grinding machines in dry phase [49].

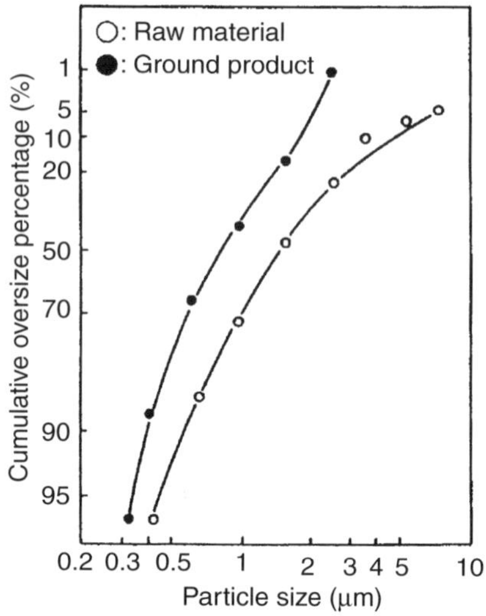

FIGURE 2.1.8 Grinding results of barium titanate by an agitated ball mill in dry phase [50].

of product per unit power consumption, and the horizontal axis shows the average particle size of the product determined by the liquid sedimentation method. The graph shows the relationship for the average particle sizes and the amount of ground powder for various grinding machines. Clearly, the frictional grinding mill called Angmill has the highest grinding capacity among three grinding machines of impact type for the size range below 3 μm of the product average size.

Figure 2.1.8 shows the particle size distribution for barium titanate after continuous grinding by an agitated ball mill in dry phase [50]. The average product size reached 0.8 μm after grinding. Fairly narrow particle size distribution could be achieved without any classification. This result demonstrates that dry-phase processing can produce submicron powder in continuous operation.

Figure 2.1.9 shows the change of average particle size with processing time [51]. These experiments used an agitated ball mill with a horizontal axis, and media balls of various sizes were used to grind the raw powder for functional ceramics. The product particle size reached under 1 μm, and decreased further with milling time. The product particle size also depended on the diameter of the media ball, and decreased with decreasing ball diameter.

However, there was a lower limit in the particle size at prolonged milling time, and the product particle size might even increase after a certain milling time. In that case, the minimum particle size was defined as the equilibrium particle size [52]. The grinding limit has been noted in many systems and is

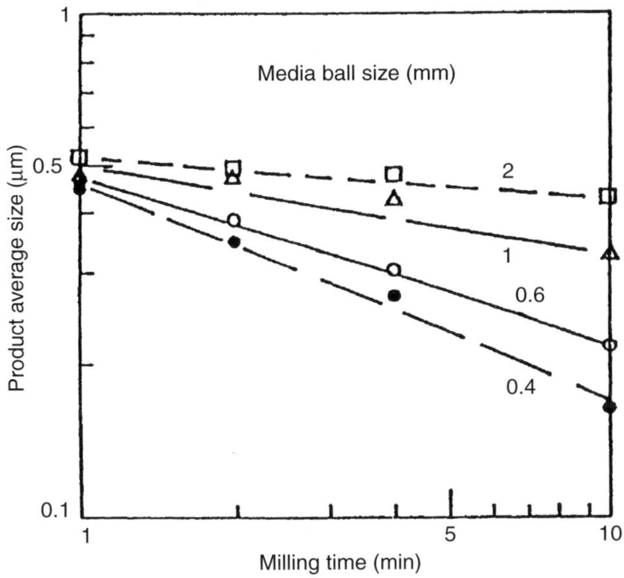

FIGURE 2.1.9 Average particle size of ceramic powder against milling time with continuous agitated ball mill of horizontal bar [51].

known to be governed by many factors, that are classified into the following categories [52]:

1. Machine problems
   (i) escape of particles from grinding zone. This occurs in impact mills and jet mills. Rumpf concluded through theoretical calculation that the lower limit should be 0.1–1 μm [53].
   (ii) Coating of media and cushioning of powder.
2. Powder size problems
   (i) Limit of size by limited brittleness. Schoenert suggested that solid particles deform plastically for size under 1 μm [54].
   (ii) Agglomeration and recombination of fine powder particles that are newly produced.

Negative grinding was proposed by Theimer [55] and Huettig [56]. The excessive force onto the product increases the particle size. Figure 2.1.10 shows the relationship between the equilibrium diameter of ground powder and the maximum centrifugal force acting onto a single media ball in a planetary ball mill [57]. A straight relationship is noted between the equilibrium size and the force in a logarithmic scale. Clearly, the particle size increases with increasing

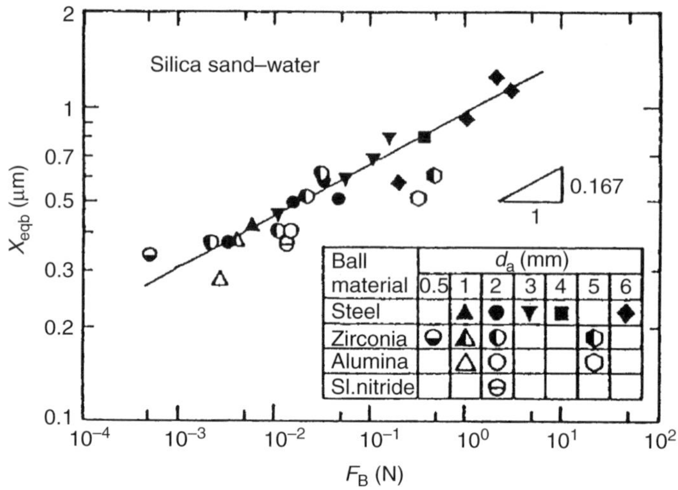

FIGURE 2.1.10 Relationship between the maximum centrifugal force F_B acted onto single media ball and the equilibrium diameter of ground powder [57].

force. The excessive force on the powder is negative in the production of fine product.

2.1.4.2.1 Quality Control of Ground Powder

Quality control is essential for commercial production of raw materials with consistent properties. Details for the characterization of powder is presented elsewhere in this textbook. Important factors to be considered include the particle size and impurities.

Accurate determination of particle size is very important in the product. Various kinds of particle size analyzers are used for this purpose. General feature of powders corresponding to size range 5–95% can be specified easily through the analysis. However, it is often difficult to examine the particles of extreme size. Coarse particles in raw powder materials are known to have a very detrimental effect on the strength of the ceramics [58]. However, it is extremely difficult to determine their amount accurately, since their content in the raw ceramics powder is usually very low. An appropriate method for measurement is lacking.

Figure 2.1.11 shows the size distribution of silicon nitride powder ground by ball milling [59]. In this experiment, media balls with 5-mm diameter were used, and the X-ray sedimentation method was used to measure the particle size distribution. The raw powder apparently has particles with size over 40 μm.

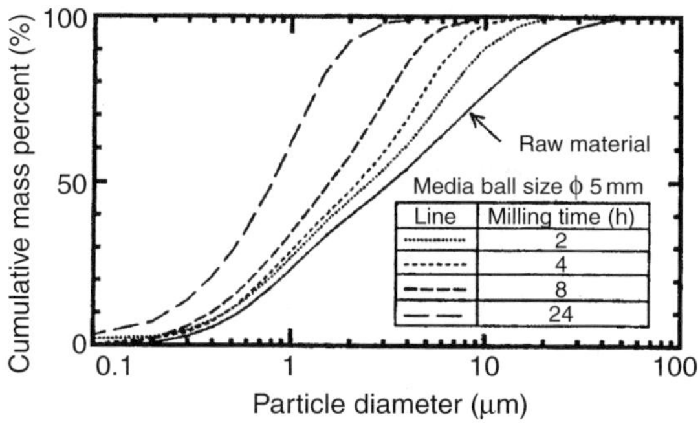

FIGURE 2.1.11 Particle size distribution of the powder ground with 5-mm media balls [59].

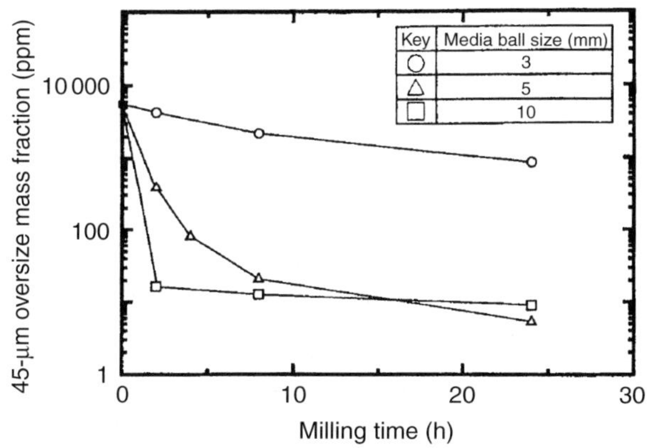

FIGURE 2.1.12 Change of 45-μm oversize mass fraction of ground powder with milling time [59].

No particles with more than about 20 μm in diameter were found after 2 h of grinding. Figure 2.1.12 shows the mass fraction for the coarse particles with size over 45 μm in the ground powder against the milling time [59]. The fraction was measured by sieving, and all samples of a batch were sieved to measure the amount of coarse powder in the ground powder. Clearly, a small amount of coarse particles are present in the ground powder. Conventional particle size analyzers are not able to identify these large particles of extremely small concentration. A new method must be developed for measuring large particles with high accuracy for precisely controlling the particle size distribution.

Contamination is always present in ground powder and affects the quality of final products, and must be minimized. The contaminants come from the materials of the grinding machine and the media due to their abrasion. Impaction

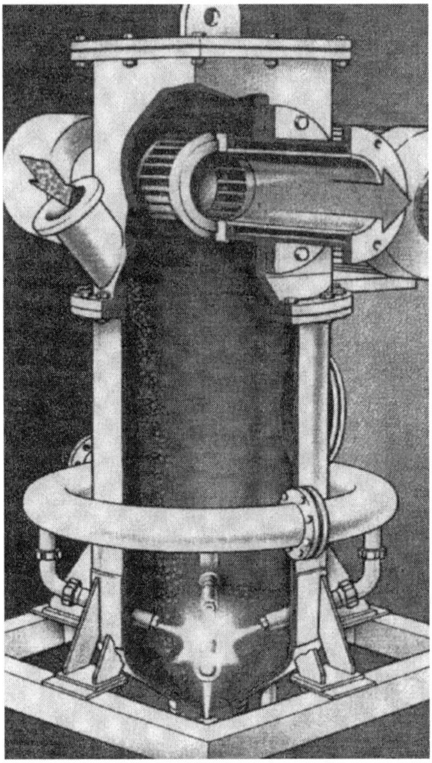

FIGURE 2.1.13 Jet mill of fluidized nozzle type [60].

blade, media balls and inner walls of the grinding machine are potential sources of metallic impurities. There are several methods to reduce the contamination caused by the abrasion. One common approach is to use hard-wear-resistant materials. Another approach is the use of polyurethane for wear protection in agitated ball mills. One of the interesting solutions is to grind powder by mutual impaction or attrition. A example is shown in Figure 2.1.13, which shows a jet mill of fluidized nozzle type [60]. Powder particles collide at high speed with each other by the air jet from a nozzle. The low probability of collision with the inner walls of the mill reduces the abrasion of the grinding machine. Another case is to use the grinding mechanism as shown in Table 2.1.8 for frictional grinding mills. A compressive force is applied on the powder bed in this apparatus, and the friction force is generated between the particles in the powder bed. Abrasion from the wall of the vessel is eliminated.

Mechanochemical phenomena should be considered in the ground powder of extremely small size. It may have a positive effect as noted in the

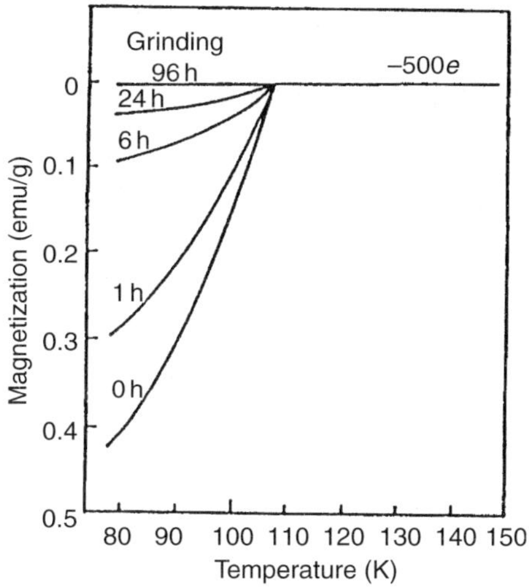

FIGURE 2.1.14 Change of magnetization with grinding time of Bi-based superconductive powder [61].

reduced sintering temperature of ceramic parts. It can be also detrimental to the properties of ceramics. For example, Figure 2.1.14 shows the relationship between the magnetization of the Bi-based superconducting powder and its critical temperature [61]. An agitated wet ball milling was used in the experiments, and the grinding time was changed. The magnetization decreased with increasing grinding time. The mechanism of mechanochemical effect often remains unclear. Attention has to be focused when submicron power is fabricated by grinding techniques.

2.1.4.2.2 Grinding Aids and Chemically Assisted Comminution

Grinding aids are defined as the additives added into the powder material during the grinding process to improve the grinding efficiency as well as the quality of the product. A number of scientific papers have been published in the fields of cement industry and pharmaceutics. Grinding aids are used in many industries. Grinding aids include gas, liquid and solid materials. The following mechanisms are proposed, but ambiguities are still present regarding their detailed functions [62]:

2.1.4.2.2.1 *Reduction of strength in solids*

Several phenomena have been proposed in this subject. An old example is the Rehbinder effect of grinding aids [63]. Recently, another case was pointed out in inorganic powder. It was proposed by Ikazaki *et al.* and is called “chemically assisted comminution” [64]. In a typical example, soda glass powder was found to be more efficiently ground when it was immersed in molten lithium nitrate before grinding. The origin of this interesting result is ascribed to the ion-exchange between lithium and sodium ions during immersion of the soda glass powder in molten lithium nitrate. Cracks were formed in the soda glass powder by the size difference for these ions, reducing the strength of the glass particles.

2.1.4.2.2.2 *Control of the mechanical properties of powder material*

Wet grinding uses liquids as effective grinding aids to disperse powder materials in the mills. Many kinds of dispersants are used to reduce the apparent viscosity of the slurry. In dry grinding, various kinds of additives such as ultra-fine powders or lubricants are used to improve the dispersibility of the powder materials.

The interaction between particles should affect the process of grinding. There is an interesting example for grinding graphite powder in different atmospheres. Figure 2.1.15 shows the relationship between the specific surface area of the graphite powder against the power consumption per ground powder

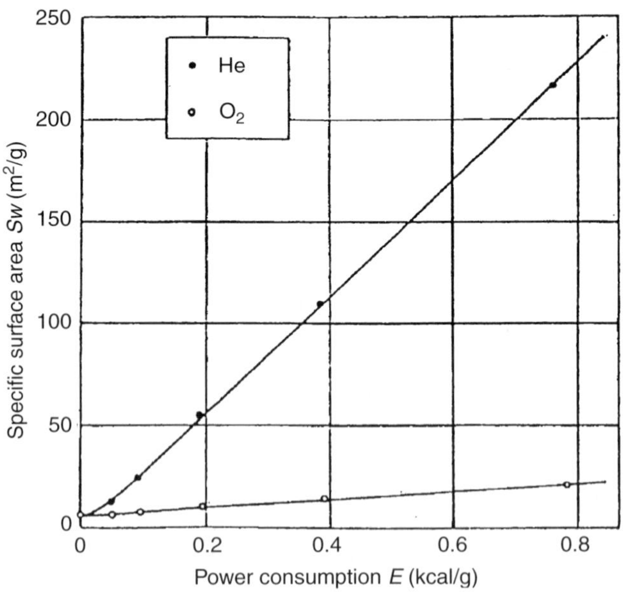

FIGURE 2.1.15 Effect of ambient gas conditions on the grinding results of graphite powder by a ball mill [65].

weight [65]. A ball mill was used in this experiment, and two kinds of gases were selected as grinding aids. The grinding rate in helium gas is much higher than that in oxygen gas. The results are explained by the difference of dynamic friction coefficients of graphite in these gases. The friction coefficient in helium gas is larger than that in oxygen gas [66]. The high friction coefficient of the powder accelerates the grinding of the powder bed between media balls.

2.1.5 MIXING OF POWDERS

2.1.5.1 Introduction

Mixing is an important unit operation to produce a homogeneous mixture of more than two kinds of particular components in dry or wet phase. Mixing is often used as a pre-treatment operation prior to the subsequent grinding, granulation, forming or coating process, etc. The degree of mixing state required depends on the application of the resultant mixture. For a majority of applications, a random powder mixture is the final target. In this mixture, various kinds of powder particles are randomly distributed three-dimensionally in the final product. The aim of the mixing operation also includes the achievement of an ordered mixture for certain applications. An example is the coating of fine particles in which the surface of the core particles is covered by a powder of a different kind in an orderly fashion. In this section, the mixing mechanism and mixing apparatus are introduced.

2.1.5.2 Mixing Mechanism

Mixing of powders is basically defined as the exchange of the particle location in a powder bed. In an mixer, three mechanisms work under an applied external force to move powder particles.

2.1.5.2.1 Convection Mixing

In convection mixing, a large bulk movement of particles such as circulation flow occurs in the mixer by the rotation of the mixing vessels or mixing bars such as puddles. It produces a macroscopic mixing of powders, and provides a high mixing rate. This mixing mechanism is typical and predominant in the initial stage of the mixing process. This mixing mechanism is very effective for quick tumbling mixing of a whole product in a batch mixing process, and is distinguished from microscopic mixing processes.

2.1.5.2.2 Shearing Mixing

Shearing mixing results from the velocity distribution of powder migration in a powder bed. This mixing mechanism includes slip and impaction between particles, and disintegration of powders by compressive and/or tensile stress, which is generated in the zone between the inner wall and the edge of the stirring bar. It contributes to both the overall mixing in the vessel and the semi-microscopic mixing.

2.1.5.2.3 Diffusive Mixing

In diffusive mixing, moving particles exchange the position with the neighboring particle randomly, and eventually form a homogeneous mixture after the exchange is repeated many times. The local mixing achieved by the exchange of position between neighboring particles is defined as diffusive mixing. It involves a random walk of particles, and is analogous to the atomic and molecular diffusion in liquid and solid phases. Particle mixing along the rolling axis in a rotary type mixer belongs to the diffusive mixing type. The mixing rate in this mechanism is extremely low compared to the those of convection mixing and shearing mixing, but the diffusive mixing is crucial for the completion of microscopic mixing of relevant powder components. Virtually any mixing process involves all the mixing mechanisms explained above, and it is vary rare that only a single mixing mechanism will be occurring.

2.1.5.3 Classification of Mixers

Table 2.1.9 shows the overview of existing mixers and their mixing mechanisms in dry phase. Many types of mixing machines have been developed and are available. High-shearing mixers provide very intensive mixing. This type of mixer also allows the fabrication of composite particles, as will be explained later. Grinding machines shown in Table 2.1.8 are commonly used for mixing powder in the wet phase. Ball mills and agitated ball mills are used to mix and grind a variety of powder materials for ceramic processing in preparing slurries.

Intensive mixing in particular grinding machines may sometimes reduce the homogeneity of powder mixture. An example is shown in Figure 2.1.16, which shows the change of size distribution with milling time for mixed powders of Y_2O_3, $BaCO_3$, and CuO (Y:Ba:Cu = 1:2:3) [67]. Ball milling in the dry phase was used in this experiment, and the particle size was measured by the laser diffraction method. The particle size of the mixed powder decreases with milling time in the beginning. It shows a grinding limit after milling for 90 min, and the particle size increases considerably after a milling time of 300 min. A separate

TABLE 2.1.9 Classification and Mechanism of Existing Mixers

Mixers		Batch type	Continuous type	Mixing mechanism
Rotary type	Horizontal cylinder	○	○	
	Inclined cylinder	○		
	V-type	○		Convective mixing
	Double cone	○		
	Continuous V-type	○	○	
Rotary type with impeller	Horizontal cylinder	○	○	
	Inclined cylinder	○		Shearing mixing
	V-type	○		
	Double cone	○		
Horizontal agitation	Ribbon	○	○	
	Screw	○	○	Convective mixing
	Pin or rod	○	○	
Vertical agitation	Ribbon	○		
	Screw	○		
	Rotating disc		○	Convective mixing
	High-speed flow type	○		
	Muller		○	
Fluidized mixing		○		Convective mixing
Motionless mixer			○	Convective mixing
High-shearing mixer		○	○	Diffusive mixing or composite formation process

analysis with SEM showed that the agglomeration of powder particles was responsible for the increased particle size during milling.

Figure 2.1.17 shows the variation of AC susceptibility of calcined powders with temperature. The powders were obtained by the calcination of powder mixtures processed by different milling time as shown in Figure 2.1.16. In Figure 2.1.17, the susceptibility of powder samples becomes higher with the milling time up to 90 min of milling. The data of the powder after 60 min of milling time without ball media are also plotted as a reference, and it has apparently lower property than those of powder mixed with media balls. However, the susceptibility becomes lower when the milling time reaches 300 min. The reason is explained by the change of particle size distribution of mixed powder

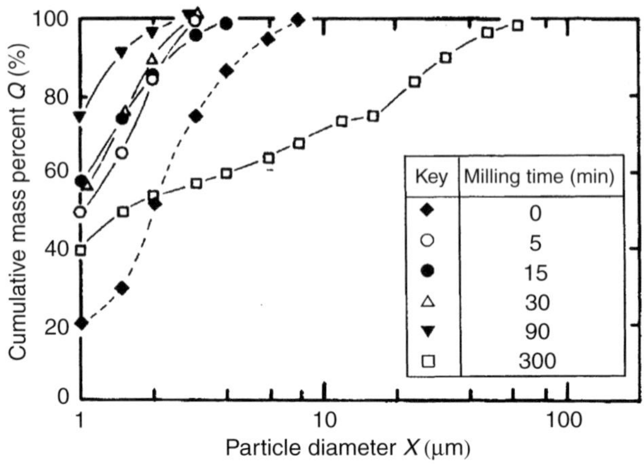

FIGURE 2.1.16 Change of size distribution of mixed powder with ball milling time in dry phase [67].

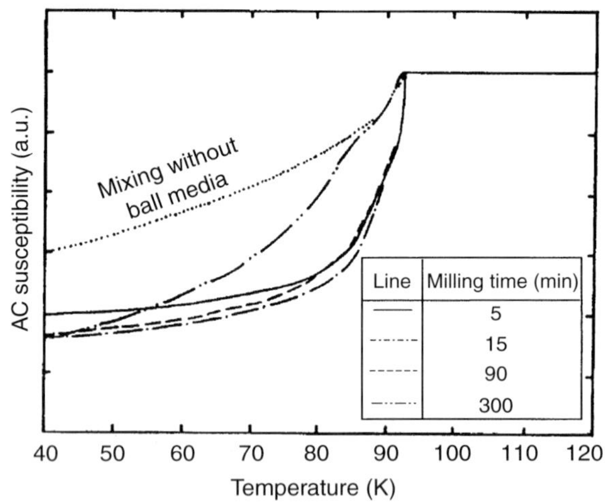

FIGURE 2.1.17 Variation of AC susceptibility of calcined powders with temperature [67].

with milling time as shown in Figure 2.1.16. The agglomeration obtained in milling for 300 min appears to be responsible for inhomogeneity of the mixed powder and also for the degradation of the superconductive properties. Apparently, there is an optimum grinding time, and excessively long grinding causes inhomogeneity due to agglomeration and also increase contamination.

2.1.5.4 Ordered Mixing

The objective of mixing powder materials is to blend different powder components and to achieve homogeneity in the mixture for a majority of cases. Recently, there has been strong interest for achieving a precisely controlled mixing state. Ordered mixtures as shown in Figure 2.1.18 are generated when fine particles can adhere onto the surface of core particles [68]. This particle structure has already been used in a wide range of industries. Recently, an ordered mixture was also used as the first step for producing composite particles, in which fine particles are fixed or embedded onto the surface of core particles. This type of composite particle structure is classified into type (a) as shown in Figure 2.1.19 [69]. The composite particle structure has a very high potential for a wide variety of applications. For example, such composite particles can be used for surface modification technology for powder materials [69].

This mixing technique can be also applied for fabricating composite particles as shown in type (b) of Figure 2.1.19 [69]. Composite particles are easily fabricated by making use of a high-shearing mixer or fine grinding machines. With these machines, fine particles are easily fixed onto the surface of core particles by strong mechanical action. Table 2.1.10 shows examples of machines used to make composite particles. Type I denotes fine grinding machines. Almost all fine grinding machines have the potential to make composite particles.

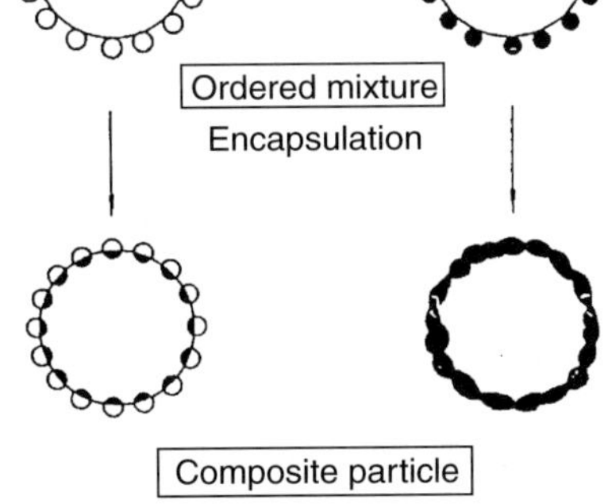

FIGURE 2.1.18 Ordered mixture and composite particle [68].

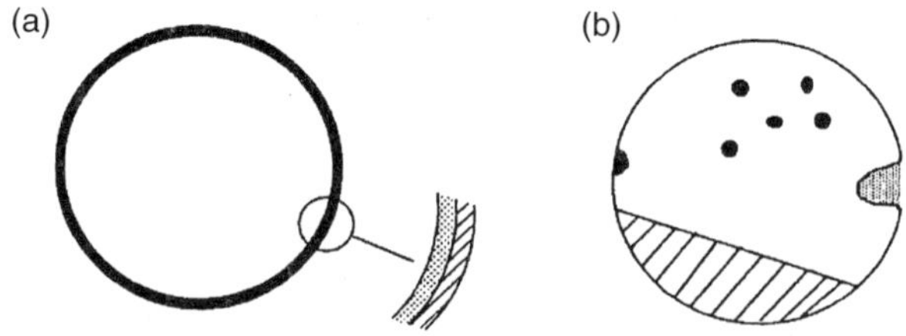

FIGURE 2.1.19 Classification of composite particles [69].

TABLE 2.1.10 Machines Used to Fabricate Composite Particles

	Type	Advanced machines for fabricating composite particles
Fine grinding mill	Impaction-type	Hybridization System (Nara Machinery Co.)
	Pin mill, disc mill	
	Centrifugal classifying type	KOSMOS (Kawasaki Heavy Industry Co.)
	Frictional grinding mill	Mechanofusion System (Hosokawa Micron Corp.)
	Ball mill	
	Tumbling	
	Vibration	
	Planetary	
	Centrifugal fluidizing (CF mill)	
	Agitated ball mill	
	Mixing vessel type	
	Jet mill	
Others	Mortar	
	Cylindrical vessel type with rotating disc	Mechanomill (Okada Seiko Co.)
	Elliptical vessel type with high-speed elliptical rotor	Theta-composer (Tokuju Corp.)

So far, several advanced machines based on the fine grinding principle have been developed in Japan as indicated in Table 2.1.10. Besides this, type II shows machines with other principles such as shearing mixer. Mixers of the high-energy type may also be used for making composite particles. The more interesting point is that the composite particles are produced by the mechanical method in dry phase. In other words, it is very simple to produce composite particles [69]. Generally, mixing operation has a high potential to be used as powder processing machines including the fabrication of composite particles. Wider applications can be expected in this area in the future.

2.1.6 CLASSIFICATION OF POWDERS

2.1.6.1 Introduction

Classification is widely used for the separation of materials in different classes, such as chemical component, particle size, particle shape, color, density, etc. In the operation called sizing or size classification, the particles are classified into several classes of particle size. This section focuses on the size classification techniques. Size classification can be conducted both in dry and wet phases, and includes sieving machines, pneumatic classifiers and wet classifiers. Their details are explained next.

2.1.6.2 Sieving Machine

Sieving is widely used in ceramics manufacturing for preparing raw powder materials, suspensions and granules. Examples include the separation of coarse particles in a suspension, and adjustment of size distribution for the granules made by spray drying. This technique is applicable even for the mixture of powder containing various different kinds of materials, since what governs the classification is simply the opening size of the sieve. The opening size of the sieve is defined by ISO standards [70]. The minimum opening size reaches down to 3 μm. Electroformed sieves have very accurate opening size and are used in a smaller size region.

Sieving machines are classified into three types, flat screen, vibrating screen and pneumatic dispersion types. The pneumatic dispersion type is particularly effective in sieving fine powders.

2.1.6.3 Pneumatic Classifier

Pneumatic classification is preferentially applied in the classification of small particles, for which classification by sieving is difficult. The machines are classified into gravitational classifiers, inertial classifiers and centrifugal classifiers. Gravitational classifiers apply the gravitational force and the fluid drag force, which act simultaneously and exert opposite effects on the particles. The principle of separation is based on the increased gravitational force relative to fluid drag force with increasing particle size. The direction of movement is opposite for fine and coarse particles under these forces, and fine particles are separated from coarse particles. The zigzag classifier shown in Figure 2.1.20 is a typical example of this type. The particles are well dispersed in this classifier, and

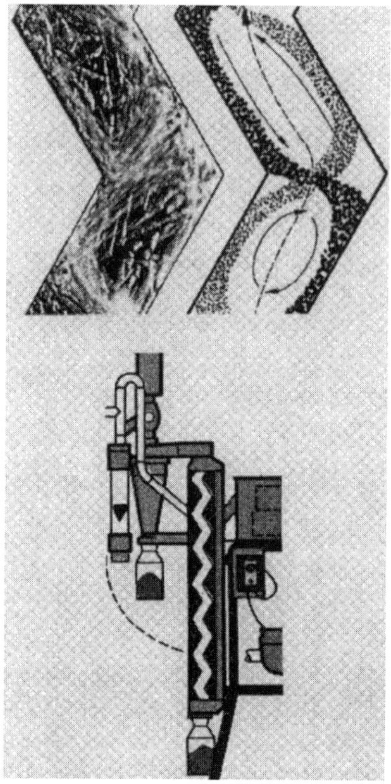

FIGURE 2.1.20 Gravitational classifier (zigzag classifier) [71].

collide with the zigzag-shaped wall. Coarse particles fall down by gravitational force, while fine ones move upwards in the airflow. The cut point is adjustable at any value between 0.1 and 6 mm by controlling the flow rate of air for various materials [71].

Inertial classifiers use inertial force and fluid drag force acting on the particles. The cut size of the powder is smaller than that of a gravitational classifier. Centrifugal classifiers use centrifugal force and fluid drag force acting on the particles. Small cut size of separation is achieved in this method, since the sedimentation velocity in a centrifugal field is several hundred times larger than that in the gravitational field. A rotation stream is needed to generate a centrifugal field. There are free vortex type and forced vortex type for generating the rotation stream. Figure 2.1.21 shows an example of a centrifugal classifier. In this classifier, the coarse particles are rejected from the rotating separation rotor towards the outside of the rotor by the centrifugal force and fall down to the coarse outlet at the bottom. The fine particles pass through the separation

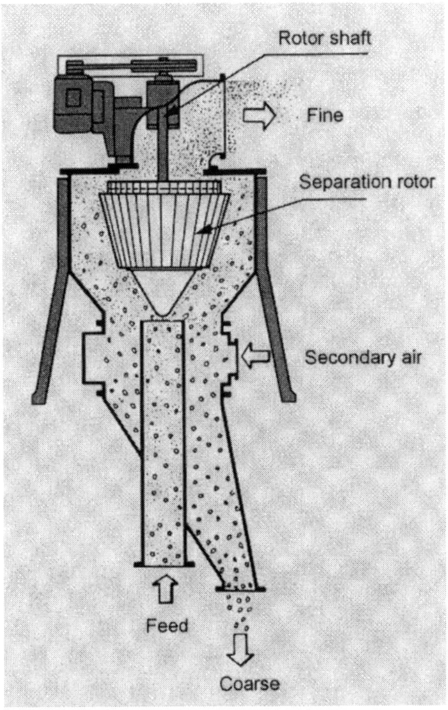

FIGURE 2.1.21 Centrifugal classifier (micron separator) [72].

rotor together with the air stream by the drag force and go to the fine outlet at the top [72].

A well-designed dispersion technique is crucial in the classification. The fine powder must be well dispersed before entering the classifier. A difficult part is the very rapid re-agglomeration of particles in an air stream after dispersion. Primary particles readily form agglomerates when they collide with each other. A very important point is immediate classification after dispersing powder materials in dry classification.

2.1.6.4 Wet Classifier

Wet classifiers comprise mainly gravitational classifiers and centrifugal classifiers. The gravitational classifiers utilize the particle size dependence of sedimentation velocity in slurries. The apparatus has a simple structure, large

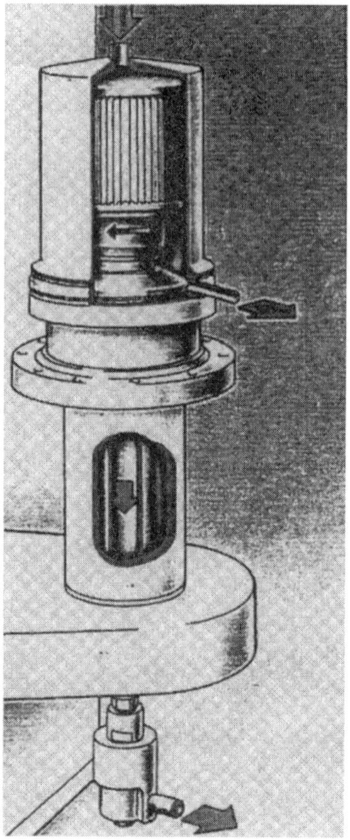

FIGURE 2.1.22 Centrifugal wet classifier (Hydroplex) [73].

scale and large cut size of about 50–200 μm, including hydro separators, spiral classifiers and rake classifiers.

The centrifugal classifiers separate particles by the balance among the drag force of a liquid stream and the centrifugal force generated by a free vortex or a forced vortex. The free vortex type includes the liquid cyclone. The forced vortex type has a more complicated configuration compared to the free vortex type, and is still actively developed for further improvement. Figure 2.1.22 shows an example of a new wet classifier, called Hydroplex [73]. This classifier uses the principle of counter-flow, in which the particles receive the centrifugal force induced by a rotating deflector wheel and the drag force from the liquid stream. The operating range of feed particle sizes is below 200 μm, whereby cut points down to the fineness of approximately 4 μm can be achieved in the wet phase.

REFERENCES

1. Lange, L. L. (1989). *J. Am. Ceram. Soc.* **72**: 3–15.
2. Uematsu, K., Ito, H., Zhang, Y., and Uchida, N. (1995). *Ceram. Trans.* **54**: 83–89.
3. Uematsu, K. (1996). *Powder Technol.* **88**: 291–298.
4. Handwerker, C. A., Moris, P. A., and Coble, R. L. (1989). *J. Am. Ceram. Soc.* **72**: 130–136.
5. Bae, S., and Baik, S. (1994). *J. Am. Ceram. Soc.* **77**: 2499–2504.
6. Echeverria, L. M. (1990). *Ceram. Trans.* **12**: 649–658.
7. Coble, R. L., and Gupta, T. K. (1967). *Sintering and Related Phenomena*, p. 423, New York: Gordon and Breach.
8. Reed, J. S. (1995). *Principles of Ceramics Processing*, 2nd edn, p. 215, New York: Wiley.
9. Kingery, W. D., Bowen, H. K., and Uhlmann, D. R. (1976). *Introduction to Ceramics*, 2nd edn, New York: Wiley.
10. Tsubaki, J. (1994). *Seramikkusu* **29**: 28–30.
11. Reed, J. S. (1995). *Principles of Ceramics Processing*, 2nd edn, p. 492, New York: Wiley.
12. Shui, A., Kato, Z., Tanaka, S., Uchida, N., and Uematsu, K. (2001). *Am. Ceram. Soc. Bull.* **80**: 29–32.
13. Lange, F. F., Davis, B. I., and Aksay, I. A. (1983). *J. Am. Ceram. Soc.* **66**: 407–408.
14. Zheng, J., and Reed, J. S. (1989). *J. Am. Ceram. Soc.* **72**: 810–817.
15. Cutler, I. B. (1978). Active powders, *Ceramic Processing before Firing*, pp. 21–29, Onoda, G. Y., and Hench, L. L., eds., New York: Wiley.
16. Uematsu, K., Miyashita, M., Sekiguchi, M., Kim, J.-Y., Uchida, N., and Saito, K. (1991). *J. Am. Ceram. Soc.* **74**: 2170–2174.
17. Rhodes, W. H., and Wench, B. J. (1973). *J. Am. Ceram. Soc.* **56**: 495.
18. Dynys, F. W., and Holloran, J. W. (1984). *J. Am. Ceram. Soc.* **67**: 596–601.
19. Yamada, K., Umezaki, H., and Furubayashi, T. (1987). Present situation of alumina powder. *Bull. Ceram. Soc. Jpn* **22**: 15–22.
20. Takeuchi, Y., and Kameda, I. (1994). Alumina powder for new ceramics. *Bull. Ceram. Soc. Jpn* **29**: 659–662.
21. Nakano, K. (2001). Alumina powders and their prices. *Bull. Ceram. Soc. Jpn* **36**: 248–253.
22. Mussler, B. H. *et al.* (2001). Materials review. *Am. Ceram. Soc. Bull.* **80** (8): 49–112.
23. Homepage of Alcan Inc.
24. Homepage of Alcoa Inc.
25. Homepage of Taimei Cemicals Co., Ltd.
26. Suzuki, T. (1987). Present situation of zirconia powder. *Bull. Ceram. Soc. Jpn* **22**: 22–28.
27. Arakawa, T., Ogai, M., and Fukuda, H. (1994). Powder for toughened zirconia ceramics. *Bull. Ceram. Soc. Jpn* **29**: 663–665.
28. Fukuda, H. (2001). Zirconia. *Bull. Ceram. Soc. Jpn* **36**: 254–256.
29. Homepage of Sumitomo Osaka Cement Co., Ltd.
30. Homepage of Saint-Gobain Ceramic Materials K. K.
31. Homepage of TOSOH Corporation.
32. Arakawa, T., Ohno, K., and Ueda, K. (1987). Present situation of silicon nitride powder. *Bull. Ceram. Soc. Jpn* **22**: 34–39.
33. Ametani, K., Nakajima, Y., and Isozaki, K. (1994). Silicon nitride powder by direct nitridation of metal silicon. *Bull. Ceram. Soc. Jpn* **29**: 673–675.
34. Kohtoku, Y. (1994). Silicon nitride powder by imide decomposition process. *Bull. Ceram. Soc. Jpn* **29**: 676–678.
35. Kawasaki, T. (2001). Silicon nitride powder. *Bull. Ceram. Soc. Jpn* **36**: 260–261.
36. Matthew, D. V. (1999). Advanced Materials & Powders Handbook. *Am. Ceram. Soc. Bull.* **78** (6): 69–81.

37. Kuramoto, N. (1987). Aluminum nitride powder and its situation. *Bull. Ceram. Soc. Jpn* **22**: 29–34.
38. Taniguchi, H., and Kuramoto, N. (1991). Recent progress in aluminum nitride powder. *Bull. Ceram. Soc. Jpn* **26**: 733–737.
39. Taniguchi, H. (1994). Aluminum nitride powder. *Bull. Ceram. Soc. Jpn* **29**: 679–681.
40. Nagano, Y. (2001). Features of aluminum-nitride and its applications. *Bull. Ceram. Soc. Jpn* **36**: 262–264.
41. Homepage of Starck, H. C. GmbH & Co. KG.
42. Homepage of Tokuyama Corporation.
43. Heuer, A. H., Fryburg, G. A., Ogbuji, L. U., Mitchell, T. E., and Shinozaki, S. (1978). $\beta \rightarrow \alpha$ Transformation in polucrystalline SiC: I, microstructural aspects. *J. Am. Ceram. Soc.* **61**: 406–412.
44. Yamamoto, M. (1987). Present situation of SiC powder. *Bull. Ceram. Soc. Jpn* **22**: 46–51.
45. Takahashi, K. (2001). Silicon carbide. *Bull. Ceram. Soc. Jpn* **36**: 257–259.
46. Homepage of Yakushima denko Co., Ltd.
47. Homepage of Pacific Rundum Co., Ltd.
48. Yokoyama, T., Urayama, K., Naito, M., Kato, M., and Yokoyama, T. (1987). The Angmill Mechanofusion System and its applications. *KONA* **5**: 59–68.
49. Yokoyama, T., Urayama, K., Naito, M., Kato, M., and Yokoyama, T. (1988). Ultra-fine grinding by a new attrition type mill and its application to the particle reformation. Proc. 2. "Japan–Soviet Symposium on Mechanochemistry," pp. 281–290, March, Tokyo.
50. Ishikawa (1992). Proc. 15. "Special Seminar of Powder Technology, E1-1," Assoc. Powder Process Ind.& Eng., Japan, eds, October, Tokyo.
51. Kugimiya, K. (1992). Ultra-fine grinding in nano-powder production process. *Funsai (The Micromeritics)* **36**: 177–180.
52. Jimbo, G. (1992). Chemical engineering analysis of fine grinding phenomenon and process. *J. Chem. Eng. Jpn* **25**: 117–127.
53. Rumpf, H. (1959). Beanspruchungstheorie der Prallzerkleinerung *Chemie-Ing-Techn.* **39**: 323–337.
54. Schoenert, K., and Steier, K. (1971). Die Grenze der Zerkleinerung bei kleiner Korngroessen. *Chemie-Ing-Techn.* **43**: 773–777.
55. Theimer, O. (1952). Ueber die Auswertung von Zermahlungs-experimenten mit Hilfe von Zermahlungs-funktionen, *Kolloid Z.* **128**: 68–74.
56. Huettig, G. F. (1953). Zur Kinetik der Zermahlungsvorgaenge. *Z. Fuer Elektrochemie* **57**: 534–539.
57. Yokoyama, T., Taniyama, Y., Jimbo, G., and Zhao, Q. Q. (1991). The grinding equilibrium size of in-water grinding of silica sand by a planetary ball mill. *J. Powder Technol. Jpn* **28**: 751–758.
58. Lange, F. F. (1989). Powder processing science and technology for increased reliability. *J. Am. Ceram. Soc.* **72**: 3–15.
59. Naito, M., Hotta, T., Hayakawa, O., Shinohara, N., and Uematsu, K. (1998). Ball milling conditions of a very small amount of large particles in silicon nitride powder. *J. Ceram. Soc. Jpn* **106**: 811–814.
60. Hosokawa Alpine Aktiengesellschaft Catalogue "Fluidized opposed jet mills."
61. Awano, M., Kani, K., Kodama, Y., Takagi, H., and Kuwahara, Y. (1990). Mechanochemical effects of grinding on the Bi–Pb–Sr–Ca–Cu–O superconductor. *JJAP* **29**: L254–L257.
62. Naito, M., and Jimbo, G. (1985). Effect of additives on the powder mechanical properties and its application for mechanical processing. *Funsai* **29**: 104–114.
63. Rehbinder, P., and Kalinkovskaya, N. (1932). *J. Tech. Phys. (USSR)* **2**: 726–755.

64. Ikazaki, F., Kamiya, K., Uchida, K., Kawai, A., Yoda, S., and Gotoh, A. (1996). Chemically assisted dry comminution of an inorganic powder. *Adv. Powder Technol.* 7: 111–120.
65. Uehara, A., and Jimbo, G. (1980). Effect of gas atmosphere on ball mill grinding of graphite. *Kagaku Kogaku Ronbunshu* 6: 101–103.
66. Rowe, G. W. (1960). Some observations on the frictional behavior of boron nitride and of graphite. *Wear* 3: 274–285.
67. Yokoyama, T., Kondo, A., Naito, M., Yoshikawa, M., Kamiya, H., and Jimbo, G. (1990). Effect of vibration ball milling on the superconducting properties of a YBaCu-system composite oxide. *J. Ceram. Soc. Jpn* **98**: 1097–1102.
68. Koishi, M. (1987). Biryushi Sekkei, Kogyo Chousakai.
69. Naito, M., Kondo, A., and Yokoyama, T. (1993). Applications of comminution techniques for the surface modification of powder materials. *ISIJ Int.* **33**: 915–924.
70. ISO2194 (1991). Industrial screen-woven wire cloth, perforated plate and electroformed sheet – Designation and nominal sizes of openings.
71. *Alpine Handbook Mechanical Processing Technology*, 94.
72. Hosokawa Micron Corporation Catalogue "Classifier series".
73. Hosokawa Alpine Aktiengesellschaft Catalogue "Hydroplex Wet Processing Systems".

Handbook of Advanced Ceramics
S. Sōmiya *et al.* (Eds.)

CHAPTER 3

3.1 Forming and Drying

YUNPENG YANG, ERIC LAARZ, SUNITA KAUSHIK,
ERIK MUELLER and WOLFGANG SIGMUND
Department of Materials Science and Engineering, University of Florida, Gainesville, FL 32611-6400, USA

3.1.1 INTRODUCTION

Ceramic processing is an exciting field that requires the combination of knowledge from several scientific disciplines such as surface chemistry, solution thermodynamics, polymer physics, quantum mechanics, and colloid science. This chapter will introduce the fundamentals of some of these sciences with a specific view for ceramic processing and highlight recent findings. Furthermore, it will give some practical guidance on forming and drying by summarizing the published literature. The intention of this chapter is not to be complete since each of the scientific fields has a vast body of literature. We will refer to review articles and books that will allow for further reading.

Due to the high melting temperatures exceeding most other materials ceramic materials are typically not melt processed as is for metals or polymers. Therefore, 90% of the world markets in ceramics are materials derived from powders. Before sintering, these powders have to be handled and processed. Since most of the desired properties of advanced ceramics have microstructures of micron and sub-micron size ceramic powders with particle sizes in the sub-micron range are applied. As soon as the particle diameter drops below 100 μm surface forces are being about equal as the gravitational pull on the particles, yielding agglomerates for particles that are smaller than 100 μm. For particles sizes of 1 μm to 1 nm, surface forces dominate their behavior in processing being 3–10 orders of magnitude larger than gravitational forces. Therefore, a full understanding of surface forces or colloidal forces is required in order to be able to control ceramic powder processing on an advanced level. The science of colloids especially the origin of surface forces and their quantitative measurement will be presented in the beginning. It should be stressed that even for processes that use dry pressing of granules the particles had been processed in slurry at some point (Fig. 3.1.1).

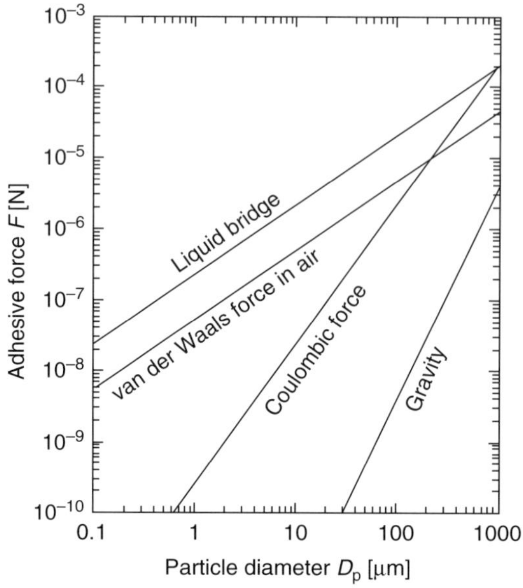

FIGURE 3.1.1 Comparison of adhesive surface forces depending on particle size [1].

After the introduction to colloid science for ceramic processing, ceramic forming techniques will be presented. First the casting techniques can be divided into two categories, that is, drained casting and constant volume casting. Drained casting techniques solidify slurry by removal of the dispersing media that goes along with shrinkage of the consolidating body, examples are slip casting or pressure slip casting. Constant volume techniques solidify the slurry without removal of the dispersing medium.

The next steps in processing are drying and binder burnout. Here, we additionally introduce a section on green body characterization since there has been tremendous progress in this field. Furthermore, sintering has only limited capability of healing flaws caused in processing before firing. Therefore, finding flaws and relating them to causes of the processing steps is most important to eliminate them from the processing cycle.

3.1.2 FUNDAMENTALS OF POWDER PROCESSING

3.1.2.1 General Concepts and Theories in Colloidal Processing

Due to the size range of advanced ceramic powder particles, that is, sub-micron and nowadays, nanometer size it is essential to understand the sciences behind particulate behavior of these objects. Objects that have at least one dimension in the range of 1 nm to 1 μm are considered colloids. The term comes from the Greek word for "glue" since these materials have the tendency to stick.

In ceramic processing these surface forces that cause sticking have to be controlled. In order to produce slurry with desirable properties, a number of suspension parameters must be carefully considered. Stability is the most desired property in a ceramic suspension. Interparticle forces play an important role in deciding the stability of the slurry. Particles suspended in a liquid medium will spontaneously agglomerate due to the attractive van der Waals forces between the surfaces. To prevent this spontaneous flocculation, modifications to the particle surface have to be made in order to produce a repulsive potential that can counteract the attractive van der Waals potential. High solids' loading is another desirable property in slurry. Generally, it is preferable to have as high a solids content as possible and yet have a low enough viscosity for the slurry to be fluid for filling molds or spray drying.

3.1.2.1.1 Origin of van der Waals Forces

Attractive forces between atoms and molecules were discovered by van der Waals when he had to make changes (add a finite volume and add attractive forces) to the ideal gas law (Boyle's law) in order to make it work for real gases over a broader temperature range. Since then several physicists explained the origin of these attractive forces. Typically three fundamental origins are described, the so-called Keesom forces, the Debye forces and the London or dispersion forces. The first two require at least one permanent dipole and the last depends only on transient dipoles inducing dipoles in neighboring atoms or molecules. Hamaker was the first to describe these energies more quantitatively in 1937. He showed that the energy of van der Waals attraction (E_{vdW}) equals a materials constant (A) times a distance (d) and body geometry dependent function $G(d)$.

$$E_{vdW} = A \times G(d) \tag{1}$$

Nowadays, A is referred to as Hamaker constant. Although this material-specific Hamaker constant is fundamentally important the number of materials where these numbers are known for are very limited. This has to do with the fact that the theoretical and experimental access to Hamaker constants still proves difficult. We refer the interested reader to two excellent review papers on Hamaker constants and van der Waals forces in ceramic processing by L. Bergström and R. French [2, 3].

3.1.2.1.2 Stabilization of Colloidal Suspensions

Dispersing particles by making them mutually repulsive is called "stabilization". It can be achieved by segregating either charged ions or polar molecules in solution preferentially to the particle surface (electric double layer stabilization), by adsorbing polymer chains onto the particles (steric stabilization) or by a combination of the two mechanisms (electrosteric stabilization).

3.1.2.1.2.1 *Electric double layer forces*

The surface charge of a colloidal particle in a suspension forms the basis of electric double layer stabilization. The magnitude of surface charge determines the magnitude of repulsion between the particles and thus the stability of the slurry. Oppositely charged ions (counterions) and oppositely charged ends of polar molecules in solution get attracted to this charge on the surface of the particle, resulting in the formation of a "diffuse double layer" of ions. Guoy and Chapman developed this theoretical concept of the double layer in the early twentieth century. For a particle surrounded by the diffuse double layer the shear slippage will occur at some distance from its surface. The electric potential at this shear slippage plane is called the zeta potential. The zeta potential is the potential at the surface of the electrokinetic unit moving through the liquid medium [4].

The main mechanism by which oxide particles develop a surface charge is by adsorption of ions from solution. The surfaces of the oxide particles are normally hydrated in water. Adsorption of H^+ ions produces a positively charged oxide surface at low pH, whereas at high pH the adsorption of OH^- ions produces a negatively charged oxide surface. The pH at which the surface is neither negatively nor positively charged is known as the point of zero charge (Fig. 3.1.2).

The inner and the outer Helmholtz layer together form the Stern layer. The value of the zeta potential is believed to be close to the potential at the Stern layer. Hence, it is more practical to measure the zeta potential. The pH value at which the zeta potential is zero is called the iso-electric point (IEP). The IEPs of various compounds are given in Table 3.1.1.

Theoretically, the relationship between surface charge density σ_0 and surface potential ϕ_0 of a particle dispersed in a liquid is given by [4]

$$\sigma_0 = (8\varepsilon\varepsilon_0 RTc)^{1/2} \sinh\left(\frac{zF\phi_0}{2RT}\right) \tag{2}$$

where ε is the dielectric constant of the medium, ε_0 the permittivity of free space, R the gas constant, F the Faraday constant, c the concentration of the electrolyte ions and z the valence of the electrolyte ions. For low potentials (i.e. when ϕ_0 is less than ~50 mV), the above equation becomes

$$\sigma_0 = \varepsilon\varepsilon_0 \kappa \phi_0 \tag{3}$$

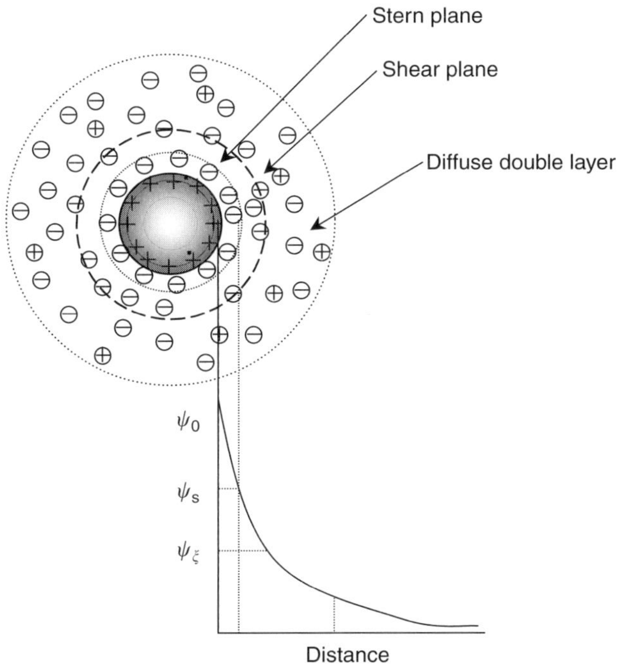

FIGURE 3.1.2 Electric double layer. The electric potential as a function of distance from the particle surface. The distance equal to $1/\kappa$ is the Debye length. ψ_s = surface potential; ψ_ζ = zeta potential; ψ_0 = zero potential.

TABLE 3.1.1 Nominal Iso-Electric Points of Oxides

Material	Iso-electric point
α-Alumina (Al_2O_3)	9–9.5
Silica (SiO_2)	3–4
Zinc oxide (ZnO)	9
Titania (TiO_2)	4–5
Calcium carbonate ($CaCO_3$)	9–10
Lead oxide (PbO)	10
Magnesia (MgO)	12–13
Zirconia (ZrO_2)	4–5
Stannic oxide (SnO_2)	4–7

where $1/\kappa$ is referred to as the thickness of the double layer or the Debye length and is defined by

$$\kappa^2 = \frac{2cz^2F^2}{\varepsilon\varepsilon_0 RT} \tag{4}$$

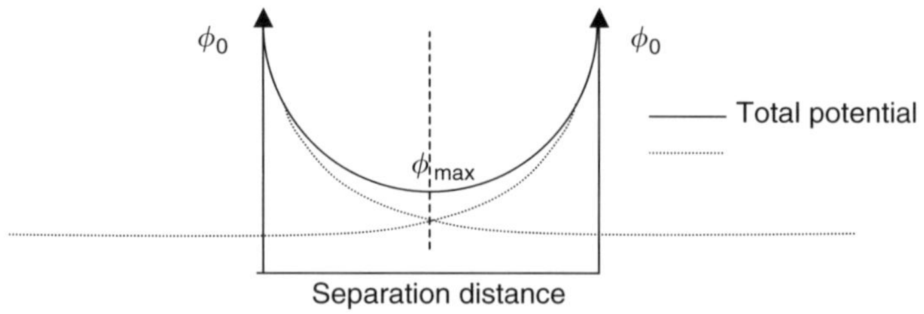

FIGURE 3.1.3 Overlap of electric double layers due to approach of two parallel surfaces, resulting in an increase in the ionic concentration, which results in repulsion between them.

3.1.2.1.2.2 *Repulsion between double layers*

Considerable importance has to be given to the double layers around colloidal particles when they approach each other. When two particles with like charges approach each other, their double layers start overlapping as soon as there is considerable interpenetration of the two diffuse layers (Fig. 3.1.3).

The overlap of the ionic clouds causes a local increase of the ionic strength. Therefore, the amount of water in this region is depleted and an osmotic pressure is developed. The diffusion of water into this region to equilibrate the concentrations causes the particles to repel. This repulsion is the reason for the apparent stability of the suspension.

3.1.2.1.2.3 *Total potential energy of interaction*

The repulsive forces acting between two particles were calculated by two Russian scientists, Derjaguin and Landau, as well as by two Dutch scientists, Verwey and Overbeek, almost at the same time in the 1940s. Hence the theory of interaction between electric double layers is referred to as DLVO theory (Fig. 3.1.4).

The DLVO theory states that the sum of the expressions for the attractive potential energy, V_A, and the repulsive potential energy, V_R gives the total interaction energy V_T

$$V_T = V_A + V_R \tag{5}$$

3.1.2.1.2.4 *Electric double layer stabilization*

As the double layers of two approaching colloidal particles with the same charge start to overlap, the repulsion between them opposes the van der Waals attraction. If the repulsive potential is large enough to overcome the attractive van der Waals potential, it leads to electric double layer stabilization.

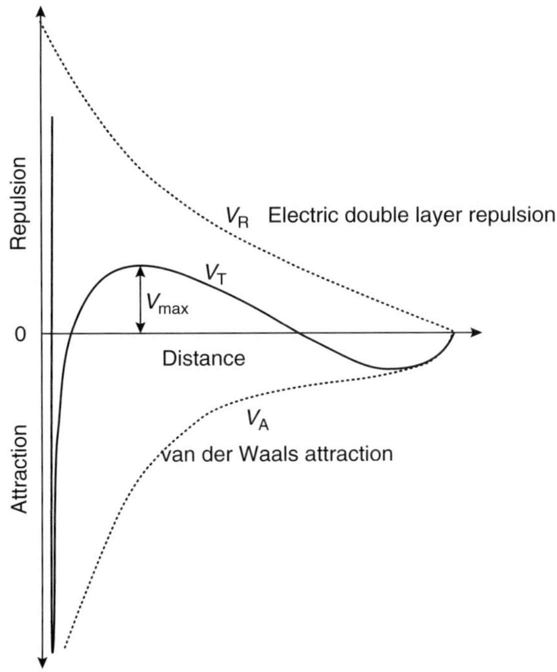

FIGURE 3.1.4 Potential energy between two particles in a liquid medium due to van der Waals attraction and electric double layer repulsion.

3.1.2.1.2.5 *Steric stabilization*

Steric stabilization is based mainly on two mechanisms. When two particles with adsorbed polymers approach each other, the number on conformations that the polymer can assume decreases due to the presence of the other particle. This results in a loss of conformational entropy. Secondly, the concentration of the polymer increases in the region of overlap. This results in osmotic repulsion between the particles in the suspension, which leads to stabilization of the suspension.

3.1.2.1.2.6 *Electrosteric stabilization*

Stabilization of slurry can also be achieved by a combination of electrostatic and steric forces. This can be done in three ways (i.e. with the presence of a charged particle and a neutral polymer molecule, in the presence of a neutral particle and a charged polymer molecule, or in the presence of a charged particle and a charged polymer molecule). The presence of charged particles produces

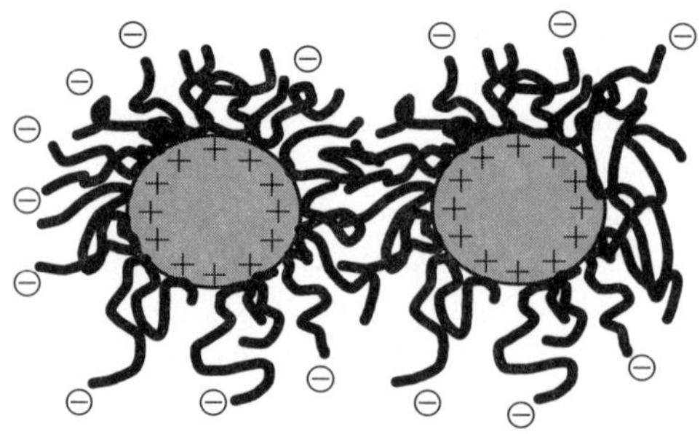

FIGURE 3.1.5 Schematic of electrosteric stabilization.

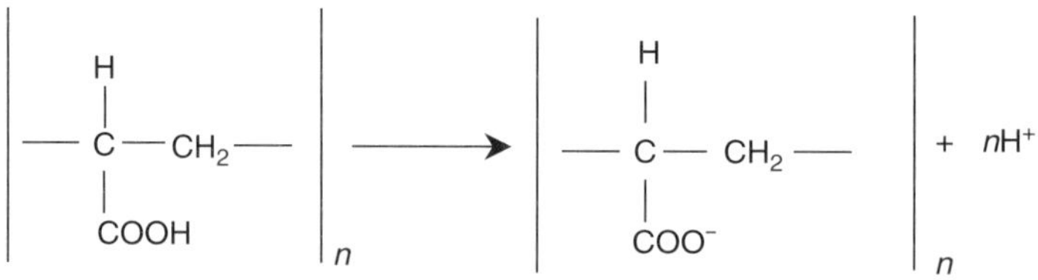

FIGURE 3.1.6 Dissociation of acrylic acid.

long-range repulsive potential and polymer molecules produce short-range repulsive potential as the particles approach one another (Fig. 3.1.5).

Polymer molecules have ionizable groups, which ionize in polar solvents, which give rise to a charged polymer. These polymers are called polyelectrolytes.

Figure 3.1.6 shows a schematic diagram of the dissociation of poly (acrylic acid) segments. The solvent conditions decide the fraction of dissociated functional groups. As the fraction of the dissociated functional groups, α, increases, the charge on the polymer varies from zero to highly negative or positive.

3.1.2.1.2.7 *Charge regulation*

At high concentrations, when the separation distances between the particles are very small (i.e. when it approaches 0), the repulsive pressure between the particles approaches infinity. Thus, the surface concentration of ions, which is proportional to the repulsive pressure P, also rises dramatically and approaches infinity [5, 6].

$$P = -\frac{\rho_s kT}{D} \tag{6}$$

$$\rho_s = \frac{2\sigma}{ze} \tag{7}$$

where σ is the surface charge density, ρ_s the surface concentration of ions, and D the separation distance between particles.

However, the infinite pressure is based on the assumption that, the surface charge remains constant. In practice, this is not so. When two particles are forced close to each other, the counterions are forced to adsorb onto their original sites. Hence, as D approaches 0, σ also falls. Hence, the surface potential remains constant. This effect is known as charge regulation. However, if there is no binding, the surface charge density remains constant and at the smallest possible separation distance, the counterion concentration approaches a constant value $2\sigma/eD$.

Thus, the charge regulation effect can be summarized as:

- The constant charge EDL repulsion is always higher than the constant potential EDL repulsion.
- Real surfaces show an intermediate behavior between constant charge and constant potential [6].

3.1.2.2 Coagulation

Most colloids are thermally unstable. They remain suspended only when the potential barrier is $\gg kT$. If the potential barrier is very small (i.e. $\ll kT$), particles come together and adhere. If this association is irreversible, the effect is called coagulation. If the adhesion between the particles is reversible, it is called flocculation. A *coagulate is* defined as a mass of particles adhered to each other due to van der Waals attraction formed in an aqueous suspension.

3.1.2.2.1 Kinetics of Coagulation

Particles in a suspension do not have a definite kinetic energy. At any given point in time, some particles may have higher energy as compared to others. The rate of coagulation is one of the most important characteristics of a colloidal suspension. The effect of DLVO theory on the kinetic behavior of particles in a suspension has been extensively studied [7]. The stability ratio W is given by

$$W = \frac{\text{Number of collisions between particles}}{\text{Number of collisions that lead to coagulation}} \tag{8}$$

and, the rate of slow coagulation is given by

$$R_s = W * R_f \tag{9}$$

where R_s is the rate of slow coagulation (when $V_T \gg kT$) and R_f the rate of fast coagulation (when $V_T \ll kT$).

3.1.2.2.2 Rate of Fast Coagulation

The rate of coagulation in the absence of repulsive barrier was addressed by von Smoluchowski [8]. He calculated the flux of a particle, which would come in contact with a fixed particle. The rate of fast coagulation is given as

$$R_f = 8\pi D a v_0^2 \tag{10}$$

where $D = kT/6\pi a\eta$ (m^2/s) (assuming uniform spheres), a is the particle radius (m), v_0 the initial particle concentration (number m^3), and η the viscosity of the dispersing medium (Pa s).

The time for the bulk concentration of the particles to reduce by half is given by

$$t^{1/2} = \frac{3\eta}{4kTv_0} \tag{11}$$

Smaller particles have faster coagulation rates as compared to larger particles.

3.1.2.2.3 Rate of Slow Coagulation

In the presence of energy barrier $\gg kT$, the rate of coagulation becomes orders of magnitude smaller than that calculated by Eq. 10. This potential energy barrier can be considered as activation energy for a coagulation event to take place. The equation for the rate of slow coagulation can be written as

$$R_s = \frac{8\pi D v_0}{\int_{2a}^{\infty}\left(\exp(V/kT)/r^2\right)\mathrm{d}r} \tag{12}$$

where V_T is the magnitude of the potential energy barrier and r the Distance from the particle center.

Substituting the values of R_s and R_f in Eq. 9, the value of stability ratio becomes

$$W = \frac{R_s}{R_f} = 2a\int_{2a}^{\infty} \exp\left(\frac{V_T/kT}{r^2}\right)\mathrm{d}r \tag{13}$$

This equation is solved numerically or graphically.

3.1.2.3 Rheological Properties of Suspensions

Rheology is the science of the deformation and flow of matter [6]. Rheology plays a very important role in the processing of the colloidal suspensions. Knowledge of rheological behavior can help characterize and optimize the state of dispersion especially at high solids loading. This chapter discusses the fundamentals in and techniques useful for rheology.

Colloidal processing requires the suspension to have as high a solids loading and as low a viscosity as possible. Contrary to this the viscosity of ceramic suspension increases with the addition of more particles to the system. In order to achieve maximum solids loading with minimum viscosity, it is necessary to study the rheological behavior of the slurry so that its properties can be manipulated in order to achieve the above.

For liquids and suspensions, a characteristic property is used to describe the flow behavior, that is, *viscosity*. Viscosity measurements can be used as a method of analysis for determining the optimum solids loading of a colloidal system. By controlling the viscosity, it is possible to achieve optimum green body properties after forming. Viscosity can be defined as [9] (Fig. 3.1.7)

$$\eta = \frac{\tau}{\dot{\gamma}} \tag{14}$$

where τ is the shear stress, $\dot{\gamma}$ the shear rate, and η the shear viscosity.

Fluids are characterized by the way in which the viscosity changes with shear rate at a constant temperature. If the viscosity of the suspension is independent of the shear rate the fluid is said to exhibit Newtonian behavior (e.g. water). However, not all fluids exhibit Newtonian behavior. Colloidal suspensions are more often non-Newtonian in nature. Fluids in which the viscosity rises with the increase in shear rate are called shear-thickening fluids. This behavior is

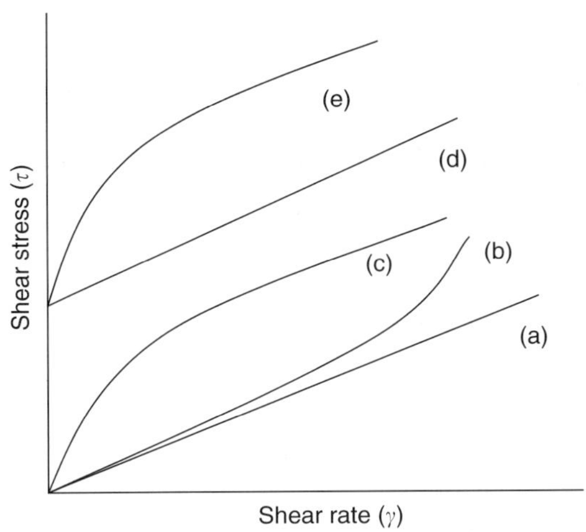

FIGURE 3.1.7 Typical rheological behavior of colloidal suspensions: (a) Newtonian; (b) shear thickening; (c) shear thinning/pseudoplastic; (d) Bingham; (e) shear thinning with yield stress.

exhibited by suspensions with a very high solids loading, which will eventually result in dilatancy of the slurry at very high strain rates. In some fluids, however, the viscosity decreases with increase in shear rate. These are called shear thinning or pseudoplastic fluids.

3.1.2.4 Parameters Affecting the Viscosity of a Suspension

3.1.2.4.1 Influence of Solids Loading on the Viscosity (Assuming Hard Sphere System)

For dispersions of non-interacting particles, the viscosity depends on the volume fraction of solids added to it. Increase in solids loading will cause an increase in the viscosity. Einstein proposed a method to describe this phenomenon. It is given by

$$\eta = \eta_s(1 + k_1\phi) \tag{15}$$

where η is the viscosity of the suspensions, η_s the viscosity of the suspending medium, k_1 the Einstein coefficient, which is equal to 2.5 for spheres, and ϕ the volume fraction of solids in the slurry.

This equation does not take into account the influence of particle size. Hence, this theory holds good only when there is no interaction between the suspended particles. Thus, this equation is valid only for dilute suspensions. When the system involves interaction between the particles (i.e. in case of concentrated suspensions) the situation becomes more complex and then the contributions due to particle interactions must also be included. This can be accounted for by including higher-order terms of ϕ. Thus, the extended Einstein equation is given as

$$\eta = \eta_s(1 + k_1\phi + k_2\phi^2 + \cdots) \tag{16}$$

A convenient way of expressing the viscosity of slurry is by dividing the slurry viscosity η_s, by the viscosity of the suspending media, η_l. This is a dimensionless quantity called the relative viscosity η_r. Increasing the volume fraction of the solids causes an increase in viscosity higher than that predicted by the power law expansion.

Thus, the effect of volume fraction of solids on the viscosity behavior is best studied in relation to the maximum packing fraction. With the addition of more and more particles, a stage is reached when the particles are jam-packed resulting in a continuous three-dimensional contact between the particles throughout the system. This leads to immobilization of the fluid. At this point, when no flow takes place, the suspension is said to have maximum packing fraction ϕ_m.

In Refs [10, 11], a model for hard spheres is introduced. The equation for this model is given by

$$\eta_r = \left(1 - \frac{\phi}{\phi_m}\right)^{-[\eta]\phi_m} \tag{17}$$

where $[\eta]$ is the intrinsic viscosity.

However, this model applies only to monosize spheres. It does not take into consideration the particle size distribution of the powders. For data analysis of soft-sphere systems typically the high shear rate viscosities are taken into consideration for the calculations. This is because, at high shear rates the viscosity is dominated by the hydrodynamic force acting on the suspension.

3.1.2.4.2 Effect of Particle Size Distributions on the Viscosity of a Suspension

Most commercial ceramic powders have a continuous particle size distribution. The effect of particle size distribution on the viscosity becomes pronounced only at very high solids loading close to the maximum packing fraction of the suspension. Typically, the viscosity of polydisperse systems is lower than monodisperse systems at the same volume fraction, as the interstitials between the larger particles can be occupied by the smaller particles present [12]. Also, as the Krieger–Dougherty equation (17) suggests, the viscosity of monodisperse suspensions reaches a very high value at low solids loading. However, there is an optimum ratio in which the different particle sizes must be present. If a dispersion comprises of a higher fraction of the smaller particles, it may exhibit a lower maximum solids loading, thixotropy or plasticity. This occurs because of the increase in specific surface area able to bind more water and dispersant which would otherwise be available as dispersion liquid to yield a flowable dispersion. In case of dispersions with a high percentage of large particles, dilatancy is observed which obstructs flow. This lowers the maximum solids loading. When a high percentage of medium size particles are present in the dispersion, the particle size distribution becomes very narrow. The dispersion behavior approaches that of monodisperse systems. This again lowers the value of maximum solids loading due to trapping of water in the interstices of the large particles, which would otherwise be available for flowable dispersions [13] (Fig. 3.1.8).

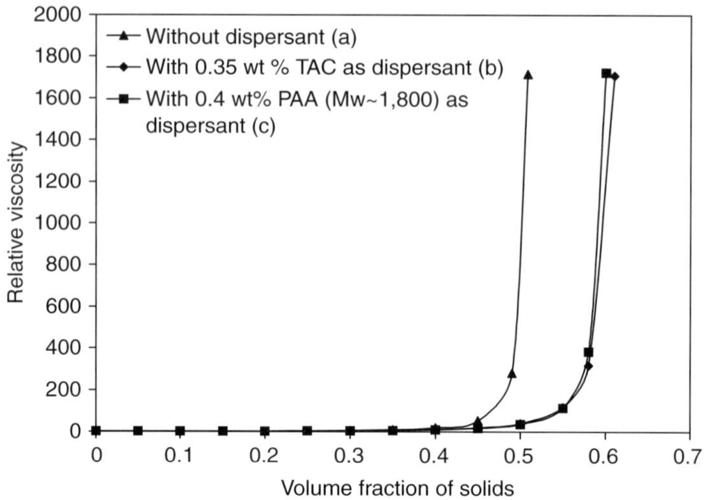

FIGURE 3.1.8 Dougherty-Krieger plots for alumina slurries: (a) stabilized by electric double layer; (b) electric double layer with short range repulsion from citrate anions; (c) electrosteric stabilization caused by poly (acrylic acid). Major differences in maximum solids loading are caused by agglomeration and subsequent solvent trapping.

3.1.2.4.3 Variation of Viscosity with pH

The pH of the suspension has a significant effect on the suspension viscosity. In case of oxides, the viscosity also indirectly varies with the zeta potential [9]. Viscosity of alumina slurry increases with increase in pH up to a certain value. Beyond a certain pH value the viscosity is inversely proportional to the suspension pH. The pH at which the viscosity reached a maximum value is the IEP of the suspension (Fig. 3.1.9).

3.1.2.5 Rheology of Sterically Stabilized Systems

In case of dispersions, which are stabilized by the addition of a polymer as a deflocculant, its rheology depends on three factors. They are

- The Flory–Huggins parameter χ. If the value of this parameter is less than 0.5, then the repulsion between the suspended particles will be strong enough for them to remain dispersed.
- The thickness of the layer of polymer adsorbed on the particles. The adsorbed layer thickness should be large enough to overcome the attractive van der Waals forces in order to prevent the particles from sticking

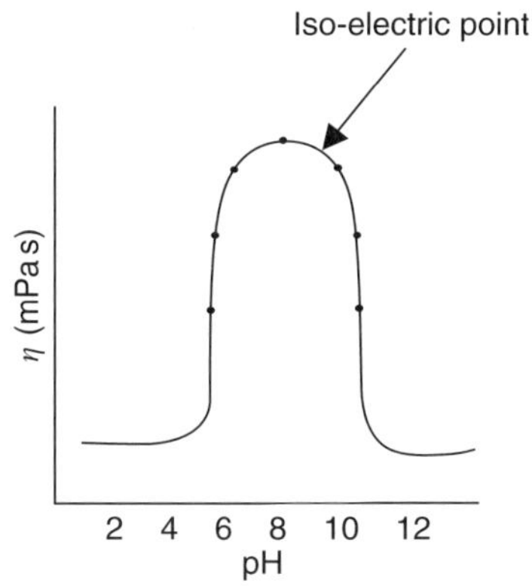

FIGURE 3.1.9 Schematic of effect of pH on the viscosity of alumina suspension.

to each other. The overlapping of the polymer layers is crucial as it is the prime cause of repulsion.

- The ratio of particle diameter to the adsorbed layer thickness [13]. The thickness of the adsorbed layer decreases with the increase in particle size. Hence, dispersions of small particles, due to large thickness of adsorbed dispersant layer tend to act like soft spheres.

3.1.2.6 Rheology of Electrosterically Stabilized System

Polyelectrolytes are widely used as dispersants for highly concentrated slurries (>50 vol%). They combine the principles of both electric double layer and steric stabilization. The effectiveness of electrosteric stabilization depends on pH and ionic strength as discussed in Chapter 2. At low solids loading (~20 vol%), the viscosity is relatively low and is affected very little by changes in viscosity. As the solids loading increases, the pH has a very significant effect on the viscosity. The amount of polyelectrolyte added also has a profound effect on the rheology of the suspension. The amount should be the optimum amount just enough to saturate the surface. Addition of higher than optimum amounts of polyelectrolytes results in excess amount of the polymer in the solution. The presence of the free polyelectrolyte in the solution can result in depletion flocculation, in case of high solids loading suspensions.

Conformation of the adsorbed polyelectrolyte also plays an important role in the rheological behavior of electrosterically stabilized slurries. The conformation in turn depends on the pH of the suspension. A detailed study of the adsorption behavior of poly electrolytes on Al_2O_3 [14, 15] shows that the amount of polyelectrolyte adsorbed on the particle increases with a decrease in the pH value. When the pH increases to a value greater than the point of zero charge, the fraction of the polyelectrolyte dissociated is ~1. Hence, the negative charges in the polyelectrolyte repel each other. This reduces the loop formation in the conformation of the polyelectrolyte adsorbed on the particle surface. The polymer thus adsorbs in a extended conformation and the repulsion is short range. In this case, the stabilization is mainly due to the electrostatic forces. Conversely, at low pH values, the amount of polyelectrolyte dissociated and adsorbed on the particles surface decreases. As a result of this, the polyelectrolyte becomes uncharged. This favors the formation of the loop conformation of the polyelectrolyte segments. With thicker layer of adsorbed polymer with chains extending into the solution, the contribution of the steric forces towards stabilization is dominant [16].

3.1.2.7 Processing Additives

To overcome attractive van der Waals forces requires that additives have to be used in ceramic processing. Additives can have a wide variety and always have an impact on the particle–dispersing medium or particle–particle interfaces. Commonly, both organic and inorganic additives are used to improve the properties of the final ceramic green bodies and the sintered products [16]. For example, polymeric additives are used many times since they provide superior mechanical properties for handling. At the same time, polymeric compounds introduce specific problems that are related to their size.

In the dry powder direct compaction process, fine ceramic raw powders are processed into granules by adding organic binders to control the packing uniformity of the consolidated body, and the granule is made up of hundreds of fine raw powder particles. In theory, these granules are assumed to be perfectly homogenous distributions of particles and binder. In reality, granules might be hollow or show strong gradients of binder and fine particles enriched on the surface.

When dispersing ceramic powders in the solvent, dispersants are introduced to make a highly loaded suspension stable by adjusting the interparticle forces. While in tape casting or injection molding, the selection of suitable additives is one of the most vital steps. The following section will give a brief description of the commonly used additives according to their serving functions.

3.1.2.7.1 Dispersants or Deflocculants

In wet powder processing, a number of fundamental interactions can be used to alter the interparticle forces [16–19]. These forces include attractive van der Waals forces, repulsive EDL forces, attractive or repulsive steric forces, and attractive capillary forces. EDL forces arise when solute ions or charge carrier function groups are attracted to or dissociation of soluble ions from particle surfaces occurs. Steric forces are developed by macromolecules that are attached to the surface of the particle. While charged macromolecules, that is, polyelectrolytes produce repulsive electrosteric forces [20]. Combination of the attractive van der Waals interaction and the repulsive double-layer repulsion is the foundation of the well-known DLVO theory, which provides an overall net interaction energy [18].

The attractive forces can be altered by changing the dispersing medium. The strongest impact on the van der Waals interaction range can be achieved by switching from high dielectric solvents to low dielectric properties.

The general explanation of the particle surface charge mechanism in the polar solvents is attributed to the accumulation of ionic species at the interface [16, 20]. The relationship between suspension microstructure, interparticle potential energy and the suspension structure can be schematically shown in Figure 3.1.10 [16].

Dispersants can be low molecular weight organic chemicals that carry functional groups for anchoring on surfaces and a buoy moiety. The buoy part is best if additional highly polar groups are available especially carboxylic acid groups as those develop negative surface charges in protic dispersing media

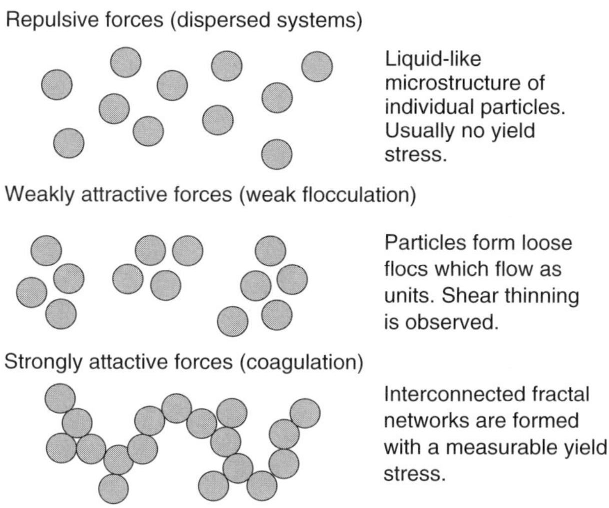

FIGURE 3.1.10 Schematic of the connection between suspension microstructure, interparticle potential energy and the suspension structure (from Ref. [16]).

above pH 3. Typical examples are citric acid or maleic acid. Other organic dispersants include polymers and poly electrolytes with molecular weights from 2000 for highly charged poly electrolytes and 100 000 for polymers. The size and the required charge of a successful dispersant can be predicted based on the particle size, the dispersing medium, that is, the main parameters controlling the van der Waals forces. In general, the repulsive barrier has to be close to or above $-0.5\,kT$ in order to have the Brownian energy to move the particles out of the attraction to achieve free flow.

Typical polar or dissociable functional groups for dispersant are hydroxyl (—OH), carboxyl (—COOH), sulfonate ($-SO_3^-$), sulfate ($-OSO_3^-$), ammonium ($-NH_4^+$), amino ($-NH_2$), imino (—NH—), and poly oxyethylene ($-CH_2\ CH_2O-$) groups (Fig. 3.1.11). Steric stabilization is the main contribution from these non-ionic macromolecules in the solid/solvent systems. The higher efficiency of the steric stabilization is realized by non-ionic macromolecules as compared to low molecular weight compounds [19]. Some commonly used dispersants are shown in Table 3.1.2.

The effect of dispersants can be strongly influenced by the solubility of the ceramic material. The solubility of ceramic materials is finite and covers many orders of magnitude depending on the material and the pH of the dispersing media. Increased solubility will cause an increase in counterions which will compress the electric double layer. Under these conditions

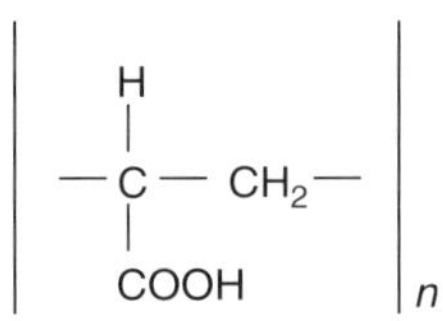

FIGURE 3.1.11 Schematic of polyacrylic acid.

TABLE 3.1.2 Commonly Used Dispersants in Ceramic Processing

Low molecular weight	High molecular weight
Sodium borate	Poly(acrylic acid) (PAA)
Sodium carbonate	Poly(methacrylic acid) (PMAA)
Sodium pyrophosphate	Ammonium polyacrylate
Sodium silicate	Sodium polyacrylate
Citric acid	Polyisobutene
Ammonium citrate	Menhaden fish oil
Sodium citrate	Phosphate ester
Sodium tartrate	Sodium polysulfonate
Sodium succinate	
Glyceryl trioleate	

a purely EDL-stabilized system will collapse. Sterically stabilized systems will also change since the solubility and conformation of polymers depends on ionic strength. Polymers might be salted out, that is, the suspension will coagulate. These mechanisms sometimes occur unwanted in processing or may be especially designed as part of a process as in the direct casting or constant volume techniques. An excellent source for understanding the solubility of ceramic materials is Mesmer's book *The Hydrolysis of Cations* [21].

3.1.2.7.2 Binders

Dry ceramic particle networks are very brittle and have very low strength, typically less than megapascals in compression. The only way to improve handling is to introduce materials that offer a toughening and strength increase, but also can be burned out. Therefore, polymeric materials are introduced. Typically, the addition of binder to the ceramic/solvent system is necessary to provide high strength to the green tape in the tape casting process after the solvent has evaporated [23]. During the granulation of fine powders, such as in spray drying, organic binders provide sufficient strength to bind hundreds of fine particles to form large ($>50\,\mu m$) agglomerates and a free-flow powder [17]. The binder is a temporary bonding media for the powder particles and should give the green body a good handling and storage characteristics but should not result in cracks, pinholes and other defects in the unfired and fired parts. Table 3.1.3 gives typical recipes for spray granule additives, and Table 3.1.4 shows some typical data for industrially produced granules. In some colloidal processing, for example, tape casting, organic binder increases the viscosity and causes the change of the flow characteristics of the slurry. Moreover, binder, dispersant, and other additives in the ceramic mixture must be compatible with each other so as to avoid phase separation during the evaporation of the dispersing medium. A high molecular weight and a low glass transition temperature are advantageous. The binder must decompose into volatile products without residues of ash and carbon during the burnout process [22]. Table 3.1.5 lists the properties of some commonly used binder materials.

TABLE 3.1.3 Typical Recipes for Spray Granule Additives

Additive component	Vol.%	Type
Poly(vinyl alcohol) (PVA)	83.1	Binder
Glycerin	8.3	Plastisizer
Ethylene glycol	4.1	Lubricant
Dispersant	4.1	Dispersant
Non-ionic wetting agent	0.2	Dispersant
Defoamer	0.2	Defoamer

TABLE 3.1.4 Typical Data for Industrially Produced Granules

Granules	Primary particle size (μm)	Organic additive (vol. %)	Granule size (μm)	Granule density (%)	Tap density (%)
Al_2O_3 (substrate)	0.7	3.6	92	54	32
Al_2O_3 (spark plugs)	2.0	13.3	186	55	34
ZrO_2 (sensors)	1.0	10.5	75	55	37
Mn–Zn ferrites	0.7	10.0	53	55	32
SiC	0.3	20.4	174	45	29

TABLE 3.1.5 Commonly Used Binders in Ceramic Processing

Binders	Soluble in water	Soluble in organic solvents	Viscosity grade (effect to change the solvent viscosity)	Biodegradable
Poly(vinyl alcohol)	✓		Low–medium	
Poly(acrylic acid)	✓		Medium–high	
Polyacrylamide	✓		High	
Poly(ethylene oxide)	✓		Low–medium	
Polyethylenimine	✓		Low–medium	
Poly(methylacrylic acid)	✓		Medium–high	
Polyvinylpyrrolidone	✓		Low	
Methylcellulose	✓		Low–high	✓
Hydroxypropylmethylcellulose	✓		Low–high	✓
Hydroxyethylcellulose	✓		Low–high	✓
Sodium carboxymethylcellulose	✓		Low–high	✓
Ethyl cellulose		✓	Medium–high	
Starches	✓		Low–medium	✓
Dextrins	✓		Very low	✓
Sodium alginate	✓		High	✓
Ammonium alginate	✓		Very high	✓
Poly(vinyl butyral)		✓	Medium–high	
Poly(vinyl formol)		✓	Medium–high	
Poly(methyl methacrylate)		✓	Medium–high	
Polysilazane		✓	Medium–high	
Gum arabic	✓		Very low	✓
Guar gum	✓		Very high	✓

3.1.2.7.3 Plasticizers

Most binders require the addition of plasticizers to improve the flexibility and workability of the dried green body, such as tape. The choice of plasticizer is affected by compatibility, efficiency and stability. External plasticizing is

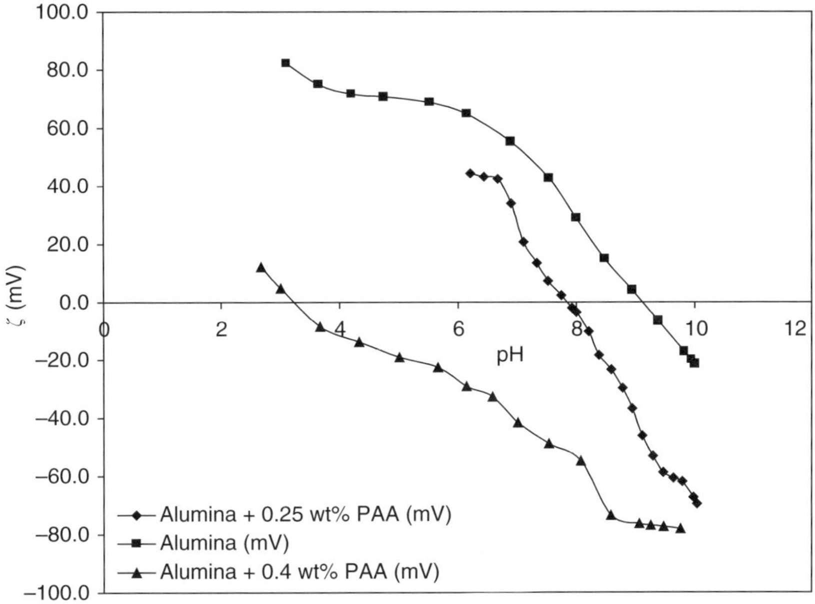

FIGURE 3.1.12 Shift in the isoelectric point of RC-LSC3DBM due to adsorption of polyacrylic acid $M_W \sim 1800$ on the surface of alumina particles. The amounts of PAA added are 0.25 and 0.4 wt% of alumina powder.

widely used, which means that binder polymers are mixed with low molecular weight plasticizers (M_W~300–400) or with oligomers in the M_W range of 3000–4000 [6]. Most widely used plasticizers are either glycols or phthalates such as polyethylene glycol or dioctylphthalate [23]. The principal function of plasticizers is to increase the chain mobility of binder molecules. With the optimal concentration of plasticizer, the intermolecular forces between polymer chains are weakened and the interchain distance between polymers, the intermolecular mobility is increased. Therefore, plasticization yields a decrease of the glass transition temperature of the softened polymer. Plasticizers are usually lower molecular weight organic compounds with a high boiling point, and they must be soluble in the same solvent used to process the binders (Figs 3.1.12 and 3.1.13).

3.1.2.7.4 Lubricants

Lubricants are commonly used in extrusion and injection molding to reduce friction between the particles or between the particles and the die wall. Stearic

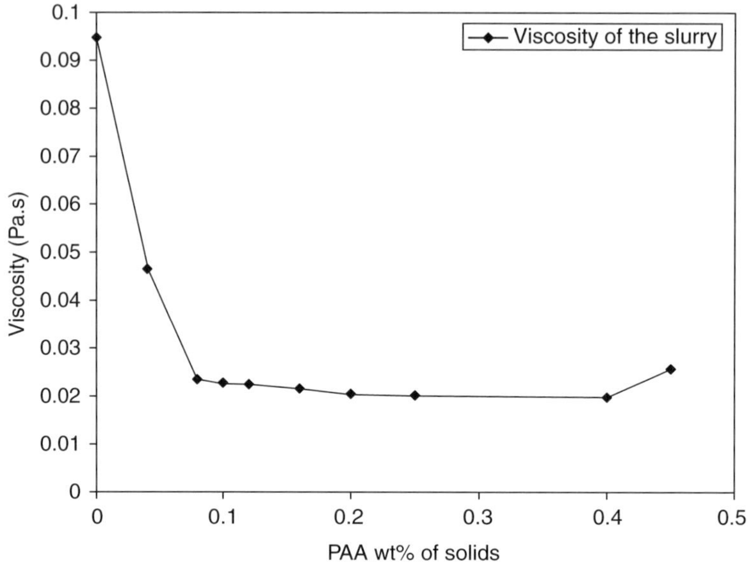

FIGURE 3.1.13 The amount of PAA $M_w \sim 1800$ required for lowest viscosity of aqueous slurry is in between 0.1 to 0.4 wt% for the here presented alumina powder at pH 9.

acid, stearates, and various waxy substances are commonly used lubricants. Table 3.1.6 shows some examples of the combination of the organic additives used in industrial pressing powders [4, 9, 23–26].

3.1.3 FORMING TECHNIQUES

3.1.3.1 Casting Techniques

3.1.3.1.1 Drained Casting

The drained casting techniques rely on phenomena that lead to separation of the liquid and solid constituents of a ceramic suspension on a macroscopic scale. The common goal with all approaches is to consolidate particles into a green body with desired shape and a mechanical strength sufficient for mold removal and handling.

Various techniques for liquid phase removal have been developed in the past. In slip casting, the solvent is drained by flow through porous walls of the mold. Capillary forces that depend on the pore size, a, and the surface tension at the liquid–vapor interface, γ_{LV}, give rise to a pressure gradient in the liquid that

TABLE 3.1.6 Some Examples of the Combination of the Organic Additives Used in Industrial Pressing Powders

Product	Solvent	Binder	Plasticizer	Others
Alumina	Acetone	Polyvinyl alcohol (low viscosity)	Polyethylene glycol (PEG) ($M_w \sim 400$)	Mg stearate (lubricant)
	Trichloroethylene (TCE)/ethanol	Polyvinyl butyral (PVB)	Poly(alkylene glycols)	Menhaden fish oil (dispersant)
Alumina substrates	Water	Polyethylene glycol ($M_w \sim 20\,000$)	None	Talc, clay (lubricant)
Alumina spark plug insulation	Acetone	Microcrystalline wax emulsion	KOH+tannic acid	Wax, talc, clay (lubricant)
TiO_2	Toluol	Ethyl cellulose Methylabietate	Diethyl oxalate Triethylene glycol Dihydroabietate	
Al_2O_3/ZrO_2	TCE/ethanol	Polyvinyl butyral	Polyethylene glycol/ phthalates mixture	Glyceryl trioleate (dispersant)
Mullite/SiC_w	Methyl ethyl ketone (MEK)/TCE	Poly(ethyl methacrylate) Poly(methyl acrylate) Poly(methacrylic acid)	Benzyl butyl phthalate (BBP)	
Mn–Zn ferrites	MEK/TCE	Polyvinyl alcohol (low viscosity)	Polyethylene glycol ($M_w \sim 400$)	Zn stearate (lubricant)
Ba titanate	MEK/ethanol	Acrylic binder	PEG/BBP	Phosphate ester (dispersant)
Ceramic tile	Water	Clay	Water	Talc, clay (lubricant)
Hotel china	Water	Clay, polysaccharide	Water	Clay (lubricant)
Refractories	Water	Ca/Na, lignosulfonate	Water	Stearate (lubricant)
$La_xSr_{1-x}MnO_3$	Ethanol/toluol	PVB	PEG/octyl phthalate	Glyceryl trioleate

drives the liquid flow

$$P_C = -\frac{2\gamma_{LV}}{a} \tag{18}$$

Often the growth rate of the slip-cast layer perpendicular to the mold surface exhibits a parabolic decrease with time. This makes slip casting a comparably slow technique and less suitable if thick consolidated layers are required. Osmotic consolidation exploits the osmotic pressure in a ceramic suspension that is in contact with a polymer solution via a semi-permeable membrane. Here, the difference in chemical potentials forces the solvent to flow through the membrane into the polymer solution. The resulting osmotic pressure can be defined as

$$\Pi = \frac{\mu_S - \mu_P}{V_m} \tag{19}$$

where μ_0 is the chemical potential of the solvent in the ceramic suspension, μ_P the chemical potential in the polymer solution, and V_m the molar volume of the solvent. Compared to the capillary pressures in slip casting, osmotic pressures are significantly higher. Alternatively, an external, mechanical pressure can be applied to the ceramic suspension, which drives the solvent through a filter membrane. This technique is known as filter pressing. Consolidation rates are comparably high for filter pressing because the employed pressures are usually about one or two orders of magnitude larger than typically encountered in slip casting. The flux of liquid through the porous particle layer is a function of the pressure gradient ∇P, the permeability D, and the liquid viscosity η according to Darcy's law

$$J = -\frac{D}{\eta}\nabla P \tag{20}$$

In tape casting, a ceramic suspension is deposited on a flat carrier substrate and the solvent is removed simply by evaporation. The rate of solvent removal is therefore mainly dependent on temperature, convective and diffusive solvent transport at the liquid–vapor interface, and internal capillary pressures in the porous layers of the cast tape.

Macroscopic separation of solid and liquid phases in a ceramic suspension can also be achieved with the help of body forces acting on the dispersed particles. Analogous to particle sedimentation caused by gravitational forces, centrifugal forces can be used to consolidate particles in a rotating mold. The settling velocity of a single spherical particle in the centrifugal force field can be approximated by a modified Stokes equation for particle movement in dilute suspensions

$$v = \frac{2\Delta\rho R^2\omega^2 z}{9\eta} \tag{21}$$

where $\Delta\rho$ is the density difference between solid and liquid, R the particle radius, ω the angular velocity, and z the distance from the rotation axis. One advantage of this method is the linear relation between cast layer growth and casting time, because no liquid has to be forced through a porous layer as in slip casting. Like in slip casting, cast bodies made by centrifugal casting can suffer from inherent density gradients. An electric field can be used in a similar manner to accelerate particles suspended in a liquid if the particles carry a surface charge. The corresponding technique used to consolidate particle into dense layers is called electrophoretic deposition. Similar to centrifugal casting, the accelerating force is balanced by viscous forces resulting in a steady state particle velocity

$$v = \frac{\varepsilon_r \varepsilon_0 \zeta E}{f_H \eta} \tag{22}$$

with ε_r being the relative dielectric constant, ε_0 the permittivity of free space, ζ the particle's zeta-potential, E the applied electric field strength, and f_H the Henry constant, which ranges from 1 to 1.5 depending on the particle size and ionic strength.

In suspensions with purely repulsive particle interactions, the consolidation pressure has to exceed the osmotic pressure of the ceramic suspension in order to achieve dense particle packing. The osmotic pressure of a hard-sphere system, Π, is a function of the solids volume fraction, Φ, namely

$$\Pi(\Phi) = \frac{RT\Phi(1 + \Phi + \Phi^2 - \Phi^3)}{V(\Phi_{max} - \Phi)^3} \tag{23}$$

where R is the gas constant, T the temperature, V the molar volume of the colloid phase, and Φ_{max} the maximum solids loading [27].

In suspensions with attractive interparticle forces, the applied pressure for consolidation has to exceed the yield stress of the space filling particle network. The relation between the yield stress, τ_Y, and the solids volume fraction can often be described by an empirical power law of the type

$$\tau_Y \propto \Phi^n \tag{24}$$

where n is a constant fitting parameter.

3.1.3.1.1.1 *Tape casting*

Liquid casting techniques allow for easy manipulation of the shape of a ceramic before sintering. Wet casting techniques involve molding a slurry—a mixture of one or more ceramic powders in a liquid, which can be either water or an organic solvent. Often other additives must be added to enhance the properties of the slurry during the casting process. Tape casting uses these principles of wet casting originally used for drain casting in order to make thin ceramic tapes. Tape casting is a method for making thin, flat ceramics as small as a few microns.

Introduction and background

The main advantage of tape casting is the ease of use for producing thin, flat ceramic and metallic parts. These parts cannot be pressed. Only recent developments in extrusion allow similar qualities. Tape casting involves spreading ceramic slurry onto a table where the liquid is removed. Tapes can be cast from 1 μm to 1 mm thickness, depending on the desired application. The original purpose of this practice was for applications in electronics and radio, which is still the chief market today. Tape-cast products include dielectric capacitors, piezoelectric actuators, along with thick and thin films.

The patent for tape-cast ceramics was granted to Glenn Howatt in 1952, although the methods for this process had been researched extensively during the Second World War [28]. In 1961, a patent was awarded to the American Lava Corp. who developed a method to continuously cast ceramic tapes. This paved the way for large-scale operations and in-line processing. In the late 1950s and early 1960s, Warren Gyurk developed a way to use ceramic tapes to make multilayered laminates, such as with today's multilayered capacitors [29].

Today, the tape casting industry is looking for technological advances using the original techniques of Howatt and Gyurk. Powder particle sizes, which were once several microns, are now in the submicron range. This has produced considerable research into various slurry additives used to optimize the properties of the slurry. Thinner films are being developed, which means more refinements are being looked into modern processing equipment and materials. With these advances, many exciting, new technological frontiers will be reached in tape casting over the next number of years.

Slurry composition

Although shaping, binder burnout, and sintering are important stages in determining the final properties of a tape-cast part, the original slurry composition is the most important influence on the final product. The variables of a slurry for a tape casting run must be in the optimized for viscosity, particle size, and miscibility. In general, the desirable viscosity is 3000–5000 mPa s with a powder particle size of 0.5–2 μm. However, these properties vary depending on the main powder, casting speed and thickness, and final application.

Solvents/dispersing media

A common problem with almost all powders is that they are agglomerated. Therefore, the particles must be dispersed. In order for this to occur, and to make a slurry, a solvent and a dispersant are added. The solvent is the medium for the slurry that gives it flow characteristics. A solvent must:

- dissolve other organic additives,
- have a low viscosity,
- be inclined to not foam,
- have a low boiling point,
- be non-reactive with the powder,
- be non-volatile and non-toxic.

The right solvent for a slurry has to meet all of the previous characteristics [23, 28] for the given powder. For example, an alumina system will use trichloroethylene as a solvent, whereas a $BaTiO_3$ system has mostly an ethylene system. These two systems are shown in Table 3.1.7. These are based on the colloidal chemistry of the individual powder and the other additives. Although water-based systems are currently being researched, water has a lower evaporation rate and higher polarity, and therefore the majority of tape casting systems still prefer non-aqueous systems. Generally, blends of an alcohol and an organic solvent are used. This allows the other additives to be dissolved in the solvent and keep the viscosity low during milling. Novel developments in green processing will allow avoiding organic solvents in the future.

Dispersants

The dispersant, also called a surfactant or deflocculant, are molecules or polymer chains adsorbed onto the powder particles. Dispersants are important to add to a slurry system so that the cast tape can be uniform and homogenous. Dispersion also helps provide uniform packing in the green body yielding uniform shrinkage during drying, binder burnout and sintering. Like with other additives, dispersants are system specific. However, Menhaden fish oil was found to be useful for a variety of non-polar systems. Menhaden fish oil is

TABLE 3.1.7 Compositions for Typical Tape Casting Slurries

Function	Composition [23]	Vol %	Composition [30]	Vol %
Powder	Alumina	27	Barium titanate	66
Solvent	Trichloroethylene	42	Ethanol	10
	Ethyl alcohol	16	Methylethyl ketone (MEK)	10
Dispersant	Menhaden fish oil	1.8	Menhaden fish oil	0.6
Binder	Polyvinyl butyral	4.4	Acryloid B-7 MEK	8.8
Plasticizer	Polyethylene glycol	4.8	Polyethylene glycol	2.6
	Octyl phthalate	4.0	Butyl benzyl phthalate	2.6
Homogenizer			Cyclohexanone	0.4

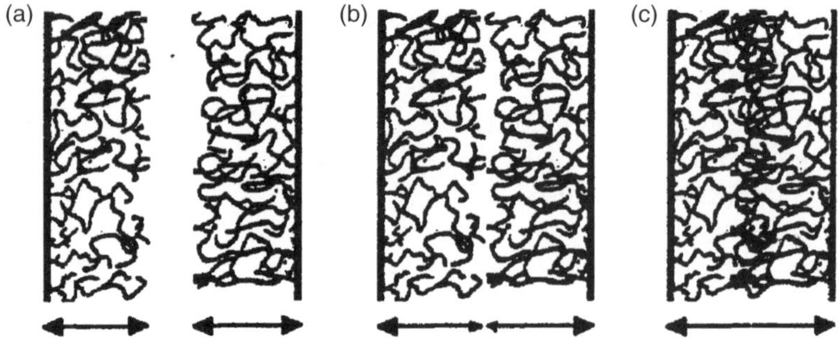

FIGURE 3.1.14 Steric interaction in polymer coated surfaces: (a) no interaction; (b) little repulsion; (c) penetration of polymer layers—strong repulsion.

highly purified non-drying oil derived from *Brevoortia tyrannus*, a herring like fish living off the Eastern coast of the United States. The oil is transformed into a useful dispersant for ceramic processing through auto-oxidation. This process causes a poorly controlled polymerization of the highly unsaturated triglycerides. The auto-oxidation is stopped after a certain viscosity has been reached. Menhaden fish oil for ceramic processing is a polydisperse polymeric (dispersity above 40, standard synthetic polymers are around 2) mixture having high molecular weight fraction above 100 000.

It still contains fatty acids, non-reacted triglycerides, and an intermediate molecular weight fraction. This multitude causes problems in processing due to the non-reproducibility of the oil and therefore its colloidal stabilizing ability. Synthetic substitutes have been developed. As with all dispersants complete surface coverage of the powder is required yielding low amounts of typically 1 wt% of dispersant (Fig. 3.1.14).

Binders

The purpose of the binder is to provide strength to the dried ceramic tape. This is important especially in storing and machining the dry tape. Since there is no other material group available to yield these properties organic polymers are introduced. The strength of polymers is derived from the size, that is, the molecular weight. The solubility of polymers in the dispersing medium and the viscosity of the polymer solution limit the size of the polymer. Therefore, binders are typically polymers with a molecular weight around 100 000. Since the binder is another polymer in the slip, the surface chemistry has to be tailored as well as the hierarchy in processing has to be kept in order to achieve the very well-defined nanostructure. Dispersants have to be added first to the slurry, then the binder. If they are added at the same time or reversed unfavorable nanostructures on the surfaces are achieved that lead to flocculation and high viscosities.

The binder should:

- form a strong, cohesive tape when dried,
- be miscible or emulsifiable in the solvent,
- be inexpensive,
- readily burnout without leaving a carbon residue.

The polymeric properties [30] of the binder are important to allow the ceramic to maintain its strength and shape until firing.

Plasticizers

Most synthetic polymers have glass transition temperatures above room temperature. The brittle behavior of glasses is not useful for ceramic green tapes therefore the glass transition temperature of the polymers has to be lowered. This is most easily done by adding solvents with a low vapor pressure, or in other words, a high boiling point with typically above 250°C. These solvents are referred to as plasticizers. Adding too much plasticizer degrades the strength of the binder, which will also cause problems during machining. Therefore, the binder and plasticizer ratio much be carefully monitored to provide optimal properties for the tape. Generally, two different plasticizers are added to slurry although one suffices for the green tape. This tradition of using two might have a positive impact on drying and burnout.

Other additives

Several other additives can be used to enhance the final product. Wetting agents, or homogenizers, increase the solubility of the many liquids in the slurry. These wetting agents also contribute to high-quality tape surfaces. Cyclohexanone is a common homogenizer. Defoaming agents, such as dimethylsilicones, are used to prevent foaming and allow trapped air bubbles to escape easier on degassing. They commonly have to be used in aqueous slurries or when sub-micron powders are used. They work by removing surplus surfactants from the bubble surface, thus increasing the bubbles interfacial energy and making it unstable. However, the complexity of having many organic additives makes the prediction of the usefulness of the defoamer difficult (Figs 3.1.15–3.1.17).

Other tape casting techniques

> *Waterfall technique*: Analogous to the glazing of ceramic tile, this casting technique involves using a pump to pour a curtain of slurry onto a substrate, which is then dried. The slurry "curtain" or waterfall must be kept

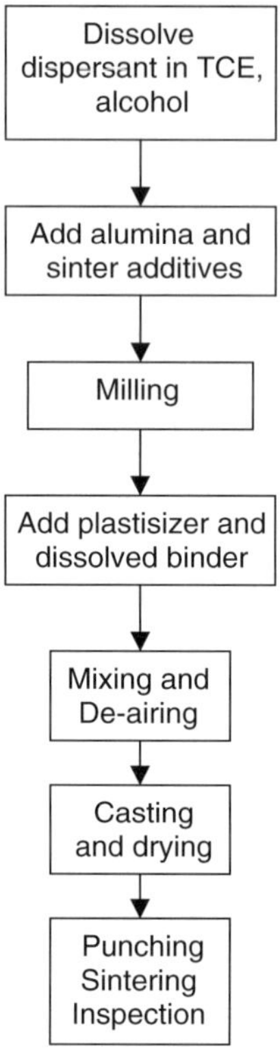

FIGURE 3.1.15 Flowchart for tape casting alumina substrates.

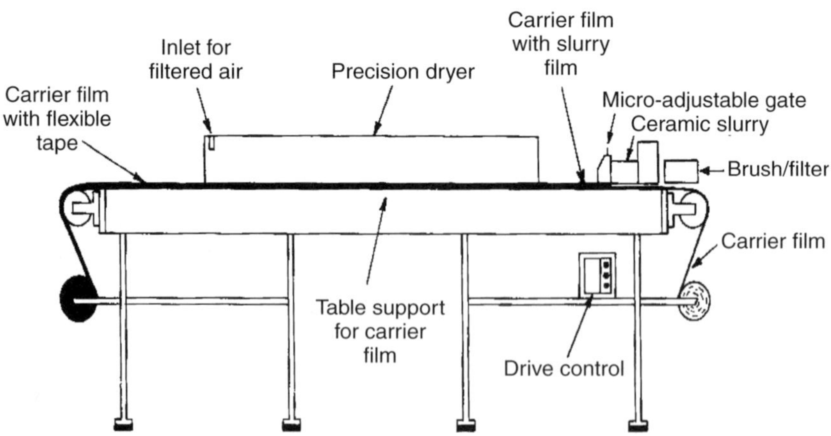

FIGURE 3.1.16 Diagram of a continuous casting machine (from Ref. [9]). This material is used by permission of John Wiley & Sons, Inc.

FIGURE 3.1.17 Laboratory scale tape caster for batch casting of ceramic tapes.

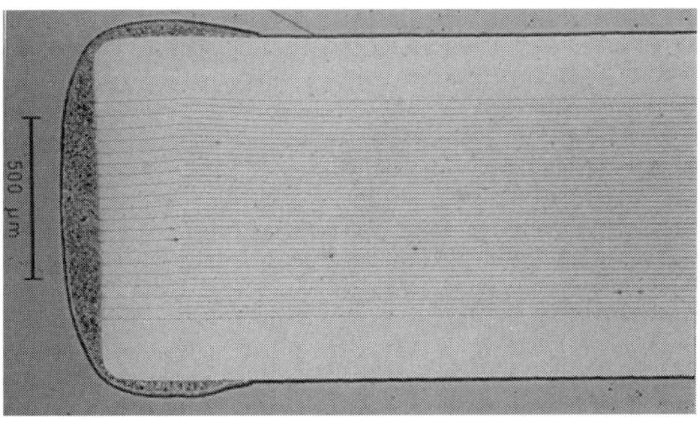

FIGURE 3.1.18 Polished long-transverse section of a multilayer capacitor.

thin and uniform to avoid splashing or other non-uniformities. This technique is used for making porous electrodes for fuel cells and multilayer ceramic capacitors [31] (Figs 3.1.18 and 3.1.19).

Paper casting: In this process, a low-ash paper is passed through slurry. The slurry sticks to the paper, which is passed through a dryer and rolled onto a reel (Fig. 3.1.20). The paper is removed during BBO. This process is used for honeycomb structures such as heat exchangers and catalytic converters [31].

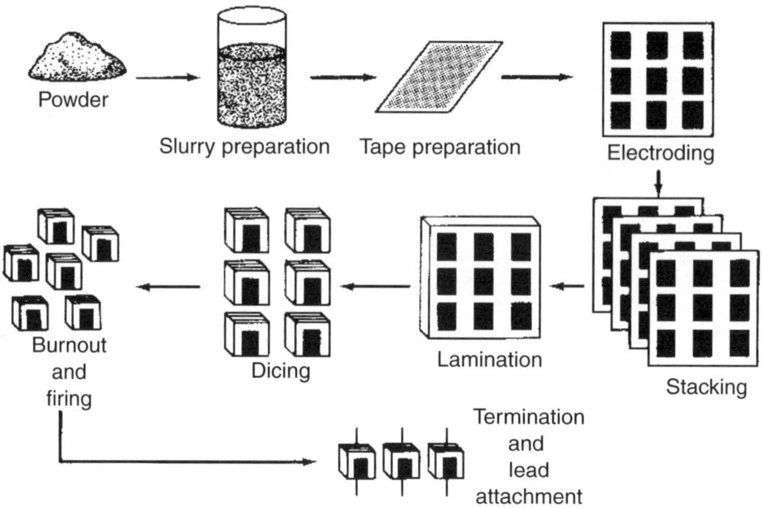

FIGURE 3.1.19 Steps for fabricating mutlilayer capacitors using tape casting (from Ref. [31]).

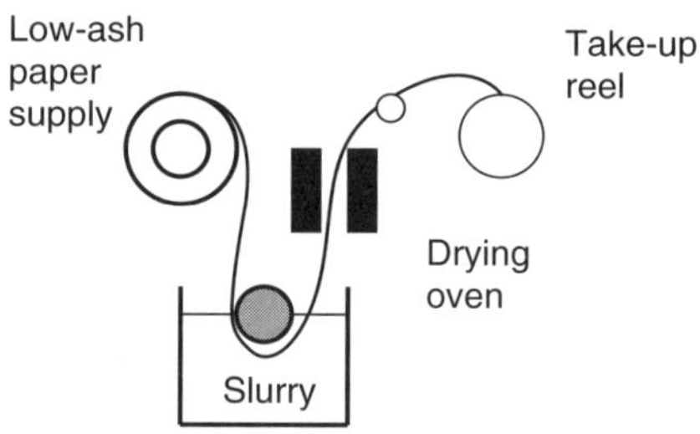

FIGURE 3.1.20 Schematic showing the paper-casting process.

Common problems and defects: All of the variables involved in tape casting must be carefully monitored and controlled to prevent defects in the final product, such as cracks, non-uniformity, delamination, and surface roughness. The majority of these problems can be traced back to the slurry composition. A common problem during casting is excessive sticking of the tape to the belt. Adding more dispersant and plasticizer will alleviate this problem. On the other hand, adding too much of both will cause the tape to release early from the belt and curl up. A weak tape is a sign of too little binder, just as a brittle tape is deficient in plasticizer. Even the smallest deviation in additive composition can greatly change how the tape and the final product turn out.

3.1.3.1.2 Constant Volume Techniques

3.1.3.1.2.1 *Gel casting*

Wet powder processing is conducted by dispersing ceramic powders in the solvent to form a highly loaded, while stable, suspension for the following consolidation. It is known that hard agglomerates limit uniform consolidation of green compacts of ceramics and results in defects in the final sintered parts. Since there are always soft agglomerates or hard agglomerates in the raw powders, de-agglomeration is needed to break them into individual particles. The aggregate structures of fine powders can be dispersed by adjusting pH and/or many types of dispersants in suspension, and sometime the efficiency can be enhanced by ball mill. Normally dispersant and other organic additives may be required to generate repulsive energy barrier to avoid flocculation. The suspension can then be cast into either a porous mold (slip casting) to consolidate the slip or a non-porous mold (constant volume casting) to induce flocculation by adjusting process parameters. After the wet green body is formed, it is demolded and drying is followed to get the dry green body. The up-to-date ceramic forming techniques through wet powder processing will be discussed in the following section.

3.1.3.1.2.2 *Extrusion*

Extrusion is a manufacturing technique for the production of long pieces of material with a common cross-section. The material is forced through a carefully shaped nozzle at high pressure to form a continuous length body. Extrusion is important in the manufacture of components from metals to ceramics and plastics. Highly ductile metals such as aluminum are extruded cold (at room temperature) to form hollow bars and tubes for low-strength applications such as window frames. Other metals such as copper and brass are extruded hot to form water pipes and pipe fittings. In the ceramics industry, extrusion is used to form square-section bars which can be cut into bricks, while viscous alumina slips are extruded into complex, thin-walled honeycombs, used as support media for catalysts. Thermoplastics become viscous on heating and can be extruded in many forms [4, 9].

In the manufacturing of ceramic green bodies by conventional extrusion, a ceramic paste is prepared by mixing ceramic powder with polymers and other additives first. Then a rotating screw, together with heaters, heats the material and continuously pushes this paste through a die with the shape of the profile to form a green ceramic string. After the die, the profile is cooled by air or water and cut into desired lengths. The advantages of the extrusion process are that profiles of all shapes can be made and production volumes are normally high. Water (traditional ceramics) or organic materials (technical ceramics) are added

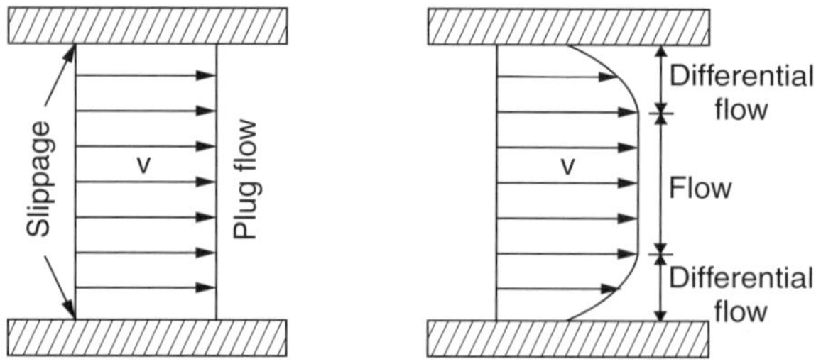

FIGURE 3.1.21 Cross-section velocity profiles for (left) plug flow with slippage at wall and (right) differential laminar flow near wall (from Ref. [9]). This material is used by permission of John Wiley & Sons, Inc.

to provide adequate plasticity for forming. The major difficulty is removing the organic material prior to firing, without causing cracks or distortions.

During extrusion, the driving force produced by the auger or piston must exceed the resistive force of the plastic material and the die–wall friction. The pressure motivating flow is highest in the barrel and decreases along the axis of the extruder. The particular flow behavior depends on the geometry of the die and the flow properties of the materials. Figure 3.1.21 shows the cross-section velocity profiles during extrusion.

In conventional extrusion, the fabricated object takes the cross-sectional shape of the extrusion die and can be no smaller than the die orifice. Use of co-extrusion for microfabrication (MFCX), which is a new method to produce axis-symmetric ceramic objects with micrometer-size feature in two dimensions, has also been reported. The MFCX process uses simple round or square dies for size reduction. Shaping is done by using a shaped object of two or more materials as the extrudate. One material is the primary ceramic plastic compound, the ceramic which is to be microfabricated. The second material is a fugitive substance, whose function is to fill space and have similar flow behavior as the primary material during extrusion. These are extruded together, or co-extruded, to reduce the cross-sectional dimensions. The fugitive material is later removed. This method can be used to produce several shapes in alumina and PbO-containing ferric ceramics with carbon black, removed by oxidation, as fugitive. Composites with layers as fine as 8 μm have been made with this process [32–34].

Figure 3.1.22 shows schematically the geometry of the ceramic composites that can be microfabricated by co-extrusion.

3.1.3.1.2.3 *Injection molding*

Injection molding techniques are frequently employed to produce ceramic parts with complex shapes that are difficult or even impossible to realize with

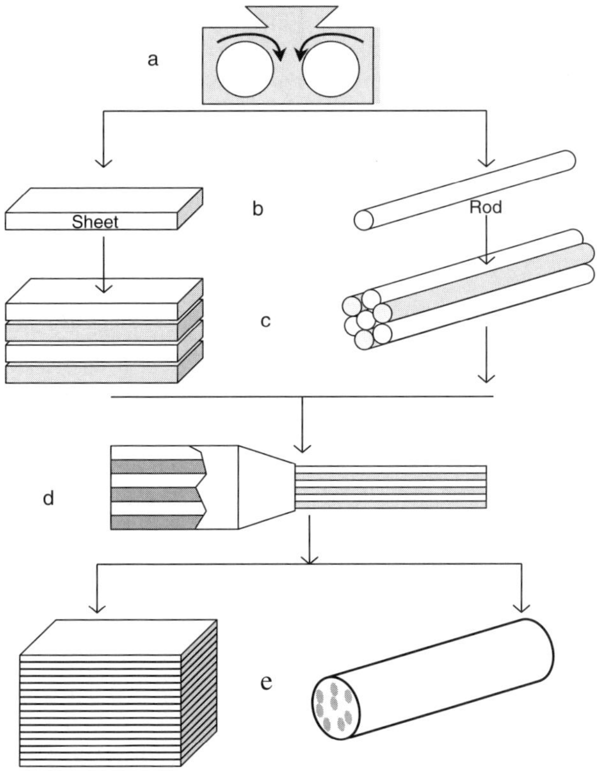

FIGURE 3.1.22 Piezoelectric ceramic/polymer composite formed by co-extrusion: (a) compounding powder; (b) performing; (c) lamination; (d) extrusion; and (e) 1–3 composite.

many other forming techniques. Moreover, when using injection molding, it is comparably easy to switch the production to a different product simply by exchanging the mold and feedstock.

Basically, the feed material consists of a mixture of one or more ceramic powders and a polymer or wax as a binder phase. Additionally, dispersants and plastisizers may be added to enhance flowability of the mix and to prevent agglomeration of the particulate phase. The feed material is fed into a heated chamber in order to soften or liquify the binder, which is necessary for optimum mold filling. The filling of the mold is usually accomplished by the action of a plunger or a screw that forces the feed material through a nozzle into the mold cavity. After mold filling is completed, the material is allowed to cool down until the binder solidifies. Then, the mold can be opened and the ceramic part ejected.

The factors that may limit the use of injection molding techniques are related to cost competitiveness and difficulties to produce large parts. The design of an injection molding machine has to take account of the high pressures created during an injection cycle. Moreover, precise control of the mass of the injected material, of the materials temperature, and of the position of the plunger or screw is essential for the manufacturing of high-quality parts. Wear of the surfaces that come in contact with the highly abrasive feed material is another factor that increases the costs of injection molding processes. Binder formulation is crucial for the quality of injection molded parts [4].

Suitable rheological properties of the feed material are required during mixing and injection. This means that the injected material must have a sufficiently low viscosity at relevant shear rates and temperatures. Accordingly, not only the binder viscosity must be low but also the solids fraction of the feedstock needs to be well dispersed in the binder matrix phase. Furthermore, viscoelastic properties of the feed material have to be considered; for example, time and temperature-dependent relaxation of shear stresses after mold filling might lead to unwanted dimensional changes. Binder properties are also of considerable importance in subsequent processing stages; for example, burn-out behavior of the binder can pose limitations on the geometry and the dimensions of the ceramic part [35].

3.1.3.1.2.4 *Pressing (die, CIP)*

Dry powder processing involves the direct compaction of either the raw fine particles or the granular powders in a rigid die. The commonly used methods are die pressing, cold isostatic pressing (CIP), hot pressing (HP), and hot isostatic pressing (HIP). Die pressing is widely used to produce simple-shaped engineering parts because it can be easily automated. HP and HIP can make the forming of the green body and sintering possible at the same time, but they require specific apparatus. These consolidation techniques will be described in the following sections.

Die pressing

Die pressing is conducted by applying stresses directly in a vertical direction to the granules of ceramic powders that are loaded in a mold to generate loose bonding between powders so that the compact can have some mechanical strength to be handled. During die pressing, the shape of the ceramic powder granule changes (illustrated schematically in Figures 3.1.23 and 3.1.24), which comprises three steps: (a) packing of granules; (b) deformation of granules; and (c) sliding in between the granules so that the size of the voids between the deformed granules can be decreased.

When the stress or force is applied only from one side, for example, from the upper side of the mold (Fig. 3.1.25), this die pressing is referred to as uniaxial pressing. The advantage of this process is fast and easy automation,

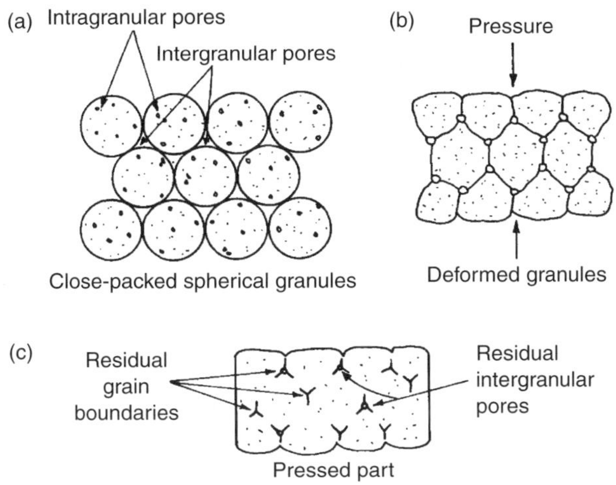

FIGURE 3.1.23 Change of granule shape on pressing: (a) packing of granules; (b) deformation of granules; (c) sliding in between the granules (from Ref. [9]). This material is used by permission of John Wiley & Sons, Inc.

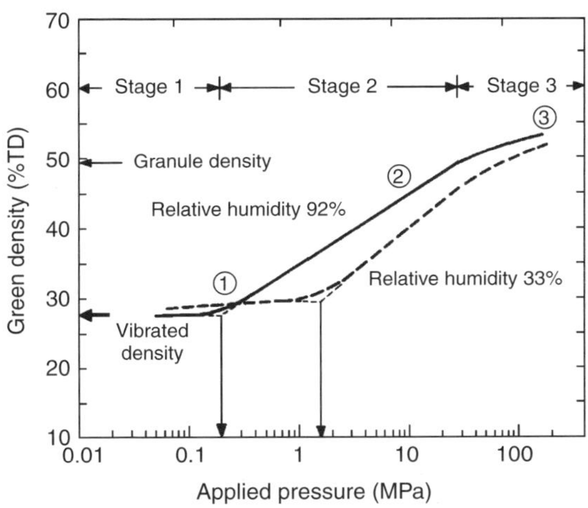

FIGURE 3.1.24 Compaction states while pressing: (1) packing of granules; (2) deformation of granules; (3) sliding in between the granules (from Ref. [9]). This material is used by permission of John Wiley & Sons, Inc.

but the shortcoming is the stress gradient that may result in density gradient in the compact green body. Sometimes lamination may also occur because of the inhomogeneous distribution of the applied stress, and some typical flaws in pressed parts.

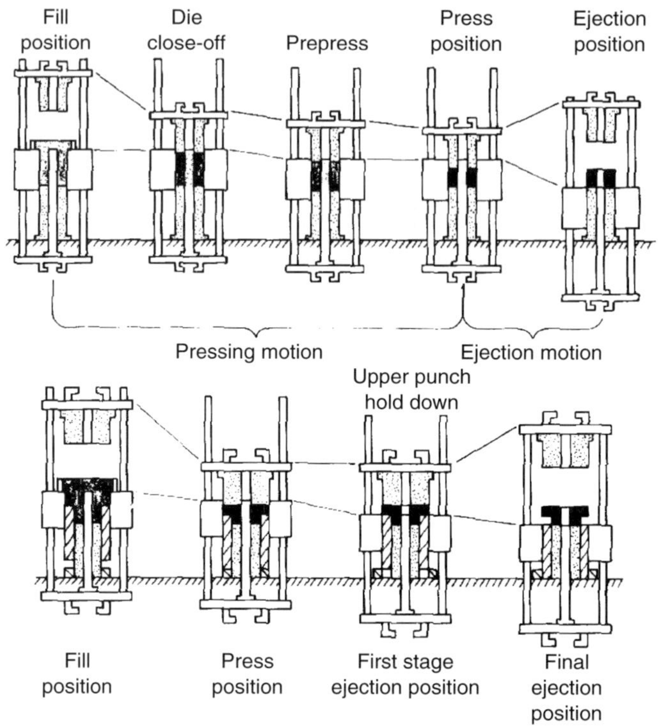

FIGURE 3.1.25 Press and ejection movements for different press matrices (from Ref. [9]). This material is used by permission of John Wiley & Sons, Inc.

To minimize the above problem, the stress can be applied from both upper and bottom directions, this is called biaxial pressing. For very large parts, there is still a problem of density gradient, especially in the areas along walls of the mold because of frictional resistance.

Cold isostatic pressing

CIP utilizes the hydrodynamic forces through liquid media to apply the stress to the powder compact in all directions, therefore, it can be used to form parts with only one elongated dimension, relatively complex, or those that may not easily be die-pressed. Typically, CIP may be divided into wet-bag and dry-bag type CIP. In the wet-bag process (Fig. 3.1.26), the powder is filled into a mold first and then the stress or force is applied in all directions. Operating pressures up to 200 MPa are commonly employed and wet bag presses operating to 500 MPa are used for pressing large blocks. This requires that the rubber mold filled with ceramic powders be immersed in a liquid media. Then the force applied to the liquid media can be passed on to the rubber mold and the force is the

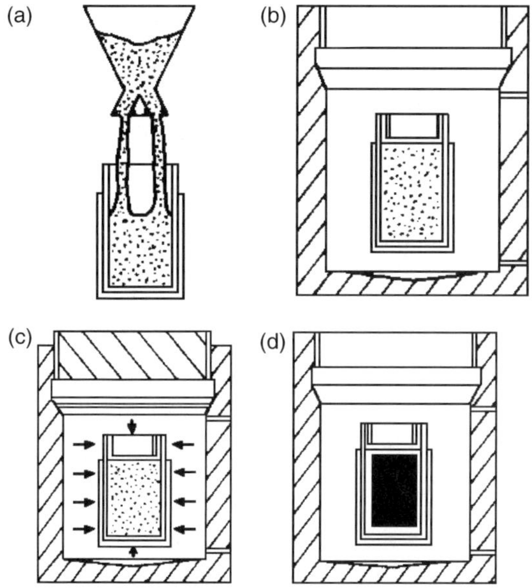

FIGURE 3.1.26 Schematic of the CIP process: (a) telling of powder; (b) installation of mold; (c) isotactic pressing (d) removal of mold (from Ref. [9]). This material is used by permission of John Wiley & Sons, Inc.

same in all directions. Therefore, the ceramic compact green body will have less density gradient. But this method encounters problems when the part is too complicated and/or the surface finish is not good. As a variation of the wet-bag CIP, in the dry-bag processing, the forces are only applied radially through the pressurized liquid medium between a flexible mold and a rigid shell, as is shown schematically in Figure 3.1.27. Therefore, this process can be used to automatically produce ceramic parts.

Hot pressing

Compared with CIP, HP is more complex and requires special apparatus (Fig. 3.1.28). It renders the compaction of powders and the sintering of green body at the same time. A mechanical pressure (> 20 MPa) increases the sintering driving pressure for densification by acting against the internal pore pressure without increasing the driving force for grain growth. Therefore, sintering temperature can be reduced and a relatively finer microstructure and higher density can be obtained. HP is commonly used to sinter a high-purity product that requires very high sintering temperatures under normal atmospheric pressure and sintering aids are thought to cause drastic influence on the mechanical properties under the actual serving environment. Another variation of HP is HIP, which uses hot pressurized gas in the range of 10–200 MPa as

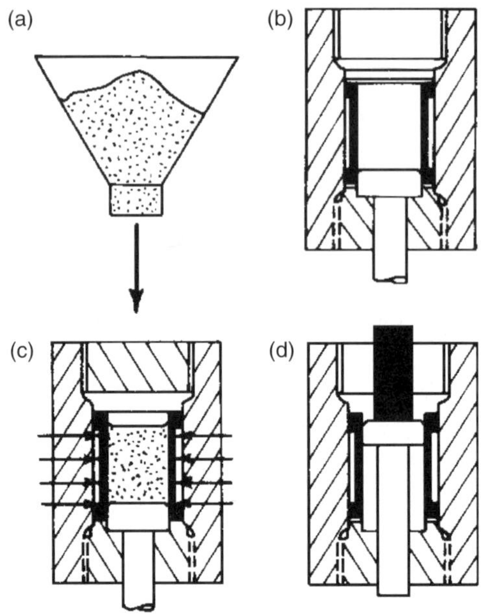

FIGURE 3.1.27 Dry-bag pseudo-isopressing—simplified automatic variation of CIP: (a) powder filling; (b) open mold; (c) pseudo-isotactic pressing; (d) ejection of part after decompression (from Ref. [6]). This material is used by permission of John Wiley & Sons, Inc.

the medium to apply the external pressure to the sintering compacts. HP or HIP can also be used as a post-sintering technique to reduce the size of the closed pores. Similar to the CIP technique, the shape of the product should not be very complicated. The size of the part is also limited because of the dimension of the chamber inside the apparatus.

3.1.4 DRYING

A crucial task in the drying process of ceramic green bodies is the minimization of non-uniform drying stresses in order to prevent cracking, non-uniform deformation, or the formation of density gradients in the green body structure. The development of non-uniform drying stresses has been related to pressure gradients in the pore liquid and in the vapor phase of unsaturated pore space, and to temperature gradients that cause differential thermal expansion [36, 37]. The evolution of the pressure and temperature distribution in a porous body is dependent on several transport phenomena such as diffusion of solvent molecules, liquid flow, and the conduction of heat. Interparticle

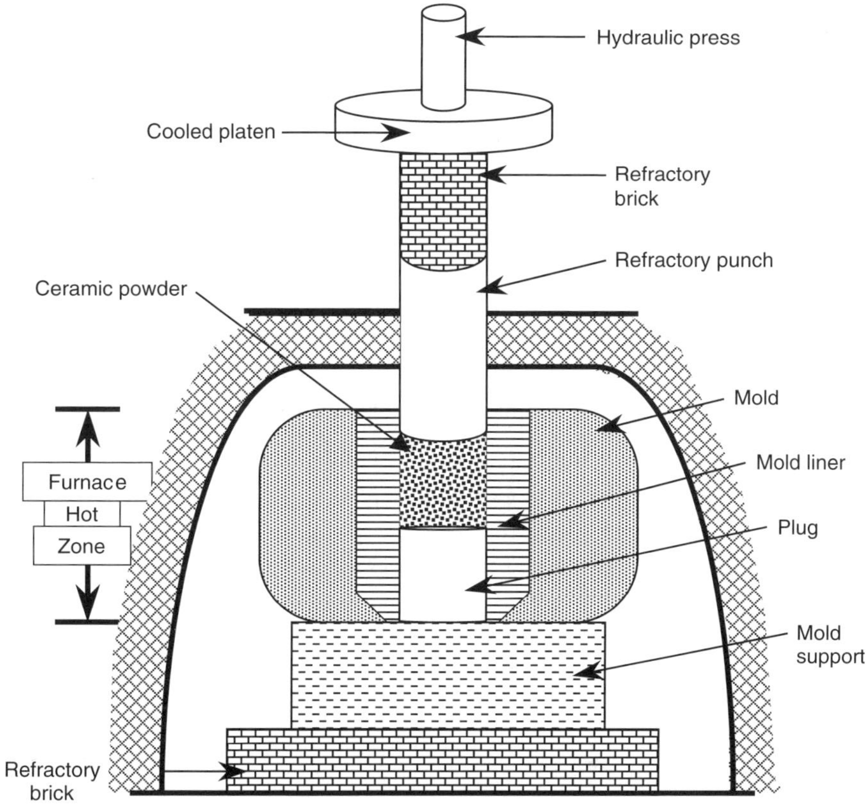

FIGURE 3.1.28 Schematic drawing of a hot press furnace (axial pressing in a furnace at up to 2000°C).

forces, friction forces, and the resulting viscoelastic properties of the particle network determine deformation and particle rearrangement in response to the mechanical drying stresses.

In ceramic processing, the shrinkage of a green part during drying is usually caused by capillary phenomena due to the small size of the pores. In some cases, other driving forces may become important, for example, osmotic forces in mixed solvents or in the presence of solutes. Capillary forces originate from the change in free surface energy, ΔE, when a solvent is spread on a ceramic surface

$$\Delta E = \gamma_{SL} + \gamma_{LV} - \gamma_{SV} \tag{25}$$

where γ_{SL}, γ_{LV}, and γ_{SV} are the specific energies of the solid–liquid, liquid–vapor, and solid–vapor interfaces, respectively. The force balance at the three-phase line is described by Young's equation [34]

$$\gamma_{SV} = \gamma_{SL} + \gamma_{LV} \cos(\theta) \tag{26}$$

where θ is the contact angle. For $\theta > 90°$, the solvent does not spread on the solid surface, whereas spontaneous wetting and spreading occurs for $\theta = 0°$ where $\Delta E < 0$. The wetting of the pore walls by the solvent leads to a tensile stress in the liquid that can be described as a negative pressure. The pressure depends not only on the interfacial specific energies but also on the pore size. For a cylindrical capillary with radius, a, the capillary pressure, P_C, can be written as

$$P_C = -\frac{2(\gamma_{SV} - \gamma_{SL})}{a} = -\frac{2\gamma_{LV} \cos(\theta)}{a} \tag{27}$$

For a contact angle $\theta = 0°$, the curvature of the liquid–vapor interface equals the capillary radius. The pressure difference across the liquid–vapor interface of a cylindrical pore is $\Delta P = \gamma_{LV}/a$ as derived from the Laplace equation

$$\Delta P = \gamma_{LV} \left(\frac{1}{R_1} + \frac{1}{R_2} \right) \tag{28}$$

where R_1 and R_2 are the principal radii of the curved liquid–vapor interface.

Due to the capillary pressure, a tensile stress develops in the liquid, which is equivalent to a compressive force acting on the particle network of the green body. Thus, solvent is forced from the interior of the green body to the exterior surface as the particle network contracts. Accordingly, the volume change of the green body equals the solvent volume that has evaporated at the green body surface. The contraction of the particle network is characteristic for the first stage of the drying process. It is called constant rate period (CRP) because the green body weight decreases at a constant rate. The further the particle network is compressed, the stiffer it becomes. The end of the CRP is marked by a maximum in capillary pressure and, thus, cracking of the green body is most likely to occur at this point of the drying process. The stresses that may lead to cracking originate from local differences in capillary pressure and the corresponding differential shrinkage. Such differential strains are inevitable near the drying front where the local tension in the liquid is higher than in the saturated interior pore space. For example, during drying of a flat plate, the pressure raises in a parabolic fashion from the midplane towards the outer surfaces of the plate [39]. As is apparent from Eq. 27 [20, 40], the magnitude of the capillary pressure depends on the pore size. Therefore, drying stresses

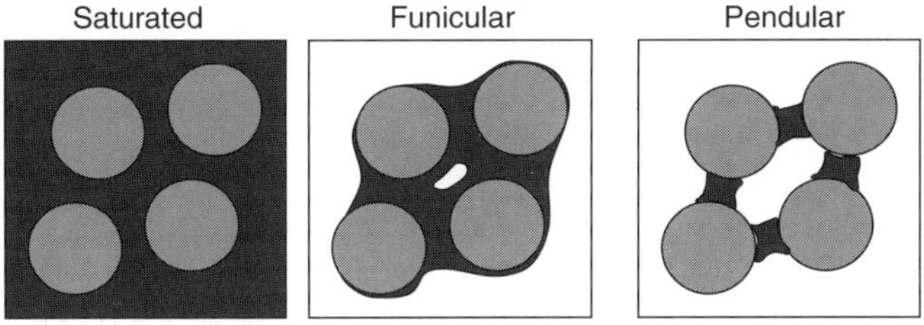

FIGURE 3.1.29 Schematic drawing of the solvent distribution in a porous body during different drying stages.

are more severe in gels with pore sizes in the nanometer range as compared to conventional ceramics with micron-sized particles. The magnitude of drying stresses is also expected to increase with the size of the green body and the drying rate [37].

At the end of the CRP, sometimes called leatherhard point or critical point, the network compressive strength exceeds the capillary pressure. Hence, further contraction is impossible and pores are partially drained, that is, the porous body is no longer saturated with solvent and enters a funicular state (Fig. 3.1.29). The term falling rate period (FRP) is used for this stage of the drying process because the rate of weight loss is decreased as compared to the CRP. During the FRP the drying front moves gradually into the interior of the green body.

Although the pores are becoming partially drained, the pore walls are still covered by a contiguous liquid film. The partial vapor pressure, p_V, of the solvent at the curved liquid–vapor interface is related to the tension, P, in the liquid or the interfacial energy, γ_{LV}, by

$$p_V = p_0 \exp\left(-\frac{PV_m}{R_g T}\right) = p_0 \exp\left(-\frac{2V_m \gamma_{LV} \cos\theta}{R_g T a}\right) \tag{29}$$

where p_0 is the equilibrium vapor pressure over a flat liquid–vapor interface, V_m the molar volume of the liquid, R_g the gas constant, and T the temperature. Evaporation takes place at the exterior green body surface when the ambient partial pressure is below p_V. The solvent is transported by liquid flow along the pressure gradient towards the exterior surface; transport via vapor diffusion through the drained pores is comparably small at this stage. The pressure driven flux of liquid through the porous body is described by an empirical expression known as Darcy's law

$$J = -\frac{D}{\eta}\nabla P \tag{30}$$

with D being the permeability of the porous structure and η the solvent viscosity. The lower the permeability, the higher is the resistance to liquid flow in the particle network and, thus, the greater is the pressure gradient needed to drain the porous body. The solvent evaporation at the surface is endothermal and leads to a lower temperature at the surface relative to both the ambient temperature and the temperature in the interior of the green body. This leads to a corresponding temperature dependent decrease in liquid surface tension and increase in vapor pressure at the surface, which further promotes the flow of liquid from the interior to the surface. The rate at which the drying front moves inside the porous body has been found to be inversely proportional to the square root of time [41].

The funicular state cannot be maintained any longer when the distance between the drying front and the green body surface is so large that the pressure gradient is insufficient to drive a liquid flow that balances the solvent evaporation at the surface. At this point a transition from the funicular to the pendular state (Fig. 3.1.29) takes place. In the pendular state, totally drained pores appear that are not connected by a liquid film to the liquid in the surrounding pores. Consequently, evaporation of solvent inside the porous network becomes large enough that diffusion of vapor through the empty pore space becomes the rate determining transport mechansim for solvent molecules. Fick's law links the diffusive flux, J_{D}, to the concentration gradient, ∇C, in the vapor phase

$$J_{\mathrm{D}} = -D_{\mathrm{C}} \nabla C \tag{31}$$

with D_{C} being the diffusion coefficient.

Willett *et al.* [42] have measured the capillary forces induced by pendular liquid bridges between sperical particles and showed that gravitational distortion and disjoining pressures have negligible effect on the measured attractive forces. Additionally, they developed a scaling method that could be used to establish closed-form expressions that accurately describe the measured force–distance curves.

It is apparent from the above discussion that both mass and heat transfer mechanisms are operative simultaneously during the drying process. The complex interplay of these processes determines the kinetics of the overall drying process. Moreover, the control of the drying conditions which affect the drying kinetics is not only decisive for the drying time but is also vital for any effort to minimize drying stresses and the risk of cracking and non-uniform deformation [36, 37]. The work of Ghosal *et al.* [43, 44] is one example that illustrates how modeling of the transport processes can be employed to predict and control appropriate drying conditions in industrial ceramic processing. These authors developed a one-dimensional physical model for the drying of gelcast ceramic parts. By identifying the rate-limiting mechanisms for unidirectional solvent removal at all stages in the drying process they could apply comparably simple theoretical expressions to describe the drying process as a function of several controlled drying parameters, namely the partial solvent pressure, the drying temperature and the thickness of the gel-cast part.

3.1.5 GREEN BODY CHARACTERIZATION

In the past few years, the innovative use of new techniques for the characterization of ceramic green bodies allowed a significantly improved description of green body microstructures. Techniques like immersion microscopy and X-ray computed tomography have helped to gain a better understanding of microstructure evolution in the powder compaction process and they proved to aid in the identification of the origin and nature of strength-limiting defects. The main body of information available in the literature deals with the characterization of spray-dried granules and dry-pressed compacts, thus reflecting the dominating role of pressing techniques in the industrial manufacturing of ceramic parts.

3.1.5.1 MORPHOLOGY

Characteristic features of green body morphologies are typically in the 0.1–100 μm size range, which makes scanning electron microscopy (SEM) a versatile technique for morphology studies. The morphology of spray-dried granules and fracture surfaces of compacted green bodies are typical examples. Clearly, granule morphology is of importance in dry pressing operations since it will strongly influence particle packing during compaction and, thus, the size and number of pores in the green compact. SEM studies on granule shape and stereological evaluation of granule cross-sections have been used to investigate the correlation between suspension characteristics of a spray-dried slurry and the morphology of the resulting granules [45, 46]. It has been found that spray drying of flocculated, colloidally unstable powder suspensions will in general result in the formation of rather uniform and approximately spherical granules that do not contain big internal pores. In contrast, at identical spray-drying conditions irregularly shaped granules with large dimples or internal pores are formed when well dispersed, colloidally stable suspensions are spray dried. Additional results concerning the influence of spray-drying conditions suggest that the colloidal stability of spray-dried suspensions has a greater influence on the granule morphology than a moderate variation of spray drying parameters such as inlet temperature of the drying chamber or flow rate of the feed slurry [47]. However, as discussed below, compacting and sintering different batches of spray-dried granules with practically identical morphology can still lead to considerable variations in sintered strength if they differ with respect to other granule properties (e.g. granule strength and binder distribution).

3.1.5.2 Mechanical Properties

Sufficient green body strength is required to prevent fracture or cracking during handling and drying. In some cases, the green strength may even be high enough to allow green machining, which is convenient for dimension control or the production of more complex shapes. In the case of dry pressing methods, however, the granule strength needs to be optimized in a way that both handling without granule fracture and substantial deformation or fracture during compaction is possible. In the past, mechanical characterization of green bodies has been limited almost exclusively to compression tests of single granules, granular beds, or dry green compacts. The compressive load required to crush a single granule, P_{max}, is determined in a diametral compression test and, assuming that the granule is spherical, can be related to the granule tensile strength, σ_T, by the equation

$$\sigma_T = \frac{2.8P_{max}}{\pi d^2} \tag{32}$$

where d is the granule diameter [44]. A similar equation can be derived for diametral compression testing of cylindrical powder compacts

$$\sigma_T = \frac{2P_{max}}{\pi dl} \tag{33}$$

where d is the sample diameter and l is the sample length.

Similar to the analysis of mechanical properties of sintered ceramic bodies, statistical analysis of experimentally obtained strength data represents a very useful and important approach for the evaluation of the influence of processing parameters and green body microstructure on the green strength [49]. The commonly used two-parameter Weibull model can be expressed as

$$\ln \ln \left(\frac{1}{1 - P_F} \right) = m \ln \left(\frac{\sigma}{\sigma_0} \right) + C \tag{34}$$

with the probability P_F of specimen failure, the Weibull modulus m, the experimentally determined stress σ, a normalizing stress σ_0, and a constant factor C [50]. The slope of a linear fit to the strength data in a $\ln\ln(1/(1 - P_F))$ ln versus σ plot delivers the Weibull modulus, which is often used as a measure of sample reliability. As discussed below in more detail, combining Weibull analysis and characterization of sample microstructure provides a powerful tool for identification of the nature and the formation mechanisms of fracture origins in ceramic green bodies.

Carnheim and Green [51] have recently evaluated several theoretical models for the description of elastic properties of granular compacts. If a number of simplifying assumptions are valid, all theoretical models can be represented by the same equation

$$\frac{E}{E_0} = \beta\phi^z \tag{35}$$

where ϕ is the particle packing fraction, β a constant prefactor, E the Young modulus of the compact, E_0 the Young modulus of a single particle, and z the density exponent [50]. The most important assumption included in this model is a constant a/D ratio, that is, the ratio of average particle diameter, D, and average diameter of the contact area between adjacent particles, a, is assumed to be independent of the applied compaction stress. Fitting Eq. 35 to experimental data gave density exponents much higher than predicted by the theoretical models, which led the authors to the conclusion that the a/D ratio varies with compression load [51]. Experimental data [51] and theoretical analysis [52] showed that such deviations from theoretical predictions occur in systems containing a compliant binder phase which can significantly influence the stiffness of a granular compact structure depending on the applied compaction stresses. Moreover, mechanical stresses in granular compacts are transmitted along force-chains [53] that comprise only a fraction of the total particle number and whose structure changes continuously in response to the increasing applied compaction stress. The theoretical models of the elastic compact behavior, however, are based on assumptions involving a static compact structure with all particles participating in the stress transfer.

Experimental compaction curves, that is, plots of overall sample density versus applied pressure, have been used frequently to describe the compaction behavior of granular beds in a die compaction process. A characteristic compaction curve is depicted schematically in Figure 3.1.30.

The compaction curve displays two distinctly different compaction stages; both show a linear $P - \rho$ relationship but with a much stronger P dependence of σ in stage II. The transition between stage I and II is sometimes referred to

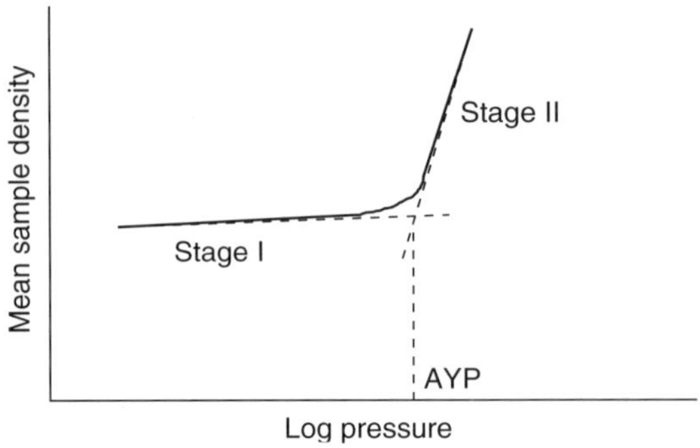

FIGURE 3.1.30 Typical compaction curve as observed in die compaction of granular powders.

as the apparent yield point (AYP) and can be estimated as the crossover point of the slopes of the two linear parts of the curve.

3.1.5.3 Microstructure

Information about the microstructure of a ceramic in the green state is of great value since inhomogeneities in the green body microstructure often persist in the sintering stage and are likely to become the origin of strength-limiting defects in the sintered body [16]. Many different types of microstructural defects may form in a ceramic processing scheme prior to sintering. In the case of advanced ceramics, pores are considered to be mainly responsible for the degradation of fracture strength. The green body microstructure is most often described in terms of overall sample density or density distributions. High green densities are advantageous for optimized densification at low sintering temperatures and a homogeneous density distribution minimizes the deteriorating effects of differential sintering. Additionally, a detailed description of defect populations present in the microstructure is desirable. This is related to the brittle nature of ceramic green bodies; their fracture mechanics are governed by parameters such as defect shape and defect size distribution rather than bulk densities [50].

Mercury intrusion porosimetry can be used to determine the volume fraction of interconnected internal pores in dry powder compacts and the corresponding green body bulk density. Assuming a cylindrical pore shape with radius r, the relationship between pore radius and external pressure can be described by the Young–Laplace equation

$$\Delta P = \frac{-2\gamma \cos\theta}{r} \tag{36}$$

where γ is the surface tension of the liquid and θ is the contact angle between the liquid and a pore wall.

For the investigation of highly compacted green bodies with a low pore number density and few interconnected pores immersion microscopy is preferred over mercury intrusion porosimetry. Immersion microscopy is an optical method for direct observation of green body microstructures [16]. Immersion of the specimen in a liquid with a refractive index equal to that of the powder particles eliminates reflection of light at the particle/liquid interface. The fraction of light, R, that is reflected at the interface depends on the matching of the refractive indices of the particle, n_1, and the immersion liquid, n_2, as expressed by the equation

$$R = \frac{\left((n_1/n_2) - 1\right)^2}{\left((n_1/n_2) + 1\right)^2} \tag{37}$$

The value of the relative refractive index $n' = n_1/n_2$ required for optimum performance depends on the features that are investigated. A value of $n' = 1$ is desirable when secondary phases, for example, binders, or particle orientation (by means of cross-polarized light) are in the focus of the investigation. A value of $n' = 1.02$–1.04 produces better optical contrast and is preferable for the detection of microstructural defects, such as pores and cracks [54].

Immersion microscopy has been used extensively to show the close relation between green body microstructure and mechanical strength of the sintered ceramic [Shinohara *et al.* 1995, 47, 55]. For example, correlating Weibull statistics of strength data with statistical data on pore size and number density of samples made at different process conditions illustrated with good accuracy that the relative differences in mean strength between samples where matched quantitatively by the relative difference in number density of the largest pores in these samples [47]. Clearly, this approach provides an excellent method to investigate the influence of processing parameters on the mechanical properties of green or sintered ceramic bodies. The immersion microscopy technique was further used to show the effect of granule properties on the microstructure of die-pressed granular compacts [45–47, 54–57]. Spray-dried granules under load may exhibit brittle fracture or plastic deformation depending on the granule density, binder content, binder distribution, and binder structure. In general, granules with a higher fracture strength are more likely to survive die pressing at high compaction stresses, which leads to an increased number of large, residual voids in the green compact that are not eliminated during sintering. In contrast, weak granules facilitate compression to high green densities, more uniform density distributions and smaller number densities of large pores. In principle, the influence of granule properties on the porosity in the green compact vanishes at sufficiently high compaction stresses. In practice, however, uniaxial die-pressing at too high pressures results in internal cracks due to lamination [58] or requires a costly subsequent high-pressure CIP treatment [59]. The compaction behavior of spray-dried granules is also strongly influenced by the processing conditions during pressing. The relative humidity in particular affects the granule fracture strength and the intergranular and granule–die wall friction, all of which play an eminent role in the evolution of density profiles in the compact [56, 57, 60]. At low relative humidity, granules are more brittle which leads to lower overall densities but also to a more homogeneous density distribution. At high relative humidity the opposite trends are observed.

X-ray computed tomography has proven to be a powerful tool to determine internal density distributions of green bodies [60]. Density profiles can be generated with this technique by passing a finely collimated X-ray beam on a horizontal plane through the specimen while the material- and density-dependent X-ray transmission is recorded with a detector. In this process, the sample is continuously rotated and translates slowly past the detector in order to construct a two- or three-dimensional representation of the internal density distribution [61, 62]. Deis and Lannutti [60] used X-ray computed tomography to give a detailed picture of the evolution of localized high- and low-density regions during uniaxial die-pressing. They could also show that the stage I–II transition in the compaction curve (Fig. 3.1.35) is not attributed to two independent asynchronous densification mechanisms; particle rearrangement mechanisms and particle deformation (or fracture) mechanisms are both active at all compression stages but the relative importance of these two types of densification mechanisms varies substantially with respect to the spatial location in the die cavity [63]. During stage I, local high-density centers near the advancing ram grow until they finally join and form a layer of uniform density adjacent to the advancing ram, which marks the transition point between stage I and II. During stage II this high density layer expands toward the stationary ram in a wavelike manner. Based on these findings, the stage I–II transition was explained by a change in structural force transmission patterns. Compaction forces are transmitted at the granule–granule contact points, but during stage I, due to the irregular particle packing at low densities, a significant part of the applied pressure is directed toward the die wall and dissipated by intergranular friction. With increasing pressure, the density of the layer adjacent to the advancing ram increases and becomes more uniform due to deformation and fracture of granules. Accordingly, the intergranular contact area increases gradually and at the beginning of stage II the major part of the applied pressure is transmitted in the direction of the advancing ram, thus causing the much more efficient densification characteristic for stage II.

3.1.6 REMOVAL OF ORGANIC ADDITIVES

The risk of an uncontrolled release of large volumes of organic decomposition products at sintering temperatures makes it necessary to burn out the organic content at lower temperatures prior to sintering. The formation of excessive amounts of volatile decomposition products is unwanted because they create internal pressure gradients that lead to cracking, if they exceed the green body strength. Solvent extraction, that is, immersion of the green body in a fluid that dissolves the polymer, is an alternative method to remove the organic content, but suffers from problems with comparably high costs and environmental hazards [64].

With temperatures of about 300–700°C in the burn-out step, one can achieve slower decomposition kinetics and a better control of the debinding process, thus reducing the risk for crack formation in the green body. It is desirable to operate at temperatures where the partial pressure of volatiles formed by polymer decomposition is below 1 atm [36]. At higher temperatures, the local pressure is higher in the pores than at the external surface, which results in tensile stresses at the surface. Alternatively, it is possible to increase the ambient pressure to reduce the partial pressure of the volatiles [65].

In a ceramic green body, polymeric binders usually constitute the largest fraction of the organic phase compared to other organic components such as dispersants or plasticizers. There are several pathways for the decomposition of organic additives, depending on the chemical structure of the polymeric components, the burn-out temperature, and the composition of the solid and the gas phase. The reaction products that form in the decomposition process may be solid or volatile and the polymer degradation may be caused by thermal activation or by oxidative attack.

Analogous to green body drying, the pore size distribution in the green body has a large effect on the transport of reactants and reaction products. Similarly, the heat transfer is also affected by the pore dimensions, in addition to the influence of material dependent heat conductivities of the solid and the gas phase. Inappropriate heating conditions can lead to cracking induced by internal temperature gradients and the corresponding thermal stresses in the green body. Additional factors influencing the evolution of temperature gradients are endothermal or exothermal reactions involved in the polymer degradation process. As a consequence of the limited burn-out temperature relatively longer times are often needed to achieve complete removal of the organic green body content. In the case of forming techniques requiring a high binder content (e.g. extrusion and injection moulding) debinding times of up to several days may be necessary.

Polymer removal proceeds via chain scission induced by either thermal or oxidative degradation, followed by evaporation of the low-molecular-weight fraction of the chain scission products [36]. Slow diffusive transport of oxygen through the pore space is often a rate determining process in oxidative polymer degradation. The kinetics of oxidative attack can be enhanced by the catalytic activity of metals, present in the ceramic body, for example, in form of metal oxides. Cyclization or cross-linking reactions on the other hand slow down polymer removal due to the resulting increase in polymer molecular weight. Most often it is required that polymeric additives depolymerize into volatile low-molecular-weight products rather than solid carbonaceous end-products that will remain in the green body after burn-out. In an oxidizing atmosphere, it is possible to remove residual carbon at temperatures below 800°C. In a reducing atmosphere, however, carbon will not be removed even at high sintering temperatures.

Lu and Lannutti [66] have investigated the effect of binder removal on the green body dimensional tolerances with X-ray computed tomography and *in situ* laser dilatometry. They showed that during debinding capillary force gradients cause the liquid binder to migrate to the high-density regions. This gives rise to a higher relative densification in the low-density regions. Therefore, green bodies with pronounced density gradients exhibit differential dimensional changes during binder burn-out. In contrast, samples with negligible density gradients showed uniform dimensional changes. It can be expected that the local redistribution of liquid binder is affected by the melt viscosity of the binder and the surface tension at the liquid–vapor interface. Based on these findings, it can be also be assumed that the total debinding time is determined by the binder removal from the high-density regions.

REFERENCES

1. Gotoh, K., and Higashitani, K., eds. (1997). *Powder Technology Handbook*. New York: Marcel Dekker, Inc.
2. Bergstrom, L. (1997). Hamaker constants of inorganic materials. *Advances in Colloid and Interface Science* 70: 125.
3. Roger H. French (2000). Origins and applications of London dispersion forces and Hamaker constants in ceramics. *J. Am. Ceram. Soc.* 83 (9): 2117–2146.
4. Rahaman, M. N., and NetLibrary Inc. (1995). *Ceramic Processing and Sintering*, New York: Marcel Dekker.
5. Hunter, R. J., and White, L. R. (1987). Foundations of colloid science. Oxford Oxfordshire New York: Clarendon Press; Oxford University Press.
6. Israelachvili, J. N. (1985). *Intermolecular and Surface Forces*. London: Academic Press.
7. Napper, D. H., and Hunter, R. J. (1972). Hydrosols in M.T.P. *Int. Rev. Sci. Phys. Chem. Ser. I.* 7: 241.
8. Smoluchowski Von, M. (1916). *Z. Phys.* 17: 557, 585.
9. Reed, J. S. (1995). *Principles of Ceramics Processing*, 2nd edn, New York: Wiley, pp. xxii, 658.

10. Krieger, I. M., and Dougherty, T. J. (1959). A mechanism for non-newtonian flow in suspensions of rigid spheres. *Transactions of the Society of Rheology* **3**: 137–152.
11. Krieger, I. M. (1972). Rheology of monodisperse lattices. *Advances in Colloid and Interface Science* **3**: 111–136.
12. Conley, J. R. F. (1996). Practical Dispersion: A Guide to Understanding and Formulating Slurries, New York: VCH, xviii, 446.
13. Dobiáa, B., Qiu, X., and Rybinski, W. V. (1999). Solid–liquid dispersions. *Surfactant Science Series*, Vol. 81, pp. vii, 562, New York: Marcel Dekker.
14. Cesarano, J., III, and Aksay, I. A. (1988). Processing of highly concentrated aqueous α-alumina suspensions stabilized with polyelectrolytes. *J. Am. Ceram. Soc.* **71**: 1062–1067.
15. Cesarano, J., Aksay, I. A., and Bleier, A. (1988). Stability of aqueous alpha-Al_2O_3 suspensions with poly (methacrylic acid) poly-electrolyte. *J. Am. Ceram Soc.* **71** (4): 250–255.
16. Sigmund, W. M., Bell, N. S., and Bergstrom, L. (2000). Novel powder-processing methods for advanced ceramics. *J. Am. Ceram. Soc.* **83**: 1557–1574.
17. Lange, F. F. (1989). Powder processing science and technology for increasing reliability. *J. Am. Ceram. Soc.* **72**: 3–15.
18. Horn, R. G. (1990). Surface forces and their action in ceramic materials. *J. Am. Ceram. Soc.* **73**: 1117–1135.
19. Lewis, J. A. (2000). Colloidal processing of ceramics. *J. Am. Ceram. Soc.* **83**: 2341–2359.
20. Myers, D. (1989). *Surfaces, Interfaces, and Colloids—Principles and Applications*, pp. 286–296. New York: VCH.
21. Baes, C. F., and Mesmer, R. (1986). *The Hydrolysis of Cations*. Malabar, Florida: Krieger Publishing Comp.
22. Roosen, A. (1988). *Ceram. Trans.* Vol. 1, Part B., Ceramic Powder Science, p. 675.
23. Böhnlein-Mauß, Sigmund, W., Wegner, W., Meyer, W. H., Heßel, F., Seitz, K., and Roosen, A. (1992). The function of polymers in the tape casting of alumina. *Adv. Mater.* **4**: 73–81.
24. Mistler, R. E., Runk, R., and Shanefield, D. (1978). Tapecasting of ceramics, *Ceramic Processing before Firing*, pp. 441–448, Onoda, G., and Hench, L., eds., New York: Wiley.
25. Paul, C. W., and Cotts, P. M. (1986). *Macromolecules* **19**: 692–699.
26. Napper, D. H. (1983). *Polymeric Stabilization of Colloidal Dispersions*. London: Academic Press, pp. xvi, 428.
27. Guo, J., and Lewis, J. A. (1999). Aggregation effects on compressive flow properties and drying behavior of colloidal silica suspensions. *J. Am. Ceram. Soc.* **82**: 2345–2358.
28. Mistler, R. E. (1998). Tape casting: past, present, future. *Am. Ceram. Soc. Bull.* 77.
29. Shanefield, D. J. (1995). *Organic Additives and Ceramic Processing*, Boston, MA: Kluwer Academic.
30. Onoda, G. Y., Jr., and Hench, L. L. (1978). *Ceramic Processing before Firing*, New York: John Wiley and Sons.
31. Richerson, D. W. (1992). *Modern Ceramic Engineering*, New York: Marcel Dekker.
32. Greenwood, R., Kendall, K., and Bellon, O. (2001). A method for making alumina fibres by co-extrusion of an alumina and starch paste. *J. Eur. Ceram. Soc.* **21**: 507–513.
33. Von Hoy, C., Barda, A., Griffith, M., and Halloran, J. W. (1998). Microfabrication of ceramics by co-extrusion. *J. Am. Ceram. Soc.* **81**: 152–158.
34. Wright, J. F., Jr, and Reed, J. S. (2001). Polymer-plasticized ceramic extrusion. Part 1. *Am. Ceram. Soc. Bull.* **80**: 31–35.
35. Mutsuddy, B. C., and Ford, R. C. (1995). *Ceramic Injection Molding; Materials Technology Series*, London: Chapman & Hall.
36. Ring, T. A. (1996). *Fundamentals of Ceramic Powder Processing and Synthesis*, San Diego, CA: Academic Press.
37. Brinker, C. J., and Scherer, G. W. (1990). *Sol–Gel Science*, San Diego, CA: Academic Press.

38. Young, T. (1855). *Miscellaneous Works*. Vol. 1, Peacock, G., ed., London: Murray.
39. Scherer, G. W. (1989). *J. Non-Cryst. Solids* **109**: 171–182.
40. Bast, R. (1993). Organic additives for ceramics—an overview. *Br. Ceram. Trans.* **92**: 217–218.
41. Shaw, T. M. (1987). *Phys. Rev. Lett.* **59**: 1671–1674.
42. Willett, C. D., Adams, M. J., Johnson, S. A., and Seville, J. P. K. (2000). Capillary bridges between two spherical particles. *Langmuir* **16**: 9396–9405.
43. Ghosal, S., Emami-Naeini, A., Harn, Y. P., Draskovich, B. S., and Pollinger, J. P. (1999). A physical model for the drying of gelcast ceramics. *J. Am. Ceram. Soc.* **82**: 513–520.
44. Harn, Y. P., Ghosal, S., Aral, G., Emami-Naeini, A., Draskovich, B., and Maxey, L. C. (1997). Real-time model-based control systems design for gelcast drying process, *Proceedings of the 1997 IEEE International Conference on Control Applications*, pp. 271–276, Piscataway: NJ: Institute of Electrical and Electronic Engineers.
45. Takahashi, H., Shinohara, N., Okumiya, M., Uematsu, K., JunIchiro, T., Iwamoto, Y., and Kamiya, H. (1995). Influence of slurry flocculation on the character and compaction of spray-dried silicon nitride granules. *J. Am. Ceram. Soc.* **78**: 903–908.
46. Walker, W. J., Jr, and Reed, J. S. (1999). Influence of slurry parameters on the characteristics of spray-dried granules. *J. Am. Ceram. Soc.* **82**: 1711–1719.
47. Hotta, T., Nakahira, K., Naito, M., Shinohara, N., Okumiya, M., and Uematsu, K. (2001). Origin of the strength change of silicon nitride ceramics with the alteration of spray drying conditions. *J. Eur. Ceram. Soc.* **21**: 603–610.
48. Hiramatsu, Y., and Oka, Y. (1966). Determination of the tensile strength of rock by a compression test of an irregular test piece. *Int. J. Rock. Mech. Mining Sci.* **3**: 89–99.
49. Bortzmeyer, D., Langguth, G., and Orange, G. (1993). Fracture mechanics of green products. *J. Eur. Ceram. Soc.* **11**: 9–16.
50. Quinn, G. D. (1991). Strength and proof testing, *Engineered Materials Handbook*, Vol. 4, Ceramics and Glasses, Metals Park, OH: ASM International.
51. Carnheim, T. J., and Green, D. J. (2001). Mechanical properties of dry-pressed alumina green bodies. *J. Am. Ceram. Soc.* **84**: 1405–1410.
52. Elata, D., and Dvorkin, J. (1996). Pressure sensitivity of cemented granular materials. *Mech. Mater.* **23**: 147–154.
53. Kuhn, L. T., McMeeking, R. M., and Lange, F. F. (1991). A model for powder consolidation. *J. Am. Ceram. Soc.* **74**: 682–685.
54. Uematsu, K. (1996). Immersion microscopy for detailed characterization of defects in ceramic powders and green bodies. *Powder Technol.* **88**: 291–298.
55. Takahashi, H., Shinohara, N., Uematsu, K., and JunIchiro, T. (1996). Influence of granule character and compaction on the mechanical properties of sintered silicon nitride. *J. Am. Ceram. Soc.* **79**: 843–848.
56. Shinohara, N., Okumiya, M., Hotta, T., Nakahira, K., Naito, M., and Uematsu, K. (1999). Seasonal variation of microstructure and sintered strength of dry-pressed alumina. *J. Am. Ceram. Soc.* **82**: 3441–3446.
57. Naito, M., Hotta, T., Abe, H., Shinohara, N., Cho, Y.-I., Okumiya, M., and Uematsu, K. (2001). Relevance of the fracture strength to process-related defects in alumina ceramics. *Mater. Trans.* **42**: 114–119.
58. Özkan, N., and Briscoe, B. J. (1997). Characterization of die-pressed green compacts. *J. Eur. Ceram. Soc.* **17**: 697–711.
59. Shin, D.-W., Yoon, D.-H., Kim, C.-J., Chung, Y.-S., and Auh, K. H. (1998). Effect of the granule properties and compaction pressure on the green and sintered densities of Al_2O_3/15 vol%ZrO_2. *J. Ceram. Soc. Japan* **106**: 363–368.

60. Deis, T. A., and Lannutti, J. J. (1998). X-ray computed tomography for evaluation of density gradient formation during the compaction of spray-dried granules. *J. Am. Ceram. Soc.* **81**: 1237–1247.
61. Phillips, D. H., and Lannutti, J. J. (1993). X-ray computed tomography for the testing and evaluation of ceramic processes. *Am. Ceram. Soc. Bull.* **72**: 69–75.
62. Copley, D. C., Eberhard, J. W., and Mohr, G. A. (1994). Computed tomography. Part I. Introduction and industrial applications. *JOM* 14–26.
63. Kong, C. M., and Lannutti, J. J. (2000). Localized densification during the compaction of alumina granules: the stage I–II transition. *J. Am. Ceram. Soc.* **83**: 685–690.
64. Chartier, T., Ferrato, M., and Baumard, J. F. (1995). Supercritical debinding of injection moulded ceramics. *J. Am. Ceram. Soc.* **78**: 1787–1792.
65. Tsai, D. S. (1991). *AIChE J.* **37**: 547–554.
66. Lu, P. K., and Lannutti, J. J. (2000). Effect of density gradients on dimensional tolerance during binder removal. *J. Am. Ceram. Soc.* **83**: 2536–2542.

Handbook of Advanced Ceramics
S. Sōmiya *et al.* (Eds.)

CHAPTER 4

4.1 Sintering of Ceramics

LUTGARD C. DE JONGHE[1] and MOHAMED N. RAHAMAN[2]

[1] *Department of Materials Science and Engineering, University of California at Berkeley, and the Lawrence Berkeley National Laboratory, Materials Sciences Division, Berkeley, CA 94720-1760, USA*
[2] *Department of Ceramic Engineering, University of Missouri-Rolla, Rolla, MO 65409-0330, USA*

4.1.1 INTRODUCTION

Because of the high melting point of the raw materials, the fabrication of ceramics commonly includes a heat treatment step in which a powder, already formed into a required shape, is converted into a dense solid. This step is referred to *sintering* (or *firing*). In general, a ceramist, wishing to produce a material with particular properties, must identify the required microstructure and then design processing conditions that will produce this required microstructure. The objective of sintering studies is therefore commonly to understand how the processing variables such as temperature, particle size, applied pressure, particle packing, composition and sintering atmosphere influence the microstructure that is produced. Several texts have been published in recent years that provide a detailed treatment of the theory, principles, and practice of sintering [1–3]. In the present article, we outline the basic principles of sintering and how they are applied practically to the production of ceramics with controlled microstructures.

There are four categories of sintering which are dependent on the composition being fired and on the extent to which second phases are formed during the heat treatment:

Solid-state sintering: the shaped green body is heated to a temperature that is typically 0.5–0.9 of the melting point. No liquid is present and atomic diffusion in the solid state produces joining of the particles and reduction of the porosity.

Liquid-phase sintering: a small amount of liquid, typically less than a few volume percent of the original solid mixture, is present at the sintering temperature. The liquid volume is insufficient to fill the pore space so that additional processes are required to achieve full densification. Liquid-phase sintering is an important method for the industrial fabrication of many ceramics. While the liquid phase can allow enhanced densification at lower temperatures, a difficulty, particularly in the case of ceramics for structural applications, is that the maximum use temperature of the material may be controlled by the softening temperature of the solidified liquid phase.

Vitrification: for a mixture where a relatively large volume of liquid (typically greater than ~25% of the original solid volume) is formed on heating, the liquid volume is sufficient to fill the volume of the remaining pores. Thus, a dense product can be achieved by the formation of the liquid, flow of the liquid into the pores, and either crystallization or vitrification (glass formation) of the liquid on cooling. Vitrification is particularly important for the production of ceramics from naturally available raw materials such as clays.

Viscous sintering: a consolidated mass of glass particles is heated near to, or above its softening temperature. Densification occurs by viscous flow of the glass under the influence of surface tension.

Under some conditions, densification achieved by any of the four categories of sintering might be inadequate. A solution to this problem is the application of an external pressure during heating, giving the method of *pressure sintering* (of which hot pressing and hot isostatic pressing are common examples). Pressure sintering has the disadvantage of increased fabrication costs but is effective when a microstructure of high density and fine grain size must be guaranteed. To distinguish it from pressure sintering, sintering performed without the application of an external pressure is commonly referred to as *conventional sintering*. A list of some ceramic compositions and the way they are sintered is given in Table 4.1.1.

4.1.2 SOLID-STATE SINTERING

4.1.2.1 Driving Forces

It is generally accepted that the reduction in the surface free energy of a sintering compact, due to the elimination of internal surface area associated with the pores, provides the driving force for sintering. When compared to other processes (e.g. chemical reactions), the decrease in the surface free energy during sintering is rather small (~100 J/mol for particles with an initial diameter of 1 μm) but the distance that matter has to be transported is also small (of the order of the particle size), so that sintering occurs at a reasonable rate at sufficiently high temperatures.

TABLE 4.1.1 Selected Ceramic Compositions and the Sintering Process Used for their Densification

Composition	Sintering process	Application
Al_2O_3	Solid-state sintering with MgO additive	Sodium vapor arc lamp tubes
	Liquid-phase sintering with a silicate glass	Furnace tubes; refractories
MgO	Liquid-phase sintering with a silicate glass	Refractories
Si_3N_4	Liquid-phase sintering with oxide additives (e.g. Al_2O_3 and Y_2O_3) under nitrogen gas pressure or under an externally applied pressure	High-temperature structural ceramics
SiC	Solid-state sintering with B and C additives; liquid-phase sintering with Al, B and C or oxide additives	High-temperature structural ceramics
ZnO	Liquid-phase sintering with Bi_2O_3 and other oxide additives	Electrical varistors
$BaTiO_3$	Liquid-phase sintering with TiO_2-rich liquid	Capacitor dielectrics; thermistors
$Pb(Zr,Ti)O_3$ (PZT), $(Pb,La)(Zr,T)O_3$ (PLZT)	Sintering with a lead-rich liquid phase; hot pressing	Piezoelectric actuators and electro-optic devices
$ZrO_2/(3\text{–}10\ \text{mol\%}\ Y_2O_3)$	Solid-state sintering	Electrical conducting oxide for fuel cells
Mn–Zn and Ni–Zn ferrites	Solid-state sintering under a controlled oxygen atmosphere	Soft ferrites for magnetic applications
Porcelain	Vitrification	Electrical insulators; tableware
SiO_2 gel	Viscous sintering	Optical devices

The specific energy and curvature of the particle surfaces provide an effective stress on the atoms under the surface. For a curved surface with principal radii of curvature r_1 and r_2, this stress is given by the equation of Young and Laplace:

$$\sigma = \gamma_{sv}\left(\frac{1}{r_1} + \frac{1}{r_2}\right) \tag{1}$$

where γ_{sv} is the specific surface energy. The *diffusion potential*, μ, which drives matter transport is in this case found by equating the mechanical work performed by the stress to the thermodynamic work required for the reduction of the surface free energy. A commonly used relation is

$$\mu = \sigma \Omega \tag{2}$$

where Ω is the atomic or molar volume. The equation for μ is actually more complex for polycrystalline ceramics where the pores are in contact with the grain boundaries [4].[1] For example, in the final stage of sintering where the pores are assumed to be spherical, one expression is [5]:

$$\mu = \Omega \left(\frac{2\gamma_{gb}}{G} + \frac{2\gamma_{sv}}{r} \right) \tag{3}$$

where γ_{gb} is the specific energy of the grain boundary, G is the diameter of the grains, and r is the radius of the pores. According to Eq. 3, the chemical potential consists of two contributions, one attributed to the pores and the other attributed to the boundaries.

A further development is to relate the chemical potential to an externally applied stress. The driving force for sintering is then defined in terms of a *sintering stress*, Σ, which is the *equivalent externally applied stress* that has the same effects on sintering as the curved surfaces of pores and grain boundaries [4]. The formulation of the driving force in terms of a fictitious externally applied stress is advantageous in the analysis of sintering where mechanical stress effects arise (e.g. pressure sintering and constrained sintering). It also provides a conceptual basis for designing experiments to measure the sintering stress (see Section 4.1.2.7).

4.1.2.2 Stages of Sintering

The microstructure of a powder compact, consisting initially of discrete particles, evolves continuously during sintering. However, it is sometimes convenient to divide the process into three idealized stages defined in terms of the microstructure, to force correspondence between simple, established sintering models. The *initial stage* would begin as soon as some degree of atomic mobility is achieved and, during this stage, sharply concave necks form between the individual particles. The amount of densification is small, typically the first 5% of linear shrinkage, and it can be considerably lower if coarsening mechanisms are very active. In the *intermediate stage*, the high curvatures of the initial stage

[1] Cannon, R. M. On the effects of dihedral angle and pressure on the driving forces for pore growth or shrinkage, unpublished work.

have been moderated and the microstructure consists of a three-dimensional interpenetrating network of solid particles and continuous, channel-like pores. This stage is considered valid to ~5–10% porosity and therefore covers most of the densification. Grain growth (coarsening) starts to become significant. As sintering proceeds, the channel-like pores break down into isolated, closed voids, which marks the start of the *final stage*. Grain growth can be more extensive in the final stage and difficulties are commonly encountered in the removal of the last few percent of porosity.

4.1.2.3 Mechanisms of Sintering

Sintering of crystalline materials can occur by several mechanisms (i.e. atomic transport paths and their associated sources and sinks): vapor transport (evaporation/condensation), surface diffusion, lattice (volume) diffusion, grain boundary diffusion, and dislocation motion. Figure 4.1.1 shows a schematic representation of the matter transport paths for two sintering particles. A distinction is commonly made between *densifying* and *non-densifying* mechanisms. Vapor transport, surface diffusion, and lattice diffusion from the particle surfaces to the neck lead to neck growth and coarsening of the particles without densification. Grain boundary diffusion and lattice diffusion from the grain boundary to the neck are the most important densifying mechanisms

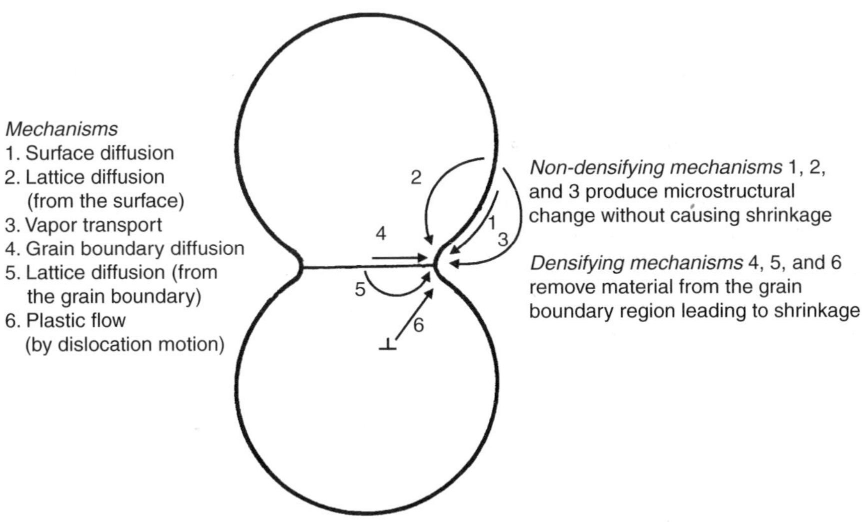

FIGURE 4.1.1 Schematic representation of the sintering mechanisms for a system of two particles.

in polycrystalline ceramics. Diffusion from the grain boundaries to the pores permits neck growth as well as shrinkage (densification). Plastic flow by dislocation motion can cause neck growth and densification through deformation (creep) of the particles in response to the sintering stress. Plastic flow is more common in the sintering of metal powders. For glass powders, which cannot have grain boundaries, densification and neck growth occurs by viscous flow, involving deformation of the particles.

In addition to the alternative mechanisms, there are additional complications arising from the diffusion of the different *ionic species* making up the compound. To preserve charge neutrality of the local composition, the flux of the different ionic species will be coupled [6–8]. A further complication arises because each ionic species can diffuse along different paths. For an oxide with the formula M_xO_y, the effective (*ambipolar*) diffusion coefficient is given by [9]:

$$\bar{D} = \frac{(x+y)[D_l^M + \pi\delta_{gb}D_{gb}^M/G][D_l^O + \pi\delta_{gb}D_{gb}^O/G]}{y[D_l^M + \pi\delta_{gb}D_{gb}^M/G] + x[D_l^O + \pi\delta_{gb}D_{gb}^O/G]} \tag{4}$$

where D is the diffusion coefficient, G is the grain size, δ is the thickness of the grain boundary, the subscripts l and gb refer to lattice and grain boundary diffusion, respectively, and the superscripts M and O refer to the metal and oxygen ions, respectively. According to Eq. 4, whatever the relative magnitudes of the four diffusivities, it is the *slowest diffusing species along its fastest path* that controls the rate. Another complicating factor is that the rate-controlling mechanism for a given material can change with changing conditions of the process variables such as temperature, grain size, and composition [10].

A complicating factor that has been exploited effectively in solid-state sintering is that aliovalent dopants and impurities alter the point defect concentration [11, 12], thereby altering the diffusion coefficients and the sintering rates. For example, the addition of TiO_2 to Al_2O_3 influences the sintering rate in a way that could be explained in terms of the effect of TiO_2 on the concentration of Al vacancies [13]. In general, the effects of dopants on the sintering rates cannot be easily explained because of the multiplicity of the dopant role [14, 15]: for example, influence on more than one diffusion process, segregation at the grain boundaries, influence on surface and grain boundary energies, and the formation of second phases.

4.1.2.4 Competition between Densification and Coarsening

The various sintering mechanisms do not operate independently. Vapor transport and surface diffusion compete with the densifying mechanism. They

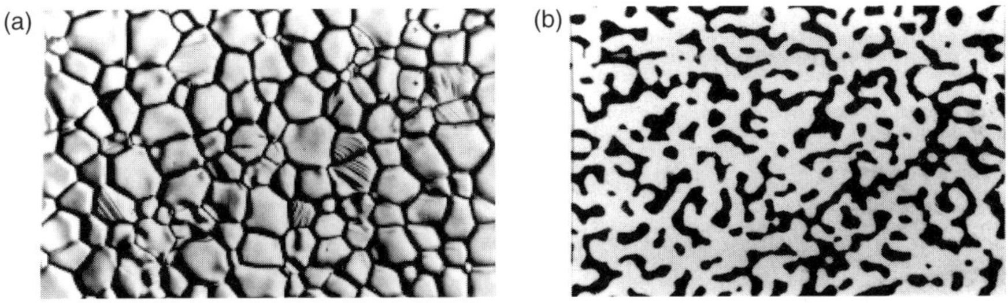

FIGURE 4.1.2 (a) The surface of an Al_2O_3 ceramic from which all porosity has been removed during sintering in which the densifying processes dominate. (b) The sintering of silicon where coarsening mechanisms dominate results in formation of a continuous network of solid material (white) and porosity (black). The microstructural change is not accompanied by any densification. (From Brook, R. J. (1991). *Concise Encyclopedia of Advanced Ceramics*. With kind permission of Elsevier, Oxford.)

lead to coarsening of the microstructure and a reduction of the driving force for sintering, so that a significant reduction in the densification rate can result. Sintering is, therefore, said to involve a competition between densification and coarsening. The production of ceramics with high density (Fig. 4.1.2a) would require choosing the sintering conditions so that the non-densifying mechanisms are not very active (see Section 4.1.7). When coarsening mechanisms dominate, the production of a highly porous body is favored (Fig. 4.1.2b).

4.1.2.5 Effects of Grain Boundaries

In the sintering of polycrystalline materials, part of the energy decrease due to elimination of internal surface area associated with the pores goes into creating new grain boundary area. The grains also have a tendency to grow, by migration of the boundaries, to reduce the energy associated with the grain boundaries, thereby leading to an increase in the diffusion distance. The presence of the grain boundaries also dictates the equilibrium shape of the pores at the intersection with the boundaries. At equilibrium, the chemical potential of the atoms in the pore surface must be the same everywhere, which is equivalent to saying that the curvature of the pore surface is the same everywhere. This means that the pore surface must consist of circular arcs in two-dimensional models, and of spherical caps in a very limited number of three-dimensional geometries, for isotropic solids. There must also be a balance of forces at the junction between the grain boundary and the pore surface. Ignoring possible torque terms, as

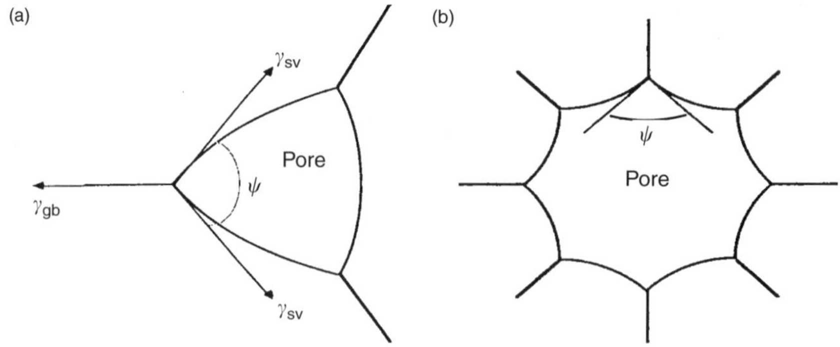

FIGURE 4.1.3 Pore shape and pore stability are determined by the dihedral angle and the pore coordination number: (a) The pore with the concave surfaces will shrink while (b) the pore with the convex surfaces will grow (or become metastable).

shown in Figure 4.1.3a this balance of forces leads to

$$\cos\left(\frac{\psi}{2}\right) = \frac{\gamma_{gb}}{2\gamma_{sv}} \tag{5}$$

where ψ is the dihedral angle, γ_{gb} is the specific grain boundary energy, and γ_{sv} is the specific surface energy.

The pore surface in Figure 4.1.3a is concave (negative curvature) and during sintering, the surface will move towards its center of curvature, so that the pore will shrink. However, as shown in Figure 4.1.3b, when the pore is surrounded by a large number of grains (i.e. the *pore coordination number* is large), the balance of forces at the grain boundary/pore junction dictates that the pore surface will become convex (positive curvature). In this case the pore will grow or become metastable (R. M. Cannon, unpublished work). Whether or not pores can shrink in a powder compact has been analyzed in thermodynamic terms by Kingery and François [16]. The balance between the reduction in pore surface area and the increase in the grain boundary area leads to a criterion that prescribes the maximum pore coordination number that will permit a pore to shrink. As shown in Figure 4.1.4, for a given dihedral angle (e.g. 120°), pores with a coordination number less than a certain critical value, ($N_c = 12$), will shrink while pores with $N > N_c$ will grow. The large pores in a powder compact will have large values of N and will therefore be difficult to shrink, leading to residual porosity in the sintered material. The importance of controlling particle packing for the achievement of high density is therefore clear.

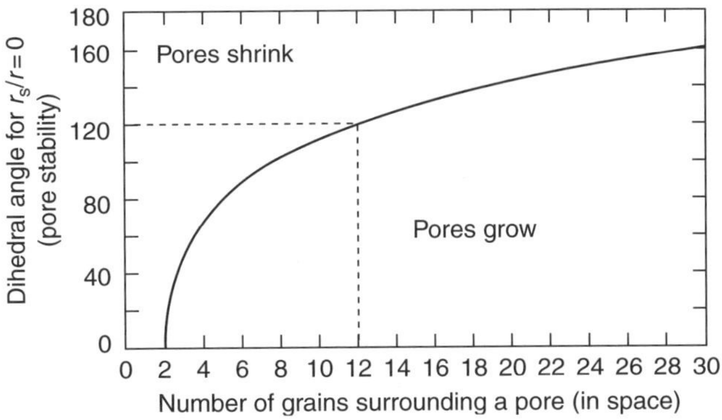

FIGURE 4.1.4 Conditions for pore stability as a function of pore coordination number and dihedral angle. For a dihedral angle of 120°, pores surrounded by more than 12 grains will grow (from Ref. [16]).

4.1.2.6 Sintering Models

Matter transport during sintering is driven by chemical potential gradients arising from differences in the curvature of the free surfaces in the compact of particles. When compared to atoms under a flat surface at the same temperature and composition, the atoms under a convex surface have a higher chemical potential while the atoms under a concave surface have a lower chemical potential. This gradient in the chemical potential produces a driving force for matter transport from the (convex) particle surfaces and from the grain boundaries between the particles to the (concave) necks between the particles (Figure 4.1.1). Mathematical models have been developed to analyze the rate of matter transport along these paths. The models can be classified into three types: (a) analytical models; (b) scaling laws; and (c) numerical models.

4.1.2.6.1 Analytical Models

The analytical models assume a simplified, highly idealized geometry for each stage of sintering and, for each mechanism, the diffusional flux equations are solved analytically to provide equations for the sintering kinetics. They have often been criticized for their simplified and generalized assumptions about the microstructure of the sintering compact (e.g. monodisperse particles, uniform packing, generalized grain and pore shapes, absence of grain growth). However, they are useful for providing an excellent *qualitative* understanding of sintering in terms of the driving force, the mechanisms and the influence of primary variables (e.g. particle size, temperature, applied pressure, heating rate, atmosphere).

For the two-sphere geometry assumed for the initial stage (Fig. 4.1.1), equations for neck growth (and shrinkage for the densifying mechanisms) have been derived for each mechanism and are summarized elsewhere [2, 17, 18]. The initial-stage equations played a key role in the early development of sintering theory but their usefulness in practical sintering is quite limited. For the intermediate and final stages of sintering, Coble [19–21] assumed the grain shape to be a tetrakaidecahedron and derived equations for conventional sintering by lattice and grain boundary diffusion, and for sintering with an applied pressure. The most useful feature of the sintering equations is the predicted dependence of the densification rate on the primary processing variables. For any stage of the sintering process, the model predictions for the densification rate, $\dot{\varepsilon}_\rho = (1/\rho)\,\mathrm{d}\rho/\mathrm{d}t$, can be expressed in general form [1, 2]:

$$\dot{\varepsilon}_\rho = \left(\frac{1}{\rho}\right)\frac{\mathrm{d}\rho}{\mathrm{d}t} = \frac{HD\Omega\phi^{(m+1)/2}}{G^m kT}(p_\mathrm{a} + \Sigma) \qquad (6)$$

where H is a numerical constant that depends on the assumed geometry of the model and on the sintering mechanism, Ω is the atomic volume of the rate controlling species, G is the grain (or particle) size, k is the Boltzmann constant, and T is the absolute temperature. The diffusion coefficient D and the grain size exponent, m, take on their appropriate values for a given mechanism: $D = D_\mathrm{l}$ and $m = 2$ for lattice diffusion, where D_l is the lattice diffusion coefficient; $D = D_\mathrm{gb}\delta_\mathrm{gb}$ and $m = 3$ for grain boundary diffusion, where δ_gb is the grain boundary thickness. The term Σ is the sintering stress, defined earlier as the *equivalent externally applied stress* that has the same effects on sintering as the curved surfaces of pores and grain boundaries. The term ϕ, referred to as the *stress intensification factor*, is a geometrical parameter that relates the externally applied stress to the actual stress on the grain boundaries (see Sections 4.1.2.7 and 4.1.2.8). For diffusional mass transport, Eq. 6 predicts that for an equivalent microstructure, the densification rate increases almost exponentially with T (due to the strong Arrhenius dependence of D on T), increases linearly with the applied pressure, p_a, and has a strong dependence on G. For reasonably homogeneous particle packing and under conditions when grain growth is limited (e.g. during hot pressing under a significant stress), these predictions have been verified by experiments. In conventional sintering, while the benefits of small particle size are significant, the measured densification rates are considerably lower than the predicted values because of inhomogeneous particle packing and the inevitable coarsening processes (see Section 4.1.7.3).

4.1.2.6.2 Herring's Scaling Law

Herrings scaling law [22] does not assume a specific geometrical model. Instead, it assumes that the geometrical changes remain similar. The law considers the effect of change of scale (e.g. particle size) on the rate of matter transport for individual mechanisms. The times for equivalent geometric change can be expressed as:

$$\frac{t_2}{t_1} = \left(\frac{G_{2,0}}{G_{1,0}}\right)^m \tag{7}$$

where the subscripts 1 and 2 represent two different powders with initial particle size $G_{1,0}$ and $G_{2,0}$, respectively and m is an integer that depends on the mechanism of matter transport: $m = 4$ for surface diffusion and for grain boundary diffusion, $m = 2$ for vapor transport, and $m = 3$ for lattice diffusion. For an equivalent geometrical change, the law states that the rate of sintering varies as $1/G^m$. Since the densification rate varies as $1/G^4$ for grain boundary diffusion and as $1/G^3$ for lattice diffusion, the grain boundary path is favored for fine particles. The rate of coarsening by surface diffusion also varies as $1/G^4$ so that grain boundary diffusion and surface diffusion are expected to be in competition during the sintering of fine particles. Because of the general approach used in its derivation, the scaling law might be expected to have some advantage over the analytical models that assume a specific geometry. A problem is that the requirement of geometrically similar microstructural change is not easy to achieve for real powder compacts and the law has not found any significant applicability in sintering [23].

4.1.2.6.3 Numerical Modeling

In practice, more than one mechanism operates during sintering. While sintering with *concurrent mechanisms* has been treated by the analytical models [24], the problem is better solved by numerical simulations that allow the use of more realistic geometries [25, 26]. In particular, finite-element modeling has been used very effectively in the analysis of a variety of sintering problems [27–29]. The calculations can be fairly complex, which may be a reason why the approach has not yet achieved much popularity.

4.1.2.6.4 Sintering Maps

Sintering maps have been constructed to show the dominant mechanisms as functions of the temperature and density [30, 31]. The construction employs the sintering equations derived in the analytical models and data for the material parameters in the equations. Because of the inadequacy of the database and the drastic simplifications of the models, the applicability of the maps is limited.

Nevertheless, they have proved useful in visualizing conceptual relationships between the various mechanisms and changes in the sintering behavior under different sintering conditions.

4.1.2.7 The Sintering Stress

The formulation of the driving force for sintering, the *sintering stress*, in terms of an equivalent external stress, as outlined earlier, provides a conceptual basis for designing experiments to measure the sintering stress. Gregg and Rhines [32] used a *zero creep* experiment in which the externally applied stress required to counteract the shrinkage of the compact due to sintering was determined. De Jonghe and coworkers [33–37] developed the technique of loading dilatometry, in which a small, controlled uniaxial stress is applied to the sintering compact. Measurement of the time-dependent axial and radial strains allows the determination of the volumetric (densification) strain rate and the creep strain rate. Venkatachari and Raj [38] used a sinter-forging technique in which considerably higher uniaxial stresses were applied.

In the loading dilatometry experiments, the hydrostatic component of the applied uniaxial stress is small compared to the sintering stress, so that the sintering mechanism of the loaded powder compact is expected to be identical to that for a freely sintered compact. Equation 6 can be written as [1, 2]

$$\dot{\varepsilon}_\rho = \frac{HD\Omega\phi^{(m+1)/2}}{G^m kT}\Sigma \tag{8}$$

Assuming the same mechanism for creep under the low uniaxial stress, p_z, an equation for the creep rate is

$$\dot{\varepsilon}_c = \frac{H'D\Omega\phi^{(m+1)/2}}{G^m kT}p_z \tag{9}$$

where H' is a numerical constant. The ratio of the densification rate to the creep rate can be found from Eqs 8 and 9 as

$$\frac{\dot{\varepsilon}_\rho}{\dot{\varepsilon}_c} = \frac{F\Sigma}{p_z} \tag{10}$$

where F is a constant. Measurement of the densification rate and the creep rate at a fixed value of p_z for identical samples indicates that the ratio $\dot{\varepsilon}_\rho/\dot{\varepsilon}_c$ is fairly constant over a wide density range for several materials (Fig. 4.1.5).

Making reasonable assumptions for the value of F, the value of Σ can be determined from Eq. 6 using the measured values of $\dot{\varepsilon}_\rho/\dot{\varepsilon}_c$ and p_z [34]. The sintering stress determined in this way is shown in Figure 4.1.6 for CdO. It is of

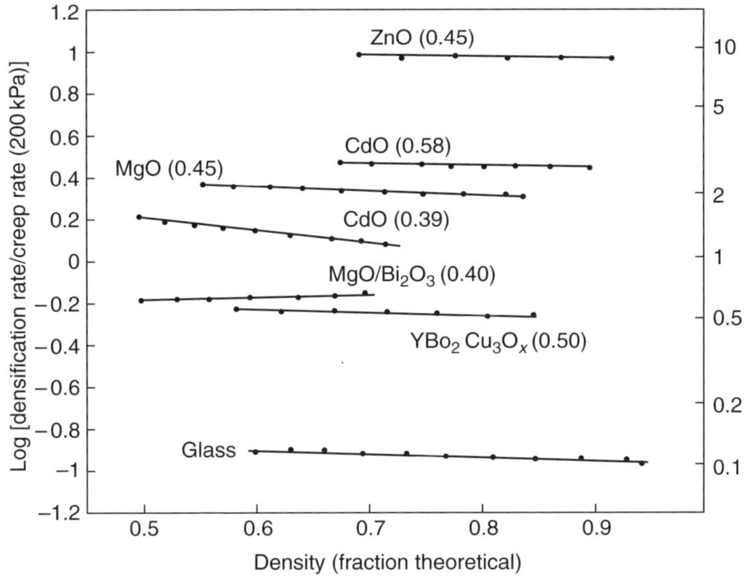

FIGURE 4.1.5 The ratio of the densification rate to the creep rate at a constant stress (0.2 MPa) for various ceramics and for glass. The numbers in parentheses are the initial densities (from Ref. [36]).

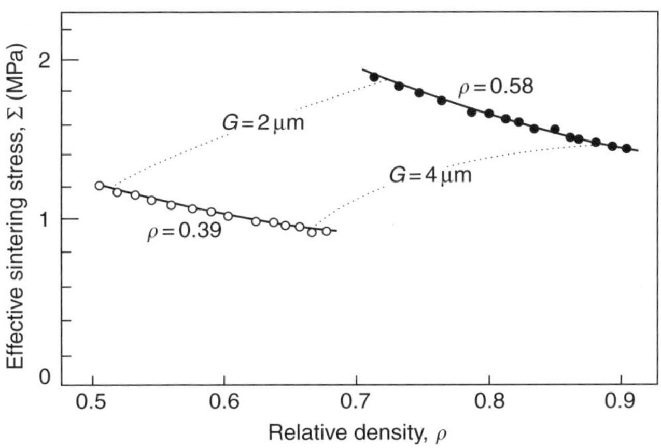

FIGURE 4.1.6 Sintering stress versus relative density for CdO powder compacts of two different initial densities ρ (from Ref. [34]).

the order of 1–2 MPa and, for a given green density, appears to be relatively constant over a large part of the densification process. Based on the sintering model of Coble [21], it might be expected that Σ would increase with densification after the initial stage of sintering. The observed decrease may have its origins in processes such as differential densification and coarsening that are not taken into account in the models. In fact, the sintering stress of Cu powder compacts measured by Gregg and Rhines [32] increased with densification up to ~95% of the theoretical density, after which it decreased due presumably to coarsening [39]. A decrease in Σ with initial density, seen in Figure 4.1.6, has also been observed for ZnO [37]. At a given density, Σ is also expected to vary inversely as the particle (or grain) size but this has not been studied in experiments of the type described here.

4.1.2.8 The Stress Intensification Factor

Solid-state sintering involves atomic transport from the grain boundaries to fill up the pores and is dependent on the *effective stress* on the grain boundaries. We therefore require a way to relate the externally applied stress, σ_a, to the effective stress on the grain boundary, σ_e. This has been commonly achieved (see Eq. 6) by using a parameter, ϕ, referred to as the *stress intensification factor*, that depends on the geometry of the porous compact:

$$\sigma_e = \sigma_a \phi \tag{11}$$

The stress intensification factor is equal to the ratio of the total cross-sectional area (of the solid phase plus the pores), A_t, to the effective cross-sectional area (of the solid-phase only), A_e. For a polycrystalline material, ϕ is equal to the cross-sectional area of the grain, A_g, divided by the grain boundary area, A_{gb}:

$$\phi = \frac{A_t}{A_e} = \frac{A_g}{A_{gb}} \tag{12}$$

If the pores are spherical and randomly distributed through the porous compact, such as a glass in the final stage of sintering, ϕ takes a simple form [21]:

$$\phi = 1/\rho \tag{13}$$

where ρ is the relative density of the compact. In general, the relationship may be expected to be complex. However, computer calculations by Beere [40] on the equilibrium shapes of pores in a porous compact yield results that can be well fitted to simple analytical expressions. One expression is [41]

$$\phi = \exp[a(1 - \rho)] \tag{14}$$

where a is a constant that depends on the dihedral angle of the pores (Fig. 4.1.2). Data for the creep rates of some porous compacts measured in the loading dilatometry experiments described earlier [34, 35] yield values for ϕ that can be well fitted by Eq. 14. An example of such data is shown in Figure 4.1.7 for CdO.

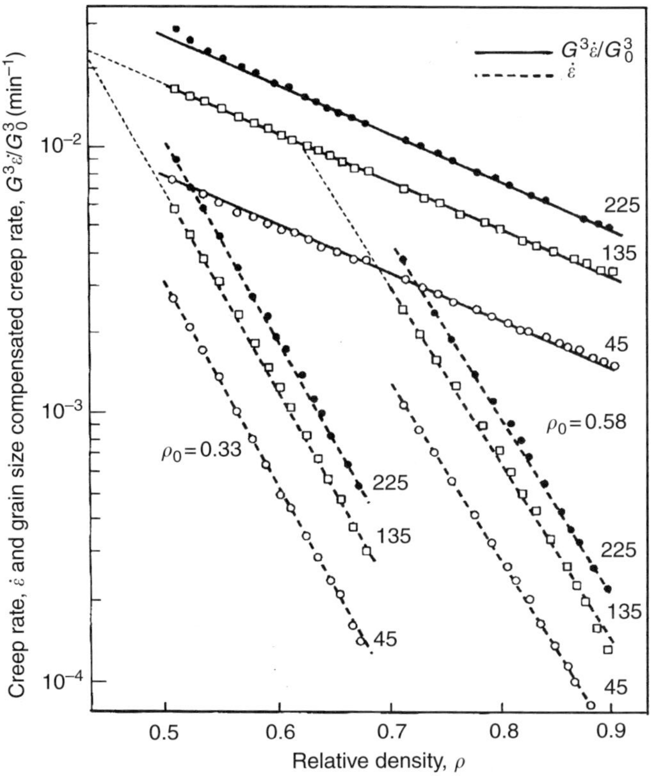

FIGURE 4.1.7 Dependence of the creep rate and the grain size compensated creep rate of CdO powder compacts on relative density for compacts with initial densities of 0.39 and 0.58 subjected to uniaxial stress shown in kPa (from Ref. [34]).

4.1.3 GRAIN GROWTH AND COARSENING

Grain growth is the term used to describe the increase in the average grain size of a polycrystalline material. The grain boundary is a region with a complex structure, about 1–2 unit cells wide, between two crystalline domains. It has a higher energy than the bulk crystal and, hence, a reduction in the grain boundary area will reduce the free energy of the system. The grain boundary moves by diffusion of atoms (ions) from one side of the boundary to the other so that atoms previously aligned with one grain becomes aligned with the other, causing it to grow at the expense of its neighbor. Growth occurs in such a way that the convex grain loses atoms while the concave grain gains atoms with the result that the boundary moves towards its center of curvature.

The term *coarsening* is frequently used to describe the process in porous ceramics whereby the increase in the average grain size is accompanied by an increase in the average pore size. Coarsening, as we have observed, reduces the driving force for sintering and increases the diffusion distance for matter transport, thereby reducing the rate of sintering. The suppression of coarsening mechanisms therefore forms a key requirement for the achievement of high density. Another requirement is that the microstructure be stabilized such that the pores and the grain boundaries remain attached. This second requirement depends primarily on the ability to reduce the intrinsic (pore-free) mobility of the grain boundaries. An understanding of grain growth in dense ceramics therefore forms a key step towards controlling grain growth during the sintering of powder compacts.

Coarsening of precipitates or particles in a solid, liquid, or gaseous medium is commonly referred to as *Ostwald ripening*. It is especially important in liquid-phase sintering where matter transport driven by chemical potential differences causes the smaller grains to become smaller and the larger particles to grow, resulting in an increase in the average grain size. Many features of coarsening (pore growth and grain growth) in solid-state sintering are also shared by the Ostwald ripening process.

Grain growth in ceramics is generally divided into two types: (i) *normal* grain growth and (ii) *abnormal* grain growth, which is sometimes referred to as *exaggerated grain growth* or *discontinuous grain growth*. In normal grain growth, the average grain size increases with time but the grain size distribution remains self-similar (invariant in time). Abnormal grain growth is the process whereby a few large grains grow rapidly at the expense of the smaller grains, giving a bimodal grain size distribution. *Anisotropic grain growth* is a type of abnormal grain growth in which the abnormal grains grow in an elongated manner, commonly with faceted, straight sides. In porous ceramics, abnormal grain growth is accompanied by breakaway of the boundaries from the pores and, as outlined earlier, must be avoided if high densities are to be achieved.

4.1.3.1 Normal Grain Growth

Normal grain growth in pure, dense, single-phase materials has been analyzed by a number of different approaches [42]. In one of the earliest models, Burke and Turnbull [43] analyzed the diffusion of atoms across an isolated boundary under the driving force of the pressure difference across the curved boundary. *Mean field theories* consider the change in size of an isolated grain embedded in a matrix that represent the average effect of the while array of grains [44–47]. As discussed by Smith [48, 49] during grain growth in real systems, certain topological requirements of a space-filling array of grains must be balanced with the requirements of interfacial tension. These topological requirements are ignored in the model of Burke and Turnbull and in the mean field theories but are considered in the model of Rhines and Craig [50]. More recently, the use of computer simulations has provided a valuable technique for the analysis of grain growth. The simulations of Srolovitz *et al.* [51, 52], employing a Monte Carlo method that allows topological requirements to be taken into account, show a remarkable ability to provide realistic pictures of grain growth and to provide a good fit to some experimental data (Fig. 4.1.8).

The grain growth models predict a kinetic equation of the form

$$G^m = G_0^m + Kt \tag{15}$$

where G is the grain size at time t, G_0 is the grain size at $t = 0$, and K is a temperature-dependent rate constant obeying the Arrhenius equation. For the Burke and Turnbull model and for the mean field theories, $m = 2$ (i.e. parabolic growth kinetics) while the topological model of Rhines and Craig predicts $m = 3$. A non-integral value of $m = 2.44$ has been obtained in the computer simulation depicted in Figure 4.1.8. In practice, grain growth data for dense ceramics yield m values ranging from 2 to 4.

In the model of Burke and Turnbull, the rate constant K for a pure, pore-free material is related to the *intrinsic* mobility of the boundary, M_b, by [2]

$$K = 2\alpha M_b \gamma_{gb} \tag{16}$$

where α is a numerical constant that depends on the shape of the grains and γ_{gb} is the grain boundary energy. The mobility is related to the diffusion coefficient for the atoms across the boundary, D_a, by

$$M_b = \frac{D_a}{kT}\left(\frac{\Omega}{\delta_{gb}}\right) \tag{17}$$

where k is the Boltzmann constant, T is the absolute temperature, Ω is the atomic volume, and δ_{gb} is the grain boundary thickness.

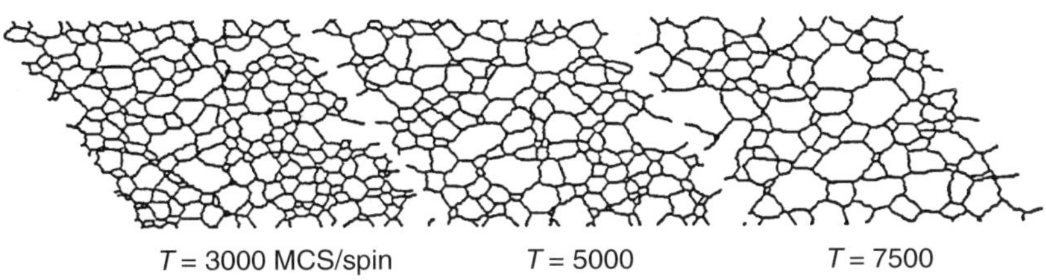

FIGURE 4.1.8 Grain growth in a dense polycrystalline solid determined by computer simulation employing a Monte Carlo method (from Ref. [51], with kind permission of Elsevier, Oxford).

The grain growth models based on the evolution of a single, identified grain do not consider the evolution of the *distribution* of grain sizes with time. As a first approximation, the grain growth rate depends on the difference between the grain size at the mode and at the mean of the distribution. Practically, this means that the rate of grain growth will be higher for a highly skewed distribution (i.e. a distribution with a wide distribution of grain sizes in which the mean grain size is significantly greater than the mode). The rate of grain growth should also be zero for a symmetrical distribution, although it would become broader and shallower with time. A monosize or narrow initial distribution, as may be produced from particles with the same size, will first broaden without an increase in the average grain size. After some incubation time, grain growth takes off, with the average grain size increasing with time.

4.1.3.2 Abnormal Grain Growth

Microstructures of polycrystalline ceramics that have been heated for some time at a sufficiently high temperature often show very large (abnormal) grains in a matrix of finer grains (Fig. 4.1.9). It is important to understand and to be able to control abnormal grain growth for two main reasons [53]. First, the occurrence of abnormal grain growth during sintering limits the attainment of high density. For example, the pores trapped in the abnormal grain shown in Figure 4.1.9 are difficult to remove. Second, the large abnormal grains are commonly detrimental to the properties of the material. An important goal in practical sintering is therefore the suppression of abnormal grain growth. As described later, this is commonly achieved through the use of dopants.

Considerable effort has been devoted to the experimental and theoretical study of abnormal grain growth in ceramics [45, 54–57]. While some insight into the problem has been achieved, the process is not well understood. Earlier explanations considered that the grain size distribution of the starting material was the major factor leading to abnormal grain growth. They were based on the Hillert theory of grain growth [45], which predicted that any grain with a size greater than twice the average grain size would be predisposed to growing abnormally. This explanation is not supported by recent computer simulations [58] and theoretical analysis [59], which show that although the large grains grow, they do not outstrip the normal grains. The normal grains grow at a faster relative rate so that the large (abnormal) grains eventually return to the normal size distribution. The size effect is therefore not a sufficient criterion for abnormal grain growth. Inhomogeneities in chemical composition, liquid phases, and particle packing have long been suggested as possible causes of abnormal grain growth.

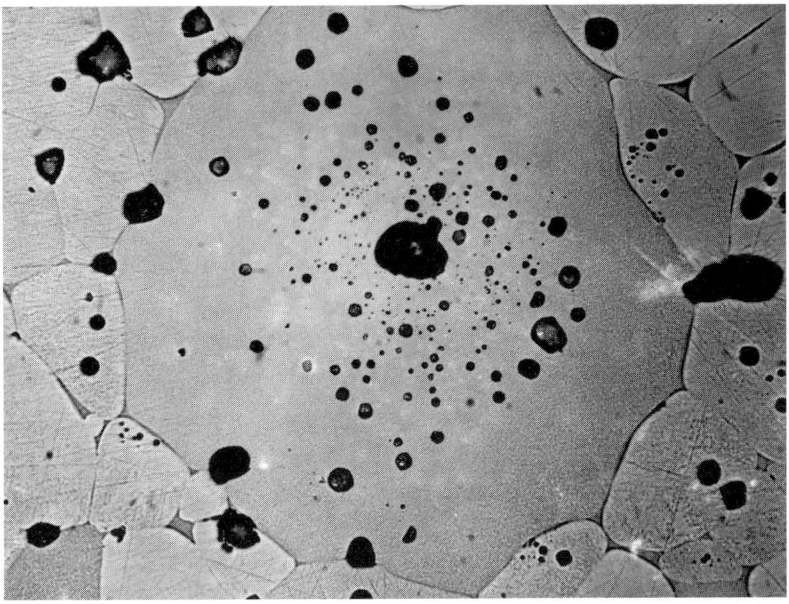

FIGURE 4.1.9 Abnormal grain growth in an MgO ceramic. The pores trapped inside the abnormal grain are difficult to remove, thereby leading to residual porosity in the sintered material (magnification = 700).

Computer simulations and theoretical analysis indicate that true, abnormally rapid grain growth can occur if the boundary of the abnormal grain has a higher mobility or a lower energy than the surrounding matrix grains [60–62]. Grain orientation and grain boundary structure are two factors that can lead to variations in the mobility and energy of grain boundaries. Boundaries of large abnormal grains often show faceting on low-index crystallographic orientations. Frequently, the growth planes are those with low surface energy. Low-energy boundaries often have a lower mobility. Growth that is slow normal to and rapid parallel to the low-energy surfaces often lead to plate-like abnormal grains.

Abnormal (anisotropic) grain growth in Al_2O_3 has received considerable attention, largely based on attempts to understand the role of MgO in suppressing abnormal grain growth in this system [63]. While the single most important role of MgO is the reduction of the boundary mobility [64], it is also recognized that MgO plays an important role in reducing the anisotropies in the surface and grain boundary energies and in the boundary mobilities [65–67]. Doping of Al_2O_3 with TiO_2 has been observed to enhance the formation of elongated, faceted grains [68, 69].

While the suppression of abnormal grain growth forms a key goal in sintering, the ability to exploit abnormal grain growth in a *controlled* manner can

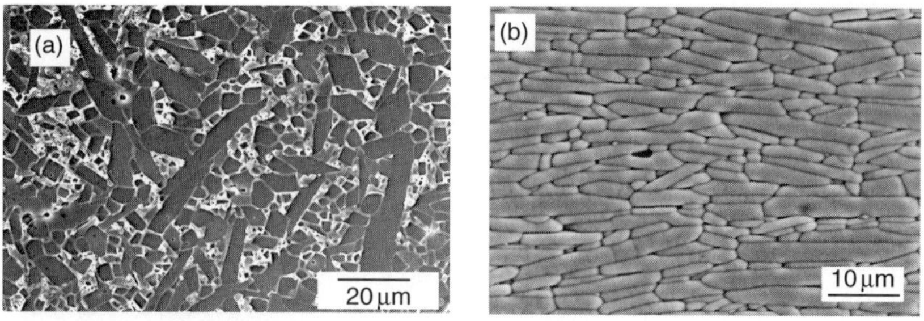

FIGURE 4.1.10 Exploitation of abnormal (anisotropic) grain growth for the production of (a) self-reinforced mullite with enhanced fracture toughness and (b) bismuth titanate ceramic with anisotropic dielectric properties.

provide significant property advantages in several ceramic systems. The growth of a controlled distribution of elongated (abnormal) grains in a matrix of finer grains (Fig. 4.1.10a) has been employed to produce a significant enhancement of the fracture toughness of silicon nitride [70], silicon carbide [71], and mullite [72]. Alignment of the abnormally growing grains (Fig. 4.1.10b) by the process of templated grain growth [73], hot-pressing, or sinter forging has been used to achieve anisotropic dielectric properties in layered-structured ferroelectrics such as bismuth titanate [74, 75]. Abnormal grain growth has been used effectively for many years to produce some single-crystal ferrites by seeding a fine-grained material with a large single crystal. The method has been used more recently to grow barium titanate [76] and lead-magnesium niobate [77] single crystals from polycrystalline materials.

4.1.3.3 Ostwald Ripening

The basic theory of Ostwald ripening was developed by Greenwood [78], Lifshitz and Slyozov [79], and Wagner [80], and is generally referred to as the LSW theory. The theory applies strictly to a dilute dispersion of particles in a solid, liquid, or gaseous medium. The characteristic feature of coarsening is that at any given time, particles smaller than a critical size will dissolve, surrounding themselves with a zone of excess solute that will find its way to particles larger than the critical size. The rate-limiting step can be either diffusion between the particles or the deposition or dissolution of atoms at the particle surfaces (reaction controlled growth). In either case, the theory predicts that at some later time, a *self-similar* (steady-state) particle size distribution is approached asymptotically in which the growth of the critical radius, r^*, follows a simple law of the form given earlier by Eq. 15 for normal grain

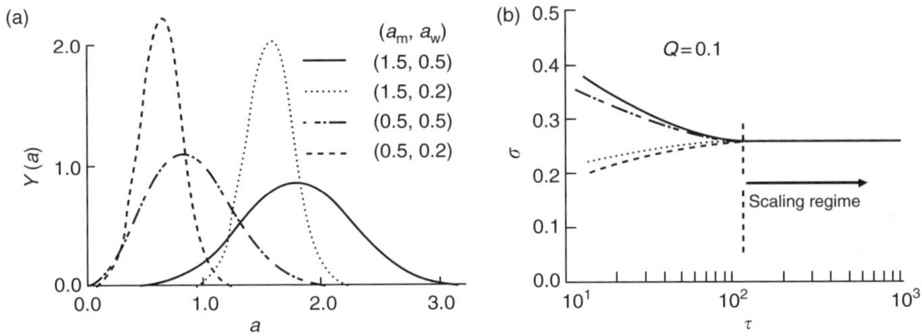

FIGURE 4.1.11 Effect of initial particle size distribution (a) on the time variation of the standard deviations of the distribution and the rate-coarsening rate (b) (after Ref. [84], with kind permission of Elsevier, Oxford).

growth:

$$(r^*)^m = (r_0^*)^m + Kt \tag{18}$$

where K is a temperature-dependent rate constant, t is the time, and m is an exponent that depends on the rate-controlling mechanism: $m = 3$ for diffusion control and $m = 2$ for surface reaction control.

The LSW theory has been the subject of considerable study and revision [81]. A key factor is the interaction between the particles when the dispersion of particles in the medium is no longer dilute. Ardell [82] found that for diffusion control, the exponent $m = 3$ is still valid but the self-similar size distribution approaches that for reaction control (i.e. it is broader and more symmetrical) when the volume fraction of particles is greater than ~0.2. Another important factor is the approach to the steady-state LSW distribution. As shown in Figure 4.1.11, for a modified Gaussian distribution of initial particle sizes with varying maximum value, a_m, and width, a_w, an analysis by Enomoto *et al.* [83, 84] indicate that for distributions with a broad width, the standard deviation of the distribution function decreases to the characteristic time-invariant value, while for systems with a narrow width, the standard deviation increases to the time-invariant value. This behavior during the transient regime indicates that the self-similar state acts as a strong attraction for the evolution of the particle size distribution.

4.1.3.4 Control of Grain Growth

The most effective approach for inhibiting grain growth involves the use of additives (*dopants*) that are incorporated into the powder to form a solid solution [14, 15, 53]. The concentration of the dopant (the *solute*) is often believed to

TABLE 4.1.2 Examples of Dopants Used for Grain Growth Control in Some Common Ceramics

Host	Dopant	Concentration (at %)	Reference
Al_2O_3	MgO	0.025	[63, 93]
$BaTiO_3$	Nb; Co	0.5–1.0	[94, 95]
ZnO	Al	0.02	[96]
CeO_2	Y; Nd; Ca	3–5	[91, 92]
Y_2O_3	Th	5–10	[97]
SiC	(B + C)	0.3B + 0.5C	[98]

be well below its solid solubility limit in the ceramic (the *host*) but this is not clear in some systems. Examples of systems where the dopant approach has been used successfully are given in Table 4.1.2. The effectiveness of the dopant in suppressing grain growth is often found to depend on its ability to segregate at the grain boundaries. Segregation may occur right at the disordered "core" of the grain boundary or in an adjacent space-charge layer [85–89].

The inhibition of grain growth is believed to occur by a mechanism of *solute drag*. In this mechanism, a strong interaction occurs between the segregated solute and the grain boundary so that the solute must be carried along with the moving boundary. Solute diffusion across the boundary is assumed to be slower than that of the host atoms and therefore becomes rate controlling. In the solute drag theory put forward by Cahn [90], the boundary mobility of the doped material is given by

$$M_b' = \frac{D_b \Omega}{4kT\delta_{gb} Q C_0} \tag{19}$$

where D_b is the diffusion coefficient for the solute atoms across the boundary of width δ_{gb}, Ω is the atomic volume of the host atoms, k is the Boltzmann constant, T is the absolute temperature, Q is a partition coefficient for the solute distribution between the boundary and the lattice, and C_0 is the concentration of the solute atoms in the lattice. According to Eq. 19, the boundary mobility is directly proportional to the diffusivity of the solute across the boundary and inversely proportional to the segregated concentration of the solute (QC_0). This indicates that aliovalent solutes with larger ionic radii than the host would be effective for suppressing grain growth. While the influence of trivalent cations on the grain growth of CeO_2 may be consistent with this prediction [91, 92], in general the choice of an effective dopant is complicated because of the multiplicity of the dopant role.

At low driving force or low boundary velocity [99], the mobility of the boundary is limited by solute drag and is given by Eq. 19. With increasing velocity, the boundary breaks away from the solute and has a mobility given by the intrinsic value (Eq. 17). Observations indicate that the boundary motion is not uniform but that it starts and stops as if making transitions between the solute-drag and intrinsic regimes or between different solute drag regimes [100].

Another effective approach for grain growth control, but one that is less widely used, involves the use of fine, inert, precipitates at the grain boundaries. An example is the use of ZrO_2 particles to control the grain growth of Al_2O_3 [101]. The precipitates inhibit grain growth by a pinning mechanism suggested by Zener in a communication to Smith [48]. According to Zener, the matrix reaches a limiting grain size given by

$$G_{\max} = \frac{2\alpha r}{3f} \tag{20}$$

where α is a geometrical constant that depends on the shape of the matrix grains, r is the radius of the precipitates (assumed to be spherical), and f is the volume fraction of the precipitates. Recent work by Stearns and Harmer [102, 103] indicates that the pinning particles are not distributed as randomly as the Zener model assumes. By incorporating the observed fraction of pinning particles at the grain boundaries into the Zener model, a modified relationship was developed to describe the grain growth data of Al_2O_3 containing fine SiC particles.

4.1.3.5 Grain Growth and Pore Evolution in Porous Compacts

In porous powder compacts, grain growth is very limited in the initial stage of sintering but coarsening of the microstructure can occur by surface diffusion and vapor transport. The mechanism most commonly cited for coarsening in this early stage of sintering is that suggested by Greskovitch and Lay [104]. It assumes that the coarsening processes (surface diffusion; vapor transport) lead to rounding of the particle shapes and growth of the necks between the particles, thereby producing the structural changes for boundary migration to be energetically favorable. The grain growth rate increases in the intermediate stage and migration of the boundaries leads to coalescence of the pores so that the average size of the pores also increases. Grain growth is most pronounced in the final stage when the pores pinch off and become isolated. Normal grain growth has been analyzed using an idealized model consisting of a nearly spherical pore on an isolated grain boundary. The kinetics of normal grain growth can be described by Eq. 15 where value of the exponent m (2–4) now depends on mechanism of the pore motion and the extent of the drag produced by the pores on the boundaries [53]. Several mechanisms have the same value of m so that an unambiguous determination of the grain growth mechanism from experimental data is difficult.

An understanding of pore evolution during sintering is important for a broader understanding of the microstructural evolution. However, very few investigations have been devoted to pore evolution due largely to the difficulty of quantitative analysis and the time-consuming nature of detailed stereological characterization. On the basis of stereological observations, Rhines and DeHoff [105] showed that the pore network changes by collapse of the network and the reforming of a new network with lower connectivity. The variation of the open and closed porosity during the sintering of a heterogeneously packed UO_2 powder was studied by Coleman and Beeré [106]. The results showed that the volume of closed porosity started to increase when the open porosity had decreased to 15 vol%. For a homogeneously packed powder, it may be expected that the formation of closed porosity would occur more suddenly and when the volume of open porosity had decreased to a value of less than 10%.

4.1.3.6 Interactions between Pores and Grain Boundaries

The attainment of high density during sintering, as outlined earlier, depends on the ability to stabilize the microstructure such that the pores and the grain boundaries remain attached. For an idealized geometry of an isolated, spherical pore at a grain boundary, typical of an idealized microstructure in the final stage of sintering, the interaction between the pore and the boundary has been used to determine the conditions for pore attachment and breakaway. A migrating boundary will apply a force on any pore situated at the boundary and will try to drag it along [107, 108]. As illustrated in Figure 4.1.12, the force causes the pore to change its shape. The leading surface of the pore becomes less strongly curved than the trailing surface. The difference in curvature causes a flux of matter from the leading surface to the trailing surface. The result is that the pore attempts to move with the boundary.

For each of the mechanisms indicated in Figure 4.1.12, analysis of the atomic flux from the leading surface of the pore to the trailing surface yields a pore mobility, M_p, of the form [53]:

$$M_p = \frac{AD\Omega}{kTr^n} \tag{21}$$

where A is a constant, Ω is the atomic volume, k is the Boltzmann constant, T is the absolute temperature, r is the radius of the pore, D is the appropriate

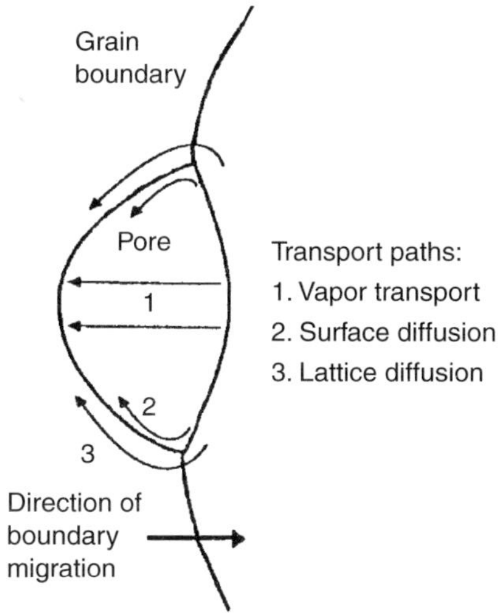

FIGURE 4.1.12 Possible transport paths for a pore moving with a grain boundary.

diffusion coefficient, and m is an exponent that depends on the mechanism: $D = D_s$ and $m = 4$ for surface diffusion; $D = D_l$ and $m = 3$ for lattice diffusion; $D = D_g$ and $m = 3$ for vapor transport. According to Eq. 21, fine pores are highly mobile and are better able to remain attached to the boundary. In practice, strategies for avoiding breakaway of the boundary (abnormal grain growth) are based mainly on suppressing grain growth (reducing M_b) through the use of dopants.

4.1.3.7 Microstructural Maps

Brook [109] developed a method to visualize the interaction between the pore and the boundary during sintering. Two situations can be visualized: (i) the pore remains attached to the migrating boundary (referred to as *attachment*) or (ii) the boundary breaks away, leaving the pore behind (*separation*). The conditions under which each situation becomes important are determined as functions of the grain size and the pore size. The results are represented on a diagram, commonly referred to as a *microstructural map*, of the grain size versus the pore size (Fig. 4.1.13). With the pores attached to the boundary, the mobility of the boundary can be controlled by pore drag (*pore control*) or by the intrinsic boundary parameters (*boundary control*). Pore–boundary separation is predicted to occur only for certain ratios of grain to pore size. Brook also considered

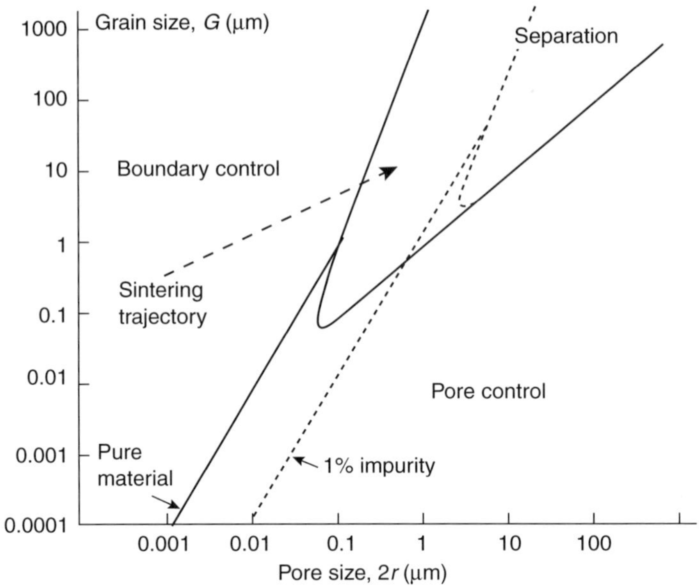

FIGURE 4.1.13 Grain size versus pore size map showing the conditions for pore attachment and pore separation. A trajectory for densification accompanied by coarsening of the microstructure is shown. If the separation region is not intersected, then the attainment of complete densification is possible (from Ref. [109], with kind permission of Elsevier, Oxford).

the effect of a segregated impurity (solute) on the boundary mobility. The mobility limited by solute drag is lower than the intrinsic mobility, allowing larger pores to migrate with the boundary before separation. The separation region is therefore shifted to larger grain and pore sizes. In the sintering of ceramics where the densification is accompanied by coarsening, both the average grain size and the average pore size increase, as shown by the trajectory on Figure 4.1.13.

A modification of the grain size versus pore size map was used by Berry and Harmer [110] to show the sintering trajectories for Al_2O_3 on a *grain size* versus *density* map (Fig. 4.1.14). The influence of dopants and other variables can now be discussed in terms of their effects on the separation region and the sintering trajectory. For the achievement of high density during sintering, the separation region must be avoided. According to Figures 4.1.13 and 4.1.14, the use of a dopant that reduces the boundary mobility has the effect of shifting the separation region to higher grain sizes, thereby making it easier for the trajectory to bypass the separation region. Flattening of the grain size versus density trajectory by enhancing the ratio of the densification rate to the grain growth rate is also effective. This can be achieved by the use of a dopant that lowers the boundary mobility or by selecting a heating schedule that enhances the densifying mechanisms relative to the coarsening mechanisms (see Section 7).

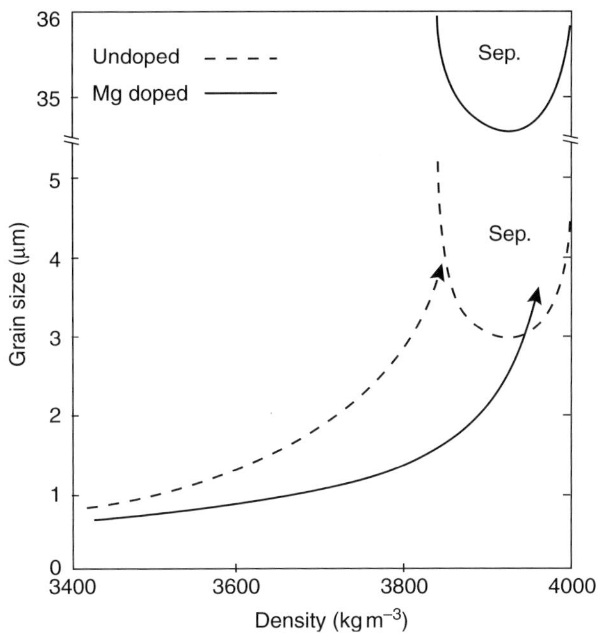

FIGURE 4.1.14 Grain size versus density map for undoped Al_2O_3 and for Al_2O_3 doped with 250 ppm MgO. The depicted trajectory for the doped Al_2O_3 is the result of raising the surface diffusivity by a factor of 4, reducing the lattice diffusion coefficient by a factor of 2, and reducing the grain boundary mobility by a factor of 34 (courtesy of M. P. Harmer).

While the microstructural maps provide a valuable method for representing the interaction between densification and coarsening, the idealized geometry of the assumed microstructure (a spherical pore on an isolated boundary of uniform curvature) means that the maps have limited applicability to actual final-stage microstructure. Furthermore, factors such as grain size distribution, pore size distribution, number of pores, and the dihedral angle will affect the simple relationships between pore size and boundary curvature assumed for separation [108, 111, 112].

4.1.4 LIQUID-PHASE SINTERING

In many sintering processes, the intervention of a liquid phase is commonly sought to assist in the densification. The purpose is usually to enhance densification rates, achieve accelerated grain growth, or to produce specific grain boundary properties. In these materials, the distribution of the liquid and of the

TABLE 4.1.3 Some Liquid-Phase Sintering Systems

System	Reference
Alumina + $Mg_3Si_4O_{10}(OH)_2$ (talc)	[113]
$ZnO + Bi_2O_3$	[114]
$SiC + Y_2O_3 + Al_2O_3$	[115]
Si_3N_4 + MgO, Sialons	[116, 117]
$BaTiO_3 + B_2O_3$	[118]

resulting heterophases after densification is of critical importance to the properties of the sintered body. Typically, the amount of liquid involved is small, a few volume percent, which can make precise control over the composition of the liquid difficult. Examples of theses are silicon ceramics, such as SiC and Si_3N_4, where the native silicon oxide film on the starting powders is only known approximately, while the silica participates in forming the liquid phase together with the intentionally introduced additives. In some instances, such as the densification of alumina, the amount of liquid phase is quite small, and so difficult to detect that numerous studies that were believed to be solid-state sintering actually involved liquid silicate phases. The solidified liquid phase could only be revealed by careful high-resolution transmission electron microscopy. Some examples of common liquid-phase sintering systems are listed in Table 4.1.3.

A related process is *activated sintering*. In this case, minor amounts of additives that segregate strongly to the grain boundaries can enhance grain boundary transport rates significantly, accelerating densification even below any eutectic temperatures in the additive/host system. Examples are found in the sintering of ZnO with Bi_2O_3 additive [119], and Al_2O_3 with CaO [120]. However, in many other cases there is no clear difference in the principles involved in activated sintering [121] and in liquid-phase sintering, other than the activated system involving the addition of less than 1% additive for which the resulting liquid grain boundary film can be difficult to detect. Clearly, the enhanced densification rates achieved by liquid-forming additives are only of interest if the properties of the material remain within useful bounds. For example, liquid-forming additives tend to degrade high-temperature properties such as creep and fatigue resistance, because of persistent intergranular films of nanometer thickness. An example of such a film is shown in Figure 4.1.15.

Liquid-phase sintering is generally regarded as proceeding in a sequence of dominant stages: (a) melting of the liquid-forming additive and redistribution of the liquid; (b) rearrangement of the majority solid phases driven by capillary stress gradients; (c) densification and shape accommodation of the solid phase involving solution-precipitation; (d) final densification driven by residual porosity in the liquid phase (Fig. 4.1.16).

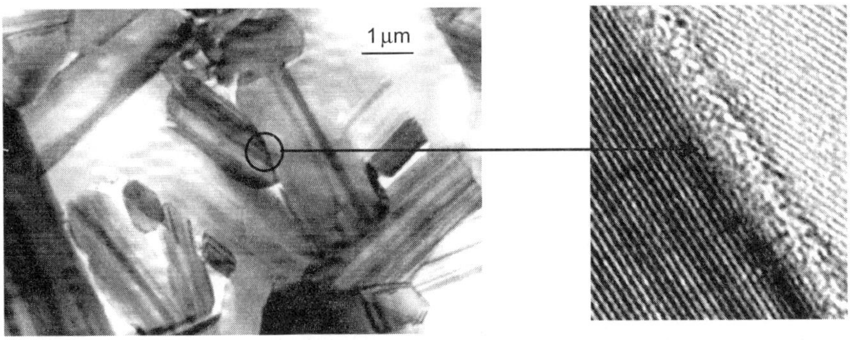

FIGURE 4.1.15 Persistent amorphous grain boundary phase, approximately 1-nm thick, in SiC containing Al, B and C as sintering additives (from Ref. [122]).

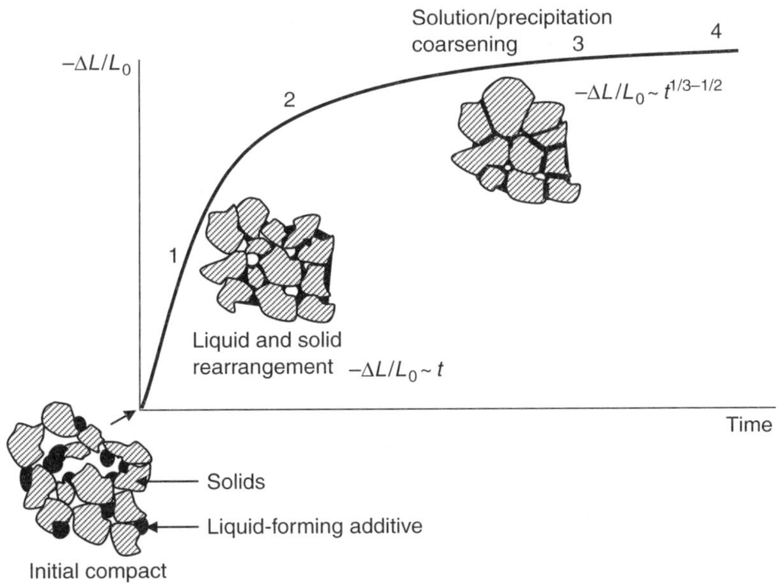

FIGURE 4.1.16 Schematic evolution of powder compact during liquid-phase sintering. The various stages are overlapping significantly.

4.1.4.1 Liquid-Phase Redistribution

Capillary stress gradients can be significant when submicron liquid-phase surface curvatures are present. For a surface energy of about 1 J/m^2 and a radius of curvature of 0.5 μm, pressure differentials within the sample can reach several MPa, sufficient to cause rapid fluid flow and particle rearrangement when the compact's effective viscosity is still relatively low. The pressure gradients will cause fluid to flow from regions with large liquid pores to regions of smaller liquid pores, and as a consequence from the surface to the interior of the compact. This principle leads to a sequential pore filling [123]. However, in the final stages of sintering continued densification of the solid particle skeletal network can eventually lead to partial expulsion of the liquid phase. One difficulty in modeling liquid redistribution is that local liquid surface curvature depends sensitively on local geometry, so quantitative statements can only be made for highly idealized microstructures [124]. Although these are highly idealized rigid particle arrays, such models can provide further insight in the process of liquid redistribution.

Rearrangement is typically fast, but kinetics are only of interest to the extent that, together with grain growth, they are an integral part of microstructure development. Once the rearrangement stage is over, models that do not incorporate simultaneous grain growth lead to shrinkage rates proportional to time$^{1/3}$ when diffusion through the liquid is rate controlling or time$^{1/2}$ when liquid/solid phase boundary reactions are rate controlling. These models further assume that the grain boundary liquid phase does not change composition or structure. This is not always the case. There are now several examples where the composition of the grain boundary film is found to be different from that of the liquid in the adjacent triple junction or even from boundary to boundary [125, 126], as shown in Figure 4.1.17. It is quite likely that such is more the rule than the exception.

A second-stage rearrangement has also been recognized. This refers to the instance in which the liquid phase is thought to penetrate the grain boundaries of dense, multigrain particles. A minimum requirement for partial or complete grain boundary penetration is that the dihedral angle is near zero (see Figure 4.1.18).

The dihedral angle, ψ, is related to the solid/liquid interfacial energy, γ_{sl} and the grain boundary energy γ_{gb} by the vector force balance, similar to Eq. 5:

$$2\gamma_{sl}\cos(\psi/2) = \gamma_{gb} \tag{22}$$

In the event that sufficient liquid is present and interfacial energies are fairly uniform, this may lead to a microstructure in which individual, rounded grains are dispersed in the liquid. Such a microstructure is commonly seen in metallic systems, as shown in Figure 4.1.19, and in some ceramic systems such as MgO/silicate [127]. In many ceramic systems, angular or ideomorphic grain shapes are more commonly found as a result of liquid-phase sintering.

While liquid redistribution is relatively easy, it is remains desirable to start with a compact that is as homogeneous as practical, for example, by using powders in which the particles are pre-coated with the liquid-forming phase: the

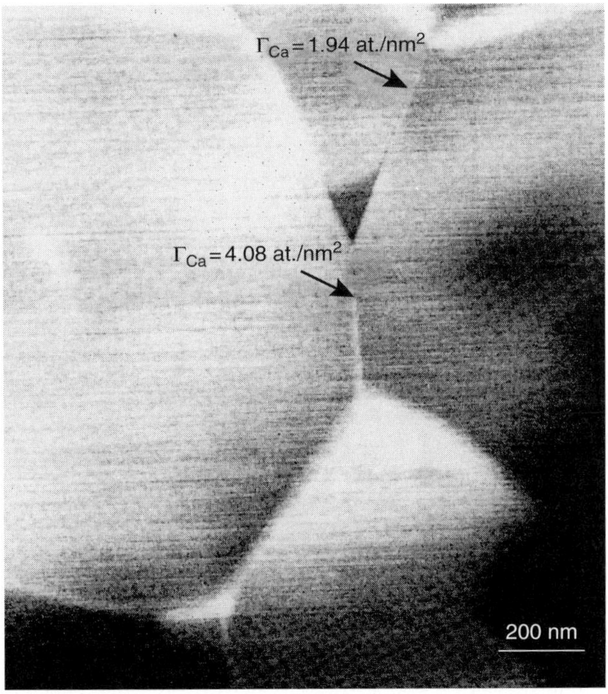

FIGURE 4.1.17 STEM image showing two neighboring grain boundary films with remarkably different calcium excess in calcia-doped silicon nitride (from Ref. [125]).

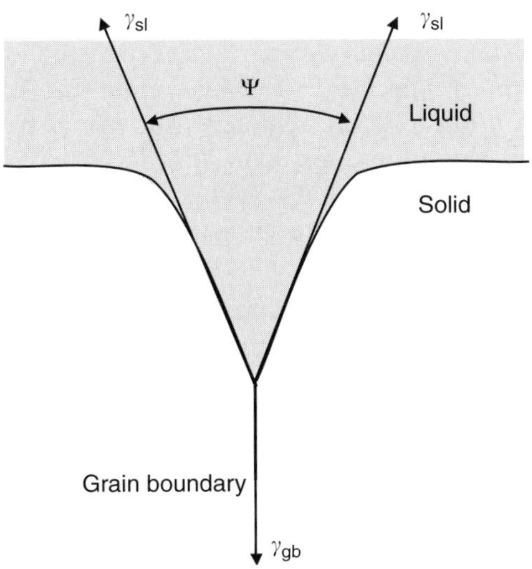

FIGURE 4.1.18 Dihedral angle for liquid at a grain boundary.

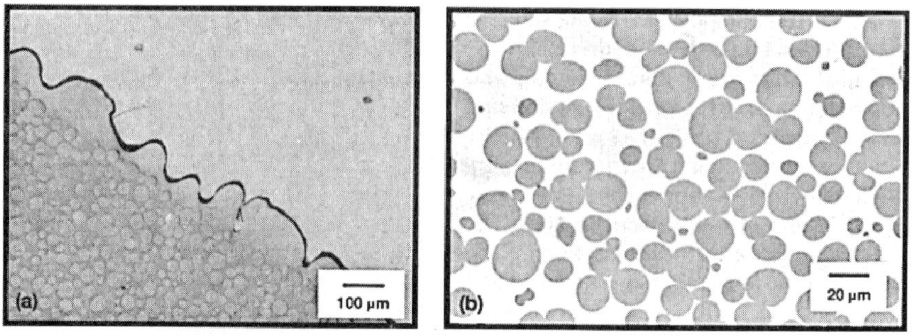

FIGURE 4.1.19 Micrographs of 78W–15.4Ni–6.6Fe after liquid-phase sintering for 1 min in microgravity at 1507°C: (a) low magnification view of sample edge and wavy solidified liquid; (b) high magnification view of sample interior showing solid grain agglomeration (courtesy R. M. German).

same concerns about homogeneity of starting compact applies to liquid-phase sintering as it does to solid-state sintering. Methods whereby powders can be coated include fluidized bed vapor deposition, and chemical precipitation from aqueous suspensions [128].

4.1.4.2 Effects of Gravity

The relatively low effective viscosity of powder compacts containing a liquid phase leads to ready creep, and the weight of the sample itself can cause significant deformation. Judicious support for large pieces is thus necessary, and has been longtime practice [129]. Additional problems are the possibility of liquid redistribution. This is most likely for liquids such as molten metals, or when a significant amount of liquid is present (e.g. more than 5 vol%) resulting in possible particle settling or liquid drainage. The effects have been examined under microgravity conditions. An interesting result of these studies is that solids suspended in a liquid still tend to agglomerate, as shown in Figure 4.1.19. Since this is a metallic system, surface charge effects, as in colloidal suspensions, cannot be active. It is possible that the metal particles are propelled, colliding and joining, as a result of uneven dissolution/precipitation rates along their surfaces. It is possible that liquid films that may be detected by high-resolution transmission electron microscopy remain at the contact areas.

4.1.4.3 Grain Boundary Films

Initially, when there is a significant amount of liquid between the grains, the thickness of the grain boundary is kinetically determined. The thinning rate,

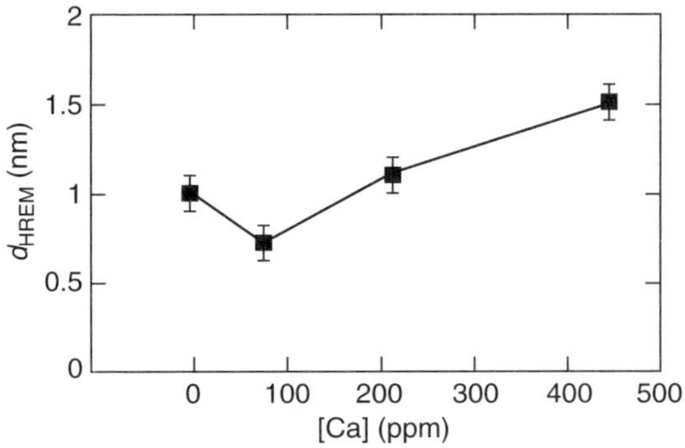

FIGURE 4.1.20 Equilibrium grain boundary thickness measured by HREM (d_{HREM}) plotted against bulk calcium doping level in silicon nitride (from Ref. [133]).

a consequence of capillary viscous flow, of the liquid film at a grain boundary of fixed cross-section has been calculated by Lange [130] as proportional to d^5, where d is the instantaneous film thickness. This predicts an infinite time for the film thickness to reach zero. However, when the film gets very thin, that is, in the nanometer range, effects other than viscous flow become dominant, including charge interactions between surfaces [131], as well as structural and chemical forces [132]. Compositional effects on what appears to be an equilibrium grain boundary film thickness have been the subject of careful investigation. Some of the results are shown in Figure 4.1.20.

4.1.4.4 Effective Stress

Concepts that have been developed in soil mechanics [134, 135] can also be applied to liquid-phase sintering. Considering a cross-section that runs through the particle–particle contacts, the particle network, or skeleton, will carry an effective stress given by

$$\sigma_y^{\text{eff}} = \left(\sum F_{iy}\right)/a \tag{23}$$

where F_{iy} is the component of the force in the y direction at the ith particle–particle contact, and a is the external cross-section of the sample. In the absence of any externally applied stress and pressure gradients, the effective stress is then

$$\sigma_y^{\text{eff}} = p(1-a) \tag{24}$$

where p is the capillary pressure in the liquid phase.

When a significant amount of liquid is present between the particles, the effective stress will be less than the equilibrium one, because the effective stress causes liquid to flow out of the contact areas. Once the liquid films become very thin, kinetic or other factors [136] obstruct fluid flow, and the effective stress will approach the equilibrium one. This effective stress is the driving force for solution precipitation mechanisms, together with Ostwald ripening processes. The effective stress concept is similar to the sintering stress multiplied by the stress concentration factor discussed earlier.

The capillary pressure, p, follows from the surface curvature of the liquid phase, which is determined by the particle size, the solids volume, and the liquid volume. In simple, idealized geometries the equilibrium curvature can be readily computed from geometrical considerations. Additionally, computational programs are now available to calculate such curvatures for a number of more complex geometries. In practice, the capillary pressure is not readily determined because of the distribution of particle sizes, complex packing geometries, and distribution of pore sizes.

In the first stage of consolidation, densification is rapid due to rearrangement of the particle skeleton, rapid dissolution of small network particles, and removal of liquid between network particles. Rearrangement of the initial particle network may occur in as little as a few minutes. Computational approaches to rearrangement have recently been made [137]. Once a quasi-steady-state grain boundary film is established, densification proceeds similar to solid-state densification at a comparatively enhanced rate. Rapid transport in the liquid phase makes grain growth more prominent in liquid-phase sintering. Inclusion of a dispersion of inert particles can assist in limiting grain growth.

4.1.4.5 Role of the Wetting Angle

The wetting angle, θ, for the liquid/solid/vapor system depends on the various interfacial energies, and is usually referred to a droplet of liquid on a flat solid surface. While this does not represent the full equilibrium, it is often cited as such, and is given by

$$\cos\theta = (\gamma_{sv} - \gamma_{sl})/\gamma_{lv} \tag{25}$$

where γ_{sv}, γ_{sl}, and γ_{lv} are the specific energies of the solid/vapor, solid/liquid, and liquid/vapor interfaces, respectively. No correction for surface curvature is considered, although the surface will support a curvature-dependent stress. The corrections are generally considered negligible. The γ_{lv} values of many inorganic melts, such as silicates, are often between 0.1 and 0.5 J/m^2, with 300 mJ/m^2 cited as a most common value for molten silicates [138], with surface tension for liquids, liquid metals, and metal oxides ranging as high as 2 J/m^2. The change of surface tension with temperature for melts such as silicates has been less well established. There are some indications in the older literature that the surface tension of a number of glasses would increase at lower temperatures [138, p. 208], and this might account for phenomena such as dewetting at lower temperatures. Compositional changes also modify the surface tension; as an example, the surface tension of Fe_xO–SiO_2 melts round 1300°C decreasing approximately linearly, from about 550 mJ/m^2 at zero wt% SiO_2 to 350 mJ/m^2 at 45 wt% SiO_2 [139].

The geometries that have been considered so far for model determination of the wetting geometry ignore the complication that for effective liquid-phase sintering, the particles have to be partially soluble in the liquid phase, leading to detailed wetting geometries or liquid distribution that requires modification of the simple two-dimensional Young's equation. These effects, including transient ridge formation near the triple junction, have been examined by Cannon *et al.* [140]. Figure 4.1.21 shows some of the configurational changes when solubility of the solid in the liquid is taken into account.

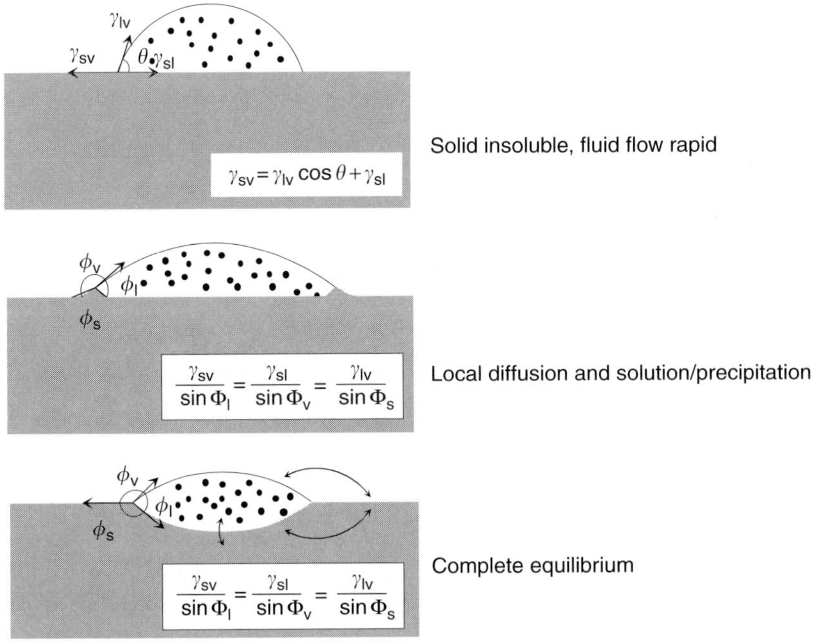

FIGURE 4.1.21 Shape changes associated with partial solubility of the solid in the liquid [140] (courtesy R. M. Cannon).

During fast heat-up and melting, where the rapid densification occurs, it may be difficult to draw a clear distinction between reactive wetting and non-reactive wetting. Reactive wetting can then be taken as referring to the case where the liquid and the solid react to form a new solid phase.

4.1.4.6 Relative Amount of Liquid Phase

The distribution of the liquid phase depends on the dihedral angles and the relative amount of liquid phase, as well as on the porosity. For contact angles larger than zero, it becomes possible to have the liquid phase actually produce repulsive forces rather than attractive forces between the particles. For such a case, further densification can only proceed by grain growth. This is readily appreciated by considering the neutral effect case as a function of liquid volume,[2] as illustrated approximately in Figure 4.1.22. Calculations of capillary forces between unequal particles were discussed by Huang *et al.* [141]. A more detailed example using the *surface evolver* can be found in the article by Carter [142].

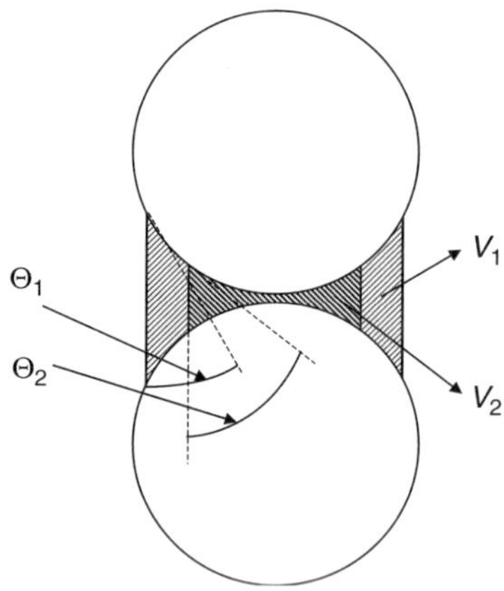

FIGURE 4.1.22 Two-dimensional illustration of the dependence of the neutral volume of liquid phase for a two-particle geometry. At fixed particle separation the neutral conditions will depend on the equilibrium contact angles Θ_i, and the corresponding liquid volumes V_i. With decreasing contact angle, the neutral liquid volume increases. The exact neutral condition in three dimensions still involves some liquid surface curvature when viewed in a two-dimensional section. A liquid volume in excess of the neutral volume produces a repulsive condition between the particles.

[2]The actual neutral case still involves some liquid surface curvature in three dimension.

If sufficient liquid is present, initial rearrangement leads to a fully dense material. The relative amount of liquid and solid at this condition depends on the rearranged density of the solids. In the event that the solids could rearrange to approach a dense random packing, then approximately 35 vol% of liquid would fill all the voids space without further solid skeleton densification. Such large volume fractions of liquid are often used in porcelains and in cemented carbides. In the case of clayware and porcelains, the liquid phases are molten silicates that remain as glass after cooling [143]. This gives the ceramic ware glassy appearance, and such ceramics are referred to as vitrified. The precise phase compositions in vitrified ceramics such as triaxial whiteware (clay–feldspar–silica compositions) can, in principle, be determined from the corresponding phase diagrams, although such phase diagrams are often incomplete and difficult to interpret if more than two components need to be considered [144]. In this regard, it is interesting to note that even the "simple" binary system alumina–silica, the subject of numerous investigations, remains a subject of study. One reason is that equilibrium is difficult to determine.

4.1.5 REACTION SINTERING

Reaction (or reactive) sintering is a process whereby new phases are formed during the firing process. Many ceramic systems involve some degree of reaction, for example, silicon ceramics where a small amount of additives are included to react with native oxide. In other systems, second phases are present only during a part of the densification process, and are absorbed in the matrix phase. This is considered transient second-phase sintering, and numerous examples can be found in the literature [145, 146]. Reactive sintering is defined here as the class of synthesis processes in which the major phase results from the reaction of at least two component powders. Some examples of reactive sintering are listed in Table 4.1.4.

The reactions, which are either solid/solid or solid/liquid, typically involve energy changes that are considerably larger than the changes in surface energies,

TABLE 4.1.4 Some Reaction Sintering Systems

Reactant powders	Resulting ceramic	Reference
$3Al_20_3 + 2ZrSiO_4$	$3Al_2O_3 + ZrO_2$	[147]
$Si_3N_4 + AlN + Al_2O_3$	Sialon	[148]
$SiO_2 + Al_2O_3$	Mullite	[149]
$ZnO + Al_2O_3$	$ZnAl_2O_4$	[150]

FIGURE 4.1.23 Scanning electron micrograph of calcium hexa-aluminate (CA_6 arrowed) formed during the densification of alumina powder mixed with 5 wt% particulate CaO, after 20 h at 1330°C. The plate-shaped CA_6 phase replicates the porous alumina matrix during solid-state reaction/grain growth (from Ref. [120]).

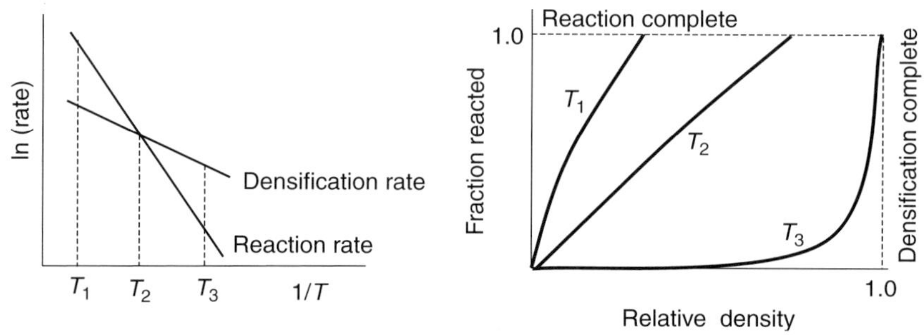

FIGURE 4.1.24 Schematic relationship between densification rate and reaction rate as a function of temperature T, and the resulting process trajectory. If the reaction rate is fast compared to densification, complete reaction can occur with limited densification (after Ref. [147]).

and as a result can lead to microstructures that inhibit densification [120]. An example is shown in Figure 4.1.23.

Qualitatively, it is convenient to consider the reaction and the densification as two separate processes, and put forth a conceptual diagram that illustrates some issues. Such diagrams have been proposed by Yangyun and Brook [147] as shown in Figure 4.1.24. These conceptual arguments lead to the conclusion that the best process is one where densification is completed before reaction can interfere, assuming that there are no deleterious molar volume differences between products and reactants.

An interesting exploitation involving this concept is the reaction sintering of silica-coated alumina and silicon carbide particles. This *transient viscous sintering* process, developed by Sacks *et al.* [151], involves rapid viscous densification due to the persistent presence of the inter-particle silica glass layer, while the subsequent reaction produces dense mullites or mullite–SiC composites.

An example of reaction sintering in which the reaction is delayed until after completion of densification can be found in the work of Claussen *et al.* [152] for the formation of a zirconia–mullite composite from zircon and alumina powders, using a dual-stage processing. In this case, it was possible to achieve nearly full densification at around 1450°C, while the reaction between zircon and alumina only initiated and proceeded to completion at about 1600°C. The resulting ceramic is a nanosize dispersion of monoclinic and tetragonal zirconia in a mullite matrix.

Since energy changes are much larger for chemical reaction than for surface area changes, it would be interesting if reaction energy could be persuaded to drive densification. There is, however, no evidence that the free energy of reaction can act directly as a driving force for densification.

4.1.5.1 Influence of Processing Parameters

The temperature T_2, in Figure 4.1.23, where reaction and densification proceed approximately at the same rates, can be affected by the choice of particle sizes, by the transport mechanisms involved in product formation, and by the detailed distribution of the reactant particles. Smaller starting powder sizes, R, accelerate densification with respect to reaction rates, and thus can assist in reaching higher sintered densities. This follows from the general dependence of densification rates on $1/R^m$ (where $m = 3$ or 4) while the reaction rates [153, 154] depend roughly on $1/R^n$ (where $n = 1$ or 2). Apparent activation energies, as implied in Figure 4.1.23, may, however, also be affected by changes in the particle size, so it may not always be possible to achieve full densification prior to reaction simply by reducing the particle sizes of the starting powders.

The importance of the reactant powder distribution and molar volume changes was evident in studies on reactive sintering of ZnO–Fe_2O_3 and ZnO–Al_2O_3 powders. For ZnO–Fe_2O_3, the spinel-phase $ZnFe_2O_4$ forms at nearly constant molar volume, with inter-diffusion of Zn and Fe ions through the spinel product phase. In this case, reaction sintering proceeds without difficulty, as evident in Figure 4.1.25 for well-mixed powders [155].

For ZnO–Al_2O_3 a significant molar volume change accompanies the formation of zinc aluminate, and the system is far more sensitive to reactant powder distribution. Significant differences in densification behavior were found for

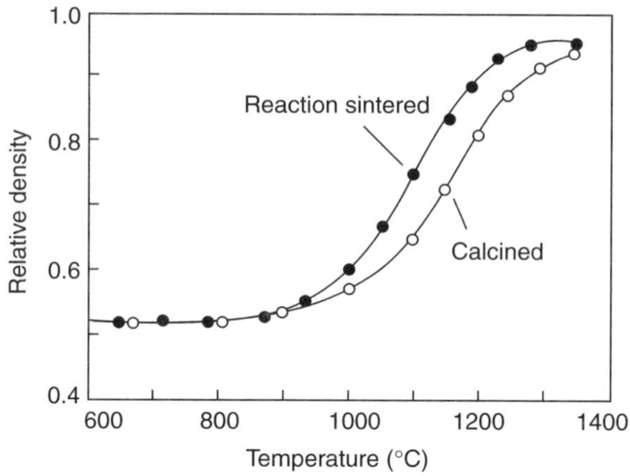

FIGURE 4.1.25 Relative density versus temperature for sintering of ZnO–Fe_2O_3 powder mixtures and for pre-calcined powders of $ZnFe_2O_4$ (from Ref. [155]).

mixed powders compared to ZnO-coated alumina powders prepared by chemical precipitation, as shown in Figure 4.1.26. The expansion produced as a result of the reaction is evident.

4.1.6 CONSTRAINED SINTERING

Densification can be opposed when the sintering matrix has to maintain strain compatibility with a dimensionally invariant phase. Such invariant phases are either dispersions of a second phase or rigid substrates. Variants of these constraints are cases where the second phase has a different densification rate from the matrix. The differential densification rates lead to back stresses in the surrounding densifying matrix that can produce varying degrees of damage. In general, the damage takes the form of higher localized porosity, or more severely, of formation of crack-like defects [157] or film delamination.

4.1.6.1 Dispersed Inert Phases

For the case of dispersed inert second phases, an important issue is whether the dispersion forms a continuous skeleton or is agglomerated to some degree. Various models have been proposed to analyze the behavior of well-dispersed systems. In that case, a concentric sphere model is appropriate, in which the

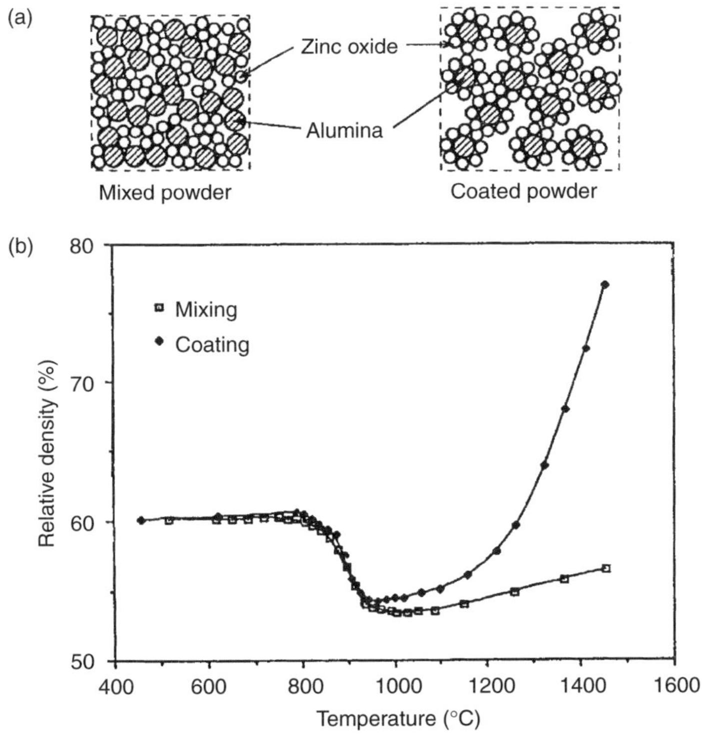

FIGURE 4.1.26 Effect of reactant powder distribution on densification for reaction sintering of $ZnO–Al_2O_3$. Considerably more densification results for coated powders (from Ref. [156]).

invariant phase is contained as a rigid sphere surrounded by a uniform shell of densifying material. The effects of the back stresses caused by the presence of the rigid particle is then analyzed using a viscoelastic description of the sintering process. However, the system is basically near the viscous limit, predicting relatively small back stresses. Scherer's analysis [158] leads to a linear densification rate of the composite, $\dot{\varepsilon}_c$, that follows from the free (i.e. unconstrained) densification rate of an identical matrix, $\dot{\varepsilon}_{fm}$, predicted by

$$\dot{\varepsilon}_c = [(1-f)K_{cs}/4G_m]\dot{\varepsilon}_{fm} \qquad (26)$$

where f is the volume fraction of the inert inclusions, and K_{cs} is related to the shear viscosity, G_m, and the bulk viscosity, K_m, of the matrix is given by $K_{cs} = 1/(1/4G_m + f/3K_m)$.

Another way of expressing the results is to invoke the densification rate of the composite as predicted by a simple, linear rule of mixture, which for an inert and uniformly dispersed inclusion phase would be

$$\dot{\varepsilon}_c^{rm} = (1 - f)\dot{\varepsilon}_{fm} \tag{27}$$

Combining Eqs 26 and 27 leads to

$$\dot{\varepsilon}_c/\dot{\varepsilon}_c^{rm} = 1/[1 + f(4G_m/3K_m)] \tag{28}$$

The theory predicts relatively small effects, with a general absence of stored elastic strain, contrary to conclusions based on other viscoelastic models [159, 160]. Viscoelastic sintering models in which large stored elastics strains arise are, however, not representative of densification driven by surface energy [161]. For volume fractions up to about 15% the Scherer treatment predicts the composite experimental densification rate fairly well for glass matrix systems, but significantly underestimates the effects of rigid inclusions at higher volume fractions. The agreement is less satisfactory for ceramic matrix systems. The most likely reasons are that compact/matrix particle packing is modified near inert inclusion in the powder compaction stage and, more importantly, that formation of rigid skeletal inclusion networks tend to form effectively obstructing densification [128, 162, 163]. This latter point is strongly supported by the significant difficulties in densification that are encountered when attempting to make fiber- or whisker-containing ceramic matrix composites by powder compaction routes.

4.1.6.2 Supported Films

Stresses that arise in films densifying on rigid substrates were analyzed by Scherer and Garino [164]. When particulate films on rigid substrates densify, a back stress will be generated by the invariant substrate, opposing densification parallel to the substrate surface. The stress is effectively a tensile stress that maintains strain compatibility with the substrate, subjecting the densifying film to simultaneous tensile creep. These stresses can be substantial. If the film is sufficiently thin, the stress distribution normal to the film surface can be considered uniform. The interface adhesion or shear strength must be sufficiently high to support the force balance. For a fixed interface strength a critical film thickness results beyond which the interface can no longer support the creep stress, resulting in interface failure, such as delamination. Typically, for sintering of particulate films on rigid substrates, this critical thickness is less than 40 μm, but can be much less when large densification strains or weak interfaces are involved [165]. For the sintering of sol–gel films, for example, the critical thickness is commonly less than 0.5 μm. An example of a cracked and delaminated sol–gel film after sintering is shown in Figure 4.1.27.

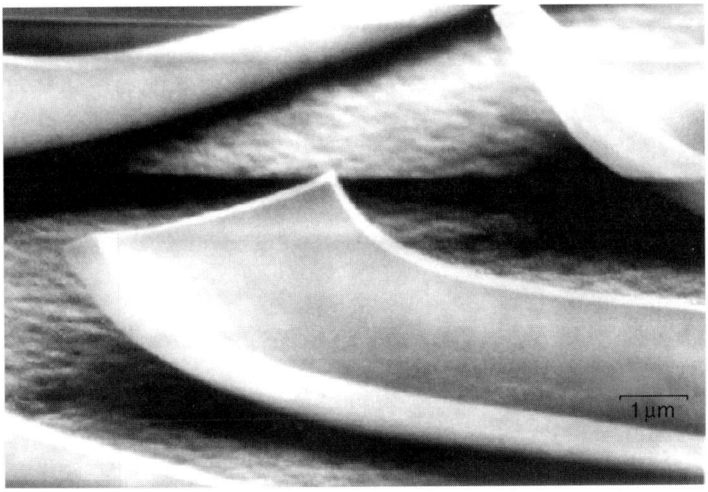

FIGURE 4.1.27 Delamination of a sintered zirconia thin film produced by sol–gel methods on a nanoporous (0.02 μm) alumina substrate (from Ref. [166]).

Failure by cracking can also be initiated in the film itself. Elastic film cracking occurs most readily during the drying stage, if the films are deposited by sol–gel or particulate colloidal methods. Using simple fracture mechanics concepts, a crack stability analysis again leads to a critical film thickness, in which the fracture toughness of the film is a determining factor. The treatments generally lead to a critical film thickness that is proportional to the square of the film fracture toughness. The fracture toughness of the films can, however, vary widely depending on their structure, and a critical thickness is therefore difficult to predict quantitatively. In general, a higher critical thickness results from better interface adhesion, reduction of strain-to-full-density (i.e. high green density), lowering of the creep viscosity relative to the densification viscosity, careful drying prior to densification, a uniform green structure, and maintaining low densification rates at low sintered densities. A factor that must also be considered for thin polycrystalline films is their dimensional stability. Instabilities enter when film grain sizes are larger than the film thickness, when grain boundary grooving can reach the film/substrate interface [167].

4.1.7 SINTERING PRACTICE

A wide variety of techniques have been developed to obtain dense ceramics with a desired microstructure and phase composition. In general, these methods involve a combination of a heating schedule and applied pressure. Heating schedules can be simple, as in isothermal sintering, or have a complex temperature–time relationship, as in rate-controlled sintering, while pressure may be applied either uniaxially, with or without a die, or by the surrounding gas. Control of the sintering atmosphere is also important, and precise control of oxygen [168] or nitrogen partial pressure as a function of temperature may in some cases be beneficial or essential [169]. Insoluble gases trapped in closed pores may obstruct final stages of densification or lead to post-densification swelling and, in these cases, a change of sintering atmosphere or vacuum sintering is indicated [170]. Atmosphere protection may also be provided by packing the sample in an appropriate powder [171]. The practice of sintering further includes a control of the particle characteristics, green compact structure, and consideration of chemistry as a function of processing conditions.

4.1.7.1 Heating Schedules

A general heating schedule is shown in Figure 4.1.28. Binder burnout, removal of volatiles such as water, and conversion of additives such as organometallics or polymers take place. Typically, the first hold temperature is, at most, a few hundred degrees centigrade. The heat-up rate should be carefully controlled, since it is quite possible that rapid heating causes boiling and evaporation of organic additives leading to specimen bloating or even shattering. Stage 2 can be included to promote chemical homogenization or reaction of powder components. Stage 3 represents the heat-up to the isothermal sintering stage 4, during which the majority of the densification and microstructure development takes place, which is then followed by a cool-down. An additional hold in stage 5,

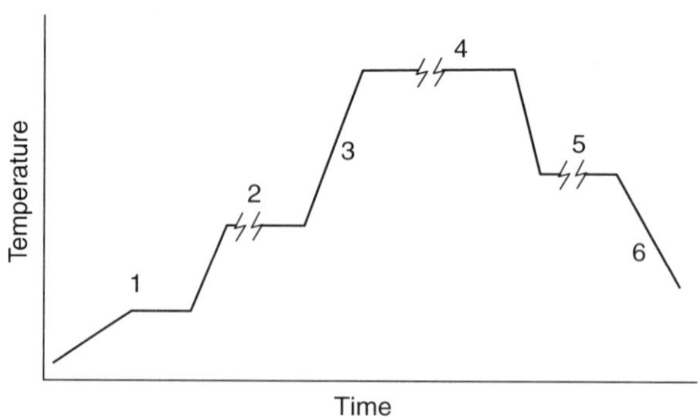

FIGURE 4.1.28 Generalized heating schedule.

prior to the final cool-down of stage 6, may also be included to relieve internal stresses or allow for precipitation or other reactions.

4.1.7.1.1 Isothermal Sintering

In isothermal sintering, perhaps better called isothermal-stage sintering, the temperature is increased monotonically to a sintering hold temperature and lowered to room temperature afterwards. The holding time is long compared to the heat-up time. This is the most common heating schedule. The heat-up times are limited by the sample size and by the thermal characteristics of the furnace. Heat-up times for large bodies can stretch over many hours, to avoid temperature gradients that could lead to cracking or to the formation of an outer dense layer on an incompletely densified core, as would result from differential densification.

During the heat-up phase of the isothermal sintering, significant densification and microstructural changes can take place, which all too often are ignored in sintering studies. Also, as a general rule, densification processes should be carried out significantly above the intended service temperature.

4.1.7.1.2 Constant Heating Rate Sintering

In this case, the sample is heated to a specified temperature at a controlled heating rate, and immediately cooled. Constant heating rate experiments may actually be simpler to analyze theoretically [24, 172] than the "isothermal" ones, because strict isothermal sintering is not possible. In practice, the use of constant heating rates is again limited by the sample size. High constant heating rates are then most useful in laboratory studies. Manipulation of the heating rate can be useful in the case where reactions take place during densification. For example, Boccaccini *et al.* [173] studied densification and crystallization of a glass powder in which densification was inhibited by crystallization at 1 K/min, but densification was complete prior to crystallization at 15 K/min.

Chu *et al.* [174] studied the microstructure and densification kinetics of ZnO over a wide range of constant heating rates. In general, higher heating rates lead to a finer grain size. The relative density versus temperature clustered in a fairly narrow band, for heating rates between 0.5 and 15 K/min. An interesting observation is that the derivative of the strain with respect to temperature for this range of constant heating rates falls on the same master curve. Lange [175] made a similar observation for alumina.

4.1.7.1.3 Multi-stage Sintering

Multi-stage sintering is frequently used in practice, sometimes introducing extended temperature plateaus or more complex temperature–time sequences

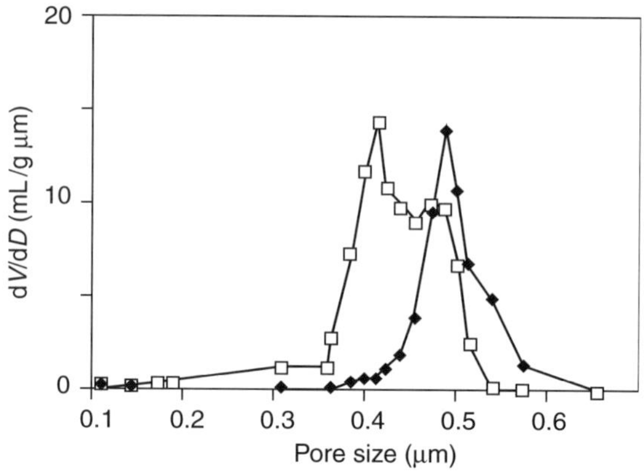

FIGURE 4.1.29 Pore size distribution narrowing after 50 h at 800°C for an alumina powder compact, determined by mercury porosimetry (from Ref. [177]).

in the heating schedule, with the purpose of achieving either specific chemical or microstructural features. An example of staged sintering, involving two separate peak temperatures in which the temperature is reduced between the stages is in the sintering of the ion-conducting ceramic sodium beta alumina [176]. Optimum results with respect to microstructure and strength were achieved in this case by reducing the sintering temperature by about 150°C after reaching the first peak temperature (about 1500°C).

A reduction in grain size and an increase in uniformity were also found by Lin *et al.* [177] and by Chu *et al.* [178] for MgO, Al_2O_3, and ZnO using a two-stage sintering technique. The first stage consisted of an extended hold at a temperature where shrinkage was 0.5–1% after 48 h. This temperature is readily found by running a dilatometery trace at constant heating rate. At the holding temperatures limited coarsening occurs that appears to have the effect of homogenizing the porous powder compact. Figure 4.1.29 shows the change in pore size distribution after limited coarsening for an alumina powder (AKP-50; Sumitomo America Corp.). Excessive coarsening should be avoided.

4.1.7.1.4 Rate-Controlled Sintering

In rate-controlled sintering, the densification rate is coupled with the temperature control of the sintering furnace in such a way as to keep the densification rate constant or limited [179, 180]. The result is a fairly complicated temperature history that at times approaches the staged sintering processes. While the underlying mechanisms are not fully understood, beneficial effects on microstructure have been reported in several instances, including where chemical reactions occur during densification [181, 182]. Manipulation of the sintering temperature is necessarily limited by the thermal impedance of the sintering furnace and by the size of the sample.

4.1.7.1.5 Fast-Firing

Ratio-controlled sintering and fast-firing are based on the principle of increasing the densification rate relative to the coarsening rate [183]. In ratio-controlled sintering, a temperature is selected that depends on the relative activation energies of these competing processes. This is shown schematically in Figure 4.1.30. The preferred sintering temperature is one where the densification rate is higher than the coarsening rate [184].

The most common situation is one where at higher temperatures densification rates are faster than coarsening rates. This leads to the process of fast-firing, in which the sample is taken to the high sintering temperature as rapidly as the system allows. The process can be particularly effective for thin walled tubes, for example, in rapid zone sintering where the ceramic piece is moved through a hot zone.

4.1.7.1.6 Microwave Sintering

Microwave heating is effective for heating more complex ceramic forms very rapidly [185]. Heating rates in excess of 1000°C/min can be achieved. In this method, the ceramic body, usually contained in non-absorbing or weakly absorbing insulation such as loose, non-conducting powder, is placed within a microwave cavity. A schematic of the method is shown in Figure 4.1.31.

The method is quite straightforward. It is even possible to use a simple consumer microwave oven to achieve densification of ceramic powder compacts

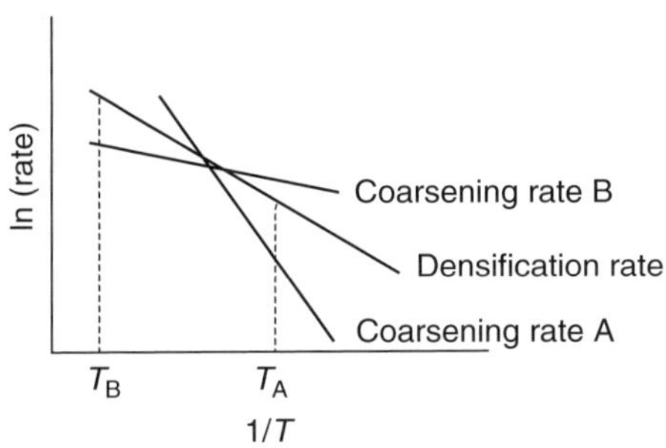

FIGURE 4.1.30 Possible relation of coarsening rates to densification rate. The preferred sintering temperatures T_A or T_B correspond to the coarsening rates A or B.

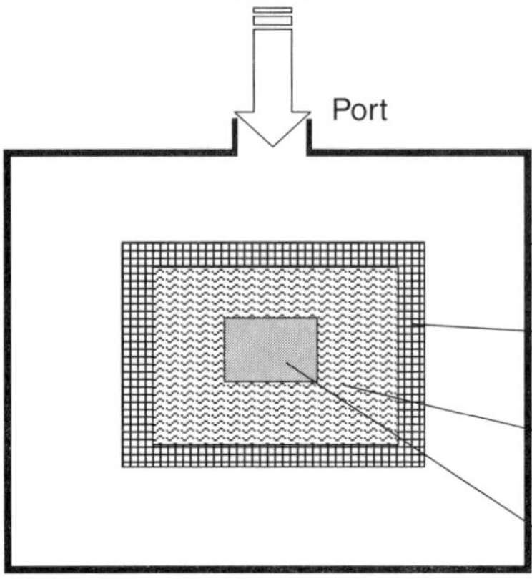

FIGURE 4.1.31 Schematic microwave sintering setup.

when properly insulated [186]. The shape of the ceramic body affects the local heating rates significantly and achieving sufficiently uniform heating can be difficult [187]. The microwave frequency plays a significant role in the temperature gradients that can develop within the ceramic body. High frequencies tend to heat the exterior of the sample more than the interior, and a combination of frequencies ranging between about 2.5 and 85 GHz has been proposed as providing more uniform heating [188]. Continuous microwave sintering has also been reported, in which samples are passed through the apparatus [189].

A variant of microwave sintering is one where the microwaves ignite a plasma that surrounds the ceramic part. Very high sintering rates can be achieved this way. The schematic of the method is shown in Figure 4.1.32. Johnson *et al.* [190] achieved full densification for tubular specimens by translating the tube at a rate of 25 mm/min through a plasma of about 50 mm in height. It is, as yet, not clear if the plasma itself enhances the densification rates in addition to the high temperature created by the plasma.

4.1.7.1.7 Plasma-Assisted Sintering

Several attempts have been made at increasing the heating rates, causing a need for various superlatives such as "superfast" or "ultrafast" sintering, by passing a DC current pulse through a powder compact contained in a graphite die, under an applied pressure of 30–50 MPa. This method has been called spark plasma sintering (SPS). A schematic of the method is shown in Figure 4.1.33. Specimen temperatures are difficult to assess in this method, and are usually

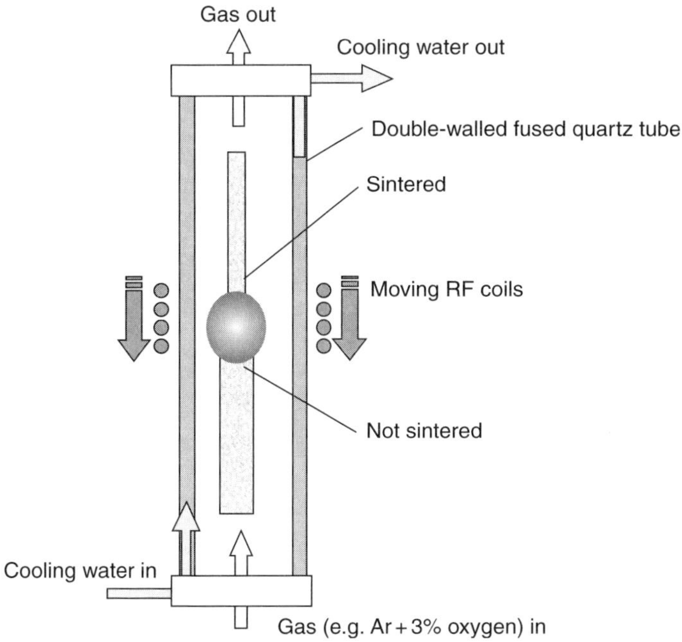

FIGURE 4.1.32 Schematic of plasma sintering setup.

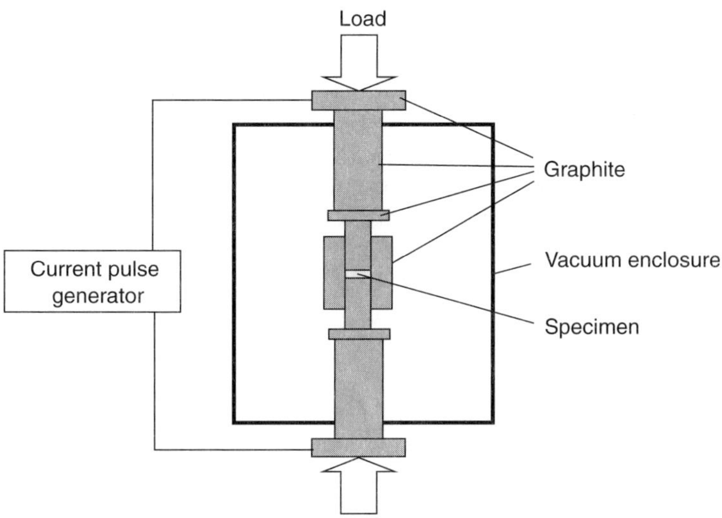

FIGURE 4.1.33 Schematic of a spark plasma sintering apparatus.

measured by optical pyrometry on the graphite die wall. Both the die and the specimen are heated by the current pulse. Heating rates of 600 K/min and more have been reported. The high heating rates are thought to be caused, in part, by spark discharges generated in the voids between the particles. Remarkably high densification rates may be achieved under such conditions, with minimal grain growth. This approach is particularly useful in producing dense ceramic bodies from nanosized powders [191, 192]. The specimen shapes that can be prepared by SPS are limited to simple slabs that can be contained in the compression die. A related method is a thermal explosive forming, in which a reactive mixture of components, for example, Ti and C, is heated up in a die, and then ignited by passing an electrical current pulse. The process may perhaps be regarded as a self-propagating high-temperature synthesis under pressure. Formation of dense boride, nitride, and carbide composites has been reported [193].

4.1.7.2 Pressure-Assisted Sintering

4.1.7.2.1 Hot pressing

Hot pressing is a convenient laboratory method for preparing dense samples [194]. Heat and pressure are applied to a sample contained typically in a high-strength graphite die, at applied pressures around 50 MPa. The die wall thickness is approximately equal to the sample diameter if high-strength graphite [195] is used. Other die material such as aluminum oxide, silicon carbide, or dies confined by an outer metal or fiber-wound mantle are also used in rare cases. Reaction of the graphite die with the sample can be decreased by spray coating the inside of the die with boron nitride, if the processing temperature remains below about 1350°C. Above that temperature the boron nitride can react with graphite. Boron nitride sprays are commercially available [196]. Lining of the graphite die with graphite foil is also useful for prolonging the die life. Multiple plate-shaped samples may be hot pressed together, using graphite spacers.

An interesting variant allowing for shape flexibility was developed by Lange *et al.* [197]. In this method, the sample is packed in coarse powder in the hot pressing die, developing a roughly isostatic pressure on the work piece. A schematic of the standard process is shown in Figure 4.1.34. More complex systems have also been developed, for example, a multi-bore machine for hot pressing nuclear reactor pellets [198].

In a typical hot pressing process, a moderate pressure is applied form the start, 10–20 MPa. Upon reaching the sintering temperature the pressure is increased to maximum. An applied pressure is usually maintained during the cool-down period as well. A schematic of the thermomechanical treatment is shown in Figure 4.1.35.

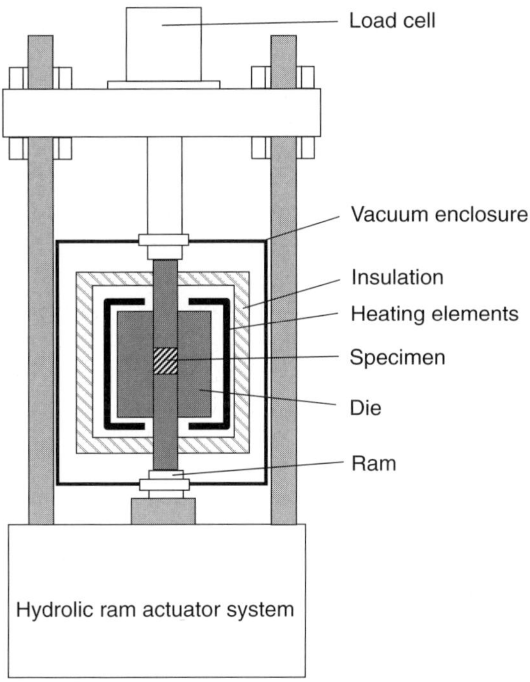

FIGURE 4.1.34 Schematic of the hot pressing process.

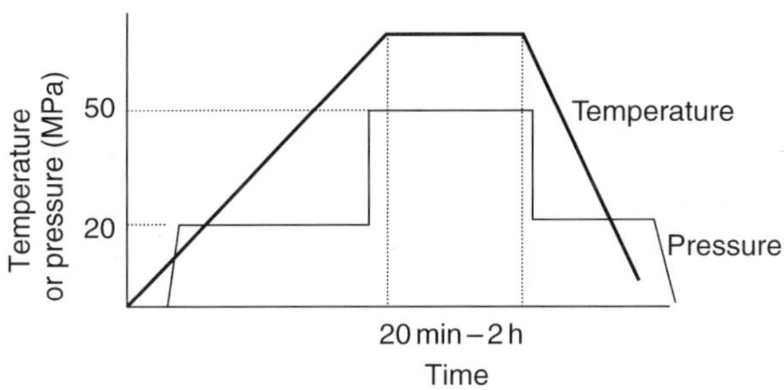

FIGURE 4.1.35 Schematic temperature/pressure schedule for hot pressing with a high-strength graphite die.

If no reliable temperature measurement is available, it is possible to follow the ram displacement as the temperature is increased. An appropriate sintering temperature is then reached when the rate of the ram displacement rapidly increases to indicate a projected full densification in 20–30 min. Hot pressing is less effective for very fine powder that is well compacted prior to hot pressing. For submicron powders the intrinsic sintering stress can be significantly in excess of what the hot-press die can tolerate, and the applied stress then appears less effective. Benefits may still be derived from the applied pressure when it can assist in the rearrangement process or in the collapsing of large pores. Some guidance for selecting the appropriate hot pressing temperature may be obtained form pressure sintering maps [199], although trial-and-error is more reliable. In case full densification is difficult to achieve, additives can be used similar to the free sintering method.

As a result of the significant uniaxial strain in hot pressing, texture may develop in the finished part. Trapping of insoluble gases in residual pores may cause swelling of the ceramic at elevated service temperatures. This can be avoided by hot pressing in vacuum.

Reactive hot pressing has also been used successfully [200, 201]. Densification is more readily achieved in this case compared to unconfined sintering. Related to reactive hot pressing is hydrothermal hot pressing. In this method, powders are compacted under hydrothermal conditions. The powder is essentially compacted inside an autoclave at temperatures between 100 and 350°C. The ram design should allow for fluid to leave the sample. Materials such as hydroxyapatite [202] and amorphous titania [203] have been compacted this way.

4.1.7.2.2 Sinter Forging

Sinter forging is similar to hot pressing, but without confining the sample in a die, as shown in Figure 4.1.36. Uniaxial strains are, as a result, significantly larger than in die hot pressing [204]. For fine powders use can be made of possible superplastic deformation modes during sinter forging. During the process, the strain rate has to be limited to avoid damaging the sample [205].

4.1.7.2.3 Hot Isostatic Pressing (HIPing)

A schematic of the equipment is shown in Figure 4.1.37 [206]. In this method, developed around 1955, the pre-consolidated powder is tightly enclosed in a glass or metal container, sealed under vacuum and positioned in the pressure vessel. Alternatively, the sample can be pre-densified to closed porosity by traditional sintering, after which a canister is not needed in subsequent HIPing. A compressor introduces inert gas pressure, and the sample is heated to the sintering temperature, which may be up to 2000°C. During this time the gas pressure rises further to as much as 30 000 psi, and the container collapses around the sample, transmitting the isostatic pressure to the sample. Commercial hot

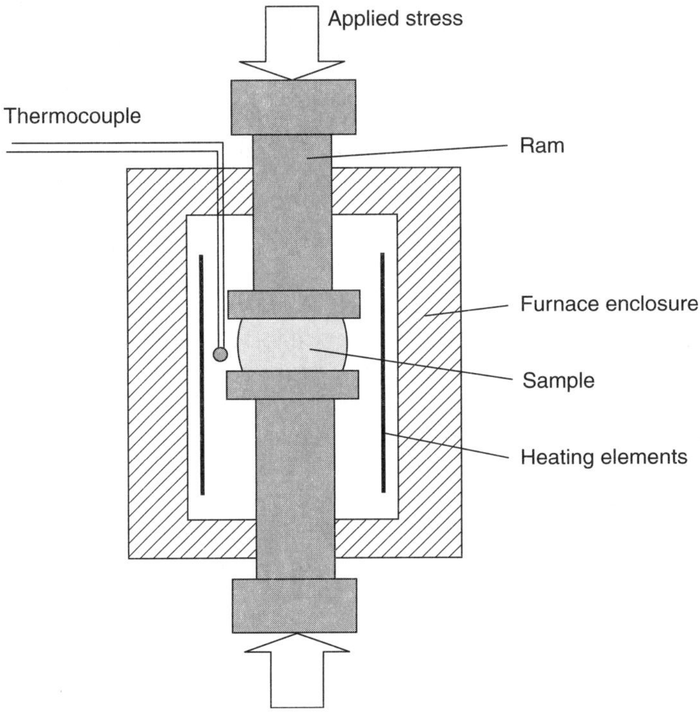

FIGURE 4.1.36 Schematic setup for sinter forging.

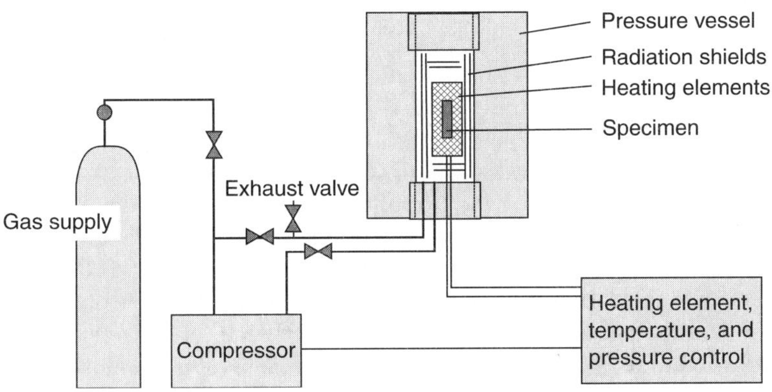

FIGURE 4.1.37 Schematic of hot isostatic pressing equipment.

isostatic presses may have internal chamber diameters approaching 1 m. Heating elements are typically graphite, molybdenum, tungsten, or tantalum. The quality of the ceramics produced by hot isostatic pressing is perhaps the highest obtainable by any other pressure method, since externally heated dies cannot withstand the pressure that can be applied by the HIP. HIPing is also used to recondition parts in which internal damage, such as resulting from fatigue or creep, may have accumulated during service.

4.1.7.3 Particle and Green Compact Characteristics

While the sintering behavior of real powders is considerably more complex than that assumed in the models, the sintering theory clearly indicates the key parameters that must be controlled to optimize sintering. The key factors are the particle size of the powder and the particle packing in the green body but other characteristics, such as size distribution, shape and structure of the particles, can also exert a significant influence.

4.1.7.3.1 Particle Size

The matter transport rates for the solid-state sintering mechanisms depend strongly on the particle size of the powder (Eqs 6 and 7). Assuming sintering conditions (e.g. the heating schedule) where the densifying mechanisms dominate, then the rate of densification is predicted to vary as $1/G^m$, where G is the particle size, and the exponent $m = 4$ for grain boundary diffusion and $m = 3$ for lattice diffusion. Thus, reduction of the particle size provides an important method for speeding up sintering. This is illustrated by the data shown in Figure 4.1.38 for CeO_2 powder compacts with three different particle sizes [207, 208]. Because of packing difficulties, the observed sintering rate is lower than that predicted by the models but the enhancement of the rate with decreasing particle size is still considerable.

A consequence of the improvement in the sintering rate is the use of lower sintering temperatures. Nanoscale particles (less than 50–100 nm in size) exhibit large reductions in the sintering temperature. From the data in Figure 4.1.38, isothermal sintering of the 10-nm particles can be achieved at less than $\sim$1150°C, compared to a sintering temperature of 1500–1600°C for the 1 μm particles. Unfortunately, nanoscale particles are prone to packing problems so that the green density is commonly lower than that for coarser powders. The shrinkage during sintering can therefore be large. Contamination can also be a problem due to the large surface area of the powder. The removal of surface impurities, such as hydroxyl groups, from the surface of nanoscale particles can

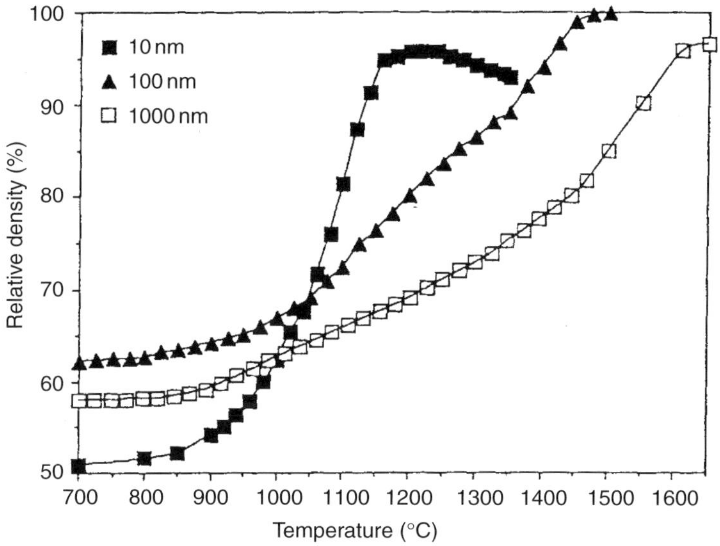

FIGURE 4.1.38 Effect of particle size on the sintering of CeO_2 powder compacts during constant heating rate sintering (10°C/min) in an oxygen atmosphere (from Ref. [207]).

also be difficult. Decomposition during sintering can lead to trapped gases in the pores that can limit the final density, as seen in Figure 4.1.38 for the 10-nm particles.

4.1.7.3.2 Particle Size Distribution

The solid-state sintering models generally assume that the particles are monodisperse but a particle size distribution can have significant effects on sintering. Coble [209] modeled the initial-stage sintering of a linear array of particles with different sizes and found that an equivalent particle size, considerably smaller than a conventional average size, can be used to account for the shrinkage. The results also show that the sintering rates of binary mixtures of particles are intermediate between those for the end-member sizes. A simple rule of mixtures has, in fact, been proposed to describe the sintering data of binary mixtures of alumina powders where the coarser phase is relatively inactive [210, 211]. Commonly, differential densification between the coarser phase and the finer phase coupled with interactions between the particles in the coarser phase can severely inhibit densification (see Chapter 6).

The use of mixtures of discrete particle sizes or a wide distribution of particle size can result in an increase in the packing density (because the finer particles fit into the interstices of the larger particles), so that the shrinkage required for

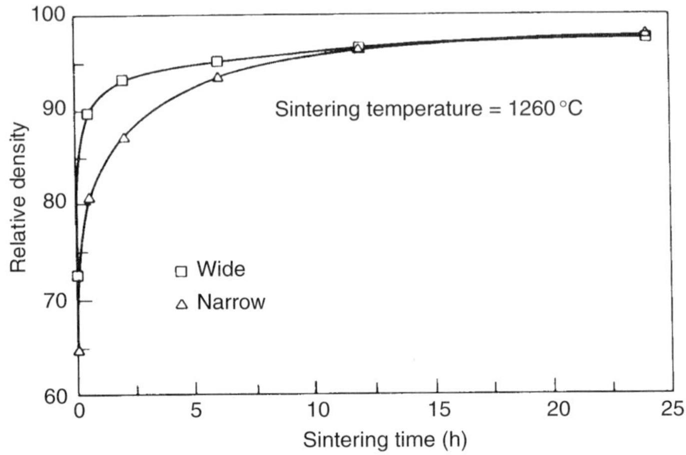

FIGURE 4.1.39 Effect of particle size distribution on the sintering of slip cast Al_2O_3 powder compacts prepared from wide and narrow size distribution powders having the same median particle size (from Ref. [228], with kind permission of American Ceramic Society).

complete densification is reduced. This reduction in the shrinkage is important in industrial sintering of large objects. As shown in Figure 4.1.39, increasing the width of the particle size distribution has a benefit in the early stages of sintering due to the presence of the fine powders with the higher driving force for sintering [212–214]. The behavior in the later stages of sintering depends critically on the particle packing of the green body. For heterogeneous packing, an increase in the width of the particle size distribution is expected to enhance the detrimental effects of differential densification and grain growth (due to the enhanced driving force arising from the size difference) so that the final density may decrease. However, if the particle packing is homogeneous, with small pores of narrow size distribution, a high final density can be achieved regardless of the width of the initial particle size distribution (Fig. 4.1.39), emphasizing the importance of the pore structure in sintering.

4.1.7.3.3 Particle Shape and Particle Structure

Particle shape influences sintering primarily through its effect on the packing of the green body. Deviations from the spherical or equiaxial shape lead to a reduction in the packing density and packing uniformity, resulting in a reduction of densification. Compacts of acicular (or elongated) particles can, however, be sintered to high density if the particles are aligned and the packing is homogeneous, as demonstrated for acicular Fe_2O_3 particles [215]. Some solids maintain or develop faceted shapes during sintering [216–219] but the effects of shape are often complicated by other factors (e.g. packing, influence of composition).

Particles used in sintering are commonly dense, discrete units that can either be polycrystalline or a single crystal. Polycrystalline particles of Y_2O_3-stabilized ZrO_2 have been observed to sinter faster than single-crystal particles and this has been attributed to the larger number of grain boundaries contributing to mass transport [220]. A problem is that the single crystals and the polycrystalline particles have different Y_2O_3 content so that the structural effects cannot be clearly separated from compositional effects. Spherical, monodisperse particles synthesized by the Stober process consist of an aggregate of much finer particles (Fig. 4.1.40). The influence of this particle substructure on sintering has been well characterized by Edelson and Glaeser [221, 222]. For TiO_2, heating to 700°C produces sintering of the fine particles in the spheres and conversion to rutile so that the spheres consist of one to three grains by the time the spheres start to sinter at ~1000°C.

4.1.7.3.4 Particle Packing

For enhanced sintering rates and the attainment of high density, the particles must be homogeneously packed with a high packing density. These

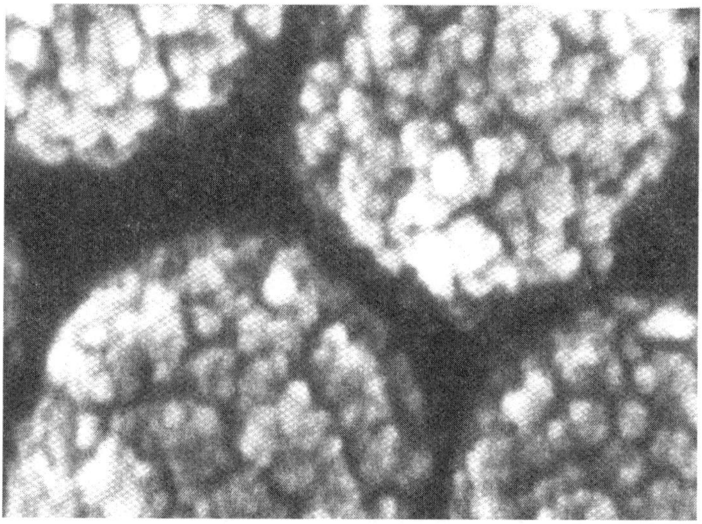

FIGURE 4.1.40 High-resolution SEM micrographs showing the substructure of monodisperse TiO_2 particles prepared by the Stober process. The spherical particles (~0.35 μm) consist of an aggregate of much finer particles (~10 nm) (courtesy of A. M. Glaeser).

packing characteristics produce fine, uniform pores with a low pore coordination number corresponding to the shrinking pore in Figure 4.1.3. Furthermore, the number of particle contacts is maximized, thereby providing many grain boundaries and short diffusion paths for rapid matter transport into the fine pores. Rhodes [223] provided a clear demonstration of these principles in a study of the sintering of Y_2O_3-stabilized ZrO_2. Using a fine powder (crystallite size ~10 nm) synthesized by the hydrolysis of alkoxides, he prepared a suspension and allowed the agglomerates to settle. The fine particles in the supernatant were then used to prepare compacts by gravitational settling in a centrifuge. After drying the compact had a density of 0.74 of the theoretical density. As shown in Figure 4.1.41, sintering for 1 h at 1100°C produced almost complete

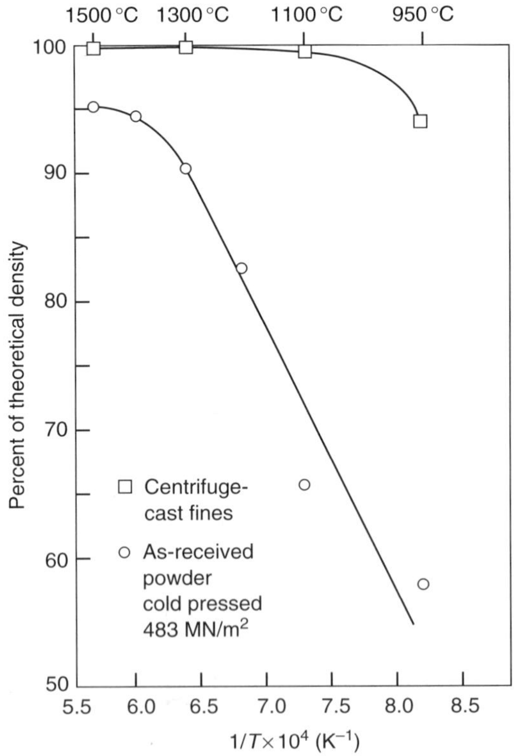

FIGURE 4.1.41 Effect of particle packing on the sintering of Y_2O_3-stabilized powder compacts after 1 h at various temperatures. The compact with the more homogeneous microstructure (formed by centrifugal casting of fine, agglomerate-free particles) reaches a significantly higher density than a compact formed by die-pressing the agglomerated, as-received powder (from Ref. [223], with kind permission of American Ceramic Society).

densification. In comparison, compacts prepared by die-pressing of as-received, agglomerated powder reached a density of only 0.95 of the theoretical after sintering for 1 h at 1500°C.

Barringer and Bowen [224, 225] developed a processing approach based on the uniform packing of monodisperse, spherical particles. While rapid sintering was achieved at considerably lower temperatures, the compacts could not be sintered to full density (the residual porosity was ~5%). A problem is that the uniform consolidation of monodispere particles leads to the formation of small regions with three-dimensional, *ordered* packing (typical of the packing in crystals), referred to as *domains*, which are separated by packing flaws (voids) at the domain boundaries. The faster sintering of the ordered regions, when compared to the domain boundaries, leads to differential stresses that cause enlargement of the voids. These voids correspond to the growing or stable pores in Figure 4.1.3.

Liniger and Raj [226] have suggested that dense, *random* packing of particles with a distribution of sizes may provide an alternative ideal packing geometry for enhanced sintering. When compared to the ordered packing of monodisperse particles, the random structure would be less densely packed than the domains but the fluctuations in density should be less severe, leading to homogeneous sintering and the reduction of flaw generation due to differential sintering. Homogeneously packed Al_2O_3 green bodies formed by colloidal processing of fine powders have, in fact, been sintered to almost full density at temperatures as low as ~1200°C [227, 228].

It is important to appreciate the implications of deviating from the ideal structure because most practical sintering, particularly for industrial applications, are not performed with an ideal system of homogeneous, densely packed, fine particles. A key problem is *differential densification*, which may lead to enlargement of the voids or even the generation of crack-like voids in the less dense regions (Figure 4.1.42). The large voids correspond to the growing or stable pore in Figure 4.1.3 so that they limit densification. The effects of differential densification on particle rearrangement in two-dimensional packing of spheres have been well characterized by Weiser and De Jonghe [229] and by Exner [230]. Another factor is that the denser regions can support grain growth: enhanced normal grain growth starts at an earlier point during sintering, thereby increasing the probability for the initiation of abnormal grain growth.

A relevant question is whether the effects of differential densification can be corrected or reversed during sintering. Lange [231] has proposed that a limited amount of normal grain growth is beneficial for sintering in that it reduces the pore coordination number of the pores. This means that grain growth can convert the growing pore with a high coordination number in Figure 4.1.3 to a shrinking pore with low coordination number. Support for Lange's suggestion

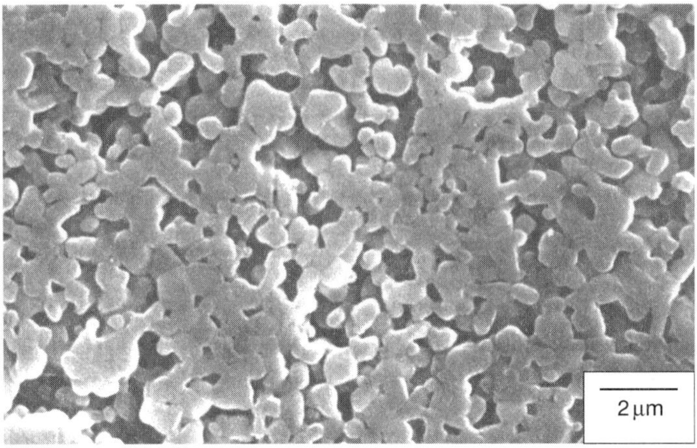

FIGURE 4.1.42 Differential densification during the sintering of an Al_2O_3 powder compact formed by die-pressing, leading to the development of denser and less dense regions.

was observed in the sintering studies of Lin *et al.* [232] on heterogeneously packed MgO powder. Grain growth can, however, provide only a short-term benefit because the increase in the diffusion distance eventually produces a significant reduction in the overall sintering rate [233]. For nanoscale CeO_2 powders, Chen [234] has suggested a mechanism of coarsening coupled with repacking of the porous network to explain the sintering of compacts with low green density to almost full density.

As described earlier, De Jonghe and coworkers [177, 178] have demonstrated that a two-step sintering approach, which involves an initial pre-coarsening step prior to densification, leads to some improvement of the microstructural homogeneity during sintering and to corresponding microstructural benefits in the sintered material.

While the effects of heterogeneous packing can be alleviated to some extent during sintering, the benefits may not be as compelling as may be required in most systems so that a more realistic approach would be based on optimizing the particle packing in the green body.

4.1.7.3.5 Effect of Green Density on Sintering

The green density of the powder compact is an important parameter in sintering. Higher green density leads a decrease in the shrinkage during sintering (due to the reduced pore volume) and to an increase in the sintering stress due to the reduced pore size [37]. Several studies have shown a correlation between the sintered density and the green density: the sintered density is observed to decrease with green density for green density values below 55–60% of the theoretical value [235–237].

Commonly, compacts with green densities lower than 40–45% of the theoretical value are difficult to sinter to high density. Increasing green density is found to delay the onset of enhanced grain growth during the later stages of sintering [238]. These trends in the data can be explained in terms of the higher probability for the occurrence of pores with large coordination numbers with decreasing green density coupled with the enhancement of differential densification. In general, the benefits of higher green density for sintering are clear, so optimization of the green density would form a useful approach. However, in some systems a lower green density does not necessarily influence the ability to attain a high sintered density, as exemplified by the data in Figure 4.1.43 for ZnO. Even with a low green density, compacts of some nanoscale powders can also be sintered to high density; as described earlier, CeO_2 powders with a green density as low as 20% which have been sintered to almost full density [239].

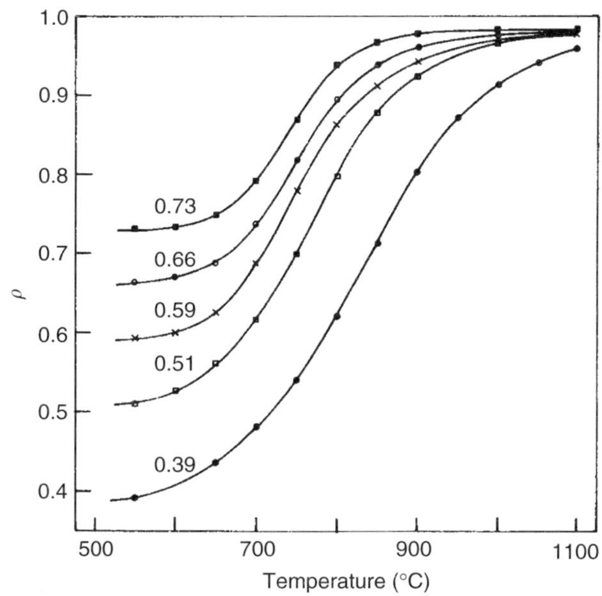

FIGURE 4.1.43 Effect of green density on the sintering of ZnO powder compacts during constant heating rate sintering (5°C/min) to 1100°C showing that for this powder, the final density is almost independent of the green density for a wide range of green density values (0.39–0.73) (from Ref. [37]).

4.1.7.4 Chemistry Considerations

The sintering atmosphere can have a decisive effect on the microstructural evolution, particularly in the case of advanced ceramics that must meet exacting property requirements. Atmosphere is not directly considered in the sintering models but the theories can provide a basis for understanding the phenomena arising from the atmosphere conditions. Both physical phenomena (e.g. the atmospheric gas trapped in the pores) and chemical phenomena (e.g. volatility, ionic oxidation state and defect chemistry) are involved.

4.1.7.4.1 Gases in Pores

In the final stage of sintering, gas from the atmosphere (or from volatile species in the solid) is trapped in the pores when the pores pinch off and become isolated. Further sintering is influenced by the solubility of the gas atoms in the solid, as has been demonstrated by Coble [170] in the sintering of Al_2O_3. Coble found that MgO-doped Al_2O_3 can be sintered to theoretical density in O_2 or H_2, which can diffuse out to the surface of the solid, but not in air, N_2, He, or Ar, which have a limited solubility in Al_2O_3.

If it has a high solubility in the solid, the gas does not have a significant influence on sintering. For an insoluble gas, application of the gas laws shows that sintering stops when a limiting porosity, P_L, is reached, as given by

$$P_L = P_0 \left(\frac{p_0 r_0}{2\gamma_{sv}} \right)^{2/3} \tag{29}$$

where p_0 is the pressure of the sintering atmosphere, P_0 and r_0 are the porosity and radius, respectively, of the pores (assumed spherical in shape) when they become isolated, and γ_{sv} is the specific energy of the solid/vapor interface. According to Eq. 29, the limiting density is controlled essentially by the pressure of the gaseous atmosphere and by the radius of the pores when they become isolated. If practical, sintering in vacuum and homogeneous packing of fine particles (giving fine pores) can improve the final density.

When coarsening occurs with a slightly soluble or insoluble gas trapped in the pores, pore coalescence leads to a phenomenon referred to as *bloating*, whereby the density of the body starts to decrease, as illustrated in Figure 4.1.44. Bloating is also a common occurrence when relatively dense ceramics, previously hot pressed in a carbon die (or furnace), are annealed in an oxidizing atmosphere [240]. Oxidation of carbonaceous impurities leads to the generation of a gas that provides the pressure for the expansion.

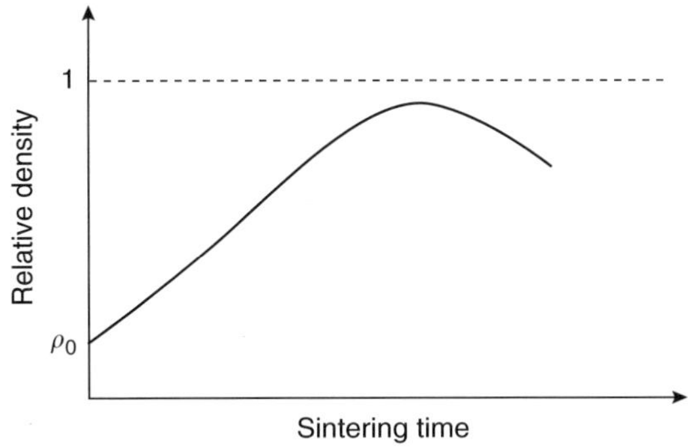

FIGURE 4.1.44 Schematic diagram illustrating the phenomenon of bloating produced by coarsening with gas-filled pores. The density decreases with time after reaching a maximum value.

4.1.7.4.2 Vapor transport

The heating schedule, as described earlier, provides an important process parameter for controlling the relative rates of the densifying and coarsening mechanisms. Readey and coworkers [242–244] have demonstrated how enhanced vapor transport, produced by sintering in a reactive atmosphere, can provide another key process variable for microstructure control. For example, vapor transport in Fe_2O_3 can be produced by the introduction of HCl gas into the atmosphere:

$$Fe_2O_3\ (s) + 6HCl\ (g) \rightarrow Fe_2\ Cl_6\ (g) + 3H_2O\ (g) \tag{30}$$

Densification decreases as the HCl pressure and the amount of vapor transport increases, and stops when the HCl pressure reaches values greater than 10^{-2} MPa. As shown in Figure 4.1.45, the decrease in densification is caused by particle coarsening. Water vapor in the atmosphere has also been shown to produce modifications in the densification/coarsening ratio [245–247].

4.1.7.4.3 Volatilization and Decomposition

For some powder compositions (e.g. sodium β-alumina and lead-based ferroelectric ceramics), evaporation of volatile components can occur during sintering, thereby making it difficult to control the composition and the microstructure of the sintered material. Evaporation of PbO in lead-based electroceramics can be represented as [248, 249].

$$PbO\ (s) \rightleftarrows PbO\ (g) \rightleftarrows Pb\ (g) + \tfrac{1}{2}O_2 \tag{31}$$

Lead is poisonous so that, in addition to controlling lead loss, the evaporated lead must be contained. In practice, this is achieved by surrounding the sample

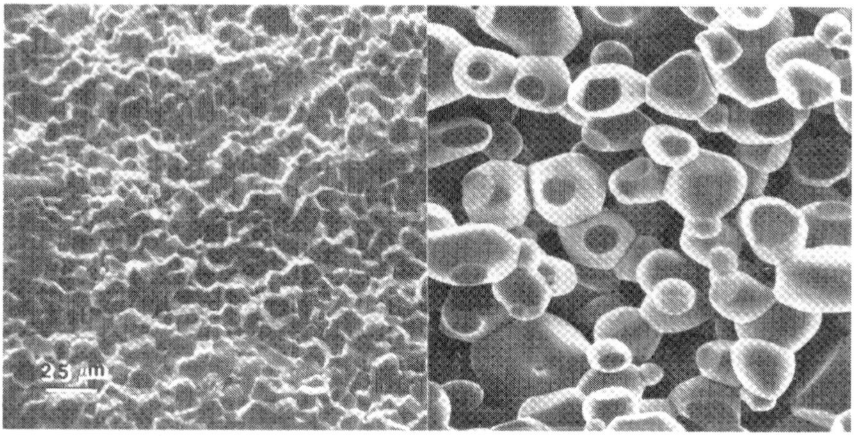

FIGURE 4.1.45 Scanning electron micrographs of fractured surfaces of Fe_2O_3 powder compacts sintered for 5 h in air (left) and in 10% HCl (right) (courtesy D. W. Readey).

with lead-based powder compositions (e.g. $PbZrO_3$ and PbO for lead zirconate–titanate samples) to provide a positive vapor pressure in a closed alumina crucible [171].

Silicon nitride shows significant decomposition at the high temperatures (1700–1800°C) required for its densification. The decomposition reaction can be written as

$$Si_3N_4 \rightleftarrows 3Si\ (g) + 2N_2\ (g) \tag{32}$$

The vapor pressure of the N_2 gas generated by the decomposition is ~0.1 atm at the sintering temperature so that considerable weight loss can occur if the decomposition is not controlled. One solution is to surround the sample with a powder of the same composition in a closed graphite crucible under N_2 gas at normal atmospheric pressure. A better method involves raising the N_2 gas pressure in the sintering atmosphere (to values of 10–20 atm or higher) so that the reaction in Eq. 32 is driven to the left [250].

4.1.7.4.4 Oxidation State

The sintering atmosphere influences the oxidation state of certain cations, particularly those of the transition elements (e.g. chromium), and control of the oxidation state has been shown to have a significant effect on the densification of chromium-containing oxides [168, 251–253]. Figure 4.1.46 shows the porosity of three chromites as a function of oxygen partial pressure in the sintering

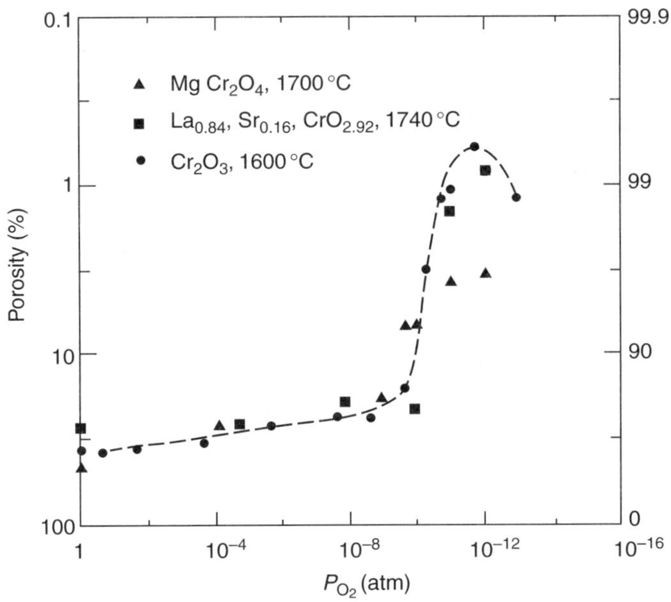

FIGURE 4.1.46 Dependence of porosity (or density) of sintered chromites on oxygen partial pressure in the sintering atmosphere (courtesy of H. U. Anderson).

atmosphere after sintering for 1 h at temperatures ranging from 1600–1740°C. The chromites show little densification if the oxygen partial pressure is greater than $\sim 10^{-8}$ atm. However, densities of ~99% of the theoretical value can be obtained by sintering in an oxygen partial pressure of $\sim 10^{-12}$ when the Cr ion is stabilized in its trivalent state (as Cr_2O_3). At higher oxygen partial pressure, Cr_2O_3 becomes unstable and vaporizes as CrO_2 or CrO_3. The high vapor pressure enhances evaporation/condensation processes, leading to neck growth and coarsening but little densification. The dramatic effects of oxidation state control observed for the chromites have not been repeated in other ceramic systems so that the approach sees limited applicability.

When compared to the undoped material, the use of TiO_2 as a dopant [254] allows the densification of Cr_2O_3 at a higher oxygen partial pressure (10^{-1} atm) and at a lower temperature (1300°C). It was suggested that the Ti^{4+} ions replace the chromium ions that are in an oxidation state greater than 3, thereby suppressing the formation of the volatile CrO_2 and CrO_3. The use of the TiO_2 dopant has also been claimed to produce chromium vacancies, leading to an enhancement of the sintering rate.

Atmosphere control is particularly important in the sintering of ferrites for magnetic devices, such as manganese zinc ferrite and nickel zinc ferrite [255].

In addition to limiting the evaporation of zinc, the atmospheric conditions must also give the right concentration of ferrous iron in the sintered ferrite for achieving the desired magnetic properties. This is commonly done by first sintering in an atmosphere of high p_{O_2} (0.3–1.0 atm) to minimize the evaporation of zinc, followed by annealing at a lower temperature in a low p_{O_2} atmosphere (10–100 ppm) to give the desired concentration of ferrous iron.

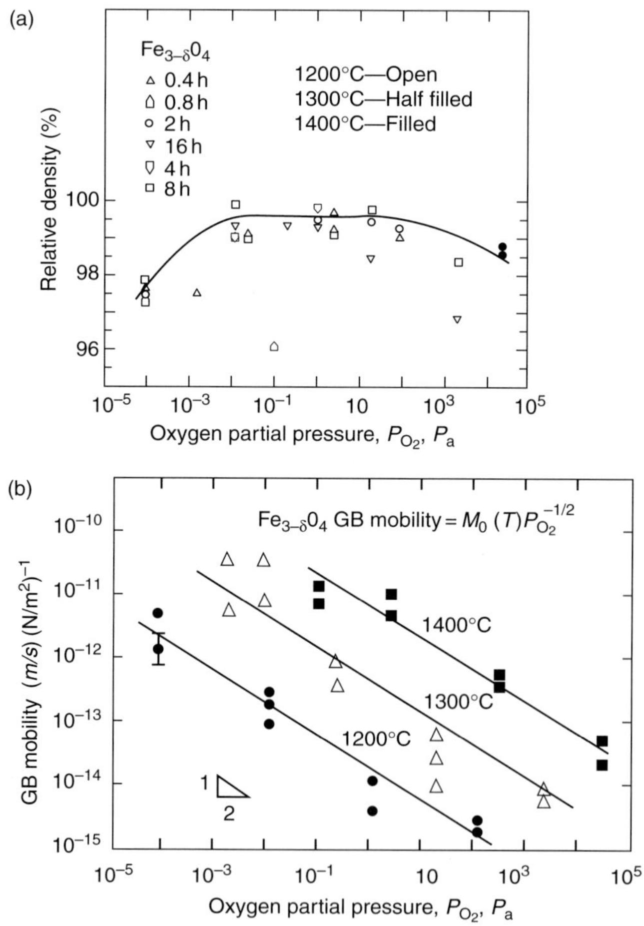

FIGURE 4.1.47 Data for (a) the relative density and (b) the boundary mobility as a function of the oxygen partial pressure in the atmosphere during the sintering of Fe_3O_4 powder compacts under the conditions of time and temperature indicated (from Ref. [257], with kind permission of American Ceramic Society).

4.1.7.4.5 Defect Chemistry and Stoichiometry

The sintering behavior of the highly stoichiometric oxides (e.g. Al_2O_3 and ZrO_2) shows almost no dependence on the oxygen partial pressure (p_{O_2}) in the atmosphere. For other ceramic oxides, particularly the transition metal oxides, the p_{O_2} can change the concentration and type of lattice defects and, hence, the stoichiometry of the compound [11, 12]. The sintering of these oxides would therefore be expected to depend significantly on the p_{O_2} of the atmosphere. Because they have a significantly larger ionic radius than most cations, the oxygen ions are often less mobile. Thus, an atmosphere that produces an increase in the oxygen vacancy concentration may enhance sintering. In practice, it is often difficult to separate the influence of defect chemistry and stoichiometry from the influence of changes in the non-densifying mechanisms (e.g. vapor transport and surface diffusion) brought about by the change in the atmosphere.

The influence of stoichiometry has been well demonstrated by Reijnen [256] in the sintering of $NiFe_2O_4$ and $NiAl_2O_4$. Densification is severely retarded for powder compositions with an excess of Fe or Al. This behavior was explained in terms of a reduction in the oxygen vacancy concentration. Figure 4.1.47 shows the densification and grain growth behavior of Fe_3O_4 as a function of p_{O_2} at three different temperatures [257]. While the type and concentration of the lattice defects may be expected to vary over the wide p_{O_2} range, the sintered density is insensitive to the atmosphere. However, the p_{O_2} has a significant effect on the grain growth: high p_{O_2} leads to small grain size whereas the grain size is much larger at low p_{O_2}. While the data cannot be fully explained, it has been suggested that oxygen lattice diffusion controls the pore mobility. Oxygen lattice diffusion is faster at lower p_{O_2}, so an increase in the pore mobility will give a faster grain growth rate if the boundary mobility is controlled by the pore motion (pore control in Figure 4.1.12).

REFERENCES

1. Evans, J. W., and De Jonghe, L. C. (1991). *The Production of Inorganic Materials*, New York: Macmillan.
2. Rahaman, M. N. (1995). *Ceramic Processing and Sintering*, New York: Marcel Dekker.
3. German, R. M. (1996). *Sintering Theory and Practice*, New York: Wiley.
4. De Jonghe, L. C., and Rahaman, M. N. (1988). Sintering stress of homogeneous and heterogeneous powder compacts. *Acta Metall.* **36**: 223–229.
5. Raj, R. (1987). Analysis of the sintering pressure. *J. Am. Ceram. Soc.* **70**: C210–C211.
6. Howard, R. E., and Lidiard, A. B. (1964). Matter transport in solids. *Rept. Progr. Phys.* **27**: 161–240.
7. Ruoff, A. L. (1965). Mass transfer problems in ionic crystals with charge neutrality. *J. Appl. Phys.* **36**: 2903–2907.

8. Readey, D. W. (1966). Mass transport and sintering in impure ionic solids. *J. Am. Ceram. Soc.* **49**: 366–369.
9. Gordon, R. S. (1973). Mass transport in the diffusional creep of ionic solids. *J. Am. Ceram. Soc.* **56**: 147–152.
10. Coble, R. L., and Cannon, R. M. (1978). Current paradigms in powder processing. *Mater. Sci. Res.* **11**: 151–170.
11. Kroeger, F. A. (1964). *The Chemistry of Imperfect Crystals*, Amsterdam: North Holland Publishing Co.
12. Kofstad, P. (1972). *Nonstoichiometry, Diffusion, and Electrical Conductivity in Binary Metal Oxides*, New York: Wiley.
13. Brook, R. J. (1972). Effect of TiO_2 on the initial sintering of Al_2O_3. *J. Am. Ceram. Soc.* **55**: 114–115.
14. Harmer, M. P. (1984). Use of solid solution sintering additives in ceramic processing. *Adv. Ceram.* **10**: 679–691.
15. Brook, R. J. (1988). Additives and the sintering of ceramics. *Sci. Sintering* **26**: 115–118.
16. Kingery, W. D., and François, B. (1967). Sintering of crystalline oxides, I. Interactions between grain boundaries and pores, *Sintering and Related Phenomena*, pp. 471–498, Kuczynski, G. C., Hooton, N. A., and Gibbon, G. F., eds., New York: Gordon & Breach.
17. Exner, H. E. (1979). Principles of single phase sintering. *Rev. Powder Metall. Phys. Ceram.* **1**: 1–251.
18. Coblenz, W. S., Dynys, J. M., Cannon, R. M., and Coble, R. L. (1980). Initial stage sintering models: a critical analysis and assessment, *Sintering Processes*, pp. 141–157, Kuczynski, G. C., ed., New York: Plenum.
19. Coble, R. L. (1961). Sintering of crystalline solids, I. Intermediate and final stage diffusion models. *J. Appl. Phys.* **32**: 787–792.
20. Coble, R. L. (1967). Mechanisms of densification during hot pressing, *Sintering and Related Phenomena*, pp. 329–347, Kuczynski, G. C., Hooton, N. A., and Gibbon, G. F., eds., New York: Gordon & Breach.
21. Coble, R. L. (1970). Diffusion models for hot pressing with surface energy and pressure effects as driving forces. *J. Appl. Phys.* **41**: 4798–4807.
22. Herring, C. (1950). Effect of change of scale on sintering phenomena. *J. Appl. Phys.* **21**: 301–303.
23. Song, H., Coble, R. L., and Brook, R. J. (1984). The applicability of Herring's scaling law to the sintering of powders. *Mater. Sci. Res.* **16**: 63–79.
24. Johnson, D. L. (1969). New method of obtaining volume, grain boundary, and surface diffusion coefficients from sintering data. *J Appl. Phys.* **40**: 192–200.
25. Bross, P., and Exner, H. E. (1979). Computer simulation of sintering processes. *Acta Metall.* **27**: 1013–1020.
26. Svoboda, J., and Riedel, H. (1995). New solution describing the formation of interparticle necks in solid-state sintering. *Acta Metall. Mater.* **43**: 1–10.
27. Jagota, A., and Dawson, P. R. (1988). Micromechanical modeling of powder compacts—I. Unit problems for sintering and traction-induced deformation. *Acta Metall.* **36**: 2551–2561.
28. Jagota, A., and Dawson, P. R. (1990). Simulation of viscous sintering of two particles. *J. Am. Ceram. Soc.* **73**: 173–177.
29. Jagota, A. (1994). Simulation of the viscous sintering of coated particles. *J. Am. Ceram. Soc.* **77**: 2237–2239.
30. Ashby, M. F. (1974). A first report on sintering diagrams. *Acta Metall.* **22**: 275–289.
31. Swinkels, F. B., and Ashby, M. F. (1981). A second report on sintering diagrams. *Acta Metall.* **29**: 259–281.

32. Gregg, R. A., and Rhines, F. N. (1973). Surface tension and the sintering force in copper. *Metall. Trans.* **4**: 1365–1374.
33. De Jonghe, L. C., and Rahaman, M. N. (1984). A loading dilatometer. *Rev. Sci. Instrum.* **55**: 2007–2010.
34. Rahaman, M. N., De Jonghe, L. C., and Brook, R. J. (1986). Effect of shear stress on sintering. *J. Am. Ceram. Soc.* **69**: 53–58.
35. Rahaman, M. N., and De Jonghe, L. C. (1987). Creep-sintering of ZnO. *J. Mater. Sci.* **22**: 4326–4330.
36. Chu, M.-Y., De Jonghe, L. C., and Rahaman, M. N. (1989). Effect of temperature on the densification/creep viscosity during sintering. *Acta Metall.* **37**: 1415–1420.
37. Rahaman, M. N., De Jonghe, L. C., and Chu, M.-Y. (1991). Effect of green density on densification and creep during sintering. *J. Am. Ceram. Soc.* **74**: 514–519.
38. Venkatachari, K. R., and Raj, R. (1986). Shear deformation and densification of powder compacts. *J. Am. Ceram. Soc.* **69**: 499–506.
39. Aigeltinger, E. H. (1975). Relating microstructure and sintering force. *Int. J. Powder Met. Powder Technol.* **11**: 195–203.
40. Beere, W. (1975). A unifying theory of the stability of penetrating liquid phases and sintering pores. *Acta Metall.* **23**: 131–138.
41. Vieira, J. M., and Brook, R. J. (1984). Kinetics of hot-pressing: the semi-logarithmic law. *J. Am. Ceram. Soc.* **67**: 245–249.
42. Atkinson, H. V. (1988). Theories of normal grain growth in pure single phase systems. *Acta Metall.* **36**: 469–491.
43. Burke, J. E., and Turnbull, D. (1952). *Progr. Metal Phys.* **3**: 220.
44. Feltham, P. (1957). Grain growth in metals. *Acta Metall.* **5**: 97–105.
45. Hillert, M. (1965). On the theory of normal and abnormal grain growth. *Acta Metall.* **13**: 227–238.
46. Louat, N. P. (1974). On the theory of normal grain growth. *Acta Metall.* **22**: 721–724.
47. Hunderi, O., and Ryun, N. (1980). The kinetics of normal grain growth. *J. Mater. Sci.* **15**: 1104–1108.
48. Smith, C. S. (1948). Grains, phases, and interfaces: an interpretation of microstructure. *Trans. AIME* **175**: 15–51.
49. Smith, C. S. (1952). Grain shapes and other metallurgical applications of topology, *Metal Interfaces*, pp. 65–113, Cleveland, OH: ASM.
50. Rhines, F. N., and Craig, K. R. (1974). Mechanism of steady-state grain growth in aluminum. *Metall. Trans.* **5**: 413–425.
51. Anderson, M. P., Srolovitz, D. J., Grest, G. S., and Sahni, P. S. (1984). Computer simulation of grain growth—I. Kinetics. *Acta Metall.* **32**: 783–791.
52. Srolovitz, D. J., Anderson, M. P., Sahni, P. S., and Grest, G. S. (1984). Computer simulation of grain growth—II. Grain size distribution, topology, and local dynamics. *Acta Metall.* **32**: 793–802.
53. Brook, R. J. (1976). Controlled grain growth, *Treatise on Materials Science and Technology*, pp. 331–364, Wang, F. F. Y., ed., New York: Academic Press.
54. Chol, G. R. (1971). Influence of milled powder particle size distribution on the microstructure and electrical properties of sintered Mn–Zn ferrites. *J. Am. Ceram. Soc.* **54**: 34–39.
55. Monahan, R. D., and Halloran, J. W. (1979). Single crystal boundary migration in hot-pressed aluminum oxide. *J. Am. Ceram. Soc.* **62**: 564–567.
56. Rios, P. R. (1992). Abnormal grain growth in pure materials. *Acta Metall.* **40**: 2765–2768.
57. Rios, P. R. (1994). Abnormal grain growth in materials containing particles. *Acta Metall.* **42**: 839–843.

58. Srolovitz, D. J., Grest, G. S., and Anderson, M. P. (1985). Computer simulation of grain growth—II. Grain size distribution, topology, and local dynamics. *Acta Metall.* **33**: 2233–2247.
59. Thompson, C. V., Frost, H. J., and Spaepen, F. (1987). The relative rates of secondary and normal grain growth. *Acta Metall.* **35**: 887–890.
60. Rollet, A. D., Srolovitz, D. J., and Anderson, M. P. (1989). Simulation and theory of abnormal grain growth—anisotropic grain boundary energies and mobilities. *Acta Metall.* **37**: 1227–1240.
61. Yang, W., Chen, L., and Messing, G. L. (1995). Computer simulation of anisotropic grain growth. *Mater. Sci. Eng. A* **195**: 179–185.
62. Kunaver, U., and Kolar, D. Three-dimensional computer simulation of anisotropic grain growth in ceramics. *Acta Mater.* **46**: 4629–4640.
63. Harmer, M. P. (1990). A history of the role of MgO in the sintering of α-Al_2O_3. *Ceram. Trans.* **7**: 13–49.
64. Berry, K. A., and Harmer, M. P. (1986). Effect of MgO solute on microstructure development in Al_2O_3. *J. Am. Ceram. Soc.* **69**: 143–149.
65. Handwerker, C. A., Dynys, J. M., Cannon, R. M., and Coble, R. L. (1990). Dihedral angles in magnesia and alumina: distributions from surface thermal grooves. *J. Am. Ceram. Soc.* **73**: 1371–1377.
66. Kaysser, W. A., Sprissler, M., Handwerker, C. A., and Blendell, J. E. (1987). Effect of liquid phase on the morphology of grain growth in alumina. *J. Am. Ceram. Soc.* **70**: 339–343.
67. Song, H., and Coble, R. L. (1990). Origin and growth kinetics of platelike abnormal grains in liquid-phase-sintered alumina. *J. Am. Ceram. Soc.* **73**: 2077–2085.
68. Horn, D. S., and Messing, G. L. (1995). Anisotropic grain growth in TiO_2-doped alumina. *Mater. Sci. Eng. A* **195**: 169–178.
69. Powers, J. D., and Glaeser, A. M. (1996). Titanium effects on sintering and grain growth of alumina, *Sintering Technology*, pp. 333–340, German, R. M., Messing, G. L., and Cornwall, R. M., eds., New York: Marcel Dekker.
70. Becher, P. F. (1991). Microstructural design of toughened ceramics. *J. Am. Ceram. Soc.* **74**: 255–269.
71. Cao, J. J., Moberly-Chan, J. J., De Jonghe, L. C., Gilbert, C. J., and Ritchie, R. O. (1996). In situ toughened silicon carbide with Al–B–C additions. *J. Am. Ceram. Soc.* **79**: 461–469.
72. Huang, T., Rahaman, M. N., Mah, T.-I., and Parthasarathay, T. A. (2000). Effect of SiO_2 and Y_2O_3 additives on the anisotropic grain growth of dense mullite. *J. Mater. Res.* **15**: 718–726.
73. Seabaugh, M. M., Kerscht, I. H., and Messing G. L. (1997). Texture development by templated grain growth in liquid-phase-sintered α-alumina. *J. Am. Ceram. Soc.* **80**: 1181–1188.
74. Horn, J. A., Zhang, S. C., Selvaraj, U., Messing, G. L., and Trolier-McKinstry, S. (1999). Templated grain growth of textured bismuth titanate. *J. Am. Ceram. Soc.* **82**: 921–926.
75. Patwardhan, J. S. (2000). Texture and grain alignment in piezoelectric bismuth titanate ceramics, MS Thesis, University of Missouri-Rolla, Rolla, MO.
76. Yamamoto, T., and Sakuma, T. (1994). Fabrication of barium titanate single crystals by solid state grain growth. *J. Am. Ceram. Soc.* **77**: 1107–1109.
77. Li, T., Scotch, A. M., Chan, H. M., Harmer, M. P., Park, S., and Shrout, T. R. (1998). Single crystals of $Pb(Mg_{1/3}Nb_{2/3})O_3$-35 mol% $PbTiO_3$ from polycrystalline precursors. *J. Am. Ceram. Soc.* **81**: 244–248.
78. Greenwood, G. W. (1956). The growth of dispersed precipitates in solutions. *Acta Metall.* **4**: 243–248.
79. Lifshitsz, I. M., and Slyozov, V. V. (1961). The kinetics of precipitation from supersaturated solutions. *Phys. Chem. Solids* **19**: 35–50.

80. Wagner, C. (1961). Theory of Ostwald ripening. *Z. Electrochem.* **65**: 581–591.
81. Fischmeister, H., and Grimvall, G. (1973). Ostwald ripening—a survey. *Mater. Sci. Res.* **6**: 119–149.
82. Ardell, A. J. (1972). The effect of volume fraction on particle coarsening: theoretical considerations. *Acta Metall.* **20**: 61–71.
83. Enomoto, Y., Kawasaki, K., and Tokuyama, M. (1987). Computer modeling of Ostwald ripening. *Acta Metall.* **35**: 904–913.
84. Enomoto, Y., Kawasaki, K., and Tokuyama, M. (1987). The time dependent behavior of the Ostwald ripening for the finite volume fraction. *Acta Metall.* **35**: 915–922.
85. Kingery, W. D. (1974). Plausible concepts necessary and sufficient for interpretation of ceramic grain boundary phenomena: I, grain-boundary characteristics, structure and electrostatic potential. *J. Am. Ceram. Soc.* **57**: 1–8.
86. Kingery, W. D. (1974). Plausible concepts necessary and sufficient for interpretation of ceramic grain boundary phenomena: II, solute segregation, grain boundary diffusion, and general discussion. *J. Am. Ceram. Soc.* **57**: 74–83.
87. Chiang, Y.-M., Henriksen, A. F., Finello, D., and Kingery, W. D. (1981). *J. Am. Ceram. Soc.* **64**: 383.
88. Ikeda, J. A. S., and Chiang, Y.-M. (1993). Space charge segregation at grain boundaries in titanium dioxide: I, relationship between lattice defect chemistry and space charge potential. *J. Am. Ceram. Soc.* **76**: 2437–2446.
89. Ikeda, J. A. S., Chiang, Y.-M., Garratt-Reed, A. J., and Vander Sande, J. B. (1993). Space charge segregation at grain boundaries in titanium dioxide: II, model experiments. *J. Am. Ceram. Soc.* **76**: 2447–2459.
90. Cahn, J. W. (1962). The impurity drag effect in grain boundary motion. *Acta Metall.* **10**: 789–798.
91. Chen, P.-L., and Chen, I.-W. (1994). Role of defect interaction in boundary mobility and cation diffusivity of CeO_2. *J. Am. Ceram. Soc.* **77**: 2289–2297.
92. Rahaman, M. N., and Zhou, Y.-C. (1995). Effect of dopants on the sintering of ultra-fine CeO_2 powder. *J. Eur. Ceram. Soc.* **15**: 939–950.
93. Coble, R. L. (1962). Transparent alumina and method of preparation, US Patent # 3, 026, 210.
94. Brook, R. J., Tuan, W. H., and Xue, L. A. (1988). Critical issues and future directions in sintering science. *Ceram. Trans.* **1B**: 811–823.
95. Rahaman, M. N., and Manalert, R. (1998). Grain boundary mobility of $BaTiO_3$ doped with aliovalent cations. *J. Eur. Ceram. Soc.* **18**: 1063–1071.
96. Rahaman, M. N., De Jonghe, L. C., Voight, J. A., and Tuttle, B. A. (1990). Low temperature sintering of zinc oxide varistors. *J. Mater. Sci.* **25**: 737–742.
97. Jorgensen, P. J., and Anderson, R. C. (1967). Grain-boundary segregation and final-stage sintering of Y_2O_3. *J. Am. Ceram. Soc.* **50**: 553–558.
98. Prochazka, S., and Scanlan, R. M. (1975). Effect of boron and carbon on sintering of SiC. *J. Am. Ceram. Soc.* **58**: 72.
99. Yan, M. F., Cannon, R. M., and Bowen, H. K. (1977). Grain boundary migration in ceramics, *Ceramic Microstructures '76*, pp. 276–307, Fulrath, R. M., and Pask, J. A., eds., Boulder, CO: Westview Press.
100. Glaeser, A. M., Bowen, H. K., and Cannon, R. M. (1986). Grain-boundary migration in LiF: I, mobility measurements. *J. Am. Ceram. Soc.* **69**: 119–125.
101. Lange, F. F., and Hirlinger, M. M. (1984). Hindrance of grain growth in Al_2O_3 by ZrO_2 inclusions. *J. Am. Ceram. Soc.* **67**: 164–168.
102. Stearns, L. C., and Harmer, M. P. (1996). Particle-inhibited grain growth in Al_2O_3-SiC: I, experimental results. *J. Am. Ceram. Soc.* **79**: 3013–3019.

103. Stearns, L. C., and Harmer, M. P. (1996). Particle-inhibited grain growth in Al_2O_3-SiC: II, equilibrium and kinetic analyses. *J. Am. Ceram. Soc.* **79**: 3020–3028.
104. Greskovich, C., and Lay, K. W. (1972). Grain growth in very porous Al_2O_3 compacts. *J. Am. Ceram. Soc.* **55**: 142–146.
105. Rhines, F., and DeHoff, R. (1983). Channel network decay in sintering. *Mater. Sci. Res.* **16**: 49–61.
106. Coleman, S. C., and Beeré, W. B. (1975). The sintering of open and closed porosity in UO_2. *Philos. Mag.* **31**: 1403–1413.
107. Shewmon, P. G. (1964). *Trans. AIME* **230**: 1134.
108. Hsueh, C. H., Evans, A. G., and Coble, R. L. (1982). Microstructure development during final/initial stage sintering: I, pore/grain boundary separation. *Acta Metall.* **30**: 1269–1279.
109. Brook, R. J. (1969). Pore–grain boundary interactions and grain growth. *J. Am. Ceram. Soc.* **52**: 56–57.
110. Berry, K. A., and Harmer, M. P. (1986). Effect of MgO solute on microstructure development in Al_2O_3. *J. Am. Ceram. Soc.* **69**: 143–149.
111. Handwerker, C. A., Cannon, R. M., and Coble, R. L. (1984). Final stage sintering of MgO. *Adv. Ceram.* **10**: 619–631.
112. Carpay, F. M. A. (1977). Discontinuous grain growth and pore drag. *J. Am. Ceram. Soc.* **60**: 82–83.
113. Kim, H. Y., Lee, J.-A., and Kim, J. J. (2000). Densification behavior of fine alumina and coarse alumina compacts during liquid phase sintering with the addition of talc. *J. Am. Ceram. Soc.* **83**: 3128–3134.
114. Elfwing, M., Osterlund, R., and Olsson, E. (2000). Differences in wetting characteristics of Bi_2O_3 polymorphs in ZnO Varistor Materials. *J. Am. Ceram. Soc.* **83**: 2311–2314.
115. Kim, D.-H., and Kim, C. H. (1990). Toughening behavior of silicon carbide with additions of yttria and alumina. *J. Am. Ceram. Soc.* **73**: 1431–1434.
116. Terwilliger, G. F., and Lange, F. F. (1975). Pressureless sintering of silicon nitride. *J. Mater. Sci.* **10**: 1169–1174.
117. Riley, F. L. (2000). Silicon nitride and related materials. *J. Am. Ceram. Soc.* **83**: 245–265.
118. Rhim, S. M., Bak, H., Hong, S., Kim, O. K. (2000). Effects of heating rate on the sintering behavior and the dielectric properties of $Ba_{0.7}Sr_{0.3}TiO_3$ ceramics prepared by boron-containing liquid-phase sintering. *J. Am. Ceram. Soc.* **83**: 3009–3013.
119. Luo, J., Wang, H., and Chiang, Y.-M. (1999). Origin of solid-state activated sintering in Bi_2O_3-doped ZnO. *J. Am. Ceram. Soc.* **82**: 916.
120. Wu, S. J., De Jonghe, L. C., and Rahaman, M. N. (1985). Subeutectic densification and second phase formation in Al_2O_3–CaO. *J. Am. Ceram. Soc.* **68**: 385–388.
121. German, R. M. (1984). *Powder Metallurgy Science*, p. 182, Princeton, NJ: Metal Powder Industries Federation.
122. Zhang, X. F., Sixta, M., and De Jonghe, L. C. (2000). Grain boundary evolution in hot-pressed ABC–SiC. *J. Am. Ceram. Soc.* **83**: 2813–2820.
123. Kwon, O. J., and Yoon, D. N. (1981). Closure of isolated pores in liquid phase sintering of W–Ni. *Int. J. Powder Technol.* **17**: 127–133.
124. Shaw, T. M. (1986). Liquid redistribution during liquid phase sintering. *J. Am. Ceram. Soc.* **69**: 27–34.
125. Gu, H., Pan, X. Q., Cannon, R. M., and Ruhle, M. (1998). Dopant distribution in grain-boundary films in calcia-doped silicon nitride ceramics. *J. Am. Ceram. Soc.* **81**: 3125–3135.
126. Brydson, R., Chen, S.-C., Riley, F. L., Milne, S. J., Pan, X., and Rühle, M. (1998). Microstructure and chemistry of intergranular glassy films in liquid-phase-sintered alumina. *J. Am. Ceram. Soc.* **81**: 369–379.

127. Buist, D. S., Jackson, B., Stephenson, I. M., Ford, W. F., and White, J. (1965). The kinetics of grain growth in two-phase (solid–liquid) systems. *Trans. Br. Ceram. Soc.* **64**: 173–209.
128. Kapolnek, S., and De Jonghe, L. C. (1991). Particulate composites from coated powders. *J. Eur. Ceram. Soc.* 7: 345–351.
129. Norton, F. H. (1974). *Elements of Ceramics*, Reading, MA: Addison-Wesley.
130. Lange, F. F. (1982). Liquid phase sintering: are liquids squeezed out from between compressed particles? *J. Am. Ceram. Soc.* **65**: C23.
131. Clarke, D. R., Shaw, T. M., Phillips, A. P., and Horn, R. G. (1993). Possible electrical double layer contribution of the equilibrium thickness of intergranular glass films in polycrystalline ceramics. *J. Am. Ceram. Soc.* **76**: 1201–1204.
132. Wang, H., and Chiang, Y.-M. (1998). Thermodynamic stability of intergranular amorphous films in bismuth-doped zinc oxide. *J. Am. Ceram. Soc.* **81**: 89–96.
133. Tanaka, I., Kleebe, H.-J, Cinibulk, M., Bruley, J., Clarke, D., and Ruhle, M. (1994). Calcium concentration dependence of the intergranular film thickness in silicon nitride. *J. Am. Ceram. Soc.* 77: 911–914.
134. Wu, T. H. (1976). *Soil Mechanics*, Boston, MA: Allyn and Bacon, Inc.
135. Bridgewater, J. (1994). Mixing and segregation in particle flow, *Granular Matter, An Interdisciplinary Approach*, Metha, A., ed., New York: Springer-Verlag.
136. Clarke, D. R. (1987). On the equilibrium thickness of intergranular glass phases in ceramic materials. *J. Am. Ceram. Soc.* **70**: 15–22.
137. Lee, S.-M., Chaix, J.-M., Martin, C. L., Allibert, C. H., and Kang, S. J. (1999). Computer simulation of particle rearrangement in the presence of liquid. *Met. Mater.* **5**: 197–204.
138. Morey, G. W. (1938). *Properties of Glass*, p. 211, New York: Reinhold Publ. Co.
139. Mills, K. C., and Keene, B. J. (1987). Physical properties of BOS slags. *Int. Mater. Rev.* **32**: 1–120.
140. Cannon, R. M., Saiz, E., Tomsia, A. P., and Carter, W. C. (1995). Reactive wetting taxonomy. *Mat. Res. Soc. Symp. Proc.* **357**: 279–292.
141. Huang, K. S., German, R. M., and Lenel, F. V. (1986). Capillary forces in the early stages of liquid phase sintering. *Rev. Powder Metall. Phys. Ceram.* **3**: 113–164.
142. Carter, W. C. (1996). Surface evolver as a tool for materials science research, *Mathematics of Microstructural Evolution*, Chen, L., Fulz, B., Cahn, J., Manning, J., Morral, J., and Simmons, J., eds., Warrendale, PA: TMS.
143. Cambier, F., and Leriche, A. (1996). Vitrification, *Materials Science and Technology*, Vol. 17B, pp. 123–144, Cahn, R. W., Haasen, P., and Kramer, E. J., eds., New York: VCH.
144. Bergeron, C. G., and Risbud, S. H. (1984). *Introduction to Phase Equilibria in Ceramics*, Columbus, OH: American Ceramic Soc.
145. De Jonghe, L. C., and Goo, E. (1978). Transient eutectics in sintering of sodium beta alumina, *Processing of Crystalline Ceramics*, pp. 433–441, Palmour III, H., Davis, R. F., and Hare, R. T., eds., New York: Plenum Press.
146. Chick, L. A., Liu, J., Stevenson, J. W., Armstrong, T. R., McCready, D. E., Maupin, G. D., Coffey, G. W., and Coyle, C. A. (1997). Phase transitions and transient liquid-phase sintering in calcium-substituted lanthanum chromite. *J. Am. Ceram. Soc.* **80**: 2109–2120.
147. Yangyun, S., and Brook, R. J. (1985). Preparation of zirconia toughened ceramics by reaction sintering. *Sci. Sintering* **17**: 35–47.
148. Jack, K. H., and Wilson, W. I. (1972). *Nat. Phys. Sci.* **238**: 28–34.
149. Bartsch, M., Saruhan, B., Schmucker, M., and Schneider, H. (1999). Novel low-temperature processing route of dense mullite ceramics by reaction sintering of amorphous SiO_2-coated gamma-Al_2O_3 particle nanocomposites. *J. Am. Ceram. Soc.* **82**: 1388–1392.
150. Leblud, C., Anseau, M. R., di Rupo, E., Cambier, F., and Fierens, P. (1981). Reaction sintering of $ZnO–Al_2O_3$ mixtures. *J. Mater. Sci. Lett.* **16**: 539–544.

151. Sacks, M. D., Bozkurt, N., and Scheiffele, G. (1991). Fabrication of mullite and mullite–matrix composites by transient viscous sintering of composite powders. *J. Am. Ceram. Soc.* **74**: 2428–2437.
152. Claussen, N., and Jahn, J. (1980). Mechanical properties of sintered, in situ-reacted mullite–zirconia composites. *J. Am. Ceram. Soc.* **63**: 228–229.
153. Koga, N., and Criado, J. M. (1998). Kinetic analyses of solid-state reactions with a particle size distribution. *J. Am. Ceram. Soc.* **81**: 2901–2909.
154. Hulbert, S. F., and Popovich, M. J. (1969). Kinetics and mechanisms of the reaction between TiO_2 and $SrCO_3$. *Mater. Sci. Res.* **4**: 442.
155. Rahaman, M. N., and De Jonghe, L. C. (1993). Reaction sintering of zinc ferrite during constant rates of heating. *J. Am. Ceram. Soc.* **76**: 1739–1744.
156. Hong, W. S., De Jonghe, L. C., Yang, X., and Rahaman. M. N. (1995). Reaction sintering of $ZnO–Al_2O_3$. *J. Am. Ceram. Soc.* **78**: 3217–3224.
157. Evans, A. G. (1982). Considerations of inhomogeneity effects in sintering. *J. Am. Ceram. Soc.* **65**: 497–501.
158. Scherer, G. W. (1987). Sintering with rigid inclusions. *J. Am. Ceram. Soc.* **70**: 719–725.
159. Raj, R., and Bordia, R. K. (1984). Sintering behavior of bimodal powder compacts. *Acta Metall.* **32**: 1003–1019.
160. Hsueh, C.-H., Evans, A. G., Cannon, R. M., and Brook, R. J. (1986). Viscoelastic stresses and sintering damage in heterogeneous powder compacts. *Acta Metall.* **34**: 927–936.
161. Bordia, R. K., and Scherer, G. W. (1988). On constrained sintering: parts 1, 2, and 3. *Acta Metall.* **36**: 2393–2416.
162. Sudre, O., Bao, G., Fan, B., Lange, F. F., and Evans, A. G. (1992). Effect of inclusions on densification: II, numerical model. *J. Am. Ceram. Soc.* **75**: 525–531.
163. Hu, C.-H., and Rahaman, M. N. (1992). Factors controlling the sintering of ceramic particulate composites: II, coated inclusion particles. *J. Am. Ceram. Soc.* **75**: 2066–2070.
164. Scherer, G. W., and Garino, T. J. (1985). Viscous sintering on a rigid substrate. *J. Am. Ceram. Soc.* **68**: 216–220.
165. Hu, M., Thouless, M., and Evans, A. G. (1988). The decohesion of thin films from brittle substrates. *Acta Metall.* **36**: 1301–1308.
166. Kueper, T. W. (1992). Sol–Gel ceramic electrolyte films on porous substrates, PhD Thesis, University of California, Berkeley, CA.
167. Miller, K. T., Lange, F. F., and Marshall, D. B. (1990). The instability of polycrystalline films: experiment and theory. *J. Mater. Res.* **5**: 151–157.
168. Anderson, H. U. (1974). Influence of oxygen activity on the sintering of $MgCr_2O_4$. *J. Am. Ceram. Soc.* **57**: 34–37.
169. Hwang, S.-L., Becher, P. F., and Lin, H.-T. (1997). Desintering process in the gas-pressure sintering of silicon nitride. *J. Am. Ceram. Soc.* **80**: 329–335.
170. Coble, R. L. (1962). Sintering of alumina: effect of atmosphere. *J. Am. Ceram. Soc.* **45**: 123–127.
171. Snow, G. S. (1973). Improvements in atmosphere sintering of transparent PLZT ceramics. *J. Am. Ceram. Soc.* **56**: 479–480.
172. Woolfrey, J. L., and Bannister, M. J. (1972). Nonisothermal sintering techniques for studying initial stage sintering. *J. Am. Ceram. Soc.* **55**: 390–397.
173. Boccaccini, A. R., Stumpfe, W., Taplin, D. M. R., and Ponton, C. B. (1996). Densification and crystallization of glass powder compacts during constant heating rate sintering. *Mater. Sci. Eng. A* **219**: 26–31.
174. Chu, M-Y., Rahaman, M. N., and Brook, R. J. (1991). Effect of heating rate on sintering and coarsening. *J. Am. Ceram. Soc.* **74**: 1217–1225.

175. Lange, F. F. (1989). Approach to reliable powder processing, *Ceramic Trans.*, 1, *Ceramic Powder Science IIB*, pp. 1069–1083, Messing, G. L., Fuller, E. R., and Hausner, H., eds., American Ceramic Society. Westerville, OH.
176. Duncan, J. H., and Budgen, W. G. (1981). Two-peak firing of beta-alumina. *Proc. Br. Ceram. Soc.* **31**: 221–228.
177. Lin, F. J. T., De Jonghe, L. C., and Rahaman, M. N. (1997). Microstructure refinement of sintered alumina by a two-step sintering technique. *J. Am. Ceram. Soc.* **80**: 2269–2278.
178. Chu, M.-Y., De Jonghe, L. C., Lin, M. F. K., and Lin, F. J. T. (1991). Pre-coarsening to improve microstructure and sintering of powder compacts. *J. Am. Ceram. Soc.* **74**: 2902–2911.
179. Huckabee, M. L., Hare, T. M., and Palmour III, H. (1978). Rate controlled sintering as a processing methods, *Processing of Crystalline Ceramics*, pp. 205–215, Palmour III, H., Davis, R. F., and Hare, R. T., eds., New York: Plenum Press.
180. Huckabee, M. L., Paisley, M. J., Russell, R. L., and Palmour III, H. (1994). RCS-taking the mystery out of densification profiles. *J. Am. Ceram. Soc.* **73**: 82–89.
181. Agarwal, G., Speyer, R. F., and Hackenberger, W. S. (1996). Microstructural development of ZnO using a rate-controlled sintering dilatometer. *J. Mat. Res.* **11**: 671–678.
182. Ragulya, A. V. (1998). Rate-controlled synthesis and sintering of nanocrystalline barium titanate powder. *Nanostruct. Mater.* **10**: 349–355.
183. Pickup, H., Eisele, U., Gilbart, E., and Brook, R. J. (1985). Analysis of coarsening and densification kinetics during the heat treatment of nitrogen ceramics, *Non-Oxide Technical and Engineering Ceramics*, pp. 41–51, Hampshire, S., ed., New York: Elsevier Applied Science.
184. Brook, R. J. (1982). Fabrication principles for the production of ceramics with superior mechanical properties. *Proc. Br. Ceram. Soc.* **32**: 7–15.
185. Johnson, D. L. (1984). Ultra-rapid Sintering. *Mater. Sci. Res.* **10**: 243–247.
186. Fang, Y., Agrawal, D. K., Roy, D., and Roy, R. (1994). Microwave sintering of hydroxyapatite ceramics. *J. Mater. Res.* **9**: 180–189.
187. Birnboim, A., and Carmel, Y. (1999). Simulation of microwave sintering of ceramic bodies with complex geometry. *J. Am. Ceram. Soc.* **82**: 3024–3033.
188. Birnboim, A. *et al.* (1998). Comparative study of microwave sintering of zinc oxide at 2.45, 30, and 83 GHz. *J. Am. Ceram. Soc.* **81**: 1493–2001.
189. Cheng, J., Agrawal, D., Roy, R., and Jayan, P. S. (2000). Continuous microwave sintering of alumina abrasive grits. *J. Mater. Process. Technol.* **108**: 26–32.
190. Henrichsen, M., Hwang, J., Dravid, V. P., and Johnson, D. L. (2000). Ultrarapid phase conversion in beta-alumina tubes. *J. Am. Ceram. Soc.* **83**: 2861–2869.
191. Gao, L., Shen, Z., Miyamoto, H., and Nygren, M. (1999). Superfast densification of oxide/oxide ceramic composites. *J. Am. Ceram. Soc.* **82**: 1061–1069.
192. Takeuchi, T., Tabuchi, M., and Kageyama, H. (1999). Preparation of dense $BaTiO_3$ ceramics with submicrometer grains by spark plasma sintering. *J. Am. Ceram.* **82**: 939–943.
193. Gutmanas, E., and Gotman, I. (1999). Dense high-temperature ceramics by thermal explosion under pressure. *J. Eur. Ceram. Soc.* **19**: 2381–2390.
194. Vasilos, T., and Spriggs, R. M. (1966). Pressure sintering of ceramics, *Progress in Ceramic Science*, Vol. 4, pp. 97–107, Burke, J., ed., New York: Pergamon Press.
195. POCO Graphite Inc, P.O. Box 2121, Decatur, TX 76234, USA.
196. Carborundum Boron Nitride, 168 Creekside Drive, Amherst, NY 14228, USA.
197. Lange, F. F., and Terwilliger, G. R. (1973). The powder vehicle hot-pressing technique. *Bull. Am. Ceram. Soc.* **52**: 563–569.
198. Rigby, F. (1981). Development of hot pressing techniques at Springfields Nuclear Laboratories. *Proc. Br. Ceram. Soc.* **31**: 249–257.

199. Wilkinson, D. S., and Ashby, M. F. (1975). The development of pressure sintering maps. *Mater. Sci. Res.* Vol. 10, *Sintering and Catalysis*, pp. 473–478, Kuczynski, G., ed., New York: Plenum Press.
200. Zhang, G.-J., Deng, Z.-Y., Kondo, N., Yang, J.-F., Ohji, and T. (2000). Reactive hot pressing of ZrB_2–SiC composites. *J. Am. Ceram. Soc.* **83**: 2330–2338.
201. Wen, G., Li, S. B., Zhang, B. S., and Guo, Z. X. (2000). Processing of *in situ* toughened B–W–C composites by reaction hot pressing of B_4C and WC. *Scripta Mater.* **43**: 853–858.
202. Hosoi, K., Hashida, T., Takahashi, H., Yamasaki, N., and Korenaga, T. (1996). New processing technique for hydroxyapatite ceramics by the hydrothermal hot-pressing method. *J. Am. Ceram. Soc.* **79**: 2771–2779.
203. Yanagisawa, K., Ioku, K., and Yamasaki, N. (1997). Formation of anatase porous ceramics by hydrothermal hot-pressing of amorphous titania spheres. *J. Am. Ceram. Soc.* **80**: 1303–1306.
204. He, Y. J., Winnubst, A. J. A., Verweij, A., and Burggraaf, A. J. (1994). Sinter-forging of zirconia toughened alumina. *J. Mater. Sci.* **29**: 6505–6512.
205. Kondo, N., Suzuki, Y., and Ohji, T. (1999). Superplastic sinter-forging of silicon nitride with anisotropic microstructure formation. *J. Am. Ceram. Soc.* **82**: 1067–1073.
206. Saller, H., Paprocki, S., Dayton, R., and Hodge, E. (1964). A method of bonding. Can. Patent # 680160.
207. Zhou, Y.-C., and Rahaman, M. N. (1993). Hydrothermal synthesis and sintering of ultra-fine CeO_2 powders. *J. Mater. Res.* **8**: 1680–1686.
208. Rahaman, M. N., and Zhou, Y.-C. (1995). Effect of dopants on the sintering of ultra-fine CeO_2 powder. *J. Eur. Ceram. Soc.* **15**: 939–950.
209. Coble, R. L. (1973). Effects of particle size distribution in initial stage sintering. *J. Am. Ceram. Soc.* **56**: 461–466.
210. Onoda, G. Y., and Messing, G. L. (1978). Packing and sintering relations for binary powders. *Mater. Sci. Res.* **11**: 99–122.
211. Smith, J. P., and Messing, G. L. (1984). Sintering of bimodally distributed alumina powders. *J. Am. Ceram. Soc.* **67**: 238–242.
212. Patterson, B. R., and Benson, L. A. (1983). The effect of powder size distribution on sintering. *Progr. Powder Met.* **39**: 215–230.
213. German, R. M. (1992). Sintering densification for powder mixtures of varying distribution widths. *Acta Metall. Mater.* **40**: 2085–2089.
214. Chapell, J. S., Ring, T. A., and Birchall, J. D. (1986). Particle size distribution effects on sintering rates. *J. Appl. Phys.* **60**: 383–391.
215. Yamaguchi T., and Kosha, H. (1981). Sintering of acicular Fe_2O_3 particles. *J. Am. Ceram. Soc.* **64**: C84–C85.
216. Mikijelj, B., and Whittemore, O. J. (1987). Grain cuboidization during sintering of MgO–$MgCl_2$ (1%). *Am. Ceram. Soc. Bull.* **66**: 809–812.
217. Kumar, P., and Johnson, D. L. (1974). Sintering of CoO. *J. Am. Ceram. Soc.* **57**: 62–64.
218. Thompson, R. J., and Munir, Z. A. (1982). Influence of particle size on the sintering kinetics of ultrapure sodium chloride. *J. Am. Ceram. Soc.* **65**: 312–316.
219. Wang, Z. Y., Harmer, M. P., and Chou, Y. T. (1986). Pore-boundary configurations in lithium fluoride. *J. Am. Ceram. Soc.* **69**: 735–740.
220. Slamovich, E. B., and Lange, F. F. (1990). Densification behavior of single-crystal and polycrystalline spherical particles of zirconia. *J. Am. Ceram. Soc.* **73**: 3368–3375.
221. Edelson, L. H., and Glaeser, A. M. (1988). Role of particle substructure in the sintering of monosized titania. *J. Am. Ceram. Soc.* **71**: 225–235.
222. Edelson, L. H., and Glaeser, A. M. (1988). Effects of thermal pretreatment on coarsening of nominally monodispersed titania. *J. Am. Ceram. Soc.* **71**: C198–C201.

223. Rhodes, W. W. (1981). Agglomerate and particle size effects on sintering of yttria-stabilized zirconia. *J. Am. Ceram. Soc.* **64**: 19–22.
224. Barringer, E. A., and Bowen, H. K. (1982). Formation, packing, and sintering of monodisperse TiO_2 powders. *J. Am. Ceram. Soc.* **65**: C199–C201.
225. Barringer, E. A., and Bowen, H. K. (1988). Effect of particle packing on the sintered microstructure. *Appl. Phys. A* **45**: 271–275.
226. Liniger, E., and Raj, R. (1987). Packing and sintering of two-dimensional structures made from bimodal particle size distributions. *J. Am. Ceram. Soc.* **70**: 843–849.
227. Yeh, T. S., and Sacks, M. D. (1990). Effect of green microstructure on sintering of alumina. *Ceram. Trans.* **7**: 309–331.
228. Cesarano III, J., and Aksay, I. A. (1988). Processing of highly concentrated aqueous α-alumina suspensions stabilized with polyelectrolytes. *J. Am. Ceram. Soc.* **71**: 1062–1067.
229. Weiser, M. W., and De Jonghe, L. C. (1986). Rearrangement during sintering in two-dimensional arrays. *J. Am. Ceram. Soc.* **69**: 822–826.
230. Exner, H. E. (1979). Principles of single phase sintering. *Rev. Powder Metall. Phys. Ceram.* **1**: 1–251.
231. Lange, F. F. (1984). Sinterability of agglomerated powders. *J. Am. Ceram. Soc.* **67**: 83–89.
232. Lin, M., De Jonghe, L. C., and Rahaman, M. N. (1987). Creep-sintering of MgO powder compacts. *J. Am. Ceram. Soc.* **70**: 360–366.
233. Harmer, M. P., and Zhao, J. (1987). Effect of pores on microstructure development, *Ceramic Micorstructures '86, Mater. Sci. Res.*, Vol. 21, pp. 455–464, Pask, J. A., and Evans, A. G., eds., New York: Plenum Press.
234. Chen, P. L., and Chen, I.-W. (1997). Sintering of fine oxide powders: II, sintering mechanisms. *J. Am. Ceram. Soc.* **80**: 637–645.
235. Rahaman, M. N., De Jonghe, L. C., and Chu, M.-Y. (1991). Effect of green density on densification and creep during sintering. *J. Am. Ceram. Soc.* **74**: 514–519.
236. Bruch, C. A. (1962). Sintering kinetics for the high density alumina process. *Am. Ceram. Soc. Bull.* **41**: 799–806.
237. Palmour, H., Geho, M., Russell, R. L., and Hare, T. M. (1992). Study of D_0 effects on subsequent densification behavior in spinel and alumina ceramics, *Sintering '91*, pp. 37–44, Chaklader, A. C. D., and Lund, J. A., eds., Brookfield, VT: Trans Tech.
238. Mansour, N. A. L., and White, J. (1963). Causes of residual porosity in sintered uranium dioxide. *Powder Metall.* **6**: 108–121.
239. Occhionero, M. A., and Halloran, J. W. (1984). The influence of green density on sintering, *Mater. Sci. Res.* **16**: 89–102.
240. Bennison, S. J., and Harmer, M. P. (1985). Swelling of hot-pressed alumina. *J. Am. Ceram. Soc.* **68**: 591–597.
241. Readey, D. W. (1990). Vapor transport and sintering. *Ceram. Trans.* **7**: 86–110.
242. Quadir, T., and Readey, D. W. (1989). Microstructure development of zinc oxide in hydrogen. *J. Am. Ceram. Soc.* **72**: 297–302.
243. Lee, J., and Readey, D. W. (1984). Microstructure development of Fe_2O_3 in HCl vapor. *Mater. Sci. Res.* **16**: 145–157.
244. Readey, M. J., and Readey, D. W. (1987). Sintering TiO_2 in HCl atmospheres. *J. Am. Ceram. Soc.* **70**: C358–C361.
245. Thompson, A. M., and Harmer, M. P. (1993). Influence of atmosphere on the final stage sintering kinetics of ultra-high-purity alumina. *J. Am. Ceram. Soc.* **76**: 2248–2256.
246. Anderson, P. J., and Morgan, P. L. (1964). Effect of water vapor on the sintering of MgO. *Trans. Faraday Soc.* **60**: 930–937.
247. Wong, B., and Pask, J. A. (1979). Experimental analysis of sintering of MgO powder compacts. *J. Am. Ceram. Soc.* **62**: 141–146.

248. Northrop, D. A. (1967). Vaporization of lead zirconate-lead titanate materials. *J. Am. Ceram. Soc.* **50**: 441–445.
249. Northrop, D. A. (1968). Vaporization of lead zirconate-lead titanate materials: II, hot-pressed compositions near theoretical density. *J. Am. Ceram. Soc.* **51**: 357–361.
250. Mitomo, M. (1976). Pressure sintering of Si_3N_4. *J. Mater. Sci.* **11**: 1103–1107.
251. Ownby, P. D., and Jungquist, G. E. (1972). Final sintering of Cr_2O_3. *J. Am. Ceram. Soc.* **55**: 433–436.
252. Ownby, P. D. (1973). Oxidation state control of volatile species in sintering. *Mater. Sci. Res.* **6**: 431–437.
253. Anderson, H. U. (1978). Fabrication and property control of $LaCrO_3$ based oxides. *Mater. Sci. Res.* **11**: 469–477.
254. Callister, W. D., Johnson, M. L., Cutler, I. B., and Ure, R. W., Jr. (1979). Sintering chromium oxide with the aid of TiO_2. *J. Am. Ceram. Soc.* **62**: 208–211.
255. Reynolds III, T. (1976). Firing, *Treatise on Materials Science and Technology, Vol. 9*, pp. 199–215, Wang, F. F. Y., ed., New York: Academic Press.
256. Reijnen, P. (1969). Nonstoichiometry and sintering of ionic solids, *Reactivity of Solids*, pp. 99–114, Mitchell, J. W., DeVries, R. C., Roberts, R. W., and Cannon, P., eds., New York: Wiley.
257. Yan, M. F. (1980). Grain growth in Fe_3O_4. *J. Am. Ceram. Soc.* **63**: 443–447.

Handbook of Advanced Ceramics
S. Sōmiya *et al.* (Eds.)

5.1 Precursor-Derived Ceramics

MARKUS WEINMANN and FRITZ ALDINGER
Max-Planck-Institut für Metallforschung, Heisenbergstrasse 3, D-70569 Stuttgart, Germany

5.1.1 INTRODUCTION

The condensation of organometallic compounds into merely inorganic materials by proper thermal treatment under controlled atmosphere is a fairly new and simple process of producing new types of ceramics. It allows for an easy control of elemental composition, chemical homogeneity as well as atomic architecture. Moreover, it provides a means for controlling and adjusting design and microstructure shaping of ceramic materials in a way not known from other techniques. In recent years, this process scheme has therefore created substantial interest, both scientifically and for practical purposes.

The overall process for the formation of precursor-derived ceramics (PDCs) is outlined in Figure 5.1.1.

It consists of three major steps. First, suitable monomers are transformed into polymeric macromolecules referred to as precursors, which are subsequently cross-linked at moderate temperatures into preceramic networks with defined rheological behavior in order to provide proper processing capabilities. The networks are finally transformed into inorganic materials by a thermal heat treatment at temperatures usually ranging from 1000 to 1400°C.

The general idea behind the process concept is to generate preferred structural features in the organometallic precursors (this can mostly be realized by comparably simple preparative organometallic chemistry) and to subsequently transform the precursors into ceramics under retention of the specially designed building blocks. Thermolysis must thereby include a series of controllable condensation steps, which are aimed for the specific architecture of the desired ceramic material. As a consequence of being mostly reaction controlled, precursor thermolysis is especially valuable for the densification of covalently bonded inorganics.

Conventionally, low diffusion capability of non-oxide ceramics even at very high temperatures is overcome by the use of sintering aids. Such additives are

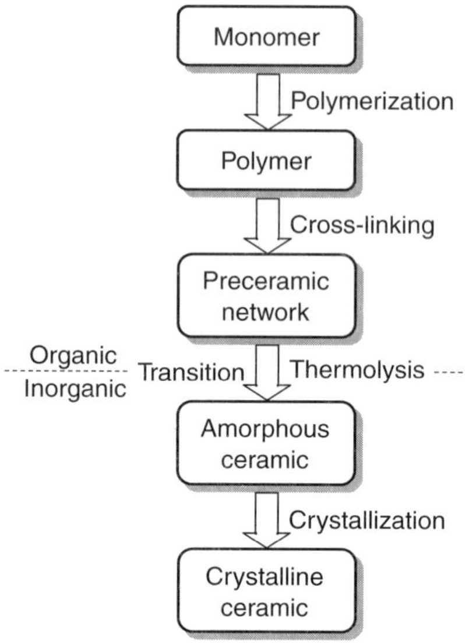

FIGURE 5.1.1 Flow diagram for the preparation of precursor-derived ceramics.

mostly of oxide type and therefore degrading the otherwise unique properties of non-oxide ceramics such as thermal, chemical, and mechanical stability over wide temperature ranges. In contrast, thermolysis of ceramic precursors provides a means to realize the production of non-oxide ceramic, thereby retaining their excellent properties without compromise.

One remarkable consequence of the low atomic mobility is that as-thermolyzed PDCs can be formed, which remain amorphous up to rather high temperatures. The reason for this is that substantial thermal activation is required for nucleation, grain growth, and the formation of crystalline phases. Crystallization thereby very much depends on chemical/elemental composition and microstructure of the materials and the atmosphere applied. Under certain conditions, crystallization can be retarded even up to 1800°C providing novel glass-like structural materials with superb thermomechanical properties. However, crystalline materials are also of great interest since their microstructure formation can be controlled during devitrification to much extend providing, for example, a means for stabilizing nano-sized morphologies.

This chapter intends to review the fundamentals of synthetic approaches and the processing of non-oxide silicon-based PDCs, to reveal possibilities of microstructure development and to give hints on their potential application. It has been shown that the basis for microstructure design of PDCs is the type of macromolecule, that is, its chemical composition and molecular structure. Therefore, first, synthetic procedures, which deliver different types of organometallic, silicon-based non-oxide ceramic precursors, are evaluated. Next, thermolysis reactions will be considered, since the structure of the ceramic materials is also widely determined by the structure of the molecular precursors. Further, phase reactions controlling the high-temperature stability of these materials will be discussed and finally recent trends in technologically relevant applications be reported.

5.1.2 PRECURSOR SYNTHESIS

5.1.2.1 Precursors to Silicon Carbide

Silicon carbide is the only (solid) binary compound of silicon and carbon. It is used for multiple purposes. However, the most important feature is its outstanding mechanical behavior. Silicon carbide is one of the most frequently used ceramic hard materials [1].

Silicon carbide has an extremely low self-diffusion coefficient, even at very high temperature. Processing, that is, compaction of silicon carbide powders thus requires sintering aids, which may—to a certain extent—unfavorably determine the properties of the processed ceramic materials [2–5].

In contrast, precursor thermolysis of suitable organosilicon compounds is a method that delivers sinter-aid-free silicon carbide. As will be shown in subsequent sections, suitable precursors for this purpose are, for example, polysilanes, polycarbosilanes, polyalkenylsilanes, polyalkynylsilanes, or their mixtures (co-polymers or polymer blends). In the following, the most important of the above-mentioned organosilicon polymers will be reviewed chronologically, starting with polysilanes, which are composed of alternating $[SiR_2]$ units, followed by polysilaethylenes, which are composed of $[SiR_2{-}CR_2]$ building blocks. Finally, polycarbosilanes with alternating $[Si{-}C{-}C]$ units will be discussed.

5.1.2.1.1 Polysilanes

Polysilanes are well-studied polymers [6–8]. However, although many research groups all over the world heavily investigated synthetic approaches, especially with the goal to obtaining polysilanes as precursors to phase-pure silicon carbide, there is only a very limited number of procedures described in the literature that deliver near-stoichiometric silicon carbide.

Polysilanes, also referred to as poly(silylene)s, usually have a linear polymer backbone consisting entirely of silicon atoms to which two organic substituents

R are bonded. When sufficiently purified, poly(diorganosilane)s, $[SiR^1R^2]_n$ (R^1, R^2 = singly bonded organic group) are stable in oxidative and hydrolytic environments and thus can be handled easily without requiring inert gas techniques. Polysilanes are often insoluble, not meltable, and intractable, which impedes their applicability in many practical processes. Chemical and physicochemical properties are strongly influenced by the molecular weight and the nature of the silicon-bonded substituents.

Polysilanes were first synthesized in the early 1920s by Kipping and Sands [6] by a Wurtz–Fittig-type dehalogenation reaction of dichlorodiphenylsilane with molten potassium in xylene delivering poly(diphenylsilane):

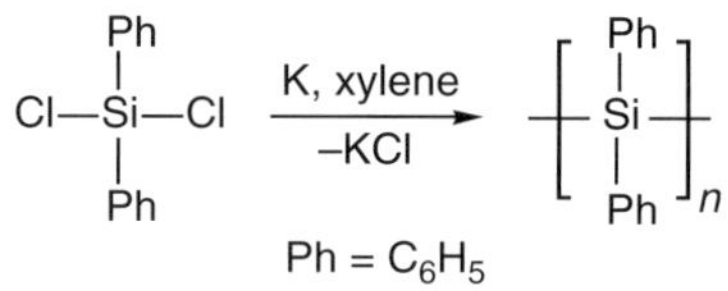

(1)

Burkhardt, in 1949 published the first clear description of the synthesis of a polysilane with a potential for technological application [9]. Dichlorodimethylsilane was reacted with sodium in a Wurtz–Fittig-type reaction in benzene analogous to the above-mentioned procedure:

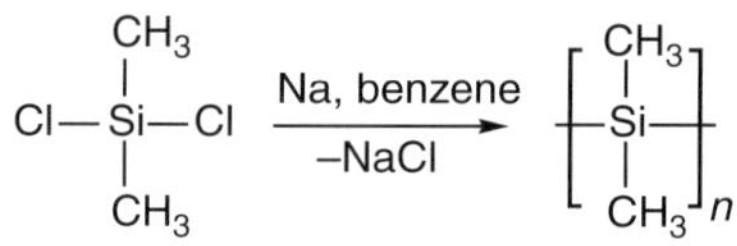

(2)

Poly(dimethysilane) is a white powder, which is poorly soluble in organic solvents and does not melt without decomposition. It will be highlighted below that it represents the first precursor for the production of refractory silicon carbide fibers.

A large number of polysilanes with different silicon-bonded substituents and molecular weight have now been synthesized. Product yields are frequently unsatisfactory because of (i) the low solubility of the polymers in organic solvents and (ii) the expandable processing (purification) including hydrolysis in order to remove excess alkali metal. A further essential drawback of the Wurtz-coupling is the limited functional group tolerance, which allows only for the synthesis of polysilanes with more or less "inert" side groups such as alkyl or aryl groups.

The mechanism of the Wurtz-type coupling of chlorosilanes appears to be very complex. Alkali metal surface reactions are probable but dehalogenation reactions in solution have also to be considered. Moreover, mechanisms based on silyl radicals are discussed as well as silyl anions, which appear to be key species in the chain growth of polysilanes.

An alternative procedure for the synthesis of polysilanes, which delivers products in high purity and higher yields, is the dehydrocoupling of silanes in the presence of suitable catalysts such as η^2-alkynyl titanocene or -zirconocene as described by Chang and Corey [10]:

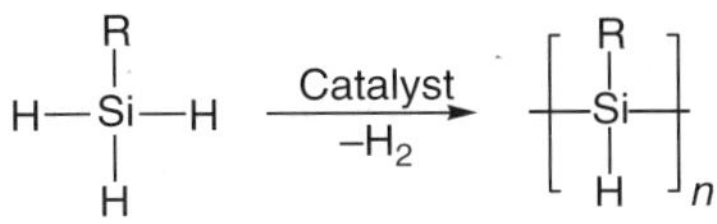

Catalysts:
η^2-Alkynyl-titanocenes,
η^2-Alkynyl-zirconocenes

(3)

In contrast to the Wurtz–Fittig reaction, no solid by-products are formed. The only by-product is hydrogen. As a consequence, purification steps such as hydrolysis and filtration are not required. The authors also suggested a conclusive mechanism of the silane polymerization by dehydrocoupling.

By elimination of the side-on coordinated alkynyl ligand a highly reactive 14VE (VE = valence electron) metallocene species forms which inserts into a Si—H bond of the silane, thus forming a silylmetallocene. This species most likely eliminates hydrogen whereby a metalla-silanediyl intermediate forms, which then adds a silane molecule. The disilylated metallocene rearranges by a hydrogen shift to the metal center and Si—M bond cleavage (see Scheme 5.1.1).

Woo *et al.* [11] investigated dehydrocoupling reactions of silanes using zirconocene and hafnocene hydride complexes as catalysts and postulated alternative mechanisms. In contrast to the above, they suggested a mechanism that is based on two σ-bond metathesis reactions that pass through four-center transition states (see Scheme 5.1.2).

5.1.2.1.2 Polycarbosilanes

In contrast to polysilanes, polymer backbones in polycarbosilanes are not exclusively composed of silicon atoms. Here, aliphatic units link the silicon atoms. Frequently, also unsaturated units as well as aromatic ring systems served as building blocks for the formation of polycarbosilanes.

Polycarbosilanes of the general structure $[R_2Si{-}CH_2]_n$ have been synthesized by a variety of reaction pathways. The most important and intensely investigated access to polycarbosilanes certainly is the Yajima process [12]. This process was attractive as it thermally converts insoluble and unmeltable poly(dimethylsilane) into soluble and processable

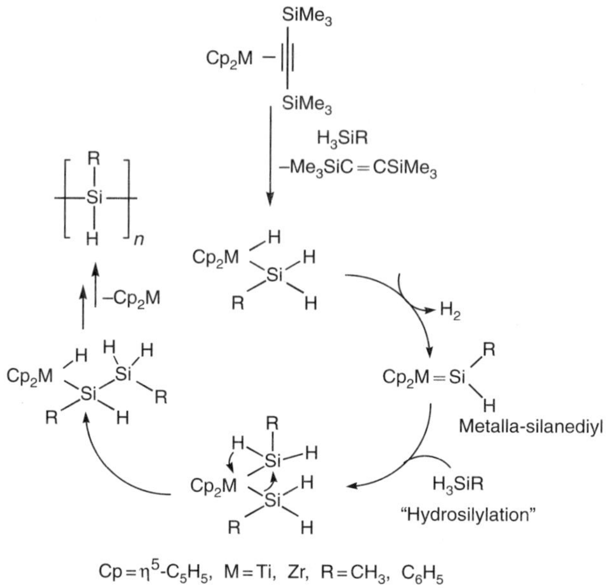

SCHEME 5.1.1 Suggested mechanism for the formation of polysilanes by metallocene-catalyzed dehydrogenative coupling of alkylsilanes, $RSiH_3$, via metalla-silanediyl intermediates [10].

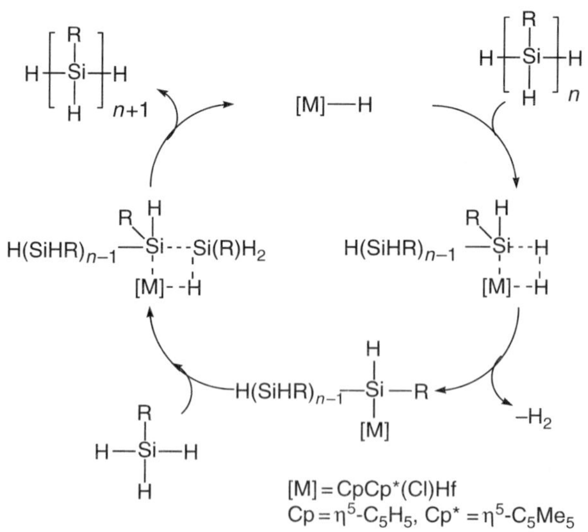

SCHEME 5.1.2 Proposed mechanism for the formation of polysilanes by hafnocene-catalyzed dehydrogenative coupling of alkylsilanes, $RSiH_3$, via metalla-silacyclobutane transition state [11].

poly(methylsilylene-methylene), $[HSi(CH_3){-}CH_2]_n$. This precursor can be either cast or spun. Silicon carbide fibers obtained from this precursor are commercially available under the trade names Nicalon™ or High Nicalon™, depending on the processing applied (compare Section 5.1.5.4)

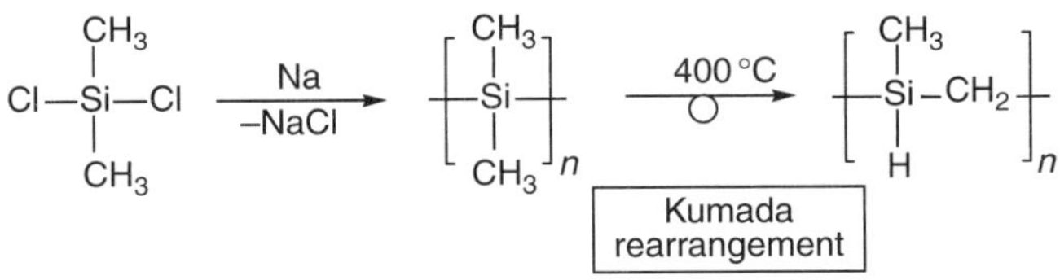

(4)

The key step in the Yajima process is a rearrangement of the polymer backbone involving pendent methyl groups, referred to as Kumada rearrangement or Kumada reaction, which occurs at around 400°C. It is a methylene migration from one of the silicon-bonded methyl groups attached to the polymer backbone to the polymer chain. The process is radical-induced.

The mechanism including the single key steps in the Kumada rearrangement of poly(dimethylsilane) is shown in Scheme 5.1.3. Initially, Si—Si bonds in the polymer are homolytically cleaved forming silyl radicals, followed by

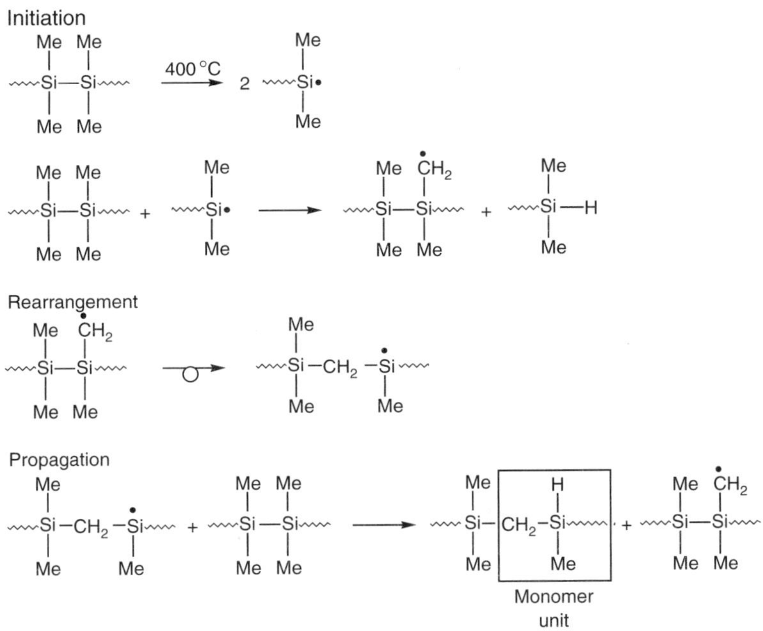

SCHEME 5.1.3 Key steps in the Kumada rearrangement of poly(dimethylsilane) [12].

H-radical migration from a methyl group to the silyl radical (initiation). The methylene radical that is thereby generated inserts into a Si—Si bond whereby a new silyl radical forms (rearrangement). Propagation proceeds by a repeated H-radical migration to the silicon radical with the formation of a methylene radical. Termination most probably takes place by radical recombination.

Polycarbosilanes with C1 building blocks can also be obtained by Grignard reactions of chloromethyl–trichlorosilane (Cl_3Si-CH_2Cl) in diethyl ether solution [13]. Following treatment of the as-obtained poly(dichlorosilylene-methylene) with $LiAlH_4$ delivers branched polycarbosilane with an idealized structure, $[H_2Si-CH_2]_n$. Due to its compositional analogy to polyethylene, it is also referred to as polysilaethylene. Thermolysis of polysilaethylene delivers near-stoichiometric SiC:

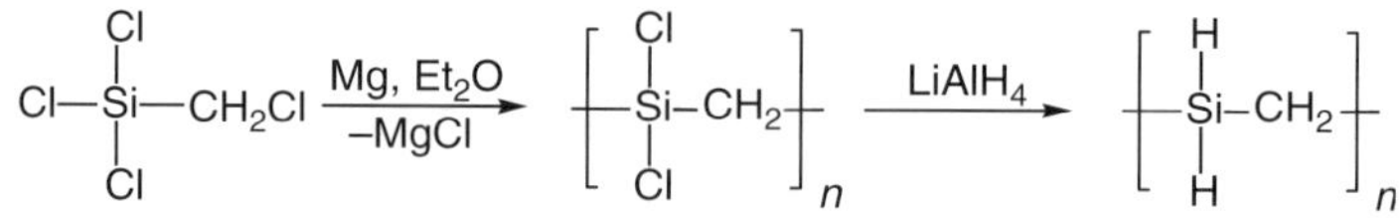

(5)

The disadvantages of this reaction pathway are long reaction times that are required for obtaining sufficient polymer yields. Changing the solvent towards a more polar one, such as tetrahydrofuran, strongly influences the selectivity of the reaction. If starting from chloromethyldialkyl- or diarylchlorosilanes, according to reaction pathway (a), 1,3-disilacyclobutane derivatives are obtained (see Scheme 5.1.4).

The transformation into linear polymers can be achieved by ring-opening polymerization (ROP) reactions using Pt-group metal complexes, such as

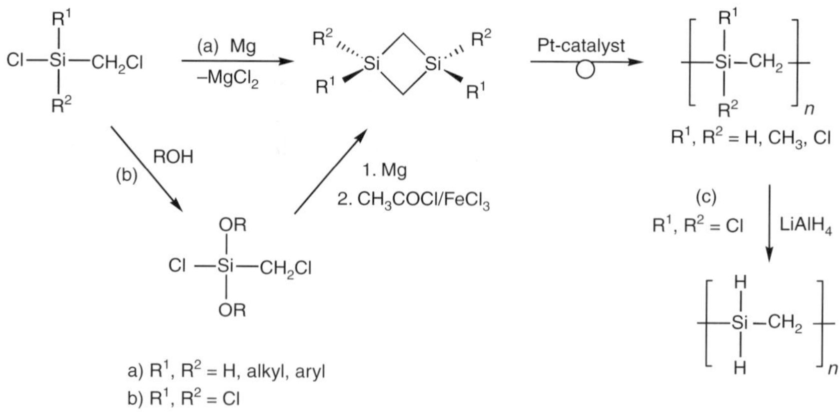

SCHEME 5.1.4 Synthesis of polycarbosilanes by Pt-catalyzed ring opening polymerization of 1,3-disilacyclobutane derivatives [15].

chloroplatinic acid as catalysts [14]. Interrante and Wu [15] developed a modified multi-step polysilaethylene synthesis starting from methyltrichlorosilane, $Cl_3Si{-}CH_3$, which is a cheap, commercially available compound. Halogenation using chlorine in the presence of UV light delivers chloromethyl–trichlorosilane, $Cl_3Si{-}CH_2Cl$ (this reaction is not shown in Scheme 5.1.4. $Cl_3Si{-}CH_2Cl$ could not directly be converted into 1,1,3,3-tetrachloro-1,3-disilacyclobutane. Rather, two of the three silicon-bonded chlorine atoms were initially substituted by ethoxy (—OEt) groups, which could selectively be realized in alcoholysis reactions prior to the reductive coupling. The ethoxy groups in the target compound were subsequently replaced with chlorine atoms by a treatment with acetic chloride acid in the presence of $FeCl_3$. Pt-catalyzed ROP and subsequent reduction with $LiAlH_4$ then delivered linear polysilaethylene in approximately 60% yield.

Procopio and Berry published a procedure for the synthesis of oligosilaethylene by dehydrogenative coupling reactions (see Scheme 5.1.5).

Oligomeric carbosilanes were obtained from trimethylsilane under mild conditions in the presence of $H_3Ru(SiMe_3)(PMe_3)_3$, which was *in situ* obtained from $H_2Ru(PMe_3)_4$ and $HSiMe_3$. The proposed mechanism involves the

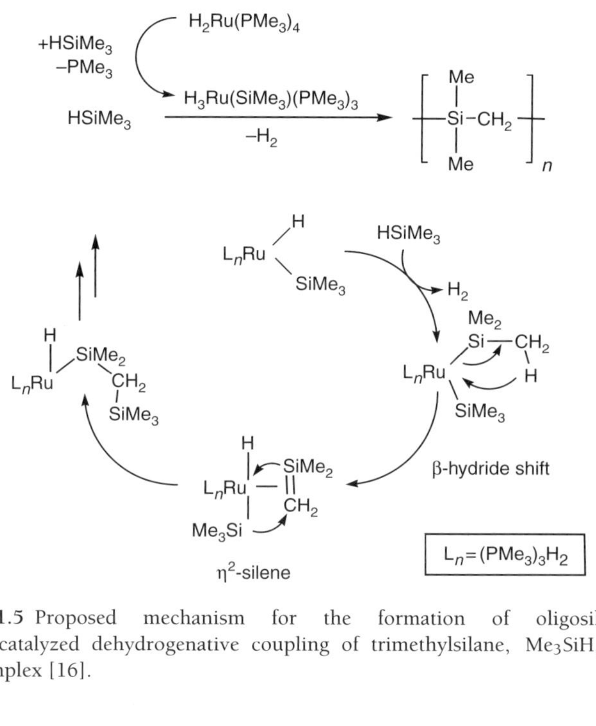

SCHEME 5.1.5 Proposed mechanism for the formation of oligosilaethylene by $H_2Ru(PMe_3)_4$-catalyzed dehydrogenative coupling of trimethylsilane, Me_3SiH, via η^2-silene ruthenium complex [16].

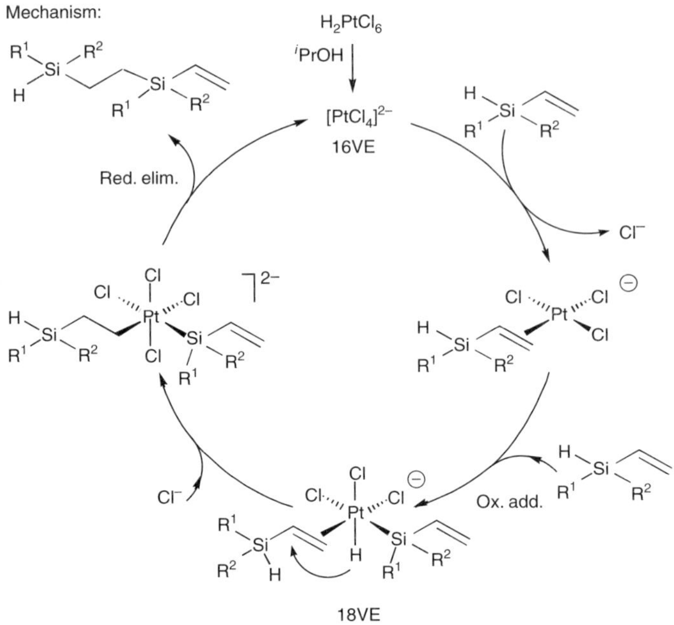

SCHEME 5.1.6 Proposed mechanism for the formation of polycarbosilanes by $PtCl_4^{2-}$ catalyzed hydrosilylation of vinylsilanes [19].

formation of a disilyl ruthenium complex, which upon a β-hydride shift transforms into a η^2-silylene species. (Analogous to ethylene, this is the monomer unit of poly(silylene-methylene)!) Insertion of the silylene into the remaining Ru—Si bond results in formation of a singly silylated ruthenium complex and chain growth by one $SiMe_2CH_2$ unit.

Polycarbosilanes with aliphatic C_2 building blocks in the polymer backbone are precursors to SiC, which have reached remarkable attractiveness. Most of the publications in this field trace back to work of Corriu *et al.* [17, 18] who have applied hydrosilylation of vinylhydridosilanes, $(R^1)(R^2)SiH(HC{=}CH_2)$ (R^1, R^2=H, Cl, alkyl, aryl, NR_2), for the preparation of linear polycarbosilanes of the general type $[(R^1)(R^2)Si{-}C_2H_4]_n$ [17, 18]. Even though this type of hydrosilylation does not occur regioselectively, formation of the β-addition product ($C_2H_4{=}CH_2CH_2$) is strongly preferred (see Scheme 5.1.6).

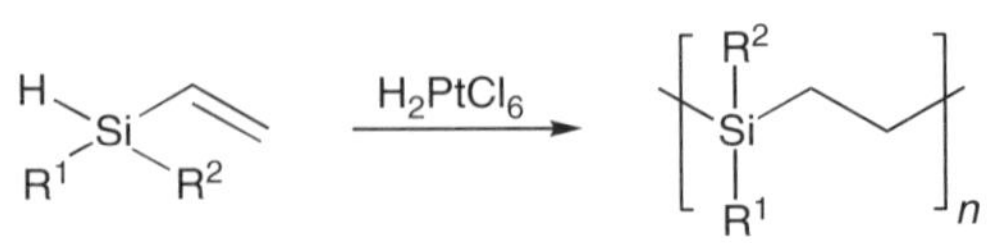

(6)

Polymerization by hydrosilylation of vinylsilanes requires a catalyst such as H_2PtCl_6 in 2-propanol solution, also known as Speyers' reagent [19]. In 2-propanol solution, $PtCl_6^{2-}$ is reduced to $PtCl_4^{2-}$, which is the catalytically active species.

The mechanism involves 16VE and 18VE species. Initially, one of the chloride ligands in 16VE fragment $PtCl_4^{2-}$ is replaced with a vinylhydridosilane ligand, thus forming a η^2-olefin $PtCl_3^-$-complex. In a second step, a vinylhydridosilane molecule adds oxidatively to the complex whereby an 18VE octahedral complex forms, followed by an insertion of the η^2-coordinated olefin into a platinum–hydrogen σ-bond. The vacancy, which forms upon this migration, is occupied by a chloride ligand that adds to the complex. Finally, the carbosilane is eliminated reductively with regeneration of the catalytic species $PtCl_4^{2-}$.

Similarly, $PtCl_4^{2-}$-catalyzed hydrosilylation of alkynylhydridosilanes of the general type $(R^1)(R^2)SiH(C{\equiv}CH)$ is a synthetic pathway for obtaining linear polycarbosilanes in which alternating silicon atoms and olefinic C_2 units constitute the polymer backbone [20, 21]:

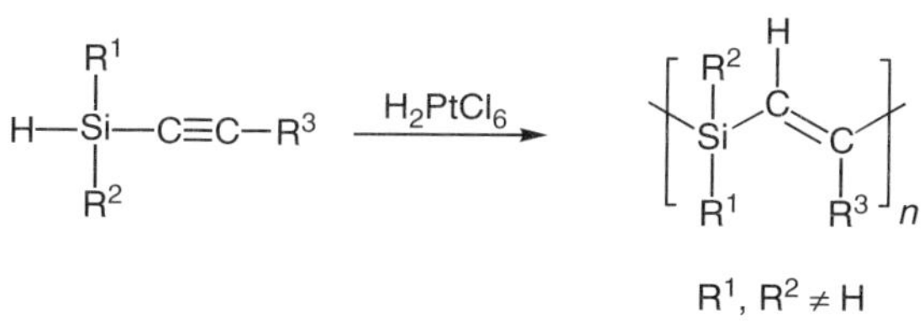

(7)

Formation of linear poly(alkynylsilanes) of this type is only observed if $R^1, R^2 \neq H$. The use of dihydridoalkynylsilanes, $(R)SiH_2(C{\equiv}CH)$, as starting compounds, in contrast, delivers highly cross-linked, glass-like polycarbosilanes. The reason, therefore, is that the olefin units, which are generated by the hydrosilylation of the alkynyl units, may react further and transform into aliphatic C_2-building blocks:

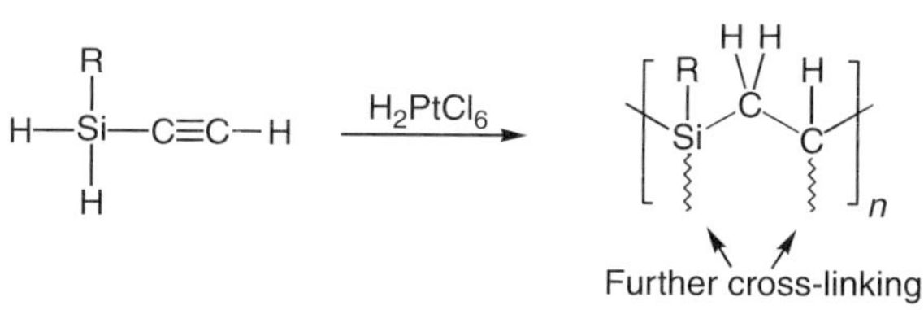

(8)

Polycarbosilanes of the general type $[(R^1)(R^2)Si{-}C{\equiv}C]_n$, in which silicon atoms are linked via C≡C building blocks, are not accessible by hydrosilylation reactions. In contrast, metathesis reactions of dichlorosilanes with dialkalimetal acetylides or the respective Grignard reagents deliver this type of polymers in good yields [20]:

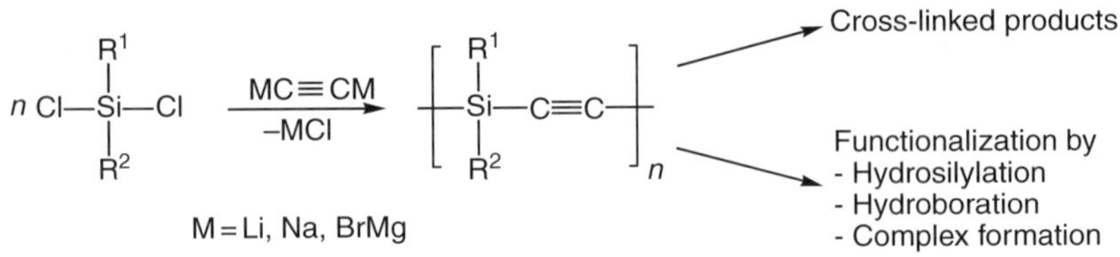

(9)

The alkynyl unit represents a highly reactive site within the polymer backbone. This is an important issue with respect to obtaining highly cross-linked precursors for ceramics in order to prevent polymer degradation and thus volatilization of low-weight molecular species during thermolysis. Moreover, the alkynyl group can be functionalized by for example, hydroboration, hydrosilylation, or attachment of transition metal complex fragments by coordination, providing an access for the synthesis of multi-component ceramics. In addition, poly(butadiynylsilanes), $[(R^1)(R^2)Si{-}C{\equiv}C{-}C{\equiv}C]_n$, obtained from 1,4-dilithio-1,3-butadiyne proved to be precursors with interpenetrating networks to silicon carbide/metal carbide (M = Ti, Nb, Ta) nano-composites [22]. Preparation of branched polymers by regiospecific hydrosilylation of poly(butadiynylsilanes) was also reported [23].

5.1.2.2 Precursors to Silicon Nitride

For the preparation of silicon nitride-based PDCs, oligo- or polysilazanes, $[(R^1)(R^2)Si{-}N(R^3)]_n$ (R = H, alkyl, aryl, amine), or poly(silylcarbodiimide)s, $[(R^1)(R^2)Si{-}N{=}C{=}N]_n$, have been studied intensively during the last 30 years, resulting in an immense number of structurally different precursors that were published. In general, such precursors deliver silicon nitride/silicon carbide composites if R^1, R^2, and/or R^3 contain carbon atoms. The formation of phase-pure silicon nitride can only be realized by thermolysis of perhydridopolysilazane in which R^1, R^2, and R^3 are hydrogen or by performing the polymer-to-ceramic conversion in an ammonia atmosphere. The latter, however, is limited to powders or parts with comparably small dimensions and/or cross-sections.

5.1.2.2.1 Carbon-Free Polysilazanes

Narsavage *et al.* [24] published the polymer-to-ceramic conversion of tetrakis(ethylamino)silane in an inert gas atmosphere (Ar), from which they expected phase-pure silicon nitride. The idea behind this concept was that silicon is exclusively bonded to nitrogen. It was expected that Si_3N_4 formation by volatilization of the nitrogen-bonded ethyl groups takes place during thermolysis. However, carbon was not completely degassed and the resultant material was an amorphous silicon carbonitride.

Consequently, the only remaining possibility for obtaining carbon-free Si_3N_4 is thermolysis of polysilazanes, containing only Si, N, and H. First attempts for the preparation of such polymers were reported already in 1885 by Schutzenberger and Colson [25] by ammonolysis of tetrachlorosilane according to Eq. 10. Polymer formation was accompanied with precipitation of ammonium chloride as a by-product, which was removed by filtration. Glemser and Naumann [26] performed more detailed investigations on this system in 1959:

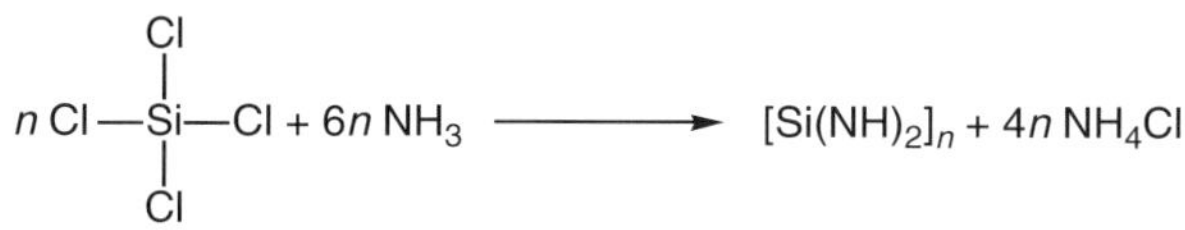

(10)

Silicon diimide, $Si(NH)_2$, is a colorless solid that is insoluble and infusible. With increasing temperature it decomposes with elimination of ammonia to form amorphous Si_3N_4 at 1000°C. Further heat treatment to 1400–1500°C results in the crystallization of α-Si_3N_4.

A processable Si_3N_4 precursor, perhydridopolysilazane, can be obtained by ammonolysis of dichlorosilane in polar solvents such as diethylether or dichloromethane [27] (see Scheme 5.1.7).

It is composed of linear and cyclic motifs but rapidly aging by losing hydrogen and further cross-linking thus increasing its viscosity gradually from becoming oily over waxy to glassy. Thermolysis to 1050°C releases a ceramic material in 70% yield, which according to X-ray investigations consists of α-Si_3N_4, β-Si_3N_4, and elemental silicon. Excess Si could be removed almost completely

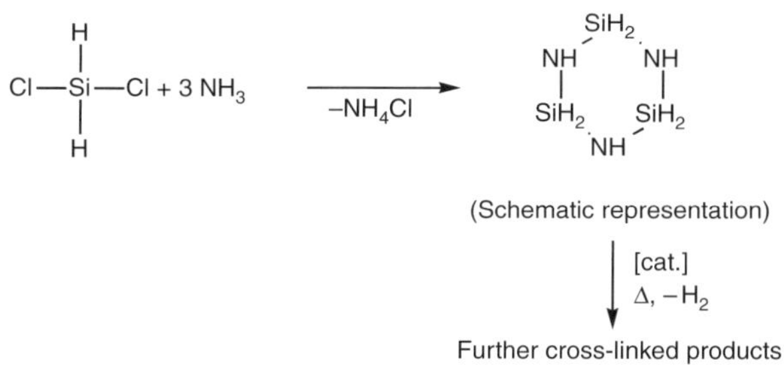

SCHEME 5.1.7 Synthesis of processable precursors to phase-pure silicon nitride by ammonolysis of dichlorosilane and subsequent cross-linking [27].

by performing thermolysis in an ammonia atmosphere. However, a critical issue with respect to transferring this process into a technical scale is the use of H_2SiCl_2, which is a highly flammable gas and which can disproportionate with the formation of SiH_4 and $SiCl_4$. A modified process, allowing for a safer handling of the highly reactive starting compounds, was developed at Tonen Company. Prior to ammonolysis, H_2SiCl_2 was modified by reaction with pyridine, which resulted in the formation of $H_2SiCl_2 \cdot (NC_5H_5)_2$ [28].

Blanchard and Schwab [29] modified the process developed by Seyferth *et al.* shown in Scheme 5.1.7. They avoided formation of free silicon by synthesizing a perhydridopolysilazane, which delivers stoichiometric silicon nitride even after thermolysis in an inert gas atmosphere, using co-ammonolysis of dichlorosilane and trichlorosilane.

5.1.2.3 PRECURSORS TO TERNARY Si—C—N CERAMICS

5.1.2.3.1 Carbon-Containing Polysilazanes

While the synthesis of perhydridopolysilazanes is limited to a small number of suitable starting compounds such as H_2SiCl_2, $HSiCl_3$, $SiCl_4$ and NH_3 or $H_2N{-}NH_2$ as silicon and nitrogen sources, respectively, the chemistry of carbon-containing polysilazanes of the general type $[(R^1)(R^2)Si{-}N(R^3)]_n$ (R^1, R^2, R^3 = H, alkyl, aryl, etc.) is a very multifaceted field. First publications in this field by Rochow appeared already in 1966 [30].

In contrast to H_2SiCl_2 and $HSiCl_3$, functionalized chlorosilanes of the general type $(R^1)(R^2)SiCl_2$ (R^1, R^2 = H, alkyl, aryl, halide, etc.), which are, in general, used in the synthesis of Si—C—N polymers, are mostly cheap and commercially available. Therefore, much effort has been made within the last 30 years towards specifically designing polysilazanes for numerous applications. In contrast to perhydridopolysilazanes, which deliver amorphous Si—N ceramics, carbon-containing polysilazanes after thermolysis to ~1100°C release amorphous Si—C—N ceramics. Interestingly, the latter materials possess improved high-temperature mechanical properties compared to silicon nitride derived from perhydridopolysilazanes. A particularly critical issue is "free", that is, excess silicon. The reason is that silicon segregation is known to lower the crystallization temperature of silicon nitride-based materials drastically [31]. In carbon containing Si—C—N materials, which usually contain "free" carbon, silicon segregation is suppressed because of the formation of silicon carbide.

There is a remarkable number of synthetic pathways described in the literature that deliver polysilazanes. The desired application, the choice of the chemical composition and molecular structure of the monomers, reaction conditions, and eventually further cross-linking finally determine the synthetic approach to the precursors. The procedure used most frequently for the synthesis of polysilazanes is ammonolysis or aminolysis of chlorosilanes.

Ammonolysis of chlorosilanes in general proceeds according to:

$$R_{4-x}SiCl_x + \frac{3}{2}\times NH_3 \longrightarrow \frac{1}{n}[R_{4-x}Si(NH)_{x/2}]_n + x\,NH_4Cl \qquad (11)$$
$$x = 1\text{–}3$$

The single steps are (i) initial substitution of Cl atoms with NH_2 units followed by (ii) a condensation step upon which Si—NH—Si linkages are built. Depending on the number of chlorine atoms per silicon atom and the nature of the silicon-bonded groups R, monomeric, dimeric, cyclic, short-chain linear, or highly cross-linked polysilazanes form.

Ammonolysis of monochlorosilanes R_3SiCl results in monomeric species. In the case of "small" silicon-bonded substituents R(R=H, CH_3), disilazanes $R_3Si—NH—SiR_3$ form. Certainly, if silicon-bonded substituents are sufficiently small and reaction temperatures are high, these species condense further, whereby tertiary amines bearing Si_3N motifs form [32]. In contrast, bulky substituents R, such as isopropyl, isobutyl, or mesityl, inhibit the above-mentioned condensation step due to steric effects, thus causing the formation of monomer silylamines R_3SiNH_2 [33, 34].

Hexamethyldisilazane, $(H_3C)_3Si—NH—Si(CH_3)_3$ (HMDS), synthesized by ammonolysis of trimethylchlorosilane in a large scale, is industrially by far the most important monomeric silazane (cost <US$10/kg). Since it is a volatile liquid (boiling point: 125°C), it cannot be directly used for the solid-state thermolysis process. However, it is an important single source precursor for the preparation of Si—C—N coatings by CVD processes [35] and a very important source for the synthesis of polysilazanes by *trans*-amination (redistribution, metathesis) reactions [36].

Ammonolysis of dichlorosilanes, R_2SiCl_2, delivers oligo- or polysilazanes of the general type $[R_2Si—NH]_n$, presuming sufficiently high reaction temperatures and sterically less demanding silicon-bonded substituents R such as hydrogen, methyl, ethyl, or phenyl. Starting from dichlorosilanes with sterically demanding substituents R such as *i*-propyl, *t*-butyl, or mesityl delivers silyldiamines, $R_2Si(NH_2)_2$, according to the above-described ammonolysis of monochlorosilanes R_3SiCl.

Usually, product mixtures are obtained that consist of six and eight-membered rings (trimers and tetramers, respectively) besides low molecular ($M < 2000$) chain molecules. The low molecular weights of the tri- and tetrameric species result in low boiling temperatures and cause their volatilization during thermolysis whereas the chain molecules frequently degrade during

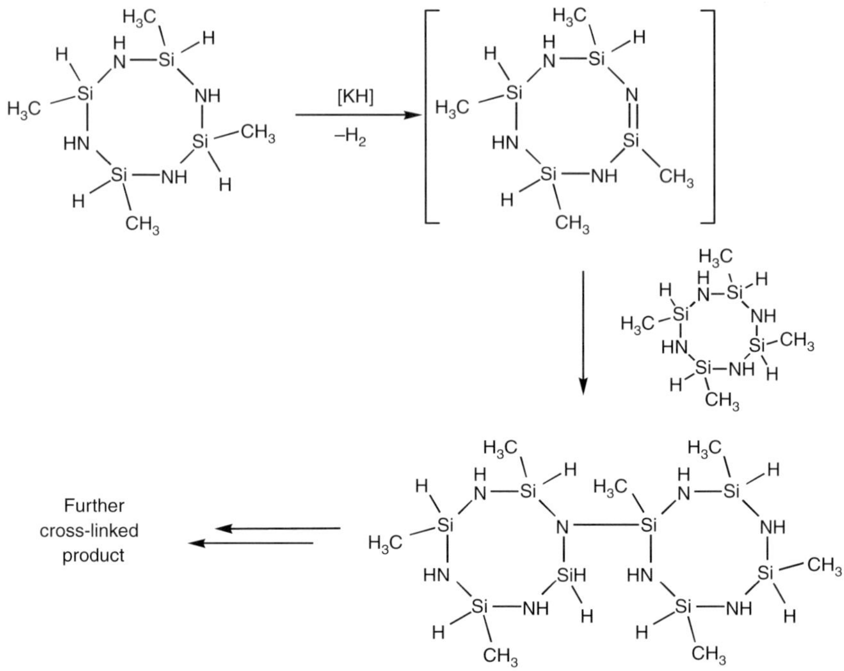

SCHEME 5.1.8 Cross-linking of cyclotetra(methylsilazane) with catalytic amounts of KH. By increasing the cross-link density, ceramic yields could be increased from below 30% to above 80% [38, 39].

the heat treatment with volatilization of low molecular gas species [37]. As a consequence, ceramic yields of polysilazanes are frequently unsatisfactory.

Seyferth and Wiseman [38, 39] published a method for increasing ceramic yields of cyclic silazanes bearing Si—H and N—H units considerably by cross-linking with catalytic amounts of KH. In the case of cyclotetra(methylsilazane), the ceramic yield could be increased from below 30% to approximately 84%. The authors proposed a mechanism that involves silylene-imine motifs as key species [39] (see Scheme 5.1.8).

The highly reactive silylene-imine intermediate, which could not be isolated, intermolecularly adds Si—H in a hydrosilylation-type reaction. The as-formed species can again be deprotonated and react further. However, there are also mechanisms desirable excluding silylene-imine formation, which were postulated for the dehydrocoupling of boron-modified silanes using *n*-butyl lithium as a catalyst [40]. Here, *n*-butyl lithium initially deprotonated the silazane N-atoms resulting in highly nucleophilic amides. Subsequent intermolecular substitution of a silicon-bonded hydride by an amide results in the formation of a new Si—N bond. The hydride that is released again deprotonates a silazane N-atom that closes the catalytic cycle.

A possibility for the synthesis of highly branched high ceramic yield polysilazanes is the attachment of reactive sites, for example, vinyl groups to the polymer backbone. These allow for further cross-linking reactions during thermolysis and thus very efficiently avoid polymer backbone degradation. Vinyl-substituted oligo- or polysilazanes can be obtained by ammonolysis of vinyldichlorosilanes $(H_2C{=}CH)Si(R)Cl_2$ ($R{=}H, CH_3$) according to:

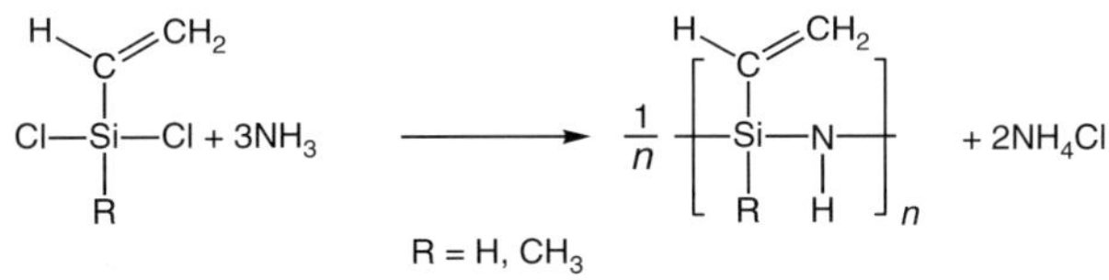

(12)

Depending on the silicon-bonded unit R and reaction conditions applied, $[(H_2C{=}CH)Si(R){-}NH]_n$ can cross-link by three different reaction pathways (compare Scheme 5.1.9): (i) olefin polymerization; (ii) hydrosilylation; and (iii) dehydrocoupling (the latter reactions only take place if R=H). It should be noted that both olefin polymerization and hydrosilylation take place without formation of by-products. This is an important issue that makes vinyl-substituted polysilazanes candidates for the production of PDC matrix composites [41].

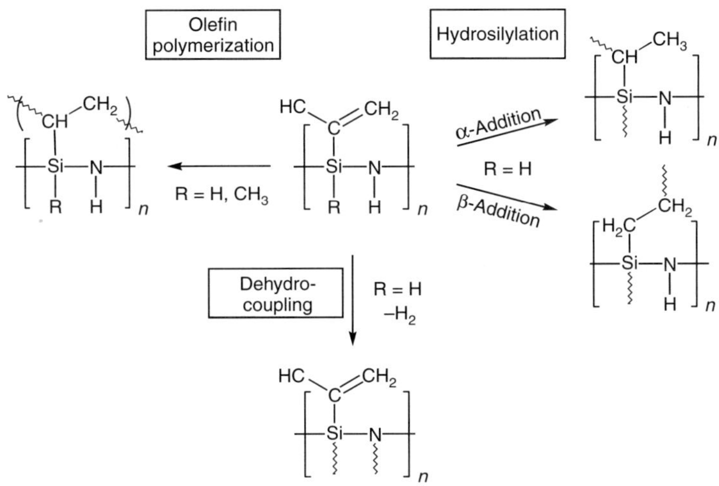

SCHEME 5.1.9 Possible cross-linking reactions of poly(vinylsilazanes).

NMR spectroscopy and chemical analysis suggest that hydrosilylation is the main reaction in the thermally induced cross-linking of $[(H_2C{=}CH)SiH{-}NH]_n$ [42]. Addition of catalytic amounts of Speyers' reagent can accelerate it. Thermally induced *trans*-amination (not shown in Scheme 5.1.9) in contrast, which would be accompanied with ammonia evaporation, occurs to a minor extend. By adding catalytic amounts of potassium hydride, dehydrocoupling reactions become favored. Cross-linking in this case proceeds by Si—N bond formation with elimination of hydrogen. Possible mechanisms for catalysis, hydrosilylation and dehydrocoupling were already mentioned earlier and are not discussed in detail here. In contrast, potassium hydride promoted anionic polyaddition reaction of the vinyl groups (anionic olefin polymerization) was not observed.

Due to the absence of Si—H motifs in $[(H_2C{=}CH)Si(CH_3){-}NH]_n$ neither hydrosilylation nor dehydrocoupling can be considered for thermal cross-linking. However, based on the structural characterization by solid-state ^{13}C- and ^{29}Si—MAS NMR, FT-IR, and elemental analysis, Bill *et al.* [43] showed that between 250 and 350°C—which is by ~120°C higher than the temperature required for hydrosilylation reactions of $[(H_2C{=}CH)SiH{-}NH]_n$—the vinyl groups are transformed into aliphatic hydrocarbons. More detailed information on thermolysis reactions of polysilazanes are provided in Section 5.1.3.2.

Very important silazane-based precursors are co-polysilazanes obtained by co-ammonolysis or aminolysis of chlorosilane mixtures. Polymerization of such mixtures allows for the synthesis of modified polymers with tunable properties such as solubility, rheology, and (latent) reactivity, which are difficult to achieve when using single-source precursors. Originally, co-polysilazanes were developed to improve the processing of the precursors for silicon nitride fibers, that is, to adjust proper rheology and latent reactivity. Moreover, the importance of chemical constitution, that is, phase composition was recognized [44]. A main aim in this regard was to decrease the amount of free carbon in Si—C—N ceramics in order to shift decomposition temperatures towards higher values.

A commercially available precursor in this series is NCP200, a "polyhydridomethylsilazane", which is obtained by co-ammonolysis of dichlorodimethylsilane and dichloromethylsilane [45]. Because of commercial availability, low cost, good processability, and high ceramic yield of ~80% (this value, however, depends on the processing), this precursor could successfully be applied for the preparation of dense silicon carbonitride ceramics [46]. Structurally comparable is NN710, a "polyperhydridomethylsilazane", which can be synthesized by co-ammonolysis of dichloromethylsilane and dichlorosilane [47].

Recently, a precursor which is derived from this system and which delivers stoichiometric silicon carbide/silicon nitride was published [48] (see Scheme 5.1.10).

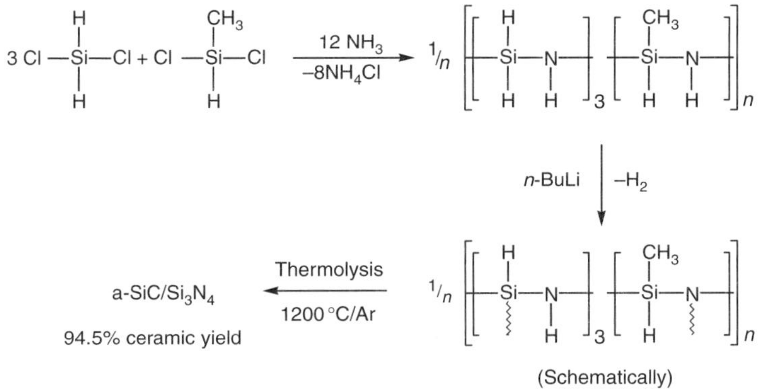

SCHEME 5.1.10 Synthesis of a precursor for stoichiometric SiC/Si_3N_4 [48]. A key step is subsequent cross-linking of the initially obtained precursor in order to avoid thermal degradation with volatilization of low molecular weight species other than hydrogen.

The polymer was obtained by co-ammonolysis of H_2SiCl_2 and $H_3CSiHCl_2$ in a 3:1 molar ratio and subsequent base-catalyzed cross-linking using *n*-BuLi as a catalyst. While the as-obtained polymer is a viscous oil, which rapidly ages, the cross-linked precursor is a hard glass-like infusible solid, which is insoluble in all common organic solvents. In TGA-MS investigations, it was shown that this type of polymer could be pyrolyzed without formation of gaseous by-products other than hydrogen, thus delivering amorphous SiC/Si_3N_4 with a ceramic yield of 94.5%.

According to the procedures described above, ammonolysis of trichlorosilanes $RSiCl_3$ usually results in the formation of $[(R)Si(NH)_{1.5}]_n$. Such poly(silsesquiazane)s (these are polymeric silazanes in which three silicon-bonded substituents are nitrogen atoms and in which N:Si = 1.5) were first published in 1967 by Andrianov and Kotrelev [49] who performed ammonolysis of methyltrichlorosilane:

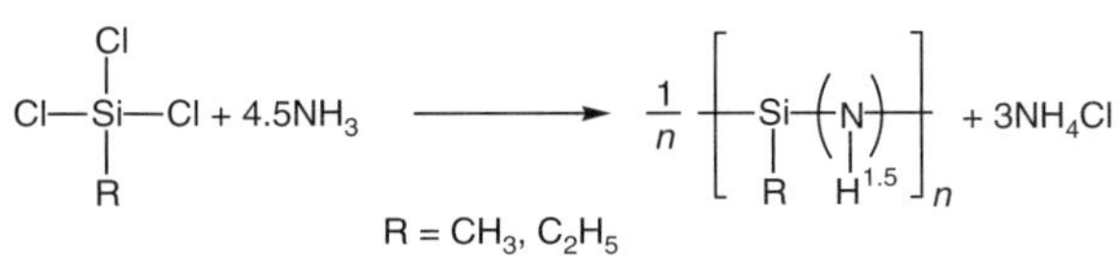

(13)

In contrast to poly(silsesquioxane)s, $[RSiO_{1.5}]_n$, which can be obtained by hydrolysis of trichloro- or trialcoxysilanes under water-starved conditions and which are structurally well investigated [50–52], there are not much details known about the molecular structure of poly(silsesquiazane)s. However, similar to poly(silsesquioxane)s they are obtained as a mixture of molecules with cage structures as well as oligomeric and highly branched polymeric products.

Recently, the structure of a hexameric silsesquiazane, $(H_3C)_6Si_6(NH)_9$, which was obtained from sodium amide and $(H_3C)SiCl_3$ in *n*-hexane solution at −78°C, was reported. Sodium amide was used instead of ammonia and thus formation of ammonium chloride precipitate avoided. Single crystal X-ray analysis displayed a cage-type structure of $(H_3C)_6Si_6(NH)_9$ in which two six-membered ring systems, $[(H_3C)Si(NH)]_3$, with chair conformation are linked via the silicon centers with each one NH unit [53].

Polysilazanes are also accessible by transition metal-catalyzed dehydrocoupling of silanes and ammonia or amines. In contrast to the above-described ammonolysis and aminolysis of chlorosilanes, no solid by-products form in the dehydrocoupling reactions; the only by-product is hydrogen, which evaporates during the reaction. In this regard, it has to be mentioned that dehydrocoupling of silanes and ammonia allows for the preparation of highly cross-linked polysilazanes that due to their comparably low solubility cannot be obtained using the conventional salt elimination process.

Laine and Blum, who found that ruthenium catalysts such as $Ru_3(CO)_{12}$ are most suitable, first reported catalytic synthesis of polysilazanes in 1986 [54–57]:

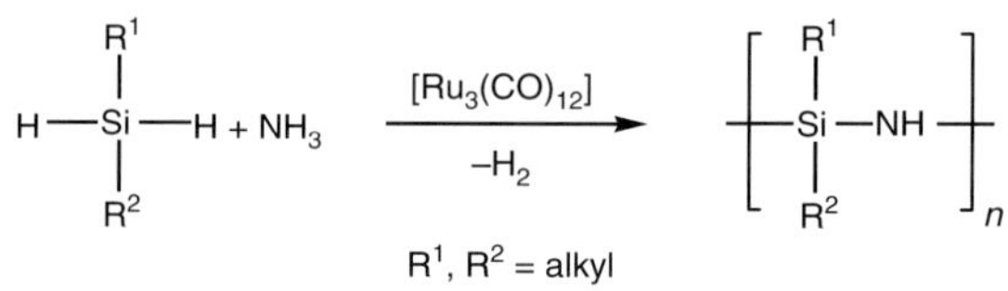

(14)

In addition, the authors demonstrated that the transition metal complex noted above is also useful for cross-linking low molecular weight oligosilazanes such as trimethylcyclotrisilazane, $[(H_3C)SiH{-}NH]_3$, or tetramethylcyclotetrasilazane, $[(H_3C)SiH{-}NH]_4$.

Liu and Harrod [58] reported on the dehydrocoupling of ammonia and less reactive tertiary silanes such as diphenylmethylsilane using dimethyltitanocene as a catalyst. The authors observed that primary silanes ($PhSiH_3$) surprisingly reacted retarded even though the opposite was expected. As an explanation for this phenomenon, they referred to the possibility of homo coupling reaction of the silanes (Si—Si linking), resulting in disilanes or trisilanes, which possess lower reactivity toward cross-dehydrocoupling with ammonia than secondary or tertiary silanes R_2SiH_2 or R_3SiH, respectively.

An alternative method for the synthesis of polysilazanes, which also avoids the formation of solid by-products during the polymerization step, are redistribution reactions of chlorosilanes with hexamethyldisilazane,

HMDS [36]:

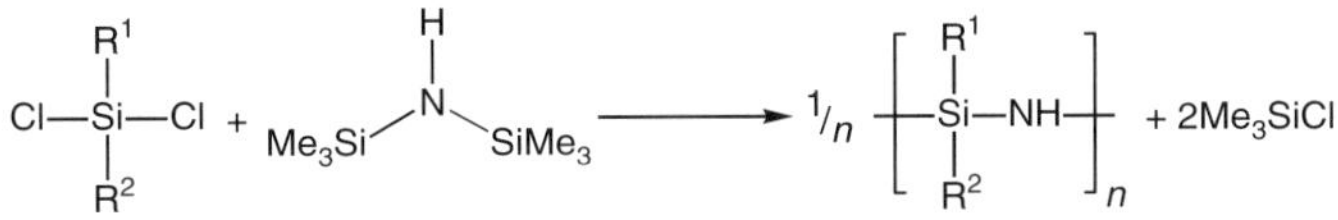

(15)

The reaction can be performed in solution or without solvent using excess HMDS. Both HMDS and trimethylchlorosilane that form are volatile liquids, which can easily be removed from the crude reaction mixture by a distillation step under reduced pressure.

As pointed out above, the molecular weight of polysilazanes can be increased by a cross-linking step, which occurs either by *trans*-amination or dehydrocoupling reactions. In both cases polymers are obtained with retention of the silazane ring structures.

The conversion of low molecular weight cyclic silazanes, for example, cyclodi-, tri-, or tetrasilazanes, by ROP in contrast, results in the loss of cyclic structural motifs. The idea behind this process is to obtain linear high molecular weight polysilazanes with proper rheological properties, as they are required for polymer fiber spinning.

Ring-opening of cyclosilazanes can be induced by treatment with catalytic amounts of a strong base such as MeLi, *n*-BuLi, or *t*-BuLi [59], strong acidic compounds such as triflic acid methylester, $F_3CSO_3CH_3$ [60–63], or by transition metal catalysts such as $Ru_3(CO)_{12}$ [54]. For example, Suom and coworkers [64] investigated in detail $F_3CSO_3CH_3$– and organo alkali salt-induced ROP of cyclodisilazanes covering small substituents in dichloromethane solution:

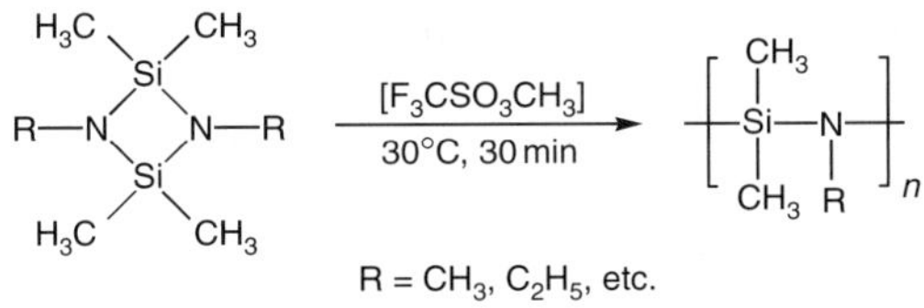

(16)

They observed that four-membered ring systems, which have higher tension, react faster and more readily than cyclotrisilazanes, independent if basic or acidic conditions were applied. Remarkably, polysilazanes with molecular weights of up to 18 000 using acidic conditions and up to 100 000 using basic conditions could be obtained by this method. Another remarkable issue using ROP is the ability to obtaining copolymers if starting from cyclodisilazanes with differently substituted silicon atoms:

(17)

5.1.2.3.2 Poly(silylcarbodiimide)s

Disilylated carbodiimides are known since the early 1960s and mainly Ebsworth and Mays, as well as Pump and Wannagat, investigated the synthesis and reaction behavior of the monomeric compounds [65–70]. However, because of their volatility di-silylated carbodiimides cannot be used directly as precursors for ceramics.

In 1964, Pump and Rochow [71] reported on poly(silylcarbodiimide)s for the first time, which were obtained by the reaction of dichlorosilanes with disilver cyanamide:

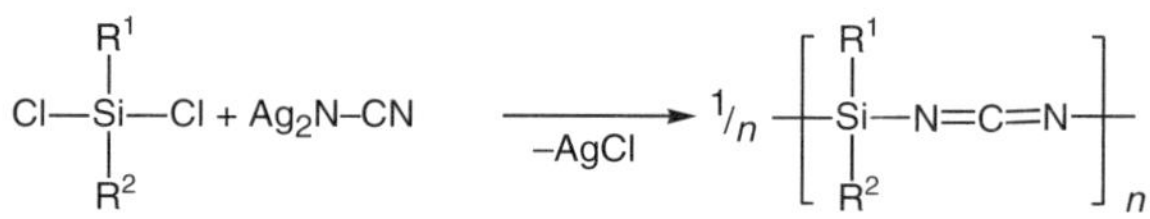

(18)

However, for a transformation into a technical scale and the application of the polymers as a preceramic material this approach is by far too expensive.

A patent of Klebe and Murray in 1968 described a modified procedure for the synthesis of poly(silylcarbodiimide)s [72]. In a *trans*-silylation reaction starting from bis(triorganosilyl)carbodiimides (usually bis(trimethylsilyl)carbodiimide) and di- or trichlorosilanes, the precursors could be obtained in simple procedures and high yields:

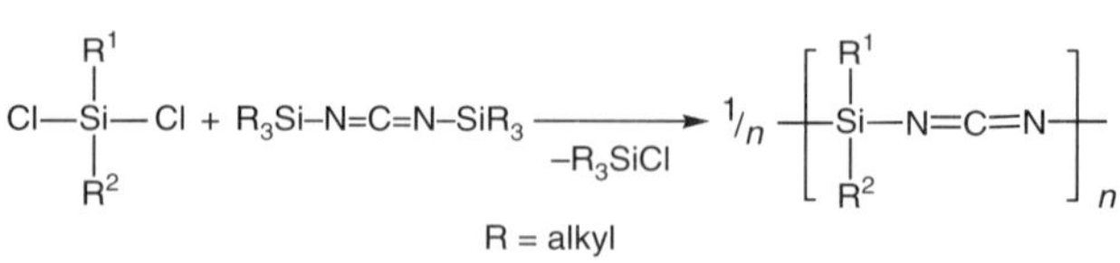

(19)

However, besides the simplified processing—no solid by-products formed in this reaction—a main step forward was that the costs for precursor synthesis could be reduced tremendously.

In the beginning of the 1990s (more than 20 years after the appearance of the patent of Klebe and Murray!), poly(silylcarbodiimide)s were rediscovered and their applicability as preceramics studied in much detail. The synthetic

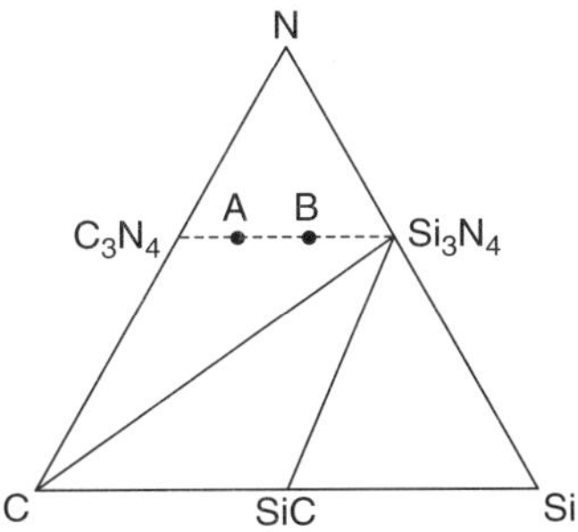

FIGURE 5.1.2 Ternary Si—C—N phase diagram (T < 1440°C, 1 atm N_2). A and B: compositions SiC_2N_4 and Si_2CN_4, respectively, the first crystalline ternary phases in the Si—C—N system obtained from poly(silylcarbodiimide) $[Si(NCN)_2]_n$ [81].

approach was improved significantly by developing a non-oxide sol—gel process, and the molecular structure of cyclo-tetra(dimethylsilylcarbodiimide) using single crystal X-ray diffraction (XRD) determined [73–82]. Moreover, the polymer-to-ceramic transformation was investigated, especially considering intermediates that form upon the thermal treatment. In the case of the thermolysis of highly cross-linked poly(silylcarbodiimide), which was obtained from tetrachlorosilane and bis(trimethylsilyl)carbodiimide in the presence of pyridine (pyridine was added to accelerate the *trans*-amination reaction) according to Eq. 20, the formation of the first crystalline phases in the Si—C—N system was thereby observed (compare Figure 5.1.2):

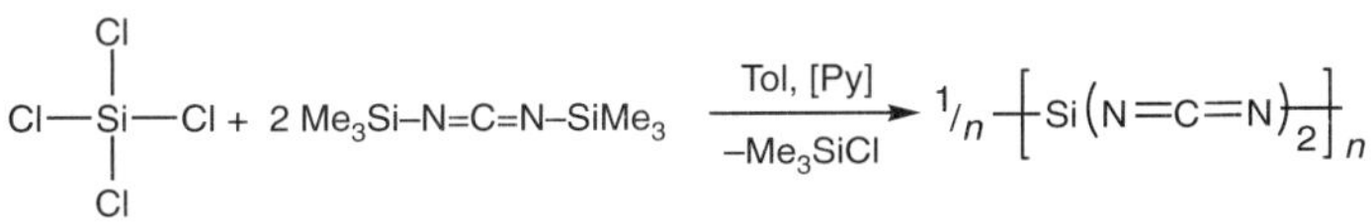

(20)

β-SiC_2N_4 (space group *Pn*3*m*) crystallizes above 400°C. It possesses an anticuprite structure, which can be described as two interpenetrating crystobalite lattices. Above 920°C it decomposes with the loss of cyanogene and nitrogen to form Si_2CN_4 (space group *Aba*2) [81].

5.1.2.4 Precursors to Quaternary Si—B—C—N Ceramics

Organometallic polymers containing the elements silicon, boron, carbon, nitrogen, and hydrogen are a comparably young class of preceramic polymers. However, the exceptional high-temperature stability of their derived ceramics, which is superior to those noted in the previous sections, coupled with a high resistance towards oxidation, that is, self-healing capabilities under oxidative conditions, have prompted considerable interest in these new materials. As a consequence, the number of publications dealing with synthetic approaches as well as materials properties steadily increased within the last decade.

Takamizawa *et al.*, in 1986, first described a method for the preparation of an inorganic fiber containing silicon, carbon, boron, and nitrogen. For this purpose, polydimethylsilane, $[(H_3C)_2Si]_n$, and B-trimethyl-N-triphenylborazine, $[B(CH_3){-}N(C_6H_5)]_3$, were mixed and reacted at elevated temperature. In a similar manner, the authors reacted block-copolymers carrying $Si(CH_3)_2$ and $Si(C_6H_5)_2$ units in the polymer backbone with the above noted borazine derivative [83, 84]. Such polymers were spun and rendered infusible by either subsequent curing in oxidative atmosphere or by irradiation with actinic rays. Investigations in order to determine the molecular structure of the precursors were not performed.

5.1.2.4.1 Borazine-Based Si—B—C—N Precursors

The chemistry of borazine and its derivatives is an important field in inorganic chemistry. It is thus natural that scientists dealt with potential precursors to Si—B—C—N ceramics long before this topic became relevant for materials science.

Such an example is the work published by Noeth in the early 1960s [85]. He investigated the reaction of trichloroborane or alkyldichloroborane with (trialkylsilyl)dialkylamines, $R_3SiNR'_2$, and hexamethyldisilazane, both important starting compounds in the synthesis of Si—B—C—N polymers. It turned out that chloroboranes readily react with $R_3SiNR'_2$ under Si—N bond cleavage and B—N bond formation. For example, (trimethylsilyl)dimethylamine, $Me_3Si{-}NMe_2$, reacted with trichloroborane to form dimethylaminodichloroborane and trimethylchlorosilane:

$$Me_3Si{-}NMe_2 + BCl_3 \longrightarrow Me_2N{-}BCl_2 + Me_3SiCl \qquad (21)$$

Using hexamethyldisilazane, $(Me_3Si)_2NH$, instead of $Me_3Si{-}NMe_2$ and alkyldichloro-borane, $RBCl_2$, resulted in the formation of monomeric B-alkylated borazine derivatives:

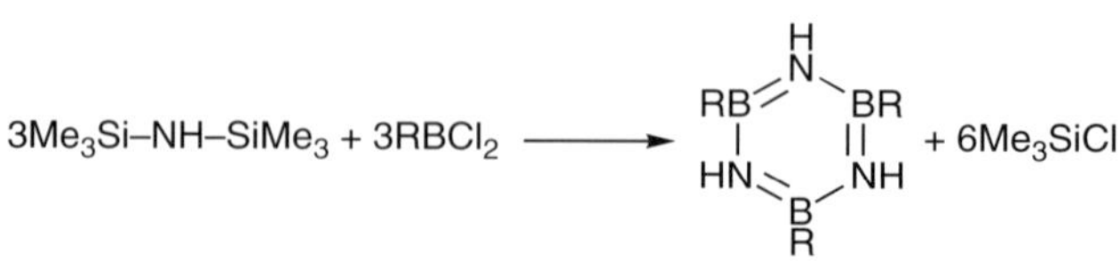

(22)

Highly cross-linked borazine-based precursors to Si—B—C—N ceramics with comparably low silicon content were synthesized by Narula *et al.* [86] starting from B-chloroborazine derivatives and hexamethyldisilazane:

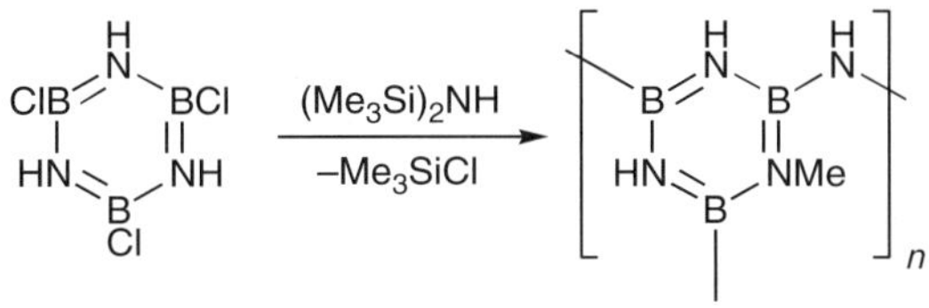

(23)

Even though not evident form the molecular structure depicted above, silicon and carbon are present in the oligomer-capping groups.

Seyferth and Plenio published further borazine-based polymers in 1990. They reported on the synthesis of "borasilazanes" as polymeric precursors for borosilicon nitride [87, 88] (see Scheme 5.1.11).

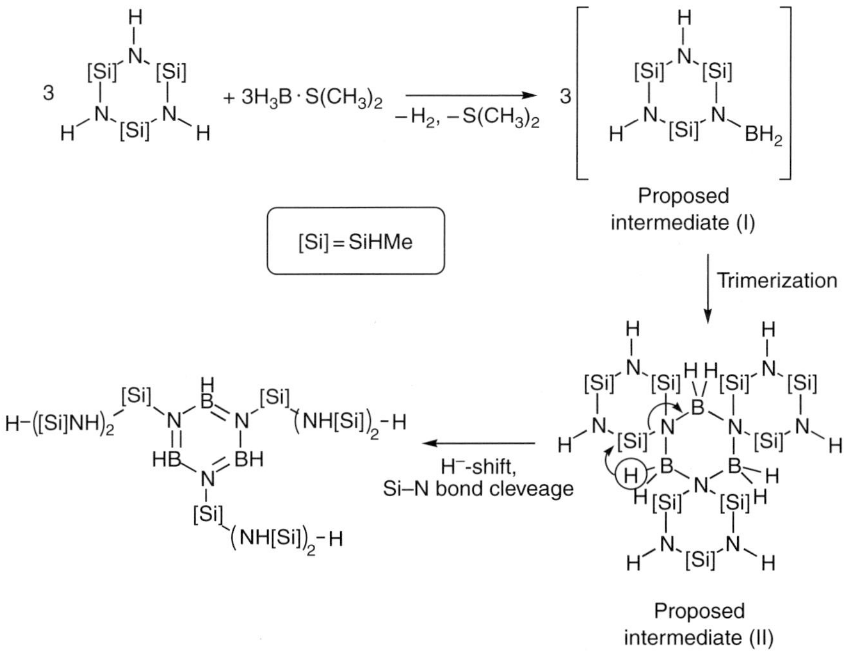

SCHEME 5.1.11 Synthesis of borazine-based polyborosilazanes via dehydrogenative coupling of borane dimethyl sulfide and cyclic silazanes [87, 88].

Initially, borane adds to the cyclic silazane with the formation of a silazane-borane adduct (not shown) and elimination of dimethyl sulfide. Subsequent dehydrocoupling (intermediate I) and trimerization result in the formation of a borazane (proposed intermediate II in Scheme 5.1.11) in which silazane rings are maintained. A shift of a boron-bonded hydride to silicon, accompanied with a ring opening reaction of the silazane motif finally releases the precursor.

Blum and Laine [89, 90] independently published a similar procedure starting from poly-N-methylsilazane, $[SiH_2{-}N(CH_3)]_n$, and ammonia or amine borane complexes, $BH_3 \cdot NH_3$ or $BH_3 \cdot N(CH_3)_3$, respectively, in the presence of catalytic amounts of $Ru_3(CO)_{12}$. They obtained viscous products that were not characterized by spectroscopic methods in more detail. Ceramic yields were in the range of 50%.

A series of Si—B—C—N precursors, which were obtained by dehydrogenative, and/or dehydrosilylative coupling reactions of monomeric or oligomeric silazanes with borazine as the boron source, was published by Sneddon *et al.* [91–97].

For example, hydridopolysilazane was reported to react smoothly with liquid borazine in different stoichiometry at temperatures between 50 and 90°C to give borazine-substituted polysilazanes in high yields. Spectroscopic investigations indicated that the polysilazane used retained its backbone but was substituted with pendent borazine substituents by means of a borazine-boron to polymer-nitrogen linking, resulting from both, hydrogen, and trimethylsilane elimination [91, 92]:

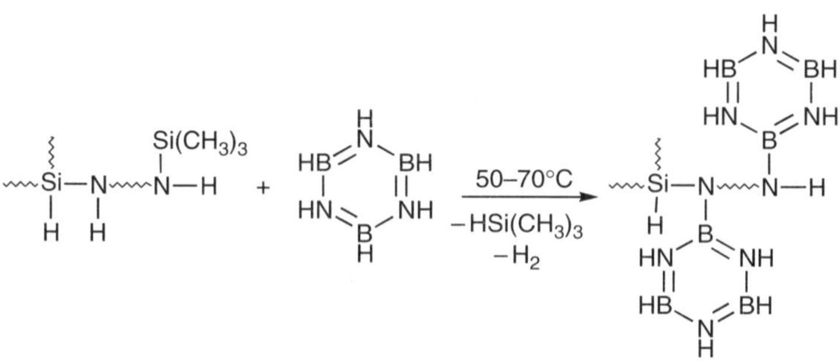

(24)

However, molecular weight and distribution studies suggested an inhomogeneous distribution of borazine throughout the molecular weight distribution.

The evidence for dehydrosilylative coupling reactions was confirmed by reaction of tris(trimethylsilylamino)silane with borazine [93, 94]:

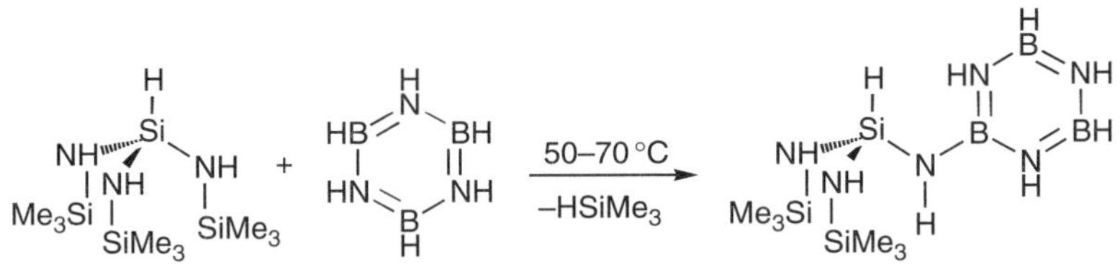

(25)

This finding was traced back to the observation that after vacuum fractionation of the volatile polymerization by-products, 0.25 mol of trimethylsilane per mole of borazine could be isolated. However, polymerization again occurred due to B—H and N—H dehydrocoupling reactions. Moreover, formation of borazylene motifs was discussed, and the following idealized product structure suggested [93]:

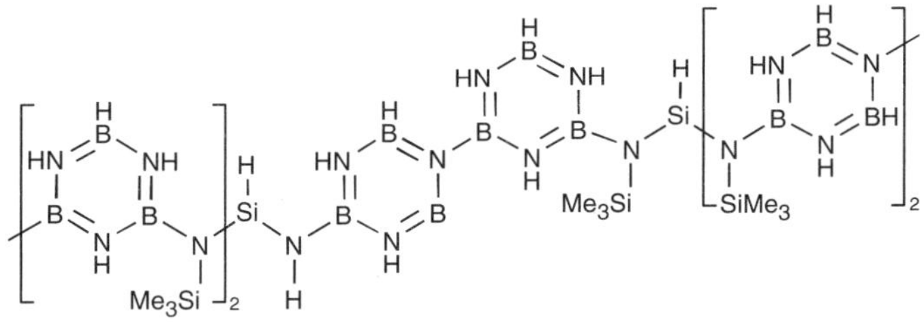

(26)

Depending on the silazane to borazine stoichiometry, ceramic yields between 38 and 42% were obtained.

Treatment of borazine with cyclic silazanes such as hexamethylcyclotrisilazane also appears to be complex. The main reaction, however, is a dehydrocoupling of borazine-boron atoms with the nitrogen atoms of the silazanes according to

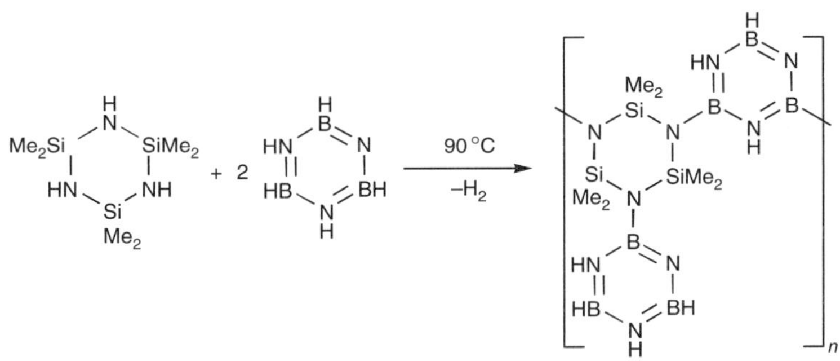

(27)

whereas ring opening reactions with Si—N bond cleavage may also appear, but less likely. Thermolysis up to 1400°C delivers ceramics with 55–62% yield, depending on the borazine to silazane ratio used.

A crucial issue in the synthesis of borazine-based polymers is the lack of selectivity in the polymerization reactions. This fact causes difficulties in controlling structure and properties of the precursors and the ceramic materials. Especially if a precursor is designed for drawing fibers, latent reactivity must—though it is absolutely required for obtaining high ceramic yields and rendering green fibers infusible—be adjusted very carefully. The pendent borazine units in the above noted precursors possess both B—H and N—H functions, which upon a thermal treatment may react further and increase molecular weight and viscosity of the polymeric precursors. Consequently, they are not suitable for processes that require stable melt viscosities, as, for example, in the generation of polymer fibers by melt-spinning [97].

As a possible approach out of this dilemma, Sneddon *et al.* designed mono- and difunctional borazine derivatives such as B-diethylborazine (DEB) or B-monoethylborazine (MEB), which can be prepared by metal-catalyzed hydroboration of ethylene by borazine [95]. In contrast to "pure" borazine, these species have only one or respectively two reactive B—H units. The modification by dehydrocoupling of polysilazanes such as hydridopolysilazane (HPZ) with DEB resulted in a precursor with pendent borazine units (see Scheme 5.1.12).

As a side reaction, Si—N bond cleaving of HPZ with formation of DEB—$NHSiMe_3$ was observed. However, the polymeric precursor that had

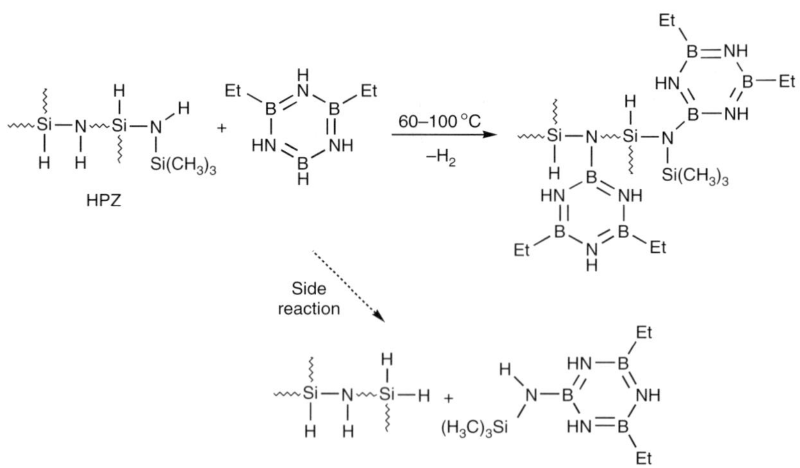

SCHEME 5.1.12 Synthesis of borazine-based Si—B—C—N precursors by dehydrogenative coupling of mono-functional borazine derivatives and perhydridopolysilazane [95].

a sufficient glass transition temperature and an onset of weight loss >200°C could be melt spun to Si—B—C—N—H green fibers, which were subsequently transformed into ceramic fibers by thermolysis up to 1400°C.

5.1.2.4.2 Si—B—C—N Precursors from Ammonolysis and Aminolysis Reactions of Boron-Modified Chlorosilanes

Totally different approaches to Si—B—C—N ceramic precursors, which do not make use of borazine derivatives and which—with one exception—do not produce borazine motifs during the polymerization step were published by Jansen *et al.* [98–102] and by Riedel *et al.* [103–109].

Baldus and Jansen [98, 99] reported on ammonolysis and aminolysis reactions of the single-source precursor trichlorosilyl-amino-dichloroborane (TADB), which they obtained starting from hexamethyldisilazane (HMDS) in a two-step reaction with (i) tetrachlorosilane and (ii) trichloroborane:

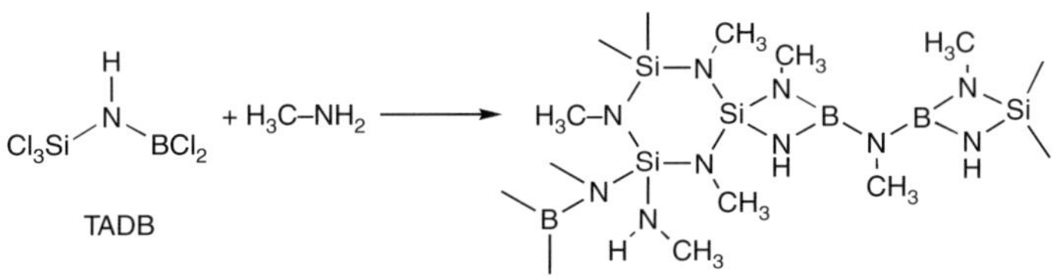

(28)

IR and NMR data of the meltable and soluble polyborosilazane, which was obtained in 80% yield, as well as elemental analysis, suggested the above polymer structure in which $Si_3(NCH_3)_3$ six-membered ring systems are connected via HN—B and $N(CH_3)B$—units. Because of the lack of resonance signals at ~30 ppm in the ^{11}B-NMR spectrum, the existence of borazine-like structures was excluded. The as-obtained polymer could be transformed into Si—B—C—N—H green fibers, which upon thermolysis delivered ceramic fibers with extraordinary high-temperature properties (compare Section 5.1.5.4).

Hydroboration of $(H_2C{=}CH)(CH_3)SiCl_2$ with borane Lewis base adducts such as borane dimethyl sulfide, $H_3B{\cdot}SMe_2$, results in the formation of tris(methyldichlorosilyl-ethyl)borane [110, 111]. Subsequent ammonolysis as reported by Riedel *et al.* results in a Si—B—C—N ceramic precursor, which is composed of silazane chains that are cross-linked via C—B—C bridges [106]:

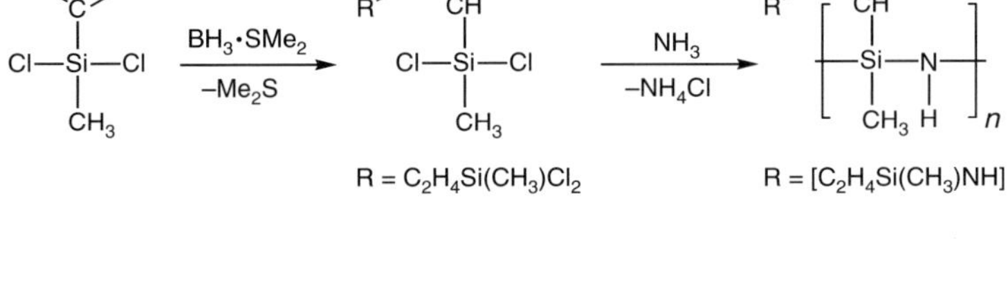

(29)

A crucial subject is the filtration process, which has to be applied for removing the solid by-product ammonium chloride. This separation can be extremely time-intensive since the precipitate partly appears colloidally distributed within the product solution. Moreover, ceramic yields are comparably low at around 50%. It will be shown below that structurally similar precursors can be obtained by a simple modification of the reaction coordinate.

Using $H_2BCl{\cdot}SMe_2$ or $HBCl_2{\cdot}SMe_2$ as hydroboration reagents instead of $H_3B{\cdot}SMe_2$, results in the formation of bis(methyldichlorosilyl-ethyl)chloroborane and (methyldichlorosilyl-ethyl)dichloroborane [111], which after ammonolysis release more highly cross-linked precursors, where the B:Si ratio increases from 1:2 to 1:1 compared to 1:3 in BH_3-derived precursors. Parallel to that, ceramic yields increased from 50% over 56–76%. Interestingly, polymers derived from the dichloroborane derivative $H_3CSiCl_2{-}C_2H_4{-}BCl_2{\cdot}SMe_2$ possess molecular structures that are composed of six- and eight-membered silazane ring systems, Si—B—N heterocycles, and borazine units [112, 113]:

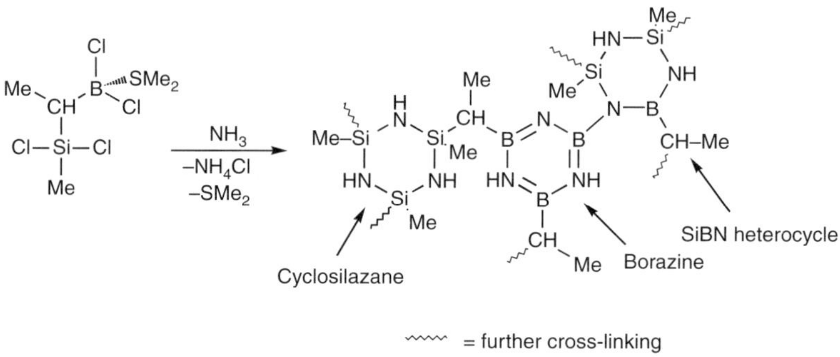

(30)

An approach that allows for increasing the ceramic yields in the above systems starts from vinyldichlorosilane or vinyltrichlorosilane instead of methyldichlorosilane [114–118]:

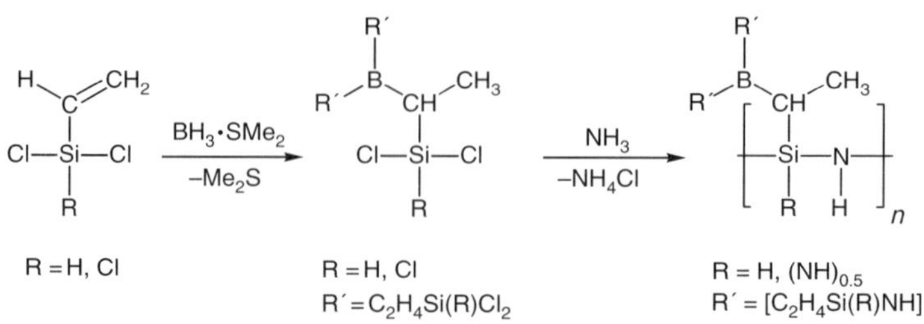

(31)

In contrast to the methyl-substituted derivative, these species provide latent reactivity that enable for further cross-linking during thermolysis. As a result, ceramic yields are above 80%. However, a remarkable disadvantage is the decreased solubility of these precursors compared to that developed by Riedel *et al.* Consequently, the processing is difficult and polymer yields are low. In the case of the highly cross-linked precursor with $R = (NH)_{0.5}$ the polymer yield is even below 5%, which is a result of its high cross-link density along with its low solubility.

However, both polymer yields and processing were significantly improved by switching the above reaction sequence, that is, by first performing ammonolysis followed by hydroboration [114–118]:

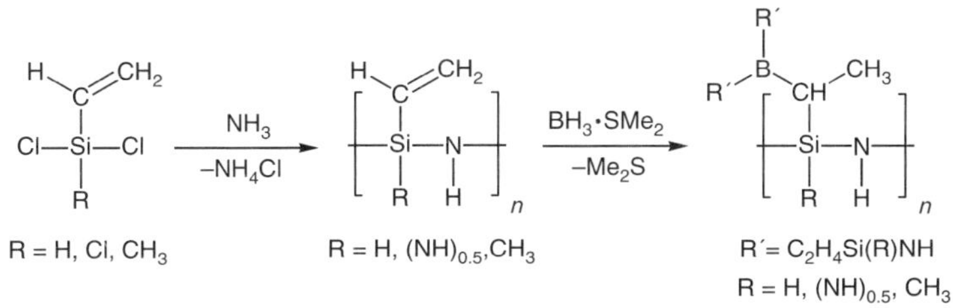

(32)

Ammonolysis of vinyl-substituted chlorosilanes $(H_2C{=}CH)Si(R)Cl_2$ ($R = H, Cl, CH_3$) delivers the respective poly(vinylsilazane)s, which are colorless liquids. Subsequent hydroboration then releases the polymeric precursors in quantitative yields. Whereas the methyl- and the hydrogen-substituted polymers possess sufficient solubility to remain in solution, the more highly cross-linked precursor $[B(C_2H_4Si(NH)_{1.5})_3]_n$ precipitated as a fine-grained solid from the reaction during borane addition. Remarkably, the work-up of Si—B—C—N precursors obtained in this way is rather simple and requires only for the removal of all volatile components by vacuum evaporation. According to this procedure, boron-modified polysilazanes with different boron contents were synthesized and the high-temperature properties of their derived materials evaluated (compare Section 5.1.4.4, Figure 5.1.18) [119, 120].

5.1.2.4.3 Si—B—C—N Precursors from Boron-Modified Hydridosilanes

An alternative method for the formation of Si—B—C—N polymers as precursors to ultra-high-temperature ceramics is a dehydrocoupling of ammonia or alkyl amines with tris(hydridosilylethyl)boranes, $B(C_2H_4Si(CH_3)_nH_{3-n})_3$ ($n = 0$–2; $C_2H_4 = CH_2CH_2, CHCH_3$) [40, 121, 122] (see Scheme 5.1.13).

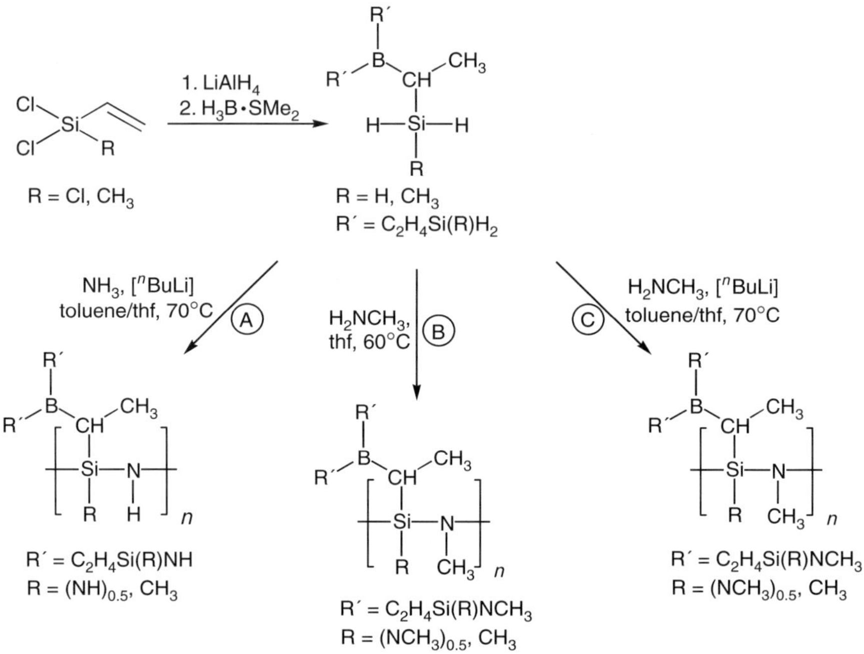

SCHEME 5.1.13 Dehydrocoupling reactions of tris(hydridosilylethyl)boranes with ammonia and methylamine [40, 121, 122].

It was observed that ammonolysis of $B(C_2H_4Si(CH_3)_nH_{3-n})_3$ (route A) does not occur directly and requires basic catalysts such as *n*-butyl lithium similar to a method described by Seyferth and Wiseman [38], in which potassium hydride was used for the cross-linking of cyclic silazanes.

A possible mechanism for this base-catalyzed dehydrocoupling was given [40] (see Scheme 5.1.14).

It was suspected that *n*-butyl lithium initially deprotonates ammonia with the formation of lithium amide and evaporation of *n*-butane (1). The more nucleophilic amide then replaces a silicon-bonded hydride that subsequently deprotonates ammonia with the evolution of molecular hydrogen (2 → 1). The proposed silyldiamine or -triamine was not stable under the reaction conditions applied and by fast elimination of ammonia (3) the polymeric precursors formed.

A precursor system that was designed especially for the production of fiber-reinforced high temperature Si—B—C—N ceramic matrix composites was also published [123–125]. Glass-like polymeric precursors were obtained by a thermally induced hydrosilylation reaction of oligo(vinylsilazane) using the

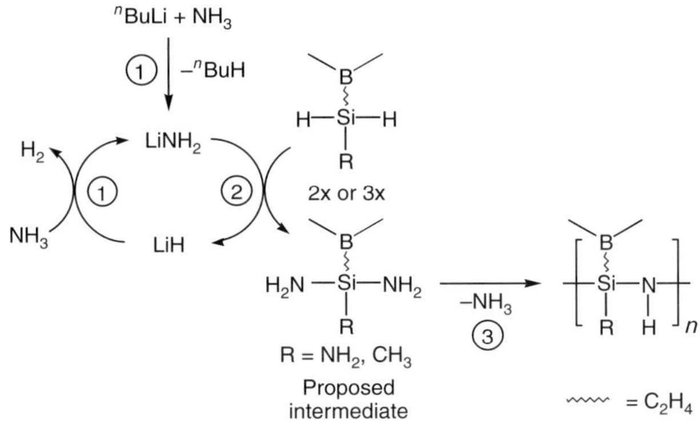

SCHEME 5.1.14 Proposed mechanism for the formation of boron-modified polysilazanes by base-catalyzed dehydrogenative coupling of tris(hydridosilylethyl)boranes with ammonia [40].

above–noted tris(hydridosilylethyl)boranes:

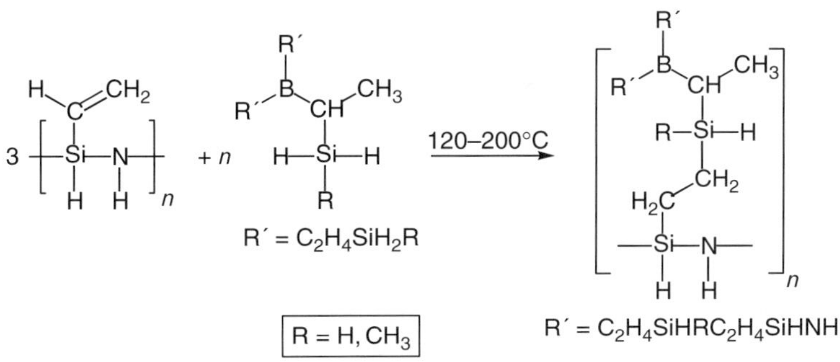

(33)

Remarkably, neither solvents nor catalysts were required for this process by-product formation was not observed. For the cross-linking step, the starting compounds were mixed at room temperature and the resulting colorless reaction mixture carefully degassed in vacuum. During a stepwise heating of the mixtures to 120°C within 2 h and then to 180–200°C within 4–5 h, a remarkable viscosity increase was observed, which resulted in the formation of the transparent, colorless, and glass-like solid.

Preliminary experiments have proven the applicability of the silane/silazane mixture for the preparation of fiber–matrix composites in the resin-transfer-molding (RTM) process [126]. Prerequisites for this process, which the precursor mixture fulfills are, for example, low viscosity (fluidity) of the starting compounds to allow for sufficient penetration, adequate vapor pressure, appropriate chemistry (no solvents or by-products), controllable reactivity during cross-linking, and high thermolysis conversion yields/low shrinkage during thermolysis. For further details see Section 5.1.5.5.

5.1.2.4.4 Si—B—C—N Precursors Based on Poly(silylcarbodiimides)

According to the procedure for the synthesis of poly(silylcarbodiimide)s as precursors to ternary Si—C—N ceramics described earlier, their boron-modified analoga are precursors to Si—B—C—N materials. They are accessible by different methods: First by the hydroboration of vinyl-substituted poly(silylcarbodiimide)s with borane dimethyl sulfide. The polymeric starting compounds can be obtained either from the respective vinyl-substituted chlorosilanes with cyan amide, $H_2N{-}CN$, in the presence of pyridine or from vinyl-substituted chlorosilanes and bis(trimethylsilyl)carbodiimide, $Me_3Si{-}N{=}C{=}N{-}SiMe_3$ [127–129]:

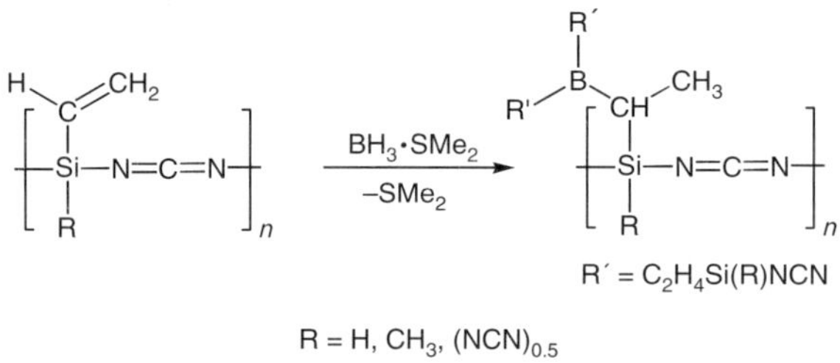

(34)

IR and NMR results pointed out that boron is attached exclusively to the silicon-bonded vinyl units. An addition to the carbodiimide unit, as it was reported for alkyl alumination reactions of monomeric silylcarbodiimides, was not observed [130].

Alternatively, boron-modified poly(silylcarbodiimide)s were obtained in a non-oxide sol–gel process involving tris(chlorosilylethyl)boranes that were reacted with excess bis(trimethylsilyl)carbodiimide [127–129]:

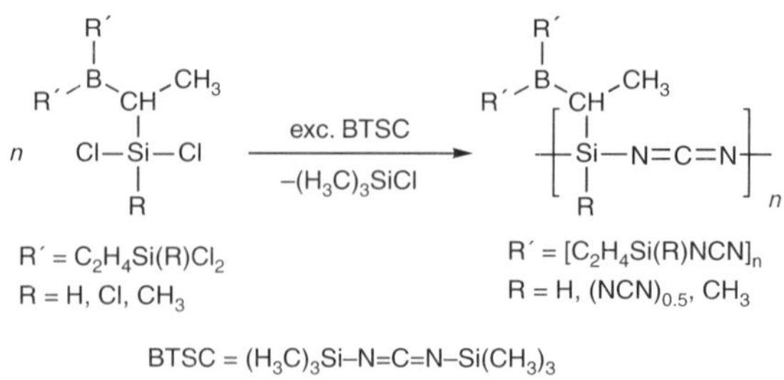

(35)

Depending on the silicon-bonded substituents R, the viscosity of the mixture (sol) increased considerably within 30–60 min ($R = Cl$, H) or within

1 day ($R = CH_3$). After approximately 3 days, the as-obtained reaction mixture, which was then transformed into a gel, began to age and separated from excess bis(trimethylsilyl)carbodiimide and the by-product chlorotrimethylsilane. Purification was preformed by removing the volatile components in high vacuum.

Remarkably, this synthetic pathway is rather cheap and easy to perform. Excess starting compound as well as the by-product chlorotrimethylsilane can easily be recycled. The precursors are obtained in quantitative yields and without the use of solvents. Moreover, time-intensive processing steps are not essentially necessary.

Unfortunately, the high-temperature properties of ceramics derived from such boron-modified poly(silylcarbodiimide)s were not superior over those obtained from boron-free Si—C—N precursors. This was traced back to the chemical composition of the materials, that is, their nitrogen content that was supposed to be too high. However, by the above method it was not possible to reduce the nitrogen content in these materials. Therefore, a synthetic pathway was developed that allows for the synthesis of boron-modified poly(silylcarbodiimide)s with adjustable nitrogen contents while the boron and silicon concentrations were kept almost constant [131, 132] (see Scheme 5.1.15).

Dehydrocoupling was performed in a one-pot reaction by dissolving the silanes in tetrahydrofuran and subsequent mixing with different amounts of cyan amide. A catalyst as required for the dehydrocoupling of tris(hydridosilylethyl)boranes and ammonia or amines that was described above was not used. To ensure quantitative turnovers, the reaction mixtures were refluxed for 12 h whereby strong hydrogen evolution and precipitation of the precursors occurred.

The above examples show that preparative organometallic chemistry allows for the production of a wide variety of molecular precursors for PDCs. Nevertheless, there is still need for further development. Future work will focus on precursors that release phase-pure ceramics as well as composites with tunable properties.

Besides synthetic approaches, the thermal conversion of the precursors also referred to as thermolysis or pyrolysis is a key step in the formation of precursor-derived ceramics.

5.1.3 THERMOLYSIS

The thermally induced degradation of precursors is a rather complex process, which involves numerous chemical reactions in the solid state as well as in the

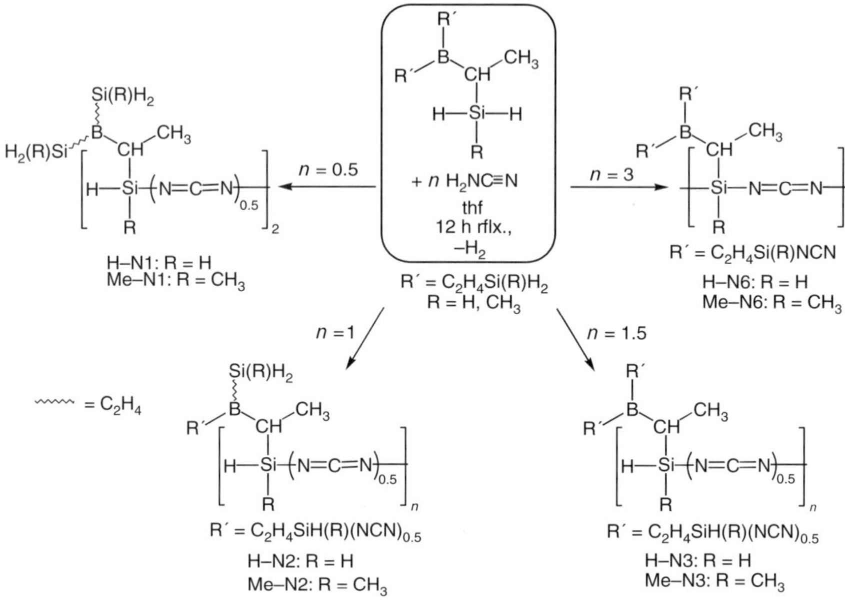

SCHEME 5.1.15 Synthesis of boron-modified poly(silylcarbodiimides) by a dehydrogenative coupling of tris(hydridosilylethyl)boranes, $B(C_2H_4Si(R)H_2)_3$, with cyanamide, $H_2N{-}CN$. The molar ratio of the starting compounds was 1:0.5 to 1:3. Indexes H and Me differentiate between polymers derived from $B(C_2H_4SiH_3)_3$ and $B(C_2H_4Si(CH_3)H_2)_3$, respectively, whereas the indexes N1—N6 give the numbers of nitrogen atoms per boron atom. Chemical compositions ranging from ($C_{6.5}H_{20}NSi_3B$) (H—N1) to ($C_9H_{15}N_6Si_3B$) (H—N6) and ($C_{9.5}H_{26}NSi_3B$) (Me—N1) to ($C_{12}H_{21}N_6Si_3B$) (Me—N6) could be realized by this procedure [132].

vapor phase. Due to limitations of monitoring such reactions, there is still a lack of understanding thermolysis mechanisms in detail. However, since thermolysis significantly influences both composition and microstructure of PDCs, an understanding of the single steps that occur during the heat treatment is of considerable interest. Accordingly, thermolysis of different types of precursors has been studied [37, 41, 43, 133–150] and reviewed [44, 82, 151] extensively.

As noted in previous sections, preceramic polymers are composed of a polymer backbone to which either single atoms or molecular units are attached. These attachments do not only influence physical–chemical properties such as solubility, viscosity, melting, etc., and thus directly determine precursor processability. With respect to thermolysis, it is important that pendent groups may also provide functionality and as a consequence latent reactivity. The latter is known to be a key issue for obtaining high polymer-to-ceramic conversion yields (high "ceramic yields"), since it provides a means for further cross-linking, thus inhibiting polymer backbone degradation and volatilization of low weight molecular backbone components. In contrast, pendent, non-functional groups are eliminated as gaseous by-products, which do not only cause a mass loss upon the thermal treatment, but which are also responsible for retention of impurities or creation of gas-generated pores and cracks. As a logical consequence, molecular precursors for ceramics should be designed in a way that exclusively molecular hydrogen is split-off during the heat treta-ment, which would provide the benefit of an easy outgassing because of its high mobility in any ceramic material. Unfortunately, high ceramic yields and good processability are often contradictory. Therefore, compromises are necessary in the precursor design, usually mandating processing with "low-ceramic-yield" precursors in order to obtain higher-quality ceramic products [151].

A relevant aspect in precursor thermolysis, which until now limits their applicability to the formation of ceramics in small dimensions (thin films, fibers), is the change in density and volume with increasing temperature. Even if ceramic yields are high, dramatic volume shrinkage occurs during thermolysis, which is due to the increase in density. Whereas the density of polymeric precursors is typically in the range of $1\,g/cm^3$, the respective values for fully thermolyzed amorphous ceramics is usually above $2.1\,g/cm^3$. This means that the increasing density of a specimen during thermolysis causes volume shrinkage to below 50% of the initial value. Considering the loss of gaseous by-products during thermolysis, this value is even smaller, for example, in the case of a precursor with 60% ceramic yield, it is below 30%. Considering an increase in density up to more than $3.0\,g/cm^3$ due to the crystallization of the amorphous material at elevated temperature (compare Section 5.1.5.2), the whole situation worsens. Near net shape manufacturing of polymer-derived ceramics is thus difficult to perform and mostly requires either passive or reactive filler materials to inhibit shape distortion [152, 153].

The complexity of the thermolysis process requires for the combination of contrasting investigation tools such as thermal and elemental analysis as well as spectroscopic and diffraction methods. Common methods are, for example, simultaneous thermogravimetric analysis (TGA) in which mass changes during the heat treatment are monitored combined with differential thermal analysis (DTA) to receive valuable information on energetic changes during the polymer-to-ceramic conversion. Elemental analysis of the thermolysis intermediates is essential, since it is the only method that allows for an examination of the compositional changes of the precursors that occur during the polymer-to-ceramic conversion. Structural characterization of thermolysis intermediates provides distinct information on chemical reactions that occur upon ceramization. For this purpose, solid-state MAS NMR (Magic Angle Spinning Nuclear Magnetic Resonance, in the case of silicon containing precursors especially ^{29}Si-NMR [154]) and FT-IR (Fourier Transform Infrared

Spectroscopy) are state-of-the-art. Both are non-destructive methods. NMR spectroscopy represents a tool to probe the local environment of single NMR-active nuclei in amorphous insoluble materials. IR spectroscopy, additionally allows for a rapid determination of characteristic vibrational frequencies of structural units, that is, stretching vibrations of O—H, N—H, C—H, Si—H, C=N, C=C, N=C=N, etc., in the precursor or the thermolysis intermediates. A helpful, supplementary method in this regard is Raman spectroscopy.

5.1.3.1 Thermolysis of Precursors to Silicon Carbide

Among numerous precursors that are described in the literature for the formation of silicon carbide ceramics, there is a small number of polymers to silicon carbide that reached commercial interest. In this regard, the most important systems are poly(dimethylsilane), which is historically seen as the first commercial precursor to SiC/C composites and poly(methylsilane), which upon thermolysis delivers near-stoichiometric silicon carbide.

As noted in Section 5.1.2.1.2, poly(dimethylsilane), $[Si(Me)_2]_n$, can be thermally rearranged to yield poly(methyl-silylenemethylene), $[Si(Me)(H)CH_2]_n$. In contrast to $[Si(Me)_2]_n$, the latter is a soluble and meltable polymer. $[Si(Me)(H)CH_2]_n$ was first applied in 1975 to produce SiC-containing fibers in the so-called Yajima process [12, 155–157]. The procedure is schematically presented in Figure 5.1.38 in Section 5.1.5.4.

Poly(methyl-silylenemethylene) has a molecular weight (MW) of 1200. It can be melt spun and delivers green fibers with poor tensile strengths of around 5^{-10} MPa. Even though the precursor is obtained in a rather simple procedure and in high yields it suffers from the lack of latent reactivity, that is, the ability to undergo thermally induced cross-linking reactions after the spinning process. As a consequence, fiber integrity is lost in the thermolysis step. To maintain fiber integrity, it is thus necessary to subsequently cure the green fibers. In this step, the fiber surface is chemically modified and the fiber rendered infusible. Curing can be performed by different procedures: According to Scheme 5.1.16, heating of the fibers in air results in a partial oxidation of surface Si—H (rather than C—H) groups with formation of Si—OH units. A subsequent condensation step produces a thin Si—C—O film on the surface of the fibers [151].

This silicon oxycarbide film efficiently inhibits melting of the fiber during thermolysis. Oxygen, which is introduced in this step, cannot be removed during the heat treatment and remains in the fiber, limiting the maximum application temperature to approximately 1150°C. Above this temperature, decomposition with formation of volatile species such as CO and SiO as

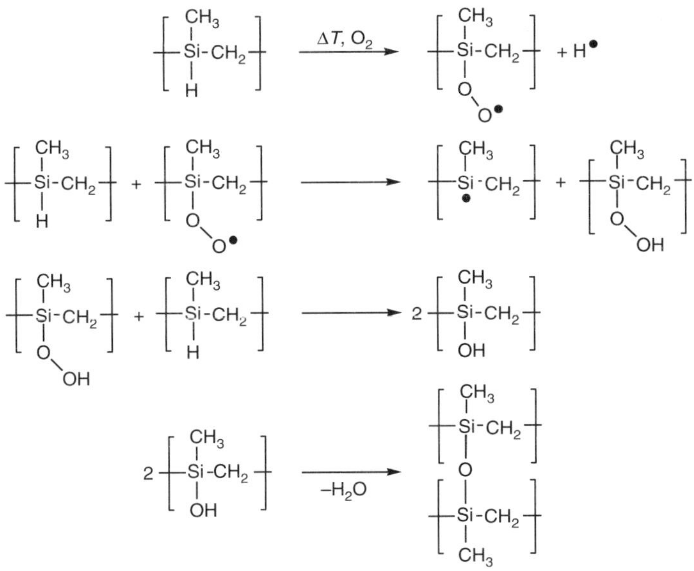

SCHEME 5.1.16 Proposed mechanism for the surface oxidation in the curing step of silicon carbide fibers [151].

well as crystallization occurs, leaving pores and large crystallites behind, which substantially worsen mechanical properties. Alternatively, electron beam irradiation can serve for fiber curing. Compared to green fiber curing in air, it is much more expensive and more difficult to perform but no oxygen is introduced into the green fiber. Consequently, thermolysis delivers oxygen-free SiC/C fibers with enhanced mechanical properties and higher maximum application temperature.

Thermolysis of the green fibers has to be performed in an inert atmosphere. Yajima *et al.* [157–159] reported on thermolysis of Nicalon^{TM} green fibers in vacuum at finally 1500°C to obtain 7-μm thick SiC fibers that exhibit a density of ~2.5 g/cm^3, tensile strengths of 1.5–3 GPa and a Youngs modulus of 170–200 GPa, depending on the processing applied. In contrast, Hi-Nicalon^{TM} fibers exhibit a density of 2.74–3.1 g/cm^3, tensile strengths of 2.8 GPa, and elastic moduli of 270–420 GPa, depending on their (poly-) crystallinity. Moreover, their maximum application temperature is above 1400°C compared to 1150°C for that of Nicalon^{TM} fibers. For further details consider Section 5.1.5.4.

Thermogravimetric analysis of the thermolysis step, which released ceramics in 60% yield, indicated that the thermal decomposition of the polymer proceeds gradually above 300°C and is completed at 800°C. However, the pyrolytic degradation of the polymer does not take place “optimally” according to

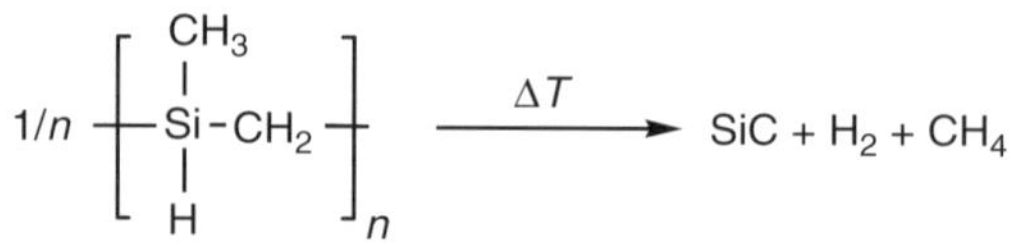

(36)

since almost half of “excess” carbon, which is present in the precursor as methyl groups, is retained in the ceramic fibers. Consequently, no phase-pure SiC but SiC/C composite fibers without defined microstructures are obtained.

To obtain phase-pure SiC, alternative improved precursors were developed. The idea behind these investigations was to design polymers in which the final elemental composition of the ceramic, that is, the stoichiometry (1:1) of the constituting elements silicon and carbon is already pre-formed and that do not produce gaseous by-products upon thermolysis other than hydrogen. Polymers, which fulfill these requirements, are polymethylsilane, $[SiMeH]_n$ [10], and polysilaethylene, $[H_2SiCH_2]_n$ [13,15], which can be obtained according to procedures described in Section 5.1.2.1. $[H_2SiCH_2]_n$ can also be generated by a Kumada rearrangement of $[SiMeH]_n$. However, although the precursors exhibit a 1:1 Si:C ratio, it is difficult to obtain stoichiometric or near-stoichiometric SiC. The reasons are that (i) Kumada rearrangement is a radical reaction, which takes place at elevated temperature and thus low selectivity and (ii) that depolymerization of the precursors during thermolysis occurs along with the volatilization of carbon and/or silicon species. IR and NMR investigations of the thermolysis intermediates provide information on chemical reactions that occur during the heat treatment. In the case of $[Si(Me)(H)CH_2]_n$, at least four different steps can be distinguished. Below 400℃, low-weight oligomers evaporate. Up to 550℃, molecular weight increases due to the formation of new SiC bonds that can be traced back to a dehydrocoupling of Si—H and C—H units [134, 160]. In contrast, there is no evidence for the formation of Si—Si linking. Further heating to 800℃ results in the degradation of residual Si—H, CH_3, and Si—CH_2—Si units with evaporation of low molecular weight organic species, that is, molecular hydrogen and methane. This stage corresponds to the organic-to-ceramic transition and releases an inorganic covalent SiC network. Additionally, a phase of free carbon forms, which is evident from ^{13}C MAS NMR investigations [160]. Above 800°C, only very small weight loss is observed, which is due to the evaporation of residual hydrogen. Heating to temperatures exceeding 1100°C finally results in the crystallization of the predominantly amorphous ceramic. Details on this topic are presented in Section 5.1.4.

5.1.3.2 Thermolysis of Precursors to Silicon Carbide/Nitride Ceramics

Due to the structural diversity of Si—C—N precursors compared to SiC polymers, thermolysis appears much more complex. Whereas for SiC usually polysilanes or polycarbosilanes serve as preceramics, polysilazanes or poly(silylcarbodiimide)s are used as precursors for ternary Si—C—N ceramics.

The ternary phase diagram in Figure 5.1.3 distinguishes different types of compositions in the Si—C—N system.

Poly(silsesquiazane)s [161] or poly(silylcarbodiimide)s [75, 77, 78, 162], both comparably rich in nitrogen, usually deliver ceramics with compositions located on the tie line C—Si_3N_4 (A). In contrast, those derived from polysilazanes with the general structure $[(R^1)(R^2)Si{-}NH]_n$ (R^1, R^2 = single-bonded organic unit) [27, 37, 41, 42, 135, 138, 163–167] result in ceramic compositions located in the three-phase field SiC/Si_3N_4/C (B). Laine *et al.* published an exception from this general rule. They obtained Si_3N_4/C (type A) compositions by thermolysis of $[H_2Si{-}N(CH_3)]_n$ rather than silicon nitride and silicon carbide-based composites [145]. Compositions located in the three-phase field SiC/Si_3N_4/Si (C) were recently obtained by co-pyrolysis of polycarbosilane PCS and perhydridopolysilazane PHPS [168–170].

Even though a comparison of the elemental composition of the precursors and their derived ceramics gives first hints for possible reactions that occur during the heat treatment, it is inevitable to structurally characterize thermolysis intermediates accurately. In general, spectroscopic methods such as NMR and IR spectroscopy, elemental analysis as well as (mass spectra-coupled) thermogravimetric analysis (TGA) are considered for this purpose.

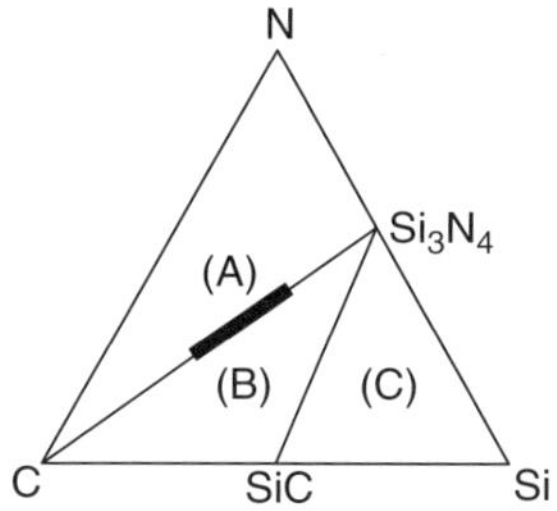

FIGURE 5.1.3 Ternary Si—C—N phase diagram ($T < 1440$°C, 1 atm N_2): (A) compositions on the tie-line C—Si_3N_4; (B, C) compositions in the three phase fields SiC/Si_3N_4/C or SiC/Si_3N_4/Si, respectively.

5.1.3.2.1 Thermolysis of Polysilazanes

Most of the work in this field concentrated on thermolysis of polysilazanes. It turned out that the molecular structure, that is, the polymer backbone, pendent functional groups, and cross-linking of the precursors to a significant extent influence the polymer-to-ceramic conversion. As mentioned in Section 5.1.2.2, ammonolysis of dichlorosilanes usually delivers six- or eight-membered cyclosilazanes $[SiR^1R^2{-}NH]_n$ ($n = 3, 4$). Because of their low molecular weight, as-obtained polymers are mostly unsuitable as precursors for ceramics. In the worst case, thermolysis of such volatile molecules results in a 100% mass loss. Moreover, cyclosilazanes are generally liquids, which have poor rheological properties, disqualifying them as precursors for ceramic fibers or bulk materials.

A procedure for increasing ceramic yields of low-weight polysilazanes is a catalyst-induced cross-linking by dehydrocoupling of Si—H and N—H units. For example, Seyferth *et al.* described cross-linking reactions of cyclotetrasilazane, yielding a highly branched polymer, which is soluble in tetrahydrofuran [38, 39]. Details on the procedure including a proposed mechanism are presented in Section 5.1.2.3. The increased cross-linking was directly mirrored in increased ceramic yields from below 30% to approximately 84%.

The incorporation of reactive sites in oligomeric silazanes, for example, vinyl groups, also results in increased ceramic yields. This is directly reflected by a comparison of ceramic yields of $[SiMeH{-}NH]_n$ (20%) [165] with that of $[SiViH{-}NH]_n$ (Vi = $HC{=}CH_2$, 83%) [42] or $[SiMe_2{-}NH]_n$ (50%) and $[SiViMe{-}NH]_n$ (76%) [41], as determined by TGA.

Detailed investigations on the thermolysis of vinyl-substituted oligosilazanes of the general type $[SiViR{-}NH]_n$ and the influence of the vinyl group in cross-linking reactions were published by Bill *et al.* (R = Me, $(NH)_{0.5}$) [43, 171] and by Mocaer *et al.* (R = H) [144].

Based on solid-state ^{1}H-, ^{13}C- and ^{29}Si-NMR and FT-IR investigations, Bill *et al.* suggested conclusive thermolysis mechanisms.

For example, cross-linking reactions involving Si—C coupling accompanied by the elimination of hydrogen (Eq. 37) or methane (Eq. 38), as shown below, were considered:

(37)

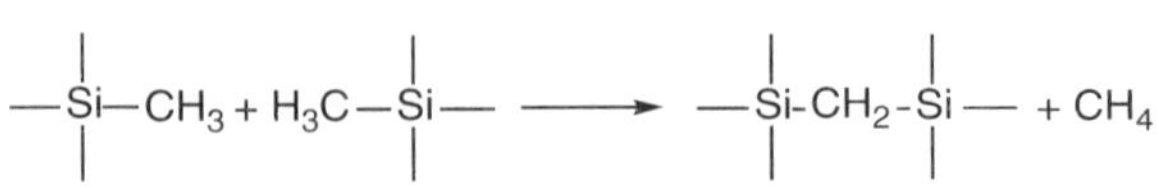

(38)

They occur at around 500°C and most likely follow radical mechanisms. Further increasing the temperature finally results in the formation of CSi_4 tetrahedra:

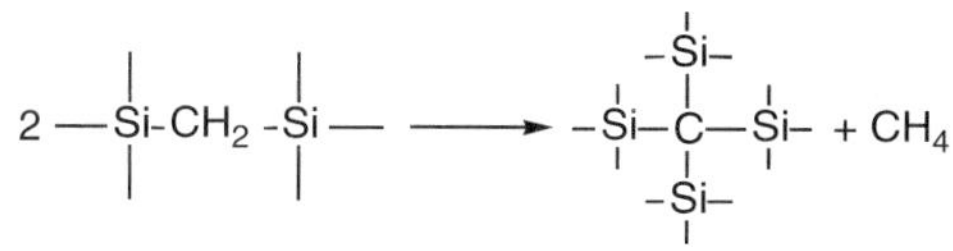

(39)

Vinyl groups that are attached to the silazane skeleton enable cross-linking by olefin polymerization at low temperatures, usually around 300°C. It will be discussed below (Figure 5.1.4) that ^{13}C-NMR spectroscopy with cross-polarization and magic angle spinning (CP-MAS) is an ideal tool to monitor the progress in such reactions. The sp^2-carbon atoms of the vinyl groups generate NMR signals at approximately 130–140 ppm. During the cross-linking they are transformed into sp^3-carbon atoms, which generate resonance signals at around 10–30 ppm:

(40)

Raising the temperature to higher values results in Si—C bond cleavage and β-hydride shift with formation of Si—H bonds and hydrogenated sp^2-carbon atoms. Under the given conditions, the latter eliminate hydrogen and transform into sp^2-carbon:

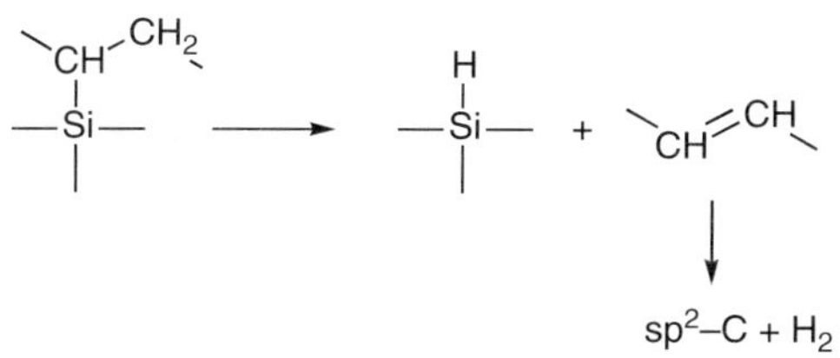

(41)

Si—H units play an important role in that they present reactive sites for hydrosilylation and/or dehydrocoupling. The former additionally requires C=C or C≡C units in the polymer and mostly takes place below 200–250°C. Si—N bond formation by dehydrocoupling of Si—H and N—H units is important, since in combination with the above discussed C=C polymerization and subsequent Si—C bond cleavage it sufficiently explains the transformation of Si—C units, present in the precursor, into Si—N (finally SiN_4) units.

(42)

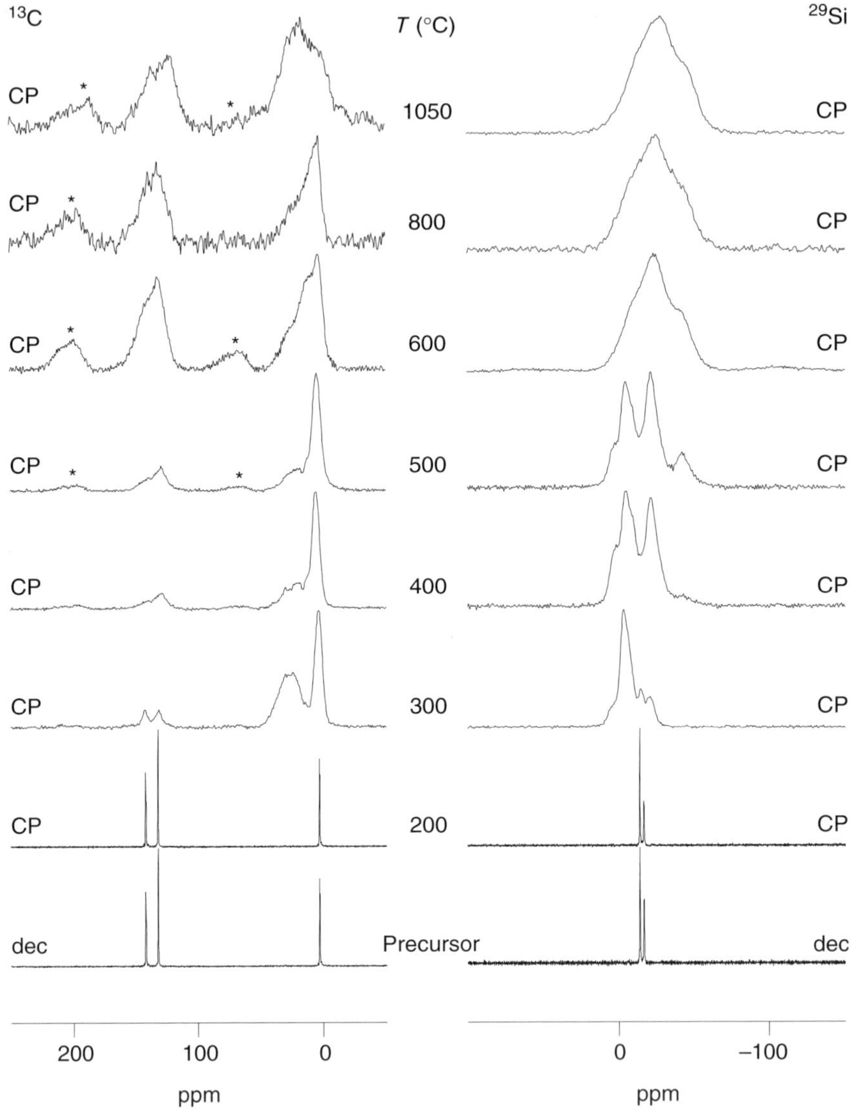

FIGURE 5.1.4 Experimental ^{13}C- (left) and ^{29}Si- (right) NMR spectra of poly(methylvinylsilazane), $[SiViMe{-}NH]_n$ at various stages of the thermolysis process. All spectra were obtained under magic angle spinning conditions using the cross-polarization technique [172].

In the pyrolytic degradation of oligo- or polysilazanes of the type $[SiR^1R^2\text{-}NH]_n$, N—H bonds are frequently involved in *trans*-amination reactions. These take place at temperatures around 200–500°C and result in the formation of NSi_3 sites among gaseous ammonia, which is eliminated:

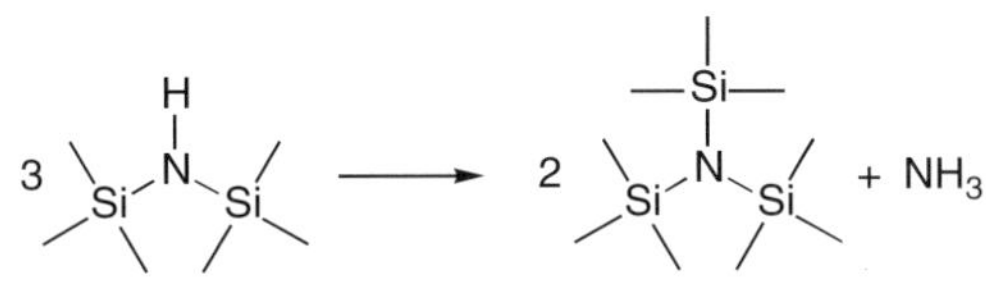

(43)

The loss of ammonia leads to a reduction of the nitrogen content in the ceramic materials. The tendency for *trans*-amination reactions is especially pronounced for poly(silsesquiazane)s of the general type $[SiR(NH)_{1.5}]_n$, since these polymers are rich in nitrogen. Their composition—hydrogen not taken into account—is located in the three-phase field $C\text{—}Si_3N_4\text{—}N$ (compare Figure 5.1.3) whereas that of the derived ceramic lies on the tie line $C\text{—}Si_3N_4$, as will be discussed below.

Finally, Si—N coupling with elimination of methane is frequently considered. This reaction requires for a cleavage of thermodynamically relatively stable Si—C bonds and thus takes place exclusively at elevated temperature:

(44)

In the following, the pyrolytic degradation of two structurally different oligosilazanes, which yield ceramics with different chemical composition, is discussed in detail. For this purpose, poly(methylvinylsilazane) (PMVS), $[SiViMe\text{—}NH]_n$, and nitrogen-rich poly(vinylsilsesquiazane), $[SiVi(NH)_{1.5}]_n$ (PNVS), are considered. Both polymers do not only differ in their chemical composition but also in their cross-link density. The first is a linear oligomer (or cyclomer) whereas the second constitutes a highly branched network.

Reliable information on the pyrolysis mechanism is received especially from NMR spectra, which for this intention have to be analyzed in detail. In the following, such a spectral breakdown is represented as an example for PMVS [172].

The ^{13}C CP MAS NMR spectrum of as-synthesized PMVS precursor shows NMR signals at around 130 and 140 ppm, which are generated by the olefin carbon atoms and one at 0 ppm, which is due to the silicon-bonded methyl group. The appearance of the spectrum does not change after heating the polymer to 200°C, indicating that in this temperature range no chemical modifications of the precursor involving carbon atoms take place. On heating the sample to 300°C, drastic changes are observed. The signals at 130 and 140 ppm significantly decrease in intensity whereas a broad resonance appears at ~25 ppm.

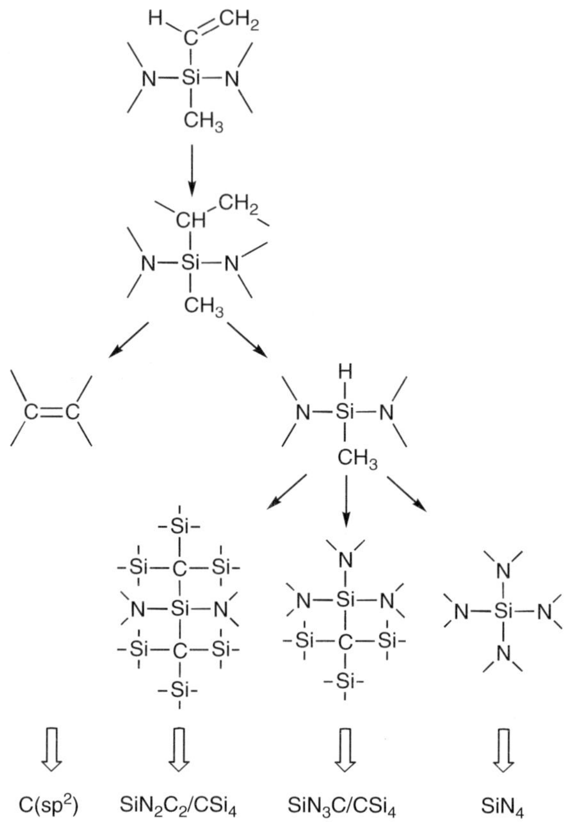

SCHEME 5.1.17 Proposed thermolysis intermediates and structural ceramic units that evolve during the polymer-to-ceramic conversion of poly(methylvinylsilazane), $[SiViMe{-}NH]_n$ [172].

This observation points to an olefin polymerization of the pendent vinyl groups in the precursor according to Eq. 40. In Scheme 5.1.17, this reaction indicates the onset of the thermal degradation of the polymer. In the ^{13}C-NMR spectrum of the sample heated to 400°C, the latter signal decreases in intensity whereas a new broadened signal at around 130 ppm is found. This points to a Si—C bond cleavage and the formation of sp^2 carbon atoms as pointed out in Eq. 41. Further raising the temperature to 500°C causes only little changes in the appearance of the spectrum. However, signals appear broader and the resonance of the sp^2 carbon atoms is somewhat more intensive. Remarkably, the signal of the methyl groups still remains unchanged. At 600°C, again drastic changes are observed. The intensity of the signal at 130 ppm strongly increases whereas that at 0 ppm significantly broadens. This temperature marks the decomposition of residual organic functions, that is, the silicon-bonded methyl groups. It is accompanied with the loss of hydrogen and/or methane and the formation of Si—C bonds (compare Eqs. 37–39). Moreover, methane that is split off partly decomposes to graphitic carbon and molecular hydrogen.

Further information on the thermal degradation of PMVS is received from ^{29}Si-NMR spectroscopy. The spectrum of the precursor displayed in Figure 5.1.4 shows two sharp lines at −14 and −17 ppm, which point to a mixture of cyclic (trimeric and tetrameric) species. Heating to 200°C causes no changes in the signal appearance whereas raising the temperature to 300°C results in a low field shift of the signals to around −3 ppm. This is in accordance with the transformation of the sp^2-hybridized vinyl carbon atoms to sp^3-hydrocarbons, which was also concluded from the respective ^{13}C-NMR spectra. At 400°C, a new signal at −21 ppm evolves. This value is typical for $C(sp^3)SiN_2H$ sites and clearly points to a rearrangement with Si—C bond cleavage and Si—H bond formation, as suggested in Eq. 41. However, it is also possible that in this stage $C(sp^3)SiN_3$ sites form by a dehydrocoupling reaction of Si—H and N—H units. They appear in a comparable chemical shift range as $C(sp^3)SiN_2H$ sites. At 500°C, a new signal at −45 ppm emerges, which can unequivocally be assigned to SiN_4 sites, most probably due to reactions according to Eq. 44. Above 500°C, the NMR signals broaden and overlap substantially. However, it can doubtless be concluded that SiC_2N_2, $SiCN_3$, and SiN_4 sites generate the broad signal observed at and above 600°C.

According to the observations in the ^{13}C- and ^{29}Si-NMR spectra of the pyrolysis intermediates of PMVS and the proposed pyrolysis mechanism in Scheme 5.1.17, a ceramic evolves, which is composed of sp^2-carbon and an amorphous Si—C—N phase, in which silicon is bonded in $SiC_{4-x}N_x$ ($x = 2$–4) environments. This is in accordance with the finding of the elemental analysis, which confirm that the elemental composition of PMVS-derived ceramic is located in the three-phase field C—SiC—Si_3N_4 (compare Figure 5.1.3).

The lack of aliphatic silicon-bonded hydrocarbons and the higher cross-linking in PNVS, in contrast, result in a ceramic material with different composition. Chemical reactions that take place during the heat treatment are depicted in Scheme 5.1.18. They are basically comparable with those of PMVS. This is reflected by a similar onset of the decomposition temperature, marked by a polymerization of the silicon-bonded vinyl groups at around 300°C.

During the further heat treatment, the aliphatic hydrocarbon chains are split off with the formation of $SiHN_3$ and SiN_4 sites among sp^2-carbon that segregates. Subsequent dehydrocoupling of Si—H and N—H units as well as *trans*-amination reactions result in the formation of a ceramic material, which is composed of sp^2-carbon and silicon nitride. This is nicely reflected in both ^{13}C- and ^{29}Si-NMR spectra of final thermolysis products (not shown here), which give no evidence for any Si—C units [171, 172]. Consequently, compositions of

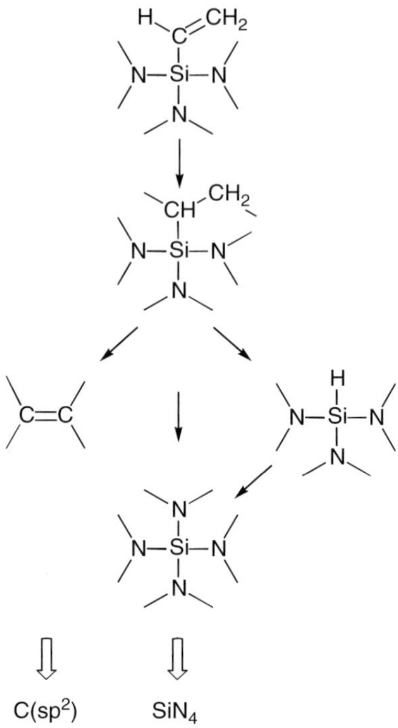

SCHEME 5.1.18 Proposed thermolysis intermediates and structural ceramic units that evolve during the polymer-to-ceramic conversion of nitrogen-rich poly(vinylsilsesquiazane) $[SiVi(NH)_{1.5}]_n$ (PNVS) [171].

ceramics obtained from PNVS are located on the tie line $C—Si_3N_4$, which is indeed verified by elemental analysis.

Recently, amorphous ceramics with compositions along the tie line $SiC—Si_3N_4$ could be obtained [48]. For the development of such materials the above-discussed findings were considered carefully. The synthesis of the precursor was discussed in detail in Section 5.1.2.3 (compare Scheme 5.1.10).

In this regard, the main objective was to design a precursor in which the final elemental composition of the ceramic, that is, the Si_4N_4C stoichiometry of the constituting elements is already pre-formed and which does not suffer from an elimination of gaseous thermolysis by-products other than hydrogen. Therefore, the following idealized thermolysis reactions were considered, presuming only hydrogen evolution occurs upon heat treatment. Accordingly, thermolysis of perhydridopolysilazanes, $[SiH_2—NH]_n$ (sum formula $Si_4N_4H_{12}$ for $n = 4$), results in a composite $Si_3N_4 + Si$ whereas thermolysis of poly(methylsilazane), $[H_3CSiH—NH]_n$ (sum formula $Si_4N_4C_4H_{20}$ for $n = 4$), delivers a material with a chemical composition Si_3N_4 + SiC + 3C. It was thus concluded that copolymers composed of (SiH_2—NH) and (H_3CSiH—NH) building blocks in a 3:1 ratio such as $[(SiH_2-NH)_3(H_3CSiH-NH)]_n$ (sum formula $Si_4N_4CH_{14}$) should deliver Si_3N_4 + SiC ceramics without "free" carbon or silicon. Consequently, precursor synthesis was performed by co-ammonolysis of H_2SiCl_2 and $H_3CSiHCl_2$ in a 3:1 ratio at −10°C using tetrahydrofuran as solvent according to Scheme 5.1.10.

To avoid depolymerization and volatilization of low molecular species during the polymer-to-ceramic conversion, the as-obtained precursor was further cross-linked by base-catalyzed dehydrocoupling according to a method described earlier by Seyferth *et al.* using 0.1 mol% of *n*-butyl lithium as a catalyst. The theoretical ceramic yield of the cross-linked precursor, presuming only hydrogen evaporation, was expected to be around 95%.

The polymer-to-ceramic conversion was monitored by means of mass spectra-coupled simultaneous thermogravimetric analysis (TGA-MS), as shown in Figure 5.1.5.

Thermolysis up to 1200°C takes place in two-step decomposition in the 200–900°C range. The ceramic yield of 94.5% is the highest value observed for Si—C—N precursors so far. Thermolysis of a precursor not cross-linked in contrast, delivers ceramics in only 78% yield (M. Weinmann, unpublished results). Accordingly, the high ceramic yield of the cross-linked precursor

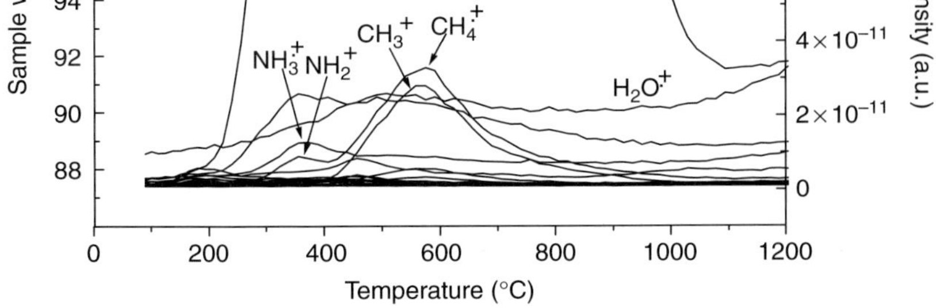

FIGURE 5.1.5 Mass spectra-coupled simultaneous thermogravimetric analysis of cross-linked $[(SiH_2-NH)_3(H_3CSiH-NH)]_n$. Fragments ranging from $m/z = 2$ to $m/z = 84$ are considered. Heating rate: 5°C/min; atmosphere: flowing Ar [48].

is mainly a consequence of the high cross-link density of the polymer combined with sufficient latent reactivity due to the presence of both Si—H and N—H units. These enable thermally induced dehydrocoupling reactions according to Eq. 42 and efficiently avoid depolymerization and volatilization of low molecular species during the heat treatment. The observed mass loss of 5.5% corresponds to the hydrogen content of the precursor determined by elemental analysis and points to the fact that exclusively hydrogen evaporates upon thermolysis. This assumption was confirmed by the results of the mass spectroscopic investigations. Hydrogen elimination starts at approximately 200°C and continues up to 1000°C. Above 1000°C the concentration of hydrogen increases again, parallel with that of a species with $m/z = 18$. The latter corresponds to water, which in this temperature range, is eliminated from the alumina crucible in which the measurement was performed. The amount of gaseous species below 1000°C with a higher m/z ratio than 2, however, is low, suggesting that the Si:C:N ratio is retained during the polymer-to-ceramic transformation. However, a very small amount of ammonia and methane and their corresponding fragments can be detected unequivocally, though in low intensity, at ~380 and ~580°C, respectively. A precise comparison of the relative intensities of methane and ammonia with that of hydrogen is not yet possible, since the large amount of hydrogen exceeded the detector's capacity.

5.1.3.2.2 Thermolysis of Polysilylcarbodiimides

In contrast to the Si—N skeleton in polysilazanes, which is retained during the thermal degradation of the polymers, the Si—N=C=N backbone in poly(silylcarbodiimide)s is decomposed quantitatively. Neither findings in NMR nor IR investigations give a hint for the presence of such structural units in the amorphous ceramics. The only exception are two metastable crystalline phases, SiC_2N_4 and Si_2CN_4 (compare Figure 5.1.2), which were published by Riedel *et al.* [81]. As already pointed out in Section 5.1.2.3, they were obtained by a reaction of tetrachlorosilane and bis(trimethylsilyl)carbodiimide and subsequent heating to 400 and 920°C, respectively.

Basically, there are two different degradation mechanisms that additionally have to be considered in the thermolysis of poly(silylcarbodiimide)s. The first mechanism involves elimination of acetonitrile, H_3C—CN, or HCN from poly(silylcarbodiimide)s exhibiting (NCN)Si—CH_3 or (NCN)Si—H units, respectively:

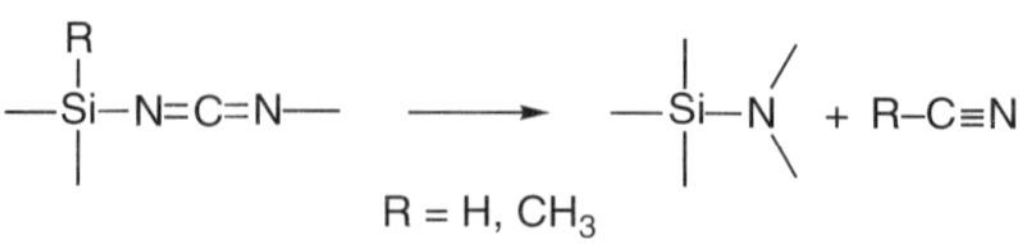

(45)

Both CH_3CN and HCN elimination result in the formation of novel SiN motifs. Acetonitrile that splits of subsequently decomposes with formation of sp^2-carbon. Simultaneous elimination of cyanogene, N≡C—C≡N, and nitrogen takes place at elevated temperatures.

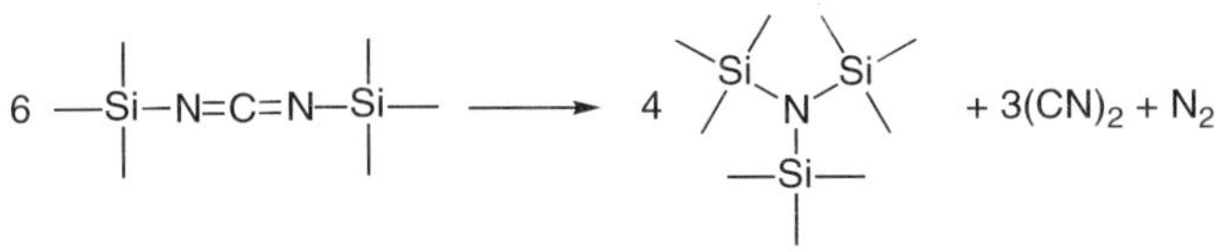

(46)

According to the formerly mentioned degradation, Si—N—Si linkages form. In addition, this type of reaction is responsible for the thermal degradation of SiC_2N_4 to Si_2CN_4, as already mentioned above (Figure 5.1.2 [81]).

Even though structurally different, polysilazanes and poly(silylcarbodiimide)s follow comparable degradation routes. As an example, the progress in the thermal decomposition of poly(methylvinylsilylcarbodiimide), $[SiViMe—NCN]_n$ (PMVSC), which was investigated by solid-state ^{13}C- and ^{29}Si-MAS NMR is discussed (Figure 5.1.6, Scheme 5.1.19) and compared with that of poly(methylvinylsilazane) $[SiViMe—NH]_n$ (see Scheme 5.1.17) [150].

Similar to the ^{13}C CP MAS NMR spectrum of PMVS, PMVSC shows NMR signals at around 135 ppm, which are generated by the vinyl carbon atoms as well as a line at 0 ppm, which is due to the silicon-bonded methyl group. An additional signal at 125 ppm can be attributed to the carbodiimide carbon atom, which vanishes after heating to 200°C. According to the thermolysis behavior of PMVS, further heating to 300°C causes polymerization of the pendent vinyl groups and results in the appearance of a broad resonance at around 20–25 ppm, which can be attributed to aliphatic carbon atoms. The latter disappears after heating to 400°C. Moreover, the line width of the N=C=N resonance at 125 ppm increases considerably, most probably because of the formation of "free" carbon, which also shows resonance in this chemical shift range. At 500°C, signals broaden further and the low-field signal increases in intensity, indicating further formation of graphite. The absence of C—H resonance signals after heating to 600°C points to the fact that the thermal degradation of the precursor is almost completed. Similar insights are obtained from ^{29}Si MAS NMR spectroscopy. Olefin polymerization at temperatures up to 300°C causes a low-field shift of the ^{29}Si signal from −35 to −20 ppm, which is a consequence of the transformation of $SiC(sp^2)C(sp^3)(NCN)_2$ sites into $SiC(sp^3)_2(NCN)_2$ motifs. Further heating causes multiple reactions, resulting in the formation of at least four differentiable ^{29}Si sites. According to Scheme 5.1.19, formation of $SiN_2C(sp^3)(NCN)$ ($\delta = -32$ ppm) and $SiN_3C(sp^3)$ ($\delta = -16$ ppm) sites is preferred. Above 500°C, all low-field signals vanish whereas a broad signal forms that can unequivocally be assigned to SiN_4 sites. In conclusion, a comparison

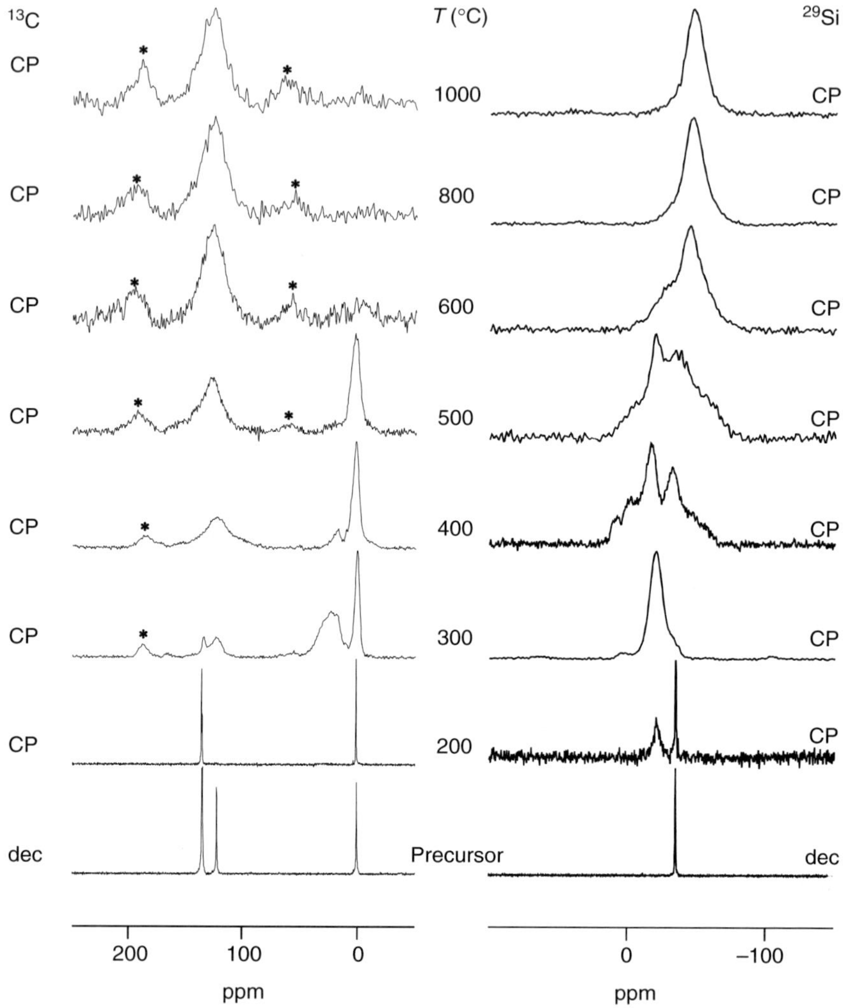

FIGURE 5.1.6 Experimental ^{13}C- (left) and ^{29}Si- (right) NMR spectra of poly(methylvinylsilylcarbodiimide), $[SiViMe{-}NCN]_n$ at various stages of the thermolysis process. All spectra were obtained under magic angle spinning (MAS) conditions using the cross-polarization technique [150, 172].

of $[SiViMe{-}NH]_n$ (PMVS) and $[SiViMe{-}NCN]_n$ (PMVSC) clearly shows that the polymer structure and chemical composition significantly influence the polymer-to-ceramic conversion and therefore the chemical composition of the PDCs. Whereas PMVS delivers ceramics with a composition located in the

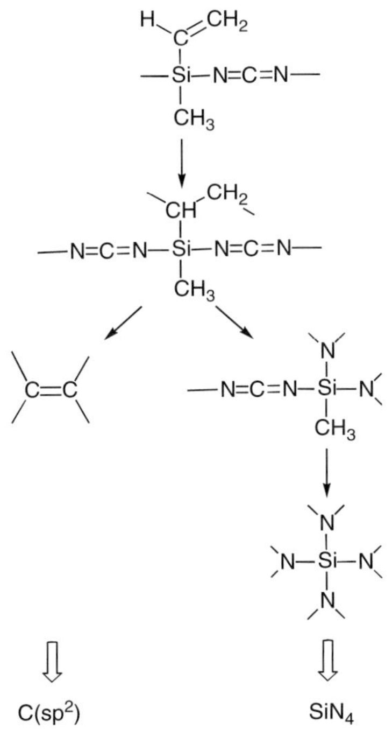

SCHEME 5.1.19 Proposed thermolysis intermediates and structural ceramic units that evolve during the polymer-to-ceramic conversion of poly(methylvinylsilylcarbodiimide), $[SiViMe{-}NCN]_n$ [150, 172].

three-phase field $C{-}SiC{-}Si_3N_4$ (Figure 5.1.3), thermolysis of PMVSC results in a material with a composition located on the tie line $C{-}Si_3N_4$.

5.1.3.3 Thermolysis of Precursors to Si—B—C—N Ceramics

Until now, not much is published on the thermolysis behavior of precursors for Si—B—C—N ceramics. So far, only precursors of the general structure $[B(C_2H_4SiRX{-})_3]_n$ (R = single bonded organic ligand; X = NH, NCN) have been investigated [171–173].

The extension of the ternary Si—C—N system with boron results in additional reactions during the thermal degradation of the polymeric precursors. In this regard, rearrangements that involve B—C bond cleavages and B—N bond formations have to be considered especially.

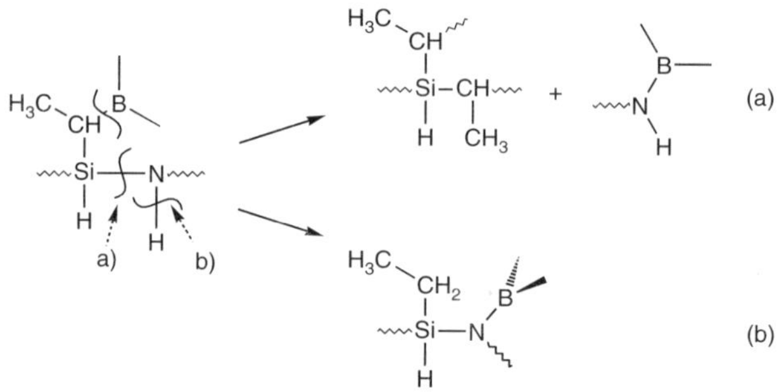

SCHEME 5.1.20 Possible polymer skeleton rearrangements during thermolysis of boron-modified polysilazanes $[B(C_2H_4SiHNH-)_3]_n$ (compare Eq. 32) due to B—C bond cleavage: (a) including Si—N bond cleavage and Si—C bond formation; (b) without Si—N bond cleavage.

In this regard, two different reactions must be taken into account: B—C bond cleavage and B—N bond formation with (Scheme 5.1.20, reaction pathway a) or without (Scheme 5.1.20, reaction pathway b) Si—N bond cleavage. In the first case (a), $Si-CH(CH_3)-Si$ motifs form whereas simultaneous B—C and N—H bond cleavage (b) result in the formation of pendent $Si-CH_2-CH_3$ units.

Such reactions are best monitored by ^{11}B, ^{15}N, or ^{29}Si MAS NMR spectroscopy, whereas ^{13}C MAS NMR does not allow for a distinct differentiation of individual resonance signals, because of a significant signal overlap of all aliphatic carbon atoms. Unfortunately, the natural abundance of ^{15}N is too low for obtaining NMR spectra with sufficient signal-to-noise ratio. For performing ^{15}N-NMR spectroscopy, samples must therefore, be synthesized enriched in ^{15}N.

As an example, the single steps in the thermal degradation of boron-containing $[B(C_2H_4SiH-NH)_3]_n$, which can be obtained from the above-mentioned PHVS $[SiViH-NH]_n$ in a hydroboration reaction using borane dimethyl sulfide [114], will be discussed on the basis of ^{13}C, ^{29}Si (Fig. 5.1.7) and ^{11}B MAS NMR spectroscopy (left spectrum in Figure 5.1.8). Additionally, the results of ^{15}N MAS NMR investigations (Figure 5.1.8, right spectrum) will be reviewed.

The ^{13}C-NMR spectrum of the polymeric precursor $[B(C_2H_4SiH-NH)_3]_n$ (Figure 5.1.7, bottom left), exhibits two strong signals at 12 and 28 ppm, which refer to $B-C_2H_4-Si$ groups along with two very weak resonance signals at 124 and 139 ppm. The latter correspond to remaining C=C units that due to steric effects did not react during the hydroboration of PHVS. After heating to 200°C, the weak olefin signals observed in the precursor vanished completely. This points to the fact that increasing the temperature enables quantitative hydroboration of the C=C units. Upon heating to 400°C, the signal at 28 ppm (aliphatic

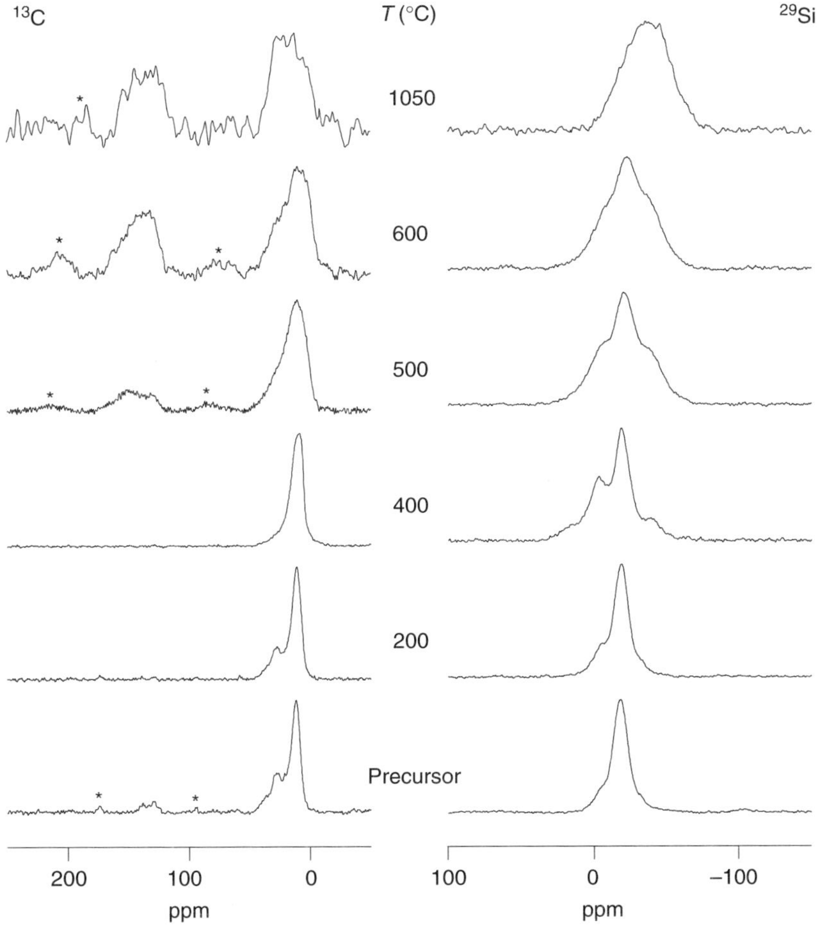

FIGURE 5.1.7 Experimental ^{13}C- (left) and ^{29}Si- (right) NMR spectra of boron-containing $[B(C_2H_4SiH{-}NH)_3]_n$ [114], which was obtained by hydroboration of PHVS $[SiViH{-}NH]_n$ [42], at various stages of the thermolysis process. All spectra were obtained under magic angle spinning conditions using the cross-polarization technique [172, 173].

$CHCH_3$) disappeared whereas aliphatic signals at 12 ppm (generated by CH_2 and $CHCH_3$) are still present. Further heating to 600°C results in a remarkable signal broadening and the appearance of a new, very broad resonance signal at around 130 ppm, which can be attributed to graphite-like domains. At higher temperature, the intensity of the latter spectral component increases further at the expense of the signal intensity of the aliphatic carbon atoms, which is now centered at around 15 ppm, a value typical for CSi_4 environments.

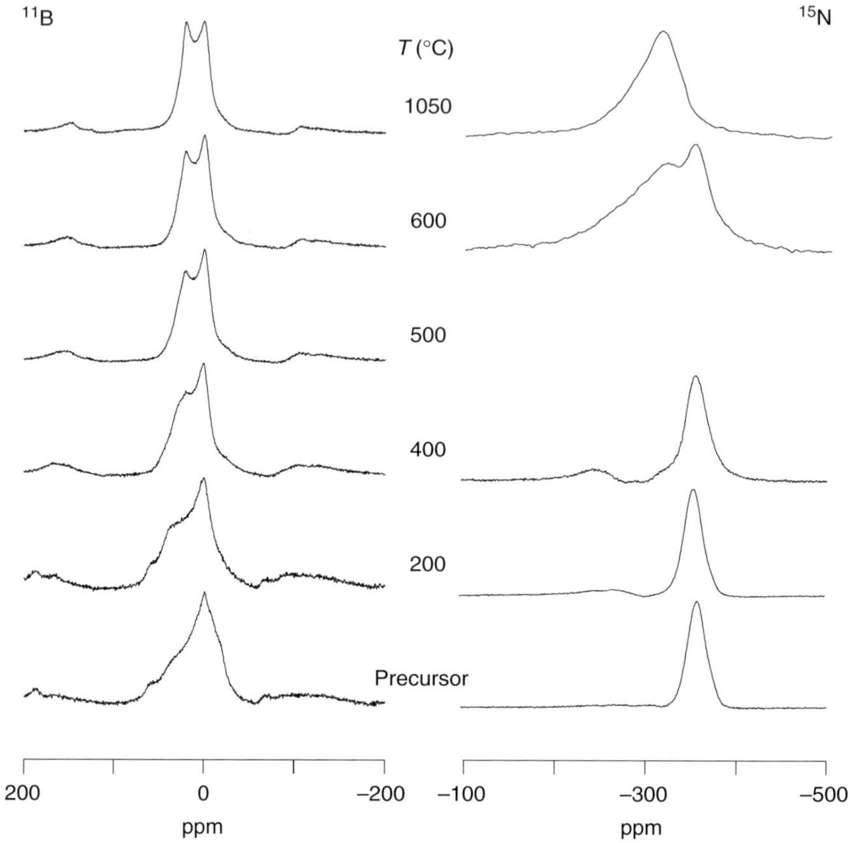

FIGURE 5.1.8 Experimental ^{11}B- (left) and ^{15}N- (right) NMR spectra of boron-containing $[B(C_2H_4SiH{-}NH)_3]_n$ [114], which was obtained by hydroboration of PHVS $[SiViH{-}NH]_n$ [42], at various stages of the thermolysis process. All spectra were obtained under magic angle spinning conditions [172, 173].

The ^{29}Si-NMR spectrum of the precursor in which silicon has a (SiHC(sp^3)N$_2$ environment) exhibits a broad line centered at −13 ppm, which is in accordance with theoretical calculations. Upon heating to 400°C, two new signals appear. One of these signals with very low intensity is observed high-field-shifted to −37 ppm. It refers to SiC(sp^3)N$_3$ sites that must have formed in dehydrocoupling reactions involving Si—H and N—H units. In contrast, the second signal is observed at lower field (−4 ppm), corresponding to SiHC$_2$(sp^3)N units that most likely have formed by a polymer skeleton rearrangement including Si—C bond formation by B—C and Si—N bond cleavage according to reaction pathway (a) in Scheme 5.1.20. Further raising the temperature results in a broad resonance signal centered at 30 ppm, which is caused by additional dehydrogenative Si—N coupling, finally resulting in the formation of a mixture of $SiC(sp^3)_2N_2$ and $SiC(sp^3)N_3$ sites. FT-IR spectra of the thermolysis intermediates confirm the above assumption. Due to the quantitative dehydrogenative Si—N coupling, N—H and Si—H stretching vibrations, which are observed in the IR spectra of the polymeric precursors and which appear as very strong signals at around 3380 and 2124 cm^{-1}, respectively, completely disappear above 600°C.

Accordingly, ^{11}B and ^{15}N MAS NMR spectra show pronounced changes in the 400–600°C range. Figure 5.1.8 on the left shows the experimental ^{11}B-NMR spectra that refer to the excitation of the central transition ($m_I = -\frac{1}{2}$ to $m_I = +\frac{1}{2}$).

Owing to the quadrupolar moment of the boron nucleus, the signals of all ^{11}B-NMR spectra are significantly broadened (a second-order broadening, which cannot be eliminated by fast rotation at the magic angle).

It can be seen clearly that the resonance signals in the ^{11}B-NMR spectra of the polymeric precursor and the thermolysis intermediates below 500°C are centered at around 0 ppm. They appear broadened, featureless, and without any fine structure. Their line shapes point to a significant heterogeneity in the local environment of the boron nuclei, presumably due to the presence of a mixture of trigonal and tetrahedral coordination spheres. However, exceeding 500°C results in a low-field shift to 15 ppm and the formation of a signal with distinct special features (quadrupolar coupling constant = 2.8–2.9 MHz), which are typical for trigonal-coordinated boron nuclei and which is almost identical with the spectrum of h-BN. This observation confirms the above-proposed rearrangement (Scheme 5.1.20) of the precursor at 500°C, which is accompanied by a B—C bond cleavage and B—N bond formation.

^{15}N-NMR spectra of ^{15}N-enriched samples are given in the right column of Figure 5.1.8. Whereas the NMR spectrum of the precursor exhibits a line at around −360 ppm arising form NSi_2H units, the 400°C sample exhibits an additional broad signal at −245 ppm. The latter can be attributed to NHB_2 and NHBSi units, which are generated by the above-discussed rearrangement. After heating the sample to 600°C, a very broad signal is seen in the ^{15}N-NMR spectrum, which can be traced back to a superposition of NHB_2, NHBSi, NSi_3, and NSi_2H sites. Finally, after heating to 1050°C, only one signal remains at around −320 ppm, which clearly reflects the formation of an amorphous ceramic network including NB_3 and NSi_3 structural units.

Remarkably, the ceramic yield of $[B(C_2H_4SiH{-}NH)_3]_n$ is above 85%, which is an exceptionally high value. Figure 5.1.9 shows the result of thermogravimetric investigations of the precursor.

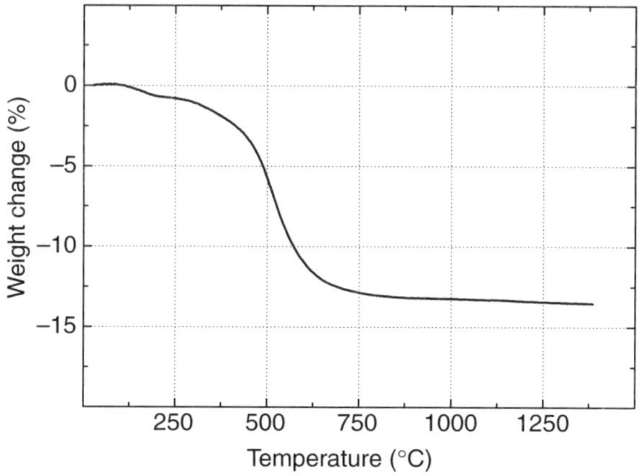

FIGURE 5.1.9 Thermogravimetric analysis of $[B(C_2H_4SiH{-}NH)_3]_n$ [114], which was obtained by hydroboration of PHVS $[SiViH{-}NH]_n$ [42]; heating rate: 2°C/min; atmosphere: flowing argon.

Thermal degradation of the precursor occurs in one step and starts at around 280–300°C (the insignificant mass loss below 200°C is due to the evaporation of residual solvent that was used in the synthesis of the polymer). This is exactly the temperature range in which the onset of the B—C bond cleavage followed by the above-mentioned rearrangement (compare Scheme 5.1.20) was observed by solid-state MAS NMR spectroscopy. The mass loss of 15% is due to the volatilization of both methane, which is eliminated from $Si{-}CH(CH_3){-}Si$ motifs, and hydrogen. The latter evaporates due to elimination from hydrocarbons and dehydrogenative Si—N coupling.

5.1.4 HIGH-TEMPERATURE PROPERTIES

5.1.4.1 General Comments

Choosing a material basis for high-temperature structure materials suffers from a principal dilemma. Oxide materials show creep affinity at high temperatures, limiting their application in energy facilities and aerospace industry [174]. Non-oxide materials, which possess increased creep stability at high temperatures, are not thermodynamically stable in air and react with oxygen and/or moisture [175–177]. Those that form protective oxide layers either contain oxide-type grain boundary phases promoting creep or increased internal oxygen diffusion at higher temperatures. Promising candidates with oxidation resistance and mechanical stability at high temperatures are covalent ceramics, which are prepared without sintering additives. According to the earlier sections, such materials can be obtained by thermolysis of suitable organometallic precursors in an inert gas atmosphere [101, 178–182]. Due to low self-diffusion coefficients in covalent ceramics, conventional sintering of powder compacts without additives is not suitable for the production of composites based on covalent ceramics. Multinary non-oxide ceramics are of special interest since these materials combine properties of the constituting binary phases. In this context, amorphous Si—B—C—N ceramics obtained from organosilicon precursors became of fundamental interest, since these materials recently demonstrated a short-time thermal stability up to 2000°C without mass loss and only minor crystallization as will be discussed at the end of this section. These attributes, coupled with a high resistance toward oxidative attack at elevated temperatures, give rise to extensive studies.

5.1.4.2 Thermodynamic Aspects

A major aspect in understanding high-temperature behavior of precursor-derived amorphous ceramic composites is thermodynamics, that is, the consideration of phase reactions of the constituting thermodynamically stable unary, binary, or ternary phases. Solid-state reactions of such phases may in the worst case result in the decomposition of the ceramic composites.

For example, the stable and metastable solid phases of the quaternary system Si—B—C—N are shown in Figure 5.1.10.

Aside from the pure components silicon, boron, carbon, and nitrogen, a number of well-characterized binary phases such as SiC, Si_3N_4, BN, $B_{4+\delta}C$ as well as intermetallic silicon boron phases are known. Besides that, several metastable binary and ternary phases have been revealed in the subsystems. For example, in the binary C—N system, β-C_3N_4 has been predicted by Liu and Cohen [183]. In the system Si—C—N, β-SiC_2N_4, which decomposes above 900°C, and Si_2CN_4 (see also Sections 5.1.2.3 and 5.1.3.2), which is stable up to around 1000°C, have been reported [81]. Crystalline, ternary B—N—C phases are to our knowledge not known so far. However, because of the almost similar lattice parameters of h-BN and graphite (both are layered materials with sheets of trigonally coordinated atoms) they build solid solutions. Stoichiometric (but not phase pure) compositions of BCN shown in Figure 5.1.10 have usually been obtained by reactions of trichloroborane with N-containing hydrocarbons such as acetonitrile, polyacrylnitrile, CVD processes using boranes, ammonia, and hydrocarbons or by thermolysis of borane–pyridine complexes [184–192].

Moreover, several amorphous compositions have been disclosed in as-thermolyzed PDCs, for example, amorphous carbon, amorphous silicon nitride, and amorphous silicon carbonitride [43, 193–197]. A set of consistent

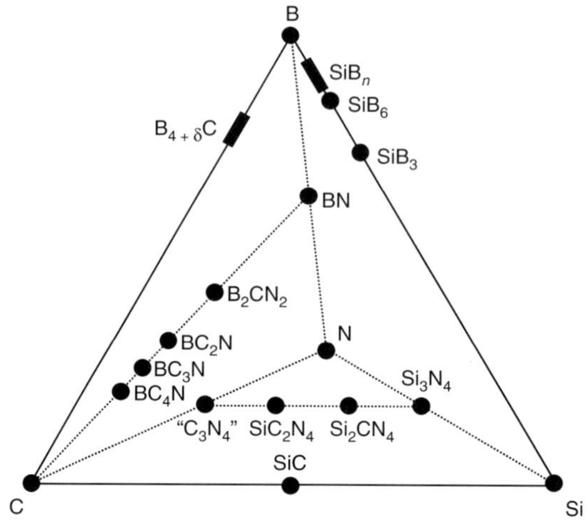

FIGURE 5.1.10 Si—B—C—N concentration tetrahedron including stable and metastable binary and ternary solid phases in the quaternary Si—B—C—N system [199, 200]. Note that compositions BC_xN are stoichiometric but not crystalline; C_3N_4 was calculated in detail [183] but so far has not bar isolated in crystalline form.

thermodynamic data of the stable phases of the quaternary Si—B—C—N system has been developed [198–200], which enables a detailed description of phase equilibria.

With respect to possible applications of precursor-derived non-oxide ceramics, Si—C—N is the most important subsystem in the quaternary system. It was stated above that such ternary compositions can be obtained by thermolysis of polysilazanes or poly(silylcarbodiimide)s at 1000–1400°C. As-obtained Si—C—N materials are usually amorphous. In other words, they are in a non-equilibrium state, where phase configuration cannot be reproduced by equilibrium calculations. However, spectroscopic methods such as solid-state MAS-NMR, and IR spectroscopy in combination with wide-angle scattering reveal a partial segregation of the amorphous ceramics into structural units that refer to thermodynamically stable phases. At higher temperature, segregation proceeds and the amorphous state is transformed into crystalline composites, consisting of Si_3N_4, SiC, and/or C, depending on the molecular structure and composition of the precursor and the conditions applied during thermolysis.

An important tool for predicting thermally induced decomposition reactions are calculations of phase equilibria by the *CalPhaD* (*Cal*culation of *Pha*se *D*iagrams) approach using the set of thermodynamic data mentioned above.

A number of thermodynamic calculations in the SiC—N system have already been published [200–204]. It turned out that decomposition of such ceramics

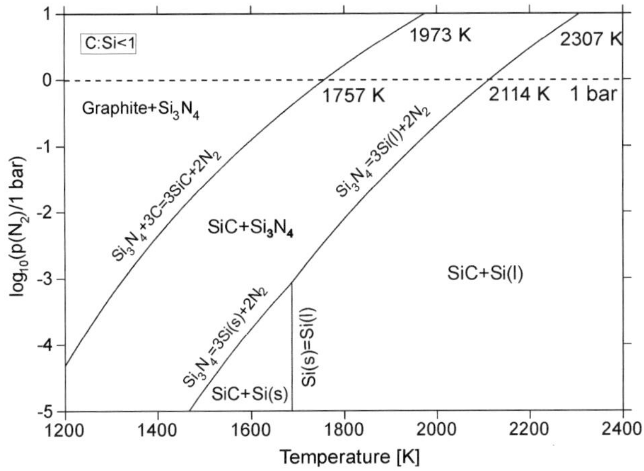

FIGURE 5.1.11 Partial pressure diagram for C/SiC/Si_3N_4 composites. Of major interest is the dependency of the decomposition temperature of Si_3N_4 and C with formation of SiC and N_2. At 1 bar $p(N_2)$ this temperature is 1484°C (1757 K), whereas at 10 bar $p(N_2)$ this value is shifted to 1700°C (1973 K). Assuming 10^{-4} bar $p(N_2)$ in an argon atmosphere, the decomposition of such a composite material should occur at 954°C (1227 K) [204].

is to a significant extent controlled by decomposition reactions of silicon nitride either by a carbothermal reduction according to

$$Si_3N_4 + 3C \xrightarrow{1484^\circ C} 3SiC + 2N_2 \quad (1013\,\text{mbar } N_2) \tag{47}$$

or by decomposition into the elements:

$$Si_3N_4 \xrightarrow{1841^\circ C} 3Si + 2N_2 \quad (1013\,\text{mbar } N_2) \tag{48}$$

It can be concluded from Figure 5.1.11 that the onset of the above decomposition reactions strongly depends on the nitrogen partial pressure. In 1 bar N_2 atmosphere the above reactions take place at 1484 and 1841°C, respectively.

Increasing the nitrogen partial pressure to 10 bar results in a shift of the decomposition temperature due to carbothermal reduction to 1700°C whereas at 1 mbar one would—considering the results of the *CalPhaD* computations—expect decomposition already at around 1050°C.

5.1.4.3 High-Temperature Properties of Si—C—N Ceramics

As an example to illustrate the influence of chemical composition on high-temperature properties, the decomposition behavior of two different Si—C—N ceramics obtained by thermolysis of commercially available poly(vinylsilazane) VT50 [43] and polyperhydridosilazane NCP200 [45], which possess different chemical compositions, is briefly discussed.

In Figure 5.1.12, calculated isothermal sections in the ternary Si—C—N system are shown: (a) represents phase equilibria between 1414 and 1484°C. In this temperature range, three three-phase fields exist: C/Si_3N_4/Gas, C/Si_3N_4/SiC, and Si_3N_4/SiC/liquid. Below 1414°C (melting point of silicon), the latter is composed of Si_3N_4/SiC/Si. The composition of VT50 ceramic (symbolized by a triangle) is located on the tie line Si_3N_4—C. Presuming quantitative segregation into the pure (crystalline) phases, the sample is composed of Si_3N_4 and graphite. In contrast, the composition of NCP200 ceramic is located within the three-phase field C/Si_3N_4/SiC. Accordingly, a fully crystalline sample would be composed of SiC, Si_3N_4, and graphite.

CalPhaD predicts that raising the temperatures to above 1484°C (Figure 5.1.12b) initiates the carbothermal reduction of Si_3N_4 according to Eq. 47. The expected quantitative loss of nitrogen in the case of VT50-derived ceramic should result in the formation of a SiC/C composite. In contrast, thermally induced degradation of ceramics obtained from NCP200 (symbolized by a rhombus) should deliver a composition, which is located on the tie line Si_3N_4/SiC. The calculations are confirmed by X-ray diffraction of annealed ceramic samples. A comparison of X-ray diffraction patterns of VT50- and NCP200-derived ceramics after annealing at 1800°C in a nitrogen atmosphere (5 h, 1 bar) is shown in Figure 5.1.13.

Whereas in the XRD pattern of annealed VT50 ceramic only β-SiC reflections appear, the pattern of annealed NCP200-derived ceramic exhibits both α-Si_3N_4 and β-SiC reflections. Raising the temperature to 2000°C (not shown in the figure) results, as expected, in the dissociation of α-Si_3N_4 into the elements. After cooling to room temperature, XRD patterns of the respective sample now clearly show α-Si reflections besides those of β-SiC.

The contrasting crystallization of VT50 and NCP200 ceramics is also reflected in their different mass-loss behavior, which was investigated by TGA mass-loss accompanied degradation is high-temperature TGA.

Figure 5.1.14 shows a comparison of the TGA curves of VT50 (a) and NCP200 ceramics (b) [205]. It is evident that VT50 decomposes in a one-step reaction with the loss of 30% of its original sample weight. The determined value exactly fits with the nitrogen content of the ceramic determined by elemental analysis and confirms the results of the *CalPhaD* calculations shown in Figure 5.1.12. Moreover, the experimentally observed decomposition at around 1500°C, perfectly fits the thermodynamically calculated value (1484°C). The corresponding DTA curve displays a strong endothermic peak, centered at around 1665°C.

In contrast, NCP200-derived ceramics decompose in a two-step reaction. Compared to the result of the calculations, the onset of the first step, which

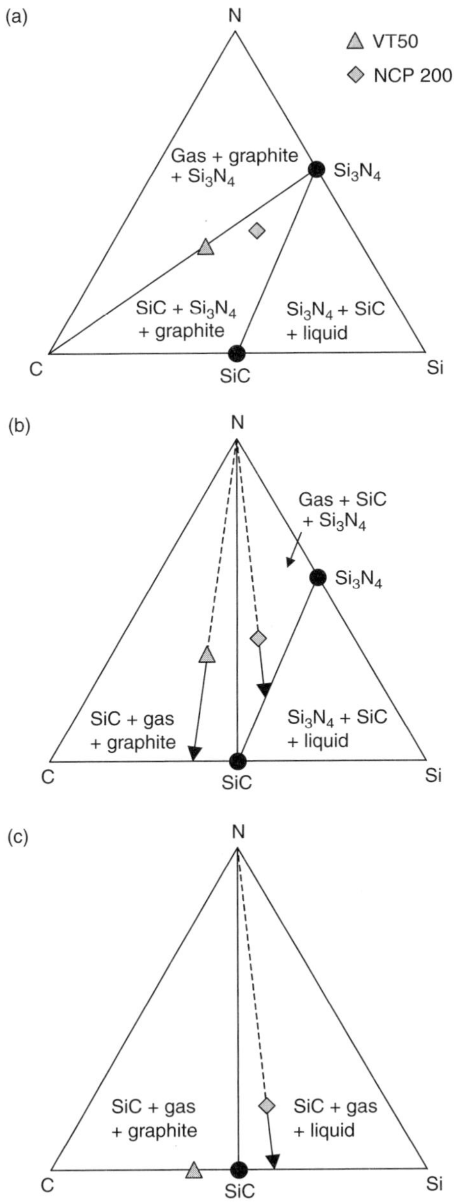

FIGURE 5.1.12 Ternary Si/C/N phase diagrams including compositions of ceramics derived from VT50 (triangles) and NCP200 (rhombs) at 1 atm N_2 an various temperatures: (a) $1414 < T < 1484°C$; (b) $1484 < T < 1841°C$; (c) $T > 1841°C$. Decompositions of as-obtained VT50 ceramic above 1484°C occurs in one step delivering SiC/C composites, whereas decomposition of NCP200-derived ceramic proceeds in two steps [43].

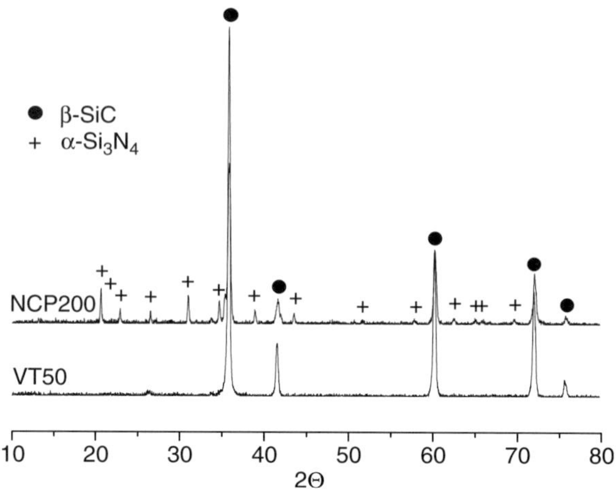

FIGURE 5.1.13 XRD patterns of VT50- and NCP200-derived ceramics after annealing to 1800°C in 0.1 MPa N_2 for 5 h [205].

is due to the carbothermal reduction of Si_3N_4, is at around 1550°C and thus shifted by 60°C to higher temperature. The second step is caused by Si_3N_4 dissociation into the elements and starts at around 1800°C. The mass loss of the individual decomposition steps is 12 and 15 mass%, respectively. The sum (27 mass%) exactly matches the nitrogen content of the ceramic material. In addition, the values of the mass loss of the single steps are purported by the amount of “free carbon”.

It is interesting to note that indeed all as-thermolyzed ceramics with an overall composition within the three-phase field C/Si_3N_4/SiC decompose like VT50- or NCP200-derived ceramics in one- or two-step reactions, only depending on whether their C:Si ratio is >1 or <1, respectively.

In this context, phase fraction diagrams are very suitable to quantitatively predict phase reactions. As an example, Figure 5.1.15 shows such a diagram of NCP200-derived ceramic. The total pressure was considered to be 1 bar.

As-pyrolyzed NCP200-derived ceramic is composed of 8 mass% of graphite, 19 mass% of SiC, and 73 mass% of Si_3N_4. Due to the carbothermal reduction of Si_3N_4 at 1484°C, graphite is quantitatively consumed and bonded as SiC. Correspondingly, it is expected that the amount of Si_3N_4 is reduced to 38 mass% with the formation of 12 mass% of gas (N_2). The second decomposition step at 1841°C consumes residual Si_3N_4. As a consequence, an additional 15 mass% of gas forms along with 22% of a liquid (Si). As expected, the amount of SiC is not affected by this reaction.

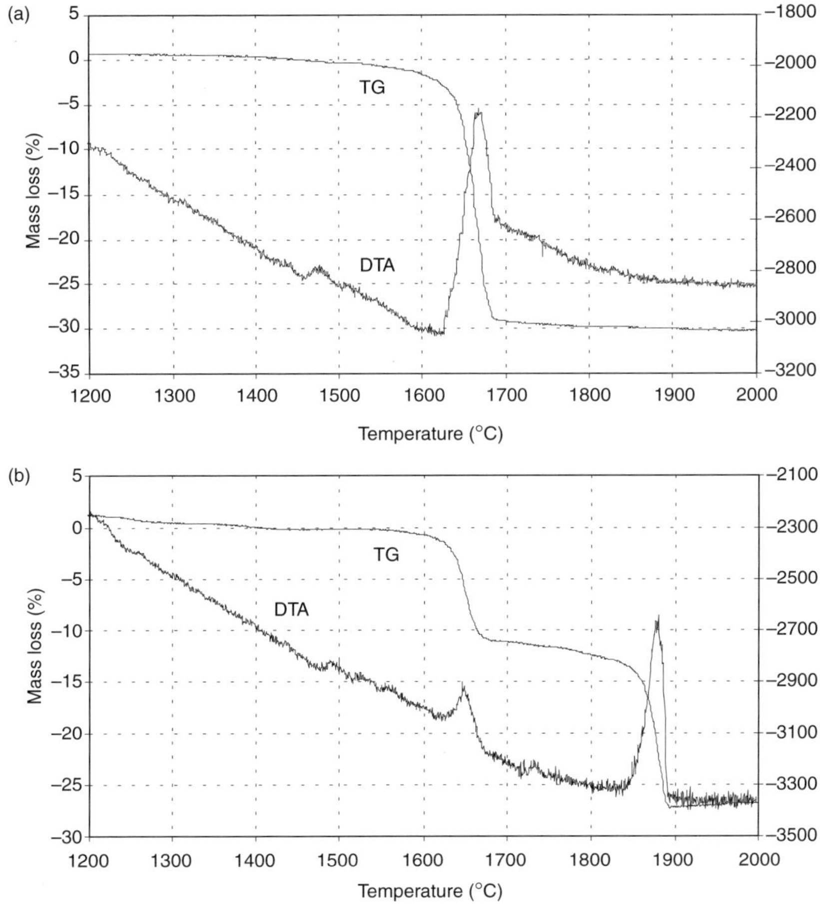

FIGURE 5.1.14 High-temperature TGA/DTA investigations of VT50- and NCP200-derived ceramics in 0.1 MPa N_2. Heating rate: 5°C/min [205].

5.1.4.4 High-Temperature Properties of Si—B—C—N Ceramics

In the mid-1990s, it was observed that organometallic polymers containing the elements silicon, boron, carbon, nitrogen, and hydrogen are precursors for silicon nitride, silicon carbide, and boron nitride-based Si—B—C—N ceramics, which possess extraordinary high-temperature stability, coupled with a high resistance towards oxidation [101, 106, 114].

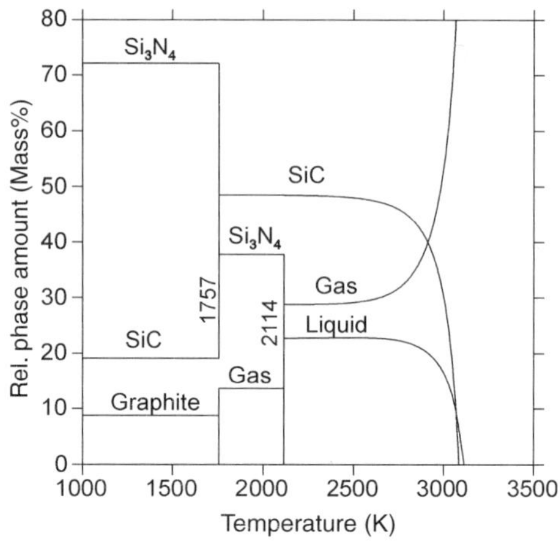

FIGURE 5.1.15 Calculated Si—C—N phase fraction diagram (p_{tot} = 1 bar) for NCP200-derived ceramics [204, 205]. Carbothermal reduction of Si_3N_4 at 1484°C (1757 K) results in the formation of SiC and N_2 (gas). Decomposition of remaining Si_3N_4 into the elements takes place at 1841°C.

Results of detailed investigations of the high-temperature properties of such materials by high-temperature TGA, XRD, and TEM of thermally annealed samples will be shown below. For a first comparison of the thermal properties of Si—C—N and Si—B—C—N materials, a HT-TGA of a material derived from boron-free poly(vinylsilazane) (PVSZ) and from a boron-modified polysilazane in an argon atmosphere is shown in Figure 5.1.16.

It is clearly seen that PVSZ-derived ceramics start to decompose below 1500°C. As has been mentioned, this breakdown is primarily a consequence of the decomposition reaction of Si—N units, which are present in the amorphous material together with "free" carbon causing the formation of silicon carbide and the elimination of nitrogen. Si—B—C—N ceramics obtained from $[B(C_2H_4SiH{-}NH)_3]_n$ [114], in contrast, resist thermal degradation up to 1950°C. This example nicely shows that by modification with around 5 mass% of boron the onset of decomposition in Si—C—N ceramics can be shifted from around 1500°C, by roughly 450°C, to 1950°C.

From the thermodynamic point of view, the unusual high-temperature stability of the Si—B—C—N ceramic cannot be understood. According to the calculated phase fraction diagram shown in Figure 5.1.17 (composition Si: 24.0, B: 8.0, C: 44.0, N: 24.0 atom%), the Si—B—C—N ceramic should at temperatures below 1484°C be composed of BN (16 mol%), Si_3N_4 (28 mol%), SiC (24 mol%), and 32 atom% of graphite [114].

Analogous to the above-mentioned Si—C—N ceramic, it is expected that in the case of the quaternary material, Si_3N_4 also reacts above 1484°C with

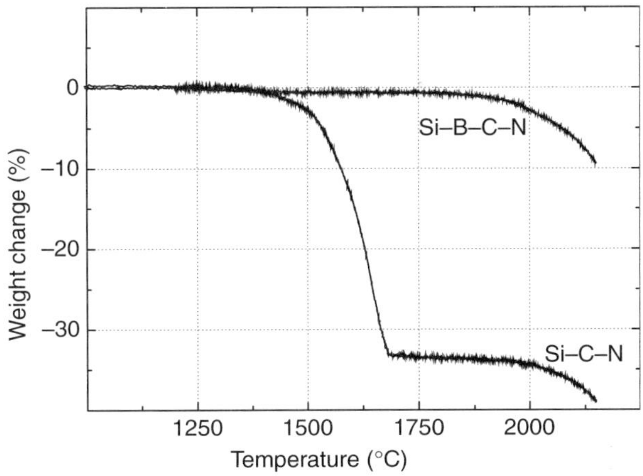

FIGURE 5.1.16 Comparison of the thermal behavior of ternary Si—C—N and quaternary Si—B—C—N ceramics [114] by high-temperature TGA. Heating rate: 5°C, 1 bar argon.

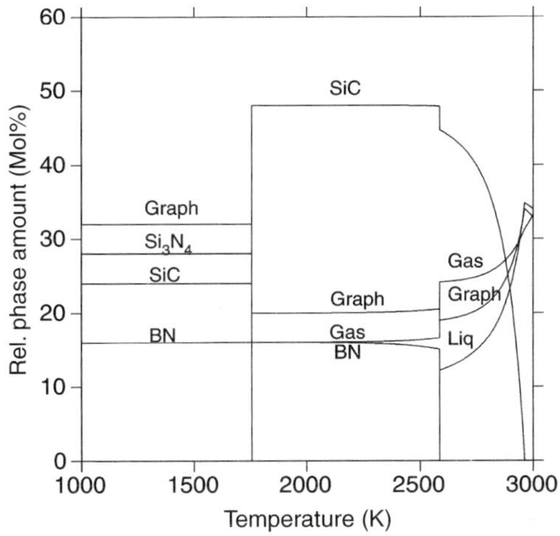

FIGURE 5.1.17 Calculated Si—B—C—N phase fraction diagram (p_{tot} = 1 bar) for ceramics derived from $[B(C_2H_4SiH{-}NH)_3]_n$. Chemical composition: Si: 24.0, B: 8.0, C: 44.0, N: 24.0 atom%. It is expected that due to the carbothermal reduction of Si_3N_4 decomposition of the ceramic composite occurs at 1484°C (1757 K), resulting in the formation of SiC and N_2 (gas) [114].

graphite according to Eq. 47. Consequently, the original sample weight should decrease due to the loss of gaseous nitrogen. The amount of free carbon in this case decreases from 32.0 to 20.0 atom% whereas the amount of silicon carbide increases from 24.0 to 48.0 mol%. However, from the high-temperature TGA

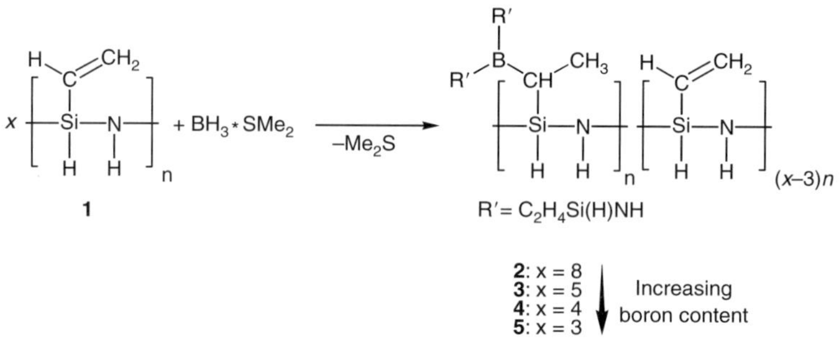

SCHEME 5.1.21 Synthesis of Si—B—C—N polymers with adjustable boron concentration [119].

investigations (compare Figure 5.1.16) it can be assumed that in contrast to the ternary ceramics the thermodynamically expected decomposition reaction does not take place in the case of the Si—B—C—N ceramic. In addition, boron, which is present as boron nitride, does not directly participate in any decomposition reactions below 2000°C.

Systematic investigations on the connection of chemical composition and thermal stability of precursor-derived Si—B—C—N ceramics reveal that both the boron and nitrogen content significantly influence thermal properties. By stepwise increasing the amount of boron in Si—B—C—N precursor-derived materials, Müller *et al.* [119, 120] showed that at least 4 wt% of boron are needed to efficiently avoid thermally induced decomposition. For example, a series of Si—B—C—N ceramics obtained from poly(vinylsilazane) (PVS), which was hydroborated with different amounts of borane dimethyl sulfide, is discussed in more detail (see Scheme 5.1.21).

Table 5.1.1 gives the chemical composition of the ceramics derived from the above precursors in wt%; values in parenthesis are atom%. Phase compositions, which were obtained using the *CalPhaD* calculations on the basis of the results of the chemical compositions obtained by elemental analysis, are given in Table 5.1.2. Oxygen values are below 1.5% and therefore not considered in the thermodynamic calculations.

From the chemical composition of the ceramics, which was determined by elemental analysis, it is obvious that the increasing boron concentration in the precursors is directly reflected in an increasing amount of boron in the ceramic materials. It is highest in **5c**, which was derived from quantitatively hydroborated poly(vinylsilazane) **5** (compare Scheme 5.1.21) and decreases in the row **4c**>**3c**>**2c**. Accordingly, the concentration of Si, C, and N increase in that order.

The results of the thermodynamic calculations in Table 5.1.2 point out that with increasing boron content the phase fraction of boron nitride increases from 7.3 mol% in **2c** to 18.2% in boron-rich **5c**. Consequently, the amount of

TABLE 5.1.1 Composition of Ceramic Materials 1c–5c (wt%, atom% in brackets) Obtained from Precursors Shown in Scheme 5.1.21 [120]

Ceramic	1c	2c	3c	4c	5c
Si	43.0 (26.8)	42.9 (26.0)	43.0 (26.0)	42.7 (25.3)	41.6 (24.7)
C	29.2 (42.5)	29.6 (41.9)	29.0 (41.0)	29.0 (40.2)	27.6 (38.4)
N	24.3 (30.3)	22.3 (27.1)	22.1 (26.8)	22.1 (26.3)	22.3 (26.6)
O	0.4 (0.4)	1.3 (1.4)	1.1 (1.2)	1.3 (1.4)	1.3 (1.4)
B	—	2.3 (3.6)	3.2 (5.0)	4.5 (6.9)	5.8 (9.0)
Empirical formula*	$SiC_{1.6}N_{1.1}B_{0.00}$	$SiC_{1.6}N_{1.0}B_{0.14}$	$SiC_{1.6}N_{1.0}B_{0.19}$	$SiC_{1.6}N_{1.0}B_{0.27}$	$SiC_{1.6}N_{1.1}B_{0.36}$

*O-content is neglected.

TABLE 5.1.2 Calculated Relative Phase Fractions of Crystalline 1c–5c Materials (atom%, O-content is neglected) and Predicted (Δm_p) and Observed Mass Loss (Δm_o at 1870°C) due to Si_3N_4 Decomposition [120]

Ceramic	1c	2c	3c	4c	5c
BN	0	7.3	10.2	14.0	18.2
Si_3N_4	53.3	41.6	38.5	34.2	31.3
SiC	8.1	17.0	19.6	22.2	23.3
C	38.6	34.0	31.7	29.5	27.2
Δm_p	25%	20%	18%	17%	15%
Δm_o	24%	19%	13%	8%	0.8%

nitrogen available for the formation of silicon nitride is reduced. This has two consequences: the silicon nitride phase fraction decreases from 41.6 mol% in **2c** to 31.3 mol% in **5c**, whereas the silicon carbide phase fraction increases from 17.0 to 23.3%. The relative increase of silicon carbide again results in a decrease of the amount of graphite. In conclusion, increasing boron concentration results in increasing boron nitride and silicon carbide but decreasing silicon nitride and graphite phase fraction and vice versa.

According to Scheme 5.1.21 and Tables 5.1.1 and 5.1.2, the total mass loss due to the carbothermal reduction of silicon nitride increases with increasing phase amount of silicon nitride in the row **1c** > **2c** > **3c** > **4c** > **5c**. This conclusion is confirmed by high- temperature TGA as shown in Figure 5.1.18.

As expected, boron-free **1c** starts to decompose at around 1550°C and loses 24% (=Δm_o) of its original sample weight. This is in accordance with the thermodynamically calculated (predicted) value of 25% (Δm_p) given in Table 5.1.1. The incorporation of even small amounts of boron such as in **2c** results in a shift of the onset of decomposition by 80 to 1630°C. However, the total mass

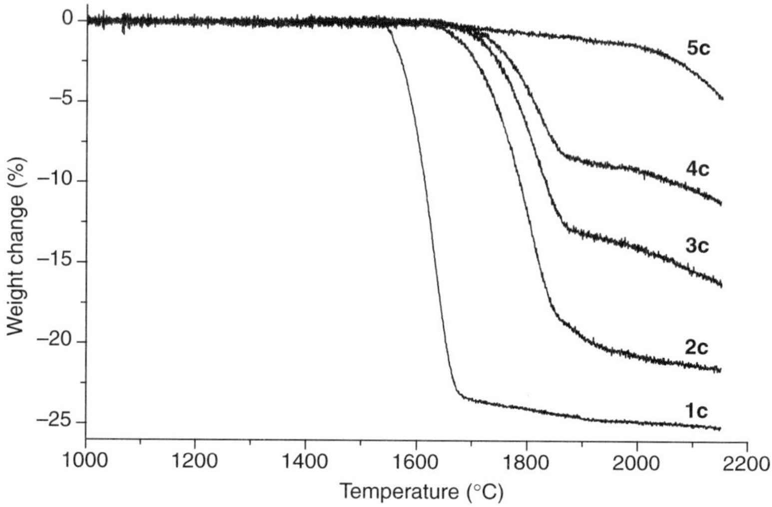

FIGURE 5.1.18 Comparison of the thermal behavior of quaternary Si—B—C—N ceramics with different boron concentrations by high-temperature TGA. Heating rate: 5°C/min, 1 bar argon. For details concerning the molecular structure of the polymeric precursors and the elemental and phase composition of the derived ceramics consider Scheme 5.1.21 and Tables 5.1.1 and 5.1.2, respectively [120].

loss of 19% corresponds to the expected value (Δm_p=20%). Further increased boron contents in **3c** and **4c** result in an additional shift of the decomposition temperature to around 1700°C. Moreover, a significant deviation of found (**3c**: $\Delta m_o = 13\%$, **4c**: $\Delta m_o = 8\%$) and calculated (**3c**: $\Delta m_p = 18\%$, **4c**: $\Delta m_p = 17\%$) mass losses are observed. **5c**, which possesses the highest boron content, shows no signs of decomposition in the expected temperature range. Thermal degradation is significantly retarded and starts around 2000°C. At this temperature carbothermal reduction of boron nitride is expected:

$$4BN + C \longrightarrow B_4C + 2N_2 \qquad (49)$$

Analogous to the carbothermal reduction of silicon nitride (compare Eq. 47), the onset of BN decomposition is also depending on the nitrogen partial pressure (see Figure 5.1.19).

The different behavior in the high-temperature TGA shown in Figure 5.1.18 is mirrored in the crystallization/phase composition of the materials after an additional heat treatment at elevated temperature. For example, after heating to 1800°C (1 bar argon) for 5 h, the XRD pattern of **2c** and **3c** correspond to the diffraction patterns of silicon carbide. This is in agreement with the high-temperature TGA investigation that indicated quantitative (**2c**) or almost

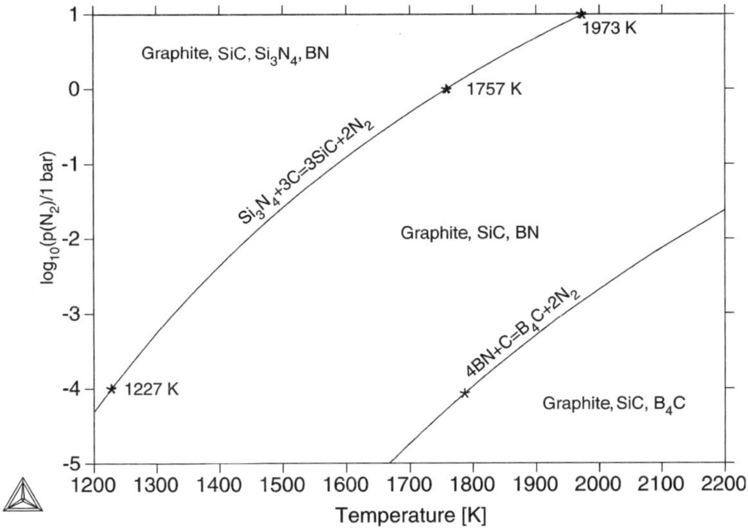

FIGURE 5.1.19 Partial pressure diagram for C/SiC/Si_3N_4/BN composites. In contrast to the ternary C/SiC/Si_3N_4 system (compare Figure 5.1.11), additional carbothermal reduction of boron nitride must be considered in the high-temperature stability of the ceramic composites [114, 205].

quantitative thermal degradation (**3c**) of silicon nitride at this temperature (see Figure 5.1.20).

In the case of the boron-richer materials **4c** and **5c**, β-Si_3N_4 reflections are distinctly detected besides SiC peaks. The line widths in both XRD patterns suggest a nanocrystalline arrangement. At 2000°C, finally, the thermodynamically expected cleavage of Si—N bond proceeds in the case of **4c**, resulting in the absence of Si_3N_4 reflections in the XRD diagram. In contrast, the XRD pattern of **5c** which by high-temperature TGA proved to be the only thermally stable composite at this temperature, still displays β-Si_3N_4 reflections. In addition, all samples heat-treated to 2000°C exhibit an additional broad reflection at 26.6°, which is characteristic for graphitic carbon.

Investigations on the influence of the nitrogen content on the thermal stability on Si—B—C—N ceramics were published recently [131]. Therefore, boron-modified poly(silylcarbodiimide)s with adjustable nitrogen content were synthesized by a dehydrocoupling reaction of tris(hydridosilylethyl)boranes, $B(C_2H_4SiRH_2)_3$ ($R = H, CH_3$), and cyan amide (compare Scheme 5.1.15) [132]. Investigations of the polymer-to-ceramic conversion of the precursors in an Ar atmosphere revealed that they deliver Si—B—C—N ceramics in 65–85% yield, depending on the elemental composition and the molecular structure, that is, the cross-linking of the polymers. For example, TGA of a series of polymers

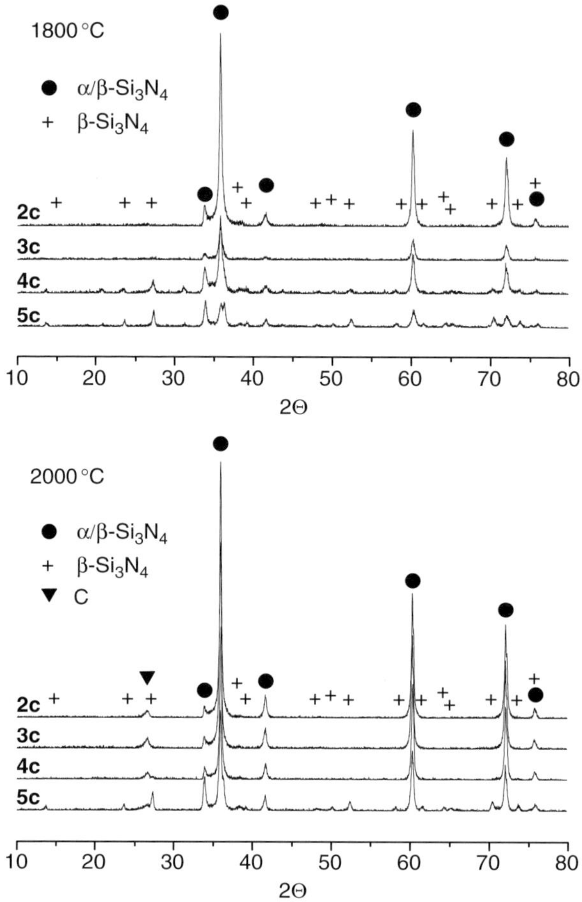

FIGURE 5.1.20 XRD patterns of the ceramics obtained from **1c–5c** (compare Scheme 5.1.21) after annealing to 1800°C (top) and 2000°C (bottom) in 1 bar argon for each 5 h [119]. It is evident that after heating to 1800°C in the case of less thermally stable materials **2c** and **3c** (compare Figure 5.1.18) only reflections of silicon carbide are observed, whereas in the case of ceramics derived from **4c** and **5c** additional silicon nitride reflections are found. In the case of **4c** the latter vanish after annealing the material at 2000°C [120].

obtained by the reaction of $B(C_2H_4SiH_3)_3$ [121] with different amounts of $H_2N{-}CN$ is shown (see Figure 5.1.21). The calculated chemical composition of the precursors (idealized sum formula) is given in Table 5.1.3.

Thermolysis of H—N1 to H—N6 releases ceramics in 75–84% yield. The polymers with the lowest degree of cross-linking, H—N2 and H—N3, have the highest ceramic yields of approx. 84%. The more highly cross-linked precursors,

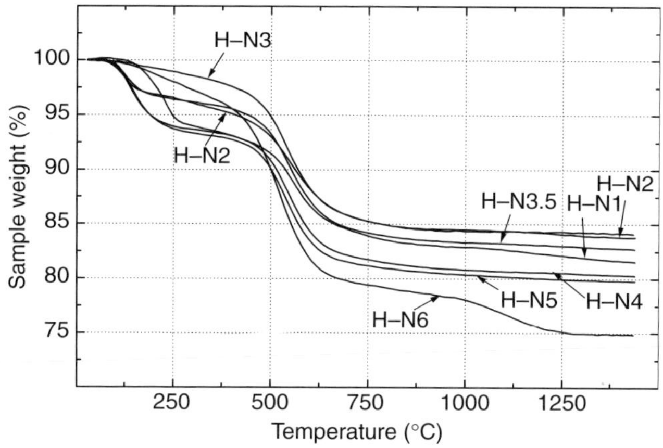

FIGURE 5.1.21 Thermogravimetric analysis of H—N1 to H—N6 (compare Scheme 5.1.15). Heating rate 2°C/min, argon atmosphere [132].

TABLE 5.1.3 Idealized sum formulas of boron-modified silylcarbodiimide H—N1 and poly(silylcarbodiimide)s H—N2 to H—N6 obtained from $B(C_2H_4{-}SiH_3)_3$ and $H_2N{-}C \equiv N$ [132]

	Silane:cyan amide	Sum formula (monomer unit)
H—N1	1:0.5	$C_{6.5}H_{20}NSi_3B$
H—N2	1:1	$C_7H_{19}N_2Si_3B$
H—N3	1:1.5	$C_{7.5}H_{18}N_3Si_3B$
H—N4	1:2	$C_8H_{17}N_4Si_3B$
H—N5	1:2.5	$C_{8.5}H_{16}N_5Si_3B$
H—N6	1:3	$C_9H_{15}N_6Si_3B$

H—N4, and H—N5, deliver ceramics in 81 and 80% yield, respectively, whereas the most highly cross-linked precursor H—N6 delivers ceramics in 75% yield. Mass spectra coupled TGA revealed that silicon, boron, or nitrogen-containing species did not evaporate during thermolysis. It can thus be concluded that the Si:B:N ratio in the precursors was maintained in the ceramics.

High-temperature properties of ceramics derived from H—N2 to H—N6 can clearly be correlated with their nitrogen content (see Figure 5.1.22 and Table 5.1.4).

Ceramics obtained from H—N5 and H—N6, which possess the highest nitrogen content among the materials investigated, decompose earliest at around 1670°C and finally lose 25 and 30% of their original sample weight. The decomposition temperature of H—N4 is shifted by 50 to 1720°C, the final mass loss of less than 10% is significantly lower than that of H—N5 and H—N6. H—N2,

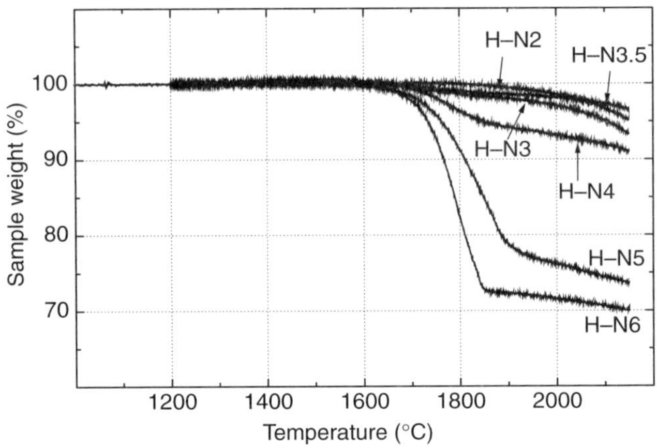

FIGURE 5.1.22 Comparison of the thermal behavior of quaternary Si—B—C—N ceramics derived from H—N2 to H—N5 with different nitrogen concentrations by high-temperature TGA. Heating rate: $T < 1200$ 10°C/min, $T > 1200$ 2°C/min, 1 bar argon atmosphere. For details concerning the molecular structure of the polymeric precursors and the elemental and phase composition of the derived ceramics see Scheme 5.1.15 and Table 5.1.4, respectively [131].

TABLE 5.1.4 Composition of ceramic materials H—N2 to H—N6 [(a) mass% and (b) calculated phase fraction in mol%]

(a) (mass%)	Si	B	C	N
H—N2	37.9	6.0	37.2	18.9
H—N3	37.8	5.6	36.1	20.5
H—N3.5	37.2	4.8	35.7	22.3
H—N4	35.5	4.8	35.1	24.6
H—N5	33.1	4.5	35.3	27.1
H—N6	31.5	4.9	32.2	31.4
(b) (mol%)	BN	Si_3N_4	SiC	C
H—N2	17.5	21.9	23.7	36.9
H—N3	16.7	23.4	23.2	36.7
H—N3.5	14.0	31.8	14.6	39.6
H—N4	13.9	36.0	8.7	41.4
H—N5	12.9	41.1	1.2	44.8
H—N6	14.0	48.2	−6.8	44.6

H—N3, and H—N3.5 behave like **5c** ceramics (compare Figure 5.1.18). None of these materials decompose below 1950°C.

The difference in the thermal mass loss behavior is again reflected in the different crystallization behavior of the ceramic materials. For example, the

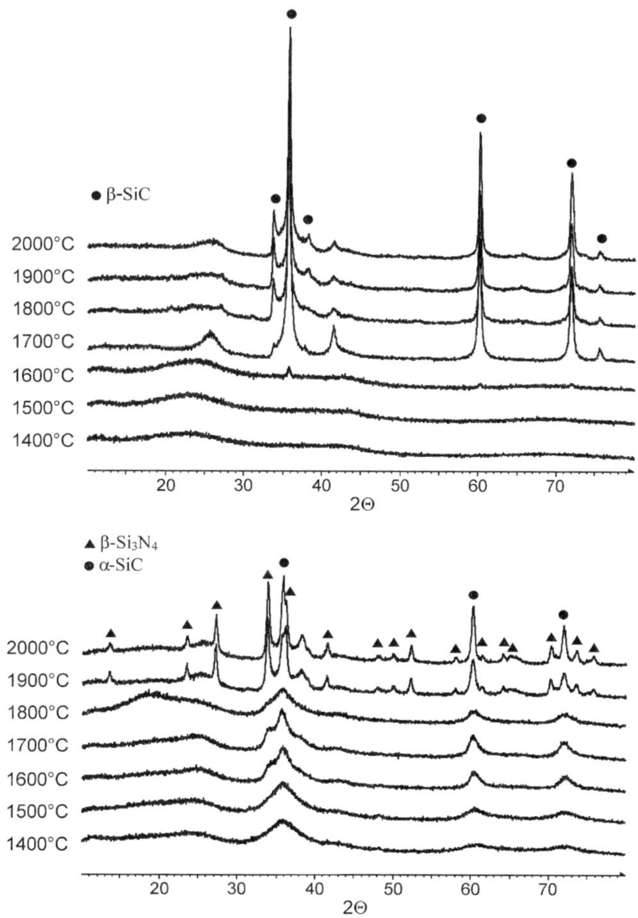

FIGURE 5.1.23 XRD patterns of the ceramics obtained from H—N2 (top) and H—N5 (bottom) annealed at 1400–2000°C (100°C steps, 1 bar N_2) each for 3 h [131].

phase evolution observed by XRD of H—N2 and H—N5, which were annealed under similar conditions, is shown in Figure 5.1.23.

As-obtained ceramics (1400°C) are predominantly amorphous. However, in the patterns of H—N2, ceramic, broadened reflections at $2\Theta=36°$ and 61° point to the formation of nano-crystalline silicon carbide. The appearance of this XRD pattern remains unchanged, even after heating the sample to 1800°C, indicating that there is no further crystallization up to this temperature. After annealing to 1900°C, reflections of both α-SiC und β-Si_3N_4 appear. Remarkably, the latter do not decrease in intensity even after heating the samples to

2000°C. The crystallization behavior of H—N5 ceramic is significantly different and reflects the decomposition behavior observed by the TGA shown in Figure 5.1.22. No crystallization is observed below 1600°C. Above this temperature, only β-SiC forms whereas there is no evidence for the crystallization of silicon nitride. This finding again indicates decomposition of Si—N units due to a carbothermal reduction either before or during crystallization.

The above examples clearly state that under certain conditions Si—B—C—N ceramics are thermally much more stable than thermodynamics predict. In order to understand in detail the thermal stability of such ceramics and to find an universal relationship between the thermal stability and the compositions of the materials, a number of different types of Si—B—C—N ceramics were investigated in detail by means of TGA and XRD. Additionally, spectroscopic and microscopic methods were incorporated to obtain information on the atomic structure of the amorphous and nano-crystalline assemblies.

Thermodynamic calculations reveal that almost all compositions of Si—B—C—N ceramics investigated so far are located in the four-phase equilibrium field Si_3N_4 + SiC + C + BN (compare Figure 5.1.10). Thermally stable compositions in this equilibrium field are located close to the three-phase equilibrium SiC + C + BN whereas thermally less stable compositions are located close to the three-phase equilibrium Si_3N_4 + C + BN (Si_3N_4-enriched area). This fact is illustrated in Figure 5.1.24 schematically, showing a calculated isothermal section in the Si—B—C—N system at a constant boron content of 10 at%.

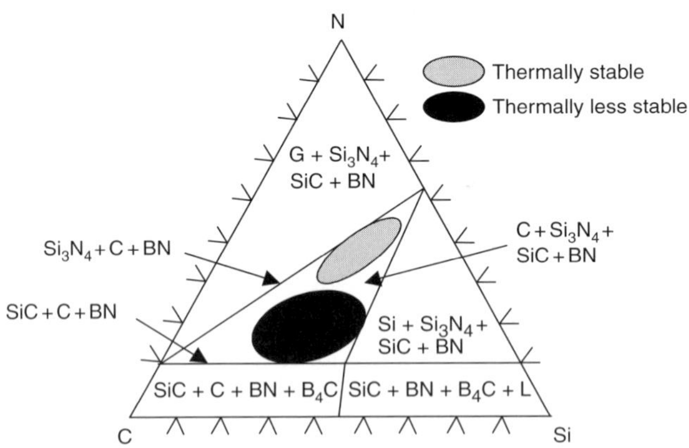

FIGURE 5.1.24 Calculated phase equilibria in the quaternary Si—B—C—N system at 10 at% boron and a total pressure of 1 bar. Thermally stable Si—B—C—N ceramics with compositions in the four-phase system C/Si_3N_4/SiC/BN are usually poor in nitrogen, whereas thermally less stable materials are often located close to the three-phase field system C/Si_3N_4/BN (on the tie-line C—Si_3N_4).

Structural analysis of as-obtained amorphous ceramics by means of small and wide angle scattering as well as by FT-IR and solid-state NMR spectroscopy reveal a homogeneously ordered atomic array originating from the polymer structure. Remarkably, the structural units of the thermodynamically stable phases are already preformed within the amorphous states on an atomic and a medium-range scale. Crystallization of these amorphous ceramics initially yields metastable amorphous phases $Si_{3+y/4}C_yN_{4-y}$ ($Y = 0$–4) and BN/C. The former is related to the composition line Si_3N_4—SiC (see Figure 5.1.10). As already shown in the X-ray studies of annealed H—N2 and H—N5 in Figure 5.1.23, crystallization of such composites delivers both silicon nitride and silicon carbide.

Detailed studies on the phase evolution of amorphous Si—B—C—N ceramics by solid-state NMR using MAS were published by Schuhmacher *et al.* [172, 173]. They compared the crystallization behavior of high-temperature stable Si—B—C—N ceramic, which was obtained from $[B(C_2H_4SiH{-}NH)_3]_n$ (MW33, [114]), with that of a less stable material, derived from nitrogen-rich boron-modified poly(silsesquiazane) $[B(C_2H_4Si(NH)_{1.5})_3]_n$ (MW60, [114]).

The representative ^{29}Si MAS NMR spectra given in Figure 5.1.25 cover the temperature range from 1050 to 2000°C. After thermolysis to 1050°C,

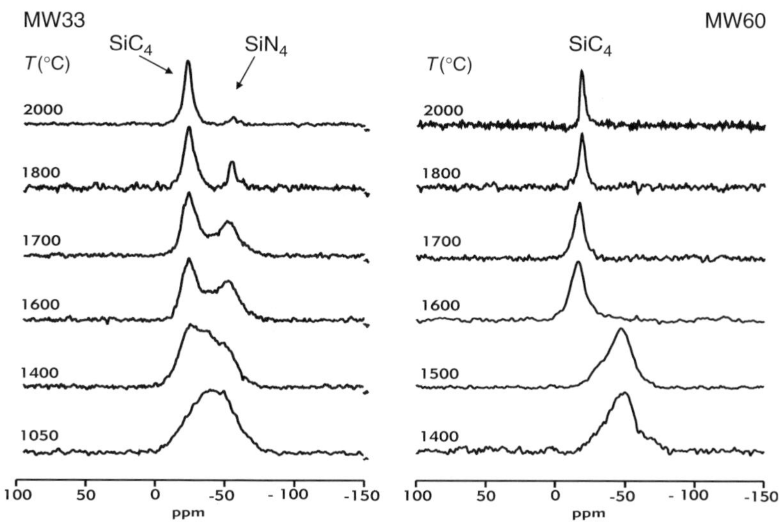

FIGURE 5.1.25 Experimental ^{29}Si-NMR spectra of ceramics derived from boron-modified polysilazanes $[B(C_2H_4SiH{-}NH)_3]_n$ MW33 (left) [114] and boron-modified poly(silsesquiazane) $[B(C_2H_4Si(NH)_{1.5})_3]_n$ MW60 (right) [114] after annealing up to finally 2000°C (atmosphere: 1 bar nitrogen). All spectra were obtained under MAS conditions and single-pulse excitation [172, 173].

^{29}Si-NMR spectra of both ceramics appear almost similarly (the NMR spectrum of the 1050°C MW60 sample is not shown). They are each characterized by a broad resonance centered at around −40 ppm. Apart from a line narrowing in the spectrum of MW60-derived ceramic, further heat treatment to 1400°C causes only minor changes in the signal appearance. The chemical shift range in combination with the broadened resonance signals at and below this temperature point to the presence of a mixture of SiC_xN_{4-x} sites ($x = 0$–2), suggesting a fully amorphous structure. After heating to 1600°C significant progress in the phase evolution of both materials occurs. In the case of MW33-derived material two new resonance signals centered at −19 and −49 ppm evolve, which indicate the formation of SiC_4 and SiN_4 sites, respectively. However, a broad signal, which appears in between, still points to a remarkable amount of SiC_xN_{4-x} sites. In contrast, annealing of MW60 ceramic to 1600°C causes the disappearance of high-field-shifted signals. The only resonance signal in the ^{29}Si-NMR spectrum of MW60 ceramic is found at −19 ppm and suggests quantitative degradation of Si—N motifs due to a carbothermal reduction with formation of silicon carbide.

Heating MW33 ceramics further results in additional de-mixing. As a consequence, only two, well-separated signals remain after heating to 1800°C, which exhibit considerably smaller line widths than those obtained from ceramics annealed at lower temperatures. It was claimed that this is a consequence of the presence of SiC and Si_3N_4 crystallites. Obviously, all structural components with silicon in mixed coordination, such as SiC_xN_{4-x} sites are disintegrated at this temperature, as also found by X-ray studies on the same samples.

Even though obtained from structurally different precursors, the crystallization behavior of ceramics obtained from MW33 and MW60, shown in Figure 5.1.26, is comparable with that obtained from boron-modified polysilylcarbodiimides depicted in Figure 5.1.23. Ceramics with "high" nitrogen contents decompose around 1550–1600°C with crystallization of silicon carbide, whereas in the case of ceramics with "low" nitrogen content neither thermal degradation nor crystallization is observed below 1750°C. At and above this temperature such materials transform into nanocrystalline silicon nitride/silicon carbide composites.

Investigations of the structural evolution of Si—B—C—N ceramics by high-resolution transmission electron microscopy (HR TEM) were also the subject of recent publications. For example, Jalowiecki *et al.* [196] investigated by HR TEM the microstructure of a boron-doped silicon carbonitride composite, which was derived from a boron-doped polyhydridomethylsilazane. As-obtained amorphous material was subsequently annealed at 1800°C for 50 h in an argon atmosphere. HR TEM investigations revealed that the annealed ceramic possessed a quite unusual microstructure. It was segregated into a turbostratic BNC_x phase that occurred along grain boundaries of nano-sized silicon carbide and silicon nitride crystals as shown in Figure 5.1.27.

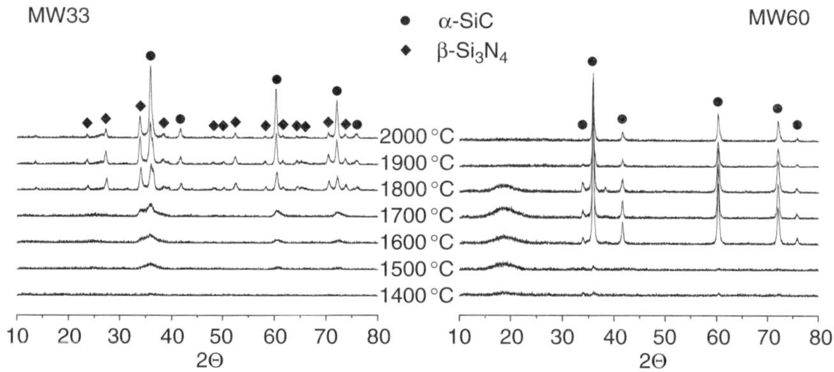

FIGURE 5.1.26 XRD patterns of the ceramics obtained from boron-modified polysilazanes $[B(C_2H_4SiH{-}NH)_3]_n$ (MW33, left) and $[B(C_2H_4Si(NH)_{1.5})_3]_n$ (MW60, right) after pyrolysis at 1400 and annealing at 1500–2000°C (100°C steps, 1 bar nitrogen) each for 3 h. In the case of the MW33-derived ceramics, α/β-SiC (•) and β-Si_3N_4 (◆) reflections are observed; in the case of MW60 ceramic, only α/β-SiC (•) reflections are found [114].

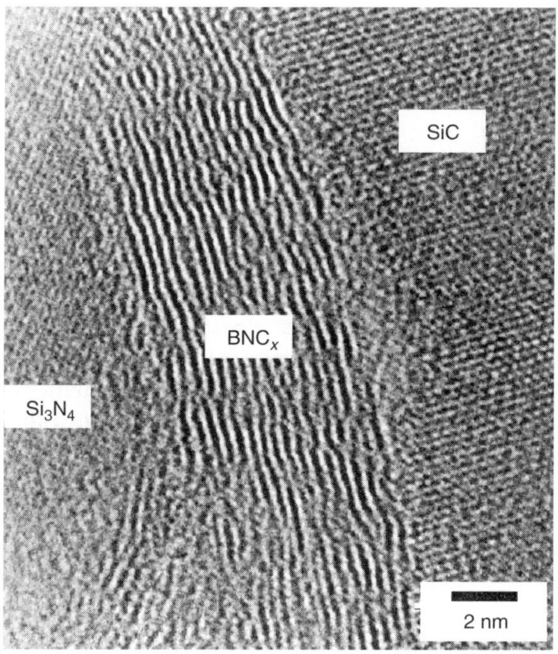

FIGURE 5.1.27 High-resolution TEM image of Si—B—C—N ceramics obtained from a boron-doped poly(methylsilazane) after annealing to 1800°C for 50 h in a nitrogen atmosphere. It is composed of SiC and Si_3N_4 nano-crystals that are separated by a turbostratic BNC_x phase [196].

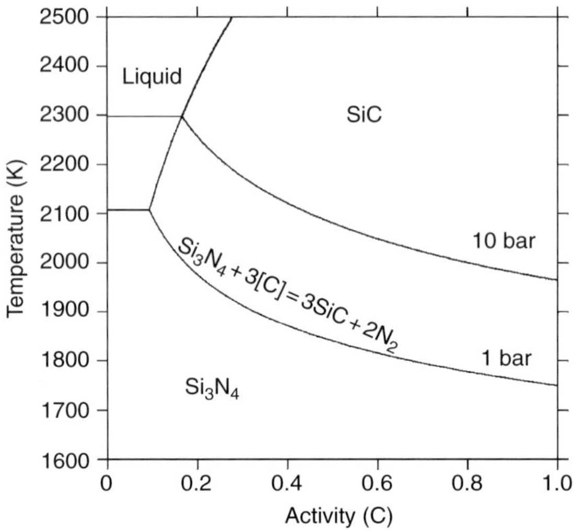

FIGURE 5.1.28 Calculated phase diagram displaying the relationship of decomposition temperature of silicon nitride due to carbothermal reduction versus nitrogen partial pressure and carbon activity. At 1 bar N_2 and $a_C = 1$ decomposition occurs at 1484°C. Assuming $a_C = 0.2$ results in a shift of this value to ~1700°C. Shifting the nitrogen partial pressure at $a_C = 0.2$–10 bar results in a further increase of the silicon nitride decomposition temperature to ~1980°C [205].

It was supposed that the BNC_x phase inhibits diffusion processes and consequently retards crystal growth. Moreover, it binds free carbon, thus decreasing the carbon activity. The connection of carbon activity and decomposition temperature of silicon nitride due to the reaction with free carbon, which was calculated thermodynamically [203], is given in Figure 5.1.28.

Obviously, a decreasing carbon activity shifts the onset of the carbothermal reduction of silicon nitride according to Eq. 47 to higher temperatures (at 1 bar N_2, this reaction takes place at 1484°C). Presuming a carbon activity of 0.5 results in a shift by 90°C, whereas an activity of 0.1 would result in a shift of the decomposition temperature of around 330°C. Apparently, a further decrease of the carbon activity does not result in a further shift of the onset of Si_3N_4 decomposition to higher temperature since silicon nitride decomposition into the elements (at 1848°C), which is indicated by the horizontal line, becomes the dominant reaction.

A quantitative encapsulation of the silicon nitride grains by the BNC_x grain boundary phase presumably prevents the release of nitrogen that forms by the silicon nitride degradation [204]. The correlating nitrogen partial pressure within this "shell" can thus be allowed to rise to higher values than atmospheric pressure. According to Figure 5.1.11, the onset of both carbothermal reduction of silicon nitride and decomposition into the elements is directly connected to the nitrogen partial pressure. Whereas at 1 bar the carbothermal reduction of silicon nitride occurs at 1484°C, a decomposition temperature of 1700°C can be calculated for a nitrogen pressure of 10 bar.

By combining the topics discussed above, the unusual high-temperature stability of Si—B—C—N ceramics can be understood unequivocally. The BNC_x grain boundary phase decreases the carbon activity and, due to an encapsulation effect, increases the internal nitrogen partial pressure considerably. Both effects result in a shift of the onset of the carbothermal reduction of silicon nitride to higher values, that is, to 2030°C, at 10 bar nitrogen pressure and $a_C < 0.2$ (compare Figure 5.1.28, upper line). However, as mentioned above, a stabilization of silicon nitride is only possible if the amount of boron (i.e. the boron nitride phase fraction) exceeds a certain value, which was empirically found to be at around 8 at%. This amount approximately corresponds to 16 mol% of BN in the final ceramic, if derived from boron-modified polysilazanes, $[B(C_2H_4SiR{-}NH)_3]_n$, or boron-modified poly(silylcarbodiimide)s, $[B(C_2H_4SiRN{=}C{=}N)_3]_n$. If the boron content in the materials is lower, there is obviously no sufficient dissolution of free carbon within the turbostratic BN phase. The second critical point is the relative phase amount of Si_3N_4. Even though there is still a lack of more detailed studies, it can be concluded that the relative phase amount of silicon nitride should not exceed 30 mol%. If the respective value is higher, an efficient encapsulation of silicon nitride crystals that evolve during a heat treatment is not possible. Very recent papers report in detail on the high-temperature mass stability of Si—B—C—N ceramics by considering relative ratios of $(SiC + BN + C)/Si_3N_4$, $Si_3N_4/(Si_3N_4 + SiC)$, and BN/(BN + C) [40, 119, 120, 131, 132]. However, this is a subject that needs more detailed investigations on the development of new synthetic approaches to novel type of precursors for Si—B—C—N ceramics, which allow for the preparation of ceramics with adjustable chemical compositions.

5.1.5 APPLICATIONS

5.1.5.1 General Comments

The wide variety of available organometallic polymers, suitable as preceramic materials, offers exceptional opportunities for the development of PDCs. According to Figure 5.1.29, bulk parts, fibers, coatings, infiltrated media, fiber–matrix composites, and near-net shape manufactured ceramic components are available [181, 182, 206–210].

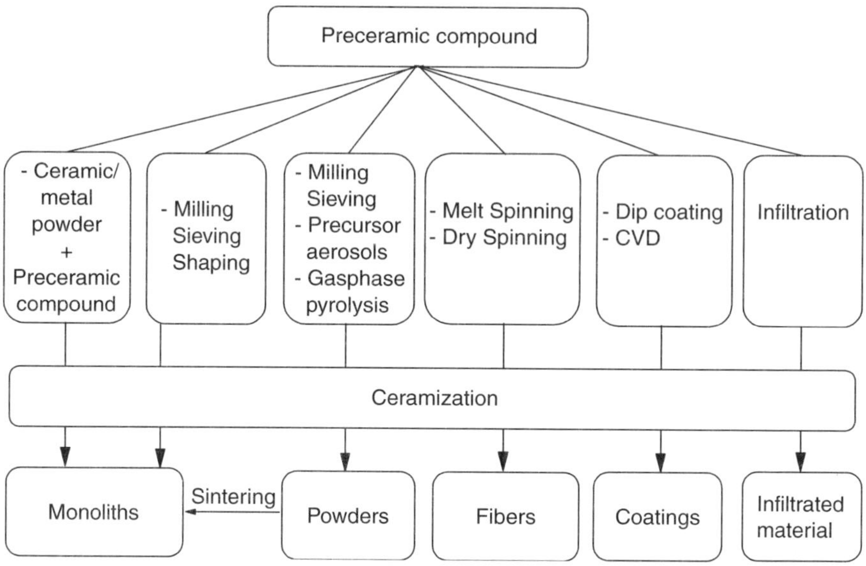

FIGURE 5.1.29 Preparation of ceramic materials by precursor processing.

The preforms—also referred to as green parts—can be produced using techniques well known from polymer process engineering. The different approaches and techniques illustrated in Figure 5.1.29 will be discussed in the following sections topologically.

5.1.5.2 Bulk Ceramics

Bulk ceramics can be obtained by polymer powder compaction and subsequent ceramization according to the flow scheme shown in Figure 5.1.30. In contrast to conventional sintering processes, no sinter additives are required for the densification. Powder compaction is usually achieved by cold, warm, or hot isostatic or uniaxial pressing, depending on the physical/chemical properties of the precursor. A key step is the initial transformation of volatile or liquid precursors into highly cross-linked preceramic infusible networks.

The polymer powder compaction process for the preparation of PDC bulk parts by cold isostatic pressing (CIP) was developed by Riedel *et al.* [46, 211, 212] in the early 1990s. The final density of ceramic materials obtained by this procedure was in the range of maximum 90%. For obtaining dense monolithic bodies, however, warm pressing of suitable (non-volatile) precursors is the state-of-the-art. Warm uniaxial pressing as described by Seitz *et al.* [213]

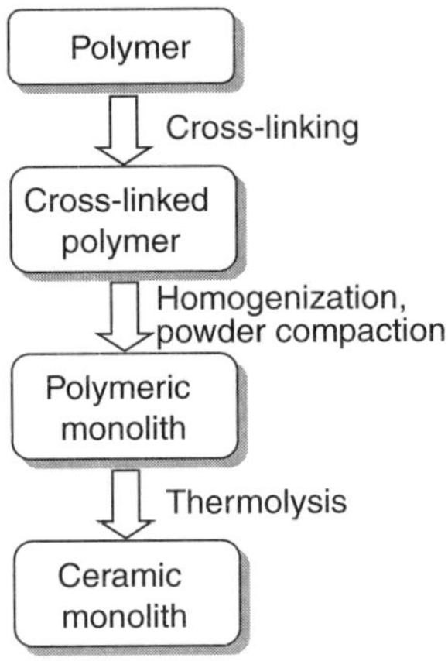

FIGURE 5.1.30 Flow scheme for the preparation of precursor-derived bulk ceramics.

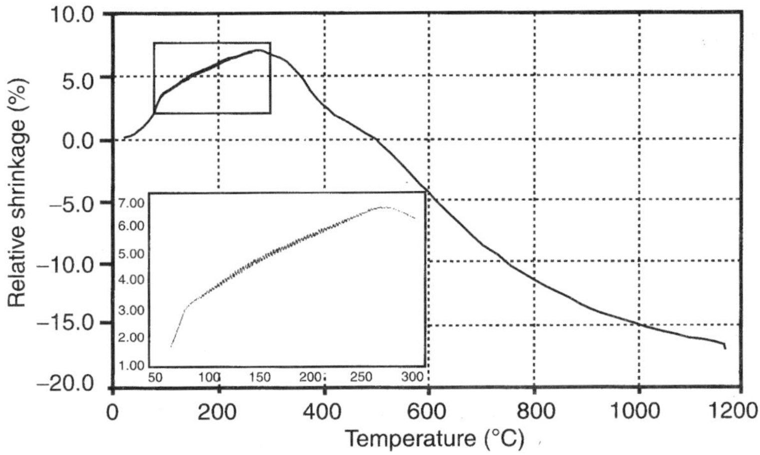

FIGURE 5.1.31 TM analysis of cold isostatically pressed green bodies obtained from poly(vinylsilazane) [213].

of poly(vinylsilazane), VT50, delivered glassy green bodies with 7.5% of open porosity. Subsequent thermolysis released ceramic bulk parts with relative densities of around 97%.

Optimized conditions for this process were determined by means of thermo-mechanical analysis (TMA). Figure 5.1.31 shows such a diagram obtained from a cold isostatically pressed poly(vinylsilazane) body (precursor: VT50, Hoechst AG) [213].

TMA investigations give information on the shrinkage (or expansion) of materials upon heat treatment. If it is performed in an oscillating modus, it additionally allows for the determination of the softening temperature of the polymeric precursor (compare clip in Figure 5.1.31). This value is of major importance since an increasing temperature results in a decrease in viscosity of the polymeric precursor, which enables a facilitated powder compaction. On the other hand, the maximum temperature applied for the powder compaction must be considered carefully. Exceeding this value, which refers to the maximum of the TMA curve, results in shrinkage of the green part due to a beginning ceramization during compression, causing the formation of shrinkage cracks that usually appear parallel to the pressing direction. As a consequence, shape integrity is lost and in the worst case the monoliths break into pieces.

Haug *et al.* [214] investigated in detail plastic forming of boron-modified poly(silylcarbodiimide)s for the production of Si—B—C—N ceramic monoliths and correlated the conditions applied for the powder compaction with the open porosity of the derived bulk parts. It turned out that the open porosity of bulk ceramics obtained from green bodies prepared at 100°C/20 MPa could be decreased from about 20 to 12% by increasing the applied pressure to 43 MPa. If the same polymer was plastically formed at 120°C the porosity decreased from 6.6% at 20 MPa to 5% if 43 MPa was applied for the densification of the polymeric powder (see Figure 5.1.32).

It is obvious that the temperature applied during plastic forming of the precursors more significantly influences the open porosity than increasing pressure. However, in the above case, further raising the temperature did not result in an additional reduction of the open porosity, since at and above 150°C shape integrity was totally lost as evident from Figure 5.1.33 [214].

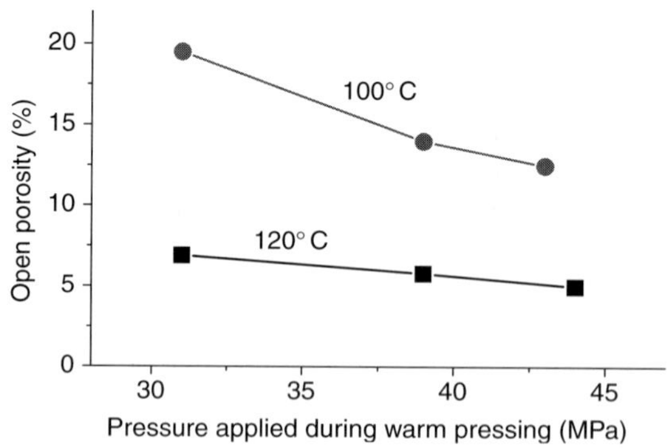

FIGURE 5.1.32 Open porosity of ceramics derived from boron-modified poly(silylcarbodiimide) $[B(C_2H_4Si(CH_3)NCN)_3]_n$ [127]. Powder compaction was performed by plastic forming of the polymer powder at different temperatures and pressures and subsequent pyrolysis at 1400°C for 2 h [214].

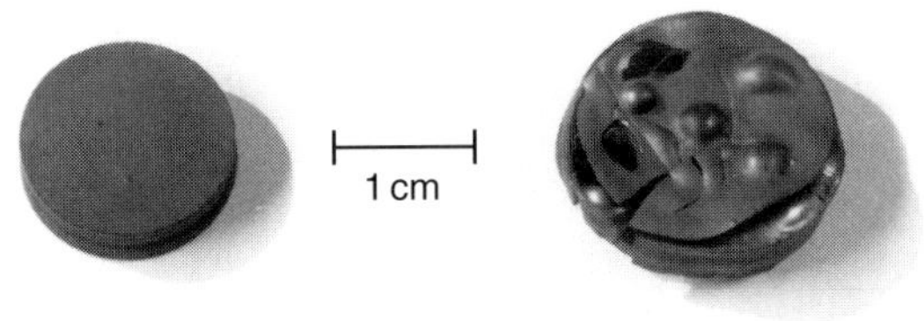

FIGURE 5.1.33 Limits in the densification of bulk ceramics derived from boron-modified poly(silylcarbodiimide) $[B(C_2H_4Si(CH_3)NCN)_3]_n$ [127] by means of plastic forming. After polymer powder compaction at 140°C sufficient open porosity is provided for releasing gaseous thermolysis by-products whereas after compaction at 150°C the green body is too dense to efficiently release such gases [214].

The absence of low-melting grain boundary phases in precursor-derived materials leads to extraordinary high-temperature mechanical properties that can be very different from those of conventionally sintered ceramics.

Recent studies of the high-temperature mechanical properties of as-thermolyzed Si—C—N and Si—B—C—N materials have shown that the amorphous state reveals an outstanding mechanical stability, even at rather high temperatures [215–217]. However, the deformation mechanisms are not yet well understood.

Christ *et al.* characterized amorphous Si—B—C—N ceramics using isothermal compression creep testing in the temperature range of 1200–1500°C [215, 216]. It was found that the deformation rate contains a stress-independent section as well as a stress-dependent component that is proportional to the applied stress, indicating that this portion of the deformation mechanism is based on viscous flow.

As shown in Figure 5.1.34 for Si—B—C—N ceramics derived from $[B(C_2H_4SiH{-}NH)_3]_n$ (MW33) and $[B(C_2H_4SiMe{-}NH)_3]_n$ (T2-1), a continuous decrease in the deformation rate from $\sim 10^{-6}$ to $10^{-8}\,s^{-1}$ with time was found, which was mainly explained with a reduction of free volume in the amorphous material. Stationary creep was not observed, even after 450 h, indicating that the materials improve their creep stability during the creep tests.

It is obvious and interesting to note that the curves of both materials can be divided into two ranges. The first one referred to as range I up to $\sim 2 \times 10^{-5}$ is similar to primary creep in conventional materials, which often can be described using the Norton power law.[1] Range II is characterized by an accelerated decrease of the deformation rate as compared to range I.

[1] Primary creep $\dot{\varepsilon} = A\sigma^n \exp(-Q/RT)t^{-c}$ where σ is the applied stress, n the stress exponent, Q the activation energy, R the gas constant, T the absolute temperature, t the time, and c the time exponent. A is a constant.

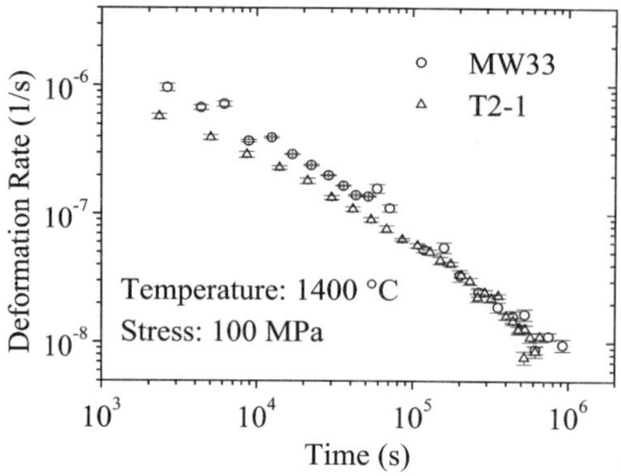

FIGURE 5.1.34 Deformation rates of amorphous ceramics derived from $[B(C_2H_4SiH{-}NH)_3]_n$ (MW33) [114] and $[B(C_2H_4SiMe{-}NH)_3]_n$ (T2-1) [106] during isothermal creep testing at 1400°C under a load of 100 MPa [215]. Atmosphere: air.

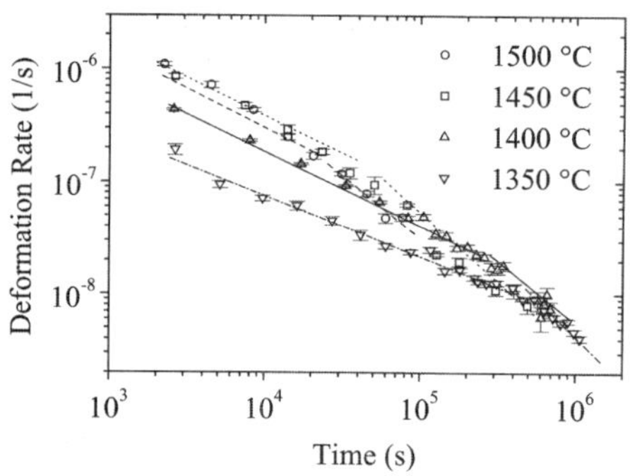

FIGURE 5.1.35 Deformation rates of $[B(C_2H_4SiMe{-}NH)_3]_n$ (T2-1) [106] derived amorphous ceramics during isothermal creep testing at various temperatures under a load of 50 MPa [215]. Atmosphere: air.

Christ *et al.* [215, 216] as well as Thurn *et al.* [218, 219] also observed that the temperature dependence of the creep behavior of both Si—C—N and Si—B—C—N ceramics is comparably low and qualitatively the same for all temperatures. For example, the influence of the temperature on the deformation rate of T2-1-derived Si—B—C—N ceramic is shown in Figure 5.1.35.

Among the investigated samples, deformation rate was lowest at 1350°C and highest at 1500°C. The range I to range II transition occurred earlier at elevated temperatures. This shift was explained by a stress-independent reduction of free volume, which proceeds diffusion-controlled and thus accelerates with increasing temperature.

All creep tests mentioned above were performed in air and thus represent long-time oxidation experiments. Once tested, the specimen usually possessed porous oxide surface layers consisting of silicon and oxygen as determined by EDX and WDX. Although the thickness of such oxide layers increased with testing time a saturation value was reached. Whereas after short-time exposures, even up to 1700°C, the thickness of the surface layers was below 2 μm [114], the thickness of such layers was around 50 μm after annealing in air for 300 h at 1400°C and did not increase even after annealing for 500 h.

5.1.5.3 Ceramic Coatings

Precursor-derived ceramic coatings can protect sensitive substrates from hydrolysis, oxidation, corrosion, and erosion. In contrast to chemical and physical vapor deposition, which are expensive procedures for obtaining ceramic coatings, the formation of ceramic coatings by a dip-coating process is a rather simple and comparably economic process [220, 221]. The substrate is introduced into a solution of the precursor, pulled out, dried, and subsequently thermolyzed. The thickness of the adherent polymer layer can be varied by changing parameters such as pull-out velocity, solution viscosity, polymer concentration, and the number of the coating—thermolysis cycles. High velocity, viscosity, and polymer concentration result in an increase of the polymer (and thus ceramic) layers in thickness and vice versa. Unfortunately, with increasing thickness, coatings tend to crack during thermolysis. This is mainly a consequence of the loss of gaseous by-products during the heat treatment and an increase in density during the polymer-to-ceramic transformation. Frequently, filler materials, either reactive (e.g. metal powders) or passive (SiC, Si_3N_4, etc.) [152, 153] are added to the polymer solution or melt to minimize such volume changes and densification-induced stresses that are associated with the polymer-to-ceramic conversion of the coatings. Moreover, incorporation of fillers allows for specifically influencing chemical, mechanical, and physical properties of the ceramic coatings.

A multifaceted field, which is currently under investigation, is the protection of carbon-fiber-reinforced ceramic composites from oxidation, hydrolysis, and erosion by PDC coatings. Because of their high maximum application temperature and low density, carbon-fiber-reinforced composites are very promising materials for realizing mechanically chargeable, light-weight structures with a high potential for aerospace applications [222]. However, under the harsh re-entry conditions, carbon fibers in such composites are oxidized and erode.

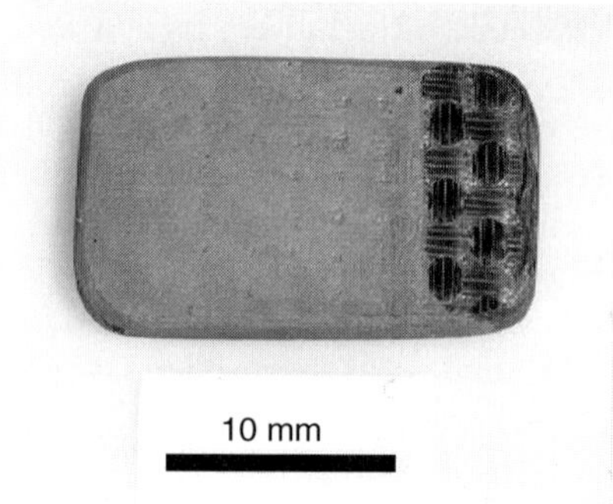

FIGURE 5.1.36 Carbon-fiber-reinforced silicon carbide composite covered with a ceramic Si—B—C—N coating. The coating was obtained by a dip-coating process using a slurry of 42 vol.% of silicon in a toluene solution of $[B(C_2H_4SiMeNH)_3]_n$ [106] and subsequent thermolysis [220, 221].

As a consequence, shape integrity of the exposed components is lost, which in the worst case results in the total damage of the composite.

Figure 5.1.36 shows such a carbon-fiber-reinforced silicon carbide substrate coated with a polymer-derived Si—B—C—N ceramic after one coating–thermolysis cycle. The precursor was obtained by ammonolysis of $B(C_2H_4SiMeCl_2)_3$ according to Ref. [106]. Additionally, 42 vol.% of Si powder was added to the polymer solution to avoid crack formation during the polymer-to-ceramic conversion [220, 221].

It is evident that the ceramic coating homogenously covered the surface of the substrate. The right part in the image in which the fiber structures are visible represents the uncoated substrate, which was fixed to the dip-coating equipment. Figure 5.1.37 displays a comparison by TGA of the oxidation behavior of the pure and the coated composite at 1400°C.

It is evident that the ceramic coating efficiently prevented the substrate from degradation. Whereas an exposure of the uncoated substrate to air after 500 min resulted in the loss of approximately 55% of the original sample weight, there was nearly no mass change observed during the oxidation of the protected substrate. The observed 55% mass loss roughly corresponded to the amount of carbon fibers in the substrate and was due to a total "fiber-burn-out".

5.1.5.4 Ceramic Fibers

The fabrication of high-modulus refractory ceramic fibers requires controllable polymer rheology and adjustable polymer reactivity. Precursor fibers are obtained by either melt or dry spinning. Different principles for fiber processes

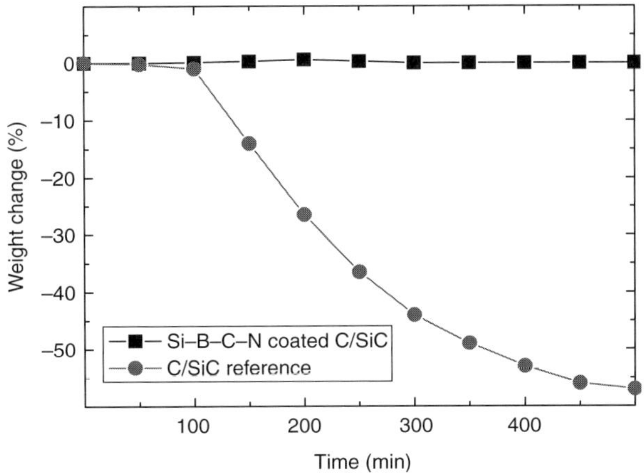

FIGURE 5.1.37 Comparison of the oxidation behavior at 1400°C in air of C/SiC composites with (squares) and without (circles) Si—B—C—N coating. The mass loss of around 55% in the case of the uncoated CMC corresponds to the amount of carbon fibers in the material [220, 221].

such as extrusion, down drawing from preforms, or up drawing from precursor melt or solutions are known [223].

Once spun, the precursor fiber is rendered infusible by subsequent curing, which represents an additional chemical surface reaction. Without this step, fiber integrity is lost because of melting and/or creep of the green fiber before full transformation into the ceramic state.

The first non-oxide silicon-based ceramic fibers were obtained from poly(dimethylsilane) by the Yajima process [12] (see Figure 5.1.38).

Synthesis of polydimethylsilane through alkali-metal-promoted dehalocoupling of dichlorodimethlysilane and its thermal (Kumada) rearrangement into processable poly(methylsilylene-methylene), $[HSi(CH_3){-}CH_2]_n$, also referred to as polysilapropene, were already discussed in detail in Section 5.1.4.1.2. Polysilapropene fibers were obtained by melt spinning in an inert gas atmosphere such as nitrogen or argon at around 300°C.

According to Figure 5.1.38, curing of the green fibers was performed by either controlled surface oxidation in air at approximately 150°C, which resulted in the oxidation of Si—H units accompanied with the formation of Si—O—Si linkages on the fiber surface. Heating such partially oxidized fibers in an inert gas atmosphere to 1200°C delivered Si—C—O fibers, which are commercially available under the trademark Nicalon™ [224]. As evidenced from (high-resolution) transmission electron microscopy and X-ray photoelectron spectra, such fibers are composed of free carbon, nanocrystalline β-silicon carbide, and an amorphous non-stoichiometric Si—C—O matrix. Heating the fibers to temperatures exceeding the final thermolysis temperature results in the

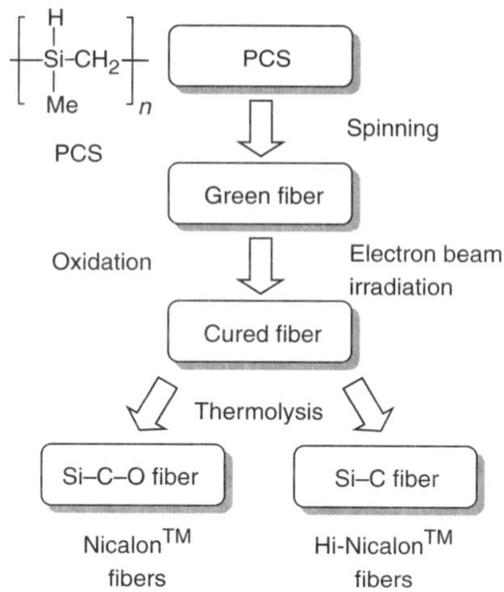

FIGURE 5.1.38 Flow scheme of the preparation of SiC Nicalon™ and Hi-Nicalon™ fibers [12].

decomposition of the ternary amorphous matrix phase by a weight loss caused by the elimination of SiO and CO as well as further crystallization. A further crucial subject is the hydrogen content of the ceramic fibers, which is progressively reduced when heating the fibers to 1200–1300°C or higher. All these effects are accompanied by a coarsening of the fiber morphology and result in a dramatic deterioration of the high-temperature mechanical properties.

Alternatively, electron beam (2 MeV) or γ-ray (^{60}Co) irradiation served for the chemical modification of the fiber surfaces, by which contamination of the fibers with oxygen could almost be avoided. The mechanism behind theses processes is based on radical reactions. Noticeably, free radicals, which remain in the green fibers, have to be removed by additional annealing steps under inert atmosphere to reduce the reactivity of the extremely sensitive radical species in order to avoid oxidation and/or hydrolysis when exposing the fibers to air. Commercially available silicon carbide fibers, which are obtained by melt spinning and subsequent electron beam curing, are, for example, Hi-Nicalon™ fibers [224]. Even though oxygen-free, these fibers also crystallize at comparably low temperatures (below 1300–1400°C). In contrast to Si—C—O fibers, formation of nanocrystalline β-SiC and random segregation of carbon occur, whereas there is no evidence for an amorphous Si—C—O phase.

Thermodynamic calculations predict that in an inert atmosphere oxygen-free SiC/C fibers are thermally stable up to approximately 2500°C. Unfortunately, such calculations only consider phase transformation or decomposition reactions that involve evaporation of gaseous species whereas crystal growth cannot be taken into account. The latter, which is driven by a reduction of the surface energy, results in grain coarsening with increasing temperature. Both silicon carbide grain coarsening and the increase of the size of the carbon domains negatively influence mechanical properties of the fibers.

A critical aspect is the oxidation behavior of SiC fibers [225]. At increased temperature, either passive or active oxidation occurs, depending on the oxygen partial pressure. Passive oxidation is accompanied with the elimination of CO or CO_2 and appears at lower temperatures and high oxygen pressure. According to Eq. 50 or Eq. 51, it results in the conversion of (surface-near) silicon carbide into silica, which forms a protective layer on the ceramic fiber:

$$SiC + 2O_2 \longrightarrow SiO_2 + CO_2 \qquad (50)$$

$$SiC + 1.5O_2 \longrightarrow SiO_2 + CO \qquad (51)$$

Even though carbon-containing species are evaporated in these reactions, the ceramic fibers gain in weight.

In contrast, oxidation at low oxygen (partial) pressure according to Eq. 52

$$SiC + O_2 \longrightarrow SiO + CO \qquad (52)$$

results in the volatilization of carbon and silicon-containing species. It is consequently accompanied with a weight loss and in the worst case results in the total damage of the ceramic fiber.

Recent publications also report on the synthesis and high-temperature properties of silicon nitride-based ceramic fibers obtained from polysilazanes (see Section 5.1.4.3) or polyborosilazanes (see Section 5.1.4.4). In previous sections, it was discussed that the derived ternary or quaternary ceramics resist crystallization up to temperatures of 1500°C and approximately 1800°C, respectively. This is mainly a consequence of the random amorphous ceramic network in which silicon atoms are present in mixed coordination spheres.

Even though numerous Si—C—N precursors have been published since the mid-1970s, there are only a very limited number of procedures described in the literature that so far deal with the conversion of such polymers into ceramic fibers [226–234]. For example, Si—C—N green fibers were obtained by melt spinning of a polymeric silazane HPZ, which was synthesized by reacting trichlorosilane with hexamethyldisilazane [228]. The green fibers were chemically cured by exposure to a stream of argon/trichlorosilane at moderate temperature and subsequently thermolyzed at 1200°C. Alternatively, curing using trichloroborane served for rendering polymeric Si—C—N fibers infusible [234]. In contrast to as-obtained Nicalon™ fibers, which consist of free carbon, nanocrystalline β-silicon carbide and an amorphous non-stoichiometric Si—C—O matrix, HPZ-derived as-thermolyzed Si—C—N fibers are composed of an amorphous non-stoichiometric $SiN_xC_{1-(3x/4)}$ and free carbon. Remarkably, grain growth in Si—C—N fibers proceeds retarded compared to Nicalon™ fibers, but above 1500°C in an Ar atmosphere, degradation by carbothermal reduction of Si—N units leads to a total breakdown of the fiber.

Baldus and Jansen obtained non-oxide silicon nitride-based ceramic fibers, which do not decompose below 1800°C and which additionally possess superior oxidation resistance [98–101]. According to Eq. 28 (compare Section 5.1.2.4.2), trichlorosilyl-amino-dichloroborane, TADB, was reacted with methylamine to deliver a polyborosilazane, which after cross-linking appeared glass-like but still soluble and fusible. The as-obtained precursor could be melt spun at 100–120°C using a multifilament spinneret, which delivers 200 single fibers. Subsequent curing and continuous heating to 1500°C released almost oxygen-free (<1 mass%) ceramic fibers with an approximate elemental composition close to $SiBN_3C$. Oxidation experiments in air at 1500°C for 50 h resulted in the formation of an amorphous 1.7-μm Si—C—O layer, which protected the fiber from further oxidation. Tensile strengths of such oxidized fibers were comparable with those of the as-obtained oxygen-free fibers.

5.1.5.5 Fiber-Reinforced Ceramics

Fiber-reinforced ceramic matrix composites (CMCs) have been shown to exhibit extraordinary thermo-mechanical properties combined with low density. They are consequently highly interesting candidates for manifold applications under severe conditions, especially in aerospace [222].

CMCs can be obtained by different methods. The most established procedures are liquid-phase routes, gas-phase routes, or their combination, referred to as hybrid processes [222].

Liquid-phase routes occur by infiltration of liquids such as melts, solutions, or slurries (which will not be discussed here in detail) into a fiber preform. The first one, usually a reactive melt infiltration, requires matrix materials with sufficiently low melting point, such as silicon (1410°C) or aluminum (660°C). During the infiltration step, the molten elements react either with the fiber preform, for example carbonaceous fiber prepregs, or the atmosphere. Even though these methods allow for a rapid preparation of near-net-shape CMCs, there are some drawbacks, which have to be considered carefully. For example, the preparation of carbon-reinforced SiC composites is performed by capillary infiltration of liquid silicon. Molten silicon readily wets carbon and under the applied conditions reacts—at the expense of the carbon fibers—to SiC, even in the presence of amorphous carbon. To avoid fiber degradation, it is essential to protect the fibers, for example, by depositing boron nitride coatings, which is an additional time-intensive and expensive procedure.

The preparation of Si—C-based CMCs with inert fibers such as Nicalon is also possible. However, in this case it is inevitable to carefully adjust the ratio of amorphous carbon in the prepreg and silicon that is infiltrated in order to avoid segregation of residual non-reacted elemental silicon, since such segregation usually worsens thermo-mechanical properties of the CMC dramatically.

Precursor-derived CMCs, which can be obtained by polymer impregnation and pyrolysis process, PIP, do not suffer from these drawbacks. Although rather simple, the process cannot be applied for every kind of polymeric precursor. Prerequisites for an effective polymer infiltration and loading of the preforms are sufficient fluidity of the polymer solution to allow for an efficient penetration into the preform, adequate solubility, and thus concentration of the polymer solution to prevent extensive shrinkage during the drying step, appropriate cross-linking chemistry of the polymeric precursor, and high polymer-to-ceramic conversion yields. Even if a precursor fulfills all requirements, a large number of infiltration/thermolysis cycles is always necessary to obtain with maximum density CMCs.

Recent papers describe the fabrication of ultra-high-temperature CMCs with precursor-derived Si—B—C—N ceramic matrix using two-component precursors. In contrast to the above-mentioned system, this route avoids the use of solvents [123–125]. According to Eq. 33 (Section 5.1.2.4.3), the polymeric matrix is obtained by a thermally induced hydrosilylation of oligo(vinylsilazane) with tris(hydridosilylethyl)boranes, $B[C_2H_4Si(CH_3)_nH_{3-n}]_3$ ($n = 0$–2). Cross-linking of the precursor mixture occurs quantitatively without requiring catalysts and without the formation of by-products. The physical/chemical properties of the starting compounds and the straightforward synthetic procedure—the starting compounds were simply mixed and reacted at moderate temperature—allowed for the preparation of fiber-reinforced Si—B—C—N green parts using the resin transfer molding (RTM) process. The single steps of the RTM process, as it was performed for the preparation of ceramic fiber–matrix composites, is shown schematically in Figure 5.1.39.

Since the monomeric starting compounds are extremely sensitive to oxygen and/or moisture, a special equipment, which is shown in Figure 5.1.40, was developed in order to allow for a manipulation of all substances under inert gas conditions.

The equipment consists of a Schlenk flask that serves as a reservoir for the precursor mixture and a brass or stainless steel mold, which contain the preforms. Flask and mold are linked via an interconnection glass tube. The valves attached allow separate evacuation of the two main parts. Initially, the mold is filled with the fiber material and the whole system thoroughly evacuated. The central valve is closed and the Schlenk flask flooded with inert gas. The reaction mixture is introduced and the Schlenk flask evacuated again. After the central valve is opened, the precursor mixture is injected into the mold by slow introduction of argon through the valve attached to the Schlenk tube. After injection,

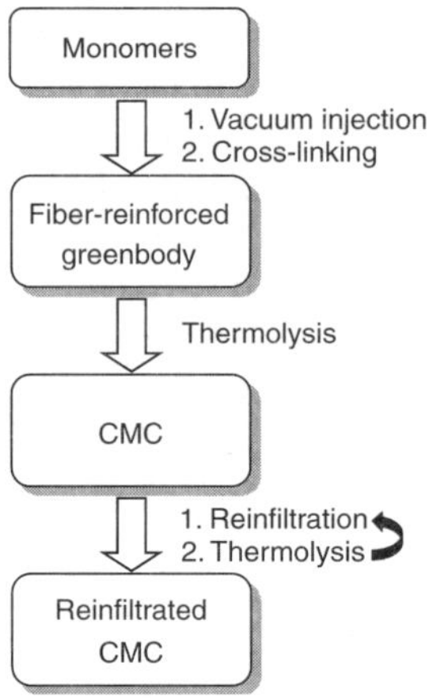

FIGURE 5.1.39 Flow scheme of the preparation of fiber-reinforced ceramic–matrix composites by vacuum injection method. Repetition of the reinfiltration–thermolysis process is required to obtain dense CMCs [123–125].

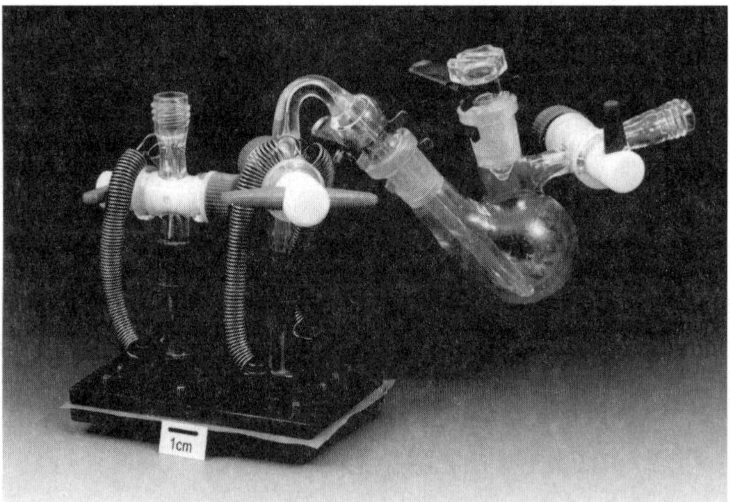

FIGURE 5.1.40 RTM vacuum injection equipment consisting of a brass mold (or alternatively stainless steel mold) containing fiber fabrics or three-dimensional weave and a Schlenk flask that serves as reservoir for the polymeric precursor [123–125].

the central valve is closed again and the brass mold heated within an oil bath to cross-link the precursors. The fiber-reinforced green body is removed from the brass mold within a glovebox. Subsequent thermolysis of the bulk part in Al_2O_3 Schlenk tubes in an argon atmosphere up to 1400°C delivered fiber-reinforced ceramic composites shown in Figure 5.1.41.

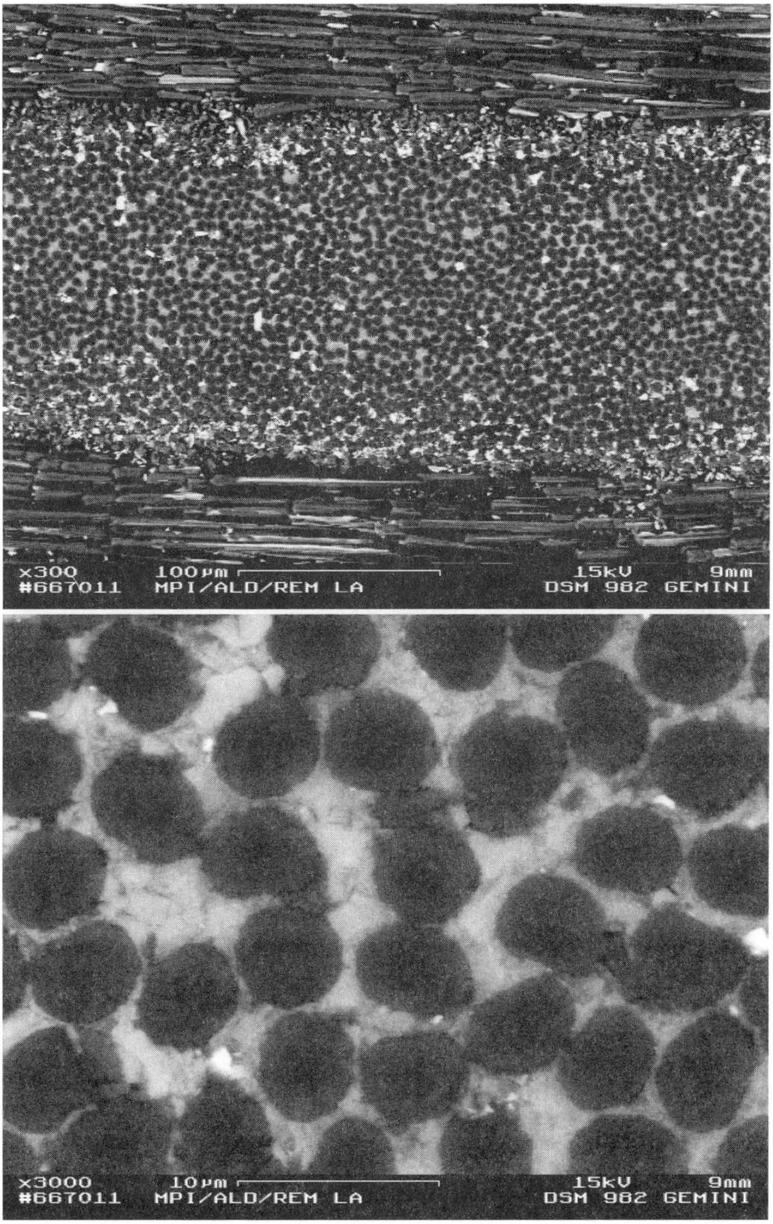

FIGURE 5.1.41 Scanning electron micrographs of a cross-section of the fiber-reinforced ceramic obtained by RTM process of carbon woven fabrics using Si—B—C—N two-component precursor. Top: 300× magnification, bottom: 3000× magnification. In the upper part of the top image a fiber bundle (dark gray) parallel to the view is seen, followed by a fiber bundle with perpendicular orientation in the center of the image. The image at 3000× magnification shows fibers well embedded by the Si—B—C—N matrix. White spots in both images are due to impurities caused by the preparation of the samples [123–125].

Thermolysis of the green parts delivered fiber-reinforced ceramics with an amorphous Si—B—C—N matrix. Depending on the starting compounds used, the materials were thermally stable in inert gas atmosphere up to 2000°C. Transformation of the amorphous materials into crystalline SiC/Si_3N_4 composites, which was performed by an additional heat treatment, occurred around 1750°C. Moreover, the materials possessed a superior oxidation resistance. TGA in air pointed out that even though passive oxidation starts at 1200°C it proceeds retarded. At 1600°C the mass gain was still below 0.5%.

REFERENCES

1. Riedel, R. (2000). *Handbook of Ceramic Hard Materials*, Germany: Wiley-VCH, and literature cited therein.
2. Suzuki, K., and Sasaki, M. (1987). In *Fundamental Structural Ceramics*, p. 75, Somiya, S., and Bradt, R. C., eds., Tokyo: Terra Scientific.
3. Tanaka, H. (1991). In *Silicon Carbide Ceramics, Vol. 1*, p. 213, Somiya, S., and Inomata, Y., eds., Barking: Elsevier.
4. Edington, J. W., Rowcliffe, D. J., and Hawshall, J. L. (1975). *Powder Metall. Int.* **7**: 82.
5. Narushima, T., Goto, T., Hirai, T., and Iguchi, Y. (1997). *Mater. Trans. JIM* **38**: 821.
6. Kipping, F. S., and Sands, J. E. (1921). *J. Chem. Soc.* **119**: 830.
7. Miller, R. D., Hofer, D., Rabolt, J., and Fickes, G. N. (1985). *J. Am. Chem. Soc.* **107**: 2172.
8. Miller, R. D. (1989). *Angew. Chem. Adv. Mater.* **101**: 1773.
9. Burkhardt, C. A. (1949). *J. Am. Chem. Soc.* **71**: 963.
10. Chang, L. S., and Corey, J. Y. (1989). *Organometallics* **8**: 1885.
11. Woo, H.-G., Walzer, J. F., and Don Tilley, T. (1992). *J. Am. Chem. Soc.* **114**: 7047.
12. Yajima, S., Hayashi, J., and Omori, M. (1975). *Chem. Lett.* 931.
13. Whitmarsh, C. K., and Interrante, L. V. (1991). *Organometallics* **10**: 1336.
14. Smith Jr., T. L. (1986). U.S. Patent 4,631,179.
15. Wu, H.-J., and Interante, L. V. (1992). *Macromolecules* **25**: 1840.
16. Procopio, L. J., and Berry, D. H. (1991). *J. Am. Chem. Soc.* **113**: 4039.
17. Boury, B., Carpenter, L., and Corriu, R. J. P. (1990). *Angew. Chem.* **102**: 818.
18. Boury, B., Corriu, R. J. P., Leclercq, D., Mutin, P., Planeix, J.-M., and Vioux, A. (1991). *Organometallics* **10**: 1457.
19. Suam, J. C., and Speier, J. L. (1958). *J. Am. Chem. Soc.* **80**: 4104.
20. Lay, U. (1991). PhD Thesis, University of Heidelberg, Germany.
21. Weinmann, M. (1992). Diploma Thesis, University of Heidelberg, Germany.
22. Corriu, R. J. P., Gerbier, P., Guérin, C., and Henner, B. (2000). *Chem. Mater.* **12**: 805.
23. Kunai, A., Toyoda, E., Nagamoto, I., Horio, T., and Ishikawa, M. (1996). *Organometallics* **15**: 75.
24. Narsavage, D. M., Interrante, L. V., Marchetti, P. S., and Maciel, G. E. (1991). *Chem. Mater.* **3**: 721.
25. Schutzenberger, H., and Colson, C. R. (1885). *R. Acad. Sci.* **92**: 1508.
26. Glemser, O., and Naumann, P. (1959). *Z. Anorg. Allg. Chem.* **298**: 134.
27. Seyferth, D., Wiseman, G. H., and Prud'homme, C. (1983). *J. Am. Ceram. Soc.* **66**: C-13.
28. Arai, M., Sakurada, S., Isoda, T., and Tomizawa, T. (1987). *Polym. Prep.* **28**: 407.

29. Blanchard, C. R., and Schwab, S. T. (1994). *J. Am. Ceram. Soc.* **77**: 1729.
30. Rochow, E. G. (1966). *Pure Appl. Chem.* **13**: 247.
31. Funayama, O., Arai, M., Tashiro, Y., Aoki, H., Suzuki, T., Tamura, K., Kaya, H., Nishii, H., and Isoda, T. (1990). *Nippon Seramikkusu Kyokai Gakujutsu Ronbunshi* **98**: 104; Chem. Abstr. 112 (1990) 103719.
32. Wannagat, U. (1967). *Fortschr. Chem. Forsch.* **9**: 102.
33. Bailey, D. L., Sommer, L. H., and Whitmore, F. C. (1948). *J. Am. Chem. Soc.* **70**: 435.
34. Wannagat, U., and Brandmair, F. (1955). *Z. Anorg. Allg. Chem.* **280**: 223.
35. Wilden, J., Wank, A., Asmann, M., Heberlein, J. V., Boulos, M. I., and Gitzhofer, F. (2001). *Appl. Organomet. Chem.* **15**: 841.
36. Mooser, J. P., Nöth, H., and Tinhof, W. (1974). *Z. Naturforsch.* **29b**: 166.
37. Blum, Y. D., Schwartz, K. B., and Laine, R. M. (1989). *J. Mater. Sci.* **24**: 1707.
38. Seyferth, D., and Wiseman, G. H. (1984). *J. Am. Ceram. Soc.* **67**: C-132.
39. Seyferth, D. (1988). In *Transformation of Organometallics into Common and Exotic Materials: Design and Activation*, NATO ASI Ser. E: Appl. Sci., Vol. 141, p. 133, Laine, R. M., ed., Dordrecht: Kluwer Publ.
40. Weinmann, M., Nast, S., Berger, F., Müller, K., and Aldinger, F. (2001). *Appl. Organomet. Chem.* **15**: 867.
41. Lücke, J., Hacker, J., Suttor, D., and Ziegler, G. (1997). *Appl. Organomet. Chem.* **11**: 181.
42. Chong Kwet Yive, N. S., Corriu, R. J., Leclerq, D., Mutin, P. H., and Vioux, A. (1991). *New J. Chem.* **15**: 85.
43. Bill, J., Seitz, H., Thurn, G., Dürr, H., Canel, J., Janos, B. Z., Jalowiecki, A., Sauter, A., Schempp, S., Lamparter, H. P., Mayer, J., and Aldinger, F. (1998). *Phys. Stat. Sol.* **166A**: 269.
44. Laine, R. M., and Sellinger, A. (1998). In *The Chemistry of Organic Silicon Compounds*, Vol. 2, Rappoport, Z., and Apeloig, Y., eds., London: J. Wiley & Sons.
45. NCP 200, Nichimen Inc.; Chisso, Japan. For detailed investigations see also Ref. [43].
46. Konetschny, C., Galusek, D., Reschke, S., Fasel, C., and Riedel, R. (1999). *J. Eur. Ceram. Soc.* **19**: 2789.
47. NN710, Tonen Corp., Japan.
48. Weinmann, M., Zern, A., and Aldinger, F. (2001). *Adv. Mater.* **13**: 1704.
49. Andrianov, K. A., and Kotrelev, G. V. (1967). *J. Organomet. Chem.* **7**: 217.
50. Voronkov, M. G., and Lavent'yev, V. I. (1982). *Top. Curr. Chem.* **102**: 199.
51. Day, V. W., Klemperer, W. G., Mainz, V. V., and Millar, D. M. (1985). *J. Am. Chem. Soc.* **107**: 8262.
52. Calzaferri, G., Herren, D., and Imhof, R. (1991). *Helv. Chim. Acta* **74**: 1278.
53. Räke, B., Roesky, H. W., Usón, I., and Müller, P. (1998). *Angew. Chem.* **110**: 1508.
54. Blum, Y. D., and Laine, R. M. (1986). *Organometallics* **5**: 2081.
55. Laine, R. M., Blum, Y. D., Chow, A., Hamlin, R. D., Schwartz, K. B., and Rowcliffe, D. J. (1987). *Polym. Prep.* **28**: 393.
56. Youngdahl, K. A., Laine, R. M., Kennish, R. A., Cronin, T. R., and Balavoine, G. G. (1988). *Mater. Res. Soc. Symp. Proc.* **121**: 489.
57. Blum, Y. D., Schwartz, K. B., Crawford, E. J., and Hamlin, R. D. (1988). *Mater. Res. Soc. Symp. Proc.* **121**: 565.
58. Liu, H. Q., and Harrod, J. F. (1992). *Organometallics* **11**: 822.
59. Seyferth, D., and Schwark, J. M. (1989). *Organometallics* **8**: 1980.
60. Duguet, E., Schappacher, M., and Soum, A. (1992). *Macromolecules* **25**: 4835.
61. Bruzaud, S., and Soum, A. (1996). *Macromol. Chem. Phys.* **197**: 2379.
62. Bruzaud, S., Mingotaud, A.-F., and Soum, A. (1996). *Macromol. Chem. Phys.* **198**: 1873.
63. Cazalis, C., Mingotaud, A.-F., and Soum, A. (1996). *Macromol. Chem. Phys.* **198**: 3441.

64. Bouquey, M., Brochon, C., Bruzaud, S., Mingotaud, A.-F., Schappacher, M., and Soum, A. (1996). *J. Organomet. Chem.* **521**: 21.
65. Ebsworth, E. A., and Mays, M. J. (1961). *J. Chem. Soc.* 4879.
66. Ebsworth, E. A., and Mays, M. J. (1962). *Angew. Chem.* **74**: 117.
67. Pump, J., and Wannagat, U. (1962). *Angew. Chem.* **74**: 117.
68. Pump, J., and Wannagat, U. (1962). *Ann. Chem.* **652**: 21.
69. Birkofer, L., Ritter, A., and Richter, P. (1962). *Tetrahedron Lett.* **5**: 195.
70. Pump, J., Rochow, E. G., and Wannagat, U. (1963). *Monatsh. Chem.* **94**: 588.
71. Pump, J., and Rochow, E. G. (1964). *Z. Anorg. Allg. Chem.* **330**: 101.
72. Klebe, J. F., and Murray, J. G. (1968). US Patent 3,352,799.
73. Kienzle, A., Obermeyer, A., Riedel, R., Aldinger, F., and Simon, A. (1993). *Chem. Ber.* **126**: 2569.
74. Obermeyer, A., Kienzle, A., Weidlein, J., Riedel, R., and Simon, A. (1994). *Z. Anorg. Allg. Chem.* **620**: 1357.
75. Gabriel, A. O., and Riedel, R. (1997). *Angew. Chem. Int. Ed. Engl.* **36**: 384.
76. Gabriel, A. O., and Riedel, R. (1997). "Proceedings of the 21st Annual Cocoa Beach Conference and Exposition", The American Ceramic Society 1997, ISSN 0196-6219, Florida USA, Vol. 18, p. 713.
77. Gabriel, A. O., Riedel, R., Storck, S., and Maier, W. F. (1997). *Appl. Organomet. Chem.* **11**: 833.
78. Riedel, R. and Gabriel, A. O. (1999). *Adv. Mater.* **11**: 207.
79. Gabriel, A. O., Riedel, R., Dreßler, W., Reichert, S., Gervais, C., Maquet, J., and Babonneau, F. (1999). *Chem. Mater.* **11**: 412.
80. Balan, C., Völger, K. W., Kroke, E., and Riedel, R. (1997). *Macromolecules* **33**: 3404.
81. Riedel, R., Greiner, A., Miehe, G., Dreßler, W., Fueß, H., Bill, J., and Aldinger, F. (1997). *Angew. Chem. Int. Ed. Engl.* **36**: 603.
82. Kroke, E., Li, Y.-L., Konetschny, C., Lecomte, E., Fasel, C., and Riedel, R. (2000). *Mater. Sci. Eng.* **R26**: 97.
83. Takamizawa, M., Kobayashi, T., Hayashida, A., Takeda, Y., and Joetsu, N. Ger. Offen. DE 344 430 6.
84. Takamizawa, M., Kobayashi, T., Hayashida, A., and Takeda, Y. (1986). US Pat. 4,604,367.
85. Noeth, H. (1961). *Z. Naturforsch. B* **16**: 618.
86. Narula, C. K., Schaeffer, R., and Paine, R. T. (1987). *J. Am. Chem. Soc.* **1009**: 5556.
87. Seyferth, D., and Plenio, H. (1990). *J. Am. Ceram. Soc.* **73**: 2131.
88. Seyferth, D., Plenio, H., Rees Jr., W. S., and Büchner, K. (1991). *Silicon Ceramics with a Dash of Boron in Frontiers of Organosilicon Chemistry*, p. 15, Cambridge: The Royal Society of Chemistry.
89. Blum, Y. D., and Laine, R. M. (1989). US Patent 4,801,439.
90. Blum, Y. D., and Laine, R. M. (1991). US Patent 5,017,529.
91. Su, K., Remsen, E. E., Zank, G. A., and Sneddon, L. G. (1993). *Polym. Prep.* **34**: 334.
92. Su, K., Remsen, E. E., Zank, G. A., and Sneddon, L. G. (1993). *Chem. Mater.* **5**: 547.
93. Wideman, T., Su, K., Remsen, E. E., Zank, G. A., and Sneddon, L. G. (1995). *Chem. Mater.* **7**: 2203.
94. Wideman, T., Su, K., Remsen, E. E., Zank, G. A., and Sneddon, L. G. (1996). *Mater. Res. Soc. Symp. Proc.* **410**: 185.
95. Wideman, T., Cortez, E., Remsen, E. E., Carroll, P. J., and Sneddon, L. G. (1997). *Chem. Mater.* **9**: 2218.
96. Wideman, T., Remsen, E. E., Cortez, E., Chlanda, V. L., and Sneddon, L. G. (1998). *Chem. Mater.* **10**: 412.

97. Wideman, T., Fazen, P. J., Su, K., Remsen, E. E., Zank, G. A., and Sneddon, L. G. (1998). *Appl. Organomet. Chem.* **10**: 412.
98. Jansen, M., and Baldus, H.-P. (1992). Ger. Offen. DE 41 07 108 A1.
99. Baldus, H.-P., Wagner, O., and Jansen, M. (1992). *Mater. Res. Soc. Symp. Proc.* **271**: 821.
100. Baldus, H.-P., Jansen, M., and Wagner, O. (1994). *Key Eng. Mater.* **89–91**: 75.
101. Baldus, H.-P., and Jansen, M. (1997). *Angew. Chem.* **109**: 338; *Angew. Chem. Int. Ed. Engl.* **36**: 328.
102. Baldus, P., Jansen, M., and Sporn, D. (1999). *Science* **285**: 699.
103. Riedel, R., Kienzle, A., Petzow, G., Brück, M., and Vaahs, T. (1994). Ger. Offen. DE 43 20 783 A1.
104. Riedel, R., Kienzle, A., Petzow, G., Brück, M., and Vaahs, T. (1994). Ger. Offen. DE 43 20 786 A1.
105. Kienzle, A. (1994). PhD Thesis, Universität Stuttgart 1994.
106. Riedel, R., Kienzle, A., Dressler, W., Ruwisch, L., Bill, J., and Aldinger, F. (1996). *Nature* **382**: 796.
107. Riedel, R., Bill, J., and Kienzle, A. (1996). *Appl. Organomet. Chem.* **10**: 241.
108. Riedel, R., Ruwisch, L. M., An, L., and Raj, R. (1998). *J. Am. Ceram. Soc.* **81**: 3341.
109. Jansen, M., and Jäschke, T. (1999). *Z. Anorg. Allg. Chem.* **625**: 1957.
110. Jones, P. R., and Myers, J. K. (1972). *J. Organomet. Chem.* **34**: C9.
111. Ruwisch, L. M., Dürichen, P., and Riedel, R. (2000). *Polyhedron* **19**: 323.
112. Ruwisch, L. M. (1998). PhD Thesis, Technische Universität Darmstadt.
113. Riedel, R., and Ruwisch, L. M. (1998). In *McGraw-Hill Yearbook of Science & Technology 1999*, p. 70, McGraw Hill, New York.
114. Weinmann, M., Schuhmacher, J., Kummer, H., Prinz, S., Peng, J., Seifert, H. J., Christ, M., Müller, K., Bill, J., and Aldinger, F. (2000). *Chem. Mater.* **12**: 623.
115. Aldinger, F., Weinmann, M., and Bill, J. (1998). *Pure Appl. Chem.* **70**: 439.
116. Weinmann, M. (1999), in *Precursor-Derived Ceramics*, p. 83, Bill, J., Wakai, F., and Aldinger, F., eds., Weinheim: Wiley-VCH.
117. Weinmann, M., and Aldinger, F. (1999), in "New Properties from Atomic Level Processing"—Proceedings of the Symposium on International Joint Project Ceramics Superplasticity, Tokyo, p. 16.
118. Weinmann, M., Seifert, H. J., and Aldinger, F. (2000), in *Contemporary Boron Chemistry*, p. 88, Davidson, M. G., Hughes, A. K., Marder, T. B., and Wade, K., eds., Cambridge: The Royal Society of Chemistry.
119. Müller, A., Gerstel, P., Weinmann, M., Bill, J., and Aldinger, F. (2000). *J. Eur. Ceram. Soc.* **20**: 2655.
120. Müller, A., Gerstel, P., Weinmann, M., Bill, J., and Aldinger, F. (2001). *J. Eur. Ceram. Soc.* **21**: 2171.
121. Weinmann, M., Kamphowe, T. W., Fischer, P., and Aldinger, F. (1999). *J. Organomet. Chem.* **592**: 115.
122. Weinmann, M., Bill, J., and Aldinger, F. (1999). Ger. Offen. DE 197 41 459 A 1.
123. Weinmann, M., Kamphowe, T. W., Schuhmacher, J., Müller, K., and Aldinger, F. (2000). *Chem. Mater.* **12**: 2112.
124. Weinmann, M., Kamphowe, T. W., Lee, S.-H., and Aldinger, F. (2001). In *Verbundwerkstoffe und Werkstoffverbunde*, p. 268, Wielage, B., and Leonhardt, G., eds., Weinheim: Wiley-VCH.
125. Weinmann, M., Kamphowe, T. W., Bill, J., and Aldinger, F. (1999). Ger. Offen. DE 197 41 460 A 1.
126. See, for example, Generazio, E. R. (1995). *Advanced Ceramic Matrix Composites: Design Approaches, Testing & Life Prediction Methods*, Lanchaster: Technomic Pub Co.

127. Weinmann, M., Haug, R., Bill, J., Aldinger, F., Schuhmacher, J., and Müller, K. (1997). *J. Organomet. Chem.* **541**: 345.
128. Weinmann, M., Haug, R., Bill, J., De Guire, M., and Aldinger, F. (1998). *Appl. Organomet. Chem.* **12**: 725.
129. Schuhmacher, J., Müller, K., Weinmann, M., Bill, J., and Aldinger, F. (1999). In "Proc. Werkstoffwoche 1998, Band VII, Keramik/Simulation", p. 321, Heinrich, J., Ziegler, G., Hermel, W., and Riedel, H., eds., Weinheim: Wiley-VCH.
130. Lechler, R., Hausen, H.-D., and Weidlein, J. (1989). *J. Organomet. Chem.* **359**: 1.
131. Weinmann, M., Zern, A., Hörz, M., Berger, F., Müller, K., and Aldinger, F. (2002). *J. Met. Nano Mater.* **386–388**: 335.
132. Weinmann, M., Hörz, M., Berger, F., Müller, A., Müller, K., and Aldinger, F. (2002). *J. Organomet. Chem.* **659**: 29.
133. Schwartz, K. B., Rowcliffe, D. J., Blum, Y. D., and Laine, R. M. (1986). *Mater. Res. Soc. Symp. Proc.* **73**: 407.
134. Soraru, G. D., Babonneau, F., and Mackenzie, J. D. (1990). *J. Mater. Sci.* **25**: 3886.
135. Lavedrine, A., Bahloul, D., Goursat, P., Chong, Kwet Yive, N. S., Corriu, R. J., Leclercq, D., Mutin, P. H., and Vioux, A. (1991). *J. Eur. Ceram. Soc.* **8**: 221.
136. Boury, B., Corriu, R. J., and Douglas, W. E. (1991). *Chem. Mater.* **3**: 487.
137. Schmidt, W. R., Interrante, L. V., Doremus, R. H., Trout, T. K., Marchetti, P. S., and Maciel, G. E. (1991). *Chem. Mater.* **3**: 257.
138. Chong Kwet Yive, N. S., Corriu, R. J., Leclercq, D., Mutin, P. H., and Vioux, A. (1992). *Chem. Mater.* **4**: 141.
139. Han, H. N., Lindquist, D. A., Haggerty, J. S., and Seyferth, D. (1992). *Chem. Mater.* **4**: 705.
140. Corriu, R. J., Leclercq, D., Mutin, P. H., and Vioux, A. (1992). *Chem. Mater.* **4** 711.
141. Schmidt, W. R., Marchetti, P. S., Interrante, L. V., Hurley Jr., W. J., Lewis, R. H., Doremus, R. H., and Maciel, G. E. (1992). *Chem. Mater.* **4**: 937.
142. Chong Kwet Yive, N. S., Corriu, R. J., Leclercq, D., Mutin, P. H., and Vioux, A. (1992). *Chem. Mater.* **4**: 1263.
143. Brodie, N., Majoral, J.-P., and Disson, J.-P. (1993). *Inorg. Chem.* **32**: 4646.
144. Mocaer, D., Pailler, R., Naslain, R., Richard, C., Pillot, J. P., Dunoguès, J., Gérardin, C., and Taulelle, F. (1993). *J. Mater. Sci.* **28**: 2615.
145. Laine, R. M., Babonneau, F., Blowhowiak, K. Y., Kennish, R. A., Rahn, J. A., Exarhos, G. J., and Waldner, K. (1995). *J. Am. Ceram. Soc.* **78**: 137.
146. Babonneau, F., Maquet, J., Bonhomme, C., Richter, R., Roewer, G., and Bahloul, D. (1996). *Chem. Mater.* **8**: 1415.
147. Bahloul, D., Pereira, M., and Gérardin, C. (1997). *J. Mater. Chem.* **7**: 109.
148. Gérardin, C., Taulelle, F., and Bahloul, D. (1997). *J. Mater. Chem.* **7**: 117.
149. Liu, Q., Wu, H.-J., Lewis, R., Maciel, G. E., and Interrante, L. V. (1999). *Chem. Mater.* **11**: 2038.
150. Schuhmacher, J., Weinmann, M., Bill, J., Aldinger, F., and Müller, K. (1998). *Chem. Mater.* **10**: 3913.
151. Laine, R. M., and Babonneau, F. (1993). *Chem. Mater.* **5**: 260.
152. Greil, P. (1998). *J. Eur. Ceram. Soc.* **18**: 1905.
153. Greil, P. (1995). *J. Am. Ceram. Soc.* **78**: 835.
154. Williams, E. A. (1989). NMR spectroscopy of organosilicon compounds, in *The Chemistry of Organic Silicon Compounds*, Patai, S., and Rappoport, Z., eds., London: John Wiley & Sons.
155. Yajima, S., Hayashi, J., and Omori, M. (1975). *Chem. Lett.* 1209.
156. Yajima, S. (1983). *Ceram. Bull.* **62**: 893.
157. Yajima, S., Hayashi, J., Omori, M., and Okamura, K. (1976). *Nature* **261**: 683.
158. Yajima, S., Hayashi, J., Omori, M., and Okamura, K. (1976). *J. Am. Ceram. Soc.* **59**: 324.

159. Yajima, S., Iwai, T., Yamamura, T., Okamura, K., and Hasegawa, Y. (1981). *J. Mater. Sci.* **16**: 1349.
160. Soraru, G. D., Babonneau, F., and Mackenzie, J. D. (1988). *J. Non-Cryst. Solids* **106**: 256.
161. Polysilazanes with general structure $[RSi(NH)_{1.5}]_n$, for example, commercially available poly(vinylsilsesquiazane) VT50, Hoechst AG, Germany.
162. Riedel, R., Kroke, E., Greiner, A., Gabriel, A. O., Ruwisch, L., and Nicolich, J. (1998). *Chem. Mater.* **10**: 2964.
163. Popper, P. (1967). *Br. Ceram. Res. Assoc. Spec. Publ.* **57**: 1.
164. Seyferth, D., Wiseman, G. H., and Prud'homme, C. (1984). *Mater. Sci. Res.* **17**: 263.
165. Seyferth, D., and Wiseman, G. H. (1984). *Am. Chem. Soc. Polym., Div. Polym. Prep.* **25**: 10.
166. Vaahs, T., Brück, M., and Böcker, W. D. (1992). *Adv. Mater.* **4**: 224.
167. Bahloul, D., Pereira, M., Goursat, P., Choong Kwet Yive, N. S., and Corriu, R. J. P. (1993). *J. Am. Ceram. Soc.* **76**: 1156.
168. PCS is a commercially available product of Nippon Carbon Co. Ltd., Type S, Tokyo, Japan.
169. PHPS is a commercially available product of Tonen Co., N-N110, Tokyo, Japan.
170. Iwamoto, Y., Völger, W., Kroke, E., Riedel, R., Saitou, T., and Matsunaga, K. (2001). *J. Am. Ceram. Soc.*, **84**: 2170.
171. Bill, J., Kamphowe, T. W., Müller, A., Wichmann, T., Zern, A., Jalowiecki, A., Mayer, J., Weinmann, M., Schuhmacher, J., Müller, K., Peng, J., Seifert, H. J., and Aldinger, F. (2001). *Appl. Organomet. Chem.* **15**: 777.
172. Schuhmacher, J. (2000). PhD Thesis, Universität Stuttgart, Germany.
173. Schuhmacher, J., Berger, F., Weinmann, M., Bill, J., Aldinger, F., and Müller, K. (2001). *Appl. Organomet. Chem.* **15**: 809.
174. Evans, A. G. (1996). in *Ceramic and Ceramic Composites as High-temperature Structural Materials: Challenges and Opportunities in High-temperature Structural Materials*, Cahn, R. W., Evans, A. G., and McLean, M., eds., London: Chapman & Hall.
175. Gogotsi, Y. G., and Lavrenko, V. A. (1992). *Corrosion of High-Performance Ceramics*, Berlin: Springer-Verlag.
176. Jorgensen, P. J., Wadsworth, M. E., and Cutler, I. B. (1959). *J. Am. Ceram. Soc.* **42**: 613.
177. Franz, I., and Langheinrich, W. (1971). *Solid State Electron.* **14**: 499.
178. Peuckert, M., Vaahs, T., and Brück, M. (1990). *Adv. Mater.* **2**: 398.
179. Toreki, W. (1991). *Polym. News* **16**: 1.
180. Birot, M., Pillot, J.-P., and Dunoguès, J. (1995). *Chem. Rev.* **95**: 1443.
181. Bill, J., and Aldinger, F. (1995). *Adv. Mater.* **7**: 775.
182. Riedel, R. (1996). In *Materials Science and Technology, Chapter 11, Advanced Ceramics from Inorganic Polymers*, p. 1, Cahn, R. W., Haasen, P., and Kramer, E. J., eds., Weinheim: VCH.
183. Liu, A. Y., and Cohen, M. L. (1989). *Science* **245**: 841.
184. Sasaki, T., and Barlett, N. (1989). "Proc. 197th ACS National Meeting", p. 46.
185. Kouvetakis, J., Sasaki, T., Shen, C., Hagiwara, R., Lerner, M., Krishnan, K. M., and Barlett, N. (1989). *Synth. Met.* **34**: 1.
186. Kawaguchi, M., and Kawashima, T. (1993). *J. Chem. Soc. Chem. Commun.* 1133.
187. Kawaguchi, M., Kawashima, T., and Nakajima, T. (1993). *Denki Kagaku.* **61**: 1403.
188. Maya, L. (1988). *J. Am. Ceram. Soc.* **71**: 1104.
189. Maya, L. (1988). *J. Electrochem. Soc.* **135**: 1278.
190. Riedel, R., Bill, J., and Passing, G. (1991). *Adv. Mater.* **3**: 551.
191. Bill, J., Riedel, R., and Passing, G. (1992). *Z. Anorg. Allg. Chem.* **610**: 83.
192. Bill, J., Friess, F., and Riedel, R. (1992). *Eur. J. Solid State Inorg. Chem.* **610**: 83.
193. Dürr, J., Schempp, S., Lamparter, H.-P., Bill, J., Steeb, S., and Aldinger, F. (1997). *Solid State Ionics* **101–103**: 1041.

194. Dürr, J., Lamparter, H.-P., Bill, J., Steeb, S., and Aldinger, F. (1988). *J. Non-Cryst. Solids* **232–234**: 155.
195. Schempp, S., Dürr, J., Lamparter, H.-P., Bill, J., and Aldinger, F. (1998). *Z. Naturf.* **53a**: 127.
196. Jalowiecki, A., Bill, J., Aldinger, F., and Mayer, J. (1996). *Composites Part A* **27**: 717.
197. Schuhmacher, J., Müller, K., Weinmann, M., Bill, J., and Aldinger, F. (1998). "Proceedings Werkstoffwoche 98, Symposium 9b: Physik und Chemie der Keramik", Munich, Germany (in German).
198. Kasper, B. (1996). PhD Thesis, Universität Stuttgart, Germany.
199. Seifert, H. J., and Aldinger, F. (1996). *Z. Metallkd.* **87**: 841.
200. Seifert, H. J., and Aldinger, F. (1999). In *Precursor-derived Ceramics*, p. 165, Bill, J., Wakai, F., Aldinger, F., eds., Weinheim: Wiley-VCH.
201. Weiss, J., Lukas, H.-L., Lorenz, J., Petzow, G., and Krieg, H. (1981). *Calphad* **5**: 125.
202. Seifert, H. J., Peng, J., and Aldinger, F. (2001). *J. Alloys Compounds* **320**: 251.
203. Peng, J., Seifert, H. J., and Aldinger, F. (1999). In *Ceramics—Processing, Reliability, Tribology and Wear*, "Proc. EUROMAT '99", Vol. 12, p. 120, Müller, G. ed., Weinheim: Wiley-VCH.
204. Seifert, H. J., Lukas, H.-L., and Aldinger, F. (1998). *Ber. Bunsenges. Phys. Chem.* **102**: 1309.
205. Peng, J. (2002). PhD Thesis, Universität Stuttgart, Germany.
206. Schubert, U., and Hüsing, N. (2000). *Synthesis of Inorganic Materials*, Weinheim: Wiley-VCH.
207. Greil, P. (2000). *Adv. Eng. Mater.* **2**: 339.
208. Bill, J., Wakai, F., and Aldinger, F. (1999). *Precursor-Derived Ceramics: Synthesis, Structure and High-Temperature Mechanical Properties*, Weinheim: Wiley-VCH.
209. Riedel, R. (1995). *Naturwissenschaften* **82**: 12.
210. Bill, J., and Aldinger, F. (1996). *Z. Metallkd.* **87**: 827.
211. Riedel, R., Pasing, G., Schönfelder, H., and Brook, R. J. (1992). *Nature* **355**: 714.
212. Riedel, R., Seher, M., Mayer, J., and Szabó, D.-V. (1995). *J. Eur. Ceram. Soc.* **15**: 703.
213. Seitz, J., and Bill, J. (1996). *J. Mater. Sci. Lett.* **15**: 391.
214. Haug, R., Weinmann, M., Bill, J., and Aldinger, F. (1999). *J. Eur. Ceram. Soc.* **19**: 1.
215. Christ, M., Thurn, G., Weinmann, M., Bill, J., and Aldinger, F. (2000). *J. Am. Ceram. Soc.* **83**: 3025.
216. Christ, M. (2001). PhD Thesis, Universität Stuttgart, Germany.
217. Zimmermann, A., Bauer, A., Christ, M., Cai, Y., and Aldinger, F. (2002). *Acta Materialica* **50**: 1187.
218. Thurn, G., Canel, J., Bill, J., and Aldinger, F. (1999). *J. Eur. Ceram. Soc.* **19**: 2317.
219. Thurn, G., and Aldinger, F. (1999), in *Precursor-Derived Ceramics*, p. 83, Bill, J., Wakai, F., Aldinger, F., eds., Weinheim: Wiley-VCH.
220. Bill, J., and Heimann, D. (1996). *J. Eur. Ceram. Soc.* **16**: 1115.
221. Heimann, D. (1997). PhD Thesis, Universität Stuttgart, Germany.
222. Krenkel, W., Naslain, R., and Schneider, H. (2001). *High Temperature Ceramic Matrix Composites*, Weinheim: Wiley-VCH and references cited therein.
223. Wallenberger, F. T. (2000). *Advanced Inorganic Fibers: Processes, Structures, Properties, Applications*, Dordrecht: Kluwer Academic Publishers, and references cited therein.
224. Nicalon^{TM} and Hi-Nicalon^{TM} are trademarks of Nippon Carbon Co., Tokyo, Japan.
225. Livage, J., Sanchez, C., and Babonneau, F. (1998). In *Chemistry of Advanced Materials*, Interrante, L. V., and Hampden- Smith, M. J., eds., New York: Wiley-VCH.
226. Verbeck, W. (1973). Ger. Offen. DE 22 18 960.
227. Nakaido, Y., Otani, Y., Kozakai, N., and Otani, S. (1987). *Chem. Lett.* 705.
228. LeGrow, G. E., Lim, T. F., Lipowitz, J., and Reaoch, R. S. (1987). *Am. Ceram. Soc. Bull.* **66**: 363.
229. Vaahs, T., Brück, M., and Böcker, W. D. (1992). *Adv. Mater.* **4**: 224.

230. Jones, R. E., Rabe, J. A., and Peterson, A. L. (1992). *Can. Chem. News* 9.
231. Delverdier, O., Monthioux, M., Mocaer, D., and Pailler, R. (1994). *J. Eur. Ceram. Soc.* **14**: 313.
232. Lu, L., Feng, C. X., and Song, Y. C. (1998). *J. Mater. Sci. Lett.* **17**: 481.
233. Motz, G., Hacker, J., and Ziegler, G. (2001). *Ceram. Eng. & Sci. Proc.* **21**: 307.
234. Hacker, J., Motz, G., and Ziegler, G. (2001). In *Ceramic Materials and Components for Engines*, Heinrich, J. G., and Aldinger, F., eds., Weinheim: Wiley-VCH.

Handbook of Advanced Ceramics
S. Sōmiya *et al.* (Eds.)

5.2 Pyrolysis of Active and Passive Filler-loaded Preceramic Polymers

PETER GREIL
Department of Materials Science, University of Erlangen-Nuernberg, Martensstrasse 5, D-91058 Erlangen, Germany

5.2.1 INTRODUCTION

Organometallic polymer precursors offer unique opportunities for manufacturing of ceramic components using versatile plastic shaping techniques established in engineering plastics industry such as casting, pressing, injection molding, or extrusion. Pyrolysis of silicon-containing polymers such as $[(R)(R')SiX]_n$ and $[(R')SiX_{1.5}]_n$ with R, R′ = H, alkyl, aryl, arenyl, etc. groups, which can be a poly(silane) X = SiR_2, -(carbosilane) X = CH_2, -(silazane) X = NH, (siloxane) X = O or more complex copolymers or mixtures thereof, has been effective in the preparation of ceramic materials in the system Si—C—N—O at temperatures significantly lower than conventionally sintered materials [1–7]. Due to the presence of four potential bridging sites per Si, the actual composition and physical form of the polymer can be quite variable and the structure of the pyrolysis residue can be amorphous, partly or fully (nano) crystalline depending on the heat treatment conditions [8, 9]. A variety of preceramic polymers have been used to serve as advanced binder systems for processing of ceramic and metallic powder systems or, more recently, to manufacture monolith and composite components. In contrast to conventional fugitive hydrocarbon binders, such as polyethylene or polystyrene, which have to be removed completely leaving a substantial amount of porosity, *low-loss* preceramic binders offer the potential of high-density green compacts. The cure of the preceramic polymer binders provides a significantly higher green strength compared to conventional polymeric binders, their high ceramic yield of up to 90 wt% reduces the volume of decomposition products which have to be released, and rheology can be varied in a wider range allowing a variety of processing techniques for shaping the articles. Si_3N_4 bodies were prepared with polysilazane binders [10] and Al_2O_3/SiC-nanocomposites were obtained using polycarbosilane as a source for SiC nanoparticles [11]. Polysilazanes and polysiloxanes served as binders for sintered SiC monoliths because they afford stable SiC/C ceramic chars at temperatures below the sintering temperature of SiC [12]. Preceramic polymers have been applied as low-loss binders in powder metallurgy for *in situ* formation of dispersed ceramic phases in the metal matrix [13].

Attempts have been made to manufacture parts entirely of preceramic polymers [14–17]. High levels of uncontrolled shrinkage, pore and flaw formation, as well as high costs have limited application to ceramic fibers, coatings, and foams [18–21]. Using inert fillers dispersed in the polymer phase ceramic monoliths and fiber-reinforced composite bodies were perepared which, however, required multiple infiltration processes to reach acceptable mechanical properties [22–25]. Active filler controlled reaction pyrolysis was shown to be successful in manufacturing ceramic components of variable composition and high dimensional accuracy [26–35]. During pyrolysis in an inert atmosphere (Ar), the cured polymer precursor decomposes to an amorphous ceramic residue with a composition in the system Si—O—C—N—B. The solid residue and gaseous decomposition products (hydrocarbons C_xH_y) may react with fillers based on metallics or intermetallics such as Ti, Mo, W, Nb, V, B, $CrSi_2$, $MoSi_2$, etc. to form carbide, nitride, oxide, boride or mixed phases thereof.

In the following, the fundamental aspects of filler to polymer interaction during manufacturing of ceramics from filler-loaded preceramic polymers will be discussed. Both active as well as inert fillers may be used to reduce overall dimensional changes associated with the thermal degradation of the polymer phase. Additionally, the fillers have a strong impact on resulting properties such as mechanical, electrical, and thermal behavior.

5.2.2 PRECERAMIC POLYMER PYROLYSIS

A large variety of presursors for ceramic products in the compositional systems Si—N, Si—C—N, Si—B—N, Si—B—N—C, Si—Ti—C, Si—O—C, Si—C—Al—O—N have been synthesized, most of them an a laboratory scale (for review, see, e.g. Refs [2, 3, 6, 7, 14, 16, 19, 20, 36–40]). Figure 5.2.1 represents a scheme of major silicon-containing preceramic polymer systems.

After shaping, the low viscous preceramic polymer has to be cured to form a non-meltable thermoset. Curing may be achieved by thermal activated condensation or hydrosilation typically in the temperature range of 150–300°C. Above 400°C, rearrangement and radical reactions result in cleavage of chemical bonding and release of organic functional groups. The organic–inorganic transition

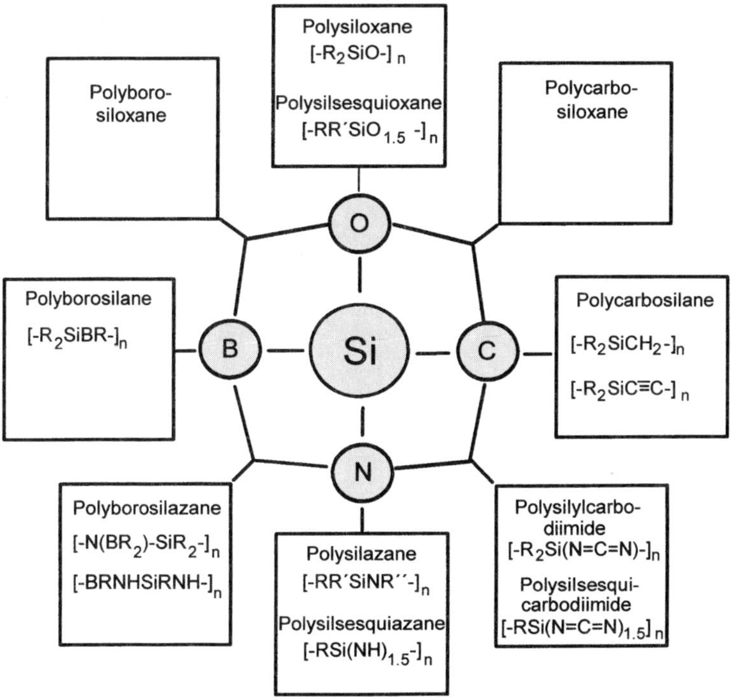

FIGURE 5.2.1 Silicon-containing preceramic polymers used for ceramic synthesis.

which is almost terminated at 800–1000°C is associated with a weight loss of typically 10–30 wt%. Decomposition reactions result in the release of gaseous decomposition products, R′, and in the formation of an amorphous, inorganic ceramic residue

$$[R_2SiO_{1.5}]_n \longrightarrow Si{-}O{-}C_{am} + R' \longrightarrow SiC + SiO_2 + C \quad (1a)$$

$$[R_2SiNH]_n \longrightarrow Si{-}N({-}C)_{am} + R' \longrightarrow Si_3N_4 \quad (1b)$$

$$[R_2SiCH_2]_n \longrightarrow Si{-}C_{am} + R' \longrightarrow SiC \quad (1c)$$

$$[R_2SiBNHR_3]_n \longrightarrow Si{-}B{-}N{-}C_{am} + R' \longrightarrow SiBN_3C \quad (1d)$$

where R=H, CH_3, C_2H_5, C_6H_5, CH=CH_2, etc. At temperatures above 400°C an open-pore channel network is formed that, upon further heating up to 800–1000°C can diminish (transient porosity). Depending on the polymer composition precipitation of excessive carbon in the form of turbostratic carbon clusters and ribbons [9] and crystallization of SiC, Si_3N_4, SiO_2, B_4C, BN, etc. occurs at different temperatures: 1000–1200°C in Si–C [41] but >1600°C in Si—B—N—C [42]. Amorphous Si—B—N—C ceramics were found to exhibit exceptional creep resistance even at temperatures up to 1500°C [43].

Compared to oxygen-poor preceramic polymers in the system Si—B—N—C, which are particularly interesting for applications at very high temperatures up to 1500°C (aerospace) ceramics derived from Si—O—C(—N—B) precursors are limited to applications at temperatures below approximately 1200°C. However, they offer an industrial availability, good handling and storage behavior, and can easily be manufactured in industrial processes. Their molecular architecture can be modified in a very wide range and hence the processing behavior can be tailored. Thermal degradation of poly(alkylsiloxane) involves the following basic steps:

$$\text{Curing:} \quad [RR'SiO_{2/2}]_m \longrightarrow [RSiO_{3/2}]_n + \text{gas } (H_2O, ROH, \text{etc.}) \qquad (2a)$$

$$\text{Pyrolysis:} \quad [RSiO_{3/2}]_n \longrightarrow Si{-}O{-}C_{am} + \text{gas } (C_xH_y, CO_2, \text{etc.}) \qquad (2b)$$

$$\text{Crystallization:} \quad Si{-}O{-}C_{am} \longrightarrow Si\text{–}O\text{–}C + C_{turbostratic} \rightarrow SiC + SiO_2 + C \qquad (2c)$$

In the amorphous structure of $Si{-}O{-}C_{am}$, part of the carbon is bonded to silicon-linking tetrahedral $SiCO_{3/2}$ and $SiC_2O_{2/2}$ species [44], and excessive carbon forms clusters of trubostratic carbon (basic structural units, BSUs) at lower temperatures and graphitic carbon ribbons at higher temperatures [9].

5.2.3 FILLER-LOADED PRECERAMIC POLYMERS

Basic limitations for the production of bulk components from polymeric precursor systems result from the extraordinary high volume shrinkage and a pronounced density increase following the polymer-to-ceramic conversion. For example, density typically increases by a factor of 2–3 from the precursor (1 g/cm^3) to the ceramic residue (SiO_2, 2.2–2.6 g/cm^3; Si_3N_4 and SiC, 3.0–3.2 g/cm^3), and volume shrinkage may exceed 50% [30]. Volume change upon polymer to ceramic conversion can be significantly reduced by loading the polymer precursor with inert or reactive filler particles. In the case of inert fillers, the fraction of polymer is decreased and accordingly the volume shrinkage is reduced. In the case of reactive fillers volume shrinkage of the polymer phase may be compensated by an equivalent expansion of the filler upon reaction with solid or gaseous decomposition products of the precursor [45]. Reduction of shrinkage is governed by an increase of ceramic yield of the polymer phase, formation of a rigid network of the filler reaction product and by expansion of the filler reaction product. Reactivity of the filler may involve complete conversion of the filler particle or be restricted to the

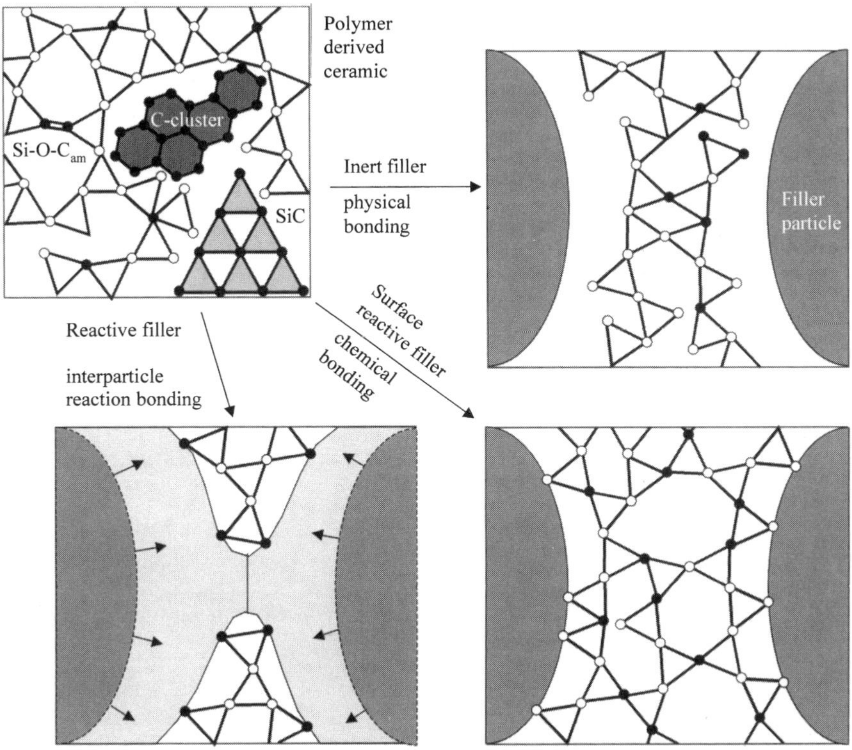

FIGURE 5.2.2 Scheme of polymer/filler interaction during heating of a polysiloxane – inert filler mixture.

surface making a chemical (primary) bonding to the polymer phase possible. Figure 5.2.2 shows a scheme of the filler – polymer interaction for polysiloxane as the polymer precursor system. While physical bonding is dominated by residual stresses due to polymer shrinkage and thermal expansion misfit, surface reaction involves the formation of primary bonds via O-bridges, for example. The filler reaction accounts for the formation of solid interparticle bonding and the generation of an interconnected filler network microstructure.

5.2.3.1 Control of Dimensional Change upon Pyrolysis

Reaction bonding of an inert or reactive filler powder by a preceramic polymer involves a series of complex structural decomposition and rearrangement

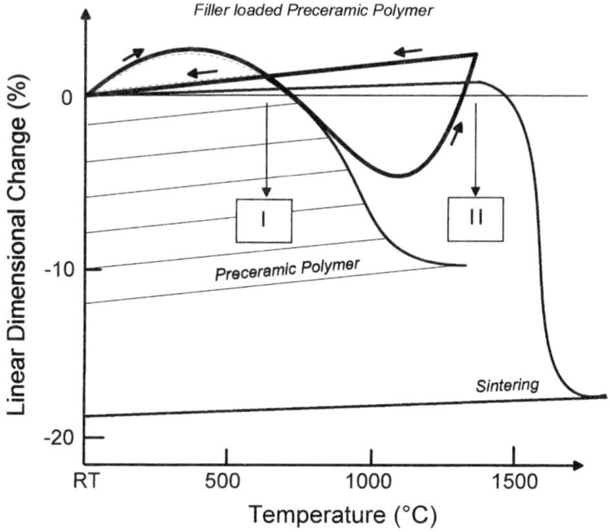

FIGURE 5.2.3 Dimensional changes during pyrolysis of a preceramic polymer compared to powder sintering. I and II denote critical temperatures to achieve near net shape manufacturing with filler loaded preceramic polymers: (I) partial pyrolysis with inert fillers; (II) full pyrolysis with reactive fillers.

reactions associated with pronounced density and volume changes which give rise for a characteristic expansion–shrinkage behavior during heating (Fig. 5.2.3). In the case of the filler reaction product having a greater specific volume compared to the starting filler phase, expansion of the filler may be used for partial compensation of the polymer to ceramic shrinkage (II in Fig. 5.2.3) [45]. Although total dimensional changes could be reduced to a linear shrinkage of less than 1–5% [46], dimensional accuracy for many future applications such as in microtechnical devices, hybrid composite components, etc. is still not sufficient. Furthermore, during heating extended volume changes due to selective shrinkage and expansion processes result in transitional strains which may cause deformation or even cracking of components. Reduction of dimensional scattering may be obtained by loading the polymer with non-reacting fillers. Dimensional accuracy is achieved by balancing thermal expansion and polymer shrinkage during low temperature treatment, for example, partial pyrolysis (I in Fig. 5.2.3). An example for I is the partial pyrolysis of poly(methylsiloxane) filled with SiC powder at temperatures below 800°C

$$[CH_3SiO_{1.5}]_n + SiC_{Filler} \xrightarrow{<800^\circ C} Si{-}O{-}C{-}H_{amorphous} + SiC + gas \qquad (3)$$

whereas pyrolysis of same polymer filled with boron powder in nitrogen atmosphere at temperatures above 1400°C represents II

$$[CH_3SiO_{1.5}]_n + B_{Filler} + N_2 \xrightarrow{>1400°C} SiC + SiO_2 + C + BN + gas \quad (4)$$

5.2.3.2 Active Fillers

During heating, a transient open pore system is generated at 400–800°C where the preceramic polymer reaches a maximum in the decomposition rate. Pore diameters of up to 500 μm were observed which provide the transportation path for decomposition species leaving the component [46]. Simultaneously gaseous species from the polymer as well as from a reactive atmosphere (e.g. N_2, CH_4, etc.) may be easily transported to the surface of the active filler. The filler particle reaction can be described according to the shrinking core model where the reaction starts at the particle surface and continues from the surface to the core until the whole initial particle volume is consumed [30]. Typical grain sizes of reactive filler powders are 1–10 μm. Thus, dispersion of reactive metals, intermetallics or ceramic phases such as Ti, Cr, V, Si, B, $CrSi_2$, $MoSi_2$, AlN, etc. results in a shape-invariant precursor–ceramic conversion. Based on the polymer/active-filler approach a variety of ceramic microcomposite materials with matrices having a composition in the system Si–C–N–O and a variety of filler reaction product phases have been prepared such as from polysilane/B and Si powders [28], polycarbosilane/nano-Co particles [47], polysilazane/Si powders [33], polysiloxane/Si and Al powders [29, 48], and polysiloxane/Ti, Nb, Mo, B, $CrSi_2$, $MoSi_2$, powders [34, 49–51].

Assuming isotropic dimensional change, strain associated with reaction and pyrolysis, ε, can be expressed by the total volume change $\psi_{total}(\equiv (V_{product}/V_{educt}) - 1)$

$$\varepsilon = (\psi_{total} + 1)^{1/3} - 1 \quad (5)$$

For the total volume change ψ_{total} an expression was derived which takes into account the filler volume fraction, V_{filler}, and the specific volume changes upon pyrolysis and reaction of the polymer, $\psi_{polymer}$, and the filler phase, ψ_{filler} [45]

$$\psi_{total} = \left(1 - \frac{V_{filler}}{V_{filler}^{max}}\right)(\psi_{polymer} - 1) + V_{filler}(\psi_{filler} - 1) + V_{pore} \quad (6)$$

V_{filler}^{max} is a critical filler volume fraction which corresponds to the maximum packing density of the filler powder (corresponding to the tapping density ≈0.5)

TABLE 5.2.1 Specific Molar Volume Changes of Potential Reactive Filler Systems. Reactants are Solid Carbon, Gaseous Hydrocarbons or Reactive Gas Atmospheres

Reactant	C(s)	C(g)	N_2(g)	O_2(g)
Ti	−24	+14	+8	+78
Zr	−21	+9	+3	+40
Mo	−5	+22	–	–
Al	+9	+53	+26	+28
B	−7	+20	+142	–
Si	−30	+7	+13	+88
$MoSi_2$	−4	+48	–	–
$CrSi_2$	−7	+54	+23	+69
$TiSi_2$	−10	+47	+53	+171

and V_{pore} is the residual porosity. Some of important filler reaction schemes (n corresponds to the major valency state of the metal filler) are:

Carburization	solid C	$F + xC$	$\longrightarrow FC_x$
	gaseous C	$F + xCH_4$	$\longrightarrow FC_x + 2xH_2$
Oxidation	solid SiO_2	$4/nF + SiO_2$	$\longrightarrow F_{4/n}O_2 + Si$
	gaseous O_2	$4/nF + O_2$	$\longrightarrow F_{4/n}O_x$
Nitridation	gaseous N_2	$6/nF + N_2$	$\longrightarrow F_{6/n}N_2$

Table 5.2.1 gives typical values of the molar volume change of the filler ψ_{filler} associated with these reactions. Equation 6 can be used to select appropriate filler volume fractions to achieve zero shrinkage upon polymer pyrolysis ($\psi_{total} = 0$)

$$V_{filler} = \frac{(\psi_{polymer} - 1) - V_{pore}}{\left((\psi_{pore} - 1)/V_{filler}^{max}\right) - \left(\psi_{filler} - 1\right)} \tag{7}$$

For typical values of $\psi_{polymer} = 0.25$–0.5 and $V_{pore} = 0.2$–0.3, filler volume fractions of $V_{filler} = 0.15$–0.3 are necessary to achieve zero shrinkage (net-shape processing) when a reactive filler with $\psi_{filler} = 1.2$–1.5 (e.g. Cr, Nb, Al, B, $CrSi_2$, etc.) is used [46]. Figure 5.2.4 shows critical volume fractions of a selection of active fillers to achieve zero shrinkage as calculated from Eq. 7. Generally, zero shrinkage could only be obtained in the presence of a certain amount of porosity, V_{pore}. Figure 5.2.5 summarizes experimental data of linear shrinkage and porosity observed in a variety of examined polysiloxane/filler systems [46].

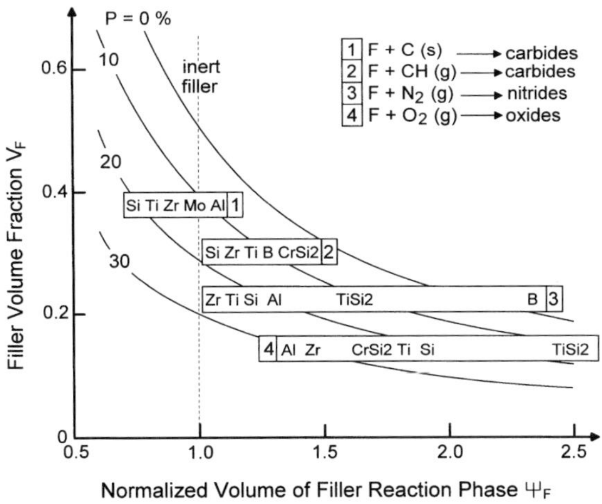

FIGURE 5.2.4 Filler volume fractions needed to achieve net shape pyrolysis as a function of the filler expansion upon reaction (normalized volume of filler reaction phase).

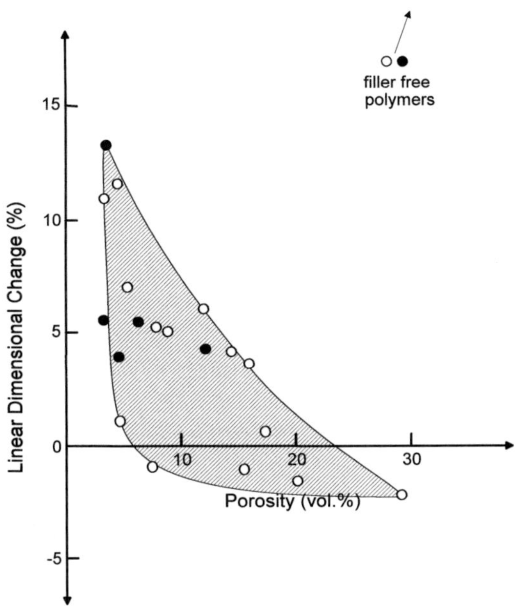

FIGURE 5.2.5 Linear dimensional change and porosity of reactive filler loaded polymethyl- (o) and polyphenyl- (•) siloxanes pyrolyzed at 1200–1400°C in nitrogen.

5.2.3.3 Passive Fillers

In the case of passive fillers, which do not react with the polymer upon pyrolysis, dimensional change will be dominated by the thermally induced structural rearrangment and decomposition reactions in the polymer. While thermal expansion of both the polymer and the filler phases will dominate initial expansion behavior below the onset of polymer decomposition, shrinkage of the polymer phase above approximately 200–300°C will reduce total expansion and finally turn it to a pronounced shrinkage at temperatures above 500–700°C. Thus, the total dimensional change during heating, $\varepsilon(T)_{\text{total}}$, may be expressed by assuming superposition of expansion, $\varepsilon(T)_{\text{expansion}}$, and shrinkage, $\varepsilon(T)_{\text{shrinkage}}$, processes

$$\varepsilon(T)_{\text{total}} = \varepsilon(T)_{\text{expansion}} + \varepsilon(T)_{\text{shrinkage}} \tag{8}$$

Thermally induced expansion $\varepsilon(T)_{\text{expansion}}$ during heating the polymer/filler mixture from T_{start} to the temperature T_{end} is a function of the temperature dependent coefficient of thermal expansion of the composite, $\alpha(T)_{\text{composite}}$,

$$\varepsilon(T)_{\text{expansion}} = \int_{T_{\text{start}}}^{T_{\text{end}}} \alpha(T)_{\text{composite}}\, \mathrm{d}T \tag{9}$$

Due to the complex reaction process taking place during heating and the unknown bulk and shear moduli a simple rule of mixture was applied, at least to a first approxiamtion, to describe the mean thermal expansion of the polymer/filler mixture

$$\alpha(T)_{\text{composite}} \approx \alpha(T)_{\text{polymer}} V_{\text{polymer}} + \alpha(T)_{\text{filler}}(1 - V_{\text{polymer}}) \tag{10}$$

V_{polymer} denotes the volume fraction of polymer phase and $(1-V_{\text{polymer}})$, the volume fraction of filler in a two-phase mixture. From room temperature to 600°C, the thermal expansion coefficient of a non-reacting filler, $\alpha(T)_{\text{filler}}$, is supposed to be nearly independent on temperature when no phase transitions occur. Thermal expansion coefficient of the polymer phase, however, exhibits an irreversible change from the high level of the polymer ($\alpha_{\text{polymer}} > 300 \times 10^{-6}\,\text{K}^{-1}$ at room temperature) down to the extremely small value of the amorphous Si–O–C pyrolysis product ($\alpha_{\text{SiOC}} < 1 \times 10^{-6}\,\text{K}^{-1}$) which is similar to the values for amorphous SiO_2–glass (Fig. 5.2.6). Sigmoidal shape of the α versus T curve may well be expressed by a relation of the following type:

$$\alpha(T) \approx \alpha_{\text{polymer,max}} \left[\exp\left(\frac{(T - T_0)R}{W^*}\right) + 1\right]^{-1} + \alpha_{\text{SiOC,min}} \tag{11}$$

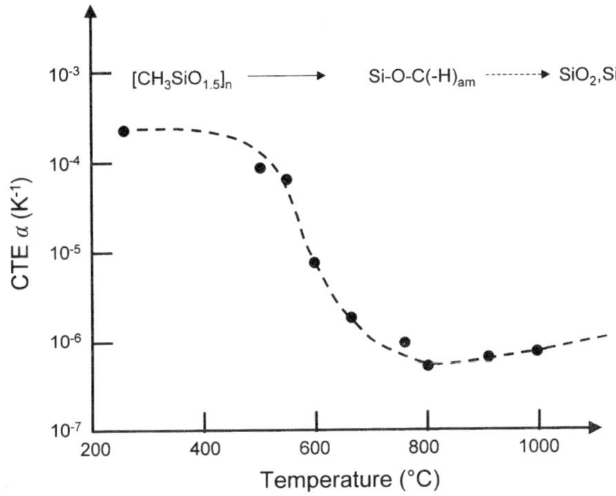

FIGURE 5.2.6 Change of coefficient of linear thermal expansion of polymethylsiloxane during heating in argon atmosphere.

where $\alpha_{\text{polymer,max}}$ and $\alpha_{\text{SiOC,min}}$ are the upper ($300 \times 10^{-6}\,\text{K}^{-1}$) and lower ($5 \times 10^{-7}\,\text{K}^{-1}$) boundary values, respectively. Taking R as the gas constant (8.31451 J/mol) and W^* as a characteristic energy (200 J/mol), the exponential term represents an energy ratio which describes the sigmoidal change of $\alpha(T)$ around a characteristic temperature T_0 (570°C). Although W^* and T_0 are fit parameters, both parameters may be considered as characteristic values for the structural changes which occur in the polymer phase upon thermal degradation.

Due to the density change from the polymer to the ceramic phase, a tremendous irreversible volume shrinkage is usually found during heating which may attain values up to 80 vol.%. Similarly to the description of the change of the coefficient of thermal expansion the sigmoidal curve of linear shrinkage may be approximated by

$$\varepsilon(T)^*_{\text{shrinkage}} \approx \varepsilon_{\text{shrinkage,max}} \left[\left(\exp\left(\frac{(T - T_0)R}{W^{**}} \right) + 1 \right)^{-1} - 1 \right] \tag{12}$$

and

$$\varepsilon(T)_{\text{shrinkage}} = \varepsilon(T)^*_{\text{shrinkage}} V_{\text{polymer}} \tag{13}$$

$\varepsilon_{\text{shrinakge,max}}$ is the total linear shrinkage associated with the polymer to ceramic conversion at temperatures above 1200°C (0.27 for the polymethylsiloxane examined). W^{**} and T_0 have similar meanings as in Eq. 11. Using Eqs 11–13, dimensional change upon heating of a polymer/inert filler mixture can be simulated. A variety of ceramic powders can be selected as inert fillers, which allow tailoring of thermal expansion of the resulting reaction product. Selection of optimized heat treatment conditions for achieving both zero shrinkage and a thermal expansion similar to steel ($13 \times 10^{-6}\,K^{-1}$) results in the two requirements

$$\varepsilon(T_c) = \alpha(T_c)_{composite} T_c \quad (14)$$

and

$$\alpha(T_c)_{composite} = \alpha_{steel}\ (T_c) \quad (15)$$

Solving Eqs 14 and 15 simultaneously (MATHCAD 6.0) finally results in a ε versus T relation the parameter of which, for example, $\alpha(rt)_{Polymer} = 320 \times 10^{-6}\,K^{-1}$, $\alpha(800°C)_{SiOC} = 5 \times 10^{-7}\,K^{-1}$, $\alpha_{filler} = 8 \times 10^{-6}\,K^{-1}$, $V_{polymer} = 0.68$, $T_0 = 610°C$, $\varepsilon_{polymer} = 0.293$, $W^* = 210\,J\,mol$, $W^{**} = 1100\,J\,mol$, were adjusted to fit the experimental curve and to yield a critical temperature $T_c = 675°C$ (Fig. 5.2.7). It may be seen that experimental heating was finished at 630°C which gave rise for a residual expansion of approximately 0.5%. Following the simulation results, a peak temperature of 675°C was selected to manufacture complex shaped components (vacuum pump rotor) which showed a linear deviation of the final dimension of less than 0.05% from the initial geometry [35].

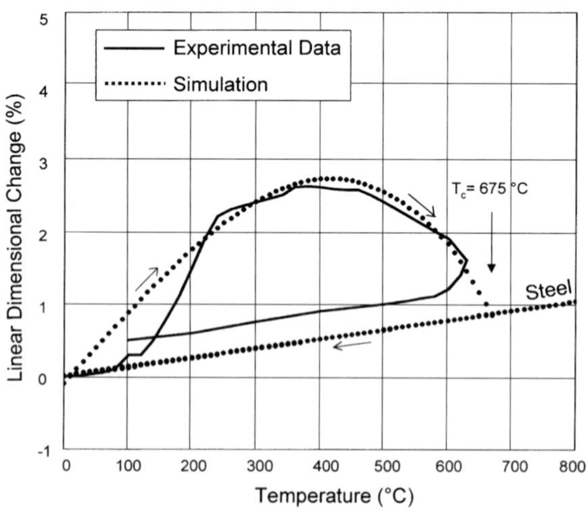

FIGURE 5.2.7 Simulation of dimensional change according to Eqs 8–15 to achieve net shape heat treatment and a CTE similar to steel.

5.2.4 MATERIALS AND PROPERTIES

A large variety of microstructures and materials can be generated by combining different preceramic polymers with active and passive fillers (Fig. 5.2.8). Composite materials from reactive filler containing polymer mixtures are characterized by a three-dimensional interconnected network of the filler reaction phase (carbide, nitride, oxide, etc.) embedded in a nanostructured polymer-derived Si—O—C—N matrix. Passive fillers can be dispersed as nanoparticles along grain boundaries (intergranular) or inside the grains (intragranular). Precipitation of carbon may result in the formation of percolation network structures and controlled nucleation of pores in the low-viscosity polymer may yield low-density ceramic foams.

The material properties of the pyrolyzed ceramic microcomposites are dominated by the interaction of the various interpenetrating networks in the hierarchical microstructure, for example, the filler network and the precipitation network of turbostratic carbon in the polymer-derived bonding phase. Generally, the material properties may be distinguished into properties varying continously with the volume fraction of filler phase and into properties, which exhibit a non-linear change at a critical percolation threshold. Figures 5.2.9 and 5.2.10 show the fracture strength and toughness, respectively, of various

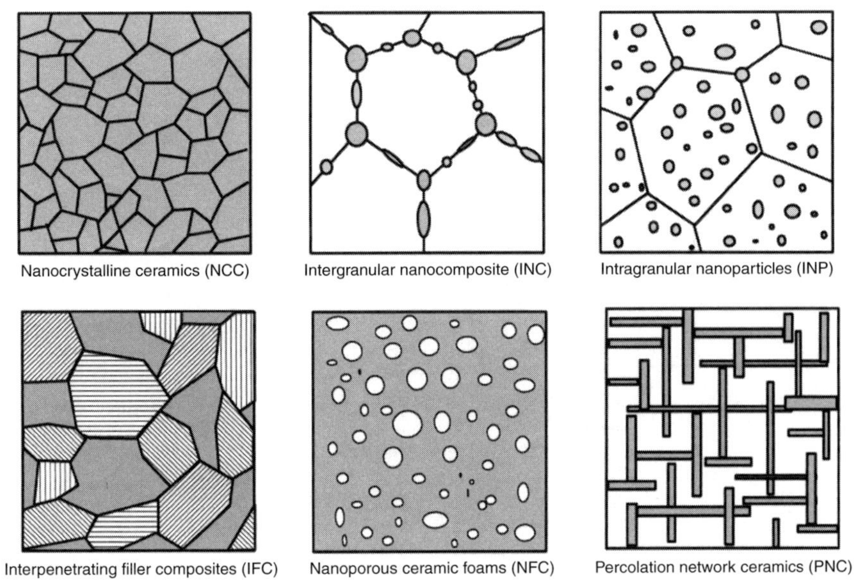

FIGURE 5.2.8 Ceramic microstructures to be obtained from preceramic polymer/filler systems.

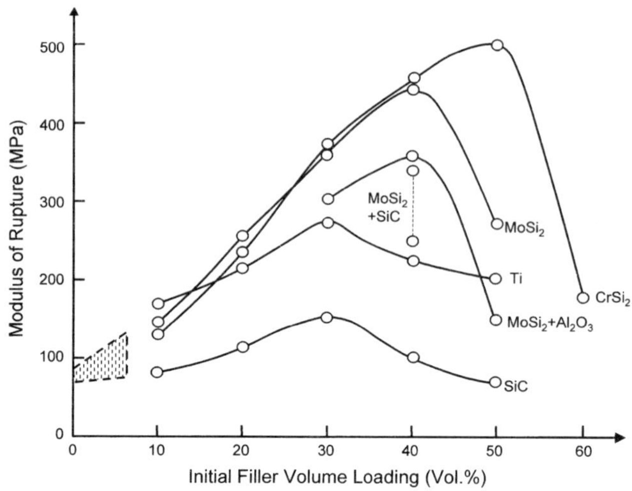

FIGURE 5.2.9 Fracture strength of pyrolyzed polysiloxane/filler systems as a function of filler loading.

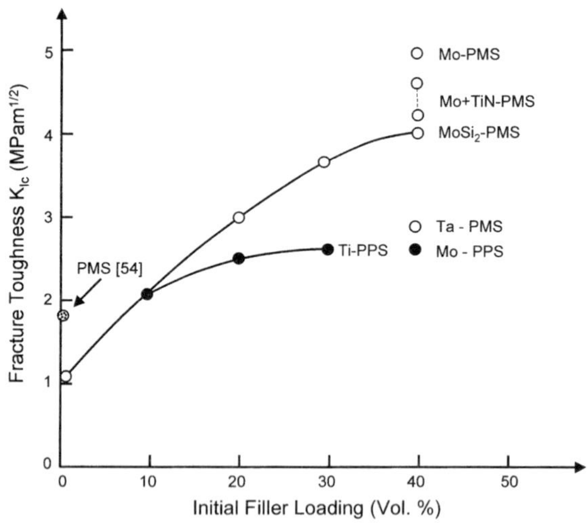

FIGURE 5.2.10 Fracture toughness of pyrolyzed polysiloxane/filler systems as a function of filler loading.

polysiloxane/filler-derived ceramics as a function of initial filler volume fraction. Modulus of rupture (MOR) increases almost linearly with increasing filler content up to a maximum typically at 40–50 vol%. At higher filler fractions, formation of large filler particle clusters as well as excessive porosity results in a pronounced strength loss. Fracture toughness, K_{Ic}, was found to vary in the range of 2–5 MPa $m^{1/2}$. Evaluation of thoughness and strength as a function of filler volume fraction and particle size according to the model of Hasselmann and Fulrath [52] indicated a distinct correlation between mean particle distance and effective crack length. It was found that in the filler loading range $V_{filler} = 0.1–0.5$, strength increased with decreasing normalized interparticle distance (ratio of mean interparticle distance, d_{ip}, to the particle size, d_p) from $10d_p$ at $V_{filler} = 0.1$ to less than $2d_p$ at $V_{filler} = 0.5$ according to $MOR \propto (d_{ip}/d_p)^{-1/2}$ [53].

In addition to the variation of filler volume fraction and particle size total porosity, V_{pore}, has a distinct effect on strength. Extrapolation of strength versus residual porosity using the relation $MOR = MOR_0 \exp\ (-5V_{pore})$ indicates a strength for the dense material, MOR_0, of approximately 650 MPa [35]. An ultimate strength of even 790 MPa was measured when a filler-free polymethylsiloxane, $[CH_3SiO_{1.5}]$, was pyrolyzed under controlled conditions to yield a highly dense Si—O—C fiber [54].

The coefficient of thermal expansion (CTE) is governed by the Si—O—C matrix expansion and the filler loading (Fig. 5.2.11). For inorganic Si—O—C pyrolysis residue from polymethylsiloxane a CTE ranging from approximately $3.1 \times 10^{-6}\ K^{-1}$ [44] to $<1 \times 10^{-6}\ K^{-1}$ (own measruements) was found. Accordingly, loading the polymer with fillers which have a higher CTE allows tailoring of the composite behavior. Partially pyrolyzed polysiloxane, which has a CTE in

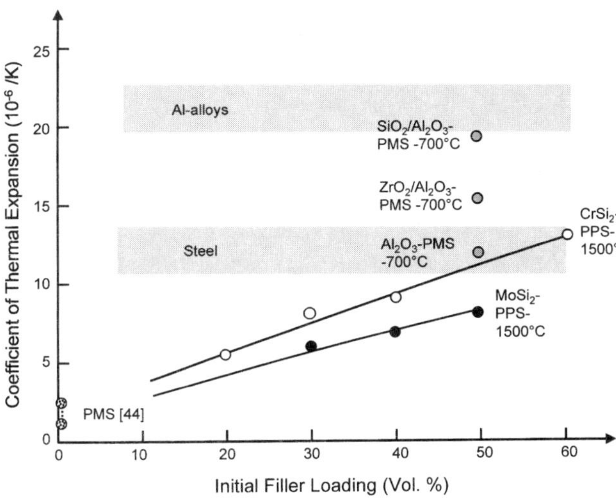

FIGURE 5.2.11 Coefficient of thermal expansion of polysiloxane/filler systems after pyrolysis at different temperatures.

the transition region between metal–organic polymer (high CTE) and inorganic amorphous Si—O—C network (Fig. 5.2.6) is of particular interest for design of materials with a high CTE. Due to a thermal expansion similar to ferritic steel (CTE 11–13 × 10^{-6} K^{-1}), substitution of thermally loaded metallic components is facilitated.

Precipitation of an interconnecting carbon network or a filler network can cause non linear property changes such as electrical insulator-conductor transition when a percolation threshold is exceeded (Fig. 5.2.12). Below the percolation threshold, the electrical conductivity in filler-free Si—O—C pyrolysis residue can be described according to Motts mechanism of variable-range-hopping of localized charge carriers. Above the percolation threshold regular electron band conduction due to intrinsic conductivity of turbostratic carbon precipitated at elevated temperatures (above approximately 800°C) predominates [55]. Percolation threshold was found to depend on the shape of the high conducting cluster particles and typically varies between 0.01 and 0.6 in composite materials [56]. For spherical particles, values of 0.2–0.3 were derived from site percolation models [57]. Significantly lower values can be expected for elongated, fiber-like particles as it was shown by Monte Carlo studies [58]. TEM

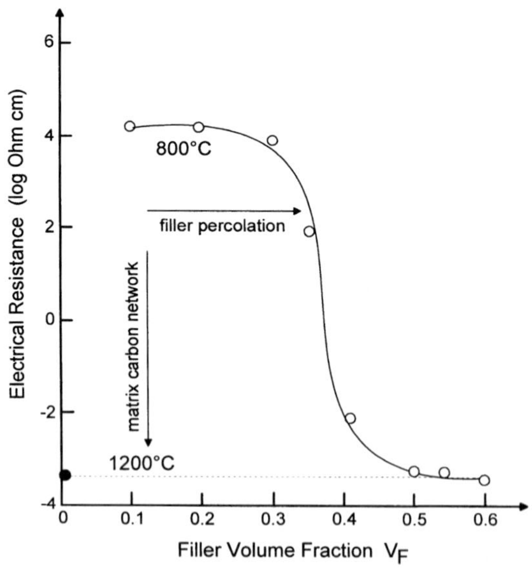

FIGURE 5.2.12 Electrical dc-resistance of $MoSi_2$-containing polysiloxane after different pyrolysis temepratures. At 800°C, $MoSi_2$ behaves as an inert filler dispersed in an insulating Si—O—C—H matrix (filler percolation network), at 1200°C precipitation of carbon dominates resistance behavior (carbon percolation network).

investigations of pyrolyzed Si—O—C ceramics have shown a ribbon-like carbon morphology [55] for which percolation thresholds of 0.1–0.2 were calculated using general effective media (GEM) theory [59]. Adding filler materials with high conductivity such as metals or intermetallics offers the possibility to obtain composite ceramics of high conductivity even at low heat treatment temperatures. For example, polymethoxysilane loaded with 40–60 vol% of MoSi (mean particle size $2\,\mu m, \sigma \approx 500\,\Omega^{-1}\,cm^{-1}$) was processed into heating elements which obtained a conductivity of $0.01\,\Omega^{-1}\,cm^{-1}$ at 600°C and $10^4\,\Omega^{-1}\,cm^{-1}$ at 1400°C [60].

5.2.5 APPLICATIONS

Filler-loaded preceramic polymers offer a great variety of chemical and structural modifications which allow tailoring of properties in a wide range. Figure 5.2.13 summarizes potential applications based on powders, fibers, infiltrants, coatings, foams and bulk materials derived from polymer/filler-systems.

Low-viscosity polymer/filler suspensions can be used as infiltrant or fiber coating to generate the matrix in fiber reinforced ceramic composite materials. A ceramic matrix for carbon-fiber-reinforced ceramic matrix composites was formed through pyrolysis of polysiloxane/boron mixtures without multiple

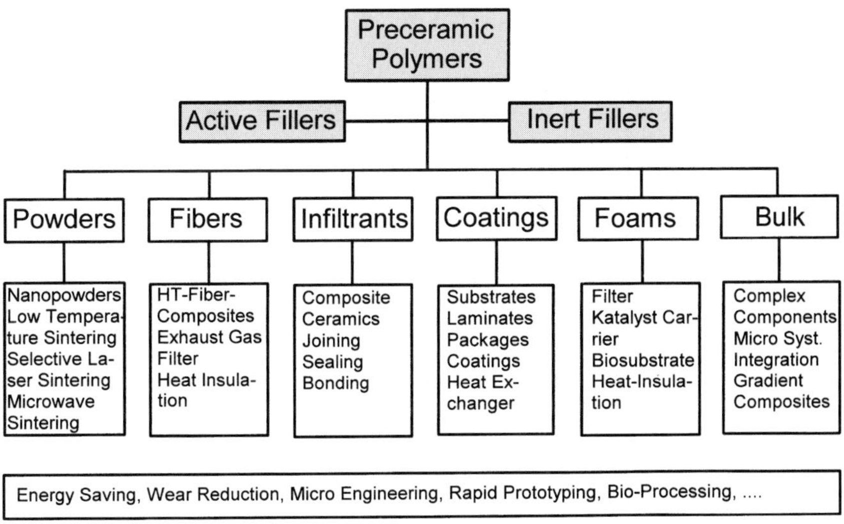

FIGURE 5.2.13 Survey of potential fields of application for filler-loaded polymer derived-ceramics.

infiltration steps [51]. At 1300°C, SiC, B_4C, BN and SiOC were compatible with the C-fiber resulting in a tensile strength of 250 MPa. While oxygen-containing Si—O—C matrix materials derived from low-cost polysiloxanes are restricted to service temperatures below 1200°C, oxygen-poor Si—B—C—N systems derived from polysilazanes and borosilazanes offer a great potential for significantly higher application temperatures. High-temperature resistant and damage tolerant light weight structures from polymer-derived ceramic matrix composites with C-, SiC- and SiBNC-fibers are expected to find increasing use in space and aircraft structures (engine components, thermal protection devices), energy technology (combustion chamber of gas turbines, hot gas tubes, exhaust purification filters), and transportation technology (brakes).

Polymer-derived amorphous and crystalline coatings are of particular interest for protecting carbon, porous non-oxide ceramics, and refractory metals against corrosion, oxidation and wear. Polymethoxysilane loaded with $MoSi_2$ powder served as a self-healing oxidation protection system on porous recrystallized SiC components for high-temperature furnace components [61]. Open, as well as closed cell foams with porosities up to 90% and pore diameters ranging from 200 to 400 μm were prepared from polysiloxane polymers by physical and chemical foaming reactions [21]. Tailored foams that contain functional fillers to provide certain chemical and physical properties can be used as insulation devices (closed cells) or as high-temperature resistant filter device (open cells) in metallurgy, energy and environmental technologies. The open cell foams are of particular interest to serve as a susbtrate for interpenetrating phase composites materials [62], which are infiltrated with an organic, inorganic or metallic melt. Due to the rheological properties of polymer-filler suspensions which can be varied in a wide range, tape casting can be used to manufacture substrates and laminated structures. Selective laser pyrolysis of polymer-filler layers offers novel possibilities to fabricate three-dimensional components according to the solid free form technologies [63].

5.2.6 CONCLUSIONS

Incorporation of fillers offers the possibility to improve the mechanical as well as the physical and chemical properties of polymer-derived ceramics. Preceramic polymers containing active fillers can be transformed to ceramic microcomposites with reduced shrinkage during pyrolysis. Volume invariant polymer–ceramic conversion is attributed to the chemical reaction of the reactive filler with solid or gaseous decomposition products of the polymer resulting in the formation of a network of the filler reaction phase. Non-linear property changes occur when critical percolation limits for network formation are exceeded. Reduction of the pyrolysis temperature results in partially converted organic–inorganic hybrid materials. Compared to high temperature polymers, such novel materials have an improved temperature stability whereas compared to conventional ceramics they offer better geometrical accuracy at significantly lower manufacturing temperatures. Tailoring the hierarchical microstructures offers new possibilities for polymer-derived ceramics with improved properties for application in the fields of machinary, electronics, and medicine.

ACKNOWLEDGMENTS

Financial support from Deutsche Forschungsgemeinschaft (DFG) is gratefully acknowledged. The author thanks P. Buhler for providing results of thermal expansion measurement.

REFERENCES

1. Wills, R. R., Markle, R. A., and Mukherjee, S. P. (1983). Siloxanes, silanes, and silazanes in the preparation of ceramics and glasses. *Bull. Am. Ceram. Soc.* **62**: 904.
2. Wynne, K. J., and Rice, R. W. (1984). Ceramics via polymer pyrolysis, *Ann. Rev. Mater. Sci.* **14**: 297.
3. Pouskouleli, G. (1989). Metallorganic compounds as preceramic materials I. Non-oxide ceramics. Part II. Oxide ceramics, *Ceram. Int.* **15**: 213, 255.
4. Maya, L. (1990). Derivatization to preformed polymers to produce ceramic precursors. *J. Am. Ceram. Soc.* **73**: 2714.
5. Lavendrine, A., Bahloul, D., Goursat, P., Choong Kwet Yive, N., Corriu, R., Leclerq, D., Mutin, H., and Vioux, A. (1991). Pyrolysis of polyvinylsilazane precursors to silicon carbonitride. *J. Eur. Ceram. Soc.* **8**: 221.
6. Bill, J., and Aldinger, F. (1995). Precursor-derived covalent ceramics, *Adv. Mater.* **7**: 775.
7. Corriu, R. J. (2000). Ceramics and nanostructures from molecular precursors. *Angew. Chem. Ind. Ed.* **39**: 1376.
8. Soraru, G. D., Babonneau, F., and Mackenzie, J. D. (1990). Structural evolutions from polycarbosilane to SiC ceramics. *J. Mater. Sci.* **25**: 3866.
9. Monthioux, M., and Delverdier, O. (1996). Thermal behavior of (organosilicon) polymer derived ceramics. V: Main facts and trends. *J. Eur. Ceram. Soc.* **16**: 721.
10. Schwartz, K. B., Rowcliffe, D. W., and Blum, Y. D. (1988). Microstructural development in Si_3N_4/polysilazane bodies during heating. *Adv. Ceram. Mater.* **3**: 320.
11. Su, B., and Sternitzke, M. (1995). A novel processing route for alumina/SiC-Nanocomposites by Si-polymer pyrolysis, *Fourth Euro Ceramics*, Vol. 4, p. 109, Bellosi, A., ed., Gruppo Editiorale Faenca Editirice, Faneza/I.
12. Atwell, W. H., Burns, G. T., and Zank, G. A. (1991). Silicon carbide preceramic polymers as binders for ceramic powders, *Inorganic and Organometallic Oligomers and Polymers*, p. 147, Harrod, J. F., and Laine, R. M., eds., Dordrecht: Kluwer.

13. Seyferth, S., and Czubarow, P. (1994). Application of preceramic polymers in powder metallurgy: their use as low-loss binders and for the in situ formation of dispersed ceramic phases in the metal matrix. *Chem. Mater.* **6**: 10.
14. Walker, B. E., Rice, R. W., Becher, P. F., Bender, B. A., and Coblenz, W. S. (1983). Preparation and properties of monolithic and composite ceramics produced by polymer pyrolysis. *Bull. Am. Ceram. Soc.* **62**: 916.
15. Riedel, R., Passing, G., Schönfelder, H., and Brook, R. J. (1992). Synthesis of dense silicon-based ceramics at low temperatures. *Nature* **355**: 355.
16. Riedel, R., Seher, M., Mayer, J., and Szabo, D. V. (1995). Polymer-derived Si-based bulk ceramics. Part I: Preparation, processing and properties. *J. Eur. Ceram. Soc.* **15**: 703.
17. Seitz, J., and Bill, J. (1996). Production of compact polysilazane-derived Si/C/N-ceramics by plastic forming. *J. Mater. Sci. Lett.* **15**: 391.
18. Cooke, T. F. (1991). Inorganic fibers—a literature review. *J. Am. Ceram. Soc.* **74**: 2959.
19. Lipowitz, J. (1991). Polymer-derived ceramic fibers. *Bull. Am. Ceram. Soc.* **70**: 1888.
20. Colombo, P., Paulson, T. E., and Pantano, C. G. (1997). Synthesis of silicon carbide thin films with polycarbosilane (PCS). *J. Am. Ceram. Soc.* **80**: 2333.
21. Colombo, P., and Modesti, M. (1999). Silicon oxycarbide ceramic foams from a preceramic polymer. *J. Am. Ceram. Soc.* **82**: 573.
22. Semen, J., and Loop, J. G. (1990). Structural ceramics derived from a preceramic polymer. *Ceram. Eng. Sci. Proc.* **11**: 1387.
23. Hurwitz, F. I., Gyekenyesi, J. Z., Conroy, P. J., and Rivera, A. L. (1990). Nicalon/siliconoxycarbide ceramic composites. *Ceram. Eng. Sci. Proc.* **11**: 931.
24. Sato, K., Suzuki, T., Funayama, O., Isoda, T., and Itoh, T. (1992). Preparation of carbon fiber reinforced composite by impregnation with perhydropolysilazane followed by pressureless firing. *Ceram. Eng. Sci. Proc.* **13**: 614.
25. Schwab, S. T., Graef, R. C., Blanchard, C. R., Pan, Y. M., Davidson, D. L., Maciel, G. E., Hawkins, B. L., Dec, S. F., Davis, M. F., and Lewis, R. (1994). Preceramic polysilazanes for structural applications. Ceram. *Eng. Sci. Proc.* **15**: 1093.
26. Jamet, J., Spann, J. R., Rice, R. W., Lewis, D., and Coblenz, W. S. (1984). Ceramic-fiber composite processing via polymer-filler matrices. *Ceram. Eng. Sci. Proc.* **5**: 677.
27. Seyferth, D., Bryson, N., Workman, D. P., and Sobon, C. A. (1991). Preceramic polymers as 'reagents' in the preparation of ceramics. *J. Am. Ceram. Soc.* **74**: 2687.
28. Jiang, Z., and Rhine, W. E. (1993). Preparation of nanophase SiC/Si_3N_4, SiC/B_4C and stochiometric SiC from mixtures of a vinylic polysilane and metal powders. *Mater. Res. Soc. Symp. Proc.* **286**: 437.
29. Colombo, P., Abdirashid, M. O., Gugielmi, M., Eposit, L. M. D., and Agostini, L. (1994). Preparation of ceramic composites by active-filler-controlled-polymer pyrolysis. *Mater. Res. Soc. Symp. Proc.* **346**: 403.
30. Greil, P. (1995). Active-filler-controlled pyrolysis of preceramic polymers. *J. Am. Ceram. Soc.* **78**: 835.
31. Yu, S. H., Riman, R. E., Danforth, S. C., and Leung, R. Y. (1995). Pyrolysis of titanium-metal-filled poly(siloxane) preceramic polymers: effect of atmosphere on pyrolysis product chemistry. *J. Am. Ceram. Soc.* **78**: 1818.
32. Walter, S., Suttor, D., Erny, T., Hahn, B., and Greil, P. (1996). Injection moulding of polysiloxane/filler mixtures for oxycarbide ceramic composites. *J. Eur. Ceram. Soc.* **16**: 387.
33. Lighfoot, A., Haggerty, J. S., and Rhine, W. E. (1996). Ceramic materices and monoliths synthesized from a combination of reactive polymers, fillers and atmospheres. *Ceram. Eng. Sci. Proc.* **17**: 653.
34. Dernovsek, O., Bressiani, J. C., Bressiani, A. H. A., Acchar, W., and Greil, P. (2000). Reaction bonded niobium carbide ceramics from polymer-filler mixtures. *J. Mater. Sci.* **35**: 2201.

35. Greil, P. (2000). Polymer derived engineering ceramics. *Adv. Eng. Ceram.* **2**: 339.
36. Peuckert, M., Vaahs, T., and Brück, M. (1990). Ceramics from organometallic polymers, *Adv. Mater.* **2**: 398.
37. Schmidt, W. R., Hurley, W. J., Doremus, R. H., Interrante, L. V., and Marchetti, P. S. (1991). Novel polymeric precursors to Si–C–Al–O–N ceramic composites. *Ceram. Trans. Adv. Compos. Ceram.* **19**: 19.
38. Ishikawa, T., Yamamura, T., and Okamura, K. (1992). Production mechanism of polytitano carbosilane and its conversion of the polymer into inorganic materials. *J. Mater. Sci.* **27**: 6627.
39. Baldus, H. P., and Jansen, K. (1997). Novel High Performance Ceramics—amorphous inorganic networks from molecular precursors. *Angew. Chem. Int. Ed.* **36**: 328.
40. Livage, J., Sanchez, C., and Babonneau, F. (1998). Molecular precursor routes to inorganic solids, *Chemistry of Advanced Materials*, p. 48, Interrante, L. V., and Hamden-Smith, M. J., eds., New York: Wiley-VCH.
41. Bouillon, E., Langlais, F., Pailler, R., Naslain, R., Cruege, F., Sarthou, J. C., Delpuech, A., Laffon, C., Lagarde, P., Monthioux, M., and Oberlin, A. (1991). Conversion mechanism of a polycarbosilane precursor into an SiC-based ceramic material. *J. Mater. Sci.* **26**: 1333.
42. Baldus, H. P., Wagner, O., and Jansen, M. (1992). Synthesis of advanced ceramics in the system Si–B–N and Si–B–N–C employing novel precursor compounds. *Mater. Res. Soc. Symp. Proc.* **271**: 821.
43. Christ, M., Thurn, G., Weinmann, M., Bill, J., and Aldinger, F. (2000). High-temperature mechanical properties of Si–B–C–N-precursor-derived amorphous ceramics and the applicability of deformation models developed for metallic glasses. *J. Am. Ceram. Soc.* **83**: 3025.
44. Renlund, G. R., Prochazka, S., and Doremus, R. H. (1991). Silicon oxycarbide glasses. Part I. Preparation and Chemistry; Part II. Structure and properties. *J. Mater. Res.* **6**: 2716, 2723.
45. Greil, P., and Seibold, M. (1992). Modelling of dimensional changes during polymer–ceramic conversion for bulk component fabrication. *J. Mater. Sci.* **27**: 1053.
46. Greil, P. (1998). Near net shape manufacturing of polymer derived ceramics. *J. Eur. Ceram. Soc.* **18**: 1905.
47. Wang, J., Song, Y., and Feng, C. (1998). Pyrolysis of polycarbosilane/nano-cobalt particles. *Ailiao Kexue Yu Gongcheng* **16**: 69.
48. Colombo, P., Abdirashid, M. O., Maddalena, A., and Marchetti, A. (1995). Aluminium powder—preceramic polymer mixtures for the preparation of ceramic composites, *Fourth Euro Ceramics*, Vol. 4, p. 117, Bellosi, A., ed., Gruppo Editiorale Faenza Editirice, Faneza/I.
49. Erny, T., Seibold, M., Jarchow, O., and Greil, P. (1993). Microstructure development of oxycarbide composites during active-filler-controlled polymer pyrolysis. *J. Am. Ceram. Soc.* **76**: 207.
50. Seibold, M., and Greil, P. (1993). Thermodynamics and microstructural development of ceramic composite formation by active-filler-controlled pyrolysis (AFCOP), *J. Eur. Ceram. Soc.* **11**: 105.
51. Suttor, D., Erny, T., Greil, P., Goedecke, H., and Haug, T. (1997). Fiber-reinforced ceramic-matrix composites with a polysiloxane/boron-derived matrix. *J. Am. Ceram. Soc.* **80**: 1831.
52. Hasselman, D. P. H., and Fulrath, R. M. (1966). Proposed fracture theory of a dispersion strengthened glass matrix. *J. Am. Ceram. Soc.* **49**: 68.
53. Walter, S. (1999). Microstructure and mechanical properties of polymer derived oxycarbide ceramics, PhD thesis, University of Erlangen-Nuernberg, Germany (in German).
54. Renlund, G. M. (1989). Silicon-oxycarbide glasses, PhD thesis, Rensselaer Polytechnic Institute, NY.
55. Cordelair, J., and Greil, P. (2000). Electrical conductivity measurements as a Microprobe for structure transitions in polysiloxane derived Si–O–C ceramics. *J. Eur. Ceram. Soc.* **20**: 1947.

56. McLachlan, D. S., Blaskiewicz, M., and Newnham, R. E. (1990). Electrical resistivity of composites. *J. Am. Ceram. Soc.* **73**: 2187.
57. Lux, F. (1993). Models proposed to explain the electrical conductivity of mixtures made of conductive and insulating materials. *J. Mater. Sci.* **28**: 285.
58. De Bondt, A., Froyen, L., and Deruyettere, A. (1992). Electrical conductivity of composites: a percolation approach. *J. Mater. Sci.* **27**: 1983.
59. Stauffer, D. (1985). *An Introduction to Percolation Theory*, London: Taylor and Francis.
60. Cordelair, J. (1999). Electrical properties of polymer/filler-derived ceramics, PhD thesis, University of Erlangen-Nuernberg, Germany (in German).
61. Mueller, C. (1998). Oxidation protection systems for porous ceramics, PhD thesis, University of Erlangen-Nuernberg, Germany (in German).
62. Clarke, D. R. (1992). Interpenetrating phase composites. *J. Am. Ceram. Soc.* **75**: 739.
63. Jacubenas, K., and Marcus, H. L. (1995). Silicon carbide from laser pyrolysis of polycarbosilane. *J. Am. Ceram. Soc.* **78**: 2263.

Handbook of Advanced Ceramics
S. Sōmiya *et al.* (Eds.)

5.3 Template-Induced Deposition of Ceramic Materials from Aqueous Solution

J. BILL, R. HOFFMANN, T. FUCHS and F. ALDINGER
Max-Planck-Institut für Metallforschung and Institut für Nichtmetallische Anorganische Materialien, Universität Stuttgart, Pulvermetallurgisches Laboratorium, Heisenbergstrasse 3, D-70569 Stuttgart, Germany

5.3.1 INTRODUCTION

The huge potential of functional and structural properties of ceramics provides a powerful source for the realization of many advanced technologies. Therefore, these materials play a key role within areas such as energy supply, information technology, transportation systems as well as medical or manufacturing technology. The trends that can be identified with respect to the further evolution of these technologies are growing materials complexity because of increased functionality, integration of different materials into multifunctional components, and miniaturization of devices as well as the exploitation of nanosize effects [1]. Owing to these requirements, there is a need for adequate processing techniques for the manufacture of complex as well as miniaturized components and devices. In this connection, multidisciplinary research is gaining importance within the field of materials science involving chemistry, physics, and increasingly, biology.

Conventionally, ceramics are produced by powder compaction and sintering [2–5], or techniques like chemical and physical vapor phase deposition as well as molecular beam epitaxy (MBE) [6]. All of them need more or less high temperatures and sophisticated production equipment (Fig. 5.3.1) which often cause technical limitations and make manufacturing eventually costly.

Typical technical limitations are, for example, complexity of shapes, materials integration, shrinkage and restrictions in the type of substrate materials.

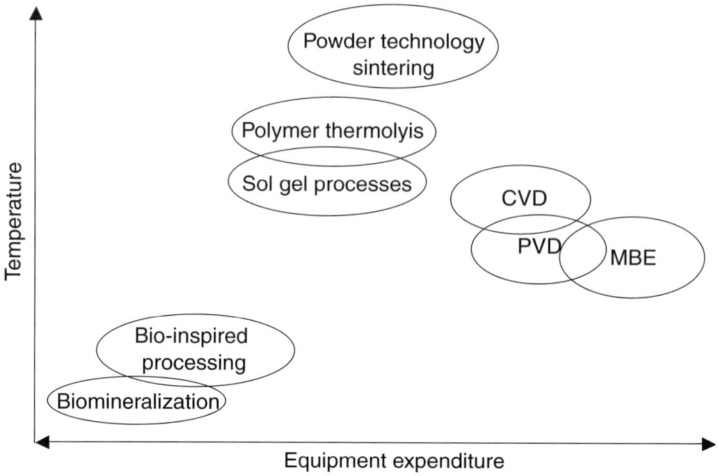

FIGURE 5.3.1 Methods of material synthesis classified by the processing temperature and the equipment expenditure.

Consequently, there is a search for new pathways that enable ways to overcome these limitations and to process the creation of complex shaped and structured materials at low temperatures in combination with low-cost equipment.

In order to decrease the processing temperatures and expenditure, in recent times, sol–gel processes for the preparation of oxide-based materials as well as the solid state thermolysis of preceramic polymers have been applied in the case of non-oxide ceramics [7, 8]. In both cases, molecular precursors are synthesized and subsequently transformed into ceramics by heat treatment at moderate temperatures. Basically, the synthesis and transformation step allows an "atomic mixing" and an architectural design on an atomic scale, leading to micro- or nanostructures with higher homogeneity than in the case of conventional powder technology processes where powder mixing is applied. Furthermore, synthesis and crosslinking steps allow to vary the specific properties of the precursors such as composition, solubility, fusibility or viscosity extensively, providing a versatility in shaping capabilities similar to that enjoyed with standard polymers. Accordingly, ceramic monoliths, films, fibers and powders can be prepared.

In recent times, processes of biomineralization [9–11] have created much interest since nature demonstrates how to produce complex materials and shapes with high functionality at ambient conditions with respect to temperature, pressure and atmosphere, that is, by simple means not achieved in conventional processing. Frequently, biomineralization leads to the formation

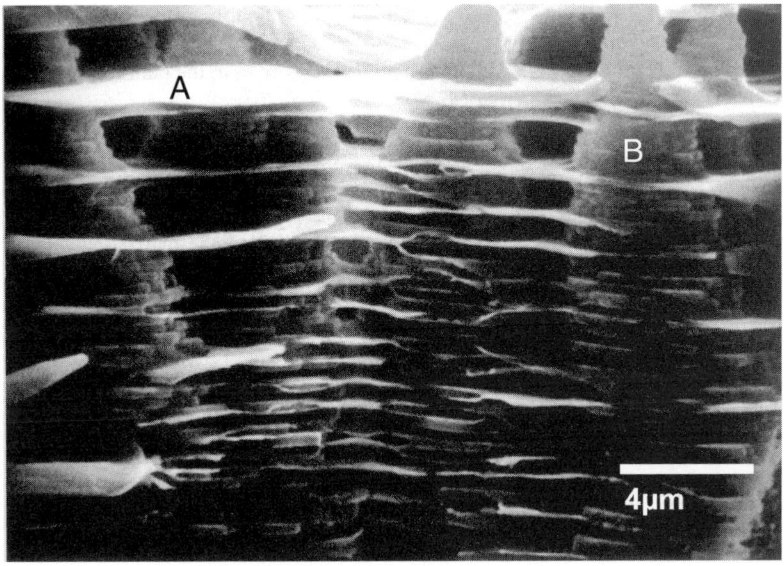

FIGURE 5.3.2 Microstructure of nacre (mother-of-pearl) consisting of protein-based sheets (A) and multilayers of aragonite ($CaCO_3$) (B) [12].

of highly complex structures that are found, for instance, within bones as well as sea or egg shells. An example is given in Figure 5.3.2, which shows the microstructure of nacre (mother-of-pearl) [12].

The material can be considered an organic/inorganic hybrid material consisting of stacked disks made of $CaCO_3$ (aragonite) that exhibit a size in the range of 5 μm and are embedded in protein-based interfaces.

According to Mann [13] biomineralization, basically, can be divided into four steps:

1. *Supramolecular preorganization*: Construction of an organized reaction environment prior to mineralization by the self-assembly of enclosed protein cages and lipid vesicles or by the facilitated construction of extended protein–polysaccharide networks.
2. *Interfacial molecular recognition*: Controlled nucleation of inorganic clusters, where the preformed organic supramolecular systems provide a framework for the assembly of the inorganic phase.
3. *Vectorial regulation of crystallization*: Assembly of the mineral phase through crystal growth and termination enabling the formation of unusual and elaborate textures and shapes by cellular intervention.
4. *Cellular processing*: Formation of higher-order architectures involving larger scale cellular activity.

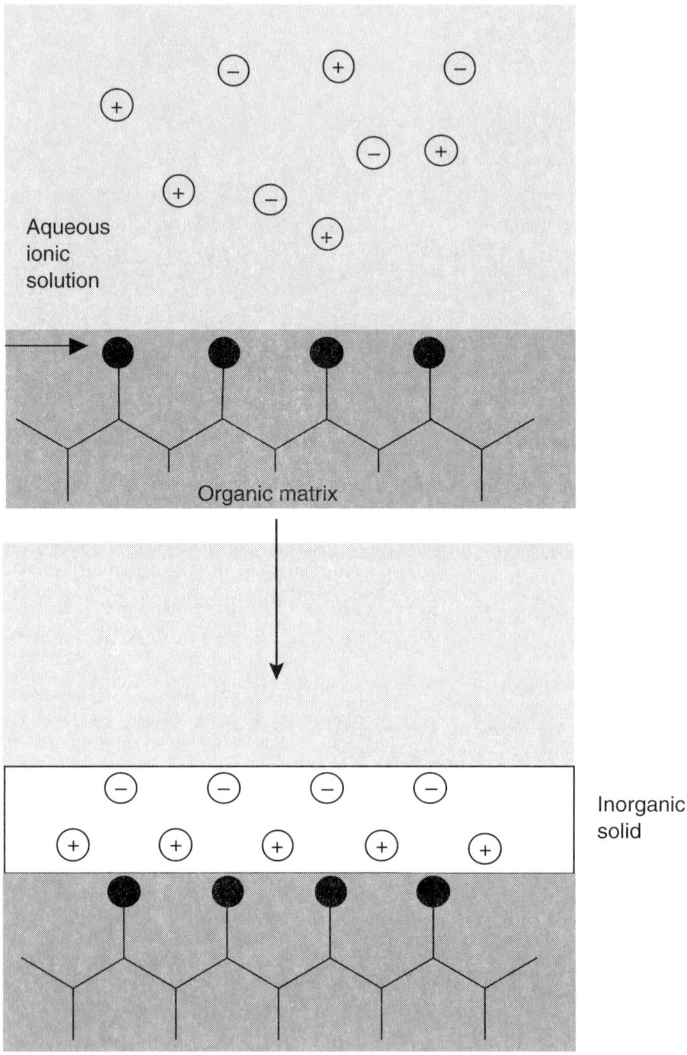

FIGURE 5.3.3 Schematic representation of the deposition of an inorganic ionic phase on an organic template.

The main principle that contributes to the formation of such highly complex architectures according to steps 2 and 3 is shown in Figure 5.3.3.

The principle involves an organic matrix acting as a template for the deposition and an inorganic solid from an aqueous ionic solution at ambient temperature. The corresponding activation energy ΔG^* is related to the following parameters [13, 14]:

$$\Delta G^* \propto \frac{\sigma^3}{(kT \ln S)^2} \tag{1}$$

First of all, ΔG^* depends on the degree of supersaturation $S = c_L/c_S$, (c_L/c_S is the ratio of the concentration of the precipitating ionic species in the solution and the ionic concentration at equilibrium) and the temperature T (k represents Boltzmann's constant). The main role of the organic matrix is to decrease the activation energy of nucleation. Such an effect is achieved by heterogeneous nucleation connected with a low interfacial energy between the organic matrix and the inorganic phase which contributes to a decrease of the net interfacial energy σ of the system. As a result, the organic phase acts as a template for the deposition of the inorganic material.

This principle is adopted in the case of the deposition of inorganic solids induced by self-assembled monolayers (SAMs) [15–17] (Fig. 5.3.4).

According to this bio-inspired concept, an organically modified surface is prepared by the covalent attachment of bifunctional surfactants to a substrate. The surface group of the long-chain hydrocarbon molecules supports the deposition of an inorganic phase if substrates containing such a SAM are put into an appropriate solution containing suitable molecular precursors.

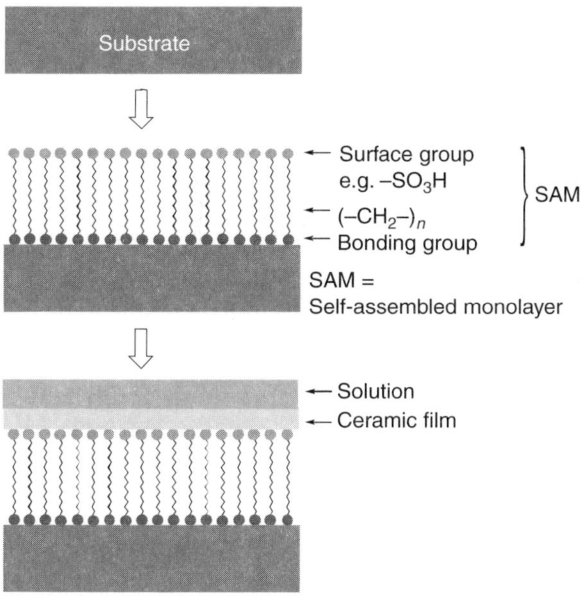

FIGURE 5.3.4 Preparation of ceramic films using organic self-assembled monolayers.

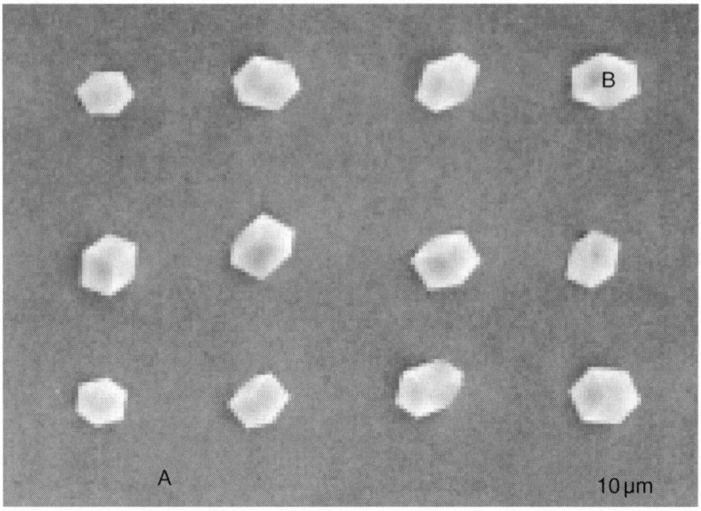

FIGURE 5.3.5 Ordered two-dimensional array of $CaCO_3$ (calcite) crystals (B) deposited on a patterned SAM terminated by $-CO_2H$ and $-CH_3$ sites (Ag substrate). $-CH_3$ sites are present within region A, $-CO_2H$ sites within region B. The crystals are formed by diffusion of carbon dioxide and ammonia vapors into a $CaCl_2$ solution (deposition time: 30 min) [19].

As an example, Figure 5.3.5 shows $CaCO_3$ crystals deposited on a silver substrate which was previously modified with a SAM [18–20].

In this case, patterned SAMs prepared via microcontact printing were employed. The organically patterned surface consists of a carboxylic acid ($-CO_2H$) and a methyl-terminated SAM. Deposition of $CaCO_3$ only occurs at CO_2H sites yielding an oriented growth of calcite crystals from the (012) crystallographic plane (space group $R\bar{3}c$) and thus a micropatterned surface.

Another example is the deposition of goethite FeO(OH) [21–27], which exhibits technical relevance, for example, as a precursor for magnetic materials employed in magnetic storage devices. The TEM dark-field image in Figure 5.3.6 shows a cross-section of such a film deposited on a silicon substrate organically modified with a SAM that is terminated by $-SO_3H$ groups from an aqueous solution of iron(III) nitrate at 70°C [23, 25].

The FeO(OH) films possess a thickness of around 4 μm. The organic monolayer induces heterogeneous nucleation resulting in columnar structures of orthorombic crystals with an orientation of the crystallographic [001] direction parallel to the substrate surface.

In recent years, a variety of further inorganic materials have been deposited including Y_2O_3 [28, 29], TiO_2 [30–41], ZrO_2 [28, 39, 42, 43], ZrO_2/Y_2O_3 [28], V_2O_5 [40], Fe_2O_3 [21], Fe_3O_4 [21, 44], ZnO [45–48], SnO_2 [49–51], PbS [52–54], CdS [55], and ZnS [56]. In the following, the template-induced

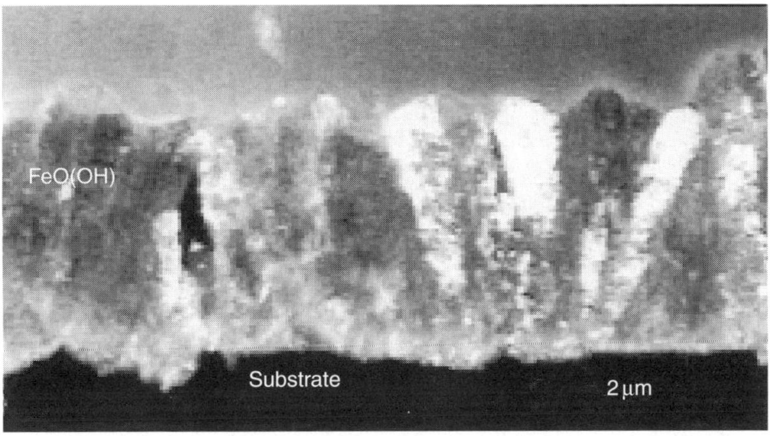

FIGURE 5.3.6 TEM dark field image of a FeO(OH) film deposited on a SAM terminated with $-SO_3H$ sites (Si substrate) from an aqueous solution of iron(III) nitrate at 70°C (deposition time: 90 min) [23].

formation of inorganic films is treated by means of the technically important systems ZrO_2, TiO_2 and ZnO, for example, applied for the production of solid-oxide fuel cells [57], dielectrics [58], and surface acoustic wave transducers [59], respectively. Static as well as liquid flow deposition are treated. Furthermore, the addition of ligands for complexation as well as of water-soluble copolymers to the aqueous deposition solution in order to control film formation is presented.

5.3.2 SAM FORMATION

An SAM is a highly ordered array of hydrocarbon chains that are covalently bonded to the substrate surface [60]. It has been demonstrated that a chain length of at least nine carbon atoms is necessary to provide sufficient van der Waals bonds necessary for the formation of an SAM. On the contrary, chain lengths above 20 carbon atoms were found to exhibit a decreased solubility in organic solvents. Highly ordered and stable SAMs are formed by chains that consist of 16 atoms [28, 29].

The preparation of siloxane-anchored SAMs on silicon substrates is shown in Figure 5.3.7.

For the formation of the SAM, the substrate is activated by "piranha" solution (H_2SO_4/H_2O_2) leaving behind a hydrolyzed layer of amorphous silicon oxide. Trichlorosilyl functional groups ($-SiCl_3$) then provide reactive sites for the

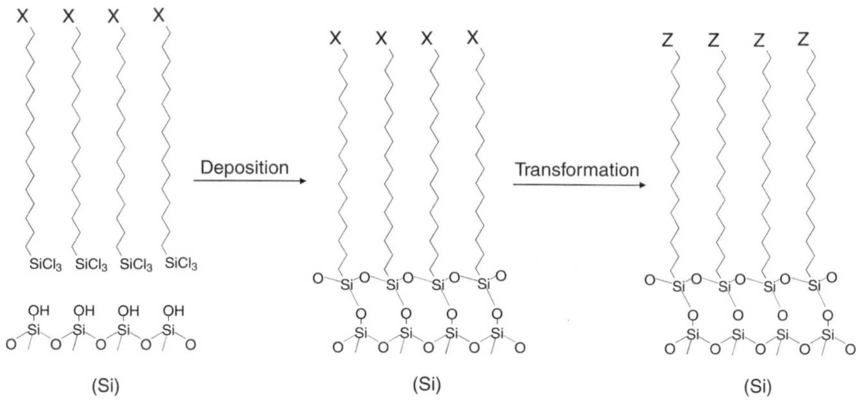

FIGURE 5.3.7 Formation of siloxane-anchored SAMs on silicon substrates.

attachment of the hydrocarbon molecules to the substrate, resulting in SAM formation. Besides crosslinking, van der Waals interactions between the chains also contribute to the stabilization of the SAM. After SAM formation the terminating functional groups X of the SAM molecules can be further modified and transformed. Such *in situ* transformation steps are especially applied if the terminating group X is not chemically compatible with the $SiCl_3$ units as in the case of the preparation of sulfonate-functionalized surfaces ($Z = SO_3H$) which are observed by oxidation of thioacetate as precursor functionality ($X = SCOCH_3$) with oxone ($2\ KHSO_5 \cdot KHSO_4 \cdot K_2SO_4$) [61].

5.3.3 OXIDE DEPOSITION

For the deposition of an oxide film by immersing a SAM-modified substrate into an appropriate precursor solution, basically, two mechanisms have been considered (Fig. 5.3.8).

On the one hand, dissolved ionic species may attach to the SAM-modified substrate according to an ion-by-ion growth mechanism of the film. On the other hand, colloidal clusters may be formed in the solution by hydrolysis and condensation reactions of the dissolved species and subsequently attracted to the functional groups of the organic monolayer by electrostatic interactions. The former mechanism is connected with heterogeneous nucleation and may lead to directed growth of nuclei of the inorganic phase as demonstrated by the examples given in Figures 5.3.5 and 5.3.6. The latter mechanism involves homogeneous nucleation in solution and as a result polycrystalline films with random crystal orientation may be observed in this case. The hydrolysis and

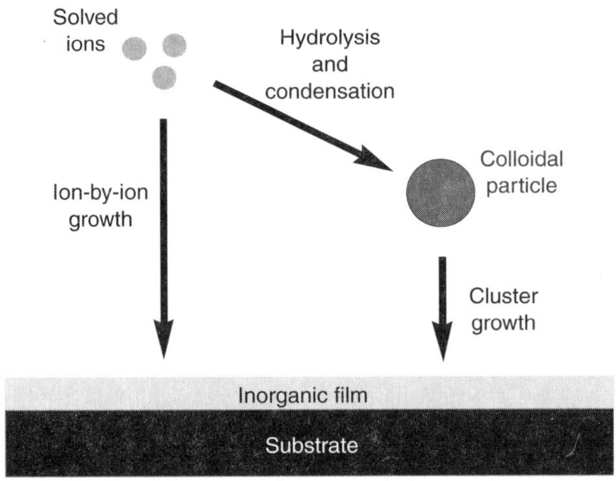

FIGURE 5.3.8 Basic mechanisms for the formation of inorganic films from aqueous inorganic solutions induced by self-assembled monolayers.

condensation reactions which provide the base for the formation of the colloidal particles are especially observed in the case of highly charged transition metal ions like Zr^{4+} [28] or Ti^{4+} [30]. Basically, the formation of the colloidal particles can be divided into the following steps [62]:

- formation of the zero-charge precursor $[M(OH)_z(OH_2)_{N-z}]^0$;
- nucleation;
- growth of nuclei;
- aging of particles.

The first step is combined with a deprotonation of the metal cations which are solvated by dipolar water molecules:

$$\left[M(OH_2)_N\right]^{z+} + hH_2O \rightarrow \left[M(OH)_h(OH_2)_{N-h}\right]^{(z-h)+} + hH_3O^+ \qquad (2)$$

This reaction depends on the acid concentration of the solution and can also be influenced by the complexation with anionic species, providing a means to control the precipitation of the inorganic solid from solution:

$$\left[M(OH_2)_N\right]^{z+} + X^- \rightarrow \left[M(OH_2)_{N-1}X\right]^{(z-1)+} + H_2O \qquad (3)$$

The creation and growth of the oxide-based nuclei by further condensation of the hydrolyzed metal ions involves olation:

$$M\text{–}OH + M\text{–}OH_2 \rightarrow M\text{–}OH\text{–}M + H_2O \qquad (4)$$

and oxolation

$$M\text{–}OH + HO\text{–}M \rightarrow M\text{–}O\text{–}M + H_2O \qquad (5)$$

Further aging of the solution involves processes like Ostwald ripening and agglomeration of the formed particles.

5.3.3.1 Deposition of Zirconia Films

Zirconia thin films can be deposited from hydrochloric acid-containing aqueous solutions of zirconium sulfate on silicon substrates which are modified with a sulfonate-functionalized SAM at 70°C [28]. No film growth is observed on bare silicon substrates under otherwise same conditions, demonstrating the essential function of the SAM. The film thickness determined by TEM as a function of the deposition time is shown in Figure 5.3.9.

As can be seen, the growth rate of the films decreases continuously. Deposition for 24 h results in a film thickness of around 40 nm. Obviously, aging of the deposition solution and the reduction of the concentration of the precipitating species slow down the film formation.

Typical micrographs obtained by transmission electron microscopy (TEM) are shown in Figure 5.3.10.

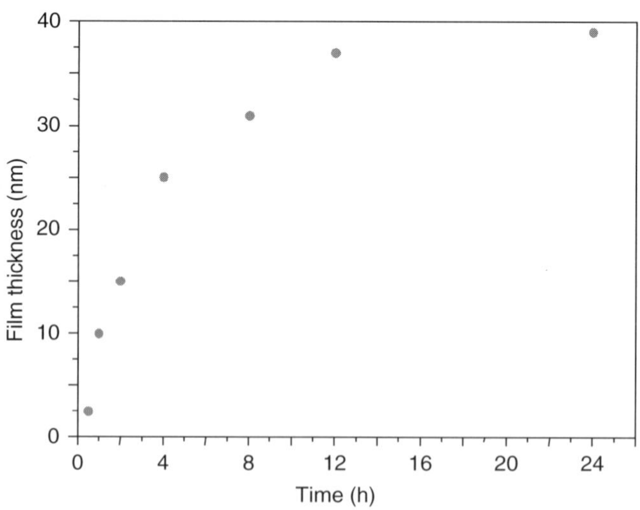

FIGURE 5.3.9 Evolution of the film thickness of zirconia-based films deposited from a 4 mM aqueous acidic solution of $Zr(SO_4)_2{\cdot}4H_2O$ and 0.4 M HCl on silicon substrates modified with sulfonate-functionalized SAMs at 70°C [28].

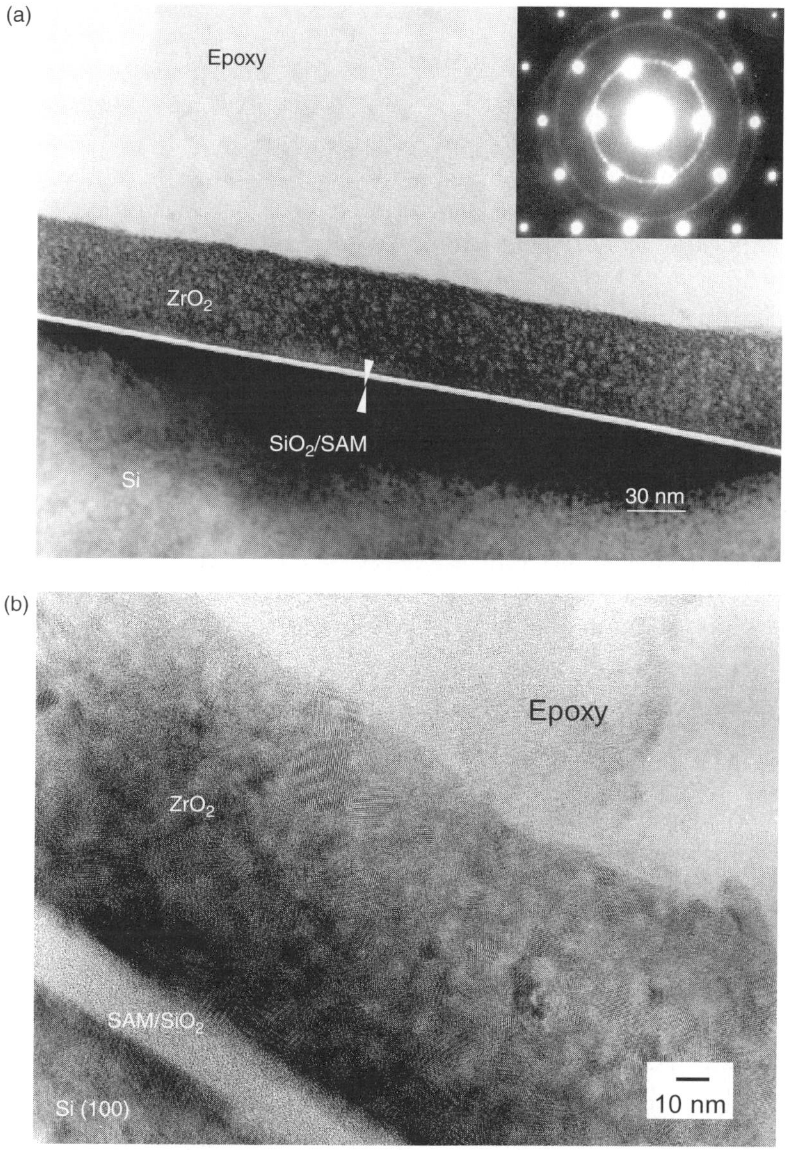

FIGURE 5.3.10 TEM [39] (a) and HRTEM [28] (b) image (cross-section) of a zirconia-based film deposited during four successive immersions, each in a freshly prepared 4 mM aqueous acidic solution of zirconium sulfate (0.4 M HCl), for 1 h each (deposition temperature: 70°C). Silicon substrates modified with sulfonate-functionalized SAMs were employed. The upper part of the image traces back to the TEM preparation of the sample.

The zirconia film in Figure 5.3.10 exhibits a thickness of around 60 nm. It consists of polycrystalline tetragonal zirconia as revealed by electron diffraction. Analytical TEM observations show that the tetragonal nanocrystals are embedded into an amorphous basic zirconium sulfate matrix. The amount of this amorphous grain boundary phase is significantly reduced in the case of the film prepared by four successive immersion steps as shown by high-resolution TEM (Fig. 5.3.10b). As can be seen, the crystal size is in the range of 5–10 nm.

The polycrystalline random nature of the film indicates cluster growth via the deposition of colloidal nanoparticles [28]. Moreover, if the pH conditions (<1) that are chosen for film deposition are taken into account this growth model is supported by considering the acidity of the sulfonate group that refers to a deprotonated negatively charged substrate surface and the isoelectric points of basic zirconium sulfate as well as of zirconia particles that suggest positively charged colloidal particles. X-ray small-angle scattering studies of the solutions also provide experimental evidence for the presence of 2 nm particles within the starting solution [39]. Furthermore, the suggested mechanism is supported and quantified by Derjaguin–Landau–Verwey–Overbeck (DLVO) calculations, which shows attractive electrostatic as well as van der Waals interactions between the precipitate and the SAM [35]. These calculations predict that film growth continues due to van der Waals attraction of additional particles to the already deposited film. As can be seen from Figure 5.3.11a, the topography of the surface includes a nanoscale background roughness in accordance with the suggested deposition mechanism of colloidal nanoparticles.

In addition, a micron-scale roughness is detected consisting of protrusions and depressions that can be attributed to the attachment or detachment of larger particles or agglomerates formed within the solution during aging. Atomic force microscopy (AFM) investigations demonstrate that such defects can be minimized by carrying out four successive immersions, each in a freshly prepared aliquot of solution for 1 h (Fig. 5.3.11b).

5.3.3.2 Deposition of Titania Films

A thin film made of titania obtained via the deposition on SAMs can be seen in Figure 5.3.12.

The deposition on a sulfonate-functionalized SAM was carried out using a 0.5 M aqueous solution of $TiCl_4$ in 6 M HCl at 80°C. Lower HCl concentrations lead to immediate bulk precipitation within the solution. With higher concentrations no film formation is observed [32]. The process leads to a polycrystalline film that consists of nanocrystals with a size in the range of 2–4 nm, which are embedded into an amorphous matrix. Furthermore, a layer

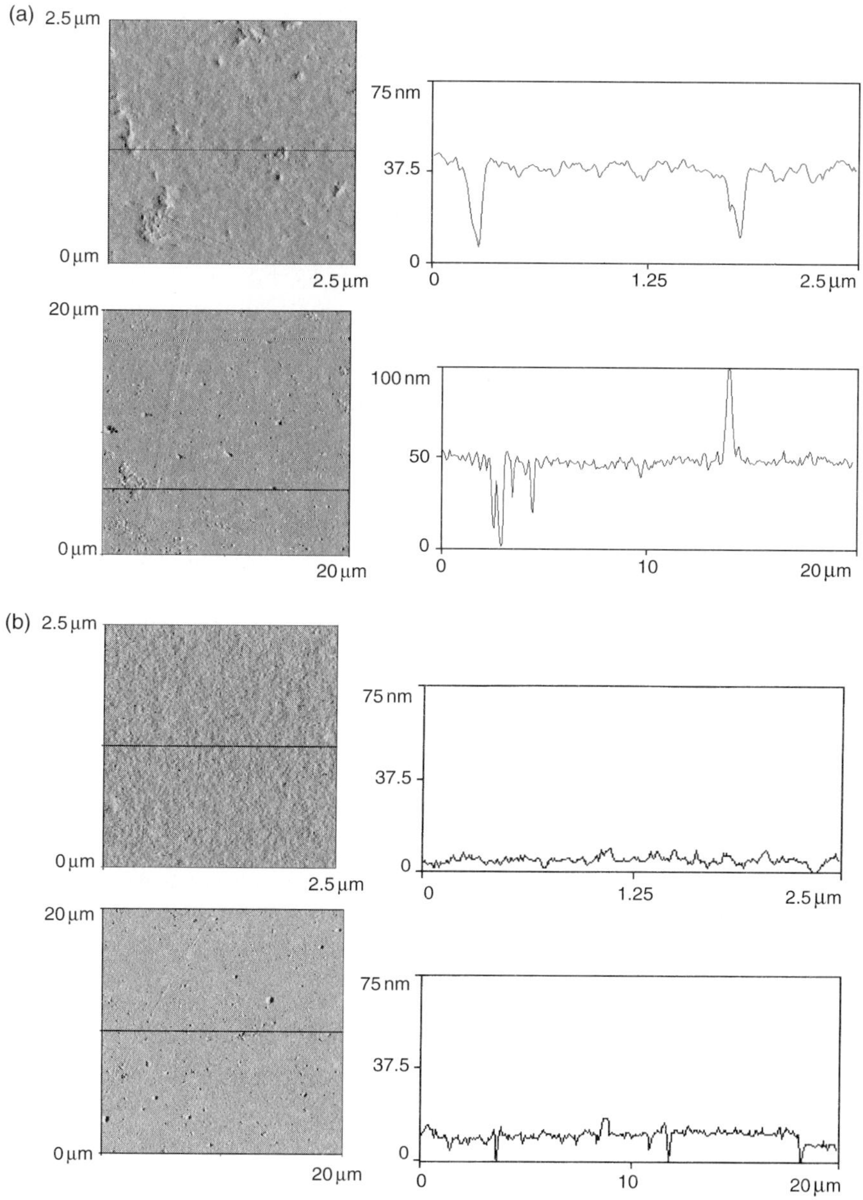

FIGURE 5.3.11 AFM images and corresponding line profiles of zirconia-based films deposited from a 4 mM aqueous acidic solution of zirconium sulfate (0.4 M HCl) on silicon substrates modified with sulfonate-functionalized SAMs at 70°C. Deposition time: 6 h (a) and 4 × 1 h (b) [39].

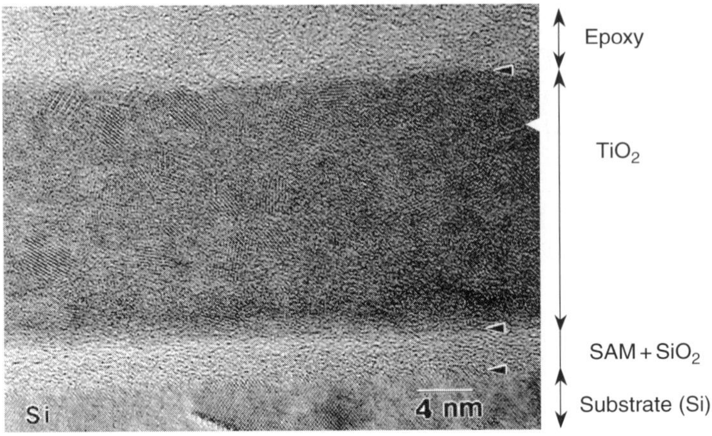

FIGURE 5.3.12 TEM image of a TiO_2 film deposited on a SAM with a sulfonate end group (Si substrate) from a 0.5 M aqueous solution of $TiCl_4$ in 6 M HCl at 80°C (deposition time: 2 h) [30].

between the silicon substrate and the TiO_2 film with a thickness of 4 nm can be seen. This layer can be attributed to the SAM and the native oxide layer. Ellipsometric results show that it consists of a 2.5 nm thick SAM layer and 1.5 nm thick SiO_2 layer.

The topography of the surface determined by AFM measurements (Fig. 5.3.13) exhibits a nano-scale background roughness as well as micron- and submicron-scale protrusions and depressions.

The nano-scale roughness is visible even within the smoothest regions, indicating that this roughness is a specific feature of the film formation process by the deposition of colloidal nanosized particles consistent with the cluster growth mechanism described above (Fig. 5.3.10). In accordance with these observations the selected pH conditions suggest positively charged titania particles within the solution and a negatively charged SAM-terminated substrate surface similar to the conditions described in Section 5.3.3.1 for the zirconia film formation. Furthermore, a quantitative treatment of the attractive interaction between the titania particles and the surface by DLVO calculations is valid in this case [35]. Electron diffraction reveals that the film consists of metastable anatase whereas the thermodynamically stable phase rutile cannot be identified. The formation of a metastable phase was also observed in the case of tetragonal zirconia films (Section 5.3.3.1) and demonstrates another main important feature of this low-temperature route. Possible mechanisms [62] for the evolution of anatase and rutile in solution are schematically shown in Figure 5.3.14. At first, the olation of the zero charge precursor $[Ti(OH)_4(OH_2)_2]^0$ creates octahedral units, which are initially formed by hydrolysis of $TiCl_4$ and subsequent deprotonation reactions (Eq. 2). Condensation of this precursor

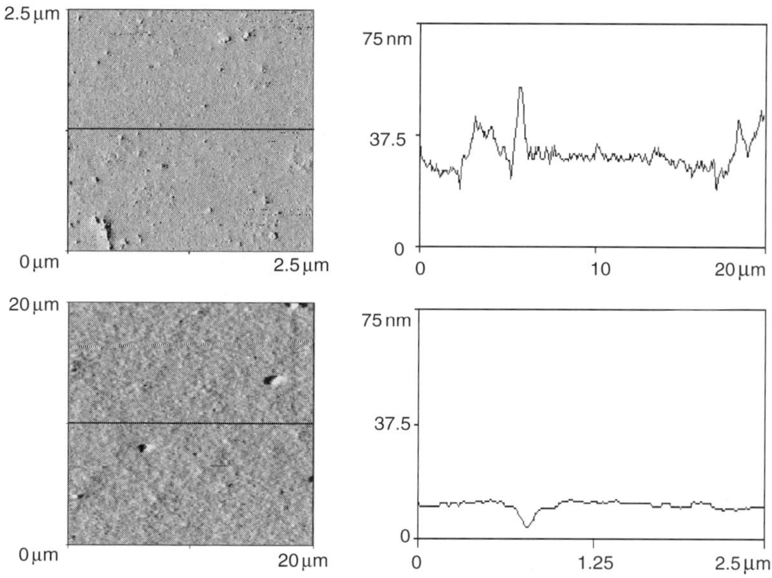

FIGURE 5.3.13 AFM images and corresponding line profiles of an as-deposited TiO_2 film on a sulfonate-terminated SAM [39].

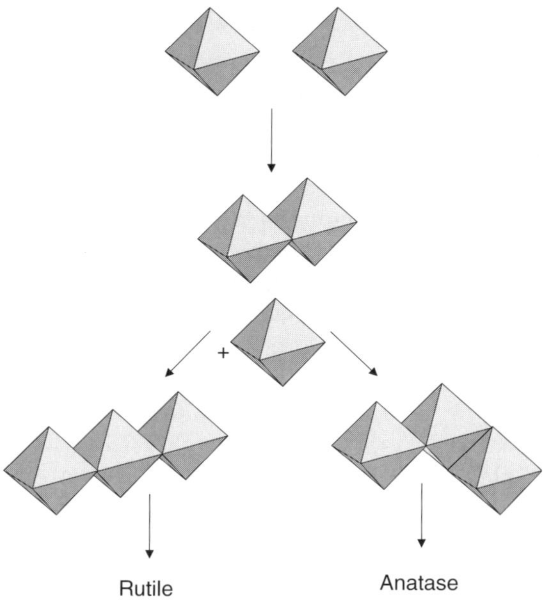

FIGURE 5.3.14 Possible formation of rutile and anatase from octahedral structural units.

results in a dimer $[Ti_2(OH)_8(OH_2)_2]^0$ in which both octahedra share an edge that corresponds to a double hydroxo bridge:

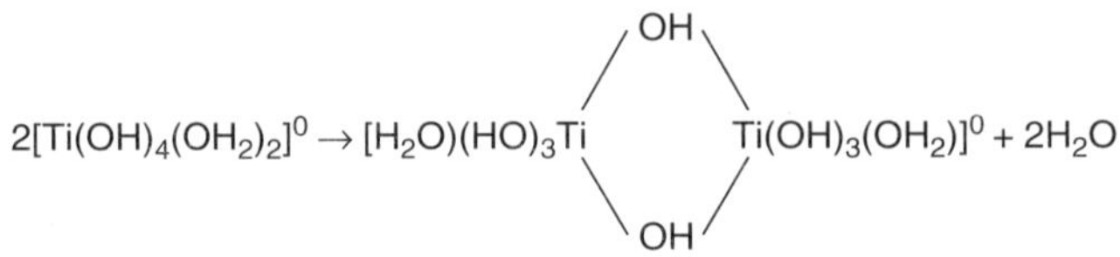

(6)

The formation of the titania phases that consist of TiO_6 octahedra involve further growth of this structural unit by oxolation with additional octahedral units. According to this mechanism the probability of a connection through edge-sharing in the same plane, leading to straight small chains is much smaller than the probability of edge sharing in different planes yielding curved or bent chains (Fig. 5.3.14).

As a result, the formation of anatase is favored under any experimental conditions according to this explanation based on statistical arguments. Furthermore, it is known that rutile is subsequently formed via a solution–dissolution process (Ostwald ripening) of anatase. This phase transformation is not observed in the present case, indicating that the metastable anatas phase is “trapped” within the inorganic film prior to the phase transformation within the solution into rutile.

Anatase is maintained if heat treatment of the films up to 600°C in air is applied whereas rutile is only barely detectable [30]. Annealing at 200, 400 and 600°C for 2 h results in average crystal diameters of 5, 10 and 25 nm, respectively. This indicates a coarsening of the crystal size with increasing temperature compared with the crystal size found within the as-deposited films (2–4 nm). The thickness of the SAM/SiO_2 layer remains unchanged at 4 nm (1.5 nm SiO_2/2.5 nm SAM). However, the SAM/SiO_2 interface was not found any more after this treatment [30]. At 600°C, the thickness is determined to be 10 nm, indicating some oxidation of the Si substrate. No loss of the structural integrity of the films was observed [30].

Further experiments that concern the thermal stability of the SAM/SiO_2 layers by annealing in nitrogen showed that these layers are preserved up to 300°C (2 h). At 400°C, the thickness of the layer decreased from 4 nm (1.5 nm SiO_2/2.5 nm SAM) to 2 nm, indicating the decomposition of the SAM. Debonding of the oxide film after this treatment is not observed [63].

According to these results, template deposition provides unique opportunities for the preparation of homogeneous thin films at low temperatures. Basically, the deposition process requires only minor equipment costs and is suitable for the line-of-sight deposition also of complex-shaped as well as of temperature-sensitive materials. Further efforts to reduce the acidity of the deposition medium which might not be compatible with certain substrate materials like plastics have been carried out applying the stabilization of solutions against bulk precipitation by complexation of titanium with ligands. Baskaran *et al.* used solutions of titanium lactate to deposit TiO_2 films on plastic surfaces modified with sulfonated SAMs [34]. As far as the above-mentioned aqueous solution of $TiCl_4$ is concerned, a decrease of the acid concentration leads to a spontaneous precipitation. This solution can be stabilized by the addition of hydrogen peroxide, yielding the peroxo complex $Ti(O_2)^{2+}$ by side-on coordination of O_2^{2-} to Ti^{4+}. If this second approach is applied, titania films can be deposited from aqueous solutions of $Ti(O_2)^{2+}$ (10 mM, prepared from $TiCl_4$ and aqueous H_2O_2) at 80°C [41]. In this case, the concentration of the added HCl is 0.14 mol/l and thus drastically reduced compared to the conditions (6 M) mentioned above. The film formation from this solution at 80°C applying $-SO_3H$, –OH and $-NH_2$ functionalized SAMs monitored by SEM can be seen in Figure 5.3.15.

Following the above-mentioned argument that positively charged colloidal particles of titania are attracted by the SAM, the highly acidic $-SO_3H$ functionality provides the highest net attraction. As a result, the formation of a uniform film made of densely packed particles is observed in the case of sulfonate-functionalized SAMs after 4 as well as after 6 h deposition time. On the contrary, only growth of islands is detected on hydroxyl- and amino-terminated SAMs, indicating heterogeneous nucleation. The islands exhibit a size in the range between 70 and 200 nm after 4 h which obviously grow together if the deposition time is expanded to 6 h, resulting in an inhomogeneous surface topography.

A cross-sectional TEM micrograph of a film deposited on sulfonate functionalities for 6 h is shown in Figure 5.3.16, providing evidence that in spite of the drastically reduced acidic conditions, a homogeneous titania film is observed.

The TEM image reveals a film thickness in the range between 55 and 75 nm. The film is free of pores and consists of nanocrystals. Electron diffraction exhibits the presence of anatase.

The topography of the film surface investigated by AFM is visualized in Figure 5.3.17.

The image reveals a dense packing of agglomerates that extend from 50 to 100 nm. The height varies in the range 50–60 nm (statistical roughness parameter: rms = 15–18 nm). These results are observed for the whole film, indicating a very homogenous topography of the film. On the contrary, larger micro-sized inhomogeneities that trace back to precipitates from solution are not detected.

The thickness of the titania films depends on the acidity of the solution and can be enhanced by decreasing the HCl concentration (Fig. 5.3.18).

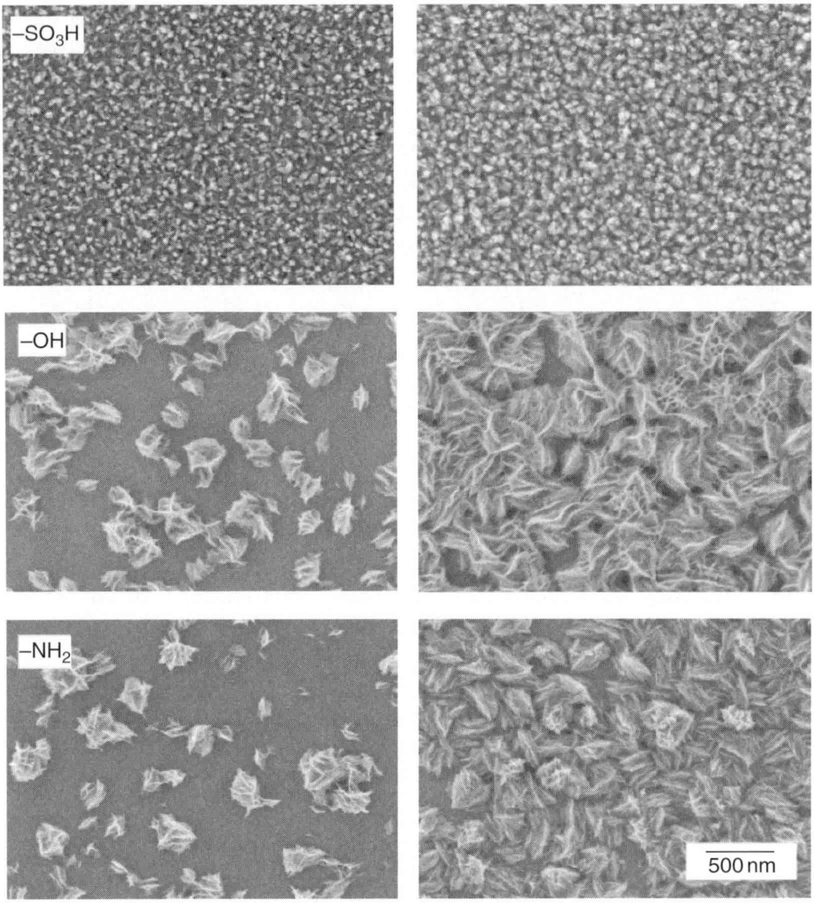

FIGURE 5.3.15 SEM images of the deposition of titania thin films after 4 h (left) and 6 h (right) on sulfonate-, hydroxyl- and amine-functionalized SAMs (Si substrate) from aqueous solutions of $Ti(O_2)^{2+}$ at 80°C [41].

The diagram reveals a nearly linear dependence between the acid concentration and the film thickness, which can be enhanced up to 170 nm by decreasing the acid concentration to 0.12 mol/l. Further decrease of the concentration leads to precipitation within the solution.

As already described by means of the zirconia films the thickness of the produced films is limited by aging of the used solutions due to a decrease of the concentration of the film-forming species in the solution. Another effect that traces back to aging is the formation of large particles and agglomerates which deposit on the substrate surface. In order to prevent such effects and

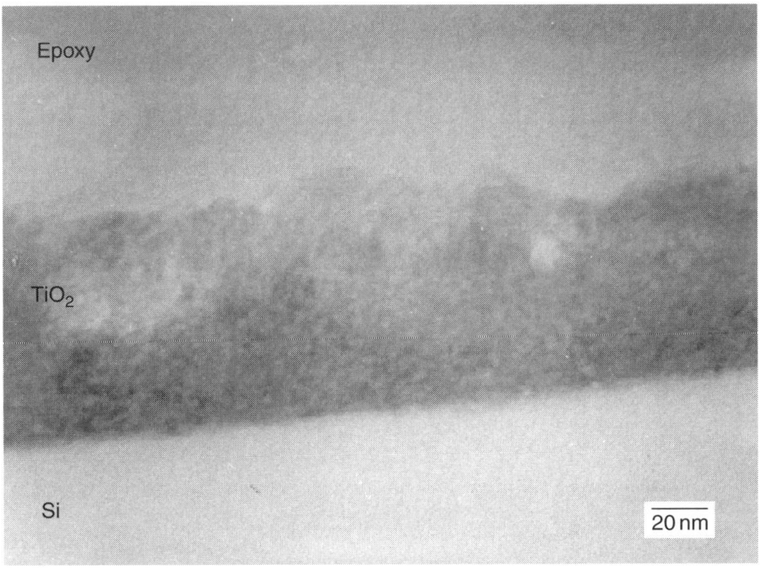

FIGURE 5.3.16 TEM image of a titania film obtained by deposition from an aqueous solution of $Ti(O_2)^{2+}$ at 80°C on a sulfonated SAM (Si substrate; deposition time, 6 h) [41].

FIGURE 5.3.17 AFM image of a titania film obtained by deposition from an aqueous solution of $Ti(O_2)^{2+}$ at 80°C on a sulfonated SAM (Si substrate; deposition time, 6 h) [41].

to provide nanoparticles to the growing film, continuous flow techniques have been developed which apply continuous flowing of a fresh precursor solution over SAMs [49, 50]. In this connection, nanocrystalline films made of SnO_2 were prepared, using sulfonate-functionalized SAMs and an aqueous

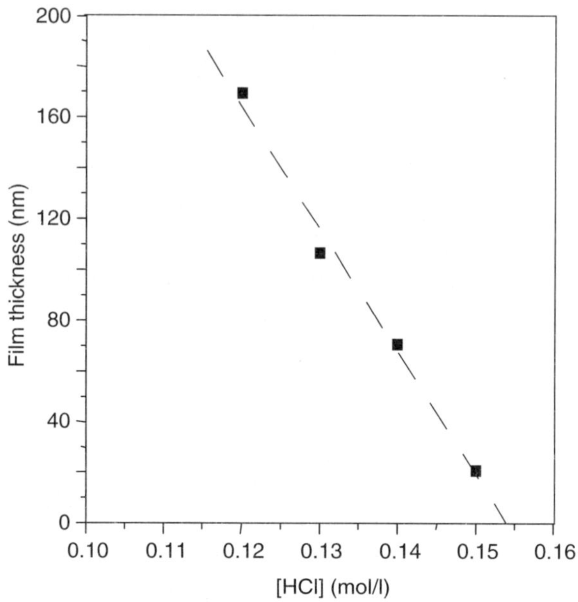

FIGURE 5.3.18 Thickness versus HCl concentration of titania films deposited from solutions of $Ti(O_2)^{2+}$ on a sulfonated SAM (Si substrate) at 80°C, 4 h, determined by ellipsometry [41].

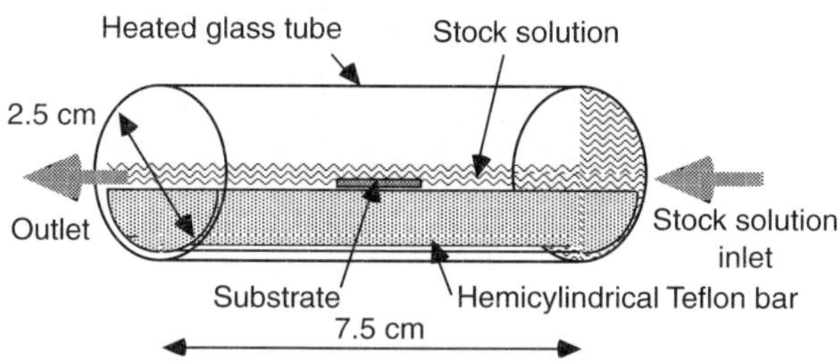

FIGURE 5.3.19 Experimental set-up used for the preparation of ceramic films on SAM-functionalized substrates [49, 50].

hydrochloric acid-containing $SnCl_4$ solution at 80°C for 15 h. The flow rate was 2.8 ml/min. After a deposition time of 15 h, the film thickness was about 1 μm. The experimental set-up can be seen in Figure 5.3.19.

The deposition chamber consists of a glass tube that contains a hemicylindrical teflon bar in order to reduce the chamber volume, to improve the profile of the liquid inside and to reduce the size of stagnant regions near both ends.

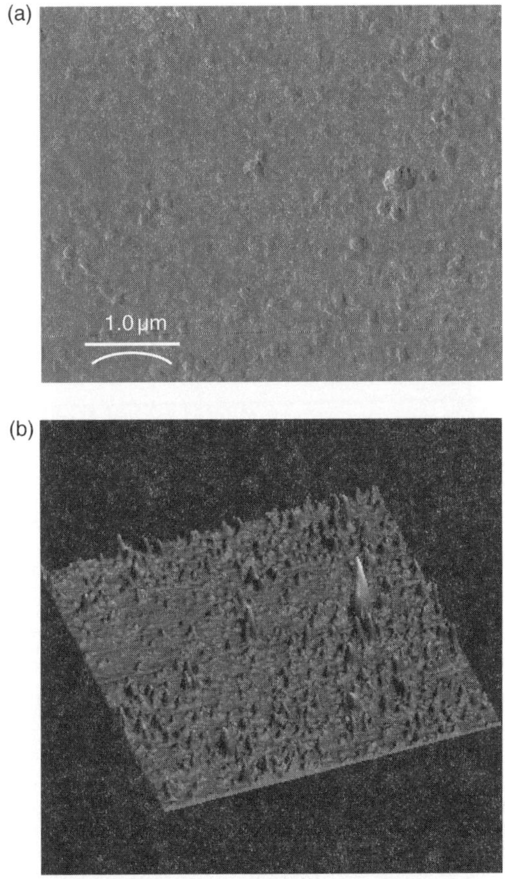

FIGURE 5.3.20 AFM image of a substrate modified with a sulfonate-functionalized SAM (Si substrate) after 1 h deposition at 80°C by the continuous-flow technique (solution: 10 mM $[TiO_2]^{2+}$, 0.12 N HCl) (a) and three-dimensional presentation (b) [64].

The solution within the deposition chamber is continuously renewed by a peristaltic pump.

This technique can also be used for the preparation of titania films from the above-mentioned solutions, containing $Ti(O_2)^{2+}$ cations [64]. As can be seen from the AFM image in Figure 5.3.20, after 1 h island-like precipitations are observed, which indicate heterogeneous nucleation at the initial stage of the film formation.

The lateral dimension of the nucleated areas is determined to be about 100 nm, the height in the range 10–20 nm. If the deposition time is increased

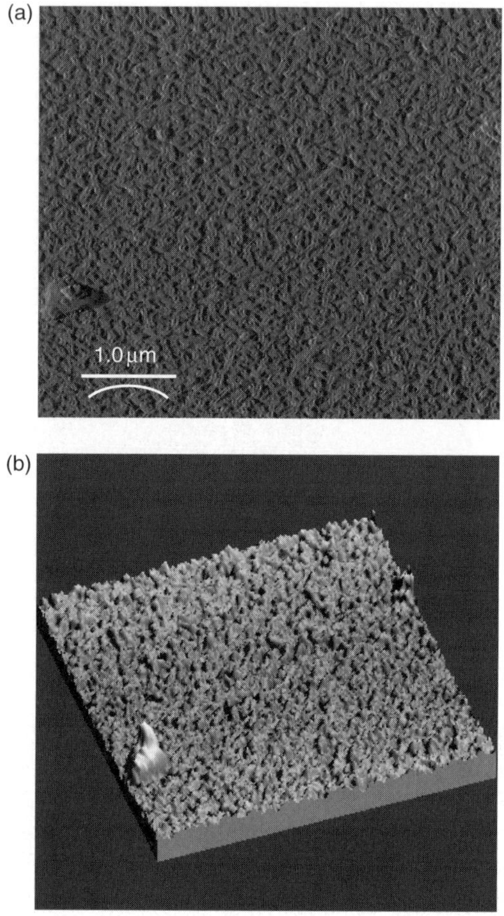

FIGURE 5.3.21 AFM image of titania films after 6 h deposition on a sulfonate-functionalized SAM (Si substrate) by the continuous-flow technique applying acidic aqueous solutions (0.12 N HCl) of 10 mM $[TiO_2]^{2+}$ at 80°C (a) and three-dimensional presentation (b) [64].

to 6 h, the islands grow together resulting in a continuous film that covers the whole substrate surface (Fig. 5.3.21).

The rms roughness is determined to be around 12 nm. A TEM image of a film obtained after 2 h is shown in Figure 5.3.22.

The film consists of extremely fine nanosized particles and exhibits a thickness of about 60 nm. A TEM image of the film observed after 6 h deposition time is shown in Figure 5.3.23.

The image reveals a thickness in the range 350–400 nm. The films exhibit a polycrystalline structure. Electron diffraction reveals the presence of anatase.

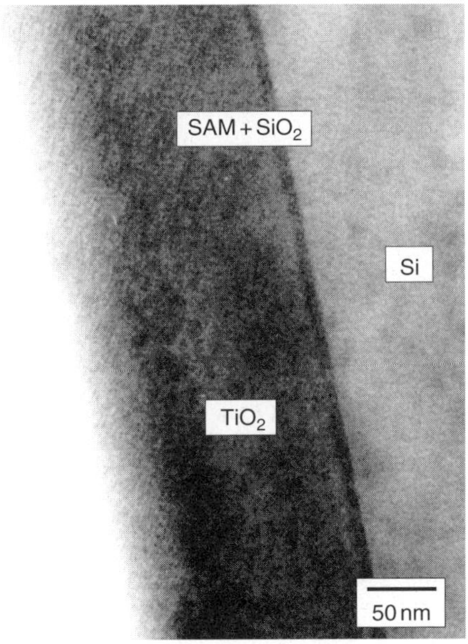

FIGURE 5.3.22 TEM image of a titania film deposited on a sulfonate-functionalized SAM (Si substrate) by the continuous flow technique and static deposition applying acidic aqueous solutions (0.12 N HCl) of 10 mM $Ti(O_2)^{2+}$ at 80°C (deposition time: 2 h) [64].

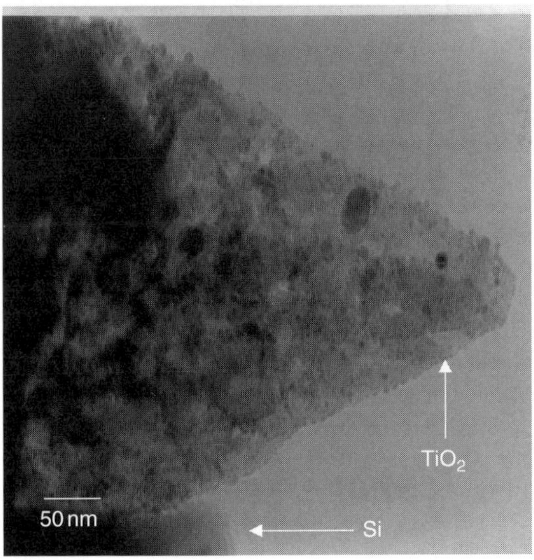

FIGURE 5.3.23 TEM image of a titania film deposited on a sulfonate-functionalized SAM (Si substrate) by the continuous-flow technique for 6 h applying acidic aqueous solutions (0.12 N HCl) of 10 mM $[TiO_2]^{2+}$ at 80°C [64].

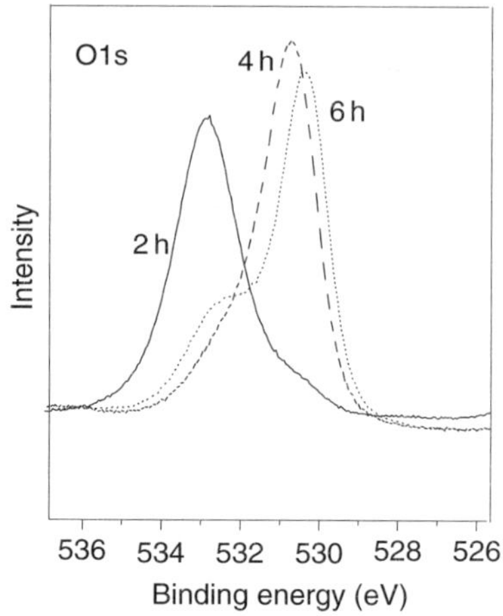

FIGURE 5.3.24 XPS spectra of a titania film deposited on a sulfonate-functionalized SAM (Si substrate) by the continuous-flow technique, applying acidic aqueous solutions (0.12 N HCl) of 10 mM $Ti(O_2)^{2+}$ at 80°C for 2, 4 and 6 h [64].

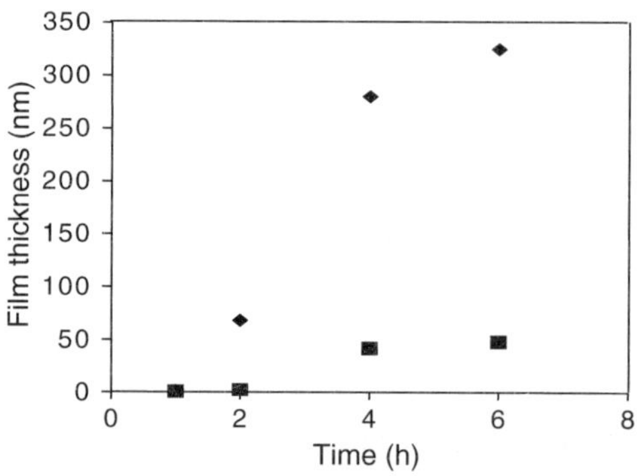

FIGURE 5.3.25 Time dependence of the thickness of titania films deposited on a sulfonate-functionalized SAM (Si substrate) by the continuous flow technique (♦) applying acidic aqueous solutions (0.12 N HCl) of 10 mM TiO_2^{2+} at 80°C [64]. The values obtained by static deposition (■) are also shown.

Furthermore, the grain size compared to the structural features of the film deposited after 2 h (Fig. 5.3.25) is increased which indicates grain growth within the film during deposition. These variations within the film are also in accordance with results obtained by X-ray photoelectron spectroscopy (XPS) (Fig. 5.3.24).

The oxygen signal is shifted with increasing deposition time. This shift is typically observed in the case of the deposition of oxides from aqueous solution and can be related to a transformation of hydrate—via hydroxide—into oxide-based species. The investigation of the dependence of film thickness on the deposition time by ellipsometry is shown in Figure 5.3.25 and compared with the values obtained by static deposition.

As can be seen from the diagram, film growth is observed after an induction time of 1 h where initial precipitation is detected by AFM (see Fig. 5.3.22). After 6 h, a thickness in the range 300–350 nm is achieved in accordance with the TEM observations whereas static deposition only leads to films with a thickness of about 60 nm.

As a result, the continuous flow technique allows the preparation of films with a thickness substantially increased as compared to the static deposition described above.

5.3.3.3 Zinc Oxide Films

Owing to the tendency to form rather elongated zincite crystals [65], ZnO films deposited from aqueous solutions of zinc salts contain micron-sized crystals and exhibit a rather high surface roughness. Furthermore, their adhesion to the substrate is difficult to control [66, 67]. Recently, it has been shown that these problems can be eliminated by the addition of water-soluble additives. In this connection it is well known that the size of zincite microcrystals that are typically observed during thermohydrolysis of zinc salts up to 100°C can be controlled and reduced by the addition of water-soluble copolymers [68]. As a result, polymethacrylic acid partially grafted with polyethyleneoxide side chains has been employed successfully yielding adherent and uniform films [48]. Silicon that was initially oxidized with piranha solution and subsequently organically modified at the surface with γ-mercaptopropyltrimethoxysilane $Si(OCH_3)_3(CH_2)_3{-}SH$ was used as the substrate material. Beside the copolymer the aqueous deposition solution contains $ZnCl_2$, $Zn(CH_3COO)_2$ as well as hexamethylenetetraamine (HMTA) which serves to increase the pH of the solution. As can be seen from Fig. 5.3.26, a smooth homogeneous films is obtained by this process.

The images exhibit canyon-like features, attributable to film formation via the initial formation of islands at the surface which subsequently grow together.

Micron-sized crystals of zincite at the surface cannot be detected revealing the important role of the copolymer. XPS investigations show that the copolymer is present within the films. Furthermore, the results reveal that the organic phase is removed by annealing at 500°C resulting in nearly phase-pure ZnO films. The growth rate estimated by ellipsometric data is 5–20 nm/h.

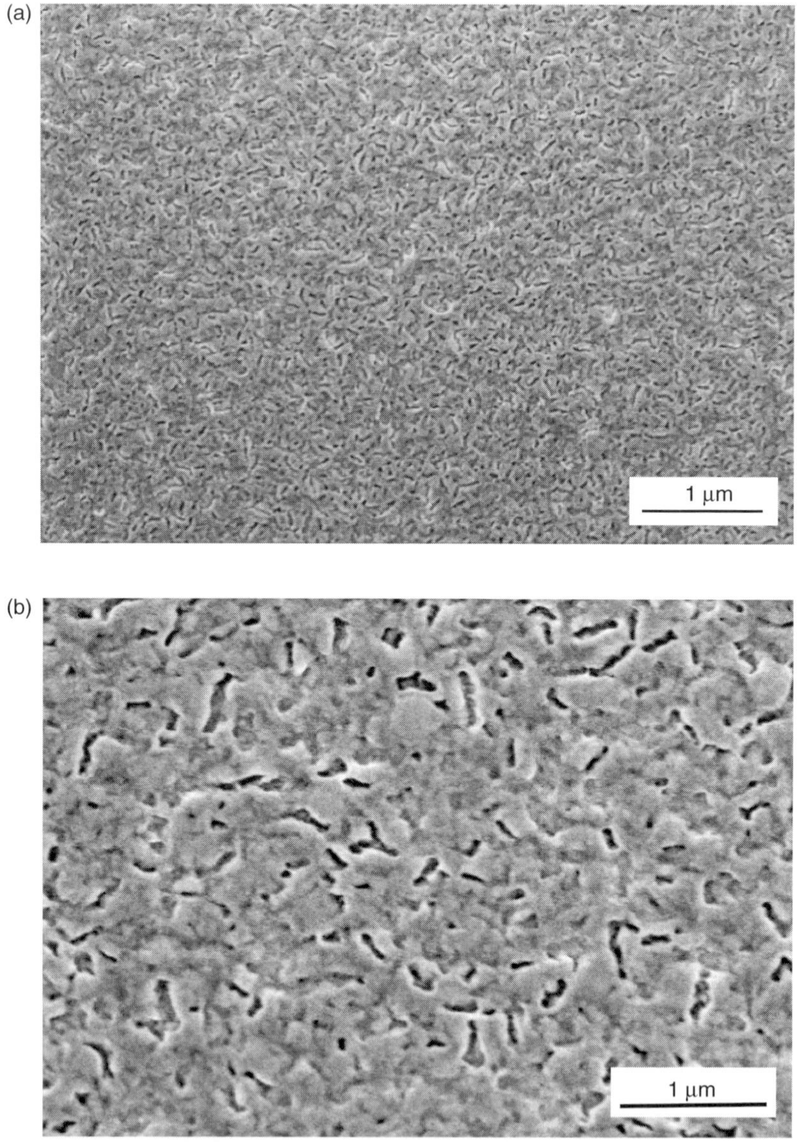

FIGURE 5.3.26 SEM image of a film deposited from an aqueous solution, containing $ZnCl_2$, $Zn(CH_3COO)_2$ and HMTA as well as a copolymer that consists of polymethacrylic acid partially grafted with polyethylenoxide side chains (deposition time: 6 h). The surface of the applied substrate material Si was initially modified with γ-mercaptopropyltrimethoxysilane $Si(OCH_3)_3(CH_2)_3{-}SH$ [48].

5.3.4 CONCLUSIONS

In summary, the bio-inspired processing of template-induced deposition provides a powerful means for the evolution of inorganic materials at low temperatures. Although this technique is still in its infancy, it seems to be very promising for the production of oxide films for practical uses. Besides further unary compounds, multicomponent oxides like yttria-stabilized zirconia, barium titanate and PZT $Pb(Ti,Zr)O_3$ are of special interest to be processed by this technique.

ACKNOWLEDGEMENTS

This project was supported by the Deutsche Forschungsgemeinschaft and the BMBF through grant 03C0294C/8. We also thank Prof. Mark De Guire for helpful discussions and for reviewing the manuscript.

REFERENCES

1. Setter, N., and Waser, R. (2000). Electroceramic materials. *Acta Mater.* **48**: 151–178.
2. Petzow, G. (1988). The optimisation of the microstructure of high technology ceramics. *Pract. Met.* **25**: 53–73.
3. Motzfeld, K. (1993). Silicon carbide: synthesis, structure and sintering, Proceedings of the "International Conference on Engineering Ceramics '92", pp. 7–41, Haviar, M., ed., Bratislava: Reproprint.
4. Omori, M., and Takei, H. (1982). Pressureless sintering of SiC. *J. Am. Ceram. Soc.* **65**: C-92.
5. Rixecker, G., Biswas, K., Wiedmann, I., and Aldinger, F. (2000). Liquid-phase sintered SiC ceramics with oxynitride additives. *J. Ceram. Process. Res.* **1**: 12–19.
6. Ohring, M. (2000). *The Materials Science of Thin Films*, Boston, MA: Academic Press.
7. Segal, D. (1989). *Chemical Synthesis of Advanced Ceramic Materials*, Cambridge: Cambridge University Press.
8. Narula, C. K. (1995). *Ceramic Precursor Technology and its Applications*, New York: Marcel Dekker.
9. Lowenstam, H. A., and Weiner, S. (1989). *On Biomineralization*, New York: Oxford University Press.
10. Mann, S., Webb, J., and Williams, R. J. P. (1989). *Biomineralization*, Weinheim: VCH Verlagsgesellschaft.
11. Baeuerlein, E. (2000). *Biomineralization*, Weinheim: Wiley-VCH.
12. Ball, P. (1997). *Made to Measure*, p. 193, Princeton, NJ: Princeton University Press.
13. Mann, S. (1998). *Biomimetic Materials Chemistry*, New York: VCH Publishers.
14. Calvert, P., and Rieke, P. (1996). Biomimetic mineralization in and on polymers. *Chem. Mater.* **8**: 1715–1727.

15. Heuer, A. H., Fink, D. J., Laraia, V. J., Arias, J. L., Calvert, P. D., Kendall, K., Messing, G. L., Blackwell, J., Rieke, P. C., Thompson, D. H., Wheeler, A. P., Veis, A., and Caplan, A. I. (1992). Innovative materials processing strategies: a biomimetic approach. *Science* **255**: 1098–1105.
16. Bunker, B. C., Rieke, P. C., Tarasevich, B. J., Campbell, A. A., Fryxell, G. E., Graff, G. L., Song, L., Liu, J., Virden, J. W., and McVay, G. L. (1994). Ceramic thin-film formation on functionalized interfaces through biomimetic processing. *Science* **264**: 48–55.
17. Niesen, T. P., and DeGuire, M. R. (2001). Review: deposition of ceramic thin films at low temperatures from aqueous solution. *J. Electroceramics* **6**: 169–207.
18. Aizenberg, J., Black, A. J., and Whitesides, G. M. (1999). Control of crystal nucleation by patterned self-assembled monolayers. *Nature* **398**: 495–498.
19. Aizenberg, J., Black, A. J., and Whitesides, G. M. (1999). Oriented growth of calcite controlled by self-assembled monolayers of functionalized alkanethiols supported on gold and silver. *J. Am. Chem. Soc.* **121**: 4500–4509.
20. Aizenberg, J. (2000). Patterned crystallization of calcite in vivo and in vitro. *J. Cryst. Growth* **211**: 143–148.
21. Maiti, M. (1994). Synthesis of iron oxide thin films on organic templates on silicon, M.Sc. thesis, Case Western Reserve University, OH.
22. Rieke, P. C., Marsh, B. D., Wood, L. L., Tarasevich, B. J., Liu, J., Song, L., and Fryxell, G. E. (1995). Aqueous solution deposition kinetics of iron oxyhydroxide on sulfonic acid terminated self-assembled monolayers. *Langmuir* **11**: 318–326.
23. Tarasevich, B. J., and Rieke, P. C. (1996). Nucleation and growth of oriented ceramic films onto organic interfaces. *Chem. Mater.* **8**: 292–300.
24. Rieke, P. C., Wiecek, R., Marsh, B. D., Wood, L. L., Liu, J., Song, L., Fryxell, G. E., and Tarasevich, B. J. (1996). Interfacial free energy of nucleation for iron oxyhydroxide on mixed self-assembled monolayers. *Langmuir* **12**: 4266–4271.
25. Liu, J., Kim, A. Y., Wang, L. Q., Palmer, B. J., Chen, Y. L., Bruinsma, P., Bunker, B. C., Exharos, G. J., Graff, G. L., Rieke, P. C., Fryxell, G. E., Virden, J. W., Tarasevich, B. J., and Chick, L. A. (1996). Self-assembly in the synthesis of ceramic materials and composites. *Adv. Colloids. Interf. Sci.* **69**: 131–180.
26. Nagtegaal, M., Stroeve, P., and Tremel, W. (1998). Growth of FeO(OH) crystals on self-assembled monolayers on gold. *Thin Solid Films* **327–329**: 571–575.
27. Nagtegaal, M., Stroeve, P., Ensling, J., Gütlich, P., Schurrer, M., Voit, H., Flath, J., Käshammer, J., Knoll, W., and Tremel, W. (1999). Soft-chemical growth of gamma-FeO(OH) films on self-assembled monolayers of substituted alkylthiols on gold(111). *Chem. Eur. J.* **5**: 1331–1337.
28. Agarwal, M., DeGuire, M. R., and Heuer, A. H. (1997). Synthesis of ZrO_2 and Y_2O_3-doped ZrO_2 thin films using self-assembled monolayers. *J. Am. Ceram. Soc.* **80**: 2967–2981.
29. Agarwal, M., DeGuire, M. R., and Heuer, A. H. (1997). Synthesis of yttrium oxide thin films with and without the use of organic self assembled monolayers. *Appl. Phys. Lett.* **71**: 891–893.
30. Shin, H., Collins, R. J., DeGuire, M. R., Heuer, A. H., and Sukenik, C. N. (1995). Synthesis and characterization of TiO_2 thin films on organic self-assembled monolayers: Part I. Film formation from aqueous solutions. *J. Mater. Res.* **10**: 692–698.
31. Collins, R. J., Shin, H., DeGuire, M. R., Heuer, A. H., and Sukenik, C. N. (1996). Low temperature deposition of patterned TiO_2 thin films using photopatterned self-assembled monolayers. *Appl. Phys. Lett.* **69**: 860–862.
32. Shin, H. (1996). "Deposition mechanism and electrical properties of TiO_2 thin films an self-assembled organic monolayers on Si, PhD thesis, Case Western Reserve University, OH.

33. Huang, D., Xiao, Z. D., Gu, J. H., Huang, N. P., and Yuan, C. W. (1997). TiO_2 thin films formation on industrial glass through self-assembly processing. *Thin Solid Films* 305: 110–115.
34. Baskaran, S., Song, L., Liu, J., Chen, Y. L., and Graff, G. L. (1998). Titanium oxide thin films on organic interfaces through biomimetic processing. *J. Am. Ceram. Soc.* **81**: 401–408.
35. Shin, H., Agarwal, M., DeGuire, M. R., and Heuer, A. H. (1998). Deposition mechanism of oxide thin films on self-assembled organic monolayers. *Acta Mater.* **46**: 801–815.
36. Shin, H., DeGuire, M. R., and Heuer, A. H. (1998). Electrical properties of TiO_2 thin films on self-assembled organic monolayers on Si. *J. Appl. Phys.* **83**: 3311–3317.
37. Xiao, Z., Su, L., Gu, N., and Wei, Y. (1998). The growth of TiO_2 thin films on mixed self-assembly monolayers from solution. *Thin Solid Films* **333**: 25–28.
38. Xiao, Z., Xu, M., Gu, J., Huang, D., and Lu, Z. (1998). Ti-based oxide thin films deposited from solution on self-assembled monolayers. *Mater. Chem. Phys.* **52**: 170–174.
39. Niesen, T. P., DeGuire, M. R., Bill, J., Aldinger, F., Rühle, M., Fischer, A., Jentoft, F. C., and Schlögl, R. (1999). Atomic force microscopic studies of oxide thin films on organic self-assembled monolayers. *J. Mater. Res.* **14**: 2464–2475.
40. Niesen, T. P., Wolff, J., Bill, J., DeGuire, M. R., and Aldinger, F. (1999). Synthesis and characterization of titania and vanadia thin films at organic self-assembled monolayers, *Organic–Inorganic Hybrid Materials II*, Vol. 576, pp. 197–202, Klein, L. C., Francis, L. F., DeGuire, M. R., and Mark, J. E., eds., Warrendale, PA: Materials Research Society.
41. Niesen, T. P., Bill, J., and Aldinger, F. (2001). Deposition of titania thin films by a peroxide route on different functionalized organic self-assembled monolayers. *Chem. Mater.* 13: 1552–1559.
42. Sampathkumaran, U., DeGuire, M. R., Heuer, A. H., Niesen, T., Bill, J., and Aldinger, F. (1999). Formation of self-assembled monolayers and ceramic films on semiconductor and oxide substrates, Proceedings of "The Symposium on Innovative Processing and Synthesis of Ceramics, Glasses, and Composites II", Vol. 94, pp. 307–318, Bansal, N. P., and Singh, J. P., eds., Westerville, OH: American Ceramic Society.
43. Fischer, A., Jentoft, F. C., Weinberg, G., Schlögl, R., Niesen, T. P., Bill, J., Aldinger, F., DeGuire, M. R., and Rühle, M. (1999). Characterization of thin films containing zirconium, oxygen, and sulfur by scanning electron and atomic force microscopy. *J. Mater. Res.* **14**: 3725–3733.
44. Palacin, S., Hidber, P. C., Bourgoin, J. P., Miramond, C., Fermon, C., and Whitesides, G. M. (1996). Patterning with magnetic materials at the micron scale. *Chem. Mater.* **8**: 1316–1325.
45. DeGuire, M. R., Niesen, T. P., Supothina, S., Wolff, J., Bill, J., Sukenik, C. N., Aldinger, F., Heuer, A. H., and Rühle, M. (1998). Synthesis of oxide and non-oxide inorganic materials at organic surfaces. *Z. Metallk.* **89**: 758–766.
46. DeGuire, M. R., Niesen, T. P., Wolff, J., Supothina, S., Bill, J., Aldinger, F., and Rühle, M. (1999). Synthesis of oxide and non-oxide inorganic materials at organic surfaces, *Precursor-derived Ceramics*, Proceedings of the "International Workshop on Grain Boundary Dynamics of Precursor-derived Ceramics", 10–16 November, 1996, Schloß Ringberg, Bavaria, Germany, pp. 143–161, Bill, J., Wakai, F., and Aldinger, F., eds., Weinheim: Wiley-VCH.
47. Kovtyukhova, N. I., Buzaneva, E. V., Waraksa, C. C., Martin, B. R., and Mallouk, T. E. (2000). Surface sol–gel synthesis of ultrathin semiconductor films. *Chem. Mater.* **12**: 383–389.
48. Hoffmann, R., Fuchs, T., Niesen, T. P., Bill, J., and Aldinger, F. (2002). *Surf. Interf. Anal.* **34**: 708–711.
49. Supothina, S., DeGuire, M. R., Heuer, A. H., Niesen, T. P., Bill, J., and Aldinger, F. (1999). Deposition of tin(IV) oxide ceramic films on organic self-assembled monolayers, *Organic–Inorganic Hybrid Materials II*, Vol. 576, pp. 203–208, Klein, L. C., Francis, L. F., DeGuire, M. R., and Mark, J. E., eds., Warrendale, PA: Materials Research Society.

50. Supothina, S., and DeGuire, M. R. (2000). Characterization of SnO_2 thin films grown from aqueous solutions. *Thin Solid Films* **371**: 1–9.
51. Sampathkumaran, U., Supothina, S., Wang, R., and DeGuire, M. R. (2000). Ceramic thin films on organic self-assembled monolayers: synthesis and the mechanism of formation. *Mineralization in Natural and Synthetic Biominerals*, Vol. 599, pp. 177–188, Li, P., Calvert, P., Levy, R. J., Kokubo, T., and Schied, C. R., eds., Warrendale, PA: Materials Research Society.
52. Meldrum, F. C., Flath, J., and Knoll, W. (1997). Chemical deposition of PbS on self-assembled monolayers of 16-mercaptohexadecanoic acid. *Langmuir* **13**: 2033–2049.
53. Meldrum, F. C., Flath, J., and Knoll, W. (1999). Formation of patterned PbS and ZnS films on self-assembled monolayers. *Thin Solid Films* **348**: 188–195.
54. Meldrum, F., Flath, J., and Knoll, W. (1999). Chemical deposition of PbS on a series of omega-functionalised self-assembled monolayers. *J. Mater. Chem.* **9**: 711–723.
55. Hwang, Y. K., Woo, S. Y., Lee, J. H., Jung, D. Y., and Kwon, Y. U. (2000). Micropatterned CdS thin films by selective solution deposition using microcontact printing techniques. *Chem. Mater.* **12**: 2059–2063.
56. Flath, J., Meldrum, F. C., and Knoll, W. (1998). Nucleation and growth of semiconductor particles on self-assembled monolayers by chemical solution deposition. *Thin Solid Films* **327–329**: 506–509.
57. Minh, N. Q. (1993). Ceramic fuel cells. *J. Am. Ceram. Soc.* **76**: 563–588.
58. Fuyuki, T., and Matsunami, H. (1986). Electronic properties of the interface between Si and TiO_2 deposited at very low temperatures. *Jpn. J. Appl. Phys.* **9**: 1288–1291.
59. Foster, N. F., and Rozgonyi, G. A. (1966). Zinc-oxide film microwave acoustic transducers. *Appl. Phys. Lett.* **8**: 221–223.
60. Ulman, A. (1996). Formation and Structure of Self-Assembled Monolayers. *Chem. Rev.* **96**: 1533–1554. Ulman, A. (1991). "An Introduction to Ultrathin Organic Films." Academic Press, San Diego.
61. Collins, R. J., and Sukenik, C. N. (1995). Sulfonate-functionalized, siloxane-anchored, self-assembled monolayers. *Langmuir* **11**: 2322–2324.
62. Jolivet, J. P. (2000). *Metal Oxide Chemistry and Synthesis*, New York: John Wiley.
63. Shin, H., Wang, Y., Sampathkumaran, U., DeGuire, M. R., Heuer, A. H., and Sukenik, C. N. (1999). Pyrolysis of self-assembled monolayers on inorganic substrates. *J. Mater. Res.* **14**: 2116–2123.
64. Fuchs, T., Hoffmann, R., Niesen, T., Tew, H., Bill, J., and Aldinger, F. (2002). *J. Mater. Chem.* **12**: 1597–1601.
65. Andrès-Vergès, M., Mifsud, A., and Serna, C. J. (1990). Formation of rod-like zinc oxide microcrystals in homogeneous solutions. *J. Chem. Soc. Faraday Trans.* **86**: 959–963.
66. Ito, K., and Nakamura, K. (1996). Preparation of ZnO thin films using the liquid film method. *Thin Solid Films* **286**: 35–36.
67. Saeed, T., and O'Brien, P. (1995). Deposition and characterisation of ZnO thin films grown by chemical bath deposition. *Thin Solid Films* **271**: 35–38.
68. Öner, M., Norwig, J., Meyer, W. H., and Wegner, G. (1998). Control of ZnO crystallization by a PEO-b-PMAA diblock copolymer. *Chem. Mater.* **10**: 460–463.
69. Xiao, Z., Gu, J., Huang, D., Lu, Z., and Wei, Y. (1998). The deposition of TiO_2 thin films on self-assembly monolayers studied by X-ray photoelectron spectroscopy. *Appl. Surf. Sci.* **125**: 85–92.

Handbook of Advanced Ceramics
S. Sōmiya *et al.* (Eds.)

5.4 Reaction Casting of Ceramic/Metal and Ceramic/Intermetallic Composites

KEN H. SANDHAGE[1] **and NILS CLAUSSEN**[2]
[1]*Department of Materials Science and Engineering, The Ohio State University, Columbus, OH, USA*
[2]*Advanced Ceramics Group, Technische Universitat Hamburg-Harburg, Hamburg, Germany*

5.4.1 INTRODUCTION

Ceramic/metal and ceramic/intermetallic composites can exhibit unique and attractive combinations of mechanical, thermal, and electrical properties. For example, such composites can possess higher hardness, wear resistance, stiffness, and creep resistance (particularly at elevated temperatures) than monolithic metals or intermetallic compounds, along with higher fracture strength, toughness, thermal shock resistance, and electrical conductivity than many monolithic ceramics [1–9]. Alternately, such composites may possess similar properties, but at significantly reduced weight relative to monolithic materials. As a result, appreciable worldwide activity has been, and is being, conducted to develop ceramic/metal or ceramic/intermetallic components for a variety of applications in the automotive, aircraft, energy production, and basic materials processing industries [10–14].

The fabrication of dense, ceramic-rich composites of complex shape often involves steps that are time consuming, labor intensive, or that involve high applied temperatures and/or pressures (e.g., hot uniaxial or isostatic pressing) [15–17]. Flaws generated during composite processing, such as distortions in dimensions and shape, have also resulted in low yields or have necessitated the use of an expensive final machining step. Such technical difficulties and relatively high costs encountered in conventional processing need to be avoided in order to increase the commercial viability of ceramic/metal or ceramic/intermetallic components.

Casting techniques are commonly used in metallurgical processing to rapidly and economically produce dense components with complex and near net shapes. Ceramic/metal composites may also be produced by the casting of: (i) a semi-solid metal mixture with ceramic powder into a die [18] or (ii) molten metal into a porous ceramic-bearing preform [4, 7–9, 15, 19, 20]. However, these methods are not well suited to the fabrication of high-melting composites, owing to the complications and costs associated with the casting of refractory metals. These processes are also not capable of producing composites with very high ceramic contents (e.g., composites with discontinuous metal). Such limitations can be avoided with "reactive casting" approaches that utilize chemical reactions between low-melting, easily-cast liquids and porous, shaped, ceramic-bearing preforms to produce higher-melting ceramic/metal or ceramic/intermetallic composites. The purpose of this chapter is to describe two reactive casting methods that are capable of producing dense, near net-shaped ceramic/metal and ceramic/intermetallic composites at modest temperatures: the Infiltration Alumina–Aluminide Alloy (i-3A) process and the Displacive Compensation of Porosity (DCP) process.

5.4.2 THE INFILTRATION ALUMINA–ALUMINIDE-ALLOY PROCESS

In the i-3A process, interpenetrating composites of aluminum oxide and aluminide compounds are produced by the reactive casting of molten aluminum (or an aluminum alloy) into porous, shaped oxide-bearing preforms [21–24]. Aluminum and aluminum-based alloys possess several characteristics that make these metals attractive for reactive casting: (i) low melting points; (ii) the ability to form high-melting alloys or intermetallic compounds; and (iii) high affinities for oxygen. The latter feature allows aluminum-rich liquids to undergo displacement (oxidation–reduction) reactions with porous preforms containing a variety of oxide phases. The aluminothermic reduction of a reactant oxide results in the formation of aluminum oxide and the liberation of a metallic element that can then undergo further reaction with excess molten aluminum to produce a higher-melting alloy or intermetallic compound. Several examples of displacement reactions used to produce alumina/aluminide composites at temperatures just above (727°C) or well above (1027°C) the melting point of aluminum are shown in Table 5.4.1 [21–25]. In addition to being thermodynamically favored, these reactions can also be quite exothermic. Indeed, the heat released per mole of alumina formed within several of these reactions is not far from that of the well-known thermite reaction (reaction 7 in Table 5.4.1).

By carefully controlling the preform phase content and microstructure in the i-3A process, dense, near net-shaped alumina/aluminide composites with tailored microstructures can be produced. In order for the final composite to

TABLE 5.4.1 Thermodynamic Data for Displacement Reactions between Molten Aluminum and Several Solid Oxides [26]

Displacement reaction[a]	ΔH°_{rxn} (kJ/mol O)[b]		ΔG°_{rxn} (kJ/mol O)[b]	
	727°C	1027°C	727°C	1027°C
(1) $13\{Al\} + 3TiO_2(r) = 2Al_2O_3 + 3TiAl_3$	−185.508	−186.381	−132.907	−116.981
(2) $7\{Al\} + 3TiO_2(r) = 2Al_2O_3 + 3TiAl$	−138.151	−138.556	−106.602	−97.068
(3) $5\{Al\} + 3NiAl_2O_4 = 4Al_2O_3 + 3NiAl$	−115.963	−116.640	−101.075	−96.511
(4) $5\{Al\} + 3NiO = Al_2O_3 + 3NiAl$	−465.120	−467.592	−414.081	−398.409
(5) $3\{Al\} + 3NiO = Al_2O_3 + Ni_3Al$	−387.185	−388.067	−351.955	−341.250
(6) $10\{Al\} + 3Nb_2O_5 = 5Al_2O_3 + 6Nb$	−188.303	−189.099	−161.288	−153.073
(7) $6\{Al\} + 3Fe_2O_3 = 3Al_2O_3 + 6Fe$	−443.530	−441.130	−400.321	−387.857

[a]The braces refer to species present in a liquid state (i.e. as a pure liquid or in a liquid solution). $TiO_2(r)$ refers to rutile.
[b]The thermodynamic values are expressed on a per mole of O basis (e.g. the values shown for reaction 1 were obtained by dividing ΔG° and ΔH° values per mole of this reaction by a factor of 6).

retain the shape and dimensions of the preform, the preform must remain rigid during reactive casting of the Al-bearing liquid. Rigid preforms can be produced by introducing inert Al_2O_3 along with the reactant oxide into the preform, and then lightly sintering the preform in order to build up necks between the Al_2O_3 particles. The resulting rigid alumina network can then inhibit compaction or distortion of the preform due to capillary pressure generated during infiltration. In addition to rigidity, the preform must possess proper phase and pore contents in order to generate a fully reacted composite with the desired amount and type of aluminide phase. For example, consider the following reactions:

$$3TiO_2(s) + 13\{Al\} + xAl_2O_3(s) = (2+x)Al_2O_3(s) + 3TiAl_3(s) \quad (1)$$

$$3TiO_2(s) + 7\{Al\} + xAl_2O_3(s) = (2+x)Al_2O_3(s) + 3TiAl(s) \quad (2)$$

For the case where $x = 7$ in reaction 1, a preform comprising of 24.0 vol% TiO_2 and 76.0 vol% Al_2O_3 is converted into a composite product consisting of 33.3 vol% $TiAl_3$ and 66.7 vol% Al_2O_3 (the molar volumes of TiO_2(rutile), Al_2O_3, TiAl, and $TiAl_3$ are 18.80, 25.58, 19.28, and 38.39 cm^3/mol, respectively [27]). Because the volume of the solid products (9 mol of Al_2O_3 and 3 mol of $TiAl_3 = 345.4\ cm^3$) is 46.7% larger than the volume of the solid reactants (7 mol of Al_2O_3 and 3 mol of $TiO_2 = 235.5\ cm^3$), the rigid TiO_2–Al_2O_3 preform should possess a porosity of 46.7% in this case to accommodate the reaction-induced increase in solid volume without a change in external dimensions. (Given that the molar volume of Al(l) at 900°C is 11.6 cm^3/mol, a $3TiO_2{:}7Al_2O_3$ preform with an open porosity in excess of 39.0% possesses sufficient pore volume for the Al(l) required for such a reaction [28].) If the value of x is changed to 2 and the preform possesses a porosity of 43.0%, then reactive casting of molten Al into such a preform should yield a composite of 36.1 vol% TiAl and 63.9 vol% Al_2O_3 (via reaction 2). By adjusting the relative amounts of aluminum oxide, reactant oxide, and porosity of the preform, interpenetrating alumina/aluminide composites can be produced with a range of phase contents and compositions.

The rate of reactive casting by the i-3A process can be enhanced with the use of modest applied pressures. For example, Wagner *et al.* [22] examined the gas pressure infiltration and squeeze casting of molten Al into porous, shaped, rigid TiO_2/Al_2O_3 preforms. Schematics of the gas pressure infiltration and squeeze casting processes are shown in Figure 5.4.1a and b, respectively. For gas pressure infiltration, the preforms were suspended in a bed of aluminum chips by an Al_2O_3 holder. The system was evacuated and heated at a rate of 10°C/min to 850°C. The pressure was then increased by introducing argon at

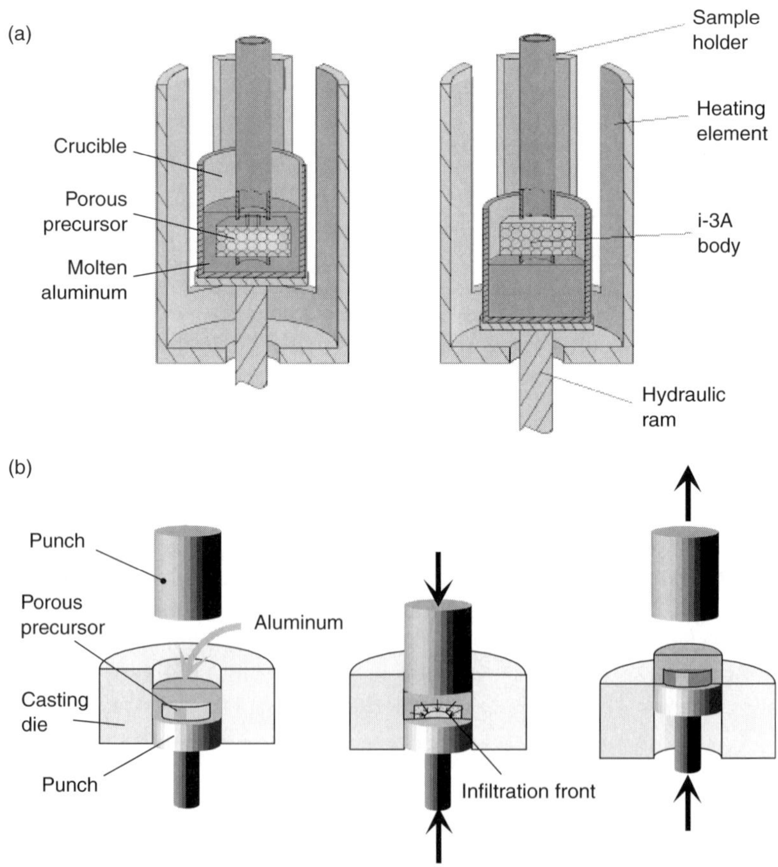

FIGURE 5.4.1 Schematics of the: (a) gas pressure infiltration and (b) squeeze casting processes used by Wagner *et al.* [22] (reprinted with permission from Elsevier Science).

a rate of 2 MPa/min to a peak pressure of 12 MPa. After holding for 10 min at this peak pressure, the furnace was cooled to 660°C and the bath was lowered so as to remove the specimen. For squeeze casting, the shaped, porous preforms were pre-heated to 900°C and inserted into a die that was heated to 300°C. Molten aluminum was then poured into the die and an upper ram was lowered to rapidly squeeze the aluminum into the preform (i.e., within 1 s). The phase content of the specimens after squeeze casting was found to be quite different than after gas pressure infiltration [22]. In the latter case, specimens removed from the furnace were completely reacted and consisted of an interpenetrating mixture of 58.1 vol% Al_2O_3 and 36.0 vol% $TiAl_3$, along with a 5.9 vol% Al (the presence of excess Al was due to the increase in the preform porosity resulting from the conversion of anatase to rutile prior to reactive infiltration). The microstructure of gas-pressure-infiltrated $Al_2O_3/TiAl_3$-bearing composite is shown in Figure 5.4.2a. The microstructure of a dense $Al_2O_3/NbAl_3/Nb_2Al$ composite produced by a similar gas pressure infiltration process (e.g. at peak casting conditions of 12 MPa and 800°C for 20 min) is also shown in Figure 5.4.2b [23]. After squeeze casting, no aluminide formation was detected. This was attributed to the short exposure time of the titania in the preform to molten aluminum during squeeze casting. Additional annealing of the squeeze-cast specimens at 640°C for 4 h resulted in complete conversion of the TiO_2 into a mixture of Al_2O_3 and $TiAl_3$ with a microstructure similar to that shown in Figure 5.4.2a. Because the porous preforms contained a rigid network of 70 vol% Al_2O_3 (i.e. the preforms were partially sintered at 1250°C for 30 min), no distortions in shape or changes in dimension were detected upon reactive casting (i.e. near net-shape processing). The near net-shaped, $Al_2O_3/TiAl_3$-bearing composites exhibited fracture strengths (four-point bending) of 494–542 MPa, fracture toughness values (indentation based) of 5.0–8.6 MPa $m^{1/2}$, and hardness values of 5.65–7.4 GPa [22].

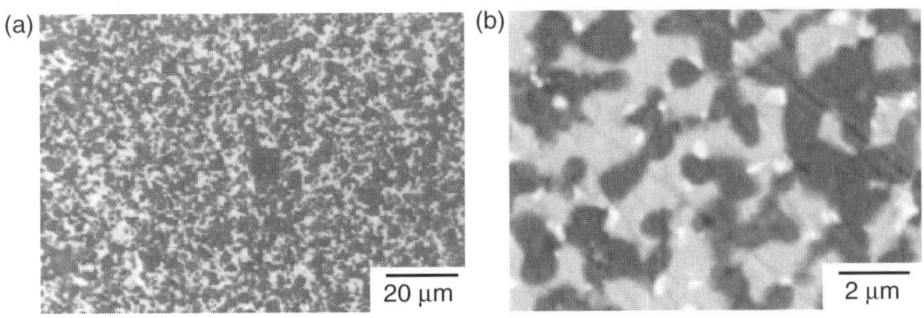

FIGURE 5.4.2 (a) Optical micrograph of a polished cross-section of an i-3A-derived composite with interpenetrating Al_2O_3 (dark grey) and $TiAl_3$ (light grey) phases, and (b) secondary electron image of a polished cross-section of an i-3A-derived composite with interpenetrating Al_2O_3 (dark grey) and $NbAl_3$ (light grey) phases, with a smaller amount of Nb_2Al (white). (Reprinted with permission from Elsevier Science and Wiley/VCH Verlag.)

The rate of reaction during squeeze casting can be enhanced by increasing the exposed surface area of the reactant oxide in the preform. Beyer *et al.* [24] used the amount of reactant oxide within the preform (50–100 vol% TiO_2) and the preform sintering conditions (no sintering versus sintering at 900–1150°C) to alter the reactant surface area. The preforms were then heated to 650°C and squeeze cast with an Al–Si alloy liquid (German standard G AlSi12(Cu) preheated to 750°C) at a peak pressure of 39 MPa in a die heated to 350°C. Preforms of TiO_2/Al_2O_3 sintered at $\geq$1025°C exhibited no apparent reaction with the molten aluminum during casting, whereas unsintered preforms completely reacted into a mixture of Al_2O_3, $TiAl_3$, and Ti_5Si_3 (i.e. no residual Al or TiO_2 were detected by X-ray diffraction analyses, and no melting endotherms were observed by differential thermal analyses at <900°C). The extent of reaction during casting of sintered preforms was also enhanced by increasing the amount of the titania reactant. However, 100% titania preforms sintered at $\leq$900°C exhibited appreciable thermal runaway during squeeze casting that, in turn, caused macroscopic cracking and extensive phase coarsening. These authors concluded that, with proper control of the reactant content, alumina content, and rigidity of the preform (i.e. the extent of sintering), dense, fully reacted composites that retain the shape and dimensions of the starting preform can be directly produced upon squeeze casting [24].

Recent work by Kumar *et al.* [29] has demonstrated that metal–alloy-rich (75 vol% metal) composites can also be produced by the reactive squeeze casting of Al-rich liquids into porous preforms. These authors produced dense composites of a Ni-based alloy (76.9 at% Ni, 10.5 at% Al, 8.2 at% Co, 4.4 at% Cr) reinforced with Al_2O_3. An Al–Si liquid (preheated to 700°C) was squeeze cast at a pressure of 39 MPa into a porous preform (preheated to 600°C) comprising of an intimate mixture of Cr_2O_3, Ni, and Co. The infiltrated preform was then annealed for 5 h at 1000°C. The resulting metal–matrix composite was free of residual Al (as determined by the absence of melting endotherms at <1000°C) and did not exhibit detectable cracking upon Vickers indentation at loads up to 200 N.

5.4.3 THE DISPLACIVE COMPENSATION OF POROSITY PROCESS

The ceramic product of the displacement reactions discussed above, α-Al_2O_3, is a close-packed-type oxide with a modest molar volume, as shown in Table 5.4.2. Indeed, when expressed on a per-mole-of-oxygen basis, alumina is among the most compact of oxides. As a result, aluminothermic displacement reactions tend to produce less oxide than is consumed (see the third column in

TABLE 5.4.2 Molar Volumes of Several Oxide Compounds

Oxide	Molar volume[a] (cm^3/mol oxide)	Volume per mole O V_{mO} (cm^3/mol O)	% Volume difference of α-Al_2O_3 with oxide
α-Al_2O_3	25.58	8.53	
TiO_2(rutile)	18.80	9.40	−9.3
$NiAl_2O_4$	39.25	9.81	−13.0
$MgAl_2O_4$	39.76	9.94	−14.2
Fe_2O_3	30.31	10.10	−15.5
$Al_6Si_2O_{13}$	134.40	10.34	−17.5
ZrO_2 (monoclinic)	21.19	10.59	−19.5
NiO	10.97	10.97	−22.2
Fe_3O_4	44.56	11.14	−23.4
MgO	11.25	11.25	−24.2
Nb_2O_5	58.69	11.74	−27.3
SiO_2 (cristobalite)	25.78	12.89	−33.8

[a]Values are based on room temperature crystallographic data [27].

Table 5.4.2). Consequently, such reactions are inherently incapable of producing ceramic-rich composites with discontinuous metal or intermetallic phases (e.g. electrically insulating composites).

In the DCP process, a reactive liquid is cast into a porous, shaped, rigid, ceramic-bearing preform. As for the i-3A process, the infiltrating liquid undergoes a displacement reaction with a ceramic phase in the preform. Unlike the i-3A process, however, displacement reactions are chosen that yield a larger volume of ceramic than is consumed [30–34]. Because the preform is rigid, this reaction-induced increase in solid ceramic volume acts to fill the pores in the preform (i.e. reaction-induced densification). If the metallic product of the reaction is solid at the reaction temperature, then dense, ceramic/metal composites can be fabricated. If, on the other hand, the metallic product is molten, then this metallic liquid will be extruded out of the specimen as the pores become filled with the more voluminous ceramic product, so that a composite with a very high ceramic content can be produced (this latter process is also referred to as PRIMA-DCP, where PRIMA refers to pressureless reversible infiltration of molten alloys). Hence, by varying the melt composition, the preform porosity, and the preform phase content, dense near net-shaped composites with a range of compositions and phase contents (e.g. continuous to discontinuous metal reinforcements) can be produced. Several examples of thermodynamically favored DCP reactions are shown in Table 5.4.3.

Kumar and Sandhage [30] used the DCP process to fabricate lightweight, near net-shaped MgO/Mg–Al alloy composites. Rigid, porous alumina preforms

TABLE 5.4.3 Thermodynamic and Volume Change Data for Several DCP-like Displacement Reactions [26, 27]

	Displacement reaction[a]	ΔG°_{rxn} [1027°C] (kJ/mol O or C)	% Change in ceramic volume[b]	% Change in solid volume[c]
(1)	$3\{Mg\} + Al_2O_3 = 3MgO + 2\{Al\}$	−37.421	+31.9%	+31.9%
(2)	$3\{Mg\} + 4Al_2O_3 = 3MgAl_2O_4 + 2\{Al\}$	−16.010	+16.6%	+16.6%
(3)	$\{Mg\} + NiAl_2O_4 = MgAl_2O_4 + Ni$	−87.174	+1.3%	+18.1%
(4)	$\{Zr\} + WC = ZrC + W$	−150.623	+24.5%	+100.9%
(5)	$\{Hf\} + WC = HfC + W$	−207.114	+20.1%	+96.6%

[a]The braces refer to species present in a liquid state (as a pure liquid or in a liquid solution).
[b]% Change in ceramic volume = the reaction-induced change in ceramic volume (e.g. in reaction 1, this value is $100 \cdot \{3V_m[MgO] - V_m[Al_2O_3]\}/V_m[Al_2O_3] = 100 \cdot \{3(11.25) - 25.58\}/25.58 = +31.9\%$).
[c]% Change in solid volume = the reaction-induced change in total solid volume (e.g. in reaction 4, this value is $100 \cdot \{V_m[ZrC] + V_m[W] - V_m[WC]\}/V_m[WC] = 100 \cdot \{15.56 + 9.55 - 12.50\}/12.50 = +100.9\%$).

were exposed to an excess of molten magnesium at 1000°C and ambient pressure. The Mg(l) wet, infiltrated, and reacted with the Al_2O_3 (via net reaction 1 in Table 5.4.3) to yield dense MgO/Mg–Al composites. Because the rigid Al_2O_3 network was converted into a more voluminous, rigid MgO network, the preform shape and dimensions were retained (to within a few percent) by the final composites. By varying the preform porosity from 47 to 29%, composites with MgO contents ranging from 70 to 86 vol%, respectively, could be produced. Composites with ≥83 vol% MgO were found to be electrically insulating; that is, sufficient Mg–Al liquid was extruded from these specimens during the pore-filling reaction, that the remaining metallic phase in the composite was discontinuous. The ability to produce such ceramic-rich (discontinuous metal) composites well below conventional ceramic sintering temperatures is a unique feature of the DCP process. As the preform porosity increased, the time required for complete consumption of the alumina decreased, which was attributed to a decrease in the amount of solid reactant and an increase in liquid/solid interfacial area [30]. The conversion of Al_2O_3 into MgO was found to occur via the formation of an intermediate solid product, $MgAl_2O_4$ (spinel). The spinel formed as a continuous, adherent layer on alumina, with magnesia present as a continuous layer on the spinel. Indeed, these layers formed in an epitaxial manner, with good alignment of the close-packed-type oxygen planes (and close-packed directions within those planes) in the MgO, $MgAl_2O_4$, and Al_2O_3 layers [31]. Such epitaxy suggests that reactive casting may be used to convert porous preforms with oriented ceramic grains into dense, textured ceramic composites. The high relative density of the MgO/Mg–Al composites (≥98% of theoretical) resulted in good resistance to hydration. After exposure for 24 h at 71°C (ASTM hydration test C492), the composites exhibited negligible weight change (≤0.04%). The average density, fracture strength (four-point bending), and fracture toughness (indentation) values of composites with 74 ± 4 vol% MgO were 3.1 g/cm^3, 244 MPa, and 5.4 MPa m$^{1/2}$, respectively [32]. Such composites are attractive for applications that require lightweight materials with enhanced stiffness relative to Mg–Al alloys and enhanced fracture strength and toughness relative to MgO. Dense, near net-shaped, MgO-rich composites with discontinuous Mg–Al may also be attractive for electrical insulation.

Rogers *et al.* [32] used the DCP process to produce co-continuous $MgAl_2O_4$/Fe–Ni–Al heat resistant alloy composites by the pressureless reactive casting of Mg–Al liquids at 900°C into porous preforms comprised of a mixture of Fe and $NiAl_2O_4$. The following net reaction was used in this case:

$$4/3\{Mg_{3/4}Al_{1/4}\} + xFe(s) + NiAl_2O_4 = MgAl_2O_4 + y[Fe_{x/y}Ni_{1/y}Al_{1/3y}] \quad (3)$$

where $\{Mg_{3/4}Al_{1/4}\}$ refers to a Mg–Al liquid solution, $[Fe_{x/y}Ni_{1/y}Al_{1/3y}]$ refers to a solid Fe–Ni–Al alloy, $y = x + 4/3$, and $x = 4$–8. In this reaction, the magnesium present in the liquid undergoes a selective displacement reaction with the nickel oxide in nickel aluminate. The displaced nickel can then react with solid iron and liquid aluminum to produce a solid Fe–Ni–Al alloy. The low-melting Mg–Al liquid solution is thereby converted into a higher-melting composite. The products of this net reaction possess a volume that is larger than the solid reactants (e.g. for $x = 4$, the reaction results in a 15% increase in volume [27, 35], so that the pores in a rigid Fe/$NiAl_2O_4$ preform can be filled as the reaction proceeds. The microstructure of a DCP-derived, $MgAl_2O_4$/Fe–Ni–Al-bearing composite is shown in Figure 5.4.3a. Microprobe

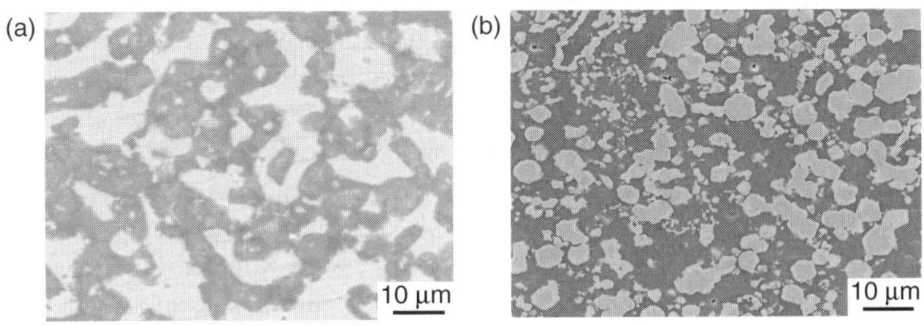

FIGURE 5.4.3 (a) Backscattered electron image of a polished cross-section of a co-continuous $MgAl_2O_4$/Fe–Ni–Al alloy-bearing composite (oxide = grey phase) produced at 900°C, and (b) secondary electron image of a polished cross-section of a ZrC/WC/W-bearing composite (dark grey phase = ZrC, light particles = WC core with a W skin) produced at 1300°C by the DCP method. (Reprinted with permission of the American Ceramic Society.)

analyses indicated that the metal alloy possessed a composition of 82.0 at% Fe, 12.8 at% Al, 3.9 at% Ni, and 1.3 at% Mg, whereas the spinel was comprised of 50.1 mol% MgO, 1.6 mol% NiO, 37.8 mol% Al_2O_3, and 10.5 mol% Fe_2O_3. Such composites possessed average fracture strengths and toughnesses (both via four-point bending) of 380 MPa and 11.3 MPa $m^{1/2}$, respectively [32].

Non-oxide-based composites have also been synthesized by the DCP route. Dickerson *et al.* [33, 34] have fabricated ZrC/W-bearing composites by the reactive casting of a Zr_2Cu liquid into porous WC preforms. The net displacement reaction in this case was:

$$0.5\{Zr_2Cu\} + WC = ZrC + W + 0.5\{Cu\} \quad (4)$$

Zr_2Cu melts congruently at 1025°C, which is more than 800°C lower than the melting point of pure Zr [36]. This liquid was found to wet and pressureless infiltrate into porous WC at 1200–1300°C. As indicated in Table 5.4.3, the combined molar volumes of ZrC and W are about twice (100.9%) the molar volume of WC [27]. Hence, a rigid preform with $\geq$50% open porosity can accommodate the increase in solid volume associated with complete conversion of WC into ZrC and W. The residual molten copper generated upon removal of zirconium from the Zr_2Cu liquid cannot form stable compounds with ZrC or W. Hence, copper is an ideal transient element for such a DCP route, in that it dramatically lowers the temperature required for reactive infiltration and can then be extruded as a noble liquid back out of the specimen as the pores are filled with the solid products, ZrC and W. The presence of small amounts of trapped copper in the final composite have essentially no effect on the refractoriness of the resulting composite (i.e. the melting point of tungsten is reduced by only about 8°C in the presence of copper [37]. The microstructure of a DCP-derived, ZrC/WC/W-bearing composite produced at a peak temperature of 1300°C for 1 h is shown in Figure 5.4.3b. By varying the preform porosity and the amount of inert W introduced into the preform (i.e. W/WC preforms), Dickerson *et al.* [33, 34] have produced dense composites with carbide (ZrC + WC) contents ranging from 17.2 to 71.8 vol%. As shown in Figure 5.4.4, a DCP-derived, ZrC/W-bearing composite can retain the shape and dimensions of a porous WC-bearing preform to within a few percent. ZrC/W-bearing composites can be attractive for very high temperature applications (e.g. in excess of 2500°C) where excellent resistance to creep, erosion, or thermal cycling are required (the average linear thermal expansion coefficients for polycrystalline ZrC and W are quite similar from room temperature to 2700°C) [38–40]. ZrC/W composites have also been found to exhibit high fracture strengths at elevated temperatures (e.g. 979 MPa at 1200°C for hot pressed 20 vol% ZrC/80 vol% W composites [41]). The DCP route allows for the fabrication of dense, near net-shaped ZrC/W composites at temperatures far below the ultimate use temperature and with much more modest

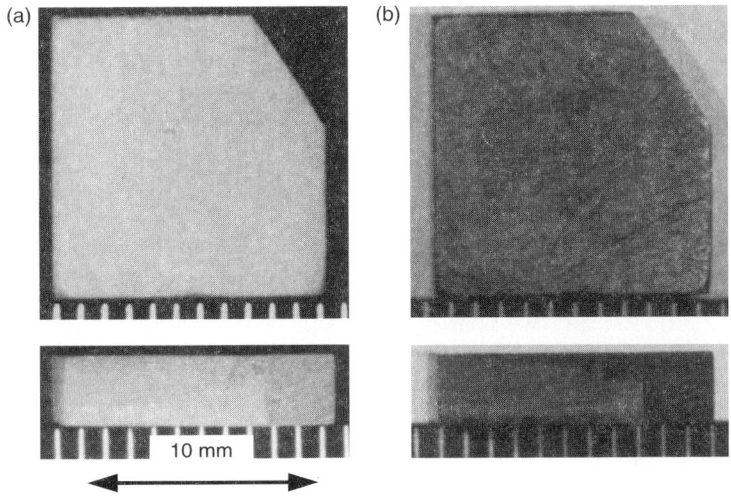

FIGURE 5.4.4 Optical images of (a) a porous WC preform (top-down and side views) and (b) the same specimen after conversion at 1300°C into a dense ZrC/W-bearing composite by the DCP method (one corner was cut off the preform in order to allow for unambiguous identification of each corner after conversion).

processing conditions than required with conventional processing (e.g. vacuum hot pressing at 2000°C at 20 MPa [41]).

5.4.4 SUMMARY

The reactive casting of ceramic/metal and ceramic/intermetallic composites can be an attractive, economical means of fabricating ceramic/metal and ceramic/intermetallic composites. Low-melting metallic liquids can be rapidly cast into porous preforms and then converted by displacement (oxidation–reduction) reactions into higher-melting composites. By controlling the liquid composition and the preform chemistry and microstructure, a wide variety of ceramic–matrix, metal–matrix, or co-continuous ceramic/metal or ceramic/intermetallic composites can be synthesized. Rigid preforms can be converted into dense composites that retain the preform shape and dimensions to within a few percent or less. With reactive casting approaches, such as the i-3A and DCP methods, costly and labor-intensive processing steps (e.g. hot pressing, extensive machining) that have inhibited the commercial application of ceramic-bearing composites can be avoided.

REFERENCES

1. Rodel, J., Prielipp, H., Claussen, N., Sternitzke, M., Alexander, K. B., Becher, P. F., and Schneibel, J. H. (1995). Ni_3Al/Al_2O_3 composites with interpenetrating networks. *Scripta Met. Mater.* **33**: 843–848.
2. Daehn, G. S., Starck, B., Xu, L., Elfishawy, K. F., Ringnalda, J., and Fraser, H. L. (1996). Elastic and plastic behavior of a co-continuous alumina/aluminum composite. *Acta Mater.* **44**: 249–261.
3. Ewsuk, K. G., Glass, S. J., Loehman, R. E., and Tomsia, A. P., and Fahrenholtz, W. G. (1996). Microstructure and properties of Al_2O_3–Al(Si) and Al_2O_3–Al(Si)–Si composites formed by in situ reaction of Al with aluminosilicate ceramics. *Met. Mater. Trans. A* **27**: 2122–2129.
4. Prielipp, H., Knechtel, M., Claussen, N., Streiffer, S. K., Mullejans, H., Ruhle, M., and Rodel, J. (1995). Strength and fracture toughness of aluminum/alumina composites with interpenetrating networks. *Mater. Sci. Eng. A* **197**: 19–30.
5. Breslin, M. C., Ringnalda, J., Xu, L., Fuller, M., Seeger, J., Daehn, G. S., Otani, T., and Fraser, H. L. (1995). Processing, microstructure and properties of co-continuous alumina/aluminum composites. *Mater. Sci. Eng. A* **195**: 113–119.
6. Schon, S., Prielipp, H., Janssen, R., Rodel, J., and Claussen, N. (1994). Effect of microstructural scale on thermal shock resistance of aluminum-reinforced alumina. *J. Am. Ceram. Soc.* **77**: 701–704.
7. Knechtel, M., Prielipp, H., Mullejans, H., Claussen, N., and Rodel, J. (1994). Mechanical properties of Al/Al_2O_3 and Cu/Al_2O_3 composites with interpenetrating networks. *Scripta Metall. Mater.* **31**: 1085–1090.
8. Travitsky, N. A., and Claussen, N. (1992). Microstructure and properties of metal infiltrated RBSN composites. *J. Eur. Ceram. Soc.* **9**: 61–65.
9. Aghajanian, M. K., Rocazella, M. A., Burke, J. T., and Keck, S. D. (1991). The fabrication of metal matrix composites by a pressureless infiltration technique. *J. Mater. Sci.* **26**: 447–454.
10. Southam, R. E., Reinicke-Murmann, J., Kreuter, P., and Rogers, G. (1990). Analysis of potential improvements in engine behavior due to ceramic valve train components, Paper No. 900452, SAE Technical Paper Series, Warrendale, PA: SAE International.
11. Sheppard, L. M. (1990). Automotive performance accelerates with ceramics. *Am. Ceram. Soc. Bull.* **69**: 1012–1021.
12. Breslin, M. C., Xu, L., Daehn, G. S., and Fraser, H. L. (1993). Applications of co-continuous ceramic composite materials for automotive components, Paper No. 930184, SAE Technical Paper Series, Warrendale, PA: SAE International.
13. Smith, V., Deckman, B., and Brueck, D. (1994). Advanced ceramics: where do we go from here. *Am. Ceram. Soc. Bull.* **73**: 49–52.
14. Asnani, M., Southam, R. E., and Wills, R. R. (1987). Contact fatigue damage in ceramic and metal valve tips, Paper No. 870419, SAE Technical Paper Series, Warrendale, PA: SAE International.
15. Cornie, J. A., Chiang, Y-M., Uhlmann, D. R., Mortensen, A., and Collins, J. M. (1986). Processing of metal and ceramic matrix composites. *Am. Ceram. Soc. Bull.* **65**: 293–304.
16. Rice, R. W. (1990). Processing, microstructures, performance, and economics of ceramic composites, *Metal and Ceramic Matrix Composites: Processing, Modeling, and Mechanical Properties*, pp. 159–166, Bhagat, R. B., Clauer, A. H., Kumar, P., and Ritter, A. M., eds., Warrendale, PA: The Minerals, Metals and Materials Society.
17. Leipold, M. H. (1976). Hot pressing. *Treatise on Materials Science and Technology, Vol. 9, Ceramic Fabrication Processes*, pp. 99–134, Wang, F. F. Y., ed., New York: Academic Press.

18. Mehrabian, R., Riek, R. G., and Flemings, M. C. (1974). Preparation and casting of metal-particulate non-metal composites. *Metall. Trans.* **5**: 1899–1905.
19. Gonzalez, E. J., and Trumble, K. P. (1996). Spontaneous infiltration of alumina by copper–oxygen alloys. *J. Am. Ceram. Soc.* **79**: 114–120.
20. Toy, C., and Scott, W. D. (1990). Ceramic–metal composite produced by melt infiltration. *J. Am. Ceram. Soc.* **73**: 97–101.
21. Claussen, N. (1999). Processing of advanced ceramic composites: issues of producibility, affordability, reliability, and tailorability. *Br. Ceram. Trans.* **98**: 256–257.
22. Wagner, F., Garcia, D. E., Krupp, A., and Claussen, N. (1999). Interpenetrating Al_2O_3–$TiAl_3$ alloys produced by reactive infiltration. *J. Eur. Ceram. Soc.* **19**: 2449–2453.
23. Scheu, C., Dehm, G., Kaplan, W. D., Wagner, F., and Claussen, N. (1998). Microstructure and phase evolution of niobium–aluminide–alumina composites prepared by melt-infiltration. *Phys. Stat. Sol. (a)* **166**: 241–255.
24. Beyer, P., Janssen, R., and Claussen, N. (2000). Synthesis of aluminide–alumina composites by reactive squeeze casting. *Adv. Eng. Mater.* **2**: 734–747.
25. Fahrenholtz, W. G., Ewsuk, K. G., Loehman, R. E., and Tomsia, A. P. (1996). Formation of structural intermetallics by reactive metal penetration of Ti and Ni oxides and aluminates. *Met. Mater. Trans. A* **27**: 2100–2104.
26. Barin, I. (1989). *Thermochemical Data of Pure Substances*, Parts I and II, pp. 42, 48, 61, 71, 72, 666, 868, 869, 1067, 1069, 1359, 1546, 1642, 1714, Weinheim: VCH Verlagsgesellschaft.
27. JCPDS Card Files. Card numbers 4-806, 4-835, 4-850, 5-678, 6-696, 9-97, 10-339, 21-1152, 21-1276, 25-1047, 33-664, 35-784, 35-789, 37-1449, 39-1491, 43-1042, 43-1484, 44-1188, 45-946. Joint Committee for Powder Diffraction Data, Swarthmore, PA (now International Center for Diffraction Data, Newtown Square, PA).
28. Brandes, E. A. (1983). *Smithells Metals Reference Book*, 6th edn, Ch. 14, p. 7, London: Butterworth & Co.
29. Kumar, P., Travitsky, N. A., Beyer, P., Sandhage, K. H., Janssen, R., and Claussen, N. (2001). Reactive casting of ceramic composites (R-3C). *Scripta Mater.* **44**: 751–757.
30. Kumar, P., and Sandhage, K. H. (1999). The displacive compensation of porosity (DCP) method for fabricating dense, shaped, high-ceramic-bearing bodies at modest temperatures. *J. Mater. Sci.* **34**: 5757–5769.
31. Kumar, P., Dregia, S. A., and Sandhage, K. H. (1999). Epitaxial growth of magnesia and spinel on sapphire during incongruent reduction in molten magnesium. *J. Mater. Res.* **14**: 3312–3318.
32. Rogers, K. A., Kumar, P., Citak, R., and Sandhage, K. H. (1999). The fabrication of dense, shaped ceramic/metal composites at $\leq$1000°C by the displacive compensation of porosity method. *J. Am. Ceram. Soc.* **82**: 757–760.
33. Dickerson, M. B., Snyder, R. L., and Sandhage, K. H. (2002). Dense, near net-shaped, carbide/refractory metal composites at modest temperatures by the displacive compensation of porosity (DCP) method. *J. Am. Ceram. Soc.* **85**: 730–732.
34. Dickerson, M. B., Snyder, R. L., and Sandhage, K. H. (2001). Low-temperature fabrication of dense, near net-shaped tungsten/zirconium carbide composites with tailored phase contents by the PRIMA-DCP process. *Ceram. Eng. Sci. Proc.* **22**: 97–107.
35. Pearson, W. B. (1958). *A Handbook of Lattice Spacings and Structures of Metals and Alloys*, pp. 347–351, New York: Pergamon Press.
36. Kneller, E., Khan, Y., and Gorres, U. (1986). The alloy system copper–zirconium. Part I. Phase diagram and structural relations. *Z. Metallkunde* **77**: 43–48.
37. Subramanian, P. R., and Laughlin, D. E. (1991). Cu–W (copper–tungsten), *Phase Diagrams of Binary Tungsten Alloys*, pp. 76–79, Nagender Naidu, S. V., and Rama Rao, P., eds., Calcutta: Indian Institute of Metals.

38. Storms, E. K. (1967). The zirconium–zirconium carbide system, *The Refractory Carbides*, pp. 18–35, New York: Academic Press.
39. Touloukian, Y. S., Kirby, R. K., Taylor, R. E., and Desai, P. D. (1975). *Thermophysical Properties of Matter: Thermal Expansion of Metallic Elements and Alloys*, Vol. 12, p. 354, New York: Plenum Press.
40. Touloukian, Y. S., Kirby, R. K., Taylor, R. E., and Lee, T. Y. R. (1977). *Thermophysical Properties of Matter: Thermal Expansion of NonMetallic Solids*, Vol. 13, p. 926, New York: Plenum Press.
41. Song, G. M., Zhou, Y., Wang, Y. J., and Lei, T. C. (1998). Elevated temperature strength of a 20 vol% ZrC/W composite. *J. Mater. Sci. Lett.* **17**: 1739–1741.

Handbook of Advanced Ceramics
S. Sōmiya *et al.* (Eds.)

5.5 Novel Colloidal Forming Technique: Direct Coagulation Casting

B. BALZER and L. J. GAUCKLER
ETH Zurich, Institute for Nonmetallic Materials, CH-8092 Zurich, Switzerland

5.5.1 INTRODUCTION

One of the most promising strategies for producing high-quality ceramics from powders is the wet-processing route. Less and smaller defects can be expected in the final microstructure due to better control of particle–particle interactions and increased homogeneity of particle packing in the wet stage compared to dry processing. Many attempts have been made in recent years to improve ceramic processing by new forming techniques. This can be achieved by either consolidating the dispersing medium or by coagulating the suspended particles. Consolidation of the dispersing medium is applied in gel-casting [1–3] where monomers in the suspension are polymerized or polymer gelation is induced at higher temperature [4]. Polymerization or gelation can be induced by UV radiation [5], by heat [6], or by a catalyst [2]. In freeze-casting and Quickset [7–10], the dispersing medium is frozen and removed after demolding by sublimation.

In flocculation and coagulation of a densely packed suspension, the interparticle forces are controlled in order to accomplish a liquid-to-solid transition. Consolidation may be induced either by applying heat, changing pH, or by increasing the ionic strength of the suspension. Bergström [11] and Napper [12] used a sterically stabilized suspension and destabilized it by heating to form a rigid green body. Lange [13], Velamakanni and Coworkers [14, 15] and Franks *et al.* [16] coagulate suspensions by adding salt and then increase the solids loading by pressure filtration or centrifugation. Pujar *et al.* [17] proposed a reaction injection molding method using heterocoagulation of silica–alumina mixtures.

Recently, we introduced a new approach based on the idea to destabilize a high solids loading aqueous suspension by a time-delayed, enzymatically catalyzed internal reaction [18–21]. The suspension, containing powder,

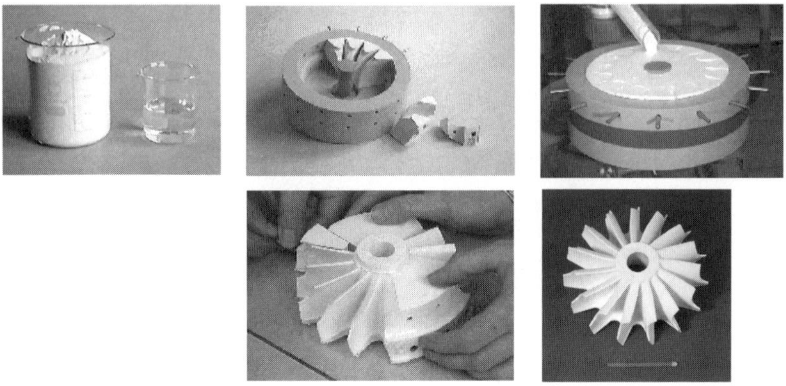

FIGURE 5.5.1 Processing complex-shaped parts by DCC: (upper row from left) starting materials, mold, casting of the suspension, (lower row from left) demolding of the wet green body, sintered turbo rotor.

a water-soluble substrate, a catalyst, and optionally polymer molecules, is cast into a mold where the products of the substrate decomposition initiate the coagulation, changing either the pH of the suspension or increasing the ionic strength. This method, called direct coagulation casting (DCC) is suitable for ceramic parts even with complex geometry (Fig. 5.5.1).

DCC combines the advantages of stabilized aqueous suspensions (cheap, harmless, and easily removable liquid; high solids content, and good flow properties) with a homogeneously induced coagulation: instead of adding the coagulating agents from outside, they are produced inside the suspension, which leads to homogeneous and strong green bodies. The coagulation rate is controlled by substrate and enzyme concentration and temperature. The amount of organics can be kept very low, therefore no special binder-removing step is necessary.

5.5.2 BASICS

In water, metal oxide surfaces are hydrolyzed:

$$\mathrm{MeO} + \mathrm{H_2O} \leftrightarrow \mathrm{Me(OH)_2}$$

Depending on the pH of the liquid, the neutral particle surface will react further, forming either positive or negative surface charges [22]:

$$\text{At acidic pH:} \quad \mathrm{Me{-}OH} + \mathrm{H_3O^+} \leftrightarrow \mathrm{Me{-}OH_2^+} + \mathrm{H_2O}$$

$$\text{At alkaline pH:} \quad \mathrm{Me{-}OH} + \mathrm{OH^-} \leftrightarrow \mathrm{MeO^-} + \mathrm{H_2O}$$

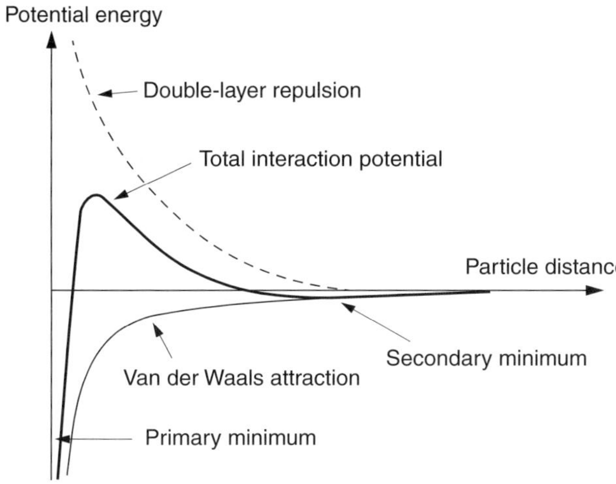

FIGURE 5.5.2 Interaction between suspended particles (electrostatic stabilization) according to DLVO theory.

When no adsorbing compounds are present, the pH therefore controls the surface charge and the surface potential Ψ_0. The pH at which the net surface charge of the suspended material is zero is called the isoelectric point (IEP). It is not a characteristic of the pure compound alone, but depends on the suspension properties as well—it can, for instance, be shifted by the presence of adsorbants that change the chemical nature of the particle surface.

According to DLVO theory [22], the net force between particles in an electrostatically stabilized suspension is the sum of the attractive Van der Waals forces and the electrostatic repulsion caused by the double layer of ions around the particle (see Figure 5.5.2).

Depending on the nature of the two contradicting forces, the resulting potential curve can show a high energy barrier preventing the particles from touching (stabilized suspension) or a barrier low enough to allow the particles to reach the primary minimum (coagulation). Furthermore, secondary minima can exist, mostly shallow ones that lead to weakly attractive forces. The double layer responsible for the electrostatic repulsion consists of the rather tightly bound Stern layer of counterions and the diffuse double layer in which counterions and co-ions diffuse freely (see Figure 5.5.3).

Within the Stern layer, the potential Ψ (V) drops linearly from Ψ_0 to Ψ_S. In the diffuse double layer, the potential decreases approximately according to:

$$\Psi = \Psi_S e^{-\kappa(d-d_S)} \tag{1}$$

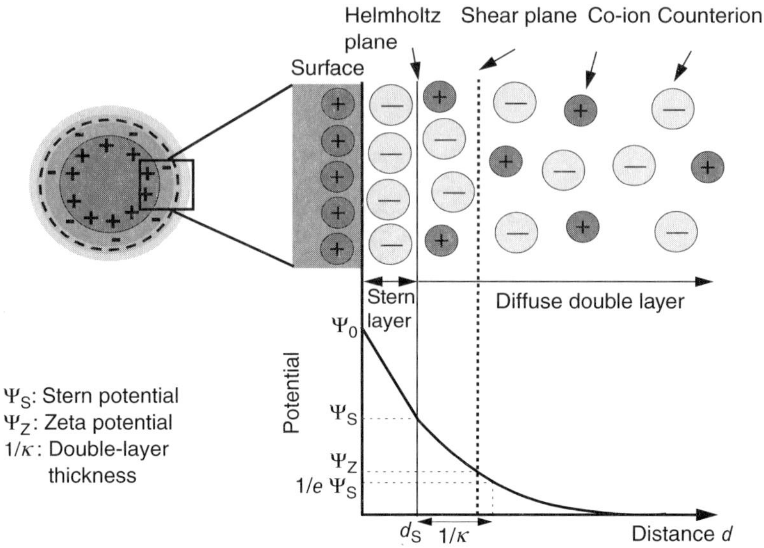

FIGURE 5.5.3 Stern model of the double layer around a positively charged particle surface (after Brinker and Scherer [22]).

Ψ_S being the Stern potential, d the distance from the particle surface, and d_S the thickness of the Stern layer. κ is the reciprocal of the double layer thickness (or Debye–Hückel length), which is defined as the distance from the Helmholtz plane at which Ψ has dropped to 1/e of Ψ_S. It is given by

$$1/\kappa = \sqrt{\frac{\varepsilon\varepsilon_0 RT}{F^2 \sum_i c_i z_i^2}} \tag{2}$$

where ε_0 (A s/V m) is the permittivity of vacuum, ε is the dielectric constant of the dispersing liquid, F (C/mol) is Faraday's constant, R (J/K mol) the gas constant, c_i (mol/l), and z_i (—) are the concentrations and valences of the ions present in the liquid. The latter add up to the ionic strength I (mol/l) of the liquid:

$$I = \frac{1}{2}\sum_i c_i z_i^2 \tag{3}$$

The higher the ionic strength, the quicker the potential drops with increasing distance from the particle surface, the thinner therefore the double layer.

In summary, the double-layer repulsion of given particles in an electrostatically stabilized suspension depends on the pH, the kind and concentration of dissolved salts, and the possible presence of adsorbing compounds. For destabilizing the suspension, there are two possibilities: shifting the pH to the IEP (i.e. decreasing the surface potential to zero) and increasing the ionic strength (i.e. shrinking the double layer, thus reducing the range of the double-layer repulsion).

5.5.2.1 Possible Reactions For Coagulating Suspensions *In Situ*

The most obvious way to change the net potential energy curve from repulsive to attractive is to add a pH-shifting agent or a high concentration of salt. Both lead to inhomogeneous coagulation since the added compound starts to produce its effects before it could be homogeneously distributed throughout the suspension. A better way would, therefore, be to produce this agent inside the suspension by reactions that can be delayed and controlled from outside. In principle, there are two families of useful reactions for shifting the pH:

- thermally activated reactions, in particular hydrolyses, and
- enzyme-catalyzed reactions.

Some possible reactions are listed in Table 5.5.1. More information on the different agents and their effects can be found in Ref. [23].

Since nearly all of these reactions produce ions as well, they increase the ionic strength as a side-effect to the pH shift, and which of the two mechanisms leads to coagulation in the end depends mainly on the starting pH and the concentration of the agents.

We will now concentrate on the reaction that has proved most useful for coagulating aqueous alumina suspensions: the hydrolysis of urea. In principle,

TABLE 5.5.1 Possible Reaction for Shifting the pH *In Situ*

	pH shift
Thermally activated reactions	
Hydrolysis of urea (T > 80°C)	$3 \rightarrow 7$
Hydrolysis of formamide (T > 60°C)	$3 \rightarrow 7$ or $12 \rightarrow 7$
Hydrolysis of esters	$11 \rightarrow$ acidic pH, depending on pK_S of acid
Hydrolysis of lactones	$9 \rightarrow$ acidic pH, depending on pK_S of acid
Enzyme-catalyzed reactions	
Hydrolysis of urea by urease	$4 \rightarrow 9$ or $12 \rightarrow 9$
Hydrolysis of amides by amidase	$3 \rightarrow 7$ or $11 \rightarrow 7$
Hydrolysis of esters by esterase	$10 \rightarrow 5$
Oxidation of glucose by glucoseoxidase	$10 \rightarrow 4$

it can be activated thermally, but this has some drawbacks: first, the whole procedure is more difficult because the bodies will extend and have to be protected against drying; second, at higher temperature, the developing gases are less soluble in water, and instead of dissolving and influencing the pH, part of them just produce bubbles in the green body. The enzymatic catalysis, however, activates the reaction already at room temperature, and a final pH of 9 can be reached—exactly the IEP of alumina. Furthermore, urease is an enzyme readily available commercially, and by adjusting its concentration, the kinetics of the coagulation process can be controlled in a wide range.

5.5.2.2 The Urea–Urease System

The hydrolysis of urea, either at temperatures above 80°C or catalyzed by urease, leads to the formation of ammonia and carbon dioxide, which in water react further to the main products NH_4^+, HCO_3^- and a small amount of CO_3^{2-}. The buffer pH of the reaction products is around 9.2:

$$
\begin{aligned}
CO(NH_2)_2 + H_2O &\leftrightarrow CO_2 + 2NH_3 \\
NH_3 + H_2O &\leftrightarrow NH_4^+ + OH^- \\
CO_2 + 2H_2O &\leftrightarrow HCO_3^- + H_3O^+ \\
HCO_3^- + H_2O &\leftrightarrow CO_3^2- + H_3O^+
\end{aligned}
$$

The commercially available urease is produced from jack beans and usually sold not as a pure enzyme, but in preparations containing buffers, stabilizing agents, etc. Its molar mass is given as 480 000 g/mol [24]. The effective amount of enzyme, which means the activity, is measured in units; 1 unit is defined as the amount of urease necessary to produce 1 μmol NH_3 per minute (at pH 7 and 25°C).

The urease activity depends on pH and temperature. The optimum conditions are a neutral pH and a temperature around 45°C [25], but even at room temperature a sufficient activity can be measured. Below 10°C, the enzyme works only very slowly, therefore cooling can be used to delay the beginning of the coagulation until the enzyme is added to the suspension, homogeneously distributed, and the suspension cast into molds.

5.5.2.2.1 Enzyme Kinetics

The rate of enzyme-catalyzed reactions is proportional to the enzyme concentration. At low substrate concentrations (in this case, substrate means urea), it increases linearly with the substrate concentration, whereas above a certain high substrate concentration, it asymptotically approaches a maximum value. These observations led Michaelis and Menten to propose the following reaction mechanism [26]:

$$E + S \underset{k_{-1}}{\overset{k_1}{\rightleftarrows}} ES \xrightarrow{k_2} P + E$$

E represents the enzyme, S the substrate, ES the enzyme–substrate complex (intermediate product), P the product and k the rate constants of the respective steps.

The formation rate of the product is given by

$$r = \frac{d[P]}{dt} = \frac{k_2[E]_0[S]}{K_M + [S]} \quad (4)$$

[X] being the momentary concentration of substance X, $[E]_0$ the initial enzyme concentration and

$$K_M = \frac{k_{-1} + k_2}{k_1} \quad (5)$$

the Michaelis constant.

Equation 4 is called the Michaelis–Menten equation. $[S] \ll K_M$ gives proportionality between r and [S], whereas with $[S] \gg K_M$, the reaction rate is constant: $r_{max} = k_2[E]_0$. For urease, K_M should be between 1.4 and 30 mmol/l [25,27].

If the pH is to be shifted from around 4 to 9 in a high solids loading alumina suspension, the minimum necessary urea concentration depends mainly on the powder surface that has to be neutralized. Experiments with different urea concentrations (see Figure 5.5.4) show that, for a powder with $d_{50} = 0.5\,\mu m$

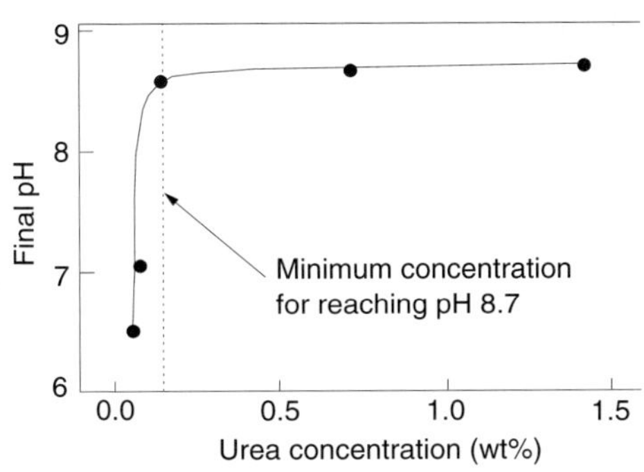

FIGURE 5.5.4 Final pH reachable with different urea concentrations in a 60 vol% alumina suspension (2 units urease per g Al_2O_3, 25°C, starting pH: 4.5).

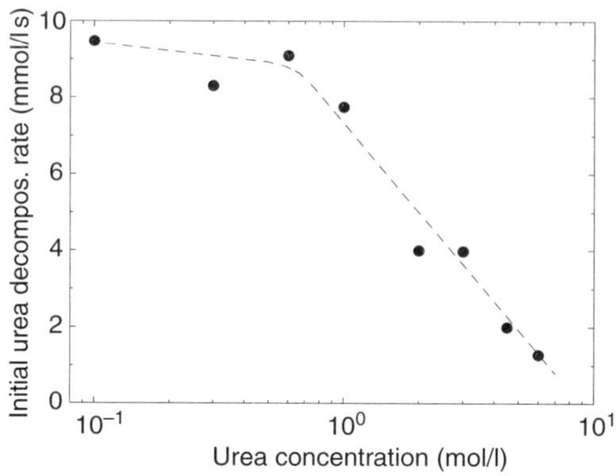

FIGURE 5.5.5 Initial urea decomposition rate (measured via conductivity) in solutions containing 30 units/ml urease and different urea concentrations (no powder!).

(BET surface $9.5\,m^2/g$), at least 0.17 wt% urea (based on the alumina weight) are needed to adjust a final pH of 8.7, which is sufficient to coagulate the suspension.

At low urea concentrations, the kinetics of the urea–urease system follows the Michaelis–Menten principles. If the urea concentration is too high, however (as necessary for coagulating alumina suspensions via increase of ionic strength, see Section 5.5.3.2), the Michaelis–Menten equation ceases to be valid, because urease is inhibited by urea at high concentrations, possibly because two urea molecules dock at the active sites of the enzyme where normally one urea and one water molecule should sit and react with each other [27, 28]. Figure 5.5.5 shows that this effect becomes perceptable around 1 mol/l urea.

As can be seen from the data shown in this section, the kinetics of the enzymatically catalyzed urea decomposition depends on various parameters and can be controlled within a wide range, corresponding to the respective coagulation mechanism. In the following chapters, we will show the connection between different coagulation mechanisms, their kinetics, and the resulting properties of the wet green bodies.

5.5.3 COAGULATION OF ALUMINA SUSPENSIONS BY DIFFERENT METHODS

5.5.3.1 pH Shift (ΔpH)

As mentioned before, alumina particles in water have an IEP of about 9. If they are stabilized around pH 4 by adding an acid, the urea decomposition can

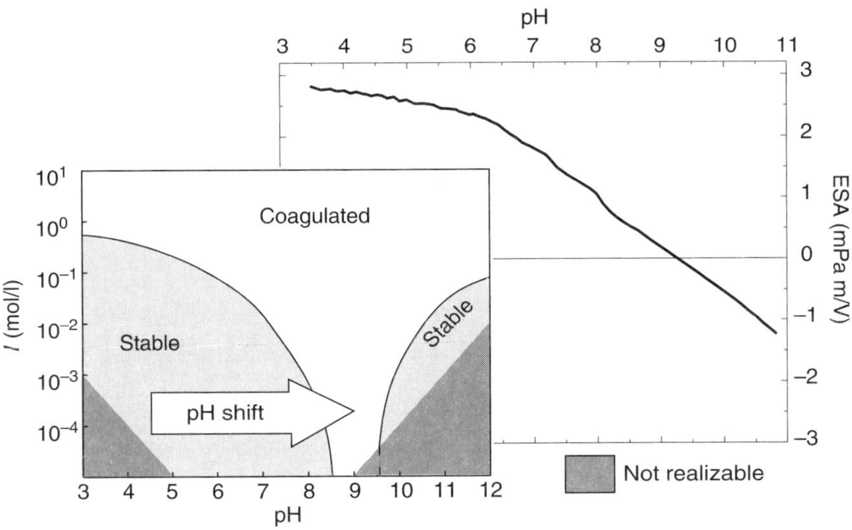

FIGURE 5.5.6 ESA (electrokinetic sonic amplitude, representing Zeta potential) signal versus pH for pure Al_2O_3 and stability diagram (schematic) of an acid-stabilized alumina suspension.

be used to shift the pH towards 9, thus reducing the repulsive double-layer potential to zero and leading to coagulation (see Figure 5.5.6). In practice, DCC samples are prepared as follows: the alumina powder is added stepwise to the water containing hydrochloric acid and urea, then the suspension is ball-milled for about 24 h and subsequently cooled with ice. The urease, dissolved in water, is added to the cooled suspension that can then be cast into the molds and allowed to warm up to room temperature where the coagulation will occur.

The experimental results given in the following were measured on suspensions or bodies made of Al_2O_3 powders with d_{50} between 0.5 and 0.6 μm.

The transition from a viscous liquid to a stiff wet green body can best be followed by oscillatory rheology. Figure 5.5.7 shows a time-dependent rheological oscillation measurement (details described in Ref. [29]) of the ΔpH coagulation of an alumina suspension. After reaching room temperature, the pH changes quickly from 4.5 to 7, whereupon the storage modulus G' and the loss modulus G'' increase sharply. In this example, the coagulation is finished within 1 h. The body starts to solidify around pH 7, and once pH 9 is reached, G' and G'' do not change any more, even though the urea decomposition may take a little more time to be completed. The change of the mechanical properties may be explained using the DLVO theory. The neutralization of the surface charge produces a deep primary minimum in the interparticle potential curves (see Figure 5.5.8). Therefore, the particles stick to each other upon first touch, and their arrangement from the stabilized suspension is practically frozen in

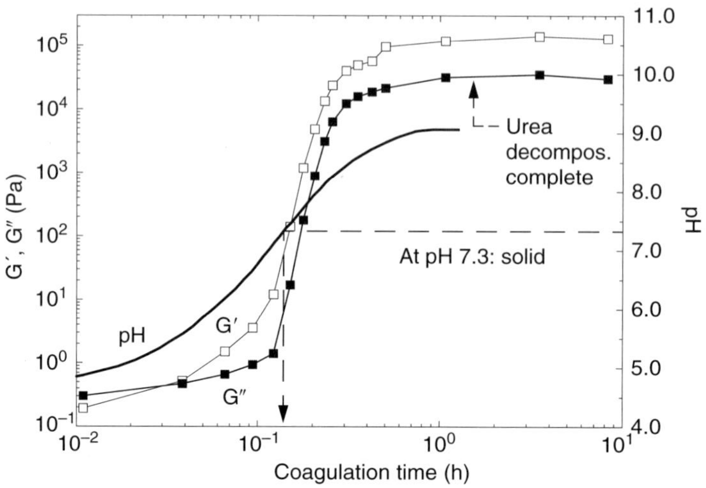

FIGURE 5.5.7 G′, G″ and pH versus time for ΔpH coagulation (54 vol%, 0.2 wt% urea, room temperature).

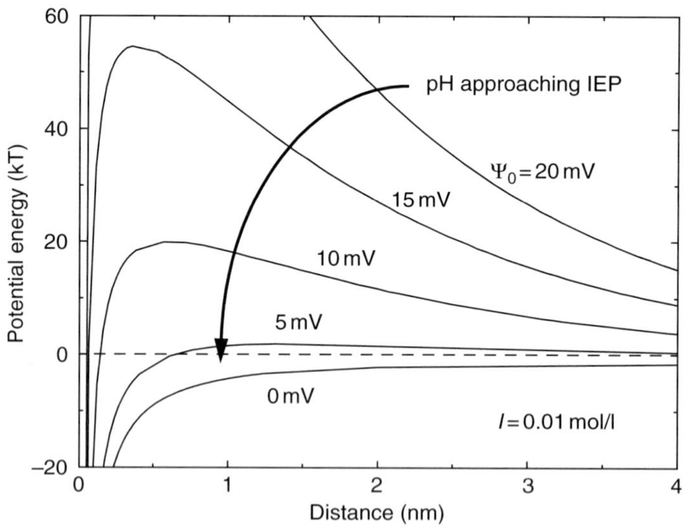

FIGURE 5.5.8 Interparticle potential energy for slurries coagulated via pH shift, calculated with Stabil 4.5 for Windows (R.V. Linhart and J.H. Adair, University of Florida, FL, USA). Hamaker Constant of alumina in water: $A = 5.32 \times 10^{-20}$ J.

(rapid coagulation), as shown by Wyss *et al.* [30]. The assumption is that this leads to a rather low medium number of nearest neighbors and a homogeneous microstructure.

5.5.3.2 Increase of Ionic Strength (ΔI)

As an example for a ΔI coagulation, an alumina suspension modified by an adsorbant can be used. Citrate ions adsorb on the alumina surface and shift the IEP to values as low as pH 3.4 for a monolayer adsorption [31]. If diammonium citrate is used as dispersant, the suspension is at the same time buffered at about pH 9 where it is stable (see Figure 5.5.9). In this case, the same enzymatic reaction as above can be used to increase the ionic strength and compress the double layer without changing the pH.

The minimum urea concentration needed for ΔI coagulation is about 1.5–2 mol/l (for pH shift, it was around 0.2 mol/l). Such high concentrations lead to inhibition of the urease as mentioned in the end of Section 5.5.2.2; therefore, somewhat higher urease concentrations are necessary, and the coagulation is still considerably prolonged. The time required to completely decompose the urea at room temperature range from some hours to more than 3 days, depending on the conditions [32].

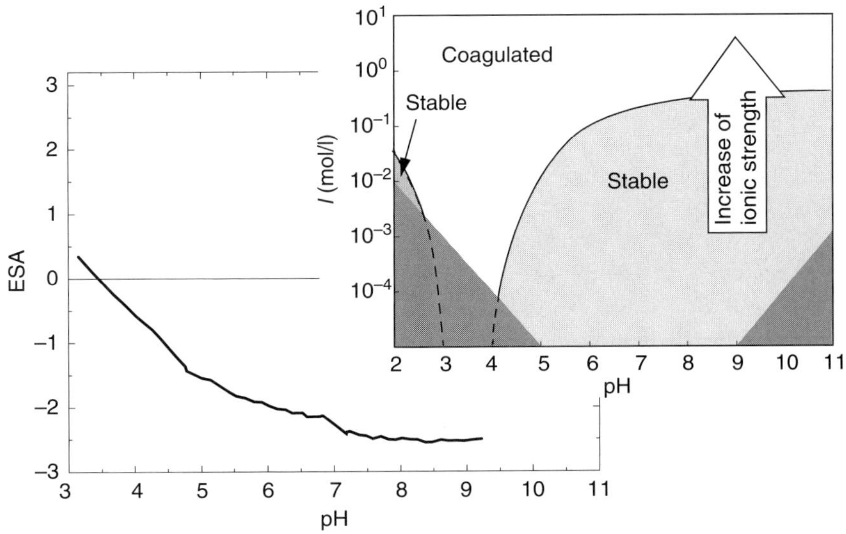

FIGURE 5.5.9 ESA (electrokinetic sonic amplitude, representing Zeta potential) signal versus pH for citrate-coated Al_2O_3 and schematic stability diagram.

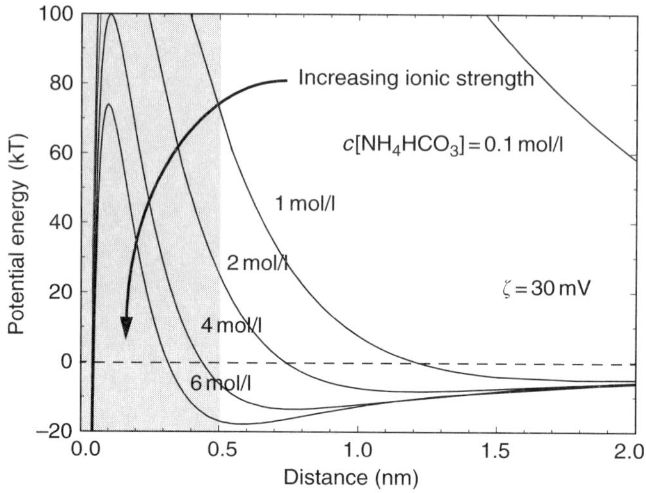

FIGURE 5.5.10 Interparticle potential energy for Δ*I* coagulation (grey area: 0.5-nm thick citrate layer), calculated as in Figure 5.5.8.

The increasing ionic strength leads to the development of a secondary minimum in the potential energy curves (see Figure 5.5.10). The citrate layer on the particles is about 0.5 nm thick [31], and the primary minimum lies within it; therefore, the particles cannot reach this minimum. In the shallow secondary minimum, however, the attractive forces are weak, and the particles can still move after the first touch, due to their thermal energy of about 5 kT. Thus, rearrangement can take place with time, and in the resulting network, particles have a higher average number of nearest neighbors, compared to a ΔpH network. Due to the rearrangement, shrinkage is observed in Δ*I* bodies, but not in ΔpH bodies [33].

In Figure 5.5.11, the coagulation of a 54 vol% alumina suspension via Δ*I* was followed by a rheological measurement. The process is slower than the ΔpH coagulation, but, even after all the urea is decomposed and the final ionic strength reached, G' (meaning the elastic properties) still increases with time, whereas G'' (viscous properties) decreases, indicating that rearrangement processes take place [29].

In diffusing-wave spectroscopy (DWS) measurements, the dynamics of ΔpH and Δ*I* systems were compared during aggregation [30]. The resulting elastic properties are in good agreement with the macroscopic rheological measurements. The DWS measurements confirm differences in microstructure between bodies produced via the two different methods: Δ*I*-destabilized systems are more inhomogeneous (on the length scale of a few particle

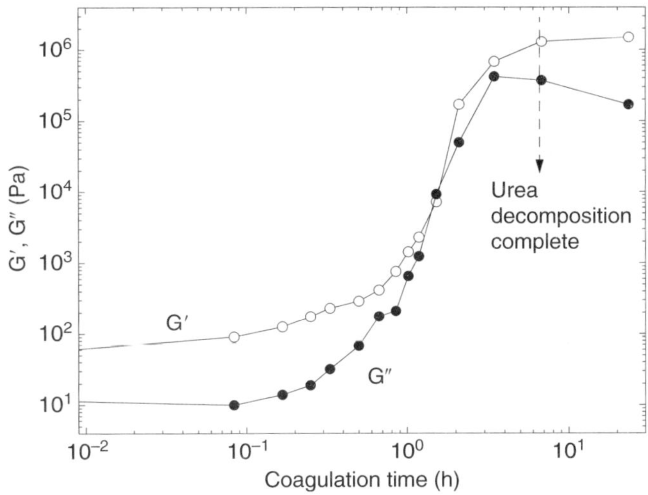

FIGURE 5.5.11 G′ and G″ versus time for ΔI coagulation (54 vol%, 1.6 mol/l, room temperature).

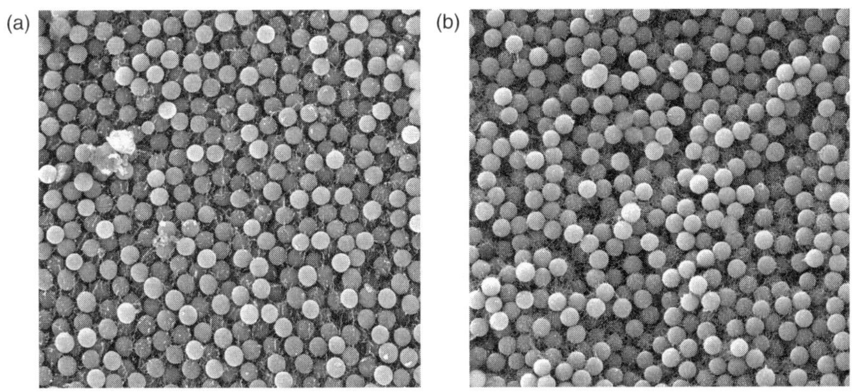

FIGURE 5.5.12 Cryo-SEM picture of (a) a stable and (b) a ΔI coagulated silica suspension (40 vol%), picture length about 10 μm [34].

diameters) than ΔpH systems. These differences can also be observed in Cryo-SEM images of high-pressure frozen (wet) samples (see Figure 5.5.12 and Ref. [34].

The consequences for the strength of the wet green bodies will be discussed in the next section.

5.5.3.3 Comparison of the Properties of ΔpH and Δ*I* Coagulated Wet Green Bodies

Since a freshly demolded wet green body has to withstand compression and shear forces during transport and handling even before it is dried, its strength (called "wet green strength" in the following) has to be as high as possible. The wet green strength of cylinders (see mold in Figure 5.5.13a) can be tested by compression as pictured in Figure 5.5.13b. Figure 5.5.13c shows a resulting stress–strain curve from which the wet green strength σ_c can be determined.

We have seen from the rheological measurements that ΔI coagulated suspensions need more time for solidification. In the compression tests, this becomes more evident (see Figure 5.5.14): after 2 h, ΔpH bodies have reached their final strength, whereas ΔI bodies cannot even be demolded. After 3 days, however, the ΔI bodies have obtained a strength five times higher than that of the ΔpH bodies, although their solids content is a little lower in this example.

It was already explained that in ΔpH bodies the arbitrary particle arrangement of the stable suspension is frozen in, whereas the particle network in ΔI bodies is more inhomogeneous due to the particle rearrangement during coagulation. This inhomogeneity now leads to areas of higher density.

In general, a very inhomogeneous distribution of stresses is observed for granular materials [35, 36]. A small percentage of the particles, arranged in the so-called force chains, bears the substantial part of the load. Though the length scale of these inhomogeneities in the force-distribution is longer than the length scale of structural inhomogeneities found in ΔI bodies, we assume that force chains are formed in the denser areas of the particle arrangement, where the

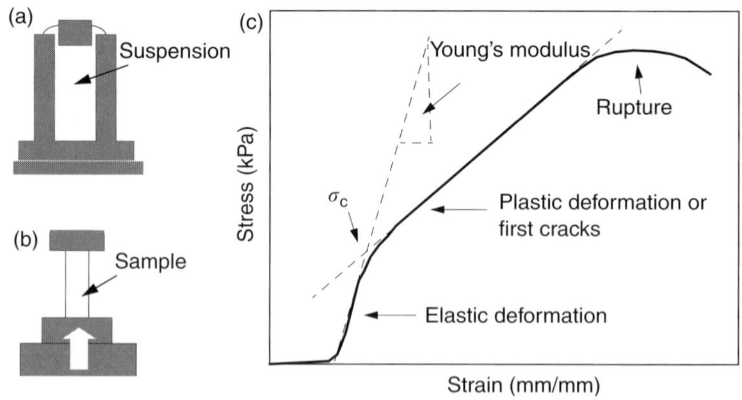

FIGURE 5.5.13 (a) Casting mold. (b) Compression test. (c) Schematic stress–strain curve.

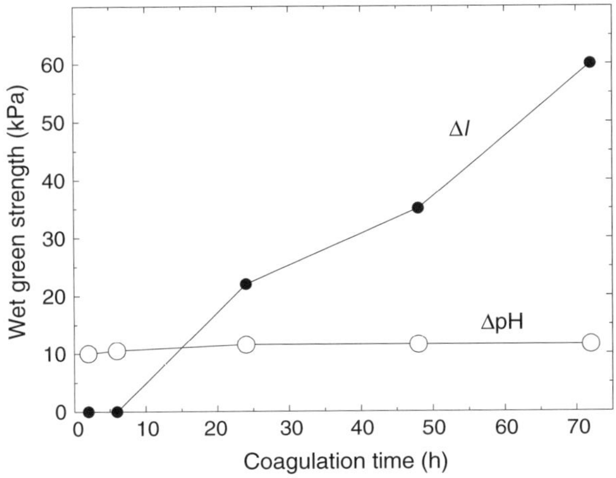

FIGURE 5.5.14 Wet green strength versus coagulation time for alumina cylinders coagulated via ΔpH (59 vol%) and Δ*I* (58 vol%).

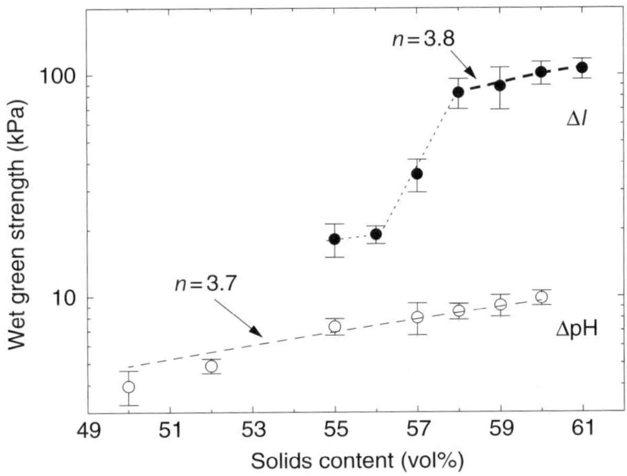

FIGURE 5.5.15 Wet green strength versus solids content for alumina cylinders coagulated via pH shift and increase of ionic strength (coagulation time: 3 days).

coordination number is higher. Consequently, structural inhomogeneities as observed in Δ*I* bodies would increase the mechanical strength of the particle network.

If the strength values for different solids contents (Figure 5.5.15) are compared, we see that even the dependence of the wet green strength σ on the solids content Φ is different for the two coagulation methods. The ΔpH data can be fitted by a power law derived from percolation theory, as, for instance, used by Yanez *et al.* [37]:

$$\sigma \propto (\Phi - \Phi_c)^n \tag{6}$$

where Φ_c is the percolation threshold (here: $\Phi_c = 0.16$, after Zallen [38]). n was found to be 3.7, which corresponds well with the exponents between 3 and 5 usually given for similar systems [39, 40]. For ΔI, the strength values at $\Phi = 58$ vol% and above obey the same law with $n = 3.8$, but below, there is a shift to another dependency, which is yet to be investigated further. One could assume that, in a ΔI sample at solids loadings below 58 vol%, the particle density is too low to form dense force chains that are still able to construct a strong network throughout the whole body; therefore, the wet green strength drops to values relatively near those of the ΔpH body.

5.5.4 DCC OF PARTICLE SYSTEMS WITH POLYMER BINDERS

Polymer binders are used in ceramic forming to increase the strength of the green bodies. This effect can be exploited in DCC for increasing the wet green body strength as well. If binders are to be used in DCC they have to meet three basic requirements:

1. the addition of the binder must not increase the suspension viscosity substantially;
2. the binder has to react to a change in pH or ionic strength;
3. the swelling, gelling, etc., of the binder has to occur time-delayed, otherwise its coagulation would be quicker than that of the powder.

A class of substances fulfilling all these requirements are polyelectrolytes on the basis of polyacrylic acid. As an example, we will present the use and effects of alkali-soluble swellable thickeners (ASTs), in this case two commercially available thickeners, Acusol 820 and 830 (more details in Ref. [41]).

ASTs are cross-linked copolymers consisting of hydrophobic (e.g. ethylene, butadiene, styrene) and hydrophilic monomers (carbonic acids as acrylic, methacrylic, fumaric or maleinic acid). At low pH, the carboxyl groups are undissociated, which, together with the hydrophobic components, makes the polymer insoluble. At high pH, the carboxyl groups deprotonate, the now negatively charged groups repel each other, their hydratization improves, together resulting in the polymer particle growing and taking up water, that is, swelling (see Figure 5.5.16).

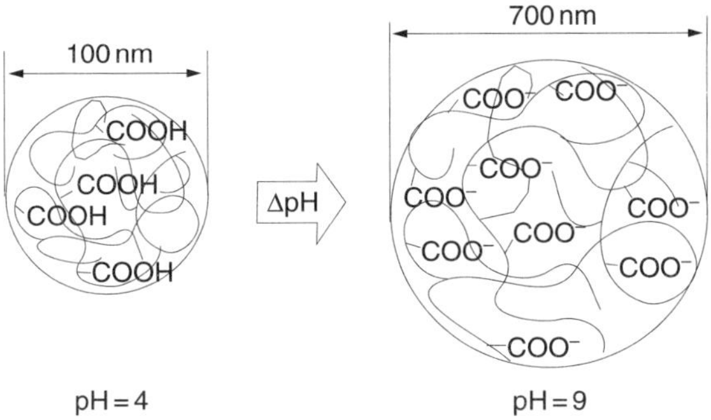

FIGURE 5.5.16 Swelling of AST molecules upon the transition acidic (insoluble) to alkaline (deprotonated, soluble).

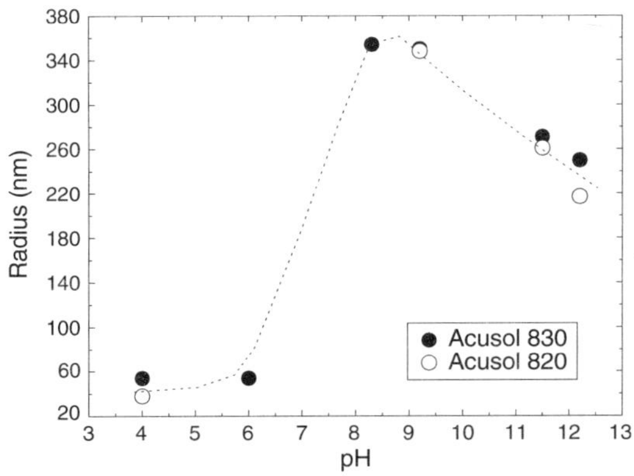

FIGURE 5.5.17 Radius of Acusol molecules in dependence of pH, measured by light scattering (data from Ref. [41]).

Figure 5.5.17 shows how the radius of Acusol particles changes with pH. The fact that the maximum size (seven times the original radius!) is reached just around pH 9 makes Acusol an excellent binder for the coagulation of alumina via DCC (ΔpH).

On the other hand, the negative charges are screened by counterions dissolved in the solution; therefore, a high ionic strength reduces the diameter of the swollen polymer "spheres", which makes Acusol unsuitable as binder for the ΔI coagulation, also because the starting pH of 9 is in the wrong region as well.

For increasing the green strength of wet alumina bodies in the DCC process, only very small amounts of ASTs are necessary (see Figure 5.5.18, the polymer amount is calculated based on the alumina weight). Whereas binderless samples

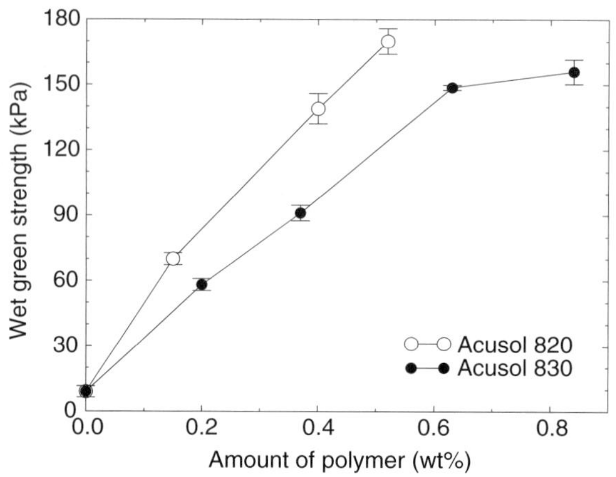

FIGURE 5.5.18 Wet green strength of Al_2O_3 bodies (ΔpH, 57 vol%) with different Acusol contents (data from Ref. [41]).

reach strength values around 10 kPa, up to 170 kPa can be found with Acusol addition. The maximum possible polymer content is found around 0.5–1 wt% since too many swelling polymer particles can induce cracks in the powder network. In addition, too strong a swelling makes the body volume increase, thus the bodies cannot be demolded.

Our explanation for this extraordinary increase in strength caused by Acusol is just an assumption that could not be proven yet. The experimental results [41] suggest that, first, the strength added by the polymer gel is far too small to add significantly to the strength of the particle network. Second, there is no perceptible bridging involved. On the other hand, the swelling of the polymer particles seems to be the crucial factor in the strengthening effect. The swelling polymer cannot fill every small water "isle" between the alumina particles, but it can try to suck the water out of it. This could draw the particles closer together, decreasing the surface-to-surface distance that increases, on the one hand, the local solids loading and, on the other hand, the bond strength between the particles.

5.5.5 OTHER CERAMIC MATERIALS

Of course, the DCC method is not restricted to alumina only. Two other ceramic compounds that were already coagulated via DCC shall be mentioned here as examples: SiC and Si_3N_4. Silicon carbide (IEP around 2–3), with boron and carbon black as sintering additives, could be stabilized at pH 11 using tetramethyl ammonium hydroxide (TMAH) and coagulated by increasing the ionic strength with the urea/urease system [42, 43]. Wet green strength values up to 50 kPa could be achieved depending on urease concentration and coagulation time.

A mixture of α-Si_3N_4 with Al_2O_3 and Y_2O_3 as sintering aids can be coagulated via DCC as well, though the processing is more complicated [44]. Since the IEPs of the three powders are very different, all the surfaces have to be modified in order to process them together. This can be done with citric acid for the two oxides and with $Al(OH)_3$ for Si_3N_4. Afterwards, the pH is adjusted around 9.5 with TMAH, and the ionic strength is increased as above, giving wet green bodies with strength values up to 200 kPa (58 vol%) that can be sintered to 97% of the theoretical density.

5.5.6 SUMMARY AND OUTLOOK

The DCC method allows to control the interparticle forces in a suspension via an internal chemical reaction. It can be used for producing ceramic components with complicated shapes (see Figure 5.5.19), homogeneous microstructures, high strength, and reliability [45]. This ceramic forming technique might also find applications in microfabrication of very small ceramic components and structures used for sensors and actuators. Small ceramic dots, lines, and three-dimensional objects made from ceramic suspensions may become useful in MEMS technology (see Figure 5.5.20).

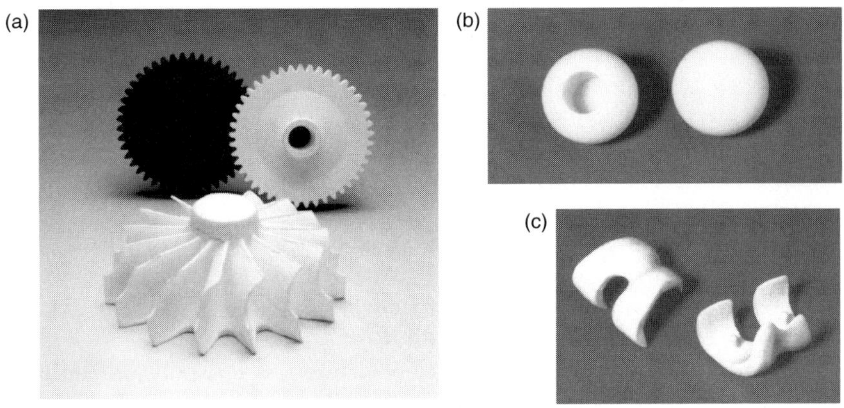

FIGURE 5.5.19 Ceramic parts produced via DCC: (a) SiC and Al_2O_3 cog-wheels and Al_2O_3 turbo rotor; (b) Al_2O_3 hip joint spheres; (c) Al_2O_3 knee implants.

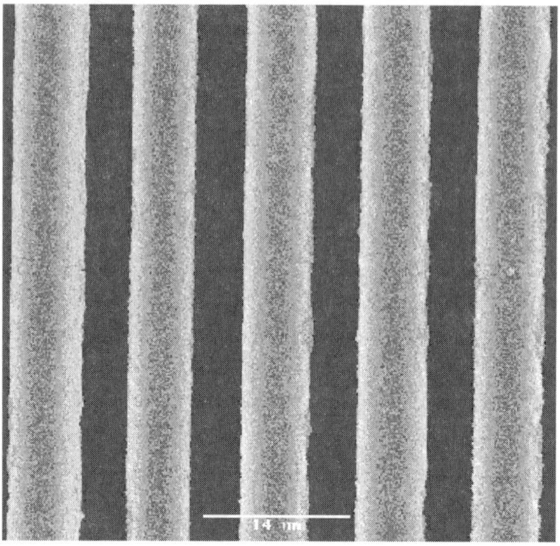

FIGURE 5.5.20 Alumina lines (7 μm in thickness) on a Si substrate, from Schönholzer *et al.* [52].

Besides triggering the sol–gel transition of electrostatically stabilized suspensions, DCC can also be used to change the conformation of polymers in suspensions and thereby change the mechanical properties from a viscous liquid to a viscoelastic gel. Examples are ASTs [41], polysaccharides, and alginates [46]. Another class are polymer emulsions containing two oligomers, one in the hydrophilic phase and the other in the hydrophobic phase. Upon changing the pH by an internal reaction, both react forming a three-dimensional cross-linked polymer network in the aqueous phase [47]. These polymers can also be combined with particle suspensions in order to create new processing methods for new materials.

Besides application-oriented topics, a method like DCC can offer new possibilities to solve basic scientific questions that are still open today. The mechanical properties of particle gel networks are highly dependent on the chemical pathway used to induce the sol–gel transition. It has been shown earlier that changing the pH produces comparatively weak networks whereas destabilizing a particle suspension via increasing the salt content results in particle networks with elastic moduli and yield strength one order of magnitude higher [32]. This is also observed in geological sediments.

There are two possible reasons for this puzzling observation. The different sol–gel kinetics might lead to different microstructures of the final gels, characterized by different coordination numbers of the particles in the network [48]. This was also found in computer simulations using Brownian dynamics

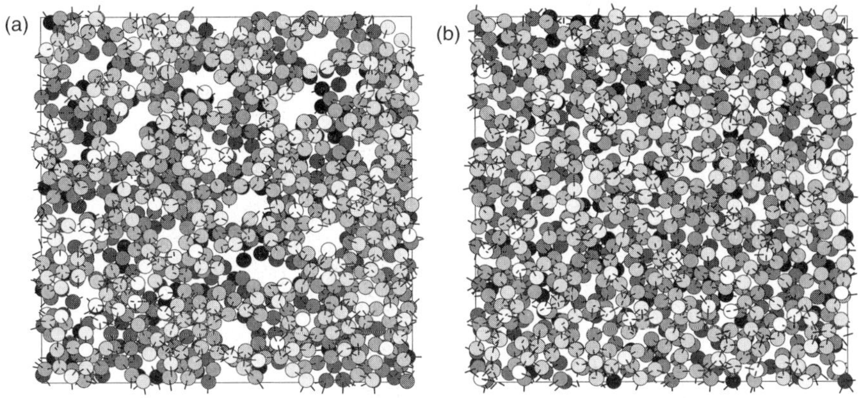

FIGURE 5.5.21 Simulated structures for 40 vol% coagulated particle suspensions (particles drawn in reduced size), from Ref. [41, p. 341]: (a) delayed coagulation due to energy barrier in interaction potential; rearrangement; (b) rapid switch from repulsive to purely attractive interaction potential, starting configuration is frozen in.

of monosized suspensions. Figure 5.5.21 illustrates the different microstructures resulting from different coagulation kinetics: on the one hand, a delayed coagulation due to an energy barrier in the interaction potential leads to a particle rearrangement (Figure 5.5.21a), on the other hand, a rapid switch from a repulsive to a purely attractive interaction potential preserves the random order of the particles (Figure 5.5.21b) [49]. Transferred to our system, case (a) can be seen as similar to the ΔI coagulation, case (b) as corresponding to ΔpH coagulation (see interparticle potential curves in Sections 5.5.3.1 and 5.5.3.2). The resulting local inhomogeneities in case (a) might strengthen the network, as discussed in Section 5.5.3.3.

Another possibility is that, due to the different chemical pathways, the bonds between the coagulated particles become different in strength. It is well known that the attractive interparticle potential in coagulated suspensions becomes much stronger upon decreasing the number of residual water layers between particles in contact [50]. It has recently been shown by new light scattering techniques such as diffusing wave spectroscopy that, after gelling, the microstructure of a polystyrene network remains unchanged whereas the dynamics show still considerable stiffening of the system during ageing [51]. This would support the hypothesis that, upon coagulating a suspension via increase of the ionic strength, the bonds between particles are strengthened with time.

A deeper insight into the structure and the mechanics of gel networks is yet to be acquired. Temperature-induced or enzymatically catalyzed internal reactions, as outlined in this article, can essentially contribute to the clarification of such problems.

REFERENCES

1. Janney, M. A. (1990). Method for forming ceramic powder into complex shape. US Patent 4.894.194.
2. Young, A. C., Omatete, O. O., Janney, M. A., and Menchhofer, P. A. (1991). Gelcasting of Alumina. *J. Am. Ceram. Soc.* **74**: 612–618.
3. Omatete, O. O., Strehlow, R. A., and Walls, C. A. (1992). Drying of gelcast ceramics. in *Ceramic Transactions, Vol. 26:* Forming Science and Technology for Ceramics, pp. 101–107, Cima M. J., ed., Westerville, OH: American Ceramic Society.
4. Sarkar, N., and Greminger Jr., G. K. (1983). Methylcellulose polymers as multifunctional processing aids in ceramics. *Am. Ceram. Soc. Bull.* **62**: 1280–1288.
5. Landham, R. R., Nahass, P., Leung, D. K., Ungureit, M., Rhine, W. E., Bowen, H. K., and Calvert, P. D. (1987). Potential use of polymerizable solvents and dispersants for tape casting of ceramics. *Am. Ceram. Soc. Bull.* **66**: 1513–1516.
6. Tormey, E. S., Pober, R. L., Bowen, H. K., and Calvert, P. D. (1984). Tape casting—future developments, in *Advances in Ceramics, Vol. 9: Forming of Ceramics*, pp. 140–149, Mangels J. A., and Messing G.L., eds., Columbus, OH: American Ceramic Society.
7. Cowan, R. E. (1976), in *Treatise on Materials Science and Technology, Vol. 9, Ceramic Fabrication Processes*, pp. 153–171, Wang F. F. Y., ed., New York: Academic Press.
8. Bollman, H. (1957). Unique new forming technique. *Ceram. Age*, **70**: 36–38.
9. Grala, E. M. (1957). Further investigation of the feasibility of the freeze casting method for forming fuel-size infiltrated titanium carbide turbine blades. NACA (Nat. Adv. Comm. Aeronaut.) Tech. Notes 3769; Ceram. Abstr., p. 108.
10. Maxwell, W. A., Gurnick, R. S., and Francisco, A. C. (1954). Preliminary investigation of the freeze casting method for forming refractory powders. NACA (Nat. Adv. Comm. Aeronaut.) Res. Memo E 53L21, p. 19.
11. Bergström, L. (1994). Method for forming ceramic powders by temperature induced flocculation, US Patent 5.340.532.
12. Napper, D.H. (1977). Steric stabilization. *J. Colloid Interface Sci.* **58**: 390–399.
13. Lange, F. F. (1986). Forming a ceramic by flocculation and centrifugal casting US Patent 4,626,808.
14. Velamakanni, B. V., Chang, J. C., Lange, F. F., and Pearson, D. S. (1990). New method for efficient colloidal particle packing via modulation of repulsive lubricating hydration forces. *Langmuir* **6**: 1323–1325.
15. Velamakanni, B. V., and Lange, F. F. (1993). Method for preparation of dense ceramic products, US Patent 5.188.780.
16. Franks, G. V., Velamakanni, B. V., and Lange, F. F. (1995). Vibraforming and in situ flocculation of consolidated, coagulated alumina slurries. *J. Am. Ceram. Soc.* **78**: 1324–1328.
17. Pujar, V. V., Cawley, J. D., Hu, P., and Lee, L. J. (1992). Reaction injection moulding of silica–alumina mixtures using heterocoagulation. *Mater Res. Soc. Symp. Proc.* **249**: 317–322.
18. Graule, T. J., Baader, F. H., and Gauckler, L. J. (1994). Shaping of ceramic green compacts direct from suspension by enzyme catalyzed reactions. *cfi/Ber. DKG* **71**: 317–323.
19. Graule, T. J., Baader, F. H., and Gauckler, L. J. (1994). Enzyme catalysis of ceramic forming. *J. Mater. Educ.* **16**: 243–267.

20. Graule, T. J., Baader, F. H., and Gauckler, L. J. (1995). Direct coagulation casting—a new green shaping technique, in *Ceramics: Charting the Future*, pp. 1601–1616, Vincenzini P., ed., Techna Srl.
21. Graule, T. J., Baader, F. H., and Gauckler, L. J. (1995). Casting uniform ceramics with direct coagulation. *Chemtech* **25**: 31–37.
22. Brinker, C. J., and Scherer, G. W. (1990). *Sol–Gel Science*, p. 240ff, New York: Academic Press.
23. Gauckler, L. J., Graule, T. J., and Baader, F. H. (1999). Ceramic forming using enzyme catalyzed reactions. *Mater. Chem. Phys.* **61**: 78–102.
24. Kistiakowsky, G. B., and Rosenberg, A. J. (1952). The kinetics of urea hydrolysis by urease. *J. Am. Chem. Soc.* **74**: 5020–5025.
25. Schmidt-Steffen, A. (1986). Untersuchung der Reaktion von Harnstoff mit an Membranen kovalent gebundener Urease (Investigation of the reaction of urea with urease bound covalently to membranes). PhD Thesis, Universität Essen.
26. Drauz, K., and Waldmann, H. (eds.) (1995). *Enzyme Catalysis in Organic Synthesis*, Vol. I, p. 113, Weinheim: VCH Verlagsgesellschaft mbH.
27. Rachhpal-Singh, and Nye, P. H. (1984). The effect of soil pH and high urea concentrations on urease activity in soil. *J. Soil Sci.* **35**: 519–527.
28. Grunwald, P. (1974). Physikalische und chemische Eigenschaften von freier und an aktivem Aluminiumoxid gebundener Urease (Physical and chemical properties of free urease and urease bound to active alumina). PhD Thesis, Universität Hamburg.
29. Balzer, B., Hruschka, M. K. M., and Gauckler, L. J. (2001). In-situ rheological investigation of the coagulation in aqueous alumina suspensions. *J. Am. Ceram. Soc.* **84**: 1733–1739.
30. Wyss, H. M., Romer, S., Scheffold, F., Schurtenberger, P., and Gauckler, L. J. (2001). Diffusing-wave spectroscopy in concentrated alumina suspensions during gelation. *J. Colloid Interface Sci.* **240**: 89–97.
31. Hidber, P. C., Graule, T. J., and Gauckler, L. J. (1996). Citric acid—a dispersant for aqueous alumina suspensions. *J. Am. Ceram. Soc.* **79**: 1857–1867.
32. Balzer, B., Hruschka, M. K. M., and Gauckler, L. J. (1999). Coagulation kinetics and mechanical behavior of wet alumina green bodies produced via DCC. *J. Colloid Interface Sci.* **216**: 379–386.
33. Yang, J., Huang, Y., Meier, L. P., Wyss, H. M., Tervoort, E., and Gauckler, L. J. (2002). Direct coagulation casting via increasing ionic strength. *Key Eng. Mater.* **224–226**: 631–636.
34. Wyss, H. M., Hütter, M., Müller, M., Meier, L. P., and Gauckler, L. J. (2002). Quantification of structure in stable and gelated suspensions from cryo-SEM. *J. Colloid Interface Sci.* **248**: 340–346.
35. Mueth, D. M., Jaeger, H. M., and Nagel, S. R. (1998). Force distribution in a granular medium. *Phys. Rev. E* **57**: 3164–3169.
36. Peterson, I. (1997). Dry sand, wet sand. Digging into the physics of sandpiles and sand castles. *Sci. News* **152**: 186–187.
37. Yanez, J. A., Laarz, E., and Bergström, L. (1999). Viscoelastic properties of particle gels. *J. Colloid Interface Sci.* **209**: 162–172.
38. Zallen, R. (1983). *The Physics of Amorphous Solids*, p. 170, New York, Wiley.
39. Yanez, J. A., Shikata, T., Lange, F. F., and Pearson, D. S. (1996). Shear modulus and yield stress measurements of attractive alumina particle networks in aqueous slurries. *J. Am. Ceram. Soc.* **79**: 2917–2924.
40. Channell, G. M., and Zukoski, C. F. (1997). Shear and compressive rheology of aggregated alumina suspensions. *AIChE J.* **43**: 1700–1708.
41. Hesselbarth, D., Tervoort, E. V., Urban, C., and Gauckler, L. J. (2001). Mechanical properties of coagulated wet Particle networks with alkali swellable thickeners. *J. Am. Ceram. Soc.* **84**: 1689–1695.

42. Gauckler, L. J., Si, W., Graule, T. J., Baader, F. H., and Will, J. (1999). Enzyme catalysis of ceramic forming: alumina and silicon carbide, in *Ceramics: Getting into the 2000's—Part C*, pp. 15–40, Vincenzini P., 9th Cimtec-World Ceramics Congress.
43. Si, W., Graule, T. J., Baader, F. H., and Gauckler, L. J. (1999). Direct coagulation casting of silicon carbide components. *J. Am. Ceram. Soc.* **82**: 1129–1136.
44. Hruschka, M. K. M., Si, W., Tosatti, S., Graule, T. J., and Gauckler, L. J. (1999). Processing of β-silicon nitride from water-based α-silicon nitride, alumina, and yttria powder suspensions. *J. Am. Ceram. Soc.* **82**: 2039–2043.
45. Baader, F. H., Will, J., Tièche, D., Graule, T. J., and Gauckler, L. J. (1995). High strength alumina produced by direct coagulation casting. *Ceramic Transactions*, Vol. 51, pp. 463–467. Ceramic Processing Science and Technology, The American Ceramic Society.
46. Katsuki, H., Kawahara, A., and Ichinose, H. (1992). Preparation and some properties of porous alumina ceramics obtained by the gelatination of ammonium alginate. *J. Mater. Sci.* **27**: 6067–6070.
47. Tervoort, E. V., private communication.
48. Hütter, M. (1999). Brownian dynamics simulation of stable and of coagulating colloids in aqueous suspensions, PhD Thesis, ETH Zürich.
49. Hütter, M. (2000). Local structure evolution in particle network formation studied by Brownian dynamics simulation. *J. Colloid Interface Sci.* **231**: 337–350.
50. Shih, W. Y., Shih, W.-H., and Aksay, I. A. (1999). Elastic and yield behavior of strongly flocculated colloids. *J. Am. Ceram. Soc.* **82**: 616–624.
51. Romer, S. (2001). Aggregation and gelation of concentrated colloidal suspensions, PhD Thesis, ETH Zürich.
52. Schönholzer, U. P., Hummel, R., and Gauckler, L. J. (2000). Microfabrication of ceramics by filling of photoresist molds. *Adv. Mater.* **12**: 1261–1263.

Handbook of Advanced Ceramics
S. Sōmiya *et al.* (Eds.)

5.6 Control of Ceramic Microstructure by Templated Grain Growth

SEONGTAE KWON, EDWARD M. SABOLSKY and GARY L. MESSING
Materials Research Institute, The Pennsylvania State University, University Park, PA 16802, USA

5.6.1 INTRODUCTION

The physical properties of ceramics are strongly affected by microstructure. An example in structural ceramics is the enhanced mechanical properties of Si_3N_4 [1] and SiC [2] because of the presence of anisometric grains. The orientation of the grains also can play an important role in ceramic properties. Although most inorganic crystals exhibit anisotropy in structure, habit, and properties, the reduction in the magnitude of desired physical properties due to averaging in randomly oriented ceramics prevents the harnessing of a class of unique anisotropy dependent properties. This opportunity is most pronounced in materials with low symmetry crystal structures. Tailoring the microstructures of ceramics by aligning the grains in preferred directions can be an attractive means of modifying or improving properties of ceramics.

Techniques for inducing texture in ceramics are largely based on alignment of anisometric particles by the application of uniaxial pressure at high temperatures (i.e. hot forging) [3], recrystallization from a liquid phase or melt texturing [4, 5], or the use of less expensive, pressureless techniques based on orienting anisometric particles under the application of an anisotropic magnetic [6], electric [7], mechanical stress, or shear field during forming [8–10].

Templated grain growth (TGG) is a novel method for tailoring microstructures that circumvents many of the difficulties of the above methods by using relatively simple forming techniques. The TGG process is dependent upon the alignment of a minor portion of anisometric particles within the ceramic body during green processing. During high-temperature treatment, the oriented, anisometric particles serve as templates for the epitaxial nucleation and growth of the desired phase. Therefore, the principal physical component in TGG is the template particle, which acts as a substrate for epitaxy and as a seed for grain growth. The epitaxy dictates the crystallographic alignment of a small population of grains, which could be thought of as a population of oriented "exaggerated" grains. Thus, with further heat treatment, the volume fraction of textured materials increases.

There has been much research that demonstrates successful property modification by TGG. For example, oriented Si_3N_4 ceramics have been shown to yield interesting fracture characteristics and enhanced thermal conductivity parallel to the elongated grain direction [11, 12]. Silicon nitride with a preferred orientation of large elongated grains was fabricated by tape casting a powder mixture of α-Si_3N_4, Y_2O_3, Al_2O_3, and rod-like β-Si_3N_4 particles. Increased fracture toughness (6.8 versus 11.1 MPa $m^{1/2}$) and bending strength (1000 versus 1092 MPa) were achieved in the direction perpendicular to the grain alignment.

Tan *et al.* [13] obtained enhanced ionic conductivity perpendicular to the basal plane of textured Na^+, K^+-β''/β-Al_2O_3. They reported an approximately two order-of-magnitude difference in conductivity depending on the direction ($2.2 \times 10^{-4}\,\Omega^{-1}\,cm^{-1}$ perpendicular to the *c*-axis and $2.7 \times 10^{-6}\,\Omega^{-1}\,cm^{-1}$ parallel to *c*-axis). Such an enhancement could be useful for beta-alumina batteries.

More intensive and basic research has been carried out on TGG of Al_2O_3. Carisey *et al.* [14, 15] and Brandon *et al.* [16] fabricated a composite consisting of oriented α-Al_2O_3 platelet particles and equiaxed α-Al_2O_3 grains. They reported improved bend strength, thermal shock resistance, and inhibited crack propagation perpendicular to the basal surface of the tabular α-Al_2O_3 grains.

Most of the fundamental studies on TGG were performed on liquid-phase sintered α-Al_2O_3 systems. Seabaugh *et al.* [17] fabricated nearly 100% textured α-Al_2O_3 by tape casting and discussed the importance of various processing parameters including the presence and orientation of template particles, the presence of liquid phase at the sintering temperature, and the ratio of template to matrix grain size difference in textured α-Al_2O_3. Seabaugh *et al.* [18] also showed that texture development in TGG is controlled by the growth of the template particles, not by template rearrangement or by the preferential development of texture in the matrix. The sintering condition, liquid phase content, and template loading all affect template growth and subsequently, texture development is controlled by the initial geometry, which in turn is controlled by the template concentration.

Texturing piezoelectrics has special attraction due to the excellent properties in the (001) oriented cuts of single crystal. It was reported that ⟨001⟩ textured $Pb(Mg_{1/3}Nb_{2/3})O_3$–$32PbTiO_3$ (PMN–32PT) can be fabricated by TGG using isostructural perovskite $BaTiO_3$ [19] or $SrTiO_3$ [20] as template materials.

The textured PMN–PT showed approximately twice the strain and piezoelectric coefficient relative to random PMN–PT of the same composition. Sodium bismuth titanate, $(Na_{1/2}Bi_{1/2})TiO_3$–$BaTiO_3$, ceramics also can be textured by TGG using $SrTiO_3$ or $Bi_4Ti_3O_{12}$ templates. Because of the high degree of texture, the piezoelectric coefficient increased from 260 to $\sim$490 pC N^{-1} [21]. Strontium barium niobate ($Sr_{0.53}Ba_{0.47}Nb_2O_6$) is another example of a non-lead-based piezoelectric material, which can be textured by TGG. Duran *et al.* [22] produced highly ⟨001⟩ textured strontium barium niobate using acicular $KSr_2Nb_5O_{15}$ and V_2O_5 as templates and liquid phase, respectively. They reported that the presence of a liquid phase affects both phase formation and template growth and ~98% texture was obtained by controlling template content, the amount of liquid phase, and the sintering temperature. Strontium niobate ($Sr_2Nb_2O_7$), which holds the promise of high-temperature piezoelectricity, was textured using $Sr_2Nb_2O_7$ templates made by the molten salt method and fine matrx powder synthesized by coprecipitation and solid-state reaction [23]. Both fiber and sheet textured samples were fabricated by tape casting slurries containing 10–40 vol% templates.

In this article, the role of key factors for templated grain growth will be discussed and illustrated with examples from the literature.

5.6.2 KEY FACTORS FOR TEMPLATED GRAIN GROWTH

5.6.2.1 Template Particles

Template particles must possess a variety of properties for successful TGG. First of all, the single-crystal templates should have a suitable morphology with a high aspect ratio (e.g. >5) so that they can be mechanically aligned by shear during the green forming process, such as tape casting, extrusion, or uniaxial pressing. Thus, most templates used in TGG are platelets or whiskers.

In instances where the purpose of texturing is to access physical properties that are directly related to crystallographic orientation (like thermal conductivity, dielectricity, piezoelectricity, electrical conductivity, etc.), the template axis must match the desired crystallographic orientation. The alignment of templates is normally in one direction (fiber texture), thus textured ceramics have the benefit of grain orientation in the aligned direction only. When the templates are blade-shaped, sheet textured or biaxial ceramics can be achieved [23].

The size of templates is also an important factor. The lower size requirement is established by the driving force for template growth, which is the grain size difference between the oriented, larger grains, and the random, smaller matrix grains. Thus, the templates must be large enough so that the templated grain size must be larger than the matrix grains after the nucleation of the growing phase. The thermochemical stability also depends on the template size as discussed below. The upper size limit of templates is established by the relation between the number frequency of the templates and size. If the templates are too large, at a given weight or volume fraction of template content, the number frequency of templates can be insufficient and consequently, the growth amount required to achieve full texture becomes large.

The Hillert criteria for abnormal grain growth and Ostwald coarsening theory explains why larger grains, such as template grains, grow preferentially in a matrix of finer grains and, furthermore, demonstrate that, when there is excessive matrix coarsening or impingement, the thermodynamic driving force for template growth is significantly reduced. Thus, the uniform dispersion of templates is a key issue in TGG. The growth distance required to achieve a certain degree of texturing and the point of impingement of growing grains also depends on the template dispersion. Figure 5.6.1 illustrates the required growth distance for 100% texturing provided three dimensionally equal interparticle spacing and same growth in all direction. It should be noted that, when the ideal dispersion is achieved, the required growth distance is less than 10 μm. This calculation implies that a method to place templates in an ideal array can resolve many of the issues related to growth kinetics.

The effect of the template size is clearly observed in the fabrication of textured PMN–PT. When 100-μm scale $BaTiO_3$ templates are used, 100-μm scale grain growth was required to achieve a high texture fraction. Meanwhile, there was no need for extensive growth or assistance from a liquid phase when 5–20-μm size $SrTiO_3$ templates were used. The proper size of template can vary depending

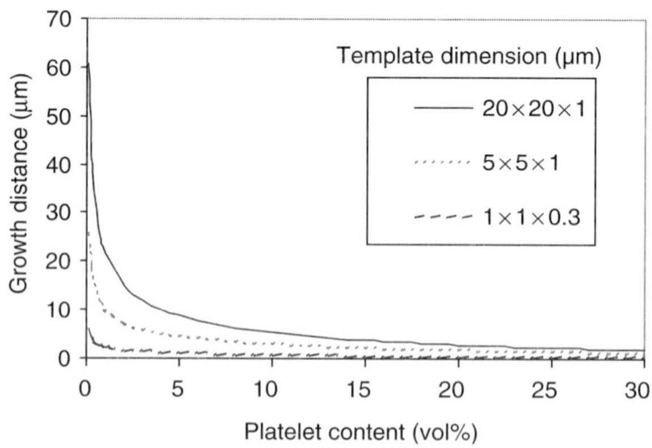

FIGURE 5.6.1 Calculated required growth distance to achieve 100% texture for various template sizes.

on the matrix size, and grain growth kinetics, but it ranges from a few to 30-μm scale. Also, the effect of size on properties, especially strength, must be considered.

There also must be epitaxial growth from the templates; thus, the lattice parameter of templates must be similar to the crystal to be grown. The chemical and thermodynamic stability of templates is also important at least during heating to the nucleation step. This means that the templates must not react or dissolve within the matrix material before nucleation and growth of stable nuclei. Since most TGG systems contain a liquid phase to enhance growth kinetics, the template must show sufficient thermodynamic stability in the presence of the liquid phase at the growth temperature. The template must not react with the liquid to form a rogue phase on the surface, which would act as a barrier to the nucleation of the desired phase. In TGG systems where the template and matrix materials are of the same composition, the solubility (S) of the material in the liquid is governed by the Gibbs–Thomson equation:

$$S = S_0 \exp\left(\frac{2\gamma\Omega}{kTr}\right) \tag{1}$$

$$S_0 = S_{T0} = S_{M0} \tag{2}$$

where S_{T0} and S_{M0} are the respective equilibrium solubilities of the template and matrix, γ is the solid–liquid surface energy, k is Boltzmann's constant, T is temperature, and r is the radius of the particle (template or matrix). The larger template particles have a lower solubility than the matrix particles due to the solubility dependence on r ($r_T \gg r_M$), which incidentally drives the Ostwald ripening process. In the case where the template and matrix are of different composition, then:

$$S_0 \neq S_{T0} \neq S_{M0} \tag{3}$$

and therefore, for TGG to occur:

$$S_{T0} \exp\left(\frac{2\gamma_T\Omega}{kTr_T}\right) \gg S_{M0} \exp\left(\frac{2\gamma_M\Omega}{kTr_M}\right) \tag{4}$$

This indicates that the TGG process is dependent on the difference in particle size between the template and matrix ($\Delta r = r_T - r_M$), the respective surface energy of the template and matrix (γ_T and γ_M), and the equilibrium solubilities of the two compositions. If the overall solubility of the template is close or below the critical solubility (S^*) of the system, then the template may not be stable.

Templates are usually synthesized by molten flux or hydrothermal crystal growth methods. The hydrothermal method is ideal for preparing low-temperature polymorphs. One of the problems associated with this method is the incorporation of OH^- ions in the crystal, which can reduce the crystal property. The templates that can be synthesized by the hydrothermal process include $SrTiO_3$, $BaTiO_3$, and $PbTiO_3$. Molten flux method produces relatively stable templates at high temperature. The size of the single crystals, however, can be sometimes too large to be used as templates in the TGG process. A variety of templates have been produced by this process including $SrTiO_3$, $BaTiO_3$, $Bi_4Ti_3O_{12}$, $Sr_3Ti_2O_7$, $KSr_2NB_5O_{15}$, and $Sr_2Nb_2O_7$.

5.6.2.2 Matrix

A fine grain, dense matrix is required since TGG is driven by the difference in surface free energies between the advancing crystal plane and the matrix grains during thermal processing. The effect of density is also important because porosity directly affects boundary mobility. Hong and Messing [24] reported a relative density of ~97% was obtained before the growth of mullite template grains started in a polycrystalline mullite matrix. Suvaci *et al.* [25] reported ~90% of theoretical density was needed before significant template growth occurred in a reaction bonded alumina oxide (RBAO) derived ceramic. Watanabe *et al.* [26] observed that growth of plate-like bismuth titanate ($Bi_4Ti_3O_{12}$) started when the sintered density was ~90% theoretical density when they started with a matrix of only plate-like particles.

Suvaci and Messing [27] proposed three stages for TGG in reaction-bonded alumina: densification, radial growth of individual template grains until template impingement, and template grain thickening. They also proposed that there is a critical size ratio between the template thickness and matrix grain size: the ratio must be greater than 1.5 for growth to continue in the absence of impingement. They also determined that the critical template concentration above which texture quality decreases due to the template–template interaction during the processing was between 5 and 10 vol% for the 20-μm size templates.

Reactive approaches for TGG, called reactive TGG or RTGG, have several advantages including the *in situ* production of fine particle size and the possible contribution of the phase transformation to the nucleation step. In RTGG, which is a combination of reactive sintering and templated grain growth, the precursor matrix transforms and densifies in a single thermal process. Seabaugh *et al.* clearly demonstrated the benefit of a reactive matrix relative to a normal matrix in texturing alumina as shown in Figure 5.6.2. It is apparent that a reactive matrix results in a much higher degree of texture in this system. The effect of a reactive matrix is also observed in piezoelectrics including PMN–PT and sodium bismuth titanate. During the TGG of PMN–PT, a reactive matrix showed much greater growth compared to an initial perovskite matrix and thus no liquid phase was required for growth [20]. The effect of a reactive matrix on formation

(a)

(b)

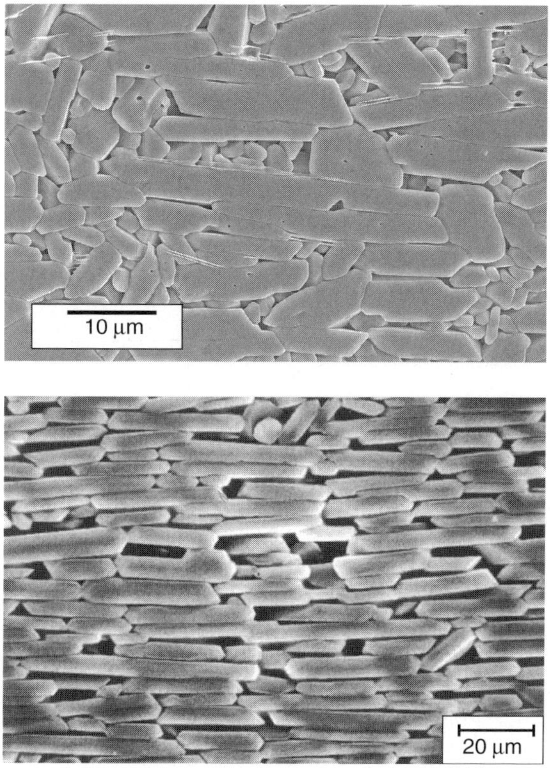

FIGURE 5.6.2 SEM micrographs of textured alumina starting with (a) α-Al_2O_3 (5 wt% templates, 5 wt% liquid phase, 1550°C, 5 h) and (b) γ-AlOOH (4 wt% templates, 5 wt% liquid phase, 1600°C, 2 h). Part (a) copyright ©2000. The American Ceramic Society.

of material in desired orientation was observed and resulted in high degree of texturing when the liquid phase exists in strontium barium niobate system [22]. Takeuchi *et al.* [28] textured $CaBiTi_4O_{15}$ (CBT) and $Na_{0.475}Ca_{0.05}Bi_{4.475}Ti_4O_{15}$ (NCBT) using RTGG and obtained 1.0 and 0.93 Lotgering factors, respectively. The piezoelectric coefficients, d_{33}, of CBT and NCBT increased from 15 to 45 pC N^{-1} and from 13 to 48 pC N^{-1} by texturing, respectively.

5.6.3 EFFECT OF LIQUID PHASE

The role of the liquid phase in TGG is significant in many ways. Liquid-phase sintering provides favorable kinetics and thermodynamic conditions for template growth. The intergranular liquid phase provides a continuous diffusion

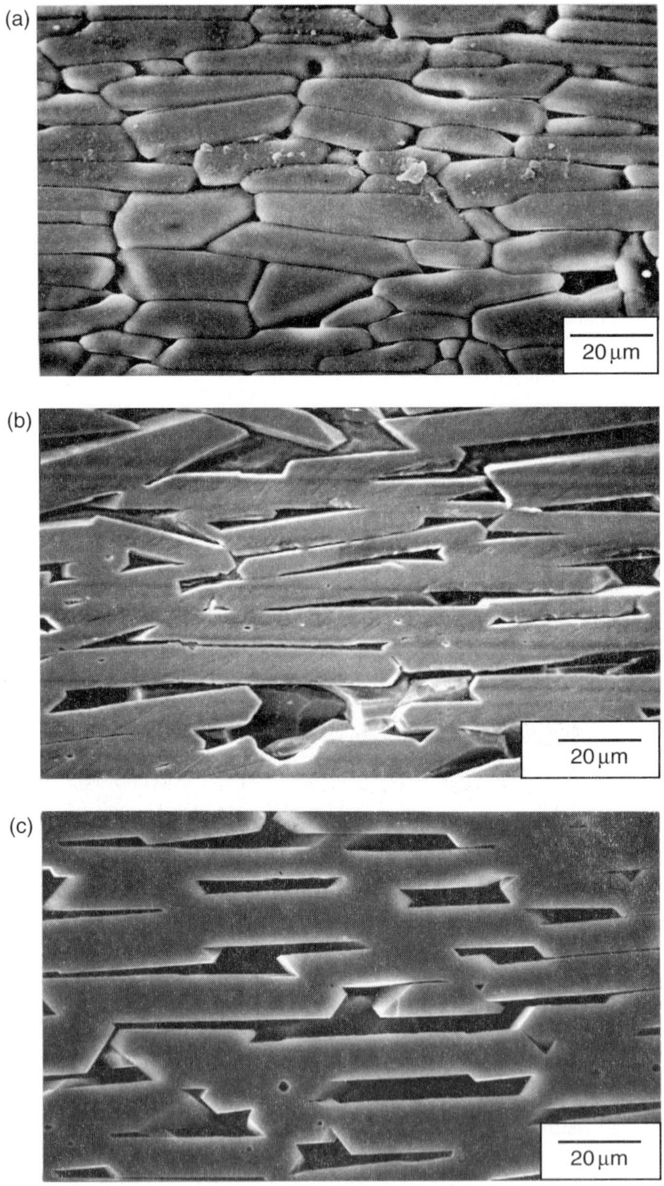

FIGURE 5.6.3 SEM micrographs of textured alumina with 4 wt% oriented template particles. All samples sintered at 1650°C, 2 h. (a) 0.5 wt% liquid phase; (b) 1 wt% liquid phase; (c) 5 wt% liquid phase. Copyright ©1997. The American Ceramic Society.

path for anisotropic growth and increases the grain growth kinetics by orders of magnitude relative to solid state diffusion. Figure 5.6.3. well illustrates the effect of the liquid-phase content on the microstructures of textured alumina. Even more importantly the liquid phase can relieve the stress of constrained sintering that results from the presence of templates, which are usually significantly larger and denser than the matrix powder. As discussed earlier, in most cases, template grain growth begins at a relative density higher than 90% , but retarded densification due to constrained sintering limits the sample density below the critical point for template growth. The textured PMN–PT with a PbO liquid phase has ~10% higher texture than when no liquid-phase former was used (Fig. 5.6.4). The enhanced kinetics resulted in a more faceted grain morphology and growth.

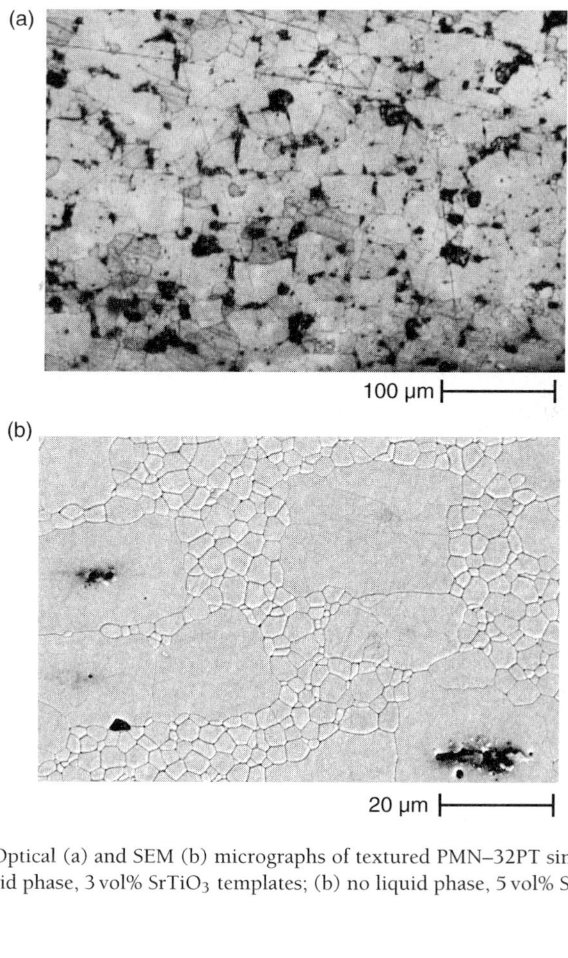

FIGURE 5.6.4 Optical (a) and SEM (b) micrographs of textured PMN–32PT sintered at 1150°C, 10 h: (a) 3% liquid phase, 3 vol% $SrTiO_3$ templates; (b) no liquid phase, 5 vol% $SrTiO_3$ templates.

5.6.4 ORIENTATION AND DISPERSION OF TEMPLATES

Anisometric templates are typically aligned by a shear force during the green-forming procedure. Thus, any type of forming process, which provides a shear, can be used for TGG including tape casting, extrusion, injection molding, roll pressing, and uniaxial pressing. It is apparent that platelet alignment is easier compared to aligning whiskers or fibers. Both cases result in fiber texture (one-dimensional alignment), but in the platelet case the rotation of template on the plane parallel to shear force during the green forming (tape-plane in tape casting case) does not affect the degree of template orientation. Meanwhile, any rotation of fiber or whisker templates from the shear force *direction* results in a decrease in the quality of template orientation. A gated doctor blade can be used to enhance the template orientation during the tape casting when the templates are fiber- or whisker-shaped [23].

Uniaxial pressing can be used to orient platy templates [13, 25]. In using platy, freeze dried β''-Al_2O_3 powder, sequential uniaxial pressing of thin layers and stacking of the layers by isostatic pressing resulted in a textured monolith. Suvaci and Messing showed that a higher compaction ratio in the uniaxial pressing of RBAO resulted in a higher degree of texturing. This is due to the fact that platelet-shaped templates, which are dispersed in a ductile matrix, such as aluminum, can be oriented by the shear force generated during the uniaxial press.

REFERENCES

1. Tani, E., Umebayashi, S., Kishi, K., Kobayashi, K., and Nishijima, M. (1986). Gas pressure sintering of Si_3N_4 with concurrent addition of Al_2O_3 and 5 wt% rare earth oxide: high fracture toughness Si_3N_4 with fiber-like structure. *Am. Ceram. Soc. Bull.* **65**: 1311–1315.
2. Becher, P. F. (1991). Microstructural design of toughened ceramics. *J. Am. Ceram. Soc.* **74**: 255–269.
3. Fuierer, P. A. (1991). Grain oriented perovskite layer structure ceramics for high temperature piezoelectric applications. PhD Thesis, The Pennsylvania State University, University Park, PA.
4. Kokubo, T., and Suzuki, N. (1982). Preparation of grain oriented ceramics of layer-structure Bi compounds by unidirectional solidification of their melts. *Yogyo-Kyokai-Shi* **90**: 633–641.
5. Monot, I., Wang, J., Desgardin, G., and Raveau, B. (1992). Crystal chemistry of the texture formation in superconductive $YBa_2Cu_3O_{7-\delta}$ ceramics. *J. Mater. Res.* **7**: 273–282.
6. Sumiya, K., Watani, S., Hayama, F., Nakamae, K., and Matsumoto, T. (1983). The orientation of magnetic particles for high density recording. *J. Magn. Magn. Mater.* **31–34**: 937–938.
7. Pridatchenko, Y. V., and Pasechnik, Z. V. (1991). Flow of dilute suspensions of deformable ellipsoidal particles in the presence of an applied electric field, unsteady orientation of particles. *Fluid Mech. Sov. Res.* **20**: 49–53.
8. Watanabe, H., Kimura, T., and Yamaguchi, T. (1989). Particle orientation during tape casting in the fabrication of grain-oriented bismuth titanate. *J. Am. Ceram. Soc.* **72**: 289–293.
9. Chou, Y-S., and Green, D. J. (1992). Silicon carbide platelet/alumina composites: I, effect of forming technique on platelet orientation. *J. Am. Ceram. Soc.* **75**: 3346–3352.

10. Wu, M., and Messing, G. L. (1994). Fabrication of oriented SiC-whisker-reinforced mullite matrix composites by tape casting. *J. Am. Ceram. Soc.* 77: 2586–2592.
11. Hirano, K., Ohashi, M., Brito, M. E., and Kanzaki, S. (1995). Processing strategy for producing highly anisotropic silicon nitride. *J. Am. Ceram. Soc.* **78**: 1687–1690.
12. Hirano, K., Watari, K., Brito, M. E., Toriyama, M., and Kanzaki, S. (1996). High thermal conductivity in silicon nitride with anisotropic microstructure. *J. Am. Ceram. Soc.* **79**: 2485–2488.
13. Tan, A., Kuo, C., and Nicholson, P. S. (1990). Preparation and characterization of textured polycrystalline Na- and K-β''/β-alumina. *Solid State Ionics* **42**: 233–241.
14. Carisey, T., Laguier-Werth, A., and Brandon, D. G. (1995). Control of texture in Al_2O_3 by gel-casting. *J. Eur. Ceram. Soc.* **15**: 1–8.
15. Carisey, T., Levin, I., and Brandon, D. G. (1995). Microstructure and mechanical properties of textured α-Al_2O_3. *J. Eur. Ceram. Soc.* **15**: 283–289.
16. Brandon, D. G., Chen, D., and Chan, H. (1995). Control of texture in monolithic alumina. *Mater. Sci. Eng. A* **195**: 189–196.
17. Seabaugh, M. M., Kerscht, I. H., and Messing, G. L. (1997). Texture development by templated grain growth in liquid-phase sintered α-alumina. *J. Am. Ceram. Soc.* **80**: 1181–1188.
18. Seabaugh, M. M., Hong, S. H., and Messing, G. L. (1998). Processing of textured ceramics by templated grain growth, in *Ceramic Microstructure: Control at the Atomic Level*, pp. 303–310, Tomsia, A. P., and Glaeser, A., eds., New York: Plenum Press.
19. Sabolsky, E. M., James, A. R., Kwon, S., Trolier-McKinstry, S., and Messing, G. L. (2001). Piezoelectric properties of ⟨001⟩ textured $Pb(Mg_{1/3}Nb_{2/3})O_3$-$PbTiO_3$. *Appl. Phys. Lett.* **78**: 2551–2553.
20. Kwon, S., Sabolsky, E. M., Messing, G. L., and Trolier-McKinstry, S. (2001) Templated grain growth of textured piezoelectric ceramics, submitted to US–Japan Conference.
21. Yilmaz, H., Messing, G. L., and Trolier-McKinstry, S. (Reactive) Templated grain growth of textured sodium bismuth titanate–barium titanate ($Na_{1/2}Bi_{1/2}TiO_3$-$BaTiO_3$) Ceramics I, in preparation.
22. Duran, C., Trolier-McKinstry, S., and Messing, G. L. (2000). Fabrication and electrical properties of textured $Sr_{0.53}Ba_{0.47}Nb_2O_6$ ceramics by templated grain growth. *J. Am. Ceram. Soc.* **83**: 2203–2213.
23. Brahmaroutu, B., Messing, G. L., and Trolier-McKinstry, S. Biaxially textured La-doped Sr2Nb207 produced by templated grain growth. *J. Mat. Res.*, in press.
24. Hong, S. H., and Messing, G. L. (1999). Development of textured mullite by templated grain growth. *J. Am. Ceram. Soc.* **82**: 867–872.
25. Suvaci, E., Seabaugh, M. M., and Messing, G. L. (1999). Reaction-based processing of textured alumina by templated grain growth. *J. Eur. Ceram. Soc.* **19**: 2465–2474.
26. Watanabe, H., Kimura, T., and Yamaguchi, T. (1991). Sintering of platelike bismuth titanate powder compacts with preferred orientation. *J. Am. Ceram. Soc.* **74**: 139–147.
27. Suvaci, E., and Messing, G. L. (2000). Critical factors in the templated grain growth of textured reaction-bonded alumina. *J. Am. Ceram. Soc.* **83**: 2041–2048.
28. Takeuchi, T., Tani, T., and Saito, Y. (1999). Piezoelectric properties of bismuth layer-structured ferroelectric ceramics with a preferred orientation processed by the reactive templated grain growth method. *Jpn. J. Appl. Phys.* **38**: 5553–5556.

Handbook of Advanced Ceramics
S. Sōmiya *et al.* (Eds.)

5.7 Hydrothermal Processing in Ceramics

SHIGEYUKI SŌMIYA
3-7-19 Seijo, Setagaya-ku, Tokyo 157-0066, Japan

5.7.1 WHAT IS HYDROTHERMAL SYNTHESIS?

The term "hydrothermal synthesis" as defined by R. Roy (personal communication) involves H_2O as a catalyst and occasionally as a component of solid phases in the synthesis at elevated temperature (<100°C) and pressure (A few atmospheres) [1]. According to Ballman and Laudise [2], hydrothermal synthesis is the use of aqueous solvents under high temperatures and high pressures to dissolve and recrystallize materials that are relatively insoluble under ordinary conditions.

There are many books and reports related to hydrothermal synthesis [2–16]. International meetings on hydrothermal reactions are also conducted. The first one was held in Japan in 1982 [17] and five more have been held till date. The other meetings were also held, for example, the International Conference on the Solvo-Thermal Reactions, which was started in 1994 in Takamatsu, Japan [18].

I prefer to use the word hydrothermal rather than solvo-thermal because the word hydrothermal has a wider meaning compared to solvo-thermal. All liquids and gases can be included in the word hydrothermal and also hydrothermal has a long history of usage compared to solvo-thermal.

5.7.2 HYDROTHERMAL PROCESSING

There are many methods for hydrothermal processing, which are shown in Table 5.7.1. Among these, some items are related mainly to geoscience. Hydrothermal processing methods in ceramics are shown in Table 5.7.2.

5.7.2.1 HYDROTHERMAL SINGLE CRYSTAL GROWTH

Chapter 6 discusses in detail the subject of crystal growth.

TABLE 5.7.1 **Hydrothermal Methods for Geosciences and Ceramics**

Hydrothermal (single) crystal growth
Hydrothermal treatment
Hydrothermal alternation
Hydrothermal dehydration
Hydrothermal extraction
Hydrothermal sintering
Hydrothermal oxidation
Hydrothermal crystallization
Hydrothermal decomposition
Hydrothermal hydrolysis–hydrothermal precipitation
Hydrothermal electrochemical
Hydrothermal mechanochemical
Reactive electrode submerged arc
Hydrothermal + ultrasonic
Hydrothermal + microwave
Hydrothermal + ultrasonic + microwave

TABLE 5.7.2 **Important Hydrothermal Processing Methods in Ceramics**

Hydrothermal (single) crystal growth
Hydrothermal dehydration
hydrothermal sintering
Reaction sintering
Hot pressing
Hot isostatic pressing
Hydrothermal precipitation–hydrothermal hydrolysis
Hydrothermal crystallization
Hydrothermal decomposition
Hydrothermal electrochemical method
Hydrothermal mechanochemical processing
Reactive electrode submerged arc process
Hydrothermal + microwave
Hydrothermal + ultrasonic
Hydrothermal + microwave + ultrasonic
Hydrothermal corrosion
Hydrothermal film
Thin
Thick
Hydrothermal phase diagrams
Cement chemistry under hydrothermal condition
Radioactive waste

5.7.2.2 Hydrothermal Fine Powder Preparations

5.7.2.2.1 References

There are many methods to produce very fine powders under hydrothermal conditions. Preparative methods for fine ceramic powders are shown in Table 5.7.3. There are many books concerning powder preparation [20–25] and hydrothermal processing [11, 26–31].

5.7.2.2.2 Characteristics of Hydrothermal Powders

Properties of hydrothermal powders are shown in Table 5.7.4 [19]. Major differences and processing between hydrothermal and other technologies are mentioned in Table 5.7.5 [19].

5.7.2.2.3 Hydrothermal Apparatus

For processing of under hydrothermal conditions, we need an apparatus containing a high-temperature–high-pressure bomb and/or an autoclave. The

TABLE 5.7.3 Preparative Methods for Fine Ceramic Powders

1. Mechanical (powder mixing)	4. Hydrothermal
Ball milling	Precipitation (coprecipitation)
Attrition milling	Crystallization
Vibration milling	Decomposition
2. Thermal decomposition	Oxidation
Heating (evaporation)	Synthesis
Spray drying	Electrochemical
Flame spraying	Mechanochemical
Plasma spraying	RESA
Vapor phase (CVD)	Hydrothermal + microwave
Freeze drying (cryochemical)	Hydrothermal + ultrasonic
Hot kerosene drying	Hydrothermal + microwave + ultrasonic
Hot petroleum drying	5. Melting and rapid quenching
Combution	
Laser beam	
Electron beam	
Sputtering	
3. Precipitation or hydrolysis	
Neutralization and precipitation	
Homogeneous precipitation	
Coprecipitation	
Salts solution	
Alkoxides	
Sol–Gel	

TABLE 5.7.4 Major Characteristics of (a) Hydrothermal Powders and (b) Ideal Powder [19]

(a) *Hydrothermal powders*
Grain size: fine grain less than 1 μm
Coagulation: no or weak agglomeration
Crystallinity: single crystal in general; depends on preparation temperature
Flow ability: forming is good
Homogeneity: good
Sinterability: good
Pores in grain: no pore
Particle size distribution: narrow
Able to synthesize low-temperature form and/or metastable form
Able to make composites like organic and inorganic mixture
Able to make a material that has a very high vapor pressure
(b) *Ideal powder*
Fine powder less than 1 μm
Soft or no agglomeration
Narrow particle size distribution
Morphology, sphere is better
Chemical composition controllable
Microstructure controllable
Uniformity
Free flowing
Less defects, dense particle
Less stress
Reactivity, sinterability
Crystallinity
Reproducibility
Process control

TABLE 5.7.5 Major Differences of Powders and Processing between Hydrothermal and the Other Technologies [19]

Powders are formed directly from solution
Powders are anhydrous, crystalline, or amorphous; it depends on production of hydrothermal powder temperature
Able to control particle size by hydrothermal temperature
Able to control particle shape by starting materials
Able to control chemical, composition, stoichiometry, etc.
Powders are highly reactive in sintering
Many cases, powders do not need calcination
Many cases, powders do not need milling process

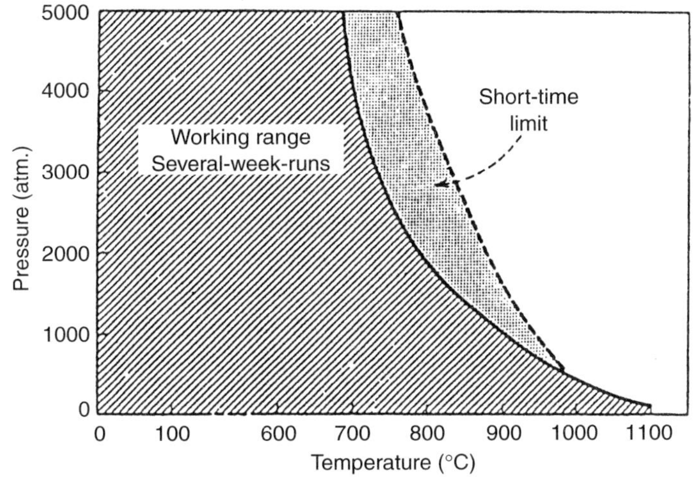

FIGURE 5.7.1 P–T limits for satellite vessel [33].

apparatus is described in Refs [2, 4, 5, 8, 32]. Many kinds of autoclaves are available to cover different *p–t* ranges and volumes (Fig. 5.7.1). In the USA, there are three companies that deal with hydrothermal apparatus:

1 *Tem-Press Division, Leco Corp.* [33]. They are the best source for specialized gases, Ar, H_2, O_2, NH_3, etc.
2 *Autoclave Engineers, Inc.* [34]. They are a moderate size company. They make a complete line of lab-scale valves, tubing, collars, all fittings for connections, etc. They also make very large autoclaves (1 m × 3 m) for chemical processes.
3 *Parr Instrument Co.* [35]. They make simple, low-pressure, low-temperature (300, 1000 bar) type of autoclaves, 50 ml to 1 l for low-temperature reactions, including vessels lined with Teflon, etc.

Kim and Stambauch [36] used continuous systems for producing zirconium oxide instead of batch autoclaves. The operating temperature was up to 180°C.

Almost all experimental and industrial hydrothermal works are operated by the batch systems.

5.7.2.2.4 Requirements of the Autoclave

Safety and corrosion of the autoclaves are very important during materials selection for a long operation. Limits of temperature and pressure and alloys for the autoclaves are shown in Tables 5.7.6 and 5.7.7 respectively [30].

TABLE 5.7.6 Limits of Pressure and Temperature of Autoclave [3]

	Pressure (MPa)	Temperature (°C)
Pyrex (5 mm inside diameter, 9 mm outside diameter)	0.6	250
Quartz (5 mm inside diameter, 9 mm outside diameter)	0.6	300
Morey type	40	400
Welded Walker–Buehler	200	480
Delta ring	230	400
Bridgman seal	370	500
Modified Bridgman seal	370	500
Cold-seal test-tube	400	200
Stellite 25	200	800
Rene 41	100	740
TZM	300	1100

TABLE 5.7.7 Alloys for Autoclaves [13]

	C	Mn	Si	Cr	Ni	Co	Mo	W	Nb	Ti	Al	Others
Low-carbon steel	0.15	0.5										
4140 Steel	0.4	0.9	0.2	1			0.2					
Stainless Steel 304	0.08	0.6	0.6	15	9							
19-9-DL	0.3	1.1	0.6	18	9		1.2	1.2	0.4	0.3		
Croloy 15-5-N	0.10	1.5	0.5	15	15		1.5	1.5	1.0			N 0.12
Timken 17-22-A	0.30	0.59	0.66	1.25	0.25		0.51				0.22	Cu 0.14
Inconel X	0.05	0.5	0.4	14.5	Bal	1.0				2.5	0.8	Fe?
Udimet 500	0.08	0.2	0.2	20	Bal	15	3.5			3	3	
Stellite 25				19–21	9–11			14–16				
Rene 25				19	Bal	11	10					
TZM							Bal			0.5		Zr 0.08

The safety of the autoclaves depends on the materials of the autoclaves, operating temperature and time. One example is shown in Figure 5.7.1. [33]. In some cases, reactor lining is important when a corrosive liquid is used. Materials used for lining are shown in Table 5.7.8 [16].

TABLE 5.7.8 Materials Used as Reactor Linings [6]

Material	T (°C)	Solutions	Remarks
Titanium	550	Chlorides	Corrosion in NaOH solution >25% In NH_4Cl solution >10% (at 400°C)
		Hydroxides	
		Sulfates	
		Sulfides	
Armco iron	450	Hydroxides	Gradual oxidation producing magnetite
Silver	600	Hydroxides	Gradual recrystallization and embrittlement, partial dissolution
Platinum	700	Hydroxides	Blackening in chlorides in the presence of sulfer ions;
		Chlorides	Partial dissolution in hydroxides
		Sulfates	
Teflon	300	Chlorides	Poor thermal conduction
		Hydroxides	
Tantalum	500	Chlorides	Begin to corrode in NH_4Cl solution 78%
Pyrex	300	Chlorides	
Copper	450	Hydroxides	Corrosion reduced in the presence of fluoride ions and organic compounds
Graphite	450	Sulfates	Pyrolytic graphite most suitable for linings
Nickel	300	Hydroxides	
Quartz	300	Chlorides	
Gold	700	Hydroxides	Partial dissolution in hydroxides
		Sulfates	

5.7.2.2.5 Autoclaves

There are many types of the autoclaves [2, 4, 8]. The most common autoclaves are shown in Figures 5.7.2 and 5.7.3 called the Morey bomb and the Tuttle tube, respectively [2, 4, 31].

5.7.2.2.6 Hydrothermal Techniques

Many hydrothermal experimental techniques are described by Ulmer [37] and Ulmer and Barnes [38].

In an ordinary case, hydrothermal experimental work uses Pt, Au and Ag tubes, mainly Pt and Au. Welding of Pt is easier than Au. Ag is the difficult to weld among the three metals. In the so-called closed system, both ends are welded. An open system is one where one end is welded and the other is open. In the latter case, the solution in the bomb can move in and put of the capsule [39].

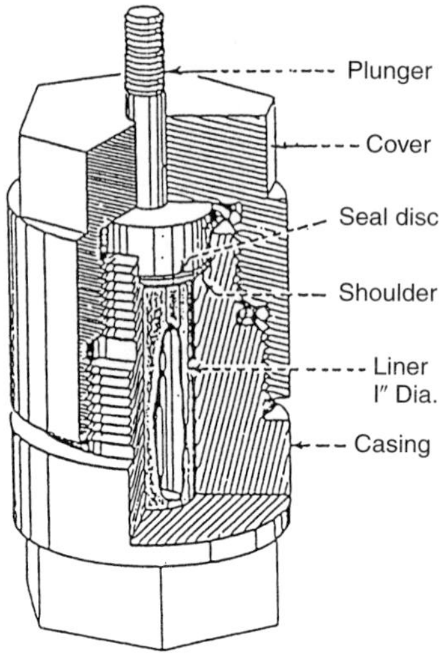

FIGURE 5.7.2 Autoclave with flat plate closure, Morey type [4].

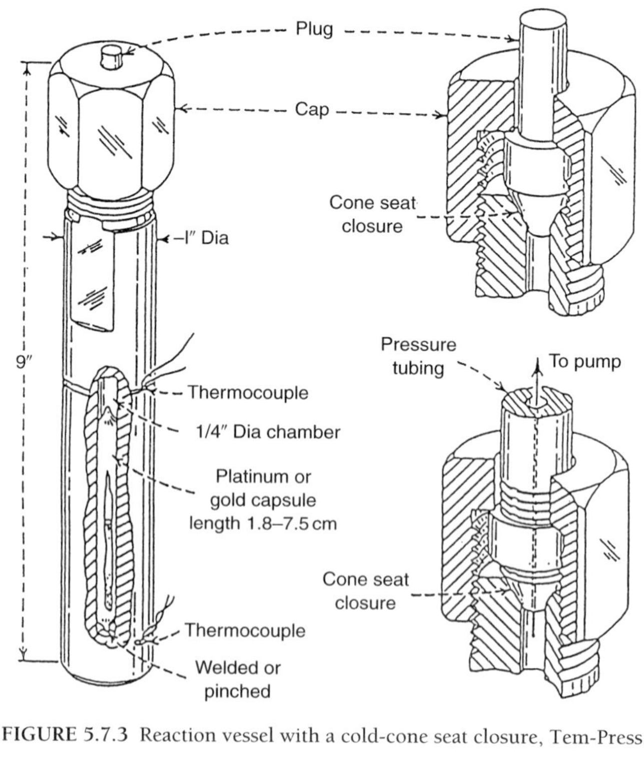

FIGURE 5.7.3 Reaction vessel with a cold-cone seat closure, Tem-Press.

5.7.3 HYDROTHERMAL POWDER PREPARATIONS

5.7.3.1 Hydrothermal Treatment [40]

Noda and coworkers [40, 41] reported an example of hydrothermal treatment to obtain more useful materials. $K_{1.7}Mg_{3.00}Al_{1.00}Si_{3.09}O_{9.52}F_{3.51}$ and $KMg_3AlSi_3O_{9.88}F_{2.24}$ were used under hydrothermal conditions at temperatures from 150 to 350°C for 1–2 days with a solution of KOH. The reaction products were hydroxyl-phlogopite and hydroxide. After hydrothermal treatment (i) the glassy phase in the samples was lesser than that the before treatment and (ii) the separation of fibrous materials in the sample was easier.

5.7.3.2 Hydrothermal Dehydration

Oota *et al.* [42] reported a hydrothermal dehydration process. The apparatus used is shown in Figure 5.7.4. At the bottom of the autoclave, dehydration reagents such as Mg, Zn, etc., were used. The starting materials were $TiO_2 \cdot nH_2O$ and NaOH along with Mg and/or Zn ribbon. The KOH solution and the powders were put into the Pt tube, leaving one side of Pt tube open. The second tube containing the dehydration reagent was placed at the bottom of the autoclave. The temperature in bomb was elevated to the desired level and kept at this temperature for 1–50 h. After the bomb was cooled to room temperature, the solid products were washed out with water into a glass beaker. The floating long fibers were removed by mechanical processes such as decantation, etc.

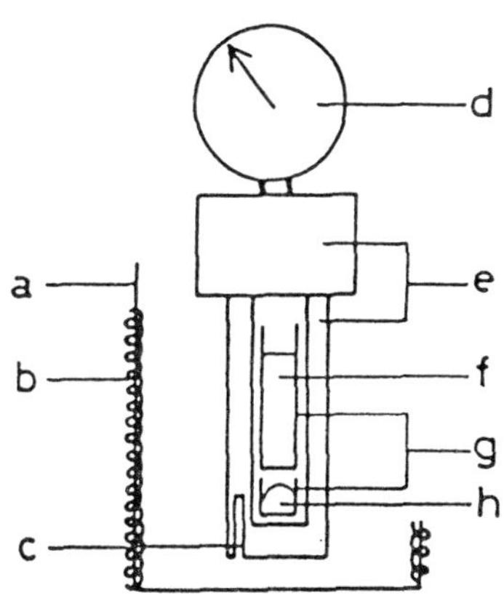

FIGURE 5.7.4 Apparatus for the hydrothermal dehydration method: (a) furnace; (b) heater; (c) thermocouple well; (d) pressure gauge; (e) autoclave; (f) nutrient; (g) Pt-tube; (h) dehydration reagent [41].

5.7.3.3 Hydrothermal Precipitation–Hydrothermal Hydrolysis

Many studies have reported hydrothermal precipitation processes [43–52], wherein zirconia, $BaTiO_3$, $PbZrO_3$, etc., were studied. A process of hydrothermal precipitation, the hydrothermal homogenous precipitation of zirconia, is shown in Figure 5.7.5. The microstructure of the powders and the sintered body are shown in Figures 5.7.6 and 5.7.7, respectively. The properties of the powders and the sintered body are shown in Table 5.7.9.

Lencka *et al.* [51] reported the hydrothermal precipitation of lead zirconate titanate solid solution. Lead acetate or nitrate as sources of Pb and either crystalline or anhydrous titanium dioxide and zirconium dioxide or hydrous titanium and zirconium oxide were used as the starting materials. Temperatures ranged from 90 to 200°C and KOH was used as a pH adjusting agent.

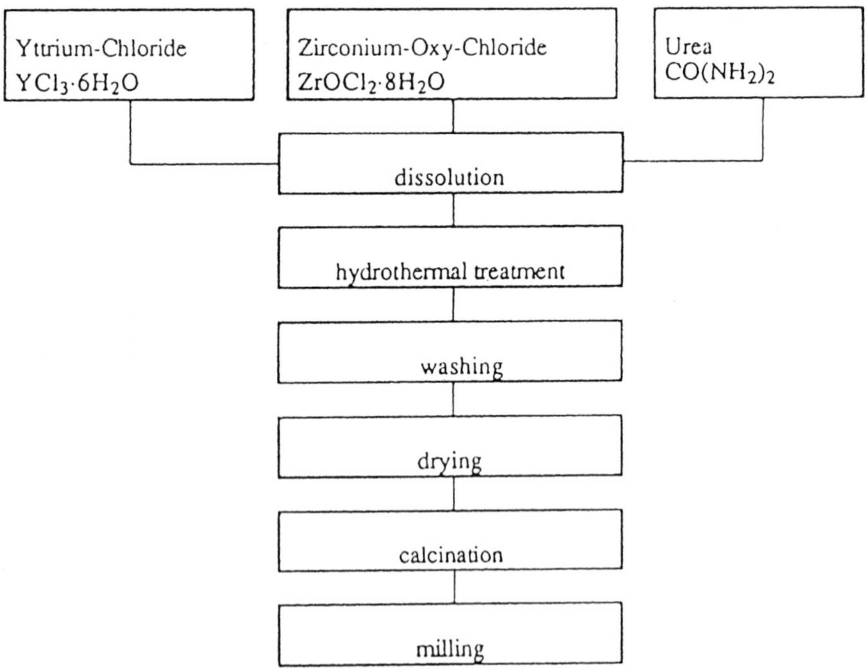

FIGURE 5.7.5 Schematic hydrothermal homogenous precipitation of zirconia by Chichibu Cement Co. Ltd.

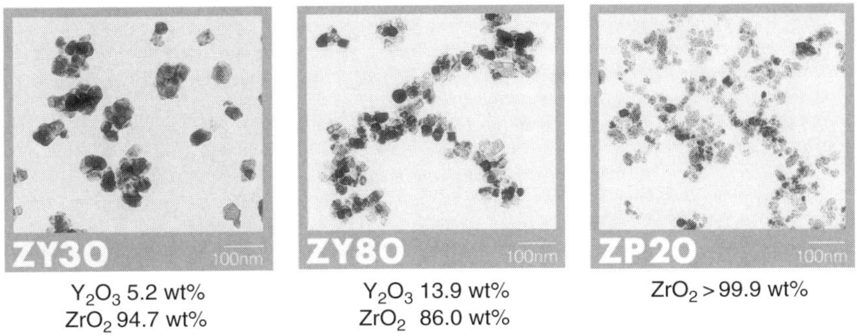

FIGURE 5.7.6 Microstructure of zirconia powders prepared by hydrothermal homogenous precipitation method.

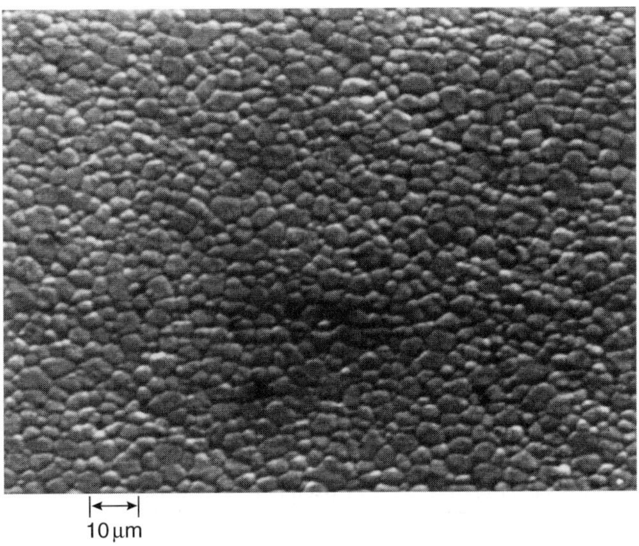

FIGURE 5.7.7 Microstructure of the sintered zirconia, Y_2O_3 3 mol%, 1400°C, 2 h (calcined at 1000°C for 1 h).

5.7.3.4 Hydrothermal Crystallization

There are many papers related to hydrothermal crystallization, that is, crystallization under hydrothermal conditions [53–58]. The starting material, amorphous hydrous zirconia was prepared from $ZrCl_4$ solution with 3 N NH_4OH, filtered, and dispersed in water. The precipitate was dried at 120°C

TABLE 5.7.9 Hydrothermal ZrO_2 Powders [19]

Powder	ZY30	ZY80	ZP20
Chemical composition			
ZrO_2 (wt%)	94.7	86.0	>9.9
Y_2O_3	5.2	13.9	—
Al_2O_3	0.010	0.010	0.005
SiO_2	0.010	0.010	0.005
Fe_2O_3	0.005	0.005	0.005
Na_2O	0.001	0.001	0.001
Cl^-	<0.01	<0.01	<0.01
Ignition loss	1.5	1.5	8.0
Crystallite size (nm)	2.2	2.2	2.0
Average particle size[a] (μm)	0.5	0.5	1.5
Specific surface area[b] (m^2/g)	20	25	95
Sintered specimens	1400°C × 2h	1500°C × 2h	
Bulk density (g/cm^3)	6.05	5.85	
Bending strength[c] (MPa)	1000	300	
Fracture toughness[d] (MPa $m^{1/2}$)	6.0	2.5	
Vicker's hardness (GPa)	12.5	11.0	
Thermal expension 20–1000°C ($\times 0^{-6}$/°C)	11.0	10.6	

[a]Photo sedimentation method.
[b]BET Method (N_2).
[c]Three-point bending method.
[d]MI method.

for 48 h. The dry powder was put into a capsule of gold and/or platinum along with the mineralizers. The Pt and/or Au tube, 27 mm ID, 0.15 mm thick, and about 30–35 mm long, was filled with the starting material and solution and treated under 100 MPa at temperature from 200 to 600°C for 24 h using the the test tube type.

The effects of mineralizers and temperatures on particle size and reaction products are shown in Table 5.7.10.

Some mineralizers produced both tetragonal and monoclinic zirconia and some mineralizers produced only monoclinic zircona fine crystals. Increasing the hydrothermal crystallization temperature to above 500°C increased the crystallite size of zirconia (Fig. 5.7.8).

Distribution of the second-phase component was very good in hydrothermal crystals of Eu_2O_3–HfO_2 cubic solid solution (Table 5.7.11) [44].

Bucko *et al.* [58] also reported hydrothermal crystallization using zirconium oxychloride water solution to precipitate zirconia gel with an ammonia water solution. The gel was washed with distilled water. Amorphous zirconia gel was hydrothermally treated at 250°C for 1–326 h in water and in NaOH solution. A 300-m^3 stainless steel autoclave lined with Teflon was used. Size of hydrothermal crystallized zirconia were 8.3 and 8.1 nm for 1 h.

TABLE 5.7.10 Phases present and Crystallite Sizes of Products Hydrothermally Crystallized at 100 MPa for 24 h [19, 54, 57]

Mineralizer	Temperature (°C)	Average crystallite size (nm)	
		Tetragonal ZrO_2	Monoclinic ZrO_2
KF (8 wt%)	200	Not detected	16
KF (8 wt%)	300	Not detected	20
NaOH (30 wt%)	300	Not detected	40
H_2O	300	15	17
LiCl (15 wt%)	300	15	19
KBr (10 wt%)	300	13	15

5.7.3.5 Hydrothermal Decomposition

There are several papers related to hydrothermal decomposition.

5.7.3.5.1 Zircon

Reynen *et al.* [59] reported hydrothermal decomposion of zircon for the production of high-purity zirconia. The hydrothermal reactions were as follows:

$$ZrSiO_4 + xCa(OH)_2 \longrightarrow ZrO_2 + xCaO \cdot SiO_2 \cdot H_2O + (x-1)H_2O$$

$$ZrSiO_4 + 2NaOH \longrightarrow Na_2ZrSiO_5 + H_2O$$

The reaction of zircon with $Ca(OH)_2$ or NaOH was described. The conditions for one of the experiments are shown below:

$ZrSiO_4$	18.43 wt%
$Ca(OH)_2$	14.9 wt%
NaOH	4.67 wt%
H_2O	62 wt%
Liquid/solid ratio	2
NaOH concentration in the solution	7 wt%
Mole ratio of $Ca(OH)_2/ZrSiO_4$	2
Temperature	350°C
Vapor pressure	170×10^5 Pa
Reaction time	8 h

After hydrothermal treatment with $Ca(OH)_2$ and NaOH, the solid phase was separated from the NaOH solution by filtration. After treatment in this process: (i) The solid was treated with sulfuric acid and evaporated to dryness. From the

(a)

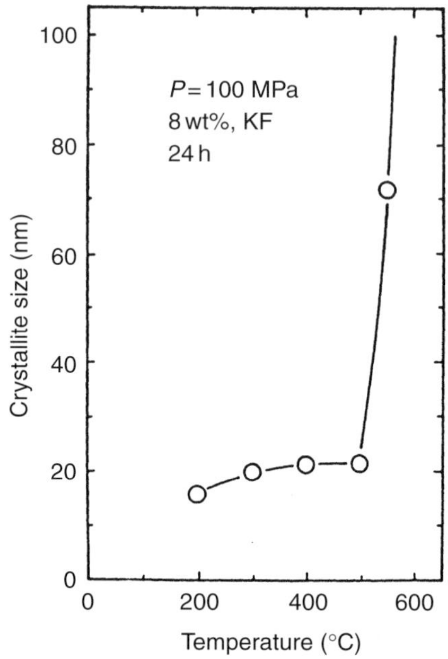

(b)

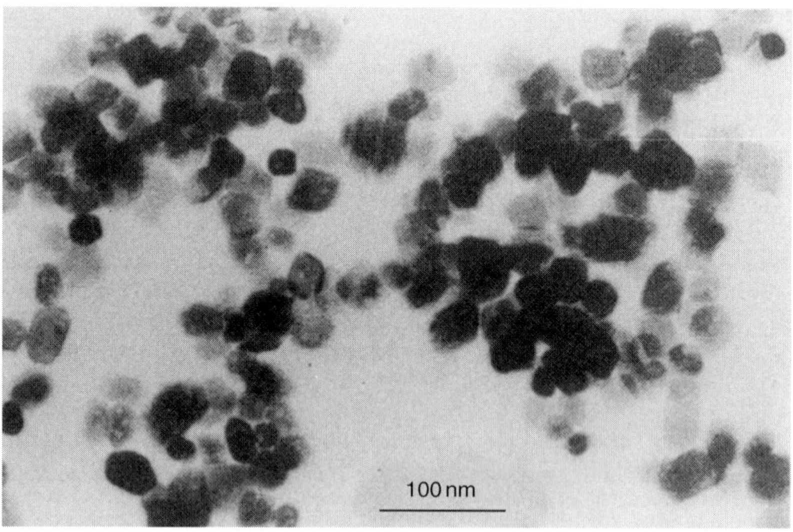

FIGURE 5.7.8 (a) Variation with temperature of crystallite size of monoclinic zirconia prepared by hydrothermal crystallization under 100 MPa for 24 h using 8 wt% KF solution. (b) TEM of zirconia (mono) powder by hydrothermal crystallization, 400°C, 100 MPa, 24 h, using 8 wt% KF solution.

TABLE 5.7.11 Quantitative EDS Analysis of 30 wt% Eu_2O_3–HfO_2 Cubic Solid Solution Particles Crystallized Hydrothermally at 300°C, 10 MPa, for 3 h from the Coprecipitate [44]

Particle no.	Composition (wt%)	
	HfO_2	Eu_2O_3
1	68.1	31.9
2	69.7	30.3
3	69.7	30.3
4	71.3	28.7
5	70.6	29.4
6	71.1	28.9
Average	70.1	29.9
Standard deviation		1.1

filtrate, a pure zirconium sulfate tetrahydrate was obtained by crystallization. (ii) The filtrate was treated with NH_4OH, then $Zr(OH)_4 \cdot xH_2O$ was obtained. Filtrations and washing are problems of this type of process.

5.7.3.5.2 Ilmenite [61, 62]

Imenite decomposed under hydrothermal conditions. For example, 50 mg of 200-mesh ilmenite powder and 60 mg of 10 M KOH were sealed in an Au capsule. Then the capsule was placed in the hydrothermal equipment. The pressure and temperature were raised to the desired value at certain time, at which the sample was quenched. The effect of pressure on hydrothermal decomposition is shown in Figure 5.7.9. The products obtained were magntite and $K_2O \cdot 4TiO_2$ in the case of ilmenite with water ratio 5:3 at 500°C, 10 M KOH, 300 kg/cm^2 for 63 h for 84 h under the same hydrothermal conditions, the product was $K_2O \cdot 6TiO_2$.

Fibrous $K_2O \cdot 6TiO_2$ was formed at 300 kg/cm^2, temperature above 600°C, 10 M KOH solution, and for 24 h. The hydrothermal reactions were as follows:

$$FeTiO_3 + H_2O \longrightarrow Fe_{3-x}O_4 + TiO_2 + H_2$$

$$nTiO_2 + KOH \longrightarrow K_2O \cdot nTiO_2 + H_2O \quad \text{where } n = 4 \text{ or } 6$$

5.7.3.5.3 Ferrite

Sisk *et al.* [63] reported the preparation of manganese zinc ferrites. The starting materials were ferric citrate, manganese acetate, and zinc acetate. The

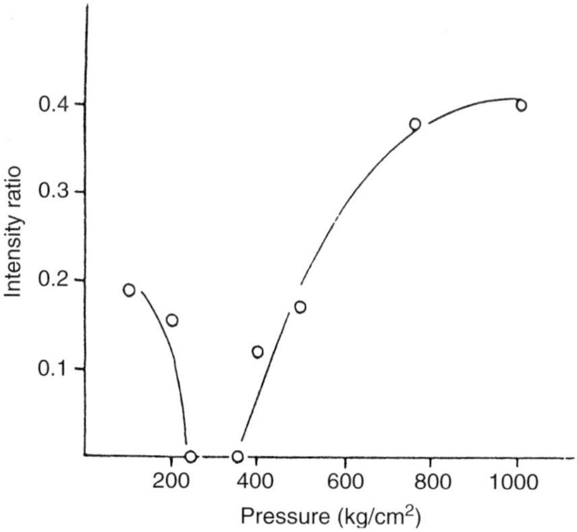

FIGURE 5.7.9 Effect of pressure on the decomposition under hydrothermal condition at 500°C, 10 M KOH mineralizer, 5:3 ilmenite to water ratio, 72 h of treatment.

mixed chemicals were put into the autoclave, heated from 130 to 200°C for 1 h. The desired powder was then obtained.

5.7.3.5.4 Organic Materials [64]

One class of organic material, polychlorinated biphenyls, was decomposed by hydrothermal treatment of 300–320°C, 180 kg/cm^2. The equipment used is shown in Figure 5.7.10. Hydrothermal decomposition is one of the methods used to decompose organic materials safety, because it is a closed method.

5.7.3.6 Hydrothermal Oxidation [65–74]

There are four types of hydrothermal oxidations: (i) direct oxidation: (ii) through hydroxide, (iii) through hydride, and (iv) mixed type.

5.7.3.6.1 Direct Oxidation

$$\text{Metal} + H_2O \longrightarrow MO + H_2$$

(a)

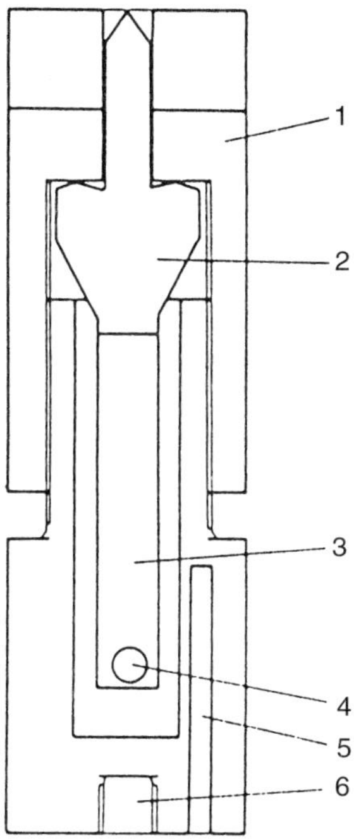

Microautoclave: (1) lid; (2) cone packing; (3) reaction chamber (60 ml); (4) ball for stirring; (5) thermocouple well; (6) socket of steel rod for taking out the autoclave

(b)

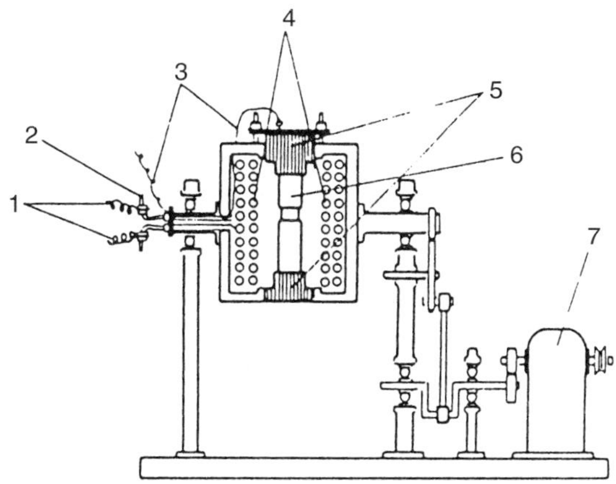

Induction heater: (1) lead wire for alternating current; (2) cooling water; (3) lead wire of thermocouple; (4) copper tube (induction coil); (5) steel blades of ferrosilicon; (6) microautoclave; (7) reduction gear

FIGURE 5.7.10 Induction heater and autoclave by Yamasaki [64].

For example,

$$2Cr + 3H_2O \longrightarrow Cr_2O_3 + 3H_2$$
$$Zn + H_2O \longrightarrow ZnO + H_2$$

5.7.3.6.2 Through Hydroxide

$$\text{Metal} + H_2O \longrightarrow M(OH)_2 \longrightarrow MO + H_2$$

For example,

$$Al + H_2O \longrightarrow Al(OH)_3 \longrightarrow Al_2O_3 + H_2$$

5.7.3.6.3 Through Hydride

$$\text{Metal} + H_2O \longrightarrow MH_2 \longrightarrow MO + H_2$$

For example,

$$Zr + H_2O \longrightarrow ZrH_x \longrightarrow ZrO_2 + H_2$$
$$Ti + H_2O \longrightarrow TiH_x \longrightarrow TiO_2 + H_2$$

5.7.3.6.4 Mixed Type

For example,

$$Al + Zr + H_2O \longrightarrow Al(OH)_3 + ZrH_x \longrightarrow Al_2O_3 + ZrO_2 + H_2$$
$$Al + Hf + H_2O \longrightarrow Al(OH)_3 + HfH_x \longrightarrow Al_2O_3 + HfO_2 + H_2$$

TEM of zirconia powder by hydrothermal oxidation at 100 MPa, 500°C for 3 h, schematic illustration of hydrothermal oxidation of zirconia powder, variation of amount of products with temperature by hydrothermal oxidation of Ti under 100 MPa for 3 h, and products and their average crystallite sizes after hydrothermal oxidation of Zr and Al metals are shown in Figures 5.7.11–5.7.13 and Table 5.7.12, respectively.

5.7.3.7 Hydrothermal Electrochemical Method [74, 75]

Yoo *et al*. reported a hydrothermal anodic oxidation. Later, they called this process the hydrothermal electrochemical method. The apparatus used is shown

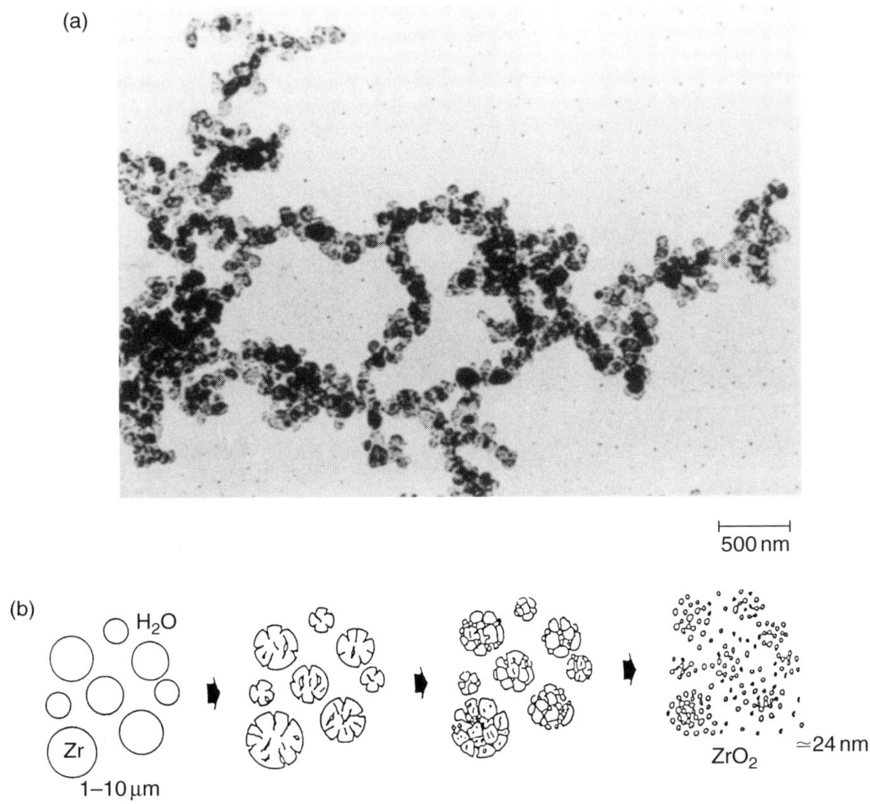

FIGURE 5.7.11 (a) TEM of zirconia powder by hydrothermal oxidation, 100 MPa, 500°C, 3 h, Zr/O = 0.33 [19]. (b) Schematic illustration of hydrothermal oxidation of Zr metal to form zirconia.

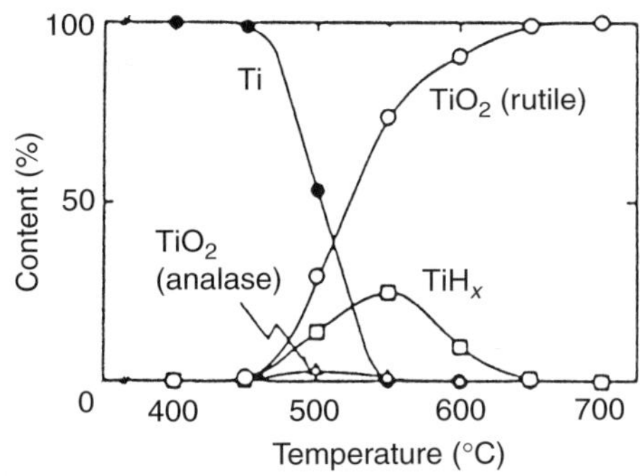

FIGURE 5.7.12 Variation of amount of products with temperature by the hydrothermal oxidation of Ti in the closed system under 100 MPa for 3 h.

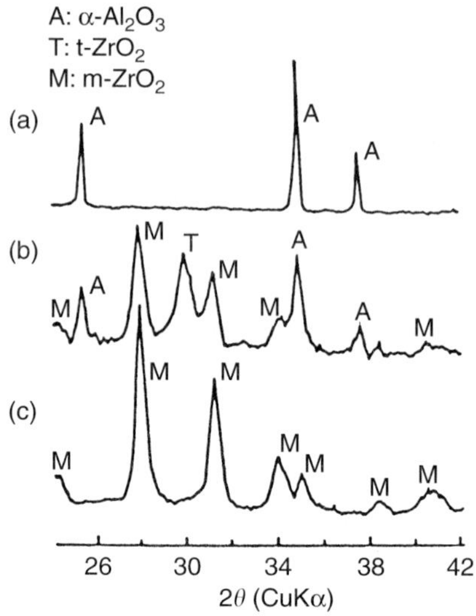

FIGURE 5.7.13 X-ray powder diffraction patterns of the products prepared at 700°C under 100 MPa for 3 h. (a) Al + H_2O; (b) $ZrAl_3$ + H_2O; (c) Zr + H_2O.

TABLE 5.7.12 Products and their Average Crystallite Size of Hydrothermal Oxidation of Metals with 100 MPa Water [73]

Starting material	Temperature (°C)	Products	Crystallite size (nm)
Zr	250–700	m-ZrO_2	23–27
Al	100–500	γ-AlOOH	—
	500–700	α-Al_2O_3	110–200

in Figure 5.7.14. In this process, Ti metal plate was used for $BaTiO_3$. $BaTiO_3$ powders were synthesized in 0.1 N $Ba(NO_3)_2$ solution. The quantity of electrolyte was 200 ml and the area of working electrode exposed to electrolyte was 2.1 cm^2 for treatment. As a counterelectrode (cathode), a platinum plate (9 mm × 30 mm × 0.15 mm) was used. The distance between the working electrode and the platinum counterelectrode was 3.0 cm. The formation diagram of the powder products by this method of titanium in $Ba(NO_3)_2$ solution and the TEM micrograph of $BaTiO_3$ powder are shown in Figure 5.7.15 and 5.7.16, respectively. $BaTiO_3$ powder preparation depended on the temperature and/or concentration of the $Ba(NO_3)_2$ solution rather than on the applied current density.

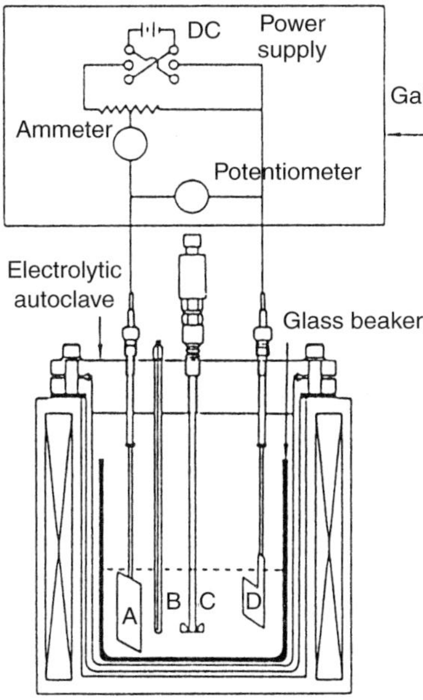

FIGURE 5.7.14 Schematic illustration of the electrochemical cell circuit arrangements for hydrothermal electrochemical method: (A) counterelectrode (Pt plate), cathode; (B) thermocouple; (C) stirrer; (D) working electrode (Ti plate), anode.

5.7.3.8 HYDROTHERMAL MECHANOCHEMICAL PROCESSING [76, 77]

Ba-ferrite was prepared by this method. The starting materials were $Ba(OH)_2$, $FeCl_3$, NaoH, etc. The conditions were 200°C, 2 MPa, for 0–24 h and 5-mm diameter stainless steel balls, numbering 200, 500 and 700, were used. The rotation rate was 10–107 rpm. Figure 5.7.17 shows the experimental apparatus. Figure 5.7.18 shows the X-ray diffraction patterns of the products. Under hydrothermal conditions, the hydrothermal mechanochemical process influenced the nucleation process in crystallization. The nucleation rate increased with increasing number of balls and/or increasing rotation speed.

5.7.3.9 REACTIVE-ELECTRODE SUBMERGED-ARC (RESA) PROCESS (FIGURE 5.7.19)

The RESA process [78, 79] is one of the new methods for making fine (10–1000 nm) oxides and non-oxide powders. This technique utilizes two metal

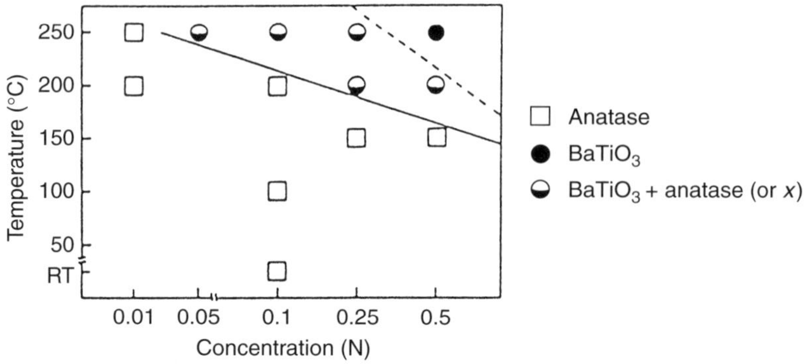

FIGURE 5.7.15 Formation diagram of powder products by anodic oxidation of titanium in $Ba(NO_3)_2$ solutions. Current density: 100 mA/cm^2.

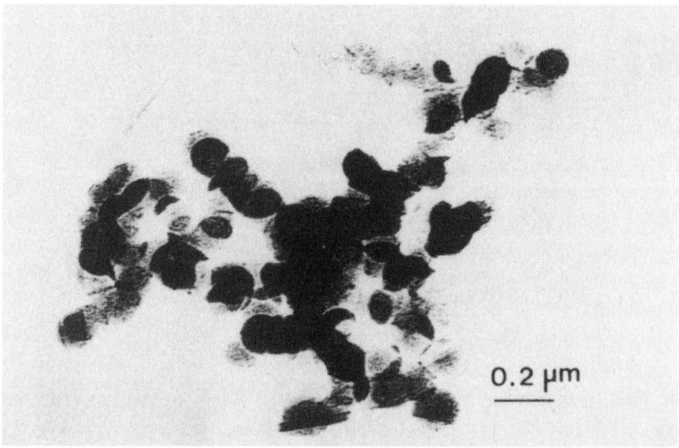

FIGURE 5.7.16 TEM micrograph of $BaTiO_3$ powder obtained after hydrothermal electrochemical method.

or conducting electrodes that are arc immersed in a suitable dielectric fluid that reacts with the metal. The spark provides an extremely high temperature for a short duration, resulting in vaporization of the electrode and the surrounding dielectric fluid and reaction within the "bubble" that forms.

The arc is struck between the two metal electrodes typically 3–6 mm in diameter that are immersed in approximately 1000 ml of liquid. In most cases, currents range from 30 to 100 A. In all cases a colloidal sol was recovered [78]. The colloids were dried and then the powders obtained.

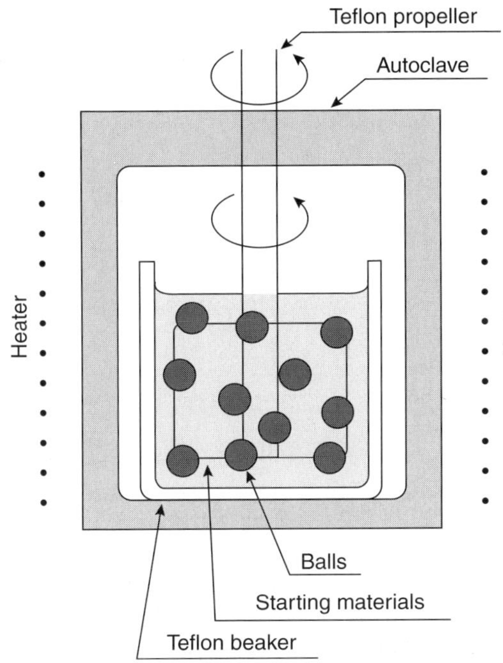

FIGURE 5.7.17 The apparatus of hydrothermal mechanochemical processing [77].

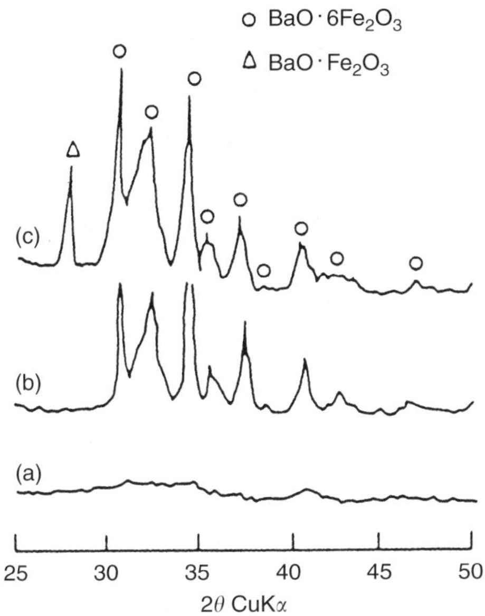

FIGURE 5.7.18 X-ray diffraction patterns of (a) starting materials, (b) material prepared at 200°C under 2 MPa for 4 h, and (c) material prepared at 200°C for 4 h, using 200 balls, 37 rpm.

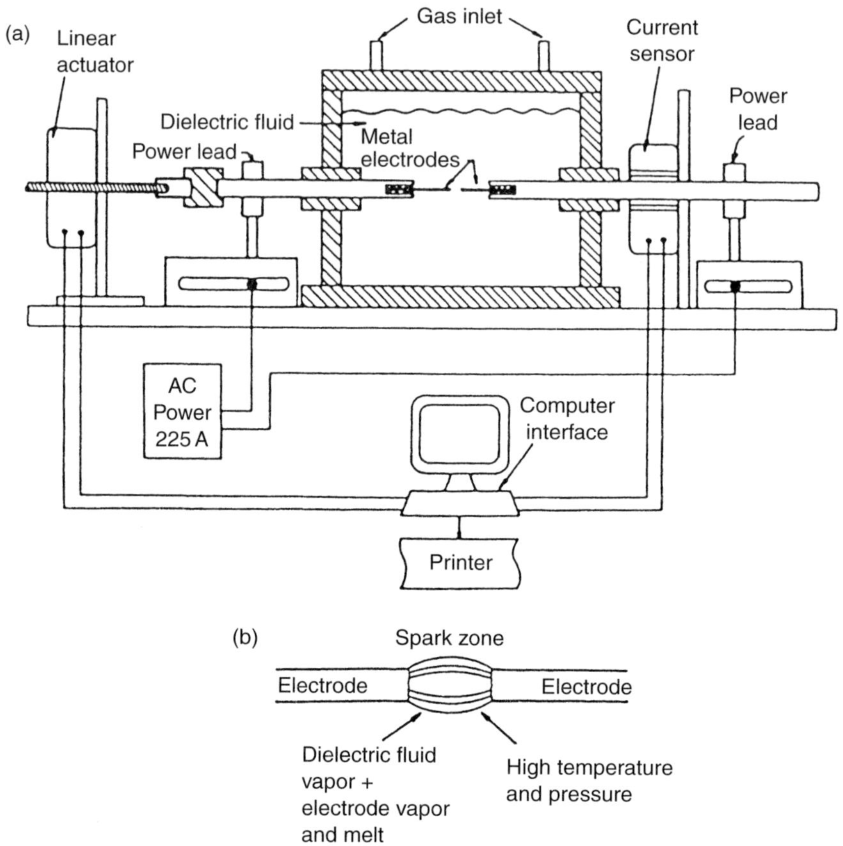

FIGURE 5.7.19 Schematic illustration of the RESA apparatus for fine powder preparation (a); and schematic of the spark between the metal electrodes submerged in a dielectric liquid (b).

5.7.3.9.1 Al

Electrodes of 1/8 in. Al rod were used. Currents ranging from 15 to 60 A at voltages ranging from 50 to 80 V were used. γ-Al_2O_3 powders, rather poorly crystallized, were obtained.

5.7.3.9.2 Zr

Pure Zr rods (2-mm diameter) and currents ranging from 40 to 60 A at voltages around 70 V were used. The product was monoclinic ZrO_2 and particles were less than 100 nm in size with an average size of 50 nm.

5.7.3.9.3 Ti

Ti rods (2-mm diameter) and currents ranging from a few ampheres to 150 A with voltages varying from a few volts to 80 V were used. The dry sol showed the poorly crystallized TiO_2 in rutile form.

5.7.3.9.4 Fe

Fe rods (2-mm diameter) gave Fe_3O_4 and Fe (α-phase).

5.7.3.9.5 W

Rods of W (1-mm diameter) with currents varying between 30 and 50 A at voltage of about 60–80 V produced the sol. The powder XRD of the dry sol showed WO_3.

5.7.3.9.6 Cr

Cr chips with currents ranging from 30 to 80 A and voltages ranging 50–70 V were used. The use of deionized water gave Cr_2O_3, Cr_3O_4, and small amounts of CrO_2; with the use of 10 vol% of H_2O_2, Cr_2O_3, Cr_3O_4, and CrO_2 appeared.

5.7.3.9.7 SiC and TiC

Two metal electrodes typically 2–6 mm in diameter were immersed in approximately 1 l of a liquid. In the case of SiC, silicone oil was used as the dielectric fluid, carbon electrodes were used as the carbon source. For TiC, TiC was formed from Ti electrodes in various organic fluids such as mineral oil and aniline. In the case of SiC, the particles were ranging from a few nanometers to 50 nm in size.

5.7.3.10 Hydrothermal-Microwave Method [80–83]

The main advantages of hydrothermal-microwave processing are as follows:

1. It increases the kinetics of reaction by one to two orders of magnitude.
2. It leads to rapid heating, which can lead to energy savings.

3. It leads to some novel phases.
4. Compounds containing Pb can be easily synthesized without damage to humans because this hydrothermal process is a closed system. Pb-vapor does not go outside the apparatus. One example of these results is shown in Table 5.7.13.

TABLE 5.7.13 X-ray Analysis of Various Powders Prepared by Hydrothermal-Microwave Technique [78–83]

	Concentration (M)				Reaction products order of abundace as determined by X-ray		
					Temp (°C)	Duration (h)	Products
$BaTiO_3$							
$Ba(No_3)_2$	$TiCl_5$	KOH					
0.35	0.33	10			109	2	$BaTiO_3$
0.35	0.33	10			144	2	$BaTiO_3$
$SrTiO_3$							
$Sr(NO_3)_2$	$TiCl_4$	KOH					
0.1334	0.13	10			164	2	$SrTiO_3$
$SrZrO_3$							
$Sr(NO_3)$	$ZrO(NO_3)$		KOH				
0.3113	0.3113		10		138	0.5	$SrZrO_3$
0.3113	0.3114		10		194	2	$SrZrO_3$
$BaZrO_3$							
$Ba(NO_3)_2$	$ZrO(NO_3)_2$		KOH				
0.3113	0.3144		10		130	2	$BaZrO_3$
0.3113	0.3144		10		194	2	$BaZrO_3$
$Pb(Zr_{0.52}Ti_{0.48})O_3$ (PZT)							
$Pb(No_3)_2$	$ZrOCl_2$		$TiCl_4$	KOH			
0.542	0.2764		0.25	10	115	0.5	PZT
0.542	0.2764		0.25	10	164	0.5	PZT
TiO_2							
$TiCl_4$	HCl						
0.5	1				164	0.5	Rutile
ZrO_2							
$ZrOCl_2$	$8H_2O$						
1.0					194	2.0	m-ZrO_2
0.5					164	2.0	Poorly crystalline m-ZrO_2

5.7.3.11 Hydrothermal + Ultrasonic

Roy and colleagues [84–87] used ultrasonic and/or hydrothermal + ultrasonic waves for materials synthesis.

Roy and colleagues also reported the synthesis of materials by using acoustic waves [84–85]. Also, Margulis [86] and Suslick [87] published similar reports.

The characteristics of reactions by ultrasonic waves are as follows:

1. acceleration of conventional chemical reaction and redox processes in aqueous solution;
2. rapid movement of fluid caused by variation of sonic pressure, which subjects the fluids to alterations of compression and rarefaction wave;
3. formation and collapse of microbubbles (cavitation), which may be responsible for most of the chemical efforts observed;
4. a large amount of vibrational energy is put into small volumes with little heating.

R. Roy (personal communication) suggested the synthesis of materials by hydrothermal-ultrasonic waves. This is a new field and is interesting.

5.7.4 HYDROTHERMAL SINTERING

There are several techniques among hydrothermal sintering processes, namely hydrothermal reaction sintering, hydrothermal hot pressing, hydrothermal hot isostatic pressing, etc.

5.7.4.1 Hydrothermal Reaction Sintering [88–95]

The first report that appeared in 1976 was by Hirano and Sōmiya for Cr_2O_3 [88]. The particle size of chromium metal was about 50 μm. The chromium metal powder and water were mixed at the molar ratio of 2:3 in a Pt capsule. The capsule used is shown in Figure 5.7.20. The capsule is hermetically sealed. The sealed capsule was placed in a test-tube-type pressure vessel and/or a high gas pressure apparatus and then subjected to the hydrothermal reaction sintering process under the desired pressure and temperature.

Above 500°C and under 1000 kg/cm^2 pressure, the reaction was

$$2Cr + 3H_2O \longrightarrow Cr_2O_3 + 3H_2$$

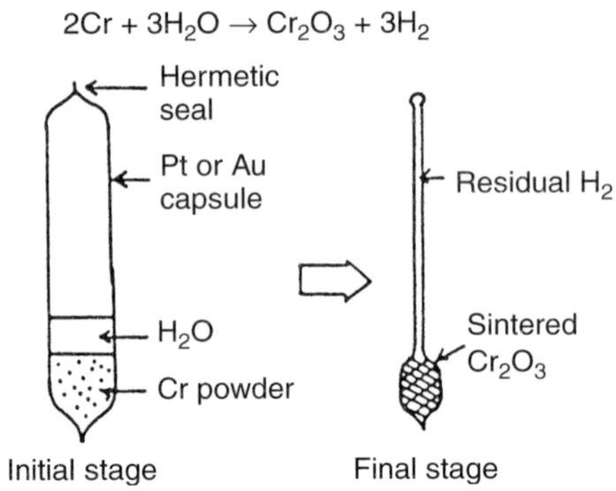

FIGURE 5.7.20 Hydrothermal reaction sintering of Cr_2O_3 [90].

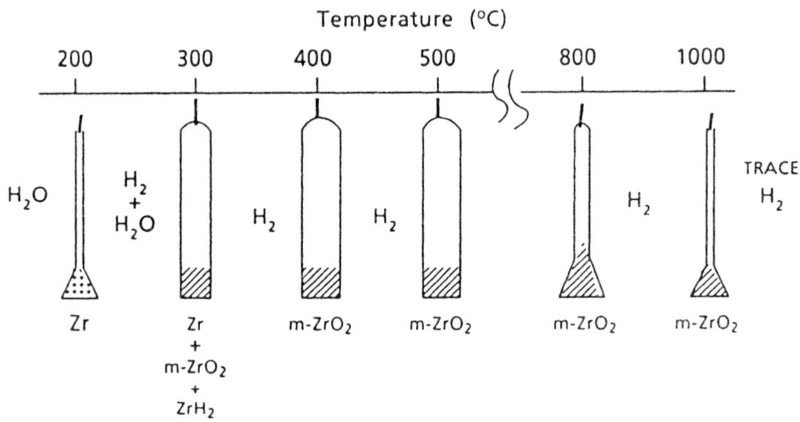

FIGURE 5.7.21 Schematic illustration of the processes in the hydrothermal reaction sintering of monoclinic ZrO_2 under 1000 kg/cm^2 in Pt capsule.

Hydrogen diffused out through the Pt capsule wall. The particle size of the starting metallic chromium powder did not influence the final density of the sintered body. Well sintered samples having density above 95% of the theoretical values were prepared above 850°C and under 1000 kg/cm^2 pressure. About 99.2% of theoretical density sample was obtained at 1000 kg/cm^2 for 3 h. The average grain size of the sintered Cr_2O_3 was 10 μm.

Coble [96] mentioned that this hydrothermal reaction sintering of Cr_2O_3 was one of the three examples wherein the particle size in the sintered body was smaller than that of the starting material.

In general, sintering is a phenomenon of grain growth, but hydrothermal reaction sintering is different. In the case of Zr, relations between temperatures and reaction products are shown in Figure 5.7.21.

5.7.4.2 Hydrothermal Hot Pressing [97–103]

Yamasaki and coworkers [97–103] have published more than 60 papers. The apparatuses they used are shown in Figures 5.7.22–5.7.24. There is a space for retreat of water. They have studied many materials such as $CaCO_3$, apatite, radioactive waste, bioactive glass-ceramics, etc.

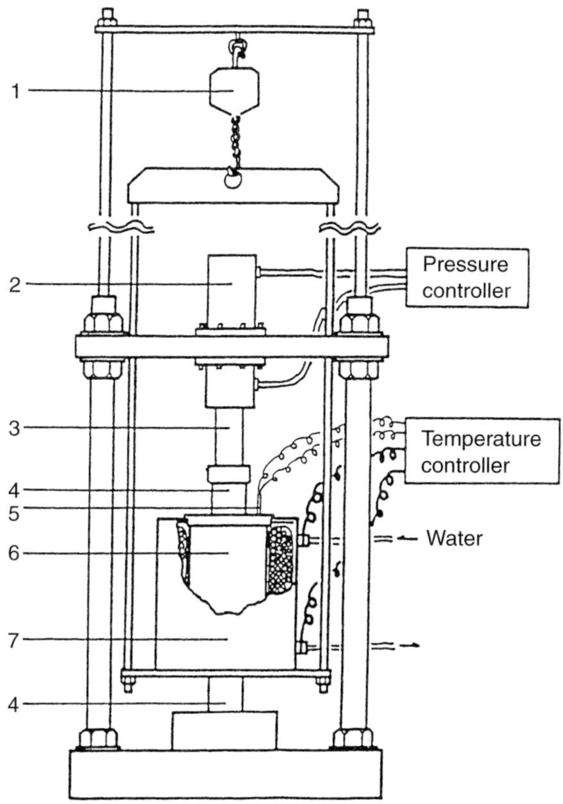

FIGURE 5.7.22 Apparatus for hydrothermal hot-pressing: (1) hoist for adjustment of furnace position'; (2) pump'; (3) ram'; (4) push rod of autoclave'; (5) thermocouple'; (6) autoclave'; (7) induction heater.

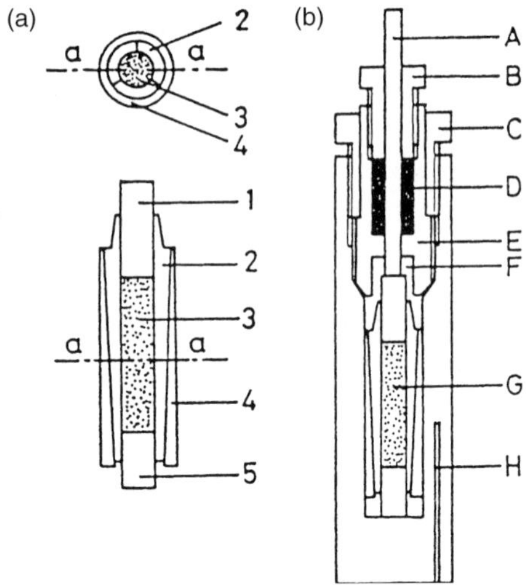

FIGURE 5.7.23 Cell and autoclave for hydrothermal hot-pressing. (a) Cell: (1,5) piston, (2) inner case, (3) sample, (4) outer case. The upper figure is a cross-section of aa. (b) Autoclave: (A) push rod, (B) nut for pushing gland packing, (C) nut for pushing cone, (D) gland packing (copper and cotton yarn), (E) cone, (F) space for water retreat, (G) sample, (H) well for thermocouple.

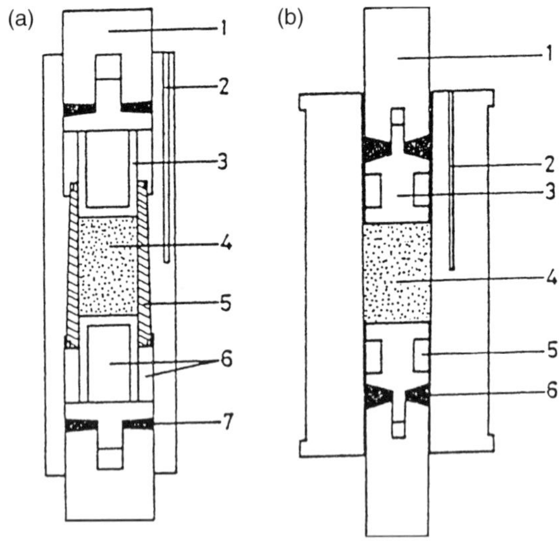

FIGURE 5.7.24 Autoclave for hydrothermal hot-pressing. (a) (1) push rod, (2) well for thermocouple, (3) piston, (4) sample, (5) cell (6) space for water retreat, (7) gland packing (Teflon containing carbon). (b) (1) push rod, (2) well for thermocouple, (3) piston, (4) sample, (5) space for water retreat, (6) gland packing (Teflon).

5.7.4.3 HYDROTHERMAL HOT ISOSTATIC PRESSING [104–106]

Hattori and coworkers [104, 105] reported the sintering of mica ceramics by this method using a borosilicate glass tube as the capsule material. The sample pellet was embeded in BN powder.

5.7.5 HYDROTHERMAL CORROSION

Recently, carbides and nitrides are being used in engine parts. Their mechanical properties and prediction of life are very important. We have to estimate the life of these when we use them for engine applications. After burning of fuel in air, we get water vapor. Therefore, reactions between carbides and nitrides and water vapor are very important when considered for application in engine parts. Recently, several papers on hydrothermal corrosion have appeared [107, 108].

Sōmiya [109] presented an overview paper of SiC and Si_3N_4 powders and sintered bodies. Reactions between water and SiC and Si_3N_4 powders are

$$SiC + 2H_2O \longrightarrow SiO_2 + CH_4$$
$$Si_3N_4 + 6H_2O \longrightarrow 3SiO_2 + 4NH_3$$

This is shown in Figure 5.7.23. Weight change of reaction sintered SiC, hot pressed Si_3N_4, and X-ray diffraction patterns of samples after hydrothermal corrosion are shown in Figures 5.7.25–5.7.29.

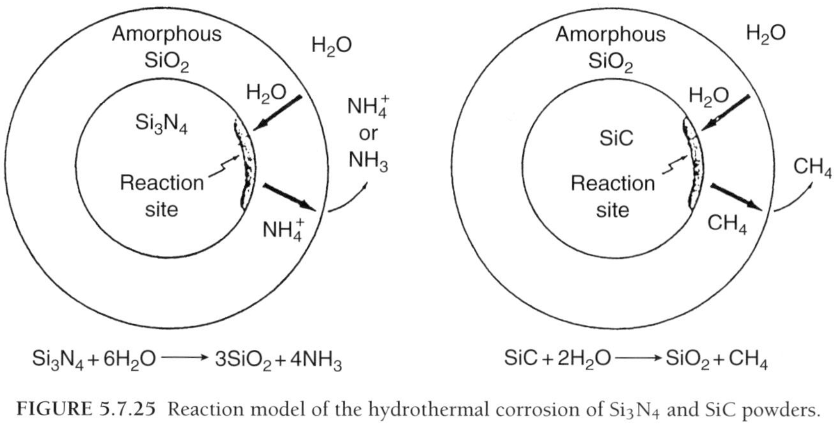

FIGURE 5.7.25 Reaction model of the hydrothermal corrosion of Si_3N_4 and SiC powders.

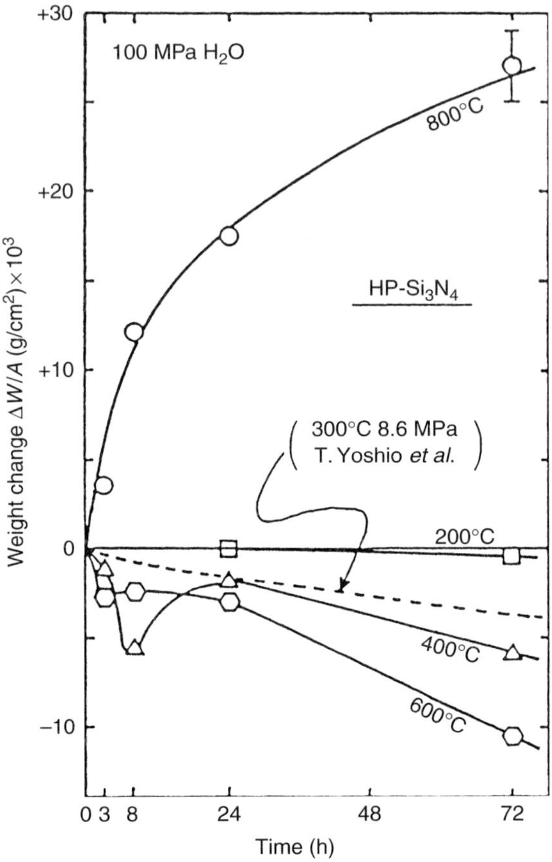

FIGURE 5.7.26 Weight change of HP-Si_3N_4 by hydrothermal corrosion under 100 MPa H_2O.

5.7.6 HYDROTHERMAL FILMS

Yoshimura and coworkers [110–115] have developed hydrothermal electrochemical processes from converting powders to films. A Ti metal plate was connected to the anode by Pt wire, then it was set in the autoclave together with a Pt plate as the cathode. It is shown in Figure 5.7.14. A direct current was passed in the electrolytic solution heated to a certain temperature under saturated vapor pressure. To obtain continuous $BaTiO_3$ films, $Ba(OH)_2$ solution was the most suitable as Ba^{2+} containing electrolytic solution under conditions of temperature above 100°C, current density of 10–100 mA/cm^2 in 0.5 N $Ba(OH)_2$ solutions provided continuous thin films showing multiple interference colors, yellow, blue, green, purple, etc.

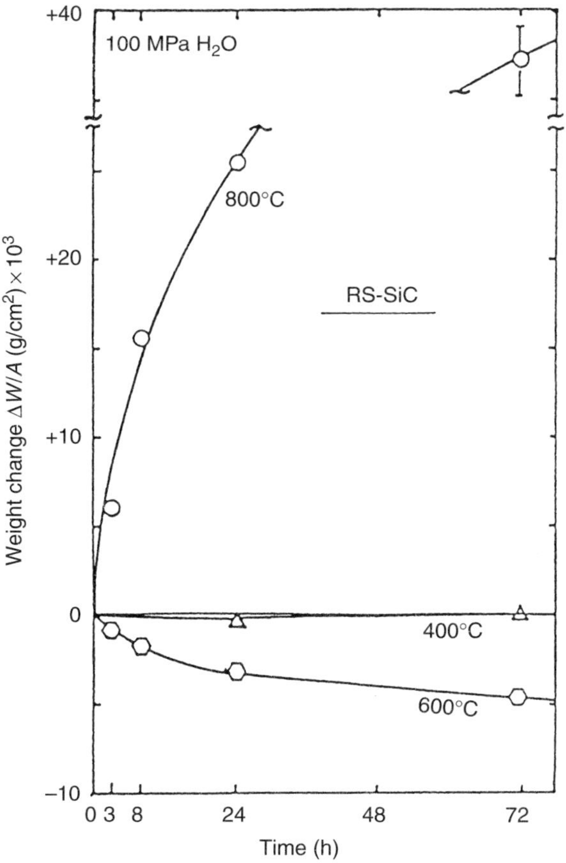

FIGURE 5.7.27 Weight change of RS-SiC by the hydrothermal corrosion under 100 MPa H_2O.

In 1990, Ishizawa *et al.* published several papers [111–115]. As for the substrate, there were several materials such as Ti metal substrate for $SrTiO_3$ [111, 112] and $BaTiO_3$ [110–113, 115], Ti deposited glass substrate for $BaTiO_3$ [115], Ti deposited flexible polymer film of polyphenylene sulfide for $BaTiO_3$ [111], etc. They were able to make thin films of $BaTiO_3$ and $SrTiO_3$ on the substrate of metal, glass, and polymer under hydrothermal electrochemical condition with electronic current or under hydrothermal condition without electronic current. Thickness by the former processing produced thicker films than the latter. In all studies, the temperature was below 200°C under the water vapor pressure at that temperature. Concentration of $Ba(OH)_2$ was 0.5 N [110], 3 N [111], 0.1–0.5 N [114, 115], etc. The electric current was 10–100 mA/cm^2

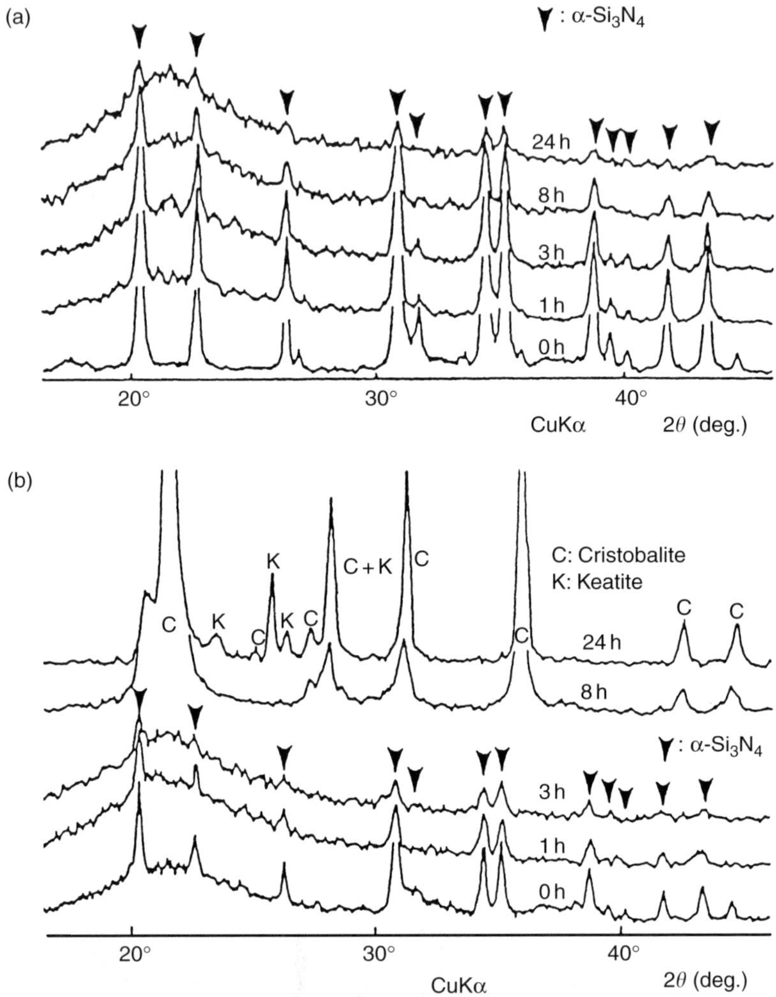

FIGURE 5.7.28 X-ray powder diffraction patterns of the products by the hydrothermal corrosion of Si_3N_4 for various times: (a) at 300°C under 100 MPa; (B) at 500°C under 100 MPa.

[110], no current [111], 10–100 mA/cm^2 [113], up to 100 mA/cm^2, up to 18 mA/cm^2, etc. The thickness of films ranged from 70 to 300 nm from 10 to 60 min at 200°C in 0.5 N $Ba(OH)_2$ with 15 mA/cm^2; 70 nm film was formed without electricity under the same conditions. Some results are shown in Figures 5.7.30 and 5.7.31.

Suchanek *et al.* [116, 117] published their report of the hydrothermal electrochemical method in a flowing solution. A schematic illustration of the flow cell for hydrothermal-electrochemical synthesis is shown in Figure 5.7.32. The

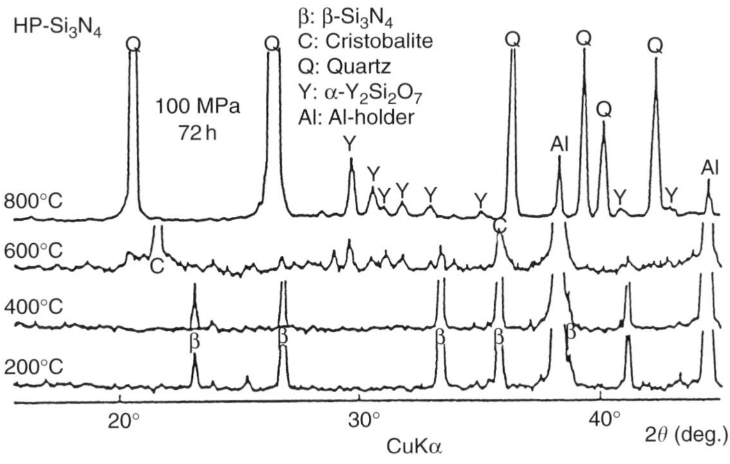

FIGURE 5.7.29 X-ray diffraction patterns of the surface of HP-Si_3N_4 after the hydrothermal corrosion at various temperatures at 100 MPa for 72 h.

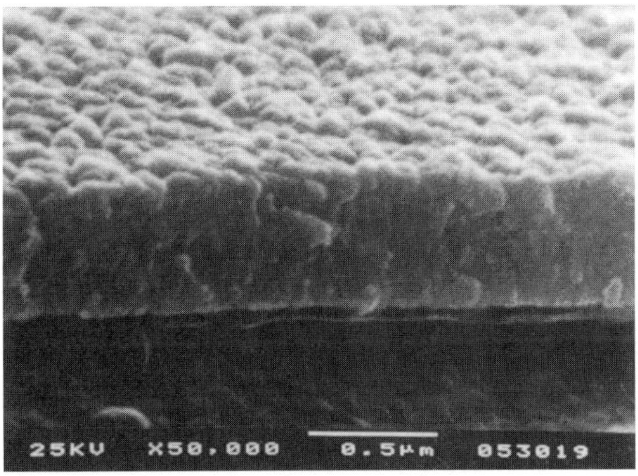

FIGURE 5.7.30 Cross-sectional scanning electron micrograph of the $BaTiO_3$ film formed at 120°C in 0.5 N $Ba(OH)_2$ for 3 h.

flow rate was 0.1–50 cm^3/min. The temperature used was up to 200°C for 1 h. The autoclave was made from two metal blocks (stainless steel SUS 304 for alkaline solutions and titanium for acidic solutions). $Ba(CH_3COO)_2$ and titanium sheet were used as sources of barium and titanium, respectively. Appropriate quantities of barium acetate were dissolved in water to yield 0.25 M solutions.

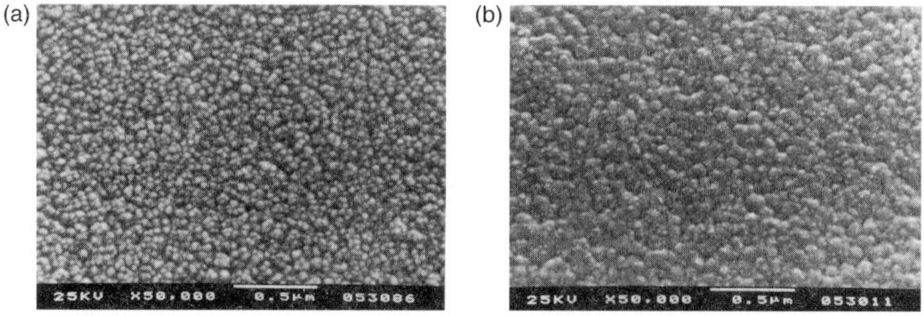

FIGURE 5.7.31 Scanning electron micrographs of the surface of $BaTiO_3$ thin films formed (a) at 150°C and (b) at 180°C, in 3 N $Ba(OH)_2$ for 1 h.

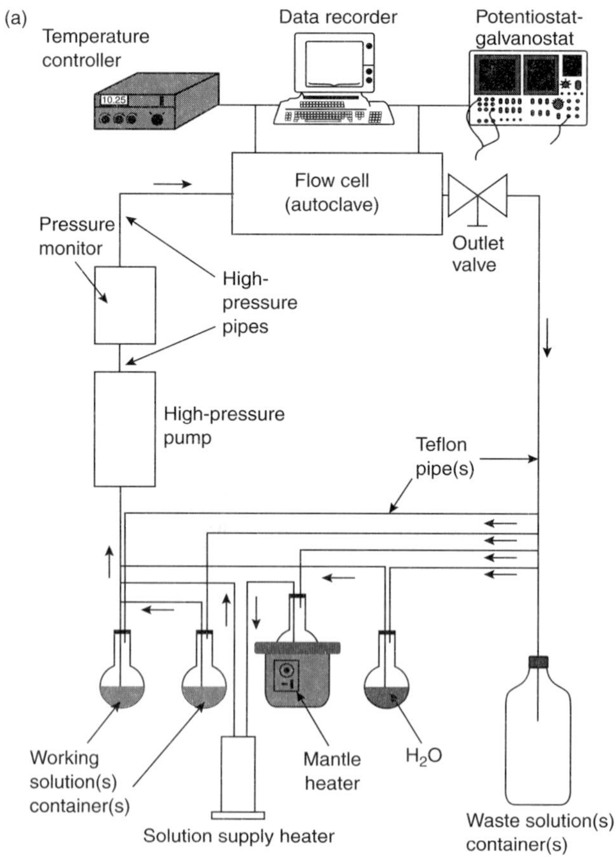

FIGURE 5.7.32 Continued.

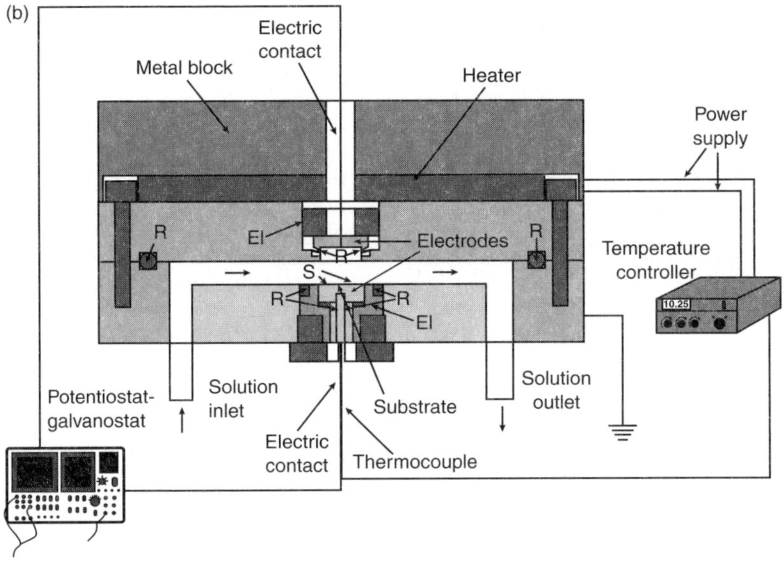

FIGURE 5.7.32 (a) Schematic diagram of the experimental setup for hydrothermal- electrochemical synthesis in a solution flow system. (b) Schematic diagram of the solution flow cell (autoclave) for hydrothermal-electrochemical synthesis: (S) screws; (R) O-ring; (EI) electric insulation. Direction of the solution flow is marked by arrows [117].

The pH of each solution was adjusted to the value of 13.0 by 1.0 N NaOH solution. During the experiments, acetate solution was constantly bubbled with argon gas. The galvanostatic conditions with constant current density was 1 mA/cm or 20 mA^{-2}. The titanium substrate served as the positive electrode (anode). The results are shown in Figure 5.7.32a and b.

Advantages of this process are (i) the possibility of recycled flow, (ii) the possibility of fabrication of multilayered materials and (iii) additional control of the film microstructure by changing the conditions of the solution flow.

REFERENCES

1. Sōmiya, S. (1994). "Advanced Materials. '93", VI, *Trans. MRS Jpn* **19B**: 1105–1110.
2. Ballman, A. A., and Laudise, R. A. (1963). Hydrothermal growth, *The Art and Science of Growing Crystals*, pp. 231–251, Gilman, J. J., ed., New York: John Wiley & Sons, Inc.
3. Roy, R., and Tuttle, O. F. (1956). Investigation under hydrothermal conditions. *Phys. Chem. Earth* **1**: 138.
4. Laudise, R. A., and Nielsen, J. W. (1961). Hydrothermal growth, *Solid State Physics*, Vol. 12, pp. 149–222, Seitz, F., and Turnbull, D., eds., New York: Academic Press.
5. Eitel, W. (1966). *Silicate Science, Vol. IV, Hydrothermal Silicate Systems*, pp. 149–408, New York: Academic Press.

6. Laudise, R. A. (1970). Hydrothermal growth, in *The Growth of Single Crystals*, pp. 275–293, New Jersey: Prentice Hall.
7. Lobachev, A. N. (1971). *Hydrothermal Synthesis of Crystals*, p. 153, New York: Consultant Bureau.
8. Chernow, A. A., Givargiew, E. I., Bagdaasarow, K. S., Kuznetsov, V. A., Demianets, L. N., and Lobachev, A. N. (1984). Growth and synthesis in hydrothermal solutions, in *Modern Crystallography III*, pp. 380–406, Vanshtein, B. K., *et al.*, eds.
9. Stambsugh, E. P., Miller, J. F., Dawson, W. J., Adair, J. H., Seedman, W. G., Muralidhara, H. S., and Kim, S. C. (1984). *Handbook No. 1, Technology for New/Improved Hydrothermal Processes to Hydrothermal Group Program Members*, p. 293, Battelle Colubus Laboratories.
10. Rabenau, A. (1985). The role of hydrothermal synthesis in preparative chemistry. *Angew. Chem. Int. Ed. Engl.* **24**: 1026–1040.
11. Dawson, W. J. (1988). Hydrothermal synthesis of advanced ceramic powders. *Am. Ceram. Soc. Bull.* **67**: 1673–1678.
12. Dawson, W. J., Wheeler, V. S., Adair, J. H., Swartz, S. L., and Stambaugh, E. P. (1987). *Handbook No. 2, Technology for New/Improved Hydrothermal Process (Phase II) to Hydrothermal Group Program Members*, Battelle Columbus Division.
13. Sōmiya, S. (1990). Hydrothermal preparation of fine powders, *Advanced Ceramics III*, pp. 207–243, Sōmiya, S., ed., London: Elsevier Applied Science.
14. Rabenau, A. (1990). *Advanced Ceramics III*, pp. 163–179, Sōmiya, S., ed., London: Elsevier Applied, Science.
15. Byrappa, K. (ed.) (1991). *Hydrothermal Growth of Crystals, Progress in Crystal Growth and Characterization of Materials*, p. 365, Oxford: Pergamon Press.
16. Byrappa, K., and Yoshimura, M. (2000). *Handbook of Hydrothermal Technology: A Technology for Crystal Growth and Materials Processing*, p. 870, Park Ridge, NJ: Noyes Publications.
17. Sōmiya, S. (ed.) (1983). "Proceedings of the First International Symposium on Hydrothermal Reactions", p. 962, Gakujutsu Bunken Fukyu Kai, Tokyo.
18. Proceedings of First International Conference on Solvo-Thermal Reactions (1994), p. 223, Shikoku National Industrial Research. Institute, Takamatsu, Japan.
19. Sōmiya, S., and Roy, R. (2000). Hydrothermal synthesis of fine oxide powders. *Bull. Mater. Sci.* **23**: 453–460.
20. Ganguli, D., and Chatterjee, M. (1997). *Ceramic Powder Preparation: A Handbook*, p. 221, Dordrecht: Kluwer Academic Publishers.
21. Messing, G. L., Mazdiyasni, K. S., MaCauly, J. W., and Haber, R. A. (eds.) (1987). *Ceramic Powder Science*, p. 825, Am. Ceram. Soc.
22. Brinker, C. J. B., Clark, D. E., and Ulrick, D. R. (eds.) (1984). Better ceramic through chemistry. "Materials Research Society Symposium Proceedings", Vol. 32, p. 394, Materials Research Society.
23. Vincenzini, P. (ed.) (1983). *Ceramic Powders*, p. 1025, Amsterdam: Elsevier Scientific Publishers.
24. Veale, C. R. (1972). *Fine Powder Preparation, Properties and Uses*, p. 147, England: Applied Science Publishers Ltd.
25. Seagal, D. (1989). *Chemical Synthesis of Advanced Ceramic Materials*, p. 182. Chapter 8, *Hydrothermal Synthesis of Ceramic Powders*, pp. 114–123. Cambridge: Cambridge University Press.
26. Johnson, Jr., D. W. (1987). *Innovation in Ceramic Powder Preparation. Advances in Ceramics, Vol. 21, Ceramic Powder Science*, pp. 1–19, Am. Ceram. Soc.
27. Stambaugh, E. P. (1986). Hydrothermal processing for advanced ceramic oxides, "Proceedings of the 1986 International Congress on Technology and Technology Exchange", pp. 71–75, October 6–8.

28. Komarneni, S., Roy, R., Breval, E., Ollinen, M., and Suwa, Y. (1986). Hydrothermal route to ultrafine powders utilizing single and di-phase gels, in *Advanced Ceramic Materials*, Vol. 1 pp. 87–92, Am. Ceram. Soc.
29. Sōmiya, S., and Yoshimura, M. (1987). Hydrothermal preparation of fine powders, *Fundamental Structural Ceramics*, pp. 11–29, Sōmiya, S., and Bradt, R. C., eds., Tokyo: Terra Scientific Publication Co.
30. Sōmiya, S. (ed.) (1989). *Hydrothermal Reactions for Materials Science and Engineering*, pp. 505, Amsterdam: Elsevier Applied Science.
31. Sōmiya, S. (1994). Hydrothermal reactions in inorganic systems. *Advanced Materials '93 VI, Trans. Mater. Res. Soc. Jpn* **19B**: 1105–1135.
32. Jacobs, P. A., and Martens, J. A. (1987). *Synthesis of High-Silica Aluminosilicate Zeolites*, pp. 3–5, Amsterdam: Elsevier Science Publishers B.V.
33. Catalogue, Tem-Pres Division, Leco Corp., 1150 Blanchard Street, Bellefonte, PA 16823, USA. Tel.: 814-355-7903; Fax: 814-355-4089.
34. Catalogue, Autoclave Engineers, Inc., 2930 W, 22nd St., Erie, PA 16512, USA. Tel.: 814-838-5700; Fax: 814-833-0145.
35. Catalogue, Parr Instrument Co., 211 53rd Street, Moline, IL 61265, USA. Tel.: 309-762-7716; Fax: 309-762-9430.
36. Kim, B. C., and Stambaugh, E. P. (1975). Hydrothermal preparation of zirconium dioxide in a continuous reactor. *J. Korean Inst. Chem. Engrs* **13**: 211–215.
37. Ulmer, G. C. (ed.) (1971). *Research Techniques for High Pressure and High Temperature*, pp. 367, Berlin: Springer-Verlag.
38. Ulmer, G. C., and Barnes, H. L. (eds.) (1987). *Hydrothermal Experimental Techniques*, p. 523, New York: John Wiley & Sons.
39. Sōmiya, S. (1984). Hydrothermal preparation and sintering of fine ceramic powders. *Mater. Res. Soc. Symp. Proc.* **24**: 255–271.
40. Noda, T., and Roy, R. (1956). OH–F exchange in fluorine phlogopipe. *Am. Mineralogist* **41**: 929–931.
41. Noda, T., Saito, H., Tata, I., Fukasawa, T., Fukase, M., and Sekine, E. (1958). Hydrothermal treatment of fluor–Phlogopite—hydrothermal treatment of aggolomerite mass of synthetic mica crystals to facilitate separation of individual crystals. *Bull. Chem. Soc. Jpn* **31**: 508–514.
42. Oota, T., Saito, H., and Yamai, I. (1979). Synthesis of potasium hexatitanate fibers by the hydrothermal dehydration method. *J. Cryst. Growth* **46**: 331–338.
43. Sōmiya, S., Yoshimura, M., Nakai, Z., Hishinuma, K., and Kumaki, T. (1987). Microstructure development of hydrothermal powders and ceramics, "Ceramic Microstructure '86. Role of Interfaces", Materials Science Research, Vol. 21, pp. 465–474, Pask, J. A., and Evance, A. G., eds., New York: Plenum Press.
44. Sōmiya, S., Yoshimura, M., Toraya, H., and Fushii, Y. (1986). Cubic Eu-doped hafnia ultrafine particles crystallized under hydrothermal condition. *Z. Anorg. Allg. Chem.* **540–541**: 251–258.
45. Nishizawa, H., Yamasaki, N., Matsuoka, K., and Mitsushio, M. (1982). Crystallization and transformation of zirconia under hydrothermal conditions. *J. Am. Ceram. Soc.* **65**: 343–346.
46. Adair, J. H., Denkewicz, R. P., Arriagada, F. J., and Osseo-Asare, K. (1988). *Preparation and In Situ Transformation in the Hydrothermal Synthesis of Crystalline Zirconium Dioxide, Ceram. Trans., Vol. 1, Ceramic Powder Science II A*, pp. 135–145. Am. Ceram. Soc.
47. Hishinuma, K., Kumaki, T., Nakai, Z., Yoshimura, M., and Sōmiya, S. (1988). *Characterization of* $Y_2O_3-ZrO_2$ *Powders Synthesized under Hydrothermal Conditions, Advances in Ceramics, Vol. 24, Science and Technology of Zirconia III*, pp. 201–209, Am. Ceram. Soc.

48. Sōmiya, S., Kumaki, T., Hishinuma, K., Nakai, Z., Akiba, T., and Suwa, Y. (1991). *Hydrothermal Preparation of* ZrO_2 *Powder, Progress in Crystal Growth and Characterization of Materials, Vol. 21, Hydrothermal Growth of Crystal*, pp. 195–198, Byrappa, K., ed., Oxford: Pergamon Press.
49. Hishinuma, K., Abe, M., Hasegawa, K., Nakai, Z., Akiba, T., and Sōmiya, S. (1993). Hydrothermal synthesis and properties of yttria: fully stabilized zirconia powders, *Science and Technology of Zirconia V*, pp. 209–214, Badwal, S. P. S, Bannister, M. J., and Hannink, R. H. J., eds., Lancaster: Technomic Publishing Co.
50. Sōmiya, S., Hishinuma, K., and Akiba, T. (1995). A new materials processing–hydrothermal processing. *Bull. Mater. Sci.* **18**: 811–818.
51. Beal, K. (1987). Precipitation of lead zirconate solid solutions under hydrothermal conditions, *Advances in Ceramics, Vol. 21. Ceramic Powder Science*, pp. 33–41, Messing, G. L., *et al.*, eds., Am. Ceram. Soc.
52. Lencka, M. M., Anderko, A., and Riman, R. E. (1995). Hydrothermal precipitation of lead titanate solid solutions: thermodynamic modeling and zirconate experimental synthesis. *J. Am. Ceram. Soc.* **78**: 2609–2618.
53. Tani, E., Yoshimura, M., and Sōmiya, S. (1981). Hydrothermal preparation of ultrafine monoclinic ZrO_2 powder. *J. Am. Ceram. Soc.*
54. Tani, E., Yoshimura, M., and Sōmiya, S. (1983). Formation of ultrafine tetragonal ZrO_2 powder under hydrothermal conditions. *J. Am. Ceram. Soc.* **66**: 11–14.
55. Nishizawa, N., Yamasaki, N., Matsuoka, K., and Mitsushio, H. (1982). Crystallization and transformation of zirconia under hydrothermal conditions. *J. Am. Ceram. Soc.* **65**: 343–346.
56. Fujiki, Y., Mitsuhashi, T., and Suzuki, Y. (1972). Hydrothermal growth and etching of ZrO_2 crystals. *J. Am. Ceram. Soc.* **55**: 223.
57. Yoshimura, M., and Sōmiya, S. (1984). *Fine Zirconia Powders by Hydrothermal Processing*, Reports R.L.E.M, Tokyo Institute of Technology, No. 9, pp. 53–64.
58. Bucko, M. M., Haberko, K., and Faryna, M. (1995). Crystallization of zirconia under hydrothermal conditions. *J. Am. Ceram. Soc.* **78**: 3397–3400.
59. Reynen, P., Bustins, H., Pavlovski, B., and von Mallinckrodt, D. (1981). Production of high-purity zirconia by hydrothermal decomposition of zircon ($ZrSiO_5$), in *Advances in Ceramics, Vol. 3, Science and Technology of Zirconia*, pp. 464–475, Heuer, A. A., and Hobbs, L. W., eds., Am. Ceram. Soc.
60. Ismail, M. G. M. U., Hirano, S., and Sōmiya, S. (1974). Decomposition of ilmenite under hydrothermal condition (preliminary work), "Proceedings of the Fourth International Conference on High Pressure" Kyoto, pp. 190–193.
61. Ismail, M. G. M. U., and Sōmiya, S. (1983). Decomposition of ilmenite under hydrothermal condition, "Proceedings of the First Symposium on Hydrothermal Reactions", pp. 669–680, Sōmiya, S., ed., Gakujutsu Bunken Fukyo Kai, Tokyo.
62. Hirano, S., Ismail, M. G. M. U., and Sōmiya, S. (1976). Decomposition of ilmenite in $KOH–H_2O$ solutions under high-pressure conditions. *J. Am. Ceram. Soc.* **58**: 277.
63. Sisk, M., Kilbride, I., and Barker, A. J. (1995). Production: of manganese zinc ferrites via the hydrothermal decomposition of metal (II) acetates and: citrates. *J. Mater. Sci. Lett.* 153–154.
64. Yamasaki, N., Yasui, T., and Matsuoka, K. (1980). Hydrothermal decomposition of polychlorinated biphenyls. *Environ. Sci. Technol.* **14**: 550–552.
65. Yoshimura, M., Kikugawa, S., and Sōmiya, S. (1981). Preparation of zircona fine powders by the reaction between zirconium metal and high temperature–high pressure solutions, "High Pressure in Research and Industry, 8th AIRAPT Conf. 9th EHPRG Conf", Sweden, August 17–22, pp. 793–796.
66. Yoshimura, M., and Sōmiya, S. (1981). Hydrothermal reaction sintering, in *Advances in Ceramics, Vol. 3. Science and Technology of Zirconia*, pp. 455–463, Am. Ceram. Soc.

67. Toraya, H., Yoshimura, M., and Sōmiya, S. (1982). Preparation of fine monoclinic hafnia powders by hydrothermal oxidation. *J. Am. Ceram. Soc.* **65**: C-172.
68. Yoshimura, M., and Sōmiya, S. (1982). Synthesis and sintering of zirconia pine powders by hydrothermal rections from zirconium metal and high temperature–high pressure solution. Sintering theory and practice. "Proceedings of the 5th International Round Conference on Sintering", pp. 417–422, Kolar, D., Pejovnik, S., and Ristic, M. M., eds., Amsterdam: Elsevier Science Publishing Co.
69. Toraya, H., Yoshimura, M., and Sōmiya, S. (1983). Hydrothermal oxidation of Hf metal chip in the preparation of monoclinic HfO powders. *J. Am. Ceram. Soc.* **66**: 148–150.
70. Toraya, H., Yoshimura, M., and Sōmiya, S. (1983). Reaction kinetics in the hydrothermal oxidation of Hf. *J. Am. Ceram. Soc.* **66**: 818–822.
71. Sōmiya, S. (1984). Hydrothermal preparation and sintering of fine ceramic powders. *Mater. Res. Soc. Sympos. Oroc.* **24**: 255–271.
72. Toraya, H., Yoshimura, M., and Sōmiya, S. (1984). Preparation of mixed fine $Al_2O_3-HfO_2$ Powders by hydrothermal oxidation, *Advances in Ceramics, Vol. 12, Science and Technology of Zirconia II*, pp. 806–815, Claussen, N., Ruhle, M., and Heuer, A., eds., Am. Ceram. Soc.
73. Sōmiya, S., Yoshimura, M., and Kikugawa, S. (1984). Preparation of zirconia–alumina fine powders by hydrothermal oxidation of Zr–Al alloys, *Emergent Process Methods for High Technology Ceramics*, Vol. 17, pp. 155–166, Davis, R. F., Palmour, H., and Porter, R. L., eds., New York: Plenum Publishing Corp.
74. Yoshimura, M., Ohira, H., and Sōmiya, S. (1987). *Formation of TiO_2 and ZnO Powders by Hydrothermal Oxidation of Ti and Zn Metals*, Rept. R.L.E.M., Tokyo Institute of Technology.
75. Yoo, S. E., Yoshimura, M., and Sōmiya, S. (1988). Preparation of $BaTiO_3$, and $LiNbO_3$, powders by hydrothermal anodic oxidation, "Sintering '87. Proceedings of the International, Institute for the Science of Sintering (IISS) Symposium", Tokyo, Japan November 4–6, 1987, pp. 108–113, Sōmiya, S., Shimada, M., Yoshimura, M., and Watanabe, R., eds., London: Elsevier Applied Science.
76. Yoo, S. E., Yoshimura, M., and Sōmiya, S. (1989). Direct preparation of $BaTiO_3$ powders from titanium metal by anodic oxidation under hydrothermal conditions. *J. Mater. Sci. Lett.* **8**: 530–532.
77. Yoshimura, M., Kubota, N., Noma, T., and Sōmiya, S. (1989). Preparation of ultrafine barium hexaferrite crystals by hydrothermal attrition mixing, *Ceramic Powder Processing Science*, pp. 205–212. Hausner, H., Messing, G. L., and Hirano, S. eds., Koln: Deutsche Keramische Gessellschaft.
78. Kumar, A., and Roy, R. (1988). RESA – A wholy new process for fine oxide powder preparation. *J. Mater. Res.* **3**: 1373–1377.
79. Kumar, A., and Roy, R. (1989). Reactive electrode submerged arc process for producing fine non-oxide powders. *J. Am. Ceram. Soc.* **72**: 354–356.
80. Komarneni, S., Roy, R., and Li, Q. H. (1992). Microwave-hydrothermal synthesis of ceramic powders. *Mater. Res. Bull.* **27**: 1393–1405.
81. Komarneni, S., Li, Q., Stefansson, and Roy, R. (1993). Microwave-hydrothermal pressing for synthesis of electronic powders. *J. Mater. Res.* **8**: 3176–9183.
82. Komarneni, S., Pidugu, R., Li, Q. H., and Roy, R. (1995). Microwave-hydrothermal processing of metal powders. *J. Mater. Res.* **10**: 1687–1692.
83. Komarneni, S. Menon, V. C., and Li, Q. H. (1996). *Synthesis of Ceramic Powders by Novel Microwave-Hydrothermal Processing, Ceram. Trans., Vol. 62. Science, Technology, and Commercialization of Powder Synthesis and Shape Forming Process*, pp. 37–46, Kingsley, J. J., Schilling, C. H., Adair, J. A., eds., Am. Ceram. Soc.
84. Roy, R., Agrawal, D. K., and Srifanth, V. (1991). Acoustic wave stimulation of low temperature ceramic reactions: the system $A1_2O_3–P_2O_5–H_2O$. *J. Mater. Res.* **6**: 2412–2416.

85. Fang, Y., Agrawal, D. K., Roy, D. M., Roy, R., and Brown, P. W. (1992). Ultrasonically accelerated synthesis of hydroxyapatite. *J. Mater. Res.* 7: 2294–2298.
86. Margulis, M. A. (1995). *Sonochemistry and Cavitation*, Translated from the Russian by Leib, G., p. 543. Amsterdam: Gordon and Breach Publishers.
87. Suslick, L. S. (1990). Sonochemistry. *Science* **247**: 1439–1445.
88. Hirano, S., and Sōmiya, S. (1976). Hydrothermal reaction sintering of pure Cr_2O_3. *J. Am. Ceram. Soc.* **59**: 534.
89. Hirano, S., and Sōmiya, S. (1976). Hydrothermal crystal growth of magnetite in the presence of hydrogen. *J. Cryst. Growth* **35**: 273–278.
90. Sōmiya, S., Hirano, S., Yoshimura, M., Itoh, S., and Kanai, H. (1979). Hydrothermal reaction sintering of Cr_2O_3 and iron oxides, "Proceedings of International Symposium on Factors in Densification and Sintering of Oxide and Non-oxide Ceramics", pp. 267–277, Sōmiya, S., and Saito, S., eds., Gakujutsu Bunken Fukyu-Kai, Tokyo.
91. Yoshimura, M., and Sōmiya, S. (1980). Fabrication of dense, nonstabilized ZrO_2 ceramics by hydrothermal reaction sintering. *Am. Ceram. Soc. Bull.* **59**: 246.
92. Yoshimura, M., and Sōmiya, S. (1981). Hydrothermal reaction sintering of monoclinic zirconia, in *Advances in Ceramics, Vol. 3, Science and Technology of Zirconia*, pp. 455–463, Am. Ceram. Soc.
93. Sōmiya, S. (1983). Hydrothermal reaction sintering of high density sintered oxides, "Proceedings of the International Symposium on Hydrothermal Reactions", pp. 887–903, Sōmiya, S., ed., Gakujutsu Bunken Fukyu Kai, Tokyo.
94. Sōmiya, S. (1984). Hydrothermal reaction sintering of oxides, in *Sintered Metal Ceramics Composites*, pp. 97–105, Upadhyaya, G. S., ed., Amsterdam: Elsevier Science Publishers.
95. Yoshimura, M., Ohira, H., and Sōmiya, S. (1987). Formation of TiO_2 and ZnO powders by hydrothermal oxidation of Ti and Zn metals, Rept. R.L.E.M., Tokyo Institute of Technology, No. 12, pp. 59–72.
96. Coble, R. L. (1982). Reactive sintering. Sintering-theory and practice. "Proceedings of the 5th International Round Table Conference on Sintering", Portoroz, Yugoslavia, September 7–10, 1981, pp. 145–151, Kolar, D., Pejovnik, S., and Ristic, M. M., eds., Materials Science Monographs, Vol. 14, Amsterdam: Elsevier Scientific Publishing Co.
97. Yanagisawa, K., Nishioka, M., and Yamasaki, N. (1985). Immobilization of radioactive waste by hydrothermal hot pressing. *Am. Ceram. Soc. Bull.* **64**: 1563–1567.
98. Yamasaki, N., Yanagisawa, K., Nishioka, M., and Kanahara, S. (1986). A hydrothermal hot-pressing method: apparatus and application. *J. Mater. Sci .Lett.* **5**: 355–356.
99. Yamasaki, N. (1987). Immobilization of high-level radioactive waste by hydrothermal hot pressing method. *J. Nucl. Sci. Technol.* **24**: 135–162.
100. Ioku, K., Kai, T., Nishida, M., Yanagisawa, K., and Yamasaki, N. (1994). Preparation of bioactive glass-ceramics by hydrothermal hot-pressing. *Phosphorus Res. Bull.* **1**: 161–166.
101. Hosoi, K., Hashida, T., Takahashi, H., Yamasaki, N., and Korenaga, T. (1996). New processing techniques for hydroyapatite ceramics by the hydrothermal hot pressing method. *J. Am. Ceram. Soc.* 79: 2771–2774.
102. Yanagisawa, K., Ioku, K., and Yamasaki, N. (1997). Formation of anatase, porous ceramics by hydrothermal hot-pressing of amorphous titania spheres. *J. Am. Ceram. Soc.* **80**: 1303–1306.
103. Hosoi, K., Hashida, T., Yamasaki, N., and Korenaga, T. (1997). Solidification behaviour of calucium carbonate via aragonite–calcite wet transformation with hydrothermal hot pressing. *J. Mater. Sci. Lett.* **16**: 382–385.
104. Hattori, T., Mohri, J., Yoshimura, M., and Sōmiya, S. (1984). Sintering of synthetic mica by a hydrothermal equipment. *Powder Metall. Int.* **16**: 131–132.
105. Hattori, T., Mohri, J., Yoshimura, M., and Sōmiya, S. (1984). Sintering of synthetic mica by hydrothermal hot isostatic process, in *Sintered Metal–Ceramic Composites*, pp. 153–155, Upadhyaya, G. S., ed., Amsterdam: Elsevier Science Publishers.

106. Tressler, R. E., and McNallan, M. (eds.) (1990). *Corrosion and Corrosion Degradation of Ceramics, Ceram. Trans*, Vol. 10, p. 494, Am. Ceram. Soc.
107. Doyama, M., Sōmiya, S., and Chang, R. P. H. (eds.) (1989). "Proceeding of MRS International Meetings on Advanced Materials. Corrosion/Coating of Advanced Materials", pp. 229–342, Materials Research Society.
108. Sōmiya, S., Doyama, M., Hasegawa, M., and Agata, Y. (eds.) (1994). *Advanced Materials '93, 1A, Trans. MRS Japan, Vol. 14A*, pp. 103–355, Amsterdam: Elsevier Science. The Netherlands.
109. Sōmiya, S. (2001). Hydrothermal corrosion of nitride and carbide of silicon. *Mater. Chem. and Phys.* **67**: 157–164.
110. Yoshimura, M., Yoo, S. E., Hayashi, M., and Ishizawa, N. (1989). Preparation of $BaTiO_3$ thin film by hydrothermal electrochemical method. *Jpn. J. Appl. Phys.* **28**: L2007–L2009.
111. Ishizawa, N., Banno, H., Hayashi, M., Yoo, S. E., and Yoshimura, M. (1990). Preparation of $BaTiO_3$ and $SrTiO_3$ polycrystalline thin films on flexible polymer film substrate by hydrothermal method. *Jpn. J. Appl. Phys.* **29**: 2467–2472.
112. Yoo, S. E., Hayashi, M., Ishizawa, N., and Yoshimura, M. (1990). Preparation of strontium titanate thin film on titanium metal substrate by hydrothermal–electrochemical method. *J. Am. Ceram. Soc.* **73**: 2561–2563.
113. Yoshimura, M., Yoo, S. E., Hayashi, M., and Ishizawa, N. (1990). In situ preparation of $BaTiO_3$ thin film by hydrothermal–electrochemical synthesis. *Ceram. Trans.* **15**: 427–436.
114. Ishizawa, N., Yoo, S. E., Hayashi, W., and Yoshimura, M. (1990). Crystalline $BaTiO_3$ thin film prepared from ti metal by hydrothermal–electrochemical method, *MRS Symposium Proceedings, Vol. 200, Ferroelectric Thin Films*, pp. 57–62, Kingon, A. I., and Myers, E. R., eds., Materials Research Society.
115. Hayashi, M., Ishizawa, N., Yoo, S. E., and Yoshimura, M. (1990). Preparation of barium titanate thin film on titanium-coposited glass substrate by hydrothermal–electrochemical method. *Nippon Seramikkusu Kyokai Gakujutsu Rombunshi* **98**: 930–933.
116. Suchanek, W., Watanabe, T., and Yoshimura, M. (1998). Preparation of $BaTiO_3$ thin films by hydrothermal–electrochemical method in the flowing solutiom. *Solid State Ionics* **109**: 65–72.
117. Suchanck, W. L., Watanabe, T., Kurai, B. S., Kumagai, N., and Yohimura, M. (1999). A solution flow system for hydrothermal–electrochemical growth of multilayered thin films. *Rev. Sci. Instrum.* **70**: 2432–2437.
118. Chien, T. A., Speck, J. S., Lange, F. F., Daykim, A. C., and Levi, C. G. (1995). Low temperature/low pressure hydrothermal synthesis of barium titanate: powder and heteroepitaxial thin film. *J. Mater. Res.* **10**: 1784–1789.
119. Lange, F. F. (1996). Chemical solution routes to single-crystal thin films. *Science* **273**: 903–909.
120. Chien, A. T., Speck, J. S., and Lange, F. F. (1997). Hydrothermal synthesis of heteroepitaxial $Pb(Zr_xTi_{1-x})O_3$ thin films at 90–150°C. *J. Mater. Res.* **12**: 1176–1178.
121. Slamovich, E. B., and Aksay, I. A. (1994). Hydrothermal processing of $BaTiO_3$/polymer film, in *MRS Symposium Proceedings, Vol. 346, Better Ceramics Through Chemistry VI*, pp. 63–68.
122. Slamovich, E. B., and Aksay, I. A. (1996). Structure evolution in hydrothermally processed (100°C) $BaTiO_3$ films. *J. Am. Ceram. Soc.* **79**: 239–247.

Handbook of Advanced Ceramics
S. Sōmiya *et al.* (Eds.)

5.8 New Thin Film Technologies

P. EHRHART and R. WASER
IFF/EKM, FZ-Jülich, D-52425 Jülich, Germany

5.8.1 INTRODUCTION

The deposition of thin film functional layers on different substrates is an essential step in many fields of modern high technology, and applications range from large area optical coatings on architectural structures to tribological coatings, ionic conductors for fuel cells and batteries, catalytic layers, the wide field of high-temperature superconductors (HTSC) and finally to the booming area of applications in microelectronics. This field is driven by the need for steady increasing miniaturization and increasing degrees of integration and includes very different electroceramic layers: high-*k* dielectric layers for dynamic random access memories (DRAMs) and gate oxides for future field effect transistors (FETs), ferroelectrics for ferroelectric RAMs (FeRAMs) and ferroelectric field effect transistors (FeFET), piezo- and pyroelectric layers, integrated optical waveguides, transparent electrodes and magnetic manganates exhibiting the giant magnetoresistance (GMR) effect. Considering this broad spectrum of applications, it is obvious that there cannot be one perfect deposition method which can be applied in all fields. In contrast, there is a wide spectrum of methods all on a high level of development and it is sometimes difficult to find the optimum choice. The most important methods in the field of interest, the deposition of ceramic materials, is summarized in Table 5.8.1 with the generally accepted subdivision into physical and chemical methods [1, 2]. Physical methods may be characterized by an adsorption and condensation process of the atoms or molecules at the substrate surface and, generally, by a locally well-defined particle source and a free flight in vacuum to the substrate. For chemical methods a chemical reaction is involved and the so-called precursor molecules dissociate at the hot substrate surface and release the atoms of interest. Some basic characteristics of the methods are summarized, however, the characterization is sometimes not very specific as the parameters depend strongly on the properties of the material of actual interest. We can restrict the discussion on

TABLE 5.8.1 Some Characteristics of Deposition Processes (after Ref. [1])

	Physical vapor deposition			Chemical deposition methods		Thermal spray
	TED/MBE	PLD	Sputtering	CSD	CVD/MOCVD	
Mechanism of production of depositing species	Thermal energy	Thermal energy	Momentum transfer	Solution of precursors	Chemical reaction	From flames or plasmas or cluster sources
Deposition rate	High, up to 750 000 Å/min	Moderate	Low, except for pure metals	Interrupted by heating	Moderate, up to 2500 Å/min	Very high
Deposition species	Atoms and ions	Atoms, ions and clusters	Atoms and ions	Precursor molecules	Precursor molecules dissociate into atoms	Droplets or clusters
Energy of deposit—species	Low 0.1–0.5 eV	Low to high	Can be high 1–100 eV	Thermal	Can be high with Plasma aid	Can be high
Throwing power						
(a) Complex shaped object	Poor, line of sight	Poor	Non-uniform thickness distribution	Poor	Good	Poor
(b) Into blind hole	Poor	Poor	Poor	Poor	Limited	Very limited

basic facts on the different methods as there are very detailed review articles and textbooks for all the deposition methods. As we want to give special emphasis to electroceramic thin films some special details necessary for the deposition of oxide films will be included.

5.8.2 PHYSICAL VAPOR DEPOSITION

Following some general trend we have subdivided this section in three parts: thermal evaporation deposition (TED) techniques with special emphasis on molecular beam epitaxy (MBE), pulsed laser deposition (PLD), and sputter deposition (SD). In a general sense, PLD is also a thermal evaporation technique, but due to the very short pulse heating it has some special characteristics. Although pulsed techniques have also been reported using electron beams, only PLD is a widely used technique at present. The distinguishing feature of sputter techniques is the plasma environment, however, using the variant of ion beam sputter deposition ultra high vacuum (UHV) conditions are also possible.

It must be kept in mind that the dividing lines are not sharp and hybrids of the techniques are used as well; this yields often different and sometimes misleading additional acronyms, which partly may also be motivated considering patent right. Ion plating is an alternative term for deposition techniques which use an ion bombardment of the film during growth; if the ions are supplied by a plasma this corresponds to bias sputtering and for evaporation under vacuum conditions this is equivalent to ion-beam assisted deposition (IBAD).

5.8.2.1 THERMAL EVAPORATION DEPOSITION/MOLECULAR BEAM EPITAXY

MBE has evolved from simple thermal evaporation techniques [2] by the application of UHV techniques to avoid disturbances by residual gases, and additionally includes different sources. The technique has been developed during the last 30 years to a high level of technical perfection [3–5] and all components are commercially available. A schematic view of a system is shown in Figure 5.8.1, which includes several independent sources in order to allow the controlled deposition of multielement compounds. The main components are summarized in Figure 5.8.2: (i) The different beam sources which must provide stable, high-purity, atomic or molecular beams impinging on the substrate; for the case of gas or vapor cells metal–organic MBE (MOMBE) systems arise,

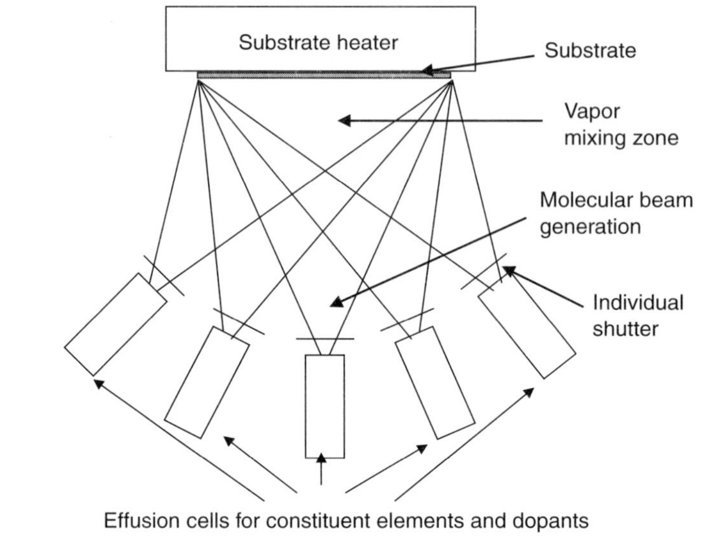

FIGURE 5.8.1 Schematic view of an MBE system for the growth of multielement compound films.

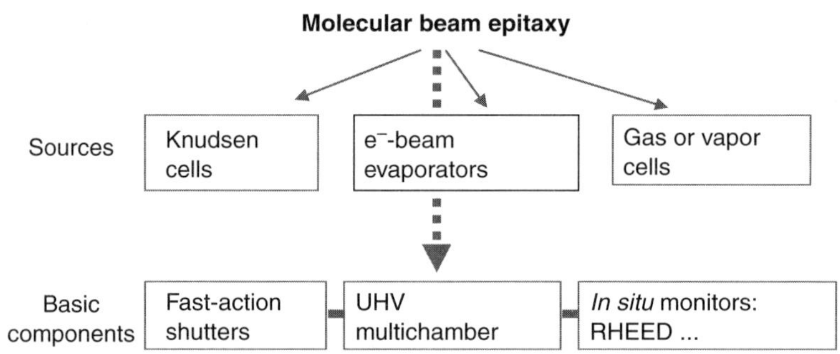

FIGURE 5.8.2 Basic components of MBE.

which also can be considered as a special case of MOCVD. (ii) The precision, fast-action shutters which are very important to control the growth of dopant profiles or multilayers. (iii) The UHV process environment is very specific for MBE and allows for the *in situ* process control by all UHV surface sensitive techniques and especially RHEED is used as a standard growth control. The oxygen partial pressure necessary for the deposition of some oxide films of interest in information technology is therefore a major problem for the UHV technique. However, several solutions have been used including differential pumping combined with a rotation of the substrate; hence, the film is temporarily exposed to oxygen whereas the sources are at high vacuum condition in order to avoid oxidation and deterioration [6].

The method allows perfect layer-by-layer growth control and rather independent control of the different deposit components and is therefore very flexible. Due to UHV conditions, very clean films can be obtained and the method therefore is ideally suited for basic investigations of growth processes. An outstanding example for the application are the feasibility studies for new epitaxial high epsilon gate oxides on silicon [7, 8].

The high flexibility is accompanied by disadvantages as a lot of parameters must be kept under control and the optimization of a process window may be time consuming. For application as a production tool, there are major drawbacks due to the generally rather slow layer-by-layer growth and the UHV techniques, which are not compatible with other components of the CMOS production lines and yield long set-up times; therefore MBE are used mainly in niche markets.

5.8.2.2 Pulsed Laser Deposition

The concept of a PLD experiment [9–11] is basically simple (Fig. 5.8.3). A short and energetic pulsed laser beam from an excimer laser is focused to a target. The pulse energy of 1 J per pulse leads to the immediate formation of a plasma

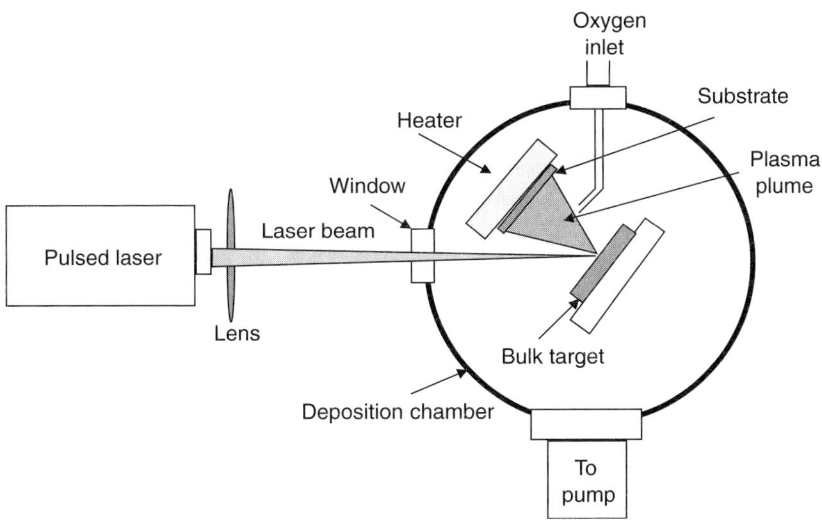

FIGURE 5.8.3 Set-up of a PLD system for oxide layers.

due to the high energy density of 3–5 J/cm^2 at the target surface. Characteristic deposition parameters are: wavelengths of the used laser beam in the ultra violet regime which yield good photon absorption within the oxide targets, and a repetition rate $\approx$50 Hz at a pulse length of 25 ns. The plasma contains energetic neutral atoms, ions and molecules and reaches the substrate surface with a broad energy distribution of 0.1 to >10 eV. On the surface of the heated substrate, which is placed, for example, on a oxygen-resistant SiC-heater, the thin film is formed. For the growth of oxide films, an ambient oxygen pressure of 0.3–1 mbar is maintained within the chamber.

Since the success of growing thin films from the high-temperature superconductors like YBaCuO, the method has been shown to be especially suited for the deposition of oxides and other multicomponent materials and has been used extensively for electro-optical thin films of $LiNbO_3$, $KNbO_3$, and $BaTiO_3$ [12]. The deposition temperature and the oxygen pressure during the growth process are used to control the crystalline perfection, the orientation and the grain size of the film.

PLD is a now a well-developed thin film preparation method and offers several advantages compared to other deposition techniques, if the formation of multicomponent thin films is considered. The most important advantage of the PLD is the stoichiometric transfer from a complex multielemental ceramic target material to the substrate without significant separation between the constituents and the rather fast deposition rates. As the base vacuum is not critical for the process, it may vary from UHV, for example, for the additional option of *in situ* growth control, up to the oxygen partial pressures required for oxide growth.

Disadvantages are the restriction to small substrate sizes which can only be overcome by mechanical scanning and the low grade of conformal deposition which is of concern for large-scale integration processes. A problem of the method is that, in addition to the energetic neutral atoms, ions and molecules, some small droplets or "boulders" may be deposited on the film. Different methods to reduce this effects have been developed, for example, time-of-flight selection of the deposits [11], as the heavier particles are slower, or the use of different off-axis geometry's [11, 13].

5.8.2.3 Sputter Deposition

Sputtering of surface atoms is known since 1852 when Sir W. R. Grove observed this effect during his investigations of plasma discharges and it was recognized soon that these sputtered atoms can be applied for the deposition of thin films. However, the large-scale technological application has developed only during the last decades [14–17] and a wide variety of technical variants of the basic sputtering process have evolved (Fig. 5.8.4).

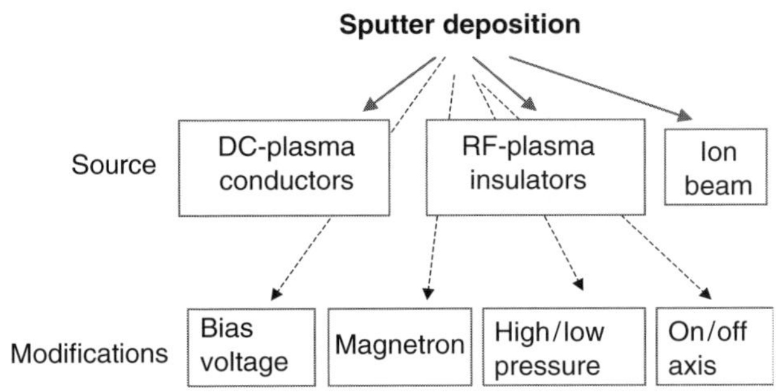

FIGURE 5.8.4 Basic modifications of sputtering systems.

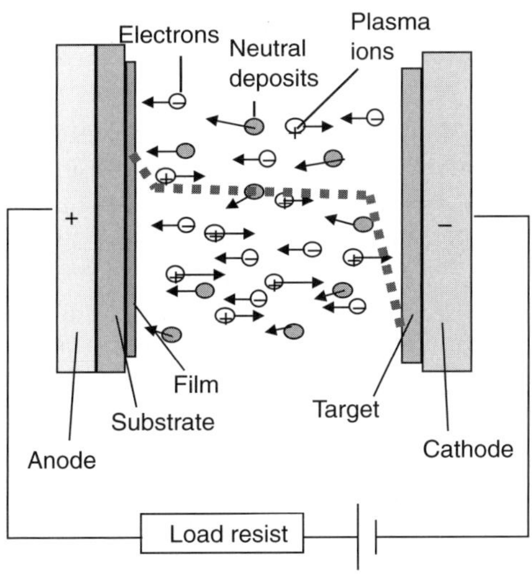

FIGURE 5.8.5 Schematics of a DC sputter system; the dotted line indicates the change of the potential between anode and cathode.

The most simple sputtering system is the so-called DC sputtering, which is schematically shown in Figure. 5.8.5. In a vacuum chamber the target material, which is eroded, is at the cathode side, and the substrate for the film is opposed at the anode side. The potential of several 100 V between these plates leads to the ignition of a plasma discharge for the typical pressures of 10^{-1}–10^{-3} mbar and the positively charged ions are accelerated to the target. These accelerated particles sputter off the deposits, which arrive to the substrate mostly as neutral atoms. In detail, the ion, which has been accelerated within the darkroom with nearly the full voltage applied, hits on the surface atoms and initiates a collision cascade which finally leads to some back-reflected atoms. These collision cascades can be simulated very reliably by "molecular dynamics" methods, and results on the sputtering yield and the dependence of the relative masses of projectile and target atom are available [14, 17]. The threshold energy for sputtering is much higher than the surface binding energy, E_b, of the atoms which is of the order of 4–8 eV. This difference can be directly understood as several collisions are necessary in order to obtain an atom in backward direction. Hence, the threshold is observed at $4E_b$ to $8E_b$ corresponding to a threshold energy of 20–50 eV. Different yields are observed for different target atoms and this difference in principle yields the deposition of a film of different stoichiometry. However, there is generally a good self regulating mechanism: due to the very low penetration depth of the sputtering process the faster eroded component is denuded after a short initial time and finally, in a quasi equilibrium, the difference in yield is compensated by the enrichment. Therefore, sputtering quite generally allows the deposition of films with the same stoichiometry as the target.

An ionization degree of less than 1% of the atoms is characteristic for a plasma and consequently a rather low sputter rate. The ionization rate can be improved by applying magnetic fields, generally using permanent magnets, which force the electrons on helical paths close to the cathode. This magnetron arrangement allows additionally a lower gas pressure, however, it has the disadvantage of a more inhomogeneous target erosion than a simple planar geometry.

DC sputtering works fine as long as the target material has a minimum electrical conductivity, however, there is electric load for insulating targets. This problem can be solved by a high-frequency plasma discharge. A typical frequency of 13.6 MHz is capacitively coupled to the target and there is only a small voltage decay across the electrode. The substrate and the deposition chamber are generally grounded. As the electrons are much faster than the ions a negative potential at the electrodes as compared to the plasma potential evolves during each cycle. With a symmetrical arrangement of cathode and anode we would obtain similar re-sputtering rates and no film growth, however, non-symmetries are introduced by the capacitive coupling of the RF and by differences in the geometry, for example, of target and grounded substrate. Nevertheless, effective deposition rates are much lower than for DC sputtering.

Compared to the other physical deposition methods the partial pressure of the sputtering gas is an additional and important process parameter which must be considered and optimized. This pressure controls the free path length of the atoms and therefore their energy, angular distribution and finally also their incorporation in the film. For metals, so-called zone diagrams for the film growth have been developed which show some systematic influence of this parameter [18].

For oxide layers, the gas pressure is even more important as negative oxygen ions are formed which can re-sputter from the film. As the re-sputtering yield also differs for different elements the stoichiometry of the growing film is changed and these changes are not compensated as it was the case for the sputter yield from the target discussed above. This re-sputtering can be reduced by high-pressure oxygen sputtering [19] as the energy of the ions is reduced by the shorter free path length.

Similar to the other deposition techniques, SD of ceramic oxide films was developed significantly along with the interest in HTSC, and sputter deposition systems are well developed and commercially available today. Advantages are large throughput, large sample sizes, good adhesion to the substrate due to the ion bombardment, often self-adjusting composition control and less severe vacuum requirements than for thermal evaporation. The systems are very flexible and can be adjusted to many specific requirements. Additional modifications are, for example, bias sputtering in order to accelerate or decelerate the ions in a controlled manner, off-axis deposition in order to obtain more homogeneous deposition, and ion beam sputtering by the use of a controlled ion gun instead of the plasma as ion source; these ion beam systems can be further developed to multiple target systems which are individually controlled.

Disadvantages of the technique are the low sputter yield for the RF sputtering of insulators and the poor conformality of the deposition.

5.8.3 CHEMICAL DEPOSITION METHODS

Chemical deposition of ceramic thin films can be conducted through the gas phase (chemical vapor deposition, CVD) or the solution phase (chemical solution deposition, CSD). In both cases, chemical precursors are employed which undergo chemical reactions for the formation of the ceramic film. For complex oxides, the precursor molecules contain typically the metal atoms, M, and organic groups, R. The introduction of such organic groups defines the subgroup of CVD, namely metal organic chemical vapor deposition (MOCVD) (or, chemically more correct, organo-metallic chemical vapor deposition, OMCVD). In most cases, such as in alkoxides, ketonates, and carboxylates, the metal atom is bound to the organic group through an oxygen atom, that is, M–O–R. This is different to precursors used for the MOCVD of compound semiconductors where the precursors are conventional metal organic compounds, that is, M–R systems. For additional liquid phase chemical techniques like electroplating or electroless plating we refer to Ref. [1, 2]. Although electroplating seems to be restricted to metals, it may also be considered for special ceramic coatings as it can be combined with anodic oxidation to form a ceramic coating, e.g., Al_2O_3.

5.8.3.1 Chemical Solution Deposition

The CSD method comprises a range of deposition techniques and of chemical routes [20, 21], the most important of which are shown in Figure 5.8.6. A generalized flowchart of the CSD of oxide thin films is shown in Fig. 5.8.7. The process starts with the preparation of a suitable coating solution from precursors according to the designated film composition and the chemical route to be used. Besides mixing, the preparation may include the addition of stabilizers, partial hydrolysis, and refluxing. The coating solution is then deposited onto substrates by *spin coating* where typically a photoresist spinner is employed and which is suitable for semiconductor wafers; *dip coating* which is often used in the optics industry for large or non-planar substrates; and *spray coating* which is based on a misting of the coating solution and deposition of the mist exploiting the gravitation or an electrostatic force [22].

The wet film may undergo drying, hydrolysis and condensation reactions depending on the chemical route. The as-deposited film possibly represents a chemical or physical network. Upon the subsequent heat treatment, a further hydrolysis and condensation and/or a pyrolysis of organic ligands may take place, again depending on the chemical route. The resulting film consists of amorphous or nanocrystalline oxides and/or carbonates. Upon further heat treatment, any carbonate will decompose and the film will crystallize through a homogeneous or a heterogeneous nucleation. Typically, the desired final film thickness is built up by multiple coating and annealing.

Depending on the type and reactivity of the precursors, the chemistry shows a wide spectrum of reaction types. On the one hand, there are the pure sol–gel reactions, that is, alkoxide precursor systems which undergo hydrolysis and condensation reactions. The formation of SiO_2 coatings starting from Si alkoxides is the classical example for this type of reaction. The condensation leads

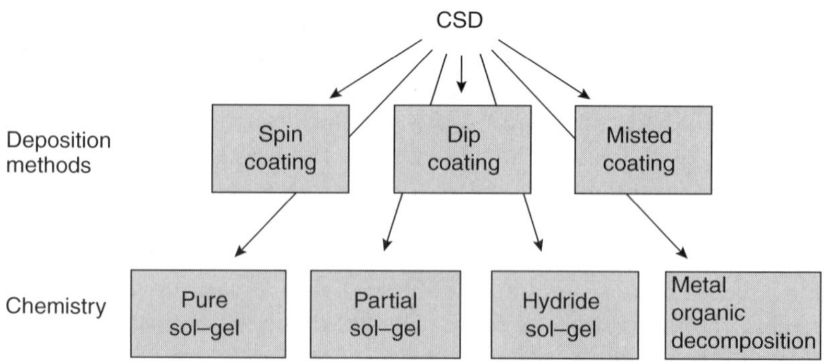

FIGURE 5.8.6 Variety of CSD techniques and chemical route types [21].

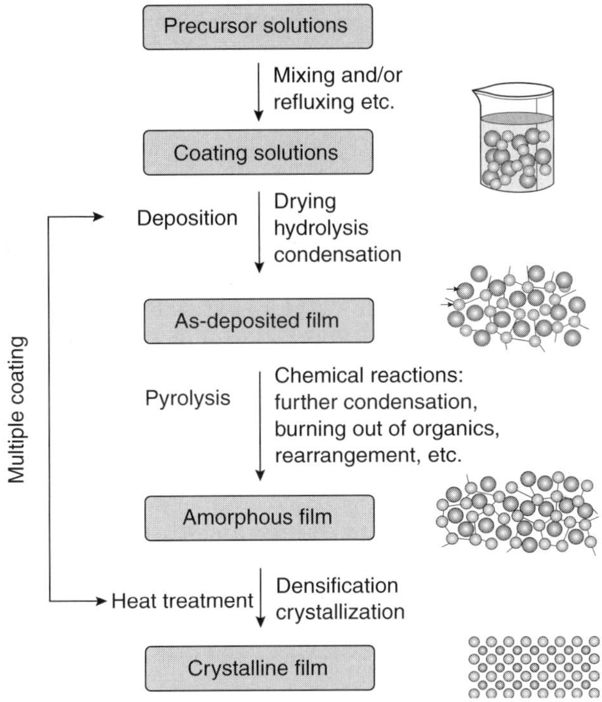

FIGURE 5.8.7 Flowchart of the CSD technique. The bars describe the states during the CSD procedure, while the arrows indicate the treatment and the internal processes [21].

to a chemical gelation in which—under appropriate reaction conditions—no pyrolysis reaction of any organic ligand occurs. On the other extreme, there is the metal organic decomposition (MOD) which typically starts from carboxylates of the cations or, in special cases, from the nitrates [23]. The carboxylates do not chemically react with water. Consequently, during the heat treatment, first the solvents are evaporated, a process which is sometimes referred to as physical gelation. Upon further heating, the carboxylates pyrolytically decompose into amorphous or nanocrystalline oxides or carbonates. There is a wide spectrum of possible reaction routes between the pure sol–gel route and the MOD route. Depending on the type of alkoxides and a possible stabilization of the precursors, there may be a partial hydrolysis and condensation while some organic ligands remain in the gelated film and undergo pyrolysis upon further heat treatment. In the synthesis of multicomponent oxide films, often hybride routes are followed, that is, there may be some precursors employed which tend to follow the sol–gel or partial sol–gel route while others undergo typical MOD reactions.

The microstructure formed during the CSD process depends strongly on the thermodynamics and kinetics of the solid state reaction from the intermediate amorphous or nanocrystalline state after pyrolysis to the final equilibrium crystalline phase. This is controlled by the chemical composition of the film system. As an example we consider the formation of perovskite structures (see Ref. [21] for review). Sr- or Ba-based oxide systems show stable intermediate carbonates whereas Pb based systems tend to form fluorite/pyrochlore nanophases. This behavior is independent of the specific precursors employed. Despite differences in the microstructure evolution, typically a heterogeneous nucleation of the perovskite at the bottom electrode (and sometimes on the film surface) is found for $Pb(Zr,Ti)O_3$ thin films. This leads to a *columnar* microstructure which is typically reported for $Pb(Zr,Ti)O_3$ films. On the other hand, in Sr or Ba titanate systems, the intermediate state after pyrolysis consists of a matrix of amorphous Sr/Ba carbonates or Sr/Ba-Ti oxocarbonates into which nanocrystalline TiO_2 particles are embedded. These particles yield a homogeneous nucleation, but there will be nuclei at the boundaries as well. For very thin coating steps, this boundary nucleation dominates and it is possible to prepare highly dense, columnar structured $SrTiO_3$, $BaTiO_3$ and $(Ba,Sr)TiO_3$ films on Pt-coated Si wafers. It is important to apply a one-step process in which the heating of the individual coating is always up to the full crystallization. In the case of $SrTiO_3$ films, an epitaxial growth of a (111) film on (111)Pt electrodes has been achieved [24].

The *advantages* of CSD are: (a) an excellent control of the film composition through the stoichiometry of the coating solution; (b) a relatively low capital investment; and (c) an easy fabrication over large areas, up to multiple square meters for dip and spray coating techniques. The *disadvantages* are: (a) some specific obstacles to achieve epitaxial films; (b) the missing opportunity to deposit atomic layer super structures; and (c) a poor step coverage for narrow sub-micron three-dimensional structures.

5.8.3.2 Chemical Vapor Deposition and Metal Organic CVD

The CVD of ceramic thin films is based on the evaporation of the precursors which are decomposing on a hot substrate in a suitable reaction chamber (Fig. 5.8.8). Within this reactor, the temperatures, the pressure of gases, reacting gases and/or inert gases, and the precursor vapor are controlled and there are excellent reviews on the methods [25–29]. The most essential requirements on the precursors are a sufficient volatility at low temperature and a sufficiently high decomposition temperature. The range between both temperatures defines the process window.

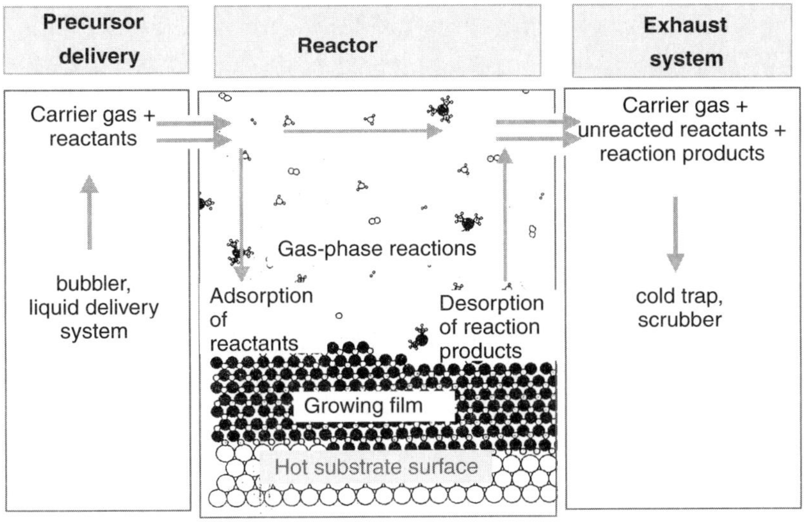

FIGURE 5.8.8 Schematics of a CVD reactor.

For many binary ceramics, as they are used in tribological coatings (carbides, nitrides, oxides, silicides and borides) or as insulators and interlayer dielectrics in CMOS technology (SiO_2, SiN_x or BPSG glasses), stable precursor molecules which are gaseous at room temperature are available for CVD. These precursors comprise halides, carbonyls, hydrides which decompose or react with, for example, NH_3, CH_4, H_2 at high temperatures. Examples are [22]:

Carbides, e.g. boron carbide:	$4BCl_3 + CH_4 + 4H_2 \rightarrow B_4C + 12HCl$
Nitrides, e.g. boron nitride:	$BCl_3 + NH_3 \rightarrow BN + 3HCl$
silicon nitride:	$3SiH_4 + 3NH_3 \rightarrow Si_3N_4 + 12H_2$
Oxides, e.g. silicon oxide:	$SiH_4 + O_2 \rightarrow SiO_2 + 2H_2$
	or by the TEOS process:
	$Si(C_2H_5O)_4 + 12O_2 \rightarrow SiO_2 + 8CO_2 + 10H_2O$

For heavy cations (especially group II elements), a reasonable volatility is often reached only by organometalic precursors, for example, by β-diketonates with several alkyl groups, such as in tetramethylheptadionates (thd). The two keto groups chelate the cation while the outer alkyl groups effectively shield any polar region of the molecule. Typically two (thd)-ligands are reacted with one cation, yielding $Ba(thd)_2$. Since the outer shell of this molecule consists entirely of alkyl groups there is only a weak van der Waals interaction between neighboring molecules and, hence, a relatively low boiling temperature. For

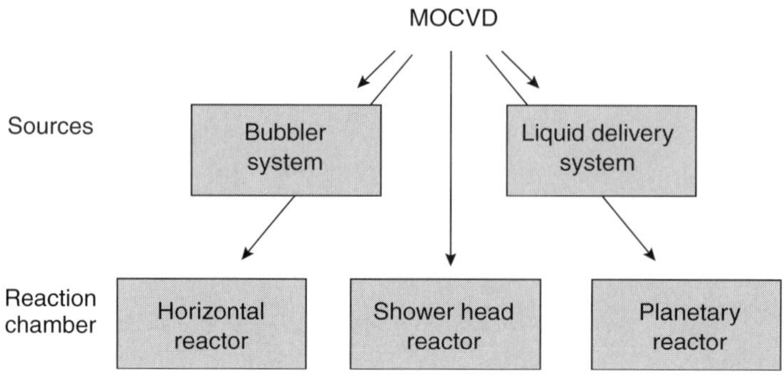

FIGURE 5.8.9 Overview of the typical precursor source systems and reactor design principles used for MOCVD of oxide thin films [21].

cations with a higher electronegativity (i.e. a lower tendency to form ionic bonds), alkyl compounds, such as $Pb(C_2H_5)_4$, and alkoxide compounds, such as $Ti(OC_3H_7)_4$, may give rise to a sufficient volatility [21].

Over the years, different techniques have been developed to evaporate and to deliver these precursors, which are liquid or solid at room temperature, into the reaction chamber (Fig. 5.8.9). The main difference is the temperature of the precursor reservoir. In the first case, the liquid, dissolved or solid precursors are brought to a designated temperature and a carrier gas flowing through or over the precursor system will pick up the precursor, at a designated rate. For liquid precursors these delivery systems are called *bubbler systems* [30]. In the latter case, liquid delivery systems [31], the precursors are mixed in a solution either as a cocktail in a batch or by mixing the single solution using liquid mass flow meters. In front of the reactor, the mixed solution is flash evaporated by contact to a high-temperature metal frit or by misting into a hot carrier gas.

The *reactor design* has to support a continuous, completely homogenous, and highly reproducible deposition reaction. For small wafers, a reactor with a horizontal gas flow is most appropriate [30]. For higher throughput, the same horizontal gas flow principle is translated into a *multiwafer planetary reactor* [32]. A system widely used for very large (e.g. 300 mm) wafers is the *showerhead design* [31]. Here, the precursor vapor is distributed over a temperature controlled plate with numerous defined holes (so-called "showerhead") which supplies the precursor vapor homogeneously over a large area. In order to reduce the thermal budget non-thermal energy deposition can be applied additionally, for example, plasma-enhanced CVD (PECVD), or photo-CVD.

There is at present only limited knowledge on the details of the chemical mechanisms involved in the MOCVD of complex oxides, that is, in the gas-phase reaction pathways and their decomposition temperatures and activation energies, and in the adsorption and surface migration processes and the incorporation efficiencies. However, some remarks on the temperature dependence of the deposition rate seem worthwhile as the reaction can quite generally be divided into two regimes:

1. *Kinetically limited growth at low temperatures*: Although the reaction kinetics includes many different process steps, the growth rate can often be described by an Arrhenius law with an effective activation energy. The exponential temperature dependence necessitates a very high temperature stability and uniformity over large wafers. Due to this temperature controlled growth the deposition rate is only weakly dependent on the flow homogeneity and this regime is best suited for conformal deposition.
2. *Transport limited regime:* At medium and higher temperatures, the major limitations of the reaction are given by the transport of the precursors and reactants to the surface and/or the nucleation sites. Hence, the deposition rate is generally less sensitive to small temperature gradients, however, the control of the flow pattern becomes important.

Major advantages of MOCVD are: (a) the process scalability to large wafer sizes and homogeneous deposition over large areas for some of the reactor designs; (b) the compatibility with semiconductor fabrication techniques; and (c) the opportunity to achieve a high-quality step coverage even for three-dimensional structures of high aspect ratio. This capability is unique for MOCVD compared to all other thin film deposition techniques and MOCVD techniques are therefore considered as the primer deposition techniques in the field of high-*k* materials like $(Ba,Sr)TiO_3$ for advanced DRAM concepts as well as for ferroelectric materials like $Pb(Zr,Ti)O_3$ for ferroelectric memories.

The major disadvantage is the requirement of a relatively lengthy adjustment of the film stoichiometry for new film compositions and the very limited *in situ* control of the processes.

5.8.4 THERMAL SPRAY AND CLUSTER BEAMS

Thermal spray is a well-established, relatively low-cost, industrial process which is widely used for the deposition of compounds, including the refractory carbides and nitrides [1, 33]. The coating material, usually in the form of a powder, is metered into a compressed gas stream and fed into the heat source where it is heated to its melting point and projected onto the substrate (Fig. 5.8.10). Many different variants of the systems are used and especially there are different heat sources. Plasma torches may be driven by DC power or

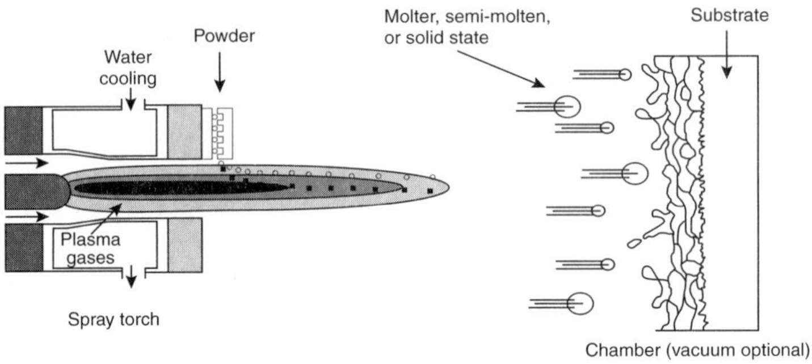

FIGURE 5.8.10 Schematics of the thermal spray process [33].

by inductively coupled RF. Alternatively detonation guns, which use the energy of continuos, controlled explosions of oxyacetylene mixtures to obtain the necessary kinetic energy or high-velocity oxy-fuel torches are used. On impact, the particles or droplets form a coating of many layers of overlapping thin lamellar particles or splats.

Almost any material that can be melted without decomposing can be deposited as a coating. In order to prevent detrimental chemical reactions, spraying can be carried out in an inert gas atmosphere. Typical thickness of the required films are from 0.05 to 0.5 mm. Porosity is generally greater than for the more continuous deposition during PVD and CVD.

The jet vapor deposition [1] and the ionized cluster beam deposition [34] may be considered as further modifications of this characteristic high-impact deposition.

5.8.5 CONCLUSIONS

In this short overview, the methods used for the deposition of thin films from ceramic materials have been introduced. Even this selection of the most basic properties of the techniques should give an impression of the wide range of possibilities which are available. All of these methods have reached very high technical standards and the necessary components are commercially available. The properties summarized in Table 5.8.1 can therefore be only a rough characterization and many modifications on the details of the techniques, which may be important, could not be discussed.

In addition, the border lines between the methods are not sharp and even hybrid methods can easily be achieved as discussed for the physical deposition methods. In addition there is an increasing trend to combinations of physical and chemical aspects which is documented by the addition "reactive" in the acronyms: reactive sputtering (using a reactive gas in the plasma), reactive evaporation (the non-metallic element is introduced as a reactive gas and the metallic component is evaporated), activated reactive evaporation (ARE), or the combination of MBE with metal-organic gas sources (MOMBE).

We have given special emphasis to deposition methods for perovskite oxide films as they are necessary for DRAM, FeRAM, FeFET, conducting oxide electrodes and optical waveguides. The deposition of high-quality films for most of the materials of interest can be achieved and has been achieved by several methods. Nevertheless, every new complex material and especially, every new material combination presents a new challenge and new ideas as well as very detailed optimization of the processes are necessary. Finally, the optimum choice of the method depends on details of the requirements and the differences between the requirements of research and production are often decisive.

APPENDIX

Acronyms and Abbreviations

ARE	activated reactive evaporation
BPSG	boron phosphor silicate glass
CMOS	complementary metal oxide silicon
CSD	chemical solution deposition
CVD	chemical vapor deposition
DC	direct current
DRAM	dynamic random access memory
FeRAM	ferroelectric random access memory
FeFET	ferroelectric field effect transistor
HTSC	high-temperature superconductor
GMR	giant magneto resistance
IBAD	ion beam assisted deposition
MBE	molecular beam epitaxy
MOCVD	metal organic chemical vapor deposition
MOD	metal organic decomposition
MOMBE	metal organic molecular beam epitaxy
PECVD	plasma enhanced CVD
PLD	pulsed laser deposition
PVD	physical vapor deposition
RF	radio frequency
RHEED	reflection high energy electron diffraction ($\approx$10 keV)

SD	sputter deposition
TED	thermal evaporation deposition
TEOS	tetraethylorthosilicate = tetraethoxysilane
UHV	ultra high vacuum ($<10^{-9}$ Torr, 1 Torr = 0.75 mbar = 0.75 h Pa)

REFERENCES

1. Bunshah, R. F. ed. (1994). *Handbook of Deposition Technologies for Films and Coatings*, 2nd edn., Park Ridge: Noyes Publ.
2. Maissel, L. I., and Glang, R., eds. (1983). *Handbook of Thin Film Technology*, New York: McGraw-Hill (reissue).
3. Parker, E. H. C., ed. (1985). *The Technology and Physics of Molecular Beam Epitaxy*, New York: Plenum Press.
4. Tsao, J. Y. (1993). *Materials Fundamentals of Molecular Beam Epitaxy*, San Diego, CA: Academic Press.
5. Herman, M. A., and Sitter, H. (1996). *Molecular Beam Epitaxy, Fundamentals and Current Status*, 2nd edn, Springer Series in Materials Science, Berlin: Springer.
6. Berberich, P., Assmann, W., Prusseit, W., Utz, B., and Kinder, H., (1993). Large area deposition of $YBa_2Cu_3O_7$ films by thermal co-evaporation. *J. Alloys Comp.* **195**: 271–274.
7. McKee, R. A., Walker, F. J., and Chisholm, M. F. (1998). Crystalline oxides on silicon: the first five monolayers. *Phys. Rev. Lett.* **81**: 3014–3017.
8. Yu, Z., Ramdani, J., Curless, J. A., Overgaard, C. D., Finder, J. M., Droopad, R., Eisenbeiser, K. W., Hallmark, J. A., Ooms, W. J., and Kaushik, V. S. (2000). Epitaxial oxide films on Si(001). *J. Vac. Sci. Technol. B* **18**: 2139–2145.
9. Hubler, G. K., ed. (1992). Pulsed laser deposition. *MRS-Bull.* **17**, no. 2.
10. Auciello, O., Kingon, A. I., Krauss, A. R., and Lichtenwalner, D. J. (1993). Physical vapor deposition of multicomponent oxide thin films: techniques, basic deposition processes and film processing–microstructure–propery relationships, *Multicomponent and Multilayered Thin Films for Advanced Microtechnologies: Techniques, Fundamentals and Devices*, pp. 151–208, Auciello, O., and Engemann, J., eds., Dordrecht: Kluwer Academic.
11. Chrisey, D. B., and Hubler, G. K. (1994). *Pulsed Laser Deposition of Thin Films*, New York: John Wiley & Sons.
12. Beckers, L., Schubert, J., Zander, W., Ziesmann, J., Eckau, A., Leinenbach, P., and Buchal, Ch. (1998). Structural and optical characterization of epitaxial waveguiding $BaTiO_3$ thin films on MgO. *J. Appl. Phys.* **83**: 3305–3310.
13. Holzapfel, B., Roas, B., Schultz, L., Bauer, P., and Saemann-Ischenko, G. (1992). Off-axis laser deposition of $YBa_2Cu_3O_{7-\delta}$ thin films. *Appl. Phys. Lett.* **61**: 3178–3180.
14. Wehner, G. H., and Anderson, G. S. (1983). The nature of physical sputtering. *Handbook of Thin Film Technology*, pp. 3-1–3-38, Maissel, L. I., and Glang, R., eds., New York: McGraw-Hill.
15. Maissel, L. (1983). Application of sputtering to the deposition of films. *Handbook of Thin Film Technology*, pp. 4-1–4-44, Maissel, L. I., and Glang, R. eds., New York: McGraw-Hill.
16. Wasa, K., and Hayakawa, S. (1992). *Handbook of Sputter Deposition Technology: Principles, Technology and Applications*, Park Ridge, NJ: Noyes Publ.
17. Rossnagel S. M. (1993). Plasma sputter deposition systems: plasma generation, basic physics and characterisation. *Multicomponent and Multilayerd Thin Films for Advanced Microtechnologies: Techniques, Fundamentals and Devices*, pp. 1–20, Auciello, O., and Engemann, J. eds., Dordrecht: Kluwer Academic.

18. Thornton J. A. (1974). Influence of apparatus geometry and deposition conditions on the structure and topography of thick sputtered coatings. *J. Vac. Sci. Technol.* **11**: 666–670.
19. Poppe, U., Klein, N., Dähne, U., Soltner, H., Jia, C. L., Kabius, B., Urban, K., Lubig, A., Schmidt, K., Hensen, S., Orbach, S., Müller, G., and Piel, H. (1992). Low-resistivity epitaxial $YBa_2Cu_3O_7$ thin films with improved microstructure and reduced microwave losses. *J. Appl. Phys.* **71**: 5572–5578.
20. Schwartz, R. W. (1997). Chemical solution deposition of perovskite thin films. *Chem. Mater.* **9**: 2325–2340.
21. Waser, R., Schneller, T., Hoffmann-Eifert, S., and Ehrhart, P. (2001). Advanced chemical deposition techniques—From research to production. *Integr. Ferroelectr.* **36**: 3–30.
22. Solayappan, N., McMillan, L. D., Paz De Araujo, C. A., and Grant, B. (1997). Second generation liquid source misted chemical deposition technology for ferroelectric thin films. *Integr. Ferroelectr.* **18**: 127–136.
23. Javoric, S., Kosec, M., and Malic, B. (2000). Water-based synthesis of $La_{0.5}Sr_{0.5}CoO_3$(LSCO) electrodes for ferroelectric thin films. *Integr. Ferroelectr.* **30**: 309–316.
24. Hoffmann, S., and Waser, R. (1999). Control of the morphology of CSD-prepared (BaSr)TiO_3 thin films. *J. Eur. Ceram. Soc.* **19**: 1339–1343.
25. Sherman, A. (1987). *Chemical Vapor Deposition for Microelectronics: Principles, Technology, and Applications*, Park Ridge, NJ: Noyes Publ.
26. Morosanu, C.E. (1990). *Thin Films by Chemical Vapor Deposition*, Amsterdam: Elsevier.
27. Hitchman, M. L., and Jensen, K. F. (1993). *Chemical Vapor Deposition, Principles and Applications*, New York: Academic Press.
28. Rees, W. S., ed. (1996). *CVD of Nonmetals*, Weinheim: VCH.
29. Pierson, H. O. (1999) *Handbook of Chemical Vapor Deposition, Principles, Technology and Applications*, 2nd edn, Park Ridge, NJ: Noyes Publ.
30. Foster, C. M. (1997). Chemical vapor deposition of ferroelectric thin films, *Thin Film Ferroelectric Materials and Devices*, pp. 167–197, Ramesh, R., ed., Boston, MA: Kluwer Acadamic.
31. Bilodeau, S. M., Carl, R., Van Buskirk, P. C., and Ward, J. (1997). MOCVD of $BaSrTiO_3$ for $\geq$1-Gbit DRAMs. *Solid State Technol.* **40**: 235–242.
32. Deschler, M., Woelk E., Schmitz, D., Strauch, G., and Juergensen, H. (1997). *Integr. Ferroel.* **18**: 119.
33. Herman, H., Sampath, S., and McCune, R. (2000). Thermal spray: Current status and future trends. *MRS Bull.* **25**, no. 7: 17–25.
34. Takagi, T. (1988). *Ionized Cluster Beam Deposition and Epitaxy*, Park Ridge, NJ: Noyes Publ.

Handbook of Advanced Ceramics
S. Sōmiya *et al.* (Eds.)

CHAPTER 6

6.1 Synthesis of Single Crystals

KIYOSHI SHIMAMURA[1] and TSUGUO FUKUDA[2]

[1]*Kagami Memorial Laboratory for Materials Science and Technology, Waseda University, Tokyo 169-0051, Japan*

[2]*Institute of Multidisciplinary Research for Advanced Materials, Tohuku University, Sendai 980-8577, Japan*

6.1.1 INTRODUCTION

No other period has utilized the "crystal growth phenomena" than today. Most electric devices are made of artificially grown single crystals such as semiconductors. Many properties of metallic materials and ceramics have been developed by paying attention to the intentional control of the crystal growth process. Furthermore, crystal growth of a highly polymerized organism compound, such as protein, under microgravity in a space laboratory, and the application for bioscience are planned. Some experiments are already in progress.

The crystal growth phenomenon itself is one that is very close to our normal daily life. For example, snow, frost and natural ice are a kind of crystal growth phenomena of water in nature. Precipitation that is observed in chemical experiments can be seen as minute crystal growth proceeded by chemical reaction. Ice is the result of "supercooling" caused by cooling water lower than the freezing point (0°C). Precipitation is generated by the "supersaturation" caused by decreasing the solubility of the target material by adding a reagent into the solution. In order to grow single crystals artificially, one of these two conditions, that is, supercooling or supersaturation, should be accomplished artificially.

Supercooling and supersaturation are realized not only in the liquid phase, such as ice formation, and in the solution phase, such as the precipitation during a chemical experiment, but also in the vapor phase. Snow is an ice crystal deposited from steam in air by supersaturation caused by cold upper air. As shown above, the combination of the crystal and liquid phase, crystal and solution phase, and crystal and vapor phase, is generally fundamental for crystal growth. The liquid phase is often called the melt phase, because it indicates the molten condition of a crystallized target material.

In practical use, the melt growth technique is the best way among the above three different combinations to achieve the highest productivity, that is, to have the highest yield of single crystals in a definite time. This is because it can be preceeded by heat, which is one of the easiest parameters to control, and it is easier to apply to large-scale operations. Hence, this technique is the most common in industry. Vapor phase crystal growth techniques have the lowest yield. However, it is a superior technique to grow two-dimensional crystals and/or high-quality crystals. From this viewpoint, vapor phase techniques are also important in industry. Solution growth techniques are recognized in between these two techniques.

When one chooses a crystal growth technique, it is fundamental to recognize whether the target material is melted congruently or incongruently. Figure 6.1.1 is a binary phase schematic diagram representing two compounds AB and A_2B with two components A and B. Compound AB melts at temperature T_1 when it is heated, whereas compound A_2B decomposes at temperature T_2 into component A and liquid, and subsequently turns into liquid when it is continuously heated. The former is called congruent melting because the composition of the solid phase and the melt phase does not change. The latter is called incongruent melting because it decomposes when it is dissolved. T_1 is called the melting point, and T_2 is called the decomposition point.

If a material melts congruently, the application of melt growth technique is useful in many cases. Figure 6.1.2 shows schematics of different types of melt growth techniques. The representative technique is a pulling technique or the Czochralski (CZ) technique. Crystals are grown from the melt directly by pulling up the seed crystal. The Bridgman technique is also a pulling technique, which enables crystal growth by cooling and solidifying the melt in a crucible.

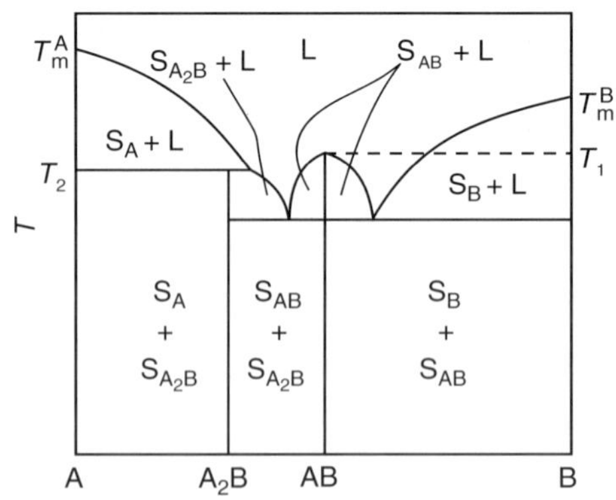

FIGURE 6.1.1 Schematic of binary phase diagram including solid phases A, B, A_2B, AB. S_{A_2B}, A_2B solid phase; S_{AB}, AB solid phase.

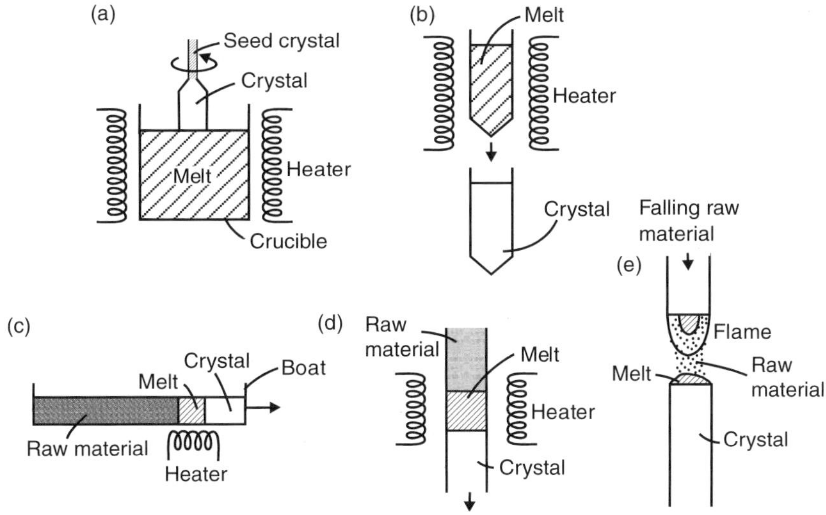

FIGURE 6.1.2 Schematics of different types of melt growth techniques.

Others are the zone melting technique, by which crystals are grown in a moving molten zone of a boat-type crucible, and the floating zone technique, in which the molten zone is kept by the surface tension in a vertically settled material without a crucible. The crystal growth is proceeded by moving the molten zone upward. The pedestal technique is a modification of the floating zone technique, where the molten zone moves downward. Crystals can also be grown by the Verneuil technique. In this technique, crystallization continues by falling liquid drops which are obtained from high melting temperature oxide powders when they pass through a flame.

In the melt growth technique, the use of a crucible is very important. In general, materials which have a high melting temperature and inactive nature are used for the crucible. To grow metal/alloy, intermetallic compounds, sulfide, halide and semiconductor crystals, ceramic series crucibles made of oxide, nitride and graphite are used. To grow oxide crystals, metal series crucibles made of gold, platinum, iridium and molybdenum are utilized. The criteria for choosing a crucible material are mainly: (a) lower melting temperature of a crucible material than that of the target crystal and (b) no chemical reaction between the crucible and melt at elevated temperatures. If a crucible can be used, the pulling techniques of CZ, Bridgman, and zone melting techniques can be applied. If a proper crucible material is not found, other techniques such as the floating zone and pedestal techniques should be considered.

The solution growth technique described below is a way to achieve crystal growth at lower temperatures with a solvent. To grow crystals that melt incongruently, the solution growth technique is preferable. In this technique, crystals are obtained from deposits in the solution in which the raw material is dissolved. The key point of this technique is to choose the proper solvent. There are two kinds of solvent. One is a liquid at room temperature such as water and organic solvents. The other is a solvent which melts only at high temperature, so-called flux. This is the reason why the solution growth technique that uses a solvent with a high melting temperature is called the flux technique. There are two solution growth techniques. One is the solvent evaporation technique and the other is the slow cooling technique. Crystal deposits are preceded by the evaporation of the solvent at a constant temperature in the former, and by the cooling solution slowly in the latter. A modification of the flux technique is the flux pulling (top-seeded solution growth, TSSG) technique. This is a technique to grow crystals by pulling up a seed crystal from the solution.

The hydrothermal technique is used to synthesize some of crystals such as quartz. This is also a kind of solution growth technique where crystals are deposited on a seed crystal from the solution under high temperature and high pressure.

In this chapter, the crystal growth methods mentioned above are described in relation to recent topical materials.

6.1.2 KYROPOLOS TECHNIQUE

The technique to pull up crystals from melt is the most popular, because relatively larger and highly perfect crystals can be grown in a short time. The CZ technique is the most representative. The Kyropolos technique is similar to the CZ technique.

A seed crystal is touched to the melt, and subsequently, a crystal is grown by slowly lowering the melt temperature. A grown crystal is taken from the crucible before it attaches to the crucible inner wall. Figure 6.1.3 shows the general concept of both techniques. The principle of the CZ technique is pulling up a crystal with rotation at constant temperature, whereas that of Kyropolos technique is slow cooling of the melt [1].

The technical progress of the Kyropolos technique has been slow, because it does not have particular advantages over the CZ technique. However, several attempts have been made by taking flux into the melt. Materials that do not melt congruently, and have a high volatile nature, or a high melting temperature, have been tried for growth using a modified Kyropolos technique (the flux Kyropolos technique). Good-quality and sized $K_3Li_2Nb_5O_{15}$ [2], $KNbO_3$ [3], $LiNdP_4O_{12}$ [4], and GeO_2 [5] single crystals that could not be attained with other techniques have been obtained in an optimized growth condition.

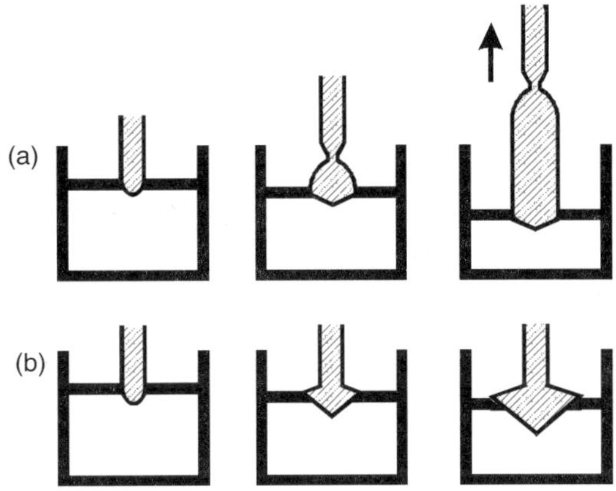

FIGURE 6.1.3 General concept of the Czochralski (a) and Kyropolos (b) techniques.

6.1.3 CZOCHRALSKI METHOD

"Crystal pulling" or the CZ technique [6] is a popular technique for crystal growth, because it can produce large, dislocation free crystals in a relatively short space of time. The main advantage of this technique is that it is a "real-time" technique whereby the size and diameter of the crystal can be controlled while it is growing. The principle of this technique is shown in Figure 6.1.4. The material to be grown is melted in a crucible. A seed crystal is then dipped into the melt and slowly withdrawn. The seed is rotated at the same time in order to attain thermal symmetry and also to stir the melt. The main requirements for growing good crystals by this technique are that the pull and rotation rates should be smooth and the temperature of the melt should be accurately controlled. The diameter of the crystal is dependent on both the temperature of the melt and the pull rate. To increase the diameter of a crystal being grown, either the power is reduced or the pull rate decreased. The opposite applies to reduce the diameter. The maximum size of crystal that can be produced by this technique is governed by the volume of the melt, the diameter of the crucible, the distance that the pull rod can move and the strength of the crystal at its minimum diameter. An additional practical limitation will be how well the seed is secured into its holder. Nothing is more depressing than seeing a freshly grown crystal becoming detached and sinking back into the melt.

Nassau and Broyer [7] have suggested that an ideal material suitable for growing by the CZ technique would have the following properties:

1. congruent melting point,
2. no destructive phase changes,

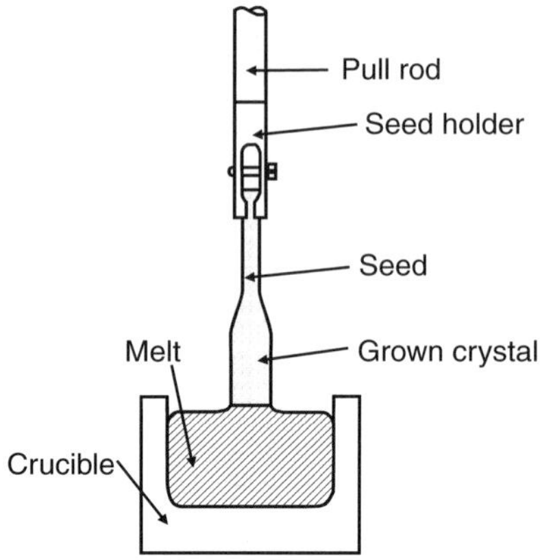

FIGURE 6.1.4 Schematic of the Czochralski technique.

3. low vapor pressure,
4. low viscosity,
5. suitable growth habit,
6. no destructive cleavage.

However, there are methods for overcoming most of the problems raised by the absence of some of these properties.

There are two basic experimental points which should be explicitly mentioned:

1. The temperature distribution in the melt should usually be such that the temperature at the solid–liquid interface is the lowest in the liquid, otherwise there is a strong possibility of spurious nucleation.
2. The surface of the melt should be free of foreign particles and films. This implies that the surface must be protected from dust particles and flakes of material condensed on the sides and top of the growth chamber, and if necessary from the atmosphere.

An important parameter governing the quality of a pulled crystal is the shape of the solid–liquid interface. The best crystals are usually grown with a flat interface because a curved one gives rise to a non-uniform impurity distribution across the face of the crystal. When an interface is convex into the melt it is possible for some low index planes to be parallel to part of the freezing isotherm and for facets to form. This formation can occur with garnets and rubies [8], but can be overcomed if a sufficiently high rotation rate is used (100–200 rpm). The shape is also governed by ambient gas pressure and the length of the crystal [9].

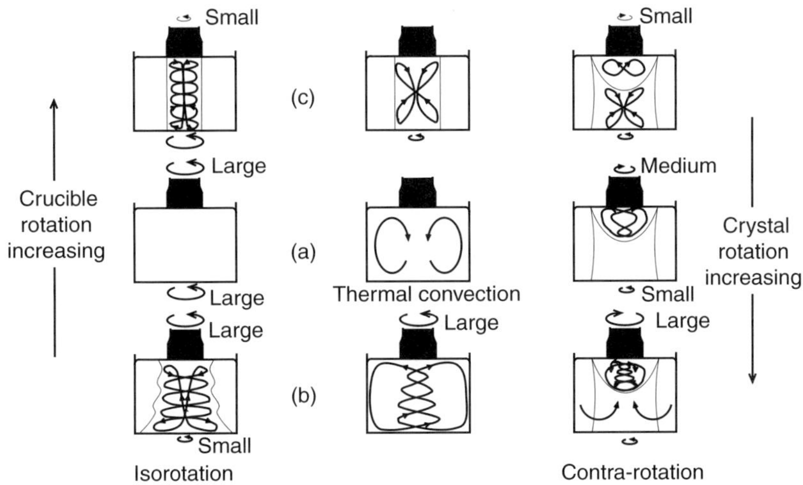

FIGURE 6.1.5 Flow patterns for different crystals and crucible rotation rates.

The interface becomes less concave towards the melt as the pressure increases and as the length of the crystal increases.

The rotation rate has other effects besides altering the shape of the interface. It also alters the flow patterns in the melt and hence the temperature and impurity distribution [10, 11]. There are four main types of flow patterns:

1. Thermal convection—this flow consists of a hot stream of liquid rising at the crucible walls and falling in at the center (Fig. 6.1.5a).
2. Forced convection due to a rotating crystal—the liquid is forced outward by centrifugal acceleration and is replaced by a central flow from the bottom of the crucible up to the center of the rotating crystal (Fig. 6.1.5b).
3. Forced convection due to a rotating crucible—here the liquid flow is the same as for thermal convection except that the upward flow is spiral in form (Fig. 6.1.5c).
4. Combined liquid rotation and thermal convection—in this case, thermal convection is discouraged by a Coriolis stabilizing force acting on particles having a horizontal velocity component.

The choice of pull rate is dependent on a number of factors:

1. diameter of the crystal required,
2. melt temperature,
3. importance of dislocation inclusions,
4. constitutional supercooling.

It is probable that as large a crystal as possible will be wanted but it should be born in mind that the larger the diameter, the more strain there will be and the higher the dislocation density. The dislocation density also increases with an increased growth rate. When a highly doped melt is used, or when pulling from a solution, the limiting factor for the pull rate may be the onset of constitutional supercooling. This can possibly be avoided if the temperature gradient in the melt can be made greater than the equilibrium liquid's temperature gradient at the interface.

Impurities, intended or otherwise, will be incorporated into a crystal while it is growing at a rate depending on the distribution coefficient, the temperature gradient and the growth rate. The distribution coefficient k is the ratio of the solute concentration in the solid (C_s) to the solute concentration in the melt (C_m)

$$C_s = k \cdot C_m.$$

As the solid–liquid interface advances, a boundary layer of increased impurity concentration builds up (Fig. 6.1.6) because the stirring rate is insufficient to remove it. As the crystal grows longer the impurity concentration in the growing crystal increases as:

$$C_s = kC_{m_0}(1 - x)^{(k-1)}$$

where C_{m_0} is the initial concentration of the solute in the melt and x is the fraction of the melt which has solidified. The impurity will often be incorporated

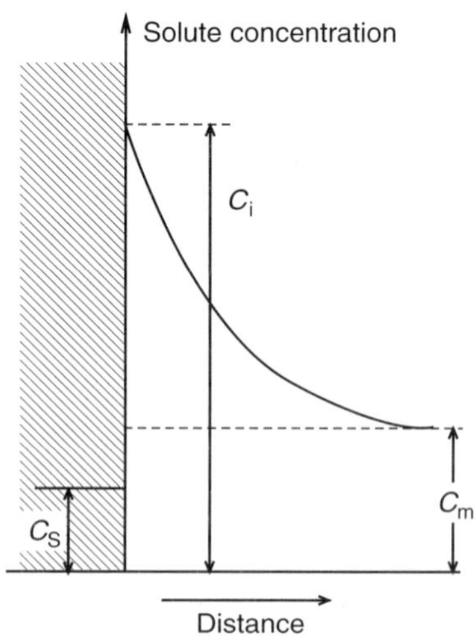

FIGURE 6.1.6 Impurity concentration at a solid–liquid interface.

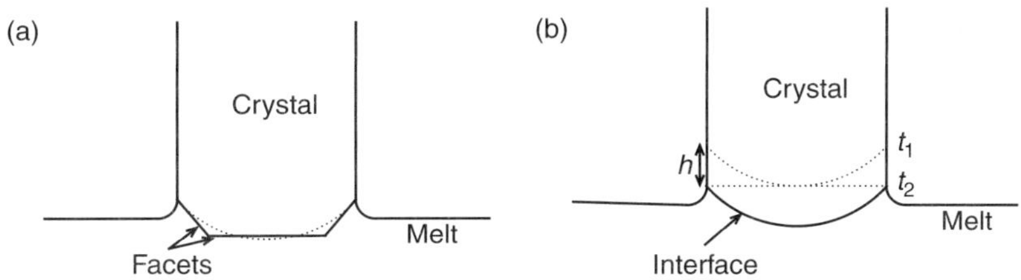

FIGURE 6.1.7 Causes of impurity distribution variation across the face of a crystal.

into the crystal in bands (thought to be caused by temperature oscillations) or be non-uniformly distributed across the face of the crystal due to different distribution coefficients on different facets (Fig. 6.1.7a) [12]. Different points on a cross-section perpendicular to the growth axis will be grown at different times if the interface is not flat (Fig. 6.1.7b), and so the impurity concentration at the center and the periphery will be different. It was shown that the ratio of the concentration at the periphery, C_p, to the concentration at the center C_c is

$$C_p/C_c = \{1 - f_v \Delta t/(V_0 - f_v t_1)\}^{(k_{eff}-1)}$$

where $\Delta t = t_2 - t_1, V_0 =$ original melt volume, $f_v =$ volume growth rate and $k_{eff} =$ effective distribution coefficient.

6.1.3.1 $LiNbO_3$ AND $LiTaO_3$ FOR SAW DEVICES

$LiNbO_3$ (LN) and $LiTaO_3$ (LT) single crystals are preferred for current electric and optical devices [13–21]. Recently, they attained a large market-size next to quartz single crystals as a material for surface acoustic wave (SAW) devices. Research to industrialize LN and LT was started in the beginning of the 1960s. In 1965, these crystals, with "cm" sizes, were grown at about the same time in the United States and Russia with the CZ technique. After this success, actual and practical investigations to grow large-size and high-quality materials for industrial purposes were started. From the beginning of the 1970s, these crystals have come into the limelight as a material for interfrequency (IF) filters for color TV. At that time, intensive research on these materials for applications in SAW devices was started. Later, a lot of effort was put into growing large-size, high-quality, and low-cost LN and LT. In the 1990s, LT became the most important candidate for the new digital communication technology including digital TV such as CATV, CS and BS, hand-held phones, etc. SAW devices made with LT are promising for next generation devices operating in the high-frequency range around 2 GHz. Currently, it seems that LT is more advantageous for advanced technologies than LN.

Although $Li_2B_4O_7$ (LBO) has been investigated as a new candidate, it has the disadvantages of deliquescence and difficulty of crystal growth compared with LT. The ideal material should have both a small frequency temperature coefficient and a large electromechanical coupling constant. This explains the rapidly increasing market for LT.

Although LN growth has been very well investigated for a long time, this is not the case with LT. Knowledge about the precise congruent melting composition and the solid solution range is lacking. However, in view of recent requirements for LT, much effort has started in the growth of LT single crystals.

The increase of diameters of LT from 2 to 4 in. was rather rapid, over almost 10 years. However, after 1985, the size has remained at 4 in. The realization of a 5-in. LT is now under investigation. Figure 6.1.8 shows LN single crystals as-grown.

For the growth of LT single crystals, Ir or Pt/Rh crucibles are normally used. Since the melting temperature of LT is reported to be 1562°C, a Pt crucible cannot be used. When LT single crystals are grown using a Pt/Rh crucible in an air atmosphere, the crystals are colored brown because of a very small amount of Rh from the crucible material. However, this coloration and the existence of Rh in LT do not affect the SAW properties. The general growth conditions that are used for production are as follows: pulling rate is 5–7 mm/h, crystal rotation speed is 8–10 rpm, the ratio between the crystal diameter and the crucible diameter is 0.6–0.7, crystal length is 80–120 mm. Although these conditions for LT are almost the same as for LN, the average temperature gradient for LT between the melt surface and 5 mm above is different from that used for LN growth (15–30°C for LT and 30–80°C for LN).

LT has a rather wide solid-solution range, and the stoichiometric composition and the congruent melt composition are different. Figure 6.1.9 shows a phase diagram for LN and LT. There have been several reports that the

FIGURE 6.1.8 View of 4-in. $LiNbO_3$ single crystals (NGK Insulators, Ltd.)

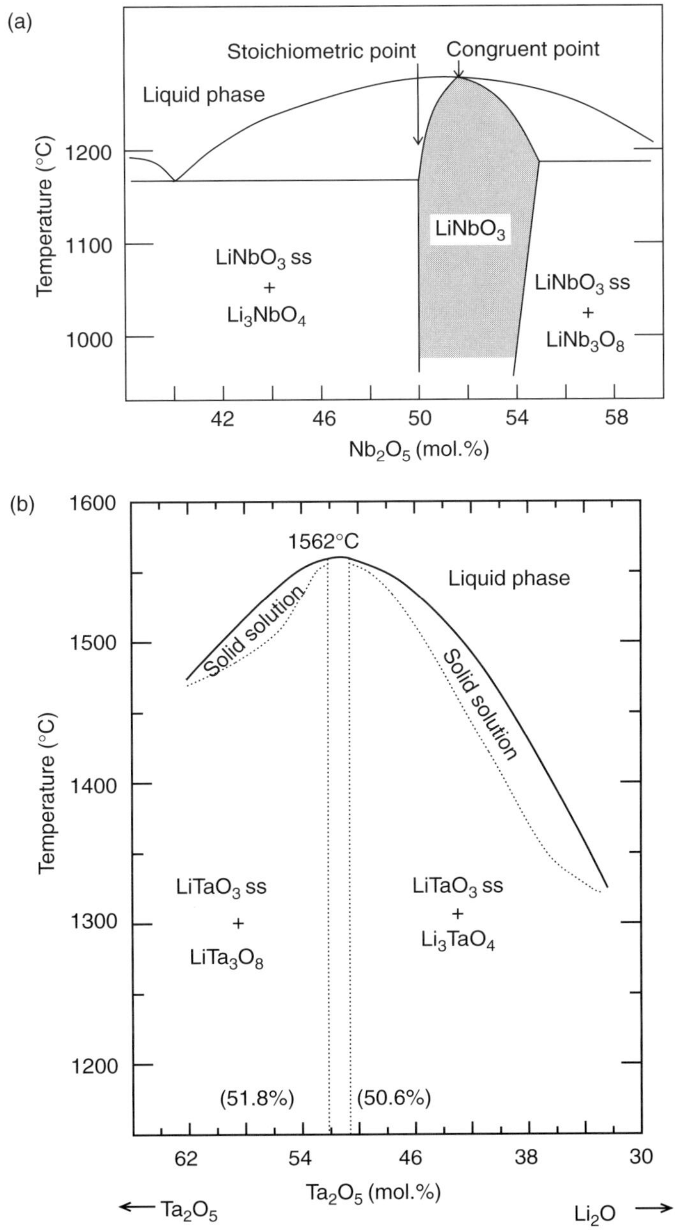

FIGURE 6.1.9 Pseudo binary phase diagram for: (a) $Li_2O-Nb_2O_5$ and (b) $Li_2O-Ta_2O_5$.

congruent melt composition of LT is around 48.7 mol.% Li_2O, and recent investigations estimate it to be 48.3–48.4 mol.% Li_2O. Knowing the exact congruent composition is very important when growing LT single crystals with high uniformity and reproducibility. One way to find out is to measure the Curie temperature (T_C), which has an almost linear dependence with the composition.

For mass production, a slightly Li_2O enriched composition is used in order to compensate for the loss during the growth process. Recent commercially available LT single crystals are guaranteed to have a compositional variation among crystal boules within ± 0.12 mol.% Li_2O. This corresponds to a T_C variation within ± 2.5°C, which satisfies the SAW velocity variation within ± 1 m/s.

Other problems observed during the growth of LT are unstable growth such as spiral phenomena and cracking. At temperature gradients smaller than 30°C/cm, spiral symptoms are sometimes observed. However, if the temperature gradient is steep, cracking occurs. Therefore, construction of the heating zone should be well designed, which in most cases is based on experience. Further investigation in this field is expected.

The required LT wafer specifications for the current SAW devices can be summarized as follows; 3–4 in. diameter, the mirror surface should be flat within 5 μm and the roughness of the back surface should be less than 0.2–2.2 μm. The roughness of back surface is also important because, if it is large, efficiency decreases based on the diffused reflection from the backside.

6.1.3.2 Twisting in $Gd_3Ga_5O_{12}$ Garnet Crystals

Gadolinium gallium garnet $Gd_3Ga_5O_{12}$ (GGG) is one of the most extensively developed garnet materials for substrate and laser applications. GGG high-quality crystals produced commercially are 4 in. in diameter and 25 cm in length [22]. GGG crystals show a tendency for spiral growth [3]. Acentric growth was also observed in other garnet crystals [22, 24]. In general, spiraling is most common in large CZ systems because of low thermal gradients at the growth interface [22]. Examples of the as-grown, twisted GGG ⟨111⟩ crystals are shown in Figures 6.1.10 and 6.1.11.

Different modifications of the CZ technique which can prevent twisting of GGG crystals were reported in Ref. [23]. The crystals were grown from ⟨111⟩ oriented seeds in a $N_2 + 2$ vol% O_2 atmosphere using iridium crucibles. Pulling and rotation rates for 1- and 2-in. crystals are 5 mm/h and 30 rpm and 2 mm/h and 10 rpm, respectively.

To test the effect of the crucible position on spiraling of GGG, the 1-in. crystals were grown using various distance between geometrical centers of crucible

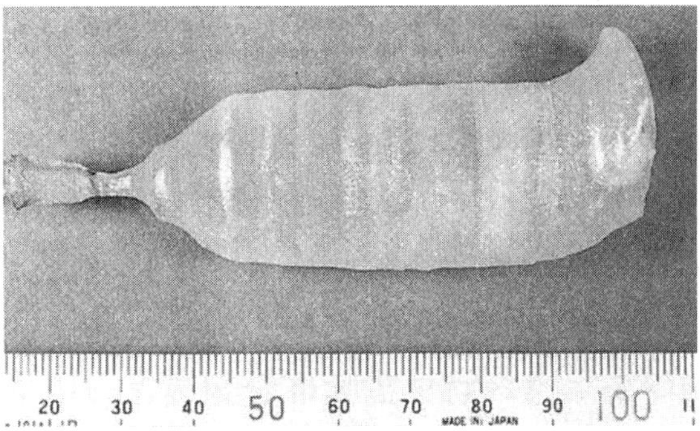

FIGURE 6.1.10 A 1-in. $Gd_3Ga_5O_{12}$ footlike twisted crystal.

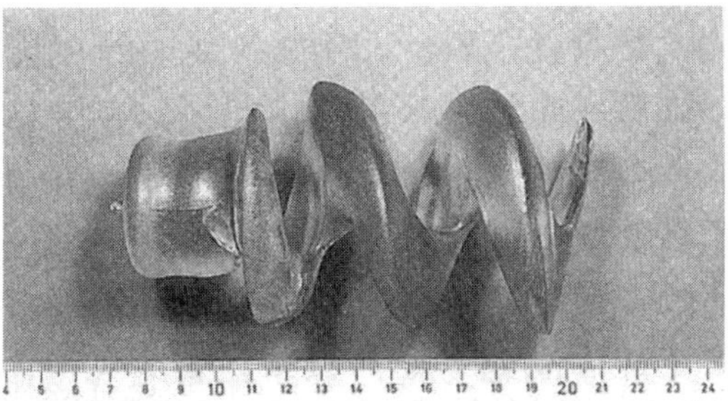

FIGURE 6.1.11 A 2-in. $Gd_3Ga_5O_{12}$ spiral twisted crystal.

and RF coil, ΔX as it is illustrated in Figure 6.1.12. It was found that if $\Delta X < 0$ (position of the coil center is lower than that of the crucible center) the spiral growth phenomena is decreased considerably. The crystals as grown in different conditions are shown in Figure 6.1.13 ($\Delta X < 0$) and Figure 6.1.14 ($\Delta X > 0$). This method resulted in successful growth of relatively small crystals only with about 1-in. diameter.

In the case of 2-in. crystals, a way related to the temperature gradient in the melt and the melt flow pattern was used to modify the growth conditions. According to Figure 6.1.15, the improvement of thermal conditions inside the heating zone was made by decreasing the amount of insulated material (ZrO_2 bubbles) under the crucible. Moreover, a hole was made under the crucible to increase heat flow in a downward direction as shown in Figure 6.1.15. The crystals grown under the conditions of cold bottom were not twisted.

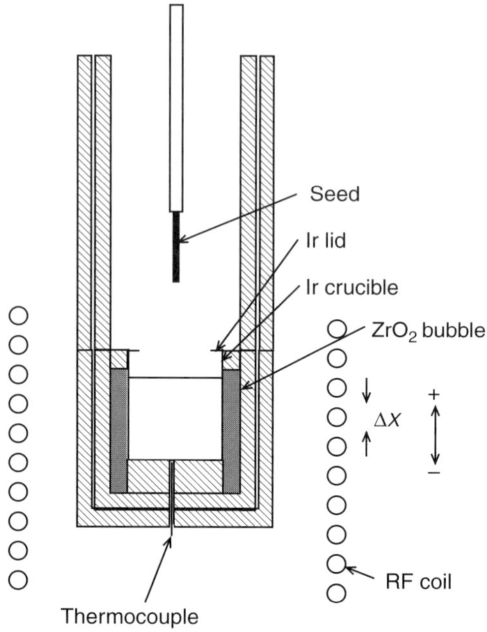

FIGURE 6.1.12 Schematic diagram of the heating zone for 1-in. $Gd_3Ga_5O_{12}$ crystal growth.

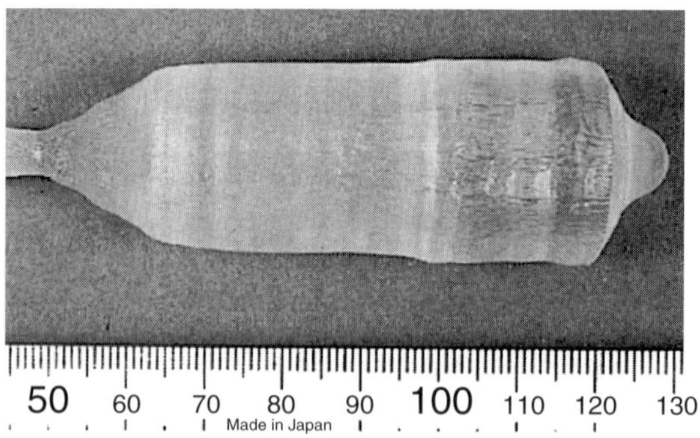

FIGURE 6.1.13 $Gd_3Ga_5O_{12}$ crystal grown with $\Delta X = -40$ mm.

To compare the intensity of natural convection, the melt flow rate was measured in the vicinity of the crucible walls. It was found that with a high thermal gradient (right in Figure 6.1.15) the rate was measured to be 21.0 ± 0.5 mm/s. However, in the opposite case (left in Figure 6.1.15) the rate observed was

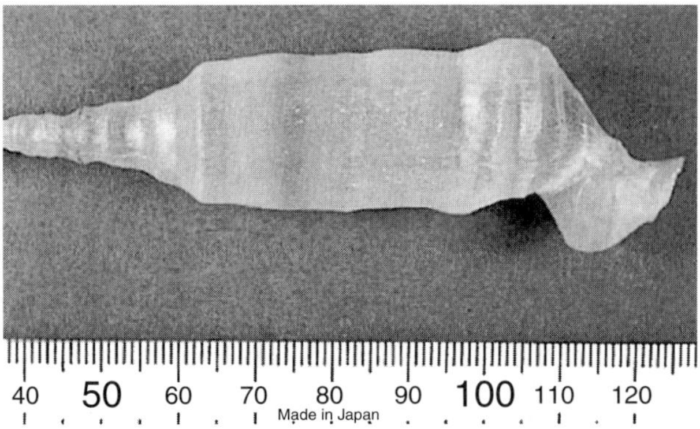

FIGURE 6.1.14 $Gd_3Ga_5O_{12}$ crystal grown with $\Delta X = +40$ mm.

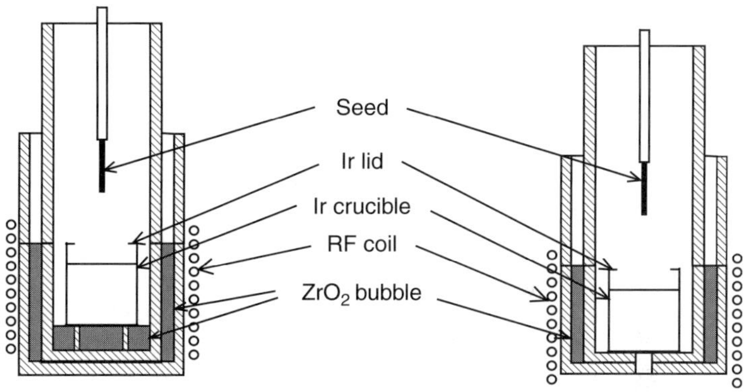

FIGURE 6.1.15 Schematic diagram of the heating zones used for the growth of 2-in. $Gd_3Ga_5O_{12}$ crystals from crucibles with hot (right) and cold (left) bottoms.

15.0 ± 2.0 mm/s. The difference of these values was assumed to follow from the difference of melt flow patterns.

There is one more method used to prevent the twisting. It is based on modification of the melt composition by doping with C^{2+} cations (C = Mg, Sr, and/or Ca). It was assumed [22] that the effect of such cations on the spiraling follows from a decrease of surface tension. The surface tension force on an iridium rod of 6 mm in diameter suspended into the doped GGG stoichiometric melts was measured. It was reported that adding 88 ppm of Ca^{2+} into the melt reduced the force to a value about 15% below the original one. However, with a corresponding addition of 160 ppm of Zr^{4+} impurity, a 33% increase of the tension force was observed.

6.1.3.3 Step Faceting in Rare-Earth Vanadate

Crystals of rare-earth orthovanadates $REVO_4$, where RE = Y, Lu, Er, Dy, and/or Gd, are well known because of their magnetic and laser properties [25–27]. These crystals are promising materials for efficient diode-laser pumped solid-state lasers and may be useful in magnetic refrigeration applications.

The crystals of this family crystallize with a tetragonal space group $I4_1/amd$ and are isomorphous with zircon $ZrSiO_4$. It is known that all these compounds melt congruently. Therefore, the CZ technique is widely used to prepare these crystals.

In spite of the fact that all these materials can be grown from stoichiometric melts as a single crystal, the wide application of orthovanadates is still limited. It follows that growth difficulties prevent production of large and high-quality crystals.

Pure $GdVO_4$ (GVO) crystals, it was reported [25], can be grown by the conventional CZ method using an RF heating furnace from a cylindrical iridium crucible with a diameter of 48 mm and a height of 48 mm. High-purity powders (99.99%) of corresponding oxides Gd_2O_3 and V_2O_5 were used as starting materials. The mixtures of stoichiometric composition were placed in the crucible and heated up to complete melting. The melting point of the GVO compound was found to be about 1780°C as it was estimated by means of an optical pyrometer.

The crystals were grown using GVO seeds under Ar + 2 vol% O_2 atmosphere with a gas flow rate of 1 *l*/min. The pulling rate was 3 mm/h and the rotation rate was 10 rpm. The crystals grown were colorless and further annealing was not necessary. The crystals had a tetragonal zircon structure as detected by X-ray diffraction powder analysis. The lattice parameters of the crystals were $a = 7.218$ Å and $c = 6.356$ Å. Variations of the lattice parameters were not detected along the growth axis.

The length of the crystals was in the range of 40–50 mm. The diameter was about 20 mm. In spite of using a diameter control system, the diameter uniformity of GVO crystals grown under such conditions was not perfect. A step faceting phenomenon was observed in the crystals grown by the conventional CZ method [26], as it is shown in Figure 6.1.16.

To prevent the formation of the facets, the crystals were also grown by the modified CZ technique where the working RF coil position was continuously changed in order to maintain its relative position to the surface of the melt in the crucible, as it is shown in Figure 6.1.17.

In this technique, the top of the RF coil was kept to be coincident with the melt surface during growth. A view of the GVO crystal grown by the modified CZ technique is given in Figure 6.1.18. It was difficult to keep a stable crystal diameter even by using this modified diameter control system. However, a

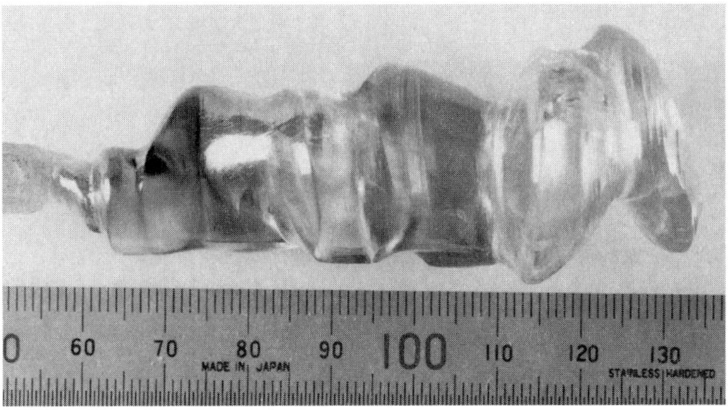

FIGURE 6.1.16 $GdVO_4$ single crystal grown with step faceting.

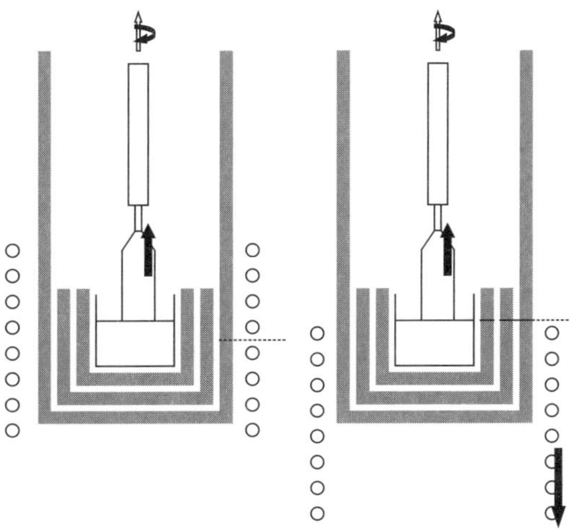

FIGURE 6.1.17 Schematic illustration of a conventional (left) and modified (right) Czochralski system.

comparison of the shapes of the crystals fabricated by both CZ systems described shows us that crystal growth conditions can be improved considerably by the technique illustrated in Figure 6.1.17 (right).

Nd-doped GVO crystals were also grown under similar conditions. In this case, 1.5 and 2.25 wt.% of Nd_2O_3 of the same purity was added into the raw materials. Different pulling rates in the range of 1–4 mm/h were applied to produce Nd : GVO crystals similar to those in Figure 6.1.19.

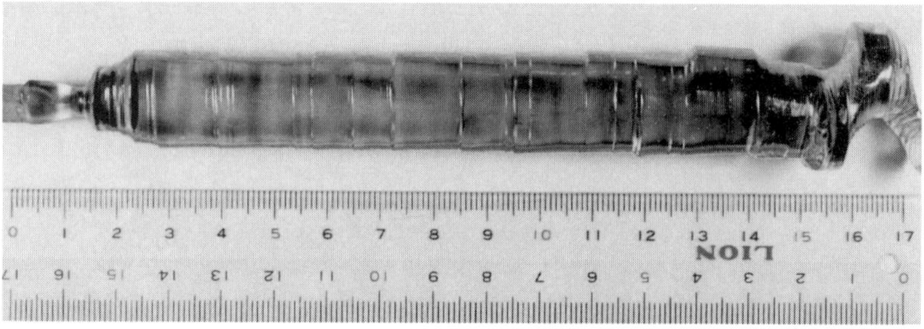

FIGURE 6.1.18 $GdVO_4$ single crystal grown with a modified Czochralski technique.

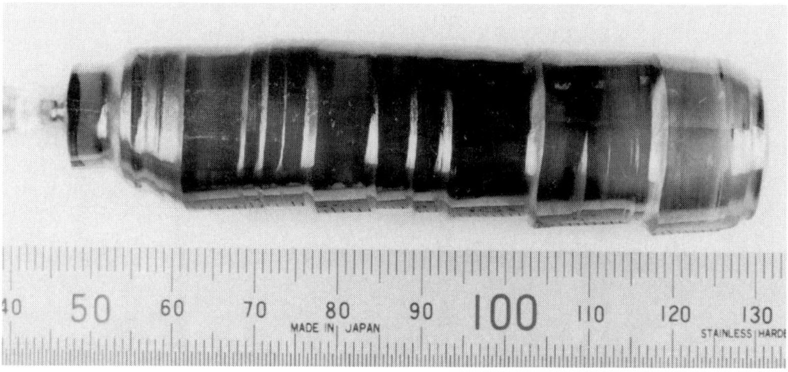

FIGURE 6.1.19 Nd : $GdVO_4$ single crystal grown with a modified Czochralski technique.

Laser diode (LD) pumped laser properties were studied using test plates (5 × 5 × 1 mm: *a*-axis elongation) cut from 1% Nd : GVO crystals. Laser oscillation was performed by pumping plates using a laser diode (500 mW) with an 808.5 nm oscillation wavelength. The results were compared with the input-output characteristics obtained for Nd : YVO_4 test samples (3 × 3 × 1 mm: *a*-axis elongation).

Slope efficiency, threshold pump power and output power of the Nd : GVO-based laser exceeded those of Nd : YVO_4-based one [26]. Therefore, it was concluded that Nd : GVO is a superior laser material.

6.1.4 EDGE-DEFINED, FILM-FED GROWTH METHOD

Edge-defined, film-fed growth (EFG) enables crystal growth by combining the crucible which contains the melt, and a die which defines the shape of the

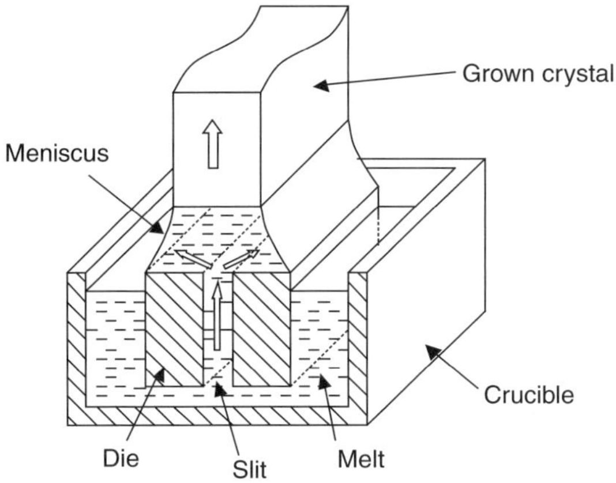

FIGURE 6.1.20 Schematic principle of crystal growth by the EFG technique.

grown crystal and supplies melt to the crystal growth interface. Figure 6.1.20 shows the principle of the EFG technique schematically. In Figure 6.1.20, the melt ascends upward through a slit by the capillary effect. At the upper end, the melt is crystallized, and the crystallized crystal is pulled upward. The characteristic point is that the shape of grown crystal is defined by the upper end shape of the die, and it is maintained constant until the end of the crystal growth. Therefore, together with a columnar crystal grown by a technique such as CZ, crystals with many different cross sectional shapes such as a square-like, plate-like (ribbon), cylindrical, and even fine diameter fiber-like, can be grown by contriving the upper end shape of a die. Furthermore, in addition to the modifying a die, by the precise control of temperature distribution during crystal growth, plural crystals with different shapes can be grown simultaneously.

On the other hand, one of the characteristics of the EFG technique is that the effective distribution coefficient of impurity is almost 1, if the equilibrium distribution coefficient of the impurity is not very small [28]. This is due to the fact that so-called back diffusion is not realistic in practical crystal growth conditions. The whole melt rising up the slit is crystallized at the upper end of a die, and impurities discharged at the crystal growth interface do not go downward toward the melt through a slit by the diffusion.

In 1971, LaBelle [29]. reported the growth of fiber-like sapphire crystals by the EFG technique [29]. Later, a crystal growth technique for plate-like and cylindrical sapphire crystals was also developed [30, 31]. Technological development to grow Si-ribbon crystals was also enthusiastically performed [32, 33].

In order to grow high-quality stable single crystals by the EFG technique, considering the conformity of the die material to the melt is important. A die should wet well with the melt, and the melt should ascend through a slit of a die by the capillary effect. A die should be stable at high temperature with no chemical reaction, and it should be easy to process. The difference in thermal conductive properties between the melt and die also largely affect the crystal growth condition.

6.1.4.1 Rutile (TiO_2) Growth by EFG Method

Rutile (TiO_2) crystals are known to be an attractive material for applications such as polarizers for optical isolators and prisms for optical communication systems [34–37]. These crystals can be grown by Verneuil [38] or the floating zone [39, 40] methods. In spite of the fact that the need for larger size and higher-quality crystals has been increasing, crystal growth of large TiO_2 crystals by the conventional CZ technique is quite difficult because of the difficulties related with diameter control. Therefore, development of other methods which can be used for growth of TiO_2 crystals seems to be important. In this subsection, application of the EFG method to the growth of shaped rutile crystals is described.

Iridium was reported to be a good crucible and die material, because no reaction of the TiO_2 melt with iridium was observed [34]. Arrangement for the determination of capillary action using a capillary-forming guide made of iridium plates is given in Figure 6.1.21. The maximum slot thickness was found to be about 5 mm with a corresponding capillary height of about 2–4 mm.

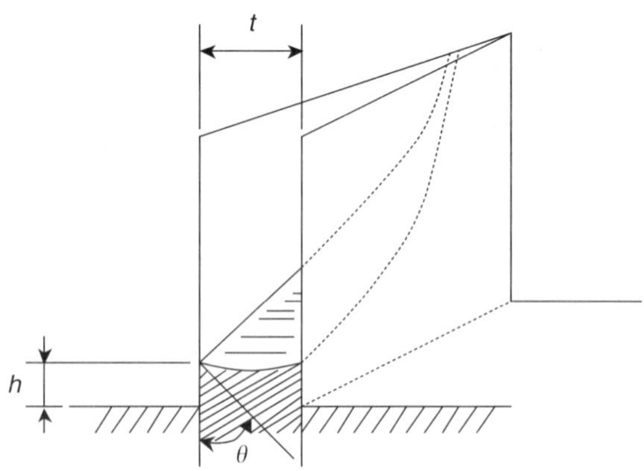

FIGURE 6.1.21 Capillary-forming guide used for the measurement of dependence of capillary height (h) on slot thickness (t).

However, with a slot thickness of about 1.5 mm, the capillary height was about 12 mm. These results show that iridium is suitable for use as a die material in the growth of shaped TiO_2 crystals by the EFG technique. The shape of the iridium die was determined according to the above results. An iridium die was mounted within a 50-ml iridium crucible which was placed in an induction heating furnace. The sizes of the die top and the slot thickness were 2 mm × 20 mm × 0.5 mm, respectively.

The crystals were grown in Ar atmosphere with a pulling rate in the range of 2–15 mm/h. The seeds were oriented toward the $z(\langle 001 \rangle)$ axis and 22.5° to the $\langle 001 \rangle$ axis.

The temperature distribution near the top of the die was controlled precisely. The temperature gradient in the vertical direction at the die top was about 150°C/cm. It was found that stability of the crystal shape depends on a corresponding horizontal temperature gradient at the die top. A view of the rutile plate single crystals grown at different temperature gradients in the horizontal direction is given in Figure 6.1.22.

With an asymmetric temperature distribution on the die top [35], the crystal also becomes asymmetric. The edge patterns of such crystals were different as shown in Figure 6.1.23.

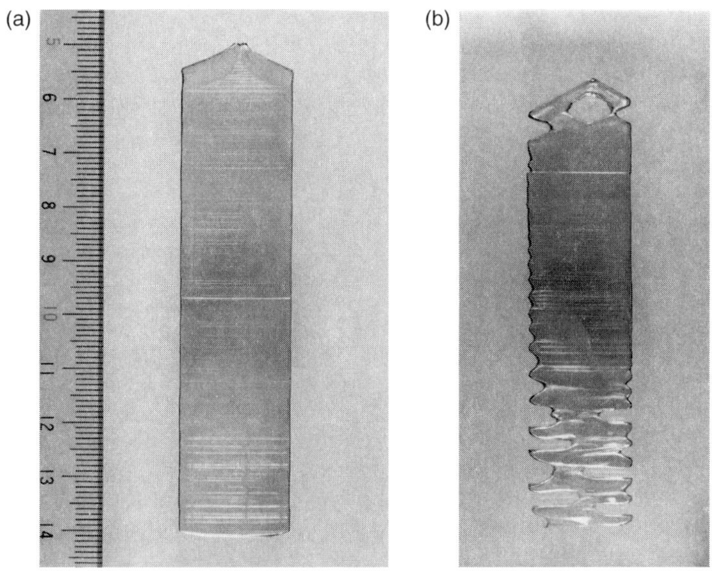

FIGURE 6.1.22 TiO_2 plate single crystals grown under 5°C/cm (a) and >5°C/cm (b) temperature gradients in the horizontal direction.

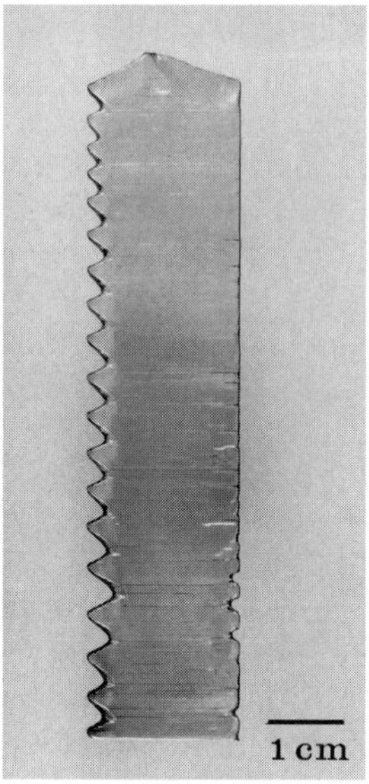

FIGURE 6.1.23 TiO_2 plate single crystals grown under an asymmetric temperature distribution along the die top.

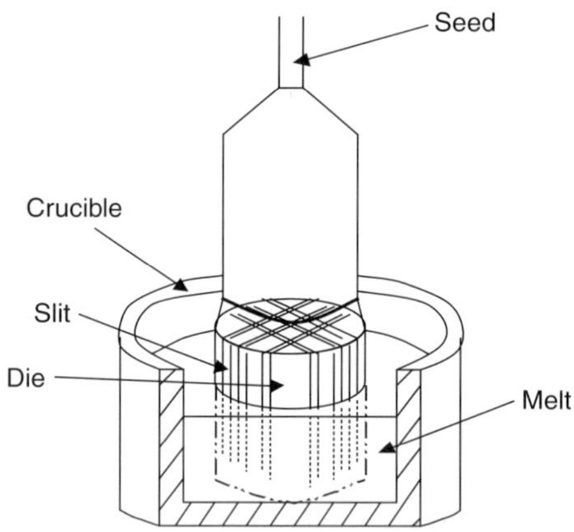

FIGURE 6.1.24 Schematic pattern of the die and crucible.

The technique described above was modified to also produce boule single crystals of TiO_2 [36, 37]. The rutile crystals were pulled from the melt on the iridium die top which was 15 mm in diameter. An arrangement of this technique is illustrated in Figure 6.1.24.

FIGURE 6.1.25 TiO_2 boule single crystal grown without seed rotation (Tokin Corporation).

The corresponding temperature distributions in the horizontal and vertical directions on the die top were 5°C/cm and 100°C/cm, respectively. A view of the crystal grown at a pulling rate of 5 mm/h is given in Figure 6.1.25.

Thus, the EFG method was found [35] to be a very attractive way to produce uniform shaped rutile crystals. The crystal growth experiments described above show that plate (ribbon) and boule single crystals can be grown by this technique.

6.1.4.2 $LiNbO_3$ Growth by Double Die EFG Method

A new crystal growth method enabling the preparation of modulation-doped crystal rods has been proposed [41]. Core-doped crystals have various advantages for optical applications. High-quality rod-like or fiber crystals doped with active laser ions only in a sharp separated inner core region have generated considerable interest. Pumping energy is absorbed only by the core part of such crystals. Therefore, the translation efficiency increases and the loss of energy by heat reduces [42].

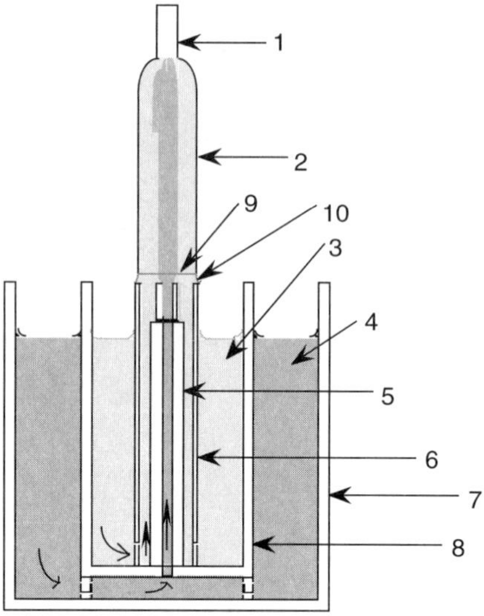

FIGURE 6.1.26 Schematic diagram of an EFG die construction used for the growth of core-doped $LiNbO_3$: (1) seed, (2) core-doped crystal, (3) undoped melt, (4) doped melt, (5) inner die, (6) outer die, (7) outer crucible, (8) inner crucible, (9) crystallization front, (10) melt meniscus.

The possibility of impurity modulation within same crystal host was shown based on growth of LN crystals core-doped with Nd^{3+} and Cr^{3+}. To obtain crystals with a modulated distribution of dopant cations, the EFG pulling technique was modified as shown in Figures 6.1.26 and 6.1.27.

The die construction was maintained in a conventional CZ system with a 30 kW RF generator used as a heat source. An outer Pt crucible with 50 mm diameter by 50 mm height and a smaller inner Pt crucible with 35 mm diameter by 45 mm height were used in the process [41, 42]. The outer crucible was filled with an Nd- and Cr-doped melt. However, the inner crucible contained the undoped melt. The melts were prepared by mixing Li_2CO_3 and Nb_2O_5 powders (99.99%) in a ratio of Li-48.6 mol.% . For each of the dopants, the starting concentration in the melt was 1 at.%.

According to Figures 6.1.26 and 6.1.27, the two contacting melts crystallized at the same rate, simultaneously leading to a core-doped single crystal formation. The crystals were grown along the $Z(c)$ axis with pulling rates in the range of 2–100 mm/h. A Pt reflector was used to decrease the axial thermal gradient over the crucible.

Figure 6.1.28 shows the cross-section of the as-grown crack-free core-doped LN crystal with a diameter of about 5 mm and a length of 60 mm.

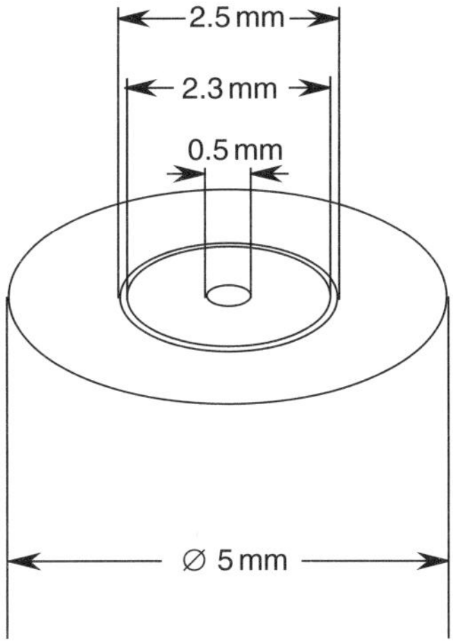

FIGURE 6.1.27 Double die construction.

FIGURE 6.1.28 Cross-section view of a Cr core-doped $LiNbO_3$ crystal (scale in mm).

Slices from the single crystalline rods with a length of about 100 mm and an outer diameter of 5 mm were cut perpendicular to the growth direction and polished to measure the dopant distribution. The composition analysis was made by the EPMA technique. The results are given in Figure 6.1.29. It was found that the concentration of Cr within the core part was considerably higher than in the periphery of the crystal. A similar core distribution tendency

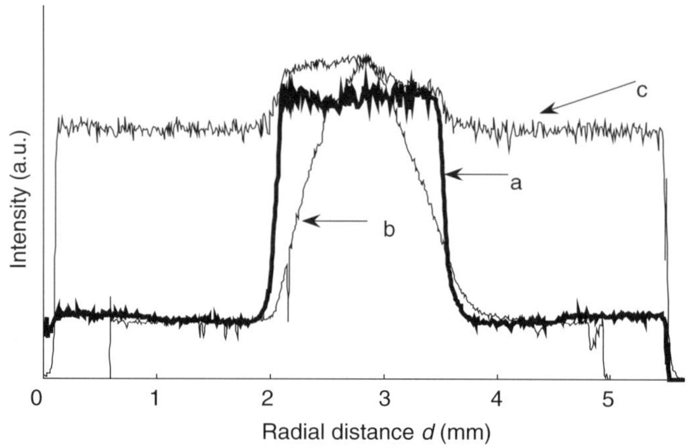

FIGURE 6.1.29 Radial distribution of Cr (a, b) and Nd (c) in core-doped $LiNbO_3$ crystals grown with a pulling rates of 4 mm/h (a, c) and 60 mm/h (b).

was observed in the case of Nd doping. However the difference between the Nd concentrations in the core and body part was not as high as supposed.

Moreover, the radial dopants distribution was reported to depend considerably on the meniscus height and the pulling rate.

It is evident that at a low meniscus height and high pulling rate, the intensity of mutual mixing between both melts decreases. Therefore, the composition profile becomes sharp. Observations indicated that if the meniscus is higher than some critical value and the crystallization rate is very low (about 2 mm/h), the dopants will be distributed uniformly along the radial direction. Therefore, a high pulling rate is preferable to obtain the crystals with a sharp concentration profile.

6.1.5 BRIDGMAN TECHNIQUE

When a crucible containing a melt of raw material to be grown is moved inside a furnace where there is a temperature gradient, the melt is crystallized uni-directionally from one end to the other. This uni-directionally crystallizing technique is called the Bridgman technique, since P. W. Bridgman was first to try this technique [43]. Figure 6.1.30 shows a schematic of the Bridgman technique. This technique is also called the Bridgman–Stockbarger technique, because D. C. Stockbarger later succeeded in growing several crystals by realizing a particularly steep temperature gradient around the crystal growth interface [44].

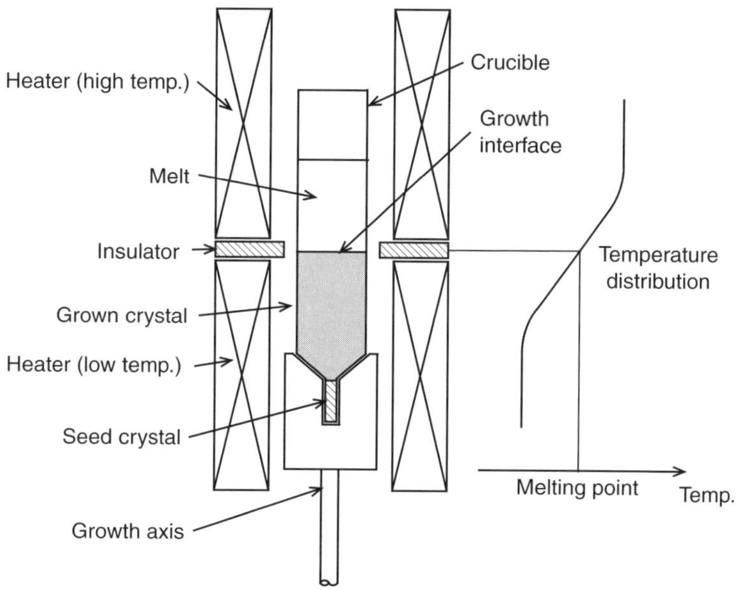

FIGURE 6.1.30 Schematic of a vertical Bridgman (VB) furnace and temperature gradient.

The technique shown in Figure 6.1.30 is called the vertical Bridgman (VB) technique since a crucible moves vertically. If uni-directional crystallization is done by changing the temperature gradient instead of moving a crucible, it is called the vertical gradient freeze (VGF) technique to distinguish it from the VB technique. Figure 6.1.31 shows a principle difference between the VB and VGF techniques. In the VB technique shown in Figure 6.1.31a, the construction of the furnace and temperature control are relatively easy, however, a mechanism is needed to move the crucible (or a furnace). Figure 6.1.31b schematically shows a case of simply lowering the temperature of the whole furnace, and this is known as the Tammann technique [45]. The VGF technique does not need a mechanism to move a crucible, however, the furnace structure and temperature control to achieve the preferable temperature gradient for crystal growth is complicated compared with that of the VB technique.

A technique to move a crucible (or a crystal growth interface) horizontally is called the horizontal Bridgman (HB) technique. The technique shown in Figure 6.1.32 is often called a boat technique, because crystal growth proceeds by moving a boat-like crucible (or a furnace) horizontally, in which the raw material is placed and melted, in a horizontal furnace with a temperature gradient. Different from the HB technique, the horizontal gradient freeze (HGF) technique changes the temperature gradient by controlling the temperature of

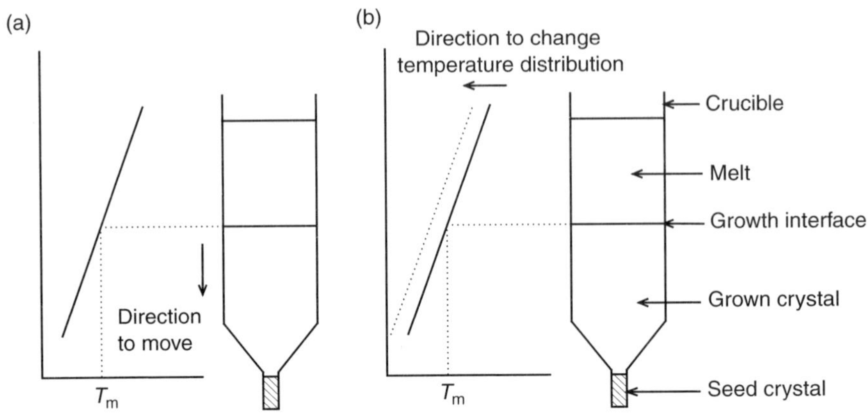

FIGURE 6.1.31 Schematic to show the difference in the (a) VB and (b) VGF techniques.

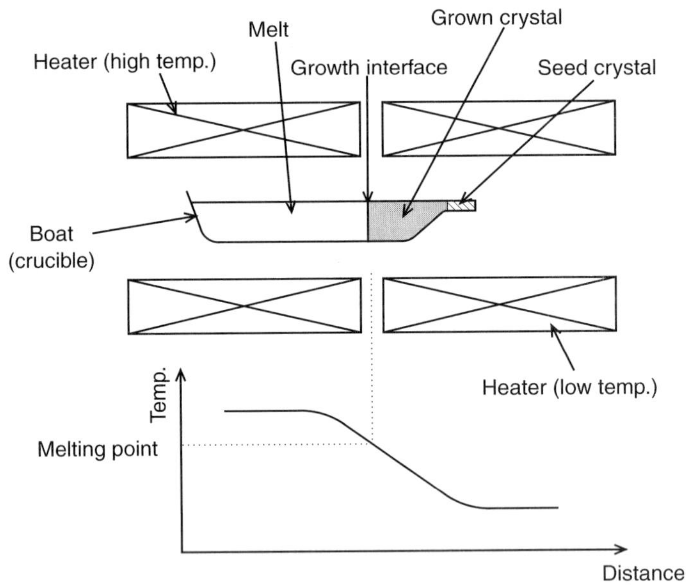

FIGURE 6.1.32 Schematic of the horizontal Bridgman (HB) furnace and temperature gradient.

a heater, maintaining a relative position between a furnace and a crucible constant. Incidentally, one HGF technique which simply lowers the temperature of the whole furnace to grow crystals horizontally is called the Kapitza technique [46]. This technique has also been used to grow several crystals.

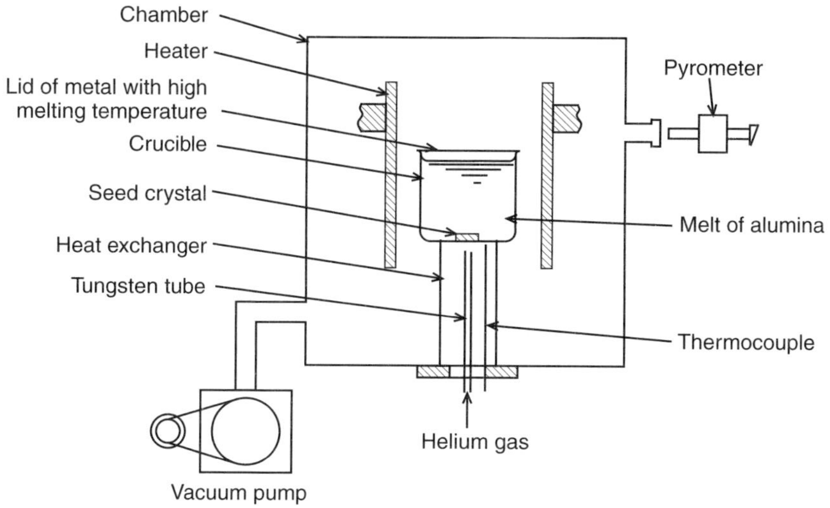

FIGURE 6.1.33 Schematic of the heat exchanger (HE) technique.

The heat exchanger (HE) technique can be called as a derivative of the VGF technique [47]. This technique was developed to grow large-size sapphire single crystals. Figure 6.1.33 shows the principle of the HE technique. The raw material is charged onto a seed crystal which is placed at the bottom of a crucible. When the raw material is melted, the bottom of the crucible corresponding to the position of a seed crystal is blown by a gas (He) for cooling in order to avoid melting the seed crystal. After seeding, the whole melt is crystallized by lowering the temperature gradually. Large sapphire crystals 300 mm in diameter have been grown. This technique has been recently applied for the growth of GaAs [48].

The crucible movement speed ideally should be as close to the deposition speed of the solid as possible. In general, it is few millimeters per hour. Different kinds of materials such as Pyrex, quartz glass, alumina, Pt, Rh, Ir and graphite are used to make crucibles. Crucible material is chosen considering the reaction with the melt, the difference of thermal expansion coefficient, and the adhering nature of the crystal to the crucible material. The shape of the crucible is also an important factor to grow better crystals. Figure 6.1.34 shows different crucible shapes to be used for the Bridgman technique. A main point is to minimize the number of nucleations. Ideally, a crystal should be grown from one nucleation. If a crystal is to be grown from a definite orientation, a seed crystal is generally fixed at the edge of the crucible.

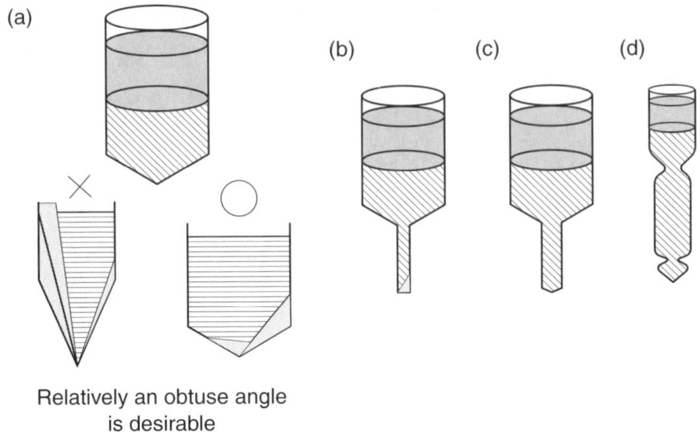

FIGURE 6.1.34 Different crucible shapes for the Bridgman technique.

6.1.5.1 $Li_2B_4O_7$ Single Crystal

Research on LBO started around 1980 in England, and, in 1981, Whatmore *et al.* reported its application for SAW devices [49]. Since then, many investigations have been done, focusing on the unique properties of LBO, such as large electro-mechanical coupling factor and zero temperature coefficient at room temperature. Recently, superior non-linear optical properties of LBO have also been reported [50]. As it can convert a wavelength of $Nd:Y_3Al_5O_{12}$ (YAG) ($\lambda = 1064$ nm) into 1/4 ($\lambda = 266$ nm) and 1/5 ($\lambda = 213$ nm), it is a promising new non-linear optical crystal to generate an ultraviolet (UV) wavelength.

LBO belongs to the tetragonal crystal system, and it is a para-dielectric piezoelectric crystal. It melts congruently with a melting temperature of 917°C. Garrett *et al.* [51] and Robertson and yong have reported its growth by the CZ technique.

Growth of LBO single crystals has several difficulties, hence, the industrial production of LBO, which requires growing large and high-quality single crystals, has been believed to be difficult. One problem is the high viscosity of its melt, which leads to bubbles forming inside crystals as a macro defect [53]. When the temperature gradient is controlled to be steep in order to reduce the formation of bubbles, grown crystals tend to crack easier because of lower thermal conductivity and anisotropy of a large thermal expansion coefficient. However, the growth of 3-in. size crystals applicable for SAW devices was

FIGURE 6.1.35 $Li_2B_4O_7$ single crystals grown by the Bridgman method (Chichibu Fuji Co., Ltd.).

reported based on the current investigations [54]. The application of the Bridgman technique was more successful in growing larger crystals with industrial quality. A recent report shows that the growth of high-quality LBO crystals with 4-in. diameter is practical [55]. Figure 6.1.35 shows LBO single crystals, 4 in. in diameter, grown by the Bridgman technique.

6.1.5.2 CaF_2 Single Crystal

Figure 6.1.36 shows a schematic illustration for the growth of CaF_2 by the VB furnace technique. The starting material is prepared from high-purity fluoride powder of CaF_2 (>99.99%). The feed material, 6.0 kg, is charged into a carbon crucible with a $\langle 111 \rangle$ oriented CaF_2 seed crystal at its bottom. The assembled crucible is sealed with a carbon lid in order to effectively prevent the volatilization of the melt during crystal growth. Vacuum treatment is performed prior to the growth. Both rotary and diffusion pumps are used to achieve a pressure of $\approx 10^{-3}$ Pa and to effectively eliminate water and oxygen traces from the growth chamber and starting material. After the furnace is heated to 1450°C, seeding is performed by adjusting the crucible to such a position that only the top of the seed is melted. The temperature gradient across the solid–liquid interface is 20–30°C/cm and the growth rate is controlled at 2.0 mm/h. Figure 6.1.37 shows an as-grown CaF_2 single crystal 6 in. in diameter.

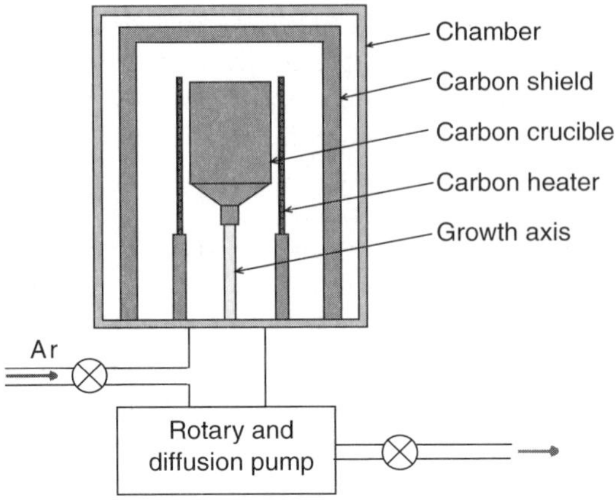

FIGURE 6.1.36 Schematic set-up for CaF_2.

FIGURE 6.1.37 CaF_2 single crystal grown by the VB technique.

6.1.6 FLOATING ZONE METHOD

The FZ method has been established as one of the single crystal growth methods. Historically, it was an application of zone refining technique reported by W. G. Pfann in 1952. The FZ method was firstly introduced by P. H. Keck around 1953 and it sets up the Si cylinder shape vertically without a boat as in the zone refining method. In a paper of Keck in 1954 [56] it was suggested that this method can be applied to various kinds of materials and is not limited to Si. Since then, the FZ method has been used for various refractory materials and has shown results in the area of physical refining, synthesizing, and growing single crystals. It was developed from the basic Keck method and, has now been expanded combining with other methods and other improvements.

Figure 6.1.38 is a schematic illustration of the FZ technique. The materials are usually put in an atmosphere completely isolated from air, and high-frequency current, an electron beam, and infrared light are used to form the floating zone. The zone is placed between two solid phases, at the top and bottom, and there is a solid–liquid–solid line-up in the vertical direction. Because the liquid phase floats on the solid phase, the name floating zone or

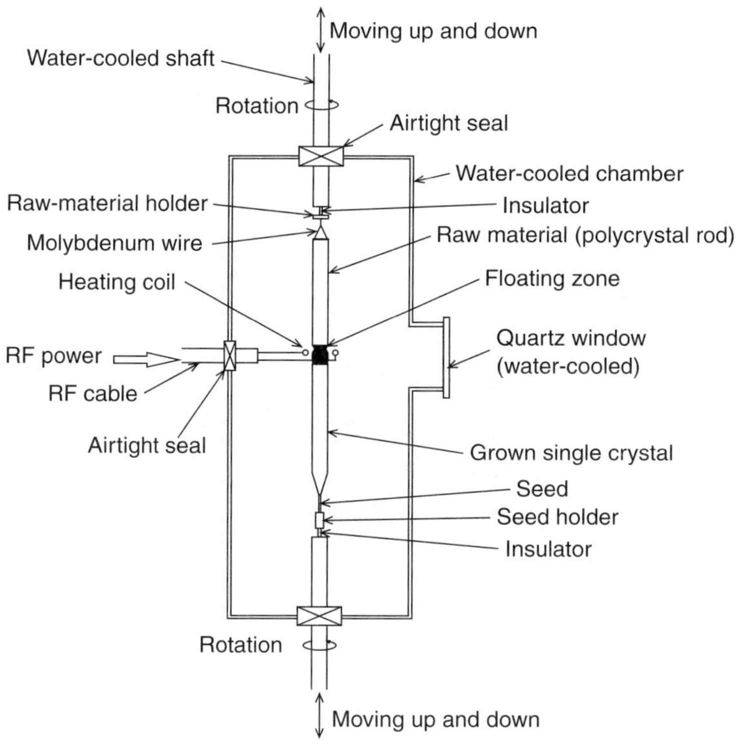

FIGURE 6.1.38 Schematic of the floating zone technique.

vertical zone was given. In the method of Keck, both up and down are acceptable movements for this floating zone, but it is necessary to specify the same direction when refining. The zone usually moves from the bottom to the top. This is because the success rate is high in this direction. Developments in the pedestal method of Dash [57] or crucible-less pulling method improved the method in moving the zone downward [58].

It is natural, for single crystals, that the form of the device is influenced by the targeted material, but, generally the following conditions apply:

1. The atmosphere must be separated from the air and a special chamber is used. In this case, vacuum, normal pressure and pressurized air or mixtured gases are used for the atmosphere. The design of the chamber, therefore, varies individually.
2. There are two mechanisms. One sets up the material and the seed in a vertical cylindrical shape and one moves and rotates them. The moving rate and the rotation rate have a range which is applicable for crystal growth of the object materials.
3. In general, high frequencies from several hundred kilohertz to several megahertz are used as energy to heat and melt. A conductor which can roll a part of a cylindrically shaped material is prepared in the insulator. The high-frequency energy is turned on to heat this conductor, and indirectly melt the material. There are methods to concentrate the electron beam to melt and to condense the arc light to form the melt in the vacuum atmosphere [59].

The above include improvements within a range where "High purity", which is a characteristic peculiar to the FZ method, is not impaired. When the device is selected and designed, a good understanding of the physical characteristics of the object material is needed. For the FZ method, the following should be considered:

1. electric conductivity or resistivity,
2. melting point,
3. Dissociation pressure of the composition at the melting point,
4. Surface tension of melt in proximity to the melting point,
5. Density (both solid phase and liquid phase),
6. thermal conductivity (in proximity to the melting point).

The above should be determined in detail, although all of them are not necessary. The chamber, the mechanism, and the heating source, etc. are decided after fully considering these points. For instance, for Si single crystals using the FZ method; the typical parameters are:

1. Electric conductivity	$10^3\ \Omega/\text{cm}$ order
2. Melting point	1418°C
3. Vapor pressure at melting point	10^{-3} Torr order
4. Surface tension	720 dyne/cm
5. Density	$\rho_l = 2.54, \rho_s = 2.33\ \text{g/cm}^3$
6. Thermal conductivity	$1.57\ \text{W/cm/cm}^3$
7. Latent heat of melting	425 cal/g

The characteristic values show that the FZ method can be easily handled and in a vacuum or inert gas, the high-frequency inducements and electron beams etc. can be used as the heat source. The Si single crystal by the FZ method can be adopted for thyristors, high-pressure diodes and power transistors. In this case, its resistivity distribution, various defects and dislocation, etc. should be controlled as well as the purity. Moreover, because the electric power needed gradually becomes large, the diameter of the crystal becomes as large as 65 and 75 mm.

Rough externals of the single crystal are shown in Figure 6.1.39, and the process to make the shape are explained in Figure 6.1.40. Compared with other materials made by the FZ method, the quantity and the quality of Si is at a highly developed industrial level. A lot of knowledge is required for the method and the devices, and many points are still not well understood.

6.1.6.1 β-Ga_2O_3 Single Crystal

β-Ga_2O_3 has been the focus of attention for gas sensor applications due to its remarkable chemical and mechanical stability at elevated temperatures, together with the change of electrical conductivity depending on the surrounding atmosphere [60]. It has been also considered for dielectric layers, surface passivation layers, as a lattice matching buffer for compounds such as GaAs and ZnO [61]. More recently, however, this material is considered as one of the most promising candidates for device applications in the UV [62, 63], since it has, at present, the shortest cutoff wavelength with a band gap of $E_g \approx 4.8$ eV [64].

For the growth and study of the basic properties of β-Ga_2O_3, the FZ technique was chosen. In order to reduce the possibility of contamination with impurities, crucible-less growth by this technique is preferable. At the same time, the composition of the crystal, from the view of deviation from stoichiometry and dopant concentration, can be controlled by the kind of gas and total pressure

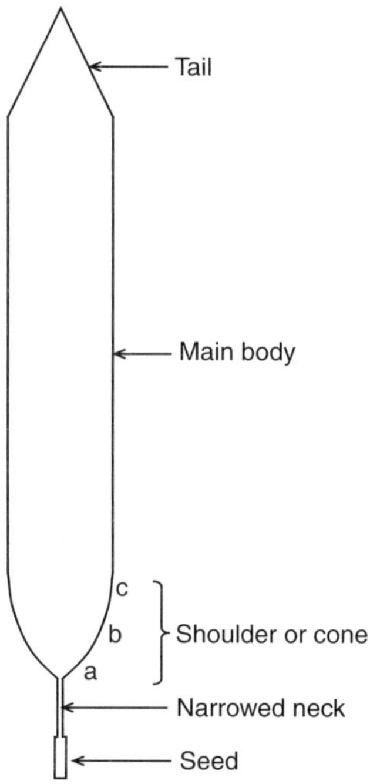

FIGURE 6.1.39 Si single crystal grown by the FZ technique.

of the growth atmosphere. There is, however, a tendency for inhomogeneous growth along the growth axis and along a cross-section. This is caused by the dynamic equilibrium with the surrounding atmosphere. Also, an increase of dopant concentration in the molten zone, caused by a relatively small segregation coefficient, enhances the dopant segregation, so that its concentration increases gradually as the crystal is growing, that is, along the growth axis.

A double ellipsoidal furnace with optical heating is used (Fig. 6.1.41). The light of two halogen lamps, 1.5 kW each, is efficiently focused in the center of the chamber after reflecting on the inside surface, which is coated by gold. The adjustable parameters for the control of the growth conditions are:

1. the growth rate,
2. the rotation speed of feed rod and seed,
3. the surrounding atmosphere, oxidative or reductive, and
4. the total pressure, 1 or 2 atm.

Growth of β-Ga_2O_3 crystals by the FZ technique consists of three steps: mixing of powders, sintering in rod-like form and crystallization in the floating zone

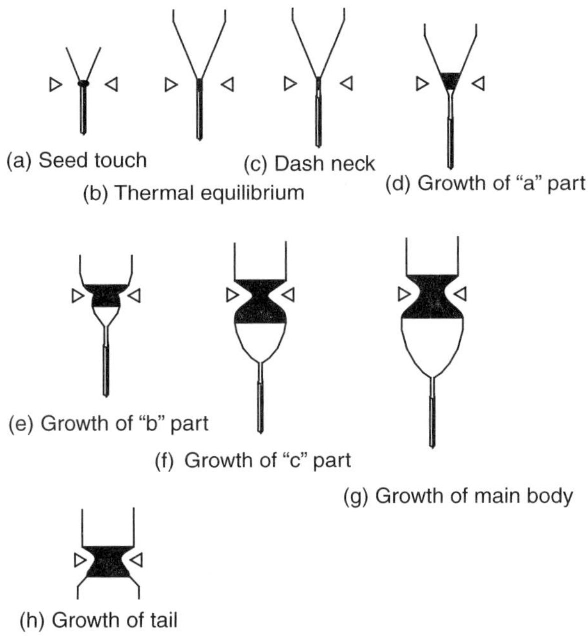

FIGURE 6.1.40 Sample of the growth process for defect-free Si single crystal by the FZ technique.

chamber. Starting materials are Ga_2O_3 powder and an oxide form of the desired dopant, in the case of doping experiments. The powders are weighed and mixed for several hours inside a polyethylene bottle by a rotary mixer, in order to obtain a homogeneous distribution. Rubber tubes are filled with the mixtures, and compressed at 500 MPa by a cold press. Subsequently, the feed rods are sintered at 1500°C for 10 h in air. Crystals have been grown with a pulling down rate of 5 mm/h, and simultaneous rotation of the feed rod and seed crystal at 20 rpm in opposite directions. The hot zone is surrounded by a quartz tube, which permits the growth under a defined atmosphere and pressure. Several gas mixtures of N_2 and O_2, ranging between 100% O_2 and 100% N_2, have been used with a constant flow rate of 500 ml/min. The total pressure is maintained at either 1 or 2 atm. A previously grown single crystal is used as a seed, with the (100) plane in the growth direction.

This technique was the most successful for the growth of high-quality single crystals of β-Ga_2O_3. Figure 6.1.42 shows as-grown β-Ga_2O_3 single crystals by the FZ technique. A controlled growth atmosphere was of vital importance for intentional modification and study of the properties of β-Ga_2O_3. According to the vaporization behavior of β-Ga_2O_3, as calculated by Lamoreaux and

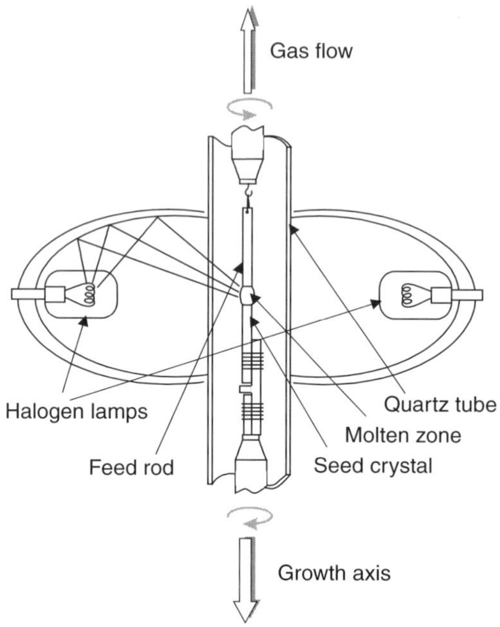

FIGURE 6.1.41 Schematic of the FZ furnace used for the growth of β-Ga_2O_3 single crystals.

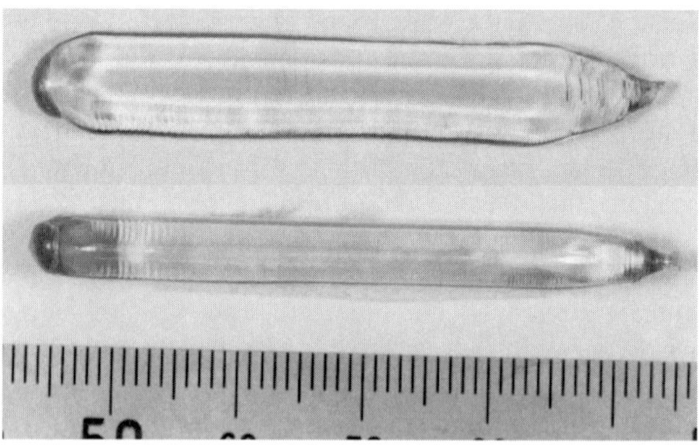

FIGURE 6.1.42 β-Ga_2O_3 single crystals grown by the FZ method.

Hildenbrand [65], the vapor pressure is reduced by several orders of magnitude by the addition of O_2 in the growth atmosphere, and the vaporization rate decreases with the increase of oxygen partial pressure. By the introduction of even a small amount of O_2 in the gas-flow, the evaporation from the floating zone could be effectively suppressed, in good agreement with Lamoreaux's estimation. The crystals had the tendency to crack, due to the complex structure and twinning affinity of β-Ga_2O_3. Further, inhomogeneities were encountered along the growth axis and perpendicular to it. These are inherent to the growth method, as it will be explained below. Crystal quality could be improved by: (i) orientation of the seed crystal; and (ii) increase of the total pressure.

By using a single crystal of β-Ga_2O_3 as a seed crystal orientated with the (100) plane in the growth axis, the crack and twinning tendency was diminished, and higher crystallinity and reproducibility was achieved. As-grown crystals exhibited two flat parallel surfaces, corresponding to the (100) cleavage plane. The sides tended to show regular striations. Crystal dimensions ranged in length of 3–5 cm, and the rectangular-like cross-section was about $5 \times 10\,mm^2$. On the other hand, it was found that the interaction between condensed phases (melt and crystal) and the surrounding atmosphere is determinant for the radial and axial inhomogeneities, since the dynamic equilibrium between each determines the diffusion processes, first from the molten zone in the surrounding atmosphere, that is, non-stochiometric evaporation, and second, from the surrounding atmosphere in the grown crystal, that is, post-growth *in situ* annealing. It was observed, that crystals grown under 1 atm tended to show a slight blue coloration gradient along the growth axis, independently of the oxygen partial pressure. This phenomenon is attributed to an *in situ* annealing, which promotes the annihilation of oxygen vacancies formed during the recrystallization process at the interface between the molten zone and the crystal. Therefore, the starting part of the crystal, which was annealed longer, is less colored or even colorless. The coloration gradient along the growth axis was clearly reduced by an increase in the total pressure, from 1 to 2 atm, when applying an oxygen concentration over 5 mol.%. This is attributed to the fact, that the total pressure as well as the oxygen concentration prevents the evaporation from the molten zone.

6.1.7 FLUX TECHNIQUE

"Flux growth" is the term most commonly used to describe the growth of crystals from molten salt solvents at high temperatures. A high temperature solvent is referred to as a flux because it permits growth to proceed at temperatures well below the melting point of the solute phase. This reduction in temperature is probably the principal advantage of flux growth over growth from the pure melt, and the reason for preferring a flux is normally one of the following:

1. The melting point of the material is too high for the available apparatus.
2. The material melts incongruently or exhibits, at some lower temperature, a phase change which leads to severe stress or fracture.

3. The material is non-stoichiometric at its melting point due to a high vapor pressure of one or more constituents.
4. Better quality crystals can be grown by this than by any other method.

Flux growth has a history of at least a 100 years and enjoyed a period of considerable activity toward the end of the nineteenth century when several attempts were made to grow crystals of gem materials, such as rubies and emeralds. This activity probably had its origins in the work of the alchemists who must have used molten salts in their attempts to make polycrystalline gold. With the development of the Verneuil furnace, interest in flux growth declined until about 1948. Renewed interest arose from the rapid advance in solid state physics, and the subject was particularly stimulated by the growth of barium titanate crystal [66], and yttrium iron garnet (YIG) [67].

In the great majority of experiments, the only satisfactory crucible material is platinum. The main disadvantages of platinum are its high cost and its softness, particularly at high temperature. Laboratory grade material, which contains a small concentration of rhodium, is preferable to pure platinum because of its greater hardness. In general, the crucible wall should be as thick as possible, since thin-walled crucibles often have a very short lifetime. The life of platinum is greater in oxidizing conditions and it is particularly important to avoid small traces of metallic lead, bismuth, iron, etc. which form low melting point eutectics with platinum. When lead-based fluxes are used, crucible life may be increased by the addition of a small quantity of PbO_2 [68]. One advantage of platinum is that it is easily welded, and sealed crucibles are frequently used for volatile fluxes.

Accurate temperature control is essential for stable growth and controllers capable of regulation to better than $\pm 0.5°C$ at the elements are available at reasonable cost from a number of manufacturers. Temperature stability at the crucible may be improved by surrounding it with ceramic material to increase the thermal capacity. A stability of $\pm 0.1°C$ at the crucible is desirable, but better regulation is often pointless because of temperature oscillation due to unstable convection.

After crystallization has taken place, it is necessary to separate the crystals from the remaining solution. The removal of the crystals is facilitated if growth occurs on a "cold finger", which may be removed from the crucible prior to cooling to room temperature. If, as in the majority of experiments, crystal growth occurs on the crucible walls, considerable time may be saved by pouring off the excess solution prior to solidification. The alternative is to cool the crucible and contents to ambient temperature and then to dissolve the solvent in some aqueous reagent in which the crystals are insoluble. This "leaching" process may take several weeks, especially if large crucibles are used.

Either method may result in stained crystals, due to thermal shock in the former case and to contraction of the solvent during solidification in the latter. Experience will indicate which method is preferable for a given system. The surface cracking which often results from hot pouring of the solution may be minimized if the crucibles are returned to the furnace as quickly as possible and cooled slowly to room temperature.

The importance of stirring in flux growth has been stressed. In practice it is not easy to find a satisfactory means of stirring a corrosive liquid at high temperatures, especially if evaporation is to be avoided. Platinum is too soft for immersed stirrers over long periods. The most satisfactory technique proposed to date is the accelerated crucible rotation technique of Scheel and Sculz-Dubois [69]. The crucible is mounted, as shown in Figure 6.1.43a, on a pedestal which is rotated at a continuously changing rate. The change in rotation rate must be smooth if nucleation is to be avoided and a typical sequence is shown in Figure 6.1.43b.

The use of stirring facilitates the restriction of nucleation by localized cooling. In Figure 6.1.43a, the central region at the base of the crucible is shown

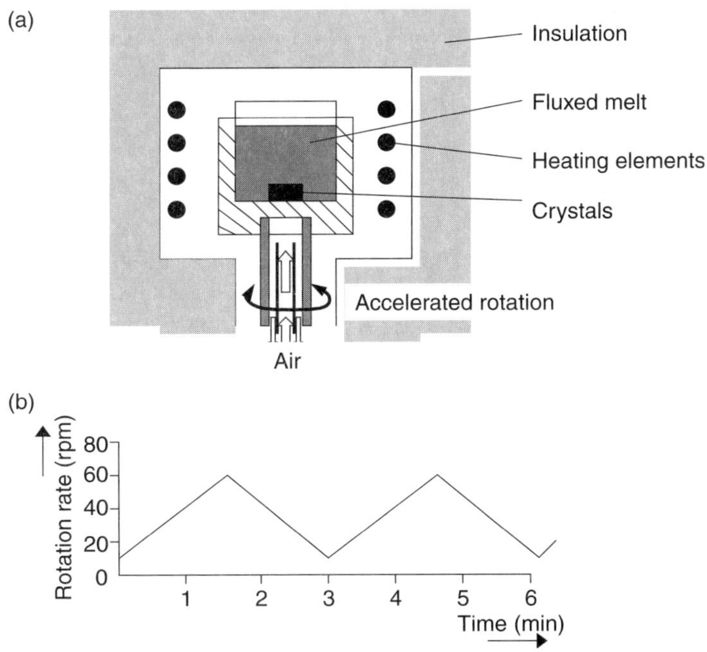

FIGURE 6.1.43 Accelerated crucible rotation technique: (a) apparatus; (b) typical variation of rotation rate.

to be cooled by a jet of air, and by this means nucleation has been restricted in a number of experiments to one or two crystals. Crystals of $GdAlO_3$ weighing over 200 g have been produced by this technique [69, 70].

6.1.7.1 Seeded Growth

Problems associated with nucleation may be eliminated by the use of a seed crystal. Seeds may be introduced by floating them on the surface of the melt, by attachment to the base of the crucible or to a platinum sinker, or by the introduction of a cooled seed holder. The main difficulty is introducing the seed when the solution is supersaturated prior to the nucleation of other crystals.

The top seeding technique, in which a seed is introduced into the top of the melt has been used with considerable success for the growth of several materials. Miller [71] grew $KNbO_3$ crystals from K_2CO_3 on a rotated seed by cooling from 1020°C at 3°C/h. Crystals up to 15 g in weight were grown in only 8 h. Von Hippel *et al.* [72] pulled $BaTiO_3$ crystals from a melt containing 50% (molar) $BaTiO_3$ and 50% TiO_2, at a rate of 6 mm/day, which is also very rapid for flux growth. In both these examples, the high growth rate is possible because of the high value of the solute concentration. Top seeded growth has also been noted for YIG [73], $Ba_2Zn_2Fe_{12}O_{22}$ [74], $Ba_xPb_{1-x}TiO_3$ solid solution [75], and GeO_2 by Goodrum [76].

6.1.7.2 Solvent Evaporation

Similar considerations of growth stability to those described for slow cooling apply when the supersaturation is produced by solvent evaporation. The main advantage of evaporation is that growth may proceed at constant temperature, so that crystals may be of more uniform composition than those obtained by slow cooling.

When doped crystals or solid solution are grown by slow cooling techniques, they will normally be inhomogeneous. For example, in the growth of crystals with $Ni_xZn_{1-x}Fe_2O_4$ compositions, if the Ni:Zn ratio in the crystal is not the same at all temperatures as that in the melt, the composition of the melt will change as growth proceeds and so this ratio will show a variation through the crystal. Growth by evaporation will yield more homogeneous crystals provided that growth occurs at a temperature where the solute composition is the same in the solid and liquid phases.

One of the disadvantages of evaporation is that a large mass of corrosive solvent is released into the atmosphere or is absorbed by the furnace brick-work. Although growth by evaporation does not require an arrangement for temperature programming, a major handicap to controlled growth is that no satisfactory means has been devised for programming of the rate of evaporation. Wood and White [77] used continuous weighing to monitor the loss in weight during crystal growth; regulation of the growth rate could be effected by changing the temperature, but then the advantage of isothermal growth is lost. Continuous regulation has not been attempted in practice, and changes in the average evaporation rate are made only by varying the size of the aperture through which evaporation occurs.

In growth by evaporation, nucleation will often occur on the surface of the melt because of the high local supersaturation created by the loss of solvent, and many highly intergrown crystals may result. However, if the temperature gradient across the crucible is arranged so that growth occurs near the crucible base, the quality of crystals grown by evaporation in PbF_2 is better [77–79].

6.1.7.3 Temperature Gradient Technique

There are two types of experiments which rely on a temperature gradient for the transport of solute from a higher to a lower temperature. In the first, the gradient is maintained across the whole fluxed melt and the solute is normally transported by convection onto a seed crystal or to grow by spontaneous nucleation. The second utilizes a thin zone of solvent, through which solute diffuse from a polycrystalline feed rod to a seed crystal.

Temperature gradient transport without the use of a seed has been applied to the growth of BeO [80] and NpO_2 [81]. The principle is illustrated in Figure 6.1.44a. Small nutrient crystals are located at the base of the crucible and are prevented by a baffle from rising into the cooler region. The solution is saturated at the highest temperature and the temperature difference across the melt, typically 5–50°C, is used to vary the supersaturation and also the rate of convective flow. Crystallization occurs at or near the surface of the crucible and will yield many intergrown crystals and the use of spontaneous nucleation is favored only for volatile fluxes which necessitate the use of sealed crucibles.

Even with volatile fluxes, seeding is possible and will normally yield crystals of improved size and quality. Reynolds and Guggenheim [82] attached seed crystals to the end of short arms on a rotating platinum rod. Crystals of cobalt ferrite 1–15 cm in the longest dimension were grown in a sodium tetraborate flux. White and Brightwell [83] used a similar seed holder (Fig. 6.1.44b) to grow large ruby crystals from PbF_2, again with the crystals well below the surface of the melt.

A combination of gradient transport with crystal pulling has been used for YIG [84–86], nickel ferrite [87], and $KTa_xNb_{1-x}O_3$ [88]. The type of apparatus

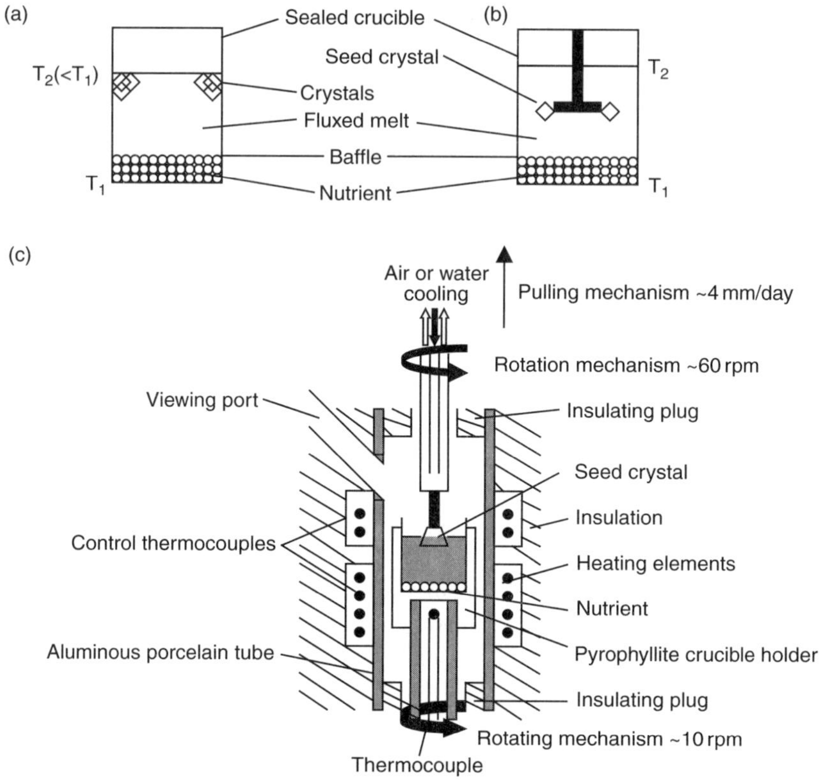

FIGURE 6.1.44 Gradient transport over the whole melt: (a) spontaneous nucleation; (b) growth on seed crystals immersed well below the melt; (c) top seeding/pulling.

used is shown diagrammatically in Figure 6.1.44c. The seed holder is rotated to increase the growth rate and to ensure a more even flow of solute across the growing face, and is withdrawn at about 0.4–4 mm/day. Crucible rotation may be used. Its main effect is to produce a more symmetrical temperature distribution in the melt. Cooling the growing crystal by a flow of air or water through the seed holder is more favorable for stable growth than the use of a submerged seed.

6.1.8 HYDROTHERMAL GROWTH TECHNIQUE

The hydrothermal technique is very important for its technological efficiency in developing bigger, purer and dislocation free single crystals. The technique is being employed to prepare many kind of materials, such as optic, magnetic, piezoelectric single crystals, and ceramics. In the hydrothermal growth method, the dissolution is carried out at the lower hotter zone (nutrient zone) and crystallization is carried out at the upper cooler zone (growth zone). The convection currents created by the temperature gradient carry the material from the nutrient zone to the growth zone. This method is used for materials with low solubility which can be increased by operating at high temperatures and consequently at high pressures, for example, ZnO, Al_2O_3 (sapphire) and SiO_2 (quartz).

The basic equipment required for hydrothermal growth of crystals are autoclaves, furnaces, thermocouples, temperature controllers and a temperature recorder. The furnaces provide the required operating temperature and temperature gradient. Nutrient material is dissolved in one part of a vertical autoclave and transported by convection to another part where the solution becomes supersaturated and deposits on a seed. For materials with a positive temperature coefficient of solubility, the nutrient is in the lower, hotter part of the autoclave and the seed in the upper, cooler part. Continuous transport of material by convection enables larger crystals to be grown than would be possible by cooling a saturated solution. Growth from an aqueous solution is typically performed in the temperature range of 300–700°C and pressure range of 500–3000 atm. The hydrothermal technique in contrast to conventional techniques offers some advantages in terms of the ability to grow:

1. Materials which decompose (e.g. VO_2) or which have a high vapor pressure near their melting points (e.g. ZnO)
2. Crystalline phases which are unstable at the melting point owing to phase transitions (e.g. α-quartz)
3. Large, good-quality crystals requiring higher perfection than that produced at the melting temperature.

There are, however, a few disadvantages too. For example, special autoclaves are needed. Unlike the CZ technique the crystal can be seen only at the end of the run, and a good quality seed of reasonable size is needed, although quality and size can be improved in successive growth runs.

Crystal growth by the hydrothermal technique has a long history, starting from Schaufhautle in 1845. He grew quartz single crystals by heating a mixture of silicic gel and water under high pressure. Many investigations were later done to develop this technique, together with the development of alloys for autoclaves which have high resistant against high temperature, high pressure and corrosion.

There has been considerable effort to industrialize this technique, consequently, mass production of quartz single crystals was enabled. The investigation for quartz single crystals continues and applications of this technique for different materials are on-going.

6.1.8.1 Quartz Single Crystal

Quartz (SiO_2) does not dissolve into most of the solution at normal temperature and pressure, however, it has a certain solubility in several aqueous solutions under high temperature and pressure. Therefore, the hydrothermal technique is most widely used for the growth of quartz single crystals.

Figure 6.1.45 shows a schematic of autoclave for the growth of quartz single crystals. Seed crystals hang at upper part of the autoclave, and natural, low-quality quartz crystals are placed at the bottom as raw material. After a certain amount of solution is poured into an autoclave, it is entirely sealed . Subsequently, when it is heated up, the expanded solution fills the autoclave and increases the inner pressure. Afterwards, maintaining the temperature of seed crystals and raw materials at 350 and 400°C, respectively, and the pressure at 1000–3000 atm, crystals continue to grow until the raw materials are used up, by the deposition on the seed crystals based on the temperature difference. Na_2CO_3 (1 N) and NaOH (0.5–1 N) are used as solutions. Figure 6.1.46 shows quartz single crystals grown this way.

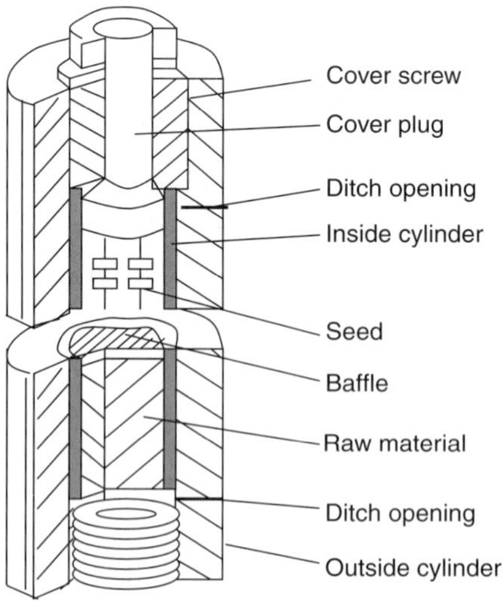

FIGURE 6.1.45 Schematic of the autoclave for the growth of quartz single crystal.

FIGURE 6.1.46 Quartz single crystals grown by the hydrothermal technique (Toyo Communication Equipment Co., Ltd.).

6.1.8.2 ZnO Single Crystal

Figure 6.1.47 shows the schematic drawing of the hydrothermal growth apparatus. Several seeds ($5 \times 5 \times 0.5\,mm^3$) suspended by a Pt wire and sintered ZnO strings as a nutrient, together with a KOH (3 M) and LiOH (1 M) aqueous solution, were put into a Pt crucible. The seed crystals and the nutrient

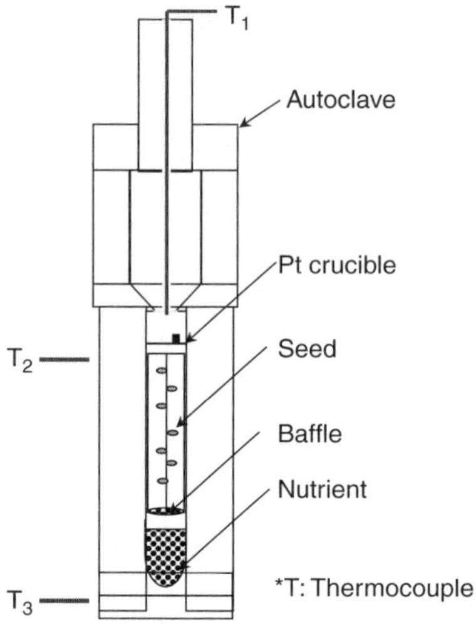

FIGURE 6.1.47 Schematic of the hydrothermal method for the growth of ZnO single crystals.

were separated by a Pt baffle. The crucible was sealed by welding and put into an autoclave. The autoclave was then put into a two-zone vertical furnace. The temperature of the autoclave was raised to about 400°C, which produced 0.1 GPa of pressure. The temperature of the seed region was set at about 10°C lower than that of the nutrient region. Growth temperature was monitored by a thermocouple (T3) inserted in the autoclave. Seed crystals grew to about 10 mm in dimension after 2 weeks. Several runs were made to obtain various single crystals (see Figure 6.1.48).

6.1.9 MICRO-PULLING DOWN TECHNIQUE

Small-diameter, long-length bulk crystals are of considerable interest for the miniaturization of multifunctional optical devices [89]. Moreover, the fiber single crystals are attractive materials for a variety of electro-optical applications because of the high surface–volume ratio and high crystal quality [90]. There are few methods which allow preparing elongated single crystals with a diameter in the range of 0.05–1.00 mm. The laser-heated pedestal method, micro-Czochralski method, and drawing down method were observed briefly in Ref. [91].

The micro-pulling down (μ-PD) technique involves growing single crystals through a micro nozzle (die) by pulling in a downward direction. A schematic illustration of the apparatus used for this process is given in Figure 6.1.49.

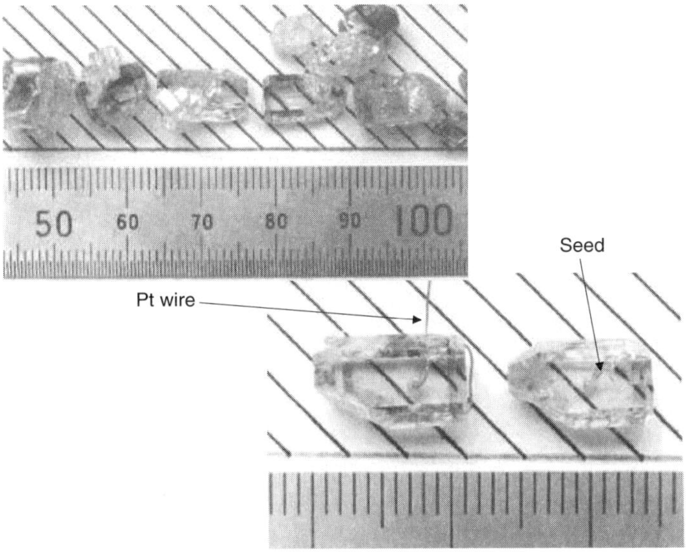

FIGURE 6.1.48 As-grown ZnO single crystals.

The growth arrangement consists of a platinum crucible heated resistively, an after-heater made from Pt wire, an annealing furnace, and a crystal lowering mechanism. High stability of the crystal growth process is ensured by high temperature gradient in the vicinity of the liquid–solid interface, which is controlled by the temperature of the crucible and the Pt after-heater. According to temperature measurements reported in Ref. [89], the axial gradient at the vicinity of crystal surface can reach 300°C/mm.

In the μ-PD technique, the solid–liquid interface is established with a narrow height meniscus (0.03–0.50 mm) and similar diameter as the crystal under the micro-nozzle. The crucible is heated by a constant current flow directly. The raw materials (mixture of oxides and corresponding salts, sintered powder, or crystal) are melted in the micro-crucible and allowed to pass through the micro-nozzle (capillary channel) of 0.3–1.2 mm diameter. The volume of the crucible is usually about 50–100 mm^3, and the amount of charged raw materials is only about 100–200 mg.

The single crystal is formed by attaching the seed to the tip of the nozzle and slowly pulling it downward. A thin platinum wire can be used instead of a seed. As a first step, polycrystalline material is crystallized on the wire, and after that, one crystallographic direction is developed by the necking process. In general the necking is not necessary in the μ-PD technique because the diameter of the crystal is small enough (0.05–1.00 mm).

6.1.9.1 $K_3Li_2Nb_5O_{15}$ AND $KNbO_3$ MICROCRYSTALS

$K_3Li_2Nb_5O_{15}$ (KLN), which has the tungsten bronze structure, is an efficient material for frequency doubling of near-infrared (Ga, Al)As diode lasers used for recording/reading data from optical disks.

Recently, the μ-PD method was developed to produce high-quality single domain KLN single crystals. A schematic of the μ-PD apparatus is shown in Figure 6.1.49. As-grown KLN microcrystals were uniform in shapes and were free from cracks, independent of the melt composition (Fig. 6.1.50). The highest growth rate resulting in crack-free single crystal was about 120 mm/h. The chemical composition in the crystal was uniform, and almost equal to that in the melt. Since incongruently melting materials such as KLN can be grown from a high growth rate, the μ-PD method is a promising growth technique.

KLN microcrystals showed strong second harmonic generation (SHG) blue light due to its superior non-linear optical properties. Figure 6.1.51 shows the blue emission obtained from KLN microcrystals. The distribution of the phase-matching wavelength along the growth axis is investigated. Deviations were about 10 nm for approximately 50 mm in length (2 nm/cm). However, SHG conversion efficiencies were quite different by the different growth rate. The

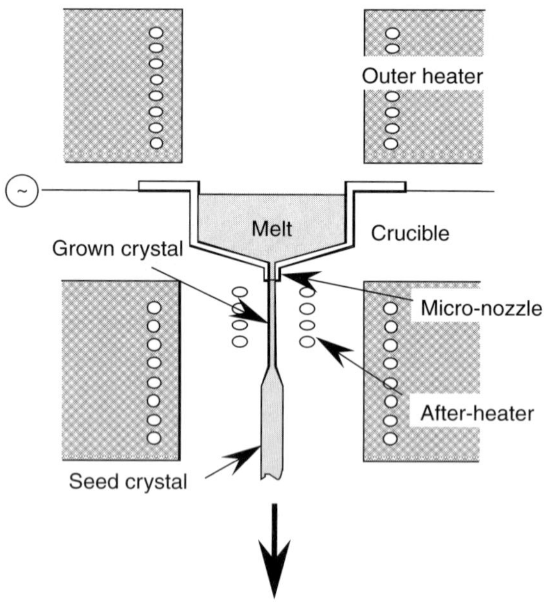

FIGURE 6.1.49 Schematic of the resistive heating-type μ-PD technique.

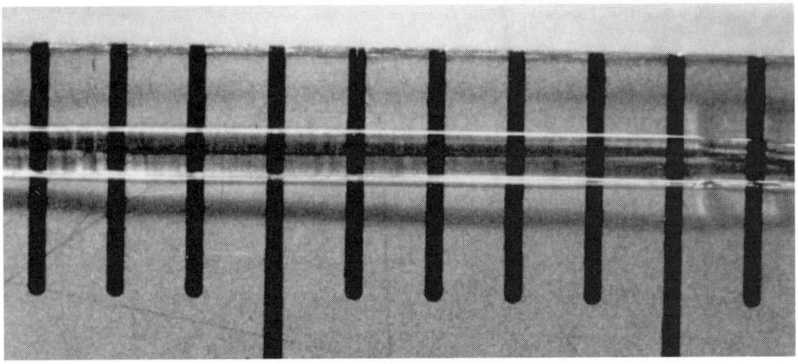

FIGURE 6.1.50 As-grown KLN microcrystal.

FIGURE 6.1.51 View of blue emission from a KLN microcrystal.

average conversion efficiency of the crystal grown at a pulling rate of 12 mm/h was 0.03%/W, while that of 80 mm/h was 0.006%/W. This may be related to the crystallinity or composition defined by the growth rate.

$KNbO_3$ (KN) is also a promising material for a SHG blue laser source, since it has a large optical non-linearity ($d_{32} = 20.5\,pm/V$). Incongruent melting and a phase transition during the cooling process are considered to be the main problems for the growth of high quality KN crystals. Figure 6.1.52 shows an as-grown KN microcrystal, grown by the μ-PD method from a melt containing

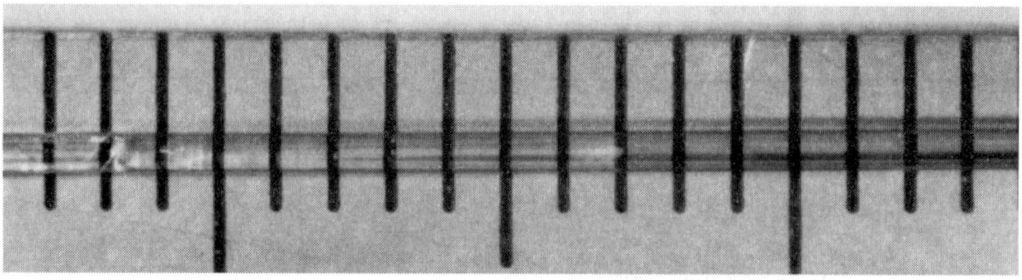

FIGURE 6.1.52 As-grown KN microcrystal.

excess K_2O as a flux. Modification of the melt composition allowed the growth of colorless, inclusion- and crack-free SHG-active single crystals with sizes up to $2 \times 2 \times 15\,mm^3$.

6.1.9.2 Micro-pulling Down Technique for High Temperature

Directionally solidified ceramic eutectics are drawing strong interest because of their high structural stability up to the melting temperature. Recently, promising results were reported for Al_2O_3/YAG [92, 93] and $Al_2O_3/GdAlO_3$ [94] systems. However, because of processing difficulties, experimental data for oxide eutectics are still limited and often uncertain.

The μ-PD method was applied for the growth of Al_2O_3/YAG eutectic fibers. The μ-PD method involves downward pulling of a crystalline fiber 0.1–2.0 mm in diameter through a capillary hole in a crucible bottom. The μ-PD method features two points important for eutectic growth: (a) a very high axial temperature gradient near the growth interface in the order of $3–5 \times 10^3\,°C/cm$. It permits high pulling rates and assures a planar interface and process stability. (b) The composition is always uniform along the fiber length, since melt convection is impossible inside the narrow capillary channel.

A high-temperature version of μ-PD employed an iridium crucible directly coupled to the RF power generator. A schematic is shown in Figure 6.1.53. A sapphire ⟨0001⟩ seed was used for the growth of the eutectic fibers. The starting materials were 4N purity Al_2O_3 and Y_2O_3 with a molar ratio of 81.3 mol.% Al_2O_3 and 18.7 mol.% Y_2O_3.

Eutectic fibers 0.25–1.00 mm in diameter and up to 500 mm in length were grown over a range of pulling rates from 0.15 to 10.00 mm/min (Fig. 6.1.54). Samples of eutectic fibers grown under various pulling rates were examined using scanning electron microscopy (SEM). Several fibers were grown with step-variable pulling rates starting from 0.15 and up to 10 mm/min. In all these

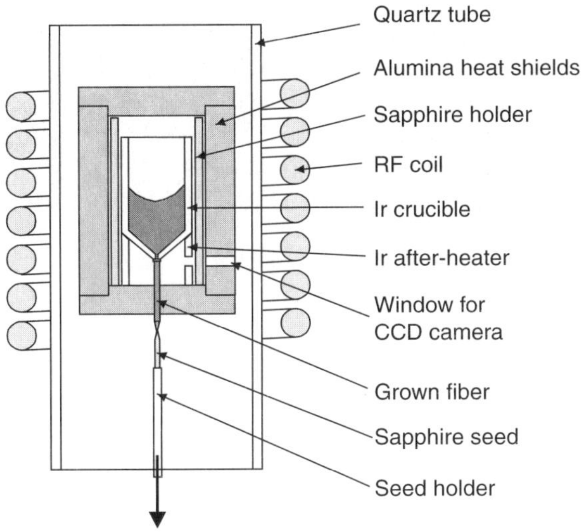

FIGURE 6.1.53 Schematic of μ-PD apparatus with RF heating system.

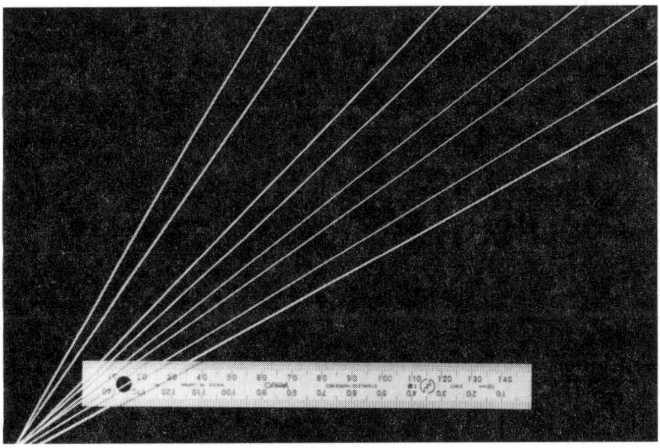

FIGURE 6.1.54 Grown eutectic fibers.

experiments, the "Chinese script" type structure was found to possess excellent reproducibility, whereas the cellular structure was not found at all (Fig. 6.1.55). The microstructure of eutectics sometimes form the structure, which resemble Chinese characters so that the microstructure is called a "Chinese-script type" microstructure. The volume fraction of YAG was 0.45 ± 0.02 for all types of structure including inter cell areas, which corresponds exactly to the theoretical

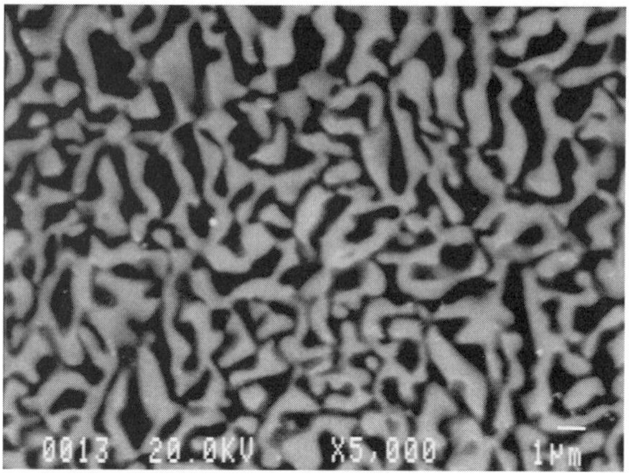

FIGURE 6.1.55 Back-scattered electron image of $Al_2O_3/Y_3Al_5O_{12}$ fibers.

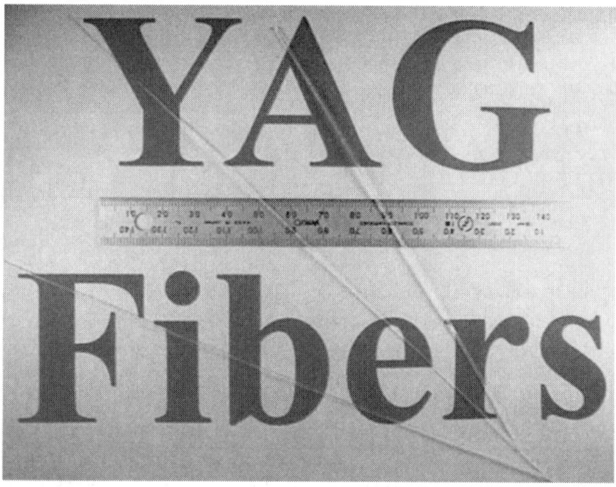

FIGURE 6.1.56 $Y_3Al_5O_{12}$ and $Nd:Y_3Al_5O_{12}$ fibers.

value for the eutectic composition (0.44). The characteristic script size was found to be very uniform for each cross-section studied. Experimentally, the generally accepted relation $\lambda - v_p^{-1/2}$, where λ is the interlaminar spacing of a conventional laminar structure and v_p is the solidification rate, also applies to the script structure of the Al_2O_3/YAG system. The proportionality constant was found to be 10, if λ has dimensions of mm and v_p, mm/s. This value is large enough to allow effective microstructure control by changing solidification rate.

The composite has been shown to have high mechanical strength up to 1800°C, excellent oxidation resistance, high thermal stability of microstructure and thermal shock stability.

Applying the same technique described above, YAG fiber crystals with 500 μm diameter have also been grown (Fig. 6.1.56). High uniformity and high transparency have been demonstrated for these fibers.

REFERENCES

1. Brice, J. C. (1973). *The Growth of Crystals from Liquid*, Amsterdam: North-Holland.
2. Fukuda, T., *et al.* (1969). *Jpn. J. Appl. Phys.* **8**: 122.
3. Fukuda, T., *et al.* (1972). *Jpn. J. Appl. Phys.* **11**: 163.
4. Yamada, T., *et al.* (1974). *J. Appl. Phys.* **45**: 5096.
5. Goodrum, J. W. (1972). *J. Cryst. Growth* **13/14**: 604.
6. Czochralski, J. (1918). *Z. Phys. Chem.* **92**: 219.
7. Nassau, K., and Broyer, A. M. (1962). *J. Appl. Phys.* **33**: 3064.
8. Cockayne, B., Chesswas, M., and Gasson, D .B. (1969). *J. Mater. Sci.* **4**: 450.
9. Arizumi, T., and Kobayashi, N. (1969). *Jpn. J. Appl. Phys.* **8**: 1091.
10. Carruthers, J. R., and Nassau, K. (1968). *J. Appl. Phys.* **39**: 5205.
11. Robertson, D. S. (1966). *Br. J. Appl. Phys.* **17**: 1047.
12. Mullin, J. B., and Hulme, K. F. (1960). *J. Phys. Chem. Solids* **17**: 1.
13. Ballman, A. A. (1965). *J. Am. Ceram. Soc.* **48**: 112.
14. Bordui, P. F., Norwood, R. G., Bird, C. D., and Calvert, G. D. (1991). *J. Cryst. Growth* **113**: 61.
15. Bordui, P. F., Norwood, R. G., Bird, C. D., and Carella, J. T. (1995). *J. Appl. Phys.* **678**: 4647.
16. Byer, R. L., Young, J. F., and Feigelson, R. S. (1970). *J. Appl. Phys.* **41**: 2320.
17. Carruthers, J. R., Pererson, G. E., and Grasso, M. (1971). *J. Appl. Phys.* **42**: 1846.
18. Fukuda, T., Matsumura, S., Yasuami, S., and Hirano, H. (1981). *Jpn. J. Appl. Phys.* **20**: 159.
19. Kozhazev, F. I., Shvartsman, S. A., Suslova, S. I., and Sokolov, V. A. (1978). *Opt. -Mekh. Prom-st.* **45**: 35.
20. Matthias, B. T., and Remeika, J. P. (1949). *Phys. Rev.* **76**: 1886.
21. Miyazawa, S., and Iwasaki, H. (1971). *J. Cryst. Growth* **10**: 276.
22. Stokowski, S. E., Randles, M. H., and Morris, R. C. (1988). *IEEE J. Quantum Electron.* **24**: 934.
23. Kumatoriya, M., and Fukuda, T. (1994). *J. Jpn. Assoc. Cryst. Growth* **21**: 144 (in Japanese).
24. Kochurikhin, V. V., Shimamura, K., and Fukuda, T. (1994). *J. Cryst. Growth* **143**: 232.
25. Kochurikhin, V. V., Shimamura, K., and Fukuda, T. (1995). *J. Cryst. Growth* **151**: 393.
26. Shimamura, K., Uda, S., Kochurikhin, V. V., Taniuchi, T., and Fukuda, T. (1996). *Jpn. J. Appl. Phys.* **35**: 1832.
27. Terada, Y., Shimamura, K., Kochurikhin, V. V., Barashov, L. V., Ivanov, M. A., and Fukuda, T. (1996). *J. Cryst. Growth* **167**: 369.
28. Chalmers, B. (1986). *J. Cryst. Growth* **82**: 70.
29. LaBelle Jr., H. E. (1971). *Mater. Res. Bull.* **6**: 581.
30. Chalmers, B., LaBelle Jr. H. E., and Mlavsky, A. I. (1972). *J. Cryst. Growth* **13/14**: 84.
31. LaBelle Jr., H. E. (1980). *J. Cryst. Growth* **50**: 8.
32. Ciszek, T. F. (1972). *Mater. Res. Bull.* **7**: 731.
33. Swartz, J. C., Surek, T., and Chalmers, B. (1975). *J. Electron. Mater.* **4**: 255.

34. Machida, H., Hoshikawa, K., and Fukuda, T. (1992). *Jpn. J. Appl. Phys.* **31**: L974.
35. Machida, H., Hoshikawa, K., and Fukuda, T. (1993). *J. Cryst. Growth* **128**: 829.
36. Machida, H., Hoshikawa, K., and Fukuda, T. (1994). *J. Cryst. Growth* **137**: 82.
37. Machida, H., Hoshikawa, K., and Fukuda, T. (1994). *J. Cryst. Growth* **137**: 516.
38. Nakamura, N., Suzuki, K., and Yazima, T. (1962). *J. Phys. Soc. Jpn.* **17**: 1806.
39. Balbashov, A. M., and Egorov, S. K. (1981). *J. Cryst. Growth* **52**: 498.
40. Hatta, K., Higuchi, M., Takahashi, J., and Kodaira, K. (1996). *J. Cryst. Growth* **163**: 279.
41. Shimamura, K., Kodama, N., and Fukuda, T. (1994). *J. Cryst. Growth* **142**: 400.
42. Rudolph, P., Shimamura, K., and Fukuda, T. (1994). *Cryst. Res. Technol.* **29**: 801.
43. Bridgman, P. W. (1925). *Proc. Am. Acad. Arts Sci.* **60**: 305.
44. Stockbargar, D. C. (1986). *Rev. Sci. Instrum.* **7**: 133.
45. Tammann, G. (1925). *Metallography*, p. 26, New York: Chemical Catalog Co.
46. Kapitza, P. (1928). *Proc. Roy. Soc.* **A119**: 358.
47. Viechnicki, D., and Schmid, F. (1974). *J. Cryst. Growth* **26**: 162.
48. Kim, T. J., Kim, H. K., Chang, S. K., Park, H. L., and Chung, C. H. (1989). *J. Cryst. Growth* **98**: 531.
49. Whatmore, R. W., Shorrrocks, N. M., O'Hara, C., Ainger, F. W., and Young, I. M. (1981). *Electron. Lett.* **17**: 11.
50. Komatsu, R., Sugawara, T., Sassa, K., Sarukura, N., Liu, Z., Izumida, S., Segawa, Y., Uda, S., Fukuda, T., and Yamanouchi, K. (1997). *Appl. Phys. Lett.* **70**: 4392.
51. Garrett, J. D., Iyer, M. N., and Greedan, J. E. (1977). *J. Cryst. Growth* **41**: 225.
52. Robertson, D. S., and Yong, Y. M. (1982). *J. Mater. Sci.* **17**: 1729.
53. Anzai, Y., Terashima, K., and Kimura, S. (1993). *J. Cryst. Growth* **134**: 235.
54. Komatsu, R., Suetsugu, T., Uda, S., and Ono, M. (1989). *Ferroelectrics* **95**: 103.
55. Fan, S., Shen, G., Wang, W., Li, J., and Le, X. (1990). *J. Cryst. Growth* **99**: 811.
56. Keck, P. H., *et al.* (1954). *Rev. Sci. Instrum.* **25**: 331.
57. Dash, W. C. (1958). *J. Appl. Phys.* **29**: 736.
58. Lenzing, J. (1975). *Solid State Technol.* 34.
59. Akashi, T., *et al.* (1969). *IEEE Trans. Magn. Magn. Mater.* **5**: 285.
60. Fleischer, M., Seth, M., Kohl, C.-D., Meixner, H. (1996). *Sensors Actuators B* **35/36**: 290.
61. Passlack, M., Schubert, E. F., Hobson, W. S., Hong, M., Moriya, N. (1995). *J. Appl. Phys.* **77**: 686.
62. Ueda, N., Hosono, H., Waseda, R., Kawazoe, H. (1997). *Appl. Phys. Lett.* **70**: 3561.
63. Orita, M., Ohta, H., Hirano, M., Hosono, H. (2000). *Appl. Phys. Lett.* **77**: 4166.
64. Tippins, H. H. (1965). *Phys. Rev. A* **140**: 316.
65. Lamoreaux, R. H., and Hildenbrand, D. L. (1987). *J. Phys. Chem. Ref. Data* **16**: 419.
66. Remeika, J. P. (1954). *J. Am. Chem. Soc.* **76**: 940.
67. Nielsen, J. W., and Dearborn, E. F. (1958). *J. Phys. Chem. Solids* **5**: 202.
68. Wanklyn, B. M. (1968). *J. Cryst. Growth* **5**: 219.
69. Scheel, H. J., and Schulz-Dubois, E. O. (1971). *J. Cryst. Growth* **8**: 304.
70. Scheel, H. J., and Elwell, D. (1972). *J. Cryst. Growth* **12**: 153.
71. Miller, C. E. (1958). *J. App. Phys.* **29**: 233.
72. Von Hippel A., *et al.* (1963). *MIT. Tech. Rep.* **178**: 44.
73. Laudise, R. A., Linares, R. C., and Dearborn, E. F. (1962). *J. App. Phys.* **33S**: 1362.
74. Aucoin, T. R., Savage, R. O., and Tauber, A. (1966). *J. App. Phys.* **37**: 2908.
75. Perry, F. W. (1967). *Crystal Growth*, p. 483, Peiser, H. S., ed., Oxford: Pergamon.
76. Goodrum, J. W. (1970). *J. Cryst. Growth* **7**: 254.
77. Wood, J. D. C., and White, E. A. D. (1968). *J. Cryst. Growth* **3/4**: 480.
78. Wang, C. C., and MacFarlane, S. H. (1968). *J. Cryst. Growth* **3/4**: 485.
79. Robertson, J. M., and Taylo, R. G. F. (1968). *J. Cryst. Growth* **2**: 171.

80. Newkirk, H. W., and Smith, D. K. (1965). *Am. Miner.* **50**: 44.
81. Finch, C. B., and Clark, G. W. (1970). *J. Cryst Growth* **6**: 245.
82. Reynolds, G. F., and Guggenheim, H. J. (1961). *J. Phys. Chem.* **65**: 1655.
83. White, E. A. D., and Brightwell, J. W. (1965). *Chem. Ind.* 1662.
84. Linares, R. C. (1968). *J. Cryst. Growth* **3/4**: 443.
85. Kestigian, M. (1967). *J. Am. Ceram Soc.* **50**: 165.
86. Bennett, G. A. (1968). *J. Cryst. Growth* **3/4**: 458.
87. Smith, S. H., and Elwell, D. (1968). *J. Cryst. Growth* **3/4**: 471.
88. Whipps, P. W. (1970). Paper presented at "Conference of British Association for Crystal Growth", Bristol.
89. Yoon, D. H., Yonenaga, I., Fukuda, T., and Ohnishi, N. (1994). *J. Cryst. Growth* **142**: 339.
90. Yoon, D. H., and Fukuda, T. (1994). *Cryst. Res. Technol.* **29**: 1119.
91. Yoon, D. H., and Fukuda, T. (1994). *J. Cryst. Growth* **144**: 201.
92. Parthasarathy, T. A., Mah, T. I., and Matson, L. E. (1993). *J. Am. Ceram. Soc.* **76**: 29.
93. Waku, Y., Ohtsubo, H., Nakagawa, N., and Kohtoku, Y. (1996). *J. Mater. Sci.* **31**: 4663.
94. Waku, Y., Nakagawa, N., Wakamoto, T., Ohtsubo, H., Shimizu, K., and Kohtoku, Y. (1997). *Nature* **389**: 49.

Handbook of Advanced Ceramics
S. Sōmiya *et al.* (Eds.)

7.1 Grinding

HITOSHI OHMORI
Institute of Physical and Chemical Research, Wako 351-0198, Japan

The grinding process in general uses grinding wheels that are rotated at high speed to grind workpieces with hard microabrasives on the grinding wheels. Grinding is mainly applied in the finishing process of precision or ultraprecision machining. Today, it has become indispensable for micrometer to nanometer level finishing, machining of hard-to-machine and brittle materials such as quenched steel and ceramics, as well as for micromachining.

7.1.1 MAIN FEATURES

The main features of grinding are as follows:

1. It can machine hard and brittle materials that cannot be machined by cutting tools. As the cutting edges of grinding wheels are made of hard abrasives such as oxides, carbides, diamond, CBN, etc., they are able to cut cemented carbide, ceramics, glass, and monocrystalline materials easily. In particular, diamond and CBN are called superabrasives, and are suitable for grinding hard-to-cut materials.
2. The required finished surface roughness and dimensional accuracy can be achieved. The microdepth of cut achieved through the development of high-precision grinding machines, small cutting edges composing the grinding wheel, and high speed rotation of the grinding wheels realize very small cutting scraps. Surface roughness and dimensional accuracy are thus better than those achieved by cutting tools.
3. Good machining efficiency: The revolution speed of the grinding wheels ranges from 600 to 4800 m/min, which is faster than cutting tools by several times to more than 10 times. For this reason, when achieving the same surface roughness as cutting, machining efficiency is excellent even though cutting scraps are microsized.
4. Self-dressing effect of cutting edges: During grinding, the abrasives on the cutting edges of the grinding wheels wear out and increase grinding

resistance, causing the abrasives to drop off and/or cracked off. Then, bond matrix is removed, and new abrasives are exposed on the new wheel surface also with chip pockets being formed at the same time.

7.1.2 GRINDING WHEEL

The grinding wheel is composed of bonding material and abrasives. Bonding materials mainly used are metal, resin, vitrified, and composite, while abrasives include oxides, carbides, diamond, and CBN, etc. These are summarized in Table 7.1.1. There are various types of grinding tool shape, mainly straight type, cup type, axial type, cutting blade type, and lapping wheel type. Figure 7.1.1 shows representative examples. The grinding wheel is composed of an abrasive layer made up of bond material and abrasives, and a wheel body. This wheel body part is mounted to the spindle of the grinding wheel via a flange, and used for grinding. Grinding wheels using metallic bond are strong and are

TABLE 7.1.1 Components of Grinding Wheel

Grinding wheel	Grain	Abrasive	Kind of abrasive
		Grain size	Mesh size of grain
	Bond	Toughness	Supporting toughness of grains
		Material	Kind of bonding material
	Pore	Matrix	Porous ratio determined by concentration of abrasives

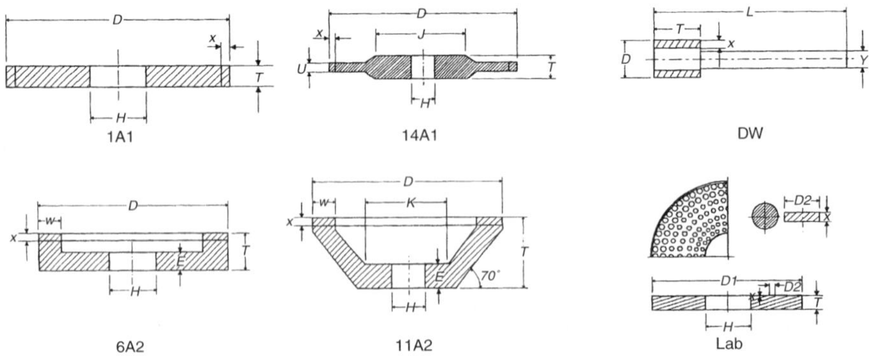

FIGURE 7.1.1 Representative examples of grinding wheel types.

FIGURE 7.1.2 Examples of sintered grinding wheels.

therefore mainly used for grinding during the forming process and for grinding hard-to-machine materials. Grinding wheels using resin bond have good elasticity. For this reason, the depth of cut of abrasives tends to be smaller than that set. This results in small surface roughness, which makes this type of grinding wheel suitable for grinding processes to obtain shiny surfaces. Vitrified bonded wheels on the other hand have high rigidity and self-dressing effect, making this type of grinding wheel indispensable for the efficient grinding of steel materials. Metal bond grinding wheels consist of several types; those that are made by sintering (Figure 7.1.2), those made by securing abrasives by electroplating, and those made by electroforming. For electroplated grinding wheels, by adopting a complex-shape wheel body, wheels with similar shapes can be made. This makes the electroplated grinding wheel suitable for forming grinding of complex shapes. Electroforming wheels are suitable for very thin cutting wheels.

7.1.3 DRESSING

Dressing is a process which removes an appropriate amount of the bonding material of the wheel to protrude the abrasives from the wheel surface to ensure good grinding performance (as well as recover grinding performance). Dressing is mainly performed by mechanical methods. The bonding material is removed by grinding soft wheels rather than the wheels used for grinding. There are two types of dressers—the grinding wheel or stick-like grindstone. Figure 7.1.3 shows dressing using a stick-like grindstone. There are also electrochemical dressing methods. One example is electrolytic dressing. As shown in

FIGURE 7.1.3 Dressing with stick-like grindstone.

FIGURE 7.1.4 Electrolytic dressing method.

Figure 7.1.4, with electrolytic dressing, the grinding wheel is taken as the positive electrode and the electrode facing the wheel work surface is taken as the negative electrode. Voltage is supplied between the two electrodes to generate electrolysis to remove the bonding material on the wheel from the reaction of the positive electrode. Since electricity must be conducted to the wheel with this method, it is limited to bonding materials with conductivity. Electrolytic dressing is mainly used for the initial dressing in ELID grinding, a precision grinding method (described later) which realizes mirror surface finishing.

7.1.4 TRUING

Truing is employed for removing burrs which generate with the spindle mounted to the grinding machine by correcting the wheel shape or for shaping the initial wheel shape into the desired shape. Truing can also be grouped into the mechanical method and electrical method. Mechanical truing is performed

FIGURE 7.1.5 Rotary truing method.

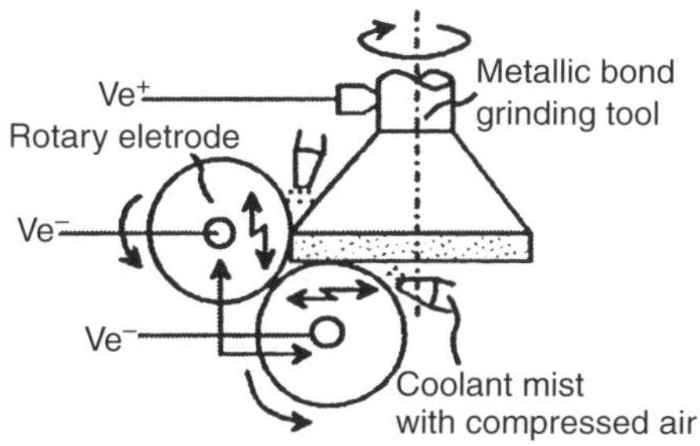

FIGURE 7.1.6 Electric discharge truing method.

FIGURE 7.1.7 Process using electric discharge truing method.

using diamond truers with single point or using rotary truers, selected according to the type, desired accuracy and shape of the grinding wheel used. Figure 7.1.5 shows the rotary truing method. One example of the electrical truing method is EDM truing. Unlike electrolysis, the bonding material is melted and sublimated by the high temperature of EDM sparks to achieve highly efficient removal. The removal efficiency, accuracy, and wheel surface conditions can be controlled by changing electrical conditions and gap. Figures 7.1.6 and 7.1.7 show the EDM truing method and the procedure.

7.1.5 GRINDING METHODS AND MACHINES

The following lists grinding methods and grinding machines available.

1. *Surface grinding* (Figs 7.1.8 and 7.1.9): Surface grinding is the process of obtaining flat surface workpieces by using a reciprocal grinding machine or rotary grinding machine mounted with straight type grinding wheels. Some of these grinding systems use cup wheels and adopt vertical grinding axis.

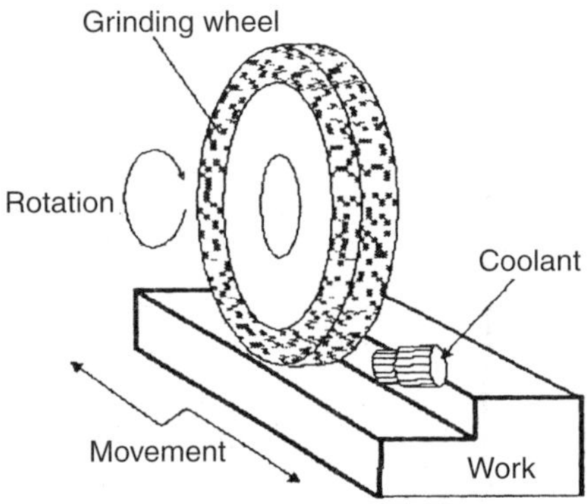

FIGURE 7.1.8 Surface grinding method.

FIGURE 7.1.9 Surface grinding machine.

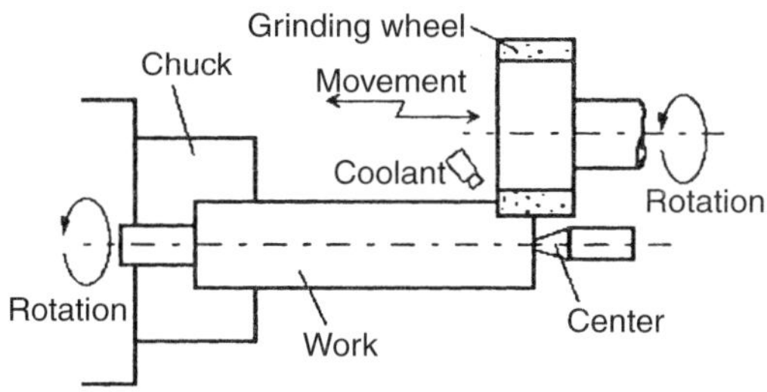

FIGURE 7.1.10 Cylindrical grinding method.

FIGURE 7.1.11 Cylindrical grinding machine.

2. *Cylindrical grinding* (Figs 7.1.10 and 7.1.11): The cylindrical grinding system is composed of an axis which holds and rotates cylindrical workpieces, and a wheel axis which grinds the outer circumference of the workpiece. The cylindrical grinding process uses the straight type or cup type wheel.
3. *Internal grinding* (Fig. 7.1.12): The internal grinding method grinds the insides of cylindrical workpieces using wheels with diameters that are smaller than the inner diameters of workpieces.
4. *Centerless grinding* (Figs 7.1.13 and 7.1.14): The centerless grinding method grinds the outside diameter of cylindrical workpieces by applying rotary drive from the frictional force with the regulating wheel, without securing the small diameter cylindrical workpiece. It can also be applied to workpieces below 100 μm.

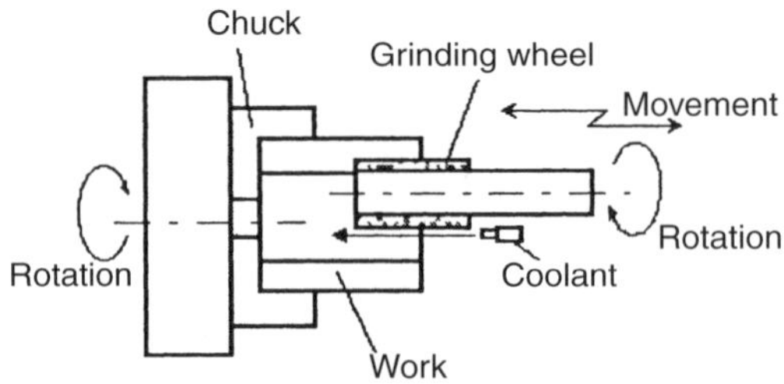

FIGURE 7.1.12 Internal grinding method.

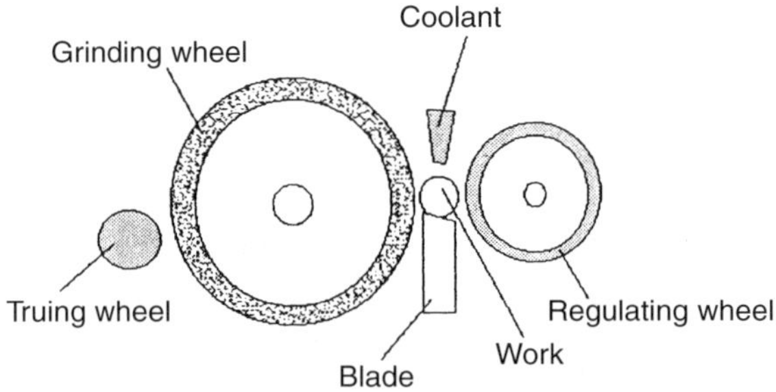

FIGURE 7.1.13 Centerless grinding method.

FIGURE 7.1.14 Centerless grinding machine.

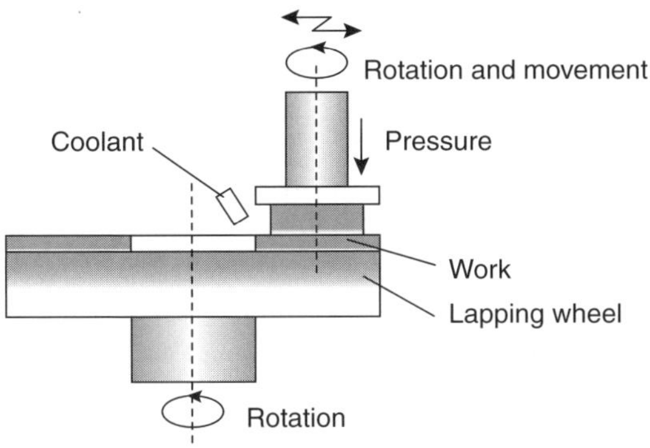

FIGURE 7.1.15 Lap grinding method.

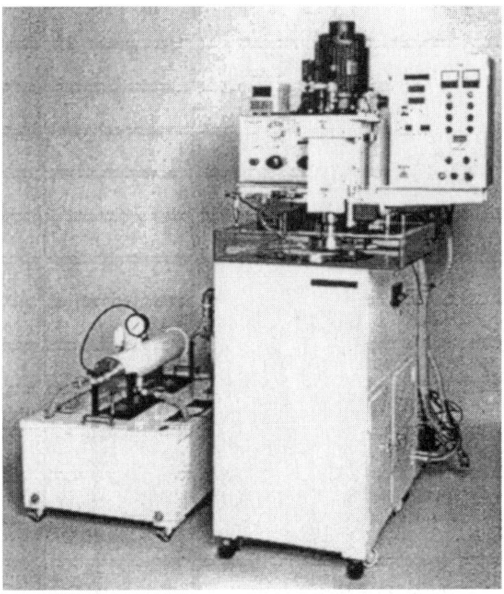

FIGURE 7.1.16 Lap grinding machine.

5. *Lap grinding* (Figs 7.1.15 and 7.1.16): Lap grinding is used to achieve flat surfaces by pressing the workpiece surface mainly with constant pressure using a lap grinding wheel.

By combining these basic grinding methods and systems or changing the axial structure, other variations can be obtained.

7.1.6 PRECISION GRINDING—MIRROR SURFACE GRINDING

Needless to say, grinding has played a major role in the precision grinding of not only machine parts, but also of optical parts and semiconductor materials. Especially in the recent years, there are increasing demands to apply diamond wheels, CBN wheels, microabrasive wheels, and precision NC grinding machine to achieve transition from conventional grinding technologies to new and higher precision technologies. ELID grinding which realizes mirror surface grinding using the above electrolytic dressing is one such example of a new precision grinding technique. ELID grinding is a grinding technique in which machining ability is maintained through the assistance of electrolytic in-process dressing. Figure 7.1.17 shows the principles of the main ELID grinding techniques. The wheel is made into the positive pole through a brush smoothly contacting the surface, and the electrode fixed below is made into the negative pole. In the clearance of approximately 0.1 mm between the two poles, electrolysis occurs through the supply of electrical current and grinding fluid. This process helps secure the protruding grains during grinding. The authors do not to intend the term "in-process" to mean "continuous", because the non-linear electrolytic phenomenon is not so much "continuous" as it is "adaptive" to the wear of the abrasives, the behavior of which is described below (Fig. 7.1.18). Figure 7.1.19 shows an example of mirror surface grinding by ELID.

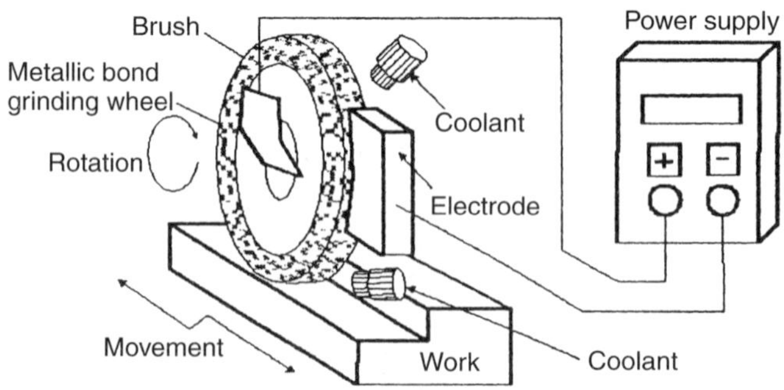

FIGURE 7.1.17 Principle of ELID grinding method.

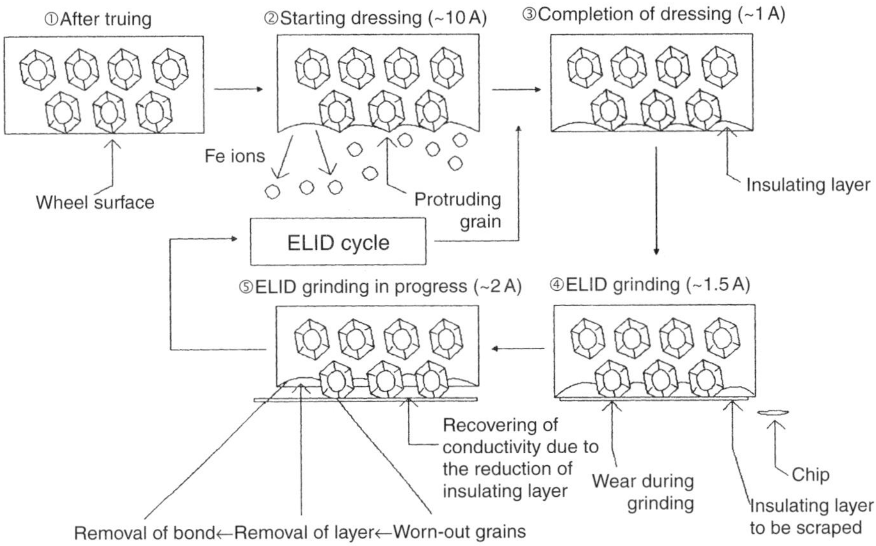

FIGURE 7.1.18 Schematic of ELID grinding mechanism.

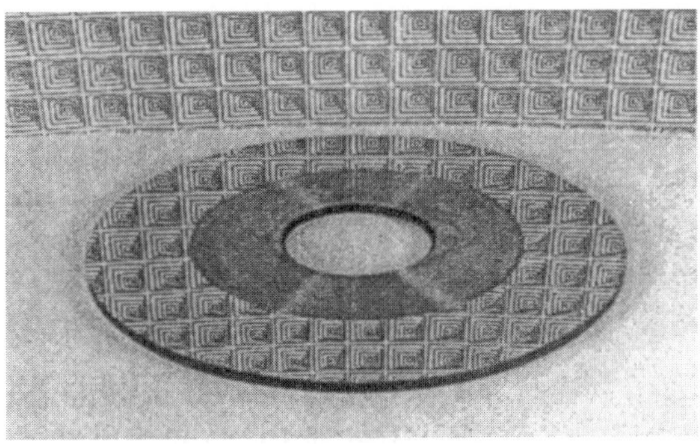

FIGURE 7.1.19 Example of ELID ground mirror-like surface.

7.1.7 ULTRAPRECISION GRINDING

Ultraprecision grinding is a grinding technique which sharply improves the profile accuracy of surfaces ground by precision grinding or mirror surface grinding. In particular, it is applied for finishing grinding processes using ultraprecision grinding machines with sharply improved driving accuracy, and can achieve submicron level grinding profile accuracy. Surfaces that can be ground—other than flat surfaces—include spherical surfaces, aspherical

TABLE 7.1.2 Main Specifications of Ultraprecision Grinding Machine

Linear axes: ultraprecision oil-hydrostatic guideway
Resolution: 10 nm
X-axis: full-stroke 1400 mm, feed-rate 10 m/min in maximum
Y-axis: full-stroke 560 mm, feed-rate 1000 mm/min in maximum
Z-axis: full-stroke 500 mm, feed-rate 1000 mm/min in maximum
Rotating axis: ultraprecision oil-hydrostatic bearing
Standard grinding wheel size: *D* 305 mm × *W* 10 mm, wheel rotation: 1800 rpm in maximum
ELID function
Open circuit voltage: 90 V, mean current: 20 A, peak current: 40 A
Duty ratio: 80% in maximum, pulse wave: square, pulse timming: 1–10 μs
NC function
Three-axes CNC unit controlable two-axes simultaneously
Capacity
Work area: 1200 mm × 500 mm, allowable weight: 1200 kg

FIGURE 7.1.20 Ultraprecision grinding machine.

surfaces, and free-form surfaces. Table 7.1.2 and Figure 7.1.20 show the specifications of the ultraprecision grinding machine and the view of the machine, respectively. Figure 7.1.21 shows the ground example produced on the machine. Control resolution below 10 nm is realized by closed loop control. With ultraprecision grinding machines with grinding spindle and hydrostatic

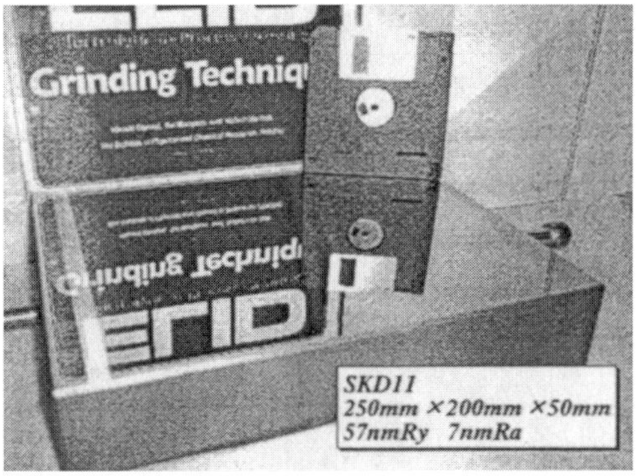

FIGURE 7.1.21 Ultraprecision ground example.

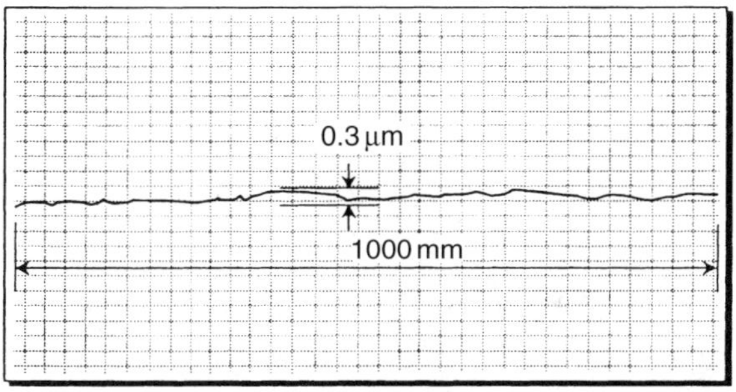

FIGURE 7.1.22 Accuracy in linearity of ultraprecision large grinding machine.

pressure guides in the table guides, very high surface accuracy of 0.3 μm in linearity can be achieved even for 1 m long grinding as shown in Figure 7.1.22. These can satisfy increasing needs for ultraprecision grinding of guides such as special optical elements and semiconductor exposure devices. In one example, we successfully achieved roughness of below 2 nm in PV by applying ELID and a special grinding wheel with 5 nm diamond abrasives. Figure 7.1.23 shows an example of measuring roughness by an atomic force microscope (AFM). As seen in this example, ultraprecision grinding will become essential for realizing low damage grinding to achieve ultraprecision and ultrasmooth features of semiconductor panels as well as effective functions.

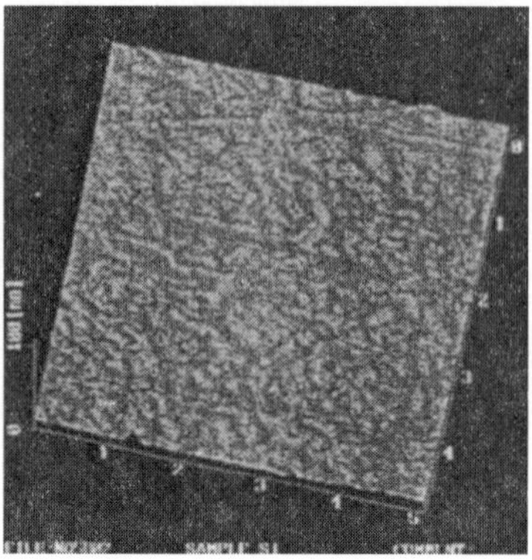

FIGURE 7.1.23 Achieved ultrasmooth ground surface.

Main body size: 580 mm × 580 mm × 580 mm

FIGURE 7.1.24 View of desktop grinding machine.

7.1.8 MICROGRINDING

Microgrinding is a method of grinding microsized workpieces. Specifically, ELID grinding is suitable for microgrinding using small workpieces, small grinding wheels and small grinding machines. Figure 7.1.24 shows a desktop grinding machine developed for small diameter cylindrical grinding. This method allows microgrinding of a micro square tool of only 50 μm on one side (Figure 7.1.25). It also enables thin and long grinding as seen in this example, as well as very thin grinding appropriate for semiconductor lamination.

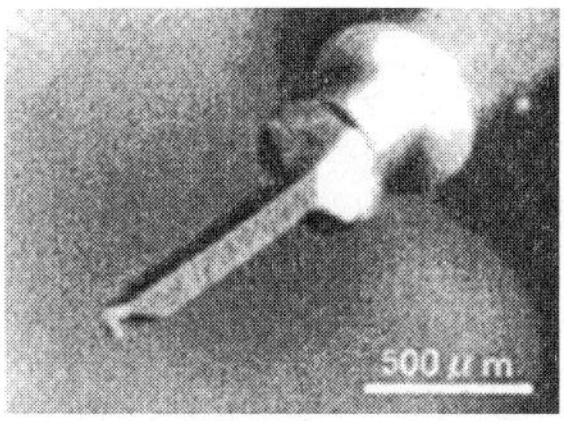

FIGURE 7.1.25 Ground microtool with 50 μm diameter.

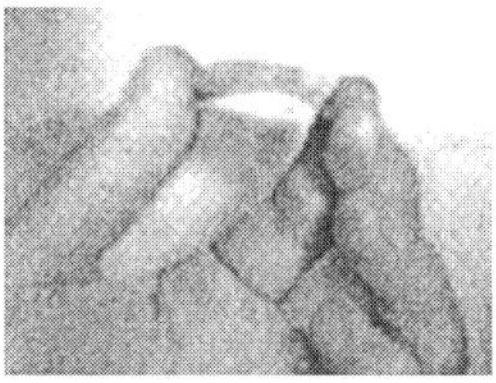

FIGURE 7.1.26 Ground silicon wafer with 30 μm thickness.

Figure 7.1.26 shows a monocrystalline silicon substrate ground to 30 μm thickness. In whichever case, the size of the ground workpiece and high quality required of the surface should be taken into consideration. Since high quality is required especially in microgrinding with a viewpoint of avoiding subsurface damage by machining on produced microcomonents, the application of ELID grinding to realize surface smoothness in the nanolevel is considered suitable.

7.1.9 EVALUATION OF GROUND SURFACE

To evaluate the ground surface, roughness measurement, flatness measurement, and profile measurement must be performed. Roughness can be measured by the contact method using a diamond stylus, non-contact method using laser, or by a method similar to AFM. Figure 7.1.27 shows a view of measurement using a diamond stylus, while Figure 7.1.28 shows a view of non-contact measurement using a laser stylus. To measure flat surfaces obtained by surface grinding, a laser interferometer as shown in Figure 7.1.29 is used. Figure 7.1.30 shows an example of flatness data obtained by the interferometer. Like roughness measurement, profile measurement can be performed by the contact probe,

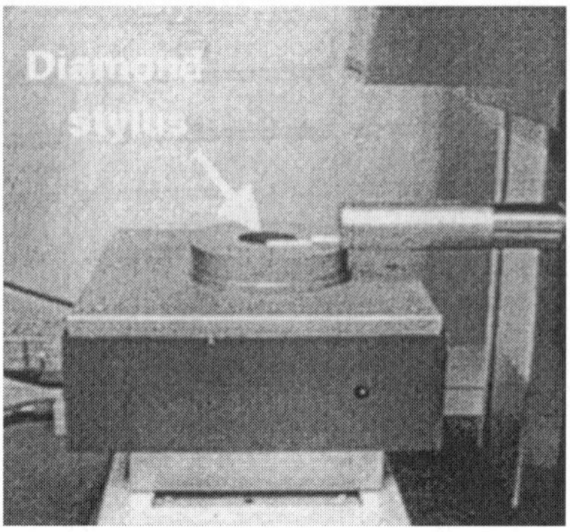

FIGURE 7.1.27 Diamond contact stylus surface measuring instrument.

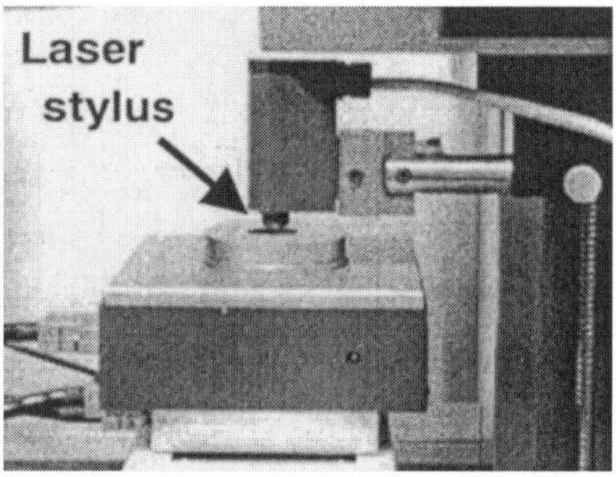

FIGURE 7.1.28 Laser stylus surface measuring instrument.

FIGURE 7.1.29 Laser interferometer.

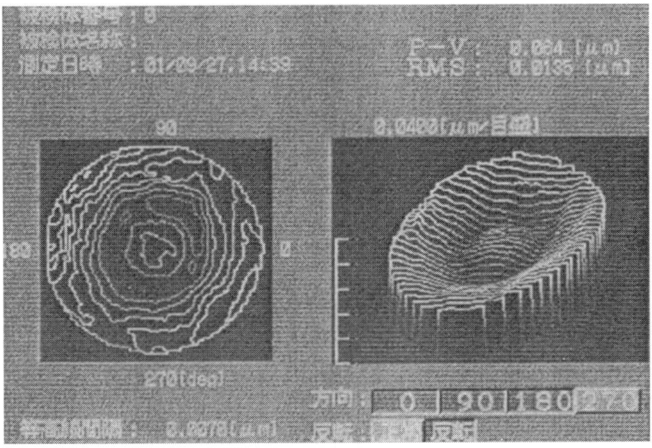

FIGURE 7.1.30 Example of flatness by laser interferometer.

FIGURE 7.1.31 On-machine contact profile measuring instrument.

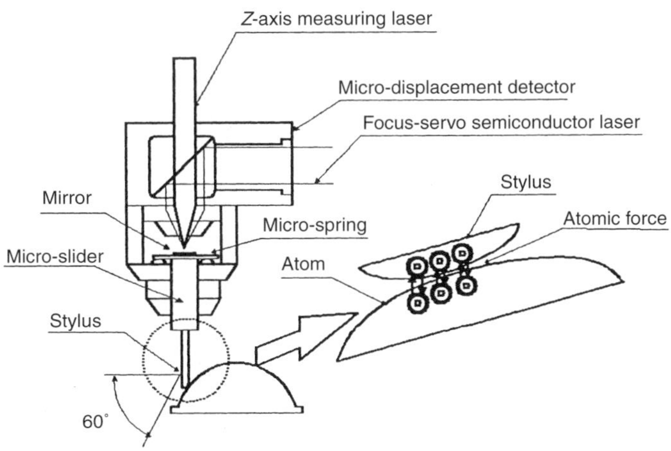

FIGURE 7.1.32 Atomic force probe profile measurement principle.

non-contact measurement using laser, and extremely low contact pressure using an atomic force probe. Figure 7.1.31 shows an example of a contact probe profile measuring unit which can be applied on the grinding machine. Figure 7.1.32 shows the principle of measurement by extremely low contact pressure using an atomic force probe.

Handbook of Advanced Ceramics
S. Sōmiya *et al.* (Eds.)

7.2 Lapping and Polishing Technology

TOSHIO KASAI
School of Engineering, Tokyo Denki University, Tokyo 101-8457, Japan

7.2.1 INTRODUCTION

Lapping, polishing and low-speed grinding occur by the sliding frictions between abrasives and work surface. Actually, the lap, polisher or abrasive stone as tool, travels across a work surface against which abrasives of slurry or stone are forced.

This processing can be applied to all solid materials for fabricating many kinds of devices, for instance, from the soft Japanese lacquer painted surface of the organic substances to hard materials such as advanced ceramics and diamonds.

The main difference between lapping, polishing and low-speed grinding is obvious in glass lens and prism manufacturing. Lapping produces a rough surface processed with coarse abrasives and a hard plate tool. In low-speed grinding, instead of lapping, coarse abrasive stones are used. On the other hand, polishing produces a mirror-like surface processed with fine abrasives and soft polishing pads.

Figure 7.2.1 shows the relationship between lapping, polishing and low-speed grinding. Each is based on a different combination of abrasives and tools. Depending on the kind of work materials, choice of abrasives and tools is not severely fixed. The appearance of finished surfaces may vary. As a matter of fact, the definition of lapping and polishing is not used with severe distinction.

In this chapter, principle of rough processing and mirror-like processing for hard and brittle materials, including ceramics, are described.

7.2.2 PROCESSING OF ROUGH SURFACES

The origin of lapping dates from tool making in the primeval age. This was inherited and developed from generation to generation to a level of mirror-polishing

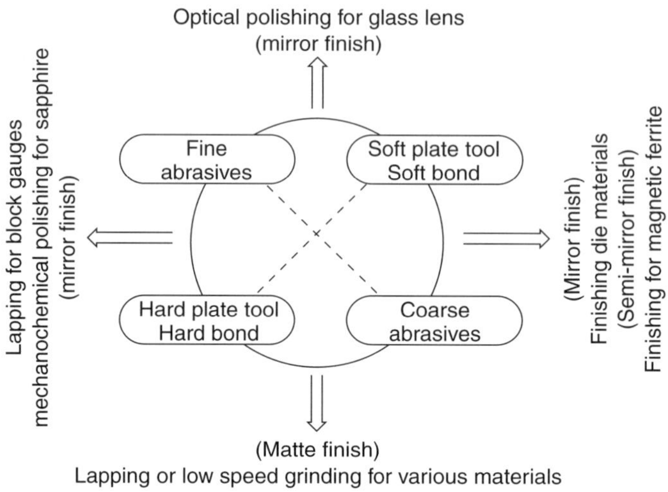

FIGURE 7.2.1 Relationship between lapping, polishing and low-speed grinding.

and advanced technology. Low-speed grinding, which used a natural abrasive stone, also has something in common with lapping.

Today, precision or ultraprecision devices with mirror-like surfaces are widely used. These are not manufactured by the mirror-polishing method alone. Prior to the polishing process, lapping or grinding is applied to obtain the surface of desired geometry and suitable surface roughness as a important technology [1].

7.2.2.1 Lapping

The main factors relating to the surface quality, accuracy and working efficiency of lapping, polishing and low-speed grinding are shown in Table 7.2.1.

7.2.2.1.1 Abrasives

In lapping, alumina or silicon carbide abrasives are usually used because of their low cost and stable supply. As another abrasives, the boron carbide abrasives and diamond powders are prepared for lapping of the hard materials. Especially, diamond powders are the hardest abrasives, but are expensive and must be carefully used. They are accordingly applied to the abrasive stone for low-speed grinding to avoid loss.

TABLE 7.2.1 Main Factors of Lapping, Polishing and Low-Speed Grinding

		Lapping	Polishing	Low-speed grinding
Finished surface		Rough	Mirror	Smooth
Abrasives	Material	Metal oxide, carbide, diamond		
	Property		Hardness, toughness, activity	
	Size	1–30 μm	<1 μm	30–100 μm
Tools	Material	Cast iron	Plastic, pitch	Metal or resin bond
	Property		Hardness, elastoplasticity, wear	
	Surface		Groove pattern, texture, shape	
Fluids			Water-type, oil-type, additives	
Machine	Type		Single-sided, Double-sided	
	Motion		Rotation, Reciprocation	
Speed (m/min)			5–30	20–20
Pressure (kPa)			1–30	10–100

Abrasives for usual lapping are classified by size and are commercially available from mesh number 300 to 8000 (approximately 50–2 μm). For the smoothing (final lapping) of glass lens and silicon wafers, alumina abrasives #1000 or #1200 are used.

The effective use of abrasive sizes is important in device fabrications. In the processing of glass lens, the mesh number of abrasives is doubled in every step to finally obtain the desired surface roughness and geometry. Such conditions are not actually used in the mass-production of glass lens since a part of the lapping process is replaced with low-speed grinding, but they become important when fabricating a special glass or crystal device. Surface irregularities and damaged layers formed during each lapping step must be removed step by step to obtain the desired accuracies of shape and size.

7.2.2.1.2 Lap

Cast iron is commonly used as a lap material because of low cost and constant accuracy on the lap surface. According to metallography, cast iron is a kind of polycrystalline metal, composed of hard pearlite iron, soft ferrite iron and free carbon. Such metal structure of lap is said to easily make fine irregularities to hold abrasives and consequently lapping advances advantageously.

As a special case, the hardened steel lap and soft plastic lap are used to satisfy the large stock removal or the small surface roughness, and glass or ceramic lap is also adopted to avoid the contamination of heavy metal ions.

7.2.2.1.3 Lapping Machines

There are three types of lapping methods, namely, hand lapping, semi-hand lapping or semi-machine lapping, and machine lapping. Hand lapping is the simplest method in which the works and immovable lap are rubbed together by the hand. In semi-hand or semi-machine lapping, the processing is advanced between the works in the hand and the rotating lap of motor drive. When all lapping operations are carried out by machine, the term machine lapping is used.

Although a part of the lapping process is replaced with the low-speed grinding in workshops producing glass lens and prisms, as mentioned above, these lapping methods have been applied to the fabrication of various devices. For instance, in semi-hand lapping, the works are manually abraded against the lap of 250 mm at 500–1000 rpm, simultaneously supplying slurry and pressing the works against the lap surface in a back and forth motion by the hand in the direction of the lap diameter or radius. Sometimes the stopping condition of lap rotation is adopted to carry out the hand lapping.

In processing aspheric lens surface, a lens lapping or polishing machine is used. This machine was improved from semi-hand lapping method. As shown in Figure 7.2.2, the work plate is pressed by a driving pin against rotating lap surface from the center of the back. The motion locus becomes uniform in work or tool surface, and the human factors in the hand lapping are suppressed.

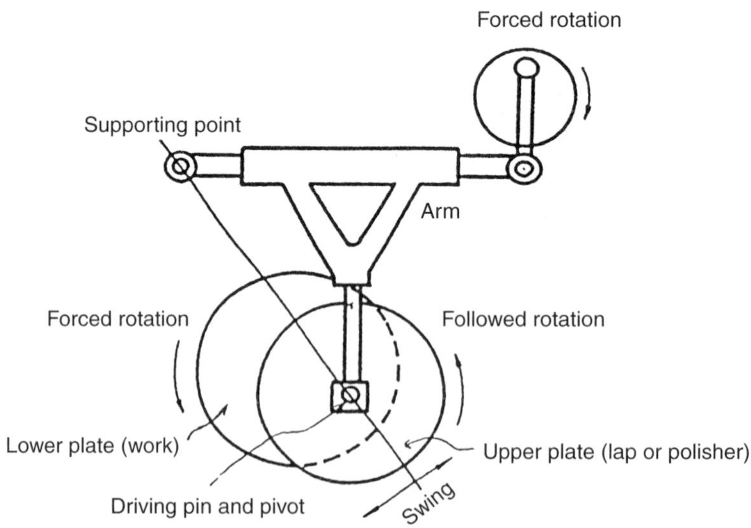

FIGURE 7.2.2 Lens lapping or polishing machine.

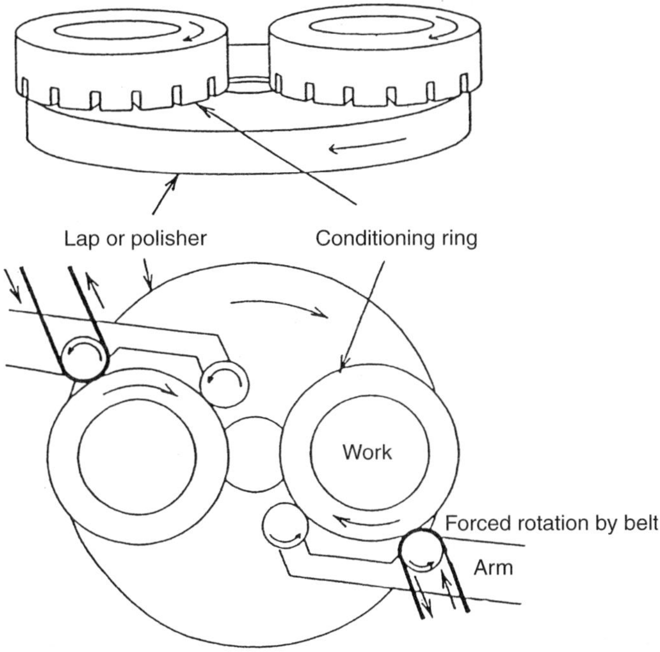

FIGURE 7.2.3 Conditioning ring-type lapping or polishing machine.

However, the subordinating work rotation to the lap rotation of motor drive is not constant, and is complex because of the reciprocating motion. The reproducibility of the high-precision processing becomes poor, though it seems that rational lapping is possible when using this machine. In other words, it is difficult to calculate the change of shape like the flatness theoretically during lapping or polishing. Therefore, special skills are required by the workers.

Conditioning ring type lapping or polishing machine as shown in Figure 7.2.3 can set the working condition with the theoretical calculation, since work and lap are in simple rotary motion. This machine has been applied to the processing of flat or long focus spherical surfaces. The adoption of conditioning ring is advantageous for obtaining accuracy. The shape accuracy of lap or polisher surface is always maintained during processing, and has yielded good results in device fabrications.

Double-sided lap or polishing machine as shown in Figure 7.2.4 is also used for simultaneous lapping of both sides the substrate type device, for example, silicon wafer, glass mask substrate, quartz oscillator substrate, ceramics plate for an electric circuit, etc. It is advantageous to process both sides of many works efficiently. However, it is necessary to correct the flatness of both lap

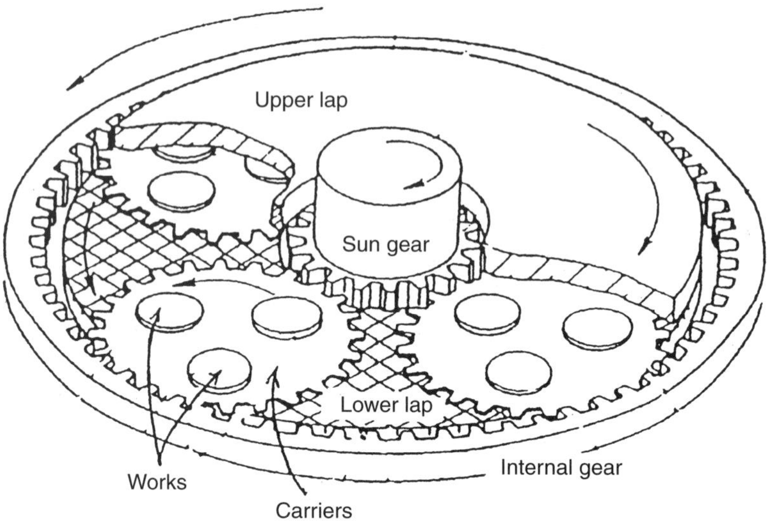

FIGURE 7.2.4 Double-sided lapping or polishing machine.

surfaces by interrupting the process, because of the undesirable lap wear and deterioration of flatness as processing advances.

In each type of machine lapping, relative speed and pressure between works and lap as a common condition are 10–50 m/min and 5–20 kPa, respectively.

7.2.2.1.4 Lapping Characteristics

Various lapping characteristics are obtained by changing the combination of work, slurry, and lap materials.

In normal lapping, the stock removal is proportional to the lapping speed, pressure and time. This relation is observed not only at stock removal but also lap wear amount, because the abrasives of free state are used. In addition, it is also applicable to polishing and common friction. The proportional constants, namely, the specific stock removals in lapping are recognized as a processing efficiency.

The result of $LiTaO_3$ single crystal lapping is shown in Figure 7.2.5. The average size of silicon carbide abrasives, processing efficiency, surface roughness and maximum depth of crack on the edge of lapped surface decrease in proportion to the decrease of the average size of silicon carbide abrasives [2].

Figure 7.2.6 shows each value in lapping with three kinds of abrasives. The processing efficiency, surface roughness and maximum depth of crack

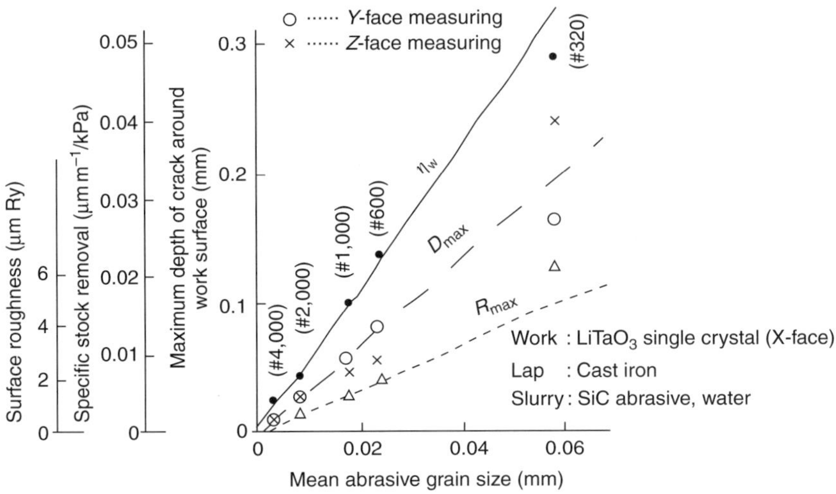

FIGURE 7.2.5 Relationship between size of abrasives, surface roughness and maximum depth of crack on edge of lapped surface.

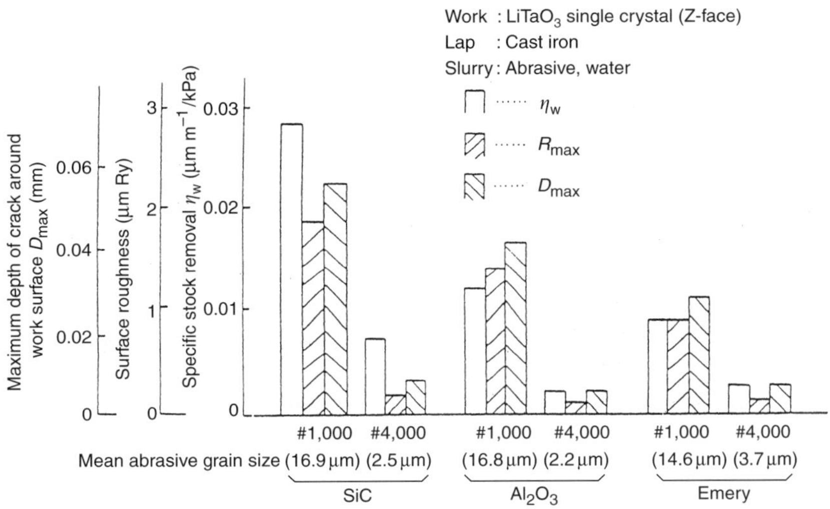

FIGURE 7.2.6 Relationship between kinds of abrasives and specific stock removal.

are shown in order of SiC > Al_2O_3 > emery, which tends to agree well with the hardness of abrasives [2, 3]. Such a relation shows the similar tendency in lapping of silicon, quartz, glass, etc. This shows evidence that the lapping advances following minute crushes by mechanical action of individual abrasive grains. However, in metal lapping, which is not accompanied by a crush, for example, aluminum alloys, copper alloys and some kinds of iron steels, such a tendency could not be observed for the different kinds of abrasives [3].

On the other hand, the cutting depth of abrasive grain is small and the crush does not occur on the work surface, when usual lapping was applied to a hard advanced ceramics. It becomes important that the hardest diamond abrasive grain is adopted in such cases. The expensive diamond abrasives are generally used as a grinding stone without using in free state.

7.2.2.2 Low-Speed Grinding

Low-speed grinding is not usual grinding but involves processing using the grinding stone, and includes horning, super finishing and sharpening the cutting edge of a blade. It has been introduced in the mass-production line of glass lens and has proved economical. In this case, various types of diamond grinding stones are applied to the smoothing of work surface prior to the mirror-polishing, instead of using free abrasives and lap.

The diamond grinding stones used in low-speed grinding, for example, are a pellet-shape of diameter 10–20 mm and thickness 3–5 mm which is made from the abrasives and bronze or resin bond material. In low-speed grinding, such pellet-shaped stones are arranged uniformly and waxed over tool plate corresponding to lap, and are used on the lapping or polishing machine under constant water or grinding fluid supply.

In glass lens manufacturing, a diamond stone of abrasive mesh number #280–320 is used instead of #1000 Al_2O_3 abrasives lapping. Considering the use of expensive diamond abrasives and economical efficiency of the processing, higher-speed and higher-pressure conditions are adopted [1].

In the processing methods using grinding stone, there is one problem that needs solving. The abrasive grain may drop out of the grinding stone surface, which can cause serious damage to the work surface, such as scratching, during processing. This requires development of a grinding stone which is able to support powerfully the abrasive grain and evolve working conditions that avoid scratching.

7.2.3 PROCESSING OF MIRROR-LIKE SURFACE

7.2.3.1 Optical Polishing

In optical polishing of glass lenses and prisms, a slurry suspended CeO_2 abrasive powders in water and a foamed polyurethane polisher having excellent wear resistance are generally used, because of the high working efficiency. The red rouge powders and pitch polisher have been used since long. However, various metal oxide abrasive powders and natural or synthetic resin polishers were examined to respond to a demand of economical efficiency and working efficiency, and the choice settled down to a combination of CeO_2 powders and polyurethane pad.

In spite of this, red rouge and pitch polisher have not completely disappeared from the special workshops engaged in production of high-precision and high-quality lenses. There is such peculiarity in the field of polishing, that an old technique of Isaac Newton' records is still left.

During polishing, the minute abrasive particles held on the polisher surface rub the work surface, and the processing advances with such an operation. The existence of chips cannot actually be confirmed there, since they dissolve in the polishing fluid.

The polishing factors related to stock removal or surface qualities are shown in Table 7.2.1.

7.2.3.1.1 Slurry and Polishing Pad

In the investigation of a suitable condition for high-quality or mirror-like surface, the first step is to select suitable abrasive powders or polishing pads. There is an example of optical polishing of $LiTaO_3$ and TeO_2 single crystals, which are a strong cleavage material and a rather soft material, respectively. The commercial abrasive powders were mixed in distilled water forming a slurry suspension, and several pitch polishers were applied. According to Nomarski microscopic and electron microscopic observation, the surface quality of 0.06 μm γ-Al_2O_3, 0.04 μm TiO_2, red rouge, ZrO_2, CeO_2, 0.3 μm α-Al_2O_3 and Cr_2O_3 deteriorated in the given order. Therefore, it is suitable for the high-quality mirror-polishing to adopt the minute abrasive powders or minutely crushable abrasive powders and the soft polisher [3].

Recently, as abrasive powders can be obtained with a particle size guaranteed by the abrasives maker, the high-quality mirror-polishing processes are carried out easily. On the other hand, even if large particles are mixed in slurry, the newly proposed bowl feed polishing method is able to avoid them. In this case, the work and polisher are submerged in the slurry, and the work is mirror-polished with the finest abrasive powders that rise up in the slurry [4].

The lapped or polished surface forms the so-called reverse surface from the tool. The form accuracy such as the flatness is also reversed from the tool surface while stock removing. When using a soft polisher, the deterioration of such accuracy often results from the wear and deformation of polisher surface. Therefore, the necessary accuracy must be always maintained on the polisher, while rough work surface is finished up to mirror-like surface. Advanced skills and significant studies are required in this area.

On the lapping or polishing machine, the need for upgradation or improvement was not felt for a long time. It looked like skilled workers were satisfied with replacing the operation by hands only with simple rotation and reciprocating motion of the machine. These have contributed to high accuracy polishing by making good use of the workers' experience and sensitivity.

However, according to the remodeled machine with extremely precise bearing, the deterioration of shape accuracy is suppressed because of the small tool wear, and float polishing was realized as a new polishing method [5]. The improvements of polishing machine are being pursue with renewed interest for making progress in ultraprecision polishing.

7.2.3.1.2 Polishing Characteristics and High-level Polishing

The relationship between polishing time, stock removal and polisher wear amount in glass polishing is shown in Figure 7.2.7. BK7 glass lapped to a mat surface was polished with CeO_2 powders and acrylic resin polisher. In the early

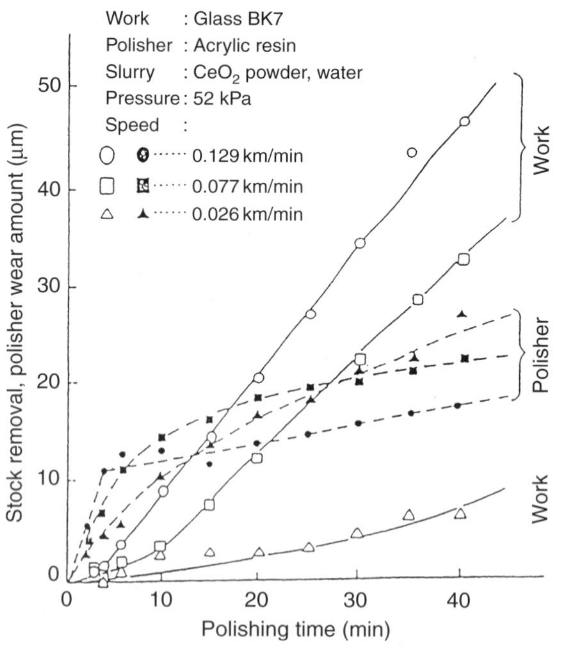

FIGURE 7.2.7 Relationship between stock removal and polisher wear in glass polishing.

stage of processing, when the work surface still remains rough, small stock removal and large wear amount of polisher are observed. However, once the surface becomes smooth, stock removal evidently increases and polisher wear amount decreases. Further, both reach a stable condition keeping proportion to the polishing time [6].

Stock removal and polisher wear amount are proportional not only to the polishing time but also to the speed and pressure of lapping. Preston's principle indicated that only stock removal during polishing with a felt polisher was proportional to the polishing speed, pressure and time [7]. Proportional constants of stock removal and tool wear indicate the specific stock removal as processing efficiency and the specific wear amount of polisher, respectively, and are important input factors for a theoretical analysis of polishing conditions [3].

When carrying out ultraprecision polishing, the following factors are necessary: (a) adoption of minute abrasive powders; (b) selection of polisher with soft surface; (c) use of ultrapure water as polishing fluid for suspension of abrasive powders and cleaning fluid of polisher, polishing jig; (d) dust-free environments.

The concept of the highest-grade mirror with optical polishing methods, namely, super smooth mirror-like surface, has been defined. Such surfaces may have a wavelike profile formed with regularly arranged atoms like the scanning tunneling microscope (STM) image of a silicon cleavage face [8]. When high-density atoms are arranged, peak and valley of the surface roughness will be one tenth or hundredth of the atomic diameter. Taking atomic terrace of crystal surface into account, it may be reasonable that the minimum surface roughness will be of atomic size.

Actual mirror-polishing advances through stock removal range from the minimum scale of an atom or molecule to the large scale of clusters of atoms. If stock removal advances along the motion track of abrasives and projecting areas of the polisher surface holding abrasives, it is possible to produce surface textures similar to the directional ridge in a cultivated field or diffraction grating. Taking the crossing of polishing motion into account, it is also possible to change it into pyramid- or corn-like arrangements. There exist invisible traces of such stock removal on polished surfaces, and there may be damaged layers from several layers to a few hundred layers under the surface. Surface quality could be improved not only by using fine abrasives but also by making the surface texture finer.

In conventional pitch polishing of glass lens, the surface roughness has remained at 5 nm Rz for a long time since it was difficult to remove the dust and large particles sufficiently. However, surface roughness of 0.1–0.3 nm Rz is obtained on optical glass when using 6 nm SiO_2 abrasive powders and a coated polisher [8]. As a matter of course, it is an advanced polishing with a clean booth and pure water equipments. The coated pitch polisher was prepared on a metal plate that was previously made to the desired shape with a 1 mm × 1 mm (width × depth) groove on 4 mm patterned grids and flatness. During the cooling process of the metal plate, which was heated to the melting point of the pitch, the thickness of the polisher becomes uniform across the surface with the aid of the pitch surface tension [9].

7.2.3.2 Chemical–Mechanical Polishing of Silicon Wafers

In the initial stages of mirror-polishing of silicon wafer, a conventional polishing technology used for the glass lens and quartz crystal oscillator plates were necessarily applied. However, it was proved to form inferior surface roughness, pull scratch and various defects including invisible damaged layer on wafers. As mentioned earlier, these defects are caused by the undesired mechanical action of the abrasive grain and the polisher. Wet-type mechanochemical polishing was newly proposed in order to satisfy LSI device fabrications and has been widely applied.

The manufacturing process is shown in Figure 7.2.8. Prior to mechanochemical polishing of the silicon wafer, the effectual deep-etching is introduced. It is very important as a pre-treatment for perfectly removing the surface defect and decreasing the surface roughness formed during lapping. After simultaneous polishing in two steps for finishing to mirror-like and defect-free surface, the final mechanochemical polishing for contamination- and haze-free silicon wafers is carried out.

In mechanochemical polishing of silicon wafers, weak alkaline slurry of colloidal SiO_2 particles of 0.01 μm abrasive grain size and soft foamed polyurethane pad are used [10]. The slurry layer is formed by the hydroplane phenomenon in the polishing motion between the wafer and the polisher. In order to thin the slurry layer, the polishing pressure is applied. On the wafer surface, formation and removal of a hydration film by slurry or soft polishing pad surface are repeated. The damage-free polishing for a final stage, as is shown in Figure 7.2.9, can be realized by keeping mechanical action less than such hydration layer thickness.

Practical use of material removing mechanism as mentioned above indicates good results in the polishing of not only semiconductor materials like silicon and germanium but also of III–V group compound semiconductor wafers such as GaAs and InP or II–VI group compound semiconductor wafers such as CdTe and ZnSe. When thin sodium hypochlorite (Antiformin, NaClO) solution and sodium bromite ($NaBrO_2$) solution are used, a reaction product film is formed and wiped off by abrasive grain or polisher surface, and the high-quality mirror-like surface is obtained.

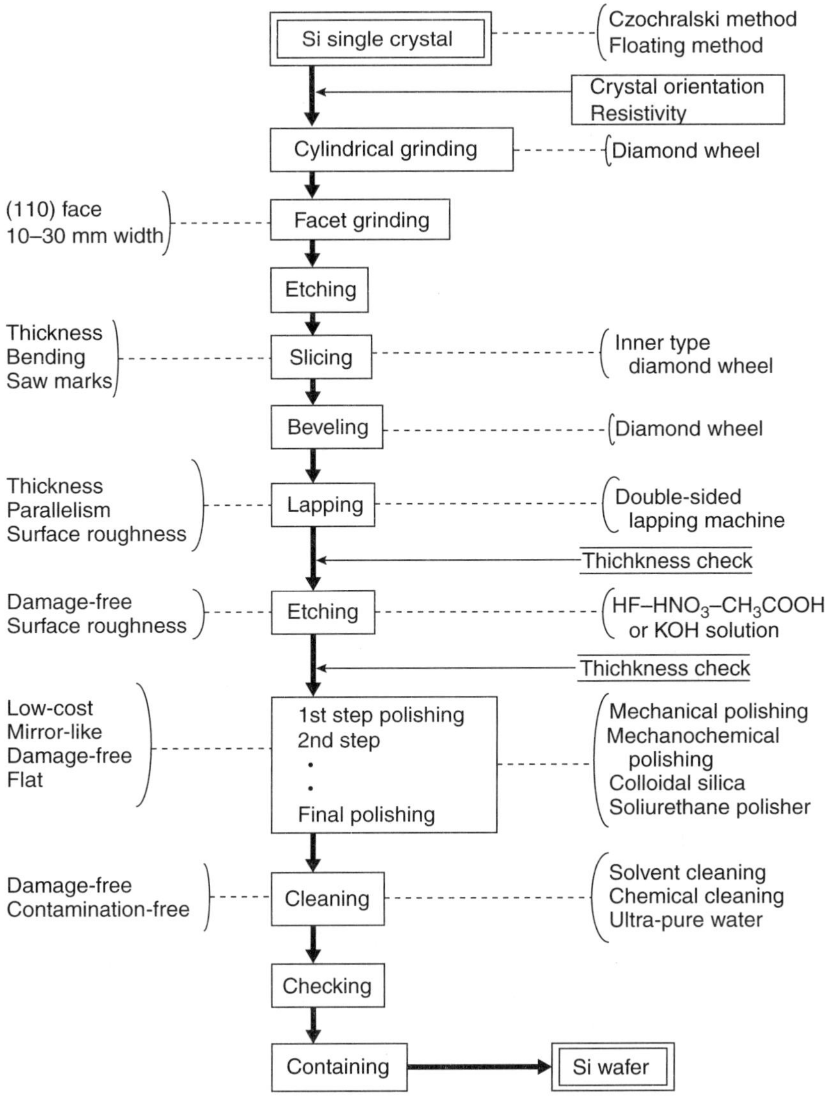

FIGURE 7.2.8 Manufacturing process of silicon wafer.

7.2.3.3 Ultraprecision Polishing

The process of mechanochemical polishing of silicon wafers has been active for upgrading various polishing methods, and the classification and analysis of

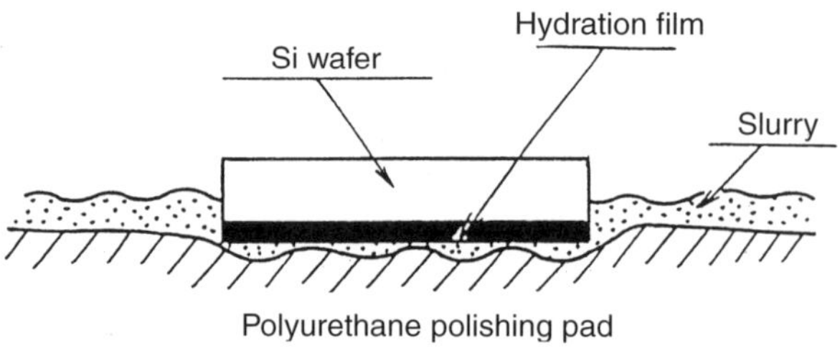

FIGURE 7.2.9 Mechanochemical polishing model of silicon wafer.

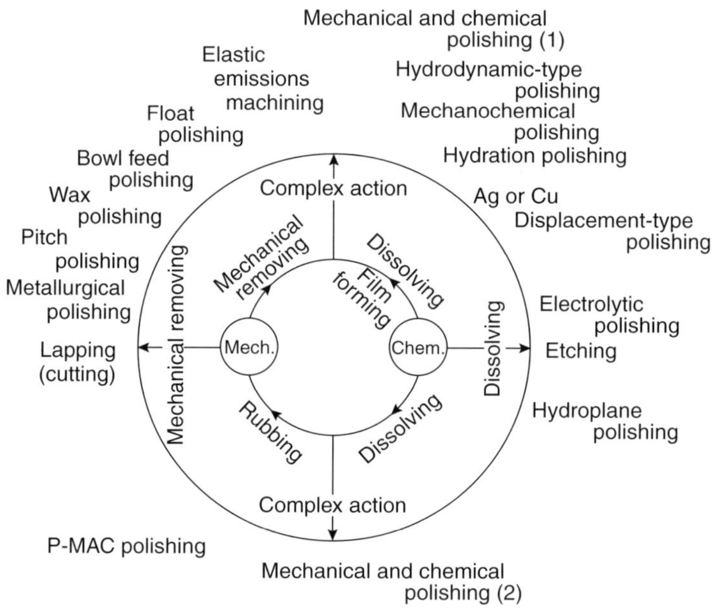

FIGURE 7.2.10 Classification of polishing methods having various complex actions.

material-removing mechanisms and methods have become the starting point for the proposed new ultraprecision polishing.

Figure 7.2.10 illustrates the material-removing mechanism in polishing. As mechanical and chemical (electric) actions, there are: (a) mechanical removal by abrasion; (b) friction to disorder the atomic arrangement; (c) etching and dissolving; and (d) formation of a thin film. The combination of these actions at different ratios constitute various polishing methods listed outside the circle shown in Figure 7.2.10. In glass polishing and mechanochemical polishing of silicon wafers, work surfaces are covered with a soft film by chemical reactions.

TABLE 7.2.2 Advanced Polishing, Conventional Lapping and Polishing

		Property of processing	Application
	(Improvement of lapping)		
	Ultraprecision low-speed grinding	Mechanically stock removing with abrasive stone and oil- or water-type solution	Block gauge End of scale Plate
	Ultraprecision lapping	With fine abrasives and metal or ceramics lap, and filtered oil- or water-type solution	
	Conventional lapping Conventional polishing	With alumina or SiC abrasives and cast iron lap, or cerium oxide powders and pitch or plastic polisher, and lens lapping or polishing machine and conditioning ring-type lapping machine	Glass lens or prism
	(Improvement of optical polishing)		Quarts plate Laser rod Opto-electronic devices X-ray mirror Large telescope mirror
Mechanical	Ultraprecision polishing [3, 8, 13]	Suppressing mechanical action and dust influence by using very fine abrasive powders, soft polisher and (ultra) pure water	
	Bowl feed polishing [4]	Submerging work and polish er in slurry to suppress mechanical action or impact and temperature rise	
	Computer aided polishing [14, 15]	Traveling polisher across on work surface by using computer	
	(Using flow effect of slurry)		X-ray mirror Optical grade dies Glass plate Ferrite
	EEM [16]	Mild stock removing action with abrasives in hydrodynamic flow of slurry around ball tool	
	Float polishing [5]	Stock removing action with abrasives in hydrodynamic flow of slurry on flat polisher	
Complex	(Removing chemical reactive products)		
	Mechanochemical polishing (wet) [10]	Growing and removing reactive products as hydration film under wet condition	Si wafer GaAs wafer
	Mechanochemical polishing (dry)[17]	Peeling off reactive product film with fine silica powder under dry friction condition	Sapphire Si–N ceramic
	Hydration polishing	Growing and wiping off hydration film with high temperature water vapor and wood or carbon tool	Sapphire ZnSe lens
	(Using chemical removal)		GaAs wafer InP wafer CdTe wafer ZnSe wafer
	Disk-type chemical polishing	Chemically removing with friction by using non-abrasive chemical solution and non-woven fiber sheet	
Chemical	Hydroplane polishing [18]	Chemical removing with non-abrasive chemical solution layer between work and polisher	
	P-MAC polishing [19]	Chemical removing by using non-abrasive chemical solution and fluorocarbon foam polisher. Working conditions change as polishing advances	

This film is mechanically removed with abrasives. These processes are repeated in order to remove surface irregularities and to gradually yield mirror-like surfaces [11, 12].

Table 7.2.2 shows various polishing methods originating from the conventional lapping and polishing methods. Such a stock-removal mechanism for making mirror-like surface is basically common to all processing methods.

REFERENCES

1. Nakamura, N. (1993). Grinding, lapping and polishing, "Optical Production Technology '93" (in Japanese), pp. 58–123. Japan Optoelectro-Mechanics Association.
2. Kasai, T., Noda, J., and Suzuki, J. (1978). Lapping characteristics of $LiTaO_3$ single crystal. Study on precision machining of opto-electronic crystals (1st report) (in Japanese) *J. Jpn. Soc. Precision Eng.* **44**: 1360–1366.
3. Kasai, T. (1979). Study on precision machining of flat planes, Electric Communication Laboratories Report No. 13634, NTT Public Corporation (in Japanese), pp. 1–268.
4. Dietz, R. W., and Bennet, J. M. (1966). Bowl feed technique for producing supersmooth, optical surface, **5**: 881.
5. Namba, Y. (1984). Mechanism of float polishing, "Technical Digest of the Topical Meeting on Science of Polishing", OSA Tub-A2.
6. Kasai, T., and Orioka, T. (1967). On the high speed polishing of glass using felt plate (5). On the polishing of glass (in Japanese), *J. Jpn. Soc. Precision Eng.* **33**: 306–311.
7. Preston, F. W., and AMICE (1927). The theory and design of plate glass polishing machines, *J. Soc. Glass Technol.* 214–256.
8. Kasai, T., Horio, K., Doy, K. T., and Kobayashi, A. (1990). Improvement of conventional polishing conditions for obtaining super smooth surfaces of glass and metal works, *Ann CIRP* **39**: 321–323.
9. Otte, G. (1965). An improved method for the production of optically flat surfaces, *J. Sci. Instrum.* **42**: 911.
10. Doy, K. T., and Watanabe, J. (1983). Effect of frictional heat on removal rate in mechanochemical polishing of crystals used in electronics, *Wear Mater. ASME* 227–234.
11. Izumitani, T., and Adachi, S. (1984). Polishing mechanism of fused silica glass, "Technical Digest of the Topical Meeting on the Science of Polishing", OSA TuB-Al-1–3.
12. Brown, N. J. (1987). Some speculations on the mechanisms of abrasive grinding and polishing. *Precision Eng.* **9**: 129.
13. Kasai, T., Horio, K., and Doy, K. T. (1992). Polishing characteristics of soft glass (laser glass), "The 3rd International Conference on Nanotechnology", Program and Abstracts, pp. 16–17.
14. Bajuk, D. J. (1976). Computer controlled generation of rotationally symmetric aspheric surfaces. *Opt. Eng.* **15**: 401.
15. Becker, K., and Beckstette, K. (1988). M400 and P400 – a pair of machines for computer-controlled fine correction of optical surfaces, "Proceedings of the International Congress for Ultraprecision Technology", Aachen, Germany, pp. 212–223.
16. Mori, Y., Yamauchi, K., and Endo, K. (1987). Elastic emission machining, *Precision Eng.* **9**: 123–128.
17. Imanaka, O., and Yasunaga, N. (1984). Mechanism of mechanochemical polishing, "Technical Digest of the Topical Meeting on Science of Polishing", OSA Tub-C2–1–4.

18. Gormley, J. V., Manfra, M. J., and Carawa, A. R. (1981). Hydroplane polishing of semiconductor crystals. *Rev. Sci. Instrum.* 52: 1256–1259.
19. Kasai, T., Matsumoto, F., and Kobayasi, A. (1988). Newly developed fully automatic polishing machines for obtainable super-smooth surfaces of compound semiconductor wafers. *Ann. CIRP* 37: 537–540.

Handbook of Advanced Ceramics
S. Sōmiya *et al.* (Eds.)

CHAPTER 8

8.1 Joining and Coating

NOBUYA IWAMOTO
Ion Engineering Laboratory, Hirakata 573-0128, Japan

8.1.1 JOINING

8.1.1.1 INTRODUCTION

Over the past several decades, worldwide researches and developments in ceramic joining have been spread out as follows: (a) fundamental studies of adhesion; (b) studies of interfacial reactions and structure; (c) development of various joining procedures; (d) joining with interlayer; (e) analysis of interfacial stress; and (f) mechanical design and analysis of joints.

Ceramic materials are being widely used as electronic, magnetic, optical, chemical, thermal, and biological components in a number of advanced technologies. Cost and difficulty in manufacturing complex components, either in one step or by joining of ceramic–metal and ceramic–ceramic, determines inhibition for more widespread use.

It is an important matter how jointed component could be maintained for longer time without problems of crack and destruction occurrence.

8.1.1.2 APPLICATIONS OF JOINED COMPONENTS

Grünling and Linsen [1] have summarized activities and trends of ceramic-to-metal joints, especially, in several international conferences and others [2–4].

In the first conference, the manufacture of Al_2O_3 cylinder joined with Al_2O_3 plate for mechanical uses and Cu joined with BeO, ZrO_2, SnO_2, or ZnO were presented [4]. As an interesting application, Al_2O_3 windows, vacuum suction parts, and vacuum tight parts were joined with 26 joining ways in the all ceramic made ZT-40 fusion reactor torus. Joinings in energy technologies such as high temperature hydrogen electrolysis, nuclear waste disposal, and heat exchangers for the solar power were developed. As other examples, the applications on oxygen partial pressure probes, ultrasonic temperature measurement equipment for

TABLE 8.1.1 Examples of Metal/Ceramic Joints and their Joining Technologies [5]. Reproduced with kind permission of Deutscher Verlag

Examples	Joining technologies	Ceramics	Metals
Transmitter valves	Metallizing and hardbrazing	Al_2O_3	FeNi, Cu
Housings for semiconductors	Matallizing and hardbrazing	Al_2O_3	FeNi, Cu
Feedthroughs	Active brazing	Al_2O_3	FeNi, Cu
Pressure sensors	Active brazing	Al_2O_3	Zr alloys
Dental implants	Diffusion bonding	Al_2O_3	Nb, Ti
Hipp implants	Form fitting	Al_2O_3	Stainless steels
Turbine blades	Form fitting	Si_3N_4	Stainless steels
Turbo charger	Active brazing	Si_3N_4	Ni alloys
Solid oxide fuel cells	Glass filler material brazing	ZrO_2	Cr alloys
Lamps	Glass filler material brazing	Glass	W, Mo
Lenses	Glueing	Glass	Steel, polymers

fast breeder reactors and solid oxide electrolyte cells and electronics and so on were reported [4].

Turwitt has given overview on joined components as shown in Table 8.1.1 [5].

8.1.1.3 Scientific Aspects of Joint Formation

Chatain *et al.* [6] have compared the calculated work of adhesion and experimentally determined tensile strength and showed good linear dependence of both values.

Li [7] has plotted these work of adhesion values as a function of electron density, at the boundary of the Wigner–Saitz cell of the corresponding metals, which is one of two parameters in the Miedema model for the heat of formation of binary alloys. He also showed the linear dependences of the work of adhesion and of the contact angle of cobalt, iron, nickel, copper, silver, and tin with various ionocovalent oxides on the free energy of the oxide formation [8].

Iwamoto [9] compared the free energy of carbide and nitride formation and the contact angle on Cu. He showed clear mutual relation among them.

8.1.1.4 Joining Processes

The word "joining" means the methods of incorporation between same or different materials to realize the constructed matters having functional characteristics. There are several possible methods of producing ceramic–ceramic and ceramic–metal joints.

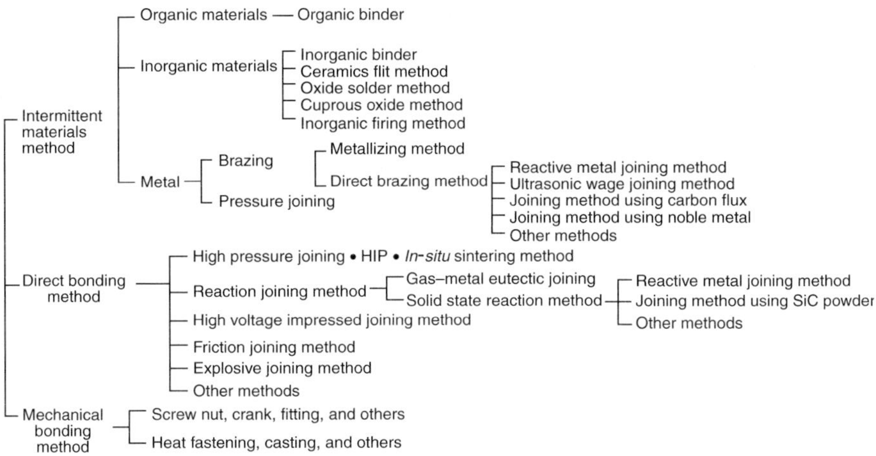

FIGURE 8.1.1 Classification of various joining procedures [10].

In Figure 8.1.1, the classification of joining procedures is given [10]. These might be categorized as either mechanical or chemical.

8.1.1.5 Mechanical Joining

Tight adhesion between ZrO_2 and steel was obtained by utilizing the difference of thermal expansion coefficient due to cooling from higher temperature under high-pressure circumstances. As other applications, ceramic armour of cylinder-head, joining between metal rotor and ceramic turbine blade with wedged metallic shin, and screw up of ceramics to leading edge of space shuttle, and high-performance heat exchanger and fastening of piston-crown in reciprocating engine were produced [11–14].

8.1.1.6 Chemical Joining

Solid state (diffusion) joining: elemental diffusion can be induced by appling pressure at high temperature, and this method has been widely applied in the field of welding. The oxide film formed on metal surface could be destroyed by applying pressure and therefore not only accelerated diffusion and reaction at the interface by exposing new metallic surface proceeded, but also thermal stress is easily released by plastic deformation.

As joining conditions [15]:

(a) Joining temperature must be settled to about 90% value of melting point of the specimen having lower one.
(b) Though, in general, ambient atmosphere is used, but joining can be done in argon or in vacuum in special cases.
(c) It becomes sufficient to give rise to good junction at contacted surface by appling only light cramping pressure.
(d) Joining time from several seconds to several hours will be used.
(e) It is necessary to give polishing ceramic having optical flatness which induces maximum joining strength.

Practical or engineering joints are made by providing enough thermal energy and pressure to promote interdiffusion and chemical bonding of the interface. Most joining methods use a liquid phase of some sort. One of the main advantages of using liquids is that they can penetrate into rough surfaces and lower the pressure required to make a strong joint.

Transient liquid phase (TLP) joining process is a branch of brazing that overcomes the presence of filler metal and is used as a sort of solid state joining. In TLP joining, the liquid interlayer resolidifies isothermally at the bonding temperature as a result of interactions between the liquid filler metal and the solid substrates.

8.1.1.7 Pressure Diffusion Joining

Though careful surface preparation is needed to obtain high-strengh bonds, joining by solid phase diffusion and reaction is advanced at elevated temperature for long times and high pressures. This procedures can be suitable for metals having low yield stress like Cu. The metals could be pressed to conform to the ceramic, and can obtain good contact between the metal and ceramic.

As for joining between same species of ceramic, for example, it is said that 0.0075–0.75 wt% MgO must be contained in Al_2O_3 to make monolithic body by applying a temperature of 1400–1500°C and a pressure of 69–138 MPa. After joining, good material can be manufactured by giving after-treatment, at a temperature of 1875°C for 10.8 ks to initiate grain growth [16].

In the case of joining between SiC and SiC, either high temperature, as high as 2100°C have been used, or to mix SiC powders at the interface to relax the difference of thermal expansion coefficients has been done.

Pressure joining of Cu/Al_2O_3 system was carried out in a vacuum furnace with an atmosphere of 10^{-6} Pa, in a temperature range of 830–1010°C, and the experimental conditions were changed to temperature of 920–1450°C in the case of Nb/Al_2O_3 joints [17].

8.1.1.8 Friction Joining

Friction joining is usually carried on metal–metal systems, and is considered as simple way joining interfaces which could be easily heated and pressed in air.

The combinations of ceramic materials, pressureless sintered silicon nitride and carbide, PSZ and Al_2O_3, and metallic materials such as Al, Al clad steel, Ti and Cu were chosen. Joinings were done under the revolution rates of 1500–6000 rpm with friction and upset pressures of 2–80 MPa, and friction times of 2–60 s. Joinings using Cu or Ti, due to low plastic deformability, did not show good result in comparison with that of Al.

In the case of Si_3N_4 and ZrO_2 to Al joints, tensile strength value of about 120 MPa was achieved without difficulty. Al/Al_2O_3 joints showed value about 60 MPa [18].

8.1.1.9 Ultrasonic Joining

Though ultrasonic welding of similar materials such as metals and plastics have been used as economical means in several industrial fields like automobile, application on compounds of dissimilar materials could not be observed.

Not only because of short joining time of less than 1.5 s, but also to low energy, so as not to raise temperature, this means has attracted interest. The joinings with various types of glass and several kinds of alloys, Ni–Co, Cr–Mo–Ti–Al and Al–Mg, were carried out with the use of aluminum interlayer.

In addition, ultrasonic joinings between the substates such as partially stabilized zirconia (ZrO_2) or alumina (Al_2O_3) and coated films of metals and brazed alloy using PVD or thermal spraying were performed. It seems, however, difficult to use this means for industrial application.

Ultrasonic bonding is a method by which two materials are joined by imposing an ultrasonic wave upon the bonding interface. It has shown that Al can be bonded to Al_2O_3, $SrO–6Fe_2O_3$, SiC, BaO, and SiO_2 [19]. It was determined that Al can be bonded to various sorts of oxides, carbides, and nitrides in air, could bond to Si_3N_4 in vacuum. Ti could be bonded to Si_3N_4 only in a high vacuum [20].

The PSZ ceramic coated with Ag–Cu–Pd braze was joined with copper sheet at a frequency of 20 kHz and with output of 2 kW. The pneumatic press of the system enabled sonotrode forces between 200 and 400 N. These joined ceramics showed good strength [20].

8.1.1.10 Microwave Joining

One of the newest techniques in joining of ceramics is microwave joining [21]. Fukushima *et al.* have reported joint strength equal to original strength for low-purity Al_2O_3 ceramics joined directly (without intermediate layer) using a single mode cavity (3 min, 0.6 MPa).

Joining of high-purity Al_2O_3 was also carried out using a lower-purity interlayer, but good joining strength was achieved using a lower-purity interlayer (70–90% strength of original value) [22].

The importance of ceramic joining arises from the fact that ceramics are difficult to machine. Since large and complex shapes are often necessary, a satisfactory joining process is desired. Ceramics would be used more widely and their fabrication costs would be reduced if techniques for joining them were simpler and more reliable.

8.1.1.11 Joining by Impressed Electric Potential

It is known that fully stabilized zirconia is an oxygen ion conductor. In an electric field above approximately 2.25 V at 900°C, the decomposition of zirconia occurs, and potential of zirconium could be increased. When additional foil of Ag 72–Cu 28 braze was inserted into Ni/ZrO_2 interface, the reaction phases like Ni_5Zr and NiZr are formed at the interface. From this effect, defect-free interfaces were obtained [23].

8.1.1.12 Reactive Metal Coat and Gas–Metal Eutectic Joining

Gas–metal eutectic joining has been widely applied, especially, to join copper and alumina. Although AlN is an attractive ceramic for electronic packing and mechanical applications because of high thermal conductivity and having matched thermal expansion coefficient with that of silicon, from a practical point of view one problem is the difficulty on bonding to metal.

The method described below was considered:

1. Oxidizing pre-treatment, AlN has lower surface energy than that of Al_2O_3, so eutectic melt does not wet the AlN surface.
2. Sputtered adherent coat of titanium and copper or other active metal on ceramic surface.

It is emphasized that a trace quantity of silicon or calcium must be contained in AlN because these elements play an important role in the minimization of thermal expansion coefficient mismatch between formed surface oxide and AlN [24].

8.1.1.13 Surface Activated Joining

Electrolytic polished single crystal of 99.999% aluminum was sputter-cleaned by Ar fast-atom beam (FAB) emitted from a saddle field FAB source at an incident angle of 45°, operated at the acceleration voltage and current of 1.5 kV and 15 mA, respectively, for 0.6 ks. Then the samples were transferred to the bonding chamber at ca 10^{-7} Pa, where a load corresponding to 40 MPa applied. In the case of Si_3N_4/Al joints, tensile strength of about 100 MPa are achieved [25].

8.1.1.14 Joining with the Use of Glassy Materials

Hot-pressed Si_3N_4 was joined using an $MgO–Al_2O_3–SiO_2$ glass [26, 27]. We applied glass solder of $CaO–SiO_2$ system, and found the formation of sialon glass at the interface between Si_3N_4 and Si_3N_4. A bonding strength of approximately 650 MPa was obtained [28, 29]. Likewise, glass solder of $Y_2O_3–SiO_2–Al_2O_3–Si_3N_4$ system was used, and a bond strength of 550 MPa was obtained [30].

8.1.1.15 Brazing

As fundamental principles of brazing, the following items must be satisfied [31], that is, aim to

1. form metallurgical bonding;
2. melt at lower temperature below the melting temperature of substrate;
3. add flux which is able to wet and protect oxidation of substrate;

4. have enough fluidity by capillary action and to give homogeneous elemental distribution of joined parts;
5. have capability not to flow out of low-melting elements in brazes;
6. proceed dissolution of parent materials;
7. have lower vapor pressure characteristics;
8. form compound having high melting temperature from the reaction with mother materials.

As for segregation after consolidation, the consideration of interfacial energy and post-diffusion is necessitated.

Families of braze alloys based on Pb, Zn, Au, Cu, and other metals and in particular eutectic compositions, for example, Au–Cu–Ti system have been developed and have found widespread use.

8.1.1.15.1 Influence of Additional Element on the Physical Properties of Cu–Ti System

Nicholas *et al.* [30] have summarized the surface energy and the solubility of Ti on various types of metal as shown in Table 8.1.2.

From Table 8.1.2 it can be noted that the element, which can not only restrict excess solubility of Ti, but also decrease surface energy, shows pronominent wetting preformance on ceramics.

As the wettability on Si_3N_4 becomes worse when excess of Ti was added in the Ag–Cu–Ti system, the Ti content has been decreased and attempts have been made to raise activity coefficient value by selecting an element, which lowers the surface energy and promotes surface segregation. The addition of In and Sn gave good result [32].

TABLE 8.1.2 Properties of Typical Elements for Brazement [32]. Reproduced with kind permission of Kluwer Academic Publishers

Element	Surface energy of molten metal (mJ/m^2)			Solubility of Ti (at%)		
	1050°C	1100°C	1150°C	1050°C	1100°C	1150°C
Cu	1364[a]	1353	1343	67	70.0	73
Sn	484	482	480	13	15.5	17
In	448	439	429	13	17.0	20
Al	810	804	798	2	3.0	5
Ag	904	897	889	4	5.5	8
Au	1160[a]	1070	1020	10[a]	7.0	12
Ga	614	609	604	11	19.0	28
Ni	1965[a]	1945[a]	1925[a]	12[a]	12.5[a]	13[a]

[a]Extracted value from melting temperature.

8.1.1.15.2 The Application of Chemical Thermodynamics to Predict Reactive Metal Brazing of Ceramics

The solvent effect arises due to the interaction between the reactive element addition and the brazing alloy [33]. Research has demonstrated that Ti is an active element in copper- and silver-based braze alloys. It was shown that the addition of Ti to silver/copper alloys can improve the wettability of many ceramics. Ag–Cu eutectic with 4.5 wt% Ti is one of the most popular active silver-base braze alloys used in metal–ceramic joints. It was, however, concluded that, Zr addition is always less effective than Ti addition for solvent metals such as Ag, Cu, Pb, and Sn.

8.1.1.15.3 Brazing Ceramics

Advances in brazing technology have given new market for the use of ceramics because of their high strength and usage at high temperature. Walz [34] approximated the development of the markets for tool tips such as saw and drill tips as well as complex shapes such as seals, gaskets, wear parts, spacers, etc., and unique applications such as satellite mirror assembly and so on.

Most structural ceramics, for example, alumina, silicon nitride, silicon carbide and zirconia, are chemically more stable than metals, and therefore they will not be wetted by traditional brazes unless the braze contains an active element.

8.1.1.15.4 Alumina

Although solid state reactions of the Ag–Cu–Ti brazes and Al_2O_3 have been pursued by many investigators, there is no established theory explaining interfacial structure. Panlasto and Kivilahti [35] have concluded that various reaction products are formed depending on the activity of titanium and oxygen. The miscibility gap in the Ag–Cu–Ti system devides the liquid braze into Ag-rich (L1) and Ti–Cu-rich (L2) liquids.

L2 liquids react with alumina forming the mixed oxide $(Ti, Al)_4Cu_2O$. Ti reacts with alumina to form phases of Ti_XO and Ti–Cu–O system in braze. The formation of Ti oxides next to the alumina is possible only in the filler alloys of highest Ag content [35]. It was also found that it provides high interfacial shear strength of the interface between brazed alumina and Ti–6Al–4V alloy sheets [36].

8.1.1.15.5 Zirconia

Mg–PSZ/nodular cast joints are aimed for utilizations at locations exposed to high temperature and wear conditions. The joints could be used under the non-stress concentrations arising from mechanical joining. Also, the joining is considered for use in automotive engine applications [37].

The brazing filler metals used were Braze 604 (Ag–30Cu–10Sn wt%) for the active-substrate process and Incusil ABA (Ag–27.5Cu–12.1In–1.2Ti wt%) for the active-filler metal process.

Voids formation was detected at brazed joints due to the formation of a non-wettable surface on the PSZ, which is presumed to be the result of Ti oxidation [38]. The wetting of partially stablized YSZ of Ag–27Cu–3Ti and Sn–10Ag–4Ti showed increment with the increase of brazing temperature and the flexual strengths became 277 and 137 MPa, respectively. The formation of TiO reaction layers was detected in both cases [39].

8.1.1.15.6 Silica

To investigate the wettability characteristics on SiO_2 is important in order to know relevance to the case of non-oxide ceramics like silicon carbide, whose surface is easily oxidized in oxygeneous atmosphere.

Wetting behavior of SiO_2 at 850°C in vacuo for Ag–28 wt% Cu Ag–35 wt% Cu–1.5 wt% Ti and, Ag–35 wt% Cu–12 wt% In–2 wt% Ti was studied. With the exception of the last mentioned braze, the wetting increased with time. The reason is the presence of extensive cracks at the surface [40].

8.1.1.15.7 Mullite

Mullite ($3Al_2O_3 \cdot 2SiO_2$) is now gaining attention as a ceramic material because of the high strength and creep resistance at elevated temperatures as Si_3N_4 and SiC. It was recently developed to induce high toughness of ZrO_2-toughened mullite.

The wetting and joining of mullite with Ag–Cu–Ti and Ag–Zr–Cu alloys was studied. Compared with the Ti-containing alloys, those with Zr were less reactive with the mullite and the Zr did not segregate as completely to the interface during heating [41].

8.1.1.15.8 Silicon Carbide

Silicon carbide ceramics having superior strength, high wear resistance and excellent corrosion-resistant and thermal characteristics have gained considerable attention. Unfortunately, silicon carbide is also brittle and has poor machinability, therefore, enormous difficulties are encountered in the fabrication of complex-shaped and large-sized components. Among the many joining technologies, brazing shows large potential to induce lesser thermal stress and distortion.

The interfacial reacted layer between Ag–Cu–2 wt%Ti braze and sintered SiC with various additives was studied. In the case of SiC containing B and C, the order of reaction layers of SiC/TiC/Ti_5Si_3/braze were formed. Likewise, the formation of same multilayer was found in the case of Al_2O_3-containing SiC. The interfacial strength between SiC containing BeO became much lower [42].

8.1.1.15.9 Cemented Carbide

Use of tungsten carbide has shown a steady growth over the years as it replaces alloy steels for the use of lathe cutting tools, wire drawing dies, forging dies, and press tools. Its economic use, however, depends on the strength and longevity of the joints between tungsten carbide and its supporting mass of steel.

These have been used for two main types of brazing alloys, silver- and copper-based. To fulfil the joint gaps which are much greater than about 0.2 mm, brazing alloy must be bonded to either side of a copper–nickel layer [43].

8.1.1.15.10 Silicon Nitride

In 1993, the international workshop on "Tailoring of High-Temperature Properties of Si_3N_4 Ceramics" was held. The history of the application of dense silicon nitride [44] and the use as turbine wheel, valve spring retainer, turbocharger turbine wheel, and also piston pin, and valves was dealt with in detail [45]. Topics related to joining applications, however, have not been treated.

The increased use of ceramics–metal joints including turbocharger rotors and ceramic caps on push rods has provided an impetus for research into techniques for joining such materials [46]. The oxidation resistance characteristics of ternary reactive-metal brazing alloys (Cu- and Ag-based) were investigated [46].

Mechanical strength of Si_3N_4/steel joined with Ag–Cu–Ti braze were studied, and optimal value like 123–158 MPa was obtained by using a sufficiently high brazing temperature [47]. The influence of cooling velocity from brazing treatment on the bending strength and thermal cycle behavior of Si_3N_4/Inconel 600 were studied by using an AE method [48].

8.1.1.15.11 Aluminum Nitride

Thick layers of copper on substrates of AlN are good supporting materials for power transisters and the hybrid mounting integratet circuit. It is very important matter to have an excellent heat contact between metal and ceramics to conduct heat energy. The bonding strength of AlN bonded with Ag–Cu–Ti braze showed the decrease with the increase of cooling rate. It was determined that the resistance against temperature cycles decrease with the length of the compound and increases with the thickness of AlN-substrate [49].

8.1.1.16 Joining with the Use of Interlayer

To make ceramics join with metals, however, there are still many problems to be solved: residual stress induced by thermal expansion mismatch, microstructure and mechanical properties of the interface and strength of ceramics itself.

With these points as background, the following procedures have been developed in order to:

1. make graded layers;
2. insert a layer having middle value of thermal expansion coefficient between metal and ceramic;
3. insert ductile layer;
4. change the shape, minimize the formation of stress;
5. form amorphous or fine-grained layer or to utilize misfit-dislocation at the epitaxial interface [50];
6. utilize phase transformation inducing volume change during cooling process;
7. use superplastic behavior of fine-grained ceramics, mainly Y-TZP;
8. insert nanocrystalline interlayer.

Tanaka [51] summarized the interlayer for effective joining of Si_3N_4 as in Table 8.1.3.

8.1.1.16.1 Selection of Interlayer

Zhou *et al.* [52] studied the influence of Cu, Kovar, Mo, and W interlayers on the magnitude and distribution of thermal stress and on the tensile strength

TABLE 8.1.3 Interlayers Used for Joining of Si_3N_4

	Relaxation of thermal expansion mismatch by using layer of middle value	Relaxation by plastic flow (ductility, superplasticity)
Metallic materials	WC–Co	Cu
	Super-Invar	Al
	Invar	Al alloy
	Koval	Ni
	Ti Alloy	Ti
	(Ti–Mo, Ti–W, etc.)	Ag–Cu
	Nb/Mo, W	
	Fe/W	and other soft metals
Composite materials	Cermet	Cermet
	($Cu–Al_2O_3$, $Ni–Al_2O_3$, etc.)	($Cu–Al_2O_3$ and others)
	Cu–C	

of brazed Si_3N_4–steel joints. As conclusion, the use of low yield strength/high thermal expansivity Cu interlayers produced higher joint strengths than low thermal expansivity Mo and W interlayers.

Furthermore, it has been clear that the use of composite Cu–W and Cu–Koval interlayer materials can give the lowest thermal stresses in Si_3N_4–steel joints.

The selection of interlayer must be determined from the viewpoints not only on thermal expansion mismatch but also on the thickness of the layer and change of interlayer mechanical properties (elastic modulus, yield strength, and work hardening rate) with temperature.

The use of more ductile and softer interlayer than nickel, for example, aluminum, is introduced to reduce tensile stress in Si_3N_4 adherent, and it is concluded that the magnitude of thermal expansion mismatch between the metal interlayer and the ceramic adherent is of minor importance for tensile residual stress reduction [53].

8.1.1.16.2 Superplastic Joining

Superplastic ceramics are considered to give effective means for structural constraction because of their deformation characteristic over 20 orders of magnitude of strain rate. Junction is achieved for lower temperatures and shorter periods of time than that for direct diffusion bonding.

Superplastically deformable yttria-stabilized tetragonal zirconia polycrystals were inserted for the use of diffusion bonding of alumina [54]. Diffusion bonding of Si_3N_4 was developed by using superplastic β-SiAlON interlayer [55].

8.1.1.16.3 Joining of Ceramics Using Nanocrystalline Interlayer

A factor to be emphasized among various joining procedures is that necessitated temperature is lower when nanocrystalline powders were insterted for successful joining of conventional alumina disks. Temperature necessitated was as low as 1250°C and a pressure of 55 MPa was used [56].

8.1.2 HEAT EXCHANGER JOINTS

Katz [57] reviewed the availability of microwave processing on ceramic sintering. Its advantages are: (a) enhanced diffusion; (b) lower processing costs because, in principle, only the ceramic is heated; and (c) improved mechanical properties of ceramics after microwave treatment because of lower grain growth kinetics.

Boch and Lequex [58] have discussed whether microwave processing can increase the sinternability of ceramics. By giving the modest configuration of furnace and sample design, microwave processing can offer the possibilities of rapid heating with appropriate power levels compared to conventional heating techniques for ceramic joining applications [59].

8.1.2.1 SiC

The use of ceramic as the fabrication material for heat exchangers adresses both disadvantages of metallic heat exchangers. First, the ceramic does not lose significant strength at the temperature in the range of 1300°C, and second, a ceramic is not attacked, or is attacked at greatly lessened rates, in comparison with that of metallic heat exchanger. Applications of ceramics as seals between tubes and tube sheet, duct linings, burner insulations, and heat exchanger/steam reformer tubes needs to withstand the corrosive combustion products ($CO_2, H_2O, N_2, O_2, HC_X, NO_X$ and SO_3) and various sorts of compounds such as halides, sulfates and so on [60].

In general, mechanical joining, brazing, and solid state bonding have been applied as joining methods for heat exchangers. Mechanical joining has advantage to be able to release stress concentrations caused by thermal cycling. It is limited, however, for use only in systems where the internal pressures and imposed loads are moderate. The main drawback of brazed joints is that brazing materials generally have a much lower service temperature. Solid state bonding gives joint irregularity, and induces failure because of different silicon content.

Joining technology using siliconized silicon carbide is considered adequate. In future, systems based on silicon nitride or silicon carbide, which are durable at high temperature and are easily repairable are wanted [61]. The application of silicon carbide as heat exchangers, has been discussed in several symposia [62, 63].

(a) Reaction bonded silicon carbide (RBSC) and sintered silicon carbide (SSC) joints: Joining of RBSC socket and SSC tube was carried at temperatures greater than 1400°C and for 3.3 ks. The assemblies have shown leak-tight behavior before and after cycling between room temperature and 1100°C [64].

(b) (Si-SiC)–(Si-SiC) joining: Microwave joining was carried on with the experimental conditions of temperature of 1250°C for 0.3 ks and a pressure of 1.2 MPa with the use of Al as interlayer. It was concluded that good joining is induced from the diffusion of Al into Si to obtain high joined fracture strength [65].

(c) SiC–SiC joining: Under the experimental conditions at the temperature approximately 1450°C and for 0.3–0.6 ks with Si as interlayer was carried out. By using Si directly as powder in an oil-based slurry or plasma-sprayed onto the surface of one of specimens, homogeneous joints were fabticated [66].

(d) RBSiC–RBSiC: The microwave heating of RBSC should result in enhanced mass transport across the interface, leading to a diffusion bonding of the pieces to be joined. This should eliminate the need to use any interlayer, thus increasing the service temperature of the joined specimens. Under a temperature of 1450–1550°C for 0.6–0.9 ks, joining was performed, and it was found that the interface region becomes dense and homogeneous [67].

(e) Hot pressed silicon carbide (HPSC–HPSC): For the SiC joining, the conditions were in the range 1190–1375°C and 0.5–2.0 MPa for 1.8 ks. The RBSC could be successfully bonded.

(f) Zirconia partially stabilized with 8 wt% Y_2O_3 (YPSZ–YPSZ) [68]: As processing parameters, temperatures of 1230–1565°C for 1.8 ks were chosen. The bonds could not be detected after joining.

8.1.3 COATINGS

Ceramics appear an surperior alternative as high-temperature materials because they have higher melting points, but they have a major drawback in that they are difficult to shape. Many coating procedures have been developed, for example, CVD, PVD, plasma spraying, electrodeposition, and sol–gel, and so on to fabricate ceramic parts.

Ceramics coatings have been widely used as oxidation-resistant materials for basal alloy, hard overlay of wear-resistant objective, rubbing seals, thermal barriers, erosion resistance, and diffusion barriers [69].

REFERENCES

1. Grünling, W., and Linssen, G. (1997). *DVS Berichte* **184**: 222–224.
2. Grünling, W., and Linssen, G. (1980). Fügen von Keramik, Glas und Metall, "1. Internationals Kolloquium", Baden-Baden, 3–5 December DVS Berichte 66. Düsseldorf: Deutscher Verlags für Schweisstechnik.
3. Grünling, W., and Linssen, G. (1985). Fügen von Keramik, Glas und Metall, "2. International Kolloquim", Bad Nauheim, 27–29 March, Fortschrittberichte von Deutscher Keramischen Gesellschaft, Band 1, Heft, Bad Honnof.
4. Grünling, W., and Linssen, G. (1989). "Third International Conference on Joining Ceramics", Glass and Metal, Bad Nauheim, Germany, April 26–28, Kraf, W., ed., Oberesel: Deutsche Gasellschaft für Metalkunde.
5. Turwitt, M. (1997). "Joining Ceramics, Glass and Metal, 5th International Conference", 12–14 May, Jena, Germany.

6. Chatain, D., Rivolett, I., and Eustathopoulos, N. (1986). *J. Chim, Phys.* **83**: 561.
7. de Boer, F. R., Boom, R., Matttens, W. C. M., Miedema, A. R., and Niessen, A. K. (1988). *Cohesion in Metals*, Amsterdam: North-Holland.
8. Li, J.-G. (1992). *J. Am. Ceram. Soc.* **75**: 3118–3126.
9. Iwamoto, N., and Suga, T. (1990). *Ceramics Joining Technologies*, pp. 30–31, Tokyo: Nikkan Kogyo Shinbun, Tokyo (in Japanese).
10. Suzumura, A. (1999). Japan Welding Society, Committee of Interfacial Joining Research, Tokyo, 3–9.
11. Hennicke, H. W., Müller-Zell, A., and Siebels, J. (1981). *Keramische Komponenten für Fahrzeug Gasturbinen II*, p. 535, Bunk, W., and Böhmer, M., eds., Berlin: Springer Verlag.
12. Rice, R. W. (1975). *Advances in Joining Technology*, p. 69, Burke, J. J., *et al.*, eds., Massachussetts: Brook Hill Publ. Co.
13. Synder, J. E., and Petrak, D. R. (1985). *Adv. Ceram.* **14**: 59.
14. Anderson, C. A. (1978). *Ceramics for High-Performance Applications – II*, p. 783, Burke, J. J., Lenoe, E. N., and Katz, R. N., eds., Massachussetts: AMMRS.
15. Bailey, F. P., and Bordige, W. E. (1981). *Surfaces and Interfaces in Ceramics and Ceramic-Metal*, p. 525, Pask, J., and Evans, A., eds., New York: Plenum Press.
16. Scott, C., and Trans, Van. B. (1984). *Amer. Ceram. Bull.* **63**: 1171.
17. Elter, D., Roeder, E., Schlicker, U., and Wagner, J. (1996). *Glastech. Ber. Glass Sci. Technol.* **69**: 326–330.
18. Zuckschwerdt, K., Greitmann, M. J., Ross, E., and Kussmaul, K. (1999). *Schweissen Schneiden* **51**: 211–212, 638–640.
19. Matsuoka, S. (1987). *J. Jpn. Soc. Technol. Plasticity*, **28**: 1186 (in Japanese).
20. Miyazawa, K., Matsuoka, S., Fujii, T., and Suga, T. (1989). *Metal–Ceramic Joints*, pp. 275–280, Iwamoto, N., and Suga, T., eds., Pittsburgh: MRS.
21. Clark, D. E., Gac, F. D., and Sutton, W. H., eds. (1991). *Microwaves: Theory and Application in Materials Processing*, Vol. 21, Westerville, OH: Am. Ceram. Soc. Inc.
22. Fukushima, H., Yamanaka, T., and Matsui, M. (1990). *J. Mater. Res.* **5**: 397–405.
23. Fritsche, B., Satir-Kolorz, A., and Feichtinger, H. K. (1997). *DVS Borichte* **184**: 92–94.
24. Chiang, W. L., Tessarotto, L. A. B., Greenhut, V. A., and Schanefield, D. J. (1991). Metal–Ceramic Joining, pp. 247–261, Kumar, P., and Gerrnhet, V. A., eds., Warrendale: The Minerals, Metals & Materials Society.
25. Suga, T. (1998). *Ceramic Microstructure: Control of Atom Level*, pp. 385–389, Tomsia, A. P., and Glaeser, A., eds., New York: Plenum Press.
26. Johnson, S. M., and Rowclitffe, D. J. (1985). *J. Am. Ceram. Soc.* **68**: 468–472.
27. Mecartney, M. L., and Sinclear, R. (1985). *J. Am. Ceram. Soc.* **68**: 472–478.
28. Iwamoto, N., Umesaki, N., and Haibara, Y. (1986). *J. Ceram. Soc. Jpn.* **94**: 880 (in Japanese).
29. Haibara, Y., Iwamoto, N., and Umesaki, N. (1989). *Ceram. Eng. Sci. Proc.* **10**: 1768.
30. Xie, R., Huang, L., and Fu, X. (1998). *J. Mater. Sci. Lett.* **17**: 761–763.
31. Schwarz, M. M. (1967). *Brazing*, p. 2, Metal Park: ASM International.
32. Nicholas, M. G., Valentine, T. M., and White, M. G. (1980). *J. Mat. Sci.* **15**: 2197.
33. Wang, G., and Lannutti, J. J. (1995). *Metall. Trans.* **26A**: 1499–1505.
34. Walz, T. (1999). *Ceram. Ind.* **2**: 47–50.
35. Paulasto, M., and Kivilahti, J. (1998). *J. Mater. Res.* **13**: 343–352.
36. Sherman, D. (1998). *Mater. Lett.* **33**: 255–260.
37. Majumdar, B. S., and Ahmad, J. (1991). *Metal-Ceramic Joining*, pp. 67–97, Kumar, P., and Greenhut, V. A., eds., Warrendale: TMS.
38. Park, E., Cawley, J. D., and Hausner, D. (1991). *Metal-Ceramic Joining*, pp. 167–180, Kumar, P., and Greenhut, V. A., eds., Warrendale: TMS.
39. Chuang, T. H., Yeh, M. S., and Chai, Y. H. (2000). *Metall. Trans.* **31A**: 1591–1597.

40. L'opez-Cuevas, J. (2000). *J. Am. Ceram. Soc.* **83**: 2913–2918.
41. Loehman, R. E., and Tomsia, A. P. (1994). *J. Am. Ceram. Soc.* 77: 271–274.
42. Kato, S., Yano, T., and Isaki, T. (1993). *J. Ceram. Soc. Jpn.* **101**: 325–330 (in Japanese).
43. Roberts, P. M. (1987). *Metal Construction* **1**: 12–18.
44. Heinlich, J. G., and Kruener, H. (1994). *Tailoring of Mechanical Properties of* Si_3N_4 *Ceramics*, pp. 19–41, Hoffmann, M. J., and Petzow, G., eds., Dordrecht: Kluwer Academic.
45. Moergenthaler, K. D., and Buehl, H. (1994). *Tailoring of Mechanical Properties of* Si_3N_4 *Ceramics*, pp. 429–441, Hoffmann, M. J., and Petzow, G., eds., Dordrecht: Kluwer Academic.
46. Kapoor, R. R., and Eagar, T. W. (1989). *J. Am. Ceram. Soc.* **72**: 448–454.
47. Santacreu, P. O., Koutny, J. L., Bartout, J. D., Bienvenu, Y., and Colin, C. (1994). *Mater. High Temp.* **12**: 293–299.
48. Iwamoto, N., and Kamai, M. (1993). *Structural Ceramics Joining II*, pp. 125–133, Moorhead, A. J., Loehman, R. E., and Johonson, S. M., eds., Westerville: American Ceramic Society.
49. Dannheim, H., Faber, M., Brunner, D., Auerswald, E., Dubek, R., and Faust, W. (1997). *Joining, Ceramics, Glass and Metal*, pp. 126–129, Turwitt, M., ed., Düsseldorf: DVS.
50. Ishida, Y., and Ichinose, H. (1986). *Zairyou Kagaku*. **172** (in Japanese).
51. Tanaka, S. (1986). F. C. Rept., 4, 13 (in Japanese).
52. Zhou, Y., Bao, F. H., Ren, J. L., and North, T. H. (1991). *Mater. Sci. Technol.* 7: 863–868.
53. Kovalev, S. P., Miranzo, P., and Osendi, M. I. (1998). *J. Am. Ceram. Soc.* **81**: 2342–2348.
54. Nagano, T., and Wakai, F. (1993). *J. Mater. Sci.* **28**: 5793.
55. Xie, R.-J., Mitomo, M., Zhan, G.-O., Huang, L.-P. and Fu, X.-R. (2001). *J. Am. Ceram. Soc.* **84**: 471–473.
56. Chain, R., and Ravi, B. G. (2000). *J. Mater. Res.* **15**: 1724–1728.
57. Katz, J. D. (1992). Microwave sintering of ceramics. *Annu. Rev. Mater. Sci.* **22**: 153–170.
58. Ph. Boch, and Lequeux, N. (1997). *Solid State Ionics* **101–103**: 1229–1233.
59. Binner, J. G. P., Davis, P. A., Cross, T. E., and Fernie, J. A. (1995). *Microwaves Theory and Application in Microwave Processing III*, pp. 335–346, Clark, D. E., Folz, D. C., and Oda, S. J., eds., Westerville: Am. Ceram. Soc. Inc.
60. Federer, J. I. (1990). *Corrosion and Corrosive Degradation of Ceramics*, Vol. 10, pp. 425–443, Westerville: Am. Ceram. Soc. Inc.
61. Bakker, W. T., and Kotschick, D. (1982). *High Temp. Technol.* 75–83.
62. Tressler, R. E., and McNallan, M., eds. (1990). *Corrosion and Corrosive Degradation of Ceramics*, Vol. 10, Westerville: Am. Ceram. Soc. Inc.
63. Foster, B. D., and Patton, J. B., eds. (1985). *Ceramics in Heat Exchanger*, Westerville: Am. Ceram. Soc. Inc.
64. Ahmad, I., Silberglitt, R., Black, W. M., Sa'adoldim, H. S., Tian, Y.-L., and Katz, J. D. (1993). *Microwaves: Theory and Application in Materials Processing II*, pp. 409–415, Clark, D. E., Tinga, W. R., and Laier, J. R., Jr., eds., Westerville, Am. Ceram. Soc. Inc.
65. Yiim, T. Y., Varadan, V. V., Varadan, V. K., and Conway, J. C. (1991). *Microwave: Theory and Application in Materials Processing*, pp. 507–514, Clark, D. E., Gac, F. D., and Sutton, W. H., eds., Westerville: Am. Ceram. Soc. Inc.
66. Silberglitt, R., Palaith, D., Black, W. M., Sa'adaldin, H. S., Katz, J. D., and Blake, R. D. (1991). *Microwave: Theory and Application in Materials Processing*, pp. 487–495, Clark, D. E., Gac, F. D., and Sutton, W. H., eds., Westerville: Am. Ceram. Soc. Inc.
67. Ahmad, I., Silberglitt, R., Black, W. M., Sa'adaldin, H. S., and Katz, J. D. (1992). *Mater. Res. Soc. Symp. Proc.* **269**: 271–276.
68. Binner, J. G. P., Fernie, J. A., and Whitaker, P. A. (1998). *J. Mater. Sci.* **33**: 3009–3013.
69. Hocking, M. G., Vasantasree, V., and Sidkly, P. S., eds. (1989). *Metallic and Ceramic Coatings*, p. 20, New York: Longman Sci. & Tech.

Handbook of Advanced Ceramics
S. Sōmiya *et al.* (Eds.)

9.1 Fracture Mechanics

T. FETT and D. MUNZ
Forschungszentrum Karlsruhe and Universität Karlsruhe, D-76021 Karlsruhe, Germany

9.1.1 INTRODUCTION

Ceramic materials fail due to the extension of processing defects such as pores, inclusions or cracks. Under rapid loading, unstable extension of these defects causes failure of the component. Under constant or cyclic loading, the defects may extend slowly until a critical size is reached, where unstable extension occurs. The subcritical extension of the defects under constant loading is called static fatigue, subcritical extension under cyclic loading is called cyclic fatigue. The defects can be described as cracks with a sharp tip and, therefore, the methods of fracture mechanics—in most cases of linear fracture mechanics—can be applied. This concept of sharp cracks has been applied successfully to describe the failure behaviour of ceramics, even in cases where the flaws are three-dimensional, e.g. pores.

The basic relations of fracture mechanics are:

- A relation between strength σ_c and flaw size a in the form of

$$\sigma_c = \frac{K_{Ic}}{\sqrt{a}Y} \tag{1}$$

  where K_{Ic} is the fracture toughness of the material and Y a constant depending on the shape of the crack.
- A relation between crack growth rate da/dt and the stress intensity factor K in the form of

$$\frac{da}{dt} = AK^n \tag{2}$$

  with the material parameters A and n.
- A relation between the crack extension during one cycle and the range of the stress intensity factor ΔK in the form of

$$\frac{da}{dN} = A(\Delta K)^n \tag{3}$$

with the material parameters A and n, which are different from the values for static fatigue.

- A relation between crack growth rate in the creep range and the loading parameter C^* in the form of

$$\frac{da}{dt} = A(C^*)^n \tag{4}$$

with the material parameters A and n.

In this chapter, the basic principles of fracture mechanics are explained, the methods to determine the material parameters are presented and details beyond the simple equations of 1–3 are described.

9.1.2 LINEAR-ELASTIC FRACTURE MECHANICS

The basics of fracture mechanics are outlined for metals in several books [1–3]. For ceramic materials a detailed overview is given in Ref. [4]. Here, a comprehensive representation of the principles will be given.

Figure 9.1.1 shows a plate of thickness B and width W containing an edge crack of depth a. Three loading modes result in high stresses at the crack tip:

mode I: tension normal to the crack plane, σ_y;
mode II: shear loading in crack direction, τ_{xy};
mode III: "out-of-plane" shear loading, τ_{yz}.

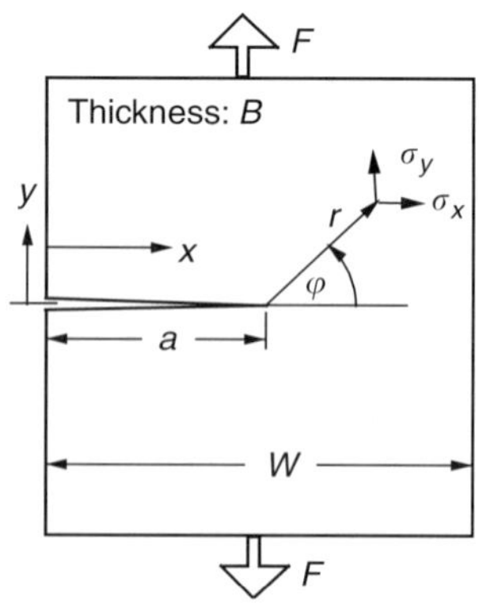

FIGURE 9.1.1 Geometrical data for an edge-cracked body (schematic representation).

Taking into consideration the singular stress term and the first regular term, the near-tip stress field can be described by

$$\sigma_{ij} = \frac{K_I}{\sqrt{2\pi r}} f_{ij}(\varphi) + \sigma_{ij,0} \tag{5}$$

where r and φ are polar coordinates with the origin at the crack tip (Fig. 9.1.1). K_I is the stress intensity factor that depends on the applied load, on the size of the crack and on the geometry of the component. The angular functions f for mode I are

$$f_{xx} = \cos\left(\frac{\varphi}{2}\right)\left[1 - \sin\left(\frac{\varphi}{2}\right)\sin\left(\frac{3\varphi}{2}\right)\right] \tag{6a}$$

$$f_{yy} = \cos\left(\frac{\varphi}{2}\right)\left[1 + \sin\left(\frac{\varphi}{2}\right)\sin\left(\frac{3\varphi}{2}\right)\right] \tag{6b}$$

$$f_{xy} = \cos\left(\frac{\varphi}{2}\right)\sin\left(\frac{\varphi}{2}\right)\cos\left(\frac{3\varphi}{2}\right) \tag{6c}$$

The first regular term in Eq. 5 represents the constant σ_x-stress contribution appearing at the crack tip, which is called the T-stress:

$$\sigma_{ij,0} = \begin{pmatrix} \sigma_{xx,0} & \sigma_{xy,0} \\ \sigma_{yx,0} & \sigma_{yy,0} \end{pmatrix} = \begin{pmatrix} T & 0 \\ 0 & 0 \end{pmatrix} \tag{7}$$

For mode II it holds that

$$f_{xx} = \sin\left(\frac{\varphi}{2}\right)\left[2 + \sin\left(\frac{\varphi}{2}\right)\sin\left(\frac{3\varphi}{2}\right)\right] \tag{8a}$$

$$f_{yy} = \sin\left(\frac{\varphi}{2}\right)\cos\left(\frac{\varphi}{2}\right)\cos\left(\frac{3\varphi}{2}\right) \tag{8b}$$

$$f_{xy} = \cos\left(\frac{\varphi}{2}\right)\left[1 - \sin\left(\frac{\varphi}{2}\right)\cos\left(\frac{3\varphi}{2}\right)\right] \tag{8c}$$

In modes II and III, the first regular stress term is zero. The stress component in thickness direction, σ_z, is given by

$$\sigma_z = \begin{cases} \nu(\sigma_y + \sigma_x) & \text{for plane strain} \\ 0 & \text{for plane stress} \end{cases} \tag{9}$$

(ν is Poisson's ratio). Plane stress conditions are fulfilled at the specimen surfaces, that is, near $z = \pm B/2$. In the centre of thick specimens, plane strain conditions occur.

The stress intensity factor can be written as

$$K_{\mathrm{I}} = \sigma\sqrt{a}Y(a/W) \tag{10}$$

In this relation, σ is a characteristic stress in the uncracked body, for example, the remote tensile stress in a tensile test or the outer fibre bending stress in a bending test. The geometric function Y depends on the crack size normalized to a characteristic dimension of the component. In the case of a curved crack front, for example, a semi-elliptical surface crack, the stress intensity factor varies along the crack front.

It follows from Eq. 5 that the stress state at the crack tip is characterized by the stress intensity factor K_{I}. Consequently, the crack propagation behaviour also depends on K_{I}.

If a component or a test specimen with a crack is loaded, K_{I} increases with increasing load until unstable crack propagation occurs at a critical value of K_{I}. This critical value is the fracture toughness K_{Ic}. It is a material property and can be determined experimentally.

The description of crack propagation behaviour with stress intensity factors is based on the consideration of stresses near the crack tip. An alternative is the consideration of energies. The crack resistance G_{Ic} is introduced, which is the energy necessary to increase the crack area by unit area.

The specific surface energy γ is the energy necessary for the generation of a unit area of new crack surface. Since two crack surfaces are created during crack extension,

$$G_{\mathrm{Ic}} = 2\gamma \tag{11}$$

is obtained.

The energy for crack extension is provided by two sources, that is, the work A done by the external forces and the elastically stored energy in the component U. The energy release rate G_{I} is the energy that is released in the case of a virtual increase of the crack by the unit area. Therefore,

$$G_{\mathrm{I}} = \frac{\mathrm{d}A}{\mathrm{d}S} - \frac{\mathrm{d}U}{\mathrm{d}S} \tag{12}$$

where S is the crack area. The elastically stored energy U enters with a negative sign, since a decrease of the elastically stored energy provides a positive energy contribution for crack extension. With increasing load G_{I} increases, too. Crack extension is initiated if $G_{\mathrm{I}} = G_{\mathrm{Ic}}$ is reached.

Further analysis is illustrated in Figure 9.1.2. For a specimen with a crack of depth a, the load F is plotted as a function of the displacements V of the points, where the load is applied. A straight line will be found for linear material behaviour. The ratio between displacement and applied load is called the

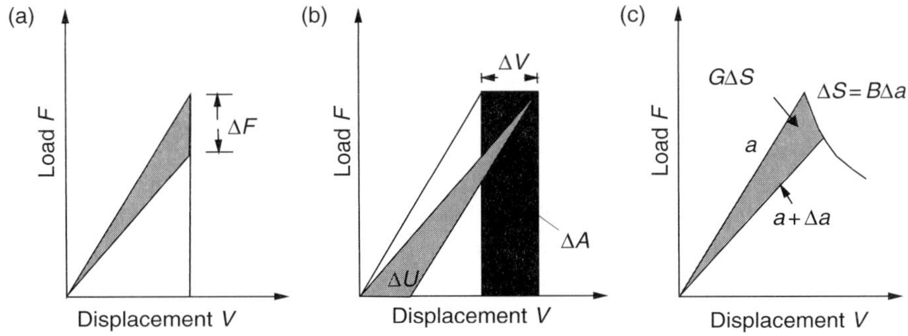

FIGURE 9.1.2 Load–displacement curve under crack extension: (a) constant displacement conditions; (b) constant load conditions; (c) general case with change in load and displacement.

compliance C

$$C = \frac{V}{F} \tag{13}$$

which depends on the crack depth a. After a crack extension Δa, the compliance has increased. In Figure 9.1.2a, the case of crack extension under constant total displacement is plotted. The load decreases by an amount of ΔF. The shaded area corresponds to the elastically stored energy released during crack propagation. The work done by the externally applied forces is zero, because the total load point displacement is kept constant. In Figure 9.1.2b, the case of crack growth under constant load F is illustrated. The displacement increases by ΔV. The work done by the external forces is

$$\Delta A = F\Delta V \tag{14}$$

The elastically stored energy in the specimen increases by the contribution ΔU. Since $\Delta A > \Delta U$, energy remains available for crack extension.

In the following considerations, the energy release rate shall be computed for the general case (Fig. 9.1.2c), where both the displacement and the load may change during crack extension.

The work done by the external load is given by

$$A = \int F\,\mathrm{d}V \tag{15}$$

and, therefore,

$$\frac{\mathrm{d}A}{\mathrm{d}S} = \frac{\mathrm{d}A}{\mathrm{d}V}\frac{\mathrm{d}V}{\mathrm{d}S} = F\frac{\mathrm{d}(CF)}{\mathrm{d}S} = F^2\frac{\mathrm{d}C}{\mathrm{d}S} + FC\frac{\mathrm{d}F}{\mathrm{d}S} \tag{16}$$

is derived. The elastically stored energy is

$$U = \tfrac{1}{2}FV = \tfrac{1}{2}F^2C \tag{17}$$

and

$$\frac{\mathrm{d}U}{\mathrm{d}S} = FC\frac{\mathrm{d}F}{\mathrm{d}S} + \frac{1}{2}F^2\frac{\mathrm{d}C}{\mathrm{d}S} \tag{18}$$

From Eqs 12, 17 and 18, it follows that

$$G_\mathrm{I} = \frac{F^2}{2}\frac{\mathrm{d}C}{\mathrm{d}S} \tag{19a}$$

For a plate of thickness B with a through-the-thickness crack, Eq. 19a can be replaced by

$$G_\mathrm{I} = \frac{F^2}{2B}\frac{\mathrm{d}C}{\mathrm{d}a} = \frac{F^2}{2BW}\frac{\mathrm{d}C}{\mathrm{d}(a/W)} \tag{19b}$$

If the compliance is known as a function of the crack area the energy release rate for the cracked component can be calculated with Eq. 19.

The loading situation may alternatively be characterized by the stress intensity factor as well as by the energy release rate. Irwin has shown that a simple relation exists between the two loading quantities, namely,

$$K_\mathrm{I}^2 = G_\mathrm{I}E' \tag{20}$$

with

$$E' = \begin{cases} E & \text{for plane stress} \\ E/(1-\nu^2) & \text{for plane strain} \end{cases} \tag{21}$$

Unstable crack propagation occurs if

$$K_\mathrm{I} = K_\mathrm{Ic} \tag{22}$$

or if

$$G_\mathrm{I} = G_\mathrm{Ic} \tag{23}$$

Equation 20 also holds for the failure condition. Consequently, we find for plane strain

$$K_\mathrm{Ic}^2 = \frac{G_\mathrm{Ic}E}{1-\nu^2} = \frac{2\gamma E}{1-\nu^2} \tag{24}$$

9.1.3 METHODS FOR THE DETERMINATION OF FRACTURE TOUGHNESS K_{Ic}

The principal procedure of fracture toughness measurements consists in the generation of a crack in a test specimen, the measurement of load at failure and the calculation of K_{Ic} from failure load or failure stress, respectively, and from crack depth using the relations

$$K_{Ic} = \sigma\sqrt{a}Y \quad \text{or} \quad K_{Ic} = \frac{F}{B\sqrt{W}}Y^* \tag{25}$$

with the crack length included in the geometric function $Y^*(a/W)$.

9.1.3.1 The Edge-Cracked Bending Bar

The specimens commonly have a thickness of $B = 3$ mm and height of $W = 4$ mm and are usually loaded in a four-point bending test arrangement (Fig. 9.1.3). The inner roller span is mostly $S_2 = 20$ mm, the outer one $S_1 = 40$ mm.

Measurements of fracture toughness for ceramic materials are often performed with specimens containing narrow saw cuts. The advantage of edge-notched bars consists in the relatively simple introduction of the notch. For a fine-grained ceramic a very narrow notch is necessary.

A procedure proposed by Nishida *et al.* [5] allows to introduce notches with notch root radii in the range of 1–10 μm. In the first step, a conventional saw cut is introduced. Then, a sharp V-shaped notch is produced by using a razor blade and diamond paste (Fig. 9.1.4). In extensive investigations the usefulness of this method was confirmed [6].

The fracture toughness can be computed from the maximum load F_{max} and the relative crack length $\alpha = a/W$:

$$K_{Ic} = \frac{F_{max}}{B\sqrt{W}} \frac{S_1 - S_2}{W} \frac{3\Gamma_M\sqrt{\alpha}}{2(1-\alpha)^{3/2}} \tag{26}$$

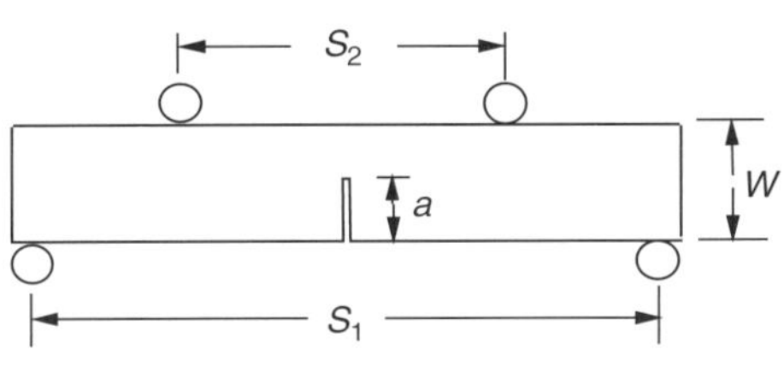

FIGURE 9.1.3 Geometric data for a four-point bending test with an edge-notched specimen.

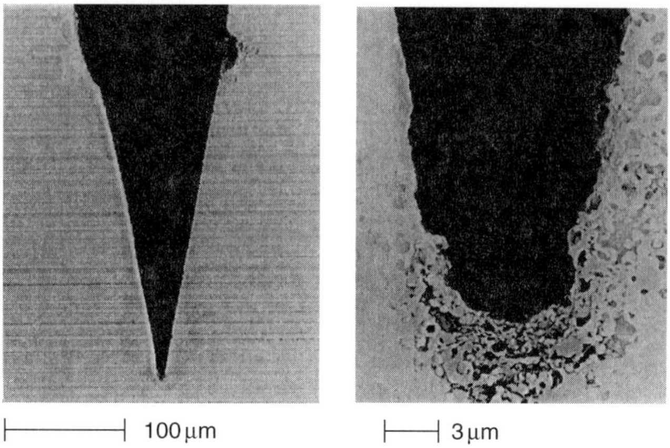

FIGURE 9.1.4 V-notch introduced into a bending bar (Si_3N_4) [6]; right: detailed view of the notch root.

with [7]

$$\Gamma_M = 1.1215\sqrt{\pi}\left[\frac{5}{8} - \frac{5}{12}\alpha + \frac{1}{8}\alpha^2 + 5\alpha^2(1-\alpha)^6 + \frac{3}{8}\exp\left(-6.1342\frac{\alpha}{1-\alpha}\right)\right] \quad (27)$$

Sometimes, three-point bending tests are also used to determine K_{Ic}. This test may be of special interest, if only a small number of specimen is available. Then, the fragments remaining from four-point bending tests can be used again. In Ref. [7], a stress intensity factor solution is given, which holds for all S_1/W ratios in the range of $S_1/W > 2$. For the special value of $S_1/W = 2.5$ (obtained, e.g. with $S_1 = 10$ mm, $W = 4$ mm), Eq. 26, with $S_2 = 0$, becomes

$$\Gamma_M = \sqrt{\pi}[0.3738\alpha + (1-\alpha)(1.044 - 2.116\alpha + 3.733\alpha^2 - 3.147\alpha^3 + 0.935\alpha^4)] \quad (28)$$

Experimental investigations on the influence of the notch width and the notch root radius ρ showed the dependency given schematically in Figure 9.1.5a. Beyond a critical notch root radius ρ_c, the measured nominal K_{Ic}-value linearly increases with $\rho^{1/2}$. Only for $\rho < \rho_c$ will a correct K_{Ic}-value be measured. The critical value ρ_c depends on the tested material. In any case, it has to be ensured that the saw cut is narrow enough. In order to obtain real toughness data from the results measured with a finite notch radius, an extrapolation technique may be applied. Basis for such consideration is a stress intensity factor solution for small cracks in front of a narrow notch.

Figure 9.1.6 illustrates a small crack of length ℓ ahead of a circular notch of radius ρ and length a_0. For an edge crack along the notch root a simple relation

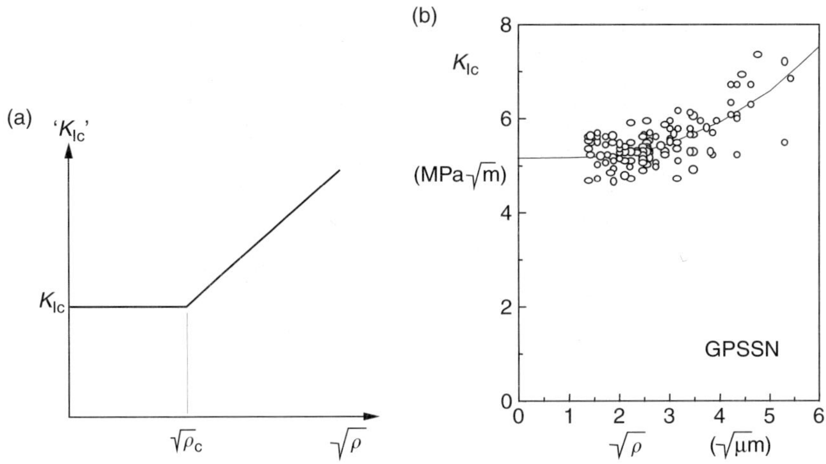

FIGURE 9.1.5 Influence of the notch root radius on the measured fracture toughness: (a) schematic representation; (b) results on silicon nitride [6, 9].

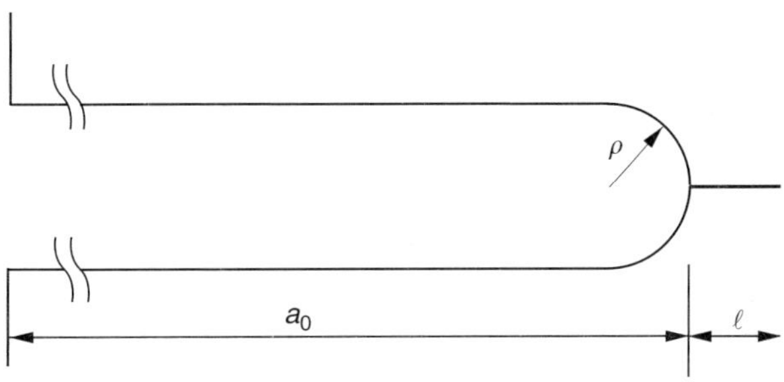

FIGURE 9.1.6 Crack in front of a narrow notch.

for the stress intensity factor K was derived based on the Boundary Collocation procedure [7]

$$K/K^* = \tanh[\beta\sqrt{\ell/\rho}], \quad \beta \cong 2.24 \tag{29}$$

In this relation, K^* is the stress intensity factor of an edge crack with the total length $a = a_0 + \ell$. This relation is applied for the evaluation of fracture toughness of ceramic materials with the natural defects at the notch root modelled by straight through-the-thickness cracks. In practical applications, crack-like defects in ceramics are described well by semi-circular or semi-elliptical cracks, but, in general, the exact geometry and size of the small crack ahead of the notch is unknown.

Damani *et al.* [8] assumed ℓ to be related to microstructural features or machining damage, for example, grain size, pores or scratches. Damani *et al.* [8] also proposed the use of the property $\beta\sqrt{\ell}$ in Eq. 29 as a free fit parameter. In this way, it can be taken into consideration that real cracks in front of a notch have a shape between an edge crack and a semi-elliptic crack. As an example, Figure 9.1.5b shows results from Ref. [9] for a silicon nitride. From these data a fit yields $K_{Ic} \cong 5.2\,\text{MPa}\sqrt{\text{m}}$.

9.1.3.2 Specimens with Chevron Notches

The most popular specimen with a chevron notch is shown in Figure 9.1.7. By two saw cuts, a triangularly shaped cross-section is produced. During loading, a sharp crack develops from the tip of the notch and the length of the crack front increases continuously. The geometrical data of the specimen and crack are:

- specimen thickness B,
- specimen width W,
- notch parameters a_0 and a_1 ($\alpha_0 = a_0/W, \alpha_1 = a_1/W$).

Between the length of the crack front b and the crack depth a the relation

$$b = B\frac{a - a_0}{a_1 - a_0} = B\frac{\alpha - \alpha_0}{\alpha_1 - \alpha_0} \tag{30}$$

holds with $\alpha = a/W$.

The normally obtained load–displacement curve is illustrated in Figure 9.1.8. The record is curved and exhibits a maximum. The non-linearity is caused by crack propagation and a related increase in compliance. At maximum load, the crack has propagated (starting from $a = a_0$) to $a = a_{max}$.

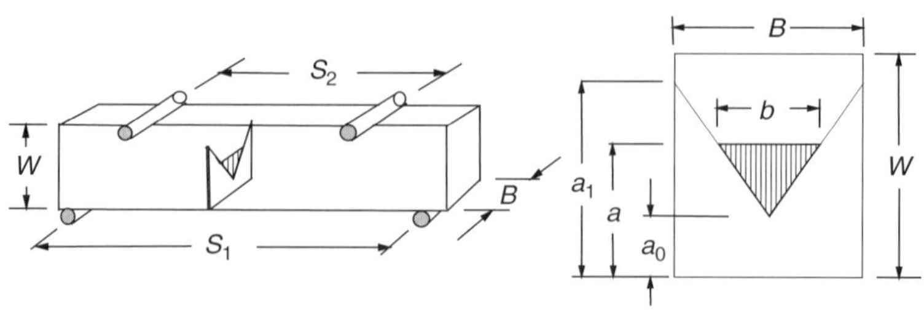

FIGURE 9.1.7 Four-point bending specimen with chevron notch.

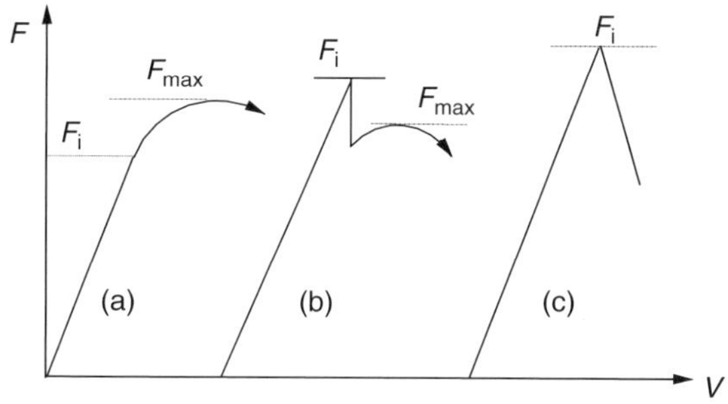

FIGURE 9.1.8 Typical load–displacement plots for chevron-notched specimens.

If C_{Ch} is the compliance of the chevron specimen, the available energy under load F is

$$\Delta P = \frac{F^2}{2W}\frac{dC_{Ch}}{d\alpha}\Delta a \tag{31}$$

Crack propagation is initiated, if ΔP equals the energy required for crack extension, namely $G_{Ic}b\Delta a$. This finally provides

$$K_{Ic} = \frac{F}{B\sqrt{W}}\left[\frac{EB}{2(1-\nu^2)}\frac{\alpha_1-\alpha_0}{\alpha-\alpha_0}\frac{dC_{Ch}}{d\alpha}\right]^{1/2} = \frac{F}{B\sqrt{W}}Y^* \tag{32}$$

The geometric function Y^* contains two terms that change with α. The derivative of the compliance with respect to α increases with α and the geometry term $(\alpha_1-\alpha_0)/(\alpha-\alpha_0)$ decreases with increasing α. If during crack extension the material resistance is constant ($K_I = K_{Ic} =$ const.), we must also have $FY^* =$ const. The minimum Y^*_m of Y^* corresponds to the maximum load F_{max}, and the fracture toughness can be expressed as

$$K_{Ic} = \frac{F_{max}}{B\sqrt{W}}Y^*_m \tag{33}$$

In order to allow K_{Ic} to be determined, the geometric function Y^*_m has to be known for different specimen geometries (W, B) and notch geometries (α_0, α_1). From compliance measurements and finite element computations, the following relation was determined for the four-point bending specimen [10]:

$$Y^*_m = (3.08 + 5.00\alpha_0 + 8.33\alpha_0^2)\left\{1 + 0.007\sqrt{\frac{S_1S_2}{W^2}}\right\}\frac{\alpha_1-\alpha_0}{1-\alpha_0}\frac{S_1-S_2}{W} \tag{34}$$

(S_1 is the outer roller distance and S_2 the inner roller distance), valid for $0.12 \le \alpha_0 \le 0.24$, $0.9 \le \alpha_1 \le 1$.

The main advantage of specimens with chevron notches is the fact that no sharp pre-crack has to be introduced. The specimen produces the sharp crack during loading. Moreover, the evaluation of fracture toughness does not need the actual crack length, since Eq. 33 only contains the maximum load F_{max} and the geometric function Y_m^*. Further test specimens with chevron notches (short rod, short bar) are addressed in Ref. [4].

In the case of a material with a rising crack resistance curve exhibiting a so-called *R*-curve behaviour (see Section 9.1.6), the resistance against crack extension increases with increasing crack extension. Consequently, the product of F and Y^* is no longer constant, but increases with crack extension. The maximum load F_{max} does not occur exactly at the minimum Y_m^* of the geometric function Y^*. Nevertheless, it can be shown that application of Eq. 33 gives a toughness value that lies on the rising *R*-curve [11]. The crack extension at F_{max}, however, depends on the specimen size. This leads to the influence of the specimen size on the measured K_{Ic}.

For an idealized notch with a disappearing notch root radius crack extension would immediately start after loading. Due to the finite notch width, crack propagation starts at a finite initial load F_i. Curve (a) in Figure 9.1.8 shows the desirable case $F_i < F_{max}$. Curve (b) in Figure 9.1.8 illustrates the case $F_i > F_{max}$ where a sudden decrease in load indicates a limited amount of unstable crack extension. Then, the load increases again and passes the maximum F_{max}. In case of curve (c) Figure 9.1.8, crack arrest is not reached and the maximum load cannot be measured. Consequently, fracture toughness cannot be determined in such a case.

9.1.3.3 Specimen with a Knoop Indentation Crack

A very simple way to produce a sharp crack is the hardness indentation test with a diamond pyramid. In the case of a Knoop pyramid, a nearly semicircular surface crack develops under load. During unloading, the damaged region closes.

In Figure 9.1.9a, the characteristic dimensions of a Knoop indentation are represented. The hardness impression on the surface has a length L and a width b. The crack profile that becomes visible on the fracture surface (see Figure 9.1.9b for a soft PZT ceramic) is characterized by the depth a and width $2c$. The depth of the impression is x and the depth of the damaged zone z_0. For some materials this quantity can be detected on the fracture surface. It was found that z_0 and b are nearly identical [12]. After unloading, the damaged region is under compressive stresses that are balanced by tensile stresses near the crack tip region.

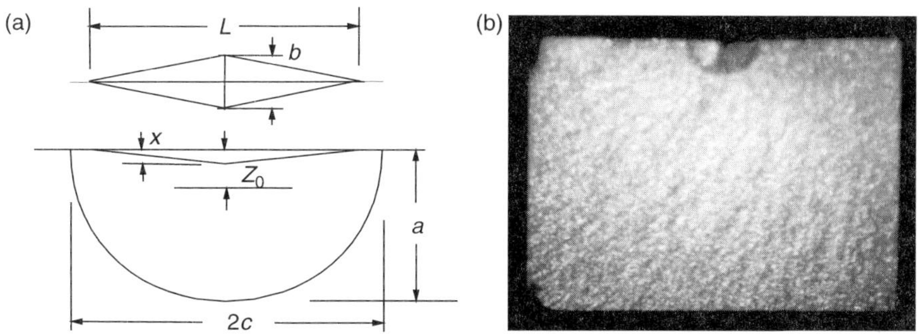

FIGURE 9.1.9 (a) Geometrical data of a Knoop indentation crack. (b) Knoop crack in a PZT specimen.

Commonly, the surface cracks are introduced into bending bars that are then broken in a four-point bending test. The computation of the fracture toughness K_{Ic} from the maximum load can be performed using the equation given by Newman and Raju [13].

The relation of Newman and Raju is appropriate for aspect ratios of $a/c < 1$:

- at the deepest point:

$$K_I = \sigma\sqrt{\pi a}\frac{M}{\Phi}H_2 \tag{35}$$

- at the specimen surface:

$$K_I = \sigma\sqrt{\pi a}\frac{M}{\Phi}H_1\sqrt{a/c}(1.1 + 0.35\alpha^2) \tag{36}$$

with the outer fibre bending stress σ

$$\sigma = \frac{3(S_1 - S_2)F}{2W^2B} \tag{37}$$

and the quantities

$$H_1 = 1 - [0.34 + 0.11(a/c)]\,\alpha \tag{38}$$

$$H_2 = 1 - \left[1.22 + 0.12(a/c)\right]\alpha + \left[0.55 - 1.05(a/c)^{0.75} + 0.47(a/c)^{1.5}\right]\alpha^2 \tag{39}$$

$$M = 1.13 - 0.09\frac{a}{c} + \left[-0.54 + \frac{0.89}{0.2 + a/c}\right]\alpha^2 + \left[0.5 - \frac{1}{0.65 + a/c} + 14(1 - a/c)^{24}\right]\alpha^4 \tag{40}$$

Φ is the complete elliptical integral of the second kind, which can be approximated by

$$\Phi \cong [1 + 1.464(a/c)^{1.65}]^{1/2} \tag{41}$$

Fracture will be caused at the point with the maximum stress intensity factor, that is, for cracks with $a/c > 0.8$ at the surface point B.

A problem in the correct evaluation of K_{Ic} is the removal of the residual stresses caused by the damaged zone directly under the Knoop indenter. The residual stresses are responsible for a positive residual stress intensity factor K_{res}, which is superimposed on the externally applied stress intensity factor and leads to a lower failure stress in the bending test. The residual stress intensity factor may be reduced during load application, but will not disappear completely before failure. Since K_{Ic} is computed with the maximum externally applied load, the resulting value will be too small. To determine correct K_{Ic} values, the residual stresses have to be removed before the bending test is carried out. This can be done by surface grinding or by annealing.

A more indirect method of toughness determination is based on crack evaluation during a Vickers indentation test. The relevant references for this test are reported in Ref. [4].

9.1.4 MODE-II AND MIXED-MODE FRACTURE TOUGHNESS

9.1.4.1 Determination of Toughness

Besides the mode-I fracture behaviour, fracture under mode II and mixed-mode loading also are of interest. The results are plotted in a mixed-mode K_I–K_{II} diagram. Three types of test specimens will be described in the following subsections.

Whereas cracks orthogonal to the specimen surfaces are applied in K_{Ic} determination, oblique edge cracks (or notches) are appropriate for mixed-mode fracture.

Figure 9.1.10a illustrates the geometrical data of a four-point bending test and Figure 9.1.11a represents the related geometric functions Y_I and Y_{II} for several values of a/W and angles β (see also Ref. [7]). Y_I and Y_{II} are defined by

$$K_I = \sigma Y_I \sqrt{a}, \quad K_{II} = \sigma Y_{II} \sqrt{a}, \quad \sigma = \frac{3}{2} \frac{S_1 - S_2}{BW^2} F \tag{42}$$

While the curves for mode-I stress intensity factors depend significantly on the relative crack depth a/W, the mode-II contributions are nearly independent of a/W. For more details see Ref. [4].

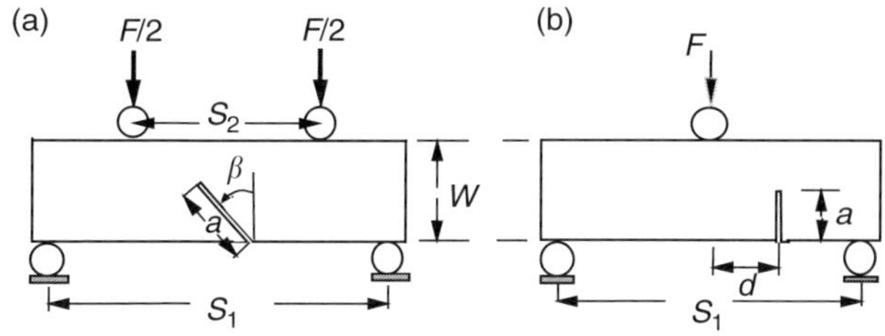

FIGURE 9.1.10 (a) Four-point bending test with an oblique notch. (b) Three-point bending test with an eccentric notch.

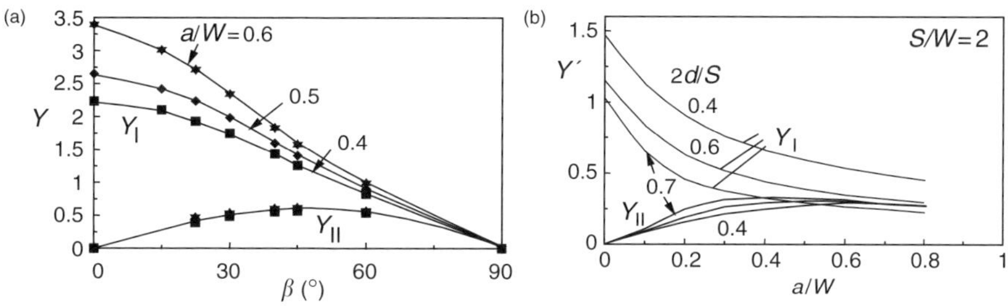

FIGURE 9.1.11 Geometric function for (a) the edge-cracked plate in bending, (b) normalized geometric functions for eccentric notches; $Y'_{II} = Y_{II}(1 - a/W)^{1/2}$, $Y'_I = Y_I(1 - a/W)^{3/2}$.

If in a three-point bending test the specimen is displaced and the notch is no longer in the centre (see Fig. 9.1.10b), a mixed-mode loading situation exists. The geometric functions of the stress intensity factors defined by Eq. 42 with $S_2 = 0$ are given in Ref. [7]. As an example, the normalized geometric functions $Y'_{II} = Y_{II}(1 - a/W)^{1/2}$ and $Y'_I = Y_I(1 - a/W)^{3/2}$ are plotted in Figure 9.1.11b.

For the determination of the pure mode-II fracture toughness K_{IIc} with edge-notched (or precracked) beams, asymmetric bending arrangements (Fig. 9.1.12) are recommended [14–16], which show a negligible mode-I contribution. The geometric functions Y_I and Y_{II}, defined by

$$K_I = \frac{F}{BW}\left(1 - \frac{d}{L}\right) Y_I\sqrt{a}, \quad K_{II} = \frac{F}{BW}\left(1 - \frac{d}{L}\right) Y_{II}\sqrt{a} \tag{43}$$

are also shown in Figure 9.1.12.

The geometric function Y_{II} is nearly independent of the value of d/W in the range of $0.375 < d/W < 0.625$. An analytical approximation can be given by

$$Y_{II} = 6.949\alpha - 9.092\alpha^2 + 25.624\alpha^3 - 46.601\,\alpha^4 + 30.261\,\alpha^5,$$
$$\alpha = a/W \tag{44}$$

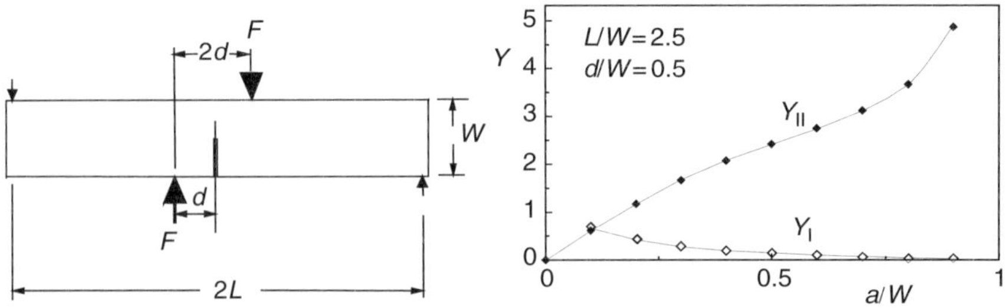

FIGURE 9.1.12 Geometrical data and geometric functions Y_I and Y_{II} of the asymmetrically loaded notched bending bar.

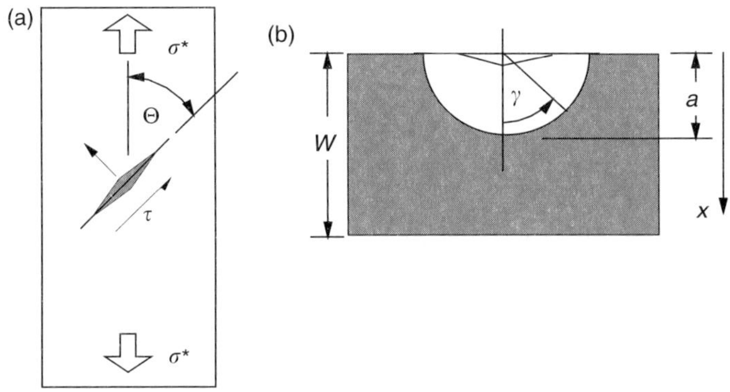

FIGURE 9.1.13 Inclined Knoop surface flaw in a bending test.

Mixed-mode studies have often been performed with controlled surface flaws in bending bars [17–19]. A single Knoop surface flaw is placed on the tensile surface at an orientation angle Θ with respect to the specimen length axis (Fig. 9.1.13). The externally applied uniaxial stress σ^* gives rise to a stress σ_n normal to the crack and a shear stress τ in the crack plane

$$\sigma_n(x) = \sigma^*(x)\sin^2\Theta, \qquad \tau(x) = \sigma^*(x)\sin\Theta\cos\Theta \tag{45}$$

In a bending test σ^* decreases linearly with the distance from the surface. In Refs [17–19], the Knoop crack was assumed to be a semicircular crack, which is nearly in agreement with experimental results, and the decrease of σ^* was neglected ($a/W \ll 1$).

The stress intensity factors used were for a crack in an infinite body, where surface effects are ignored:

$$K_{\mathrm{I}} = 2\sqrt{a/\pi}\,\sigma_{\mathrm{n}} \tag{46}$$

$$K_{\mathrm{II}} = \frac{4}{2-\nu}\sqrt{a/\pi}\,\tau\cos\gamma \tag{47}$$

$$K_{\mathrm{III}} = \frac{4(1-\nu)}{2-\nu}\sqrt{a/\pi}\,\tau\sin\gamma \tag{48}$$

with the angle γ defined in Figure 9.1.13b. The advantage of this test method is the existence of a sharp crack. Nevertheless, there are some problems that aggravate the analysis. It has to be ensured that all residual stresses generated during the indentation test have been removed by annealing or surface grinding. Since the mode-II and mode-III contributions vary along the crack-tip line, it seems impossible to determine without doubt which particular mixed-mode stress intensity factor combination was responsible for failure.

9.1.4.2 Effective Stress Intensity Factors

Under mixed-mode loading, failure of a component containing a crack occurs, if a combination of the stress intensity factors, the effective stress intensity factor K_{eff}, equals the fracture toughness K_{Ic}. The most popular failure criterion is that of the coplanar energy release rate [20]. Under plane strain conditions, it reads

$$K_{\mathrm{eff}} = \sqrt{K_{\mathrm{I}}^2 + K_{\mathrm{II}}^2 + \frac{1}{1-\nu}K_{\mathrm{III}}^2} = K_{\mathrm{Ic}} \tag{49}$$

Such a criterion makes sense only for positive K_{I}. For $K_{\mathrm{I}} < 0$, the crack faces are under compression and no singular mode-I stress field exists. Nevertheless, a mode-II stress intensity factor can occur, caused by the superimposed shear loading. Due to friction between the crack faces, K_{II} has to be calculated with an effective shear stress [21].

$$K_{\mathrm{II,eff}} = \tau_{\mathrm{eff}} Y\sqrt{a}, \qquad \tau_{\mathrm{eff}} = \begin{cases} \tau & \text{for } \sigma_{\mathrm{n}} > 0 \\ |\tau| + \mu\sigma_{\mathrm{n}} & \text{for } |\mu\sigma_{\mathrm{n}}| < |\tau| \\ 0 & \text{for } |\mu\sigma_{\mathrm{n}}| > |\tau| \end{cases} \tag{50}$$

with the geometric function Y for constant shear stresses and the friction coefficient μ. Equation 50 is only correct for cracks with completely closed crack faces and constant stresses over the crack size. This is often the case, but in the case of strongly varying stresses (occurring, e.g. near point or line contacts), this condition is not fulfilled. As outlined in Ref. [22] for edge cracks, the condition

$$K_{\mathrm{II,eff}} = K_{\mathrm{II,appl}} + \mu K_{\mathrm{I,appl}} \tag{51}$$

(where subscripts "appl" designate the applied stress intensity factors caused by the stresses in the uncracked body at the location of the crack) remains exactly valid for small (natural) cracks fully or partially closed, if $a/W \to 0$.

9.1.5 FRACTURE OF PIEZOELECTRIC CERAMICS

When applied as actuators, piezoelectric ceramics are subjected to electrical and mechanical loads. Both types of loading will influence the failure behaviour. Whereas the mechanical loading results in the stress intensity factors K_{I}, K_{II}, and K_{III}, another (electric) stress intensity factor K_{IV} appears in cracked piezoelectric components due to the electric field [23–26].

Two limit cases have to be distinguished, namely, an electrically conducting and an insulating crack. The conducting crack is assumed to be filled with a medium of high permittivity $\varepsilon_{\mathrm{r}} \to \infty$ (or has thin conducting layers on its surface). The insulating crack is filled with an idealized medium $\varepsilon_{\mathrm{r}} \to 0$. Mechanical stiffnesses are supposed to disappear in both cases.

An insulated crack in an infinite body may be considered. With the coordinate system shown in Figure 9.1.1, the dielectric displacements near the crack tip are given by

$$\begin{Bmatrix} D_x \\ D_y \end{Bmatrix} = \frac{K_{\mathrm{IV}}}{\sqrt{2\pi r}} \begin{Bmatrix} -\sin(\varphi/2) \\ \cos(\varphi/2) \end{Bmatrix} \tag{52}$$

that is, the same singularity occurs as in the case of mechanical stresses. The electric field intensity factor K_{IV} is defined as

$$K_{\mathrm{IV}} = D_y^{\infty} \sqrt{\pi a} \tag{53}$$

with the electric displacements in y-direction far away from the crack, D_y^{∞}. The energy release rate, related to K_{IV}, is

$$G_{\mathrm{IV}} = -\frac{1}{2\varepsilon_0 \varepsilon_r} K_{\mathrm{IV}}^2 \tag{54}$$

with $\varepsilon_0 = 8.86 \times 10^{-12}$ A s/V m.

Two specific experimental results may be addressed. An insulated crack in a CT specimen was investigated by Park and Sun [27]. As shown in Figure 9.1.14a, the polarization P, the applied electric field $E = U/H$ and the mechanical loading were perpendicular to the crack plane. The critical stress intensity factor K_{Ic} is plotted in Figure 1.9.14b as a function of the electric field. For positive field E a decrease in K_{Ic} and for negative field an increase of K_{Ic} was found.

The failure criterion for conducting cracks was studied by Heyer *et al.* [28, 29] in bending tests. In this case the polarization was orientated in crack plane direction. The electric boundary conditions are illustrated in Figure 9.1.15a. Figure 9.1.15b represents the results where the critical mechanical mode-I stress intensity factor K_{Ic} is plotted versus the electric intensity

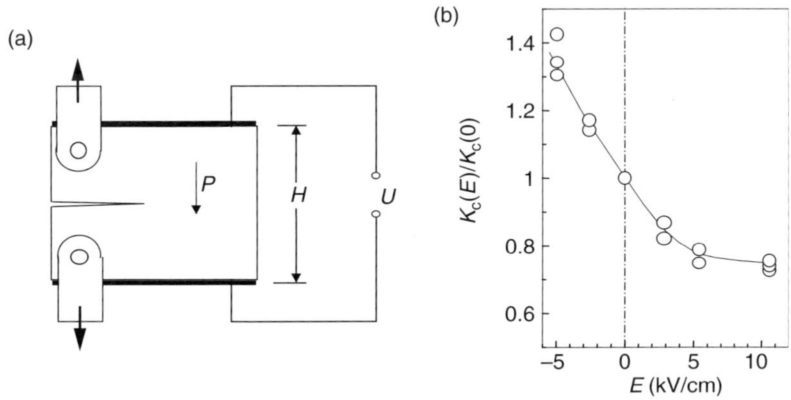

FIGURE 9.1.14 Influence of an electric field on the fracture toughness of a piezoelectric ceramic as reported by Park and Sun [27]; (a) test device; (b) fracture toughness (normalized to toughness in the absence of an electric field).

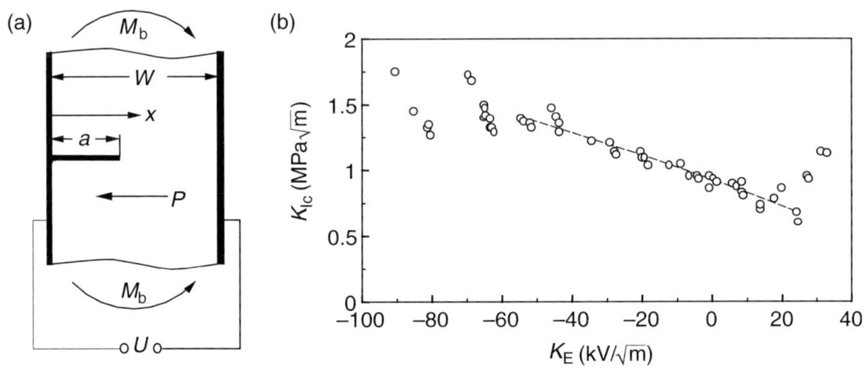

FIGURE 9.1.15 (a) Conducting crack under combined electrical and mechanical loading. (b) Fracture criterion determined by Heyer *et al.* [28, 29].

factor K_E, describing the singularity of the electric field ahead of the crack tip, and defined by the electric field in the crack plane $E(x-a)$ as

$$E = \frac{K_E}{\sqrt{2\pi(x-a)}} \tag{55}$$

The symbols in Figure 9.1.15b cover the range of crack length of $0.9 \le a \le 2.2$ mm.

9.1.6 R-CURVE BEHAVIOUR

In the case of ideally brittle materials, the fracture toughness is independent of the crack extension Δa. In this case, a so-called flat crack resistance curve (*R*-curve) is obtained. In many ceramics a different behaviour is observed. The crack growth resistance increases with increasing crack extension. The necessary energy for unit area crack extension—now called G_{IR}—increases from an initial value G_{I0}. The stress intensity factor K_{IR} increases from K_{I0}. The crack propagation behaviour is no longer described by a single value K_{Ic}, but now characterized by the $K_{IR}-\Delta a$ curve. In a first approximation it was assumed that the crack growth resistance curve is a pure material property, that is, independent of the specimen geometry, the crack size, and the loading (tension, bending, etc.).

In the interpretation of the rising crack growth resistance it is assumed that during crack extension the stress intensity factor at the crack tip K_{I0} is constant and identical to the value at the onset of crack propagation. The stress intensity factor applied during crack extension, K_R, is reduced by a shielding stress intensity factor K_{sh} (see Figure 9.1.16)

$$K_R + K_{sh} = K_{I0}, \quad K_{sh} < 0 \tag{56}$$

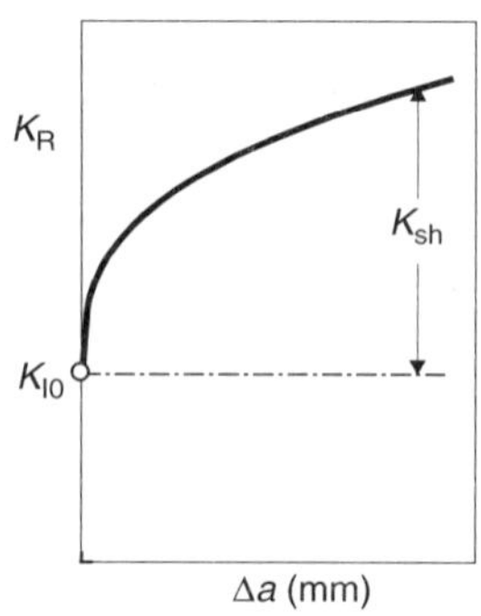

FIGURE 9.1.16 Schematic representation of an *R*-curve.

For coarse-grained materials, the shielding term is predominantly caused by crack border interactions in the wake of the growing crack (see Section 9.1.10). For materials undergoing phase transformation in front of a crack tip (mostly accompanied by a change in volume) such as zirconia, the shielding stress intensity factor is caused by residual stresses that reduce the stresses at the tip. The same holds for domain switching in PZT materials, which occurs under constant volume. In all cases K_{sh} is negative and increases with increasing crack extension. The first observed increasing R-curve was published by Hübner and Jillek [30] for alumina. Later, rising R-curves were found for many other materials. In Ref. [4], the authors presented results for a wide range of materials and covered several standard test methods. In this contribution, some recent procedures and results are shown.

9.1.7 EXPERIMENTAL RESULTS

Some experimental results are shown in Figure 9.1.17. The increase in the crack growth resistance with increasing grain size can be seen from Figure 9.1.17a for alumina. Figure 9.1.17b shows that the slope of the crack growth resistance curve is not independent of the initial crack size. Figure 9.1.17c reveals that large regions with a linear increase in crack growth resistance are possible. In Figure 9.1.17d an example is given, where after an increase in the crack growth resistance a decrease follows.

9.1.8 METHODS FOR THE DETERMINATION OF *R*-CURVES

9.1.8.1 Controlled Bending Tests

For the experimental determination of crack resistance curves for specimens with macrocracks, the actual load F and the actual crack length a have to be determined. K_R then can be calculated according to

$$K_R = \sigma\sqrt{a}\,Y(a/W) = \frac{F}{B\sqrt{W}}Y^*(a/W) \tag{57}$$

with

$$Y^* = Y\sqrt{a/W}$$

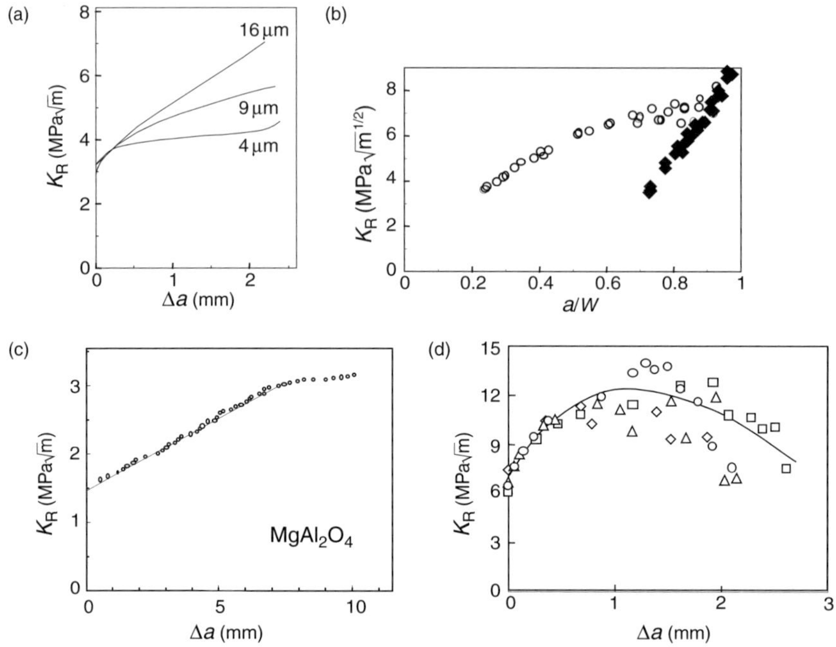

FIGURE 9.1.17 (a) Crack resistance curves for alumina with different grain sizes [31]. (b) Influence of the initial crack size on R-curves for alumina [4]. (c) $MgAl_2O_4$ spinel from DCB tests [32]. (d) R-curves of several specimens of 9 mol% Ce–TZP with a mean grain size of $d_m = 2.5\,\mu m$ [33].

The crack length can be measured optically with a microscope. An alternative is to use the compliance

$$C = \frac{V}{F} \tag{58}$$

where V is the load point displacement.

In Figure 9.1.18, a very rigid bending test device is shown [33]. R-curve experiments are carried out by three-point bending with a roller span of 40 mm. The load is recorded with a quartz load cell and the load point displacement monitored with an LVDT. The load rig is mounted horizontally on a microscope to measure the crack length, yielding an accuracy of 1 μm. This enables to compare the crack length obtained via compliance with the physical crack length visible on the surface.

Figure 9.1.18 also represents a load versus displacement curve together with partial unloadings, from which the actual compliance can be determined.

The relation between compliance and crack length can be determined experimentally for a given specimen geometry. It is also possible to calculate this

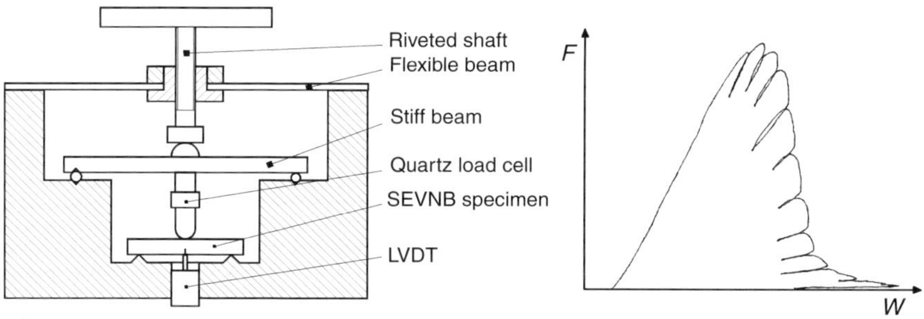

FIGURE 9.1.18 Determination of crack length via compliance: left, load rig for conducting R-curve experiments; right, load versus displacement curve for a 9 mol% Ce–TZP [33].

relation from the function Y or Y^*. From Eqs 19 and 20, it follows that

$$C = C_0 + \frac{2}{BE'} \int_0^{\alpha} Y^2(\alpha')\alpha' \mathrm{d}\alpha' = C_0 + \frac{2}{BE'} \int_0^{\alpha} Y^{*^2}(\alpha')\, \mathrm{d}\alpha' \qquad (59)$$

where Y and Y^* are the geometric functions for the stress intensity factor under the special loading situation and C_0 is the compliance of the specimen in the absence of a crack or notch. From the measured difference in compliance between an uncracked and a cracked specimen, $\Delta C = C - C_0$, $\alpha = a/W$ can be determined.

For larger crack extensions the compliance method underestimates the real crack length. This can be explained by bridging effects. A partially bridged crack leads to a higher stiffness of the cracked component than a crack without bridging interactions. The increased stiffness in the presence of bridging effects results in a lower compliance and, consequently, in a shorter crack length.

9.1.8.2 Stable Fracture Test Under Opposite Roller Loading

A recently developed test for stable crack propagation over a large range of crack extension is illustrated in Figure 9.1.19 [34]. A pre-notched bar is loaded via four rollers by the load P_{appl}, which is applied via a half-sphere to the upper rollers. As rollers, cylindrical pins (hardened steel) may be used. After a certain increase of load, the test is suspended. After unloading, the actual crack length can be measured at the side surfaces using an optical microscope. Additional application of a penetration dye allows to measure the crack depth on the fracture surface of a broken specimen. From the load P_{appl} at the moment of

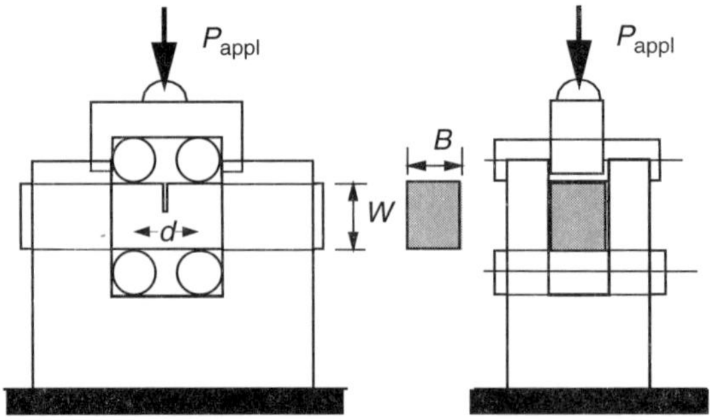

FIGURE 9.1.19 Device for a four-roller crack extension test [34].

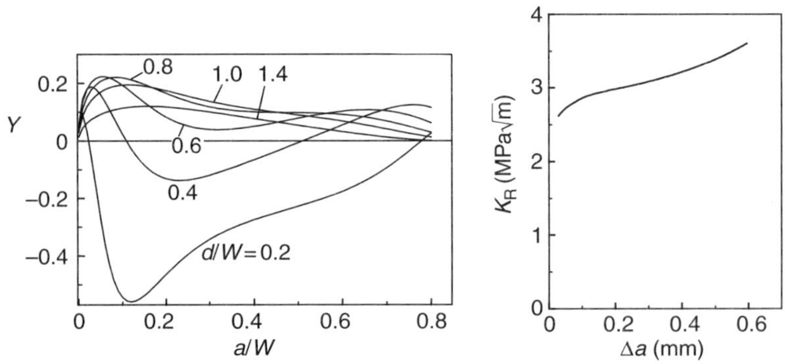

FIGURE 9.1.20 (a) Geometric function Y according to Eq. 60 for four-point loading. (b) R-curve for alumina with 4 % glass content.

unloading and the crack depth a, the mode-I stress intensity factor is obtained as

$$K_{\mathrm{I}} = \sigma^* Y \sqrt{W}, \quad \sigma^* = \frac{P_{\mathrm{appl}}}{WB} \tag{60}$$

with the geometric function Y. The mode-I stress intensity factor is shown in Figure 9.1.20a as a function of a/W for different d/W. Figure 9.1.20b shows an R-curve obtained for an alumina ceramic.

In the special case of $d/W = 1$, the geometric function Y is fitted for $\alpha = a/W \leq 0.6$ by

$$Y = 0.905\,\alpha^{1/2} - 3.358\alpha^{3/2} + 3.857\alpha^{5/2} + 1.4425\alpha^{7/2} - 3.873\alpha^{9/2} \tag{61}$$

Knowledge of the stress intensity factor solution enables to predict crack growth under increasing load. For a material without an R-curve effect, the

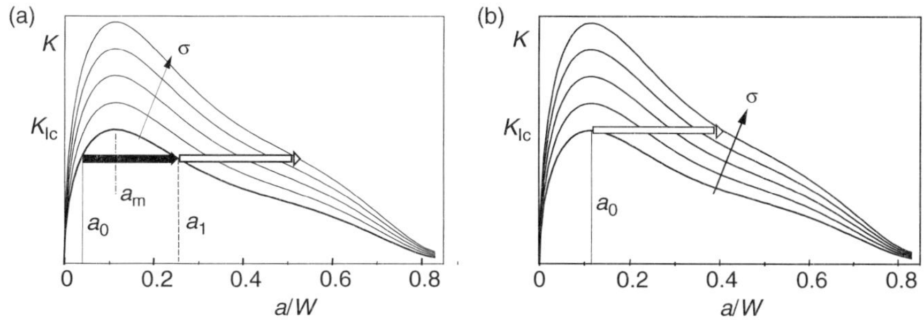

FIGURE 9.1.21 Crack growth under increasing load: (a) initial crack size $a_0 < a_m$; and (b) $a_0 = a_m$.

crack propagation phases are illustrated in Figure 9.1.21 for the special case of $d/W = 1$.

If the initial crack depth a_0 is smaller than the value a_m corresponding to the maximum stress intensity factor (Fig. 9.1.21a), crack extension is unstable up to a_1. Then, stable crack propagation takes place. For an initial crack depth $a_0 \geq a_m$, exclusively stable crack growth occurs (Fig. 9.1.21b). In the case of a sufficiently steep *R*-curve, stable crack extension appears also for $a_0 < a_m$ [34].

9.1.8.3 *R*-Curves from Strength Measurements on Knoop-Damaged Specimens

The *R*-curve in the range of short crack extension may also be determined from strength measurements on artificially damaged tensile or bending specimens [35]. Let us consider specimens with a Knoop crack of initial depth a_0 and width $2c_0$ under tension. During the increase in applied load, the crack starts to extend, if $K_A, K_B > K_{I0}$, where K_A is the stress intensity factor at the deepest point and K_B are the stress intensity factors at the surface points of the semi-ellipse. Catastrophic failure follows after a certain amount of stable crack growth. The failure condition in a strength test after stable crack growth $\Delta a = a - a_0$ in depth and $\Delta c = c - c_0$ in width direction is given as

$$K_{IA} = K_{IR}(\Delta a), \qquad K_{IB} = K_{IR}(\Delta c) \tag{62a}$$

$$\max\left(\frac{\partial(K_{IA} - K_{IR}(\Delta a))}{\partial a}, \frac{\partial(K_{IB} - K_{IR}(\Delta c))}{\partial a}\right) \geq 0 \tag{62b}$$

In the following evaluation, it is assumed that the *R*-curve is a function of Δa and independent of a_0.

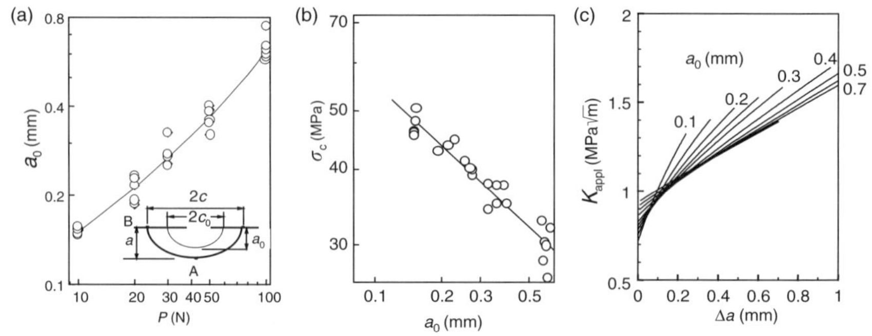

FIGURE 9.1.22 First estimation of the *R*-curve, computed for fixed $a/c = 0.95$: (a) initial crack depths; (b) applied stress intensity factors for several initial crack depths (thin curves); (c) first estimation of the *R*-curve (thick curve).

Rectangular specimens of a PZT ceramic with $3 \times 4 \times 45\,\text{mm}^3$ were damaged by Knoop indentations with different indentation loads *P*. The initial crack depth a_0 is shown in Figure 9.1.22a as a function of the indentation load *P*. After annealing above the Curie temperature, tensile tests were performed with the results plotted in Figure 9.1.22b [35].

The stress intensity factors computed with the strength as the critical stress $\sigma = \sigma_c$ must provide an envelope for the *R*-curve. For a first rough computation the aspect ratio was assumed to remain constant ($a/c = a_0/c_0 = \text{const.} = 0.95$) during crack extension. Since the geometric functions Y_A and Y_B differ slightly for this aspect ratio, the average value $Y = 1/2(Y_A + Y_B)$ was used as an approximation. With these simplifications the applied stress intensity factors for several initial crack depths were computed from

$$K_{appl} = \sigma_c \sqrt{a} Y \tag{63}$$

with σ_c taken from Figure 9.1.22b.

The resulting curves $K_{appl} = f(\sigma_c, \Delta a)$ are entered in Figure 9.1.22c with the initial crack depth as the parameter. The *R*-curve resulting from the envelope condition is introduced as a thick line. A strong increase of K_{IR} with crack extension Δa is obvious.

For further computation steps the assumption of a constant aspect ratio was dropped. The resulting higher approximations for the *R*-curve were found to be nearly identical with the simply obtained first estimation [35].

9.1.9 DETERMINATION OF K_{I0}

The increasing crack resistance (*R*-curve) behaviour of ceramic materials is of high interest for technical applications. Numerous investigations are available in literature, in which the material behaviour was studied for long crack extensions. In recent years, increasing attention was drawn to the starting point K_{I0} of an R-curve, also called "crack tip toughness" [36, 37]. Several methods for the determination of K_{I0} are mentioned below.

9.1.9.1 Extrapolation of R-Curves to $\Delta a \to 0$

In the case of a linear increase in crack resistance, the K_{I0}-value can be determined easily by extrapolating to $\Delta a = 0$. For instance, this is possible for the $MgAl_2O_4$ spinel from Figure 9.1.17c. This curve shows a pronounced initial straight line behaviour that extends over about 6 mm crack extension. We find for this range

$$K_R = K_{I0} + C\Delta a, \quad C = 0.214\,\text{MPa}\sqrt{\text{m}}/\text{mm} \tag{64}$$

and the stress intensity factor at the onset of crack propagation $K_{I0} = 1.45\,\text{MPa}\sqrt{\text{m}}$ (see Figure 9.1.17c).

9.1.9.2 K_{I0} from Near-Tip Crack Opening Displacements

A microscopic observation of the near-tip crack opening displacement during a stable crack growth test yields K_{I0} [36, 38, 39]. In Ref. [36] crack opening displacements in "Round Compact Tension" (RCT) specimens (Fig. 9.1.23a) were evaluated. Figure 9.1.23b shows the measured crack opening profile for a coarse-grained Al_2O_3 versus the distance $a-x$ from the crack tip. The near-tip COD data are related to the actual crack tip stress intensity factor K_{tip} by

$$\delta_{tip} = \sqrt{\frac{8}{\pi}}\frac{K_{tip}}{E'}\sqrt{a-x} \tag{65}$$

The dashed curve in Figure 9.1.23b represents this dependency. The analysis of the near-tip data yields $K_{I0} \cong 2.3\,\text{MPa}\sqrt{\text{m}}$.

9.1.9.3 K_{I0} from Load versus Displacement Curves

In principle, the initial stress intensity factor K_{I0} can also be determined from controlled bending fracture tests. Unfortunately, the often used evaluation of

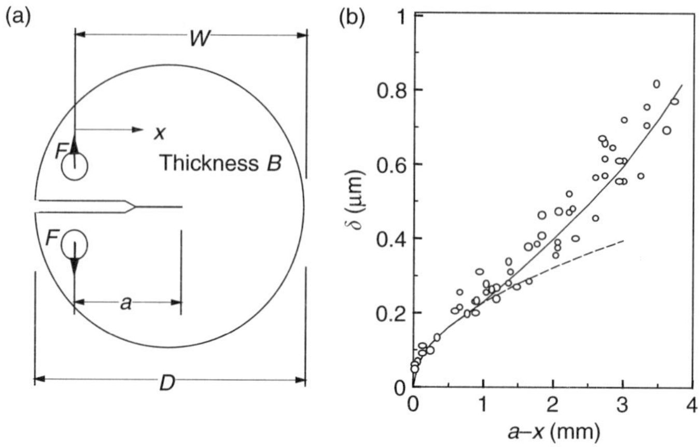

FIGURE 9.1.23 (a) RCT specimen. (b) Measured crack opening displacement, average values (solid curve), near-tip displacements (dashed curve) [39].

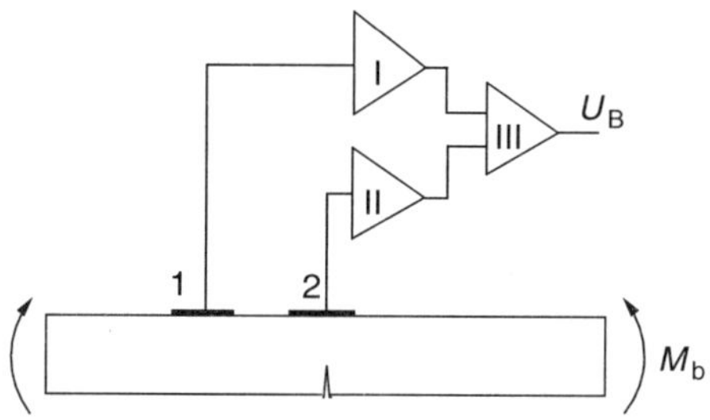

FIGURE 9.1.24 (a) Electrical equipment and amplifier arrangement. (b) Determination of the first deviation from straight-line behaviour.

the load versus displacement curve is very insensitive to small crack extensions. An improvement of the sensitivity of this procedure was proposed in Ref. [40]. Figure 9.1.24 represents the test arrangement. Two strain gauges (1) and (2) are placed at the bending bar in the compressive zone.

One of them is applied at a location remote of the notch (1) and the other one directly opposite the notch (2). Each strain gauge is the active arm of a quarter-bridge circuit. Under bending loading, the strain gauge (2) will produce a higher signal (U_2) than the other sensor (U_1) due to the higher stresses in the notched cross section. Nevertheless, the ratio of the two signals U_2/U_1 is independent of the actual load (as long as linear-elastic material behaviour is considered), but depends on changes in geometry.

Let us assume that the strains at location 2 are higher by a factor of $\lambda > 1$ than the strains at location 1, that is, $U_2/U_1 = \lambda$. In the two amplifiers I and II the strain gauge signals are differently amplified. The amplifications of amplifier I (κ_{I}) and amplifier II (κ_{II}) are conveniently chosen as

$$\kappa_{\mathrm{I}}/\kappa_{\mathrm{II}} = \lambda \tag{66}$$

Consequently, the output signals of both amplifiers are identical. The difference of the single output signals is given on the input connectors of an additional amplifier III. For this purpose the authors use the amplifier of an X–Y recorder. As long as linear material behaviour is investigated, no signal is available at the output of amplifier III. The optimum ratio λ can be found experimentally by application of a small load and by adjusting the magnification of the amplifier until the output signal vanishes. If a crack starts to grow at a certain increased load, the geometry changes and the signal of strain gauge 2 increases more than that of strain gauge 1. The consequence is a signal at the output of amplifier III, which can be recorded.

9.1.10 REASONS OF *R*-CURVE BEHAVIOUR

The most important reasons of the *R*-curve behaviour are mentioned in this section.

9.1.10.1 Bridging Effects

Coarse-grained materials show an *R*-curve effect that is caused by crack–border interactions in the wake of an advancing crack. This fact has been proved experimentally by re-notching experiments [41, 42] and by *in-situ* observations under the electron microscope [43, 44].

In Figure 9.1.25a, such a bridging interaction influenced by a friction force F is illustrated schematically. The crack face interactions localized at single grains can transfer loads that can be modelled in a more homogeneous way by the so-called bridging stresses σ_{br} that depend on the crack opening displacement δ. The bridging stresses shield the crack tip from the external loads as illustrated schematically in Figure 9.1.25b.

For bridging stresses that are caused by friction effects, Mai and Lawn [45] proposed a relation

$$\sigma_{\mathrm{br}} = \begin{cases} \sigma_0(1 - \delta/\delta_c) & \text{for } \delta/\delta_c \leq 1 \\ 0 & \text{for } \delta/\delta_c > 1 \end{cases} \tag{67a}$$

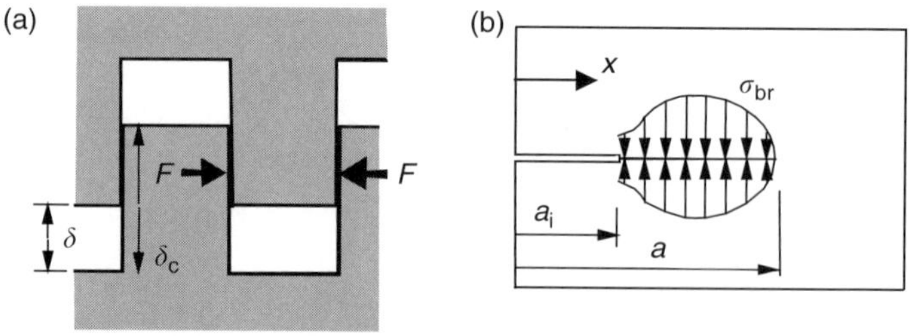

FIGURE 9.1.25 Crack surface interactions due to bridging forces (schematic): (a) local bridging events; (b) continuous bridging stress distribution.

with a characteristic displacement δ_c at which the bridging stress disappears. If the critical displacement δ_c is Γ-distributed, an averaged bridging relation is obtained [46]

$$\sigma_{br} = \sigma_0 \exp(-\delta/\delta_0) \tag{68a}$$

with a material-specific displacement δ_0.

In the case of unbroken microstructural elements, the bridging stress may increase up to a critical displacement δ_c, leading to a relation of

$$\sigma_{br} = \begin{cases} \sigma_0 \delta/\delta_c & \text{for } \delta/\delta_c \le 1 \\ 0 & \text{for } \delta/\delta_c > 1 \end{cases} \tag{67b}$$

If again the critical displacement is Γ-distributed, an averaged bridging relation is obtained:

$$\sigma_{br} = \sigma_0 \frac{\delta}{\delta_0} \exp(-\delta/\delta_0) \tag{68b}$$

The two bridging laws, Eqs 67b and 68b, are illustrated in Figure 9.1.26.

The bridging stress intensity factor results from the distribution of bridging stresses along the crack, $\sigma_{br}(x)$ in Figure 9.1.25b, as

$$K_{br} = \int_0^a h(x, a)\, \sigma_{br}(x)\, dx \tag{69}$$

where h is the fracture mechanics weight function.

9.1.10.2 Phase Transformation Effects in Zirconia Ceramics

Due to the singular stress field near a crack tip in transformation-toughened zirconia, the material undergoes a stress-induced martensitic transformation

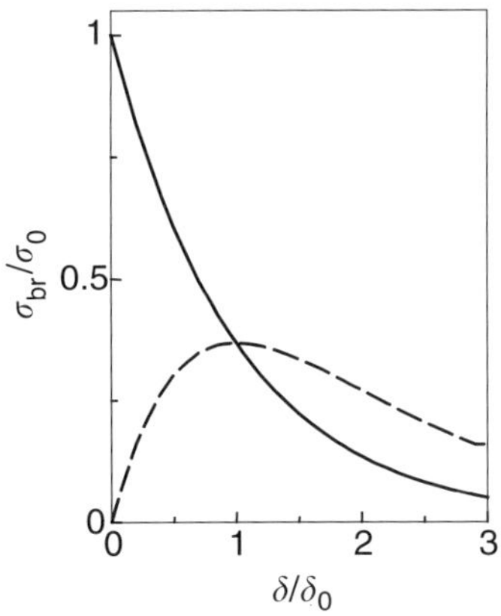

FIGURE 9.1.26 Bridging relations according to Eqs 67a (solid curve) and 68b (dashed curve).

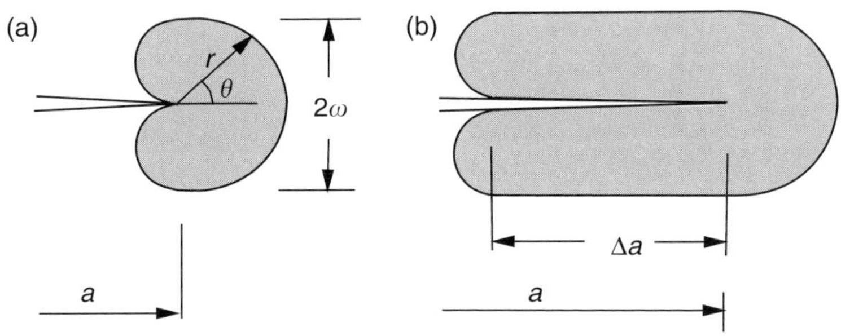

FIGURE 9.1.27 Transformation zone ahead of a crack tip: (a) without crack propagation; (b) after crack extension.

and the tetragonal phase changes to the monoclinic phase (t- to m-ZrO_2). This transformation occurs when the local stresses reach a critical value, and the result is a crack tip transformation zone.

In this section, the stress intensity factor analysis according to McMeeking and Evans [47] will be addressed briefly. Their procedure is based on an Eshelby technique [48]. Let us assume the transformation zone ahead of the crack (see Figure 9.1.27) to be removed from the original material. The transformed material is assumed to be restored to the original shape of the non-transformed material by applying surface traction **T** to the contour of the transformation zone. The zone now fits again into the body and equilibrium can be obtained by nullifying the surface traction with a layer of body forces. The surface tractions then result in a residual stress intensity factor K_{res}

$$K_{res} = \oint_{\Gamma} \mathbf{T} \cdot \mathbf{h} \, dS \qquad (70)$$

where Γ is the contour line of the transformation zone and dS is a line length increment. The vector $\mathbf{h}$ represents the weight function $\mathbf{h} = (h_y, h_x)^T$ with the components h_y and h_x.

In an early study, McMeeking and Evans [47] investigated the case that transformation was initiated when the hydrostatic tensile stress reached a characteristic value $\sigma_{hyd,c}$. For the special case of small-scale transformation conditions (transformation zone size negligible compared to crack size and component dimensions), McMeeking and Evans [47] computed the transformation zone, neglecting the perturbation of the stress field due to transformations.

The phase transformation zone for plane strain becomes

$$\omega = \frac{(1+\nu)^2}{4\sqrt{3}\pi} \left(\frac{K_I}{\sigma_{hyd,c}} \right)^2, \quad r = \frac{8}{3\sqrt{3}} \omega \cos^2(\theta/2) \qquad (71)$$

Figure 9.1.27 illustrates the transformation zone for a non-extending crack (a) and after a crack extension of Δa (b). Caused by the martensitic transformation, a volumetric expansion strain of about 4.5% occurs. These strains cause tensile stresses at some distance ahead of the crack tip and compressive stresses along the length Δa at the crack line. The compressive stresses lead to a shielding stress intensity factor that has to be overcome during crack propagation, that is, the applied stress intensity factor must be increased to maintain stable crack growth.

In the special case of a pure dilatational transformation the shielding stress intensity factor tends asymptotically to $K/p\omega^{1/2} = 0.22$. An increase of the crack growth resistance to a maximum value within a small amount of crack extension is expected.

In recent, much more complicated numerical studies (e.g. Ref. [49]), the influence of shear stresses and strains on the transformation criterion and on the zone calculation was also taken into consideration.

Compared to the nearly circular phase transformation zones that are found in MgO-doped zirconia, the zone shape of Ce-partially stabilized zirconia differs strongly. The left part of Figure 9.1.28 illustrates a typical elongated transformation zone for a zirconia (Ce–TZP) alumina composite as reported by Yu *et al.* [50]. A typical zone for a 9 mol% Ce–TZP with a mean grain size of 1.4 μm is shown in the right part of Figure 9.1.28, as reported by Rauchs *et al.* [33]. In this case, an irregular zone contour is visible.

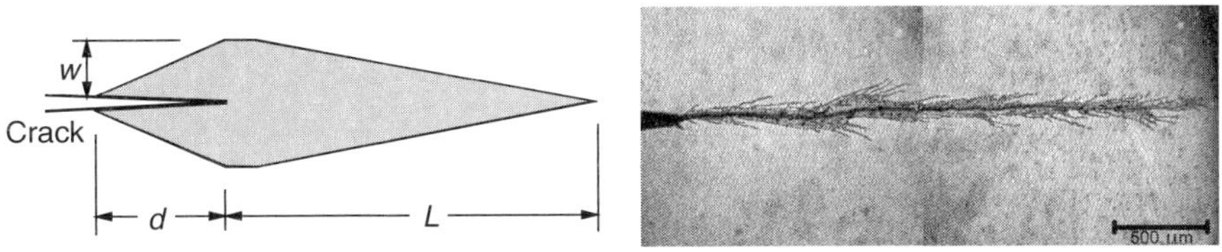

FIGURE 9.1.28 Shape of phase transformation zones in ceria-partially stabilized zirconia (Ce–TZP) alumina composites: left, schematic result obtained by Yu *et al.* [50]; right, transformation zones during crack extension for a specimen of 9 mol% Ce–TZP with a mean grain size of 1.4 μm [33].

9.1.10.3 Domain Switching in Piezoelectric Ceramics

Piezoelectric ceramics such as barium titanate or lead zirconate/titanium zirconate (PZT) are paraelectric above a critical temperature, the Curie temperature, and piezoelectric below this temperature. If during cooling the Curie temperature is reached, a spontaneous phase transformation occurs. In the case of $BaTiO_3$, this is a transformation from the cubic to the tetragonal phase. In the tetragonal state, the ratio of the lengths of the c-axis to the a-axis is about $c/a = 1.01$ at room temperature. Phase transformation is associated with an electrical polarization in the direction of the c-axis. In each grain six polarization directions are possible. The grains are subdivided into domains, where each domain has a direction of polarization. The domains are separated by domain boundaries. The overall polarization of a polycrystal after cooling below the Curie temperature is zero. Applying an electrical field leads to domain switching in the direction of the electrical field and, therefore, to a polarization of the sample. Domain switching can also be caused by mechanical stress. Applying a compressive stress in the direction of polarization leads to a switching of the domains in a direction perpendicular to the stress and, therefore, to a contraction of the specimen. Applying a tensile stress perpendicular to the orientation of polarization can lead to domain switching in the direction of the applied stress and to an elongation of the specimen. In an unpolarized specimen domain switching is also possible for those domains that are in a suitable orientation. Therefore "plastic" deformation is possible in polarized and unpolarized materials.

A simple explanation of the toughening effect may be given by the influence of domain orientation ahead of the tips of cracks [51]. Let us consider a volume element at some distance from the tip of a natural crack. In an unpoled tetragonal material we will find one-third of all domains orientated in the x-direction, one-third in the y-direction and one-third in z-direction (Fig. 9.1.29a). If a mechanical load is applied normally to the crack plane (strongly magnified

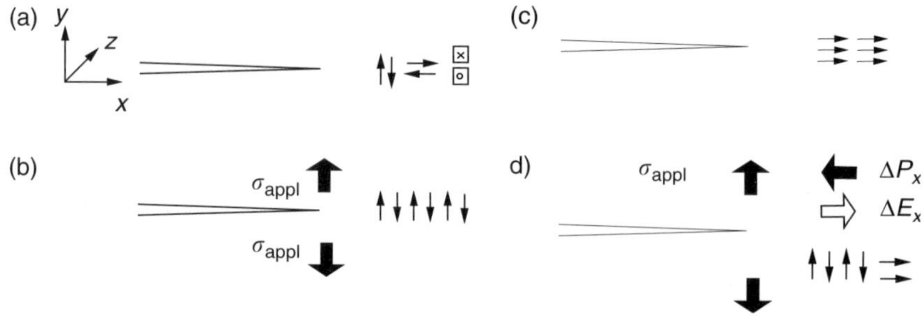

FIGURE 9.1.29 Orientation of domains ahead of a crack in an unpoled material (a) under externally applied mechanical load (b). (c) and (d) Influence of mechanical load on the domains for PZT poled in x-direction [51].

near the crack tip), most of the domains will switch into the stress direction (Fig. 9.1.29b) with antiparallel orientation sense. The increase of domains in y-direction causes a residual stress field that reduces the influence of the externally applied stress, interpreted as an increase of resistance against crack propagation [52, 53].

In the case of PZT poled perpendicularly to the stress direction, nearly all domains may be orientated in x-direction (Fig. 9.1.29c). Under externally applied tensile stresses, an orientation of the domains in the y-direction can also be observed, similar to the orientation of domains in the unpoled material (Fig. 9.1.29d). Whereas in the case of unpoled material two-third of all domains can switch into the y-direction, all domains are able to switch in the case of perpendicularly-poled PZT. From this point of view, an increased domain switching and increased crack resistance have to be expected. Due to domain switching from the poled state, depolarization ΔP_x takes place. This depolarization is strongly concentrated in the vicinity of the crack tip. The highly stressed and depoled zone is in series with the nearly unstressed bulk material that remains in the poled state. Therefore, the short-circuited electrodes—far away from the depoled zone—are without effect and the electric boundary conditions are given by a constant dielectric displacement, D = constant. Due to depoling, an electrical field ΔE_x is generated

$$D_x = \varepsilon E_x + P_x = \text{constant} \Rightarrow \Delta E_x = -\Delta P_x/\varepsilon \qquad (72)$$

(ε is the dielectric permittivity), which inhibits depoling by domain switching (Fig. 9.1.29d), that is, switching in the poled state needs a higher stress to reach the same degree of domains oriented in the y-direction.

If the PZT is poled in y-direction, the number of domains in the x- and z- directions able to switch into the y-direction is very small. Therefore, we

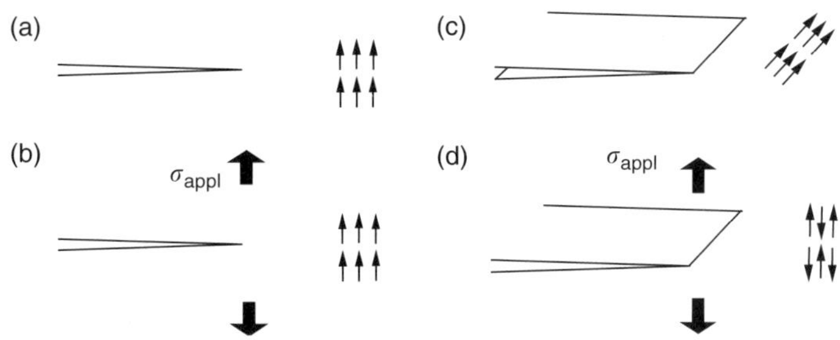

FIGURE 9.1.30 Influence of mechanical load on the domains ahead of a crack for perpendicularly-poled PZT (a, b) and PZT poled in z-direction (c, d).

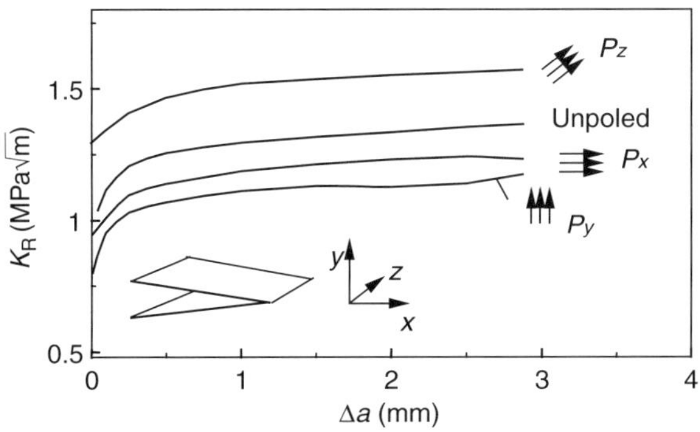

FIGURE 9.1.31 *R*-curves for a soft PZT ceramic with different poling states, compared with unpoled material [54].

do not expect any effect of reorientation (Fig. 9.1.30a and b). Hence, the lowest crack resistance is expected.

The case of a crack with the polarization being parallel to the crack tip line is shown in Figure 9.1.30c and d. In this case, all domains are able to switch into the y-direction. The highest crack resistance has to be expected.

From the previous considerations, we expect that initial polarization with its differing domain reservoir in combination with the influence of the internal electric field will lead to the following ranking of fracture toughness:

$$K_y < K_x < K_{\text{unpoled}} < K_z \tag{73}$$

where K_x, K_y, and K_z are the toughness values for poling in x-, y- and z-directions respectively. This ranking is in agreement with measurements by Lucato *et al.* [54] (see Figure 9.1.31).

9.1.10.4 Microcrack Zone and Crack Branching

Further reasons for R-curves are the energy-consuming effects that occur ahead of a crack tip also in materials without phase transformations. Ahead of a crack tip a disturbed material zone extends with increasing load. Due to the high stresses, part of the grain boundaries aligned perpendicularly to the maximum principal stress will fail and a zone of microcracks develops [55, 56]. Its shape is similar to Figure 9.1.27a. In the energy considerations the microcracks contribute to the total crack surface which becomes $\Delta S_{\mathrm{eff}} > B\Delta a$ and, consequently, increases the effective critical energy release rate. In the stress intensity factor description the microcracks reduce the effective Young's modulus ahead of the crack tip, the crack tip stresses and the effective stress intensity. From both points of view, an increased crack resistance must be concluded. An increase of the disturbed zone with increasing crack extension leads to an R-curve effect.

Also in the case of crack branching, the energy consideration yields an increase in the crack surface with increased energy consumption.

9.1.11 DETERMINATION OF BRIDGING STRESSES

The case of the computation of R-curves on the basis of known bridging stresses has been discussed in Ref. [4]. In this section, a similar problem will be treated that is based on the same basic relations.

9.1.11.1 Bridging Stresses from the *R*-Curve

Let us assume a measured R-curve that is based on bridging interactions in the crack wake. The specimen may be loaded by an external load, which leads to a stress distribution $\sigma_{\mathrm{appl}}(x)$ at the location of the crack in the uncracked component. In case the material exhibits a bridging zone with crack surface interactions, the total stress is the sum of the applied stress and the bridging stress σ_{br}, that is,

$$\sigma(x) = \sigma_{\mathrm{appl}}(x) + \sigma_{\mathrm{br}}(x). \tag{74}$$

The bridging stresses are responsible for the bridging stress intensity factor K_{br}, which is given in the representation of the fracture mechanics weight function h [57] as

$$K_{\mathrm{br}} = \int_0^a h(x, a)\sigma_{\mathrm{br}}(x)\,\mathrm{d}x \tag{75}$$

and for the applied stress intensity factor K_{appl} it holds

$$K_{\text{appl}} = \int_0^a h(x,a)\sigma_{\text{appl}}(x)\,\mathrm{d}x \tag{76}$$

The total stress intensity factor is then given as

$$K_{\text{tip}} = K_{\text{appl}} + K_{\text{br}} \tag{77}$$

As a simple example, the bridging stresses may be determined from the R-curve for $MgAl_2O_4$ spinel (Fig. 9.1.17c), which was obtained in a Double Cantilever Beam (DCB) test [32]. The weight function for the DCB specimen reads [7]

$$h = \sqrt{\frac{12}{H}}\left[\frac{a-x}{H} + 0.68\right] + \sqrt{\frac{2}{\pi(a-x)}}\exp\left(-\sqrt{12\frac{a-x}{H}}\right) \tag{78}$$

where H is the half-height of the specimen.

The total displacements of the crack surface, δ, result from the relation [4]

$$\delta(x) = \frac{1}{E'}\int_x^a h(x,a')\left[\int_0^{a'} h(x',a')[\sigma_{\text{appl}}(x') + \sigma_{\text{br}}(\delta(x'))]\,\mathrm{d}x'\right]\mathrm{d}a' \tag{79}$$

Due to the long linear range of the R-curve, a very simple attempt is possible to obtain the bridging stresses. From Eqs. 64, 75 and 78, it can be concluded that for crack extensions within the straight part of the R-curve, the distribution of the bridging stress must be reciprocal to the weight function, that is,

$$\int_0^a h(x,a)\,\sigma_{\text{br}}\,\mathrm{d}x = C(a-a_0) \quad \Leftrightarrow \quad h(x,a)\sigma_{\text{br}} = C \quad \sigma_{\text{br}} = C/h \tag{80}$$

The evaluation of bridging stresses from R-curve data is based on the solution of Eqs 75–80. The numerical strategy consists of the following steps:

1. For small displacements (corresponding to crack extensions where $\Delta a \leq 5$ mm) the bridging relation results from Eq. 80 with the displacements computed by Eq. 79. The result is shown in Figure 9.1.32a.
2. For larger displacements the function $\sigma_{\text{br}}(\delta)$ is represented by the stress values at the sampling points δ_1 and δ_2, which are interpolated by cubic splines (Fig. 9.1.32b).
3. For any chosen value of $\sigma_{\text{br}}(\delta_1)$ and $\sigma_{\text{br}}(\delta_2)$ (see Figure 9.1.32b), the bridging stresses in the whole range of $\delta \leq 4\,\mu$m were computed, the integral equation was solved and the crack profile and bridging stress distribution were obtained. From Eq. 75, the related bridging stress intensity factor,

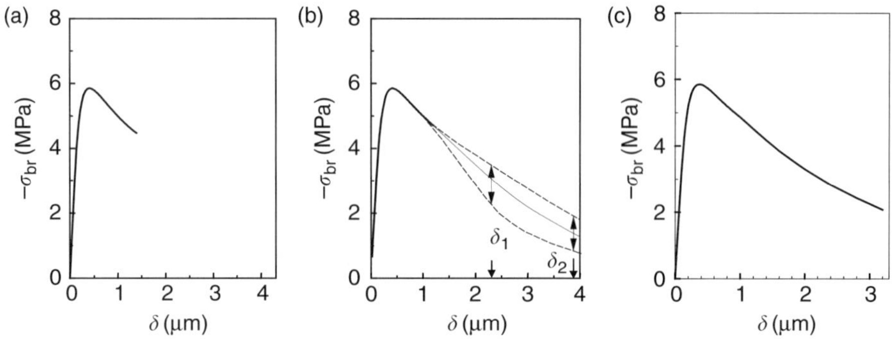

FIGURE 9.1.32 Determination of the bridging relation: (a) solution for small displacements from the linear part of the R-curve; (b) representation of the relation for large displacements by cubic splines (two sampling points); (c) final result.

K_{br}, was calculated and from Eq. 56 the R-curve values, K_R, were obtained for a number of selected Δa. The computed K_R values were compared with the experimental data. The bridging stress values at δ_1 and δ_2 were changed systematically by use of an optimization computer subroutine, until minimum deviations between the computed and measured K_R were found. The corresponding bridging stresses at δ_1 and δ_2 then establish the bridging stress relation $\sigma_{br}(\delta)$ for larger crack opening displacements (Fig. 9.1.32c).

9.1.11.2 Bridging Stresses from Crack Profiles

In order to determine the relation between bridging stresses σ_{br} and the COD δ, $\sigma_{br} = f(\delta)$, the CODs observed during stable crack propagation can be evaluated. Such evaluations were performed in Ref. [39] on coarse-grained alumina with the "Round-Compact-Tension" (RCT) specimen and by Kobayashi *et al.* [58, 59] on reinforced alumina with the "Compact-Tension" (CT) specimen. The procedure is, in principle, based on the relations of Section 9.1.11.1. In Eq. 79, now the left-hand side is known as the result of the displacement measurements. The unknown quantity is the bridging stress distribution $\sigma_{br}(x)$

$$\delta_{measured}(x) = \frac{1}{E'}\int_x^a h(x,a')\left[\int_0^{a'} h(x',a')\,[\sigma_{appl}(x') + \sigma_{br}(x')]\,dx'\right] da' \quad (81)$$

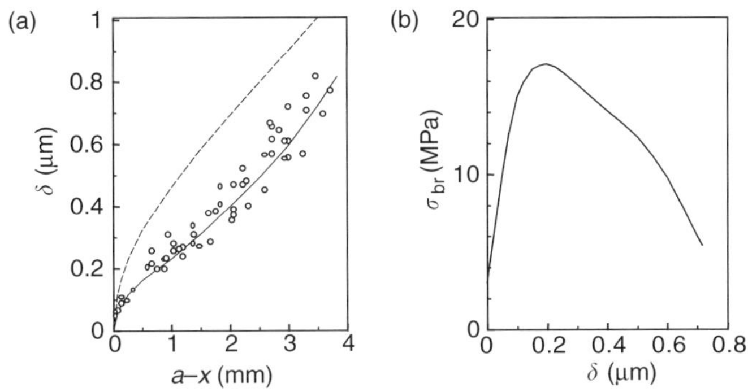

FIGURE 9.1.33 Crack profile δ for coarse-grained Al_2O_3 (symbols: Measured data, solid curve, Average dependency, dashed curve: δ_{appl}); (b) bridging stresses derived from (a) [39].

The integral equation can be solved numerically and provides the bridging relation

$$\sigma_{br} = f(\delta)$$

As an example of an application, results from Ref. [39] are reported. In Figure 9.1.33a, the crack opening displacements of Figure 9.1.23b are plotted once more. The crack opening profile for $K_{appl} = 3.75\,\mathrm{MPa}\sqrt{\mathrm{m}}$, which would occur in the absence of any bridging effect, was computed and is introduced in Figure 9.1.33a as the dashed curve. The evaluation of these data yields the bridging stress law plotted in Figure 9.1.33b.

9.1.11.3 Bridging Stresses from Post-Fracture Tensile Tests

In an effort to directly characterize the wake behaviour, Hay and White [60–62] developed the post-fracture tensile (PFT) test, which gives the crack closure stress versus the crack opening displacement relationship.

In addition to the half-thickness side groove machined into the DCB specimen (Fig. 9.1.34a) for crack guidance, two separate grooves are cut to facilitate tensile loading of the PFT specimens, as shown in Figure 9.1.34b. Next, several slices are taken from the cracked portion of the DCB specimen (indicated schematically in Figure 9.1.34a), thereby isolating the wake region. Tensile tests were performed on these specimens with the crack-face separations measured with a laser interferometric displacement gauge. In these tests, the wake stiffness and peak load capacity depend on the starting COD conditions. The PFT specimens cut from near the crack tip exhibit the greatest peak load and stiffness. From a series of load–displacement curves, the relation $\sigma_{br} = f(\delta)$ is derived.

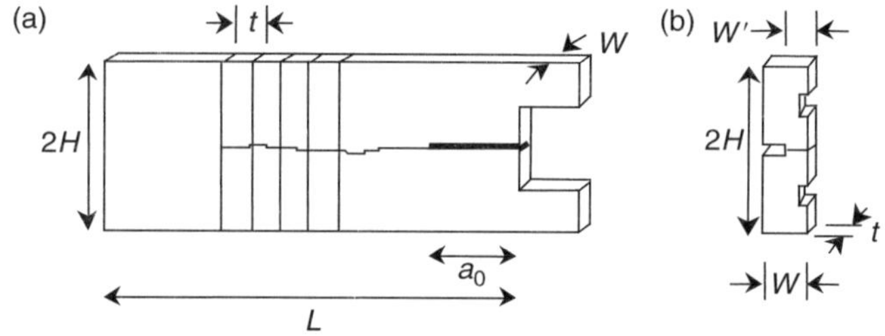

FIGURE 9.1.34 Schematic representation of the (a) double cantilever beam and (b) post-fracture tensile specimens.

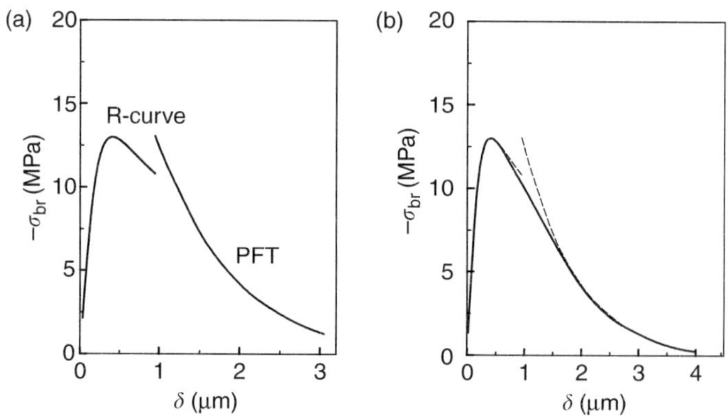

FIGURE 9.1.35 Bridging stress relation: (a) results from PFT experiments and from the evaluation of the *R*-curve; (b) interpolation curve (solid line) representing the "wide-range" bridging relation [63].

The significance of the data obtained by this procedure relies on the successful isolation of the wake without damaging the wake constituents. By using the DCB specimen, this latest incarnation of the PFT technique allows for careful control of the COD profile and for the efficient use of material.

Figure 9.1.35a shows the PFT results for a 99.7% Al_2O_3 ceramic with 12-μm mean grain size together with an evaluation of the *R*-curve using the relations of Section 9.1.11.1. By interpolation of these curves (Fig. 9.1.35b), a sufficient "wide range" bridging relation can be concluded [63].

9.1.12 SUBCRITICAL CRACK GROWTH

In the preceding sections, stable crack extension for a rising crack growth resistance curve and unstable crack extension were considered. A third type

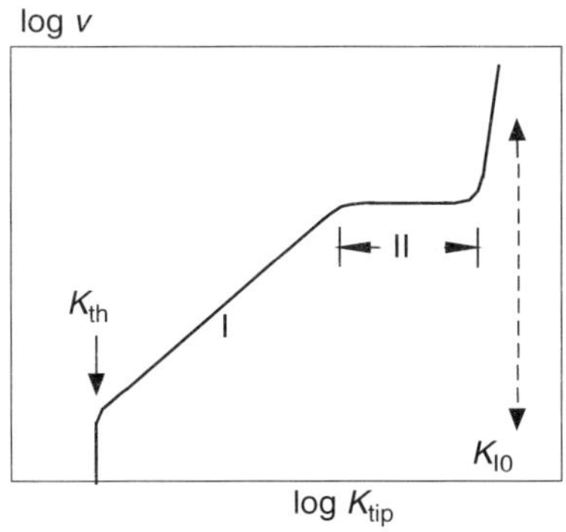

FIGURE 9.1.36 v–K curve of subcritical crack growth (schematic representation).

of crack extension is called subcritical crack growth (SCG), where a crack is growing at $K_{appl} < K_{Ic}$ or in terms of crack tip stress intensity factor $K_{tip} < K_{I0}$.

The subcritical crack growth is governed by the crack tip stress intensity factor K_{tip}. For a given material and environment there is a unique relation between the crack growth rate v and K_{tip}:

$$v = \frac{da}{dt} = f(K_{tip}) \tag{82}$$

In Figure 9.1.36, a v–K_I curve is shown in the log–log representation. At low crack growth rates an extended range (region I) occurs with a straight line. In this region, the crack growth rates fulfil a power-law relation

$$v = AK_{tip}^n = A^* \left(\frac{K_{tip}}{K_{I0}}\right)^n \tag{83}$$

with the parameters A, A^* and n depending on the material, the temperature and the environment. In seldom cases (e.g. glass) a threshold value can be detected, below which no subcritical crack growth is found [64, 65]. Sometimes, this threshold value is also denoted as K_{I0}. In order to avoid confusions and to distinguish between the onset value of stable crack extension and the threshold in SCG, the symbol K_{th} is used here. At a relative high crack growth rate, a plateau (region II) in v may occur with crack growth rates independent of K_I. Such a plateau was found for glass.

If a v–K curve is determined as shown in Section 9.1.12.1, it always has to be assessed what the relation $v(K)$ represents. In the case of small natural cracks in the order of some 10 μm, which will fail at high stresses after a very short crack extension, the difference between K_{appl} and K_{tip} may be negligible. Regarding artificial macroscopic cracks this requirement is not fulfilled for materials with an R-curve behaviour.

As has been outlined in Ref. [4] for small cracks, a representation of subcritical crack growth in terms of the applied stress intensity factor K_{appl} is approximately possible

$$v \cong AK_{appl}^{n} = A^{*}\left(\frac{K_{appl}}{K_{Ic}}\right)^{n} \tag{84}$$

where A and A^* are different from those in Eq. 83 and the exponent n is larger than that in Eq. 83. It should be emphasized once more that a description according to Eq. 84 is not possible for SCG-curves obtained with macroscopic cracks for materials with a strong R-curve behaviour.

It is common use for ceramic materials to represent subcritical crack growth by $v(K_{appl})$ curves. Therefore, in the following relations K_I is always used instead of K_{tip} and K_{appl} in order to simplify the notation.

9.1.12.1 Determination of V–K Curves with Macroscopic Cracks

Many types of specimens have been used to determine the v–K relation. As an example of a test specimen, the double cantilever beam (DCB) specimen may be addressed (Fig. 9.1.37). The specimen consists of a rectangular bar with a slot and is loaded by a pair of forces F that load the specimen in the pure mode I. If the two halves of the specimen are assumed to be clamped at the location of the crack tip, the elementary bending theory provides for the two loading point displacements

$$V = \frac{8Fa^3}{EBH^3} \tag{85}$$

and the energy release rate and stress intensity factor resulting from Eqs. 19 and 20

$$G_I = \frac{F^2}{2B}\frac{dC}{da} = \frac{12F^2a^2}{EB^2H^3}, \quad K_I = \frac{\sqrt{12}aF}{(1-\nu^2)^{1/2}BH^{3/2}} \tag{86}$$

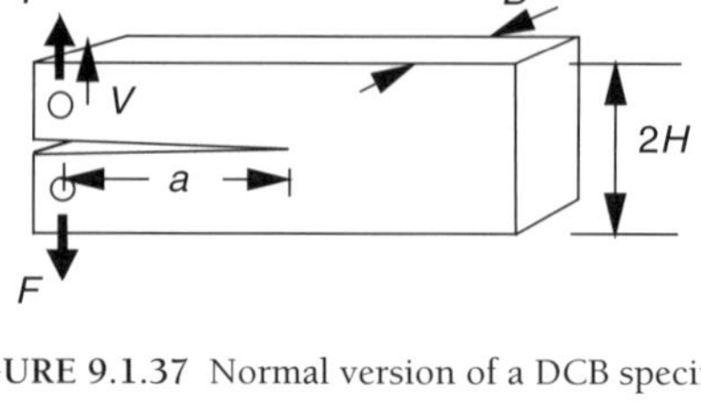

FIGURE 9.1.37 Normal version of a DCB specimen.

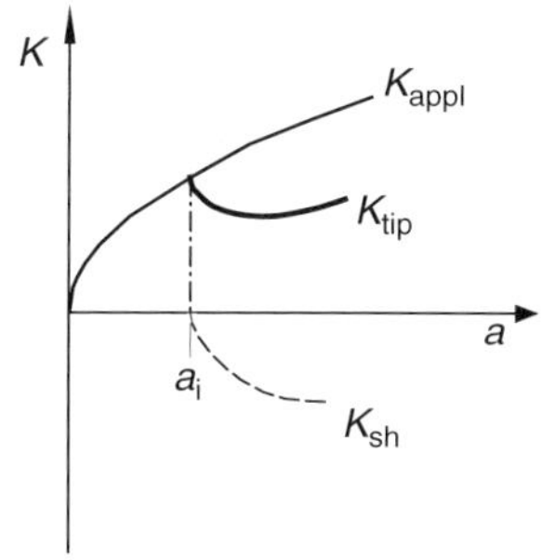

FIGURE 9.1.38 Stress intensity factor contributions in a constant load test.

If additional shear deformations are included and the condition of "fixed clamps" is modelled more accurately, it results (see, e.g. Ref. [66]) in

$$K_{\mathrm{I}} = \frac{\sqrt{12}aF}{(1-\nu^2)^{1/2}BH^{3/2}}\left[\left(1+\frac{a_0}{a}\right)^2 + \frac{1+\nu}{3}\left(\frac{H}{a}\right)^2\right]^{1/2} \tag{87}$$

with good approximation, $a_0 = 2H/3$.

Accurate measurements of the crack growth rate as a function of K_{appl} for materials with a rising R-curve showed first a decrease and after some crack extension an increase in the crack growth rate with increasing K_{appl}. This can be explained by Figure 9.1.38 where for a test with constant load K_{appl} increases according to $K_{\mathrm{appl}} = \sigma Y\sqrt{a}$. The increase in the shielding stress intensity factor is larger for small crack extensions and therefore K_{tip} first decreases corresponding to the decrease in the crack growth rate. Therefore, tests with macroscopic cracks cannot be recommended for materials with a rising crack growth resistance curve.

9.1.12.2 Evaluation of Lifetimes Under Constant Load

Under conditions of subcritical crack growth the lifetimes of a material is finite. The lifetime depends on the stress in the component, which may change with time. In cases where the simple crack growth relation 84 can be applied, the lifetime can be computed analytically.

The lifetime, t_{f}, under an arbitrary loading history, $\sigma(t)$, is given by (see, e.g. Ref. [4])

$$\int_0^{t_{\mathrm{f}}} \sigma(t)^n \,\mathrm{d}t = \int_{a_{\mathrm{i}}}^{a_{\mathrm{c}}} \frac{\mathrm{d}a}{AY^n a^{n/2}} \tag{88a}$$

If the subcritical crack extension is caused by the same flaws as the strength, the initial crack size a_i can be replaced by the strength σ_c (see Ref. [4]), leading to

$$\int_0^{t_f} \sigma(t)^n \, dt = B\sigma_c^{n-2}[1 - (\sigma_f/\sigma_c)^{n-2}] \tag{88b}$$

with a material parameter B comprising some fracture mechanical quantities

$$B = \frac{2K_{Ic}^2}{A^* Y^2 (n-2)} \tag{89}$$

In tests with constant load, $\sigma(t) = \text{const.}$, one obtains

$$t_f = B\sigma_c^{n-2}\sigma^{-n}[1 - (\sigma/\sigma_c)^{n-2}] \tag{90}$$

resulting in a strong stress dependency of the static lifetimes as a consequence of the high exponent n for ceramics. For a load sufficiently lower than the critical one, Eq. 90 is simplified to

$$t_f = B\sigma_c^{n-2}\sigma^{-n} \tag{91}$$

or in logarithmic form

$$\log t_f = -n \log \sigma + \log(B\sigma_c^{n-2}) \tag{92}$$

From this relation, the parameters n and $\log(B\sigma_c^{n-2})$ can be obtained by linear regression of $\log(t_f)$ versus $\log(\sigma)$.

As an application of this method, static lifetime measurements are reported in Figure 9.1.39 for hot-isostatically pressed Al_2O_3. They were carried out in four-point bending tests in a concentrated salt solution at 70°C [4]. In Figure 9.1.39a,

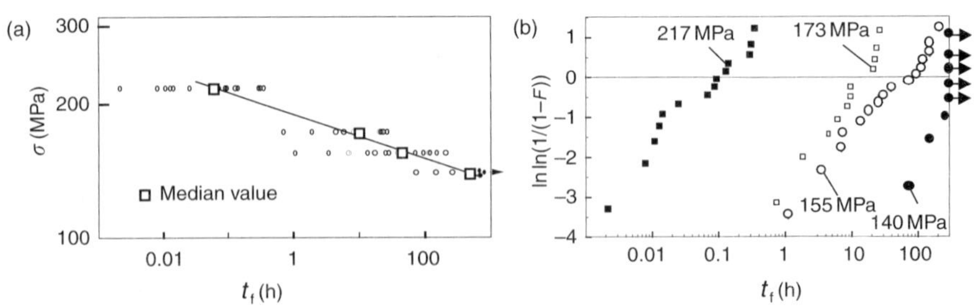

FIGURE 9.1.39 Representation of lifetime measurements on hot-isostatically pressed Al_2O_3 in concentrated salt solution at 70°C: (a) Wöhler-like representation (solid circles: suspended after 500 h); (b) Weibull representation.

the lifetime results are given and in Figure 9.1.39b in Weibull representation [67]. The data of Figure 9.1.39 result in

$$n = 20, \quad \log(B\sigma_c^{n-2}) = 45.51 \ \ (\text{for } B\sigma_c^{n-2} \text{ in MPa}^2\,\text{h})$$

The inert strength σ_c and the time to failure t_f are properties with a large amount of scatter. Both are related to the scatter of the initial flaw size a_i. The scatter of the fracture strength can be described by the Weibull distribution. The cumulative density function F for the two-parameter Weibull distribution is given by

$$F(\sigma_c) = 1 - \exp[-(\sigma_c/\sigma_0)^m] \tag{93}$$

where m and σ_0 are the Weibull parameters.

In a plot of $\ln\ln(1/(1-F))$ versus $\ln(\sigma_c)$, a straight line results and the parameter m is obtained from the slope. The scatter of lifetimes t_f is related to the scatter of strength and represented by

$$F(t_f) = 1 - \exp[-(t_f/t_0)^{m^*}] \tag{94}$$

with

$$m^* = \frac{m}{n-2} \tag{95}$$

$$t_0 = B\sigma_0^{n-2}\sigma^{-n} \tag{96}$$

One can find the value m^* as the slope in the plot $\ln\ln(1/(1-F))$ versus $\ln t_f$. Equation 95 can be used to obtain the exponent n. An analytical evaluation of the Weibull parameters has to be performed using the maximum likelihood method [68].

In order to evaluate the Weibull parameters for these lifetime distributions, the maximum likelihood method has to be applied. Due to the large scatter and the limited test duration, a number of specimens that fail during load application or survive the chosen time limit (e.g. 1000 h) will be obtained in many cases (especially for small m^*, i.e. large n and small m). In such a case, a test series that is truncated at a lower limit $t_{f,l}$ and at an upper limit $t_{f,u}$ [69] has to be evaluated.

For a number of n_l spontaneous failures with $t_f < t_{f,l}$, n_u survivals, where $t_f > t_{f,u}$, and $n_v = n - n_l - n_u$ valid test results with $t_{f,l} < t_f < t_{f,u}$, the equations for the determination of m^* and t_0 read

$$\frac{n_v}{m^*} + \sum_{i=1}^{n_v} \ln\frac{t_{fi}}{t_0} + n_l \ln\frac{t_{f,l}}{t_0} - \frac{1}{t_0^{m^*}}\left[\sum_{i=1}^{n_v}\left(t_{fi}^{m^*}\ln\frac{t_{fi}}{t_0}\right) + \frac{1}{2}n_l t_{f,l}^{m^*}\ln\frac{t_{f,l}}{t_0} + n_u t_{f,u}^{m^*}\ln\frac{t_{f,u}}{t_0}\right] = 0 \tag{97a}$$

$$t_0^{m^*} = \frac{1}{n_l + n_v}\left(\sum_{i=1}^{n_v} t_{fi}^{m^*} + n_u t_{f,u}^{m^*} + \frac{1}{2}n_l t_{f,l}^{m^*}\right) \tag{97b}$$

which have to be solved numerically.

The lifetime procedure described earlier is based on the power law relation of Eq. 84. An alternative method, also based on a statistical evaluation, was proposed in Ref. [4]. The evaluation, however, does not require the assumption of a special prescribed type of crack growth relation. The crack growth rate at the beginning of a lifetime test (subscript i) is

$$v(K_i) = -\frac{2}{t_f\sigma_c^2}\left(\frac{K_{Ic}}{Y}\right)^2 \frac{d[\log(\sigma/\sigma_c)]}{d[\log(t_f\sigma^2)]} \tag{98}$$

with the initial stress intensity factor K_i directly after load application.

In Eq. 98, the lifetime t_f *and* the inert strength σ_c of a specimen have to be known. This is, in principle, impossible, since either a strength test or a lifetime test can be performed with one specimen. Therefore, a statistical procedure has to be applied.

In a first series of tests, N samples are tested at high loading rates in an inert environment to give the distribution of σ_c. The N values of strength are arranged in increasing order. In a second series, also involving N specimens, the lifetimes t_f are measured. The results are also arranged in increasing order. The νth value of lifetime $t_{f,\nu}$ is associated with the νth value of inert bending strength $\sigma_{c,\nu}$. By plotting $\log(t_{f,\nu}\sigma^2)$ versus $\log(\sigma/\sigma_{c,\nu})$, one obtains a dependency, from which—by use of a suitable smoothing procedure—the actual derivative

$$\frac{d[\log(\sigma/\sigma_c)]}{d[\log(t_f\sigma^2)]}$$

results. By introducing this derivative into Eq. 98, the crack growth rate $v(K_{Ii})$ is obtained. The corresponding K_{Ii} follows from $K_{Ii}/K_{Ic} = \sigma/\sigma_c$. Some results at high temperature are compiled in Figure 9.1.40. Figure 9.1.40a shows results for boron-doped sintered α-silicon carbide and Figure 9.1.40b for MgO-doped HPSN.

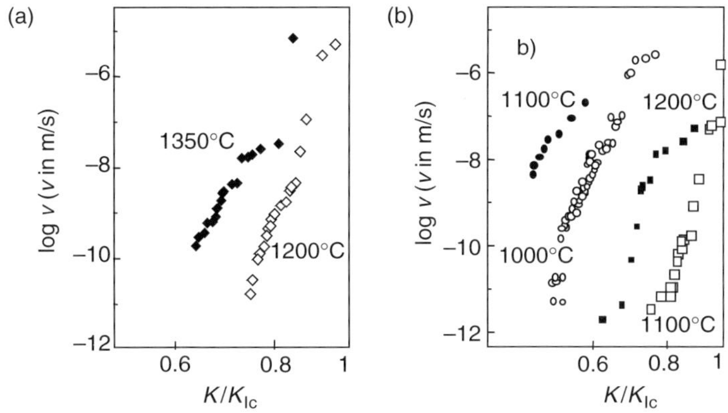

FIGURE 9.1.40 Subcritical crack growth data obtained by Eq. 95: (a) sintered α-silicon carbide (boron-doped) [70]; (b) HPSN (MgO-doped, circles; Y_2O_3-doped, squares) [71].

9.1.12.3 Dynamic Strength Test

In the dynamic strength test developed by Charles [72], the specimens are loaded with a constant loading rate $\dot{\sigma}$ until failure. The failure stress σ_f depends on the loading rate as

$$\sigma_f^{n+1} = B\sigma_c^{n-2}\dot{\sigma}(n+1)[1-(\sigma_f/\sigma_c)^{n-2}] \tag{99}$$

For very low loading rates, Eq. 99 is simplified to

$$\sigma_f^{n+1} = B\sigma_c^{n-2}\dot{\sigma}(n+1) \tag{100}$$

From measurements of strengths at different stress rates $\dot{\sigma}$, n and A of Eq. 84 can be evaluated by application of Eq. 100. In the logarithmic form this equation reads

$$\ln \sigma_f = \alpha + \beta \ln \dot{\sigma} \tag{101}$$

with

$$\alpha = \frac{1}{n+1} \ln B\sigma_c^{n-2}(n+1), \qquad \beta = \frac{1}{n+1} \tag{102}$$

A regression analysis then results in the exponent n and the lifetime-relevant quantity $B\sigma_c^{n-2}$. In this context, it should be mentioned that in the relation for predicting lifetimes under arbitrary loading the parameter $B\sigma_c^{n-2}$ can be directly introduced and an evaluation of the coefficients A and A^* of Eq. 84 is not necessary. The evaluation of the dynamic bending strengths is recommended to be performed via the median values. But sometimes also a regression of all single data points is carried out.

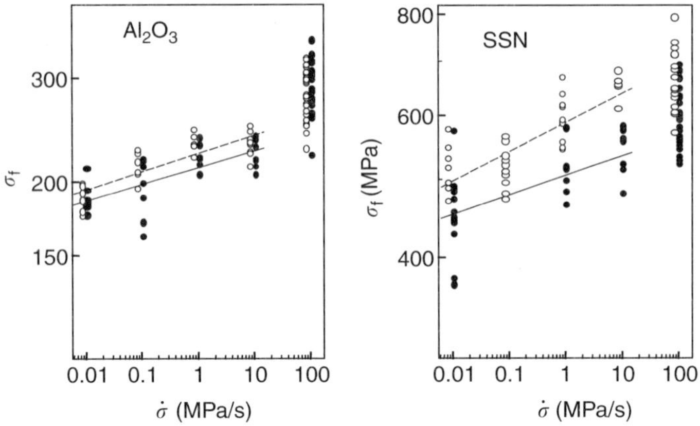

FIGURE 9.1.41 Dynamic bending strength for alumina and SSN, values at 100 MPa/s. Inert strength data obtained in air; solid and open circles symbolize different labs [73].

In the study group "Strength and Lifetimes" of the DKG/DVM (Deutsche Gesellschaft für Keramik e.V./Deutsche Gesellschaft für Materialkunde e.V.), a Round Robin was applied to dynamic bending tests including nine different ceramics [73]. As an example, Figure 9.1.41 shows the results for SSN and Al_2O_3 in water at room temperature. The different symbols indicate different labs. The data introduced at 100 MPa/s represent the inert strength measurements in air. The evaluation was performed on the basis of the median strength values and also by consideration of the single data points.

As can be seen from the data of Figure 9.1.41, large strength scatter occurs. In order to allow an assessment regarding the confidence of the determined parameters, a statistical analysis should be made. A possibility is to calculate the confidence intervals of the relevant parameters. Therefore, some statistics relations necessary for the evaluation of the measurements are reported. With the abbreviation of

$$[z] = \sum_{i=1}^{N} z_i \tag{103}$$

the estimates $\hat{\alpha}$, $\hat{\beta}$ for the parameters α and β result from N tests as

$$\hat{\beta} = \frac{[xy]N - [x][y]}{[x^2]N - [x]^2}, \qquad \hat{\alpha} = \frac{[y]}{N} - \hat{\beta}\frac{[x]}{N} \tag{104}$$

with their standard deviations

$$\varepsilon_{\hat{\alpha}} = \varepsilon\sqrt{\frac{[x^2]}{N[x^2] - [x]^2}}, \qquad \varepsilon_{\hat{\beta}} = \varepsilon\sqrt{\frac{N}{N[x^2] - [x]^2}} \tag{105}$$

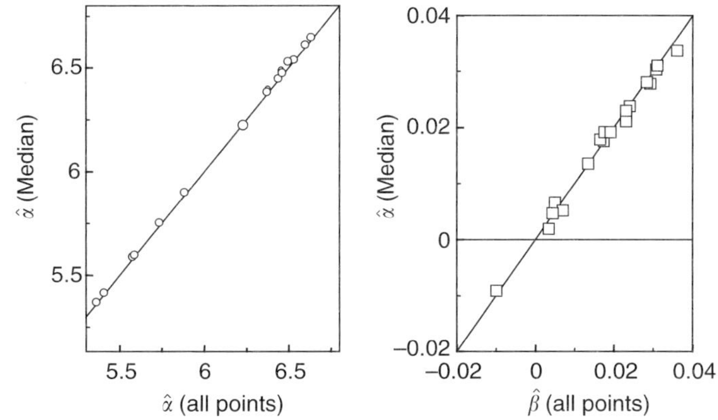

FIGURE 9.1.42 Comparison of the estimates for the regression data obtained by fitting all individual data points and by fitting the median values.

and $y = \log \sigma_f$, $x = \log \dot{\sigma}$. In Eq. 105, ε is the residual standard deviation defined by

$$\varepsilon = \sqrt{\frac{[(y - \hat{\alpha} - \hat{\beta}x)^2]}{N - 2}} \tag{106}$$

In a simplification we assume that $\hat{\alpha}$ and $\hat{\beta}$ might be normally distributed. Then, the confidence intervals to the confidence level $1 - \gamma$ can be determined easily. At a value of $\gamma = 0.1$, the true values of α and β are located with a probability of 90 % in the intervals of

$$\hat{\alpha} - t_{0.05}\varepsilon_{\hat{\alpha}} \leq \alpha \leq \hat{\alpha} + t_{0.95}\varepsilon_{\hat{\alpha}}, \qquad \hat{\beta} - t_{0.05}\varepsilon_{\hat{\beta}} \leq \beta \leq \hat{\beta} + t_{0.95}\varepsilon_{\hat{\beta}} \tag{107}$$

with the upper confidence limit factor $t_{0.05}$ determined at $\gamma/2$ and the lower confidence limit factor $t_{0.95}$ at $(1 - \gamma/2)$.

In Figure 9.1.42, the two straight-line parameters $\hat{\alpha}$ and $\hat{\beta}$ are plotted for the median value evaluation and the evaluation of all single data points. From this diagram, an excellent agreement can be seen. From these two parameters the strength $\sigma(0)$ at 1 MPa/s loading rate and the exponent n were determined.

In order to obtain an impression of the agreement between different laboratories involved in the tests (for each test a lab of a research institute and one from industry), the 90 % confidence intervals for $\sigma(0)$ and n are plotted in Figure 9.1.43. It follows from Eqs 100–102: $\log \sigma(0) = \alpha$ and $n = (1/\beta) - 1$. The horizontal and vertical dimension sizes of the rectangles represent the confidence intervals obtained by the two labs. An agreement within the 90% confidence intervals can be stated, if the rectangular areas touch the straight line with slope 1. Apart from $\sigma(0)$ for the SSN ceramic, sufficient agreement

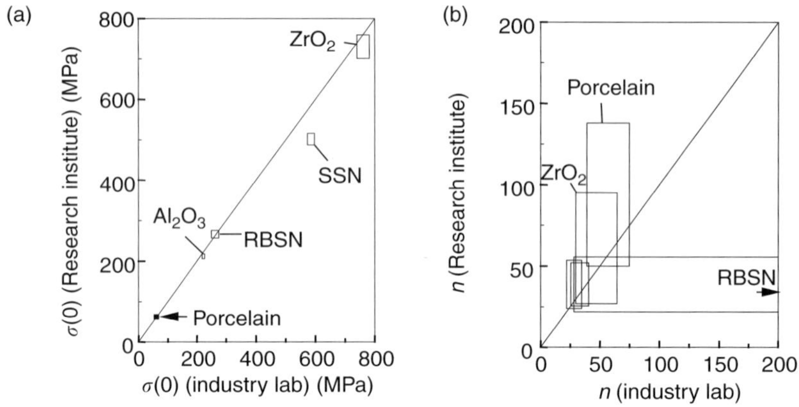

FIGURE 9.1.43 Comparison of strengths $\sigma(0)$ at a stress rate 1 MPa/s and crack growth exponents n.

was found between the labs. In contrast to this result large confidence intervals for n are visible.

9.1.12.4 Cyclic Fatigue

Under the assumption of the same mechanism being responsible for constant and cyclic loading, lifetimes can be predicted from the results obtained in tests with constant load by using Eq. 88 and introducing the relevant loading history $\sigma(t)$ under the integral sign. Unfortunately, this method is only suited for a few materials as has been confirmed by cyclic experiments on glass and porcelain [74]. In general, the crack growth rates in ceramics are larger and the lifetimes are shorter than those predicted from constant load tests.

Therefore, it is indispensable to determine the crack growth rates under cyclic loading experimentally. Similar to static fatigue tests one will find power law relations over wide regions of crack growth rates $\mathrm{d}a/\mathrm{d}N$ (N is the number of cycles). Two representations of cyclic fatigue results are well established, namely, a description in terms of the range of the stress intensity factor ΔK

$$\frac{\mathrm{d}a}{\mathrm{d}N} = A(\Delta K)^n = A^* \left(\frac{\Delta K}{K_{\mathrm{Ic}}}\right)^n \tag{108}$$

or in terms of the maximum stress intensity factor during a cycle $K_{\max}$

$$\frac{\mathrm{d}a}{\mathrm{d}N} = A(K_{\max})^n = A^* \left(\frac{K_{\max}}{K_{\mathrm{Ic}}}\right)^n \tag{109}$$

with the maximum stress intensity factor

$$K_{\max} = \sigma_{\max}\sqrt{a}Y \tag{110}$$

with the maximum stress $\sigma_{\max}$. The parameters A or A^* and n depend on the R-ratio and the frequency. The R-ratio and the range ΔK are given by

$$R = K_{\min}/K_{\max}, \quad \Delta K = K_{\max} - K_{\min} \tag{111}$$

Integration of the crack growth relations from the initial crack size a_{i} to the critical crack size a_{c} yields the number of cycles to failure N_{f}

$$N_{\mathrm{f}} = \frac{B\sigma_{\mathrm{c}}^{n-2}}{(\Delta\sigma)^n} = \frac{B\sigma_{\mathrm{c}}^{n-2}}{\sigma_{\max}^n(1-R)^n} \tag{112}$$

with B defined by Eq. 89. In a plot of $\log \Delta\sigma$ or $\log(\sigma_{\max})$ versus $\log(N_{\mathrm{f}})$ a straight line with a slope of $-1/n$ is expected.

In principle, all procedures used for static tests are also applicable for cyclic fatigue experiments. Since small cracks are of highest importance for lifetime predictions, the evaluation of crack extension parameters may be outlined here for tests with natural cracks. For this purpose let us consider a piezoelectric material. In their application as actuators, piezoelectric ceramics are subjected to changing electrical and mechanical loads. These changes can lead to an electrical and mechanical fatigue that is connected with a premature failure of the component. In Ref. [75], fatigue experiments were performed in pulsating tensile tests on an unpoled and poled commercial PZT (PIC 151, PI Ceramic). The specimens were poled parallel ($\|$) and perpendicular ($\perp$) to the tensile direction. Figure 9.1.44a shows the Weibull plots of the inert strength σ_{c}. The lifetime results obtained for $R = \sigma_{\min}/\sigma_{\max} = 0$ and a frequency of 1 Hz are summarized in Figure 9.1.44b in Weibull representation. Similar to the crack growth rate in static tests, Eq. 98, it holds for cyclic tests [76]

$$\frac{\mathrm{d}a}{\mathrm{d}N} = -\frac{2K_{\mathrm{Ic}}^2}{N_{\mathrm{f}}\sigma_{\mathrm{c}}^2 Y^2}\frac{\mathrm{d}[\log(\sigma_{\max}/\sigma_{\mathrm{c}})]}{\mathrm{d}[\log(N_{\mathrm{f}}\sigma_{\max}^2)]} \tag{113}$$

For the computation of $K_{\max}$ the proportionality between stress and the stress intensity factor can be used, that is, $K_{\max}/K_{\mathrm{Ic}} = \sigma_{\max}/\sigma_{\mathrm{c}}$.

In order to allow the determination of the logarithmic derivative in Eq. 113, an auxiliary diagram $\log(\sigma/\sigma_{\mathrm{c}})$ versus $\log(N_{\mathrm{f}}\sigma^2)$ has to be plotted. This is illustrated in Figure 9.1.45a. The derivative can be determined from a fitted curve that is not necessarily a straight line. The crack growth rates obtained with

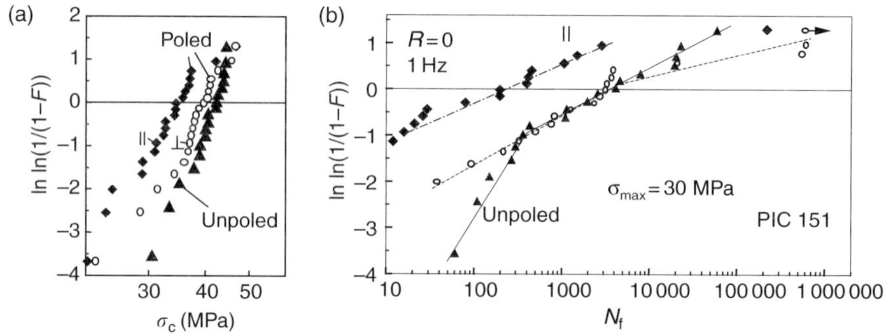

FIGURE 9.1.44 (a) Tensile strength tests on a PZT ceramic in different poling states. (b) Numbers of cycles to failure in Weibull representation (symbols as in *a*).

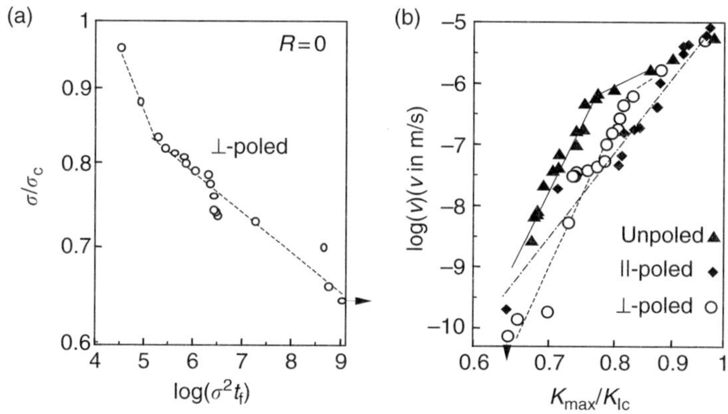

FIGURE 9.1.45 Auxiliary plot (a) and crack growth curves (b) for PZT.

Eq. 113 are entered in Figure 9.1.45b. It is obvious that crack growth rates can be described well by power law relations for the investigated materials at $v < 10^{-6}$ m/s.

To compare crack growths in tension and bending tests, Figure 9.1.46a shows bending results from Ref. [77] as straight lines and results from pulsating tensile tests for the ⊥-poled material (symbols). The results for the two different loading types (bending/tension) are in good agreement. A change of the crack growth exponent with the *R*-curve is obvious. This is in agreement with results for other ceramics (Figure 9.1.46b and c).

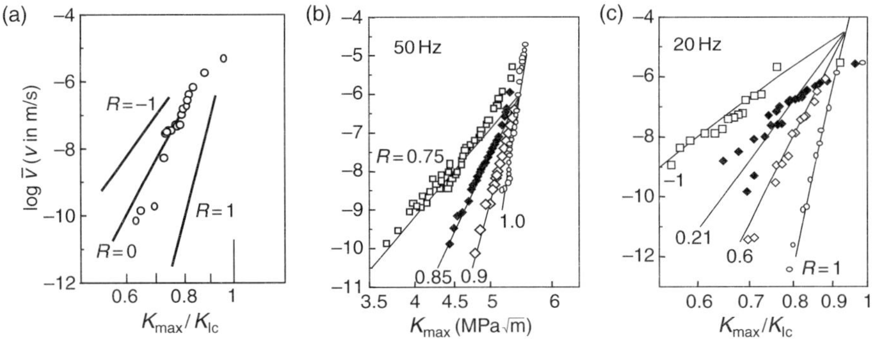

FIGURE 9.1.46 Influence of the R-ratio on crack growth rates: (a) crack growth data for ⊥-poled PZT, straight lines from pulsating bending tests [77], circles from pulsating tensile tests; (b) sintered silicon nitride (Ueno *et al.* [78]); and (c) Y-PSZ (Schmitt *et al.* [79]). Curves: fitting curves according to Eq. 109.

9.1.13 THE LOADING PARAMETER C^*

In the preceding considerations behaviour of cracks was addressed in terms of linear-elastic fracture mechanics. Whereas at low and moderate temperatures the description by stress intensity factors has to be applied, this approach may fail at high temperatures. Fracture at elevated temperatures may be caused by crack propagation that starts from existing flaws or from creep-induced damage caused by the formation and coalescence of pores. In the presence of noticeable creep, the C^*-integral proposed by Landes and Begley [80] is the crack driving parameter (for details see Riedel [81]).

The C^*-parameter is a path-independent energy-rate line integral. In the two-dimensional case (see Figure 9.1.47), it is defined by

$$C^* = \int_{\Gamma} \left[W^* dy - T_i \frac{\partial \dot{v}_i}{\partial x} \, ds \right] \tag{114}$$

$$W^* = \int_0^{\dot{\varepsilon}_{ij}} \sigma_{ij} \, d\dot{\varepsilon}_{ij}, T_i = \sigma_{ij} n_j \tag{115}$$

where W^* is the strain energy density rate, T_i is the traction vector along the contour Γ, n_j is the outward normal to Γ, v_i the displacement vector, and ds is a line length increment. The stress tensor is σ_{ij} and the strain rate tensor is $\dot{\varepsilon}_{ij}$. The integration path Γ extends from the lower crack surface counterclockwise around the crack tip and ends at the upper crack surface. A premise for the applicability of C^* as a loading parameter is the occurrence of secondary creep, which mostly is described by the power-law relation proposed by Norton [82]

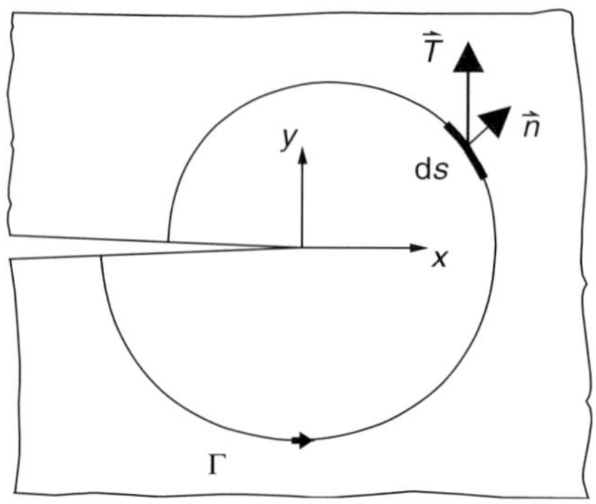

FIGURE 9.1.47 C^* as a line integral around the crack tip.

and Baily [83]

$$\dot{\varepsilon}_s = D\sigma^n \tag{116}$$

with exponents in the range $n \approx 1$–4 for ceramic materials.

The C^* integral governs the stresses and strains in front of a crack tip. Under secondary creep condition, an HRR (Hutchinson, Rice, Rosengren) stress field appears ahead of the crack tip [84, 85] and the stresses and strain rates are given by

$$\sigma_{ij} = \left(\frac{C^*}{I_n D r}\right)^{1/(n+1)} \tilde{\sigma}_{ij}(\theta), \qquad \dot{\varepsilon}_{ij} = \left(\frac{C^*}{I_n D r}\right)^{n/(n+1)} \tilde{\varepsilon}_{ij}(\theta) \tag{117}$$

where $I_n(n)$, $\tilde{\sigma}(\theta)$ and $\tilde{\varepsilon}(\theta)$ are dimensionless functions that can be taken from Ref. [86].

C^* can be expressed by the energy rate difference of two identically loaded structures containing cracks of depths a and $a + \mathrm{d}a$, respectively, that is,

$$C^* = -\frac{1}{B}\frac{\mathrm{d}U^*}{\mathrm{d}a}, \qquad U^* = \int_0^{\dot{\delta}} F\,\mathrm{d}\dot{\delta} \tag{118}$$

(B is the specimen thickness). This results directly from Eq. 114, if the outer contour of the component is used for the path Γ.

Equation 118 is the basis for a number of procedures used to determine C^*. The experimental data necessary for the evaluation of Eq. 118 are the force F, the displacement rates $\dot{\delta}$ of the load application points and the crack length a.

In the dynamic loading case with constant displacement rates $\dot{\delta}$ at the load application points, the C^* parameter can be computed as

$$C^* = \frac{F\dot{\delta}}{BW}\left[-\frac{n}{n+1}\frac{W}{F}\frac{\mathrm{d}F}{\mathrm{d}a}\right] \tag{119}$$

where F is the load and B is the thickness and W the width of the specimen.

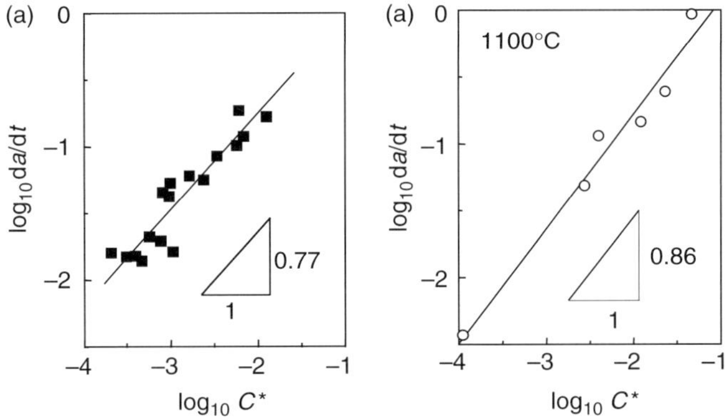

FIGURE 9.1.48 *v*–C^* curves for an alumina with glass content at 1100°C, obtained with different methods [4]: (a) d*a*/d*t* as a function of C^* from static tests; (b) d*a*/d*t* versus C^* from dynamic bending tests using the procedure of Landes and Begley [80] (d*a*/d*t* in mm h^{-1}, C^* in N/mm/h).

Similar to the description of subcritical crack growth driven by the stress intensity factor *K*, a power-law relation is also possible for creep crack growth. A relation was proposed by Riedel [81] in which the exponent is no longer a free parameter

$$\frac{da}{dt} = A(C^*)^{n/(n+1)} \tag{120}$$

Compared with the high crack growth exponents for subcritical crack growth, the creep crack growth exponents are small, namely, less than 1. Experimental results on ceramics (see Figure 9.1.48 for an Al_2O_3 ceramic with about 3% glass content) confirm Eq. 120 [4].

REFERENCES

1. Broek, D. (1986). *Elementary Engineering Fracture Mechanics*, Dordrecht, NL: Martinus Nijhoff Publishers.
2. Ewalds, H. L., and Wanhill, R. J. H. (1985). *Fracture Mechanics*, London: Edward Arnold Ltd.
3. Anderson, T. L. (1995). *Fracture Mechanics, Fundamentals and Applications*, 2nd edn., Boston: CRC Press.
4. Munz, D., and Fett, T. (1999). *Ceramics*, Heidelberg: Springer Verlag.
5. Nishida, T., Pezzotti, G., Mangialardi, T., and Paolini, A. E. (1996). Fracture mechanics evaluation of ceramics by stable crack propagation in bend bar specimens. *Fract. Mech. Ceram.* **11**: 107–114.
6. Kübler, J. (1997). Fracture toughness using the SEVNB method: preliminary results. *Ceramic Engineering & Science Proceedings*, Vol. 18, pp. 155–162.

7. Fett, T., and Munz, D. (1997). *Stress Intensity Factors and Weight Functions*, Southampton, UK: Computational Mechanics Publications.
8. Damani R., Gstrein R., and Danzer, R. (1996). Critical notch-root radius effect in SENB-S fracture toughness testing. *J. Eur. Ceram. Soc.* **16**: 695–702.
9. Rudolph, E. (1997), in *Ceramic Engineering & Science Proceedings*, vol. 18.
10. Munz, D., Shannon, J. L., and Bubsey, R. T. (1980). Fracture toughness calculation from maximum load in four point bend tests of chevron notch specimens. *Int. J. Fract.* **16**: R137–R141.
11. Shannon, J. L., and Munz, D. (1984). Specimen size and geometry effects on fracture toughness of aluminum oxide measured with short-rod and short-bar chevron-notched specimens, ASTM STP 855, pp. 270–280.
12. Lawn, B. R., and Swain, M. V. (1975). Microfracture beneath point indentation in brittle solids. *J. Mater. Sci.* **10**: 113–122.
13. Newman, J. C., and Raju, I. S. (1981). An empirical stress intensity factor equation for the surface crack. *Eng. Fract. Mech.* **15**: 185–192.
14. He, M. Y., Cao, H. C., and Evans, A. G. (1990). Mixed-mode fracture: the four-point shear specimen. *Acta Metall. Mater.* **38**: 839–846.
15. Suresh, S., Shih, C. F., Morrone, A., and O'Dowd, N. P. (1990). Mixed-mode fracture toughness of ceramic materials. *J. Am. Ceram. Soc.* **73**: 1257–1267.
16. Fett, T., Gerteisen, G., Hahnenberger, S., Martin, G., and Munz, D. (1995). Fracture tests for ceramics under mode-I, mode-II and mixed-mode loading. *J. Eur. Ceram. Soc.* **15**: 307–312.
17. Petrovic, J. J., and Mendiratta, M. G. (1976). Mixed-mode fracture from controlled surface flaws in hot-pressed Si_3N_4. *J. Am. Ceram. Soc.* **59**: 163–167.
18. Petrovic, J. J. (1985). Mixed-mode fracture of hot-pressed Si_3N_4, *J. Am. Ceram. Soc.* **68**: 348–355.
19. Marshall, D. B. (1984). Mechanisms of failure from surface flaws in mixed-mode loading. *J. Am. Ceram. Soc.* **67**: 110–116.
20. Paris, P. C., and Sih, G. C. (1965). Stress analysis of cracks, ASTM STP 381, pp. 30–80.
21. Alpa, G. (1984). On a statistical approach to brittle rupture for multiaxial states of stress. *Eng. Fract. Mech.* **19**: 881–901.
22. Fett, T. (2001). Effective mode-II stress intensity factor for partially opened natural cracks under mixed-mode loading, presented at the "10th International Congress of Fracture (ICF 10)", Honolulu, 3–7 December.
23. McMeeking, R. M. (1989). Electrostrictive stresses near crack-like flaws. *J. Appl. Math. Phys.* **40**: 615.
24. McMeeking, R. M. (1990). A J-integral for the analysis of electrically induced mechanical stress at cracks in elastic dielectrics. *Int. J. Eng. Sci.* **28**: 605–613.
25. Suo, Z. (1991). Mechanics concepts for failure in ferroelectric ceramics, *Smart Structures and Materials*, Srinvasan, A. V., ed., pp. 1–6, New York: ASME.
26. Suo, Z., Kuo, C. M., Barnett, D. M., and Willis, J. R. (1992). Fracture mechanics for piezoelectric ceramics. *J. Mech. Phys. Solids* **40**: 739–765.
27. Park, S., and Sun, C. T. (1995). Fracture criteria for piezoelectric ceramics. *J. Am. Ceram. Soc.* **78**: 1475–1480.
28. Heyer, V., Schneider, G. A., Weitzing, H., Balke, H., Drescher, J., Bahr, H.-A., and Kemmer, G. (1997). Fracture criterion for a conducting crack in piezoelectric ceramics, "Proceedings of the International Conference and Exhibition Micro Materials MICRO MAT 97", Berlin, Michel, B., and Winkler, T., eds., pp. 398–401.
29. Heyer, V., Schneider, G. A., Balke, H., Drescher, J., and Bahr, H.-A. (1998). A fracture criterion for conducting cracks in homogeneously poled piezoelectric PZT-PIC 151 ceramics. *Acta Mater.* **46**: 6615–6622.

30. Hübner, H., and Jillek, W. (1977). Subcritical crack extension and crack resistance in polycrystalline alumina. *J. Mater. Sci.* **12**: 117–125.
31. Llorca, J., and Steinbrech, R. W. (1991). Fracture of alumina: an experimental and numerical study. *J. Mater. Sci.* **26**: 6383–6390.
32. Fett, T., Munz, D., Dai, X., and White, K. W. (2000). Bridging stress relation from a combined evaluation of the *R*-curve and post-fracture tensile tests. *Int. J. Fract.* **104**: 375–385.
33. Rauchs, G., Fett, T., and Munz, D. (2002). *R*-Curve behaviour of 9Ce-TZP zirconia ceramics, *Engng. Fract. Mech.* **69**: 389–401.
34. Fett, T., Munz, D., and Thun, G. (2001). A toughness test device with opposite roller loading. *Eng. Fract. Mech.* **68**: 29–38.
35. Fett, T., and Munz, D. (2000). *R*-curve for a PZT ceramic obtained from tensile strength tests with Knoop-damaged specimens. *J. Am. Ceram. Soc.* **83**: 3199–3201.
36. Rödel, J., Kelly, J. F., and Lawn, B. R. (1990). In situ measurements of bridged crack interfaces in the scanning electron microscope. *J. Am. Ceram. Soc.* **73**: 3313–3318.
37. Seidel, J., and Rödel, J. (1997). Measurement of crack tip toughness in alumina as a function of grain size. *J. Am. Ceram. Soc.* **80**: 433–438.
38. Meschke, F., Raddatz, O., Kolleck, A., and Schneider, G. (2000). *R*-curve behaviour and crack-closure stresses in barium titanate and (Mg,Y)-PSZ ceramics. *J. Am. Ceram. Soc.* **83**: 353–361.
39. Fett, T., Munz, D., Seidel, J., Stech, M., and Rödel, J. (1996). Correlation between long and short crack *R*-curves in alumina using the crack opening displacement and fracture mechanical weight function approach. *J. Am. Ceram. Soc.* **79**: 1189–1196.
40. Fett, T., Munz, D., and Thun, G. (2000). A method to estimate crack tip toughness from bending tests on pre-notched bars. *J. Am. Ceram. Soc.* **83**: 421–423.
41. Knehans, R., and Steinbrech, R. (1982). Memory effect of crack resistance during slow crack growth in notched Al_2O_3 bend specimens. *J. Mater. Sci. Lett.* **1**: 327–329.
42. Steinbrech, R., Knehans, R., and Schaarwächter, W. (1983). Increase of crack resistance during slow crack growth in Al_2O_3 bend specimens. *J. Mater. Sci.* **18**: 265–270.
43. Frei, H., and Grathwohl, G. (1989). The fracture resistance of high performance ceramics by in situ experiments in the SEM. *Beitr. Elektronenmikroskop. Direktabb. Oberfl.* **22**: 71–78.
44. Lathabai, S., Rödel, J., and Lawn, B. R. (1991). Cyclic fatigue from frictional degradation at bridging grains in alumina. *J. Am. Ceram. Soc.* **74**: 1340–1348.
45. Mai, Y., and Lawn, B. R. (1987). Crack-interface grain bridging as a fracture resistance mechanism in ceramics: II. Theoretical fracture mechanics model. *J. Am. Ceram. Soc.* **70**: 289.
46. Fett, T., and Munz, D. (1993). Evaluation of *R*-curve effects in ceramics. *J. Mater. Sci.* **28**: 742–752.
47. McMeeking, R. M., and Evans, A. G. (1982). Mechanics of transformation-toughening in brittle materials. *J. Am. Ceram. Soc.* **65**: 242–246.
48. Eshelby, J. D. (1957). Determination of the elastic field of an ellipsoidal inclusion and related problems. *Proc. R. Soc., Lond. Ser. A* **241**: 376–396.
49. Stam, G. T. M., van der Giessen, E., and Meijers, P. (1994). Effect of transformation-induced shear strains on crack growth in zirconia-containing ceramics. *Int. J. Solids Struct.* **31**: 1923–1948.
50. Yu, C. S., Shetty, D. K., Shaw, M. C., and Marshall, D. B. (1992). Transformation zone shape effects on crack shielding in ceria-partially-stabilised zirconia (Ce-TZP)–alumina composites. *J. Am. Ceram. Soc.* **75**: 2991–2994.
51. Fett, T., Kamlah, M., Munz, D., and Thun, G. (2000). Strength of a PZT ceramic under different test conditions. *Smart Structures and Materials 2000: Active Materials: Behaviour and Mechanics*, Vol. 3992, Lynch, C. S., ed., pp. 197–208.

52. Chen, W., and Lynch, C. S. (1998). A micro-electro-mechanical model for polarization switching of ferroelectric materials. *Acta Mater.* **46**: 5303–5312.
53. Schneider, G. A., and Heyer, V. (1999). Influence of the electric field on Vickers indentation crack growth in $BaTiO_3$. *J. Eur. Ceram. Soc.* **19**: 1299–1306.
54. Lucato, S. L., Lupascu, D. C., and Rödel, J. (2000). Effect of poling direction on *R*-curve behavior in PZT. *J. Am. Ceram. Soc.* **83**: 424–426.
55. Buresch, F. E. (1978). A structure sensitive K_{Ic}-value and its dependence on grain size distribution, density, and microcracking interaction. *Fract. Mech. Ceram.* **4**: 835–847.
56. Pabst, R. F., Steeb, J., and Claussen, N. (1978). Microcracking in a process zone and its relation to continuum fracture mechanics. *Fract. Mech. Ceram.* **4**: 821–833.
57. Bueckner, H. (1979). A novel principle for the computation of stress intensity factors. *Zeit. Angew. Math. Mech. ZAMM* **50**: 529–546.
58. Yu, C. T., and Kobayashi, A. S. (1993). Fracture process zone in SiC_W/Al_2O_3. *Ceram. Eng. Sci. Proc.* **14**: 273–283.
59. Fett, T., Munz, D., Yu, C. T., and Kobayashi, A. S. (1994). Determination of bridging stresses in reinforced Al_2O_3. *J. Am. Ceram. Soc.* **77**: 3267–3269.
60. Hay, J. C., and White, K. W. (1993). Grain-bridging mechanisms in monolithic alumina and spinel. *J. Am. Ceram. Soc.* **76**: 1849–1854.
61. White, K. W., and Hay, J. C. (1994). The effect of thermoelastic anisotropy on the *R*-curve behavior of monolithic alumina. *J. Am. Ceram. Soc.* **77**: 2283–2288.
62. Hay, J. C., and White, K. W. (1995). Grain boundary phases and wake zone characterization in monolithic alumina. *J. Am. Ceram. Soc.* **78**: 1025–1032.
63. Fett, T., Munz, D., Dai, X., and White, K. W. (2000). Bridging stress relation from a combined evaluation of the *R*-curve and post-fracture tensile tests. *Int. J. Fract.* **104**: 375–385.
64. Wiederhorn, S. M., and Bolz, L. H. (1979). Stress corrosion and static fatigue of glass. *J. Am. Ceram. Soc.* **53**: 543–548.
65. Evans, A. G. (1972). A method for evaluating the time-dependent failure characteristics of brittle materials—and its application to poly-crystalline alumina. *J. Mater. Sci.* **7**: 1137–1146.
66. Freiman, S. W., Murville, D. J., and Mast, P. W. (1973). Crack propagation studies in brittle materials. *J. Mater. Sci.* **8**: 1527–1533.
67. Weibull, W. (1939). *A Statistical Theory of the Strength of Materials*, Stockholm: Ingeniörsvetenskaps-akademiens Handlingar, 151.
68. Thoman, D. R., Bain, L. J., and Antle, C. E. (1969). Inferences on the parameters of the Weibull distribution, *Technometrics* **11**: 445.
69. Lawless, J. F. (1982). *Statistical Models and Methods for Lifetime Data*, New York: Wiley.
70. Fett, T., Munz, D., Keller, K., and Himsolt, G. (1997). Subcritical crack growth in silicon carbide. *J. Mater. Sci. Lett.* **16**: 324–327.
71. Fett, T., Keller, K., and Munz, D. (1988). Determination of *v*–*K*-curves of hot-pressed silicon nitride at elevated temperatures. *Int. J. Fract.* **36**: 3–14.
72. Charles, R. J. (1958). Dynamic fatigue of glass. *J. Appl. Phys.* **29**: 1657–1661.
73. Fett, T., and Gehrke, E. (2000). A round robin on dynamic bending tests. *Ceram. Forum Int.—cfi/Berichte der Deutschen Keramischen Gesellschaft* 77: 7–11.
74. Evans, A. G., and Fuller, E. R. (1974). Crack propagation in ceramic materials under cyclic loading conditions. *Metall. Trans.* **5**: 27–33.
75. Fett, T., Munz, D., and Thun, G. (1999). Mechanical fatigue of a soft PZT ceramic under pulsating tensile loading. *J. Mater. Sci. Lett.* **18**: 1895–1898.
76. Fett, T., Martin, G., Munz, D., and Thun, G. (1991). Determination of $da/dN-\Delta K$ curves for small cracks in alumina in alternating bending tests. *J. Mater. Sci.* **26**: 3320–3328.
77. Fett, T. (1998). Mechanische Ermüdung an weicher PZT-Keramik. "30. Tagung des DVM-Arbeitskreises Bruchvorgänge", pp. 375–384.

78. Ueno, A., Kishimoto, H., Kawamoto, M., and Asakura, M. (1991). Crack propagation behavior of sintered silicon nitride under cyclic loads of high stress ratio and high frequency. *Eng. Fract. Mech.* **40**: 913–920.
79. Schmitt, R., Fett, T., and Munz, D. (1996). Cyclic fatigue of zirconia. *Fatigue Fract. Eng. Mater. Struct.* **19**: 1411–1420.
80. Landes, J. D., and Begley, J. A (1976). A fracture mechanics approach to creep crack growth. ASTM STP 590, pp. 128–148.
81. Riedel, H. (1987). *Fracture at High Temperatures*, Berlin: Springer Verlag.
82. Norton, F. H. (1929). *Creep of Steel at High Temperatures*, New York: McGraw Hill.
83. Bailey, R. W. (1929). Creep of steel under simple and compound stresses, and the use of high initial temperature in steam power plants. "Transactions of the World Power Conference", Vol. 3, p. 1089, Tokyo.
84. Hutchinson, J. W. (1968). Plastic stress and strain fields at crack tip. *J. Mech. Phys. Solids* **16**: 13–31.
85. Rice, J. R., and Rosengren, G. F. (1968). Plain strain deformation near crack tip in power law hardening material. *J. Mech. Phys. Solids* **16**: 1–12.
86. Kumar, V., German, M. D., and Shih, C. F. (1981). An engineering approach for elastic–plastic failure analysis. EPRI-Report NP-1931, Palo Alto, USA.

Handbook of Advanced Ceramics
S. Sōmiya *et al.* (Eds.)

CHAPTER 10

10.1 Testing and Evaluation

TATSUKI OHJI
AIST, Synergy Materials Research Center, National Institute of Advanced Industrial Science and Technology, Moriyama-ku, Nagoya 463-8687, Japan

10.1.1 INTRODUCTION

Because of their brittle nature, ceramics show totally different fracture behavior from that of metals and polymers. Ceramic engineers should properly understand this unique behavior when designing ceramic components for safe and reliable use. Fracture behavior of ceramics can be clarified through proper mechanical tests suitable to materials and their applications, so mechanical tests comprise a significant part of ceramic technology. Over the years, substantial progress has been made in our understanding of mechanical properties of ceramics as well as test methods needed to determine those properties.

This chapter addresses mechanical properties of ceramics; particular interest is given to fast fracture strength and time-dependent failure properties such as fatigue, slow crack growth, and creep. The testing procedure and the recent related research topics are reviewed for establishing the reliability peculiar to each of these properties. More detailed information on mechanical testing procedure of ceramics is available in other literatures [1]. Fracture toughness, which is one of the most important parameters for expressing brittle fracture of ceramics, is not discussed here in detail however, since it was discussed in Chapter 9.

10.1.2 FRACTURE STRENGTH

Fracture strength is one of the most common and widely used mechanical properties for structural ceramics because it is the most simple and important material reliability parameter. Ceramic component design demands critical investigation of fracture strength and its distribution due to the nature of probabilistic fracture of ceramics.

A number of methodologies and techniques have been developed for determining ceramics fracture strength [2–31]. Most of these techniques equate fracture strength to maximum stress (tensile or compressive) at fracture. Consequently, in order for a particular load and specimen geometry to be useful for determination of fracture strength, stress distribution must be well established. A complicating factor in determination of fracture strength is that ceramic materials strength is quite sensitive to size, shape, and surface finish. Such sensitivity is largely responsible for wide variation in strength values reported for a given material. Viable test methodology must, therefore, account for these effects.

This section describes commonly used methods, typically divided into four types: flexural tests, tensile tests, compressive tests, and biaxial tests. Recent topical research efforts on fracture strength measurements of ceramics are then introduced.

10.1.2.1 Testing Methods

10.1.2.1.1 Tensile Tests

Uniaxial tensile tests have been widely used for many years in order to obtain information on mechanical strength properties of industrial materials. However, in the field of ceramics, flexural tests are commonly used instead of tensile tests. This is because stress concentration occurring during tests cannot be relaxed by local plastic deformation in ceramics; hence, even slight eccentric loading in tensile tests causes parasitic bending stress. Susceptibility of stress concentration also causes improper fractures inside the gauge section. Especially in high-temperature measurements, the degree of difficulty increases because material and structure of the testing fixture are restricted. On the other hand, ceramics elasticity is usually maintained until fracture occurs, enabling researchers to determine strength in flexural tests by elementary stress elastic beam equations. In addition, flexural tests are superior to tensile tests due to the simple specimen geometry, low cost of specimen preparation, and adaptability to high-temperature measurements. Nevertheless, tensile tests have the following advantages. First, fracture strength and its distribution can be determined from the stress field within large effective volumes of the material. This is most useful for reliable design of large mechanical components. Second, fractures in tensile tests more frequently originate from internal natural fracture defects than those in flexural ones, resulting in more intrinsic strength. Third, even in the high-temperature region where the material shows non-elastic behavior and validity of stress calculation based on the elastic body in flexural tests is questionable, tensile tests can determine stress precisely and in a simple manner. Also, in such a region, tensile tests are most advantageous for determining stress–strain curves, and creep curves because deformation amounts can be

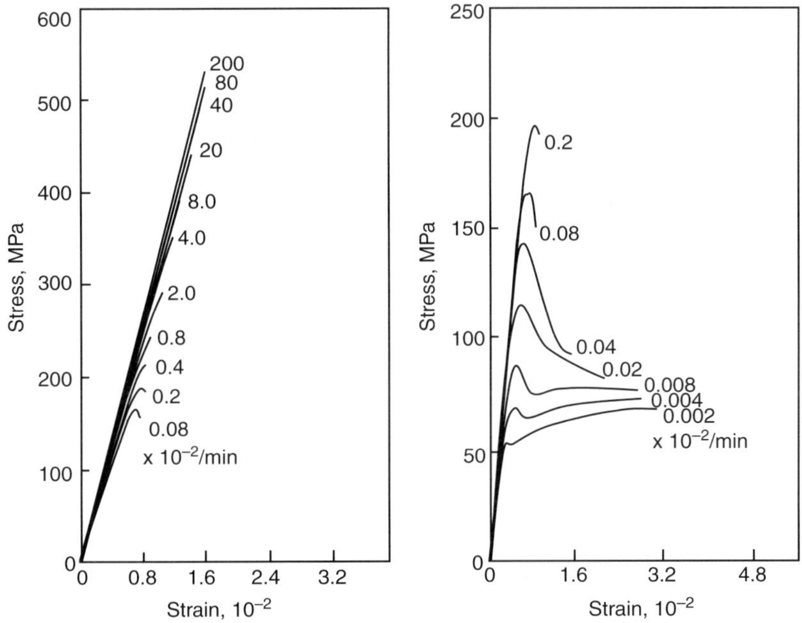

FIGURE 10.1.1 Stress–displacement curves showing effect of strain rate for a hot-pressed silicon nitride at 1260°C.

detected properly with an appropriate extensometer. Examples of non-linear stress–strain curves of silicon nitride at high temperatures in tension are shown in Figure 10.1.1.

As mentioned earlier, there are two major barriers to ceramics tensile tests. One is eccentric loading, which results in unexpected parasitic bending strain. The other is stress concentration inside the gauge section. In general cases, the latter barrier can be overcome by employing the long gauge section, which is distant from the loading points with moderate change of cross-section (this is due to St Vennant's effect). For the first barrier, there are several sources of eccentricity including off-center loading, end bending moment, end torque moment, and an imperfect or curved specimen. Difficulty in solving the first barrier exists in these sources being involved with it in the complex. Sources include an improperly aligned load train, improper specimen/grip interface, imperfect (e.g. curved or non-symmetric) specimen geometry, and so on. Examples of successful tensile testing ceramics are available in Refs [2–14].

The geometry of tensile test specimen depends on the purpose of the tensile test itself. When a tensile test is performed for knowing the tensile strength of an as-fabricated component, its dimension limits the dimensions of the specimen.

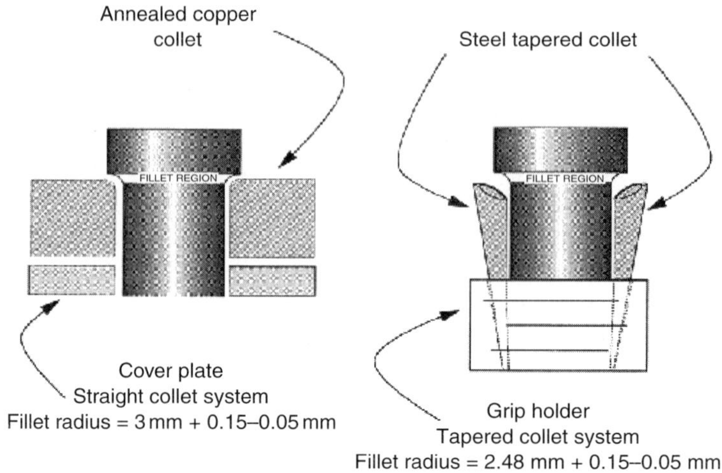

FIGURE 10.1.2 Illustration of two types of tensile specimen and grip designs [32]. Reprinted with permission of the American Ceramic Society. All rights reserved.

Gripping devices also restrict the design of the test specimen geometry. The range of possible tensile test specimen geometries is illustrated in ASTM C 1273, "Standard Test Method for Tensile Strength of Monolithic Advanced Ceramics at Ambient Temperatures."

Tennery *et al.* [32] recently reported results of cooperative work towards standardization of tensile testing ceramics conducted by 10 US laboratory groups for silicon nitride. The collaboration employed two geometries of button-head tensile specimens; 100 of a straight collet design (type A) and 50 of a tapered collet design (type B) were fractured as shown in Figure 10.1.2. The load train for type A specimens included copper collets, which transferred the load into the specimen button-head end, and that for type B specimens included steel collets, which transferred the load into the shanks of the specimen located just under the button head. These two methods for delivering the load into the specimen were selected from results of finite element analyses of a number of geometries for the specimen end and the testing machine grips necessary to hold the specimen [9]. Figure 10.1.2 also shows button-head radii for the two designs. Uniform load delivery is essential for minimizing induced bending moment during specimen loading. Relatively good agreement of results was found between the two methods.

10.1.2.1.2 Compressive Tests

Although compressive strength of ceramic materials is generally much higher than tensile strength, the dominant loading mode in many actual applications is compression. Reliable component design in these applications requires precise compressive strength measurements. Compressive strength measurements are also necessary for development of fundamental insights into micromechanical and statistical aspects of compressive failure.

Compressive failure is thought to involve coalescence of damage in the form of microcracks and microvoids that propagate by alternating growth and arrest mechanisms [15]. For ceramic materials tested at ambient temperatures, this damage is thought to be generated by localized microplasticity arising from twinning and slip [16]. Failure occurs through structural collapse when density of damage in a zone reaches a critical level. Because this size appears to depend only upon intrinsic material characteristics, compressive strength is not generally affected by specimen size.

For uniaxial testing, compressive stress is given simply by the applied load divided by the cross-sectional area. However, high stresses required for failure under compressive loading have created a number of problems in test fixture design and specimen geometries required for successful compression strength measurement. Particularly, problems related to improper alignment and load block stress concentrations can lead to generation of tensile stresses in the specimen sufficient to cause failure. Therefore, it should be noted that measured compressive strength tends to underestimate true values. Examples for successful measurements of compressive strength in ceramics are available in Refs [15–17].

10.1.2.1.3 Flexural Tests

Due to advantages including simple specimen geometry, low-cost specimen fabrication, and adaptability to high-temperature measurements, flexural tests are most widely used for measuring ceramic materials strength [18–24]. Compared to other testing methods such as tension and compression, specimens and testing fixtures for flexural tests are much simpler and, consequently, less expensive. Flexural tests, however, are disadvantageous by vulnerability of strength to surface machining, small effective volume of specimen, and other reasons. When resultant flexural strength data are used for component design, one must assume that strength-limiting flaw distribution (pores, surface flaws, agglomerates, inclusions, etc.) is equivalent to other methods. Therefore, surface machining processes for test specimens are extremely important; when machining processing causes severe surface or subsurface damage, it becomes difficult to obtain intrinsic strength. This tendency increases with higher material strength. Such materials have limited fracture origin size and require damage-less surface treatment.

Two kinds of flexure loading are commonly used: three-point loading and four-point loading (Fig. 10.1.3). In four-point flexure, pure flexure moment

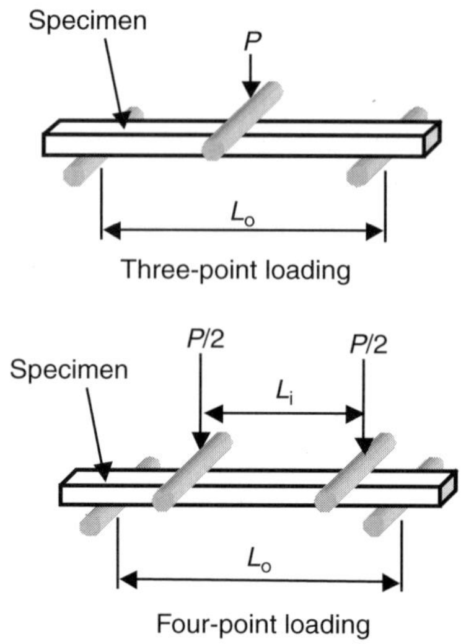

FIGURE 10.1.3 Three- and four-point loading geometries.

without shear stress is generated within an inner span and beam theory can be applied more properly for strength calculation than in three-point flexure. Also, four-point flexure has larger effective volume or effective surface than three-point flexure for identical specimen dimensions. Still, three-point loading has simpler geometry and simplifies testing fixture design. This is particularly important when adapted to high-temperature measurements or with limited available specimen size. The fracture strength, σ_f, is equated to the maximum tensile stress generated at the failure load, P, as follows:

$$\sigma_f = \frac{1.5P(L_o - L_i)}{bh^2} \tag{1}$$

where L_o and L_i are the outer and inner load spans, respectively (L_i is zero for three-point loading), and b and h are the width and height of the rectangular specimen, respectively.

10.1.2.1.4 Biaxial Tests

Strength measurements are usually performed in uniaxial loading. Most ceramic components are subjected to biaxial (or multiaxial) loading in actual practice. In addition, biaxial loading conditions are much more severe in a stress state than uniaxial conditions, leading to more reliable component design. One of the most frequently used biaxial loading techniques is biaxial flexure using thin plates as specimens. Biaxial flexure is free from the specimen edge effects that substantially affect strength measurements in three- or four-point flexural tests. This technique also can reflect flaws of any direction. In biaxial stress fields, a larger number of flaws are subjected to tensile stress than in uniaxial stress fields, resulting in more conservative strength evaluation. In addition, simplified loading geometry eases specimen preparation and lowers its cost.

Several types of loading methods have been employed to evaluate biaxial flexural strength: ball-on-ring, piston-on-ring, and ring-on-ring [25–31]. The ring can be substituted for balls placed at the circumference of the same radius since applied stresses are independent of angles between them and the number of supports for this loading geometry. The ball-on-three-ball and the piston-on-three-ball formations are, therefore, very frequently used instead, as shown in Figure 10.1.4. One advantage of the ball-on-three-ball is its tolerance of rough specimen surfaces, simplifying surface machining. The fracture strength, σ_f, can be calculated from the failure load, P, as follows:

$$\sigma_f = \frac{3P(1+\nu)}{4\pi t^2}\left[1 + 2\ln\frac{a}{b} + \left(\frac{1-\nu}{1+\nu}\right)\left(1 - \frac{b^2}{2a^2}\right)\left(\frac{a^2}{R^2}\right)\right] \tag{2}$$

where a is the radius of the support ring, R and t are the specimen radius and thickness, respectively, and ν is Poisson's ratio. b is the radius of the piston for the piston-on-ring, that of the contact area of the loading ball for the ball-on-ring, or that of the loading ring for the ring-on-ring.

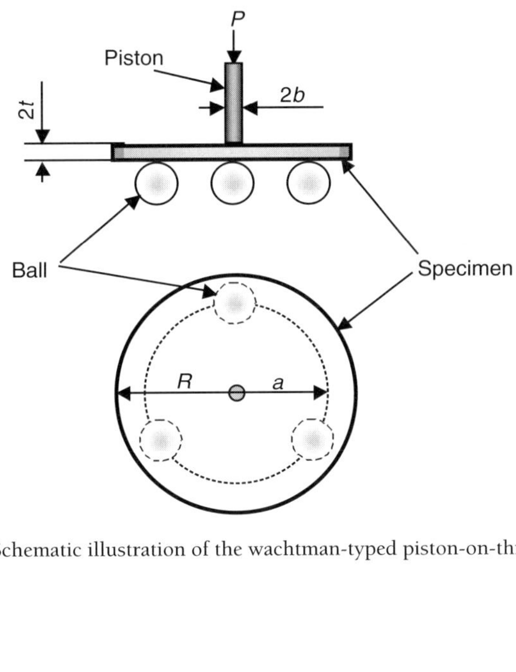

FIGURE 10.1.4 Schematic illustration of the wachtman-typed piston-on-three-ball test fixture.

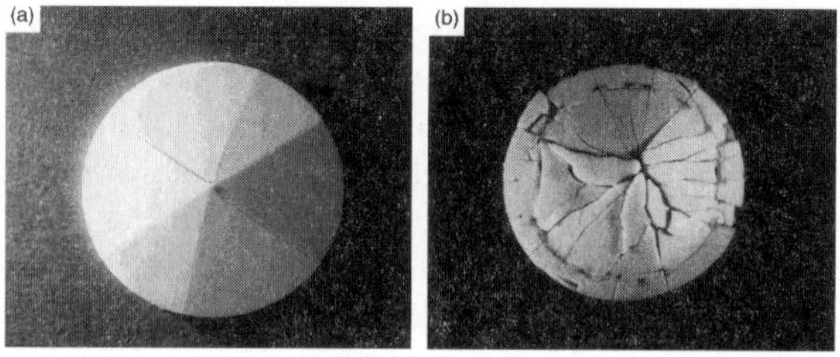

FIGURE 10.1.5 Typical fracture patterns for (a) ball-on-ring (BOR) and (b) hydraulic-burst (HB) tests [33]. Reprinted with permission of the American Ceramic Society.

Biaxial strength of ceramics is also affected by the effective volume or area of the stressed portion of the specimen. Simpatico *et al.* [33] measured biaxial strength of alumina using two tests: the hydraulic burst (HB) and the ball-on-ring (BOR). Typical fracture patterns are shown in Figure 10.1.5. The area (volume) under stress is much larger for the HE test than the BOR test. Both mean strengths and Weibull moduli from BOR tests were approximately twice the values from HE tests. The HE data can be considered as a low-probability-of-failure, low-strength tail of the BOR curve that has a lower Weibull modulus than the high stress portion.

10.1.2.2 Related Standards

In these past few decades, considerable efforts have been made to standardize fracture strength measurements for both room and high temperatures. Applicable standards for strength measurement tests follow.

ASTM, Committee C28 on "Advanced Ceramics" (USA)

- C1161-94: Standard Test Method for Flexural Strength of Advanced Ceramics at Ambient Temperature.
- C1273-95: Standard Test Method for Tensile Strength of Monolithic Advanced Ceramics at Ambient Temperatures.
- C1275-00: Standard Test Method for Monotonic Tensile Behavior of Continuous Fiber-Reinforced Advanced Ceramics with Solid Rectangular Cross-Section Specimens at Ambient Temperatures.
- C1323-96: Standard Test Method for Ultimate Strength of Advanced Ceramics with Diametrally Compressed C-Ring Specimens at Ambient Temperature.
- C1341-00: Standard Test Method for Flexural Properties of Continuous Fiber-Reinforced Advanced Ceramic Composites.

- C1358-96: Standard Test Method for Monotonic Compressive Strength Testing of Continuous Fiber-Reinforced Advanced Ceramics with Solid Rectangular Cross-Section Specimens at Ambient Temperatures.
- C1359-96: Standard Test Method for Monotonic Tensile Strength Testing of Continuous Fiber-Reinforced Advanced Ceramics With Solid Rectangular Cross-Section Specimens at Elevated Temperatures.
- C1360-96: Standard Practice for Constant-Amplitude, Axial, Tension–Tension Cyclic Fatigue of Continuous Fiber-Reinforced Advanced Ceramics at Ambient Temperatures.
- C1366-97: Standard Test Method for Tensile Strength of Monolithic Advanced Ceramics at Elevated Temperatures.
- C1424-99: Standard Test Method for Monotonic Compressive Strength of Advanced Ceramics at Ambient Temperatures.

CEN, Committee TC184 "Advanced Technical Ceramics" (Europe)

- ENV 658-1: Tensile strength of continuous fiber reinforced ceramic composites.
- ENV 658-2: Flexural strength of continuous fiber reinforced ceramic composites.
- ENV 658-3: Compressive strength of continuous fiber reinforced ceramic composites.
- ENV 820-1: Flexural strength of monolithic ceramics at high temperatures.
- ENV 820-2: Deformation of monolithic ceramics at high temperatures.
- ENV 843-1: Flexural strength of monolithic ceramics.
- ENV 843-3: Sub-critical crack growth of monolithic ceramics.

JISC, Divisional Council on Ceramics (Japan)

- R1601: Testing method for flexural strength of high-performance ceramics.
- R1604: Testing method for flexural strength of high-performance ceramics at elevated temperatures.
- R1606: Testing methods for tensile strength of high-performance ceramics at room or elevated temperatures.
- R1608: Testing method for compressive strength of high-performance ceramics.

10.1.2.3 Recent Research Topics

10.1.2.3.1 Strength of Piezoelectric Ceramics

A lead zirconate titanate (PZT) ceramic is now used for a wide variety of applications. One typical use is as an actuator. For such application, the component receives mechanical loading, so assessment of mechanical performance of the material is crucial. Recent strength measurements conducted on PZT in bending and tensile tests have shown very different results; bending strength is almost twice the tensile strength [34]. Different effects were identified as reasons for this behavior. In the case of piezoelectric materials, the stress–strain curve is non-linear because of the stress-induced change of polarization direction of domains associated with inelastic strains. This behavior complicates bending test evaluation [34]. If both different effective specimen surfaces in each loading case and influence of non-linearity and non-symmetry in stress–strain behavior are considered for stress computations of true bending stress, resultant strength values become similar. Specimens with artificial cracks are addressed in this paper to exclude any statistical effect and to determine the contribution of non-linear and non-symmetric deformation behavior. The difference between tension and bending tests also occurs in tests involving Knoop-cracked specimens [35].

10.1.2.3.2 Strength of Ceramics Thin films and Microcomponents

Thin films and minute components are now used in many technical applications including semiconductor devices, information storage media, microelectromechanical systems (MEMS), and biomedical devices [36]. Such applications are expected to expand in the future. Such small structures often have different properties than large-sized components of identical materials. Therefore, knowledge of materials properties is essential for component design and fabrication.

Currently, much attention has been paid to microcomponents, whose characteristic dimensions are in the range of 0.1–10 mm, for the new class of microfabricated machines such micro-gas turbine generator "microengines" [37]. Chen *et al.* [38] employed a biaxial flexure test to characterize single crystal silicon strength on such a scale. They used two biaxial flexure specimen types, normal planar specimens, and radius hub specimens to obtain local strength at stress concentration as shown in Figure 10.1.6. A schematic of the strength-testing apparatus for both biaxial flexure and radius hub flexure specimens is shown in Figure 10.1.7. Specimens were placed over a circular hole (7 mm in diameter) in a machined aluminum block; the fixture and load cell were placed under a modified microhardness indenter. A microscope was

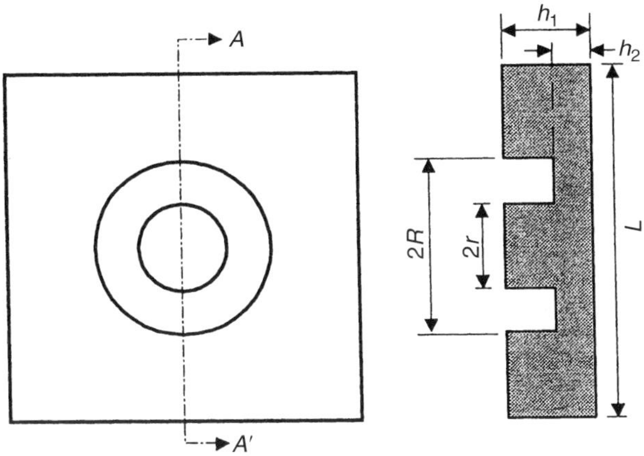

FIGURE 10.1.6 Geometry of the radiused hub flexure specimens in strength measurements of silicon on mesoscale [38]. Reprinted with permission of the American Ceramic Society. All rights reserved.

used for the X–Y stage of the indenter for accurate specimen positioning and centering. The indenter head was a steel ball bearing, 1.5 mm in diameter. Tests were performed in displacement control to cause fracture in about 30 s. Weibull reference strengths of the planar biaxial flexure specimens were in the range of 1.2–4.6 GPa. The strength of the deep reactive ion-etched (radius hub) specimens was significantly lower than that measured on planar specimens. These mechanical test results have significant implications for development of highly stressed microfabricated structures of future application.

10.1.3 FATIGUE AND SLOW CRACK GROWTH

Fatigue, or fracture after loading of some duration at a stress level less than that required for immediate fracture, is caused by several mechanisms. (While the term fatigue in the metallurgy community indicates solely cyclic fatigue, cyclic mechanical loading of a specimen, or a component, fatigue for ceramic and glass includes static and dynamic fatigue as well.) One well-known mechanism is environmentally enhanced crack growth, very frequently referred to as slow crack growth (SCG), which occurs in nearly all structural ceramics [39]. This is crack growth enhanced by some chemical reaction that substantially weakens the atomic bond at the crack tip even at stresses well below intrinsic strength.

The SCG behavior of ceramics and glass is generally expressed by a relationship between crack velocity, v, and an applied stress intensity factor, K, to

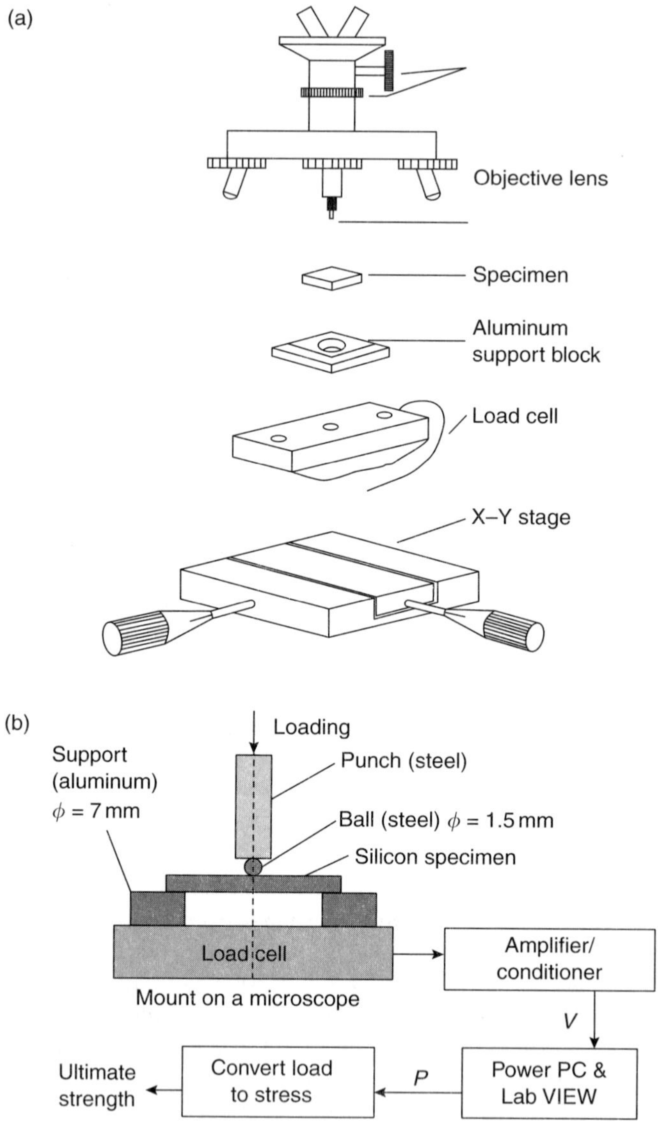

FIGURE 10.1.7 Schematic of strength test fixture (a) and the strength testing procedures (b) used in strength measurements of silicon on mesoscale [38]. Reprinted with permission of the American Ceramic Society.

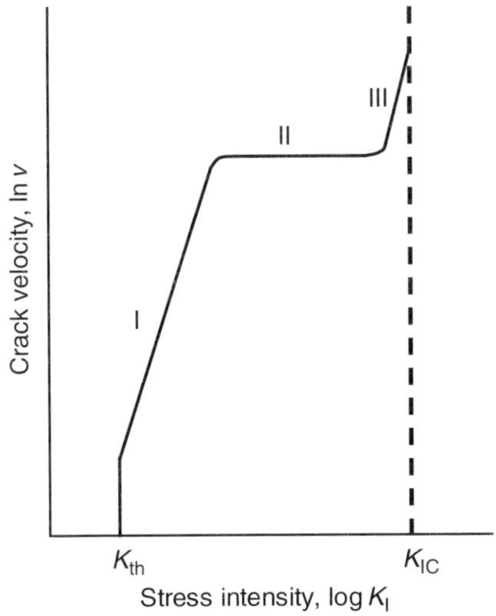

FIGURE 10.1.8 Schematic relationship between the crack velocity and applied stress intensity K during SCG.

show the so-called v–K diagram. These v–K diagrams of ceramics and glasses typically contain stages I, II, and III with threshold stress intensity below which SCG does not occur (K_{th}), as shown in Figure 10.1.8. Crack velocity at each stage of the diagram is governed by a different mechanism; the diagram varies substantially depending on materials and environments. Some cases show no threshold stress intensity while for other cases stages II and III are negligible. Once the v–K diagram is established for a particular material and condition, fracture mechanics can predict its lifetime. Since stage I crack propagation occupies, a large part of total lifetime, this stage is essential for critical lifetime prediction.

A typical example of v–K diagram exists in the classic work by Wiederhorn [40]. They measured crack growth of soda lime silica glass in inert nitrogen environments with different humidities and obtained v–K diagrams that all consist of stages I, II, and III. It is generally considered that stages I, II, and III correspond to impurity controlled activity, impurity diffusion limited, and bulk environment controlled regions, respectively. In the above crack growth study, Wiederhorn considered that moisture activity controls crack velocity at stage I, while stage II is limited by moisture diffusivity. Thus, moisture concentration is decisive and insensitive to K_I, leading to the so-called plateau region. Here, K_I is the applied stress intensity factor in mode I.

Great effort has been made to understand fundamental mechanisms determining the relationship between SCG and stress intensity factor. Although a number of mathematical expressions to describe this relationship have been proposed, the following forms are used presently:

$$v = v_0(K_I/K_{IC})^N \tag{3}$$

$$v = v_0 \exp(K_I/K_{IC})^B \tag{4}$$

Here, v_0, N, and B are constant, and K_{IC} is the critical applied stress intensity factor, or fracture toughness, of the material. Neither expression has obtained general acceptance as a "fundamental" form. Equation (3) is most commonly employed in component design due to its ease in mathematical manipulation; log(v) is linear in log(K), with N slope. Parameter N is recognized as an indicator of susceptibility to SCG; the larger the N value, the smaller the SCG. However, it is frequently observed that obtained N values do not agree between two different measurement techniques or two different specimen sizes for identical materials and environments [41]. This indicates that the power law form does not always adequately express crack velocity at small stress intensity factors. This shortcoming significantly limits potential of extrapolating the smaller velocity range.

Several techniques have been proposed for measuring a v–K diagram. Many researchers have used direct measurement of propagating cracks in a fracture mechanics test specimen [40–42]. An advantage of this method is that a v–K diagram can be drawn easily from one specimen or one test. The fracture mechanics test specimen usually has a long artificial crack for which the K value can be determined easily. However, such long artificial cracks do not always behave similarly to short natural cracks. Ceramic components are usually inspected through non-destructive tests before service so that they do not contain intolerant long cracks. Thus, it is short cracks that govern the lifetime of ceramic components in service. Another way to establish a v–K diagram is to estimate it indirectly from strength measurement results of static or dynamic fatigue tests [43–47]. Static fatigue tests involve static loading of tensile stresses lower than the fast fracture strength until failure occurs: this shows the relationship between time-to-failure and stress. In dynamic fatigue tests, tensile stress is monotonically increased at various loading rates until there is a failure to obtain a relationship between loading rate and strength. Both obtained relations lead to a v–K relationship using fracture mechanics. One advantage of these methods is that observed crack propagation behavior can be known from short flaws. This has practical importance, as mentioned above. Short cracks can occur in ceramic components as pores, inclusions, and machining surface flaws.

When K_{IC} depends on crack length a (or R-curve behavior), SCG behavior differences between short and long cracks become clear. As shown schematically in Figure 10.1.9, short crack extension starts at considerably lower stress

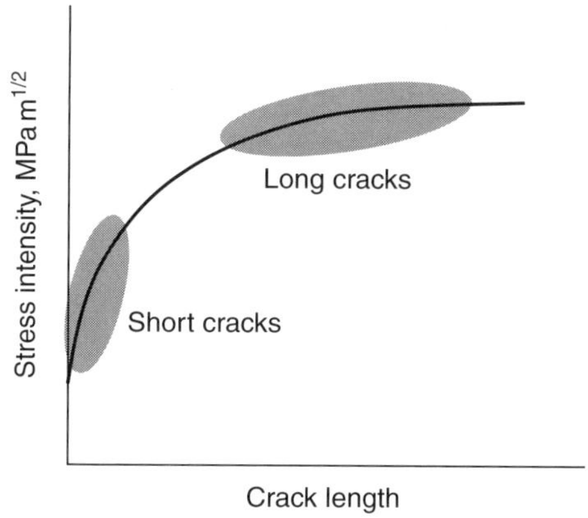

FIGURE 10.1.9 In a material with a characteristic *R*-curve behavior, short or natural cracks show a different behavior from that of long cracks.

intensities than long cracks for materials with *R*-curves; and plateau fracture toughness is not attained before catastrophic failure [48, 49]. Unlike tests where long cracks are forced to initiate from a notch tip in a controlled fashion, short cracks originate from local natural flaws that are frequently located at the surface. The short crack with the shallowest *R*-curve slope becomes unstable first (the fracture stress at this instability corresponds to flexure stress in this instance). This indicates that under equivalent loading conditions, specimens or components cannot take advantage of the full *R*-curve potential. In addition, crack resistance measurements generated from long cracks might, therefore, overestimate ceramic component toughness in real, load-bearing applications.

The relationship between material microstructure and its properties has been addressed recently; also, the *v*–*K* relationship is often characterized by information on interactions between the moving crack and the microstructure [50, 51]. Importance of fractographic analysis on specimens after the test has been recognized with strength-measuring techniques because this can show whether cracks propagate trans- or intergranularly, hence clarifying effects of grain boundary phases on microstructural crack propagation. For specimens tested at high temperatures, fractography can determine whether strength degradation arises from SCG or other mechanisms such as creep [52, 53]. Furthermore, degree of moving crack velocity can be estimated in some cases from fractography, as cracks grow trans- or intergranularly depending on their velocity [47]. Finally, direct observation of indentation cracks has been proposed recently for investigating both the *v*–*K* relationships and fundamental aspects of flaw shape evolution [54, 55].

Fatigue and slow crack growth tests described here include: (i) static and dynamic fatigue for short (or natural) cracks; (ii) cyclic fatigue for long cracks; (iii) static and dynamic fatigue for long cracks; and (iv) cyclic fatigue for long cracks. Related recent research topics are shown as well.

10.1.3.1 Testing Methods

10.1.3.1.1 Static and Dynamic Fatigue Tests Using Short Cracks

Fatigue tests for specimens containing short natural flaws can provide data closest to the real service application. The static test involves applying a mode I, plane strain static stress in an appropriate environment and measuring the time to failure. Conducting measurements at different applied stresses, the v–K curve can be established even though natural flaw sizes are unknown. The methods are essentially identical to simple strength measurement methods. Principally, the fatigue parameter N can be measured without any assumptions about flaw size or shape. However, this method is very time consuming to obtain even one data point and a number of specimens are needed for each condition to account satisfactorily for statistical variability in strength. Dynamic fatigue generates principally the same results, but testing time is shortened through testing at several constant stress rates. Any specimen configuration that is generally used for any strength test can be used for both fatigue tests. Examples of successful measurements on static and dynamic fatigue of ceramics are available for short artificial cracks [52–55] and natural cracks [57–60].

10.1.3.1.2 Cyclic Fatigue Tests Using Short Cracks

Structural ceramic components in service very often are subjected to cyclic mechanical loading. Cyclic fatigue performance of structural ceramics determines the life of components in many cases, including gas turbine rotors or blades and diesel engine valves. Therefore, characterizing and properly understanding ceramics cyclic fatigue is crucial for confident use of components subjected to cyclic loading. Many reports explain cyclic fatigue of ceramics using short or natural flaws [61–65].

Several mechanisms for cyclic fatigue of ceramics have been proposed. Most of them are related strongly to toughening mechanisms of ceramics at room temperature. They include: degradation of transformation toughening zone at the crack tip (such as in some zirconia); damage to bridging reinforcements (grains, whiskers, platelets, and brittle and/or ductile particles) over the bridged interface; cyclic loading-induced accumulation of microcracks around the crack tip; crack tip blunting/resharpening; and relaxation of any mechanically beneficial

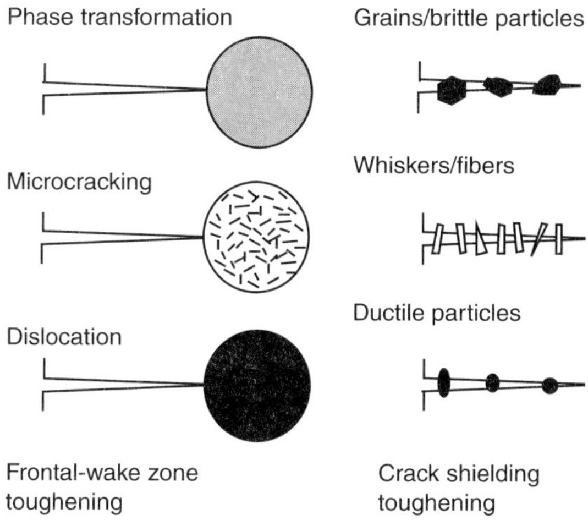

FIGURE 10.1.10 Schematic classification of primary microstructural toughening mechanisms, which are strongly involved with property degradation during cyclic loading.

residual stress due to microcracking, as shown in Figure 10.1.10 [66]. It should be noted that these degradation mechanisms are not active in static or dynamic fatigue.

In some cases, at elevated temperatures where softening of the grain boundary glassy phase can occur, other mechanisms become active. These include oxidation-induced deformation and viscoelastic effects [56, 61].

10.1.3.1.3 Static and Dynamic Fatigue Tests Using Long Cracks

Crack propagation testing using long cracks (i.e. fracture mechanics specimens) is fundamentally different from crack propagation testing with short or natural flaws (i.e. strength measurement specimens) for several reasons. Crack length being propagated may be on the order of millimeters, while crack lengths dealt with in short fatigue crack propagation were on the order of micrometers or tens or hundreds of micrometers. "Long" cracks are not easily influenced by microstructural residual stresses while "short" cracks are.

By using fracture mechanics specimens (i.e. specimens that are often used for measuring fracture toughness), one can measure directly the crack velocity v as a function of applied stress intensity K [40–42]. However, test specimens and the experimental facility are more complicated; it is probably difficult to operate at high temperature or in an extremely corrosive environment. Specimens may be loaded in load control or in displacement control and relaxation techniques may also be used. Figure 10.1.11 shows examples of frequently

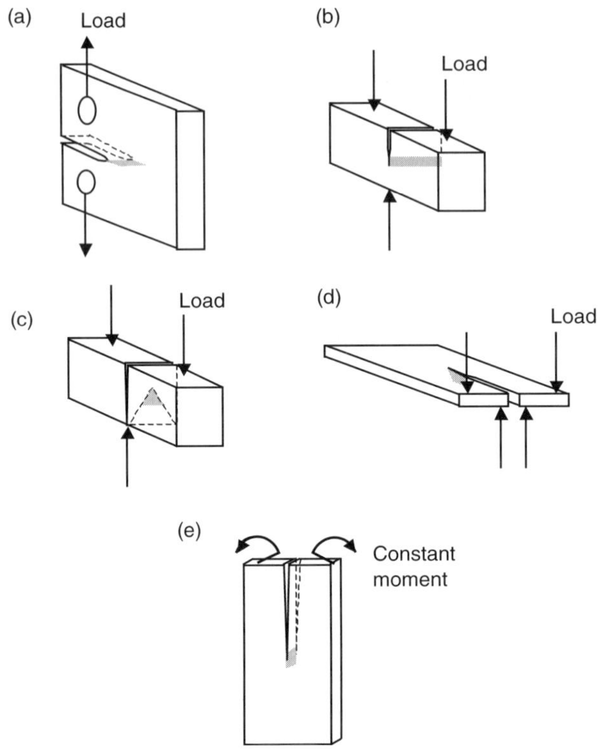

FIGURE 10.1.11 Frequently used long-crack test geometries: (a) compact tension (CT); (b) single-edge notched beam (SENB); (c) chevron notch (CN); (d) double torsion (DT); (e) double cantilever beam (DCB). Grayed area is crack.

used long crack test geometries. Test specimens must have a relatively large sharp crack that will propagate during testing. Ideally, the crack tip should be atomically sharp. However, a notch produced by a diamond saw has been used frequently, assuming the sharp crack is produced from the tip of the notch.

10.1.3.1.4 Cyclic Fatigue Tests Using Long Cracks

Also, there is substantial difference in cyclic loading between crack propagation with long cracks and that with short or natural flaws. The so-called *S–N* (stress–number of cycles) curves represent results of fatigue crack propagation testing of short cracks, while crack propagation per cycle as a function of stress intensity range or maximum stress intensity represent results of crack propagation testing with long cracks. However, toughening mechanisms, which are related

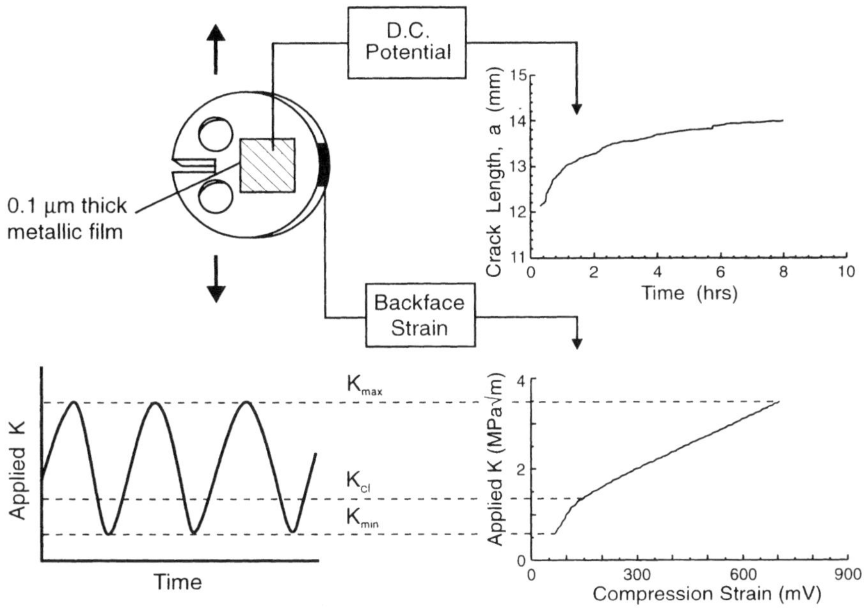

FIGURE 10.1.12 Experimental techniques used to measure cyclic fatigue-crack growth rates, showing compact-tension specimen and procedures used to monitor crack length and the stress intensity [74]. Reprinted with permission of the American Ceramic Society. All rights reserved.

to enhanced crack propagation in cyclic fatigue with short cracks, can be similarly related to that in cyclic fatigue with long cracks. There are examples of successful measurements on cyclic fatigue of ceramics with long cracks [66–74]. Figure 10.1.12 shows experimental techniques used by Ritchie and colleagues to measure cyclic fatigue crack growth rates for compact tension specimens [74].

Similarly to the v–K relationships described above, crack propagation rates or crack growth extension per unit cycle (da/dN) for cyclic fatigue testing with long cracks are also typically represented by a power-law relationship. The commonly known Paris law [75] relates da/dN to the stress intensity range ($\Delta K = K_{\mathrm{Imax}} - K_{\mathrm{Imin}}$, with K_{Imax} and K_{Imin} being maximum and minimum applied stress intensities, respectively) as

$$da/dN = C(\Delta K)^m \tag{5}$$

where C is a constant and m is a cyclic fatigue exponent. Crack propagation rate (da/dt) or v is related to da/dN as

$$da/dt = v = (da/dN)f \tag{6}$$

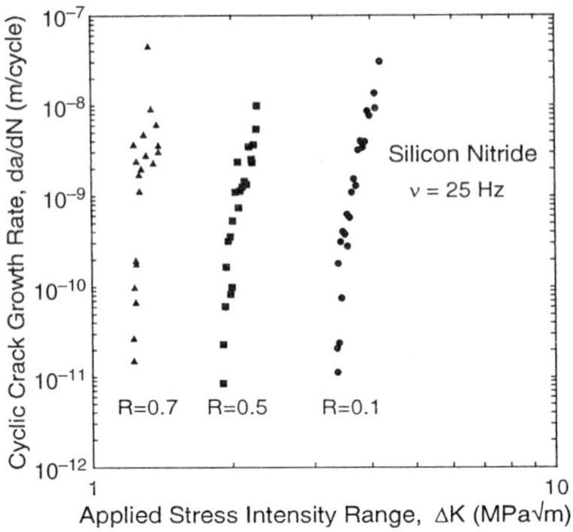

FIGURE 10.1.13 Example of graphical representation of cyclic fatigue-crack growth rates and applied stress intensity range. Influence of load ratio R on cyclic fatigue-crack growth rates, da/dN, in silicon nitride, in terms of the applied stress intensity range, ΔK [74]. Reprinted with permission of the American Ceramic Society. All rights reserved.

where f is the cyclic frequency. For determining the cyclic fatigue exponent, m, the crack growth increment per cycle da/dN, or the crack propagation rate da/dt, as a function of stress intensity range, ΔK, is used. Figure 10.1.13 shows one such graphical presentation.

10.1.3.2 Related Standards

Available standards established for fatigue tests of ceramics are shown below. It should be noted that most standards for simple strength measurements can be applied to fatigue tests.

ASTM, Committee C28 on "Advanced Ceramics" (USA)

- C1360-96: Standard Practice for Constant-Amplitude, Axial, Tension–Tension Cyclic Fatigue of Continuous Fiber-Reinforced Advanced Ceramics at Ambient Temperatures.
- C1361-96: Standard Practice for Constant-Amplitude, Axial Tension–Tension Cyclic Fatigue of Advanced Ceramics at Ambient Temperatures.
- C1368-00: Standard Test Method for Determination of Slow Crack Growth Parameters of Advanced Ceramics by Constant Stress-Rate Flexural Testing at Ambient Temperature.

- C1465-00: Standard Test Method for Determination of Slow Crack Growth Parameters of Advanced Ceramics by Constant Stress-Rate Flexural Testing at Elevated Temperatures.

CEN, Committee TC184 "Advanced Technical Ceramics" (Europe)

- ENV 843-3: Sub-critical crack growth of monolithic ceramics.

JISC, Divisional Council on Ceramics (Japan)

- R1621: Testing method for flexural fatigue of high-performance ceramics.
- R1632: Testing method for constant loading flexural rupture of high-performance ceramics.

10.1.3.3 Recent Research Topics

10.1.3.3.1 Cyclic Fatigue of *In Situ* Toughened Ceramics

Most ceramics sub-critical crack growth rates under cyclic loads exceed corresponding growth rates under sustained, quasi-static loads by many orders of magnitude at equivalent stress intensity levels [61–74]. This contrast clearly indicates that mechanical properties of ceramics are degraded by cyclic fatigue. A number of tough ceramics show crack shielding effects that substantially influence cyclic fatigue behaviors. Mechanisms for cyclic fatigue crack growth in ceramic materials are generally based on suppression of crack tip shielding in the crack wake under cyclic loading.

Gilbert *et al.* [74, 76] characterized cyclic fatigue crack propagation behavior of "*in situ* toughened" monolithic silicon nitride, in light of fatigue crack growth models for ceramics toughened by grain bridging mechanisms. Models were based on diminished crack tip shielding in the crack wake under cyclic loads due to frictional wear degradation of the grain bridging zone (Fig. 10.1.14). The notion of cyclic crack growth promoted by diminished shielding was consistent with measured (long crack) growth rates, fractography, *in situ* crack profile analyses, and measurements of back face strain compliance. Growth rates depended on maximum applied stress intensity, $K_{-\max}$, much more than on the applied stress intensity range, ΔK. Fatigue thresholds exhibited a marked dependence on the load ratio, $R = K_{-\min}/K_{-\max}$. These effects were inconsistent with traditional models of fatigue crack closure. In particular, when characterized in terms of $K_{-\max}$, growth rates below about 10^{-9} m/cycle exhibit an inverse dependence on load ratio. This observation was consistent with the grain bridging phenomenon. Specifically, with increasing R, sliding distance

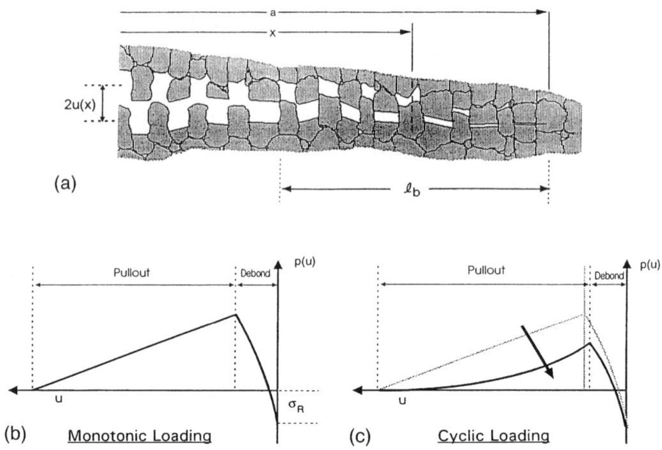

FIGURE 10.1.14 Schematic illustration of the cyclic fatigue-crack propagation involving the cyclic-loading induced degradation in crack-tip shielding by (a) grain bridging in the crack wake. The crack-opening displacement, $2u(x)$, is indicated at position x in the bridging zone. The grain-bridging stress, $p(u)$, is shown rapidly rising from the initial compressive stress, σ_R, as the grain debonds, followed by frictional sliding which gradually decreases as the grain is extracted from the matrix under (b) monotonic and (c) cyclic loading conditions [74]. Reprinted with permission of the American Ceramic Society.

between the grain bridges decreased, leading to less frictional wear and less degradation in shielding per loading cycle. Similar investigation was made on monolithic, hot-pressed alumina. Rates of sub-critical crack growth under cyclic loads were observed to be many orders of magnitude faster than corresponding growth rates under monotonic loading at equivalent stress intensity levels. Such behavior was also attributable to diminished crack tip shielding under cyclic loads. Reduced shielding also acts locally to enhance the crack tip driving force in cyclic fatigue in alumina.

The *in situ* toughening has been adapted also into silicon carbide recently. Similar crack growth resistance curves (*R*-curve) and cyclic fatigue behaviors have been found in such silicon carbide materials. Gilbert *et al.* [77] investigated *R*-curve behavior in two silicon carbide (SiC) ceramics with sharply contrasting fracture properties, using small (<400 μm) surface cracks and long (>3 mm) through-thickness cracks. The first, an *in situ* toughened material designated ABC–SiC, fails by intergranular fracture, whereas the second, a commercial SiC (Hexoloy SA), fails by transgranular cleavage. The former material, hot-pressed with aluminum, boron, and carbon additives, contained a network of plate-shaped grains [78]. An amorphous grain boundary film of ~1 nm thickness promoted debonding and crack deflection. Resultant grain bridging generated *R*-curve toughening. In contrast, no evidence of crack tip shielding is observed in the latter SiC, Hexoloy SA. Although the Hexoloy SA fails catastrophically at $<3\,\mathrm{MPa\,m^{1/2}}$, ABC–SiC exhibits much improved flaw tolerance with significant rising *R*-curve behavior and a steady-state fracture toughness of $\sim 9\,\mathrm{MPa\,m^{1/2}}$ after crack extension of $\sim 600\,\mu\mathrm{m}$. Cyclic fatigue properties of such *in situ* toughened silicon carbide were also examined using similar long (>3 mm) and small ($<200\,\mu\mathrm{m}$) cracks [79]. Crack extension was monitored optically under cyclic load. Small cracks showed high growth rates with negative dependence on the far-field driving force. This behavior was attributable to both indent induced residual stresses and relative size of cracks compared with bridging zone lengths.

At elevated temperatures also, bridging tractions developed behind a crack tip influence crack propagation substantially under cyclic loading conditions in tough monolithic ceramics. Assuming a temperature range where grain boundary phases were soft enough that bridging can occur due to a viscous layer in the boundary, McNaney *et al.* [80] developed a viscoelastic model to incorporate: microstructural details (grain shape, grain boundary thickness, and glass viscosity) along with effects of external variables (loading rate and temperature). In the model, bridging forces associated with shear resistance of the grain boundary phase were transmitted across the crack surfaces. This phenomenon is applicable to a wide range of ceramic materials at elevated temperatures. Chen *et al.* [81] examined fatigue crack growth at elevated temperatures of 25–1300°C under cyclic loading for *in situ* toughened, monolithic silicon carbide, along with roles of temperature, load ratio, cyclic frequency, and loading mode (static and cyclic). The presented extensive crack growth data were based on measurements from a crack monitoring technique using an electrical potential drop. It was found that at equivalent stress intensity levels, crack velocities under cyclic loads were significantly faster than under static loads. Fatigue thresholds were decreased with increasing temperature up to 1200°C. At 1200°C, they detected no effect of frequency in the range of 3–1000 Hz. Indeed, fractography and crack path sectioning revealed a fracture mode at 1200–1300°C that was essentially identical to that at room temperature, that is, predominant intergranular cracking with evidence of grain bridging in the crack wake. This excellent crack growth resistance was attributed to a process of grain boundary microstructural evolution at elevated temperatures, specifically involving crystallization of the amorphous grain boundary films/phases.

Excellent cyclic fatigue resistance, high strength, and high fracture toughness in the tough silicon carbide were not compromised significantly at elevated temperatures up to 1500°C [82]. Particularly, fatigue crack growth properties up to 1300°C were essentially identical to those at room temperature. Mechanistically, the cyclic fatigue crack advance mechanism was similar up to 1300°C, involving mutual competition between intergranular cracking ahead of the crack tip and interlocking grain bridging in the crack wake. Moreover, good creep resistance was observed below 1400°C, with little evidence of cavitation. Such unusually good high-temperature properties of silicon carbide were attributed to crystallization of the grain boundary amorphous phase, that can occur either *in situ*, due to prolonged thermal exposure associated with high-temperature fatigue and creep tests, or by prior heat treatment.

As stated above, cyclic fatigue growth rates in tough ceramics with rising *R*-curve behavior depend very strongly on the maximum applied stress intensity, $K_{-\max}$, and weakly on the stress intensity range, ΔK. Jacobs and Chen [83] identified mechanical equilibrium between shielding accumulation by crack growth. They further identified shielding degradation by frictional wear of sliding interfaces for steady-state cyclic fatigue in a gas pressure sintered silicon nitride. This equilibrium gave rise to a rate law for cyclic fatigue. They found that accumulation and degradation processes were the origin of strong $K_{-\max}$ dependence and weak ΔK dependence, respectively.

Chen and Engineer [84] also studied a model for fatigue crack propagation based on sliding wear of bridging grains of polycrystalline ceramics such as tough silicon nitride. Taking into account damage development and crack tip energy balance, they developed equilibrium and compatibility equations in the crack wake under monotonic and cyclic loading/unloading conditions. Fatigue was controlled by frictional resistance to reverse sliding, suggesting a mechanics similar to an elastic–plastic one of a path-dependent hardening material. The unloading crack opening displacement is more strongly dependent on $K_{\max}$ than on ΔK; such displacement caused wear on bridging grains. Incremental crack growth brought in new bridging grains with a shielding effect on the crack tip stress field. Such an effect strongly depended on $K_{\max}$, but was independent of ΔK. In the steady state, with balanced shielding accumulation and shielding degradation, the fatigue crack growth rate had the form $da/dN = A(K_{\max})^b(\Delta K)^c$, where A, b, and c are material-dependent parameters. Fatigue was predicted to have a very high b, modest c, a higher fatigue resistance for tougher ceramics, and a stronger $K_{\max}$ dependence for less tough ceramics. These predictions agreed with experimental observations.

10.1.3.3.2 Contact Fatigue

A number of structural components in practical use are very often subjected to contact fatigue or contact damage. Despite importance of contact fatigue, ceramics contact damage behavior is not understood clearly. Recently, Lawn and colleagues [85–91] studied ceramics contact damage using Hertzian contact of sphere, as shown in Figure 10.1.15. It was demonstrated that the nature of contact damage in tough ceramics, with coarse, heterogeneous microstructures, is fundamentally different from that in brittle ceramics. Generally, in fine, homogeneous structures, damage takes the form of classical macroscopic

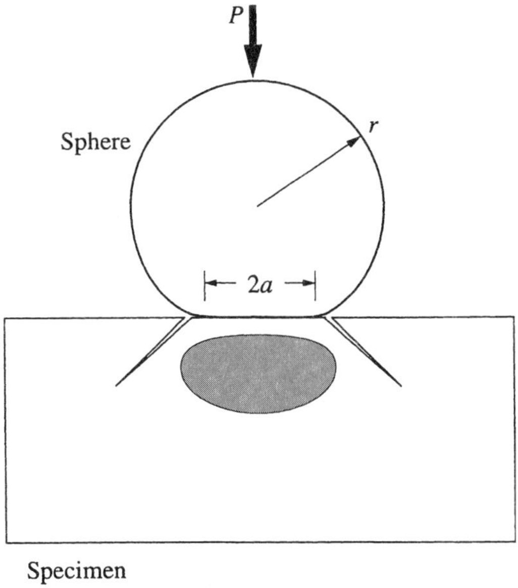

FIGURE 10.1.15 Hertzian contact of sphere on flat ceramic specimen. Beyond elastic limit, contact initiates cone fracture ("brittle mode") or subsurface deformation zone ("quasi-plastic mode") [85]. Reprinted with permission of the American Ceramic Society. All rights reserved.

cone cracks in a region of weak tension outside the contact ("brittle" mode). Strength losses from this damage mode are initially abrupt as the cone cracks pop in, but fall off only slightly thereafter with increasing numbers of cycles, indicative of time-dependent slow crack growth. In coarser, heterogeneous structures, damage manifests itself as a cloud of closed shear microcracks or "faults" at microstructurally limited weak boundaries within a subsurface shear compression deformation zone ("quasi-plastic" mode). Strength losses in initial damage stages become more gradual, corresponding to failures from individual, small microcracks, but rapidly accelerate with extensive cycling as microcracks coalesce into dangerous radial cracks. Degradation then depends more strongly on number of cycles than on integrated time in the contact, indicating true mechanical fatigue. In the toughest ceramics, the quasi-plastic mode tends to dominate all stages of the contact damage process.

For example, Lee and Lawn [92] conducted a contact fatigue study with silicon nitride. Contacts utilized WC spheres, principally in cyclic, but also in static loading. Test environments were mainly in air, but also in nitrogen and water. They employed three silicon nitrides with different microstructures for damage patterns: (i) fine (F)—almost exclusively fully developed cone cracks; (ii) medium (M)—well developed, but smaller cone cracks, plus modest subsurface quasi-plastic damage; (iii) coarse (C)—intense quasi-plastic damage

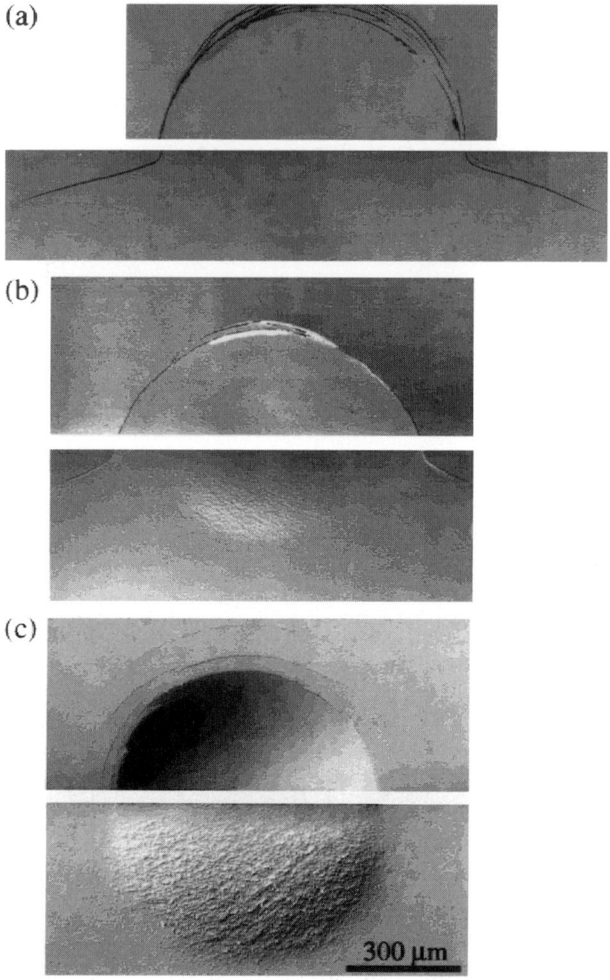

FIGURE 10.1.16 Hertzian indentation damage in silicon nitride, from tungsten carbide sphere with 1.98-mm radius and 4000-N load: (a) F, fine; (b) M, medium; and (c) C, coarse. Half-surface specimens [92]. Reprinted with permission of the American Ceramic Society. All rights reserved.

with little or no cone cracking (Fig. 10.1.16). The influence of these competing damage types on inert strength as a function of number of contacts is shown in Figure 10.1.17. In F and M microstructures, strength degradation was attributed primarily to chemically assisted slow growth of cone cracks in the presence of moisture during contact. The M material exhibited signs of enhanced failure from quasi-plastic zones at a large number of cycles.

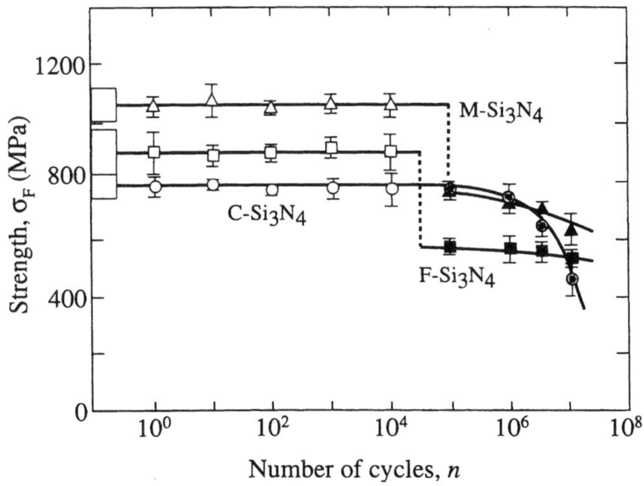

FIGURE 10.1.17 Strength of silicon nitride specimens indented with tungsten carbide sphere with 1.98-mm radius and 1000-N load, as function of number of contact cycles at 10 Hz. Data points are experimental measurements, indentation tests in air—black symbols represent failures from cone crack origins, gray symbols from quasi-plastic zones, and open symbols from other origins. Boxes at left axis represent strengths of polished, unindented specimens [92]. Reprinted with permission of the American Ceramic Society. All rights reserved.

The C material, though relatively tolerant of single cycle damage, showed accelerated strength degradation from quasi-plastic damage in cyclic loading conditions; this tendency was enhanced in water.

A model of contact damage accumulation and strength degradation from cyclic loading in heterogeneous ceramics has been also proposed by Lee *et al.* [93]. Damage evolution showed two stages: (i) cyclic contact damage at prescribed loads and environments; and (ii) strength degradation of the damaged specimens in fast fracture. They considered a quasi-plastic damage zone produced by Hertzian contact with an elastic sphere of radius r at load P over n cycles, followed by tensile loading at stress σ in flexure (Fig. 10.1.18). The damage zone consists of an array of intrinsic closed shear faults, characteristic dimension l, spacing D, with extensile wing crack c (Fig. 10.1.18). The first stage identifies damage; the second quantifies it. Contact is assumed to be essentially quasi-plastic, consisting of an array of closed frictional shear faults with attendant "extensile" or "wing" microcracks at their ends. At large numbers of cycles or loads the faults coalesce by microcrack extension, ultimately into radial cracks. Contact fatigue takes place by cyclic attrition of frictional resistance at the repeatedly sliding fault interfaces according to an empirical degradation law. They derived working relations for degraded strength in terms of numbers of cycles and contact load. They evaluated the model using data from two

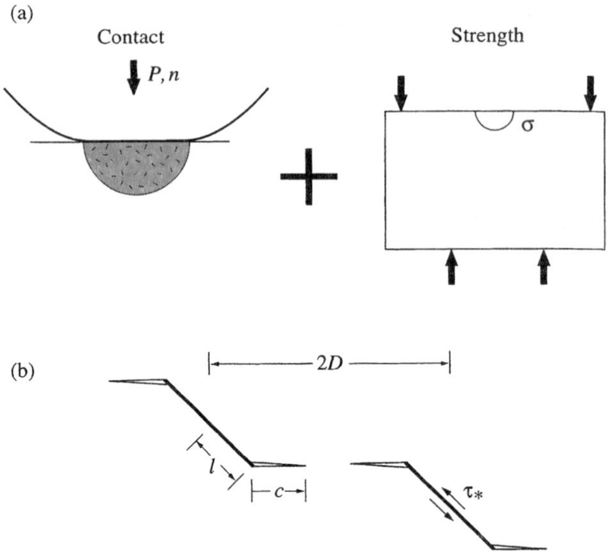

FIGURE 10.1.18 Model of cyclic strength degradation in quasi-plastic ceramic: (a) Showing two-step indentation-strength sequence. Step 1: Contact at n cycles at load P generates shear faults within quasi-plastic zone. Step 2: Application of bending stress σ takes system to fast failure (inert strength). (b) Coordinates for shear faults with extensile wing cracks. Composite cracks are treated as penny-like with virtual radius $C = c + \gamma l$, with c annular crack dimension and l microstructural fault radius, and mean separation D [93]. Reprinted with permission of the American Ceramic Society. All rights reserved.

well-studied coarse grain quasi-plastic ceramics, a micaceous glass-ceramic, and a silicon nitride. The model, though somewhat phenomenological in its treatment of micromechanical processes, well accounted for broad trends in contact fatigue strength degradation data.

Ceramics with high wear resistance are now being used for sliding components in a variety of engineering applications, including engine components (bearings, rollers, dies, tappets, valves, fuel injectors) and biomechanical components (dental restorations, hip prostheses). In such applications, scratch damage in combination with contact and wear damage are essential determinants for lifetime performance. It is known that such fatigue properties [94] and machining and wear [95, 96] are strongly affected by microstructural factors.

Pajares *et al.* [94] used Hertzian testing with spherical indenters to reveal that, in Mg–PSZ, the mode of contact damage changed dramatically with the degree of heat treatment aging: from distributed microcracking to phase transformation, and back to microcracking. On the other hand, scratch testing using a translating sharp point provides more direct information on potential processes in wear and machining operations [95–97]. Xu *et al.* [97] investigated effects of scratch interactions on material removal in alumina. Parallel scratches were introduced in aluminas with different grain sizes to investigate influence of scratch interactions on the material removal process in abrasive machining. A very small amount of material was removed when separation distance between the two parallel scratches was large or when the two scratches completely overlapped. On the other hand, at intermediate distances, the volume of material removed increased dramatically as a result of interactions between the two scratches. The maximum amount of material removed and the corresponding distance between the two scratches depended strongly on grain size and load. They showed that grain dislodgement was the predominant mechanism of material removal, irrespective of grain size.

Combining results from contact and scratch damage can provide basic information on damage modes in engineering contact and sliding applications, on material removal in wear and machining, and on flaw development in strength determination. Lee *et al.* [98] investigated contact and scratch damage in selected Mg–PSZ (peak-aged), Y–TZP, and Ce–TZP materials. Results demonstrate a need for caution in materials selection for applications in severe contact conditions. In all materials, the damage mode changed from smooth plastic deformation to limited cracking with increasing scratch load. In Mg–PSZ, plastic deformation was predominant at lower loads, with microcracking at higher loads. As for Y–TZP, plastic deformation was predominant over the range of test loads; macrocracks initiated only at relatively high loads, penetrating to a relatively large depth. The Ce–TZP showed an intermediate behavior. Bending tests on specimens subjected to scratch damage indicated a relatively high damage tolerance in Mg–PSZ and Ce–TZP. The Y–TZP showed highest initial strength, with relatively large strength loss above the critical load for macrocracking.

10.1.3.3.3 Fatigue of Piezoelectric Ceramics

Lead zirconate titanate (PZT) ceramics have potential for increased use in advanced industrial applications such as piezoelectric actuators and ultrasonic vibrators because of their quick response, compactness, and good power efficiency. When they are subjected to a powerful alternating electric field, cyclic stresses are introduced into these piezoelectric ceramics, which can lead to fracture and/or microcracking. Moreover, piezoelectric properties degradation is a serious practical concern, as this is a major cause of reduced device lifetime. However, mechanical performance and reliability of this kind of material is not well understood, though several studies have dealt with PZT durability [99–105]. Piezoelectric actuators are complex structures consisting of PZT layers, electrodes, and insulators. When they are used for automobile applications, for example, they receive severe cyclic loading operation. The lifetime of a PZT injector actuator must be more than 10^8 cycles, based on an automobile lifetime of 10^5 miles of driving. Such long-term durability should be established before practical use.

For evaluating PZT durability, a repeated durability evaluation system is very often employed to analyze changes in properties under various conditions [106, 107]. It has been suggested that the mechanism of deterioration is associated with domain switching and microcracks. Recently, high-durability PZT has been developed to prevent time-dependent deterioration of displacement, as well as failure under operation, by optimizing elemental manganese, zirconium, and antimony content [108]. The material shows high displacement and low time-dependent displacement degradation.

Another method for improving durability is forming nanocomposites with secondary phases of metallic or ceramic particles. For example, PZT nanocomposites with secondary phase dispersoids of metallic platinum showed significant improvement in fatigue resistance [109]; crack growth under electrical cyclic loading in the PZT/platinum nanocomposite was suppressed remarkably compared to that of the normal PZT material, as shown in Figure 10.1.19. Good fatigue resistance was also observed in PZT/alumina naocomposites. Tajima *et al.* [110], in an alternating electric field, evaluated crack growth behavior of PZT/alumina nanocomposites above the coercive field via optical and scanning electron microscopy. Crack extension in the 1.0 vol% alumina composite was

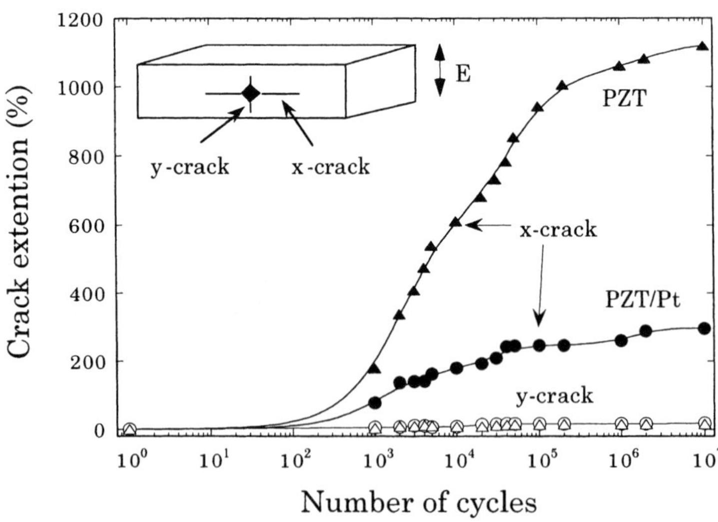

FIGURE 10.1.19 Crack extension, as a function of the number of cycles, in monolithic PZT (denoted by triangles) and the PZT/platinum nanocomposite (denoted by circles); both have been cyclically loaded at 2.6 kV/mm. Closed and open symbols, respectively, indicate *x*-crack and *y*-crack data [109]. Reprinted with permission of the American Ceramic Society. All rights reserved.

significantly less than in monolithic PZT and the 0.5 vol% alumina composite. Secondary-phase alumina dispersoids were found both at grain boundaries and within grains in composites. It was suggested that alumina dispersoids reinforce grain boundaries of the PZT matrix while acting as effective pins against microcrack propagation.

10.1.4 CREEP AND CREEP RUPTURE

Advanced ceramics have great potential for use as high-temperature materials in devices such as gas turbine blades since their heat resistance is far superior to that of currently available single crystal super-alloys. When ceramic materials are used for such applications, delayed fracture properties material should be assessed thoroughly to establish reliability. High-temperature durability of materials is very often limited by stress-enhanced cumulative damage, which is very strongly related to time-dependent deformation, or creep [111–179].

Test methods required for evaluating creep behavior are classified into three areas: flexure, tension, and compression. While all three methods involve uniaxial stress states, only flexure has a non-uniform stress state. This non-uniformity complicates creep behavior evaluation, particularly when the ceramic deforms differently in compression than in tension. Several methods have been proposed to extract uniformly stressed creep behavior from flexural tests. Still, to add sufficient reliability, such methods require at least one test in either tension or compression in addition to the flexure test. Thus, flexure tests are simple, and inexpensive; however, they should accompany complicated analysis. As ceramic components used in advanced engineering application are complicated, it is essential to collect reliable data on creep behavior to predict component durability. To do this, creep tests conducted in uniaxial tension and compression are superior to flexure tests. Particularly, tensile creep tests have become routine during the past 10 years. Principal tensile test advantages include precise determination of applied stress and strain in the uniform stress field. This section first assesses these three creep test methods, then reviews relevant research topics.

10.1.4.1 Testing Methods

10.1.4.1.1 Tensile Creep Test

There have been several testing methods for investigating ceramics, creep behaviors including flexural tests, tensile tests, and compressive tests. Among them, the tensile test gives the most advantages for obtaining creep–strain curves of materials because deformation factors such as dislocation, diffusion, and cavitation occur uniformly and throughout a gauge length area of a specimen. So, correspondence between time and deformation can be properly and easily obtained. Nevertheless, tensile tests are not so widely employed in ceramics creep measurements compared to the flexural test. This is because a stress concentration produced during the test cannot be attenuated by local plastic deformation. Hence, even slight eccentric loading causes specimen fracture around the grips and a large parasitic bending stress. Especially when there is some limit in the ceramic billet size, the specimen is required to be small and then gripped in a hot zone of a furnace (hot grip). Since ceramic materials are used for grips, their geometry and design require simplicity. During the past few decades development of tensile creep testing, such simplicity has been the subject of active development and significant progress has been made toward reliable testing [132, 140, 141, 151–160, 166–170]. Flag and specimen designs and test configuration employed by Luecke *et al.* [170] are shown in Figure 10.1.20. Recent interlaboratory verification (to which five institutes joined) on reliability of tensile creep testing silicon nitride using pin-loaded specimens and flag based, laser extensometry technique resulted in relatively uniform agreement among participating institutes [160].

10.1.4.1.2 Flexure Creep Test

Flexural creep tests have the advantage of low specimen preparation cost, easy experimental design, and minimum specimen material and preparation. Therefore, many creep studies have employed flexure tests so far [129, 138, 161–165]. There are, however, critical problems in creep testing ceramics in flexure. They include non-uniform stress distribution over the specimen cross-section, chemically and mechanically predominant surface phenomena, indirect determination of stress and strain, and small volume (surface) under maximum stress. Particularly, assumption of linear elastic behavior in flexural bars very often results in erroneous measurement results.

Two typical flexural testing methods are three- and four-point bending, in which a specimen is subjected to either static (normally dead weight) or dynamic loading. Stress distribution between the two top rollers is due to a uniformly applied moment. In three-point bending, stress distribution between two reaction points is due to both an applied moment and a shear stress at the center point of the load application.

10.1.4.1.3 Compressive Creep Test

Compressive creep testing, shown in Figure 10.1.21, is an old and very commonly used method for studying ceramic materials' creep deformation. In principle, uniform stress distribution is obtained in compressive testing, so

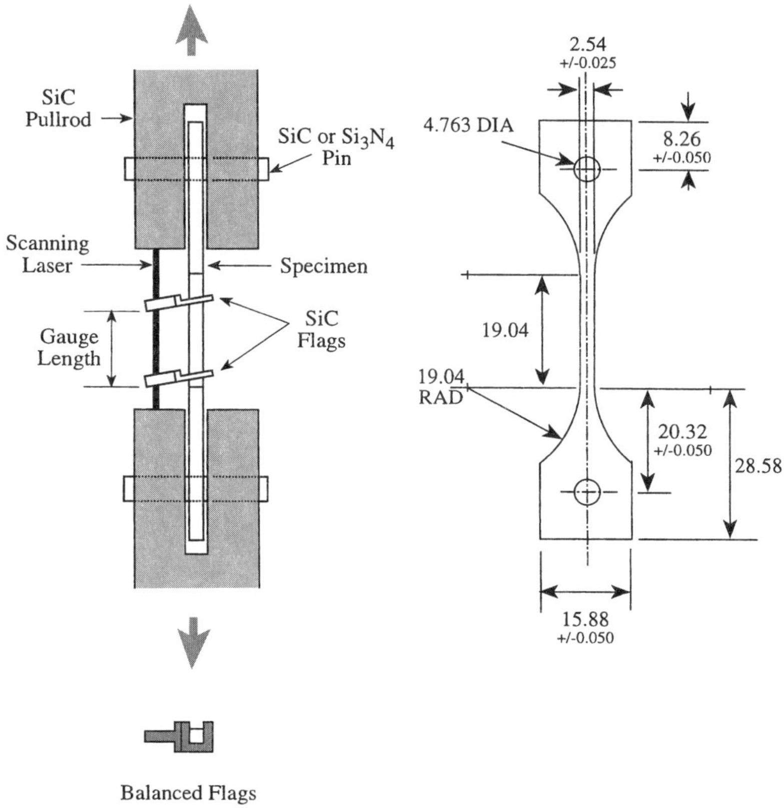

FIGURE 10.1.20 Flag and specimen designs and test configuration for NIST test technique [170]. Reprinted with permission of the American Ceramic Society.

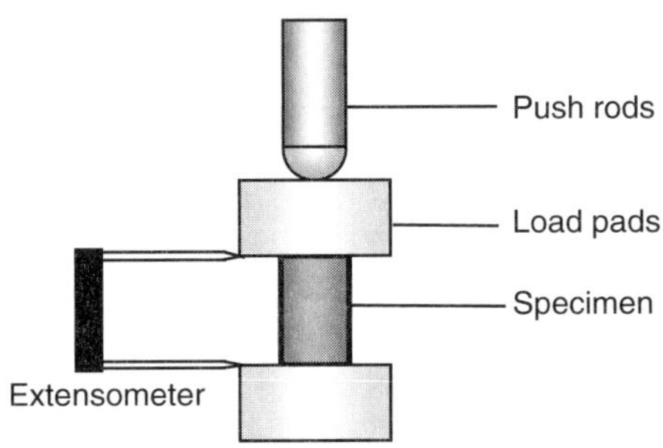

FIGURE 10.1.21 Schematic diagram of a compression test.

material behavior can be studied in a pure stress state. For a variety of reasons explained later, creep in compression can differ substantially from creep in tension. Therefore, to completely understand material creep behavior, it is important to characterize creep responses to both tension and compression. Examples of compressive creep studies of ceramics are available [171–177].

Compression test advantages include low cost of specimen preparation (compared to tensile ones), small size of specimens (allowing use of limited small batches), and uniform stress distribution (precise determination of stress–strain relationship). However, specimen alignment is very often more difficult than with flexure or tension. In addition, cavitation and microcracks effects, which greatly impact creep lifetime, cannot be evaluated in compression.

10.1.4.2 Related Standards

In recent decades, considerable efforts have been made to standardize creep measurement of ceramics. Applicable standards are listed here.

ASTM, Committee C28 on "Advanced Ceramics" (USA)

- C1291-00: Standard Test Method for Elevated Temperature Tensile Creep Strain, Creep Strain Rate, and Creep Time-to-Failure for Advanced Monolithic Ceramics.
- C1337-96: Standard Test Method for Creep and Creep Rupture of Continuous Fiber-Reinforced Ceramic Composites under Tensile Loading at Elevated Temperatures.

CEN, Committee TC184 "Advanced Technical Ceramics" (Europe)

- ENV 820-2: Deformation of monolithic ceramics at high temperatures.

JISC, Divisional Council on Ceramics (Japan)

- R1612: Testing method for bending creep of high-performance ceramics.
- R1631: Testing method for tensile creep of high-performance ceramics.

10.1.4.3 Recent Research Topics

10.1.4.3.1 Creep of Silicon Nitride

Silicon nitride, one of the most attractive candidate materials for high-temperature components, generally exhibits creep deformation at high temperatures above 1300°C. It is known that dislocation motion contributes very little to silicon nitride creep below 1700°C [180]. Also, diffusional creep involving transfer of vacancy from interfaces under tension to those under compression is unlikely to be a rate-controlling process in silicon nitride containing considerable amounts of glassy phase at the interfaces. The dominant creep mechanism for Si_3N_4 containing a grain boundary phase is grain boundary sliding (GBS). This GBS occurs by viscous flow of the grain boundary phase (which acts as a lubricant between Si_3N_4 grains) or by solution-reprecipitation of Si and N (which occur by diffusion through the grain boundary phase). In addition, cavitation at either two-grain or multiple-grain junctions is generated within Si_3N_4 containing a grain boundary phase; void nucleation and growth are controlled by the same mechanism that determines deformation. Thus, creep behavior of silicon nitride is largely dependent on characteristics and behavior of the glassy phase. These might include composition (viscosity), state of dispersion, dimensions, devitrification, and outward diffusion of the cation enhanced by oxidation. Glassy phase characteristics are affected by even a small addition of impurity [181, 182]. There have been systematic studies of the effect of small additions of fluorine or chlorine impurity on creep behavior of silicon nitrides.

Recent studies on tensile creep of several grades of silicon nitride suggest that cavitation contributes substantially to the creep strain [170, 183, 184]. Luecke *et al.* [170] reported that, during tensile creep of a HIP-ed silicon nitride, the volume fraction of cavities increases linearly with strain (Fig. 10.1.22); these cavities produce nearly all of the measured strain. Transmission electron microscopy (TEM) revealed cavities formed during creep in

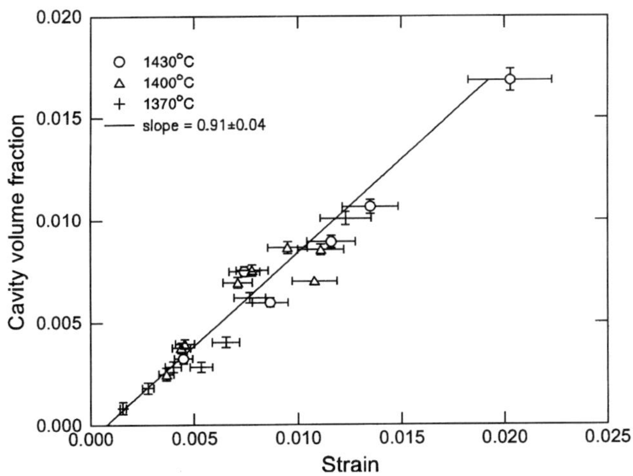

FIGURE 10.1.22 Total volume fraction of cavities in the gauge length for silicon nitride specimens tested to failure, as a function of failure strain (courtesy of William Luecke *et al.*).

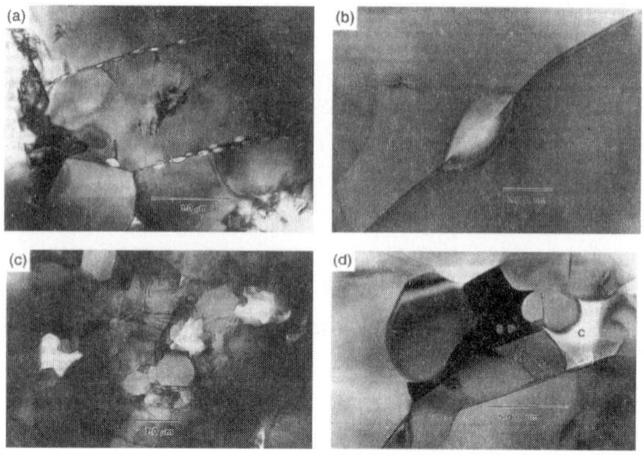

FIGURE 10.1.23 Transmission electron micrographs showing both small, lens-shaped cavities as well as the larger, more irregularly shaped cavities. (a) and (b) show the lens-shaped cavities that formed in a specimen crept at 1370°C under 125 MPa for 1521 h, while (c) and (d) show the irregularly shaped, interstitial cavities that formed in the silicate phase in a specimen deformed at 1430°C under 125 MPa for 30 h. In both micrographs the tensile axis is vertical. In (d) "C" is a cavity and "SP" is the silicate phase [170]. Reprinted with permission of the American Ceramic Society. All rights reserved.

silicate glass pockets of multigrain junctions as shown in Figure 10.1.23. They investigated cavity size distribution using small-angle X-ray scattering (SAXS) and found that cavity addition, rather than cavity growth, dominates the cavitation process (Fig. 10.1.24). It was suggested that during creep, while silicon nitride grains remain rigid, cavitation in multigrain junctions allows the silicate to flow from cavities to surrounding silicate pockets, leading to microstructure dilatation and material deformation. Grain boundary sliding accommodates this expansion and results in specimen elongation.

The tensile creep rate of silicon nitride very often increases strongly with stress. Although usual power-law functions can represent creep data, data often show curvature and systematic variations of slope with temperature and stress. Luecke and Wiederhorn [185] recently proposed a new approach to understanding silicon nitride creep, where a deformable second phase bonds a deformation-resistant major phase. Assuming that hydrostatic stresses in pockets of the second phase are normally distributed, a model was developed to accurately describe the tensile creep rate of grades of silicon nitrides. The steady creep rate, $d\varepsilon/dt$, is generally expressed by the Norton equation [186]:

$$d\varepsilon/dt = A(\sigma/\sigma_0)^n \exp(-\Delta H/RT) \quad (7)$$

where σ is the applied stress, n is the creep exponent, ΔH is the activation enthalpy, R is the gas constant, T is temperature, and A and σ_0 are constant.

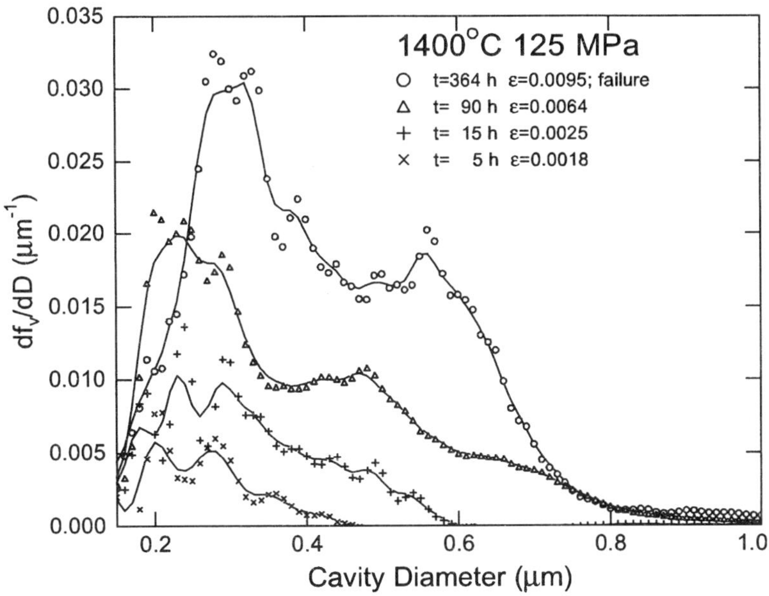

FIGURE 10.1.24 Cavity size distribution evolution as a function of strain for specimens crept at 1400°C and 125 MPa (courtesy of William Luecke *et al.*).

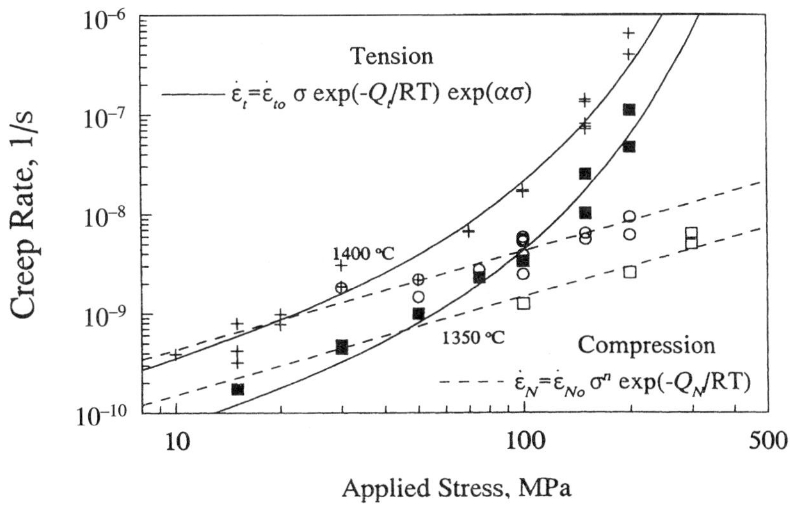

FIGURE 10.1.25 Comparison of tensile and compressive creep behavior of silicon nitride: cavitation model given by Eq. 8 is fitted creep data [188]. Reprinted with permission of the American Ceramic Society. All rights reserved.

The value of $d\varepsilon/dt$ in the developed model is given as

$$d\varepsilon/dt = B\sigma \exp(-\Delta H/RT)(\Phi^3/(1-\Phi)^2)\exp(\alpha\sigma) \tag{8}$$

where

$$\alpha = 2\sigma_c/9\sigma^2$$

σ_c is critical stress for cavity formation, and B and Φ are constants. Thus, the creep rate: increases exponentially with applied stress, is independent of silicon nitride grain size, is inversely proportional to effective viscosity of the deformable phase, and is proportional to the cube of the volume fraction of the deformable phase.

Uniaxial creep tests, such as tension and compression, are advantageous since the specimen is subject to uniform stress such that the obtained results can be analyzed in a simple manner and precisely (if tests are conducted deally). One principal difference between tensile and compressive creep tests is the degree of cavity formation; tensile creep is much more affected by cavitation. Luecke and Wiederhorn [187] have investigated silicon nitride creep in both tension and compression at 1430°C at stresses between 40 and 300 MPa. The material crept up to 100 times faster in tension than in compression. Stress dependence of the tensile strain rate was not characterized by a single stress exponent rising from about two at low stresses to about five at higher stresses. While cavitation contributed significantly to the tensile strain, the material exhibited a stress exponent of unity, and cavitation was almost completely suppressed in compression. Asymmetry in creep behavior between tension and compression is due to dilatation of the material. Yoon *et al.* [188] compared tensile and compressive creep behavior of silicon nitride in at 1350 and 1400°C in more detail. Again, over the entire investigated stress range, the tension creep rate increased linearly with stress at low stresses and exponentially at high stresses, but the compression creep rate increased linearly with stress. While the cavity volume fraction increased linearly with increased tensile creep strain, the volume fraction of cavities in compression was ~18% of that in tension at 1.8% axial strain and approached zero at strains <1%. In compression, where cavitation is suppressed, deformation occurs by solution-reprecipitation of silicon nitride. Boling-Risser *et al.* [189] also studied compression creep behavior of silicon nitride at very high temperatures. Around 1600°C, the material deformed with steady-state grain boundary sliding accommodated by solution-reprecipitation of silicon nitride through the grain boundary phase, exhibiting a stress exponent of unity.

10.1.4.3.2 Creep Rupture of Highly Heat Resistant Ceramics

In recent decades, efforts to control the grain boundary phase for improving heat resistance of silicon nitride have made significant progress. For example, some grades of silicon nitrides, SN 281 and 282 (Kyocera Corp., Kyoto, Japan), which have been recently developed for ceramic gas turbine blades [190–192], showed excellent creep resistance even at temperatures above 1400°C [193,194].

There are two regions in a delayed fracture mechanism map of silicon nitride at high temperatures; slow crack growth failure and creep damage rupture. The former is a fracture occurring when a crack grows subcritically from a pre-existing flaw and reaches critical size. This is predominant in the high-stress, short-term life region. The latter is due to formation of a macrocrack with the critical size by cavity nucleation and coalescence, which prevails in the low-stress, long-term life region. Creep life for structural ceramics is conventionally expressed as

$$t_f = C_L \sigma^{-N} \tag{9}$$

where t_f is time-to-failure, σ is applied stress, and C_L and N are constants. Exponent N determines stress dependency of creep life and is often referred to as a fatigue exponent. In Figure 10.1.26, creep lives of SN 282 are shown as functions of stress and temperature in comparison with those of conventional silicon nitrides, SN A (hot-pressed silicon nitride doped with 5 wt% yttria and 3 wt% alumina) and SN B (HIPed silicon nitride doped with 5 wt% ytterbia) [193]. Plots indicate some changes in slopes around 200–250 MPa. Fractography revealed that fracture occurred from slow crack growth in the high-stress range for SN 282, as well for SN A and SN B; the estimated fatigue exponent was 10 or higher. In marked contrast, creep damage was obvious for SN A and SN B, and the fatigue exponent was two to three in the low-stress range. Occurrence of creep damage rupture substantially limits high-temperature structural reliability of silicon nitride. In SN 282, however, creep rupture hardly occurred at stresses below 200 MPa at tested temperatures. While creep curves of conventional silicon nitrides consist of three regimes: transient, steady state, and accelerated creep regimes, these materials showed solely transient creep [193, 194]. Although high-resolution TEM study revealed that a glassy phase existed at some grain boundaries, thickness was very narrow—typically below 1 nm (Fig. 10.1.27). Glassy phase was also identified at some boundaries between silicon nitride matrix and secondary phase. In XRD analysis, there were strong peaks of $Lu_2Si_2O_7$ aside from α and β silicon nitrides (or $Lu_4Si_2N_2O_7$) [193, 194]. Excellent creep resistance was attributed to the highly refractory phase of grain boundaries and triple junctions such as $Lu_2Si_2O_7$.

Both elevated temperature strength and eutectic melting points of the Si_3N_4–Si_2N_2O–$RE_2Si_2O_7$ (RE: rare earth) system increase with decreasing rare

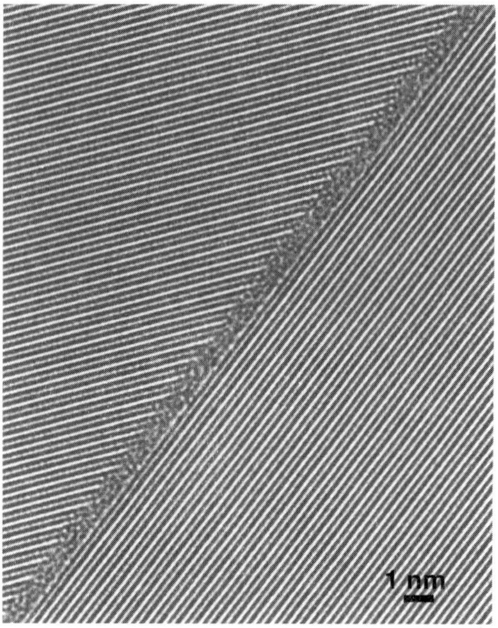

FIGURE 10.1.26 Grain boundary of SN 282. Between two silicon nitride grains [193]. Reprinted with permission of the American Ceramic Society. All rights reserved.

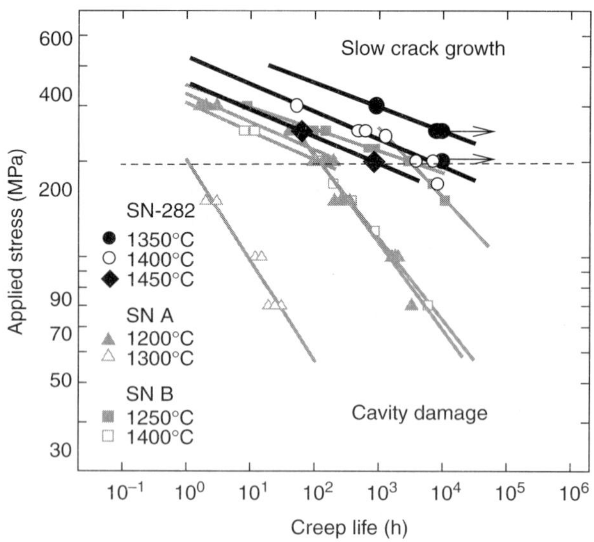

FIGURE 10.1.27 Stress dependencies of creep lives of SN 282, SN A, and SN B [193]. Reprinted with permission of the American Ceramic Society. All rights reserved.

earth ionic radii, that is, with increasing bond strength in the $RE_2Si_2O_7$ phase [195]. Since ionic radius decreases with increasing atomic number, Lu, whose atomic number is highest among lanthanoids, can give very high bonding strength of $RE_2Si_2O_7$, and a very high eutectic melting point of the Si_3N_4–Si_2N_2O–$RE_2Si_2O_7$ system. Therefore, excellent creep resistance of SN 282 observed in this study is very likely due to the $Lu_2Si_2O_7$ phase at grain boundaries and triple junctions. Both ICP and TEM studies showed that this material contains a substantial amount of the secondary phase. It is known that a large amount of secondary phase mitigates room-temperature strength of silicon nitride, but high-temperature mechanical strength is not affected by secondary phase content [195]. Another advantage of the $Lu_2Si_2O_7$ phase is good oxidation resistance. Lange *et al.* [196] revealed that materials in the Si_3N_4–Si_2N_2O–$RE_2Si_2O_7$ compatibility triangle preclude unstable compounds and are extremely resistant to oxidation. This is, for example, because both SiO_2 (oxidation product of either Si_3N_4 or Si_2N_2O) and $Y_2Si_2O_7$ can exist in equilibrium with each other and no reaction or interdiffusion between them should occur until the eutectic temperature (1660°C) is reached. Actually, it has been reported that SN 282 shows very good oxidation resistance [190–192].

Silicon carbide generally does not contain glassy phases at grain boundaries, even with doped sintering additives such as alumina. Due to this rigid interface, strength is not degraded at very high temperatures. Also, because of good high-temperature mechanical properties and good corrosion resistance, silicon carbide is one of the most important candidate materials capable of withstanding high temperatures above 1400°C. Here, we describe creep and creep rupture behavior of silicon carbide doped with 5 wt% alumina 1400°C [197]. The TEM observation revealed that there is no glassy phase at interfaces between two silicon carbide grains: even if some glassy phase were present, its thickness is on the order of atomic dimensions, as shown in Figure 10.1.28 [111]. Then, the measured creep rate of this material at 1400°C, 200 MPa, is as small as 6×10^{-12}/s. No cavity is formed during creep, though creep deformation should be controlled by grain boundary diffusion [198], and creep failure is caused by slow crack growth from a pre-existing flaw. The crack grows sub-critically along grain boundaries in a diffusive process. In this case, it is possible to apply a diffusive crack growth model [199] to results since this model is based on the following three assumptions: a crack propagates along grain boundaries by surface and grain boundary diffusion, grains on either side of the boundary behave elastically, and the crack grows along grain boundaries in a steady state at a fixed velocity. Crack velocity, v, in this model is given as

$$v = C_D[0.59K/K_G + \{0.35(K/K_G)^2 - 1\}^{1/2}]^{12} \tag{10}$$

where C_D is a constant, K is applied stress intensity, and K_G is the critical K value predicted by Griffith's theory for propagation of an atomistically sharp crack in

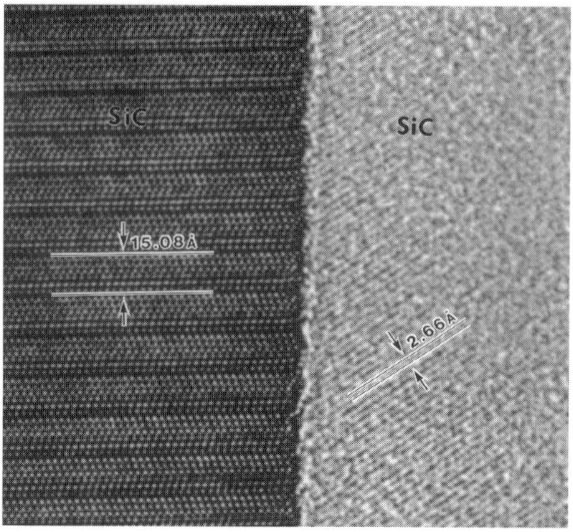

FIGURE 10.1.28 TEM image of interface between two silicon carbide grain [197]. Reprinted with permission of the American Ceramic Society. All rights reserved.

an interface. Here, K_G, related to true surface energy, is normally one order of magnitude lower than fracture toughness measured using a fracture mechanics test specimen. In this v–K relation, there is a threshold stress intensity, K_{th}, defined as $K_{th} = 1.69K_G$. Below K_{th}, applied stress is insufficient to drive the crack. Applied stress, σ, can be related to K by $K = \sigma Y \cdot a^{1/2}$ where Y is a coefficient depending on crack geometry and a is crack size. Combining this relation with Eq. 10 yields time-to-failure, t_f, as a function of σ (Fig. 10.1.29). The line is the t_f–σ curve estimated from Eq. 10. The curve agrees well with the plotted data. The curve also can predict a threshold applied stress below which the pre-existing crack does not grow. Consequently, delayed fracture does not occur. The predicted value is 165 MPa. This stress can be considered as a safety applied stress. Its value is very important in making component designs with this material. In the case of silicon nitride, short-term results obtained in the high-stress range generally tend to estimate longer creep life in the low-stress range as already described. On the other hand, silicon carbide creep fractures are controlled by diffusive crack growth from a pre-existing flaw at all stress levels due to absence of a glassy phase in it. Then, in the stress–life diagram, creep life in the low-stress range is much longer than that estimated from data in the high-stress range.

A ceramic gas turbine, where such advanced ceramics are used in high-temperature components such as turbine blades and nozzles, enables increased turbine inlet temperatures up to 1300–1400°C, resulting in substantially high thermal efficiency. An important consideration is that, even though non-oxide

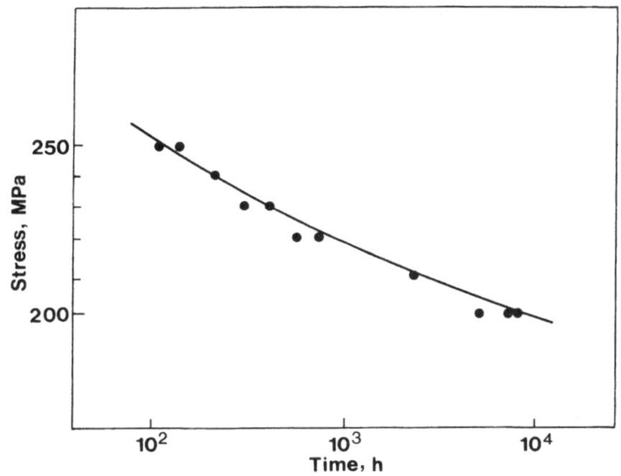

FIGURE 10.1.29 Log–log representation of time-to-failure and applied stress at 1400°C. The curve is based on Chuang's diffusive crack growth model [199]. The threshold stress was estimated at 165 MPa from the curve [197]. Reprinted with permission of the American Ceramic Society. All rights reserved.

systems such as silicon nitride and silicon carbide can be viable for structural application below 1500°C, oxide-based ceramics are needed for service above 1500°C, which is required in the near future, as pointed out by Raj [200]. Waku *et al.* [201] reported a strong ceramic composite with a different kind of microstructure made by unidirectional solidification of an $Al_2O_3/GdAlO_3$ eutectic mixture. The material showed plastic deformation at 1600°C owing to dislocation motion. High strength and resistance to brittle failure at such high temperatures indicate that this kind of oxide ceramic can be of service above 1500°C.

10.1.4.4 Dopant Effects in Creep of Alumina

It was recently discovered that creep resistance of some ceramics is improved remarkably by even small addition of dopants. For example, Cho *et al.* [202] reported that the creep rate of alumina doped with rare earths such as Y_2O_3 or La_2O_3 was reduced by 2–3 orders of magnitude. While the mechanism for enhanced creep resistance is not clearly identified, grain boundary segregation probably plays an important role. Secondary ion mass spectroscopy (SIMS) mapping [203] and scanning transmission electron microscopy (STEM) studies [204] revealed that oversized cation dopants tend to segregate to grain boundaries. It was suggested that segregation of oversized dopant ions to grain boundaries in Al_2O_3 hinders grain boundary diffusion and reduces the tensile

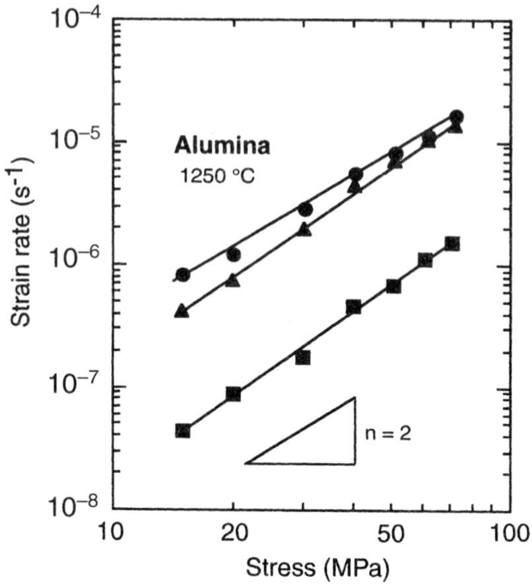

FIGURE 10.1.30 Strain rate versus stress (log–log plot) at 1250°C ((•) undoped Al_2O_3, (▲) 100-ppm-zirconium-doped Al_2O_3, and (■) 1000-ppm-zirconium-doped Al_2O_3); the strain rate of zirconium-doped Al_2O_3 was plotted at the normalized grain size of undoped Al_2O_3 (1.1 μm) [205]. Reprinted with permission of the American Ceramic Society. All rights reserved.

creep rate in this system by a few orders of magnitude. Wakai *et al.* [205] also reported the effect of zirconium segregation on hardening in tensile creep of alumina doped with 100-ppm- and 1000-ppm-zirconium. They found that doping even as little as 100 ppm of zirconium caused hardening, and the creep rate was further reduced by increasing the amount of zirconium dopant (Fig. 10.1.30). Scanning transmission electron microscopy/energy-dispersive X-ray spectroscopy study revealed that zirconium was segregated at the alumina grain boundary.

Beneficial effects of selective codoping for enhancing creep resistance of Al_2O_3 have been reported by Li *et al.* [206]. For example, the creep rate of Nd/Zr codoped alumina decreased by 2–3 orders of magnitude relative to undoped alumina (Fig. 10.1.31). It is notable that this improvement was greater than that achieved by doping with either Nd or Zr alone, indicating that incorporation of ions of differing sizes may be beneficial. Microchemical mapping using SIMS again revealed direct evidence of cosegregation of dopant ions to grain boundaries. They presumed that, in some circumstances, a combination of large and small ions can fill the grain boundary region more efficiently than can either ion individually, which inhibits diffusion and creep. Cho *et al.* [207] characterized both the effective cation and interstitial volumes at grain boundaries, inferring

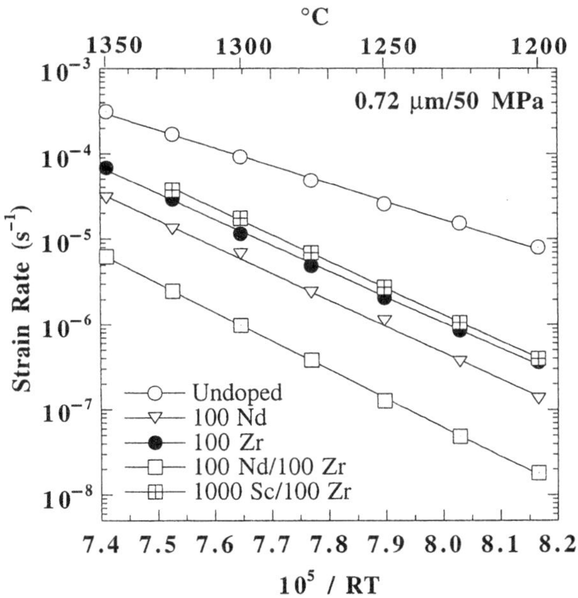

FIGURE 10.1.31 Strain rate (normalized to 0.72 μm grain size) as a function of temperature for undoped, Nd and Zr singly doped, and Nd/Zr and Sc/Zr codoped alumina [206]. Reprinted with permission of the American Ceramic Society. All rights reserved.

that relative openness of some boundary structures leads to a great accommodation of oversized ions. Boundary volume was determined by a spatially local Voronoi construction for cation (Al^{3+}) substitutional sites as well as large interstitial voids for coincidence-site lattice (CSL) boundaries with a common [0001] rotation axis. Examining spatial distribution of free volume near grain boundaries, they suggested that selective codoping can provide a more efficient means of filling available space near boundaries, thereby further enhancing creep resistance.

Yoshida *et al.* [208, 209] studied creep behavior of polycrystalline alumina with 0.05 mol% lanthanoid oxides of Y, Sm, Eu, Tm, or Lu in uniaxial compression creep testing at 1250°C. They also found improved creep resistance by doping, but the dopant effect depended on the lanthanoid type. Effects became stronger in the order Sm < Tm < Eu < Y < Lu. Each dopant cation was found to segregate in grain boundaries and was likely to suppress grain boundary diffusion. Change in chemical bonding state with doping was estimated by a first principles molecular orbital calculation using the discrete variational (DV)-X-alpha method, resulting in a good correlation between creep resistance and net charge of constituent ions. They concluded that improved creep resistance was caused by a change in the ionic bonding state in grain boundaries due to lanthanoid segregation. They also reported transient creep behavior of Lu-doped Al_2O_3 was caused by retardation of grain boundary diffusion due to Lu^{3+} ions segregation in grain boundaries.

10.1.4.5 Creep of Alumina/Silicon Carbide Nanocomposite

In the field of metals creep, it is well known that high melting oxide particles on grain boundaries significantly inhibit diffusional creep. For example, this is seen in oxide dispersion strengthened (ODS) super-alloys. It is recently known that nanometer-sized silicon carbide particle dispersion into an alumina matrix (alumina/silicon carbide nanocomposites) gives rise to significant improvements in mechanical properties, particularly in creep resistance, which is very similar to the case of metals [210–214]. For example, it was revealed that the minimum creep rate of alumina/17 vol% silicon carbide nanocomposite was about three orders of magnitude lower than that of monolithic alumina [211]. Creep life of the nanocomposite was 10 times longer and the creep strain at fracture was eight times smaller than those of the monolith at 1200°C and 50 MPa.

Another nanocomposite feature is that it tended to show almost solely transient creep until failure, while slightly accelerated and steady-state creep were observed in the monolith. The TEM study revealed that intergranular silicon carbide nanoparticles inhibited grain boundary sliding of the alumina/silicon carbide nanocomposite, as shown in Figure 10.1.32 [211]. Grain boundary sliding proceeded, and the pinning effect of the particle increased and caused transient creep. Along with grain boundary sliding, the particles plunged into adjacent alumina grains. From this process, small intergranular cavities appeared around the particles. These cavities induced intergranular crack formation, leading finally to creep failure. It was also revealed that the interface between the intergranular silicon carbide particles and the alumina matrix was much stronger than the alumina/alumina interface [212]. Rigid bonding at alumina/silicon carbide interfaces caused inhibition of vacancy nucleation and annihilation at the interface, which remarkably improved nanocomposite creep resistance.

Thompson *et al.* [213] also studied tensile creep behavior of alumina–silicon carbide nanocomposite containing 5 vol% of 0.15 μm SiC particles. For a stress level of 100 MPa in the temperature range of 1200–1300°C, SiC reduced the creep rate of alumina by 2–3 orders of magnitude (Fig. 10.1.33) In contrast to monolithic alumina, the nanocomposite exhibited no primary or secondary stages, with only tertiary creep being observed. Microstructural examination reveals extensive cavitation associated with SiC particles located at Al_2O_3 grain boundaries. Failure of the nanocomposite occurs via growth of

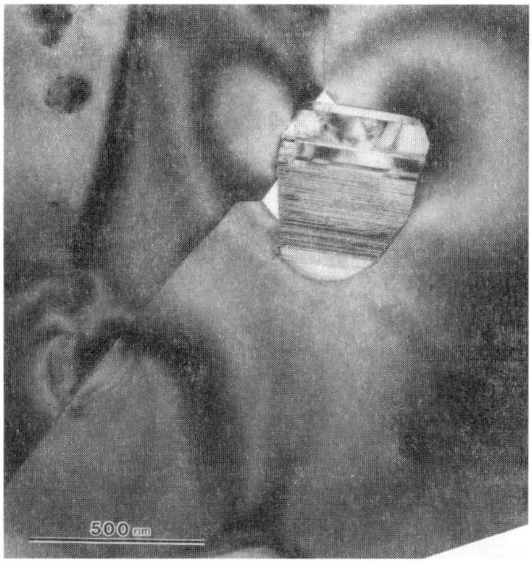

FIGURE 10.1.32 Behavior of an intergranular silicon carbide nanoparticle in alumina/silicon carbide nanocomposite after creep [211]. Reprinted with permission of the American Ceramic Society.

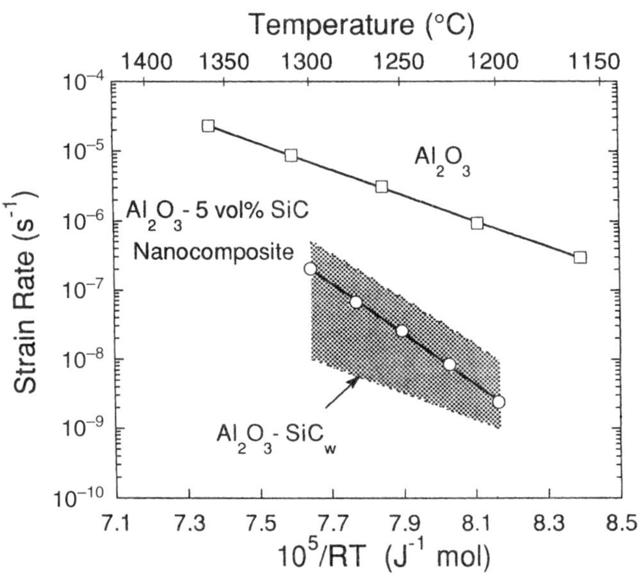

FIGURE 10.1.33 Arrhenius plot comparing the strain rate of the Al_2O_3–SiC nanocomposite at 1% total strain with fully dense Al_2O_3 of a similar grain size. Also included on this plot is a hatched band that represents flexural strain-rate data for SiC-whisker-reinfoced Al_2O_3 [213]. Reprinted with permission of the American Ceramic Society.

subcritical cracks that are nucleated preferentially at gauge corners. A modified test procedure enables creep lifetimes to be estimated and compared with creep rupture data. Several possible roles of SiC particles are considered, including (i) chemical alteration of Al_2O_3 grain boundaries, (ii) retarded diffusion along the Al_2O_3-SiC interface, and (iii) inhibition of the accommodation process (either grain boundary sliding or grain boundary migration).

Since intergranular silicon carbide nanoparticles have been found to inhibit grain boundary sliding in the creep of the alumina/silicon carbide nanocomposite, this system has a threshold stress model—a common trait with the ODS superalloys [214]. The minimum creep rates of the nanocomposite and monolith at 1200°C are shown in Figure 10.1.31 as a function of applied stress. Monolith plots well approximated one straight line. For the nanocomposite, however, minimum creep rates of 35 and 40 MPa were remarkably lower than the line extrapolated from data in the stress range above 50 MPa. These results strongly imply presence of a threshold in the creep of the nanocomposite. Threshold stress can be very crudely estimated to be in the range of 20–35 MPa. The creep lives, t_f, of the nanocomposite and monolith are shown in Figure 10.1.34, as a function of applied stress. Again, while monolith plots well followed one straight line, those of the nanocomposite deviated from a

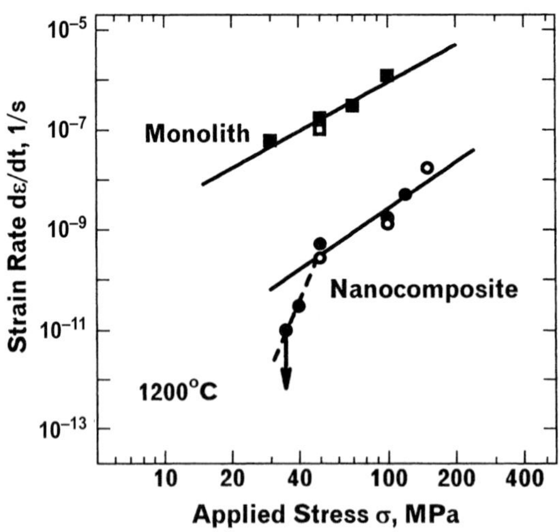

FIGURE 10.1.34 Stress dependencies of steady-state or minimum creep rates for the alumina monolith and the alumina/silicon carbide nanocomposite; the temperature is 1200°C. Closed and open symbols represent the results of the present and previous studies, respectively, and the arrow indicates that the real value is possible lower [214]. Reprinted with permission of the American Ceramic Society. All rights reserved.

straight line at stresses below 50 MPa. Creep lives of the nanocomposite at 35 and 40 MPa were substantially longer than those estimated from the higher stress range, but their measurements were interrupted after 10 000 h.

Arzt and Ashby [215] proposed a model where particles pin motion of dislocations to estimate theoretically the threshold stress. This mechanism is based on an assumption that vacancies can be absorbed and emitted only at edges of grain boundary dislocations. Thus, continuous provision of vacancies requires movement of dislocation. When a dislocation moves in the particle matrix interface, it will interact with particles whose modulus, lattice parameter, or chemical composition differ from those of the matrix. This interaction causes a pinning force on the dislocation, resulting in a threshold stress, σ_{tr}, given by

$$\sigma_{tr} = C_A G b_B / D \tag{11}$$

where C_A is a constant, G is shear modulus, b_B is Burgers' vector of grain boundary dislocation ($b_B = \sim 1.6 \times 10^{-10}$ m), and D is particle spacing on the grain boundaries. Here, C_A is 0.8 for the case of Orowan's classic calculation of the pinning of lattice dislocation by hard particles. However, at high temperatures where climb is active, C_A can be considered to vary from 0.3 to 0.8 depending on intrinsic local mobility at the particle–matrix interface (by kinetics of atom rearrangement). Then, σ_{tr}, estimated by Eq. 11, ranges from about 18 to 47 MPa. This agrees well with experimentally estimated threshold stress, 20–35 MPa. Equation 11 shows one advantage of nanoparticle dispersion. At the same volume fraction, threshold stress σ_{tr} becomes larger when particle spacing (and particle size) becomes smaller.

REFERENCES

1. Cranmer, D. C., and Richerson, D. W. (eds.) (1998). *Mechanical Testing Methodology for Ceramic Design and Reliability*, New York: Marcel Dekker, Inc.
2. Lange, F. F., and Diaz, E. S. (1978). Powder-cushion gripping to promote good alignment in tensile testing. *J. Test. Eval.* **6**: 320–323.
3. Govila, R. K. (1982). Uniaxial tensile and flexural stress rupture strength of hot-pressed silicon nitride. *J. Am. Ceram. Soc.* **65**: 15–21.
4. Sheppard, L. M. (1987). Tensile testing of ceramics. *Adv. Mater. Proc.* **131**: 11–15.
5. Seshadri, S. G., and Chia, K.-Y. (1987). Tensile testing ceramics. *J. Am. Ceram. Soc.* **70**: C242–C244.
6. Ohji, T. (1988). Towards routine tensile testing. *Int. J. High Technol. Ceram.* **4**: 211–225.
7. Vaidayanathan, R., Sankar, J., and Avva, V. S. (1988). Uniaxial tensile characteristics of silicon nitride at room temperature. *Ceram. Eng. Sci. Proc.* **9**: 1383–1392.
8. Ohji, T., Yamauchi, Y., Kanematsu, W., and Ito, S. (1990). Dependence of high temperature tensile strength for hot pressed silicon nitride. *J. Mater. Sci.* **25**: 2990–2996.
9. Jenkins, M. G., Ferber, M. K., and Martin, R. L. (1990). Evaluation of the stress state in a buttonhead, tensile specimen for ceramics. *Ceram. Eng. Sci. Proc.* **11**: 1346–1363.

10. Hedehiro, U. K., Kawamoto, H., and Ura, S. (1990). High temperature tensile tests of sintered silicon nitride. *Zairyo* **39**: 716–722.
11. Hatanaka, K., Shiota, H., and Ando, T. (1991). Tensile tet of sintered silicon nitride ceramic at elevated temperatures. *JSME Int. J. Ser 1* **34**: 351–360.
12. Hatanaka, K., Fujimitsu, T., Ando, T., and Tachibana, T. (1990). Strength evaluation of sintered silicon nitride ceramics by means of a new test system at elevated temperatures. *JSME Int. J. Ser 1* **33**: 542–550.
13. Baxter, D. F. (1991). Tensile testing at extreme temperatures. *Adv. Mater. Proc.* **139**: 22–33.
14. Ohji, T., Yamauchi, Y., Kanematsu, W., and Ito, S. (1989). Tensile rupture strength and fracture defects of sintered silicon carbide. *J. Am. Ceram. Soc.* **72**: 688–690.
15. Adams, M., and Sines, G. (1978). A statistical micromechanical theory of the compressive strength of brittle materials. *J. Am. Ceram. Soc.* **61**: 126–131.
16. Lankford, J. (1977). Compressive strength and microplasticity in polycystalline alumina. *J. Mater. Sci.* **12**: 791–796.
17. Tracy, A. (1987). A compression test for high strength ceramics. *J. Test. Eval.* **15**: 14–19.
18. Quinn, G. D., and Morrell, R. (1991). Design data for engineering ceramics: a review of the flexure test. *J. Am. Ceram. Soc.* **74**: 2037–2066.
19. Hoagland, R. G., Marschall, C. W., and Duckworth, W. H. (1976). Reduction of errors in ceramic bend tests. *J. Am. Ceram. Soc.* **59**: 189–192.
20. Newnham, R. C. (1975). Strength tests for brittle materials. *Proc. Br. Ceram. Soc.* **25**: 281–293.
21. Baratta, F. I. (1981). When is a beam a plate? *J. Am. Ceram. Soc.* **64**: C86.
22. Quinn, G. D., and Baratta, F. (1985). Flexure data can it be used for ceramics part design? *Adv. Mat. Proc.* **129**: 31–35.
23. Swank, L. R., Caverly, J. C., and Allor, R. L. (1990). Experimental errors in modulus of rupture test fixtures. *Ceram. Eng. Sci. Proc.* **11**: 1329–1345.
24. Quinn, G. D. (1990). Flexure strength of advanced structural ceramics: a round robin. *J. Am. Ceram. Soc.* **73**: 2374–2384.
25. Shetty, K., Rosenfield, A. R., McGuire, P., Bansal, B. K., and Duckworth, W. H. (1980). Biaxial flexure tests for ceramics. *Am. Ceram. Bull.* **59**: 1193–1197.
26. Wachtman, J. B. Jr., Capps, W., and Mandel, J. (1972). Biaxial flexure tests of ceramic substrates. *J. Mater. Sci.* **7**: 188–194.
27. Radford, K. C., and Lange, F. F. (1978). Loading (L) factors for the biaxial flexure test. *J. Am. Ceram. Soc.* **61**: 211–213.
28. Marshall, B. (1980). An improved biaxial flexure test for ceramics. *Am. Ceram. Bull.* **59**: 551–553.
29. Giovan, M. N., and Sines, G. (1981). Strength of a ceramic at high temperatures under biaxial and unniaxial tension. *J. Am. Ceram. Soc.* **64**: 68–73.
30. Quinn, D., and Wirth, G. (1990). Multiaxial strength and stress rupture of hot pressed silicon nitride. *J. Eu. Ceram. Soc.* **6**: 169–177.
31. Shetty, D. K., Rosenfield, A. R., Duckworth, W. H., and Held, P. R. (1983). A biaxial flexure test for evaluating ceramic strengths. *J. Am. Ceram. Soc.* **66**: 36–42.
32. Tennery, V. J., Breder, K., Ferber, M. K., and Jenkins, M. G. (2000). Tensile fracture behavior of two types of silicon nitride specimen geometries conducted by ten US groups. *J. Am. Ceram. Soc.* **83**: 1186–1191.
33. Simpatico, S., Cannon, W. R., Matthewson, M. J. (1999). Comparison of hydraulic-burst and ball-on-ring tests for measuring biaxial strength. *J. Am. Ceram. Soc.* **82**: 2737–2744.
34. Fett, T., Müller, S., Munz, D., and Thun, G. (1998). Nonsymmetry in deformation behaviour of PZT. *J. Mater. Sci. Lett.* **17**: 261–265.
35. Fett, T., Munz, D., and Thun, G. (2000). Differences in tensile and bending strength for Knoop-cracked lead zirconate titanate specimens. *J. Am. Ceram. Soc.* **83**: 669–671.

36. Kraft and Volkert, C. A. (2001). Mechanical testing of thin films and small structures. *Adv. Eng. Mater.* **3**: 99–110.
37. Epstein, A. H., and Senturia, S. D. (1997). Micro power from micro machinery. *Science* **276**: 1211.
38. Chen, K.-S., Ayon, A., and Spearing, S. M. (2000). Controlling and testing the fracture strength of silicon on the mesoscale. *J. Am. Ceram. Soc.* **83**: 1476–1484.
39. Freiman, S. W. (1988). Brittle fracture behavior of ceramics. *Ceram. Bull.* **67**: 392–402.
40. Wiederhorn, S. M. (1967). Influence of water-vapor on crack-propagation in soda-lime glass. *J. Am. Ceram. Soc.* **50**: 407–414.
41. Michaelske, T. A., Bunker, B. C., and Freiman, S. W. (1986). Stress corrosion of bulk and fibrous silica. *J. Am. Ceram. Soc.* **69**: 721–724.
42. Pletka, B. J., and Wiederhorn, S. M. (1982). A comparison of failure prediction by strength and fracture mechanics techniques. *J. Mater. Sci.* **17**: 1247–1268.
43. Ritter, J. E. Jr., and Humenik, J. N. (1979). Static and dynamic fatigue of polycrystalline alumina. *J. Mater. Sci.* **14**: 626–632.
44. Ritter, J. E., Vicedomine, M., Breder, K., and Jakus, K. (1985). Dynamic fatigue of indented soda-lime glass as a function of temperature, *J. Mater. Sci.* **20**: 2868–2872.
45. Ritter, J. E., Vicedomine, M., Breder, K., and Jakus, K. (1984). Dynamic fatigue analysis of indentation flaws using an exponential-law crack velocity function. *J. Am. Ceram. Soc.* **67**: C-198–C-199.
46. Cook, R. F., Lawn, B. R., and Fairbanks, C. J. (1985). Microstructure-strength properties in ceramics: II, fatigue relations. *J. Am. Ceram. Soc.* **68**: 616–623.
47. Zeng, K., Brede, K., and Rowcliffe, D. J. (1993). Comparison of slow crack growth behavior in alumina and SiC-whisker-reinforced alumina, *J. Am. Ceram. Soc.* **76**: 1673–1680.
48. Knehans, R., and Steinbrech, R. (1982). Memory effect of crack resistance during slow crack growth in Al_2O_3 bend specimens. *J. Mater. Sci. Lett.* **1**: 327–329.
49. Steinbrech, R., Knehans, R., and Schaarwachter, W. (1983). Increase of crack resistance during slow crack growth in Al_2O_3 bend specimens. *J. Mater. Sci.* **18**: 265–270.
50. Cook, R. F., Lawn, B. R., and Fairbanks, C. J. (1985). Microstructure-strength properties in ceramics: I effect of crack size on toughness. *J. Am. Ceram. Soc.* **68**: 604–615.
51. Cook, R. F., Fairbanks, C. J., Lawn, B. R., and Mai, Y. W. (1987). Crack resistance by interfacial bridging: its role in determining strength characteristics. *J. Mater. Res.* **2**: 345–356.
52. Wereszczak, A., Kirkland, T. P., Breder, K., Ferber, M. K., and Khandelwal, P. (1995). High temperature dynamic fatigue performance of a hot isostatically pressed silicon nitride. *Mater. Sci. Eng. A* **191**: 257–266.
53. Lin, C.-K. J., Jekins, M. G., and Ferber, M. K. (1993). Tensile dynamic and static fatigue relations for a HIPed silicon nitride at elevated temperatures. *J. Eur. Ceram. Soc.* **12**: 3–13.
54. Dwivedi, P. J., and Green, D. J. (1995). Indentation crack-shape evolution during subcritical crack growth. *J. Am. Ceram. Soc.* **78**: 1240–1246.
55. Dwivedi, P. J., and Green, D. J. (1995). Determination of subcritical crack growth parameters by in situ observation of indentation cracks. *J. Am. Ceram. Soc.* **78**: 2122–2128.
56. Lin, C.-K. J., Jekins, M. G., and Ferber, M. K. (1994). Cyclic fatigue of hot isostatically pressed silicon nitride at elevated temperatures. *J. Mater. Sci.* **29**: 3517–3526.
57. Gupta, P. K., Inniss, D., Kurkjian, C. R., and Brownlow, D. L. (1994). Determination of crack velocity as a function of stress intensity from static fatigue data. *J. Am. Ceram. Soc.* **77**: 2445–2449.
58. Mould, R. W., and Southwick, R. D. (1959). Strength and static fatigue of abraded glass under controlled ambient conditions II. *J. Am. Ceram. Soc.* **42**: 582–592.
59. Govila, R. K. (1984). Phenomenology fracture in sintered alpha silicon carbide. *J. Mater. Sci.* **19**: 2111–2120.

60. Govila, R. K. (1983). High temperature uniaxial tensile stress rupture strength of sintered alpha SiC. *J. Mater. Sci.* **18**: 1967–1976.
61. Masuda, M., Soma, T., Matsui, M., and Oda, I. (1989). Cyclic fatigue behavior of sintered Si_3N_4 at high temperatures. *J. Ceram. Soc. Jpn.* **97**: 601–607.
62. Masuda, M., Soma, T., Matsui, M., and Oda, I. (1988). Fatigue behavior of sintered Si_3N_4 under tension-compression cyclic stress. *J. Ceram. Soc. Jpn.* **96**: 275–280.
63. Lin, C.-K. J., and Socie, D. F. (1991). Static and cyclic fatigue of alumina at high temperatures. *J. Am. Ceram. Soc.* **74**: 1511–1518.
64. Masuda, M., Yamada, N., Soma, T., Matsui, M., and Oda, I. (1989). Cyclic fatigue properties of sintered Si_3N_4 at room temperature. *J. Ceram. Soc. Jpn.* **97**: 509–514.
65. Kawakubo, T., and Komeya, K. (1987). Static and cyclic fatigue behavior of a sintered silicon nitride at room temperature. *J. Am. Ceram. Soc.* **70**: 400–405.
66. Dauskardt, R. H. (1993). A frictional-wear mechanism for fatigue-crack growth in grain bridging ceramics. *Acta Metall. Mater.* **41**: 2765–2781.
67. Duskardt, R. H., James, M. R., Porter, J. R., and Ritchie, R. O. (1992). Crack fatigue-crack growth in a SiC-whisker-reinforced alumina ceramic composite: long- and small-crack behavior. *J. Am. Ceram. Soc.* **75**: 759–771.
68. Ritchie, R. O., and Dauskardt, R. H. (1991). Cyclic fatigue of ceramics: a fracture mechanics approach to subcritical crack growth and life prediction. *J. Ceram. Soc. Jpn.* **99**: 1047–1062.
69. Dauskardt, R. H., Marshall, D. B., and Ritchie, R. O. (1990). Cyclic fatigue-crack propagation in magnesia-partially-stabilized zirconia ceramics. *J. Am. Ceram. Soc.* **73**: 893–903.
70. Gilbert, J., Dauskardt, R. H., Steinbrech, R. W., Petrany, R. N., and Ritchie, R. O. (1995). Cyclic fatigue in monolithic alumina: mechanisms for crack advance promoted by frictional wear of grain bridges. *J. Mater. Sci.* **30**: 643–654.
71. Cornelissen, E., Dauskardt, R. H., Ritchie, R. O., and Thomas, G. (1994). Cyclic fatigue behavior and fracture toughness of silicon nitride ceramics sintered with rare earth oxides. *Acta Metall. Mater.* **42**: 3055–3064.
72. Jacobs, S., and Chen, I.-W. (1994). Mechanical and environmental factors in the cyclic and static fatigue of silicon nitride. *J. Am. Ceram. Soc.* **77**: 1153–1161.
73. Ewart, L., and Suresh, S. (1987). Crack propagation in ceramics under cyclic loads. *J. Mater. Sci.* **22**: 1173–1192.
74. Gilbert, C. J., Dauskardt, R. H., and Ritchie, R. O. (1995). Behavior of cyclic fatigue cracks in monolithic silicon nitride. *J. Am. Ceram. Soc.* **78**: 2291–2300.
75. Paris, P. C., and Erdogan, F. (1963). A critical analysis of crack propagation laws. *J. Basic Eng., Trans. ASME* **85**: 528–534.
76. Gilbert, C. J., Dauskardt, R. H., and Ritchie, R. O. (1997). Microstructural mechanisms of cyclic fatigue-crack propagation in grain-bridging ceramics. *Ceram. Int.* **23**: 413–418.
77. Gilbert, C. J., Cao, J. J., De Jonghe, L. C., and Ritchie, R. O. (1997). Crack-growth resistance-curve behavior in silicon carbide: small versus long cracks. *J. Am. Ceram. Soc.* **80**: 2253–2261.
78. Cao, J. J., Moberly-Chan, W. J., DeJonghe, L. C., Gilbert, C. J., and Ritchie, R. O. (1996). In situ toughened silicon carbide with Al–B–C additions. *J. Am. Ceram. Soc.* **79**: 461–469.
79. Han, Y.-S., Kim, D.-K., Gilbert, C. J., and Ritchie, R. O. (2001). Cyclic fatigue-crack propagation behavior in silicon carbide: long- and small-crack behavior. *J. Am. Ceram. Soc.* **84**: 551–554.
80. McNaney, J. M., Gilbert, C. J., and Ritchie, R. O. (1999). Effect of viscous grain bridging on cyclic fatigue-crack growth in monolithic ceramics at elevated temperatures. *Acta Mater.* **47**: 2809–2819.

81. Chen, D., Gilbert, C. J., Zhang, X. F., and Ritchie, R. O. (2000). High-temperature cyclic fatigue-crack growth behavior in an in situ toughened silicon carbide. *Acta Mater.* **48**: 659–674.
82. Chen, D., Sixta, M. E., Zhang, X. F., De Jonghe, L. C., and Ritchie, R. O. (2000). Role of the grain-boundary phase on the elevated-temperature strength, toughness, fatigue and creep resistance of silicon carbide sintered with Al, B, and C. *Acta Mater.* **48**: 4599–4608.
83. Jacobs, D. S., and Chen, I. W. (1995). Cyclic fatigue in ceramics—a balance between crack shielding accumulation and degradation. *J. Am. Ceram. Soc.* **78**: 513–520.
84. Chen, I.-W., and Engineer, M. (1999). Model for fatigue crack growth in grain-bridging ceramics. *J. Am. Ceram. Soc.* **82**: 3549–3560.
85. Lawn, R. (1998). Indentation of ceramics with spheres: a century after hertz. *J. Am. Ceram. Soc.* **81**: 1977–1994.
86. Lawn, R., Padture, N. P., Cai, H., and Guiberteau, F. (1994). Making ceramics "ductile". *Science* **263**: 1114–1116.
87. Lawn, R., Lee, S. K., Peterson, I. M., and Wuttiphan, S. (1998). A model of strength degradation from hertzian contact damage in tough ceramics. *J. Am. Ceram. Soc.* **81**: 1509–1520.
88. Kim, K., Jung, Y.-G., Peterson, I. M., and Lawn, B. R. (2000). Cyclic fatigue of intrinsically brittle ceramics in contact with spheres. *Acta Mater.* **47**: 4711–4725.
89. Lawn, B. R., Padture, N. P., Guiberteau, F., and Cai, H. (1994). A model for microcrack initiation and propagation beneath hertzian contacts in polycrystalline ceramics. *Acta Metall.* **42**: 1683–1693.
90. Padture, N. P., and Lawn, B. R. (1995). Fatigue in ceramics with interconnecting weak interfaces: a study using cyclic hertzian contacts. *Acta Metall.* **43**: 1609–1617.
91. Lawn, B. R., and Marshall, D. B. (1998). Nonlinear stress-strain curves for solids containing closed cracks with friction. *J. Mech. Phys. Solids* **46**: 85–113.
92. Lee, S. K., and Lawn, B. R. (1999). Contact fatigue in silicon nitride. *J. Am. Ceram. Soc.* **82**: 1281–1288.
93. Lee, K. S., Jung, Y.-G., Peterson, I. M., Lawn, B. R., Kim, D. K., and Lee, S. K. (2000). Model for cyclic fatigue of quasi-plastic ceramics in contact with spheres. *J. Am. Ceram. Soc.* **83**: 2255–2262.
94. Guiberteau, P. F., Lawn, B. R., and Lathabai, S. (1995). Hertzian contact damage in magnesia-partially-stabilized zirconia. *J. Am. Ceram. Soc.* **78**: 1083–1086.
95. Swain, M. V. (1979). Microfracture about scratches in brittle solids. *Proc. R. Soc. Lond. A* **366**: 575–597.
96. Xu, H. H. K., and Jahanmir, S. (1995). Microfracture and material removal in scratching of alumina. *J. Mater. Sci.* **30**: 2235–2247.
97. Xu, H. H. K., Jahanmir, S., and Wang, Y. S. (1995). Effect of grain-size on scratch interactions and material removal in alumina. *J. Am. Ceram. Soc.* **78**: 881–891.
98. Lee, S. K., Tandon, R., Readey, M. J., and Lawn, B. R. (2000). Scratch damage in zirconia ceramics. **83**: 1428–1432.
99. Larsen, K. P. K., Dormans, G. J., Taylor, D. J., and van Veldhoven, P. J. (1994). Ferroelectric properties and fatigue of $PbZr_{0.51}Ti_{0.49}O_3$ thin films of varying thickness: blocking layer model. *J. Appl. Phys.* **76**: 2405–2413.
100. Mihara, T., Watanabe, H., and Araujo, C. A. P. (1994). Polarization fatigue characteristics of sol–gel ferroelectric $Pb(Zr_{0.4}Ti_{0.6})O_3$ thin-film capacitors. *Jpn. J. Appl. Phys.* **33**: 3996–4002.
101. Jiang, Q. Y., Subbarao, E. C., and Cross, L. E. (1994). Effect of composition and temperature on electric fatigue of La-doped lead zirconate titanate ceramics. *J. Appl. Phys.* **75**: 7433–7443.
102. Tanimoto, T., and Okazaki, K. (1991). Fatigue degradation and reliability of piezoelectric ceramics. *Jpn. J. Appl. Phys.* **30**: 2410–2412.

103. Sakai, T., and Kawamoto, H. (1997). Fatigue-testing method and property changes of piezoelectric ceramics. *J. Ceram. Soc. Jpn. (Yogyo Kyokaishi)* **105**: 1032–1036.
104. Harada, A. S., Furuta, A., and Uchino, K. (1994). *Jpn. J. Appl. Phys.* **33**: 3091–3093.
105. Nagata, K., and Kinoshita, S. (1995). Relationship between lifetime of multilayer ceramic actuator and temperature. *Jpn. J. Appl. Phys.* **34**: 5266–5269.
106. Sakai, T., Ishikiriyama, M., and Shimazaki, R. (1992). Durability of piezoelectric ceramics for an actuator. *Jpn. J. Appl. Phys.* **31**: 3051–3054.
107. Sakai, T., Terai, Y., and Ishikiriyama, M. (1995). Improvement in durability of piezoelectric ceramics for actuator. *Jpn. J. Appl. Phys.* **34**: 5276–5278.
108. Sakai, T., and Kawamoto, H. (2000). Improvement in time-dependent displacement degradation of piezoelectric ceramics by manganese/indium addition. *J. Am. Ceram. Soc.* **83**: 1423–1427.
109. Hwang, H., Tajima, K., Sando, M., Toriyama, M., Niihara, K. (1998). Fatigue behavior of PZT-based nanocomposites with fine platinum particles. *J. Am. Ceram. Soc.* **81**: 3325–3328.
110. Tajima, K., Hwang, H., Sando, M., Toriyama, M., Niihara, K. (2000). Electric-field-induced crack growth behavior in PZT/Al2O3 composites. *J. Am. Ceram. Soc.* **83**: 651–653.
111. Davis, R. F., and Carter, C. H. Jr. (1998). A review of creep in silicon nitride and silicon carbide, *Advanced Ceramics*, pp. 95–125, Saito, S., ed. Oxford: Oxford University Press and Tokyo: Ohmsha Ltd.
112. Blumenthal, W., and Evans, A. G. (1984). High temperature failure of polycrystalline alumina: II, creep crack growth and blunting. *J. Am. Ceram. Soc.* **67**: 751–759.
113. Carter, C. H. Jr., and Davis, R. F. (1984). Kinetics and mechanisms of high-temperature creep in silicon carbide: I, reaction-bonded. *J. Am. Ceram. Soc.* **67**: 409–419.
114. Carter, C. H. Jr., Davis, R. F., and Bently, J. (1984). Kinetics and mechanisms of high temperature creep in silicon carbide: II, chemically vapor deposited. *J. Am. Ceram. Soc.* **67**: 732–740.
115. Chan, K. S., Lankford, J., and Page, R. A. (1984). Viscous cavity growth in ceramics under compressive loads. *Acta Metall.* **32**: 1907–1914.
116. Page, R. A., Lankford, J., and Spooner, S. (1984). Nucleation and early stage growth of creep cavities in hot pressed silicon carbide. *Acta Metall.* **32**: 1275–1914.
117. Thouless, M. D., and Evans, A. G. (1984). Nucleation of cavities during creep of liquid-phase sintered materials. *J. Am. Ceram. Soc.* **67**: 721–727.
118. Goursat, B. P., and Besson, J. L. (1985). Oxidation resistance and creep behavior of a silicon nitride ceramics densified with YO. *J. Mater. Sci.* **20**: 1156–1159.
119. Chokshi, H., and Porter, J. R. (1985). Creep deformation of an alumina matrix composite reinforced with silicon carbide whiskers. *J. Am. Ceram. Soc.* **68**: C144–C145.
120. Clarke, D. R. (1985). High temperature deformation of polycrystalline alumina containing an intergranular glass phase. *J. Mater. Sci.* **20**: 1321–1332.
121. Dalgleish, J., Slamovich, E. B., and Evans, A. G. (1985). Duality in the creep rupture of a polycrystalline alumina. *J. Am. Ceram. Soc.* **68**: 575–581.
122. Wiederhorn, S. M., Hockey, B. J., Krause, R. F., and Jakus, K. (1986). Creep and fracture of a vitreous-bonded aluminum oxide. *J. Mater. Sci.* **21**: 810–824.
123. Chokshi, H., and Porter, J. R. (1987). Cavity development during creep deformation in alumina with a bimodal grain size distribution. *J. Am. Ceram. Soc.* **70**: 197–202.
124. Dimos, and Kohlstedt, D. L. (1987). Diffusional creep and kinetic demixing in yttria-stabilized zirconia. *J. Am. Ceram. Soc.* **70**: 531–536.
125. Page, R. A., Lankford, J., Chan, K. S., Hardman-Rhyne, K. H., and Spooner, S. (1987). Creep cavitation in liquid-phase-sintered alumina. *J. Am. Ceram. Soc.* **70**: 137–145.
126. Page, R. A., and Lankford, J. (1987). The effect of cyclic creep on grain-boundary cavitation in ceramics. *J. Mater. Sci.* **22**: 137–145.

127. Carroll, F., and Tressler, R. E. (1988). Accumulation of creep damage in a siliconized silicon carbide. *J. Am. Ceram. Soc.* **71**: 472–477.
128. Chuang, T.-J., and Wiederhorn, S. M. (1988). Damage-enhanced creep in a siliconized silicon carbide: mechanics of deformation. *J. Am. Ceram. Soc.* **71**: 595–601.
129. Jakus, K., and Wiederhorn, S. M. (1988). Creep deformation of ceramics in four point bending. *J. Am. Ceram. Soc.* **71**: 832–836.
130. Lane, J. E., Carter, C. H. Jr., and Davis, R. F. (1988). Kinetics and mechanisms of high temperature creep in silicon carbide: II, sintered a-silicon carbide. *J. Am. Ceram. Soc.* **71**: 281–295.
131. Wilkinson, S. (1988). Effect of devitrification on creep deformation in glass containing ceramics. *J. Am. Ceram. Soc.* **71**: 562–565.
132. Carroll, D. F., and Tressler, R. E. (1989). Effect of creep damage on the tensile creep behavior of a siliconized silicon carbide. *J. Am. Ceram. Soc.* **72**: 49–53.
133. Chakraborty, D., and Mukhopadhyay, A. K. (1989). Creep of sintered silicon nitride. *Ceram. Int.* **15**: 237–245.
134. Page, R. A., Chan, K. S., Davidson, D. L., and Lankford, J. (1990). Micromechanics of creep-crack growth in a glass ceramic. *J. Am. Ceram. Soc.* **73**: 2977–2986.
135. Todd, J. A., and Xu, Z.-Y. (1989). The high-temperature creep deformation of Si_3N_4–$6Y_2O_3$. *J. Mater. Sci.* **24**: 4443–4452.
136. Jou, Z. C., and Virkar, A. V. (1990). High temperature creep and cavitation of polycrystalline aluminum nitride. *J. Am. Ceram. Soc.* **73**: 1928–1935.
137. Lin, T., and Becher, P. F. (1991). Creep behavior of a SiC whisker-reinforced alumina. *J. Am. Ceram. Soc.* **73**: 1379–1381.
138. Chen, C., Wiederhorn, S. M., and Chuang, T. (1991). Cavitation damage during flexural creep of Sialon-YAG ceramics. *J. Am. Ceram. Soc.* **74**: 1658–1662.
139. Chuang, T., Liu, W., and Siederhorn, S. (1991). Steady state creep behavior of SisiC C-rings. *J. Am. Ceram. Soc.* **74**: 2531–2537.
140. Hockey, J., Wiederhorn, S. M., Liu, W., Baldoni, J. G., and Buljan, S. T. (1991). Tensile creep of whisker-reinforced silicon nitride. *J. Mater. Sci.* **26**: 3931–3939.
141. Robertson, G., Wilkinson, D. S., and Caceres, C. H. (1991). Creep and creep fracture in hot-pressed alumina. *J. Am. Ceram. Soc.* **74**: 915–921.
142. Blanchard, C. R., and Page, R. A. (1992). Grain boundary sliding microdisplacement measurements during creep of AlO. *J. Am. Ceram. Soc.* **75**: 1612–1620.
143. Chadwick, M. M., Wilkinson, D. S., and Dryden, J. R. (1992). Creep due to non-Newtonian grain boundary phase. *J. Am. Ceram. Soc.* **75**: 2327–2334.
144. Corman, G. S. (1992). Creep of 6H α-SiC single crystals. *J. Am. Ceram. Soc.* **75**: 3421–3424.
145. Ferber, M. K., and Jenkins, M. G. (1992). Evaluation of the strength and creep fatigue behavior of a HIPed silicon nitride. *J. Am. Ceram. Soc.* **75**: 2453–2462.
146. Hockey, B. J., and Wiederhorn, S. M. (1992). Effect of microstructure on the creep of siliconized silicon carbide. *J. Am. Ceram. Soc.* **75**: 1822–1830.
147. Blanchard, R., and Chan, K. S. (1993). Evidence of grain-boundary-sliding-induced cavitation in ceramics under compression. *J. Am. Ceram. Soc.* **76**: 1651–1660.
148. Chan, K. S., and Page, R. A. (1993). Creep damage development in structural ceramics. *J. Am. Ceram. Soc.* **76**: 803–826.
149. Chadwick, M. M., and Wilkinson, D. S. (1993). Microstructural evolution in annealed and crept silicon nitride. *J. Am. Ceram. Soc.* **76**: 376–384.
150. Chadwick, M. M., Jupp, R. S., and Wilkinson, D. S. (1993). Creep behavior of a sintered silicon nitridek. *J. Am. Ceram. Soc.* **76**: 385–396.
151. Ohji, T., and Yamauchi, Y. (1993). Tensile creep and creep rupture behavior of monolithic and SiC whisker-reinforced silicon nitride ceramics. *J. Am. Ceram. Soc.* **76**: 3105–3112.

152. Wiederhorn, S. M., Hockey, B. J., Cranmer, D. C., and Yeckley, R. (1993). Transient creep behavior of hot isostatically pressed silicon nitride. *J. Mater. Sci.* **28**: 445–453.
153. Cinibulk, M., Thomas, G., and Johnson, S. M. (1994). Strength and creep behavior of rare-earth disilicate-silicon nitride ceramics. *J. Am. Ceram. Soc.* **75**: 2050–2055.
154. Ding, J. L., Liu, K. C., More, K. L., and Brinkman, C. R. (1994). Creep and creep rupture of an advanced silicon nitride ceramic. *J. Am. Ceram. Soc.* 77: 867–874.
155. Ferber, M. K., Jenkins, M. G., Nolan, T. A., and Yeckley, R. (1994). Comparison of the creep and creep rupture performance of two HIPed silicon nitride ceramics. *J. Am. Ceram. Soc.* 77: 657–665.
156. Menon, M. N., Fang, H. T., Wu, K. C., Jenkins, M. G., Ferber, M. K., More, K. L., Hubbard, C. M., and Nolan, T. A. (1994). Creep and stress rupture behavior of an advanced silicon nitride, Part III—stress rupture and Monkman grant relation. *J. Am. Ceram. Soc.* 77: 1235–1241.
157. Kandil, A., and Dyson, B. F. (1988). Tensile creep of ceramics: the development of a testing facility. *Int. J. High Technol. Ceram.* **4**: 243–262.
158. Carroll, D. F., Wiederhorn, S. M., and Roberts, D. E. (1989). Technique for tensile creep testing of ceramics. *J. Am. Ceram. Soc.* **72**: 1610–1614.
159. Ohji, T., and Yamauchi, Y. (1992). Long-term tensile creep for advanced ceramics. *J. Am. Ceram. Soc.* **75**: 2304–2307.
160. Luecke, W. E., and Wiederhorn, S. M. (1997). Interlaboratory verification of silicon nitride tensile creep properties. *J. Am. Ceram. Soc.* **80**: 831–838.
161. Fett, T., Keller, K., and Munz, D. (1988). An analysis of the creep of hot pressed silicon nitride in bending. *J. Mater. Sci.* **23**: 467–474.
162. Rosensfield, R., Shetty, D. K., and Duckworth, W. H. (1985). Damage analysis of creep in bending laws. *J. Am. Ceram. Soc.* **68**: 483–485.
163. Chuang, T.-J. (1986). Estimation of power-law creep parameters from bend test data. *J. Mater. Sci.* **21**: 165–175.
164. Chen, F., and Chuang, T.-J. (1990). Improved analysis for flexural creep with application to sialon ceramics. *J. Am. Ceram. Soc.* **73**: 2366–2373.
165. Krause, R. F. Jr. (1992). Observed and theoretical creep rates for an alumina ceramic and a silicon nitride ceramics in flexure. *J. Am. Ceram. Soc.* **75**: 1307–1310.
166. Lange, F. F., Diaz, E. S., and Andersson, C. L. (1979). Tensile creep testing of improved SiN. *Bull. Am. Ceram. Soc.* **58**: 845–848.
167. Luecke, W. E., and French, J. D. (1996). Sources of strain measurement uncertainty in flag-based extensonetry. *J. Am. Ceram. Soc.* **79**: 1617–1626.
168. Lin, H.-T., Becher, P. F., and Ferber, M. K. (1994). Improvement of tensile creep displacement measurements. *J. Am. Ceram. Soc.* 77: 2767–2770.
169. French, J. D., and Wiederhorn, S. M. (1996). Tensile specimens from ceramic components. *J. Am. Ceram. Soc.* **79**: 550–552.
170. Luecke, W. E., Wiederhorn, S. M., Hockey, B. J., Krausse, R. Jr., and Long, G. G. (1995). Cavitation contributes substantially to tensile creep in silicon nitride. *J. Am. Ceram. Soc.* **78**: 2085–2096.
171. Birch, J. M., Wilshire, B., Owen, D. J. R., and Shantaram, D. (1976). The influence of stress distribution on the deformation and fracture behavior of ceramic materials under compression creep conditions. *J. Mater. Sci.* **11**: 1817–1825.
172. Birch, J. M., and Wilshire, B. (1978). The compressive creep behavior of silicon nitride ceramics. *J. Mater. Sci.* **13**, 2627–2636.
173. Lange, F. F., Davis, B. I., and Clarke, D. R. (1980). Compressive creep of Si_3N_4, alloys, Part 1, effect of composition. *J. Mater. Sci.* **15**: 601–610.

174. Lange, F. F., Davis, B. I., and Clarke, D. R. (1983). Compressive creep and oxidation resistance of Si_3N_4 material fabricated in the system Si_3N_4–Si_2N_2O–$Y_2Si_2O_7$. *J. Am. Ceram. Soc.* **66**: C98–C99.
175. Tracey, A. (1987). A compression test for high strength ceramics. *J. Test. Eval.* **15**: 14–19.
176. Awaji, H., and Nagano, Y. (1993). Compressive strength of fine ceramics. *J. Ceram. Soc. Japan, Int. Edn* **101**: 945–950.
177. Debschutz, K. D., Caspers, B., Schneider, G. A., and Petzow, G. (1993). Critical evaluation of the compressive creep test. *J. Am. Ceram. Soc.* **76**: 2468–2474.
178. Wiederhorn, S. M., Roberts, D. E., Chuang, T.-J., and Chuck, L. (1988). Damage enhanced creep in siliconized silicon carbide: phenomenology. *J. Am. Ceram. Soc.* **71**: 602–608.
179. Chuang, T.-J., Liu, W., Wiederhorn, S. M. (1991). Steady-state creep behavior of SiC–SiC C-rings. *J. Am. Ceram. Soc.* **74**: 2531–2537.
180. Evans, A. G., and Sharp, J. V. (1971). Microstructural studies on silicon nitride. *J. Mater. Sci.* **6**: 1292–1302.
181. Pezzotti, G., and Ota, K. (1997). Grain-boundary sliding in fluorine-doped silicon nitride. *J. Am. Ceram. Soc.* **80**: 599–606.
182. Pezzotti, G., Ota, K., Kleebe, H.-J. (1997). Viscous slip along grain boundaries in chlorine-doped silicon nitride. *J. Am. Ceram. Soc.* **80**
183. Lofaj, F., Okada, A., Kawamoto, H. (1997). Cavitational strain contribution to tensile creep in vitreous bonded ceramics. *J. Am. Ceram. Soc.* **80**: 1619–1624.
184. Krause, R. F., Luecke, W. E., French, J. D., Hockey, B. J., and Wiederhorn, S. M. (1999). Tensile creep and rupture of silicon nitride. *J. Am. Ceram. Soc.* **82**: 1233–1240.
185. Luecke, W. E., and Wiederhorn, S. M. (1999). A new model for tensile creep of silicon nitride. *J. Am. Ceram. Soc.* **83**: 2769–2775.
186. Norton, F. H. (1929). *The Creep of Steel at High Temperatures*, New York: McGraw-Hill.
187. Luecke, W. E., and Wiederhorn, S. M. (1994). Tension compression creep asymmetry in Si_3N_4. *Key Eng. Mater.* **89**: 587–591.
188. Yoon, K. J., Wiederhorn, S. M., and Luecke, W. E. (2000). Comparison of tensile and compressive creep behavior in silicon nitride. *J. Am. Ceram. Soc.* **83**: 2017–2022.
189. Boling-Risser, M. A., Goretta, K. C., Routbort, J. L., and Faber, K. T. (2000). Effect of microstructure on high-temperature compressive creep of self-reinforced hot-pressed silicon nitride. *J. Am. Ceram. Soc.* **83**: 3065–3069.
190. Yoshida, M., Tanaka, K., Kubo, T., Terazono, H., Tsuruzono, S. Development of ceramic components for ceramic gas turbine engine (CGT302). ASME 98-GT-398.
191. Yoshida, M., Tanaka, K., Tsuruzono, S., Tatsuki, T. (1999). Development of ceramic components for ceramic gas turbine engine (CGT302), *Ceramics: Getting into the 2000s, Part D*, pp. 253–264, Edited by Vincentini, P., ed., Trieste, Italy: Elsevier Science Publishing.
192. Tanaka, K., Yoshida, M., Kubo, T., Terazono, H., Tsuruzono, S. Development and evaluation of ceramic components for small gas turbine engine. ASME 00-GT-531.
193. Ohji, T. (2001). Long-term tensile creep behavior of highly heat-resistant silicon nitride for ceramic gas turbines. *Ceram. Sci. Eng. Proc.* **23**.
194. Lofaj, F., Wiederhorn, S. M., Long, G. G., Jemian, P. R., and Ferber, M. K. (2001). Cavitation creep in the next generation silicon nitride. *Ceram. Sci. Eng. Proc.* **23**.
195. Anderson, C. A. (1981). Ceramic materials for high temperature turbines. Final Rept. US Energy Research and Development Administration, Contract No. DOF/OR 90405-TI.
196. Lange, F. F., Singhal, S. C., and Kuznicki, R. C. (1977). Phase relation and stability studies in the Si_3N_4–Si_2O_2–Y_2O_3 pseudoternary system. *J. Am. Ceram. Soc.* **60**: 249–252.
197. Ohji, T., and Yamauchi, Y. (1994). Diffusional crack growth and creep rupture of silicon carbide. *J. Am. Ceram. Soc.* **77**: 678–682.

198. Lane, J. E., Carter, C. H. Jr., and Davis, R. F. (1988). Kinetics and mechanisms of high temperature creep in silicon carbide: III, sintered α-silicon carbide. *J. Am. Ceram. Soc.* **71**: 281–295.
199. Chuang, T.-J. (1982). A diffusive crack-growth model for creep fracture. *J. Am. Ceram. Soc.* **65**: 93–103.
200. Raj, R. (1993). Fundamental research in structural ceramics for service near 2000°C. *J. Am. Ceram. Soc.* **76**: 2147–2174.
201. Waku, Y., Nakagawa, N., Wakamoto, T., Ohtsubo, H., Shimizu, K., and Kohtoku, Y. (1997). A ductile ceramic eutectic composite with high strength at 1,873 K. *Nature* **389**: 49–52.
202. Cho, J., Harmer, M. P., Chan, H. M., Rickman, J. M., and Thompson, A. M. (1997). Effect of yttrium and lanthanum on the tensile creep behavior of aluminum oxide. *J. Am. Ceram. Soc.* **80**: 1013.
203. Thompson, M., Soni, K. K., Chan, H. M., Harmer, M. P., Williams, D. B., Chabala, J. M., and Levi-Setti, R. (1997). Rare earth dopant distributions in creep-resistant Al_2O_3. *J. Am. Ceram. Soc.* **80**: 373–376.
204. Bruley, J., Cho, J., Chan, H. M., Harmer, M. P., and Rickman, J. M. (1999). Scanning transmission electron microscopy analysis of grain boundaries of creep-resistant yttrium- and lanthanum-doped alumina microstructures. *J. Am. Ceram. Soc.* **82**: 2865–2870.
205. Wakai, F., Nagano, T., and Iga, T. (1997). Hardening in creep of alumina by zirconium segregation at the grain boundary. *J. Am. Ceram. Soc.* **80**: 2361–2363.
206. Li, Y.-Z., Wang, C., Chan, H. M., Rickman, J. M., and Harmer, M. P., Chabala, J. M., Gavrilov, K. L., and Levi-Setti, R. (1999). Codoping of alumina to enhance creep resistance. *J. Am. Ceram. Soc.* **82**: 1497–1504.
207. Cho, J., Rickman, J. M., Chan, H. M., and Harmer, M. P. (2000). Modeling of grain-boundary segregation behavior in aluminum oxide. *J. Am. Ceram. Soc.* **83**: 344–352.
208. Yoshida, H., Ikuhara, Y., and Sakuma, T. (1999). High-temperature creep resistance in lanthanoid ion-doped polycrystalline Al_2O_3. *Philos. Mag. Lett.* **79**: 249–256.
209. Yoshida, H., Ikuhara, Y., and Sakuma, T. (2001). Transient creep in fine-grained polycrystalline Al_2O_3 with Lu^{3+} ion segregation at the grain boundaries. *J. Mater. Res.* **16**: 716–720.
210. Niihara, K. (1991). New design concept of structural ceramics—ceramic nanocomposite. *J. Ceram. Soc. Jpn.* **99**: 974–982.
211. Ohji, T., Nakahira, A., Hirano, T., and Niihara, K. (1994). Tensile creep behavior of alumina/silicon carbide nanocomposite. *J. Am. Ceram. Soc.* **77**: 3259–3262.
212. Ohji, T., Hirano, T., Nakahira, A., and Niihara, K. (1996). Particle/matrix interface and its role in creep inhibition alumina/silicon carbide nanocomposite. *J. Am. Ceram. Soc.* **79**: 33–45.
213. Thompson, M., Chan, H. M., and Harmer, M. P. (1997). Tensile creep of alumina–silicon carbide "nanocomposite". *J. Am. Ceram. Soc.* **80**: 2221–2228.
214. Ohji, T., Kusunose, T., and Niihara, K. (1998). Threshold stress in creep of alumina–silicon carbide nanocomposites. *J. Am. Ceram. Soc.* **81**: 2713–2716.
215. Arzt, E., and Ashby, M. F. (1982). Threshold stress in materials containing dispersed particles. *Scripta Metall.* **16**: 1285–1290.

INDEX